Vocational and Technical Resources for Community College Libraries:
Selected Materials, 1988-1994

Edited by

Mary Ann Laun
Pasadena City College

Funding and production assistance provided by

CHOICE Magazine

Association of College and Research Libraries
American Library Association

Published by the Association of College and Research Libraries, a division of the American Library Association, 50 East Huron Street, Chicago, IL 60611

Copyright © 1995 by the American Library Association. All rights reserved. Printed in the United States of America. This book, or parts thereof, may not be reproduced in any form without written permission of the publishers.

Library of Congress Cataloging-in-Publication Data

Vocational and technical resources for community college libraries:
   selected materials, 1988-1994 / edited by Mary Ann Laun.

   p. cm.
   Includes index.
   ISBN 0-8389-7775-8 (paperback)
   1. Technology—Vocational guidance—United States—Bibliography.
   2. Industrial arts—Vocational guidance—United States—Bibliography.
   3. Community college libraries—United States—Book lists. I. Laun, Mary Ann, 1950- .
T65.3.V63   1995
690'.023'73—dc20   94-46648
                                              CIP

TABLE OF CONTENTS

Introduction
vii

Acknowledgments
ix

Referees
x

Allied Health

Dental Auxiliary Careers: Dental Hygiene, Dental Assisting and Dental Lab Technology 1
Nancy Buchanan

Diagnostic Technologies 21
Isabel C. Hernandez

Dietetics and Nutrition 33
Judy Caramanica

Emergency Medical Services 40
Grace Ekins

Medical Assisting and Medical Records 47
Tori Beyer and Michelle Strazer

Medical Laboratory Technology 57
Ann Cannon

Nursing 67
Elisa Abella and Isabel C. Hernandez

Occupational Health and Safety 91
Terri Propes

Occupational Therapy Assisting 97
Nancy Seamans

Physical Therapist Assistant 103
Brenda White Turner

Respiratory Therapy Assisting 113
Sue Hollander and Mary Rose Amidjaya

Building and Construction Trades

Cabinetmaking and Woodworking 121
Patty Miller

Construction, Remodeling, and Repair 127
Barbara Edwards

Electrical Technology 141
Dan Haley

Heating, Ventilation, and Air Conditioning 150
Steven Self

Masonry 155
Steven Self

Plumbing Technology 159
Joanne Kim and Tori Beyer

Business

Accounting, Bookkeeping, and Income Tax Preparation 165
Melinda Townsel

Computers and Data Processing 173
Rebecca Kroll

Cosmetology 185
Barbara A. Heiffner

Culinary Arts/Food Service Management 190
Anne Marsh Fields

Hospitality Management 201
Julie Pinnell

Office Technologies 208
Natalie Diamond

Real Estate 213
Antoinette Byers

Small Business and Entrepreneurship 222
Barbara Alper

Travel and Tourism 226
Trudy Cleveland

Communications/Production Technologies

Journalism 233
Marit MacArthur

Public Relations 257
Jane L. Crocker

Radio and Television Production Technologies 263
Red Wassenich

Criminal Justice and Law

Court Reporting 273
Louise Treff

Law Enforcement and Criminal Justice 284
Nancy Tenhet

Legal Secretarial 293
Susan L. Brant

Paralegal 297
Susan Stussy

Education

Early Childhood Education 309
Peggy Holleman

Library Technology 328
Diane J. Turner

Engineering and Technology

Aircraft Technologies 345
Marcia Miller

Automotive Technology 351
Ellen Tiedrich

Aviation 357
Joyce Hopkins

Diesel Mechanics and Heavy Equipment Operations 362
Marcia Miller

Environmental Technologies 366
Kathy C. O'Gorman

Industrial and Mechanical Design (including CAD and CAM) 371
Debbie Bogenschutz

Laser/Electrooptics Technology 376
Debbie Bogenschutz

Small Engine Repair 380
Barbara A. Heiffner

Surveying 383
Susan L. Brant

Welding 388
Karen Fischer and Dona Mitoma

Graphic and Apparel Arts (Applied Arts)

Desktop Publishing and Printing Technologies 395
  Charles R. James

Fashion Design and Apparel Arts 411
  Dianna Thor

Interior Design 428
  Sue Swanson

Photography 443
  Kate Hickey and Robert Johnston

Visual Communication/Graphic Arts 452
  Judy Goodyear

Sciences

Agricultural Sciences and Agronomy 475
  Mary E. Coffin, Sandra J. Donovan, Crystal Havely Stratton, Debora Thomas, and Kelly Willmarth

Fire Services 491
  Terri Propes

Forestry 497
  Charlotte Cooper

Funeral Services 510
  Harolyn Cumlet and Lenora Lockett

Landscape Horticulture Technologies 517
  Connie Barber

Veterinary Technology 523
  Dr. Gail Staines

Social Services

Alcohol and Drug Abuse Counseling 533
  Pam Kessinger

Social Work and Human Services 551
  Lynn Namsick

Name Index 577
Title Index 594
Journal Index 620

Introduction

Vocational and Technical Resources for Community College Libraries presents selected annotated bibliographies for use in the selection and evaluation of collections that support vocational and technical curricular areas. The concept for this work originated in 1992 with the Book Review Committee of the Community and Junior College Libraries Section (CJCLS) of the Association of College and Research Libraries. The function of this committee is to encourage the reviewing of materials of interest to community, technical, and junior college libraries. Recognizing the acute lack of review sources for vocational and technical areas, committee members felt that a published list of resources was greatly needed to aid librarians and faculty in collection development. The committee presented the editorial concept to the CJCLS section and to *Choice* and found full support for this project.

Vocational and Technical Programs: Determination of Coverage

The major programs considered for inclusion in this publication were originally drawn from the *Chronicle Two Year College Databook 1991-1992*, and then categorized into ten discipline areas. This list was reviewed and revised by the CJCLS book review committee. In addition, a survey was sent to all CJCLS section members requesting subject category revisions and additions, and the scope of the materials to be included (audiovisual, CD-ROMs, journals, software, videos, and other). An interest survey was also included to solicit qualified bibliographers to prepare chapters. All suggestions received from the survey were carefully weighed by the CJCLS Book Review Committee.

Bibliographic Scope

These annotated bibliographies present a collection of resources, both print and nonprint, of value to libraries supporting vocational and technical curricula in community colleges or in vocational and technical institutions. The focus is on selected reference and support materials that supplement and complement the vocational curriculum in community and junior colleges. Textbooks were generally not included, except in areas where other materials were generally not available, or in areas such as nursing, where textbooks are the predominant resource published.

This bibliography serves an audience of librarians, undergraduate students, vocational and technical students, faculty, and the general public looking for information on resources relevant to these programs. The focus of this publication is on recent English language materials, published primarily in North America between 1988 and 1993. Titles released early in 1994 were also included.

These bibliographies are not meant to be exhaustive, but rather representative of the types of materials that may be used to support a curricular area. For many subject areas, availability of a work occasionally was a factor in the decision to include or exclude titles from the bibliography; some limitations faced by the bibliographers were restrictions imposed by interlibrary loan agreements on current titles, and publishers' policies concerning desk review copies. Many publishers, for example, would not release review copies except to faculty members who were considering a book for a text. In addition, publishers of nonprint materials (especially videocassettes, audiocassettes, and software), were often reluctant to "lend" review materials.

Organization and Style

This bibliography is organized by broad curricular areas, with specific chapters within the range of subject areas. For example, the area of Allied Health includes chapters such as Dental Services, Diagnostic Technologies, Emergency Medical Services, Nursing, and Physical Therapist Assistant, to name a few. Chapter titles and subheadings were determined by CJCLS members and the Book Review Committee and finalized by the General Editor, the Editor and Publisher of *Choice,* and approved by the *Choice* Editorial Board. Each chapter's print and nonprint resources are preceded by an introductory essay stating the scope of the chapter and current trends in the field, a list of accrediting and/or certifying agencies (or selected associations), and a list of selected journals.

Bibliographers and editing staff used *The Chicago Manual of Style* for the general preparation of entries and grammatical style. Some of the conventional style recommendations were adapted to meet the needs of this publication. For example, the form of the entry does not correspond to the *Chicago Manual of Style* or *AACR2R* cataloging rules. In order to facilitate indexing, and to collocate various editions and related works by editors and compilers, the personal names were moved to the top line and serve as the main access point. The form of the name reflects the author's established form of name from OCLC.

To ensure consistency of word use, the following authorities were consulted: *Webster's Third New International Dictionary of the English Language, Unabridged* (Merriam-Webster, 1986), and *12,000 Words: A Supplement to Webster's Third New International Dictionary* (Merriam-Webster, 1986). Terms not found in either of these sources due to their specialized nature or recency were verified in subject dictionaries, "new word" dictionaries, or the OCLC Authority File. In addition, an in-house list for publisher's authority names was developed to achieve a unified presentation of names and spelling.

Other elements of bibliographic description information were verified in OCLC and adjusted accordingly, and when cataloging information was not available, verification was made in *Books in Print*. Because of the intrinsic nature of the material, many resources were not listed in either source; in that case, the bibliographer was asked to reverify the bibliographic information.

Selection of Bibliographers and Referees

Selection of bibliographers occurred after an interest survey of CJCLS members was conducted. In addition, many bibliographers were referred by section members. Since community college librarians are closely involved with the instructional processes, and can draw from the program strengths and faculty in specific discipline areas, community college librarians were chosen to prepare the bibliographies. An impartial referee was desirable for the evaluation of the manuscript, and a call for referees was extended on the COMMCOLL listserv on Internet. Responses came in from throughout the United States, from the northern part of Canada to the southern states. Faculty referees were teachers in the discipline area or subject specialists and agreed to respond to a series of questions regarding the bibliography and introduction in general, as well as to specific questions on the scope, quality, and quantity of titles included. In some disciplines, referees were drawn from the Editor's local contacts and referrals from other community college librarians.

Indexes

This volume concludes with computer-generated indexes by author and by title for print, nonprint, and journal publications. The Name Index provides access to the first-named author, editor, or compiler, or the sponsoring organization. The Title Index lists the title proper of all entries for print and nonprint works. The Journal Index lists all journals cited in the "Selected Journals" section of each chapter.

Acknowledgments

This work would not have come to completion without the cooperation and efforts of many people. I am deeply grateful to all the community college librarians who gave so generously of their time and expertise to this project. Many of them worked outside the realm of their day-to-day responsibilities and strove to obtain as many relevant materials as possible, in spite of the limits of interlibrary loan and review copy policies. In addition, I would like to thank the referees for their valuable insights from the instructor's perspective; their input kept the bibliographer's work on target and relevant to the needs of students in these programs.

This project reflects the continued support of the *Choice* editorial team. Pat Sabosik, *Choice* Editor & Publisher in the planning phases of this project, and the *Choice* Editorial Board recognized the need for this project from its inception and provided valuable financial and editorial support. Francine Graf, Interim *Choice* Editor & Publisher, continued that philosophy into the progression of the project and provided many needed insights and guidance as the project moved to completion. Acknowledgment is also extended to other *Choice* staff heavily involved with this project: first and foremost, Lisa Gross, Manager, Computer Services, who finalized the page layout and entries and indexes for the entire manuscript; Beth Vanderstar, Administrative Assistant, who handled correspondence with contributors and assisted with correction; Editor Judith Douville, who reviewed and corrected formatted pages and indexes; and Clare Hoover, Production Manager, who coordinated many aspects of this project.

In addition to the bibliographers, referees, and *Choice*, the Editor relied on a local editorial team for their vital contributions:

- Tori Beyer, for her watchful proofreading and editorial assistance;
- Carol Cooper, for her keen attention to detail and style interpretations, for her editorial guidance and continued support;
- Jennifer Cooper and Susan Miali, for their painstaking corrections to manuscripts;
- Ann Pibel for her persistent and careful bibliographic verification;
- David Dowell, for his support of this project from its inception to completion;
- to the staff and resources of the Pasadena City College Library;
- Jeff Laun, for his technological insights, layout and manuscript preparation expertise, timesaving macros, and unwavering support.

And finally to my daughters, Lissa, Mindy, and Amy, I thank them for their patient acceptance and support of this effort throughout the long days and nights of editing.

Referees

In addition to the faculty members who are listed in the chapter introductions, the following faculty and subject specialists acted as referees for specific discipline areas. Their contributions and insights validated the work of the bibliographers and enhanced the manuscripts by providing a teacher's firsthand perspective.

Mr. John Babbo
Olympic College
Bremerton, Wash.
Alcohol and Drug Abuse Counseling

Dr. Daniel Bellack
Trident Technical College
Charleston, S.C.
Early Childhood Education

Dr. James Bickley
Pasadena City College
Pasadena, Calif.
Social and Human Services

Ms. Anita Bobich
Pasadena City College
Pasadena, Calif.
Dental Auxiliary Sciences

Ms. Mikki Bolliger
Pasadena City College
Pasadena, Calif.
Journalism

Ms. Joan Brandlin
Pasadena City College
Pasadena, Calif.
Dental Auxiliary Sciences

Mr. Jack Compton
College of the Canyons
Valencia, Calif.
Welding

Mr. Bob Cooper
Central Oregon Community College
Bend, Orre.
Forestry Sciences

Ms. Carlene Decker
Parkland College/Veterinary Technology
Champaign, Ill.
Veterinary Technology and Assisting

Mr. John Deeming
Central Oregon Community College
Bend, Ore.
Forestry Sciences

Ms. Joan DeFato
Los Angeles State and County Arboretum
Arcadia, Calif.
Horticulture

Dr. David Dowell
Pasadena City College
Pasadena, Calif.
Library Technician

Dr. Ahni Foley
Pasadena City College
Pasadena, Calif.
Office Technologies

Mr. Vincent Foster
Austin Community College
Austin, Tex.
Heating, Ventilating and Air Conditioning

Ms. Carolee Freer
Cypress College
Cyprus, Calif.
Court Reporting

Ms. Ann Gunkel
Cincinnati State Technical and Community College
Cincinnati, Ohio
Environmental Technologies

Mr. Richard Guymon
Maple Woods Community College
Kansas City, Mo.
Law Enforcement/Police Science

Mr. Ward Harder
Motlow State Community College
Tullahoma, Tenn.
Accounting/Bookkeeping /IncomeTax Preparation; Small Business and Entrepreneurship; Travel and Tourism

Ms. Amy Hart
Fenway Libraries Online
Boston, Mass.
Interior Design

Mr. Ted James
Pasadena City College
Pasadena, Calif.
Electrical Technology

Ms. Eleanor Kenney
Pasadena City College
Pasadena, Calif.
Emergency Medical Services

Ms. Carol Kizer
Columbus State Community College
Columbus, Ohio
Culinary Arts/Food Services; Hospitality Management

Mr. Ken Kolle
Pasadena City College
Pasadena, Calif.
Cosmetology

Mr. Stan Kong
Pasadena City College
Pasadena, Calif.
Visual Communications and Graphic Arts

Mr. Jack L'Heureux
Western Nebraska Community College
Sidney, Neb.
Aircraft Mechanics; Heavy Equipment Operation

Mr. Mike Lamb
Glasser-Miller-Lamb
Arcadia, Calif.
Funeral Services

Dr. Lily Hee Lau
Pasadena City College
Pasadena, Calif.
Nursing

Mr. Wai-Min Liu
Pasadena City College
Pasadena, Calif.
Laser/Electrooptics Technology

Mr. Dave Lucas
Cuyahoga Community College
Parma, Ohio
Respiratory Therapy

Ms. Penny McLain
Pasadena City College
Pasadena, Calif.
Diagnostic Technologies

Mr. Bret Michalski
Central Oregon Community College
Bend, Ore.
Forestry Sciences

Mr. Gene Murry
Pasadena City College
Pasadena, Calif.
Fire Science

Ms. Joyce Nakano
Pasadena City College
Pasadena, Calif.
Medical Assisting and Medical Records

Dr. Richard Niederhof
Central Oregon Community College
Bend, Ore.
Forestry Sciences

Ms. Jennifer Orsini
Pasadena City College
Pasadena, Calif.
Fashion Design and Apparel Arts

Ms. Marijean Piorkowski
Cerritos College
Norwalk, Calif.
Physical Therapy Assistant

Ms. Patricia Porterfield
St. Charles County Community College
St. Peters, Mo.
Occupational Therapy

Mr. Phil Salomon
Pasadena City College
Pasadena, Calif.
Surveying

Mr. John Schaeffer
Central Oregon Community College
Bend, Ore.
Industrial and Mechanical Design

Mr. Michael Sellmeyer
St. Charles County Community College
St. Peters, Mo.
Printing/Desktop Publishing

Dr. Brian R. Shmaefsky
Kingwood College/N. Harris Montgomery
Community College District
Kingswood, Tex.
Agricultural Sciences/Agronomy

Mr. Ron Smallwood
Northern Lights College
Fort Nelson, BC, Canada
Computers and Data Processing

Mr. Richard Smith
Gloucester County College
Sewell, N.J.
Public Relations

Ms. Sandra Smith
Gloucester County Institute of Technology
Sewell, N.J.
Automotive Technology

Ms. Sharon Thatcher
DeKalb Technical Institute
Clarkston, Ga.
Medical Lab Technology

Ms. Julia Trzk
Cuyahoga Community College
Parma, Ohio
Legal Secretarial; Paralegal

Mr. Gary Vincent
San Jacinto College
Pasadena, Tex.
Occupational Health and Safety

Ms. Yeimei Wang
Glendale Community College
Glendale, Calif.
Dietetics/Nutrition

Mr. John Wright
Metropolitan Community College
Omaha, Neb.
Aircraft Mechanics; Heavy Equipment Operation; Aviation

Ms. Debbie Zimmerman
Motlow State Community College
Tullahoma, Tenn.
Radio, Television and Studio Production Technologies

Allied Health

Dental Auxiliary Careers: Dental Hygiene, Dental Assisting and Dental Lab Technology 1
Nancy Buchanan

Diagnostic Technologies 21
Isabel C. Hernandez

Dietetics and Nutrition 33
Judy Caramanica

Emergency Medical Services 40
Grace Ekins

Medical Assisting and Medical Records 47
Tori Beyer and Michelle Strazer

Medical Laboratory Technology 57
Ann Cannon

Nursing 67
Elisa Abella and Isabel C. Hernandez

Occupational Health and Safety 91
Terri Propes

Occupational Therapy Assisting 97
Nancy Seamans

Physical Therapist Assistant 103
Brenda White Turner

Respiratory Therapy Assisting 113
Sue Hollander and Mary Rose Amidijaya

Dental Auxiliary Careers: Dental Hygiene, Dental Assisting, and Dental Lab Technology

Nancy Buchanan
Pima Community College
Tucson, Arizona

Introduction

Dentistry is the profession that maintains, improves, and corrects the health of the teeth and the supporting oral structures. Dentists are the main practitioners and have final responsibility for all dental services. People employed in the allied dental health field are called dental auxiliaries. There are three kinds of dental auxiliary careers: dental hygiene, dental assisting, and dental laboratory technology. Each handles specific functions that enable the dentist to devote more time to the specialized activities of chairside operative and restorative dentistry.

Dental Hygiene

Dental hygiene is a program of study that trains students to become licensed professionals. Hygienists work primarily under the supervision of dentists, providing preventive dental care and teaching patients how to practice good oral hygiene.

Depending on state legal restrictions, dental hygienists may provide a wide range of services. They examine the patient's teeth and mouth. They keep records of medical and dental histories, charting conditions of decay and disease for diagnosis by the dentist. They remove stains, calculus, and plaque from above and below the gumline, polish teeth, and apply fluorides and pit and fissure sealants to prevent caries. They may also take and develop X-rays, place temporary fillings and periodontal dressings, remove sutures, and polish and recontour amalgam restorations. Their duties also include infection control procedures such as sterilization and disinfection of instruments and surfaces. In some states, hygienists administer local anesthetics and nitrous oxide/oxygen analgesia. Dental hygienists also teach patients how to prevent dental problems, emphasizing good nutrition, proper brushing and flossing, and the importance of regular checkups. Some develop and promote community dental health programs. Dental hygienists must work well with others, particularly patients who may be under stress. Attentiveness to detail and manual dexterity are also important attributes for this occupation.

Most dental hygienists are employed in private dental offices. Part-time employment is common. There are also job opportunities in school systems, public health agencies, clinics, hospitals, nursing homes, health maintenance organizations, private industry, as civilian employees of the U.S. armed forces, and for those with advanced degrees, as instructors in dental hygiene schools.

Dental hygienists must be graduates of a school of dental hygiene accredited by the Commission on Dental Accreditation. To enroll in such a school, a high school diploma is a minimum requirement. Training can be obtained in a two-year certificate or associate degree program offered at a community college or a vocational/technical school. Completion of a certificate or associate degree program is sufficient to practice in a private dental office, but a bachelor's or master's degree is usually required for positions that involve teaching, research, or practice in public or school health programs. The curriculum in schools of dental hygiene consists of courses in the basic sciences, dental sciences, and liberal arts. They offer laboratory work, clinical experience, and classroom instruction.

With the exception of one state, dental hygienists must be licensed by the state in which they practice. To obtain a license, students must pass a written and clinical examination. Part of the state licensing requirements can be completed by passing the National Board of Dental Examiners written exam. Upon successful completion of license requirements, a student becomes a Registered Dental Hygienist (RDH). Continuing education is required to maintain licensure. The State Board of Dental Examiners in each state can supply licensing information.

Dental Assisting

Dental assistants perform clinical, clerical, and laboratory duties in a dental office under the supervision of dentists. In their clinical duties they help prepare the patient for the dental exam, obtain dental records, hand the dentist the proper instruments and materials, and keep patients' mouths dry and clear. Assistants also sterilize and disinfect instruments and equipment, mix compounds for cleaning or filling teeth, take and process X-rays, and prepare materials for making impressions and restorations. State laws determine which clinical tasks a dental assistant is allowed to perform.

As part of their clerical duties, dental assistants schedule appointments, keep records, process bills, receive payments, order supplies, and perform other administrative and general office tasks. Those with laboratory duties make casts of the teeth and mouth from impressions, clean and polish removable appliances, and make temporary crowns.

Most dental assistants work in private offices for one or more dentists. Others are employed in public health departments, clinics, hospitals, or dental schools; as dental office managers; or as insurance company representatives or in sales positions.

Dental assisting positions tend to be entry-level. Many assistants learn their skills on the job in positions requiring little or no experience and no education beyond high school. Others, however, go on to receive training in trade schools, technical institutes, and community colleges that offer dental assisting programs. Armed forces schools also train dental assistants. Most programs take one year or less and lead to a certificate or diploma.

Community colleges award two-year associate degrees. The American Dental Association has an accredited correspondence course, as does the University of Kentucky College of Dentistry. Accredited dental assisting programs include classroom, laboratory, and preclinical instruction in skills and theory. Students gain clinical experience by working in dental schools and local dental offices and clinics that are affiliated with programs. Further education is required for advancement to positions in dental assisting education. Dental assistants who wish to become hygienists must enroll in a dental hygiene program. High school students may prepare for this career by taking courses in general science, biology, health, chemistry, and office practices. Dentists recruit assistants who are reliable, able to follow directions, work well with others, and have manual dexterity.

Some states offer licensure to qualified dental assistants with extended training and education. Upon passing the licensure exam, they become Registered Dental Assistants. Graduates of accredited programs may wish to obtain certification from the Dental Assisting National Board by passing an examination to become Certified Dental Assistants (CDA), but this is not always required for employment. To keep CDA credentials, individuals must undergo periodic retesting or acquire more education.

Dental Laboratory Technology

Dental lab technicians are never seen by most patients, but patients benefit immensely from their behind-the-scenes activities. These highly skilled craftpersons work in laboratories, making and repairing dental appliances according to dentists' and orthodontists' prescriptions and impressions.

Dental lab technicians construct or fabricate artificial teeth, fixed bridges, removable dentures and partials, crowns, inlays, and orthodontic appliances. They use specialized equipment and precision instruments on a range of materials; e.g., gold, silver, stainless steel, porcelain, plastics, and acrylics. The work is very delicate and time-consuming. They may specialize in one of five areas: orthodontic appliances, crowns and bridges, complete dentures, partial dentures, or ceramics.

Most dental lab technicians work in commercial dental laboratories. Others may be employed by private dental practices, federal and state health agencies, hospitals, and clinics. Some work as sales representatives or in product development for dental companies. For most technicians, opening one's own laboratory is the route to higher earnings in this field. They may also aspire to become clinical instructors.

There are two ways of becoming a Dental Lab technician, and both require a high school diploma. An apprenticeship program involves on-the-job training in a commercial dental lab. Apprenticeships last three to four years and the trainee receives a minimal salary. The academic program involves completing a two-year certificate or associate degree program in a community college, vocational/technical school, or trade school. Training is also available through the U.S. armed forces. Programs provide classroom instruction in dental materials science, oral anatomy, fabrication procedures, ethics, and related subjects. Students are given supervised practical experience in the school or in an associated dental laboratory. Useful high school preparation includes courses in chemistry, shop, mechanical drawing, art, and ceramics. Students should have a high degree of manual dexterity, a good sense of color perception, artistic talent, and an affinity for accurate, detailed work.

Dental lab technicians with appropriate training and experience can become Certified Dental Technicians (CDT). Although this is not mandatory for employment, it is recommended. Certification is offered by the National Board for Certification in Dental Technology. For initial certification, candidates must pass a basic, comprehensive written examination. Upon successful completion of the examination, the candidate then takes a written and practical examination in one of five lab specialties. Specific continuing education requirements must be met to maintain certification status.

This bibliography has been written for students, instructors, and librarians in community colleges, vocational/technical institutes, and trade schools, as well as for those working in the field. Some resources listed in this bibliography are appropriate for more than one field of study; the following codes, which appear after each annotation, indicate the applicability of each resource for different fields of study: DH (Dental Hygiene), DA (Dental Assisting), DL (Dental Lab Technology), and ALL (general works).

Accreditation

Commission on Dental Accreditation (in conjunction with the American Dental Association)
444 N. Michigan Ave., Suite 3400
Chicago, Ill. 60611

Selected Associations

✦ Dental Hygienist

American Dental Hygienists' Association
444 N. Michigan Ave., Suite 3400
Chicago, Ill. 60611

✦ Dental Assistant

American Dental Assistants Association
919 N. Michigan Ave., Suite 3400
Chicago, Ill. 60611

National Dental Assistants Association
5506 Connecticut Ave. NW, Suite 24
Washington, D.C. 20015

Dental Assisting National Board
216 E. Ontario St.
Chicago, Ill. 60611

✦ Dental Lab Technician

National Board for Certification in Dental Technology
555 E. Braddock Rd.
Alexandria, Va. 22305

National Association of Dental Laboratories
3801 Mt. Vernon Ave.
Alexandria, Va. 22305

Selected Journals

Access.
American Dental Hygienists Association. Monthly (except May-June and September-October issues). ISSN 1050-0758

Dental Abstracts.
Mosby. Monthly. ISSN 0011-8486

Dental Assistant Journal.
American Dental Hygienists Association. Quarterly.
ISSN 0011-8508

Dental Clinics of North America.
Saunders. Quarterly. ISSN 0011-8532

Dental Economics.
PennWell Pub. Co. Monthly. ISSN 0011-8583

Dental Lab Products.
Medec Communications. Bimonthly. ISSN 0146-9738

Dental Teamwork.
American Dental Association. Bimonthly. ISSN 0895-318X

Dentistry Today.
Enterprise Communications. Nine times per year. ISSN 8750-2186

General Dentistry.
Academy of General Dentistry. Bimonthly. ISSN 0363-6771

Journal of Dental Education.
American Association of Dental Schools. Monthly.
ISSN 0022-0337

Journal of Dental Hygiene: JDH.
American Dental Hygienists Association. Ten times per year.
ISSN 1043-254X

Journal of Periodontology.
American Academy of Periodontology. Monthly. ISSN 0022-3492

Journal of Prosthetic Dentistry.
Mosby. Monthly. ISSN 0022-3913

Journal of Public Health Dentistry.
American Association of Dental Schools. Quarterly.
ISSN 0022-4006

Journal of the American College of Dentists.
The College. Quarterly. ISSN 0002-7979

Journal of the American Dental Association.
The Association. Monthly. ISSN 0002-8177

LMT: Lab Management Today.
Dental Lab Publications. Monthly. ISSN 1058-7845

RDH: [The National Magazine for Dental Hygiene Professionals].
Stevens Pub. Co. Bimonthly. ISSN 0279-7720

Trends and Techniques in the Contemporary Dental Laboratory.
National Association of Dental Laboratories. Bimonthly.
ISSN 0746-8962

Many state dental associations publish journals that provide valuable articles and perspectives from within the state.

Print Materials

✦ Reference Works

American Dental Association
CDT-1: Current Dental Terminology, A User's Manual
Chicago, American Dental Association, 1991. 108p.
   This handbook of dental procedure codes incorporates the seventh revision of the American Dental Association's *Code on Dental Procedures and Nomenclature.* Spiral bound. ALL

Freiberg, Marcos A.
Bilingual Dictionary of Dental Terms: Spanish-English=Diccionario Bilingüe de Términos Odontológicos: Inglés-Español
San Francisco, Ism Press, 1990. 111p. ISBN 0910383227
   An Argentine dentist now living in the United States has compiled a dictionary of technical terms in Spanish and English for dental professionals. Provides pronunciation guides. ALL

Glossary of Prosthodontic Terms
6th ed. St. Louis, Mo., Mosby, 1994. 72p.
   This standard lexicon for prosthetic dentistry has been prepared by the Academy of Prosthodontics and the Education Committee of the National Association of Dental Laboratories. (Available from Mosby's Journal Reprint Department.) DL

Jablonski, Stanley
Jablonski's Dictionary of Dentistry
Malabar, Fla., Krieger Publishing, 1992. 887p. ISBN 0894644777
   Defines terminology in all specialties of dentistry and allied fields of science, technology, and health care, including dental practice management and health insurance. Includes terms no longer in use but of historical interest. ALL

Santos, Virginia
Spanish for Dentistry
Virginia Santos, 1991. 87p.
   Presents Spanish phrases and sentences that dental professionals might need to communicate with Spanish-speaking patients. It is divided into sections such as: how to take a case history, chairside procedures, specialties, and preventive dentistry. DH DA

Silverman, Sol
Color Atlas of Oral Manifestations of AIDS
Philadelphia, Decker, 1989. 113p. ISBN 1556641990
   With an emphasis on photographs, this work is designed to give clinicians a visual approach to diagnosis and treatment of oral lesions associated with AIDS. It provides an overview of the epidemiology of AIDS, provides information on the biology of the infection, and describes progression of the disease. ALL

Williams, David F., ed.
Concise Encyclopedia of Medical and Dental Materials
New York, Pergamon Press, 1990. 412p. ISBN 0080361943
   Sixty technical articles cover broad topics in the field of medical and dental materials science and are accompanied by extensive cross-references and bibliographies. Includes a detailed subject index. ALL

Wischnitzer, Saul
Barron's Guide to Medical and Dental Schools
6th ed. New York, Barron's, 1993. 374p. ISBN 0812016319
   This handbook helps students choose and gain admission to the medical or dental school of their choice. Includes career guidance suggestions, facts and statistics on medical and dental careers, the latest admissions data, full-length model Medical College Admission Test and Dental Admission Test questions, profiles of the schools, and information on how to choose a specialty. ALL

Zwemer, Thomas J., ed.
Boucher's Clinical Dental Terminology: A Glossary of Accepted Terms in All Disciplines of Dentistry
4th ed. St. Louis, Mo., Mosby, 1993. 433p. ISBN 0801667062

Presents the current consensus on the meaning and proper usage of dental terminology. Appendices include abbreviations, codes on dental procedures and nomenclature, and medical tables. ALL

✦ General Works

Anderson, Pauline Carter, and Martha Burkard
The Dental Assistant
6th ed. Delmar Publishers, 1995, 768p. ISBN 0827352816

A text and workbook for the entry-level dental assistant. Emphasis is on the rationale and development of manipulative skills, with a solid foundation of theoretical knowledge. Introduces terminology, facts, and methods as they apply to clinical or lab procedures. Includes review exercises.

Andujo, Emily, ed.
Appleton and Lange's Review for the Dental Assistant
3rd ed. Norwalk, Conn., Appleton and Lange, 1992. 270p. ISBN 0838501354

Provides approximately 1,200 questions in the Dental Assisting National Board Certification Examination format, with keyed answers. Easy to read chapters offer a course synopsis covering the basics of dental, clinical, and behavioral sciences. Provides detailed information on the exam and tips for effective test-taking. DA

Aoshima, Hitoshi
Collection of Ceramic Works: A Communication Tool for the Dental Office and Laboratory
Chicago, Quintessence Publishing, 1992. 91p. ISBN 4874173861

This visual guide gives dental technicians a blueprint of the characteristics intrinsic to each tooth. This information is used to determine the proper tooth shape and color and is also helpful in illustrating prospective treatment to a patient. DH DL

Ash, Major M.
Oral Pathology: An Introduction to General and Oral Pathology for Hygienists
6th ed. Philadelphia, Lea and Febiger, 1992. 321p.
ISBN 0812114345

This extensively revised and rewritten edition of the classic text takes into account new information about disease patterns, immune response, and aging. Ash provides a broad background in oral pathology for the dental hygienist. DH

Ash, Major M.
Wheeler's Dental Anatomy, Physiology, and Occlusion
7th ed. Philadelphia, Saunders, 1993. 478p. ISBN 0721643744

This revision of a classic text reflects changes in recent literature. Introduces the orofacial complex, beginning with nomenclature, then leads into tooth development, dental physiology, and provides detailed descriptions of each tooth group. DH

Avery, James K.
Essentials of Oral Histology and Embryology: A Clinical Approach
St. Louis, Mo., Mosby, 1992. 224p. ISBN 1556641885

Avery's concise textbook and atlas of developmental and structural microscopic anatomy covers cell, tissue, and organ development and function. Electron microscopic photographs assist with learning and clarify content. Includes glossary and index. DH

Barton, Roger E., Stephen R. Matteson, and Richard E. Richardson, eds.
Dental Assistant
6th ed. Philadelphia, Lea and Febiger, 1988. 655p.
ISBN 0812111419

Prepared by the faculty of the University of North Carolina School of Dentistry, this edition has three new chapters: stress management for the dental patient; the elderly and handicapped patient; and hospital dentistry. Stresses the basic operations of dentistry, the dental office, and the general duties and deportment of the dental assistant. Includes review questions at the end of each chapter. DA

Bell, Welden E.
Orofacial Pains: Classification, Diagnosis, Management
4th ed. Chicago, Year Book Medical Publishers, 1989. 448p.
ISBN 0815106572

Familiarizes students with knowledge of pain mechanisms and how to use that knowledge at a clinical level. Reviews characteristics of orofacial pain syndromes and the problems caused by chronic pain. DH

Berkovitz, B. K. B., G. R. Holland, and B. J. Moxham
Color Atlas and Textbook of Oral Anatomy, Histology, and Embryology
2nd ed. St. Louis, Mo., Mosby, 1992. 328p. ISBN 0815106971

Considered to be one of the most useful anatomical textbooks for students. This is an extensive atlas of oral histology, presenting detailed information on every topic related to this discipline. Brief text accompanies many illustrations. DH

Berns, Joel M.
Understanding Periodontal Diseases
2nd ed. Carol Stream, Ill., Quintessence Books, 1993. 73p.
ISBN 0867152397

A patient education resource, this updated edition impresses upon patients the importance of regular dental checkups. It discusses periodontal diseases: what they are, how they start, how they can be prevented, and how they are treated. Includes color drawings, large type, and clear language. DA DH

Brand, Richard W., and Donald E. Isselhard
Anatomy of Orofacial Structures
4th ed. St. Louis, Mo., Mosby, 1990. 496p. ISBN 0801635055

Written for the beginner studying anatomy of the oral cavity, this illustrated text covers oral histology and embryology, head and neck anatomy, and dental anatomy. Contains a set of removable flash cards of the teeth. Includes tests and references, a glossary, and an index. DH

Burt, Brian A., and Stephen A. Eklund
Dentistry, Dental Practice, and the Community
4th ed. Philadelphia, Saunders, 1992. 339p. ISBN 0721631959
   This new edition of a classic public health text provides substantial revisions and new material on infection control, pit and fissure sealants, smokeless tobacco, and a chapter on oral health care in Canada. Includes introductions and summaries for each chapter. ALL

Charbeneau, Gerald T.
Principles and Practice of Operative Dentistry
3rd ed. Philadelphia, Lea and Febiger, 1988. 496p. ISBN 0812111354
   Charbeneau brings together the traditional subjects of operative dentistry from a conservative point of view, within the concepts of occlusal function and periodontal health. Emphasizes disease prevention as the ultimate goal. Includes glossary and appendix with quality evaluation criteria. DA

Chiche, Gerard J., and Alain Pinault
Esthetics of Anterior Fixed Prosthodontics
Chicago, Quintessence Publishing, 1994. 202p. ISBN 0867152583
   Presents the principles and methods of achieving the best possible esthetic results. Teaches how the dental technician can work effectively with the restorative dentist and the orthodontist. Superior color illustrations. DL

Chilo, V., Monica E. Strong, and G. Borea
Life Threatening Emergencies in Dentistry
[Italy?], Ishiyaku EuroAmerica, 1988. 187p. ISBN 0912791691
   When an unexpected emergency occurs in the dental chair, it is difficult to consult complex medical texts. This book is a succinct, easy-to-use compilation of medical information that is helpful when identifying and handling difficulties that require emergency treatment, e.g., epileptic seizures, cardiac arrest, and pregnancy problems. DA DH

Collins, W. J. N., and T. F. Walsh
Handbook for Dental Hygienists
3rd ed. Boston, Mass., Butterworth-Heinemann, 1991. 350p. ISBN 0723609802
   This concise but comprehensive textbook covering current practice is designed for students as well as those currently working in the field. DH

Cottone, James A., G. T. Terezhalmy, and John A. Molinari
Practical Infection Control in Dentistry
Philadelphia, Lea and Febiger, 1991. 286p. ISBN 0812113268
   Cottone's purpose is to aid in the development of a single set of policies and procedures that can be used in infection control with all patients. Diseases discussed include viral hepatitis, HIV infection, and AIDS. Effective infection control practices are thoroughly covered. ALL

Cowan, Fred F.
Dental Pharmacology
2nd ed. Philadelphia, Lea and Febiger, 1992. 446p. ISBN 0812113853
   Presents a concise summary of all the significant drug classes used as adjuncts in dental care, including anesthetics, anti-infectives, analgesics, sedatives, and chemotherapeutic agents. A new section focuses on drug abuse, emergency drugs, and mercury toxicity. DA DH

Craig, Robert G., ed.
Restorative Dental Materials
9th ed. St. Louis, Mo., Mosby, 1993. 581p. ISBN 0801668727
   Provides students with the fundamental and applied information necessary to understand the basis for the laboratory and clinical use of materials. Discusses the effect of manipulation and technical procedures on materials properties. This edition updates the chapter on biocompatibility of materials and contains the latest information on toxicity of mercury and amalgam. DL

Craig, Robert G., William J. O'Brien, and John M. Powers
Dental Materials: Properties and Manipulation
5th ed. St. Louis, Mo., Mosby, 1992. 328p. ISBN 0801610753
   This text provides information on the use and handling of dental materials for bridges, crowns, and amalgams. Presents an in-depth treatment of the principles that control the properties and performance of dental products. Provides section summaries and self-tests. DL

Dale, Barry G., and Kenneth W. Aschheim, eds.
Esthetic Dentistry: A Clinical Approach to Techniques and Materials
Philadelphia, Lea and Febiger, 1993. 510p. ISBN 0812114671
   Contributors from a variety of specialties detail the clinical aspects of cosmetic dentistry. Covers such diverse topics as dental photography, color science, psychology, jurisprudence, marketing, custom staining, and advanced technologies. ALL

Darby, Michele Leonardi, ed.
Mosby's Comprehensive Review of Dental Hygiene
3rd ed. St. Louis, Mo., Mosby, 1994. 864p. ISBN 0801679656
   Selectively reviews the current body of knowledge in dental hygiene in an easy-to-use, outlined, question-and-answer format. Prepares the reader to take the National Board Dental Hygiene Examination and has a "simulated" sample exam with rationales for right and wrong answers. DH

Dawson, Peter E.
Evaluation, Diagnosis, and Treatment of Occlusal Problems
2nd ed. St. Louis, Mo., Mosby, 1989. 633p. ISBN 0801627885
   Interest in occlusion and the temporomandibular joint is on the increase. This text simplifies a complex subject by attempting to explain how the masticatory system works. Discusses diagnostic methods and controversial views, evaluated pro and con. Includes guidelines for treatment of every type of occlusal problem. DH

DeBiase, Christina B.
Dental Health Education: Theory and Practice
Philadelphia, Lea and Febiger, 1991. 314p. ISBN 0812113667
   Presents a comprehensive view of health care issues that affect the population through various stages of life and their relation to oral health. Among the issues addressed are: prenatal fluorides, teething, child abuse, eating disorders, designer orthodontic appliances, smokeless tobacco, AIDS, Alzheimer's disease, and cancer care. DH

De Lyre, Wolf R., and Orlen N. Johnson
Essentials of Dental Radiography for Dental Assistants and Hygienists
4th ed. Norwalk, Conn., Appleton and Lange, 1990. 446p.
ISBN 0838524680
    This in-depth introduction to dental radiography covers theories, basic principles, and the entire spectrum of common procedures. The fourth edition reflects technique and equipment changes and the greater concern with quality control, sterilization, and protection from exposure to hepatitis, AIDS, and other diseases. Includes review questions and glossary. DA DH

Denissen, Harry, and others
Atlas of Porcelain Restorations
Padua, Piccin, 1990. 94p. ISBN 8829908312
    This atlas details the indications, preparations, lab work, and esthetic results for porcelain veneers, crowns and veneer bridges, as well as porcelain-metal bridges. DA DL

Dental Hygiene Employment Reference Guide
Chicago, American Dental Hygienists' Association, 1992. 55p.
    Features job interview preparation tips, advice on negotiating for a better position, and samples of employment agreements. DH

Dental Hygienist As Change Agent: A Curriculum Module
Chicago, American Dental Hygienists' Association, 1992. 54p.
    Presents the essentials of the legislative environment as it relates to dental hygiene practice. Explains how dental hygienists can influence the regulatory system in shaping their political future. DH

Dionne, Raymond A., and James C. Phero
Management of Pain and Anxiety in Dental Practice
New York, Elsevier, 1993. 420p. ISBN 0444817492
    The latest research, advancements, and techniques in the management of pain in dentistry and oral surgery. Includes patient evaluation and monitoring, pharmacological management of acute pain, and pharmacological management of anxiety. DA DH

Doherty, S. Adele, and C. Ray Bennett, eds.
Substance Abuse in Dentistry: A Proviso for a Drug-Free Profession
Brentwood, Tenn., D.S.H. Pub. Co., 1991. 97p.
ISBN 0963051105
    Proceedings of a conference that addressed the issue of substance abuse by dental health professionals, which is believed to be greater than in the general population. Encourages the American Dental Association and other dental groups to take a leading role in addressing abuse among colleagues, and provide help in achieving recovery and preventing relapse. ALL

DuBrul, E. Lloyd
Sicher and DuBrul's Oral Anatomy
8th ed. St. Louis, Mo., Ishiyaku EuroAmerica, 1988. 356p.
ISBN 091279142X
    This definitive textbook for this specialized field of anatomy details the components that make up the head and neck. This edition contains recent relevant research. Part One covers anatomy and functional analysis; Part Two addresses anatomy and clinical application. DH

Ehrlich, Ann B.
Business Administration for the Dental Assistant
4th ed. Champaign, Ill., Colwell Systems, 1991. 194p.
ISBN 0940012332
    This combination textbook and workbook outlines the responsibilities of a dental assistant, such as appointment control, records management, purchasing, and inventory control. A forms packet gives students experience working with the dental forms they will use on a daily basis. DA

Ehrlich, Ann B.
Fundamentals I: Introduction to Dental Terminology, Charting, and Procedures
3rd ed. Champaign, Ill., Colwell Systems, 1993. 175p.
ISBN 0940012375
    Dental terminology chapters include a word mastery list, the eight dental specialties, and a word recognition list. Dental charting chapters cover the Universal Numbering System, charting terms and symbols, treatment codes and terms, and ADA procedure codes. The last section classifies cavities. Includes a post-test and illustrations. DA DH

Ehrlich, Ann B.
Fundamentals II: Infection Control, Local Anesthesia and Oral Surgery
2nd ed. Champaign, Ill., Colwell Systems, 1988. 170p.
ISBN 0940012294
    Provides up-to-date information on the most current CDC and ADA guidelines. Includes information on disease transmission, infection control, operatory cleanup, sterilization, local anesthesia, and oral surgery. Includes post-test and illustrations. DA DH

Ehrlich, Ann B., and Hazel O. Torres
Essentials of Dental Assisting
Philadelphia, Saunders, 1992. ISBN 0721632629
    This combination text and how-to workbook is intended for the entry-level dental assistant. Easy to read and extensively illustrated, this work provides criterion sheets for self-evaluation, and a self-testing exercise after each chapter. Covers the fundamentals of dental assisting in addition to disease transmission, infection control, and hazards management based on current OSHA guidelines. DA

Employee Rights Handbook for Dental Hygienists
Chicago, American Dental Hygienists' Association, 1993. 35p.
    Designed to give dental hygienists insight into their rights as employees, this handbook includes topics such as federal antidiscrimination and wage laws, employee benefits, employment contracts, wrongful discharge, and independent contracting. DH

Employment Savvy: ADHA's Quick Reference Guide to Successful Job Hunting
Chicago, American Dental Hygienists' Association, 1989. 11p.
    This helpful booklet outlines tips on writing effective résumés, maximizing interview opportunities, negotiating for new jobs, and making the right career moves. DH

Ethics Conference I: Professional Accountability, Dental Hygiene in the 90's
Chicago, American Dental Hygienists' Association, 1991. 133p. and 5 videocassettes

The proceedings of a conference held during the ADHA's annual session address the revision of the Code of Ethics. DH

Expanding Dental Practice with Computer Technology
Princeton, N.J., Princeton Dental Resource Center, 1992. 47p.

Summarizes and highlights state-of-the-art developments in computer technologies related to dental practices. This is one of a series of monographs that provides practicing professionals with information on technology and trends in dentistry. ALL

Finkbeiner, Betty Ladley, and Jerry Crowe Patt
Practice Management for the Dental Team
3rd ed. St. Louis, Mo., Mosby, 1991. 375p. ISBN 0801662346

Finkbeiner presents a business practice manual for the dental office assistant. Spiral bound. DA DH

Foster, Malcolm S.
Protecting Our Children's Teeth: A Guide to Quality Dental Care from Infancy through Age Twelve
New York, Insight Books, 1992. 248p. ISBN 0306441225

Written by a pediatric dentist, this consumer-oriented book provides clear, concise descriptions of the development of children's teeth, as well as advice about baby teeth, and proper brushing techniques. DA DH

Freedman, George A.
Color Atlas of Porcelain Laminate Veneers
St. Louis, Mo., Ishiyaku EuroAmerica, 1990. 237p.
ISBN 0912791527

Discusses techniques in design, manufacture, and clinical application of porcelain veneers including topics such as fusion, preparations, clinical cases, inlays, and color. DA DL

Frommer, Herbert H.
Radiology for Dental Auxiliaries
5th ed. St. Louis, Mo., Mosby, 1992. 351p. ISBN 0801617014

This new edition of a standard text incorporates new chapters on infection control, panoramic radiography, and legal considerations. Include self-study and examination questions. DA DH

Garber, David A., and Ronald E. Goldstein
Porcelain and Composite Inlays and Onlays: Esthetic Posterior Restorations
Chicago, Quintessence Publishing, 1994. 159p. ISBN 0867151714

Garber presents a step-by-step manual on indications, preparation, placement, and laboratory fabrication of indirect porcelain and composite inlays and onlays. Clarifies the fundamental techniques, alternative techniques, and future trends such as computer-aided design, computer-aided manufacturing, and other computer modalities. DL

Garber, David A., and Ronald E. Goldstein
Porcelain Laminate Veneers
Chicago, Quintessence Publishing, 1988. 136p. ISBN 0867151943

A how-to book on the preparation, fabrication, and placement of porcelain laminate veneers. Describes the techniques, materials, and equipment necessary to deliver this esthetic or cosmetic treatment. A mini-atlas photo series gives step-by-step details on procedures. Includes sections on porcelain technology, laboratory procedures, and alternative therapies. DL

Gardner, Martha
Basic Anatomy of the Head and Neck
Philadelphia, Lea and Febiger, 1992. 202p. ISBN 0812114485

Teaches students a combination of regional detail and systemic overviews for a more complete understanding of head and neck anatomy. Emphasizes pathways through which vessels and nerves approach oral structures. Drawings. DH

Goldstein, Ronald E.
Change Your Smile
2nd. rev. ed. Chicago, Quintessence Publishing, 1988. 254p.
ISBN 0867151870

Goldstein covers cosmetic dentistry for the consumer including various facets of beautifying the teeth, whether it be repairing, straightening, or reshaping. Provides suggestions on analyzing a smile, coping with stains, closing gaps, and rejuvenating an aging smile. Success stories are used as examples. ALL

Haga, Michio
Techniques for Porcelain Laminate Veneers
St. Louis, Mo., Ishiyaku EuroAmerica, 1990. 46p.
ISBN 0912791918

Haga's concise, inexpensive guide to porcelain veneer technology is written for the dental technician. Many illustrations. DL

Haring, Joen Iannucci, and Laura Jansen Lind
Radiographic Interpretation for the Dental Hygienist
Philadelphia, Saunders, 1993. 197p. ISBN 0721637043

This textbook covers film mounting, viewing, descriptive terminology, identifying restorations and foreign objects, and relevant facts about dental and periodontal disease. DH

Harris, Norman O., and Arden G. Christen
Primary Preventive Dentistry
3rd ed. Norwalk, Conn., Appleton and Lange, 1991. 580p.
ISBN 0838578985

Harris provides a comprehensive review of two infectious disease processes: dental caries and periodontal disease. This work discusses the prevention of these diseases and the clinical application of preventive methods. This edition focuses more on periodontal disease, patient education, pit and fissure sealants, tooth remineralization, and geriatric dentistry. DH DA

Harris, Ruth Roy
Dental Science in a New Age: A History of the National Institute of Dental Research
Ames, Iowa, Iowa State Univ. Press, 1992. 476p.
ISBN 0813813220

Traces the development of dentistry's body of knowledge in the United States. Makes a strong case for the importance of research, particularly in dental hygiene because research is lacking regarding instrument and equipment design, use, and maintenance. DH

Hillam, Christine, ed.
Roots of Dentistry
London, British Dental Association, 1990. 73p. ISBN 0904588254
  Surveys dental practices from ancient times to today. This lavishly illustrated book covers the history of dentures and extractions. ALL

Hoag, Philip M., and Elizabeth A. Pawlak
Essentials of Periodontics
4th ed. St. Louis, Mo., Mosby, 1990. 248p. ISBN 080162228X
  Hoag provides concise, selective coverage of the basic principles and techniques involved in periodontics. This easy-to-use reference discusses periodontal disease within the framework of ten key factors and presents information in a modified outline form. This edition adds new illustrations and revised chapters on periodontal microbiology and immunology, patient education, and surgery. Includes glossary and index. DA DH

Holroyd, Sam V., Richard L. Wynn, and Barbara Requa-Clark, eds.
Clinical Pharmacology in Dental Practice
4th ed. St. Louis, Mo., Mosby, 1988. 600p. ISBN 0801622603
  By relating pharmacology to clinical dentistry, the contributors discusses all major drug groups and how they may affect a patient's response to dental treatment. Includes new chapters on respiratory tract drugs, the pharmacological considerations of temporomandibular joints, and facial pain. Sample Board review questions and an appendix listing the two hundred most commonly prescribed drugs are also provided. DH

Howe, Geoffrey L., and F. Ivor H. Whitehead
Local Anesthesia in Dentistry
3rd ed. Boston, Wright, 1990. 129p. ISBN 0723621470
  Presents basic principles and instruction in the use of well-tried and widely accepted anesthesia techniques. The third edition incorporates changes in opinion and practice. DH

Ibsen, Olga A. C., and Joan Anderson Phelan
Oral Pathology for the Dental Hygienist
Philadelphia, Saunders, 1991. 490p. ISBN 0721629504
  This text on oral disease processes focuses on the gingiva and periodontium, since the dental hygienist is specifically involved in treating conditions of these tissues. Does not cover dental caries or periodontal disease. Includes review questions, references, glossary, illustrations, and an index. DH

Ice Pack
Chicago, American Dental Assistants Association, 1994.
  Consists of a packet of twenty-nine articles that help students prepare for the Infection Control Exam. DA

Johnson and Johnson Medical, Inc.
Infection Control Card File
Chicago, American Dental Hygienists' Association, 1994. 63 cards
  These laminated cards in a file box bring practical infection control information to dental professionals. The twenty-four topic areas include a glossary of OSHA terms, disease transmission, and elements of an infection control program. DA DH

Jong, Anthony W., ed.
Community Dental Health
3rd ed. St. Louis, Mo., Mosby, 1993. 347p. ISBN 0801663873
  This is a collection of essays by practicing professionals on concepts such as the social and financial aspects of dental care, the dental care delivery system in the United States, research in community dental health management, and ethical issues of dental care. ALL

Jordan, Ronald E.
Esthetic Composite Bonding: Techniques and Materials
2nd ed. St. Louis, Mo., Mosby, 1993. 371p. ISBN 0801669804
  Provides an in-depth review of the materials and techniques used in the field of esthetic dentistry. Each chapter ends with a summary table that categorizes available products by purpose and manufacturer. Includes over 1,100 illustrations, chiefly in color. DA

Juniper, Richard, Brian J. Parkins, and A. J. Rodesano
Emergencies in Dental Practice, Diagnosis and Management
Oxford, England, Heinemann Medical Books, 1990. 172p. ISBN 0433022663
  A small, easy-to-read book to consult when emergencies arise in the dental office. Subjects range from the fractured dental restoration and hemorrhage to the diabetic coma and cardiac arrest. The second part addresses dental emergencies in a hospital setting. Includes appendix with drug information. DA DH

Kasle, Myron J.
Atlas of Dental Radiographic Anatomy
3rd ed. Philadelphia, Saunders, 1990. 295p. ISBN 0721652921
  A supplement to radiology texts, this book enhances radiographic interpreting skills. The text begins with the basics, followed by examples of artifacts and technical errors that should be recognized, avoided, and not misread. Ultimately the reader learns how to arrive at interpretive judgments. Brief text, many illustrations. DA DH

Katsuyama, Shigeru, Teatsuya Ishikawa, and Benji Fujii
Glass Ionomer Dental Cement
English ed. St. Louis, Mo., Ishiyaku EuroAmerica, 1993. 195p. ISBN 0912791942
  The development, composition, properties, handling fundamentals, and applications of glass ionomer dental cements are discussed for use in early caries treatment. DA DL

Kawabe, Seiji
Kawabe's Complete Dentures
St. Louis, Mo., Ishiyaku EuroAmerica, 1992. 201p.
ISBN 1563860112
  This translation of a popular work by one of Japan's most prominent dentists covers the entire process of doing dentures. DL

Kendall, Bonnie L.
Opportunities in Dental Care Careers
Lincolnwood, Ill., VGM Career Horizons, 1991. 149p.
ISBN 0844285765
  Investigates the many possibilities in dental care careers, and the wide variety of specialties within each. Provides a brief history of dentistry, discusses educational requirements, and then summa-

rizes the status of dentistry. Appendices list professional associations and accredited programs. (Ed. note: Many DLT programs have lost their accreditation since this work was published.) ALL

Klatell, Jack, Andrew S. Kaplan, and Gray Williams
Mount Sinai Medical Center Family Guide to Dental Health
New York, Macmillan, 1991. 304p. ISBN 0025636758

Presents a consumer's guide to self care as well as a reference book explaining common dental problems, symptoms, and treatments. Suggests a practical program of home care from infancy to old age, including information on what to do during dental emergencies and how to tell whether a symptom is serious enough to require a trip to the dentist. ALL

Krueger, Robert F.
How to Overcome Fear of Dentistry
Cincinnati, Heritage Communications, 1988. 191p.
ISBN 0961829117

This book is written for people who are uncomfortable during dental visits or who avoid the dentist completely, but it also aids the dental staff in understanding fear from the patient's point of view. Examines sources of dental fear and phobia, relaxation techniques, the use of biofeedback, self-hypnosis, medication, and psychotherapy. Explains how to prevent dental fear in children and how to select a gentle dentist and staff. DA DH

Langlais, Robert P., and Craig S. Miller
Color Atlas of Common Oral Diseases
Philadelphia, Lea and Febiger, 1992. 167p. ISBN 0812112490

A text with high quality color photographs of common oral diseases. The broad coverage includes topics such as tooth related disorders, hypersensitivity, carcinogenesis, and oral manifestations of various diseases. DA DH

Langland, Olaf E., and others
Radiology for Dental Hygienists and Dental Assistants
Springfield, Ill., Charles C. Thomas, 1988. 237p.
ISBN 0398054703

This work aids dental auxiliaries in developing competency in the performance of dental radiographic procedures. Contents follow the *Curriculum Guidelines for Dental Auxiliary Radiology*. Includes simplified material and review questions. DA DH

Leimone, Christine A., and Ethel Earl
Dental Assisting: Basic and Dental Sciences
St. Louis, Mo., Mosby, 1988. 362p. ISBN 080162942X

Treats the sciences in greater detail than other general dental assisting textbooks. Discusses the relation of head and neck anatomy, microbiology, nutrition, oral pathology, pharmacology, and dental emergencies to dental assisting practice. DA

Leinfelder, Karl F., and Jack E. Lemons
Clinical Restorative Materials and Techniques
Philadelphia, Lea and Febiger, 1988. 359p. ISBN 0812110730

Fundamental knowledge about the properties and use of dental materials has been combined with information about clinical applications. The first group of topics deals with materials used by the dentist and dental assistant directly in the treatment of patients. The second group discusses materials used in the dental laboratory. DA DL

Levison, H.
Textbook for Dental Nurses
7th ed. Boston, Mass., Blackwell Scientific, 1991. 371p.
ISBN 0632029560

Written for students preparing to take the British version of the dental assisting and dental hygiene exams, this is also a good basic text for dental auxiliaries. The seventh edition has more information on health and safety at work, cross infection, and current trends in dental practice. DA DH

Malamed, Stanley F.
Handbook of Local Anesthesia
3rd ed. St. Louis, Mo., Mosby, 1990. 332p. ISBN 0801630762

The third edition continues to provide knowledge and understanding of the field of local anesthesia in dentistry. Includes new information on current and anticipated developments and focuses on advances in the techniques of electronic dental anesthesia (EDA). DA DH

Malamed, Stanley F.
Medical Emergencies in the Dental Office
4th ed. St. Louis, Mo., Mosby, 1993. 466p. ISBN 0801663865

This work is designed to stimulate the members of the dental profession to improve and maintain their skills in the prevention of medical emergencies and in the management of emergencies that occur. This edition suggests changes in the design of an office emergency drug and equipment kit and provides the new American Heart Association guidelines for basic and advanced life support. DA DH

Malamed, Stanley F.
Sedation: A Guide to Patient Management
2nd ed. St. Louis, Mosby, 1989. 603p. ISBN 0801632102

Presents a down-to-earth presentation of the modalities of sedation, including inhalation, intramuscular, and intravenous sedation. DH

Malone, William F. P., and David L. Koth
Tylman's Theory and Practice of Fixed Prosthodontics
8th ed. St. Louis, Mo., Ishiyaku EuroAmerica, 1989. 461p.
ISBN 0912791489

Stresses current treatment in fixed prosthodontics and related specialties. Oral diagnosis, problem solving, treatment planning, and laboratory support are dealt with as complementary procedures. Elucidates innovative dental procedures. DL

Mann, Jonathan M., D. Tarantola, and Thomas W. Netter
AIDS in the World
Cambridge, Mass., Harvard Univ. Press, 1992. 1037p.
ISBN 0674012658

This text is valuable for public health professionals who are interested in global perspectives on AIDS. It emphasizes the need for global solutions, since HIV does not respect borders or cultures. Includes appendices and references. ALL

Matteson, Stephen R., Cy Whaley, and Vickye C. Secrist
Dental Radiology
4th ed. Chapel Hill, N.C., University of North Carolina Press, 1988. 155p. ISBN 0807842052

This manual discusses techniques for taking periapical films, and issues of radiation physics, biology, and protection. It also provides a review of normal anatomy as identified on dental films. The fourth edition provides updated information on safety and new technology. Each section has a list of objectives and review examinations. DA DH

McGivney, Glen P., and Dwight J. Castleberry
McCracken's Removable Partial Prosthodontics
9th ed. St. Louis, Mo., Mosby, 1994. 550p. ISBN 0801679648

The purpose of this well-known text is to provide the information necessary to produce a partial denture that is in itself a definitive restorative entity. Features all aspects of removable partial denture fabrication. This eighth edition has a new chapter on infection control. DL

McLaughlin, Gerald
Color Atlas of Tooth Whitening
St. Louis, Mo., Ishiyaku EuroAmerica, 1991. 118p. ISBN 0912791853

In addition to state-of-the-art clinical information and explanations of applications and techniques, this guide provides information on the origins of materials and processes. DA DH

McNeil, Charles, ed.
Temporomandibular Disorders: Guidelines for Classification, Assessment, and Management
2nd ed. Chicago, Quintessence Books, 1993. 141p. ISBN 0867152532

Provides an updated version of the American Academy of Orofacial Pain's recommendations for the classification, assessment, and management of temporomandibular disorders for all dental allied health professionals. Includes a lengthy glossary. DA DH

Miles, Dale A., Margot L. Van Dis, and Thomas F. Razmus
Basic Principles of Oral and Maxillofacial Radiology
Philadelphia, Saunders, 1992. 215p. ISBN 0721634710

This text addresses the basic principles of dental radiology, including the following: the physics of radiation production, radiation biology and protection, radiographic image production, techniques for exposing intraoral and extraoral radiographs, processing, and advanced diagnostic imaging. DA DH

Miles, Dale A., and others
Radiographic Imaging for Dental Auxiliaries
2nd ed. Philadelphia, Saunders, 1993. 314p. ISBN 0721667295

This introductory text presents techniques for taking dental X-rays and information on protection for patients and operators. Beginning with basic skills, the authors advance to skills for exposing, processing, and mounting radiographs. Theory, technique, and history are also included. An appropriate text for basic certification in oral radiology. DA DH

Miller, Chris H.
Infection Control: And Management of Hazardous Materials for the Dental Team
St. Louis, Mo., Mosby, 1994. 291p. ISBN 0801669324

Covers sanitation measures in the dental office, the prevention of cross infection, safety measures when dealing with hazardous substances, and OSHA standards. ALL

Moss, Stephen J.
Growing Up Cavity Free: A Parent's Guide to Prevention
Chicago, Quintessence Publishing, 1993. 147p. ISBN 0867152567

A resource for patient consultation that offers tips to give parents so that their children will have healthy teeth. The focus of this work is prevention. It covers prenatal care to ensure the best dental outlook for newborns, and orthodontics for adolescents. DA DH

Moss-Salentijn, Letty, and Marlene Hendricks-Klyvert
Dental and Oral Tissues: An Introduction
3rd ed. Philadelphia, Lea And Febiger, 1990. 327p. ISBN 0812113209

Provides a clinical orientation in layperson's language to the nondental and dental components of the mouth. Includes appendices on histologic technique, use of the microscope, and interpretation of sectioned structures. DA DH

Muia, Paul J.
Esthetic Restorations: Improved Dentist-Laboratory Communication
Chicago, Quintessence Publishing, 1993. 257p. ISBN 0867152265

Excellence in dentistry requires flawless dentist-technician communication. This book shows what is needed for a successful restoration and how the dentist and technician can best work together. DL

Murray, John J., ed.
Prevention of Dental Disease
2nd ed. New York, Oxford Univ. Press, 1989. 503p. ISBN 0192618075

This is an adjunct text on prevention, stressing dental caries and periodontal disease. This edition added chapters on dental health education, root caries, and dental disease in handicapped persons. DA DH

Naylor, W. Patrick
Introduction to Metal Ceramic Technology
Chicago, Quintessence Publishing, 1992. 195p. ISBN 0867152370

Naylor presents introductory level, skill-oriented technical information on fabricating metal ceramic restorations, one of the most widely used restorative combinations in dentistry today. DL

Neville, Brad W., and others
Color Atlas of Clinical Oral Pathology
Philadelphia, Lea And Febiger, 1991. 385p. ISBN 081211311X

This atlas format includes over 500 illustrations, organized into twelve chapters, each examining a disease category such as cysts and tumors, infections, and the pathology of teeth. DA DH

Newman, Michael G., and Kenneth S. Kornman
Antibiotic/Antimicrobial Use in Dental Practice
Chicago, Quintessence Publishing, 1990. 260p. ISBN 0867151722

Provides current, easily accessible information on the complex world of antibiotics, antimicrobials, culture microbiology, and patient application. Includes an annotated outline with tables and appendices. The second edition added research findings since 1984. DH

Nield, Jill Shiffer, and Ginger Ann Houseman
Fundamentals of Dental Hygiene Instrumentation
2nd ed. Philadelphia, Lea And Febiger, 1988. 521p.
ISBN 0812111303

Supplies students with instruction in the basics of dental hygiene instrumentation. The focus is self-paced learning in self-contained modules. This edition includes skill-building activities that promote psychomotor development, and a section on appointment preparation with case problems for the reader to solve. Appendices identify and solve problem areas of instrumentation and provide a list of further reading. DH

Nizel, Abraham E., and Aretha S. Papas
Nutrition in Clinical Dentistry
3rd ed. Philadelphia, Saunders, 1989. 465p. ISBN 0721624235

This concise reference text discusses nutrition as it applies to dentistry and overall health. Includes topics such as basic nutritional and biochemical information; nutrition and diet-related diseases with recommendations for treatment; potential benefits of dietary counseling; the effects of diet on the health of the elderly; and the role of fluorides in dental caries prevention. DH

Noback, Charles Robert, Robert J. Demarest, and Norman L. Strominger
Human Nervous System: Introduction and Review
4th ed. Philadelphia, Lea And Febiger, 1991. 448p.
ISBN 0812113438

Written for students new to the subject and as a review for professionals. The first third of the book introduces terminology and basic material, and the balance of the book addresses special functions or specific regions of the central nervous system. DH

Pattison, Anna Matsuishi, and Gordon L. Pattison
Periodontal Instrumentation
2nd ed. Norwalk, Conn., Appleton and Lange, 1992. 485p.
ISBN 0838578047

Pattison presents an up-to-date instrumentation manual that introduces the basic clinical procedures of periodontics. Modules cover examination procedures, detection of calculus, scaling and root planing, instrument sharpening, polishing, and the periodontal patient. DH

Paul, J. Ellis
Team Dentistry: Chairside Procedures and Practice Management
London, Martin Dunitz, 1991. 215p. ISBN 0948269782

Written for dentists and staff members, this practical guide discusses how to improve clinical and administrative organization. DA DH

Pediatric Dental Care: An Update for the 90s
Evansville, Ind., Bristol-Myers-Squibb Co., 1991. 42p.

Prepared for the American Academy of Pediatric Dentistry, this work presents child dental health objectives for the year 2000. Also discussed is dental care in infancy and childhood, infant nutrition, the growth and development of the mouth, and the prevention of tooth disease. DA DH

Permar, Dorothy
Permar's Oral Embryology and Microscopic Anatomy
9th ed. Philadelphia, Lea And Febiger, 1994. 278p.
ISBN 0812116593

This ninth edition of a classic text provides comprehensive information about embryology and microscopic anatomy. Uses scanning electron micrographs to aid understanding of the formation of the face and oral cavity. DH

Perry, Dorothy A., Phyllis Beemsterboer, and Fermin A. Carranza
Techniques and Theory of Periodontal Instrumentation
Philadelphia, Saunders, 1990. 388p. ISBN 072162734X

Of interest to dental hygiene students, this work provides instruction in periodontal instrumentation, integrating information on a skill's scientific background and its use in the delivery of patient care. Major emphasis is on hand instrumentation because of its difficulty. Includes diagrams and photographs, glossary, index. DH

Phagan-Schostok, Patricia A., and Karen L. Maloney
Contemporary Dental Hygiene Practice
Chicago, Quintessence Publishing, 1988-1989. 2 v.
ISBN 0867151692 (vol.1); ISBN 0867151706 (vol. 2)

Provides a complete approach to dental hygiene education. Priority is given to providing comprehensive care. Includes topics such as meeting the needs of medically compromised patients; the fundamentals of dental hygiene therapy; the hygienist in the role of periodontal cotherapist; and dental disease control. The laboratory manual provides exercises to develop instrumentation and procedural skills. Includes glossary, illustrations, and index. DH

Phillips, Ralph W., Keith Moore, and Marjorie L. Swartz
Elements of Dental Materials: For Dental Hygienists and Dental Assistants
5th ed. Philadelphia, Saunders, 1994. 311p. ISBN 0721642985

Explains the scientific aspects of materials such as polymers, alloys, and cements and how they interact with the oral environment. The fifth edition has new chapters on dental implants and ceramics, hazardous waste disposal, disease transmission, and the safety of dental materials. The key points are easy to find with outlines, highlighting, and step-by-step lists. DA DH

Preston, Jack D., ed.
Perspectives in Dental Ceramics: Proceedings of the Fourth International Symposium on Ceramics
Chicago, Quintessence Publishing, 1988. 472p. ISBN 0867151366

Presents contributions from dentists, dental lab technicians, academicians, and scientists with skills and knowledge in dental ceramics. Includes topics such as esthetics, fabrication, and communication. DL

Provenza, D. Vincent
Fundamentals of Oral Histology and Embryology
2nd ed. Philadelphia, Lea And Febiger, 1988. 307p.
ISBN 0812110811

The aim of this revised text is to increase the scientific knowledge base of students and integrate that knowledge with clinical applications. Uses current dental terms together with standard terminology. Partial contents: the cell, basic tissues, orofacial structures, dental and paradental structures, enamel, dentin, dental pulp, temporomandibular joint. Illustrated and indexed. DA DH

Rahn, Arthur O., and Charles M. Heartwell
Textbook of Complete Dentures
5th ed. Philadelphia, Lea And Febiger, 1993. 528p.
ISBN 0812115236

Presents the various components involved in the basic principles of complete denture prosthodontics. Relates anatomy, physiology, pathology, pharmacology, and psychology with the art and mechanics involved in complete denture construction. Provides updated chapters, a revised bibliography, and new photographs. DL

Rasmussen, Richard A.
Branemark System of Oral Reconstruction: A Color Atlas
St. Louis, Mo., Ishiyaku EuroAmerica, 1992. 305p.
ISBN 1563860031

Offers an overview of osseointegration. Discusses the diagnosis, treatment planning, and basic/advanced surgical techniques. DA DH

Rateitschak, Klaus H., and others
Periodontology
2nd rev. and expanded ed. New York, Thieme Verlag, 1989. 399p.
ISBN 0865773181

Although caries are on the decline, the treatment of gingivitis and periodontitis assumes a more prominent role in dentistry. This work consists of a clinically relevant documentation of practical periodontics and related specialties, with an emphasis on prevention. Includes clinical photographs and diagrams and contains the most recent research advances. DH

Regezi, Joseph A., and James J. Sciubba
Oral Pathology: Clinical-Pathologic Correlations
2nd ed. Philadelphia, Saunders, 1993. 1 v. (various pagings).
ISBN 0721636217

Helps bridge the study of oral pathology with practical clinical considerations. Enhances diagnostic skills using disease classifications, descriptions, and photographs to facilitate identification and treatment of oral diseases. Also provides microscopic correlations, current theories on etiology and pathogenesis, and current therapies. Includes a revised bibliography. DH

Renner, Robert P., ed.
Quintessence of Dental Technology: QDT
Lombard, Ill., Quintessence Publishing, Annual. (various pagings)
ISSN 0896-6532

This annual summary of the year's latest advances in dental technology offers articles by various authors on new and international techniques. DL

Requa-Clark, Barbara, and Sam V. Holroyd
Applied Pharmacology for the Dental Hygienist
2nd ed. St. Louis, Mo., Mosby, 1989. 298p. ISBN 0801642663

Presents basic pharmacology as it pertains to dental hygiene. The drugs used in dentistry are grouped in sections and discussed in detail. Includes review questions and an appendix listing the 200 most commonly prescribed drugs. DH

Robertson, Paul B., and John S. Greenspan
Perspectives on Oral Manifestations of AIDS: Diagnosis and Management of HIV-Associated Infections
Littleton, Mass., PSG Publishing, 1988. 216p. ISBN 0884165922

In this compilation of papers and discussions from a 1988 symposium, nineteen contributors deal with the implications of HIV infection for dental practice, the effects of HIV immunosuppression, oral microorganisms and the host, approaches in managing lesions, and ethical and legal considerations for health care providers. ALL

Rudman, Jack
Certified Dental Technician (CDT)
Syosset, N.Y., National Learning Corp., 1993. 1 v. (various pagings). ISBN 0837358566

This study guide provides sample tests and answers, along with relevant study material to prepare candidates for the CDT certification exam. DL

Rudman, Jack
DHAT: Dental Hygiene Aptitude Test
Syosset, N.Y., National Learning Corp., 1992. 1 v. (various pagings)

The DHAT is administered by the American Dental Hygienists' Association to measure those skills necessary for dental hygiene practice. Most schools require the test for admission. Sample questions and answers test verbal ability, numerical ability, science background, and study and reading skills. Also includes an introduction to the profession and test-taking suggestions. DH

Rudman, Jack
National Dental Assistant Boards (NDAB)
Syosset, N.Y., National Learning Corp., 1991. 1 v. (various pagings). ISBN 0837351871

Similar in style and format to Rudman's other publications, this study guide provides sample tests and answers, along with relevant study material to prepare candidates to take the NDAB certification exam. DA

Rudman, Jack
National Dental Hygiene Boards (NDHB)
Syosset, N.Y., National Learning Corp., 1991. 1 v. (various pagings)

Similar in style and format to Rudman's other publications, this study guide provides sample tests and answers, along with relevant study material to prepare candidates to take the NDHB certification exam. DH

Runnells, Robert R.
AIDS in the Dental Office? The Story of Kimberly Bergalis and Dr. David Acer
Fruit Heights, Utah, I.C. Publications, 1993. 323p.
ISBN 0936751118

Chronicles the story of a dental visit that sparked worldwide debate on whether health care workers and patients should be required to undergo mandatory testing for the AIDS virus. The author was an expert witness in the resulting lawsuit. Concludes with a chapter on the safety of dental treatment. ALL

Sato, Sadakatsu
Eruption of Permanent Teeth: A Color Atlas
American ed. St. Louis, Mo., Ishiyaku EuroAmerica, 1990. 100p.
ISBN 0912791446

Meticulous weekly documentation of more than 600 children was done over many years to produce this amply illustrated study. DA DH

Schwarzrock, Shirley Pratt, and others
Effective Dental Assisting
7th ed. Dubuque, Iowa, William Brown, 1991. 618p.
ISBN 0697113159

Written for dental assisting students and those being trained in the office, this text aims to present every phase of dental assisting. This edition revises and expands the section on dental specialties; dental materials; dental operations with practitioners wearing gloves, glasses, and masks; and provides a new section on computers. Includes vocabulary lists at the end of each chapter and an accompanying workbook. DA

Scott, Ronald W.
Legal Aspects of Documenting Patient Care
Gaithersburg, Md., Aspen Publishers, 1994. 242p.
ISBN 0834205491

Because every component of the health care system is now subject to litigation, the author has taken a risk-management approach to patient care documentation to encourage practitioners to be objective, precise, and timely when they document treatment-related matters. DA DH

Scully, Crispian
Occupational Hazards to Dental Staff
London, British Dental Association, 1990. 338p.
ISBN 0904588270

Essential reading for all dental health care workers because it covers the potential hazards found in the workplace. ALL

Shiba, Akihiko
Conical Double-Crown Telescopic Removable Periodontic Prosthesis
St. Louis, Mo., Ishiyaku EuroAmerica, 1993. 56p.
ISBN 1563860023

Basic concepts as well as clinical applications are described for this special partial denture. This denture is used to restore phonetic function, correct occlusal abnormalities, and maintain good temporomandibular joint function. DL

Stewart, Kenneth L.
Clinical Removable Partial Prosthodontics
2nd ed. St. Louis, Mo., Ishiyaku EuroAmerica, 1992. 695p.
ISBN 0912791985

Presents a completely updated edition with special attention to clarity. New sections on attachments, implants, maxillofacial prosthetics, and laboratory procedures are included. DL

Taintor, Jerry F., and Mary Jane Taintor
Oral Report: The Consumer's Common-sense Guide to Better Dental Care
New York, Facts on File, 1988. 194p. ISBN 0816013926

These authors maintain that consumers know more about their cars than their mouths. The purpose of this book is to help foster communication between dental professionals and their patients, so that patients can possess the knowledge they need to play a more active role in their own dental health care. Provides general fee information. Includes a glossary and appendices listing dental schools and associations. ALL

Taylor, Thomas D., and William R. Laney
Dental Implants: Are They For Me?
2nd ed. Carol Stream, Ill., Quintessence Publishing, 1993. 60p.
ISBN 0867151986

This patient education resource explains what dental implants are, who should have them, how they are placed, and costs. Also provided is a special detailed section on how to clean the new teeth after dental implant therapy. Includes a typical timetable for implant treatment with a fixed prosthesis. Format includes color drawings, large type, and clear language. DA DH

Thomas, Regina Dreyer
Career Decisions for Dental Hygienists: Your Guide to Change and Opportunity
Holmdel, N.J., Career Directions Press, 1992. 199p.
ISBN 0933163037

More than twenty registered dental hygienists focus on dental hygiene as a career and the many possibilities it offers, such as hospital work, private practice, sales, working abroad, and insurance business opportunities. Includes résumé, cover letter, and interview examples and a list of funding resources. DH

Torres, Hazel O., and Ann B. Ehrlich
Modern Dental Assisting
4th ed. Philadelphia, Saunders, 1990. 980p. ISBN 072162488X

This updated, comprehensive clinical textbook/reference for dental assistants gives a complete overview of the assistant's role in modern dentistry, detailed technical and procedural material, and employment information. DA

Torres, Hazel O., and Lois Mazzuchi-Ballard
Dental Assisting Exam Preparation
Philadelphia, Saunders, 1994. 247p. ISBN 0721632955

Designed to help candidates prepare for the Dental Assisting National Board Exam or for licensure exams administered by individual states. This work provides three comprehensive tests: two on general chairside functions and one on dental radiography. A rationale follows the test questions. DA

Ubassy, Gerald
Shape and Color: The Key to Successful Ceramic Restorations
Chicago, Quintessence Publishing, 1993. 216p. ISBN 0867152079

The study of shapes and how they interact with color is the focus of this book. Stresses the need for interpersonal contact between the dentist and the lab technician. Emphasizes the importance of observing natural teeth in restoration work. Covers how to polish ceramic, create fissures, and fabricate working casts in the laboratory. DL

Universal Precautions: Employer's Compliance Manual for the Dental Office
Garden Grove, Calif., MEDCOM, 1992. 235p.

Presents a comprehensive Universal Precautions manual, which provides guidelines for establishing practices that will safeguard dental office employees and be in compliance with current OSHA requirements. ALL

Weinstein, Bruce D., ed.
Dental Ethics
Philadelphia, Lea and Febiger, 1993. 243p. ISBN 0812114442
　　Looks at important ethical issues in the dental profession today. Topics have been chosen from the Curriculum Guidelines on Ethics and Professionalism in Dentistry. Provides case histories, discussion questions and possible solutions, and suggested readings on such topics as HIV infection, addicted colleagues, and advertising. ALL

Weinstein, Philip, Tracy Getz, and Peter Milgrom
Oral Self Care: Strategies for Preventive Dentistry
3rd ed. Seattle, Wash., University of Washington Press, 1991. 230p. ISBN 1880291002
　　Serves as a guideline for creating high-quality preventive dental care. Stresses behavioral techniques with discussion and participation activities for patients. This edition has a new section on children and the elderly. Patient anxiety, fears, phobias, and pain are key elements covered in this text. DA DH

Westman, Randall Perry
Trust, AIDS, and Your Dentist
San Antonio, Tex., Sweettooth Publishing, 1993. 113p. ISBN 0963708805
　　Westman's intent is to allay people's fears of going to the dentist after the AIDS incident in Florida. This work explains the principles of sterilization and infection control to the prospective patient and provides general information on how AIDS is transmitted and lists consumer contacts such as the American Dental Association and the Centers for Disease Control. Suggests key questions to ask dentists about this issue. Includes a glossary and an index. DA DH

White, Graham E.
Osseointegrated Dental Technology
Chicago, Quintessence Publishing, 1993. 233p. ISBN 1850970319
　　Based on dental technology research and extensive clinical practice, this book explains the methods and materials necessary for making frameworks that fit well without corrective soldering. DL

Wilkins, Esther M.
Clinical Practice of the Dental Hygienist
6th ed. Philadelphia, Lea and Febiger, 1989. 802p. ISBN 0812111818
　　This work is organized to help dental hygienists proceed through logical steps in patient care. The text provides comprehensive information in a concise manner. Specific chapters are devoted to patients with oral, systemic, physical, and other problems that require special knowledge. This edition has two new chapters—one to describe debonding and one on how to care for the alcoholic patient. DH

Willett, Norman D., Robert R. White, and Samuel Rosen, eds.
Essential Dental Microbiology
Norwalk, Conn., Appleton and Lange, 1991. 406p. ISBN 0838524532
　　Written for dental students, this book provides an understanding of the characteristics of microbiology, molecular biology, and immunology, and their effects on the practice of dentistry. DH

Wilson, Thomas G., ed.
Dental Maintenance for Patients with Periodontal Disease
Chicago, Quintessence Publishing, 1989. 224p. ISBN 0867152095
　　As an addition to periodontology texts, this book is unique because it is devoted to maintenance therapy for restorative, pediatric, orthodontic, irradiated, diabetic, and temporomandibular joint patients. "At a glance" sections provide quick reference. DH

Wilson, Thomas G., Kenneth S. Kornman, and Michael G. Newman, eds.
Advances in Periodontics
Chicago, Quintessence Publishing, 1992. 383p. ISBN 0867152508
　　The editors present information on periodontology from researchers and clinicians. Essays discusses leading-edge concepts, treatments, and advanced techniques. Quick review sections at the end of each chapter provide an overview of the subject. Includes numerous graphics and color photographs. Large print. DH

Woelfel, Julian B.
Dental Anatomy: Its Relevance to Dentistry
4th ed. Philadelphia, Lea And Febiger, 1990. 438p. ISBN 0812112598
　　Woelfel demonstrates the complex, dynamic interrelationships between teeth, bones, muscles, and nerves associated with dentition. Includes new chapters on forensic dentistry and the evolution of mammalian dentition. The chapter on operative dentistry has also been updated. DA DH

Wood, Peter R.
Cross Infection Control in Dentistry: A Practical Illustrated Guide
St. Louis, Mo., Mosby, 1992. 207p. ISBN 0815194390
　　The author's stated purpose is to inform dental health care workers about the importance of implementing cross infection control in private practice. Incorporates the regulations and recommendations of the Centers for Disease Control, Environmental Protection Agency, and OSHA. Reviews procedures in a cookbook fashion. Includes a procedural guide. ALL

Woodall, Irene R., ed.
Comprehensive Dental Hygiene Care
4th ed. St. Louis, Mo., Mosby, 1993. 872p. ISBN 0801670195
　　The goal of this text is to prepare the students for the "first day" of clinical practice when a patient arrives for care. It is sequentially structured for that purpose, with a list of objectives for each chapter. Emphasizes working for and with the patient. DH

Nonprint Materials

Adkisson, Mary Ann, Richard Bebermeyer, and Robert White
Infection Control in the Dental Branch Operatory
Chapel Hill, N.C., Health Sciences Consortium, 1992.
1 videocassette (21 min.) VHS
　　After watching this video, viewers should be able to (1) identify procedures designed to reduce or eliminate infectious exposures in the dental operatory; (2) identify the field of contamination; (3) describe infection control measures to perform before, during, and after an appointment; and (4) define "universal precautions". DA DH

Basic Oral Anatomy for the Dental Health Team
Garden Grove, Calif., Medcom, 1992. 1 videocassette (13 min.) VHS

This video discusses basic oral anatomy and the use of proper technical dental terminology. Accompanied by a workbook. ALL

Bermel, Stephanie Nowysz
Nursing Bottle Caries Simulation
Iowa City, Iowa, University of Iowa, 1 videodisc, 1991. Interactive videodisc

This is a patient simulation program about a one-year old child who had tender gums and caries on his front teeth. Bermel discusses taking medical and dental histories, the physical examination, diagnosis, and management of the condition. DH

Bloodborne Pathogens
Virginia Beach, Coastal Video Communications, 1992.
1 videocassette (21 min.) VHS

This videotape provides instruction for new employees and a review for experienced employees of federal regulations and the control of bloodborne infections in an industrial setting. It is also helpful for individuals trained in first aid. ALL

Bloodborne Pathogens
Herkimer, N.Y., HCTV, 1993. 1 videocassette (60 min.) VHS

A certified OSHA instructor introduces the Pathogen Training Program, along with a discussion of HIV, AIDS, hepatitis B, and universal precautions. ALL

Blozis, George G., Joen Iannucci Haring, and Michael F. Para
AIDS: Vignettes for Dental Professionals: Four Case Presentations
Chapel Hill, N.C., Health Sciences Consortium, 1991.
2 computer disks

The user is guided through a series of decision-making situations and case studies related to managing and coping with AIDS. ALL

Borecki, Madeline
Class I Occlusal and Buccal Pit Amalgam Restoration
Chapel Hill, N.C., Health Sciences Consortium, 1989.
1 videocassette (5 min.) VHS

Designed to teach the proper technique for restoration of Class I occlusal and buccal pit preparations using amalgam as the restorative material. Uses a life-size model and allows close-up observation. DA

Borecki, Madeline, John M. Powers, and Lon T. Smith
Mixing Glass Ionomer Cements
Chapel Hill, N.C., Health Sciences Consortium, 1990.
1 videocassette (14 min.) VHS

Cements are a fundamental component of dental practice. This video introduces technicians to the precise techniques for mixing cements. The materials required for each technique are presented and the procedures are demonstrated. DA

Carroll, Kevin O.
Advanced Radiographic Techniques Part I: Occlusal and Lateral Oblique Projections
Chapel Hill, N.C., Health Sciences Consortium, 1993.
1 videocassette (20 min.) VHS

This updated and revised video demonstrates the proper procedures for making a variety of intra-oral and extra-oral radiographs, including occlusal and lateral oblique projections. For each, the author provides steps in placement, angle of projection, and use. DA DH

Chavarria, Lazaro, and Dorothy N. Bassett
Intra-Oral Radiographic Technique: Maxillary and Bite-Wing
Carrboro, N.C., Health Sciences Consortium, 1988. 1 videocassette (15 min.) VHS

Presents techniques for producing diagnostic maxillary intra-oral radiographs using film-holding devices such as the Rinn XCP instruments to reduce distortion. Also shows the technique for producing bite-wing radiographs. DA DH

Christensen, Gordon J.
Auxiliary Oriented Diagnostic Appointment
Provo, Utah, Practical Clinical Courses, 1989. 1 videocassette (55 min.) VHS

This videotape discusses the role of dental auxiliaries in data collection, radiographs, cast construction, examination of hard and soft tissues, compilation of clinical data, and patient education. DA DH

Christensen, Gordon J.
Bleaching Vital Teeth at Home and in the Dental Office
2nd ed. Provo, Utah, Practical Clinical Courses, 1990.
1 videocassette (60 min.) VHS

The many types of bleaching techniques can be confusing; this video presents the newest, most successful concepts and name brands for "at-home" and "in-office" bleaching. DA DH

Christensen, Gordon J.
Dental Cements: Selecting the Best Type
2nd ed. Provo, Utah, 1992. 1 videocassette (60 min.) VHS

There have been many new advances in cements in recent years. This video shows cement comparisons, indications, clinical mixing, and the use of the popular types of cement. DA

Christensen, Gordon J.
Fluoride: Professionally Applied and Home Use
Provo, Utah, Practical Clinical Courses, 1990. 1 videocassette (60 min.) VHS

Proper fluoride use can prevent caries in adults and children. This presentation gives a comparison of fluoride products and a demonstration of current clinical techniques, a comparison of home-use products, a method of caries prevention for radiation patients, and information on systemic fluoride supplements for children and pregnant women. DA DH

Christensen, Gordon J.
Infection Control Made Cost Effective!
Provo, Utah, Practical Clinical Courses, 1993. 1 videocassette (60 min.) VHS

Distinguishes between effective infection control procedures and products, and those that are superfluous and ineffective. Shows several ways to recover the costs of infection control. DA DH

Christensen, Gordon J.
Making Accurate, Easy, Alginate Impressions
Provo, Utah, Practical Clinical Courses, 1990. 1 videocassette (60 min.) VHS
   Directed to dental auxiliaries for improving impressions, this text includes such topics as pre-impression mouth preparation, selection of alginate brands, mixing, tray selection, placement, setting, and a critique of completed impressions. DA DH

Christensen, Gordon J., and Karen Preston
Occlusal Splints
Provo, Utah, Practical Clinical Courses, 1989. 1 videocassette (60 min.) VHS
   Teaches auxiliaries to make a simple splint and put it in the mouth. Includes rationale for splints, splint construction and placement, follow-up, and post-splint observation. DA DH

Christensen, Gordon J.
Simple Surgery for Everyday Practice
Provo, Utah, Practical Clinical Courses, 1992. 1 videocassette (60 min.) VHS
   Covers surgical procedures performed by nearly all dentists. Includes new equipment, instruments, and supplies necessary for trouble-free surgery. Live patient demonstrations show the principles steps for doing difficult extractions. DA DH

Christensen, Gordon J.
Simple Temporary Restorations for Fixed Prosthodontics
Provo, Utah, Practical Clinical Courses, 1988. 1 videocassette (58 min.) VHS
   This presentation emphasizes faster and easier procedures for temporary restorations that can be delegated to auxiliaries. Included are single crown procedures, fast multiple temporaries, comparison of various resin types, repair, and occlusal considerations. DA DH

Christensen, Gordon J.
Tooth Desensitization
Provo, Utah, Practical Clinical Courses, 1989. 1 videocassette (60 min.) VHS
   The following concepts are demonstrated on patients: home care with high-content fluoride in simple custom trays, use of desensitizing sealers, a comparison of desensitizing toothpastes, dietary considerations, and non-preparation restorations as a last resort. DA DH

DeLuca, Chester, and James A. Anderson
Diet Counseling Procedure for the Dental Clinic
Chapel Hill, N.C., Health Sciences Consortium, 1990. 1 videocassette (55 min.) VHS
   Excerpts from actual sessions with patients in a preventive dentistry clinic demonstrate the techniques used by a counselor to promote better oral and systemic health through diet modification. DH

Dental Anatomy: Mandibular Incisors
Ann Arbor, University of Michigan School of Dentistry. 1988. 1 videocassette (26 min.) VHS
   Discusses the location, morphology, terminology, and identifying characteristics of mandibular incisors. ALL

Dental Anatomy: Maxillary and Mandibular Canines
Ann Arbor, University of Michigan School of Dentistry, 1988. 1 videocassette (27 min.) VHS
   Discusses the location, morphology, terminology, and identifying characteristics of the maxillary and mandibular canines. ALL

Dental Anatomy: Maxillary Incisors
Ann Arbor, University of Michigan School of Dentistry, 1988. 1 videocassette (30 min.) VHS
   Discusses the location, morphology, terminology, and identifying characteristics of the maxillary incisors. ALL

Dental Anatomy: Maxillary Premolars
Ann Arbor, University of Michigan School of Dentistry, 1988. 1 videocassette (46 min.) VHS
   Discusses the location, morphology, terminology, and the identifying characteristics of the maxillary premolars. ALL

Dental Education Center
Toothbrushing: The Bass Method
Washington, D.C., Veteran's Administration Medical Center, 1988. 1 videocassette (9 min.) VHS
   The Bass Method of toothbrushing dislodges and removes bacterial plaque and debris from surfaces of teeth and gums using a soft, multi-tufted, blunt-tipped brush. The complete routine is described and demonstrated. DA DH

Dental Education Center
Toothbrushing: The Circular Scrub Method
Washington, D.C., Veteran's Administration Medical Center, 1988. 1 videocassette (9 min.) VHS
   The Circular Scrub Method of toothbrushing uses a soft, multi-tufted nylon brush to remove bacterial plaque and debris from the surface of the teeth and gums. The complete routine is described and demonstrated. DA DH

Dental Hygiene: A Profession of Opportunities
Chicago, American Dental Hygienists' Association, 1993. 1 videocassette (5 min.) VHS
   This discussion among six dental hygienists presents dental hygiene as a profession of opportunity and one with universal appeal. DH

Department of Veterans Affairs, and others
Infection Control in the Dental Environment
Chapel Hill, N.C., Health Sciences Consortium, 1992. 3 videocassettes (57 min.) VHS
   This three-unit, self-paced learning experience presents a rational way of managing the risk of HIV and hepatitis B infection, while continuing to provide quality patient care. ALL

East Central AIDS Education and Training Center
AIDS: Identifying Community Resources
Chapel Hill, N.C., Health Sciences Consortium, 1991. 1 videocassette (40 min.) VHS
   Provides the health care professional with insight into community resources typically available to person who is HIV positive. DA DH

East Central AIDS Education and Training Center
HIV/AIDS: Epidemiology for Primary Care Health Professionals
Chapel Hill, N.C., Health Sciences Consortium, 1991.
1 videocassette (21 min.) VHS

Explains the natural history of the human immunodeficiency virus and the syndrome of diseases that accompany it. Describes how the virus is transmitted, its impact on certain populations, and its profound impact on the medical community. ALL

East Central AIDS Education and Training Center
HIV/AIDS: Testing and Risk Assessment
Chapel Hill, N.C., Health Sciences Consortium, 1991.
1 videocassette (20 min.) VHS

HIV testing requires that the physician and other health care providers give pretest and post-test counseling to the individual. This video explains the necessity of making a complete sexual history part of a physical exam, and discusses behaviors that put a person at risk for contracting the HIV virus. ALL

Hazard Communication for the Dental Health Team
Garden Grove, Calif., Medcom, 1992. 1 videocassette (23 min.) VHS and 1 workbook.

Meets OSHA staff training requirements for the proper use of chemicals in the dental office. DA DH

Infection Control for the Dental Health Team
Garden Grove, Calif., Medcom, 1992. 1 videocassette (29 min.) VHS and 1 workbook.

Discusses infection control procedures in the dental office. ALL

Introduction to Basic Concepts in Dental Radiography
Chicago, American Dental Assistants Association, 1991.
1 videocassette (20 min.) VHS

Illustrated how dental radiography works and the procedures for exposing and processing radiographs safely and effectively. Helps to prepare assistants for taking the Dental Assisting National Board's Dental Radiation Health and Safety Exam. DA

Jones, John C., and Gordon J. Christensen
Simplified Management of Medical Emergencies
Provo, Utah, Practical Clinical Courses, 1989. 1 videocassette (58 min.) VHS

Stresses proper preparation for medical emergencies. Illustrates how to use emergency kits and how to educate staff in emergency procedures. Provides simulated emergencies such as seizures, strokes, cardiac arrest, and allergic and respiratory problems, and offers treatment recommendations. DA DH

Kirsch, A., and K. L. Ackermann
IMZ Implant System, Part I: Clinical Aspects
Carol Stream, Ill., Quintessence Publishing, 1993. 1 videocassette (40 min.) VHS

A step-by-step examination of the surgical treatment of edentulous patients receiving implant-supported restorations in the mandible. ALL

Lang, W. Paul
Pit and Fissure Sealants
Ann Arbor, University of Michigan School of Dentistry, 1988.
1 videocassette (12 min.) VHS

The objectives of this video are to present rationales for sealant use, introduce sealant materials, show application techniques, and conduct patient education in the clinical use of sealants. DA DH

Lipp, Markus D. W.
Local Anesthesia
Carol Stream, Ill., Quintessence Publishing, 1993. 1 videocassette (30 min.) VHS

Through a combination of clinical images and three-dimensional computer simulated perspectives, this work shows proper usage techniques and anesthesia's physiologic effect on nerves. The criteria for choosing anesthetics and the risk factors and complications are also addressed. DA DH

McCarthy, Frank, and J. Chen
Emergency Drugs, Devices, and Procedures
Los Angeles, USC School of Dentistry, 1989. 1 videocassette (29 min.) VHS

McCarthy (MD, DDS, DMFS) presents a rationale for minimum emergency drugs, devices, and procedures for the general dentist and specialist. This work is also useful for auxiliary personnel. DA DH

Mulligan, Nan, and others
Simulations in Medical Evaluation of Geriatric Patients [computer software]
Los Angeles, USC School of Dentistry, 1992

This interactive software provides a case study approach to medical evaluation of geriatric patients. Focus is on diagnostic skills. System requirements: IBM-PCs or compatibles. ALL

Neuendorff, G., A. Kirsch, and K. L. Ackermann
IMZ Implant System Part II: Laboratory Procedure
Carol Stream, Ill., Quintessence Publishing, 1993. 1 videocassette (30 min.) VHS

Step-by-step examination of the fabrication of an IMZ implant-supported, screw-retained restoration. DL

New Dental Assistant: Impressions and Molds
Garden Grove, Calif., Medcom, 1991. 1 videocassette (24 min.) VHS

Discusses proper techniques in making impressions and pouring dental molds. DA DH

Newsom, Jayne
Head and Neck Screening Examination Procedures for the Dental Hygienist
Chapel Hill, N.C., Health Sciences Consortium, 1992.
1 videocassette (12 min.) VHS

Demonstrates the proper techniques for conducting a head and neck dental screening examination. Topics covered include intra-oral and extra-oral examination, materials required, and infection control measures. DH

Porter, Thomas
Sign Language for the Dental Team
Chapel Hill, N.C., Health Sciences Consortium, 1990.
1 videocassette (20 min.) VHS

Introduces the viewer to the basic alphabet, the numbers 1-32, and words and phrases related to the dental environment. DA DH

Preston, Karen, and Gordon J. Christensen
Simple, Fast, High Quality Dental Radiographs (Intraoral)
Provo, Utah, Practical Clinical Courses, 1990. 1 videocassette (60 min.) VHS

Educates dental auxiliaries with a view to improving the quality of dental radiographs. Discusses types and sizes of film, patient and clinician protection, film holders, exposure times, developing and copying film, and infection control. DA DH

Prevention and Treatment Considerations for the Dental Patient with Special Needs
[s.l.], Johnson and Johnson, 1989. 40 slides

A reference tool developed by the Association of Retarded Citizens, the Academy of Dentistry for the Handicapped, and the American Dental Hygienists' Association to teach dental hygienists how to deal with special needs patients. Includes script. DA DH

Rada, Robert
Porcelain Laminate Veneers: Preparation and Placement
St. Louis, Mo., Ishiyaku EuroAmerica, 1990. 1 videocassette (25 min.) VHS

This complete presentation guides the viewer through all phases of clinical treatment, demonstrating tooth preparation, veneer evaluation and placement, finishing technique, and post-placement evaluation. DA DH

Radiographic Techniques and Safety: Introduction to X-ray and Safety Precautions
Garden Grove, Calif., MEDCOM, 1992. 1 videocassette (17 min.) VHS

Provides basic knowledge of X-rays. Discusses possible health hazards and proper precautions to take to ensure the safety of the patient and the dental health team. Includes workbook. DA DH

Radiographic Techniques and Safety: Taking Radiographs
Garden Grove, Calif., MEDCOM, 1992. 1 videocassette (24 min.) VHS

Discusses X-ray equipment, how it works, and the factors that control image formation. Includes workbook. DA DH

Radiographic Techniques and Safety: X-ray Film Processing
Garden Grove, Calif., Medcom, 1992. 1 videocassette (24 min.) VHS

Provides an understanding of how the developing process works and how to perform each step with accuracy. Includes workbook. DA DH

Richards, E. Earl, and Jessie O. Brown
Model for Oral Hygiene in Long Term Care Facilities
Chapel Hill, N.C., Health Sciences Consortium, 1990. 1 videocassette (25 min.) VHS

This videotape presents a model plan for daily oral hygiene that can be implemented in institutions that provide long-term care for the elderly. Techniques are offered for patients who can take care of themselves and also for those who are completely incapacitated. DH

Riffe, Tanya S.
Oral Hygiene Procedures for Bedridden Patients
Chapel Hill, N.C., Health Sciences Consortium, 1990. 1 videocassette (15 min.) VHS

A procedural guide for those giving basic oral hygiene care to the bedridden. Focus is on toothbrushing and flossing, as well as procedures for cleaning removable appliances. Shows adaptations for unresponsive or comatose patients. DA DH

Romano, Deborah, and Carolyn Sparks
Instrument Transfer for the Dental Assistant
Carrboro, N.C., Health Sciences Consortium, 1988. 1 videocassette (14 min.) VHS

Guidelines for efficient one- and two-handed transfer of dental instruments as applied to four grasp techniques: pen grasp, palm-thumb grasp, palm grasp, and special cases. These guidelines help to save time, reduce fatigue, and minimize eyestrain. DA

Romano, Deborah, and Carolyn L. Sparks
Suction Tip Placement
Chapel Hill, N.C., Health Sciences Consortium, 1990. 1 videocassette (6 min.) VHS

Considers the need for oral evacuation, and describes and contrasts the application of low- and high-velocity suction tips. The proper placement of the standard high-velocity evacuation tip in all sextants of the mouth is demonstrated. DA

Rudman, Jack
Dental Assistant Boards (NDAB) [computer software]
Syosset, N.Y., National Learning Corp., 1993. System requirements: IBM and compatible; all monitors including MDA, Hercules graphics, CGA, EGA, and VGA.

Provides comprehensive test preparation software for use on all IBM-compatible personal computers. Provides self-paced learning and graphics exercises to help students prepare for the NDAB exam. *Future Test Software Series.* DA

Rudman, Jack
Dental Hygienist Boards (NDHB)
Syosset, N.Y., National Learning Corp., 1993. System requirements: IBM and compatible; all monitors including MDA, Hercules graphics, CGA, EGA, and VGA.

Provides comprehensive test preparation software for all IBM-compatible personal computers. Helps students prepare for the National Dental Hygiene Board (NDHB) exam through self-paced learning and graphics. *Future Test Software Series.* DH

Skills of Daily Mouth Care: A Caregiver's Guide
Belmont, Mich., Dental Health Video Source, 1990. 1 videocassette (23 min.) VHS

Developed to educate and motivate nursing home and hospital caregivers in the skills of daily oral hygiene, this videocassette outlines methods ranging from simple daily hygiene to more involved evaluations, and the use of an ultrasonic cleaner for appliances in both sitting and supine patients. Features OSHA regulations. Comes with an instructional guide, a pretest, post-test, answers, and a caregiver's guide. DA DH

Smoot, E. Clyde, and others
Giving Your Child a Smile: Correcting Cleft Lip and Palate
Chapel Hill, N.C., Health Sciences Consortium, 1990.
1 videocassette (23 min.) VHS

Recommended for dental allied health personnel participating in the repair and rehabilitation of patients with cleft lips and palates. Introduces the viewer to the treatment plan, explains the origins of the disorder, and presents a timeline for the treatment procedure. DA DL

Tetsch, Peter
Basics of Dental Implantology: Indications, Diagnostics, Therapy, and Recall
Chicago, Quintessence Publishing, 1993. 1 videocassette (55 min.) VHS

A complete introduction to dental implantology, with an emphasis on treatment planning. ALL

Thurston, Stephen E.
As It Should Be Done: Workplace Precautions Against Bloodborne Pathogens
Butler, Pa., Applied Science Associates, 1992. 1 videocassette (24 min.) VHS

An overview of how OSHA's Standards on Bloodborne Pathogens applies in different occupations, including dentistry. Demonstrates the importance of Universal Precautions in the workplace. ALL

Universal Precautions: AIDS and Hepatitis B Prevention for the Dental Health Team
Garden Grove, Calif., Medcom, 1992. 1 videocassette (32 min.) VHS

Provides OSHA-required training and documentation of training in five lessons: HBV: Recognizing the Dangers; HIV: Recognizing the Dangers; Modes of Transmission for HBV and HIV; Personal Protective Equipment; and Safe Work Practices. Includes workbook. ALL

Visual Guide to Dental Care
Chicago, American Dental Hygienists' Association, 1993. 12p. Laminated photo

This enlarged, laminated guide simplifies the basics of dental care and is useful for patient education. Detailed photographs illustrate oral anatomy, tooth development, plaque control, tooth decay and repair, periodontal disease, tooth replacement, cosmetic dentistry, orthodontics, and wisdom teeth. DH

Weber, H., and others
Combined Fixed/Removable Prosthesis with the SAE Spark Erosion System
Chicago, Quintessence Publishing, 1993. 1 videocassette (43 min.) VHS

Shows procedures for the placement of double crowns in the mandible and the implantation of swivel latches in the maxilla with the SAE Erosion System.

Diagnostic Technologies

Isabel C. Hernandez
Miami-Dade Community College
Medical Center Campus Library
Miami, Florida

Introduction

Medical imaging is one of the fastest growing fields in allied health. Technological advances have resulted in the discovery and application of new modalities that have greatly broadened the scope of practice of radiologic technologists. Radiographers use imaging equipment to perform procedures that facilitate the diagnosis of medical problems. They prepare patients for radiologic examinations by positioning the patients to radiograph specific parts of the body and by explaining the procedure at hand. Caution is exercised to prevent unnecessary radiation exposure. Additionally, radiographers must keep patient records, adjust and maintain equipment, prepare work schedules, evaluate equipment purchases, and often manage their departments. Radiation is used not only to reproduce images of the internal areas of the body, but to treat cancer as well. Radiation therapy technologists prepare cancer patients for treatment and administer prescribed doses of ionizing radiation to specific areas of the body. Other modalities that do not involve X-rays have expanded the field, such as ultrasound and magnetic resonance scans. Sonographers use nonionizing ultrasound equipment to produce images, and often they specialize in a specific area: neurosonography (the brain), vascular sonography (the veins, etc.), echocardiography (the heart), abdominal (the liver, kidneys, spleen, and pancreas), obstetric/gynecology, or ophthalmology (the eye).

Students of radiography can become registry eligible by attending a two-year, hospital-based certificate program, a two-year associate degree program, or a four-year baccalaureate program. The Joint Review Committee on Education in Radiologic Technology (JRCERT) approves all educational programs. Applicants for a degree in radiation therapy technology can enroll in a one-, two-, or four-year program. Students of one-year programs must have graduated from an accredited program in radiography or be an allied health professional able to demonstrate competence in specific areas. The *Allied Health Education Directory* (American Medical Association, 1978-), published annually, provides a listing of all accredited programs. Applicants to a sonography program must be graduates of a two-year accredited allied health or nursing program or have equivalent education in relevant college coursework. Graduates of these programs must pass a national certification examination. Additionally, some states require students to take a state exam before they can obtain their license. Radiographers, sonographers, and radiation therapists are expected to complete continuing education courses as an annual requirement for recertification.

The American Society of Radiologic Technologists, the Society of Diagnostic Medical Sonographers, and the JRCERT are among the many professional organizations providing guidance and setting standards in the field. The American Registry of Radiologic Technologists and the American Registry of Diagnostic Sonographers are organizations that examine and certify radiologic technologists and sonographers for the many specialties.

This bibliography includes materials that support a full range of topics and modalities specific to the field of medical imaging including:
- Radiography: X-rays.
- Computerized Tomography: X-ray images digitized via computer.
- Radiation Therapy: X-rays used in the treatment of diseases.
- Magnetic Resonance Imaging: Signals made by magnetic waves and radio waves passing through tissue are interpreted by a computer to form an image.
- Fluoroscopic Imaging: X-rays used in the observation of dynamic physiological functions such as the flow of barium through the gastrointestinal tract or the injection of a contrast medium into the heart.

- Ultrasonography: ultrasound waves transmitted through various body parts and the reflections bounce off tissue to create images digitized by a computer.

The selections listed in this bibliography were made in consultation with experts in the field and a review of subject bibliographies. Greater consideration was given to titles listed in two well-established and respected biennial bibliographies: Alfred N. Brandon and Dorothy R. Hill's "Selected List of Books and Journals in Allied Health Sciences" (*Bulletin of the Medical Library Association*, July 1992, v. 80, no. 3), and "Selected List of Books and Journals for the Small Medical Library" (*Bulletin of the Medical Library Association*, April 1993, v. 81, no. 2). Additionally, the feature article published in the March 1992 issue of *Choice*, "Radiologic Technology: A Core List for Community College Libraries," by Sue P. Forrest and Kenneth Neal, was very helpful.

Acknowledgment is extended to Gregory Ferenchak, Chairperson of the Radiologic Technology Program, and Joseph Schnetzer, Chairperson of the Diagnostic Sonography Program of Miami-Dade Community College, Medical Center Campus, for their guidance and recommendations of titles for inclusion in this bibliography.

Accreditation and Certification

American Registry of Diagnostic Sonographers
2368 Victory Pkwy. No. 510
Cincinnati, Ohio 45206-2810

American Registry of Radiologic Technologists
1255 Northland Dr.
Mendota Heights, Minn. 55120

Joint Review Committee on Education in Radiologic Technology
20 N. Wacker Drive, Ste. 900
Chicago, Ill. 60606

Society of Diagnostic Medical Sonographers
12225 Grenville Ave., Ste. 434
Dallas, Tex. 75243

Selected Associations

American Society of Radiologic Technologists
15000 Central Avenue SE
Albuquerque, N.M. 87123

Society of Diagnostic Medical Sonographers
12225 Grenville Ave., Suite 434
Dallas, Tex. 75243

Selected Journals

American Journal of Roentgenology.
American Roentgen Ray Society. Monthly. ISSN 0361-803X

Applied Radiology.
Romaine Pierson Publishing. Bimonthly. ISSN 0160-9963

Canadian Journal of Medical Radiation Technology.
Canadian Association of Medical Radiation Technologists. Quarterly. ISSN 0820-5930

Diagnostic Imaging.
Miller Freeman Publications. Monthly. ISSN 0194-2514

Journal of Diagnostic Medical Sonography.
Lippincott. Bimonthly. ISSN 1055-7997

Journal of the American Society of Echocardiography.
The Society. Bimonthly. ISSN 0894-7317

Journal of X-Ray Science and Technology.
Academic Press. Quarterly. ISSN 0895-3996

Radiologic Clinics of North America.
Saunders. Bimonthly. ISSN 0033-8389

Radiologic Technology.
American Society of Radiologic Technologists. Bimonthly. ISSN 0033-8397

Radiology.
Radiological Society of North America. Twelve times a year. ISSN 0033-8419

Ultrasonic Imaging.
Academic Press. Quarterly. ISSN 0161-7346

Allied Health — Diagnostic Technologies

Print Materials

✦ Radiology

Adler, Arlene McKenna, and Richard R. Carlton, eds.
Introduction to Radiography and Patient Care
Philadelphia, Saunders, 1994. 394p. ISBN 0721634656

A textbook designed for beginning radiologic technology students to provide an introduction to the field. Topics include basic information on the profession, educational requirements, techniques, equipment, interaction with patients, ethics, and legal issues.

Andolina, Valerie, Shelly Lille, and Kathleen M. Willison
Mammographic Imaging: A Practical Guide
Philadelphia, Lippincott, 1992. 292p. ISBN 0397510969

The importance of quality performance by radiographic technologists is emphasized in relation to of the accurate diagnosis of breast cancer. Topics include the history of mammography, patient considerations, equipment, processing, and quality assurance.

Ball, John, and Tony Price
Chesney's Radiographic Imaging
5th ed. Boston, Blackwell Scientific Publications, 1989. 380p. ISBN 0632019433

The principles of medical imaging are provided in this basic text for radiographic technologists. Topics include intensifiers and television imaging, digital imaging, digital subtraction, lens and mirror optics, radiographs, and xeroradiographs. Other modalities are also discussed such as ultrasound, computerized tomography, and radionuclide imaging.

Ballinger, Phillip W.
Merrill's Atlas of Radiographic Positions and Radiologic Procedures
7th ed. St. Louis, Mo., Mosby-Yearbook, 1991. 3 v.
ISBN 0801601703 (set)

A classic text in the field that is geared to the radiologic technology student learning proper positioning skills. Very popular among students. Contains glossary, bibliography, and an extensive index. A Brandon/Hill allied health list first-purchase selection. (New edition to be published in 1995.)

Bontrager, Kenneth L.
Textbook of Radiographic Positioning and Related Anatomy
Expanded 3rd ed. St. Louis, Mo., Mosby-Yearbook, 1993. 729p. ISBN 0801605377

Geared to student radiographers, this book is simply written and focuses on physical anatomy relative to patient positioning. Designed to improve the understanding of body structures and their anatomical relationships to help develop more competent radiographers. Illustrated. A Brandon/Hill allied health list selection. (New edition to be published in 1995.)

Bouchard, Eric A.
Radiology Management: A Guide for Administrators, Supervisors, and Students
Dubuque, Iowa, Shepherd, 1994. 2 v. ISBN 1881795063 (set)

A comprehensive reference manual designed for managers of a radiology department. Organized into four major areas: people, technology, planning, and management. Contains charts, graphs, and extensive examples.

Burns, Evelyn Frank
Radiographic Imaging: A Guide for Producing Quality Radiographs
Philadelphia, Saunders, 1992. 215p. ISBN 0721632467

Designed for beginning radiography students, this work presents the factors integral to the production and evaluation of quality radiographs. Heavily illustrated, it includes X-ray reproductions, charts, and pictures. "This book is a culmination of lecture notes and lesson plans developed during 20 years of teaching radiography student." Also available: *Laboratory Manual for Radiographic Imaging* (Saunders, 1992).

Bushong, Steward C.
Radiologic Science for Technologists: Physics, Biology, and Protection
5th ed. St. Louis, Mo., Mosby, 1993. 714p. ISBN 0801664551

Mathematical and physical concepts are integrated with radiologic technology. Topics include radiography, computed tomography, magnetic resonance imaging, and ultrasound. Illustrated. A workbook is also available.

Campeau, Frances
Limited Radiography
Albany, N.Y., Delmar Publishers, 1993. 332p. ISBN 0827333358

An introductory textbook aimed at beginning radiographers. Not intended for licensed radiologic technologists, this work provides information necessary to "limited permittee" technologists who produce radiographs. A Brandon/Hill allied health list selection.

Carlton, Richard R.
Principles of Radiographic Imaging: An Art and a Science
Albany, N.Y., Delmar Publishers, 1992. 718p. ISBN 0827336055

This excellent text, written by radiography educators for beginning students, covers all the basic principles of radiography. Chapters contain an outline, objectives, a summary, review questions, and a detailed bibliography. Topics include mathematical and physical principles, the art of radiography, radiographic film, and exposure. Also discusses special imaging systems such as fluoroscopy, tomography, digital image processing, computed tomography, and magnetic resonance imaging. A Brandon/Hill allied health list first-purchase selection.

Carlton, Richard R.
Radiography Exam Review
Philadelphia, Lippincott, 1993. 517p. ISBN 0397548990

A review book that prepares students to take the examination for Registered Technologist (radiographer), given by the American Registry of Radiologic Technologists (AART). Questions require problem-solving skills and are categorized by specific content areas. Contains a posttest consisting of 200 multiple-choice questions following the review activities. A Brandon/Hill allied health list selection.

Carroll, Quinn B.
Evaluating Radiographs
Springfield, Ill., Charles C. Thomas, 1993. 357p.
ISBN 0398058784

Carroll provides instructions for radiographers on how to make the necessary adjustments in both technique and positioning to produce an optimum radiographic view when the exposure is repeated. The topics covered include general considerations, positioning quality, and technical quality. Illustrated.

Carroll, Quinn B.
Fuchs's Radiographic Exposure, Processing and Quality Control
5th ed. Springfield, Ill., Charles C. Thomas, 1993. 530p.
ISBN 0398058210

Complements radiographic physics texts by bridging the gap between theory and practice. Assumes some basic knowledge of physical principles. A Brandon/Hill allied health list selection.

Cullinan, Angeline M.
Optimizing Radiographic Positioning
Philadelphia, Lippincott, 1992. 245p. ISBN 0397510500

This text complements positioning atlases by offering suggestions on how to optimize the quality of the images for optimal diagnosis by the radiologist. Heavily illustrated. A Brandon/Hill allied health list first-purchase selection.

Cullinan, Angeline M., and John E. Cullinan
Producing Quality Radiographs
2nd ed. Philadelphia, Lippincott, 1994. 319p. ISBN 0397550316

Designed to instruct radiologic technology students on how to produce quality radiographs, with the emphasis on image quality. Written in simple terms, the concepts included are radiation physics, radiation protection, quality assurance, and radiographic equipment. Includes photographs and illustrations.

Curry, Thomas S., James E. Dowdey, and Robert C. Murry
Christensen's Physics of Diagnostic Radiology
4th ed. Philadelphia, Lea and Febiger, 1990. 522p.
ISBN 0812113101

The physical principles of radiology are covered in this basic text for radiographic technology students. A Brandon/Hill allied health list selection.

Daffner, Richard H.
Clinical Radiology: The Essentials
Baltimore, Williams and Wilkins, 1993. 391p. ISBN 0683023306

Discusses many different types of modalities including the use of X-rays, nuclear imaging, ultrasound, and magnetic resonance imaging. Trends in the field are considered, especially the use of computer technology. Chapters are arranged according to body regions, e.g., pulmonary, cardiac, breast, abdominal, gastrointestinal, urinary, obstetric, and gynecologic. Illustrated.

Dennis, Cynthia A., and Ronald L. Eisenberg
Applied Radiographic Calculations
Philadelphia, Saunders, 1993. 179p. ISBN 0721665969

Provides practicing technologists with the mathematical calculations required to produce high-quality images. Through a workbook format, the text reviews general mathematics and furnishes the basic formulas and calculations used by radiographers. Topics covered include addition, subtraction, multiplication, division, mixed numbers, fractions, decimals, ratio, proportion, algebra and geometry, exponents, scientific notation, and metric conversions.

De Vos, Dianne C.
Basic Principles of Radiographic Exposure
Philadelphia, Lea and Febiger, 1990. 147p. ISBN 0812112229

Designed to give students a clear understanding of how to formulate techniques of radiographic exposure, this work utilizes a programmed approach, from its discussions of X-ray production and the properties of X-rays to X-ray tube rating and coding charts. Chapters include laboratory experiments to aid in problem solving and comprehension.

Drafke, Michael W.
Trauma and Mobile Radiography
Philadelphia, F. A. Davis, 1990. 347p. ISBN 0803628056

Intended for trauma and mobile radiography specialties within the field of medical imaging, this book covers basic radiographic principles (e.g., anatomy, routine patient positioning, basic patient care, and image production) that must be applied in the same way as in other areas of radiography. This text presents those basic procedures, plus guidelines for producing quality images with specialized mobile units. A Brandon/Hill allied health list selection.

Ehrlich, Ruth Ann, and Ellen Doble McCloskey
Patient Care in Radiography
4th ed. St. Louis, Mo., Mosby, 1993. 311p. ISBN 0801670586

A text written for radiographers to assist them in acquiring the technical and interpersonal skills needed to provide good patient care. Topics include ethics, patient communication techniques, mobility, physical consideration of the patient, infection control measures (including universal precautions), medications, emergency situations, imaging techniques, and bedside radiography. A Brandon/Hill allied health list selection.

Eisenberg, Ronald L.
Clinical Imaging: An Atlas of Differential Diagnosis
2nd ed. Gaithersburg, Md., Aspen Publishers, 1992. 1040p.
ISBN 0834202603

Guide to radiological signs for the effective diagnosis of diseases. This excellent presentation includes every organ or body system. A Brandon/Hill medical list first-purchase selection.

Eisenberg, Ronald L., and Cynthia A. Dennis
Comprehensive Radiographic Pathology
St. Louis, Mo., Mosby, 1990. 379p. ISBN 0801661420

This textbook for radiologic technology students is a helpful reference for the clinical setting. Heavily illustrated. A Brandon/Hill allied health list first-purchase selection.

Eisenberg, Ronald L., Cynthia A. Dennis, and Chris R. May
Radiographic Positioning
Boston, Little, Brown, 1989. 384p. ISBN 0316225436

The most common 181 positions for X-raying patients are detailed with both illustrations and photographs. Includes pertinent information for density and contrast scales for each position. A Brandon/Hill allied health list selection.

Eisenberg, Ronald L.
Radiology: An Illustrated History
St. Louis, Mo., Mosby, 1992. 606p. ISBN 0801615267

Illustrated text covering over 100 years of radiology's history. Presents an overview of the development of clinical radiology by body systems with up-to-date coverage of the latest imaging modalities and therapeutic techniques. Provides information on a wide variety of topics including forensic radiology, radiology in art and archaeology, and X-rays in military medicine and dentistry.

Gurley, LaVerne Tolley, and William J. Callaway, eds.
Introduction to Radiologic Technology
3rd ed. St. Louis, Mo., Mosby-Yearbook, 1992. 349p. ISBN 0801618339

Introductory text for radiologic technologists. Provides students with information regarding the profession and basic clinical concepts. Radiation safety in the laboratory and the clinical setting is emphasized. A Brandon/Hill allied health list first-purchase selection.

Handee, William R., and E. Russell Ritenour
Medical Imaging Physics
3rd ed. St. Louis, Mo., Mosby-Yearbook, 1992. 781p. ISBN 0815142412

Discusses the physical principles behind emerging technologies in the field of medical imaging, technologies such as magnetic resonance imaging, computed tomography, digital techniques in radiography, ultrasonography, and single-photon and position tomography. The expanding role of computers in diagnostic imaging is also examined.

Hiss, Stephen S.
Understanding Radiography
3rd ed. Springfield, Ill., Charles C. Thomas, 1993. 548p. ISBN 039805827X

Presents basic knowledge of radiography for both the new student and the experienced technologist, focusing on equipment, techniques, and physical principles.

Ireland, Sandra Jones
Integrated Mathematics of Radiographic Exposure
St. Louis, Mo., Mosby, 1994. 156p. ISBN 0815148348

A reference designed to help students evaluate and solve problems with radiographic techniques and understand why the solutions work. Includes problem-solving guidelines with excellent rationales for difficult calculations. Workbook format.

Kiefer, J.
Biological Radiation Effects
New York, Springer-Verlag, 1990. 444p. ISBN 0387510893

This comprehensive text describes the effects of radiation, discussing types of radiation, physical processes, mutation, radiation damage, radiobiology, radiation protection, and biological background.

Lefave, Linda
Mammography: Pretest Self-Assessment and Review
New York, McGraw-Hill, 1993. 100p. ISBN 0070520178

Written for those preparing to take mammography technician examinations, this review book provides typical questions for practice. Each of the six chapters, arranged to coincide with the ARRT's handbook on advanced-level examinations, contains fifty multiple-choice items.

Leonard, William L.
Radiography Examination Review: 800 Multiple Choice Questions with Explanatory Answers
7th ed. New York, Medical Examination Pub. Co., 1991. 242p. ISBN 0444016228

This review book for the registry exam provides test-taking guidelines, a basic review, and practice questions. A Brandon/Hill allied health list selection.

Long, Shirley M.
Handbook of Mammography
3rd ed. Edmonton, Alberta, Canada, Mammography Consulting Services, 1994. 231p. ISBN 0969486715

This handbook, which focuses on mammographic positioning, includes information on equipment, terminology, and processing procedures. Very well organized with clear black-and-white photographs.

Mace, James D., and Nina Kowalczyk
Radiographic Pathology for Technologists
2nd ed. St. Louis, Mo., Mosby, 1992. 432p. ISBN 0801670594

A comprehensive text intended to instruct radiographers on how to produce enhanced images for optimal diagnosis. Anatomy and physiology are detailed from the point of view of radiologic science. Illustrated. A Brandon/Hill allied health list selection.

Malott, Jack C., and Joseph Fodor
The Art and Science of Medical Radiography
7th ed. St. Louis, Mo., Mosby-Yearbook, 1993. 297p. ISBN 0801663210

This text for beginning students provides a firm foundation in radiation physics and techniques. Contains new information on stereoscopic radiography, computed radiography, radiation protection, and body section radiography. A Brandon/Hill allied health list first-purchase selection.

McKinney, William E. J.
Radiographic Processing and Quality Control
Philadelphia, Lippincott, 1988. 303p. ISBN 0397509022

A text for radiologic technologists on producing quality images that covers the basic steps for film processing and tips on how to produce a good radiograph. Other topics addressed are darkroom design, the chemistry of film manufacturing, sensitometry, processing, chemistry systems, electromechanical systems, maintenance, and quality control. Contains an appendix, bibliography, and a quiz.

Parelli, R. J.
Medicolegal Issues for Radiographers
2nd ed. Dubuque, Iowa, Eastwind Publishing, 1994. 188p. ISBN 1881795071

The legal issues of radiography are presented to radiologic technology students in this excellent text. Topics include tort law, labor law, legal doctrines, ethics, licensure, certification and credentialing, health care reform, patient rights, risk management, and forensic radiology. Contains review questions, references, glossary, and index.

Plaut, Simone
Radiation Protection in the X-Ray Department
Boston, Butterworth-Heinemann, 1993. 162p. ISBN 0750606061

Intended to promote awareness of the effects of radiation on health, this concise text presents information on the risks associated with medical X-rays. Offers instruction on radiation protection in an easily accessible form, with explanations of commonly used terms.

Prue, Lucinda K.
Atlas of Mammographic Positioning
Philadelphia, Saunders, 1994. 175p. ISBN 0721636837

Written for the radiologic technologist, this text on mammography demonstrates how to position patients to produce better breast images for the radiologist.

Reeders, Jacques W. A. J., F. H. Barneveld Binkhuysen, and J. F. W. M. Bartelsman, eds
Diagnostic Imaging of AIDS
New York, Thieme Medical Publishers, 1992. 135p.
ISBN 0865774161

A well-written book on the epidemiology, pathophysiology, and prevention of AIDS as it relates to the radiology department. Imaging aspects of the disease in the nervous system, the lungs, and the gastrointestinal tract are presented. Illustrations, some in color.

Saia, D. A.
Appleton and Lange's Review for the Radiography Examination
2nd ed. Norwalk, Conn., Appleton and Lange, 1993. 264p.
ISBN 0838500587

A review book of multiple-choice questions for radiologic technology students preparing for the board certification examination. Includes test-taking strategies, practice tests with answers and explanations, and a master bibliography

Shapiro, Jacob
Radiation Protection: A Guide for Scientists and Physicians
3rd ed. Cambridge, Mass., Harvard Univ. Press, 1990. 494p.
ISBN 0674745868

An advanced text explaining the principles and practice of radiation protection for those working with radiation sources in the field of medicine. Discusses a variety of rays such as beta particles, ionizing particles, gamma rays, and neutrons. Includes information on radiation dose calculations, measurements, and controls.

Sprawls, Perry
Principles of Radiography for Technologists
Rockville, Md., Aspen Publishers, 1990. 285p. ISBN 0834200880

Provides student technologists with information on the basic principles of radiography and how to apply them to obtain better image quality. A helpful work for those preparing for examinations, and as a general reference. Companion text to *Physical Principles of Medical Imaging* (Aspen Publishers, 1993), a text used by residents in radiology. Illustrated. A Brandon/Hill allied health list first-purchase selection.

Statkiewicz-Sherer, Mary Alice
Preparation for Credentialing in Radiography
Philadelphia, Saunders, 1993. 241p. ISBN 0721632823

Designed to prepare students to take the credentialing exam in radiography, this text contains multiple-choice practice questions and a general review.

Statkiewicz-Sherer, Mary Alice, Paul J. Visconti, and E. Russell Ritenour
Radiation Protection in Medical Radiography
2nd ed. St. Louis, Mo., Mosby, 1993. 317p. ISBN 0801657504

The safe use of X-rays by health professionals is covered in this fundamental text on radiation protection. Each chapter begins with learning objectives and ends with multiple-choice review questions. Contains illustrations, photographs, and a glossary.

Thompson, Michael A., and others
Principles of Imaging Science and Protection
Philadelphia, Saunders, 1994. 522p. ISBN 0721634281

Written by radiography educators with extensive experience in the classroom, this text includes an overview of radiography and discusses mathematics and physical concepts, radiation biology, radiation protection, and other imaging modalities. Illustrated.

Torres, Lillian S.
Basic Medical Techniques and Patient Care for Radiologic Technologists
4th ed. Philadelphia, Lippincott, 1993. 271p. ISBN 0397549636

Patients' needs are taken into consideration in this radiography text, which addresses methods of patient assessment and communication skills. There is strong emphasis on patient safety. Illustrated. A Brandon/Hill allied health list first-purchase selection.

Tortorici, Marianne R.
Concepts in Medical Radiographic Imaging: Circuitry, Exposure, and Quality Control
Philadelphia, Saunders, 1992. 363p. ISBN 0721631177

This text, designed for student radiographers, discusses radiographic techniques, circuitry, and quality control. Illustrated. A laboratory manual is also available.

Towsley, Doreen, and Eric Cunningham
Biomedical Ethics for Radiographers
Dubuque, Iowa, Eastwind Publishing, 1994. 100p.
ISBN 188179508X

Contemporary ethical challenges are the focus of this text, which presents many types of ethical dilemmas encountered by providers of radiographic services. Topics include informed consent, confidentiality, death and dying issues, health care distribution, and future trends.

Watkins, Gary L., and Thomas F. Moore
Atypical Orthopedic Radiographic Procedures
St. Louis, Mo., Mosby-Yearbook, 1993. 160p. ISBN 0801662702

This reference textbook for radiographers involved with orthopedic imaging presents imaging possibilities in orthopedic procedures and positioning, with special emphasis on patient considerations. Chapters are divided into regions: upper limb, shoulder girdle, body thorax, lower limb (foot and ankle), lower limb (patella and knee), hip, pelvis, and spine.

Webb, Steve
From the Watching of Shadows: The Origins of Radiological Tomography
New York, Adam Hilger, 1990. 347p. ISBN 085274305X

Who patented techniques for tomography? When was the first patient treated? Who built the first useful equipment? Did the early pioneers know about each others' work? These are some of the questions answered in this excellent book on the history of radiology. The chronological arrangement is augmented by black-and-white illustrations, charts, and an extensive reference list.

Wentz, Gini, ed.
Mammography for Radiologic Technologists
New York, McGraw-Hill, 1992. 131p. ISBN 0071053879

The basic principles and techniques of mammography are presented in this text for radiologic technologists. Contains information on quality imaging techniques and patient positioning. Illustrated.

Wolbarst, Anthony B.
Physics of Radiology
Norwalk, Conn., Appleton and Lange, 1993. 461p.
ISBN 0838557694

Based on the science and technology that underlie diagnostic medical images, this text details information on how imaging technologies actually work. Written for the more advanced student. Contains many charts, graphs, and photographs.

Wootton, R., ed.
Radiation Protection of Patients
New York, Cambridge Univ. Press, 1993. 152p.
ISBN 0521426693

A concise text providing essential advice and guidelines for medical and allied personnel involved with radiation for diagnostic or therapeutic purposes.

✦ Computed Body Tomography

Lee, Joseph K.T., Stuart S. Sagel, and Robert J. Stanley
Computed Body Tomography with MRI Correlation
2nd ed. New York, Raven Press, 1989. 1168p. ISBN 0881673315

The application of CT and CRI to the extracranial organs of the body are treated in this comprehensive text. Presents a clear explanation of computed tomography, MRI and CT technology, and a chapter on their application to specific areas such as the neck, thorax, mediastinum, lungs, the heart, gastrointestinal tract, and spleen.

Seeram, Euclid
Computed Tomography: Physical Principles, Clinical Applications, and Quality Control
Philadelphia, Saunders, 1994. 336p. ISBN 0721667104

This text is for radiologic technology students working with computed tomography. Coverage includes historical perspectives, image processing concepts, CT mathematics and physics, instrumentation, and clinical considerations when working with both children and adults.

✦ Radiation Therapy

Ang, K. K., Johannes H. A. M. Kaanders, and Lester J. Peters
Radiotherapy for Head and Neck Cancers: Indications and Techniques
Philadelphia, Lea and Febiger, 1994. 141p. ISBN 081211678X

Radiotherapy's important role in the management of patients with head and neck neoplasms is treated in this text, which focuses on the primary cancers of the head and neck regions that are considered rare.

Bentel, Gunilla Carleson
Radiation Therapy Planning: Including Problems and Solutions
New York, McGraw-Hill, 1993. 352p. ISBN 0071053824

An overview of radiation therapy is provided in this text, which also covers historical perspectives. Bentel discusses brachytherapy, radiation therapy equipment, dose determination and calculations, and problems encountered in radiation therapy treatment planning such as positioning, documentation, and errors.

Cox, James D., ed.
Moss' Radiation Oncology: Rationale, Technique, Results
7th ed. St. Louis, Mo., Mosby, 1994. 1024p. ISBN 0801669405

Continuing the classic work by William T. Moss, this basic text on the treatment, prognosis, and complications of tumors contains the basic physical and biological concepts of radiation therapy. Information on cancers is arranged by specific areas of the body and organ systems. A Brandon/Hill first-purchase selection.

Dowd, Steven B.
Practical Radiation Protection and Applied Radiobiology
Philadelphia, Saunders, 1994. 270p. ISBN 0721649173

A text designed to teach students in radiography, nuclear medicine, and radiation therapy how to protect themselves and their patients from ionizing radiation in the clinical setting.

Halperin, Edward C., and others
Pediatric Radiation Oncology
New York, Raven Press, 1989. 434p. ISBN 0881675474

Pediatric oncology is presented in this comprehensive textbook arranged by cancer types typically found in children.

Levitt, Seymour H., Faiz M. Khan, and Roger A. Potish, eds.
Levitt and Tapley's Technological Basis of Radiation Therapy: Practical Clinical Applications
2nd ed. Philadelphia, Lea and Febiger, 1992. 414p.
ISBN 0812114663

Discusses the changing role of radiation therapy as technology progresses. Topics presented include accurate localization, adequate dosage, and daily treatment fields. Illustrated. A Brandon/Hill allied health list first-purchase selection.

Perez, Carlos A., and Luther W. Brody, eds.
Principles and Practice of Radiation Oncology
2nd ed. Philadelphia, Lippincott, 1992. 1544p. ISBN 0397511620

This comprehensive text on the use of radiation to manage patients with cancer contains an excellent historical overview. Extensive coverage of radiation physics, radiation biology, clinical treatment planning, and the use of computers in radiation therapy.

Selman, Joseph
Basic Physics of Radiation Therapy
3rd ed. Springfield, Ill., Charles C. Thomas, 1990. 749p. ISBN 0898056854

The role of physics as it applies to radiation therapy is thoroughly discussed in this volume. Topics include the mathematical basis of radiation therapy, matter and energy, the nature of radiation, high-energy units, ionizing radiation and matter, exposure and exposure rate, X-ray quality, and dose distribution. The author also wrote *Fundamentals of X-Ray and Radium Physics* (Charles C. Thomas, 1985), a standard in the field. Illustrated.

Steel, G. Gordon, ed.
Basic Clinical Radiobiology: For Radiation Oncologists
Boston, Edward Arnold, 1993. 233p. ISBN 0340601442

The biological aspects of radiotherapy are presented in this text, which focuses on significant developments in the field that have led to improvements in the radiotherapeutic management of cancer. Early chapters deal with growth and cell proliferation in tumors and normal tissues. Later chapters treat biological developments in other areas of radiotherapy.

Stryker, John A.
Clinical Oncology for Students of Radiation Therapy Technology
St. Louis, Mo., Warren H. Green, 1992. 307p. ISBN 0875274889

Geared to the radiation therapy technologist, this text presents an overview of cancer and its therapies.

Wang, C. C.
Radiation Therapy for Head and Neck Neoplasms: Indications, Techniques and Results
2nd ed. Chicago, Year Book Medical Publishers, 1990. 426p. ISBN 0815191847

A textbook on the applied anatomy and treatment techniques used in the management of head and neck cancers. Illustrated.

✦ Magnetic Resonance Imaging

Berquist, Thomas H., ed.
MRI of the Musculoskeletal System
2nd ed. New York, Raven Press, 1990. 545p. ISBN 0881676675

An atlas for radiologists presenting MRI images of the musculoskeletal system. First purchase selection on the Brandon/Hill small medical library list.

Bushong, Stewart C.
Magnetic Resonance Imaging: Physical and Biological Principles
St. Louis, Mo., Mosby, 1988. 382p. ISBN 0801618207

The principles of magnetic resonance imaging and explanations of its use in diagnosing illness are detailed in this textbook written by an established author in the field. A Brandon/Hill allied health list selection.

Edelman, Robert R., and John R. Hesselink, eds.
Clinical Magnetic Resonance Imaging
Philadelphia, Saunders, 1990. 1192p. ISBN 0721622410

Reviews the clinical applications of MR imaging for four broad anatomical regions—the central nervous system, the chest and abdomen, the pelvis, and the musculoskeletal system. Also provides MR diagnosis and interpretation for specific diseases.

Elster, Allen D.
Questions and Answers in Magnetic Resonance Imaging
St. Louis, Mo., Mosby, 1994. 278p. ISBN 080167767X

Designed to answer typical questions that students ask during the course of studies in magnetic resonance, this work is intended to supplement more comprehensive texts on the subject. Assumes a basic clinical knowledge of MR imaging.

Newhouse, Jeffrey H., and Jonathan I. Wiener
Understanding MRI
Boston, Little, Brown, 1991. 155p. ISBN 0316604747

Numerous photographs of MR images are contained in this basic text on magnetic resonance imaging for beginning students. A Brandon/Hill allied health list first-purchase selection.

Rao, Krishna C.V.G., and others
MRI and CT of the Spine
Baltimore, Williams and Wilkins, 1994. 536p. ISBN 0683071335

Selection of the ideal imaging technique to obtain the needed clinical information with the least invasive procedure and the least expense is discussed in this text. Technical considerations are also included as is discussion of specific spinal problems and complications. Numerous illustrations and black-and-white photographs are provided.

Runge, Val M.
Magnetic Resonance Imaging of the Brain
Philadelphia, Lippincott, 1994. 586p. ISBN 0397512449

This atlas on magnetic resonance imaging of the head area; the brain, skull, face, and mastoids; and neck is an excellent reference. Contains halftone MR scans and some radiographs, CT scans, and angiograms.

Stark, David D., and William G. Bradley
Magnetic Resonance Imaging
2nd ed. St. Louis, Mo., Mosby-Yearbook, 1992. 2 v. ISBN 0801649307 (set)

Magnetic resonance imaging for clinicians and students is presented in this highly recommended, comprehensive text. The inclusion of many illustrations helps make this an excellent resource. A Brandon/Hill medical list first-purchase selection.

Westbrook, Catherine
MRI in Practice
Boston, Blackwell Scientific, 1993. 286p. ISBN 0632035870

The theoretical aspects of MRI are applied to the clinical setting in this illustrated text for radiographers, MRI technologists, and radiologists.

✦ Fluoroscopic Imaging

Anderson, Charles M., Robert R. Edelman, and Patrick A. Turski, eds.
Magnetic Resonance Angiography
New York, Raven Press, 1993. 498p. ISBN 0781700949

Highly recommended for students needing basic and detailed information, this text discusses the techniques and applications of the most routine MRA procedures in the clinical setting. Includes many black-and-white photographs.

Cope, Constantin, Dana R. Burke, and Steven G. Meranze
Atlas of Interventional Radiology
Philadelphia, Lippincott, 1990. 1 v. (various pagings).
ISBN 039744656X

This atlas of the most common interventional procedures presents information in a concise manner. Describes procedural indications, contraindications, methodology, applications, reports of success, and complications and their prevention. Assumes a basic knowledge of catheterization techniques.

Grossman, William, and Donald S. Baim, eds.
Cardiac Catheterization, Angiography, and Intervention
4th ed. Philadelphia, Lea and Febiger, 1991. 698p.
ISBN 081211342X

The major techniques used in cardiac catheterization angiography and some techniques in interventional cardiology are described in this reference text. Topics include principles and techniques of cardiac catheterization, evaluation of cardiac function, and interventional techniques.

Kadir, Saadoon, ed.
Current Practice of Interventional Radiology
Philadelphia, Pa., B.C. Decker, 1991. 779p. ISBN 1556641303

Interventional radiology, including both diagnostics and techniques, is the focus of this comprehensive text. Kadir presents invasive radiologic procedures according to the regions of the body.

Lubell, David L.
Cath Lab: An Introduction
Philadelphia, Lea and Febiger, 1989. 133p. ISBN 0812112628

This text is geared to the cardiac catheterization team—those assisting the practitioner of invasive cardiology to perform the catheterization procedure. Covers general procedures, imaging techniques, and specialized techniques.

Moore, Robert J.
Imaging Principles of Cardiac Angiography
Rockville, Md., Aspen Publishers, 1990. 258p. ISBN 0834201208

Moore presents a guide to understanding how angiography images are made and shows how to achieve image quality. Clearly written, this text provides precise instructions, clinical examples, and illustrations that lead students step-by-step through the process of angiography. Also discusses issues of radiation safety and proper use of equipment.

Rosen, Robert J., and John Nosher
Angiography and Interventional Radiology
Philadelphia, Lippincott, 1991. 1 v. (various pagings).
ISBN 1563750031

This atlas on the pathological and anatomical features of angiography and interventional radiology includes basic diagnostic angiographic techniques and applications as well as interventional procedures methodology. Numerous illustrations.

Seibel, Rainer M. N., and Dietrich H. W. Gronemeyer
Interventional Computed Tomography
Boston, Blackwell Scientific Publications, 1990. 355p.
ISBN 0865421277

Focuses on the blending of interventional computed tomography with interventional radiology. Covers the entire body including the head, neck, lungs, abdomen, and skeleton. Overall arrangement is by body regions. Includes therapeutic procedures and a discussion of the trends in the field.

✦ Ultrasound

Babikian, Viken L. and Lawrence R. Wechsler
Transcranial Doppler Ultrasonography
St. Louis, Mo., Mosby-Yearbook, 1993. 323p. ISBN 1556643942

This text focuses on the clinical aspects and research applications of Doppler ultrasonography in the examination of intracranial circulation. Of particular interest is the diagnosis of blood flow velocity, as well as cerebrovascular and circulatory disorders. Illustrated.

Bushong, Stewart C., and Benjamin R. Archer
Diagnostic Ultrasound: Physics, Biology and Instrumentation
St. Louis, Mo., Mosby-Yearbook, 1991. 177p. ISBN 0801603943

This text for sonography students discusses the basic physical and biological principles of ultrasound. The author introduces basic instrumentation. A Brandon/Hill allied health list first-purchase selection.

Chervenak, Frank A., G. Isaacson, and Stuart Campbell, eds.
Ultrasound in Obstetrics and Gynecology
Boston, Little, Brown, 1993. 2 v. ISBN 0316138657 (set)

This highly recommended reference on the use of ultrasound in obstetrics and gynecology discusses theory and practice of ultrasound and its future implications. A Brandon/Hill small medical library list first-purchase selection.

Craig, Marveen
Introduction to Ultrasonography and Patient Care
Philadelphia, Saunders, 1993. 217p. ISBN 0721642292

Emphasis is placed on patient care considerations and rudiments of ultrasonography for entry-level students.

Craig, Marveen
Ultrasound Exam Review: Sonographer's Self-Assessment Guide
2nd ed. Philadelphia, Lippincott, 1994. 292p. ISBN 0397550219

This is an excellent review for the ultrasound exam. Chapters are divided into subspecialties: abdomen, neonatal neurosonography, obstetrics, gynecology, echocardiography and vascular sonography some color illustrations and answers to the practice questions.

Cudleigh, Patricia, and J. Malcolm Pearce
Obstetric Ultrasound: How, Why and When
2nd ed. New York, Churchill Livingstone, 1992. 323p.
ISBN 0443042071

This work focuses on ultrasound techniques for use during pregnancy examinations. Introduces Doppler ultrasound and transvaginal sonography. Illustrated with half-tone scans, some in color.

Diagnostic Technologies — Allied Health

Evans, D. H., and others
Doppler Ultrasound: Physics, Instrumentation, and Clinical Applications
New York, Wiley, 1989. 297p. ISBN 0471914894
   Doppler ultrasound is discussed in this advanced text that is written for practioners. Provides information on instrumentation and reviews some physics concepts.

Feigenbaum, Harvey
Echocardiography
5th ed. Philadelphia, Lea and Febiger, 1994. 695p.
ISBN 0812116925
   A standard reference on the instrumentation, examination, and clinical uses of echocardiology. Arrangement is by disease or anatomical part. Includes illustrations and extensive references.

Foley, W. Dennis, and Donald S. Emerson
Color Doppler Flow Imaging
Boston, Andover Medical Publishers, 1991. 175p.
ISBN 0962652121
   A concise text for imaging specialists that provides a clinical overview. Contains some color photographs.

Goldberg, Barry B., ed.
Textbook of Abdominal Ultrasound
Baltimore, Williams and Wilkins, Co., 1993. 527p.
ISBN 0683036246
   This text for sonographers highlights the use of ultrasound in the abdomen, retroperitoneum, and pelvic region. Focuses on the techniques and physical principles of both ultrasound and Doppler imaging. Illustrated.

Hagen-Ansert, Sandra L.
Textbook of Diagnostic Ultrasonography
3rd ed. St. Louis, Mo., Mosby, 1989. 1095p. ISBN 0801624460
   Surveys ultrasound principles, including instrumentation physics and clinical applications. This is a Brandon/Hill allied health list first-purchase selection.

Houston, Alan B., and Iain A. Simpson, eds.
Cardiac Doppler Ultrasound: A Clinical Perspective
Boston, Wright, 1988. 154p. ISBN 0723609950
   Focuses on the benefits of using Doppler ultrasound to diagnose heart conditions. Emphasizes techniques and patient care considerations within the clinical setting.

Jaffe, Richard, and Steven L. Warsof
Color Doppler Imaging in Obstetrics and Gynecology
New York, McGraw-Hill, 1992. 306p. ISBN 0071054200
   Presents an overview on the use of Doppler ultrasound in the area of obstetrics and gynecology. Discusses operating principles of this technique and its clinical and research potential. Contains numerous illustrations.

Jawad, Ibrahim A.
A Practical Guide to Echocardiography and Cardiac Doppler Ultrasound
Boston, Little, Brown, 1990. 379p. ISBN 0316458325
   This is an excellent introduction to Doppler echocardiography and ultrasound principles and techniques, especially for the diagnosis of heart conditions. A Brandon/Hill allied health list selection.

Kim, Tok-su, and Dan Eviathar Orron, eds.
Peripheral Vascular Imaging and Intervention
St. Louis, Mo., Mosby-Yearbook, 1992. 619p. ISBN 0801654831
   The focus of this work is on the peripheral vascular system. Includes a thorough discussion on the basics of peripheral arterial disease. Topics include angiographic anatomy, interpretation, cross-sectional imaging and interventional techniques (including percutaneous transluminal angioplasty, thrombolysis, transcatheter therapy, and inferior vena cava filter placement). Selected emerging modalities included are angioscopy, transluminal ultrasound, radionuclide thrombus imaging, and laser angiosurgery.

Kisslo, Joseph A., David Adams, and Robert N. Belkin
Doppler Color Flow Imaging
New York, Churchill Livingstone, 1988. 183p. ISBN 0443085633
   This advanced textbook on Doppler technology discusses techniques and uses especially as applied to cardiovascular applications.

Krebs, Carol A., Vishan L. Giyanani, and Ronald L. Eisenberg
Ultrasound Atlas of Disease Processes
Norwalk, Conn., Appleton and Lange, 1993. 432p.
ISBN 0838592457
   Written for both clinicians and students, this atlas and textbook contains information on normal sonographic anatomy, typical differences, and common abnormal conditions. Illustrated with halftone sonograms.

Kremkau, Frederick W.
Diagnostic Ultrasound: Principles and Instruments
4th ed. Philadelphia, Saunders, 1993. 411p. ISBN 0721643086
   Used as reference textbook in many sonography programs, this highly recommended, essential text discusses ultrasound concepts, instrumentation, Doppler ultrasound concepts and patient safety issues. Includes review questions and answers. A Brandon/Hill allied health list first-purchase selection.

Kremkau, Frederick W.
Doppler Ultrasound: Principles and Instruments
Philadelphia, Saunders, 1990. 286p. ISBN 0721628648
   Provides imaging technologists with an in-depth treatment of the physical properties of Doppler ultrasound. Focus is on instrumentation. Contains some color illustrations. A Brandon/Hill allied health list selection.

Kurjak, Asim, ed.
Atlas of Ultrasonography in Obstetrics and Gynecology
Park Ridge, N.J., Parthenon Pub. Group, 1992. 236p.
ISBN 1850703620
   This atlas is designed for gynecologists, obstetricians, infertility specialists, and radiologists. Includes ultrasound, color Doppler, and normal and abnormal fetal anatomy. Half-tone and color-scan illustrations are provided. Kurjak is also the editor of *Atlas of Transvaginal Color Doppler* (Parthenon Pub. Group, 1994).

Labovitz, Arthur, and George Williams
Doppler Echocardiography: The Quantitative Approach
3rd ed. Philadelphia, Lea and Febiger, 1992. 131p.
ISBN 0812114299
   This introductory text for sonographers and practicing cardiologists includes basic principles of echocardiography as well as

topics on specific disorders. Illustrated with color and half-tone scans. A Brandon/Hill allied health list selection.

Lanzer, P., and A. P. Yoganathan
Vascular Imaging by Color Doppler and Magnetic Resonance
New York, Springer-Verlag, 1991. 338p. ISBN 0387533206
   Primarily designed for radiologists, cardiologists, and vascular surgeons, this work provides a concise, state-of-the-art review of the application of new imaging modalities to cardiovascular diagnostic imaging.

Nanda, Navin C., and Reinhard Schlief, eds.
Advances in Echo Imaging Using Contrast Enhancement
Boston, Kluwer Academic Publishers, 1993. 405p.
ISBN 0792321375
   This advanced text for cardiologists focuses on echocardiography with contrast agents. Part One discuses history, basics, and safety of contrast agents; Part Two includes clinical uses of contrast agents; and Part Three deals with future perspectives.

Nanda, Navin C.
Atlas of Color Doppler Echocardiography
Philadelphia, Lea and Febiger, 1989. 544p. ISBN 0812110781
   This is an excellent reference text for physicians and technologists using color Doppler technology.

Nanda, Navin C., ed.
Doppler Echocardiography
2nd ed. Philadelphia, Lea and Febiger, 1993. 466p.
ISBN 0812115880
   Text for cardiology residents and technicians on the fundamentals and applications of Doppler techniques. Includes conventional and color Doppler, transesophageal echocardiography, and more. Color and half-tone scans.

Neilson, James P., and S. E. Chambers, eds.
Obstetric Ultrasound One
New York, Oxford Univ. Press, 1993. 298p. ISBN 0192622242
   Reviews the application of obstetric ultrasound within a historical context. In addition, pioneers in the field have contributed authoritative critiques of important new trends. Illustrated.

Obeid, Anis I.
Echocardiography in Clinical Practice
Philadelphia, Lippincott, 1992. 383p. ISBN 0397510241
   Clinicians working with echocardiography, Doppler echocardiography, and color Doppler will find this highly recommended source of value. Includes transesophageal echocardiography. A Brandon/Hill small medical library list selection.

Odwin, Charles S.
Appleton and Lange's Review for the Ultrasonography Examination
2nd ed. Norwalk, Conn., Appleton and Lange, 1993. 584p.
ISBN 083859073X
   This review book will aid students who are preparing for the American Registry of Diagnostic Medical Sonography Examination. Each chapter covers a main topic. Questions are accompanied by answers and rationales.

Pearce, J. Malcolm, ed.
Doppler Ultrasound in Perinatal Medicine
New York, Oxford Univ. Press, 1992. 346p. ISBN 0192620193
   This text, designed for obstetricians, radiographers, and ultrasonographers, focuses on the applications of Doppler ultrasound in the screening and monitoring of high-risk pregnancies. Three main sections cover basic considerations, statistical studies, and pulsed wave and color flow Doppler.

Powis, Raymond L., and Robert A. Schwartz
Practical Doppler Ultrasound for the Clinician
Baltimore, Williams and Wilkins, 1991. 187p. ISBN 0683069586
   This concise reference for imaging specialists and physicians introduces Doppler principles and technology as a diagnostics instrument.

Reece, E. Albert, Israel Goldstein, and John C. Hobbins
Fundamentals of Obstetric and Gynecologic Ultrasound
Norwalk, Conn., Appleton and Lange, 1994. 272p.
ISBN 0838592473
   Provides beginning students with the essentials of ultrasound technology for use with obstetric and gynecologic patients. Begins with basic techniques and progresses to more advanced applications. Contains ultrasonic images and illustrations.

Sanders, Roger C., ed.
Clinical Sonography: A Practical Guide
2nd ed. Boston, Little, Brown, 1991. 557p. ISBN 0316770167
   Covers a wide range of information essential to the overall understanding of sonography such as physics, instrumentation, and Doppler principles. Details the diagnostic uses of sonography in specific areas such as in fetal death and sickness, pain in upper and lower quadrants, liver function tests, and renal problems. A Brandon/Hill allied health list first-purchase selection.

Shirkhoda, Ali, and Beatrice L. Madrazo, eds.
Pelvic Ultrasound
Baltimore, Williams and Wilkins, 1993. 298p. ISBN 0683076973
   This comprehensive reference on the ultrasound of the male and female pelvis aids in the diagnosis of genital diseases. Contains current techniques such as color Doppler; excludes obstetrical ultrasound. Illustrated with color and half-tone scans.

Silverman, Norman H.
Pediatric Echocardiography
Baltimore, Williams and Wilkins, 1993. 628p. ISBN 0683077139
   This text addresses echocardiography particularly in regard to assessment of heart conditions in infants and children. Includes ultrasound, cross-sectional imaging, all Doppler modalities, and contrast echocardiography. Illustrated.

Teele, Rita L., and Jane Chrestman Share
Ultrasonography of Infants and Children
Philadelphia, Saunders, 1991. 507p. ISBN 072168775X
   A highly recommended text for sonographers concerned with the patient care needs of infants and children. The typical uses of ultrasonography in pediatrics are discussed. A Brandon/Hill small medical library selection.

Tempkin, Betty Bates
Ultrasound Scanning: Principles and Protocols
Philadelphia, Saunders, 1993. 331p. ISBN 072163706X

This informative programmed text for student sonographers focuses on scanning and image techniques, particularly in regard to specific body regions and organs. Illustrated, some with color.

Weyman, Arthur E.
Principles and Practice of Echocardiography
2nd ed. Philadelphia, Lea and Febiger, 1994. 1335p. ISBN 0812112075

This comprehensive reference presents physical principles, instrumentation, computer applications, and miscellaneous echocardiographic techniques: transesophageal, epicardial, intraoperative, and catheter-based (intravascular and intracardial).

Zwiebel, William J., ed.
Introduction to Vascular Ultrasonography
3rd ed. Philadelphia, Saunders, 1992. 423p. ISBN 0721635512

The basics of ultrasound vascular diagnosis are presented in this text for sonographers. Focus is on sonographic technique, normal features, and major pathologies. Includes duplex and color-duplex techniques. Illustrated.

✦ Continuations

The National Council on Radiation Protection publishes extensive material in the field of medical imaging. They strive to promote research in the field and to establish guidelines for the safety of both the practitioner and the patient. Monographs, reports, and proceeding are published on a continuous basis. These publications are essential to any medical imaging collection. Libraries may wish to place a standing order to receive all their publications as they are published. For a listing of titles, write to the National Council on Radiation Protection, 7910 Woodmont Ave., Suite 800, Bethesda, MD 20814.

Nonprint Materials

Euganeo, Kathleen Doran, and Allan B. Evantash
Radiographic Exposure and Technique
St. Paul, Minn., Image Premastering Services, 1992. 1 videodisk and 1 manual

Theoretical concepts of the physics and exposure principles of radiography are illustrated with animations, diagrams and radiographic examples. Images are of high quality. Students may follow along with a workbook.

Evantash, Allan B., Kathleen Evans, and Joel Kahn
An Interactive Approach to Radiographic Anatomy and Positioning
St. Paul, Minn., Image Premastering Services, 1991. 1 videodisk and 2 manuals

This interactive video provides an introductory course for radiologic technology students. Contains approximately 150 common projections, which are divided in four sections: anatomy, basic positioning, technique, and evaluation criteria. Contains radiation protection considerations and terminology. Text is enhanced with illustrations, slides, radiographs, and video sequences.

Forbes, Glenn S.
Radiologic Atlas of Brain Tumors
St. Paul, Minn., Image Premastering Services, 1991. 1 videodisk, barcode book

This reference work of approximately 1,045 cases and 10,000 images of numerous types of brain tumors is arranged according to pathology. Each case contains basic information on patient demographics, histological subtype, and imaging characteristics. Images can be viewed by selecting the appropriate bar-code.

Herrmann, Tracy
Essentials of Skull Radiography
Cincinnati, Ohio, Seven Hills Radiology, 1990. 1 videocassette (51 min.) VHS

This interactive approach to radiographic anatomy discusses positioning and pathology of the facial bones, paranasal sinuses, and the orbits.

Kettlehake, Jane E.
Fracture Radiography
Cincinnati, Ohio, Seven Hills Radiology, 1989. 1 videocassette (45 min.) VHS

Presents positioning techniques for patients with fractures and discusses the mechanism of injury and types of fractures.

Dietetics and Nutrition

Judy Caramanica
Houston Community College System
and
North Harris Montgomery Community College District
Houston, Texas

Introduction

The research and popular literature on nutrition has increased dramatically in the last few decades. This proliferation has been fueled by an increased public interest in nutrition, a rising concern for healthier lifestyles, and the promotion of physical and health education. As the United States "baby boomer" population has aged, more emphasis has been placed on nutrition to prolong one's lifespan, and prevent diseases. Another factor lies in our ethnically diverse population, which enriches the traditional American diet, but also presents issues that need to be addressed in nutrition and health care. Finally, the media, through its presentation of truths, half-truths, facts, fallacies, and fads, has stirred the public's interest in nutrition and its impact on individual nutrition needs and choices.

In vocational training, the study of nutrition is usually divided into two distinct areas of concentration: management of food services, and nutrition related to health care. Food service managers are responsible for the planning, purchasing, preparation, and serving of food and beverages in a variety of settings. These include hotel dining rooms, large institutional cafeterias, and other areas where quantities of food are involved. Several community colleges provide specialized courses in these areas. More than one hundred colleges offer a specialized four-year curriculum pertaining to hotel administration and food management. Resources for these curricular areas are presented in the *Culinary Arts/Food Services* and *Hospitality Management* chapters of this book.

The focus of this bibliography is on nutrition as it applies to health care. Students in a dietetics or nutrition vocational program are prepared to assist dietitians in maintaining all the nutritional aspects of health care. Although traditional employment has been in acute care, rehabilitation hospitals, and community agencies, nutritional professionals and paraprofessionals are now securing employment in weight control clinics, health and exercise spas, gymnasiums, community agencies concerned with nutritional education, and in developing and distributing nutritional-related programs.

Dietitians, nutritionists, and their paraprofessional partners already make up a large share of the health care professionals. Career options include employment in a full-time, part-time, or free-lance capacity. Although registered dietitians require a four-year bachelor's degree with a major in dietetics, nutrition, home economics, food science, or food service management, paraprofessional programs in community colleges and vocational/technical schools include the following options in dietetics and nutrition specialization:
- The administrative or management dietitian, who administers nutrition programs
- The clinical or therapeutic dietitian, who designs specific diets for patients in a health care facility
- The educational dietitian, who teaches principles of nutrition in educational facilities or to health care teams in hospitals or research facilities
- The community dietitian, who practices nutrition counseling at the community level
- The research dietitian, who usually has an advanced degree and may do research at a medical care center or for the community.

A program leading to a two-year associate degree as dietetic technician is offered by many community colleges and vocational-technical programs. In addition, these same institutions may also offer a one-year program leading to a degree as a dietetic assistant. Certification through an examination is offered by the American Dietetic Association for these two degrees.

Increased research has led to a proliferation of both technical and popular books on diet and nutrition. This bibliography is a representative sampling of the types of resources available. Some titles have been included to demonstrate the diversity and breadth of the topic, while others provide basic information for both the layperson and professional. Basic reference tools by health care professionals are also included.

Certification

American Dietetic Association. Commission on Dietetic Registration
216 W. Jackson Blvd., Suite 800
Chicago, Ill. 60606

Dietary Managers Association
400 E. 22nd Street
Lombard, Ill. 60148

The following associations are good sources of additional information on education, industry trends, and continuing education:

American Hotel and Motel Association
888 Seventh Ave.
New York, N.Y. 10106

Council on Hotel, Restaurant, and Institutional Education
Human Development Building, Suite S-208
University Park, Pa. 16801

Selected Journals

American Journal of Clinical Nutrition.
American Society for Clinical Nutrition.
Waverly Press. Monthly. ISSN 0002-9165

Journal of Nutrition.
Federation of American Societies for Experimental Biology.
Monthly. ISSN 0022-3166

Journal of the American Dietetic Association.
The Association. Frequency varies. ISSN 0002-8223

Nutrition Action Health Letter.
Center for Science in the Public Interest. Ten issues a year.
ISSN 0885-7792

Nutrition Reviews.
Springer-Verlag. Monthly. ISSN 0029-6643

Nutrition Today.
Williams and Wilkins. Bimonthly. ISSN 0029-666X

Proceedings of the Nutrition Society.
Cambridge Univ. Press. Three issues a year. ISSN 0029-6651

Print Materials

American Dietetic Association
Handbook of Clinical Dietetics
2nd ed. New Haven, Conn., Yale Univ. Press, 1992. 588p.
ISBN 0300052189

This handbook addresses therapeutic nutrition and nutritional intervention in disease states. It also foster "some degree of uniformity in the composition of diets by providing a comprehensively researched basis for the definitions, purposes, effects, physiology, indications, possible adverse reactions, contraindications, nutritional assessments, and quality assurance priorities" (pref.). Provides strategies for implementation and education. Includes numerous tables, references, appendices, and index.

Annual Review of Nutrition
Palo Alto, Calif., Annual Reviews, 1981- . v. ISSN 0199-9885

This publisher issues annual reviews on a variety of subjects ranging from anthropology to sociology. Volume Twelve in this series, for example, includes articles on nutrition research in India, nutrition in space, fat substitutes, and the body's processing of vitamin A. The content varies from year to year. Includes index.

Applegate, Elizabeth Ann
Power Foods: High-Performance Nutrition for High-Performance People
Emmaus, Pa., Rodale Press, 1991. 288p. ISBN 0878579672

This work is written for the athlete or person wanting to improve work performance. The first section includes chapters on beverages, carbohydrates, proteins, supplements, and fats. Each chapter discusses the body's use of these substances and which are most appropriate for an athlete. The second section describes a "Power-Eating Plan" which incorporates the power foods. Includes numerous tables and an index.

Bricklin, Mark
Prevention Magazine's Complete Nutrition Reference Handbook: Over 1,000 Foods and Meals Analyzed and Rated for Health Effect
Emmaus, Pa., Rodale Press, 1992. 596p. ISBN 0875961177

Rates prepared and natural foods and lists chief nutrients. Included are caloric, fat, sodium, protein, carbohydrate, and dietary fiber content. A description is provided for each food, including any cautions or outstanding attributes.

Burtis, Grace, Judi Davis, and Sandra Martin
Applied Nutrition and Diet Therapy
Philadelphia, Saunders, 1988. 823p. ISBN 0721612822

Authors stress a holistic approach to dietary management that offers a different focus to the subject. Each chapter has "Focus on" boxes that include information on specific topics intended to broaden the learning experience. "Fallacies and Facts" contain discussions of nutritional fads and current misinformation. Chapters are summarized and followed by a reference list.

Byrne, Kevin P.
Understanding and Managing Cholesterol: A Guide for Wellness Professionals
Champaign, Ill., Human Kinetics Books, 1991. 334p.
ISBN 0873223098

Written for the large population whose cholesterol level is in the middle range. Part One provides background information on blood cholesterol; Part Two offers other contributions to atherosclerosis; Part Three discusses theory and rationale behind dietary change; and Part Four assists health professionals in designing and implementing a cholesterol-lowering program. Includes a resource list of organizations, newsletters, community programs, and videos.

Catsberg, C. M. E., and G. J .M. Kemper-Van Dommelen
Food Handbook
New York, Halstead Press, 1989. 382p. ISBN 0747600546

Provides comprehensive information on each food category and allied products, including production, technology, additives, preservation, and quality deterioration and spoilage. Discusses legislation pertaining to each food type. Bibliography and index.

Coleman, Ellen
Eating for Endurance
Palo Alto, Calif., Bull Pub. Co., 1988. 158p. ISBN 0915950871

This work focuses on how diet affects one's ability to compete in sports. The first chapter explains how the body uses oxygen during competition, and how the body's reserve of glucose is related to food consumption. Discusses carbohydrate loading, use of caffeine, and faddism. Includes tables, appendices, and index.

Combs, Gerald F.
The Vitamins: Fundamental Aspects in Nutrition and Health
San Diego, Academic Press, 1992. 528p. ISBN 0121834905

Written as a textbook for background reading and also as a desk reference for professionals. Very technical in treatment, this work includes a chapter on each vitamin, concluding with a recommended reading list. Includes index.

Conners, C. Keith
Feeding the Brain: How Foods Affect Children
New York, Plenum Press, 1989. 277p. ISBN 0306433060

The author presents a balanced review of the effects of diet, vitamins, additives, and environmental toxins on children's behavior and functioning. A summary follows each chapter.

Drummond, Karen Eich
Nutrition for the Foodservice Professional
2nd ed. New York, Van Nostrand Reinhold, 1994. 576p.
ISBN 0442013701

Discusses nutrition relating to health, menu planning, recipe modification, foods, and their preparation in food service operations. Includes appendices and index.

Dudek, Susan G.
Nutrition Handbook for Nursing Practice
2nd ed. Philadelphia, Lippincott, 1993. 722p. ISBN 0397549288

Very comprehensive and up-to-date. The author discusses basic principles of nutrition and the nutrient requirements of the various stages of life. In the third section, "Nutrition in Clinical Practice," disorders are described and diet therapy objectives are presented. Includes several appendices and an index.

Dunne, Lavon J.
Nutrition Almanac
3rd ed. New York, McGraw-Hill, 1990. 340p. ISBN 0070349126

Discussions include detailed descriptions of nutrients, nutrients that function together, and the use of nutrients beneficial for numerous ailments. Provides information on herbs and herbal preparations. This edition includes new chapters on drinking water and its uses in the body, how pollutants in water affect the body, and waterborne diseases. Includes "Table of Food Composition," bibliography, and index.

Erdmann, Robert, and Meirion Jones
Fats, Nutrition, and Health: The Complete Guide to What Fats Are, What They Do, and What Makes Them Healthy or Harmful
Wellingborough, England, Thorsons, 1990. 144p.
ISBN 0722519621

The title is a good summary of this work's contents. Dietitian and layperson alike will find this book useful, although discussions on chemical make-up of different fats are technical. Among the appendices is a list of sources of fats and oils arranged according to their chemical composition. Includes index.

Escott-Stump, Sylvia
Nutrition and Diagnosis-Related Care
2nd ed. Philadelphia, Lea and Febiger, 1988. 618p.
ISBN 0812111613

Written as a reference tool to be used by dietetic practitioners, interns, and students to aid in determining protocols and priorities in nutritional care for numerous disorders. Each disorder listed includes a succinct description of the disease, dietary recommendations, and side effects of common drugs used for the disease. Includes an extensive bibliography and appendices.

Finn, Susan Calvert, and Linda Stern Kass
The Real Life Nutrition Book: Making the Right Food Choices Without Changing Your Life-Style
New York, Penguin Books, 1992. 389p. ISBN 0140131744

Written to simplify dietary information bombarding the consumer and to modify the resulting confusion. Offers recommendations and nutritional needs of single and combined families, and complex food choices facing today's shoppers. Provides helpful guidelines for good nutritional choices in fast food restaurants and for coping with stress. Includes various charts, appendices, an annotated list of cookbooks, and an index.

Gallagher-Allred, Charlette R.
Nutritional Care of the Terminally Ill
Rockville, Md., Aspen Publishers, 1989. 291p. ISBN 0834200600
    This author draws from ten years of experience as a volunteer consulting nutritionist in a Riverside hospice. She provides a solid text on the nutritional needs of dying persons in a palliative care setting. A fascinating, thoughtful book on a somber subject.

Garrison, Robert H., and Elizabeth Somer
The Nutrition Desk Reference
New Canaan, Conn., Keats, 1990. 306p. ISBN 0879835230
    "Combines nutrition basics and current research in a documented format for the health professional and the general public" (pref.). Information is given in a very succinct manner and includes numerous statistics, charts, diagrams, and references. Discusses the use of nutrition in prevention and treatment of many diseases including diabetes, eye disorders, kidney disorders, etc. Includes glossary and index.

Gilbert, Sara
Psychology of Dieting
New York, Routledge, 1989. 155p. ISBN 0415028442
    Discusses who diets, why, the health risks of being overweight, causes of obesity, different aspects of hunger and binge eating, and eating diseases. Offers suggestions on how to succeed in dieting. Includes notes at the end of each chapter and an index.

Gittleman, Ann Louise
Super Nutrition for Women: A Food-Wise Guide for Health, Beauty, Energy, and Immunity
New York, Bantam Books, 1991. 254p. ISBN 0553353284
    Author presents a scientific nutritional program based on a woman's body chemistry and "unique nutritional needs." Includes chapters on types of sugars and how each is used in the body, calcium inhibitors (coffee, tea, chocolate), iron requirements, minerals to avoid to make PMS less troubling, and a chapter on foods that encourage yeast infection. Includes selected references, recipes, and index.

Hamilton, Eva May Nunnelley, Eleanor Noss Whitney, and Frances Sienkiewicz Sizer
Nutrition: Concepts and Controversies
5th ed. St. Paul, Minn., West Pub. Co., 1991. 554p.
ISBN 0314810919
    Includes chapters on food technology and food safety including freezing, canning, extrusion, drying, food additives, and food safety at home. Also covers chapters on topics not ordinarily seen in nutrition books, such as nutrition and the environment, nutrition and world population, deforestation, and soil loss. Thoroughly updated from previous edition. Provides detailed appendices including an extensive "Table of Food Composition" and index.

Handbook of General and Modified Diets
3rd ed. Chester, Pa., The Center, 1989. 334p. ISBN 0917664051
    Provides numerous therapeutic diets including those for liver disease, thermal injury, cancer, renal disease, and pediatric nutrition, to list a few. Both sound nutritional principles and palatability have been taken into consideration in the construction of the diets. Compiled by the professional staff of the Department of Nutrition Services and selected medical staff members of the Crozier-Chester Medical Center.

Hendler, Sheldon Saul
The Doctors' Vitamin and Mineral Encyclopedia
New York, Simon and Schuster, 1990. 496p. ISBN 067166784X
    Provides an alphabetical guide to nutritional supplements. Covers vitamins, minerals, amino acids, nucleic acids, lipids, herbs, and miscellaneous supplements. Discusses the importance of each and whether claims concerning them are substantiated. Discussions on herbs cover uses as folk medicine. Last section offers programs for better health including "Basic Formula for Teenagers" and "Formula for Hospitalized Patients."

Institute of Medicine (U.S.). Committee on Dietary Guidelines Implementation
Improving America's Diet and Health: From Recommendations to Action
Edited by Paul R. Thomas. Washington, D.C., National Academy Press, 1991. 239p. ISBN 0309041392
    Includes strategies and actions targeted to the public and private sectors having special responsibilities for encouraging and enabling consumers to eat better. Recommendations are provided for three strategies necessary to further the implementation of dietary recommendations in the United States.

Institute of Medicine (U.S.). Committee on the Nutrition Components of Food Labeling
Nutrition Labeling: Issues and Directions for the 1990s
Edited by Donna Viola Porter and Robert O. Earl, Washington. D.C., National Academy Press, 1990. 355p. ISBN 0309043263
    This work addresses the needs of consumers to make more healthful and informed food choices. Discussion covers the background, drawbacks, omissions, and confusion surrounding nutrition labeling. Committee's recommendations are also discussed. Includes index.

Institute of Medicine (U.S.). Food and Nutrition Board
Eat For Life: The Food and Nutrition Board's Guide to Reducing Your Risk of Chronic Disease.
Edited by Catherine E. Wotek and Paul R. Thomas
Washington, D.C., National Academy Press, 1992. 179p.
ISBN 0309040493
    This publication synthesizes information and provides the nine-point dietary plan developed to reduce the risk of diet-related chronic diseases. Two previous books were published in this series: see *Diet and Health* (National Research Council, Committee on Diet and Health, 1988) and *Improving America's Diet and Health* (Institute of Medicine, Committee on Dietary Guidelines Implementation, 1991). Includes index.

Jones, Julie Miller
Food Safety
St. Paul, Minn., Eagan Press, 1992. 453p. ISBN 0962440736
    Endeavors to demonstrate the difference between toxicity and hazard as applied to foods. Covers all aspects of food safety including the food laws and their history, natural toxicity of food, and how toxicity is caused by bacteria, molds, or mycotoxins. Discussions cover a variety of topics such as food additives, food colors and flavorings, food irradiation, and pesticides. Includes glossary, list of references, and index.

Kinney, John, and others
Nutrition and Metabolism in Patient Care
Philadelphia, Saunders, 1988. 797p. ISBN 0721611567

The sixty-eight contributors to this clinical reference text are medical scientists who are noted either in the area of nutrition or metabolism. This book was written because "the nutritional condition of any given patient depends not only on the level of nutritional support, but on the metabolic changes that accompany disease and injury" (Foreword). Covers elements of nutrition, nutritional changes required because of natural body changes (aging, pregnancy, etc.), clinical conditions requiring nutritional changes, and assessment of a patient's nutritional needs.

Knight, John Barton, and Lendal Henry Kotschevar
Quantity Food Production, Planning, and Management
2nd ed. New York, Van Nostrand Reinhold, 1989. 445p. ISBN 0442240163

Covers all phases of food service operation including menu planning, preparation and cooking methods of major foods, sanitation, and equipment. Includes computer applications, appendices and index.

Kotschevar, Lendal Henry
Standards, Principles, and Techniques in Quantity Food Production
4th ed. New York, Van Nostrand Reinhold, 1988. 505p. ISBN 0442256620

Discusses the growth in the use of convenience foods, computer-operated machines, and increased use of fruits and vegetables in quantity food production. Includes chapters on menu planning, "portion costing," recipe development, the uses of computers, and scheduling. Chapters were updated to reflect the growth in popularity of convenience foods. Includes appendices, bibliography, and index.

Kwiterovich, Peter
Beyond Cholesterol: The Johns Hopkins Complete Guide for Avoiding Heart Disease
Baltimore, Johns Hopkins Univ. Press, 1989. 395p. ISBN 0801838282

The goal of this book is to enable the reader to make more informed decisions about cholesterol. Part One discusses risk of high cholesterol in adults and children; Part Two covers cholesterol and heredity; Part Three advises what to eat and why (for the reduction of cholesterol); and Part Four discusses the role of exercise and drugs. Scholarly work with each chapter ending in a summary. Includes numerous graphs, charts, recipes, glossary, and index.

Mahan, L. Kathleen, and Marian T. Arlin
Krause's Food, Nutrition, and Diet Therapy
8th ed. Philadelphia, Saunders, 1992. 933p. ISBN 0721655084

Used as a textbook and reference tool for both students and practitioners. This edition incorporates the 1989 Recommended Dietary Allowances. Includes a new chapter on "Nutrition for Athletic Training and Performance" and several other new chapters including "Nutritional Care in Pulmonary Disease," "Arthritic Disease," and "Diseases of the Nervous System." Includes extensive appendices and index.

Mayer, Jean
Diet and Nutrition Guide
New York, Pharos Books, 1990. 314p. ISBN 0886875684

This work evolved from a newspaper question-and-answer column. It contains introductory summary information followed by questions and answers. Covers numerous topics including anatomy of food labels, pesticides, "nutritional ages of man," basic components of food, weight reduction and fitness, and special diets such as those for diabetic and cancer patients. Includes a glossary of foods describing origin, uses, and nutritional value, a chart of nutritional components of fast foods, menus, and an index.

Mayo Clinic Diet Manual: A Handbook of Dietary Practices
6th ed. Philadelphia, Decker, 1988. 636p. ISBN 1556640331

Provides a comprehensive and expanded resource for healthful nutrition from infancy through adulthood, and for the evaluation and management of problems in clinical nutrition. This sixth edition has an expanded section on pediatric nutrition, and a new section on guides for diets to promoting wellness for specific disease problems. Contains tables and reading lists.

Morgan, William J.
Supervision and Management of Quantity Food Preparation: Principles and Procedures
3rd ed. Berkeley, Ca., McCutchan Pub. Corp., 1988. 498p. ISBN 0821112600

This edition contains new chapters on quality assurance and legal liabilities of food service operations. Includes appendices and index.

National Research Council (U.S.). Subcommittee on the Tenth Edition of RDAs
Recommended Dietary Allowances
Washington, D.C., National Academy Press, 1989. 284p. ISBN 0309040418

The tenth edition was revised extensively to incorporate latest scientific knowledge and interpretation concerning nutrition. RDA is given for vitamin K, which was not included before. Contains technical discussions of carbohydrates, fiber, lipids, proteins, amino acids, minerals, trace elements, water, and electrolytes. Chapters on vitamins and minerals provide recommended allowances, sources, general signs of deficiency, and toxicity levels. Includes references and index.

Poleman, Charlotte M., and Nancy J. Peckenpaugh
Nutrition Essentials and Diet Therapy
6th ed. Philadelphia, Saunders, 1991. 538p. ISBN 0721672817

One aim of this text is to teach students to differentiate between facts and fallacies. This sixth edition gives more attention to individual nutritional advice and less space to complex menus. Attention is given to food and drug interactions, not presented in previous editions. Included is a chapter on "Nutritional Care of the Developmentally Disabled," a subject not covered in most basic texts. Includes extensive appendices and index.

Quincy, Matthew
Diet Right! The Consumer's Guide to Diet and Weight Loss Programs
Berkeley, Conari Press, 1991. 121p. ISBN 0943233100

Gives an overview of major diet and weight loss programs available. Five categories are discussed: moderate calorie/grocery store food programs; package food programs; very low calorie programs (special formulated products); combination programs. Other miscellaneous programs such as the Pritikin diet are also discussed. Each program includes a description, history, what the diet consists of and how it works, approximate cost, average weight loss, personal considerations, and location.

Raso, Jack
Mystical Diets: Paranormal, Spiritual, and Occult Nutrition Practices
Buffalo, N.Y., Prometheus Books, 1993. 291p. ISBN 0879757612
Raso is assistant chief dietitian at Wyckoff Heights Medical Center in Brooklyn and is also a contributing editor to *Nutrition Forum*. His goal is to help laypersons and health professionals understand and evaluate a number of "fringe nutritional programs." Topics covered include macrobiotics, nutripathy, Gerson therapy, and "mail-order nutrition," among others.

Robinson, Corinne H., and others
Normal and Therapeutic Nutrition
17th ed. New York, Macmillan, 1990. 759p. ISBN 0024026050
This edition has extensive revision with new material including the 1989 Recommended Dietary Allowances. Contains four new chapters, including one on computer applications in clinical nutrition, and another on the nutritional and metabolic considerations involved with stress. Revised 1990 printing. Includes index.

Roth, Eli, and Sandra Streicher
Good Cholesterol, Bad Cholesterol
Rocklin, Calif., Prima Publishing and Communications, 1988. 179p. ISBN 0914629859
Two cardiovascular specialists present all aspects of cholesterol and triglycerides. Includes chapters on reading and understanding food labels, medications for controlling cholesterol, choosing foods when eating out, and low-fat recipes. Contains a bibliography, an appendix which lists fat content of food, and an index.

Rowland, I. R.
Nutrition, Toxicity, and Cancer
Boca Raton, Fla., CRC Press, 1991. 559p. ISBN 0849388120
A reference source that can be used by toxicologists, biochemists, cancer researchers, and nutritionists. This work provides practical guidance on formulating diets and designing nutritional studies in animals and humans. It also includes valuable information such as the influence of nutrition on specific biological processes such as biotransformation of foreign and indigenously produced compounds. Other topics discussed include the complex interactions between nutrition and carcinogenic processes, teratogenesis, and mutagenesis. Includes sample diets, references, and index.

Rudman, Jack
New Rudman's Questions and Answers on the OCE: Occupational Competency Examination in Quantity Food Preparation
Syosset, N.Y., National Learning Corp., 1993. 1 v. (various pagings) ISBN 0837357306
Provides examination questions covering factual and technical information on food preparation. Includes exercises in problem-solving.

Shugart, Grace Severance, and Mary Molt
Food for Fifty
9th ed. New York, Macmillan, 1993. 767p. ISBN 0024103411
In this edition, numerous recipes have been added and others have been changed to reflect current food preferences, modern eating styles, and a growing concern for nutrition. This book is divided into three sections: Part One is a guide to planning and preparing food in quantity; Part Two consists of recipes; and Part Three discusses menu planning and special food service events. Appendices and index.

Simone, Charles B.
Cancer and Nutrition: A Ten-Point Plan to Reduce Your Risk of Getting Cancer
Rev. and expanded ed. Garden City Park, N.Y., Avery Publishing Group, 1992. 338p. ISBN 0895294915
Simone is a cancer specialist who is concerned that cancer is the second most common cause of death in the United States. Approximately one-third of the book discusses causes not related to diet and nutrition. Nutritional discussions cover free radicals, food additives, pesticides, alcohol and caffeine consumption, natural food protectors such as carotene (from yellow fruits and vegetables), indoles (from cabbage family), isoflavones (from legumes), sterols (from cucumbers), sulfur (from garlic), and terpenes (from citrus fruits). Includes a diet plan to modify risks and an extensive bibliography.

Somer, Elizabeth
Nutrition for Women: The Complete Guide
New York, Holt, 1993. 457p. ISBN 0805023895
This work is written by a registered dietitian who is the editor of Health Media of America's *Nutrition Report*. It gives readable, easily understood summaries of nutritional science nutrition research pertaining to women's health. Presents dietary recommendations for many disorders including anemia, osteoporosis, skin problems, and yeast infections. Includes index, glossary, and extensive references.

Stare, Fredrick J., Robert E. Olson, and Elizabeth M. Whelan
Balanced Nutrition: Beyond the Cholesterol Scare
Holbrook, Mass., Bob Adams, 1989. 360p. ISBN 1558509208
Authors' objective is to inform the public of the nutritional threats to public health. They wrote this book because of concern over journalistic irresponsibility that presents more hype than truth about some current nutritional controversies. Stare is Professor Emeritus and founder of Harvard's Department of Nutrition. Olson is Professor of Medicine and Pharmacological Studies at S.U.N.Y., Stony Brook. Whelan is President of the American Council on Science and Health. Includes glossary, appendices, and index.

United States. Public Health Service. Office of the Surgeon General.
Surgeon General's Report on Nutrition and Health
Rocklin, Calif., Prima Publishing and Communications, 1989. 727p. ISBN 0914629964
This Surgeon General's report on nutrition and health reviews scientific evidence linking diet to several diseases including certain types of cancer, diabetes, heart disease, and obesity. Includes short history of the Federal Nutrition Policy begun in 1862. There are chapters on skeletal, dental, kidney, and gastrointestinal diseases, infections and immunity, anemia, and neurological disorders. Each

disease is discussed in terms of historical perspectives, significance for public health, and key scientific issues. Includes numerous tables and extensive bibliographies. Very comprehensive.

Vartabedian, Roy E., and Kathy Matthews
Nutripoints: The Breakthrough Point System for Optimal Nutrition
New York, Harper and Row, 1990. 450p. ISBN 0060162759

Survey of the "Nutripoints" system used to determine and compare nutritional values of foods. Each food is assigned a Nutripoint score. A high score describes a food high in important nutrients including vitamins, fiber, complex carbohydrates, and protein. This score also indicates if the food is low in fat, sugar, sodium, cholesterol, and caffeine. Includes brand name foods and fast food items. References and index.

Warner, Mickey
Recreational Foodservice Management
New York, Van Nostrand Reinhold, 1989. 481p.
ISBN 0442205961

The business of food service has emerged in its own right and with a variety of subfields. This work discusses the need for specialized education in these subfields and concentrates on recreational-food and fast-food service management, covering all aspects. Includes appendices and index.

Williams, Sue Rodwell
Basic Nutrition and Diet Therapy
9th ed. St. Louis, Mo., Mosby, 1992. 486p. ISBN 0801656869

Written with both students and practitioners in mind, this work has been thoroughly updated. Includes sections on the cultural food patterns of several groups, including Native Americans, Muslims, African Americans, and Cajuns. Includes glossary, detailed appendices, and index.

Williams, Sue Rodwell, and Bonnie S. Worthington-Roberts, eds.
Nutrition Throughout the Life Cycle
2nd. ed. St. Louis, Mo., Mosby, 1992. 498p. ISBN 0801664772

The updated edition includes a new chapter on the nutritional needs of young adults, with special focus given to young women. Details nutritional needs and nutritional problems associated with each age. Contains numerous charts, and word and phrase definitions in margins. Includes glossary and index.

Winick, Myron and others, eds.
The Columbia Encyclopedia of Nutrition
New York, Putnam, 1988. 349p. ISBN 0399132988

This book's editorial board is composed of specialists in the field and who are currently or were recently on the faculty of Columbia University College of Physicians and Surgeons. Provides concise, up-to-date synthesis of information on nutrients and diseases affected by nutrition. Includes index.

Winter, Ruth
A Consumer's Dictionary of Food Additives
3rd ed. New York, Crown, 1989. 352p. ISBN 0517572621

The focus of this work is on definitions of ingredients found in packaged foods which may be harmful or beneficial. Provides information on the sources, uses, and health considerations of each additive. Defines thousands of the chemicals that are added to the food we eat.

Zeman, Frances J.
Clinical Nutrition and Dietetics
2nd ed. New York, Macmillan, 1991. 854p. ISBN 0024315109

Integrates the theoretical basis for nutritional care with the biological sciences, biochemistry, physiology, pharmacology, immunology, and genetics. Includes a new chapter that reviews fluid, electrolyte, and acid-base balance. Also contains several chapters covering nutrition and specific diseases. Included are case studies, references, extensive appendices, and an index.

Nonprint Materials

Eat Smart
Alexandria, Va., PBS Home Video, 1991. 1 videocassette (60 min.) VHS ISBN 156111622x

A MacNeil/Lehrer report linking the American diet to heart disease and cancer. Provides a comparison of "typical" American diets to those of other nationalities that have lower incidence of disease.

Food Features Video
Tallahassee, Fla., Worth Owens Productions, 1989. 1 videocassette (94 min.) VHS

Contains four separate sections covering a different topic. Discussed are controlling sodium, controlling dietary fiber, "defensive dining," and fat detection.

Nutrition
St. Louis, Mo., Mosby, 1993. 1 videocassette (16 min.) VHS
ISBN 0801671035

Discusses physical, cultural, psychological, and socioeconomic factors associated with nutritional problems in clients.

Nutrition and Respiratory Care
Dallas, Univ. of Texas Health Science Center at Dallas, 199?. 1 videocassette (60 min.) VHS

Technical explanation of the relationships and interactions of malnutrition on ventilatory drive, respiratory muscles, lung structure, and immunity. Also discussed are the effects of nutrients on the respiratory system, particularly the nutritional support needed by the mechanically ventilated and Chronic Obstructive Pulmonary Disease patients.

Successful Aging: Overcoming Barriers to Nutrition and Health
Okemos, Mich., Dairy Council of Michigan, 1989. 2 videocassettes (164 min.) VHS

This teleconference was developed by the Michigan State University, the Dairy Council of Michigan, and the Drug Administration. Various topics concerning older adults are covered, including nutritional needs and factors influencing food intake of the elderly.

Emergency Medical Services

Grace Ekins
DeKalb College
Clarkston, Georgia

Introduction

The United States Department of Transportation's National Standard Curriculum is the basis for Emergency Medical Service (EMS) training in the United States. There are several levels of Emergency Medical Technology (EMT) training: Basic (EMT-1), Intermediate (EMT-2), and Paramedic (EMT-P). Additional certification guidelines to consider are those of the American Heart Association for Advanced Cardiac Life Support and the Prehospital Trauma and Advanced Trauma Life Support. Individual states and counties may also have their own certification guidelines.

Prior to the 1960s, ambulance drivers received very little training. Uniform standards for emergency medical services were implemented with the passage of the Highway Safety Act of 1966. Recommendations and guidelines of the Committee on Emergency Medical Services of the National Academy of Sciences and the National Research Council in 1968 led to the establishment of the National Registry of Emergency Medical Technicians (NREMT) in 1970. In 1971 the United States Department of Transportation published a national standard basic training course. The Emergency Medical Services System Act of 1973 provided funds for training.

EMT positions with fire departments offer high salaries and good benefits; however, the competition for these positions is intense. As voluntary ambulance services became salaried services, more job opportunities for EMTs have become available.

Certifications for emergency medical technicians vary by educational level and by state. After passing the required state or local exams, EMT-1 personnel may take the national exam to qualify as a member of the National Registry of Emergency Medical Technicians. This examination may also substitute for state certifications in some states, but must be requested from the appropriate state agency.

The journals, books, and nonprint materials in this bibliography focus on the training and certification of EMS personnel. Particularly useful are journals such as *JEMS* and *Emergency Medicine*, which offer up-to-date articles on state-of-the-art techniques, as well as reviews of videos and books, and advertisements for materials useful in this dynamic field. The videos listed present Emergency Medical Technology as only a visual medium can—in actual scenarios of life and death situations.

This bibliography provides an overview of materials currently available and a framework on which to build an appropriate EMS collection in a vocational program. Because of the differences in training at the various EMT levels, citations have been coded *(EMT)* or *(Paramedic level)*.

Accreditation and Certification

✦ EMTs-Basic, Intermediate, and Paramedic

National Registry of Emergency Medical Technicians
P.O. Box 29233
Columbus, Ohio 43229

✦ EMT-Paramedics

Committee on Allied Health Education and Accreditation
515 N. State Street
Chicago, Ill. 60610

National Association of Emergency Medical Technicians
9140 Ward Parkway
Kansas City, Mo. 64114

American Ambulance Association
3800 Auburn Blvd., Suite B
Sacramento, Calif. 95821

Selected Journals

Annals of Emergency Medicine.
Monthly. ISSN 0196-0644

Emergency.
Monthly. ISSN 0162-5942

Emergency Medicine.
Sixteen times a year. ISSN 0013-6654

EMS Insider.
Monthly.

JEMS: A Journal of Emergency Medical Services.
Monthly. ISSN 0197-2510

Prehospital and Disaster Medicine.
Quarterly. ISSN 1049-023X

Rescue.
Bimonthly. ISSN 1045-0246

Print Materials

Allison, E. Jackson, ed.
Advanced Life Support Skills
St. Louis, Mo., Mosby Lifeline, 1994. 274p. ISBN 0801674263
　　The author's knowledge is based on years of experience in the field and this experience is reflected in the practical approach to prehospital care. The definition, background, indications, contraindications, detailed procedure instructions, special considerations, and potential complications of each Advanced Life Support skill are covered, following the Department of Transportation curriculum. Includes skill practice sheets, review questions and answers. *(Paramedic level)*

American College of Legal Medicine
Legal Medicine: Legal Dynamics of Medical Encounters
2nd ed. St. Louis, Mo., Mosby, 1991. 670p. ISBN 0801600987
　　More than sixty distinguished contributors under the aegis of the American College of Legal Medicine cover a wide range of interrelated aspects of medicine and law. Risk managers, medical directors, and those responsible for the medical actions of others can benefit from this interpretation of the legal dynamics of medical encounters. *(Paramedic level)*

Aoki, Byron Y., and Karin McCloskey
Evaluation, Stabilization, and Transport of the Critically Ill Child
St. Louis, Mo., Mosby, 1992. 527p. ISBN 0815101147
　　Over twenty contributors provide important information dealing with the first crucial hours of treating the critically ill child. First degree principles, stabilization, management, and the transport process are reviewed. Organized by body system, this reference examines cardiopulmonary arrest, near-drowning, asphyxia, and post-resuscitation care. *(Paramedic level)*

Asken, Michael J.
Psycheresponse: Psychological Skills for Optimal Performance by Emergency Responders
Englewood Cliffs, N.J., Regents/Prentice Hall, 1993. 172p. ISBN 0893038393
　　This book provides psychological skills training to prepare EMTs for the worst situations or the "one in a hundred response." Although these skills are enhanced by experience, basic foundations are set in place so that an EMT does not gain experience at the expense of a patient. The author focuses on extraordinary situations, using "imagery" and "visualization" training approaches designed to improve concentration skills, reduce fear reactions, and stop negative afterthoughts. *(Paramedic level)*

Auerbach, Paul S.
Medicine for the Outdoors: A Guide to Emergency Medical Procedures and First Aid.
Rev. and updated. Boston, Little, Brown, 1991. 412p. ISBN 0316059323
　　This work offers brief explanations of a wide variety of medical problems encountered outdoors. It then offers understandable solutions for treating patients who are at some distance from hospitals and ambulances. *(EMT) (Paramedic level)*

Baskett, Peter J. F.
Resuscitation Handbook
2nd ed. St. Louis, Mo., Wolfe, 1993. 136p. ISBN 156375620X
　　Information is presented on both basic and highly advanced life support techniques. Algorithms guide the user through a series of symptoms to the required action. The book covers cardiac life support, trauma life support, pediatric resuscitation, and pain relief. In the techniques section, the hazards of potential complications are highlighted. *(Paramedic level)*

Bergeron, J. David
First Responder
3rd enl. ed. Englewood Cliffs, N.J., Brady, 1994. 502p.
ISBN 0893031917

This emergency medical manual is written especially for the first responder. It follows all Department of Transportation guidelines for the forty-hour course. The third edition focuses specifically on self-protection from infectious diseases during rendering of care. A workbook is available. *(EMT)*

Bledsoe, Bryan E., Robert S. Porter, and Bruce R. Shade
Brady Paramedic Emergency Care
2nd ed. Englewood Cliffs, N.J., Brady, 1994. 1015p.
ISBN 0893039799

A primary text for emergency medical personnel, covering all aspects of the field. Includes a workbook with drug cards and skill sheets. *(Paramedic level)*

Bledsoe, Bryan E.
Paramedic Pocket Reference
Englewood Cliffs, N.J., Brady, 1992. 157p. ISBN 0893038334

This handbook of medical emergencies and drugs is divided into six sections: respiratory care, emergency drugs (seventy-one are listed), intravenous fluids, advanced cardiac life support treatment algorithms, trauma emergencies, and pediatric emergencies. It is designed to assist EMTs, EMT-1s, and paramedics in the field. *(Paramedic level)*

Bosker, Gideon
The Sixty Second EMT: Rapid BLS/ALS Assessment, Diagnosis and Triage
St. Louis, Mo., Mosby, 1988. 200p. ISBN 0801611857

The author's approach is to present a case history while applying the "sixty second" approach to patient assessment. He seeks quicker, more rapid, precise, efficient, and compassionate patient assessments, diagnosis, and triage. Geared to both the EMT-A and EMT-P, the book teaches the basics of the sixty-second syncope assessment, triage, and the sixty-second neurological exam. *(EMT)* *(Paramedic level)*

Brillman, Judith C., and Ronald Quenzer, eds.
Infectious Disease in Emergency Medicine
Boston, Little, Brown, 1992. 951p. ISBN 0316108383

Covers communicable diseases that emergency medical personnel may face. Includes bibliography and index. *(Paramedic level)*

Campbell, John E.
BTLS: Basic Prehospital Trauma Care
Englewood Cliffs, N.J., Prentice Hall, 1988. 272p.
ISBN 0893031224

The BTLS, or Basic Trauma Life Support course, has become the standard prehospital course for advanced EMTs and paramedics. This book teaches the necessary skills: rapid patient assessment, resuscitation, stabilization, and transport. It focuses on what to look for in the acutely injured patient, signs and symptoms, conclusions to draw, and how to manage acute conditions. An instructor's guide is available. *(Paramedic level)*

Caroline, Nancy L.
Ambulance Calls: Review Problems for the Paramedic
3rd ed. Boston, Little, Brown, 1991. 383p. ISBN 0316128899

Designed to help the paramedic review the latest treatment algorithms. Prehospital scenarios are set up as a challenge to determine appropriate patient treatment and its progression. A series of questions is followed by the prescribed treatment for each scenario. *(Paramedic level)*

Caroline, Nancy L.
Emergency Care in the Streets
4th ed. Boston, Little, Brown, 1991. 1005p. ISBN 0316128880

This popular text for paramedic students emphasizes the fundamental concepts of advanced life-support procedures and their underlying physiology. It follows the Department of Transportation curriculum in content and sequence, and treatment recommendations reflect current medical consensus and American Hospital Association standards. Covered are medical control, stress management, trauma, hazardous materials, rescue and extrication, and geriatric emergencies. *(Paramedic level)*

Caroline, Nancy L.
Emergency Medical Treatment: A Text for EMT-As and EMT-Intermediates
3rd ed. Boston, Little, Brown, 1991. 643p. ISBN 0316128864

First published in 1982, this text covers all aspects of emergency medical treatment. A workbook is available. Includes bibliography and index. *(EMT)*

Cotton, Sherrie L.
Mosby's Paramedic Study Guide: Certification, Preparation, and Review
St. Louis, Mo., Mosby, 1989. 358p. ISBN 0801629012

This workbook assists paramedic students during and after their educational experience. It is designed to be used for notes, as a study guide, or for review. Includes questions with answers. *(Paramedic level)*

Dailey, Robert H., and others
The Airway: Emergency Management
St. Louis, Mo., Mosby, 1992. 397p. ISBN 0801612705

Management procedures for handling the first thirty to sixty minutes of airway closure are detailed. The physiologic aspects of the airway and breathing, specific airway maneuvers, how to handle special situations, drugs involved in airway management, and other topics are explained by medical doctors. *(Paramedic level)*

Dernocoeur, Kate Boyd
Streetsense: Communication, Safety, and Control
2nd ed. Englewood Cliffs, N.J., Prentice Hall, 1990. 352p.
ISBN 089303861X

"Streetsense" is the "art of applying the science of total prehospital emergency care to the unscientific environment of people" (Cover). It is an art, and is the most essential quality that any EMS provider can achieve. Includes chapters on many different topics, including communication, death and dying, special populations, fatigue, knives, guns, stress, infection, and legal risks. *(EMT)* *(Paramedic level)*

Downey, Ray
The Rescue Company
Saddle Brook, N.J., Fire Engineering Books and Videos, 1992. 328p. ISBN 091221225X

A thirty-year veteran of rescue operations describes duties and responsibilities, and provides operational guidelines for rescuers. The author comments on planning, training, operations, tools and equipment, and personnel. Included are photographs that cover urban rescues and structural collapse. *(Paramedic level)*

Edwards, David P.
Paramedic Examination Review Manual
Englewood Cliffs, N.J., Brady , 1991. 419p. ISBN 0893038059

This book provides a wide range of review materials in a single volume and gives step-by-step advice on how to prepare for the paramedic certification exam. It incorporates care methods covered by the Department of Transportation National Standard Curriculum, and presents medical terminology in concise form. *(Paramedic level)*

Emergency Care and Transportation of the Sick and Injured
5th ed. Park Ridge, Ill., American Academy of Orthopaedic Surgeons, 1992. 889p. ISBN 089203050X

Based on the Department of Transportation National Standard Curriculum, this text has chapters that begin with an overview of learning objectives, and closes with a self-help quiz. A slide set containing approximately 1500 slides is available, as are student and instructor workbooks. The first edition was the basis for the National Standard Curriculum for the EMT-A. It includes both the Department of Transportation objectives and the National Fire Protection Association standards. (Sixth edition is due to be published in spring of 1995) *(EMT)*

Gonsoulin, Sheryl M., and William Raynovich
Prehospital Drug Therapy
St. Louis, Mo., Mosby Lifeline, 1994. 318p. ISBN 0801619696

Written for paramedics, this book includes information on the basic principles of pharmacology, pharmacotherapy for patients experiencing specific emergencies, and the implications of pharmacology for paramedic practice. The approach is problem-oriented, with case studies illustrating the use of drugs in specific situations. It incorporates the new American Hospital Association Advanced Cardiac Life Support standards. *(Paramedic level)*

Grant, Harvey D., Robert H. Murray, and J. David Bergeron
Brady Emergency Care
6th ed. Englewood Cliffs, N.J., Brady, 1994. 800p.
ISBN 0893031550

This work draws its skills orientation from the 110-hour Department of Transportation course. Included are practical learning aids and skills/objectives summaries for each chapter. Illustrated with full-color drawings and four-color photographs. A self- instruction workbook is also available. *(EMT)*

Grauer, Ken and Daniel L. Cavallaro
ACLS
3rd ed. St. Louis, Mo., Mosby Lifeline, 1993. 2 v. (685p.) ISBN 0801670691 (vol. 1); ISBN 0801670705 (v. 2)

Vol. 1. *Certification Preparation.* This volume delivers practical information to help candidates pass the Advanced Cardiac Life Support (ACLS) course, with advice on how to best prepare for it. There are chapters on essential drugs and on an algorithmic approach to management.

Vol. 2. *Comprehensive Review.* Beyond the basics of ACLS testing, this volume contains practical information to further develop understanding of ACLS. There is a section on acute myocardial infarction that reviews every aspect of evaluation and management. Ten special resuscitation situations are covered. *(Paramedic level)*

Hafen, Brent Q.
Brady EMT Review Manual
2nd ed. Englewood Cliffs, N.J., Brady, 1994. 300p.
ISBN 089303200X

Review questions for Emergency Medical Technicians. *(EMT)*

Hafen, Brent Q., and Keith J. Karren
Prehospital Emergency Care and Crisis Intervention
4th ed. Englewood Cliffs, N.J., Prentice Hall, 1992. 800p.
ISBN 0893039306

This textbook aids EMTs in preparation for the Emergency Technician Course. It covers all areas from anatomy, life support, and wounds to communications, records and reports. Includes a glossary of terms. *(EMT)*

Heckman, James D.
Emergency Care and Transportation of the Sick and Injured
5th ed. Park Ridge, Ill., American Academy of Orthopaedic Surgeons, 1992. 889p. ISBN 089203050X

Based on the Department of Transportation National Standard Curriculum, this text begins with an overview of learning objectives and closes with a self-help quiz. A slide set containing approximately 1500 slides is available, as are student and instructor workbooks. The first edition was the basis for the National Standard Curriculum for the EMT-A. The 5th edition includes both the department of Transportation objectives and the National Fire Protection Association standards. *(EMT)*

Henry, Mark C., and Edward R. Stapleton
EMT: Prehospital Care
Philadelphia, Saunders, 1992. 860p. ISBN 0721613012

This textbook meets Department of Transportation standards and emphasizes an assessment-oriented approach to care. The rationale for certain actions is provided. Well illustrated, it explains how to assess a patient's condition, stabilize the patient, and initiate appropriate prehospital care swiftly and accurately. *(EMT)*

Huszar, Robert J.
Early Defibrillation
St. Louis, Mo., Mosby, 1991. 272p. ISBN 0801629276

Addresses electric countershock and cardiovascular emergencies that emergency medical personnel need to understand. *(EMT)*

Ivey, Pat
EMT: Beyond the Lights and Sirens
New York , Ivy Books, 1991. 241p. ISBN 0804107114

Pat Ivey tells her own story. She is a former teacher who became an EMT and joined the community rescue squad in Lake of the Woods, Virginia, after EMTs impressed her by spending all night looking for her son and a friend. The lost boys were found unharmed. She conveys her experiences as a student and EMT. *(EMT)*

Jacobs, Donald Trent
Patient Communication for First Responders and EMS Personnel: The First Hour of Trauma
Englewood Cliffs, N.J., Brady, 1991. 178p. ISBN 089303732X

    EMTs can help patients mobilize their own survival and healing strategies. This book describes field-tested approaches that enhance patient care through communication. *(EMT) (Paramedic level)*

Jones, Shirley A., and others
Advanced Emergency Care for Paramedic Practice
Philadelphia, Lippincott, 1992. 899p. ISBN 0397512597

    Designed for prehospital emergency care students and professionals, this textbook may be useful for education, certification, and for general reference. It is based entirely on the revised Department of Transportation curriculum, and it includes the National Registry of Emergency Medical Technicians Advanced Level Practice Examinations. *(Paramedic level)*

Kidd, J. Steven, and John D. Czajkowski
Vehicle Extrication: A Training Manual
Saddle Brook, N.J., Fire Engineering, 1991. 253p. ISBN 0878149155

    This textbook details techniques of successful automobile extrication. It is a manual intended to be used by field rescuers, rather than as a comprehensive instructor's guide. *(EMT) (Paramedic level)*

Krebs, Dennis R., Kenneth C. Henry, and Mark B. Gabriele
When Violence Erupts: A Survival Guide for Emergency Responders
St. Louis, Mo., Mosby, 1990. 228p. ISBN 0801661951

    Violence in the home and on the street has become a significant hazard to emergency response personnel, who are often under the false impression that they are viewed as "helpers." This book seeks to increase the awareness of EMTs by providing them with the education and protective skills necessary for self-defense. It covers highway incidents, approaching the motor vehicle, fleeing the armed encounter, domestic encounters, hostage situations, and bombs. *(EMT) (Paramedic level)*

Lee, Genell
Quick Emergency Care Reference
St. Louis, Mo., Mosby, 1992. 60p. ISBN 0801665841

    This small format (11 x 14 cm.) handbook has sections on assessment/intervention, electrolyte imbalances, airway obstructions, neurological care, and wounds. Dosages are included, as are symptoms of pediatric diseases, and descriptions of emergency drugs. *(Paramedic level)*

London, P. S.
Colour Atlas of Diagnosis After Recent Injury
London, Wolfe Medical Publications, 1990. 141p. ISBN 072341520X

    Illustrations of injuries in all body regions are accompanied by a discussion of evidence to enable EMTs to make an expedient diagnosis. Many of the cases are presented in photographs, which show the immediate appearance of an injury, as well as the appearance during diagnosis and after treatment. *(EMT) (Paramedic level)*

Mack, Daniel
EMT Certification Preparation and Review
St. Louis, Mo., Mosby, 1990. 190p. ISBN 0801658535

    Review questions are grouped according to subject matter in four different formats: multiple choice, matching, labeling, and discussion. A sample test with one hundred questions is also included. The questions are based on materials covered in the Emergency Medical Technician—Ambulance: National Standard Curriculum. Questions covering treatment protocols that vary from location to location are purposely avoided. *(EMT)*

Miller, C. D., and David White
EMT: Paramedic National Standards Review Self Test
Englewood Cliffs, N.J., Brady Regents/Prentice Hall, 1993. 413p. ISBN 0893039551

    This book focuses on the paramedic level of study and is designed to be used with Bryan Bledsoe's *Paramedic Emergency Care*. Answer keys for multiple-choice questions reference page numbers in Bledsoe's book. *(Paramedic level)*

Miller, C. D., and David White
EMT: Basic National Standards Review Self Test
Englewood Cliffs, N.J., Brady Regents/Prentice Hall, 1992. 239p. ISBN 089303875X

    Based on the Emergency Medical Technician National Standard Curriculum by the U.S. Department of Transportation, National Highway Traffic Safety Administration, this book is keyed to *Brady Emergency Care* by Harvey D. Grant (5th ed., 1990). Self-test format. It aids in preparation for the EMT-Basic exam and includes a section on preparation for the National Registry Exam. *(EMT)*

Miller, Robert H., and John K. Wilson
Manual of Prehospital Emergency Medicine
St. Louis, Mo., Mosby, 1992. 515p. ISBN 0801657911

    The unique two-part format of this text provides both a quick reference guide for rapid review and an in-depth study of prehospital emergency medicine. It covers a broad spectrum of emergency medical subjects, illnesses, and injuries that affect prehospital care, from pharmacological therapy to communicable diseases. Also included are subject locators and cross-references. *(Paramedic level)*

Moore, Ronald E.
Vehicle Rescue and Extrication
St. Louis, Mo., Mosby, 1991. 319p. ISBN 0801633516

    Photographs are used to highlight key information about the basic concepts of vehicular rescue. State-of-the-art tools and techniques are demonstrated, covering late model automobile structure and extrication methods, removal of accident victims from vehicles, and techniques for immobilizing injured patients. *(EMT) (Paramedic level)*

National Association of Emergency Medical Technicians.
Pre-Hospital Trauma Life Support Committee PHTLS: Pre-Hospital Trauma Life Support
Edited by Robert S. Swor. 2nd ed. Akron, Ohio, Emergency Training, 1990. 384p. ISBN 094043206

    Published in cooperation with the Committee on Trauma of the American College of Surgeons, this work is written by leading authorities. This book presents both basic and advanced explanations of prehospital trauma care and stresses the importance of considering special and specific needs of the multisystem trauma patient. *(EMT) (Paramedic level)*

Neely, Keith A.
Street Dancer
Solana, Calif., JEMS Pub., 1990. 265p. ISBN 0936174064
　　A fictional account of Neely's experiences as a Denver General paramedic, this book provides valuable insights into the "seasoning" of a veteran street medic. *(EMT) (Paramedic level)*

Nichols, David G., and others, eds.
Golden Hour: The Handbook of Advanced Pediatric Life Support
St. Louis, Mo., Mosby, 1991. 447p. ISBN 0815163959
　　This pocket-sized handbook covers only the most common critical emergency procedures. Details step-by-step instructions for both traumatic and nontraumatic emergencies, including resuscitation techniques, proven guidelines for evaluating the seriously ill or injured, effective airway management, and advanced cardiac life support. It includes a pediatric formulary for emergency drugs, with information on dosage forms, routes, and dilutions. *(Paramedic level)*

Parcel, Guy S., and Charles E. Rinear, eds.
Basic Emergency Care of the Sick and Injured
4th ed. St. Louis, Mo., Times Mirror/Mosby College Pub., 1990. 345p. ISBN 0801642671
　　This work demonstrates emergency procedures and recommendations for laypeople, trained technicians, and nursing and medical personnel. It was prepared primarily for use as a textbook for college-level first aid and emergency care courses, but is also appropriate for the training of professionals. *(EMT) (Paramedic level)*

Peto, Gloriajean and William J. Medve
EMS Driving: The Safe Way
Englewood Cliffs, N.J., Prentice Hall, 1992. 189p.
ISBN 0893038288
　　Covers aspects of ambulance driving including operational assertiveness, laws, the driving challenge, pre-driving considerations, knowledge of the vehicle, and preventive maintenance. Also includes range exercises, including basic maneuvers and emergency driving maneuvers. *(EMT) (Paramedic level)*

Phillips, Charles, Philip Froman, and Carol Hagberg
Paramedic Skills Manual
2nd ed. Englewood Cliffs, N.J., Prentice Hall, 1990. 249p.
ISBN 0893037508
　　This is a how-to student manual intended for use as part of a formal EMT paramedic course taught by a qualified instructor. It attempts to standardize the training and testing of the paramedic to provide national mobility for paramedics. *(Paramedic level)*

Politis, Jonathan F., Judith A. Creemens, and Patricia L. Tritt
The Paramedic Review Manual
3rd ed. Los Altos, Calif., NNR Pub. Co., 1992. 294p.
ISBN 0917010450
　　The purpose of this review is to prepare paramedics for the state and National Registry standardized EMT-Paramedic exams. Following a section on the design and structure of the tests are sample questions and answers with detailed explanations. *(Paramedic level)*

Ptacnik, Donald J.
The EMT Review Manual: Self-Assessment Practice Tests for Basic Life Support Skills
4th ed. Philadelphia, Saunders, 1993. 318p. ISBN 0721650430
　　The purpose of this manual is to develop the student's confidence to pass the state and National Registry standardized EMT-Paramedic exams now required in many states. Provides multiple-choice questions and answers. *(EMT)*

Shapiro, Paul D., and Mary B. Shapiro
Paramedic: The True Story of a New York Paramedic's Battles with Life and Death
New York, Bantam Books, 1991. 383p. ISBN 0553293834
　　Shapiro places the reader in the line of fire "in a world of hookers and hustlers, violence and fear, danger and death." He recounts the chaos and camaraderie of being a paramedic, or "trauma junkie." *(EMT) (Paramedic level)*

Stewart, Charles E., and Carol P. Stewart
Emergency Medical Technician Paramedic: Examination Review: 600 Questions with Explanatory Answers
New York, Medical Examination Pub. Co., 1989. 245p.
ISBN 0444014810
　　This book is designed to help the reader prepare for course examinations and state licensure examinations for EMT-P certification. It is based on experience in teaching paramedics in several states over a period of twelve years. It includes questions in broad categories with answers at the end of each section. *(Paramedic level)*

Walraven, Gail, and Miles Julihn
Paramedic Review Guide: Case Studies and Self-Assessment Questions
Englewood Cliffs, N.J., Prentice Hall, 1988. 227p.
ISBN 0893037516
　　This source of challenging self-assessment exercises is designed to correspond to the fifteen-minute module of the current Department of Transportation paramedic training curriculum. The guide contains forty-one case studies and over 600 multiple-choice self-assessment questions. *(Paramedic level)*

Worsing, Robert, ed.
Basic Rescue and Emergency Care
Park Ridge, Ill., American Academy of Orthopaedic Surgeons, 1990. 333p. ISBN 0892030402
　　Part of The American Academy of Orthopaedic Surgeons Rescue and Emergency Care Series, this book is designed to provide to fire, ambulance, law enforcement, industrial, park service, and other rescuers the fundamental skills needed to become an effective part of the modern, high-tech rescue team. *(EMT) (Paramedic level)*

Yvorra, James G., ed.
Mosby's Emergency Dictionary: Quick Reference for Emergency Responders
St. Louis, Mo., Mosby, 1989. 682p. ISBN 080163525X
　　Adapted from Mosby's *Medical and Nursing Dictionary* (2nd ed.), this reference has become the standard for definitions of

emergency response terms, and it provides clear explanations of techniques used routinely by emergency personnel. Written for first responders and EMTs, there are appendices on hazardous materials, pediatrics, Advanced Cardiac Life Support (ACLS) standards, and the Incident Command System. Includes a seventeen-page full-color atlas of the human body. *(EMT) (Paramedic level)*

Nonprint Materials

Carbusters! (Series)
Naples, Fla., American Safety Video Publishers; JEMS Communications, distributor, 1991. 6 videocassettes (210 min.) VHS

Videotapes and accompanying instructor guides form a training package designed to help rescuers reduce on-scene extrication time. Each video demonstrates specific techniques with a special emphasis on safety and coordination of the extrication effort. Contents: principles of extrication, techniques of extrication, hand tools and pneumatics, patient considerations and packaging, command philosophies and special situations, school bus rescue. *(EMT) (Paramedic level)*

CPR '93
Seattle, Wash., LifeLine Videos, 1993. 1 videocassette (35 min.) VHS

This videotape demonstrates the 1993 CPR standards for adult, child and infant CPR. *(EMT) (Paramedic level)*

EMS Training Software
Redmond, Wash., SpaceLabs Medical, First Medic Products; available through JEMS Communication, 20 computer programs

PC-based computer programs teach emergency basics to entry-level EMS personnel, with all coursework done entirely on the computer. Programs cover first-responder issues, anatomy and physiology, patient assessment, resuscitation, injuries and automated external defibrillators. A test bank program allows students to test themselves. *(EMT)*

Hands On! Skills Series
Naples, Fla., American Safety Video Publishers; Coronet/MTI Film and Video, distributors, Deerfield, Ill.), 1990. 6 videocassettes (167 min.) VHS or BETA

Designed in conjunction with George Washington University's Emergency Medical Degree Program, this how-to series combines studio demonstrations, street scenes, and extensive computer-aided graphics to illustrate basic and advanced skills. Each program features a presenter having extensive "hands on" experience. Contents: airway adjuncts, automated defibrillation, bandaging and splinting, bleeding and shock, childbirth, emergency vehicle operations. *(Paramedic level)*

Rescue
Carlsbad, Calif., Hare Publications; Springhouse Pub. Co.; also available from JEMS Communications, 1990. 1 videocassette (65 min.) VHS

Filmed for EMTs and first responders, this video has step-by-step demonstrations and reenactments of procedures and practices: patient assessment, airway management, CPR, bleeding and shock, Military Anti-Shock Trousers (MAST), pneumothorax, traction splints, oxygen therapy, and extrication and Kendrick Extrication Device (KED). *(EMT)*

Street Medicine (Video Series)
Naples, Fla., American Safety Video Publishers; available from the JEMS Bookstore, Carlsbad, Calif., 1988-1991. 9 videocassettes (388 min.) VHS

This EMS video education series attempts to close the gap between classroom learning and the streets. Competency-based programs augment the Department of Transportation curriculum for EMT and paramedic education. Lessons combine EMS authorities' material with video footage of live street action. Lesson guides are available. Contents: advanced airway management, advanced shock management, basic airway management, basic shock management, chest injuries, pediatric emergencies, prehospital burn care, the primary survey—rapid patient assessment, and, the secondary survey—multiple trauma. *(EMT) (Paramedic level)*

Medical Assisting and Medical Records

Tori Beyer
Pasadena City College
Pasadena, California
and
Michelle Strazer
College of Lake County
Grayslake, Illinois

Introduction

According to the United States Bureau of Labor and Statistics, the fields of Medical Assisting and Medical Records are expected to grow rapidly throughout the 1990s. Labor Department reports project job growth to increase by 90 percent in the field of Medical Assisting, and 75 percent in the field of Medical Records.

A predominant reason for this increase is the overall aging of America and the resultant paperwork and health care needs of an aging population. For medical assistants, a constant increase in new technology requires that there be knowledgeable personnel to utilize that technology. For records technicians, the 1991 change in Medicare reimbursement policies resulted in a need for specialists trained in medical records. Since that time, other changes in policies and records have created a high demand for trained record specialists.

Future growth is expected in the number of medical tests, treatments, and procedures available. For an Accredited Record Technician (ART), these changes mean that medical records will be increasingly scrutinized by third-party payers, courts, and consumers. The advent of computer-based patient records (CPR) will force a reevaluation of present policies and procedures, and the record technician must be prepared to make the necessary changes for utilizing that technology.

Course work for both professions includes instruction on medical terminology and diseases, anatomy and physiology, legal aspects of medical records, coding and abstraction of data, statistics, databases, and quality assurance and computers. The influx of new and improved technology will also force medical assistants and records technicians to stay abreast of a changing field. They will be expected to perform their traditional duties, and also cope with the rapidly changing technology in their field. Office functions increasingly will be computerized, and medical assistants and records technicians will play a central role in utilizing the new technology. New areas of patient care will be increasingly emphasized, such as patient teaching and holistic care. Both careers require an awareness of the changing legal and ethical considerations associated with work in these fields. Job opportunities for medical assistants and records technicians are numerous and varied. Hospitals, nursing homes, insurance providers, private medical offices, health maintenance organizations, and rehabilitation centers are just a few of the places where medical assistants and records technicians are needed.

To become certified, a student of Medical Assisting must pass a national examination which is administered and overseen by the American Association of Medical Assistants (AAMA). The examination focuses on three basic skill areas: general, administrative, and clinical. Certification for medical records technicians is usually required by most employers. To be eligible to take the examination, a medical records technician must be a graduate of a two-year associate degree program accredited by the Commission on Accreditation of Allied Health Education Programs (CAAHEP) of the American Medical Association, or a graduate of the Independent Study Program in Medical Record Technology. In addition, certification requires thirty hours of academic credit in predefined areas.

Accreditation and Certification

American Association of Medical Assistants
20 N. Wacker Dr., Suite 1575
Chicago, Ill. 60606
   Certifies medical assistants.

American Health Information Management Association
919 N. Michigan Ave., Suite 1400
Chicago, Ill. 60690
   Certifies record technicians and record administrators.

American Medical Association, Commission on Accreditation of Allied Health Education Programs (CAAHEP)
515 N. State St.
Chicago, Ill. 60610
   Certifies medical records technicans.

American Medical Technologist Association
710 Higgins Rd.
Park Ridge, Ill. 60068-5765
   Provides registration for registered medical assistants.

Selected Journals

Coding Clinic for ICD-9-CM.
American Hospital Association. Quarterly. ISSN 0742-9800

In Confidence.
American Health Information Management Association. Bimonthly.

Journal of AHIMA.
American Health Information Management Association. Monthly.
ISSN 1060-5487

Professional Medical Assistant.
American Association of Medical Assistants. Bimonthly.
ISSN 0033-0140

RMA Vital Signs.
American Medical Technologists. Quarterly. ISSN 0279-5736

Topics in Health Information Management.
Aspen Systems. Quarterly. ISSN 1065-0989

Toward an Electronic Patient Record.
Institute for Medical Records. Monthly. ISSN 1063-973X

Print Materials

✦ Clinical Procedures

Flynn, John C., ed.
Procedures in Phlebotomy
Philadelphia, Saunders, 1994. 240p. ISBN 0721646859
   This reference guide is divided into two sections: the practice of phlebotomy, and professional issues. The first section focuses on what phlebotomists do, including various techniques for drawing blood and methods for prevention control. The second section addresses professional issues such as those related to making career decisions. Included is a review test that aids students preparing for the certification exam. Index.

Frew, Mary Ann, and David R. Frew
Clinical Procedures for Medical Assisting
Philadelphia, F. A Davis, 1990. 535p. ISBN 0803638515
   This text covers the nonadministrative aspects of medical assisting, including such topics as time management and assisting with physical examinations. Each chapter offers a section, "Related Ethical and Legal Implications," which presents current issues of interest to the medical assistant. Each chapter also provides discussion questions and application exercises. One unit is dedicated to employment issues such as résumés, interviews, and entering the job market. Includes bibliographies and an extensive index.

Frew, Mary Ann, Karen Lane, and David R. Frew
Comprehensive Medical Assisting: Administrative and Clinical Procedures
3rd ed. Philadelphia, F. A. Davis, 1995. 930p. ISBN 080363871X
   Provides updated information on office technology, medical technology, and current health issues including AIDS and eating disorders. Greater emphasis has been placed on administrative procedures. This text covers administrative as well as clinical duties with chapters on management, maintaining records, financial procedures, and business machines, among others. Includes a chapter on entry into professional employment and appendices of insurance and medical law vocabulary.

Guy, Julia F.
Learning Human Anatomy: A Laboratory Text and Workbook
Norwalk, Conn., Appleton and Lange, 1992. 360p.
ISBN 0838556035
   An introductory guide for beginning students of anatomy, this text divides the study of the human body into four parts. Each part is described in detail through the use of labeled illustrations. Emphasis is on visual recognition of anatomy, so the written text is brief and concise. Each chapter provides brief methods of study and review exercises; answers appear at the end of the book. Includes a detailed index.

Keir, Lucille, Barbara A. Wise, and Connie Krebs
Medical Assisting: Administrative and Clinical Competencies
3rd ed. Albany, N.Y., Delmar Publishers, 1993. 778p.
ISBN 0827353111
   This text prepares students for entry into the medical assisting profession through the step-by-step instruction in administrative and clinical procedures. Emphasis is on safety. This new edition in-

cludes updated information on immunology and common pathologies, vocational guidance, and health care insurance. Includes glossary and index.

Kinn, Mary E., Mary Ann Woods and Eleanor F. Derge
The Medical Assistant: Administrative and Clinical
7th ed. Philadelphia, Saunders, 1993. 926p. ISBN 0721646913

Designed for two-year programs, medical offices, and independent studies, this work incorporates advances in equipment and technology and new procedures. Covers the health care team, "front office" procedures, and clinical principles and procedures. Includes new information on AIDS, a glossary, and an index.

Krebs, Connie, and Barbara A. Wise
Medical Assisting: Clinical Competencies
Albany, N.Y., Delmar Publishers, 1994. 612p. ISBN 0827349866

Provides current information on entry-level clinical procedures including anatomy, physiology, pharmacology, and emergencies. This edition is updated to provide greater detail on infectious diseases, safety guidelines, medical issues related to law and ethics, and employment. Includes a glossary and an index.

Lane, Karen, ed.
Saunders Manual of Medical Assisting Practice
Philadelphia, Saunders, 1993. 861p. ISBN 0721630634

This illustrated reference and procedural manual aids medical assistants in keeping up-to-date on diagnostic and treatment procedures, office practices, and legislative issues related to health care. Each procedure includes Universal Precaution symbols, purpose of the procedure, equipment and materials needed, step-by-step instructions, and rationales. Appendices include resources, administrative information, standards, and clinical information. Indexed.

Marshall, Jacquelyn
Clinical Microbiology
Albany, N.Y., Delmar Publishers, 1994. 160p. ISBN 0827353634

Contains sixty-two manual exercises interwoven with instructional materials that focus on pertinent microbiological issues for medical assistants. Each unit includes learning objectives, a glossary, review questions, and additional activities. Indexed. *The Clinical Laboratory Manual Series.*

Mulvihill, Mary L.
Human Diseases: A Systemic Approach
4th ed. Norwalk, Conn., Appleton and Lange, 1995. 474p. ISBN 0838539289

Provides a comprehensive approach to human disease including general mechanisms and nature of the organ systems. A review of anatomy and physiology is presented at the beginning of each chapter. This new edition expands coverage of AIDS and presents a new chapter on diet and exercise as it relates to wellness. Illustrated with photographs, line drawings, and flow charts. Includes glossary and index.

Regents/Prentice Hall Medical Assistant Kit
3rd ed. Englewood Cliffs, N.J., Regents/Prentice Hall, 1992-93. 11 v. (various pagings)

Titles in this series include *Neurology; The Respiratory System; Urology and Reproduction; Cardiology; Endocrinology; Gastroenterology; Dermatology; Bio-Organization; Orthopedics;* and *Hematology, the Lymphatic System, and the Immune System.* These titles provide a detailed overview of the body system including injuries, diseases, and disorders that affect that system. One title of the series, *Clinical Processes*, describes the responsibilities of clinical medical assistants, e.g., assisting with medical procedures, obtaining patient history, and administering medication.

Woods, Mary Ann
The Clinical Medical Assistant
Philadelphia, Saunders, 1994. 523p. ISBN 0721652352

Provides practical instruction and advice to the entry-level student as well as the practicing medical assistant. This work focuses on basic medical conditions and procedures, with an added emphasis on diseases that are diagnosed or retreated in the medical office. Includes an introduction to clinical pharmacology and information on advanced medical assisting procedures. Includes glossary and index.

Zakus, Sharron M.
Clinical Procedures for Medical Assistants
3rd ed. St. Louis, Mo., Mosby-Lifeline, 1995. 712p. ISBN 0801669839

Zakus presents an informative text that is designed to help medical assistants attain the skills and techniques required by their profession. In addition, this work functions as a reference manual on routine clinical skills and techniques, patient procedures, and patient care information. Include glossary and index.

✦ Coding

Adams, Wanda L.
Adams' Guide to Coding and Reimbursement
St. Louis, Mo., Mosby-Lifeline, 1994. 247p. ISBN 0815101368

Provides an introduction to coding including background information on *ICD-9-CM*, CPT coding, insurance companies, Medicare, claim forms, and accounts receivable. Also includes a chapter on various legal issues related to coding. Indexed.

Bowers, Charlotte R. and Marleeta K. Jones, eds.
St. Anthony's ICD-9-CM: Questions and Answers
Alexandria, Va., St. Anthony Pub., 1992. 141p. ISBN 1563290677

This reference tool addresses common coding problems regarding certain diagnoses or surgical procedures. Questions are arranged into two sections: diagnosis coding and procedural coding. Questions are then organized by code order and topic. (Codes valid for discharges on or after Oct. 1, 1991 through Sept. 30, 1992.)

Brown, Faye
ICD-9-CM Coding Handbook, with Answers
1991 rev. ed. Chicago, American Hospital Pub., 1991. 397p. ISBN 1556480717

A handbook on the statistical classification system for coding diagnoses and treatments. Elements are arranged by diseases and injuries, then followed by numerical code. The work explains basic *ICD-9-CM* coding principles and provides coding exercises useful to entry-level and experienced workers, as well as to medical records students. This work is of value for training and educational purposes.

Finnegan, Rita
Coding for Prospective Payment
Chicago, American Health Information Management Association, 1992. 337p. ISBN 0317054260

This manual focuses on orienting the student to diagnosis and coding principles. Provides coding requirements, alerts coders to concerns that may surface on drug-related group diagnosis validation reviews, and discusses the use of the *Medicare Code Editor*.

International Classification of Diseases, 9th Revision, Clinical Modification
9th ed. Ann Arbor, Mich., Commission on Professional and Hospital Activities, 1993. 3 v. ISBN 1880678578

This coding source aids the classification by diseases, diagnosis, and procedures that are required by all U.S. Public Health Service and Health Care Financing Administration programs.

Jones, Marleeta K., and others
St. Anthony's Color-Coded ICD-9-CM
Alexandria, Va., St. Anthony Pub., 1993. 1 v. (loose-leaf)
ISBN 1563290006

This inclusive volume contains official *ICD-9-CM* codes, indices, notes, footnotes, and symbols, and the American Hospital Association's Coding Clinic references. The color coding system includes symbols and footnotes that alert coders to coding and reimbursement issues. Updated periodically.

Jones, Marleeta K.
St. Anthony's DRG Optimizer: 1993
Alexandria, Va., St. Anthony Pub., 1992. 345p. ISBN 1563291185

Beginning with the admission process and ending with discharge, this book contains step-by-step process for assigning drug-related diagnosis documentation.

Jorwic, Therese M.
ICD-9-CM Coding Handbook for Physician Practices
Gaithersburg, Maryland, Aspen Pub., 1992. 1 v. (various pagings). ISBN 0834203472

A basic guide to the *ICM-9-CM Clinical Modification Diagnosis Coding* system for physicians' office personnel. This source provides the student with the fundamentals needed to use this system correctly and consistently.

Kirschner, Celeste G., and others
CPT 1994: Physician's Current Procedural Terminology
Chicago, American Medical Association, 1993. 855p.
ISBN 0899705561

This source lists descriptive terms and codes for reporting medical services and procedures performed by physicians. Provides a uniform language for communication among physicians, patients, and third parties. Revised annually.

Kotoski, Gabrielle M.
CPT Coding Made Easy
Gaithersburg, Md., Aspen Pub., 1991. 1 v. (loose-leaf)
ISBN 0834202840

Written for the nonmedical professional, this source defines medical terms, explains procedures, and gives examples of different procedures. Designed to be used with the AMA's *Physicians' CPT* (American Medical Association, 1981). Updated supplements are provided. Includes bibliographical references and index.

Morin-Spatz, Patrice
CP "Teach" Expert Coding Made Easy!
1994 version. Chicago, American Health Information Management Association, 1994. 349p. ISBN 0923369236

A how-to guide for learning the rules and regulations of *CPT-1994* (AMA, 1993). Includes details on rules of the surgery, consultations, emergency room visits, and outpatient services. A student workbook is also available with an answer key.

Puckett, Craig D., ed.
The Educational Annotation of ICD-9-CM
4th ed. Reno, Nev., Channel. Pub. 1993. 1 v. (loose-leaf)

This source provides coding information containing notations and drug-related diagnosis (DRG) information incorporated into the existing sections of the *ICD-9, Clinical Modification*.

Richard, Eugene, ed.
St. Anthony's Clinical Reference to Diagnostic Coding: Illustrations and Definitions for ICD-9-CM, Volumes 1 and 2
Alexandria, Va., St. Anthony Pub., 1992. 1 v. (loose-leaf)
ISBN 0834203170

This source is designed to translate diagnoses into correct *ICD-9-CM* codes, validate intended code sections, and improve data quality by providing code-specific clinical information. Offers hints regarding coding classification and currency of information.

Rogers, Vickie
Total Data Quality for the Coding Manager
Chicago, American Health Information Management Association, 1993. 100p. ISBN 0317054430

Provides information on developing and implementing coding audits for inpatient and outpatient coded data. Includes exercises and answer key.

Tanaka, Paul K.
Easy Coder
1994 ed. Montgomery, Ala., Unicor Medical, 1994. 503p.
ISBN 1567812015

Presents a logical, easy-to-use alphabetical format for coding that incorporates many of the basic rules within the alphabetical framework. Help text windows aid the user.

Toula, Nicholas
Basic ICD-9-CM Coding Handbook
Chicago, American Health Information Management Association, 1992. 315p. ISBN 0317054236

This basic coding workbook intended for educators, practitioners, and students includes lessons on diagnosis and procedural coding, along with information on the clinical aspects of disease. Answer key is provided.

Toula, Nicholas
ICD-9-CM Basic Coding Handbook
Chicago, American Health Information Management Association, 1994. 88p.

Contains basic rules and principles for *ICD-9-CM* coding. The handbook covers all aspects of coding from conventions used through special instructions. Lessons are arranged around each chapter, and answer key is included.

✦ Dictionaries and Terminology

Anderson, Kenneth N., Lois E. Anderson, and Walter D. Glanze
Mosby's Medical, Nursing, and Allied Health Dictionary
4th ed. St. Louis, Mo., Mosby, 1994. 1973p. ISBN 0801672252

This is a useful dictionary for nursing and allied health students. Appendix includes normal reference values, clinical calculations, units of measurement, and information on cultural differences in patients. Includes some color illustrations.

Austrin, Miriam G., and Harvey R. Austrin
Learning Medical Technology: A Worktext
7th ed. St. Louis, Mo., Mosby-Yearbook, 1991. 543p.
ISBN 0801601754

The first section of this text describes how one can learn medical terms by dividing them into roots, prefixes, and suffixes. The next two sections provide all the necessary terminology for the various body systems, highlighted with labeled illustrations. The fourth section provides updated information on the immune system and multiple-system diseases. Each chapter includes extensive review exercises, and crossword and hidden word puzzles. Indexed.

Birmingham, Jacqueline Joseph
Medical Terminology: A Self-Learning Text
2nd ed. St. Louis, Mo., Mosby, 1990. 416p. ISBN 0801602599

This introduction to basic medical terminology is intended to be used as a self-instructional manual. Chapters are brief and succinct, with review questions and answers provided. Word combining is explained, then the systems of the body are described using their unique terminology. Also included for each body system are terms related to disorders, diagnosis, and treatment of that system. Includes a bibliography and index.

Chabner, Davi-Ellen
Medical Terminology: A Short Course
Philadelphia, Saunders, 1991. 227p. ISBN 0721629393

Chabner's introduction to medical terminology requires no previous chemistry or biology background. It gives the reader basic tools for dividing words into roots, prefixes, and suffixes. The root of each medical term listed is underlined to assist the reader in associating the word with its meaning. The reader has several opportunities to practice this skill through exercises and self-tests. Glossaries of medical terms, diagnostic tests, and procedures are provided.

Cohen, Barbara Janson
Medical Terminology: An Illustrated Guide
2nd ed. Philadelphia, Pa., Lippincott, 1994. 574p.
ISBN 0397550529

An excellent terminology workbook with color illustrations. Includes topics such as "suffixes that mean condition" and "prefixes pertaining to color," etc.

Dennerll, Jean Tannis
Medical Terminology Made Easy
Albany, N.Y., Delmar Publishers, 1993. 296p. ISBN 0827352786

This guide takes a programmed learning approach to medical terminology. Students are instructed in word building instead of memorization. The text of each chapter is divided into frames, with key words missing from each frame. An answer column next to the text allows the reader to fill in the blanks then immediately check the work. Appendices include a sample patient's history and physical exam report, a list of health career opportunities, and crossword puzzles for review.

Dox, Ida, B. John Melloni, and Gilbert M. Eisner
Melloni's Illustrated Medical Dictionary
3rd ed. Scranton, Pa., HarperCollins, 1993. 533p.
ISBN 1850704791

Dox's illustrated dictionary provides more than 26,000 terms for use by health professionals.

Ehrlich, Ann B.
Medical Terminology for Health Professionals
2nd ed. Albany, N.Y., Delmar Publishers, 1993.
ISBN 0827330367

This unique text provides definitions and pronunciation for more than 1,500 medical terms, with an emphasis on medical terminology as a new language that can be acquired by learning prefixes and suffixes. Chapters are organized by body systems with separate chapters for diagnostic and imaging procedures, and mental disorders and pharmacology. An appendix provides extensive lists of prefixes, suffixes, and combined forms. Exercises accompany each chapter, and detailed labeled illustrations complement the text.

Frenay, Agnes Clare, and Rose Maureen Mahoney
Understanding Medical Terminology
9th ed. Dubuque, Iowa, Wm. C. Brown, 1993. 619p.
ISBN 0871251574

The first section of this comprehensive text presents the anatomic, diagnostic, operative, and symptomatic terms related to disorders of each of the body systems. The second section provides selected terms for various areas of medicine including oncology, gerontology, anesthesiology, and nuclear medicine. The third section contains review exercises for the whole book, answers to review questions, and an extensive index.

Hanken, Mary Alice, and Kathleen A. Waters, eds.
Glossary of Healthcare Terms
Rev. ed., 1994 ed. Chicago, American Health Information Management Association, 1994. 1 v. (various pagings)

A standard resource for students and health care professionals, this glossary includes health terms and selected statistics.

Kinn, Mary E.
Medical Terminology: Building Blocks for Health Careers
Albany, N.Y., Delmar Publishers, 1990. 467p. ISBN 0827333382

This text begins with an introduction to word combining, pronunciation, and spelling of medical terms. Body systems are discussed, terminology is defined, and phonetic spellings are provided. One chapter provides a glossary-style listing of health occupations and specialties. Appendices include lists of terms not covered in other chapters and a glossary of building blocks for word construction.

Leonard, Peggy C.
Building a Medical Vocabulary
3rd ed. Philadelphia, Saunders, 1993. 558p. ISBN 0721646905

A programmed learning approach to medical terminology that instructs in word building instead of memorization. Chapters are divided into body systems, and the text of each chapter is divided into frames with key words missing. Students can fill in the missing words, then check their work against the answers provided. Appendices include pharmacological terms and color plates of human anatomy. A list of word parts, an extensive index glossary, and a bibliography are included.

Leonard, Peggy C.
Quick and Easy Medical Terminology
Philadelphia, Saunders, 1990. 310p. ISBN 0721657265

This guide's methods for teaching include word association and programmed learning. The first section describes how to simplify learning medical terms by dividing them into roots, prefixes, and suffixes. The second section provides terminology and review exercises for each of the body systems. The third section contains a review of the entire book. Appendices provide a list of physician specialties, a glossary of word parts, and answers to exercises.

O'Toole, Marie
Miller-Keane Encyclopedia and Dictionary of Medicine, Nursing, and Allied Health
5th ed. Philadelphia, Pa., Saunders, 1992. 1780p.
ISBN 0721634567

Information is presented in clear, concise, and easy to understand terminology. Entries on diseases contain relevant information such as signs and symptoms. Includes some color illustrations.

Rice, Jane
Medical Terminology with Human Anatomy
2nd ed. Norwalk, Conn., Appleton and Lange, 1991. 381p.
ISBN 0838562035

This book begins with an introduction to word structure and combination. Each body system is discussed, with terminology defined and spelled phonetically. Separate chapters provide the latest terminology in the areas of oncology and radiology. Each body system chapter contains a "Diagnostic and Laboratory Tests" section that lists and describes tests relevant to that body system. Supplementary audiocassettes are available.

Smith, Genevieve Love, Phyllis E. Davis, and Jean Tannis Dennerll
Medical Terminology: A Programmed Text
7th ed. Revised by Jean Tannis Dennerll. Albany, N.Y., Delmar Publishers, 1995. 500p. ISBN 0827363044

Readers are instructed in word building instead of memorization. This text uses a frame approach with key words missing from each frame. An answer column next to the text allows the reader to fill in the blanks then immediately check the work. Appendices include review activities and thirteen case studies taken from actual medical records. Supplementary audiocassettes are available.

Sormunen, Carolee
Terminology for Allied Health Professionals
2nd ed. Cincinnati, South-Western Pub. Co., 1990. 563p.
ISBN 053870070X

This text-workbook is divided into body systems, with medical terms defined and pronounced for each system. A chapter on medical histories and physical examinations discusses the types of reports found in medical offices. Supplementary audiocassettes provide authentic examples of dictation to be used with listening activities in the text. Extensive review exercises are provided for each chapter.

Taber's Cyclopedic Medical Dictionary
17th ed. Philadelphia, F. A. Davis, 1993. 2590p.
ISBN 080368312X

A classic dictionary for beginning students.

Wistreich, George
Medical Terminology in Action
Dubuque, Iowa, Wm. C. Brown Publishers, 1994. 605p.
ISBN 0697108961

This text begins with an introduction to prefixes and suffixes, combining forms, and spelling of medical terms. All systems of the body are discussed; terminology is defined and spelled phonetically. Most chapters include lists of keywords, case histories, and review exercises. Numerous illustrations and an extensive index are included. A supplementary slide set is also available.

◆ Ethical and Legal Issues

Bruce, Jo Anne Czecowski
Privacy and Confidentiality of Health Care Information
2nd ed. Chicago, American Hospital Pub., 1988. 209p.
ISBN 1556480253

Discusses issues concerned with developing policies to protect the privacy and confidentiality of health care information.

Brandt, Mary
Maintenance, Disclosure, and Redisclosure of Health Information
Chicago, American Health Information Management Association, 1993. 23p.

This booklet details AHIMA's position on patient confidentiality and security of primary and secondary records. Includes guidelines for managing data related to AIDS/HIV, adoption, and treatment for alcohol and drug abuse. Contains bibliographic references.

Flight, Myrtle
Law, Liabilities, and Ethics for Medical Office Personnel
2nd ed. 1993, Albany, N.Y., Delmar Publishers. 258p.
ISBN 0827339739

This edition discusses the regulations and issues concerning ethics in the medical office. The focus of this work is to make the reader aware of important legal concepts that relate to medical care. Includes case studies, OSHA regulations, and updated information on the Americans with Disabilities Act and AIDS patient care. Includes bibliography and index.

Lewis, Marcia A.
Medical Law, Ethics, and Bioethics in the Medical Office
3rd ed. Philadelphia, F. A. Davis, 1993. 270p. ISBN 0803656246

Focus is on ethical and legal responsibilities of the medical office worker. Includes information on topics such as business management, public duties, hiring practices, genetic engineering, abortion, AIDS, and issues related to death. Appendices include addresses of Drug Enforcement Administration offices, representative codes of ethics, and sample documents (a donor card and an example of power of attorney). Indexed.

Pence, Gregory W.
Classic Cases in Medical Ethics
New York, McGraw-Hill, 1990. 397p. ISBN 0070380929

This source discusses ethical information from legal cases that caused controversy in the medical community (such as Baby M, abortion, in vitro fertilization, and mercy killing). Includes historical background details for each case.

Pozgar, George D.
Legal Aspects of Health Care Administration
5th ed. Gaithersburg, Md., Aspen Pub., 1993. 587p.
ISBN 083420360X

This source orients the health care professional to a working knowledge of health law. Provides the nonlegal student with background information as well as case decisions. Includes glossary and index.

Roach, William H., and others
Medical Records and the Law
2nd ed. Gaithersburg, Md., Aspen Pub., 1994. 346p.
ISBN 0834203170

This overview of the legal aspects of health information management and issues relating to medical records includes disposition of records in mergers and acquisitions, employer access, AIDS, and security of computer-based patient records.

✦ Office Management, Computers, and Records

Ball, Marion J., and F. Collen Morris, eds.
Aspects of the Computer-Based Patient Record
New York, Springer-Verlag, 1992. 316p. ISBN 0387977236

This source offers information about user needs, technologies, and the future of computer-based patient records (CPR).

Becklin, Karonne J., and Edith M. Sunnarborg
Medical Office Procedures
3rd ed. New York, Glencoe/McGraw-Hill, 1994. 1 v. (various pagings). ISBN 0028001206

Provides a simulation approach to medical assisting. Covers career guidance, patient records, professional activities such as office management, reports, medical meetings, administrative responsibilities, and financial responsibilities. Indexed.

Campbell, Linda, and others, eds.
Medical Transcription: Fundamentals and Practice
Englewood Cliffs, N.J., Prentice Hall Career and Technology, 1994. 381p. and cassettes. ISBN 0130164372

Presents guidelines for transcription in a variety of medical specialities such as dermatology, gastroenterology, cardiology, ophthalmology, and obstetrics and gynecology. Each chapter focuses on a medical field and provides common terminology for that field. Book chapters are related to accompanying cassettes that contain authentic dictation from doctors in each field. The cassette lessons reflect a variety of accents and dictation styles. An annotated bibliography is included, and each chapter provides exercises.

Dick, Richard S., and Elaine B. Steen, eds.
The Computer-Based Patient Record: An Essential Technology for Health Care
Washington, D.C., National Academy Press, 1991. 190p.
ISBN 0309044952

This report by the Institute of Medicine discusses the future of computer-based patient records (CPR) and their development in the health care field. The plan defines CPR as the standard for future patient records. Includes suggestions for overcoming barriers, training recommendations, and standards for data security.

Glondys, Barbara A.
Documentation Requirements for the Acute Care Patient Record
3rd ed. Chicago, American Health Information Management Association, 1993. 368p. ISBN 0317054414

This text provides guidelines for documentation in the health record along with sample forms. Checklists are also included for the medicolegal document and for medical staff orientation.

Harpole, Greg
Patient Billing: A Computerized Simulation Using Medisoft
2nd ed. New York, Glencoe/McGraw-Hill, 1995. 274p. and one computer disk ISBN 0028025571

This text/workbook contains an introduction to medical billing, a simulation, a procedures manual, and source documents. Students work through case studies and are tested on knowledge as they proceed through the *Medisoft* program, a widely used patient accounting software program. Includes index.

Huffman, Edna K.
Health Information Management
10th ed. Berwyn, Ill., Physicians' Record Co., 1994. 780p.
ISBN 0917036174

Huffman provides a basic, foundational text on medical record management. Includes bibliographical references and an index.

Humphrey, Doris
Contemporary Medical Office Procedures
Cincinnati, South-Western Pub. Co., 1990. 392p.
ISBN 0538700157

The emphasis of this detailed guide to the administrative duties of medical office workers is on new office technology and its impact on medical assisting. Brief essays by actual medical office staff introduce each chapter. Coverage includes the current medical environment, working with patients, and the role of computers in the medical office. Chapters on working with patients provide examples of effective methods of communication. Each chapter includes review exercises and references.

Johnson, Jan L.
Basic Filing for Health Information Management
Albany, N.Y., Delmar Publishers, 1994. 208p. ISBN 0827354576

Presents rules, systems, procedures and practical applications of medical filing. Uses step-by-step examples and instructions to facilitate learning. Illustrated with practice sheets, self-tests, and quizzes. Contains cards that can be taken out for hands-on exercises.

Medical Assisting and Medical Records

Johnson, Joan M., and Marc W. Johnson
Computerized Medical Office Management
Albany, N.Y., Delmar Publishers, 1994. 347p. and computer disk
ISBN 0827351127

Students utilize a simulation disk that teaches procedures in medical office work. Includes basic computer information as well as more sophisticated use of the computer. Covers patient appointments and scheduling, patient information and charts, billing and collections, insurance, and banking. Troubleshooting guide, glossary, and index.

Keir, Lucille, Barbara A. Wise, and Connie Krebs
Medical Assisting: Administrative and Clinical Competencies
2nd ed. Albany, N.Y., Delmar Publishers, 1989. 586p.
ISBN 0827335075

A comprehensive overview of all aspects of medical assisting. This second edition provides updated information on the latest procedures and equipment, plus a discussion of the latest ethical issues in health care including organ donation, artificial insemination, and issues surrounding AIDS. Every chapter is highlighted by several "Procedures" sections that provide specific instructions for the reader to practice skills covered. Includes references, a glossary, and an index.

Moss, Edna Jean
Basic Keyboarding for the Medical Office Assistant
New York, Delmar Publishers, 1995. 288p. ISBN 0827357982

Designed to aid the medical assistant in developing keyboard proficiency through lessons and drills. Includes timed drills and discussions of medical topics such as chart notes, operative reports, pathology reports, and discharge summaries. Bibliography.

Rowel, Jo Ann C.
Understanding Medical Insurance: A Step-By-Step Guide
2nd ed. Albany, N.Y., Delmar Publishers, 1994. 336p.
ISBN 0827349661

Aids students in the development of skills for medical insurance claim work. Features detailed instructions for coding diagnosis and procedures, and insurance program data sheets. Discusses current trends in the major insurance programs and includes examples of forms for various program needs. Indexed.

Worthley, John Abbott, and Philip S. DiSalvio
Managing Computers in Health Care: A Guide for Professionals
2nd ed. Ann Arbor, Mich., Health Administration Press, 1989. 343p. ISBN 0910701474

Focuses on the computer user who will be managing the office technology and dealing with computer technicians and vendors.

Zakus, Sharron M., De A. Eggers, and Margaret A. Shea
Mosby's Fundamentals of Medical Assisting: Administrative and Clinical Theory and Technique
2nd ed. St. Louis, Mo., Mosby, 1990. 702p. ISBN 0801642809

The varied responsibilities of medical assistants are covered, with an emphasis on holistic care for patients. Topics include fundamentals of health care, administrative theory and technique, and clinical theory and technique. Throughout the book are descriptions of general procedures, with the specific responsibilities of medical assistants addressed separately. Appendices include special vocabulary and sample reports. A bibliography and an extensive index are provided.

✦ Pharmacology

Asperheim, Mary Kaye
Pharmacology: An Introductory Text
7th ed. Philadelphia, Saunders, 1992. 260p. ISBN 0721637531

Discusses the drugs frequently encountered in medical assisting practice, indicating appropriate nursing implications for each, including side effects. A basic math review is also provided for dosage calculation purposes. Includes index.

Bayt, Phyllis Theiss
Administering Medications: A Competency-Based Program for Health Occupations
3rd ed. Glencoe/McGraw Hill, 1994. 456p. ISBN 0028008863

Written to meet the needs of a wide range of health professionals, this text provides a self-contained program of study to aid understanding of drug actions and procedures of administration. May be used in the classroom environment or in the office as self-study. Indexed.

Drake, Ellen and Randy Drake
Saunders Pharmaceutical Word Book, 1995
Philadelphia, Saunders, 1994. 640p. ISBN 0721641970

This dictionary of drugs will be helpful to all in the allied health professions. This easy-to-use reference provides correct spellings and capitalization of drugs, and cross-references from brand name to generic names. Provides information on common methods of administration.

Hitner, Henry, and Barbara T. Nagle
Basic Pharmacology for Health Occupations
3rd ed. New York, Glencoe/McGraw Hill, 1994. 435p.
ISBN 0028006798

Designed for allied health programs, this work presents drugs in relation to their therapeutic applications. Each section provides reviews physiology and related diseases before discussing drugs. Emphasis is placed on current drug therapy. Provides a firm foundation for additional study. Indexed.

Rice, Jane
Principles of Pharmacology for Medical Assisting
2nd ed. Albany, N.Y., Delmar Publishers, 1994. 548p.
ISBN 0827357443

Offers a complete introduction to the fundamental areas of pharmacology: mathematics, dosage calculations, administration, classification, and effect of drugs on the body systems. The second edition includes updated information on AIDS and other infectious diseases. Includes tables and practice problems with answers. Indexed.

✦ Professional Materials and Standards

Amatayakul, Margaret K., and Makhdoom A. Shah
Research Manual for the Health Information Management Profession
Chicago, American Health Information Management Association, 1992. 80p. ISBN 0317054392

This guide on research methodology, designed for faculty and students, contains all the information needed to identify, prepare, and disseminate research projects and findings in the field of medical records.

American Health Information Management Association
Professional Practice Standards for Health Information Management Services
Chicago, The Association, 1992. 110p. ISBN 0317054368
   Provides standards applying to health information management services.

American Health Information Management Association
Professional Practice Standards for Health Information Management Services in Ambulatory Care
Chicago, The Association, 1992. 182p. ISBN 031705435X
   Provides standards applying to freestanding ambulatory care centers.

American Health Information Management Association
Professional Practice Standards for Health Information Management Services in Long Term Care
Chicago, The Association, 1993. 166p. ISBN 0866883207
   Provides standards for long-term care facilities.

American Medical Record Association
Professional Practice Standards for Mental Health
Chicago, The Association, 1990. 1 v. (various pagings). ISBN 0317054384
   Provides standards for mental health, substance abuse, and mental retardation and developmental disabilities treatment centers.

Tamparo, Carol D., and Wilburta Q. Lindh
Therapeutic Communications for Allied Health Professions
Albany, N.Y., Delmar Publishers, 1992. 253p. ISBN 0827349998
   Covers basic psychology, human growth and development, and therapeutic response. Focus is on self-understanding as well as understanding others. Includes responses to clients who are experiencing depression, aggression, stress, suicidal tendencies, and abusive tendencies, to name a few.

✦ Vocational Guidance

Bailey, Susan Pritchard
Medical Record: 800 Multiple-Choice Questions with Explanatory Answers
5th ed. New York, Medical Examination Pub. Co., 1989. 279p. ISBN 0838557708
   This source prepares students for the certifying exam. Bailey presents sample questions and answers for review.

Barber, Linda George
Being a Medical Admissions Clerk
Angled Cliffs, N.J., Brady/Prentice Hall, 1994. 134p. ISBN 0893030724
   An introductory guide to the varied responsibilities of medical admissions clerks. All the basics are covered including emergency room admitting, insurance forms, medical vocabulary, and the general admitting process. The importance of interpersonal skills is also emphasized, since clerks often perform admission interviews and assist patients in the emergency room. Contains review activities at the end of each chapter and small glossaries throughout the book.

Dodson, Laurie
Being a Medical Information Coder
Englewood Cliffs, N.J., Regents/Prentice Hall, 1992. 262p. ISBN 0893030848
   A guide introducing students to the opportunities found in medical record departments throughout the health care industry.

Drafke, Michael W.
Working in Health Care: What You Need to Know to Succeed
Philadelphia, F. A. Davis, 1994. 213p. ISBN 0803628080
   An overview of the management and organizational aspects of working in health care. Topics covered include communication, management decisions, change, performance evaluations, and stress. A chapter on vocational guidance provides résumé preparation and job search information. Each chapter contains an outline, a list of learning objectives, and references.

Dreizen, LaVerne, and Thelma Audet
Medical Assistant: 800 Multiple-Choice Questions with Explanatory Answers
4th ed. East Norwalk, Conn., Medical Examination Pub. Co., 1989. ISBN 0838557724
   This guide is arranged like a certified medical assistant (CMA) examination. Multiple-choice questions are divided into four sections: fundamentals, terminology, administrative medical assisting, and clinical medical assisting. The answers also reference a page in a textbook, which students can use for further information.

Lane, Karen
The Medical Assistant Examination Guide: A Comprehensive Review for Certification
Revised reprint. Philadelphia, F. A. Davis, 1991. 192p. ISBN 0803654642
   Designed to aid medical assisting students achieve the highest possible scores on the CMA and RMA examinations, this work presents model examinations with more than 900 questions in a topically oriented format. Includes correct answers with rationales. Provides exam strategy tips and examples of applications.

Lindsey, Bonnie Joan
The Professional Medical Assistant: Clinical Practice
Albany, N.Y., Delmar Publishers, 1993. 316p. ISBN 0827341504
   Lindsey provides a comprehensive overview of the medical and clinical knowledge necessary for a career in medical assisting. Chapter topics address office surgery, pharmacology, emergencies, and health maintenance. Each chapter includes detailed instructions for procedures discussed. A chapter on professionalism addresses vocational issues, e.g., seeking the right job, changing jobs, preparing for certification, and continuing education. Includes numerous illustrations and a detailed index.

Marshall, Jacquelyn, and Kathleen Harris
Being a Medical Clerical Worker: An Introductory Core Text
Englewood Cliffs, N.J., Prentice Hall, 1990. 270p. ISBN 0893031062
   This fundamental overview of the duties of medical clerical workers covers reception techniques, office automation, health insurance, understanding medication orders, and patient records. Appendices include instructions for completing a Universal Health Insurance Claim Form, a case history example, and numerous samples of forms. Review questions for each chapter; index.

Palko, Tom, and Hilda Palko
Appleton and Lange's Review for the Medical Assistant
4th ed. Norwalk, Conn., Appleton and Lange, 1995. 299p.
ISBN 0838501974

Present a question-and-answer format for CMA and RMA certification examinations. Includes more than 1600 questions and answers with detailed explanations. Also provides information on test-taking strategies.

Rudman, Jack
Passbook for Career Opportunities Series
Syosset, N.Y. National Learning Corp, 1991- .

These workbooks prepare candidates for certification examinations. Includes questions and answers, sample forms, and discussions on patterns of questioning. Rudman also supplies information on test-taking strategies. Titles include:

Medical Assistant, 1993. 1 v. (various pagings).
ISBN 0837313651
Medical Records Assistant, 1991. 1 v. (various pagings).
ISBN 0837329523
Medical Records Clerk, 1991. 1 v. (various pagings).
ISBN 0837323096
Medical Records Technician, 1991. 1 v. (various pagings).
ISBN 0837323290

Medical Laboratory Technology

Ann Cannon
DeKalb Technical Institute
Clarkston, Georgia

Introduction

The Medical Laboratory Technology program provides educational opportunities to acquire the knowledge, skills, and attitudes necessary to succeed as a medical laboratory technician.

Medical laboratory technicians perform tests for use in treatment and diagnosis of disease. Using microscopes and other instruments, they prepare tissue samples and examine body fluids, viruses, and bacteria to help determine the causes of diseases. They perform laboratory tests such as urinalysis and blood counts as well as make quantitative and qualitative chemical and biological analyses of body specimens. All technicians work under the supervision of a medical technologist or a laboratory director. Various positions in the clinical laboratory include laboratory aide, laboratory clerical worker, medical laboratory assistant, phlebotomist, medical laboratory technician, medical laboratory technologist, laboratory manager, and pathologist. There is a growing need for medical technologists, and all branches of the American Society of Clinical Laboratory Scientists (ASCLS) are very interested in recruitment.

Community colleges and vocational schools offer several different programs for medical laboratory technician students. Generally, medical laboratory technicians have an associate degree from a college, or a diploma or certificate from a private post-secondary technical school or hospital. Graduates are eligible to take a national certification exam. Program graduates should be competent in communications, math, interpersonal relations, anatomy and physiology, and inorganic chemistry. Program graduates are eligible to work in the major medical laboratory areas of phlebotomy, urinalysis, hematology, immunology, serology, clinical chemistry, microbiology, and immunohematology.

Persons who choose medical technology as a career must meet the standards set by their profession. The various types of certification in medical technology are general, categorical, and specialist. The categorical and specialist categories are for those who wish to enter one of the specialist fields of medical technology. The general certificate MT is awarded to the medical technologist, and the general certificate, MLT, to the medical laboratory technician.

Nationally recognized accrediting agencies in the allied health field include the Council on Accreditation of Allied Health Education Programs (CAAHEP) and the Accrediting Bureau of Health Education Schools (ABHES). CAAHEP is supported by the National Accrediting Agency for Clinical Laboratory Sciences (NAACLS) and the American Medical Association.

This bibliography has been created for students of medical laboratory technology enrolled at professional schools, community colleges, vocational schools, and hospital programs, as well as those working in the field.

Accreditation and Certification

✦ Medical Laboratory Technician

American Society of Clinical Pathologists
2100 W. Harrison
Chicago, Ill. 60612

American Medical Technologists
710 Higgins Road
Park Ridge, Ill. 60068

International Society for Clinical Laboratory Technology
818 E. Olive St., Suite 918
St. Louis, Mo. 63101-1598

✦ Clinical Laboratory Technician

National Certification Agency of Medical Lab Personnel
2021 L St., NW, Suite 400
Washington, D.C. 20036

Selected Journals

Abstracts of Clinical Care Guidelines.
Mosby-Yearbook. Bimonthly. ISSN 1042-4423

Advance for Medical Technologists.
Merion Publishers. Weekly except Christmas week.
ISSN 1044-2898

Advances in the Biology of Disease.
Williams and Wilkins. Annual. ISSN 0743-5592

American Laboratory News.
International Scientific Communications. Bimonthly.
ISSN 0893-8830

Archives of Pathology and Laboratory Medicine.
American Medical Association. Monthly. ISSN 0003-9985

Clinical Chemistry (Reference Edition).
American Association of Clinical Chemistry. Monthly.
ISSN 0009-9147

Clinical Lab Letter.
Quest Pub. Co. Semimonthly (except Aug. and Dec.).
ISSN 0197-8454

Clinical Laboratory Reference.
Medical Economics, Inc. Annual. ISSN 0093-8076

Clinical Laboratory Science.
American Society of Medical Technology. Bimonthly.
ISSN 0894-959X

Clinical Pharmacology and Therapeutics.
Mosby-Yearbook. Monthly. ISSN 0009-9236

Critical Reviews in Clinical Laboratory Sciences.
CRC Press. Quarterly. ISSN 1040-8363

Journal of Clinical Investigation.
Rockefeller Univ. Press. Monthly. ISSN 0021-9738

Journal of Clinical Laboratory Analysis.
Wiley. Bimonthly. ISSN 0087-8013

Journal of Laboratory and Clinical Medicine.
Mosby-Yearbook. Monthly. ISSN 0022-2143

Labmedica.
Techcom. Six times a year. ISSN 1054-0970

Lab Report.
G and R Publications. Monthly. ISSN 1045-7313

Laboratory Equipment.
Gordon Publications. Monthly. ISSN 0023-6810

Laboratory Investigation.
Williams & Wilkins. Frequency varies. ISSN 0023-6837

Medical Laboratory Observer (MLO).
Medical Economics. Monthly. ISSN 0580-7247

MT Today.
MT Today. Weekly. ISSN 1060-7609

Print Materials

Adams, Cynthia H.
Interpersonal Skills and Health Professional Issues
Mission Hills, Calif., Glencoe Publ. Co., 1989. 326p.
ISBN 0026854821
   A clearly written and organized text focusing on the basic communication skills needed for health care workers. Helps students develop an understanding of interpersonal techniques and seeks to improve interactions with patients. Specific issues addressed are grief and loss, addiction, professional ethics, associations, laws, abuse, and discrimination. Accompanied by instructor's guide.

Addison, Lois A., and Paul M. Fischer
Office Laboratory
2nd ed. Norwalk, Conn., Appleton and Lange, 1990. 433p.
ISBN 0838572448
   This second edition introduces students to the terms, concepts, procedures and equipment used in a professional medical laboratory. Includes bibliography and index.

Alba, Augusto
Alba's Medical Technology: Board Examination Review, Volume II
7th ed. Anaheim, Calif., Berkeley Scientific Publications, 1993. 500p. ISBN 0910224145 (vol. 2)

Revised and updated, this seventh edition contains thousands of questions and answers representative of recent examinations, plus practical math for clinical chemistry and board exam tips. Features detailed outlines of every clinical laboratory subject. Extensive glossaries and index.

Anderson, Sandra K.
Computer Literacy for Health Care Professionals
Albany, N.Y., Delmar Publishers, 1992. 221p. ISBN 0827341717

This work presents a hands-on approach to the computer and applies this knowledge to the health care setting. Examples of specific medical applications are provided. Partial contents: introduction to computer literacy, fundamental components of computers, direct patient care and treatment applications, diagnostics, and communications. Appendices. Glossary.

Anderson, Shauna Christine, and Susan Cockayne
Clinical Chemistry: Concepts and Applications
Philadelphia, Saunders, 1993. 748p. ISBN 0721633722

This authoritative text discusses concepts and techniques of clinical chemistry applicable to medical laboratory technology. Delivers concrete, state-of-the-art material on the basics of clinical chemistry, the mechanisms of disease, and the correlation of laboratory data with those diseases.

Annual Review of Microbiology
Palo Alto, Calif., Annual Reviews, v.1, 1947- . ISSN 0066-4227

Published by Annual Reviews, a nonprofit scientific publisher established to promote the advancement of the sciences. A reliable compilation of diverse topics on research in microbiology. Organized by editors and editorial committees who are qualified within each discipline.

Baron, Samuel, ed.
Medical Microbiology
3rd ed. New York, Churchill Livingstone, 1991. 1340p. ISBN 0443086710

Well-balanced coverage of academic medical microbiology and clinically oriented reviews of the microbiology of organ systems. It is actually two books in one: a comprehensive textbook of microbiology, and a concise review text that is written at a level appropriate for medical students and physicians. An innovative resource that includes study reviews for each chapter and over 300 drawings and diagrams.

Becan-McBride, Kathleen, and Doris L. Ross
Essentials for the Small Laboratory and Physician's Office
Chicago, Year Book Medical Publishers, 1988. 444p. ISBN 0815105909

Covers basic laboratory diagnosis, quality control, methods of organization, and administration. Indexed. Includes bibliography.

Belsey, Richard, and others
Physician's Office Laboratory
2nd ed. Los Angeles, Practice Management Information, 1993. 468p. ISBN 1878487493

Coverage has been expanded in this new edition to include discussions of general lab techniques and practice of medical lab management, routine diagnostic tests, and methods.

Bick, Rodger L., ed.
Hematology: Clinical and Laboratory Practice
St. Louis, Mo., Mosby-Yearbook, 1993. 2 v. ISBN 0801602033

Offers a systematic, practical and complete approach to hematological diseases. A clinical hematologist discusses each disease and follows it with a complete discussion of the laboratory manifestations and diagnosis. Features 965 illustrations, and sixty-three color plates in each volume.

Brunzel, Nancy A.
Fundamentals of Urine and Body Fluid Analysis
Philadelphia, Saunders, 1994. 504p. ISBN 0721639763

Affordable, well-illustrated text featuring a host of learning aids and real-life situations students may face in their careers. Provides theory and techniques required to conduct tests on urine and body fluids.

Bryant, Neville J.
Introduction to Immunohematology
3rd ed. Philadelphia, Saunders, 1994. 530p. ISBN 072163883X

This thoroughly updated text incorporates learning aids that give students a clear grasp of the most important concepts in immunohematology.

Bryant, Neville J.
Laboratory Immunology and Serology
3rd ed. Philadelphia, Saunders, 1992. 387p. ISBN 0721642128

Extensively revised to reflect new advances and changes in the area of nonspecific and specific immunity, the immune response and in vitro antigen-antibody reactions. Features summary sections, new chapters on AIDS, Lyme disease, and syphilis. Important concepts and illustrations are highlighted with color.

Bureau of National Affairs
Medical Testing: The Complete Resource Guide to Good Laboratory Practice and HFCA Compliance
Washington, D.C., Bureau of National Affairs, 1989. 1 v. (various pagings) ISBN 1558711139

Articles by leading authorities on various topics in medical testing. New legislative and regulatory developments are analyzed with emphases on the procedures that are needed to bring laboratories into compliance and to increase their efficiency.

Calbreath, Donald F.
Clinical Chemistry: A Fundamental Textbook
Philadelphia, Saunders, 1992. 468p. ISBN 0721626211
   Well-written, clearly illustrated, and concise clinical correlations on the chemistry of assay techniques. Students will benefit from the logical flow of principles and attain a solid foundation in clinical chemistry. Although this is a textbook, few other books provide such clear, up-to-date information on these techniques. Indexed. Bibliography included.

Campbell, June Mundy
Laboratory Mathematics: Medical and Biological Applications
4th ed. St. Louis, Mo., Mosby, 1990. 446p. ISBN 0801628733
   This work contains simplified explanations of the calculations used in the clinical and general biological laboratories and is aimed at helping students develop the essential skills necessary to meet the standards for clinical performance. A good refresher for laboratory workers in specific areas.

Cella, June H.
Nurse's Manual of Laboratory Tests
Philadelphia, F. A. Davis, 1989. 524p. ISBN 0803616961
   This book is designed for students and practitioners of nursing; however, its additional features in practical use, description of specimen collections of urine and blood, and clinical applications make this a valuable resource for medical lab technicians. Indexed.

Cembrowski, George S.
Laboratory Quality Management: QC [and] QA
Chicago, ASCP Press, 1989. 264p. ISBN 089189277X
   Provides discussions of quality control practice in health care laboratories. This book is not intended for light reading but is essential for those involved in managing the quality of laboratory operations. Published by the American Society of Clinical Pathologists Press. Indexed; bibliography included.

Chabner, Davi-Ellen
Medical Terminology: A Short Course
Philadelphia, Saunders, 1991. 227p. ISBN 0721629393
   Presented in workbook-text format with easy-to-understand language. Important terms from all areas of medicine are covered, including terms from clinical procedures and laboratory tests.

Ciulla, Anna P., and Georganne K. Buescher, eds.
Medical Technology Examination Review and Study Guide
2nd ed. Norwalk, Conn., Appleton and Lange, 1992. 493p.
ISBN 0838563104
   Self-examination review guide for those preparing for the national certification exam. Entries consist of examination questions and answers on laboratory diagnosis and the medical technology field.

Davis, Brenta G., Michael L. Bishop, and Diana Mass, eds.
Clinical Laboratory Science: Strategies for Practice
Philadelphia, Lippincott, 1989. 1024p. ISBN 0397508603
   This work is based on the premise that the demands placed on the clinical laboratory are becoming more complex. Workers need to know not only the methodologies of testing, but also the rationale behind the tests. Reporting raw data without insight or understanding of the physician's needs can be counterproductive to patients. Emphasizes proper utilization of the laboratory for efficient and effective diagnosis and treatment.

Dawe, Renee A.
Math and Dosage Calculations for Health Occupations
New York, Glencoe, Macmillan/McGraw-Hill, 1993. 232p.
ISBN 0028006771
   Mathematics is presented for pharmaceutical arithmetic as it relates to dosage calculations. Can be used as a math unit in shorter programs or as a supplement or review in longer courses. Contents: common fractions, decimals, metric system of measurement, percent, proportion, system of weight and measurement, and dosage calculations.

Dawids, S. G., ed.
Test Procedures for the Blood Compatibility of Biomaterials
Boston, Kluwer Academic Pub., 1993. 684p.
ISBN 0792321073
   A collection of test descriptions used in the evaluation of biocompatibility. This work has applicability to many disciplines. Indexed.

Diagnostic Tests
Springhouse, Pa., Springhouse Corporation, 1991. 537p.
ISBN 0874343321
   The clinical procedures described and recommended in this publication are based on research and consultation with medical and nursing authorities. The work reflects current and accepted clinical practice and is designed for use in any clinical or academic setting. Partial contents: diagnostic tests, urinalysis, and magnetic resonance imaging. Includes 520 tests. Usefulness is enhanced by an easel, flip chart, and spiral binding format.

Dictionary of Clinical Tests: A Concise Guide to Tests, Scales, and Scores in Medicine
Pearl River, N.Y., Parthenon Pub. Group, 1993. 262p.
ISBN 1850704163
   Alphabetical listing of clinical tests. The authors cannot guarantee information is representative of all groups due to the fact that tests were performed in a racially controlled group.

Dorland's Hematology/Oncology Speller
Philadelphia, Saunders, 1993. 242p. ISBN 0721637507
   Furnishes references to short terms and phrases. A much-needed speller.

Dorland's Medical Speller
Philadelphia, Saunders, 1992. 1493p. ISBN 0721635997
   A definitive A-to-Z source for spelling and hyphenation of the medical words, phrases, and eponyms in current use.

Dorland's Pocket Medical Dictionary
24th ed. Philadelphia, Saunders, 1989. 669p.
ISBN 0721622028
   Provides exceptional, comprehensive, up-to-date coverage of advances in technology and medical specialties.

Finley, Paul R., ed.
Clinical Laboratory Handbook: Normal Values, General Information, SI Unit Conversion Tables
8th ed. Tucson, Ariz., University of Arizona, University Medical Center, Department of Pathology, 1992. 185p.

This handbook aids the student in the study of clinical pathology and laboratory medicine and provides a good introductory framework for the study of pathology.

Flynn, John C., ed.
Procedures in Phlebotomy
Philadelphia, Saunders, 1994. 240p. ISBN 0721646859

An all-inclusive resource, the use of which will guarantee a sound and sure background in phlebotomy. Focuses on medical careers, giving students guidance in becoming competent, successful phlebotomists in any medical laboratory. Describes blood collection methods. Includes glossary, chapter review questions.

Fong, Elizabeth
Microbiology for Health Care Careers
5th ed. Albany, N.Y., Delmar Publishers, 1994. 440p. ISBN 0827360495

This best-selling text serves as an introduction to the essentials of microbiology. It includes information on infectious diseases, achievement reviews, safety guidelines, and expanded information on the microscope. Partial contents: basic chemistry, the microscope, cellular structure, introduction to fungi, microbiological techniques, sanitary measures, and how microorganisms cause infections. Accompanied by instructor's guide, complete with answers to chapter questions.

Forbes, C. D., and William F. Jackson
Color Atlas and Text of Clinical Medicine
St. Louis, Mo., Mosby-Yearbook, 1993. 528p. ISBN 0815132719

An ideal guide to clinical medicine. Provides a unique visual and textual guide to all major medical disorders. Extensive cross-references. Indexed.

Frew, Mary Ann, and Anne Lilly
Workbook for Clinical Procedures for Medical Assisting
Philadelphia, F. A. Davis, 1990. 269p. ISBN 0803638523

This workbook provides instructional support materials based on sound educational principles and develops student skills by simulating their roles in the workplace.

Garza, Diana, and Kathleen Becan-McBride
Phlebotomy Handbook
3rd. ed. Norwalk, Conn., Appleton and Lange, 1993. 419p. ISBN 0838579000

This edition continues as a practical instruction manual to reflect the most up-to-date information on specimen collection and blood collection safety equipment and techniques. Revisions cover the latest information on quality assurance/quality control within the lab setting, safety and liability aspects as they relate to AIDS, and the latest equipment and special collection techniques.

Gelijns, Annetine, and Ethan A. Halm, eds.
Changing Economics of Medical Technology
Washington, D.C., National Academy Press, 1991. 210p. ISBN 030904491X

A book of proceedings from a December 1989 conference on the impact of regulations and reimbursement on pharmaceutical innovation.

Greene, Harry L., William P. Johnson, and Michael J. Maricic, eds.
Decision Making in Medicine
Philadelphia, B. C. Decker, 1993. 561p. ISBN 1556642261

Topics are organized by sign, symptom, problem, or laboratory abnormality. Text offers uniformity and orderliness and to the work-up of common complaints. An attempt is made to bridge the gap between the teaching of science and the practice of experienced clinicians by offering a "decision-tree" approach and an algorithm for the appropriate therapy or course of action. Indexed. Bibliographical references included.

Hales, T. B.
Hospital Laboratory: Strategy, Equipment, Management, and Economics
New York, Ellis Horwood, 1990. 123p. ISBN 0133930831

This book is designed to explain the knowledge and skills required to perform management functions in hospitals and laboratories. The author has worked in the field for twenty years, and he attempts to show how the current trend of developing hospital pathology laboratories is under review worldwide because of rapidly changing technology and inflation.

Henry, John Bernard, ed.
Clinical Diagnosis and Management by Laboratory Methods
18th ed. Philadelphia, Saunders, 1991. 1454p. ISBN 0721622127

This work equips readers with the knowledge needed to direct a good laboratory. Concentrates on medical problem-solving with emphasis on the following topics: ordering of the lab test, interpreting the lab results, preparing patients for the lab results, managing quality, and organizing a clinical lab. This edition opens with an introduction to the clinical laboratory and an explanation of the purposes of laboratory medicine.

Hoeltke, Lynn B.
Clinical Laboratory Manual Series: Phlebotomy
Albany, N.Y., Delmar Publishers, 1994. 224p. ISBN 0827355270

Part of a series designed to give medical laboratory technicians the best training in new technologies and changes in the field. The series stresses the hands-on practical approach. Partial contents: introduction to phlebotomy, phlebotomy equipment, safety in phlebotomy, phlebotomy techniques, specimen considerations, and special procedures. Appendices.

Hoeltke, Lynn B.
Complete Textbook of Phlebotomy
Albany, N.Y., Delmar Publishers, 1994. 216p. ISBN 0827362315

This text provides strong emphasis on a hands-on, practical approach to learning phlebotomy.

Isenberg, Henry D., ed.
Clinical Microbiology Procedures Handbook
Washington, D.C., American Society of Microbiology, 1992. 2 v.
ISBN 1555810381

A useful handbook that provides a compilation of pertinent methods for the diagnosis of infectious diseases and the related tasks which produce accurate, reproducible, quality-controlled results. Each section is preceded by a detailed table of contents listing procedure titles. Procedures are numbered and cross-referenced. This work is written by authors whose occupation is the everyday performance of clinical microbiological analyses "at the bench" and technical microbiologists. Looseleaf. Indexed.

Jacobs, David S., ed.
Laboratory Test Handbook
3rd ed. Hudson, Ohio, Lexi-Comp, 1994. 1513p.
ISBN 0916589137

This handbook's chapters are arranged alphabetically by major clinical laboratory disciplines. Each individual test listing provides five major types of information: test name, patient care, specimen, interpretive data, footnotes, and references. Indexed.

Kaplan, Lawrence A., and Amadeo J. Pesce, eds.
Clinical Chemistry: Theory, Analysis, and Correlation
2nd ed. St. Louis, Mo., Mosby, 1989. 1149p.
ISBN 0801627044

Serves as a teaching text for medical technologists and medical laboratory technicians. This work contains information required in a bench reference for use by clinical chemists and laboratory supervisory personnel. Editors have put together the most relevant and current information from the associated disciplines. The chapters are consistently organized. Indexed.

Kaufman, James A., ed.
Waste Disposal in Academic Institutions
Chelsea, Mich., Lewis Publishers, 1990. 192p.
ISBN 0873712560

Written specifically to present practical and cost-effective solutions for dealing with chemical waste. Provides small generators with the information needed to design and operate a waste management program that will meet current regulations. Includes fifty experiments in analytical chemistry laboratory training. Indexed.

Keir, Lucille, Barbara A. Wise, and Connie Krebs
Medical Assisting: Administrative and Clinical Competencies
3rd ed. Albany, N.Y., Delmar Publishers, 1993. 778p.
ISBN 0827353111

This easy-to-understand text prepares students for entry-level positions as medical assistants. Includes instructor's guide, bibliography, and index.

Kumar, Vinay
Basic Pathology
5th ed. Philadelphia, Saunders, 1992. 772p. ISBN 0721637329

This superb text makes basics of pathology easier for students to grasp. Offers students detailed and entertaining discussions on general pathology and specific diseases. Indexed.

Leonard, Peggy C.
Quick and Easy Medical Terminology
Philadelphia, Saunders, 1990. 310p. ISBN 0721657265

Perfect for the classroom or as a study tool. Text offers basic medical terminology.

Lindsey, Bonnie Joan, and Francis M. Rayburn
The Professional Medical Assistant: Clinical Practice
Albany, N.Y., Delmar Publishers, 1993. 316p.
ISBN 0827341504

Written in an easy-to-understand conversational style, this text introduces the student to the knowledge, skills, and procedures required of today's clinical medical assistants. Includes bibliographical references and index.

Marieb, Elaine Nicpon
Essentials of Human Anatomy and Physiology
4th ed. Redwood City, Calif., Benjamin/Cummings Pub. Co., 1994. 494p. ISBN 0805341706

Ideal for classroom teaching of anatomy and physiology, this work is designed to complement any core anatomy and physiology textbook. Updated to reflect the latest research and advances. Indexed.

Markell, Edward K.
Medical Parasitology
7th ed. Philadelphia, Saunders, 1992. 463p.
ISBN 0721634117

A classic hailed by the *Journal of the American Medical Association*. It has been updated and expanded, providing information necessary for diagnosis, treatment, and prevention of human parasitic infections. Shows specific therapies with emphasis on the indications and contraindications of specific drugs, as well as their mechanisms of action. Contains life-cycle drawings. Indexed.

Marshall, Jacquelyn
Clinical Microbiology
Albany, N.Y., Delmar Publishers, 1994. 224p.
ISBN 0827353634

Part of the *Clinical Laboratory Manual Series,* this book is designed to give medical laboratory technicians the training needed to keep abreast of the rapidly changing health care field. Includes general concepts in clinical microbiology.

Marshall, Jacquelyn
Fundamental Skills for the Clinical Laboratory Professional
Albany, N.Y., Delmar Publishers, 1993. 698p.
ISBN 0827348231

This excellent book addresses the need for a practical, skills-oriented text for the mid-level medical laboratory technician. Provides a perfect balance between theory, skills, and state-of-the-art testing procedures.

McCall, Ruth E.
Phlebotomy Essentials
Philadelphia, Lippincott, 1993. 291p. ISBN 0397549296

Intended to provide accurate, up-to-date, practical information and instruction in phlebotomy procedures and techniques. Outstanding features are: key terms, chapter objectives and questions, and suggested laboratory activities at the end of each chapter. Excellent reference for practicing phlebotomists needing to update skills or pass a national certification exam.

McFarland, Mary Brambilla
Nursing Implications of Laboratory Tests
3rd ed. Albany, N.Y., Delmar Publishers, 1994. 453p.
ISBN 0827351356

This work covers diagnosis in the laboratory, nurses' instructions related to diagnositic procedures, and patient education. Includes bibliography and index.

McClatchey, Kenneth D., ed.
Clinical Laboratory Medicine
Baltimore, Md., Williams and Wilkins, 1994. 1893p.
ISBN 0683057553

Devised primarily to help laboratory workers and students make sound decisions in laboratory testing. Indexed.

Medical Laboratory Technician
Syosset, N.Y., National Learning Corp., 1992. 1 v. (various pagings) ISBN 0837323231

One of the volumes in the *Passbook* series. This series was created to prepare applicants and candidates for testing. It is also useful for those seeking educational or career advancement. Features of the series are learning-by-doing, programmed learning, presentation of questions.

Miller, Benjamin Frank, and Claire Brackman Keane
Encyclopedia and Dictionary of Medicine, Nursing, and Allied Health
5th ed. Philadelphia, Saunders, 1992. 1780p. ISBN 0721634567

Gives students access to lucid definitions and correct spellings for words they will encounter in their health care studies. This edition is alphabetically arranged and should be used as a first-line resource. Leaders and researchers have provided "windows" that broaden the definitions and provide the reader with additional comments regarding professions and clinical topics. Can be used as a quick reference.

Morgan, Bradley J., and Joseph M. Palmisano, eds.
Medical Technologists and Technicians Career Directory
Detroit, Mich., Gale Research, 1993. 324p. ISBN 0810391546

Highlights careers in the field of medical technology. Advice from top industry professionals is given on what to expect on the job, how to get the job, career paths, training, and certification. Divided into four parts to convey the steps of a typical job search.

O'Brien, Joseph D., ed.
Medical Device Packaging Handbook
New York, Dekker, 1990. 362p. ISBN 0824776984

This medical instruments and apparatus packaging handbook encompasses product labeling, equipment, and supplies.

Pagana, Kathleen Deska
Mosby's Diagnostic and Laboratory Test Reference
St. Louis, Mo., Mosby-Yearbook, 1992. 843p. ISBN 0801637562

Comprehensive, current clinical handbook that allows easy access to 564 laboratory and diagnostic tests. Tests are listed alphabetically by name for quick reference. Indexed with bibliographical references.

Panzer, Robert J., Edgar R. Black, and Paul F. Griner, eds.
Diagnostic Strategies for Common Medical Problems
Philadelphia, American College of Physicians, 1991. 551p.
ISBN 0943126207

Revised edition of *Clinical Diagnosis and the Laboratory* (1986); however, there are few updates in the coverage. Indexed.

Paul, William E., ed.
Fundamental Immunology
3rd ed. New York, Raven Press, 1993. 1490p.
ISBN 0781700221

This third edition is not an update of the previous ones. The goal of this edition is to provide readers with an understanding of the main areas of modern immunology. Written by scientists at a time when the pace for advancement in immunology has accelerated.

Phelan, Susan, E.
Phlebotomy Techniques: A Laboratory Workbook
Chicago, American Society of Clinical Pathologists, 1992. 280p. ISBN 0891893431

Designed as a tool to teach phlebotomy techniques, this book is suitable for the classroom setting. It was written by an instructor who has several years of teaching experience with students of phlebotomy. Objective is to offer instruction in the psychomotor domain as it relates to phlebotomy practice. Students will master standard phlebotomy procedures which will ease the transition to the clinical setting.

Physician's Office Laboratory Guidelines
2nd ed. Villanova, Pa., National Committee for Clinical Laboratory Standards, 1992. ISBN 1562381598

NCCLS provides a communication forum for the development, promotion, and use of national and international standards. This work discusses laboratory procedures, bench and reference methods/evaluation, and protocols applicable to major laboratory disciplines. Loose-leaf format.

Powers, Lawrence W.
Diagnostic Hematology: Clinical and Technical Principles
St. Louis, Mo., Mosby, 1989. 555p. ISBN 0801640423

Designed as an intermediate-level presentation for clinical laboratory technologists and other laboratory scientists needing a background in blood cell morphology of function. A general synopsis is provided of laboratory procedures.

Pribor, Hugo C.
Laboratory Consultant
Philadelphia, Lea and Febiger, 1992. 780p.
ISBN 081211387X

A handbook suitable for use by clinicians, pathologists, and others responsible for interpreting laboratory findings. It is organized by laboratory test name.

Ratliff, Thomas A.
Laboratory Quality Assurance System: A Manual of Quality Procedures with Related Forms
New York, Van Nostrand Reinhold, 1993. 250p.
ISBN 0442014708
   Details fundamentals of laboratory procedures and tests. Contents: laboratory testing and quality control.

Ravel, Richard
Clinical Laboratory Medicine: Clinical Application of Laboratory Data
5th ed. Chicago, Year Book Medical Publishers, 1989. 742p.
ISBN 081517098X
   Presents the clinical symptoms of various diseases and the correct procedures to follow in diagnosing them. Background information is given to further strengthen clinical skills.

Sacher, Ronald A.
Widmann's Clinical Interpretation of Laboratory Tests
10th ed. Philadelphia, F. A. Davis, 1991. 841p.
ISBN 0803676948
   Intended for the clinician responsible for the whole patient. The reader will find a vast array of information with straightforward facts in their simplest form. Revised edition of *Clinical Interpretation of Laboratory Tests* (Widmann, 1983).

Salway, J. G., ed.
Drug-Test Interactions Handbook
New York, Raven Press, 1990. 1087p. ISBN 0881676020
   An outline of clinical chemistry and the side effects of drugs.

Schoeff, Larry E., and Robert H. Williams
Principles of Laboratory Instruments
St. Louis, Mo., Mosby-Yearbook, 1993. 473p.
ISBN 0801674891
   All clinical laboratories will benefit from this text. The invaluable information presented provides a foundation of knowledge in basic electronics and principles of instrumentation. Indexed.

Seeley, Rod R.
Anatomy and Physiology
St. Louis, Mo., Mosby-Yearbook, 1992. 980p. ISBN 0801648327
   Gives an introduction to problem-solving techniques required of future professionals. Relevant clinical examples help students learn basic anatomy and physiology. Information is easy to read. Concepts are explained simply.

Simmons, Arthur
Hematology: A Combined Theoretical and Technical Approach
Philadelphia, Saunders, 1989. 423p. ISBN 0721617166
   Superb coverage that provides students with theoretical and practical perspectives on hematology. Offers a vivid view with the use of color plates and black-and-white photos. Topics: physiology and pathophysiology of hematology, hematopoiesis, blood disorders, specimen collection, common tests, automation, and quality assurance.

Tamparo, Carol D.
Therapeutic Communications for Allied Health Professions
Albany, N.Y., Delmar Publishers, 1992. 253p.
ISBN 0827349998
   Written for students in medical assisting and other allied health programs. Text covers communications, human growth, development, and therapeutic response. Emphasis is on a solid understanding of self and others so that clients with different needs can be helped. Indexed.

Tietz, Norbert W., ed.
Applied Laboratory Medicine
Philadelphia, Saunders, 1992. 425p. ISBN 0721664741
   Presents information on seventy of the most common disorders and the clinical signs of those diseases. Provides teaching material geared toward the second-year student. The material would also be helpful in providing discussions of clinical evaluation and diagnostics.

Tietz, Norbert W., ed.
Clinical Guide to Laboratory Tests
2nd ed. Philadelphia, Saunders, 1990. 931p.
ISBN 0721624863
   An indispensable handbook for medical technology students. A reference source on common, rare, and highly specialized laboratory tests.

Tilton, Richard C., ed.
Clinical Laboratory Medicine
St. Louis, Mo., Mosby-Yearbook, 1992. 1207p.
ISBN 080165873X
   An excellent resource for laboratory specialists and students. It brings together the method selection and decision-making processes that ensure quality laboratory testing. Correlates the laboratory data with current understanding of disease pathophysiology and clinical management of patients. Indexed.

Turgeon, Mary Louise
Fundamentals of Immunohematology: Theory and Technique
Philadelphia, Lea and Febiger, 1989. 493p.
ISBN 0812112172
   Written to meet the needs of clinical laboratory students, educators, and blood bank personnel. The purpose is to integrate the basic foundations of blood banking with the technical aspects of the discipline and clinical applications.

Turgeon, Mary Louise
Immunology and Serology in Laboratory Medicine
St. Louis, Mo., Mosby, 1990. 415p. ISBN 080165131X
   Written primarily for undergraduate students in a clinical laboratory science program, this is an excellent reference for medical technologists, medical students, medical residents, nursing students, practitioners, medical assistants, and other allied health professionals. Provides a basic foundation in theoretic concepts in immunology. Contents and index included.

Volk, Wesley A.
Basic Microbiology
7th ed. New York, HarperCollins, 1992. 602p. ISBN 0060468491
   Beneficial for students who are entering careers in the health science field as well as those who merely want to learn about the world around them. This work describes the importance of microorganisms in all aspects of our lives. Half of the text is devoted to immunology and medical microbiology with the remainder of the book focusing on introductory material on bacteriology, virology, and eukaryotic microorganisms. Comprehensive coverage for students with little or a minimal background in biology or chemistry.

Walter, J. B.
Introduction to the Principles of Disease
3rd ed. Philadelphia, Saunders, 1992. 666p. ISBN 0721690823
   Pathology and clinical practice are integrated so that students can see how basic science affects patient care. A good introduction to the study of pathology.

Walters, Norma J., Barbara H. Estridge, and Anna P. Reynolds
Basic Medical Laboratory Techniques
2nd ed. Albany, N.Y., Delmar Publishers, 1990. 516p. ISBN 0827339488
   A performance-based text written primarily to introduce the prospective teacher or student to the basic principles and techniques of manual laboratory procedures. Illustrates the fundamental principles taught in most training programs. Partial contents: introduction to the medical laboratory, basic hematology, advanced hematology, and basic clinical chemistry.

Ward, Kory M., Craig A. Lehmann, and Alan M. Leiken, eds.
Clinical Laboratory Instrumentation and Automation: Principles, Applications, and Selection
Philadelphia, Saunders, 1994. 599p. ISBN 0721642187
   This work gives insights on today's technology and clinical laboratory. Offers students the latest in basic electronics, instrumentation, and automation in clinical chemistry, hematology, microbiology, immunology, and physician office laboratories.

Young, Donald S.
Effects of Drugs on Clinical Laboratory Tests
3rd ed. Washington D.C., AACC Press, 1990. ISBN 0915274531
   Covers drugs and their side effects, clinical chemistry, and laboratory diagnosis.

Zablotney, Sharon L., ed.
NCA Review for Clinical Laboratory Sciences
2nd ed. Boston, Little, Brown, 1989. 235p. ISBN 0316599256
   For self study or class use. Practice examination provides general review of diagnostic laboratories.

Zakus, Sharron M.
Mosby's Fundamentals of Medical Assisting: Administrative and Clinical Theory and Technique
St. Louis, Mo., Mosby, 1990. 702p. ISBN 0801642809
   This is a foundation text for all general, administrative, and clinical courses needed in the preparation of the medical assistant for employment. The focus is on practical aspects of the profession.

Nonprint Materials

AIDS Prevention for Laboratory Professionals
Mt. Kisco, N.Y., Guidance Associates, 1992.
1 video- cassette (69 min.) VHS
   Designed to outline specific safety procedures to follow in the laboratory to prevent accidental transmission of HIV and hepatitis B. Discusses procedures to follow when a worker is accidentally exposed to a patient's blood or other bodily fluids. Updated with 1992 OSHA guidelines.

Bloodborne Pathogens
Herkimer, N.Y., HCTV, 1993. 1 videocassette (60 min.) VHS
   Scott Millbowa, OSHA's certified instructor, presents the Pathogen Training Program. Discussion of HIV, AIDS, hepatitis, and universal precautions.

Bloodborne Pathogens
Virginia Beach, Va., Coastal Video Communications Corp., 1992. 1 videocassette (21 min.) VHS
   Developed to instruct new employees and to update experienced first responders, as well as employees trained in first aid, with current information and federal regulations. Includes the control of bloodborne infections in an industrial setting.

Chemical Safety in the Laboratory: Training for the OSHA Laboratory Standard
Van Nuys, Calif., AIMS Media, 1991. 1 videocassette (40 min.) VHS
   The main objective of this video is to give an understanding of OSHA standards. It is designed as a training program to present the fundamentals of chemical safety to all laboratory employees.

Chemical Safety in the Laboratory
Chattanooga, Tenn., Tel-A-Train, 1992. 1 videocassette (22 min.) VHS
   This program stresses the importance of OSHA's Laboratory Standard to employees and also addresses extensive employee information and training provisions of the standard. Accompanied by a viewer's guide, teacher's guide, and answer key.

Dorland's Electronic Medical Speller
27th ed. San Francisco, Reference Software International, 1992. 6 computer disks (5 1/4 in.) System requirements: Wordperfect 5.0 or 5.1; IBM-PC, DOS 2.1 or higher; Windows; hard disk with 600K free space.
   Excellent but expensive resource containing all the words and two-word forms from fifty-eight comprehensive medical, pharmacological, and specialty areas. Spell check included.

Human Body
Warren, N.J., Optical Data Corp., 1992. 3 videodiscs
   From the *Life Science Series*, this resource provides an

overview of each body system by locating and identifying each of the components while exploring the function of each component and its role in the system.

Human Body
Washington, D.C., National Geographic Society, 1992.
3 videodiscs. System requirements: Macintosh, 2 MB RAM; System 6.0.5 or higher; HyperCard 2.0 or higher.

Using video animation, medical photography, and text, the user can explore various aspects of the human body.

Infectious Medical Waste: Safe Handling and Disposal
Virginia Beach, Va., Coastal Video Communications Corp., 1990. 1 videocassette (21 min.) VHS

Describes the four types of medical waste: solid, chemical, radioactive, and infectious and lists the occupations that encounter infectious waste.

Lab Safety
Grand Rapids, Mich., Summit Training Source, 1990.
1 videocassette (12 min.) VHS

Students learn the physical and health hazards associated with chemicals; hazard identification; methods of detection and observation of hazardous chemicals, and control measures.

Lab Safety: The Chemical Hygiene Plan
Vancouver, Wash., Media Resources, 1992. 1 videocassette (30 min.) VHS

Stresses the necessity of a laboratory's having a chemical hygiene plan and describes safety rules and procedures. Includes instructor and study guides.

Laboratory Safety: Containing HIV and HBV Barriers for Your Protection
Oradell, N.J., Medical Economics, 1990. 1 videocassette (23 min.) VHS

Explains ways to contain HIV and hepatitis B viruses in the medical laboratory through the use of protective barriers and procedures.

Lab Safety: Handling Hazardous Chemicals
Virginia Beach, Va., Coastal Video Communications Corp., 1990. 1 videocassette (21 min.) VHS

Covers requirements of a company's chemical hygiene plan as outlined by OSHA. Contents: protective equipment and apparel, medical surveillance, monitoring designated areas, hazard recognition, and identification.

Medical Assisting
New York, Delmar Publishers, 1993. 12 videocassettes (276 min.) VHS

This series of twelve videos emphasizes professionalism, safety, and universal precautions throughout the administrative and clinical areas of medical assisting procedures. Each step-by-step procedure demonstrated is defined by the DACUM General Area of Competence to ensure proper procedural techniques.

Medical Waste Handling
Virginia Beach, Va., Coastal Video Communications Corp., 1993. 1 videocassette (16 min.) VHS

This program identifies various types of medical waste and their safe disposal.

Needlestick Prevention: Five Steps to Safety
Virginia Beach, Va., Coastal Video Communications Corp., 1993. 1 videocassette (14 min.) VHS

Presents five safety steps of needle use. Contents: (1) Roll up your sleeve to fight HIV; (2) A used sharp is a dangerous sharp (universal precautions); (3) Handle sharp as though your life depends on it, it does; (4) Dispose of used sharp as if your co-worker's life depends on it, it might; and (5) Watch out for the other guy.

OSHA's Bloodborne Pathogens Standard: Compliance in the Clinical Laboratory
Chicago, American Society of Clinical Pathologists, 1992.
1 videocassette and manual (ISBN 0891893490) (20 min.) VHS

This work takes a very practical approach, designed to inform clinical laboratory employees of their rights and responsibilities under OSHA's new standards on exposure to bloodborne pathogens.

Problems and Solutions: Sources of Error in Chemistry Tests
New York, ED-U-QUEST, 1990. 1 videocassette (108 min.) VHS

Students can use this video to identify specific problems in the chemistry lab. Illustrates methods to detect and avoid error.

Protecting Yourself Against Bloodborne Pathogens
Santa Barbara, Calif., Idea Bank, 1992. 1 videocassette (15 min.) VHS

Accents the importance of preventive measures. This program is intended to assist custodians who may be exposed to bloodborne pathogens in the workplace. Accompanied by employee's guide, reference guide and master for group training.

Nursing

Elisa Abella
Health Sciences Library
South Miami Hospital
Miami, Florida

and

Isabel C. Hernandez
Medical Center Campus Library
Miami-Dade Community College
Miami, Florida

Introduction

Nursing education is critical to quality health care. Nursing studies programs in colleges and universities are designed to assure enhanced collaboration between the educational and the health delivery sectors. There are many levels of professionalism within the nursing program: diploma programs, associate degree programs, baccalaureate degree programs, masters programs, and doctoral programs. A graduate of an associate degree program is eligible to take the National Council Licensure Examination (NCLEX) to become a registered nurse. Accredited programs of study in nursing have met accreditation standards set by the National League for Nursing (NLN). Accredited programs are listed in the NLN's annual publication, *State-Approved Schools of Nursing,* which lists both LPN/LVN and RN programs. There is a movement in the profession to require a bachelor's degree to be eligible to take the NCLEX examination. This debate is still unresolved and is a prevalent topic of discussion in many nursing articles today.

The American Nurses Association (ANA) established a certification program in 1973 to provide tangible recognition of professional achievement in a defined functional or clinical area of nursing. The American Nurses Credentialing Center (ANCC) bases its credentialing programs on standards set by the ANA Congress for nursing practice. The Boards on Certification of the ANCC develop the examinations, set passing scores, determine eligibility requirements for certification, and issue certificates to nurses who pass the written examination and have met the requirements of certification.

The Boards on Certification include the following:
    Board on Certification for Community Health Nursing Practice
    Board on Certification for General Nursing Practice
    Board on Certification for Maternal-Child Nursing Practice
    Board on Certification for Medical Surgical Nursing Practice
    Board on Certification for Psychiatric and Mental Health Nursing Practice
    Board on Certification for Primary Care in Adult and Family Health Nursing Practice
    Board on Certification for Gerontological Nursing Practice
    Board on Certification for Nursing Administration Practice
    Board on Certification for Nursing Continuing Education/Staff Development.

The nursing profession provides an essential service to society: the restoration and preservation of health. Nurses also are required to respond to broader responsibilities and challenges before them: adapt to changes in health care delivery, promote good health practices, implement active patient education practices, and emphasize disease prevention measures. Home-based nursing is a growing trend prompted by early patient discharge, an expanding elderly population, and the special requirements of the terminally ill. These factors, along with greater emphasis on the ambulated patient, increase the complexity of the nurse's role, and contribute to the growth of nursing specialties.

This bibliography provides libraries with a current list of resources to complement educational programs in nursing; it also provides nurses working in the clinical environment with resources they can apply in their professional practice. The community college curriculum has been carefully considered in developing this bibliography, with guidance from many established bibliographies. One of the most respected bibliographies in the field of nursing is published biennially in the March/April issue of *Nursing Outlook:* "Selected List of Nursing Books and Journals" by Alfred N. Brandon and Dorothy R. Hill. This bibliography has been in publication since 1979 and has long been considered a standard source.

In addition, Alfred N. Brandon publishes similar listings for allied health and small medical libraries. Annually in the January issue, the *American Journal of Nursing* features "Books of the Year," a list of approximately fifty of the best books published. The titles are selected by a committee of nurses having expertise in specific subject categories. In addition, the Interagency Council on Library Resources for Nursing (ICLRN) publishes "Essential Nursing References" (last published in *American Journal of Nursing,* September, 1992). This lists key publications, audiovisual resources, and computerized bibliographic databases in nursing.

The nonprint references listed have been recommended by the nursing faculty working with students in a computerized laboratory setting. Acknowledgment is extended to Diann Gregory, a faculty member in the Miami-Dade Community College Nursing Department, for her assistance in the selection and annotation of titles for the nonprint portion of this bibliography.

Accreditation and Certification

American Nurses Credentialing Center (ANCC)
600 Maryland Ave. SW, Suite 1200
Washington, D.C. 20024-2571

American Nurses Association
600 Maryland Ave. SW, Suite 1200
Washington, D.C. 20024-2571

National League for Nursing
350 Hudson St.
New York, N.Y. 10014

Selected Journals

AANA Journal.
American Association of Nurse Anesthetists. Bimonthly. ISSN 0094-6354

AAOHN Journal.
American Association of Occupational Health Nurses. Monthly. ISSN 0891-0162

AIDS Patient Care.
Mary Ann Liebert, Inc. Bimonthly. ISSN 0893-5068

American Journal of Infection Control.
Mosby. Bimonthly. ISSN 0196-6553

American Journal of Nursing.
American Journal of Nursing Company. Monthly. ISSN 0002-936X

American Nurse.
American Nurses Association. 10 times a year. ISSN 0098-1486

ANNA Journal.
American Nephrology Nurses Association. Bimonthly. ISSN 8750-0779

Advances in Nursing Science.
Aspen Publishers. Quarterly. ISSN 0161-9268

AORN Journal.
Association of Operating Room Nurses. Monthly. ISSN 0001-2092

Canadian Nurse.
Canadian Nurses Association. Monthly. ISSN 0008-4581

Cancer Nursing.
Raven Press. Bimonthly. ISSN 0162-220X

Clinical Nurse Specialist.
Williams and Wilkins. Bimonthly. ISSN 0887-6274

Computers in Nursing.
Lippincott. Bimonthly. ISSN 0736-8593

Critical Care Nursing Clinics of North America.
Saunders. Quarterly. ISSN 0899-5885

Critical Care Nursing Quarterly.
Aspen Publishers. Quarterly. ISSN 0887-9303

Geriatric Nursing.
Mosby Yearbook. Bimonthly. ISSN 0197-4572

Heart and Lung.
Mosby Yearbook. Bimonthly. ISSN 0147-9563

Image—Journal of Nursing Scholarship.
International Honor Society of Nursing. Quarterly.
ISSN 0743-5150

Journal of Cardiovascular Nursing.
Aspen Publishers. Quarterly. ISSN 0889-4655

Journal of Community Health Nursing.
L. Erlbaum Associates. Quarterly. ISSN 0737-0016

Journal of Emergency Nursing.
Mosby Yearbook. Bimonthly. ISSN 0099-1767

Journal of Neuroscience Nursing.
American Association of Neuroscience Nurses. Bimonthly.
ISSN 0888-0395

Journal of Nursing Education.
SLACK, Inc. Nine issues a year. ISSN 0148-4834

Journal of Nursing Administration.
Lippincott. Eleven issues a year. ISSN 0002-0443

Journal of Obstetric, Gynecologic and Neonatal Nursing: JOGNN.
Lippincott. Bimonthly. ISSN 0884-2175

Journal of Pediatric Health Care.
Saunders. Bimonthly. ISSN 0891-5245

Journal of Professional Nursing.
Saunders. Bimonthly. ISSN 8755-7223

Journal of Psychosocial Nursing and Mental Health Services.
SLACK, Inc. Monthly. ISSN 0279-3695

MCN, The American Journal of Maternal Child Nursing.
American Journal of Nursing Co. Bimonthly. ISSN 0361-929X

Nurse Practitioner.
Vernon Publications. Monthly. ISSN 0361-1817

Nursing and Health Care.
National League for Nursing. Ten issues a year. ISSN 0276-5284

Nursing.
Springhouse Corp. Monthly. ISSN 0360-4039

Nursing Clinics of North America.
Harcourt Brace and Co. Quarterly. ISSN 0029-6465

Nursing Diagnosis.
Nursecom. Quarterly. ISSN 1046-7459

Nursing Outlook.
Mosby Yearbook. Bimonthly. ISSN 0029-6554

Nursing Research.
American Journal of Nursing Co. Bimonthly. ISSN 0029-6562

Nursing Times.
Macmillan. Weekly. ISSN 0954-7762

Oncology Nursing Forum.
Oncology Nursing Press. Ten issues a year. ISSN 0190-535X

Pediatric Nursing.
Jannetti Publications. Bimonthly. ISSN 0097-9805

Public Health Nursing.
Blackwell Scientific Publications. Quarterly. ISSN 0737-1209

Regan Report on Nursing Law.
Medica Press. Monthly. ISSN 0034-317X

Rehabilitation Nursing.
Association of Rehabilitation Nurses. Bimonthly. ISSN 0278-4807

Research in Nursing and Health.
Wiley. Bimonthly. ISSN 0160-6891

RN.
Medical Economics Pub. Co. Monthly. ISSN 0033-7021

✦ Indexes

Cumulative Index to Nursing and Allied Health Literature.
Glendale Adventist Medical Center. Bimonthly. ISSN 0146-5554

International Nursing Index.
American Journal of Nursing. Quarterly. ISSN 0020-8124

Print Materials

✦ Reference

Anderson, Kenneth, Lois E. Anderson, and Walter D. Glanze, eds.
Mosby's Medical, Nursing and Allied Health Dictionary: Illustrated in Full Color Throughout
4th ed. St. Louis, Mo., Mosby-Yearbook, 1994. 1 v. (various pagings). ISBN 0801672252

This is a useful, illustrated dictionary for nursing students that contains normal reference values, clinical calculations, units of measurement, and information on the cultural differences of patients. A Brandon/Hill nursing list first-purchase selection and an ICLRN title.

Bolwell, Christine
Directory of Educational Software
5th ed. New York, National League for Nursing, 1993. 719p.
ISBN 0887376045

This irregularly published directory lists educational software for nursing. This is a valuable resource for faculty who are interested in locating software with multimedia and interactive learning applications.

Cohen, Barbara J.
Medical Terminology: An Illustrated Guide
2nd ed. Philadelphia, Lippincott, 1994. 574p. ISBN 0397550529

This excellent terminology workbook includes topics such as "suffixes that mean condition" and "prefixes pertaining to color." It is listed on the Brandon/Hill allied health list. Color illustrations.

Dox, Ida, Biagio John Melloni, and Gilbert M. Eisner
Melloni's Illustrated Medical Dictionary
3rd ed. New York, Parthenon Pub. Group, 1993 533p.
ISBN 1850704791

An illustrated dictionary with over 26,000 terms for health professionals. This is a Brandon/Hill nursing list selection.

Miller, Benjamin Frank, and Claire Brackman Keane
Encyclopedia and Dictionary of Medicine, Nursing, and Allied Health
5th ed. Philadelphia, Saunders, 1992. 1780p. ISBN 0721634567

Information is presented in clear, concise, and easy-to-understand terminology. Entries on diseases contain relevant information for nursing students, such as signs and symptoms. Includes some color illustrations. This is a Brandon/Hill nursing list first-purchase selection and an ICLRN recommendation.

Nursing Data Review 1993
New York, National League for Nursing, 1993. 290p.
ISBN 0887375863

This excellent reference on nursing educational statistics and faculty data also provides information on nursing schools, students and graduates. Previously published as *Nursing Data Source* (National League for Nursing, 1990). Biennial consensus of *Nurse Educators* (Nursing Digest, published annually). Recommended by the ICLRN.

Scholarships and Loans for Nursing Education
New York, National League for Nursing. Annual.
ISSN 0891-7884

A reference book on scholarships and loans available to nursing students.

State-Approved Schools of Nursing L.P.N./L.V.N.
New York, Division of Research, National League for Nursing. Annual. ISSN 0081-4423

This comprehensive directory of all licensed practical and vocational nursing programs in the United States, including the territories, is arranged alphabetically by state. Each entry includes address, telephone number, director, NLN accreditation status, state board approval, financial support, administrative control, admissions, enrollments, and graduation statistics. Recommended by the ICLRN.

State-Approved Schools of Nursing—R.N., 1993
New York, National League for Nursing. Annual.
ISSN 0081-4431

This annual publication arranged by state lists all the RN preparation programs in the United States. Entries include institution's address, telephone number, type of program, director, NLN accreditation status, state board approval, financial support, administrative control and admissions, enrollments, and graduation statistics. Recommended by the ICLRN.

Thomas, Clayton L., ed.
Taber's Cyclopedic Medical Dictionary
17th ed. Philadelphia, F. A. Davis, 1993. 2590p.
ISBN 080368312X

A classic dictionary for beginning students. This is a Brandon/Hill nursing list first-purchase selection.

♦ Administration and Management

Chenevert, Melodie
STAT: Special Techniques in Assertiveness Training for Women in the Health Professions
4th ed. St. Louis, Mo., Mosby,1994. 168p. ISBN 0801672333

Nursing faculty often recommend this excellent text to students to teach them assertiveness. Focuses on learning to stand up for oneself and for one's patients, with development of trust of one's feelings and opinions.

Cohen, Elaine L.
Nursing Care Management: From Concept to Evaluation
St. Louis, Mo., Mosby-Yearbook, 1993. 239p. ISBN 0801666988

This text focuses on case management as a nursing care delivery system and relates case management to managed care programs. Provides an informative guide to assessing an organization's readiness and capacity for case management as a means of delivering care.

Douglass, Laura Mae
The Effective Nurse: Leader and Manager
4th ed. St. Louis, Mo., Mosby, 1992. 278p. ISBN 0801663202

This excellent textbook by an authority in the field includes topics such as communication, staff leadership development, and ethical issues. Study questions follow each chapter. This is a Brandon/Hill nursing list first-purchase selection and a 1993 *AJN* "Books of the Year" title.

Hein, Eleanor C., and M. Jean Nicholson, eds.
Contemporary Leadership Behavior: Selected Readings
4th ed. Philadelphia, Lippincott, 1994. 488p. ISBN 0397550022

Presents a collection of reprints in nursing management including topics such as values, assertiveness, advocacy, whistle-blowing, political empowerment, unionism, conflict management, negotiation. This highly recommended text is particularly helpful to nursing students who need research topics.

Koch, Marylane Wade
Integrated Quality Management: The Key to Improving Nursing Care Quality
St. Louis, Mo., Mosby, 1993. 260p. ISBN 080167476X
    Incorporates quality assessment and improvement from both management and patient care aspects. Helpful in the clinical setting.

Marquis, Bessie L.
Management Decision Making for Nurses: 118 Case Studies
2nd ed. Philadelphia, Lippincott, 1994. 445p. ISBN 0397550561
    The focus of this textbook/workbook for nursing management students is on strengthening decision-making skills. Includes historical perspectives of both management and leadership concepts. Contains case studies.

Marriner-Tomey, Ann
Guide to Nursing Management
4th ed. St. Louis, Mo., Mosby-Yearbook, 1992. 498p. ISBN 0801663261
    The focus of this work is on the nurse's role in the management process: planning, organization, directing the work of staff, and evaluation of staff. This is a Brandon/Hill nursing list first- purchase selection. This author also wrote *Transformational Leadership in Nursing* (Mosby-Yearbook, 1993).

Tucker, Susan M., and others
Patient Care Standards: Nursing Process, Diagnosis, and Outcome
5th ed. St. Louis, Mo., Mosby-Yearbook, 1992. 872p. ISBN 0801662680
    This manual contains guidelines for nurses and nursing service administrators for patient care planning protocols. A Brandon/Hill nursing list first-purchase selection.

✦ Anatomy and Physiology

Guyton, Arthur C.
Textbook of Medical Physiology
8th ed. Philadelphia, Saunders, 1991. 1014p. ISBN 0721630871
    The goal of this classic text is to explain the body's homeostasis mechanisms and to discuss their functions in relation to disease. Nursing students often prefer to use this as a reference when preparing care plans. This is a Brandon/Hill medical list selection.

McCance, Kathryn L., and Sue E. Huether
Pathophysiology: The Biologic Basis for Disease in Adults and Children
2nd ed. St. Louis, Mo., Mosby, 1994. 1 v. (various pagings). ISBN 0801669022
    This is an excellent text for nursing students who need to expand knowledge of biology as it relates to specific diseases. It is also a Brandon/Hill nursing list first-purchase selection.

Marieb, Elaine Nicpon
Human Anatomy and Physiology
2nd ed. Redwood City, Calif., Benjamin/Cummings Pub. Co., 1992. 1014p. ISBN 080534120X
    Marieb presents a lab manual to support introductory courses on anatomy and physiology for nursing students. This is a Brandon/Hill nursing list selection. This author also wrote a condensed version: *Essentials of Human Anatomy and Physiology* (Benjamin/ Cummings, 1994).

Memmler, Ruth Lundeen, Dena Lin Wood, and Barbara J. Cole
Structure and Function of the Human Body
5th ed. Philadelphia, Lippincott, 1992. 314p. ISBN 0397548826
    Chapters begin with learning objectives and close with summary and review questions. Detailed illustrations aid the learning process of anatomy and physiology. This is a Brandon/Hill nursing list first-purchase selection.

Porth, Carol
Pathophysiology: Concepts of Altered Health States
4th ed. Philadelphia, Lippincott, 1994. 1270p. ISBN 039754961X
    Discusses how alterations to normal health states (i.e., motion, sensation, regulation, protection, etc.) are clinically manifested. Two other resources in this subject area are Virginia Carrieri-Kohlman and Ada M. Lindsey's *Pathophysiological Phenomena in Nursing: Human Responses to Illness* (Saunders, 1993) and Barbara L. Bullock's, *Pathophysiology: Adaptations and Alterations in Function* (Lippincott, 1992). A Brandon/Hill nursing list selection.

Solomon, Eldra Pearl
Introduction to Human Anatomy and Physiology
Philadelphia, Saunders, 1992. 314p. ISBN 0721639666
    Beginning nursing students will benefit from this introductory anatomy and physiology text. Contains some color illustrations, a glossary, review questions, chapter outlines and learning objectives. Has a companion: *Study Guide for Introduction to Human Anatomy and Physiology* (Saunders, 1992) by Eldra Pearl Solomon and Mical K. Solomon.

Tate, Philip, Rod R. Seeley, and Trent D. Stephens
Understanding the Human Body
St. Louis, Mo., Mosby, 1994. 365p. ISBN 0801671973
    This concise, introductory anatomy and physiology text is complemented by color illustrations, chapter summaries, and quizzes.

Thibodeau, Gary A., and Kevin T. Patton
Anatomy and Physiology
2nd ed. St. Louis, Mo., Mosby, 1993. 968p. ISBN 0801650054
    This standard for students of human anatomy and physiology shows how anatomical structures serve specialized functions. Homeostasis is used to illustrate how normal interactions are balanced by other forces in the body. Easy to read and illustrated with color illustrations and diagrams. Chapters include outlines, objectives, terms, review questions, and case studies. A Brandon/Hill nursing list first-purchase selection.

✦ Communication

Bradley, Jean C., and Mark A. Edenberg
Communication in the Nursing Context
3rd ed. Norwalk, Conn., Appleton and Lange, 1990. 260p.
ISBN 0838513271

Focuses on the importance of communication in nursing. Reviews many types of communication including nonverbal, nurse-client, nurse-nurse, and nurse-physician.

Fischbach, Frances Talaska
Documenting Care: Communication—The Nursing Process and Documentation Standards
Philadelphia, F. A. Davis, 1991. 677p. ISBN 0803635613

An excellent manual for nursing students on documenting and charting, presented in outline format. This is a Brandon/Hill nursing list selection.

Long, Lynette
Understanding/Responding: A Communication Manual for Nurses
2nd ed. Boston, Jones and Bartlett, 1992. 356p.
ISBN 0867204338

Long presents a handbook for nursing students on how to communicate effectively with patients. Beginning nurses often refer to this resource, which has case study illustrations. A Brandon/Hill nursing list selection.

Northouse, Peter Guy, and Laurel Lindhout
Health Communication: Strategies for Health Professionals
2nd ed. Norwalk, Conn., Appleton and Lange, 1992. 286p.
ISBN 0838536751

Presents strategies for effective communication. Discusses professional-patient, professional-professional, professional-family, and patient-family relationships as well as conflict resolution. This is a Brandon/Hill nursing list first-purchase selection.

Purtilo, Ruth B.
Health Professional and Patient Interaction
4th ed. Philadelphia, Saunders, 1990. 339p. ISBN 0721673961

This excellent resource is designed for the health professional who needs to develop communication skills for patient interactions. This is a Brandon/Hill nursing list first-purchase selection. This author also wrote *Ethical Dimensions in the Health Professions* (Saunders, 1993).

Smith, Susan
Communications in Nursing: Communicating Assertively and Responsibly in Nursing: A Guidebook
2nd ed. St. Louis, Mo., Mosby-Yearbook, 1992. 360p.
ISBN 0801663571

Smith provides an assertiveness training guide for nursing staff that focuses on overcoming communication barriers and developing rapport with patients and families. This is a Brandon/Hill nursing list selection and a 1993 *AJN* "Books of the Year" title.

Sundeen, Sandra J., and others
Nurse-Client Interaction: Implementing the Nursing Process
5th ed. St. Louis, Mo., Mosby, 1994. 403p. ISBN 0801675286

This reference text on communication discusses the behavioral aspects of the nursing process by integrating psychodynamic principles with the theoretical principles of nursing practice. Units include topics such as "Emergence of the Self" and "Stress and Adaptation." This is a Brandon/Hill nursing list first purchase selection.

✦ Community Health Nursing

Clemen-Stone, Susan, Diane Gerber Eigsti, and Sandra L. McGuire
Comprehensive Family and Community Health Nursing
3rd ed. St. Louis, Mo., Mosby-Yearbook, 1991. 899p.
ISBN 0801660688

This comprehensive textbook for nursing students discusses philosophical, economic, social, political, and technological trends that have an impact on community health nursing. Contains a historical perspective. Each chapter is followed by appendices and summaries in timetable format. 1992 *AJN* "Books of the Year" title.

Helvie, Carol O.
Community Health Nursing: Theory and Practice
New York, Springer, 1991. 470p. ISBN 0826165508

This basic text covers concepts in nursing theory that serve as a basis for practicing community health. Topics include home visits, diagnosis specified by nursing's human need theory, wellness, prevention, and codes for nurses. Includes tables and illustrations. This is a Brandon/Hill nursing list selection.

Stanhope, Marcia, and Jeanette Lancaster, eds.
Community Health Nursing: Process and Practice for Promoting Health
3rd ed. St. Louis, Mo., Mosby-Yearbook, 1992. 968p.
ISBN 0801647746

A classic resource in the field of community health nursing, this work includes topics such as AIDS, drug abuse, and adolescent pregnancy. This author also wrote *Handbook of Community and Home Health Nursing: Tools for Assessment, Intervention, and Education,* (Mosby, 1992), a Brandon/Hill nursing list first-purchase selection.

Swanson, Janice M.
Community Health Nursing: Promoting the Health of Aggregates
Philadelphia, Saunders, 1993. 757p. ISBN 0721613128

The focus of this text is on providing services to aggregates: families and schools, worksites, clinics, and community groups. Introduces elements essential to community nursing such as language variables, cultural differences and considerations, social issues, and trends. This is a Brandon/Hill nursing list first-purchase selection and a 1994 AJN "Books of the Year" title.

✦ Critical Care

Baas, Linda S., ed.
Essentials of Cardiovascular Nursing
Gaithersburg, Md., Aspen Publishers, 1991. 322p.
ISBN 0834202069

Baas presents a concise textbook for nurses in the clinical setting who care for the cardiovascular patient. Considers multiple aspects for successful recovery, hypertension, cholesterol, tobacco dependence, stress, diabetes, sexual intimacy, and physical activity. Chapters provide extensive references. This is a Brandon/ Hill nursing list selection.

Catalano, Joseph T.
Guide to ECG Analysis
Philadelphia, Lippincott, 1993. 408p. ISBN 0397550154
   This basic text assists health professionals to interpret ECGs. Includes information on the identification and description of dysrhythmia.

Clochesy, John M.
Critical Care Nursing
Philadelphia, Saunders, 1993. 1486p. ISBN 0721628567
   Clochesy's comprehensive text for nursing students in the clinical setting integrates nursing practice with ICU practice. Stresses physical and psychological support, and education of patients and their families in ethical decision-making and legal issues. This is a Brandon/Hill nursing list first-purchase selection and a 1994 *AJN* "Books of the Year" title.

Conover, Mary Boudreau
Understanding Electrocardiography: Arrhythmias and the 12-Lead ECG
6th ed. St. Louis, Mo., Mosby-Yearbook, 1992. 464p. ISBN 0801610192
   Focuses specifically on how to read electrocardiograms and emphasizes the 12-Lead ECG as a diagnostic tool. Topics are arranged by cardiac condition, i.e., arrhythmias, supraventricular ectopics, preexcitation syndromes, myocardial infarction, and chamber enlargement. Numerous illustrations. This is a Brandon/Hill nursing list selection.

Dolan, Joan T.
Critical Care Nursing: Clinical Management Through the Nursing Process
Philadelphia, F. A. Davis, 1991. 1531p. ISBN 0803626916
   This text is arranged by major organ systems: nervous, endocrine, renal, respiratory, cardiovascular, gastrointestinal, immune, and hematologic systems. Other topics include oncology, trauma, burns, shock states, and acute poisoning. Contains case studies.

Durham, Jerry D., and Felissa L. Cohen, eds.
The Person with AIDS: Nursing Perspective
2nd ed. New York, Springer, 1991. 441p. ISBN 0826156312
   An excellent resource prepared by numerous contributors for the nurse caring for the AIDS patient. Topics include AIDS etiology, HIV testing, nursing care, infection control, discharge planning, and politics. This is a Brandon/Hill nursing list selection.

Fahey, Victoria, ed.
Vascular Nursing
2nd ed. Philadelphia, Saunders, 1994. 598p. ISBN 0721665896
   Focuses on nursing care for the vascular patient. Discusses the medical and surgical aspects of nursing care and includes new topics such as intraoperative nursing care, vascular trauma, and rehabilitation. This is a Brandon/Hill nursing list first-purchase selection.

Flaskerud, Jacquelyn Haak, and Peter J. Ungvarski
HIV/AIDS: A Guide to Nursing Care
2nd ed. Philadelphia, Saunders, 1992. 515p. ISBN 0721637183

   Presents a comprehensive view of aspects of HIV/AIDS disease written for the nurse in clinical practice. Nursing interventions are organized by categories: self-care assistance, acute-care management, lifestyle alteration, health promotion, and life support. For each diagnosis, the nursing intervention is described. This is a Brandon/Hill nursing list selection and a 1993 *AJN* "Books of the Year" title.

Flynn, Janet-Beth McCann, and Nancie Pardue Bruce
Introduction to Critical Care Skills
St. Louis, Mo., Mosby-Yearbook, 1993. 524p. ISBN 0801624554
   This basic text for nursing students includes critical care procedures and patient care concepts. Reviews topics such as mechanical ventilation, defibrillation, intracranial monitoring, total parental nutrition, burn wound therapy, and organ donations. This is a Brandon/Hill nursing list selection.

Fulmer, Terry T, and Mary K. Walker, eds.
Critical Care Nursing of the Elderly
New York, Springer, 1992. 359p. ISBN 0826170501
   This book considers the special needs of the elderly in critical care situations. Topics include common geriatric problems, and legal and policy issues. Indexed. Part of the *Springer Series on Geriatric Nursing*. A 1993 *AJN* "Books of the Year" title.

Funk, Sandra G., and others, eds.
Key Aspects of Recovery
New York, Springer, 1990. 350p. ISBN 0826172903
   In this research text the editors present essays on recovery and the factors that influence or inhibit recovery. Funk also edited *Key Aspects of Comfort: Management of Pain, Fatigue, and Nausea* (Macmillan, 1994).

Guzzetta, Cathie E., and Barbara Montgomery Dossey
Cardiovascular Nursing: Holistic Practice
St. Louis, Mo., Mosby-Yearbook, 1992. 801p. ISBN 0801627842
   Guzzetta and Dossey present a comprehensive text geared to nursing students working with cardiac patients. Focus is on patient problems, and assessment, diagnosis, and treatment. This is an excellent resource for preparing care plans. A Brandon/Hill nursing list first-purchase selection.

Holloway, Nancy Meyer, ed.
Nursing the Critically Ill Adult
4th ed. Redwood City, Calif., Addison-Wesley, 1993. 700p. ISBN 0805325441
   An excellent text for nurses working with the critically ill. Well organized and specific to nursing diagnosis, this work includes nursing tips, special information highlighted in boxes, and nursing care plans. This is a Brandon/Hill nursing list selection.

Hudak, Carolyn M., Barbara M. Gallo, and Julie J. Benz
Critical Care Nursing: A Holistic Approach
6th ed. Philadelphia, Lippincott, 1994. 1079p
   This text could be used as a reference for the nurse in a clinical setting. Topics include psychosocial aspects of critically ill patients and their family, nursing care, and patient education issues. This is a Brandon/Hill nursing list first-purchase selection.

Kinney, Marguerite Rodgers, Donna Rogers Packa, and Sandra Byars Dunbar, eds.
AACN'S Clinical Reference for Critical Care Nursing
3rd ed. St. Louis, Mo., Mosby-Yearbook, 1993. 1461p.
ISBN 0801664527

Provides nursing students working in critical and intensive care settings with relevant nursing interventions. This is a Brandon/Hill nursing list selection.

Kinney, Marguerite Rodgers, and others, eds.
Comprehensive Cardiac Care
7th ed. St. Louis, Mo., Mosby-Yearbook, 1991. 492p.
ISBN 0801627702

This is a classic text for nursing students involved in the care of the cardiac patient. Provides updated information on the diagnosis and treatment of atherosclerosis, valvular heart disease, and cardiomyopathies. This is a Brandon/Hill nursing list first-purchase selection.

Neff, Janet A.
Trauma Nursing: The Art and Science
St. Louis, Mo., Mosby-Yearbook, 1993. 808p. ISBN 0801666554

A comprehensive text for nursing students to be used as a clinical reference in emergency situations. Each chapter begins and ends with a case study and review questions that are designed to aid critical thinking. Includes annotated bibliographies.

Sheehy, Susan Budassi, ed.
Emergency Nursing: Principles and Practice
3rd ed. St. Louis, Mo., Mosby-Yearbook, 1992. 733p.
ISBN 0801662486

This reference aids nurses working with emergency situations in the clinical setting. Includes a pocket guide companion. Sheehy is author of *Manual of Clinical Trauma Care: The First Hour* (Mosby, 1994). A Brandon/Hill nursing list first-purchase selection.

Starck, Patricia L., and John P. McGovern, eds.
The Hidden Dimension of Illness: Human Suffering
New York, National League for Nursing, 1992. 304p.
ISBN 088737543X

Suffering and pain are discussed from physical, psychological, sociological, and spiritual perspectives in this 1993 *AJN* "Books of the Year" title. A related text is Kathleen A. Puntillo's *Pain in the Critically Ill: Assessment and Management* (Aspen, 1991), a 1992 *AJN* "Books of the Year" title.

Swearingen, Pamela L., and Janet Hicks Keen, eds.
Manual of Critical Care: Applying Nursing Diagnoses to Adult Critical Illness
2nd ed. St. Louis, Mo., Mosby-Yearbook, 1991. 642p.
ISBN 0801650844

Offers a quick reference-format manual for students preparing care plans. Swearingen is a Brandon/Hill author, who also wrote *Manual of Medical-Surgical Nursing Care, Nursing Interventions and Collaborative Management* (Mosby, 1994).

Thelan, Lynne A., and others
Critical Care Nursing: Diagnosis and Management
2nd ed. St. Louis, Mo., Mosby-Yearbook, 1994. 978p.
ISBN 0801671671

This excellent resource focuses on critical care and its ethical, legal, and cultural issues. This is a Brandon/Hill nursing list first-purchase selection.

Trofino, Rita Bolek, ed.
Nursing Care of the Burn-Injured Patient
Philadelphia, F. A. Davis, 1991. 466p. ISBN 0803686587

This comprehensive text for the nurse working in a critical care setting with burn patients includes units on basics of burn care, care according to the severity of burns, specific anatomic care, care of the burn patients in acute care settings, and care of the special burn patient. Concise format.

Vazquez, Moya, Susan Engman Lazear, and Elaine Larson, eds.
Critical Care Nursing
2nd ed. Philadelphia, Saunders, 1992. 514p. ISBN 0721632122

Incorporates nursing diagnosis and management principles into this reference for the nurse in the clinical setting. A Brandon/Hill nursing list selection.

Workman, M. Linda
Nursing Care of the Immunocompromised Patient
Philadelphia, Saunders, 1993. 307p. ISBN 0721632130

This text for nurses working with AIDS patients stresses the immune function and AIDS therapies such as chemotherapy and radiation therapy. Contains care plans.

✦ Diet and Nutrition

Bowes, Anna De Planter, and Helen Nichols Church
Bowes and Church's Food Values of Portions Commonly Used
16th ed., revised by Jean A. Thompson Pennington. Philadelphia, Lippincott, 1994. 483p. ISBN 0397550871

This excellent reference for nursing students contains food values, vitamin and mineral content, calories, and recommended dietary allowances as revised in 1989. This is a Brandon/Hill allied health first-purchase selection.

Davis, Judi Ratliff, and Kim Sherer
Applied Nutrition and Diet Therapy for Nurses
2nd ed. Philadelphia, Saunders, 1994. 1154p. ISBN 07221667856

Davis and Sherer present a comprehensive text for nursing students addressing the special topic of dietetics. Includes information on nutritional needs, physiological aspects of the gastrointestinal tract, carbohydrates, lipids, minerals, and vitamins. This is a Brandon/Hill nursing list selection.

Dudek, Susan G.
Nutrition Handbook for Nursing Practice
2nd ed. Philadelphia, Lippincott, 1993. 722p. ISBN 0397549288

This comprehensive text addresses the nutritional aspects of nursing care and is an excellent reference when preparing care plans. Includes sample nursing diagnoses. This is a Brandon/Hill nursing list selection.

Escott-Stump, Sylvia
Nutrition and Diagnosis-Related Care
3rd ed. Lea and Febiger, 1992. 660p. ISBN 0812115562
   Focuses on the nutritional aspects of patient care and is especially helpful when preparing care plans for a specific diagnosis. This is a Brandon/Hill allied health list first-purchase selection.

Jaffe, Marie S.
Geriatric Nutrition and Diet Therapy
El Paso, Tex., Skidmore-Roth Pub., 1991. 188p.
ISBN 0944132650
   Jaffe presents a concise text in outline format on nutrition problems specific to the elderly. Marie S. Jaffe is also the author of *Home Health Nursing Care Plans* (Mosby, 1993), and *Nursing Procedures for Home Health Care* (Delmar Publishers, 1993), both Brandon/Hill nursing list first-purchase selections.

Mahan, L. Kathleen, and Marian T. Arlin, eds.
Krause's Food, Nutrition and Diet Therapy
8th ed. Philadelphia, Saunders, 1992. 933p. ISBN 0721655084
   A fundamental textbook on clinical nutrition for nursing students. This is a Brandon/Hill nursing list first-purchase selection.

Pipes, Peggy L., and Christine M. Trahms, eds.
Nutrition in Infancy and Childhood
5th ed. St. Louis, Mo., Mosby-Yearbook, 1993. 429p.
ISBN 0801665671
   The focus of this work is on nutritional needs of infants and children. Chapters provide information on growth and development, nutrient requirements of infants and children, and assessment of food intake. Includes adolescent nutrition guidelines and special diets. Summaries and review questions aid understanding.

Robinson, Corrinne H.
Basic Nutrition and Diet Therapy
7th ed. New York, Macmillan, 1993. 516p. ISBN 002402502X
   An excellent text for nursing students on the nutritional needs of patients. This author also wrote *Normal and Therapeutic Nutrition* (Macmillan, 1990). This is a Brandon/Hill nursing list selection.

Williams, Sue Rodwell
Nutrition and Diet Therapy
7th ed. St. Louis, Mo., Mosby-Yearbook, 1993. 1 v. (various pagings). ISBN 0801665655
   This basic nutrition text for nursing students covers nutrition throughout the life cycle, AIDS patients' nutritional needs, and weight management. Study guide available. This is a Brandon/Hill nursing list first-purchase selection. This author also wrote a condensed version titled *Essentials of Nutrition and Diet Therapy* (Mosby, 1994).

Worthington-Roberts, Bonnie S.
Nutrition in Pregnancy and Lactation
5th ed. St. Louis, Mo., Mosby-Yearbook, 1993. 537p.
ISBN 0801665698
   Worthington-Roberts provides an excellent resource and textbook on the nutritional needs of patients during pregnancy and lactation. This is a Brandon/Hill nursing list selection. The 1993 *AJN* "Books of the Year" work lists a related title by the author: *Nutrition During Pregnancy and Lactation: An Implementation Guide* (National Academy Press, 1992).

Zaloga, Gary P.
Nutrition in Critical Care
St. Louis, Mo., Mosby-Yearbook, 1993. 550p. ISBN 1556643977
   Nutritional support for the critically ill is the focus of this work, which covers specific topics such as burns, sepsis, gut disease, respiratory failure, and renal failure. Discusses specific feeding strategies, e.g., parenteral nutrition and enteral feeding.

✦ Geriatric Nursing

Ayer, Jennifer
Geriatric Nursing
St. Louis, Mo., Mosby-Yearbook, 1994. 300p. ISBN 0944132901
   Reviews geriatric nursing issues and includes objectives of care based on *Healthy People 2000: National Health Promotion and Disease Prevention Objectives* (U.S. Department of Health and Human Services, Public Health Service, 1992).

Birchenall, Joan M., and Mary Eileen Streight
Care of the Older Adult
3rd ed. Philadelphia, Lippincott, 1993. 422p. ISBN 0397548400
   A basic text on the psychological, emotional, social, and health care needs of the geriatric population.

Burke, Mary M.
Gerontologic Nursing: Care of the Frail Elderly
St. Louis, Mo., Mosby-Yearbook, 1992. 524p. ISBN 0801658845
   Emphasizes care of those over the age of seventy-five while stressing respect for the patient's personal integrity. This is a Brandon/Hill nursing list selection. Burke also authored *Gerontologic Nursing: Issues and Opportunities for the Twenty-First Century* (National League for Nursing Press, 1993).

Ebersole, Priscilla, and Patricia Hess
Toward Healthy Aging: Human Needs and Nursing Response
4th ed. St. Louis, Mo., Mosby-Yearbook, 1994. 876p.
ISBN 0801668166
   Provides an overview of the aging process with nursing considerations. Considers the needs of the aged in a hierarchic Maslovian manner, moving from basic concepts of gerontology to basic needs concepts with a psychosocial and physiological point of view. Illustrated. This is a Brandon/Hill nursing list first-purchase selection.

Eliopoulos, Charlotte
Gerontological Nursing
3rd ed. Philadelphia, Lippincott, 1993. 439p. ISBN 0397549741
   Focuses on health promotion and self-care, with a chapter on gerontological nursing in diverse care settings. This is a Brandon/Hill nursing list first-purchase selection. Eliopoulos, an expert in the field, has also edited *Caring for the Elderly in Diverse Care Settings* (Lippincott, 1990) and *Nursing Care Planning Guides for Long-Term Care* (Williams and Wilkins, 1990).

Hogstel, Mildred O., ed.
Nursing Care of the Older Adult
3rd ed. Albany, N.Y., Delmar Publishers, 1994. 624p.
ISBN 0827351216

Addresses the care and needs of the older adult in a variety of health care situations. This is a Brandon/Hill nursing list first-purchase selection.

Newman, Diane Kaschak
Geriatric Care Plans
Springhouse, Pa., Springhouse, 1991. 334p. ISBN 0874342635

An excellent geriatric care plan book. Another selection in this area is Marie Jaffe's *Geriatric Nursing Care Plans* (Skidmore-Roth, 1991).

Wold, Gloria
Basic Geriatric Nursing
St. Louis, Mo., Mosby-Yearbook, 1993. 378p. ISBN 0801666473

Focuses on the social, emotional, and health care needs of the elderly. Includes medications and nursing considerations.

✦ Laboratory Diagnosis

Burton, Gwendolyn R. W.
Microbiology for the Health Sciences
4th ed. Philadelphia, Lippincott, 1992. 403p. ISBN 0397548869

Burton presents this standard microbiology textbook for the health professions. Instructor's manual available. This is a Brandon/Hill nursing list selection.

Clinical Laboratory Tests: Values and Implications
Springhouse, Pa., Springhouse, 1991. 724p. ISBN 0874342708

Provides information on laboratory tests and values including ranges, with nursing implications. This is a Brandon/Hill nursing list selection.

Corbett, Jane Vincent
Laboratory Tests and Diagnostic Procedures with Nursing Diagnosis
3rd ed. Norwalk, Conn., Appleton and Lange, 1992. 750p.
ISBN 0838555942

A reference for nursing students of laboratory tests, lab values, and nursing implications. This is a Brandon/Hill nursing list first-purchase selection.

Fischbach, Frances Talaska
A Manual of Laboratory and Diagnostic Tests
4th ed. Philadelphia, Lippincott, 1992. 1020p. ISBN 0397548311

This reference handbook lists laboratory tests, procedures, normal values, and nursing implications. Includes patient preparatory procedures. A Brandon/Hill nursing list first-purchase selection.

Kee, Joyce LeFever
Laboratory and Diagnostic Tests with Nursing Implications
3rd ed. Norwalk, Conn., Appleton and Lange, 1991. 562p.
ISBN 0838555780

Kee provides a reference text in alphabetical format. Includes laboratory tests, diagnostic tests, and laboratory/diagnostic assessments of body functions. Appendices provide nursing implications and rationales. Indexed. A Brandon/Hill nursing list selection.

Pagana, Kathleen Deska
Diagnostic Testing and Nursing Implications: A Case Study Approach
4th ed. St. Louis, Mo., Mosby-Yearbook, 1994. 531p.
ISBN 0801667798

This reference manual lists diagnostic tests and explains procedures and purposes of laboratory tests, and nursing implications of such testing. A Brandon/Hill nursing list selection.

✦ Medical-Surgical Nursing

Handbook of Medical Surgical Nursing
Springhouse, Pa., Springhouse, 1994. 952p. ISBN 0874345952

Provides a synopsis for nursing students on the 320 most common problems in a medical-surgical unit. Includes nursing diagnoses and expected outcomes. A Brandon/Hill nursing list selection.

Jaffe, Marie S.
Medical-Surgical Nursing Care Plans: Nursing Diagnosis and Interventions
2nd ed. Norwalk, Conn., Appleton and Lange, 1992. 433p.
ISBN 0838562027

This excellent reference for nursing students contains seventy-five sample care plans. A Brandon/Hill nursing list selection.

Lewis, Sharon Mantik, and Idolia Cox Collier, eds.
Medical-Surgical Nursing: Assessment and Management of Clinical Problems
3rd ed. St. Louis, Mo., Mosby-Yearbook, 1992. 1879p.
ISBN 0801660394

This is a primary textbook for nursing students to consult when implementing the nursing process as a basis for the assessment of patients in the clinical setting. A Brandon/Hill nursing list selection.

Long, Barbara C., Wilma J. Phipps, and Virginia Cassemeyer, eds.
Medical-Surgical Nursing: A Nursing Process Approach
3rd ed. St. Louis, Mo., Mosby-Yearbook, 1993. 1695p.
ISBN 0801666724

A basic resource for nursing students working with medical-surgical patients. A Brandon/Hill nursing list first-purchase selection.

Smeltzer, Suzanne, and Brenda G. Bare, eds.
Brunner and Suddarth's Textbook of Medical-Surgical Nursing
7th ed. Philadelphia, Lippincott, 1992. 2012p. ISBN 0397547978

This is an excellent text for nursing students working with medical-surgical patients in the clinical setting and is often cited as a primary resource in the nursing curriculum. A Brandon/Hill nursing list first-purchase selection.

✦ Neurological Nursing

Hanak, Marcia
Rehabilitation Nursing for the Neurological Patient
New York, Springer Pub. Co., 1992. 229p. ISBN 0826176607

This reference textbook for nurses working with patients with neurological problems is arranged in three parts: wellness and patient education, nursing management, and specific disabilities. This is a Brandon/Hill nursing list first-purchase selection.

Hickey, Joanne V.
Clinical Practice of Neurological and Neurosurgical Nursing
3rd ed. Philadelphia, Lippincott, 1992. 688p. ISBN 039754222

An excellent reference for nursing students who work with neurological patients. Hickey comprehensively covers major adult neurological problems, reflects current standards of practice, and focuses on expanding basic knowledge and nursing skills. This is a Brandon/Hill nursing list first-purchase selection.

Snyder, Mariah, ed.
A Guide to Neurological and Neurosurgical Nursing
2nd ed. Albany, N.Y., Delmar Publishers, 1991. 609p. ISBN 0827343817

This handbook provides an overview of neuroscience nursing processes and diagnoses that are common to patients with neurologic problems. This is an excellent resource for nurses who need to prepare care plans. A Brandon/Hill nursing list selection.

✦ Nursing as a Profession

Annual Review of Nursing Research
11th ed., edited by Joyce J. Fitzpatrick and Joanne S. Stevenson. New York, Springer, 1993.s 335p. ISBN 0826182305

Topics include nursing practice, nursing care delivery, nursing education, nursing as a profession, family unit-focused research, alcohol/drug abuse, and patient falls in health care institutions.

Chitty, Kay Kittrell, ed.
Professional Nursing: Concepts and Challenges
Philadelphia, Saunders, 1993. 474p. ISBN 0721640613

Presents an up-to-date review of the historical and social events leading to modern nursing. Includes various definitions and philosophies of nursing theory and practice.

Ellis, Janice Rider, and Celia Love Hartley
Nursing in Today's World: Challenges, Issues and Trends
4th ed. Philadelphia, Lippincott, 1992. 542p. ISBN 0397549164

This overview of career opportunities and professional growth discusses nursing as a developing profession including historical, educational, legal, ethical, bioethical, and political perspectives, as well as future trends. This is a Brandon/Hill nursing list selection.

Kelly, Lucie Young
Dimensions of Professional Nursing
6th ed. New York, Pergamon Press, 1991. 748p. ISBN 080403034

A long-time favorite with nursing students and faculty, this work portrays nursing as a profession and presents a historical overview of the field including associations and specializations. This is a Brandon/Hill nursing list first-purchase selection. The author also wrote *The Nursing Experience: Trends, Challenges, and Transitions* (McGraw-Hill, 1992), which is also on the Brandon/Hill nursing list.

Leddy, Susan, and J. Mae Pepper
Conceptual Basis of Professional Nursing
3rd ed. Philadelphia, Lippincott, 1993. 484p. ISBN 0397549326

Focuses on current nursing theory, the process of socialization, and nursing as a profession. Includes a special chapter on the health care delivery system. A Brandon/Hill nursing list selection.

Lindberg, Janice B., Mary Love Hunter, and Ann Z. Kruszewski
Introduction to Nursing: Concepts, Issues and Opportunities
2nd ed. Philadelphia, Lippincott, 1994. 448p. ISBN 0397549865

This introductory text for beginning nursing students covers topics such as nursing history, nursing theory, health care delivery, and philosophy of nursing. A Brandon/Hill nursing list first-purchase selection.

Moloney, Margaret M.
Professionalization of Nursing: Current Issues and Trends
2nd ed. Philadelphia, Lippincott, 1992. 330p. ISBN 0397548427

Moloney examines nursing as a profession and its scientific knowledge base. This is a Brandon/Hill nursing list selection.

Zerwekh, JoAnn Graham, and Jo Carol Claborn, eds.
Nursing Today: Transition and Trends
Philadelphia, Saunders, 1994. 450p. ISBN 0721636454

Examines the transition of nursing from theory to practice and the standards of professional nursing practice.

✦ Nursing Diagnosis

Ackley, Betty J., and Gail B. Ladwig
Nursing Diagnosis Handbook: A Guide to Planning Care
St. Louis, Mo., Mosby-Yearbook, 1993. 353p. ISBN 0801677912

This nursing handbook discusses the development of nursing care plans. Nursing diagnosis is analyzed in light of the North American Nursing Diagnosis Association's approved diagnoses.

Bulechek, Gloria M., and Joanne Comi McCloskey, eds.
Nursing Interventions: Essential Nursing Treatments
2nd ed. Philadelphia, Saunders, 1992. 619p. ISBN 0721638023

Focuses on self-care assistance, acute-care management, lifestyle alteration, health promotion, and life support.

Carnevali, Doris L., and Mary Durand Thomas
Diagnostic Reasoning and Treatment Decision Making in Nursing
Philadelphia, Lippincott, 1993. 267p. ISBN 0397549210

This nursing text presents decision-making techniques and stresses the need for critical thinking skills when making clinical judgments, nursing diagnoses, prognostics, and treatment decisions. Contains exercises in diagnostic reasoning and extensive explanations.

Carpenito, Lynda Juall
Handbook of Nursing Diagnosis
5th ed. Philadelphia, Lippincott, 1993. 513p. ISBN 0397550545

Carpenito's handbook aids students in the clinical setting in the preparation of nursing care plans. This author also wrote *Nursing Care Plans and Documentation: Nursing Diagnosis and Collaborative Problems* (Lippincott, 1995).

Carpenito, Lynda Juall
Nursing Diagnosis: Application to Clinical Practice
5th ed. Philadelphia, Lippincott, 1993. 1070p. ISBN 0397550227
   Addresses nursing diagnoses specific to clinical practice. Reviews principles and concepts of nursing diagnosis including nursing assessment strategies and interventions. This is a Brandon/Hill nursing list first-purchase selection.

Carroll-Johnson, Rose Mary, and Mary Paquette, eds.
Classification of Nursing Diagnosis: Proceedings of the 10th Conference
Philadelphia, Lippincott, 1994. ISBN 0397550111
   Proceedings of the North American Nursing Diagnosis Association conference held April 25-29, 1992, in San Diego, Calif. Updates theory and practice in approved nursing diagnosis.

Doenges, Marilynn E., and Mary Frances Moorhouse
Application of Nursing Process and Nursing Diagnoses: An Interactive Text
Philadelphia, F. A. Davis, 1992. 229p. ISBN 0803626754
   This workbook is designed to help beginning nursing students develop a care plan. The authors also wrote *Nurse's Pocket Guide: Nursing Diagnoses with Interventions* (F. A. Davis, 1993), a Brandon/Hill nursing list selection.

Gettrust, Kathy V.
Nursing Diagnosis in Clinical Practice: Guides for Care Planning
Albany, N.Y., Delmar Publishers, 1992. 631p. ISBN 0827348525
   Aids nursing students in the preparation of care plans and in the formulation of nursing interventions by applying scientific rationales to NANDA approved diagnoses. A Brandon/Hill nursing list selection.

Gordon, Marjory
Nursing Diagnosis: Process and Application
St. Louis, Mo., Mosby-Yearbook, 1994. 421p. ISBN 0801660534
   This text is designed to assist nursing students in developing a nursing diagnosis as a result of assessment. Supports nursing as a critical thinking activity with clinical reasoning exercises. A Brandon/Hill nursing list selection. This author also wrote *Manual of Nursing Diagnosis, 1993-94* (Mosby, 1993).

McFarland, Gertrude K., and Elizabeth A. McFarlane, eds.
Nursing Diagnosis and Intervention: Planning for Patient Care
2nd ed. St. Louis, Mo., Mosby-Yearbook, 1993. 811p. ISBN 0801667038
   Focuses on teaching nursing students how to prepare and implement care plans, using a case study approach. Includes a thorough discussion of NANDA diagnoses, with definitions, assessment strategies, examples, rationales, and evaluations.

Neal, Margo Creighton, Mary Paquette, and Mary Mirch
Nursing Diagnosis Care Plans for Diagnosis-Related Groups
Boston, Jones and Bartlett, 1990. 420p. ISBN 0867204184
   This reference assists nursing students in preparing care plans and patient diagnostics as related to specific diagnosis-related groups.

Sparks, Sheila M.
Nursing Diagnosis Reference Manual
2nd ed. Springhouse, Pa., Springhouse, 1993. 597p. ISBN 0874345308
   This reference manual focuses on the nursing process as it relates to patient care. A Brandon/Hill nursing list selection.

✦ Nursing Ethics

Aiken, Tonia D., and Joseph T. Catalano, eds.
Legal, Ethical and Political Issues in Nursing
Philadelphia, F. A. Davis, 1994. 297p. ISBN 080360081X
   Addresses the many legal issues that have an effect on the profession such as malpractice, negligence, and ethics. Chapters treat types of law, application of ethical theories, criminal law, and the courts.

Benjamin, Martin
Ethics in Nursing
3rd ed. New York, Oxford Univ. Press, 1992. 248p. ISBN 0195067479
   Benjamin's text focuses on specific ethical issues and includes some case studies. A Brandon/Hill nursing list first-purchase selection.

Davis, Anne J.
Ethical Dilemmas and Nursing Practice
3rd ed. Norwalk, Conn., Appleton and Lange, 1991. 248p. ISBN 0838522750
   This textbook relates ethics to the clinical setting. A Brandon/Hill nursing list selection.

Pence, Terry, and Janice Cantrall, eds.
Ethics in Nursing: An Anthology
New York, National League for Nursing, 1990. 344p. ISBN 0887374611
   This anthology on nursing ethics covers excellent resource topics such as nurse-physician relations, nurses' rights, nurse-nurse relations, whistle-blowing, and patient advocacy conflicts.

✦ Nursing Fundamentals and Skills

Abrami, Patrick F., and Joyce E. Johnson, eds.
Bringing Computers to the Hospital Bedside: An Emerging Technology
New York, Springer, 1990. 150p. ISBN 0826171907
   Discusses the concept of implementing computerized patient documentation and charting in the hospitals. Hospital information systems are also reviewed.

Bates, Barbara
Guide to Physical Examination and History Taking
5th ed. Philadelphia, Lippincott, 1991. 714p. ISBN 0397547811
   This illustrated guide provides instruction on the correct way to physically assess patients. A favorite title with entry-level or first-year students.

Bolander, Verolyn Barnes, ed.
Sorenson and Luckmann's Basic Nursing: A Psychophysiologic Approach
3rd ed. Philadelphia, Saunders, 1994. 1642p. ISBN 0721640133
   This basic skills and concepts text is often cited as required reading in the nursing curriculum.

Bowers, Arden C.
Clinical Manual of Health Assessment
4th ed. St. Louis, Mo., Mosby-Yearbook, 1992. 657p. ISBN 0801608260
   Bowers' manual explicitly outlines the knowledge and skills necessary to assess patients.

Craven, Ruth F., and Constance J. Hirnle
Fundamentals of Nursing: Human Health and Function
Philadelphia, Lippincott, 1992. 1522p. ISBN 0397546696
   Provides concepts, philosophy, and skills necessary for nursing practice. This work is particularly helpful for beginning nursing students. A Brandon/Hill nursing list selection.

Earnest, Vicki Vine
Clinical Skills in Nursing Practice
2nd ed. Philadelphia, Lippincott, 1993. 1215p. ISBN 0397549261
   This basic fundamentals text for nursing students includes documentation, administration of medication, and safety issues. A Brandon/Hill nursing list selection.

Eggland, Ellen Thomas, and Denise Skelly Heinemann
Nursing Documentation: Charting, Recording and Reporting
Philadelphia, Lippincott, 1994. 260p. ISBN 0397550103
   Emphasizes the importance of accurate record keeping in the nursing profession. Contains chapters on care plans, nursing interventions, and legal aspects of record keeping.

Ellis, Janice Rider, Elizabeth Ann Nowlis, and Patricia M. Bentz
Modules for Basic Nursing Skills
5th ed. New York, Lippincott, 1992. 2 v. ISBN 0397549083 (vol. 1); ISBN 0397549091 (vol. 2)
   This programmed workbook for beginning nursing students builds basic nursing skills and reinforces fundamental concepts.

Ellis, Janice Rider, and Elizabeth Ann Nowlis
Nursing: A Human Needs Approach
5th ed. Philadelphia, Lippincott, 1994. 1 v. (various pagings). ISBN 0397550049
   Presents an overview of basic nursing concepts. Appendices contain NANDA-approved diagnoses and abbreviations. Ellis wrote *Nursing in Today's World: Challenges, Issues, Trends* (Lippincott, 1995), a Brandon/Hill nursing list selection.

Fuller, Jill, and Jennifer Schaller-Ayers
Health Assessment: A Nursing Approach
2nd ed. Philadelphia, Lippincott, 1994. 702p. ISBN 0397550030
   This beginning nursing text focuses on the skills necessary for accurate patient assessment. Topics include overview of health assessment skills such as symptom analysis and head-to-toe examination.

Fundamental Nursing Skills
Springhouse, Pa., Springhouse, 1994. 501p. ISBN 0874345960
   This concise textbook for beginning nursing students helps the reader master over 200 essential clinical procedures. Contains a section that lists commonly used equipment in clinics. A Brandon/Hill nursing list selection.

Giger, Joyce Newman, and Ruth Elaine Davidhizer, eds.
Transcultural Nursing: Assessment and Intervention
St. Louis, Mo., Mosby-Yearbook, 1991. 540p. ISBN 0801619289
   Focuses on the cultural aspects of health beliefs. Cross-cultural comparisons are a part of many nursing curriculums, and this work will provide background for students studying in this area. A Brandon/Hill nursing list selection.

Gulanick, Meg, and others
Nursing Care Plans: Nursing Diagnosis and Intervention
3rd ed. St. Louis, Mo., Mosby-Yearbook, 1994. 541p. ISBN 0801677459
   This manual for nursing students provides sample care plans. Chapters are arranged according to nursing diagnosis. A Brandon/Hill nursing list first-purchase selection.

Hannah, Kathryn J.
Introduction to Nursing Informatics
New York, Springer-Verlag, 1994. 311p. ISBN 0387979832
   Hannah's text discusses possible applications of computer technology to nursing practice, education, and research. Includes topics such as history of health care computing, confidentiality, and selection of software/hardware.

Horne, Mima M.
Fluid, Electrolyte, and Acid-Base Balance: A Case Study Approach
St. Louis, Mo., Mosby-Yearbook, 1991. 451p. ISBN 0801654793
   Horne provides information on diagnosis and interventions, using a case study approach with questions and answers. A Brandon/Hill nursing list selection.

Illustrated Manual of Nursing Practice
2nd ed. Springhouse, Pa., Springhouse, 1994. 1460p. ISBN 0874346096
   An excellent reference manual for nurses. Topics include documentation, legal issues, clinical skills, assessment, and nursing interventions. An appendix provides lab values, common drugs, and infection control measures. A Brandon/Hill nursing list selection.

Iyer, Patricia W.
Nursing Documentation: A Nursing Process Approach
St. Louis, Mo., Mosby-Yearbook, 1991. 214p. ISBN 0801623847
   Iyer's manual, geared to nursing students, emphasizes the importance aand method of accurate documentation.

Jarvis, Carolyn
Physical Examination and Health Assessment
Philadelphia, Saunders, 1992. 952p. ISBN 0721611168
   This introductory text addresses topics such as interviewing, history taking, and physical examinations. Contains full color photographs, chapter outlines, tables, transcultural considerations, NANDA nursing diagnosis boxes, and representative clinical problems.

Kozier, Barbara, and others
Techniques in Clinical Nursing
4th ed. Redwood City, Calif., Addison-Wesley Nursing, 1993. 1 v. (various pagings). ISBN 0805359508

Basic nursing procedures are illustrated with selected color illustrations. This very prolific writer also authored *Fundamentals of Nursing: Concepts, Process, and Practice* (Addison-Wesley, 1995) and *Introduction to Nursing* (Addison-Wesley, 1989). A Brandon/Hill nursing list first-purchase selection.

Metheny, Norma Milligan, ed.
Fluid and Electrolyte Balance: Nursing Considerations
2nd ed. Philadelphia, Lippincott, 1992. 372p. ISBN 0397548915

This comprehensive textbook focuses on fluid and electrolyte imbalance including fundamental concepts, systematic assessments for fluid and electrolyte disturbance, and types and applications of parenteral fluids. A Brandon/Hill nursing list first-purchase selection.

Morton, Patricia Gonce, ed.
Health Assessment in Nursing
2nd ed. Springhouse, Pa., Springhouse, 1993. 681p.
ISBN 0874344255

Addresses cultural as well as holistic aspects of patient care, encompassing physical, biological, psychosocial, and cultural dimensions. Focus is on assessment techniques and findings. Extensive illustrations reinforce the text. This is a Brandon/Hill nursing list selection.

Nursing Procedures
Springhouse, Pa., Springhouse, 1992. 772p. ISBN 0874343925

Presents 325 fundamental procedures of importance to beginning nursing students during clinicals.

Perry, Anne Griffin, and Patricia Ann Potter
Clinical Nursing Skills and Techniques
3rd ed. St. Louis, Mo., Mosby-Yearbook, 1994. 1198p.
ISBN 0801670071

This comprehensive manual for beginning nursing students focuses on fundamental skills and procedures. A Brandon/Hill nursing list first-purchase selection.

Potter, Patricia Ann
Fundamentals of Nursing: Concepts, Process and Practice
3rd ed. St. Louis, Mo., Mosby-Yearbook, 1993. 1767p.
ISBN 0801666678

Covering nursing principles and current clinical practices, this work is often used as the basic course text for nursing students. A Brandon/Hill nursing list first-purchase selection. Potter is also known for her book *Basic Nursing: Theory and Practice* (Mosby, new edition scheduled for publication in 1995).

Redman, Barbara Klug
Process of Patient Education
7th ed. St. Louis, Mo., Mosby-Yearbook, 1993. 337p.
ISBN 0801666708

This text is geared to nurses working in the clinical setting. It teaches how to provide patients and families with information needed to understand clinical problems and their prognoses. A Brandon/Hill nursing list first-purchase selection.

Seidel, Henry M., and others
Mosby-Yearbook's Guide to Physical Examination
2nd ed. St. Louis, Mo., Mosby-Yearbook, 1991. 791p.
ISBN 0801604435

This popular textbook focuses on the assessment of patients as individuals and stresses the patient interview, history taking, physical examination, and critical judgment. Includes Universal Precautions guidelines and color photographs.

Smith, Sandra Fucci
Clinical Nursing Skills: Nursing Process Model, Basic to Advanced Skills
3rd ed. Norwalk, Conn., Appleton and Lange, 1992. 986p.
ISBN 0838513352

A comprehensive textbook on nursing skills. Includes guidelines for Universal Precautions.

Spector, Rachel E.
Cultural Diversity in Health and Illness
3rd ed. Norwalk, Conn., Appleton and Lange, 1991. 345p.
ISBN 0838513964

Addresses the attitudes of health personnel and the cultural health beliefs of patients. A Brandon/Hill nursing list first-purchase selection.

Suddarth, Doris Smith
Lippincott Manual of Nursing Practice
5th ed. Philadelphia, Lippincott, 1991. 1607p. ISBN 0397547870

In this clinical reference for nursing students, emphasis is placed on nursing care procedures for specific clinical problems. This is a favorite with students.

Swartz, Mark H.
Textbook of Physical Diagnosis: History and Examination
2nd ed. Philadelphia, Saunders, 1994. 644p. ISBN 0721655300

Swartz presents a how-to handbook on diagnosis, including patient assessment.

Swearingen, Pamela L.
Photo Atlas of Nursing Procedures
2nd ed. Redwood City, Calif., Addison-Wesley Nursing, 1991. 650p. ISBN 0201132397

This essential atlas contains 1,500 black-and-white photographs. A Brandon/Hill nursing list selection.

Thompson, June M., and others
Mosby's Clinical Nursing
3rd ed. St. Louis, Mo., Mosby-Yearbook, 1993. 1714p.
ISBN 0801666961

Covers many conditions encountered in hospitals on a daily basis. Nursing practice is viewed in terms of nursing theory, nursing processes, and patient education. Four major sections discuss anatomy and physiology, preoperative nursing care, diagnostic tests, and nursing care according to nursing diagnosis. A Brandon/Hill nursing list first-purchase selection.

Tilkian, Ara G., and Mary Boudreau Gordon
Understanding Heart Sounds and Murmurs with an Introduction to Lung Sounds
3rd ed. Philadelphia, Saunders, 1993. 278p. ISBN 0721667848
    This is a particularly useful text for nursing students who need to develop assessment skills and an understanding of the various heart and lung sounds.

Timby, Barbara, Kuhn, and LuVerne Wolff Lewis
Fundamental Skills and Concepts in Patient Care
5th ed. Philadelphia, Lippincott, 1992. 808p. ISBN 0397548486
    Emphasizes holistic care, family participation, and patient teaching and includes Universal Precautions procedures. A Brandon/Hill nursing list selection.

Tucker, Susan Martin
Patient Care Standards: Nursing Process, Diagnosis, and Outcome
5th ed. St. Louis, Mo., Mosby-Yearbook, 1992. 872p. ISBN 0801662680
    This manual for nursing students contains comprehensive guidelines for preparing patient care plans. A Brandon/Hill nursing list first-purchase selection.

Weber, Janet, ed.
Nurses' Handbook of Health Assessment
2nd ed. Philadelphia, Lippincott, 1993. 457p. ISBN 0397549385
    A quick reference/pocket guide on assessment for nurses in the clinical setting. A Brandon/Hill nursing list selection.

Weldy, Norma Jean
Body Fluids and Electrolytes: A Programmed Presentation
6th ed. St. Louis, Mo., Mosby-Yearbook, 1992. 197p. ISBN 0801655773
    Weldy provides an introductory handbook for nursing students in a programmed format. Information presented begins with the most simple concepts and extends to more complicated concepts. A Brandon/Hill nursing list selection.

✦ Nursing History

Birnbach, Nettie, and Lewenson, Sandra, eds.
First Words: Selected Addresses from the National League for Nursing, 1894-1933
New York, National League for Nursing, 375p. ISBN 088737526X
    This excellent resource for advanced students meets the need for information on progressive themes in nursing throughout history. Listed in ICLRN.

Bullough, Vern L., Olga Maranjian Church, and Alice P. Stein
American Nursing: A Biographical Dictionary
New York, Garland, 1988-1992. 2 v. ISBN 082408540X (vol.1); ISBN 0824072014 (vol.2)
    An excellent reference book for nursing biography in a historical context, this work is arranged alphabetically, each entry containing a bibliography of publications by and about the subject. Includes an index by decades of birth. For each biographee, lists nursing schools attended, special interest areas or accomplishment, and states and countries of birth. Criteria for inclusion in Volume One is that the subjects had to be born before 1890 or be deceased. Volume Two subjects had to be born before 1916 or be deceased. Some black- and-white photographs in both volumes. Listed in the ICLRN.

Carnegie, Mary Elizabeth
Path We Tread: Blacks in Nursing, 1854-1990
2nd ed. New York, National League for Nursing, 1991. 301p. ISBN 0887375340
    This historical reference documents contributions of African American nurses to the nursing profession. The period covered is 1854 to 1990. A 1993 *AJN* "Books of the Year" title, and recommended by ICLRN.

Haase, Patricia T.
Associate Degree Nursing Education: An Historical Annotated Bibliography 1942-1988
Durham, Duke Univ. Press, 1990. 322p. ISBN 0822309831
    Provides a framework for research on the nursing education controversy over the two-, three-, and four-year programs. This bibliography is arranged chronologically and contains an author, title, and subject index.

Haase, Patricia T.
Origins and Rise of Associate Degree Nursing Education
Durham, Duke Univ. Press, 1990. 210p. ISBN 0822309785
    A companion volume to *Associate Degree Nursing Education: An Historical Annotated Bibliography, 1942-1988.* Outlines the historical foundations of the ADN degree and addresses the controversy surrounding basic education in nursing.

Hine, Darlene Clark
Black Women in White: Racial Conflict and Cooperation in the Nursing Profession 1890-1950
Bloomington, Indiana Univ. Press, 1989. 264p. ISBN 0253327733
    This biographical and historical text acknowledges the many contributions of African American nurses to the profession. Listed in ICLRN.

Madden, Margretta M., and Patricia Moccia
On Nursing: A Literary Celebration: An Anthology
New York, National League for Nursing, 1993. 355p. ISBN 0887375774
    A little treasure to have in any library, this anthology collects parts of works by noteworthy authors that have nursing as a theme. Contains poetry, essays, and synopses of literary works.

✦ Nursing Theory

Barnum, Barbara Stevens
Nursing Theory: Analysis, Application, Evaluation
4th ed. Philadelphia, Lippincott, 1994. 304p. ISBN 0397549423
    This text addresses the philosophy of nursing, emphasizing nursing theory development, analysis, and comparison. A Brandon/Hill nursing list first-purchase selection.

Chinn, Peggy L., and Maeona K. Kramer
Theory and Nursing: A Systematic Approach
3rd ed. St. Louis, Mo., Mosby-Yearbook, 1991. 220p.
ISBN 0801609704
   Discusses the development of nursing theory as it applies to life's situations. This is a Brandon/Hill nursing list selection.

George, Julia B., ed.
Nursing Theories: The Base for Professional Nursing Practice
3rd ed. Norwalk, Conn., Appleton and Lange, 1990. 401p.
ISBN 0838570518
   Presents the perspectives of eighteen nursing theorists. This is a Brandon/Hill nursing list selection. The fourth edition is in preparation.

Keegan, Lynn
The Nurse as Healer
Albany, N.Y., Delmar Publishers, 1994. 138p. ISBN 0827361564
   This text aids nursing students in the formulation of a personal philosophy of nursing.

Marriner-Tomey, Ann, ed.
Nursing Theorists and Their Work
3rd ed. St. Louis, Mo., Mosby-Yearbook, 1994. 530p.
ISBN 080166764X
   The theories of principal nursing theorists are analyzed by contributors. This is a Brandon/Hill nursing list first-purchase selection.

Nicoll, Leslie H., ed.
Perspectives on Nursing Theory
2nd ed. Philadelphia, Lippincott, 1992. 696p. ISBN 0397549105
   This anthology of articles on nursing theory includes current debates about the concept and progression of nursing theory, based on historical data.

Wesley, Ruby L.
Nursing Theories and Models
Springhouse, Pa., Springhouse, 1992. 144p. ISBN 0874343674
   This presentation of the work of many nursing theorists provides students with tools to analyze and develop insights about theories in the clinical setting.

Ziegler, Shirley Melat, ed.
Theory-Directed Nursing Practice
New York, Springer, 1993. 265p. ISBN 0826176305
   Applies theoretical nursing concepts to the clinical setting. Includes topics such as cognitive theory, theory of depression, family theory, and psychosocial development. A Brandon/Hill nursing list selection.

✦ Obstetrical and Maternity Nursing

Blackburn, Susan Tucker, and Donna Lee Loper
Maternal, Fetal, and Neonatal Physiology: A Clinical Perspective
Philadelphia, Saunders, 1992. 723p. ISBN 0721629369
   Covers physiological changes throughout the prenatal period. Geared for nurses in the clinical setting needing to assess and manage pregnant women, the fetus, and the neonate. This is a Brandon/Hill nursing list selection and a 1993 *AJN* "Books of the Year" title.

Bobak, Irene M.
Maternity and Gynecologic Care: The Nurse and the Family
5th ed. St. Louis, Mo., Mosby-Yearbook, 1993. 1502p.
ISBN 0801666635
   This established author in the field presents an excellent textbook for nursing students. Bobak also wrote *Essentials of Maternity Nursing* (Mosby-Yearbook, 1991), which includes clinical teaching guidelines and care plans. A Brandon/Hill nursing list first-purchase selection and a 1991 *AJN* "Books of the Year" title.

Cohen, Susan M.
Maternal, Neonatal, and Women's Health Nursing
Springhouse, Pa., Springhouse, 1991. 1201p. ISBN 0874342589
   Emphasizes the role of the nurse in obstetrical and gynecological settings. Includes topics such as trends, health care delivery, and cultural influences on health care decisions. Teaching Guides may be duplicated for patient education. A 1992 *AJN* "Books of the Year."

Dickason, Elizabeth J., Martha Olsen Schult, and Bonnie Lang Silverman
Maternal-Infant Nursing Care
2nd ed. St. Louis, Mo., Mosby-Yearbook, 1994. 875p.
ISBN 0801674085
   This standard maternity textbook for nursing students covers the entire normal childbirth cycle and discusses potential complications. Includes many illustrations. A Brandon/Hill nursing list selection.

Gorrie, Trula, Emily Slone McKinney, and Sharon Smith Murray
Foundations of Maternal-Newborn Nursing
Philadelphia, Saunders, 1994. 1041p. ISBN 0721640338
   This work provides comprehensive coverage of nursing care practices with regard to maternity nursing.

Harvey, Carol J., ed.
Critical Care Obstetrical Nursing
Gaithersburg, Md., Aspen Publishers, 1991. 220p.
ISBN 0834202352
   Harvey provides a resource for obstetrical nurses working in a critical care setting. This is a Brandon/Hill nursing list selection.

Kenner, Carole, Ann Brueggemeyer, and Laurie Porter Gunderson
Comprehensive Neonatal Nursing: A Physiologic Perspective
Philadelphia, Saunders, 1993. 1218p. ISBN 07221629924
   This comprehensive textbook covers the medical, surgical and psychosocial care of perinatal patients. A Brandon/Hill nursing list first-purchase selection and a 1994 *AJN* "Books of the Year" title.

Knuppel, Robert A., and Joan E. Drukker, eds.
High-Risk Pregnancy: A Team Approach
2nd ed. Philadelphia, Saunders, 1993. 769p. ISBN 0721634559
   Nursing students often wish to find examples of collaborative practices in textbooks. This resource provides this team approach to patient care.

Ladewig, Patricia A., Marcia L. London, and Sally B. Olds
Essentials of Maternal-Newborn Nursing
3rd ed. Redwood City, Calif., Addison-Wesley Nursing, 1994. 909p. ISBN 0805355898

This maternal and newborn nursing text reflects changes in immunization schedules specifically with regard to the hepatitis vaccine. This is a Brandon/Hill nursing list first-purchase selection.

May, Katharyn A., and Laura R. Mahlmeister
Maternal and Neonatal Nursing: Family-Care
3rd ed. Philadelphia, Lippincott, 1994. 1199p. ISBN 0397549539

Aids understanding of contemporary health care problems, with a focus on family-centered care. Includes a chapter on parenting after age thirty-five. Information is divided between the "knowledge-base" chapters and the "nursing-care" chapters. A Brandon/Hill nursing list first-purchase selection. Contains new immunization standards.

Moore, Keith L., and T.V.N. Persaud
Before We Are Born: Essentials of Embryology and Birth Defects
4th ed. Philadelphia, Saunders, 1993. 354p. ISBN 0721646654

This sourcebook presents a clear accounting of human embryology. Demonstrates the weekly progression of fetal development, both normal and abnormal. An excellent resource.

Olds, Sally B., Marcia L. London, and Patricia W. Ladewig
Maternal Newborn Nursing: A Family-Centered Approach
4th ed. Redwood City, Calif., Addison-Wesley Nursing, 1992. 1225p. ISBN 0805355804

An excellent textbook for nursing students, this work is designed to develop critical-thinking skills. Contains diagrams, charts, color photographs, and informational boxes highlighting clinical applications. A Brandon/Hill nursing list selection and a 1993 *AJN* "Books of the Year" title.

Pillitteri, Adele
Maternal and Child Health Nursing: Care of the Childbearing and Childrearing Family
Philadelphia, Lippincott, 1992. 1 v. (various pagings). ISBN 0397548621

This comprehensive text deals with both maternity and pediatric nursing concepts. The life-cycle approach includes reproduction, pregnancy through birth and postpartum, family development, and human growth and development through adolescence. A Brandon/Hill nursing list selection and a 1993 *AJN* "Books of the Year" title.

Reeder, Sharon J., Leonide L. Martin, and Deborah Koniak-Griffin
Maternity Nursing: Family, Newborn and Women's Health Care
17th ed. Philadelphia, Lippincott, 1992. 1377p. ISBN 0397548133

A standard work in maternity nursing that is often used as the primary textbook in many nursing curriculums. A Brandon/Hill nursing list selection.

✦ Oncological Nursing

Burke, Margaret Barton
Chemotherapy Care Plans: Designs for Nursing Care
Boston, Jones and Bartlett, 1992. 426p. ISBN 0867203390

This care plan book for nursing students documents care for the chemotherapy patient. A companion reference is Margaret Barton Burke's *Cancer Chemotherapy: A Nursing Process Approach* (Jones and Bartlett, 1991), which provides an overview of the principles of chemotherapy combined with nursing diagnosis. A 1992 *AJN* "Books of the Year" title.

Dow, Karen Hassey, and Laura J. Hinderley, eds.
Nursing Care in Radiation Oncology
Philadelphia, Saunders, 1992. 425p. ISBN 072164015X

Dow presents a comprehensive text for nurses caring for cancer patients, which focuses on radiation therapy and related problems.

Groenwald, Susan L., and others, eds.
Cancer Nursing: Principles and Practice
3rd ed. Boston, Jones and Bartlett, 1993. 1724p. ISBN 0867206403

This comprehensive text for the nurse working with oncology patients covers patient care and psychological and social dimensions of cancer. A Brandon/Hill nursing list first-purchase selection.

Otto, Shirley E., ed.
Oncology Nursing
2nd ed. St. Louis, Mo., Mosby-Yearbook, 1993. 802p. ISBN 0801678161

This thorough treatment dealing with cancer patients includes topics such as diagnosis, clinical management of cancer diseases, and treatment modalities. This is a Brandon/Hill nursing list first-purchase selection.

✦ Orthopedic Nursing

Maher, Ann Butler
Orthopaedic Nursing
Philadelphia, Saunders, 1994. 954p. ISBN 0721626998

Maher offers a comprehensive text on the care of patients with musculoskeletal disorders. Includes physical, sociocultural, and psychologic aspects of orthopedic nursing care. A Brandon/Hill nursing list first-purchase selection.

✦ Pediatric Nursing

Castiglia, Patricia Thorson, and Richard E. Harbin
Child Health Care: Process and Practice
Philadelphia, Lippincott, 1992. 1055p. ISBN 0397547285

This book focuses on human growth and development from conception through adolescence. Includes physiological as well as psychosocial aspects. Appendix includes growth charts, lab values, and drug dosages.

Gulanick, Meg, Michelle Knoll Puzas, and Cynthia R. Wilson
Nursing Care Plans for Newborns and Children: Acute and Critical Care
St. Louis, Mo., Mosby-Yearbook, 1992. 420p. ISBN 0801663881
   Provides sample care plans specifically developed for critically ill newborns and children.

Hazinski, Mary Fran, ed.
Nursing Care of the Critically Ill Child
2nd ed. St. Louis, Mo., Mosby-Yearbook, 1992. 1127p. ISBN 0801653126
   Integrates the psychosocial aspects of the critically ill child and the family. Includes topics such as bereavement and death, and assessment and management of pediatric pain. Provides sample care plans and charts. Brandon/Hill nursing list and 1993 *AJN* "Books of the Year" selections.

Jackson, Debra Broadwell, and Rebecca B. Saunders
Child Health Nursing: A Comprehensive Approach to the Care of Children
Philadelphia, Lippincott, 1993. 1949p. ISBN 0397547250
   This comprehensive text addresses the physical, psychosocial, and familial needs of the child, up through adolescence. A Brandon/Hill nursing list first-purchase selection.

Jaffe, Marie S.
Pediatric Nursing Care Plans
El Paso, Tex., Skidmore-Roth Pub., 1993. 416p. ISBN 0944132588
   Includes sample pediatric care plans including NANDA approved diagnoses, desired outcomes, and nursing interventions with rationales. A Brandon/Hill nursing list selection. Jaffe is also the author of *Maternal Infant Health Care Planning* (Springhouse, 1994).

Keller, Leslie, and Anne Weir
Pediatric Nursing
El Paso, Tex., Skidmore-Roth Pub., 1993. 279p. ISBN 0944132898
   This pediatric text is arranged by specific types of system disorders. Nursing care focus includes growth and development, signs and symptoms, and physiological as well as psychological issues. Appendix includes growth charts, assessment, NANDA nursing diagnoses, and medication tables.

Skale, Nedra
Manual of Pediatric Nursing Procedures
Philadelphia, Lippincott, 1992. 584p. ISBN 039754782X
   Skale's manual includes assessment, nursing diagnosis, and interventions, as well as patient education. Listed on the Brandon/Hill nursing list.

Smith, Donna Phillips, ed.
Comprehensive Child and Family Nursing Skills
St. Louis, Mo., Mosby-Yearbook, 1991. 828p. ISBN 080165209X
   This is an excellent resource for nursing students working with pediatric patients. Includes numerous skills from psychosocial care to clinical procedures such as physical assessment, visual screening, and patient education. Contains photographs, diagrams, and charts. A Brandon/Hill nursing list selection and a 1992 *AJN* "Books of the Year" title.

Speer, Kathleen Morgan
Pediatric Care Planning
2nd ed. Springhouse, Pa., Springhouse, 1994. 288p. ISBN 0874346592
   This resource is especially useful for nursing students that need to develop pediatric care plans.

Whaley, Lucille F.
Nursing Care of Infants and Children
4th ed. St. Louis, Mo., Mosby-Yearbook, 1991. 1984p. ISBN 0801653789
   This standard pediatric nursing textbook incorporates many issues facing pediatric nursing today. Includes topics such as homelessness, transcultural nursing, dual career parenting, gay families, neonatal cocaine exposure, Lyme disease, and HIV. A Brandon/Hill nursing list selection and a 1992 *AJN* "Books of the Year" title.

Wong, Donna L.
Clinical Manual of Pediatric Nursing
3rd ed. St. Louis, Mo., Mosby-Yearbook, 1990. 619p. ISBN 0801661447
   Wong presents an essential text that includes care plans, assessment strategies, and nursing interventions. The author also wrote *Whaley and Wong's Essentials of Pediatric Nursing* (Mosby-Yearbook, 1993), a 1994 *AJN* "Books of the Year" title. Both are Brandon/Hill nursing list first-purchase selections.

Yogev, Ram, and Edward Connor, eds.
Management of HIV Infection in Infants and Children
St. Louis, Mo., Mosby-Yearbook, 1992. 639p. ISBN 0801656533
   Addresses management strategies as well as controversies in HIV/AIDS treatment in infants and children.

✦ Pharmacology and Drug Therapy

Alfaro-Lefevre, Rosalinda, and others
Drug Handbook: A Nursing Process Approach
Redwood City, Calif., Addison-Wesley, 1992. 790p. ISBN 0201092786
   A concise drug handbook with nursing implications. Presented in a color-coded, quick reference format with 1,400 drugs listed.

Arcangelo, Virginia Poole
Weaver and Koehler's Programmed Mathematics of Drugs and Solutions
5th ed. Philadelphia, Lippincott, 1992. 140p. ISBN 0397549180
   This introductory workbook teaches students how to calculate drugs and solutions. Includes a pretest, basic arithmetic practice problems with reviews, and a post-test.

Baer, Charold Lee Morris, and Bradley R. Williams
Clinical Pharmacology and Nursing
2nd ed. Springhouse, Pa., Springhouse, 1992. 1386p. ISBN 0874343798
   Nursing students needing drug information for their care plans will find this work especially useful. Arranged by drug types, and includes nursing implications.

Clark, Julia B., Sherry F. Queener, and Virginia Burke Karb
Pharmacologic Basis of Nursing Practice
4th ed. St. Louis, Mo., Mosby-Yearbook, 1993. 880p.
ISBN 0801666732

This comprehensive pharmacology textbook for nursing students includes learning objectives, overviews, nursing implications summaries, chapter reviews, and informational insets.

Cornett, Emily F.
Dosages and Solutions: A Programmed Approach to Meds and Math
5th ed. Philadelphia, F. A. Davis, 1991. 273p. ISBN 0803619812

This is a favorite among nursing students as an instruction guide to calculating dosages and solutions. Previous edition was written by Dorothy M. Blume.

Daniels, Joanne M.
Clinical Calculations: A Unified Approach
3rd ed. Albany, New York, Delmar Publishers, 1994. 376p.
ISBN 0827359454

This programmed manual of dosage and solution calculation for nursing students includes information on metrics, household units, conversions, oral medications, and use of the medicine cup. Contains some illustrations. A similar text is Mary Jane Cordon's *Clinical Calculations for Nurses: With Mathematics Review* (Appleton and Lange, 1994).

Dison, Norma
Simplified Drugs and Solutions for Nurses, Including Mathematics
10th ed. St. Louis, Mo., Mosby-Yearbook, 1992. 209p.
ISBN 0801616956

Dison teaches nursing students to calculate drug dosages using a workbook format.

Gahart, Betty L.
Intravenous Medications: A Handbook for Nurses and Allied Health Professionals
St. Louis, Mo., Mosby. Annual.

Gahert's pocket-sized work is a standard reference for intravenous solutions. Appendix includes recommendations for the safe handling of cytotoxic drugs, alternative choices for infrequently used drugs, and FDA pregnancy categories. A Brandon/Hill nursing list first-purchase selection and listed in ICLRN.

Johnson, G. E.
Pharmacology and the Nursing Process
3rd ed. Philadelphia, Saunders, 1992. 799p. ISBN 0920513085

Discusses the effect of drug interactions on the physiological, psychological, and pathological body processes.

Kee, Joyce LeFever, and Evelyn R. Hayes
Pharmacology: A Nursing Process Approach
Philadelphia, Saunders, 1993. 665p. ISBN 0721636624

This nursing pharmacology textbook includes conversions, abbreviations, sample calculations, and drug information cards in tear-out sections. This author also wrote *Clinical Calculations: With Applications to General and Specialty Areas* (Saunders, 1992).

Loebl, Suzanne, George Spratto, and Adrienne L. Woods
Nurse's Drug Handbook
7th ed. Albany, N.Y., Delmar Publishers, 1994. 1420p.
ISBN 0827357109

This handbook focuses on nursing considerations of drugs. A Brandon/Hill nursing list first-purchase selection.

Lipsey, Sally I., and Donna D. Ignatavicius
Math for Nurses: A Problem Solving Approach
Philadelphia, Saunders, 1994. 389p. ISBN 0721004814

This math workbook for nursing students teaches the calculation of dosages for drugs and solutions. Contains basic math information such as fractions, multiplication, division, ratios, percentages, and conversions.

Mathewson Kuhn, Merrily
Pharmacotherapeutics: A Nursing Process Approach
3rd ed. Philadelphia, F. A. Davis, 1994. 1696p.
ISBN 0803659326

Incorporates nursing principles and considerations into pharmacology. A Brandon/Hill nursing list first-purchase selection.

McKenry, Leda M., and Evelyn Salerno
Mosby-Yearbook's Pharmacology in Nursing
18th ed. St. Louis, Mo., Mosby-Yearbook, 1991. 1220p. and pocket guide (60p.) ISBN 0801631998

This illustrated text focuses on medication administration. A Brandon/Hill nursing list first-purchase selection.

Mosby's Nursing Drug Reference
St. Louis, Mo., Mosby-Yearbook, Annual. v. ISSN 1044-8470

Provides essential data for the administration of the most common prescription and over-the-counter drugs. This is a Brandon/Hill nursing list first-purchase selection.

Nursing Student's Guide to Drugs
Springhouse, Pa., Springhouse, 1993. 1104p. ISBN 0874343895

Nursing students writing care plans and working in the clinical setting will benefit from this pharmacology text. Arranged by drug types, entries include nursing considerations such as assessment, planning, preparation and administration, monitoring, intervention, and patient teaching.

Photoguide to Drug Administration
Springhouse, Pa., Springhouse, 1992. 194p. ISBN 0874343658

This clinical text for nursing students includes specific instruction and guidelines for administering drugs. Includes numerous black- and-white photographs and illustrations.

Shannon, Margaret T., Billie Ann Wilson, and Carolyn L. Stang
Govoni and Hayes Drugs and Nursing Implications
7th ed. Norwalk, Conn., Appleton and Lange, 1992. 1135p.
ISBN 083851779X

Nursing interventions and patient education perspectives are included in this comprehensive text. It is a long-time favorite of nursing students and is often a required text in the nursing curriculum. A Brandon/Hill nursing list first-purchase selection.

Spencer, Roberta Todd, and others
Clinical Pharmacology and Nursing Management
4th ed. Philadelphia, Lippincott, 1993. 1361p. ISBN 0397549350

This comprehensive text introduces chemical agents and their applications, provides drug information on the most common drugs encountered by nurses, and emphasizes skills needed to administer medications.

✦ Psychiatric Nursing

Barry, Patricia D.
Mental Health and Mental Illness
5th ed. Philadelphia, Lippincott, 1994. 438p. ISBN 0397550138

This textbook for nursing students focuses on mental disorders and nursing considerations. A Brandon/Hill nursing list first-purchase selection.

Diagnostic and Statistical Manual of Mental Disorders: DSM-IV
4th ed. Washington, D.C., American Psychiatric Association, 1994. 886p. ISBN 0890420610

This is a standard reference text for psychiatric diagnostic criteria and definitions. This edition will have an impact on future revisions in psychiatric nursing textbooks. Also available in electronic formats from the publisher.

Fortinash, Katherine M., and Patricia A. Holoday-Worret
Psychiatric Nursing Care Plans
St. Louis, Mo., Mosby-Yearbook, 1991. 320p. ISBN 0801602521

Contains sample psychiatric nursing care plans, including rationales and nursing interventions. A Brandon/Hill nursing list selection.

Haber, Judith, and others
Comprehensive Psychiatric Nursing
4th ed. St. Louis, Mo., Mosby-Yearbook, 1992. 844p. ISBN 0801660408

Haber provides nursing students with a solid foundation for psychiatric nursing and related theoretical models. A Brandon/Hill nursing list first-purchase selection.

McFarland, Gertrude K.
Nursing Diagnoses and Process in Psychiatric Mental Health Nursing
2nd ed. Philadelphia, Lippincott, 1992. 321p. ISBN 0397547587

This text focuses on the psychosocial aspects of patient care, and it is a Brandon/Hill nursing list selection. This author also wrote *Psychiatric Mental Health Nursing: Application of the Nursing Process* (Lippincott, 1991), also a Brandon/Hill selection.

Murray, Ruth Beckman
Nursing Assessment and Health Promotion: Strategies Through the Life Span
5th ed. Norwalk, Conn., Appleton and Lange, 1993. 738p. ISBN 0838566375

Murray stresses the mental and physical needs of patients from birth to death. This is a Brandon/Hill nursing list first-purchase selection. This author also wrote *Psychiatric/Mental Health Nursing: Giving Emotional Care* (Appleton and Lange, 1991).

Rawlins, Ruth Parmelee, and Patricia Evans Heacock
Clinical Manual of Psychiatric Nursing
2nd ed. St. Louis, Mo., Mosby Yearbook, 1993. 401p. ISBN 0801663334

This concise text for nurses in the psychiatric clinical setting uses case studies to illustrate examples of patient care plans. A Brandon/ Hill nursing list first-purchase selection.

Schultz, Judith M., and Sheila Dark Videbeck
Manual of Psychiatric Nursing Care Plans
4th ed. Philadelphia, Lippincott, 1994. 470p. ISBN 0397550677

This reference for nursing students includes sample care plans for psychiatric conditions.

Shives, Louise, R.
Basic Concepts of Psychiatric-Mental Health Nursing
2nd ed. Philadelphia, Lippincott, 1990. 658p. ISBN 0397547579

Shives focuses on psychiatric conditions including diagnosis, assessment, and nursing interventions. A Brandon/Hill nursing list selection.

Stuart, Gail Wiscarz, and Sandra J. Sundeen, eds.
Principles and Practice of Psychiatric Nursing
4th ed. St. Louis, Mo., Mosby Yearbook, 1991. 1069p and pocket guide (43p.) ISBN 0801658853

This standard textbook for psychiatric nursing students includes a quick reference pocket guide: *Quick Psychopharmacology Reference (*Mosby, 1991). This is a Brandon/Hill nursing list selection.

Taylor, Cecelia Monat
Essentials of Psychiatric Nursing
14th ed. St. Louis, Mo., Mosby-Yearbook, 1994. 551p. ISBN 0801678145

This established text is useful for teaching the foundations of psychiatric nursing. A Brandon/Hill nursing list first-purchase selection.

Varcarolis, Elizabeth M.
Foundations of Psychiatric-Mental Health Nursing
2nd ed. Philadelphia, Saunders, 1994. 957p. ISBN 0721637655

This psychiatric text provides classification of mental disorders and reviews *DSM-IV* changes. Discusses historical and transcultural aspects of psychiatric nursing.

West, Patricia, and Christina L. Sieloff Evans
Psychiatric and Mental Health Nursing with Children and Adolescents
Gaithersburg, Md., Aspen Publishers, 1992. 428p. ISBN 0834202409

This text covers the emotional, interpersonal, and behavioral needs of children, adolescents, and their families and is an excellent resource for the nurse in the clinical setting. A Brandon/Hill nursing list selection and a 1993 *AJN* "Books of the Year" title.

Wilson, Holly Skodol
Psychiatric Nursing
4th ed. Redwood City, Calif., Addison-Wesley Nursing, 1992. 1004p. ISBN 0805394001

Wilson presents a foundation for psychiatric nursing including relevant information for the families of psychiatric patients. A Brandon/Hill nursing list first-purchase selection,

✦ Review Publications

AJN/Mosby-Yearbook Nursing Boards Review: For the NCLEX-RN Examination
9th ed. St. Louis, Mo., Mosby Yearbook, 1994. 687p. ISBN 0801677815

Aids students who are preparing to take the NCLEX-RN exam. Includes strategies for success, how to score the test, and reviews material. Arranged by systems such as "Nursing Care of the Client with Psychosocial and Mental Health Problems." Published approximately every two years. A Brandon/Hill nursing list first-purchase selection.

Bininger, Carol J.
American Nursing Review for NCLEX-RN
2nd ed. Springhouse, Pa., Springhouse, 1992. 658p. ISBN 0874343941

Bininger presents an outline review of nursing practice for students preparing for the NCLEX-RN exam. Includes two hundred multiple-choice questions with computerized score sheet. A Brandon/Hill nursing list selection.

Hoefler, Patricia A.
NCLEX-RN Exam: Easy Steps to Passing
2nd ed. Silver Spring, Md., Medical Education Development Services, 1992. 163p. ISBN 1565330048

This helpful text assists nursing students in the mastery of skills needed to do well on the NCLEX-RN exam. Includes chapter topics such as "Learning How to Take a Timed Exam," and "How Do I Choose Between the Two Best Options." A final exam is included as well as appendices with outlined reviews. Answers include rationales. *Medical Education Development Series.* Hoefler is also the author of *Comprehensive NCLEX-RN Review (*Mosby, 1994).

Matassarin-Jacobs, Esther
Saunders Review for NCLEX-RN
2nd ed. Philadelphia, Saunders, 1994. 784p. ISBN 0721649939

This review book prepares students for the NCLEX-RN exam. Chapter sections have pretest and answers with rationales. Contains five simulated NCLEX-RN practice tests with answers and rationales. This is a Brandon/Hill nursing list selection.

Nugent, Patricia Mary, and Barbara Ann Vitale
Test Success: Test-Taking Techniques for Beginning Nursing Students
Philadelphia, F. A. Davis, 1993. 267p. ISBN 0803665989

The authors provide tips and test-taking techniques for success in nursing exams. Includes practice questions with answers and rationales. A Brandon/Hill nursing list selection.

Saxton, Dolores F., ed.
Mosby Comprehensive Review of Nursing
St. Louis, Mo., Mosby, Published every three years.

Saxton presents multiple-choice questions and answers for board review. This is a Brandon/Hill nursing list first purchase selection. Saxton is also the author of *Mosby's Assess Test: A Practice Exam for RN Licensure* (Mosby, 1993) which contains a 375-item, multiple-choice practice exam that students can mail in for test analysis.

Sides, Maria B., and Nancy Krochek
Nurse's Guide to Successful Test-Taking
Philadelphia, Lippincott, 1994. 482p. ISBN 0397549881

Presents a systematic and efficient method of preparing for the state boards. The authors focus is on problem-solving techniques and strategies for success. Sample test questions included.

Smith, Sandra Fucci
Sandra Smith's Review for NCLEX-RN
8th ed. Los Altos, Calif., National Nursing Review, 1994. 687p. and 1 computer disk ISBN 0917010582

Smith's review book for the NCLEX exam is very popular with nursing students. Includes two different sample exams with answers.

Stein, Alice M., and Judith C. Miller, eds.
NSNA, NCLEX-RN Review
2nd ed. Albany, New York, Delmar Publishers, 1994. 722p. and 1 computer disk. ISBN 0827357206

This comprehensive nursing review for the NCLEX-RN exam explains the test plan, how the test is constructed, and how it is scoreds. It also provides tips on how to plan studies and prepare for the exam. Includes an extensive section on drugs and nursing implications, four practice tests, and an IBM-compatible disk.

Continuations

Advanced Skills
Springhouse, Pa., Springhouse.

This series addresses the need for information on more advanced nursing procedures due to growing numbers of acutely ill patients. The reference texts in this series cover the most important and most advanced procedures. Illustrated with some color photographs.

> *Deciphering Difficult ECGs.* 1993. ISBN 0874345529
> *Giving Drugs by Advanced Technique.* 1993. ISBN 0874345537
> *Mastering Advanced Assessment.* 1993. ISBN 0874345510
> *Performing Advanced Procedures.* 1993. ISBN 0874345545

Clinical Skillbuilders
Springhouse, Pa., Springhouse.

These manuals help nursing students build assessment skills in specific areas.

> *Better Documentation.* 1992. ISBN 0874344093
> *Crisis Drugs.* 1991. ISBN 0874343852
> *Danger Signs and Symptoms.* 1990. ISBN 0874343100

Diagnostic Test Implications. 1991. ISBN 0874343801
ECG Interpretation. 1990. ISBN 0874343097
Emergency Procedures. 1991. ISBN 087434378X
I.V. Therapy. 1990. ISBN 087434350X
Psychosocial Crisis. 1992. ISBN 0874344824
Rapid Assessment. 1991. ISBN 087434364X
Respiratory Support. 1991. ISBN 0874343623

Lippincott's Review Series
Philadelphia, Lippincott.

This series provides reviews of key subject areas in nursing. Each book contains a comprehensive outline review, and chapter study questions with answer keys and rationales. Students use these resources to supplement course materials and as subject reviews for the NCLEX-RN exam.

Medical-Surgical Nursing. 1992. ISBN 0397547757
Pediatric Nursing. 1992. ISBN 0397547749
Maternal-Newborn Nursing. 1992. ISBN 0397547765
Mental Health and Psychiatric Nursing. 1992. ISBN 0397547730

Mosby's Clinical Nursing Series
St. Louis, Mo., Mosby-Yearbook

Belcher, Anne E. *Cancer Nursing.* 1992. ISBN 0801618096
Brundage, Dorothy J. *Renal Disorders.* 1992. ISBN 0801616859
Canobbio, Mary M. *Cardiovascular Disorders.* 1990. ISBN 0801614058
Chipps, Esther M. *Neurologic Disorders.* 1992. ISBN 0801613728
Doughty, Dorothy Beckley. *Gastrointestinal Disorders.* 1993. ISBN 0801620961
Edge, Valerie. *Women's Health Care.* 1994. ISBN 0801680131
Gray, Mikel. *Genitourinary Disorders.* 1992. ISBN 080166876X
Grimes, Deanna E., and Richard M. Grimes. *AIDS and HIV Infection.* 1994. ISBN 0801680123
Grimes, Deanna E. *Infectious Diseases.* 1991. ISBN 0801623456
Mourad, Leona A. *Orthopedic Disorders.* 1991. ISBN 0801634385
Wilson, Susan Fickertt. *Respiratory Disorders.* 1990. ISBN 0801650879

Nurse's Photolibrary
Springhouse, Pa., Springhouse, 1993. loose-leaf.
ISBN 0874345111

This excellent updating service stresses nursing skills. Contains photographs and illustrations.

Nurse's Ready Reference
Springhouse, Pa., Springhouse.

Provides a quick-reference on clinical resources for selected disorders. These volumes are particularly helpful when working with patients in the clinical setting and in care plan development.

Diagnostic Tests. 1991. ISBN 0874343321
Alphabetical listings of laboratory tests with nursing considerations. Quick reference format.
Diseases. 1991. ISBN 0874343313
Handbook consisting of excellent overviews of diseases with nursing implications. Includes etiology, symptoms, treatment, and nursing considerations.
Medications. 1994. ISBN 0874345995
Concise handbook listing drug information with nursing implications.
Quick E.C.G. Interpretation. 1992. ISBN 0874344913
Handbook for easy interpretation of ECGs. Ready reference format.

NurseReview
Springhouse, Pa., Springhouse, 1994. annual updates loose-leaf service.

Arranged by body systems, this concise publication text is designed to keep the clinical nurse current. Contains charts and graphs. Provides extensive explanations on caring for patients with cardiac, respiratory, gastrointestinal, neurologic, vascular, genitourinary, endocrine, musculoskeletal, metabolic, hematologic, immunologic, and psychiatric problems.

Nursetest
Springhouse, Pa., Springhouse.

Like the *Lippincott's Review Series*, this series is designed to help nursing students improve their test-taking skills and increase their theoretical knowledge of nursing. Includes multiple-choice questions and answers that help review for the NCLEX-RN exam. Contains reference bibliography.

Condon, Mary Ann Blum. *Pediatric Nursing.* 1992. ISBN 0874343054
Martin, Frances L. *Medical-Surgical Nursing.* 1992. ISBN 0874343038
Olsen, June Looby. *Dosage Calculations.* 1992. ISBN 0874343011
Olsen, June Looby. *Fundamentals of Nursing.* 1992. ISBN 087434302X
Reese, Linda E. *Mental Health Psychiatric Nursing.* 1992. ISBN 0874343062
Stamps, Georgiana M. *Maternal-Newborn Nursing.* 1992. ISBN 0874343046

Real Nursing Series
Albany, N.Y., Delmar Publishers

This excellent series on timely topics of interest to more advanced nursing students presents information in a concise and organized manner. Contains extensive references and useful appendices such as agencies, bibliographies, and sources of technical information.

This series is ongoing with future topics to include critical business skills for nurses, alcohol and substance abuse, ethical dilemmas in nursing, war stories, difficult nursing decisions, gay and lesbian nursing, interventions in everyday nursing emergencies, and healing racism in nursing.

Backer, Barbara A. *To Listen, to Comfort, to Care: Reflections on Death and Dying.* 1994. ISBN 0827361785

Fiesta, Janine. *Twenty Legal Pitfalls for Nurses to Avoid.* 1994. ISBN 0827361521

Kahn, Sherry, and Nileva Saulo. *Healing Yourself: A Nurse's Guide to Self Care and Renewal.* 1994. ISBN 0827361505

Keegan, Lynn. *The Nurse as Healer.* 1994. ISBN 0827361564

Lego, Suzanne. *Fear and AIDS/HIV: Empathy and Communication.* 1994. ISBN 0827361556

Sherman, Karen M. *Communication and Image in Nursing.* 1994. ISBN 0827356897

Wolf, Zane Robinson. *Medication Errors: The Nursing Experience.* 1994. ISBN 0827362625

Zawid, Carole Israeloff. *Sexual Health: A Nurse's Guide.* 1994. ISBN 0827356854

Springhouse Notes Series
Springhouse, Pa., Springhouse.

In concise outline format, volumes in this series review nursing skills and concepts. Similar to mini-review lectures on miscellaneous topics. Students use these as additional resources for review.

Benner, Margaret P. *Mental Health and Psychiatric Nursing.* 2nd ed. 1993. ISBN 0874344875

Boyd, Mildred W. *Medical-Surgical Nursing.* 2nd ed. 1993. ISBN 0874344832

Catalano, Joseph T. *Ethical and Legal Aspects of Nursing.* 1991. ISBN 0874343151

Christ, Mary Ann. *Gerontologic Nursing.* 2nd ed. 1993. ISBN 0874344883

Conrad, Lynne Hutnik. *Maternal-Neonatal Nursing.* 2nd ed. 1993. ISBN 0874344867

Eirksson, Joann Huang. *Oncologic Nursing.* 2nd ed. 1994. ISBN 0874346134

Golub, Sharon. *Nursing Fundamentals.* 2nd ed. 1994. ISBN 0874346150

Innerarity, Sheryl A. *Fluids and Electrolytes.* 2nd ed. 1994. ISBN 0874346169

Quillman, Susan M. *Nutrition and Diet Therapy.* 2nd ed. 1994. ISBN 0874346126

Regan, Joan M. *Guide to Surviving Nursing School.* 1991. ISBN 087434316X

Selekman, Janice. *Pediatric Nursing.* 2nd ed. 1993. ISBN 0874344859

Tobler, Rita. *Fundamental Nursing Procedures.* 1992. ISBN 0874343690

Vallerand, April Hazard. *Nursing Pharmacology.* 2nd ed. 1993. ISBN 0874344840

Nonprint Materials

Auscultation of Normal Breath Sounds
La Jolla, Calif., Darox Interactive, 1986; updates ongoing; 1 videodisc, 2 computer disks, and supplementary material

Reviews the standard procedures for auscultating the posterior and anterior chest using simulated chest exams. Includes information on identification of bronchial, vesicular, and bronchovascular breath sounds. *Dxter Assessment and Intervention Series.* System requirements: IBM-PC or compatible; 640K memory and 4MB space on hard drive or floppy; IBM Infowindow display; Pioneer or Sony videodisk player. Videodisc includes *Chest Tube Therapy.*

Chest Tube Therapy
La Jolla, Calif., Darox Interactive, 1986; updates ongoing, 1 videodisc, 2 computer disks, and supplementary material

This program is designed to review intrathoracic physiology and the dynamics of chest tubes. A major component of the lesson is an exercise that tests application of the lesson principles. *Dxter Assessment and Intervention Series.* Videodisc includes *Auscultation of Normal Breath Sounds.*

Emanuelsen, Kathy Lynn
Nursing Care of the Elderly Patient with Chronic Obstructive Pulmonary Disease
New York, American Journal of Nursing, 1988. 1 videodisc; 1 user guide

This program follows a seventy-three year old man with chronic obstructive pulmonary disease (COPD) and pneumonia through the emergency room admission, and on through his stay in intensive care. Students manage the care of the client by answering questions throughout the program, and they can receive a printed copy of their scores at the end of the program.

Gilman, Barbara R., Elizabeth E. Weiner, and Jeffry S. Gordon
Managing the Experience of Labor and Delivery
Chapel Hill, N.C., Health Science Consortium, 1989. 1 videodisc; user manual

This program includes many tactile experiences. The student is led through nursing assessments at every stage of labor and delivery. This program may be used when an actual delivery has not been possible in the clinical area.

I.V. Solutions
La Jolla, Calif., Darox Interactive, 1986. Updated by supplements. 1 videodisc, 2 computer disks.

This program reviews the major types of intravenous solutions. Practical application of the information is provided through the use of mini-case studies. *Dxter Assessment and Intervention Series.*

Hall, Mary Jo Larkin
Nursing Care of Elderly Patients with Acute Cardiac Disorders
New York, American Journal of Nursing Co., 1989. 1 videodisc; 1 user guide

This program refines the student's assessment and intervention skills with a cardiac patient. Students are asked to make decisions in this program which features EKG interpretation, medication regiments, and cardioversion. Psychological aspects and family interventions are discussed.

McAfooes, Julie
Intravenous Therapy
Athens, Ohio, Fuld Institute for Technology in Nursing Education, 1990. 1 videodisc; 1 integration plan guide; 1 overview/instruction sheet

Provides in-depth instructions on preparing, starting, monitoring, and discontinuing intravenous infusions. May be used in a practice situation, as an introduction or as a review of techniques.

Ethical Dilemmas and Legal Issues in Care of the Elderly
New York, American Journal of Nursing, 1990. 1 videodisc

This is an interactive video program with a touch screen. Realistic scenarios are presented including restraining issues and resuscitation legalities. Case study approach.

Mobility Series
Athens, Ohio, Fuld Institute for Technology in Nursing Education, 1992. 1 videodisc; 1 guide

Demonstrates procedures to prevent complications involved with immobility, including protective positioning, dangling, transfer, and ambulation. Includes range of motion exercises. Provides step-by-step-instructions on moving patients. This video is best suited for a skills laboratory where the student can practice along with the program.

Novoa, Jane, and others
Art of Bedside Care: ABC's of Nursing Procedures
Miami, Fla., Miami-Dade Comm. College, 1993. 21 videocassettes

This integrated package of video programs and support materials teaches basic nursing procedures. Twenty-one videocassettes are divided into areas of care: asepsis and protection, comfort, drainage, personal hygiene, mental health, mobility, and family and child care. The programs present an overview of procedures at an elementary level, gradually progressing in complexity. Includes a facilitator's manual with suggestions for classroom applications and exercises. This title was recognized as best video in 1994 by the *American Journal of Nursing*.

O'Neill, Paula N.
Nursing Care of the Cancer Patient with Compromised Immunity: The Nursing Process
Houston, Tex., UT M.D. Anderson Cancer Center, 1989. 1 videodisc; 1 study guide; 1 user's manual

The student is introduced to the patient—a sixty-year old suffering from lymphoma who has already experienced chemotherapy. Students are able to actively collect data and work with the patient using the nursing process.

Renshaw, Sharon
Medication Administration: Module I
New York, American Journal of Nursing Co., 1990. 1 videodisc; 1 manual; and 1 user guide

This program requires students to choose equipment, sites, perform calibrations, and evaluate administration techniques. The program includes eye instillation, irrigation, and application. Contains Universal Precautions measures.

Sweeney, Mary Ann
Healthcare for Older Adults: A Continuing Education Program for Nurses
Galveston, Tex., University of Texas Medical Branch, 1990. 1 videodisc; 1 booklet

This program increases the skills and knowledge needed to care for the elderly. Covers clinical skills, assessment, nursing care, and anatomy and physiology. The humanities module includes poetry, art, and exhibits on the ages of humans. The normal aging module includes age-prevalent health problems and age-related changes. Psychological, biological, and sociological theories are discussed. Demographical data on aging are provided.

Therapeutic Communication
Athens, Ohio, Fuld Institute for Technology in Nursing Education, 1990. 1 videodisc

Helps students learn the positive and negative factors that influence effective communication. Real-life scenarios assist students in the development of techniques to refine therapeutic communication skills. Includes glossary, help section, and quizzes.

Underhill, Sandra L., and Susan L. Woods
Cardiovascular Resource Videodisc
Seattle, Wash., University of Washington, Health Sciences Center for Education Resources, 1989. 1 videodisc.

An extensive two-sided videodisc containing a large collection of material that may be used to teach cardiovascular nursing. Includes 4,000 slides on anatomy, physiology, embryology, and microscopic and gross pathology; 500 slides on arrhythmias; 200 radiographs and angiograms; 50 echocardiograms; and numerous motion sequences showing patients with signs and symptoms of cardiovascular disease for better diagnosis and management.

Occupational Health and Safety

Terri Propes
Lee Davis Library
San Jacinto College Central
Pasadena, Texas

Introduction

The Occupational Health and Safety Technology Program prepares students to work in industry as occupational health and safety technologists. Using skills gained, technologists strive to improve worker conditions through safety training, elimination of possible health and safety hazards from the workplace, and regulatory compliance. Technologists also work with management to develop loss control programs, investigate and analyze workplace accidents, and to evaluate the effectiveness of health and safety performance.

Ergonomic evaluation of work sites and job functions is becoming increasingly important in the field as is the study of toxicology in dealing with hazards in the workplace. In addition, the occupational health and safety technologist is moving into the area of employee assistance programs as employers recognize the value of such programs to the safety and well-being of their employees.

Standards for Occupational Health and Safety Technology programs are issued jointly by the American Board of Industrial Hygiene and the Board of Certified Safety Professionals.

Certification

American Board of Industrial Hygiene
4600 W. Saginaw, Suite 101
Lansing, Mich. 48917-2737

Board of Certified Safety Professionals
208 Burwash Ave.
Savoy, Ill. 61874

Selected Journals

American Industrial Hygiene Association Journal.
The Association. Monthly. ISSN 0002-8894

Industrial Hygiene News.
Rimbach Pub. Co. Seven times a year. ISSN 0147-5401

Job Safety and Health Quarterly: JS and HQ.
Occupational Safety and Health Administration. Quarterly. ISSN 1057-5820

Occupational Health and Safety.
Stevens Pub. Co. Monthly. ISSN 0362-4064

Occupational Hygiene.
Gordon and Breach Science Publishers. Quarterly.
ISSN 1061-0251

Occupational Injuries and Illnesses in the United States by Industry.
U.S. G.P.O. Four times a year. ISSN 0162-010X

Occupational Safety and Health Reporter.
Bureau of National Affairs. Weekly. ISSN 0095-3237

Occupational Safety and Health Statistics of the Federal Government.
Occupational Safety and Health Administration. Annually.
ISSN 0092-8712

OSHA Week.
Stevens Pub. Co. Weekly. ISSN 1057-1485

Professional Safety.
American Society of Safety Engineers. Monthly. ISSN 0099-0027

Safety and Health.
National Safety Council. Monthly. ISSN 0891-1797

Print Materials

Accident Prevention Manual for Business and Industry
10th ed. Chicago, National Safety Council, 1992. 2 v.
ISBN 0879121556 (Vol. 1 Administration and Programs);
ISBN 0879121564 (Vol. 2 Engineering and Technology)
Comprehensive survey of current practices in the field of industrial safety and health. Volume One covers management, regulations, and program development; Volume Two discusses technical information relating to specific industries. Includes appendices, glossary, and indexes.

Anton, Thomas J.
Occupational Safety and Health Management
2nd ed. New York, McGraw-Hill, 1989. 409p.
ISBN 0070021082
An introduction to the principles and practices of safety and health management, which also covers the history of the safety movement to hazardous waste management. Appendices and index included.

Apts, David W.
Back Injury Prevention Handbook
Boca Raton, Fla., Lewis Publishers, 1992. 186p.
ISBN 0873714199
Written by a physical therapist, this book discusses back injury prevention techniques that have proven effective in industry.

Bayer, Ronald, ed.
Health and Safety of Workers: Case Studies in the Politics of Professional Responsibility
New York, Oxford Univ. Press, 1988. 308p. ISBN 0195053656
Examination of public health policy relevant to three occupational hazards: lead poisoning, coal dust, and asbestosis. Good historical perspective on the relationship of health policy to industrial hazards.

Best's Safety Directory: Industrial Safety, Hygiene, Security
Oldwick, N.J., A. M. Best Co., Annual. 1 v. (various pagings)
ISSN 0090-7480
Seven sections, each consisting of OSHA summaries by standard number and paragraph order; safety, training and technology articles; and a buyer's guide. Indexed by company name, brand name, product name, distributors by geographical region, and advertisers' branch offices.

Blosser, Fred
Primer on Occupational Safety and Health
Washington, D.C., Bureau of National Affairs, 1992. 363p.
ISBN 0871797410
This primer on occupational safety and health regulation provides an introduction to current government programs especially OSHA and the Occupational Safety and Health Review Commission. Includes glossary, bibliography, and index.

Bokat, Stephen A., and Horace A. Thompson, eds.
Occupational Safety and Health Law
Washington, D.C., Bureau of National Affairs, 1988. 988p.
ISBN 0871795272
Comprehensive overview of OSHA law. Produced by the American Bar Association Section of Labor and Employment Law. Updated by supplements.

Brauer, Roger L.
Safety and Health for Engineers
New York, Van Nostrand Reinhold, 1990. 651p.
ISBN 0442211252
Desk reference for safety practitioners. Chapters end with classroom exercises, review questions and bibliography. Suitable for introductory course in industrial hygiene. Appendices and index included.

Burke, Mike
Applied Ergonomics Handbook
Boca Raton, Fla., Lewis Publishers, 1992. 258p.
ISBN 0873713672
Discusses prevention of musculoskeletal injury through the application of ergonomic principles to job and site analysis. Intended for professionals whose primary responsibilities are not in ergonomics. Appendices and index provided.

Campbell, Reginald L., and Roland E. Langford
Fundamentals of Hazardous Materials Incidents
Boca Raton, Fla., Lewis Publishers, 1991. 449p.
ISBN 0873713621
Training manual for handling hazardous material (HAZMAT) incidents. Identification of possible problems at hazardous waste sites and protective clothing are discussed.

Chaffin, Don B., and Gunnar B. Andersson
Occupational Biomechanics
2nd ed. New York, Wiley, 1991. 518p. ISBN 0471601349
Applies occupational biomechanics to ergonomics. The focus of this work is on the design of working environments, considering body features and task requirements. Includes appendices and index.

Chemical Safety Data Sheets
Cambridge, England, Royal Society of Chemistry, 1989-1992. 5 v.
ISBN 0851869033 (Vol. 1); ISBN 0851869130 (Vol. 2);
ISBN 0851869230 (Vol. 3); ISBN 0851863116 (Vol. 4);
ISBN 0851863213 (Vol. 5)
Substances discussed in this set include solvents, main group metals and their compounds, corrosives and irritants, toxic chemicals, and flammable chemicals. Data sheets list chemical name, physical properties, risks, safety precautions, packaging, transportation, handling and storage, chemical hazards, and fire precautions.

Cheremisinoff, Paul
Hazardous Materials Emergency Response: Pocket Handbook
Lancaster, Pa., Technomic Pub. Co., 1989. 161p.
ISBN 0877626316
Presents information necessary to perform hazardous waste cleanup within OSHA and EPA guidelines with minimal risk to personnel. Includes bibliography, appendices, and index.

Clayton, George D., and Florence E. Clayton, eds.
Patty's Industrial Hygiene and Toxicology
4th ed. New York, Wiley, 1991- . 2199p. ISBN 0471501972 (Vol. 1, part A); ISBN 0471501964 (Vol. 1, part B)

    Encyclopedic coverage of industrial hygiene. The previous edition has been expanded from twenty-seven to forty-three chapters reflecting the growing number of subjects in the field. Each chapter is written by an expert in the area.

Colvin, Raymond
Guidebook to Successful Safety Programming
Boca Raton, Fla., Lewis Publishers, 1992. 292p.
ISBN 0873714814

    Complete guide to establishing a safety program. Includes a range of information such as writing guidelines that conform to OSHA standards and off-site safety requirements. Includes case histories of successful programs and an index.

Confer, Robert G.
Workplace Health Protection: Industrial Hygiene Program Guide
Boca Raton, Fla., Lewis Publishers, 1994. 541p.
ISBN 0873713877

    Guidelines for implementing and evaluating industrial hygiene programs based on established practice and standards.

Confer, Robert G., and Thomas R. Confer, eds.
Occupational Health and Safety Terms, Definitions, and Abbreviations
Boca Raton, Fla., Lewis Publishers, 1994. 213p.
ISBN 1566700779

    Dictionary of commonly used terms and abbreviations found in journals, books, and manufacturers' literature in the field of occupational health and safety. Also includes terms and abbreviations used in government regulatory enforcement.

Corn, Jacqueline K.
Response to Occupational Health Hazards: A Historical Perspective
New York, Van Nostrand Reinhold, 1992. 182p.
ISBN 0442004885

    Explication of how social factors have had an impact on attitudes and definitions of occupational hazards. Five case studies show the relationships among public perceptions, governmental action, and technology in the history of occupational health.

Ellis, J. Nigel
Introduction to Fall Protection
Des Plaines, Ill., American Society of Safety Engineers, 1988. 141p. ISBN 0939874822

    Defines and analyzes fall hazards and the methods of fall prevention. Examines systems, training, equipment, and cost effectiveness.

Ferry, Ted S.
Modern Accident Investigation and Analysis
2nd ed. New York, Wiley, 1988. 306p. ISBN 0471624810

    Thorough introduction to the concepts and practice of accident investigation. Stresses preparation and planning, analysis, reporting, and corrective action. Appendices and index included.

Finucane, Edward W.
Definitions, Conversions, and Calculations for Occupational Safety and Health Professionals
Boca Raton, Fla., Lewis Publishers, 1993. 1 v. (various pagings)
ISBN 0873718631

    Covers virtually every definition, conversion factor, and mathematical formula likely to be encountered in the field. These sections are followed by problems and their solutions.

Fire Department Occupational Safety
2nd ed. Stillwater, Okla., Fire Protection Publications, 1991. 366p.
ISBN 0879390972

    Based on the *NFPA 1500, Standard on Fire Department Occupational Safety and Health Program*. Addresses how to set up and implement a safety program, equipment and tool maintenance, fitness and health requirements, protective equipment, training, and hazard responses.

Freedman, Warren
Law and Occupational Injury, Disease, and Death
New York, Quorum Books, 1990. 192p. ISBN 0899304109

    Covers laws and litigation pertaining to occupational injury, disease, or death. Types of work-related injury or disease are identified, as well as procedures for pursuing or defending a claim. Includes a table of cases cited and index.

Godish, Thad
Indoor Air Pollution Control
Chelsea, Mich., Lewis Publishers, 1989. 401p. ISBN 0873710983

    An overview of the literature on air pollution control followed by sections on conducting air quality investigations and methods of improving indoor air quality. Case histories are provided.

Gots, Ronald E.
Toxic Risks: Science, Regulation, and Perception
Boca Raton, Fla., Lewis Publishers, 1993. 277p.
ISBN 0873715101

    Discusses public perceptions of toxic risks and how these relate to public health concerns, policy, and scientific facts. Each chapter examines a different aspect of toxic risk and concludes with references for further reading. Includes index.

Hall, Stephen K.
Chemical Safety in the Laboratory
Boca Raton, Fla., Lewis Publishers, 1994. 242p.
ISBN 0873718968

    Chemical laboratory safety for hospitals, educational institutions and industry. The *OSHA Laboratory Standard* and the *Chemical Hygiene Plan* are discussed as well as the use of protective clothing, and emergency planning and response. Appendices and index included.

Hansen, J. Doan, ed.
Work Environment
Chelsea, Mich., Lewis Publishers, 1991-1994. 3 v.
ISBN 0873713036 (Vol. 1); ISBN 0873713923 (Vol. 2); ISBN 0873713931 (Vol. 3)

    Introduction to the basics of occupational health. Volume One covers evaluating workplace hazards, occupational health standards, controlling hazards in the work environment, personal protective equipment, and training. Volume Two addresses health care, laboratories, and biosafety. The third volume includes information on methods of identifying and correcting indoor air quality problems. Building ventilation, indoor air quality surveys, sick building syndrome, and noise abatement are examined with practical solutions outlined. Includes glossary and index.

Health and Safety in Small Industry: A Practical Guide for Managers
Boca Raton, Fla., Lewis Publishers, 1989. 204p.
ISBN 0873711955

Developed by the University of Medicine and Dentistry of New Jersey, Department of Environmental and Community Medicine. Discusses the establishment of a safety program for small companies including general principles and procedures. Five appendices list resources for assistance in establishing a program and complying with laws and regulations.

Heath, Earl D., and Ted S. Ferry
Training in the Workplace: Strategies for Improved Safety and Performance
Goshen, N.Y., Aloray, 1990. 217p. ISBN 0913690198

Practical guide to planning and implementing a safety training program. Chapters contain itemized checklists to assist in completing tasks.

Hernberg, Sven
Introduction to Occupational Epidemiology
Chelsea, Mich., Lewis Publishers, 1992. 230p. ISBN 0873716361

Introduction to occupational epidemiology for those in the fields of occupational hygiene, public health, and ergonomics. Important definitions are printed in bold type and numerous examples, charts, and graphs enhance the text. Includes chapter references and index.

Howard, Philip H., ed.
Handbook of Environmental Fate and Exposure Data for Organic Chemicals
Chelsea, Mich., Lewis Publishers, 1989-1993. 4 v.

Detailed information on the release, transportation, and degradation of chemicals in the environment and their exposure to humans and other organisms. Entries are arranged in four sections: substance identification, chemical and physical properties, environmental fate/exposure potential, and references. Includes cumulative indexes by synonyms, CAS (Chemical Abstracts Service) registry number, and chemical formula.

Kase, Donald W., and Kay J. Wiese
Safety Auditing: A Management Tool
New York, Van Nostrand Reinhold, 1990. 318p.
ISBN 0442237464

Discusses interrelationships of safety auditors and company representatives including expectations, planning, preparation, and conducting an effective safety audit.

Kletz, Trevor A.
Learning from Accidents in Industry
Boston, Butterworth-Heinemann, 1988. 158p. ISBN 0408026960

Thorough explanation of the lessons to be learned from accidents and near-misses in the workplace. Minor and major incidents are used to illustrate investigation. Identifies causal factors and methods to eliminate them. Appendices and index included.

Kohn, James P.
Behavioral Engineering Through Safety Training: The B.E.S.T. Approach
Springfield, Ill., Charles C. Thomas, 1988. 189p.
ISBN 0398054347

Discusses the use of training programs to improve occupational health and safety. Chapters begin with objectives and end with a list of references.

Konz, Stephen A.
Work Design: Industrial Ergonomics
Worthington, Ohio, Publishing Horizons, 1990. 543p.
ISBN 0942280555

A basic ergonomic primer, this book utilizes a magazine format rather than standard textbook layout. Intended as a study guide; an instructor's manual accompanies the text.

Lauwerys, Robert R., and Perrine Hoet
Industrial Chemical Exposure: Guidelines for Biological Monitoring
2nd ed. Boca Raton, Fla., Lewis Publishers, 1993. 318p.
ISBN 0873716507

Presents a brief introduction to the principles and types of biological monitoring, followed by entries on specific substances grouped by types of chemicals. Includes index.

Lewis, Richard J., Sr.
Hazardous Chemicals Desk Reference
3rd ed. New York, Van Nostrand Reinhold, 1993. 1742p.
ISBN 0442014082

Entries extracted from *Sax's Dangerous Properties of Industrial Materials*. General chemical entries have been updated.

Lewis, Richard J., Sr.
Sax's Dangerous Properties of Industrial Materials
8th ed. New York, Van Nostrand Reinhold, 1992. 3 v.
ISBN 0442011326 (set)

The first volume of this set contains a CAS Registry synonym cross-index and an extensive bibliography. Synonyms are listed in English, French, German, Dutch, Polish, Japanese, and Italian. The second and third volumes provide general chemical entries that list physical properties, animal and human toxicology, exposure limits, Department of Transportation (DOT) classification, and hazard review. Update available for 1993 (ISBN 0442016751).

Lu, Frank C., Sr.
Basic Toxicology: Fundamentals, Target Organs, and Risk Assessment
2nd ed. New York, Hemisphere Pub. Corp., 1991. 361p.
ISBN 089116894X

Includes an introduction to toxicology, toxicology of human organ systems, and the fundamentals of risk assessment. Chapter appendices and references included.

Moeller, D. W.
Environmental Health
Cambridge, Mass., Harvard Univ. Press, 1992. 332p.
ISBN 0674258584

An overview of the concepts of environmental health including those of the workplace and injury prevention. Liberally illustrated with charts and graphs, and each chapter ends with references for further reading. Includes index.

Parker, Kathryn G., and Harold R. Imbus
Cumulative Trauma Disorders: Current Issues and Ergonomic Solutions: A Systems Approach
Boca Raton, Fla., Lewis Publishers, 1992. 144p.
ISBN 0873713222

Presents ergonomic programs to eliminate cumulative trauma disorders, especially disorders of the upper extremities. Discusses OSHA guidelines, workstation design, and plant programs.

Petersen, Dan
Safety Management: A Human Approach
2nd ed. Goshen, N.Y., Aloray, 1988. 380p. ISBN 0913690120

Approaches safety management from a behavioral sciences point of view. Human factors that contribute to or decrease accidents are examined and ways to integrate them into an effective safety management program are reviewed. Appendices and index included.

Petersen, Jack E.
Industrial Health
2nd rev. ed. Edited by Leonard Wilcox. Cincinnati, Ohio, American Conference of Governmental Industrial Hygienists, 1991. 350p. ISBN 0936712910

Survey of basic industrial health and hygiene. Chapter summaries have been added to the original text.

Plog, Barbara A., ed.
Fundamentals of Industrial Hygiene
3rd ed. Chicago, National Safety Council, 1988. 915p.
ISBN 0879120827

Text for an industrial hygiene fundamentals course that is also intended as a self-instructional text for those in related professions such as safety and occupational medicine. Includes appendices and index.

Proctor, Nick H.
Proctor and Hughes' Chemical Hazards of the Workplace
3rd ed. Edited by Gloria J. Hathaway, and others. New York, Van Nostrand Reinhold, 1991. 666p. ISBN 0442004559

Summary of the toxicological and health effects of 545 commonly used chemicals. Emphasis is on signs and symptoms of overexposure. Appendices and indexes included.

Railton, W. Scott
OSHA Compliance Handbook
Rockville, Md., Government Institutes, 1992. 364p.
ISBN 0865872902

Overview of the *Occupational Safety and Health Act* for health and safety professionals. Case citations are used liberally to illustrate aspects of standards and compliance. The text of the act is included as an appendix.

Rekus, John
Complete Confined Spaces Handbook
Boca Raton, Fla., Lewis Publishers, 1995 (estimated publication date). 400p. ISBN 0873714873

Recommended procedures for entry into confined spaces. Case histories demonstrate the elements of a confined space entry program. Includes OSHA standards, glossary, and sources.

Rouf, A., and B. S. Dhillon
Safety Assessment: A Quantitative Approach
Boca Raton, Fla., Lewis Publishers, 1994. 191p.
ISBN 0873716752

Survey of the current state of quantitative safety assessment. Analysis, behavior sampling, unsafe conditions, factors contributing to accidents, and automation and safety are considered as components of assessment. Appendices included.

Royster, Judith Doswell, and Larry Royster
Hearing Conservation Programs: Practical Guidelines for Success
Chelsea, Mich., Lewis Publishers, 1990. 120p. ISBN 0873713079

Outlines the essential elements of a successful hearing conservation program. Based on practical experience and implementation of programs in industry.

Rudman, Jack
Industrial Hygienist Trainee
Syosset, N.Y., National Learning Corp., 1988. 1 v. (unpaged)
ISBN 0837330351

A series of test questions and answers for employment, licensure, or certification tests. Part of the *Passbook for Career Opportunities* series.

Schaaf, T. W. van der, D. A. Lucas, and A. R. Hale, eds.
Near Miss Reporting as a Safety Tool
Oxford, Butterworth-Heinemann, 1991. 151p. ISBN 0750611782

Describes the methods and uses of a near-miss reporting system. Organized in two sections, theoretical and case studies, this volume illustrates the value of near-miss reporting to a comprehensive safety program.

Stramler, James H.
Dictionary for Human Factors/Ergonomics
Boca Raton, Fla., CRC Press, 1993. 413p. ISBN 0849342368

Provides basic terminology in the field of ergonomics. Many definitions are supplemented with examples of usage. Includes acronyms and abbreviations with an alphabetical listing of cross-references to definitions.

Thomen, James R.
Leadership in Safety Management
New York, Wiley, 1991. 378p. ISBN 0471533262

Management-oriented guide to safety programs. Focuses on employee behavior as the major cause of industrial accidents. Presents the elements of a safety program with practical examples. Index included.

U.S. Department of Energy. Office of Environment, Safety, and Health.
Dictionary and Thesaurus of Environment, Health, and Safety
Boca Raton, Fla., Lewis Publishers, 1992. 510p.
ISBN 0873718763

Appendices include thesaurus acronyms, subject listings, and a separate shorter thesaurus for DOE vocabulary used on the Safety Performance Measurement System.

Young, Jay A., ed.
Improving Safety in the Chemical Laboratory: A Practical Guide
2nd ed. New York, Wiley, 1991. 406p. ISBN 0471530360
 This thorough guide to laboratory safety focuses on how to identify close calls and correct problems, train employees, and develop a chemical hygiene plan. Includes an index and extensive appendices on controlling hazards, regulatory requirements, and resource materials.

Occupational Therapy Assisting

Nancy Seamans
College of Health Sciences
Community Hospital of Roanoke Valley
Roanoke, Virginia

Introduction

Occupational therapy is a branch of rehabilitative medicine providing services to persons with impaired function due to physical illness or injury, congenital or developmental disorders, the aging process, or emotional disability. Professionals in occupational therapy perform evaluation and treatment. The purpose of rehabilitation is to optimize the functional abilities, prevent disability, and maintain the physical and mental health of patients. Individuals are assisted in achieving the highest level of functioning, to become self-reliant and balance all aspects of their daily lives.

Occupational therapy assistant (OTA) education provides students the opportunity to become certified as an occupational therapy assistant or COTA. COTAs work under the supervision of a registered occupational therapist (OTR). According to the U.S. Department of Labor, the projected 1990-2005 employment potential for occupational therapy assistants is expected to grow much faster than average, with an increase of 35 percent or more in positions.

Services provided by COTAs include evaluation of skills and capacities to perform daily functions; instruction in self-care, homemaking, social, and crafts skills; and maintenance of records and supplies. Treatment procedures are carried out in a wide variety of clinical settings, such as hospitals, clinics, long-term and custodial care facilities, and school districts. With appropriate supervision, COTAs also work in home-care settings.

COTA education consists of the completion of either a two-year associate degree program, or a one-year certificate program. Programs are available in both two- and four-year colleges, and universities; and they include academic, laboratory, clinical, and fieldwork components. Upon successful completion of an accredited program, graduates are eligible to take a certification examination administered by the American Occupational Therapy Association. Certification is required to practice as a COTA in most states.

This bibliography is intended for use by librarians as well as by faculty and students at institutions with OTA programs. There is limited availability of materials written specifically for OTA students or COTAs; however, much of the material referenced here is written for OTs or OTRs but has applicability to the OTA's training.

Occupational Therapy Assisting · Allied Health

Accreditation

American Occupational Therapy Association
1383 Picard Drive
P.O. Box 1725
Rockville, Md. 20849-1725

American Occupational Therapy Certification Board
4 Research Place, Suite 160
Rockville, Md. 20850-3226

Selected Journals

American Journal of Occupational Therapy: Official Publication of the American Occupational Therapy Association.
The Association. Monthly. ISSN 0272-9490

British Journal of Occupational Therapy.
College of Occupational Therapists. Monthly. ISSN 0308-0226

Canadian Journal of Occupational Therapy/Revue Canadienne d'Ergotherapie.
Canadian Association of Occupational Therapists. Five times per year. ISSN 0008-4174

Journal of Occupational Therapy Students: JOTS.
American Occupational Therapy Association. Quarterly. ISSN 1069-3262

OT Week.
American Occupational Therapy Association. Weekly. ISSN 0893-1712

Print Materials

Acquaviva, Jane D., ed.
Effective Documentation for Occupational Therapy
Rockville, Md., American Occupational Therapy Association, 1992. 293p. ISBN 0910317852

Clear and concise documentation, for purposes of both treatment and financial reimbursement is essential in occupational therapy. This volume provides guidelines on why and how to document all phases of treatment, extensive samples of forms and notes, information on how documentation is used, and the legal and ethical justifications for clarity in documentation.

Allen, Claudia Kay, Catherine A. Earhart, and Tina Blue
Occupational Treatment Goals for the Physically and Cognitively Disabled
Rockville, Md., American Occupational Therapy Association, 1992. 350p. ISBN 0910317720

Designed to assist the therapist in setting realistic, achievable treatment goals for the patient with physical and cognitive impairment. The lengthy section on activities analysis is particularly useful for the OTA student or practitioner. A study guide is also available.

Angle, Deborah K., and Julie M. Buxton
Community Living Skills Workbook for the Head Injured Adult
Gaithersburg, Md., Aspen Publishers, 1991. 361p.
ISBN 0834202107

This work focuses on assisting the traumatically brain injured (TBI) in the return to work or school. The training activities include a wide array of community reentry skills, supported by information on the role of occupational therapy. The emphasis is on practical information and guidelines for treatment.

Aquaro, Marianne and others
OT GOALS: Occupational Therapy Goals and Objectives Associated with Learning
Tucson, Ariz., Therapy Skill Builders, 1992. 185p.
ISBN 0884505448

Intended for use by the therapist working with a pediatric population in a variety of settings. Though it provides detailed guidelines for developing skills, the text also allows for therapist creativity in implementing strategies to enable children to realize specific goals and derive maximum benefit from occupational therapy sessions.

Bobath, Berta
Adult Hemiplegia: Evaluation and Treatment
3rd ed. Oxford, England, Butterworth-Heinemann, 1990. 190p.
ISBN 075060168X

Third edition of a classic work by a classic author. The Bobath Method for the treatment of individuals with paralysis is well-known in therapy circles. Bobath provides extensive guidelines for the treatment of the adult patient with hemiplegia, with emphasis on patient participation and involvement. This work is particularly useful when demonstrating how hemiplegia treatment has changed during the past several decades.

Bonder, Bette R.
Psychopathology and Function: A Guide for Occupational Therapists
Thorofare, N.J., Slack, 1991. 213p. ISBN 1556420765

Provides a foundation for understanding psychiatric diagnoses and for explaining what occupational therapists must know to work effectively with other mental health care providers. Sections on OT functions and treatment are brief but adequate. The volume's strength lies in the information presented on a variety of psychiatric disorders.

Case-Smith, Jane, ed.
Pediatric Occupational Therapy and Early Intervention
Boston, Andover Medical Pub., 1993. 389p.
ISBN 1563720264

Addresses the needs of the occupational therapist dealing with children, with emphasis on early intervention. Includes substantial information on how the families can be involved in pediatric therapy. The first five chapters are particularly beneficial for the COTA or OTA student.

COTA Supervision Information Packet
Rockville, Md., American Occupational Therapy Association, 1991. 154p. ISBN 0910317755

This reference document for supervisors of occupational therapy assistants outlines the responsibilities of the COTA. Included are discussions of role delineation, practice standards, and legal guidelines for the COTA and COTA supervisor. The questions-and-answers section is of particular benefit to the first-time supervisor.

Christiansen, Charles, and Carolyn Baum, eds.
Occupational Therapy: Overcoming Human Performance Deficits
Thorofare, N.J., Slack, 1991. 884p. ISBN 155642180X

This textbook provides the occupational therapy student with an understanding of human performance, and the roles of occupational therapy in identifying and ameliorating performance deficits. Though not an introductory text, it is a valuable and fundamental work in this field, and is well-organized and indexed.

Christiansen, Charles, ed.
Ways of Living: Self-Care Strategies for Special Needs
Rockville, Md., American Occupational Therapy Association, 1994. 495p. ISBN 1569000085

Defines self-care as a fundamental aspect of activities of daily living and tries to create a holistic view for understanding self-care interventions. This is a well-organized work with good references, illustrations, and excellent examples. The sections on assessing and teaching self-care skills are particularly well written for the student. The chapter on the self-care environment is both practical and thought-provoking.

Daniel, Marilyn S., and L. Randy Strickland
Occupational Therapy Protocol Management in Adult Physical Dysfunction
Gaithersburg, Md., Aspen Publishers, 1992. 430p. ISBN 0834203146

Addressed to the student or new practitioner, this volume provides guidelines for clinical reasoning. Discusses evaluation tools and provides assessment forms for a wide range of adult physical dysfunctions. Treatment procedures and discharge preparation guidelines are included; it is concisely written and easy to follow.

Davis, Linda J., and Martha Kirkland, eds.
The Role of Occupational Therapy with the Elderly
Rockville, Md., American Occupational Therapy Association, 1988. 449p. ISBN 0910317194

The aging of the American population is making gerontological issues an increasingly significant issue in health care education and practice. This volume, though written by and for occupational therapists, is comprehensive enough to be of value to any practitioner dealing with geriatric patients. It is designed as a foundation for education in geriatrics, and its modular format is well conceived and informative.

Diamant, Rachel B.
Positioning for Play: Home Activities for Parents of Young Children
Tucson, Ariz., Therapy Skill Builders, 1992. 213p. ISBN 0884504840

The activities in this volume are clearly illustrated and explained, and are designed to be given to parents by the therapist working with young children. The illustrations and instructions that make this a good tool for the practitioner also make it a valuable tool for OTA students.

Drake, Margaret
Crafts in Therapy and Rehabilitation
Thorofare, N.J., Slack, 1992. 211p. ISBN 1556421184

This illustrated work explains why crafts are important in occupational therapy treatment, and where and how they may best be used. Part Two explains specific crafts, including the appropriate use of each, supplies needed, therapeutic applications, and case studies. Discussion questions at the end of each unit provide additional skill-development opportunities for OT and OTA students and faculty.

Dunn, Winnie, ed.
Pediatric Occupational Therapy: Facilitating Effective Service Provision
Thorofare, N.J., Slack, 1991. 401p. ISBN 1556420145

Occupational therapy strategies for children are provided in a variety of settings. Dunn includes information for the therapist working in schools and in the community, as well as in traditional health care settings. Viewing pediatric treatment holistically, the author provides a team approach which includes the occupational therapist as part of this whole.

Glickstein, Joan K.
Therapeutic Interventions in Alzheimer's Disease: A Program of Functional Communication Skills for Activities of Daily Living
Rockville, Md., Aspen Pub., 1988. 197p. ISBN 0871898896

This volume is a particularly valuable resource for the therapist dealing with both the victim of Alzheimer's and with the patient's caregivers. A selection of brief lessons addresses daily living activities, with suggested interventions for both parties. Unit Eleven is a practical guide for helping the client and offers a list of sources for information on Alzheimer's disease and aging.

Greenberg, Naomi Schubin
Occupational Therapy Assistant Career Profile
St. Louis, Mo., Warren H. Green, 1990. 236p. ISBN 875272703

Information in this volume is useful to anyone considering a career in assistant-level occupational therapy, or to a health care provider interested in using the services of COTAs. Includes information about education, licensure, and credentials. Also discusses the varied areas of practice wherein the use of COTAs is particularly beneficial.

Hemphill, Barbara J., Cindee Quake Peterson, and Pamela Carr Werner
Rehabilitation in Mental Health: Goals and Objectives for Independent Living
Thorofare, N.J., Slack, 1992. 160p. ISBN 1556421435

The assessment and development of independent living skills for the mental health patient is the focus of this book. Discussion of preparation of individualized treatment plans is particularly informative, as is the information on documentation and goals measurement. Much of the information is presented in tabular format. Suggestions at the end of each chapter provide additional treatment alternatives.

Hertfelder, Sarah, and Carol H. Gwin, eds.
Work in Progress: Occupational Therapy in Work Programs
Rockville, Md., American Occupational Therapy Association, 1989. 291p. ISBN 0910317542

This is a review of work programming, an area of occupational therapy practice that the editors claim is experiencing a "growth spurt." Various authors discuss such occupational therapy topics as ergonomics, vocational rehabilitation, and health promotion in the workplace. Includes an introduction to the history of work.

Hirama, Haru
Occupational Therapy Assistant: A Primer
Rev. ed. Baltimore, Md., Chess Pubs., 1990. 297p.
ISBN 093527300X

A fundamental volume in OTA education and practice, this work efficiently provides basic information on everything the OTA student or practitioner needs to know. Included is the history of the OTA profession, general and specific skills needed, and practice settings.

Hopkins, Helen L., and Helen D. Smith
Willard and Spackman's Occupational Therapy
8th ed. Philadelphia, Lippincott, 1993. 948p. ISBN 039754877X

One of the "bibles" of occupational therapy, this eighth edition is a basic, comprehensive text for both the OT and the OTA. Includes chapters by sixty-five different contributors (including one OTA) from throughout the United States. This volume should be part of any OTA library.

Kiernat, Jean M., ed.
Occupational Therapy and the Older Adult: A Clinical Manual
Gaithersburg, Md., Aspen Publishers, 1991. 361p.
ISBN 0834202395

Takes a "whole person" approach to geriatric occupational therapy, with guidelines for recognizing the connection among all of the needs of the elderly. The chapter on culturally diverse elders is particularly interesting. Also noteworthy is the discussion of occupational therapy in community-based programs, including adult day care and adult home care.

Kovich, Karen M., and Diane E. Bermann
Head Injury: A Guide to Functional Outcomes in Occupational Therapy
Rockville, Md., Aspen Publishers, 1988. 236p.
ISBN 0871897628

One in an excellent series from the Rehabilitation Institute of Chicago, this volume comprehensively covers occupational therapy treatment for the head-injured patient. It is specifically written for students and therapists relatively unfamiliar with head injuries.

Lamport, Nancy K., Margaret S. Coffey, and Gayle Ilene Hersch
Activity Analysis Handbook
2nd ed. Thorofare, N.J., Slack, 1993. 196p.
ISBN 1556422156

Margaret Coffey is a certified occupational therapy assistant and her input makes the book particularly useful for the OTA student. This second edition excels in the presentation of guidelines for examining activities, for determining their therapeutic uses, and for identifying and recording outcomes. Includes student worksheets in an appendix.

Levine, Kristin Johnson
Fine Motor Dysfunction: Therapeutic Strategies in the Classroom
Tucson, Ariz., Therapy Skill Builders, 1991. 593p.
ISBN 0884503925

This volume provides the therapist with guidelines for assisting the child with any of the dysfunctions of the central nervous system, including hyperactivity, cerebral palsy, and other developmental disabilities. The intent of the activities presented are to allow the child to function in a classroom setting.

McCarthy, Karen
Activities of Daily Living: A Manual of Group Activities and Written Exercises
Framingham, Mass., Therapro, Inc., 1993. 128p.

This volume, written by a certified occupational therapy assistant, provides reproducible checklists, worksheets, activities, and plans, including discussion suggestions for use with a population needing assistance with daily living activities. Extremely simple and straight forward, but with practical guidelines for activities of daily living (ADL), this book is geared toward the COTA practitioner.

Mann, William C., and Joseph P. Lane
Assistive Technology for Persons with Disabilities: The Role of Occupational Therapy
Rockville, Md., American Occupational Therapy Association, 1991. 287p. ISBN 0910317712

Beginning with a comprehensive explanation of assistive technology, this volume covers the role of technology in improving the functioning of the disabled. The topic is particularly important as occupational therapists work with both people and technology to research, develop and maintain tools to assist those with a wide range of disabilities. Includes a discussion of the interaction with other professions as well as with the potential consumer.

Melvin, Jeanne L.
Rheumatic Disease in the Adult and Child: Occupational Therapy and Rehabilitation
3rd ed. Philadelphia, Davis, 1989. 607p. ISBN 0803661371

An increasing part of the occupational therapist's responsibility is therapy for patients with rheumatic diseases (such as arthritis, fibromyalgia, lupus, carpal tunnel syndrome, etc.) This comprehensive volume addresses all aspects of rheumatic diseases, drug therapies, assessment (with emphasis on how to assess damaged joints), activities of daily living and rehabilitation. Part Two addresses major rheumatic diseases and provides concise information on etiology, prognosis and therapy. This work also outlines the role of the OT in their treatment.

Moran, Michael L., and Mark A. Brimer
Computer Principles for Physical and Occupational Therapists
Tucson, Ariz., Therapy Skill Builders, 1994. 213p.
ISBN 0884500330

This book focuses on the increasing importance of computers in health care and seeks to provide basic computer information for occupational and physical therapists who are novices in computer basics. Though probably too basic for those who use computers regularly, there are useful sections on computer use in one's practices, and on connecting to and using on-line services.

Occupational Therapy Roles Task Force
Occupational Therapy Roles and Career Exploration and Development: A Companion Guide to the Occupational Therapy Roles Document
Rockville, Md., American Occupational Therapy Association, 1994. 1 v. (various pagings) ISBN 1569000131

Provides definitions of a variety of occupational therapy roles. Included are major functions, scope, key performance areas, qualifications and supervisory responsibilities for OTRs, COTAs, and occupational therapy educators. The companion guide outlines professional growth and development opportunities for all areas of practice. An extremely useful and concise volume for all involved in occupational therapy.

Payton, Otto D., Mark N. Ozer, and Craig E. Nelson
Patient Participation in Program Planning: A Manual for Therapists
Philadelphia, Davis, 1990. 97p. ISBN 0803668031

Written for both occupational and physical therapy students, this book is based on the assumption that a patient will be more likely to achieve goals if involved in setting them. Highly structured to ensure concept integration in practice, this text includes information on patient involvement, including planning, evaluation and implementation.

Pedretti, Lorraine Williams and Barbara Zoltan
Occupational Therapy: Practice Skills for Physical Dysfunction
3rd ed. St. Louis, Mo., Mosby, 1990. 690p. ISBN 0801638526

Though the preface to the third edition states that it is designed for students in baccalaureate and master's degree programs, it is an important work for the OTA collection as well. It provides a process approach to occupational therapy treatment of the dysfunctional adult patient, and is comprehensive and balanced without being excessively prescriptive.

Pratt, Pat Nuse and Anne S. Allen, compilers
Occupational Therapy for Children
2nd ed. St. Louis, Mo., Mosby, 1989. 656p. ISBN 0801624665

A valuable textbook resource for the OTA and OT student as well as for the practitioner working with pediatric patients. The contributors recognize the need to work with children with a variety of problems in diverse settings. The result is a well-written volume that addresses the needs of the pediatric OT patient.

Reed, Kathlyn L.
Quick Reference to Occupational Therapy
Gaithersburg, Md., Aspen Publishers, 1991. 619p. ISBN 0834202379

Provides an overview of virtually all disorders encountered by the occupational therapist. Discussion is divided into sections by major areas of disorders. Information on specific problems includes description, cause, assessment, problems, treatment/management, precautions, and prognosis and outcome. A quick reference tool with useful appendices and a thorough index.

Ross, Mildred
Integrative Group Therapy: The Structured Five-Stage Approach
2nd ed. Thorofare, N.J., Slack, 1991. 172p.
ISBN 1556420838

Intended for use with long-term care patients, this second edition provides updated theories and techniques that have evolved since publication of the first edition. The information is intended to be used with difficult-to-reach populations who do not typically respond to group therapy situations, and can be used with a range of ages.

Ryan, Sally E., ed.
Practice Issues in Occupational Therapy: Intraprofessional Team Building
Thorofare, N.J., Slack, 1993. 376p. ISBN 1556421796

Some may find this volume slightly prescriptive, but its emphasis on the team approach to treatment makes it valuable. Ryan is a COTA, as are a number of the contributing authors, and the role of the COTA is well represented, making this volume useful to the OTA student and practitioner.

Sheda, Constance and Christine Small
Developmental Motor Activities for Therapy: Instruction Sheets for Children
Tucson, Ariz., Therapy Skill Builders, 1990. 1 v. (various pagings) ISBN 0884504522

These instruction sheets are designed to illustrate methods to appropriately position physically disabled children while providing treatment. Simply drawn and basic, they would be helpful to a student or therapist just starting to work with a pediatric population. They could also be duplicated for use by family members.

Thomson, Linda Kohlman
Kohlman Evaluation of Living Skills: KELS
3rd ed. Rockville, Md., American Occupational Therapy Association, 1992. 72p. ISBN 0910317909

KELS provides the occupational therapist with a means of efficiently evaluating a patient's basic living skills. Though originally established for use with the psychiatric patient, it may also be used with other clients. Written in clear language, it is appropriate for the OT or OTA student, as well as the practitioner.

Toglia, Joan P., and Kathleen M. Golisz
Cognitive Rehabilitation: Group Games and Activities
Tucson, Ariz., Therapy Skill Builders, 1990. 383p.
ISBN 0884505634

The five sections of this volume provide a framework for group activities for those working with the cognitively impaired. The first section provides necessary fundamental information; the remainder of the book addresses specifics of group interaction, including assessment. Though the book is written for use by any cognitive rehabilitation professional, the emphasis on group activities would be of particular interest to the OT and OTA.

Wilson, A. Bennett
Wheelchairs: A Prescription Guide
2nd ed. New York, Demos Publications, 1992. 81p.
ISBN 0939957396

This brief volume describes when and where wheelchairs are indicated and how to make the necessary decisions for appropriate prescribing. Describes and illustrates the various types of chairs and their components. Includes a list of suppliers.

Nonprint Materials

Adapting the Home for the Physically Disabled
Roanoke, Va., A/V Health Services, 1991. 1 videocassette
(23 min.) VHS

Demonstrates how a home can be adapted for more efficient access and use by the physically disabled. Though some suggestions may seem obvious, this video serves as a good reminder of the obstacles faced daily by the physically disabled. Much of the information is practical and would assist the student or new therapist in recognizing when and where adaptations are necessary, and how to provide them economically.

Allen, Claudia Kay
Why Occupational Therapists Use Crafts
Colchester, Conn., S&S Worldwide, 1990. 1 videocassette
(30 min.) VHS

Crafts are a traditional part of occupational therapy and this video explains why OTs and OTAs can constructively provide low-tech treatment in an increasingly high-tech health care environment. The information assists the therapist with assessment and treatment of the cognitively disabled patient and aids in predicting success with living skills.

Heinze, Art
The Use of Upper Extremity Prostheses
Thief River Fall, Minn., Art Heinze, 1988. 1 videocassette
(49 min.) VHS

An extraordinary video by and about Art Heinze, an OTR who is also a bilateral amputee. Heinze's narrative about his personal responses to the loss of his hands, his successful rehabilitation, use of prostheses, education, marriage, and career provides the therapist with an unusual view of prosthetic patients.

Normal Infant Reflexes and Development
Tuscon, Ariz., Therapy Skill Builders, 1991. 1 videocassette
(20 min.) VHS

This video demonstrates normal infant reflexes and how they change and develop as the child matures. It would help the therapist determine what kind of development is normal and when intervention may be required. The activities presented, combined with footage of children and therapists, make this an extremely valuable teaching tool.

The Reliable Source for Occupational Therapy: An On-Line Information System for Occupational Therapy Practitioners
Rockville, Md., American Occupational Therapy Association and The American Occupational Therapy Foundation

A valuable on-line resource for both student and practitioner. Provided by the American Occupational Therapy Association, this system features a job bank, access to databases, e-mail, and a variety of occupational therapy resources. Easy to use and reasonably priced, it is accessible using a toll-free number. Requires a computer, modem and basic telecommunications software.

Physical Therapist Assistant

Brenda White Turner
Faulkner University
Montgomery, Alabama

Introduction

The Bureau of Labor Statistics reports physical therapy to be one of the fastest growing health care occupations. Increased longevity of the elderly, the baby boom generation, and interest in sports, exercise, and rehabilitation disorders are all factors contributing to this growth.

Physical therapist assistants are skilled technical health workers who perform physical therapy modalities and procedures under the direction of a registered physical therapist. They may choose to assist in a specialized number of fields (e.g., pediatric, geriatric, sports medicine, cardiopulmonary, oncological, and rehabilitative care), and work in such settings as hospitals, clinics, extended care facilities, rehabilitation centers, schools, and private practices. Students choosing this profession should possess a helpful and caring attitude toward people and a strong interest in the support of physical therapists in rehabilitation through mobilization methods. Upon receipt of a physician's prescription, physical therapists perform tests and assess needs to determine proper therapies and plan a therapeutic regime. Physical therapist assistants aid patients by assisting with therapeutic exercise routines and using modalities such as heat, light, sound, and electrical treatments. Physical therapy aides secure patients in therapy equipment, transport patients, record treatments, and perform a variety of supervised tasks.

Forty-five states and the territory of Puerto Rico offer educational opportunities in the field of physical therapy at the doctoral associate through degree level. Physical therapists must hold a bachelor's or master's degree from an accredited institution, while a licensed physical therapist assistant needs an associate degree to qualify for state licensure. Physical therapy aides generally have no formal physical therapy education but usually receive on- the-job training. Admission into most accredited programs is extremely limited. Potential students enhance their admittance by maintaining above average grades, particularly in the areas of science. Although programs vary, most college course work includes the study of chemistry, biology, anatomy, physiology, and psychology.

Accreditation

American Physical Therapy Association (APTA)
1111 N. Fairfax St.
Alexandria, Va. 22314

Selected Journals

American Geriatrics Society Journal.
Williams and Wilkins. Monthly. ISSN 0002-8614

American Journal of Sports Medicine.
American Orthopedic Society for Sports Medicine. Bimonthly. ISSN 0363-5465

American Journal of Physical Medicine and Rehabilitation.
[Association of Academic Physiatrists].
Williams and Wilkins. Bimonthly. ISSN 0894-9115

Archives of Physical Medicine and Rehabilitation.
[American Congress of Rehabilitation Medicine].
Saunders. Monthly. ISSN 0003-9993

Developmental Medicine and Child Neurology.
[American Academy for Cerebral Palsy and Developmental Medicine].
MacKeith Press. Monthly. ISSN 0012-1622

Exercise and Sport Sciences Review.
[American College of Sports Medicine].
Macmillan Press. Irregularly. ISSN 0091-6331

Infants and Young Children.
Aspen Pub. Quarterly. ISSN 0896-3746

The Journal of Bone and Joint Surgery.
The Journal. Ten times a year. ISSN 0021-9355

The Journal of Hand Therapy.
[American Society of Hand Therapists].
Hanley and Belfus. Quarterly. ISSN 0894-1130

The Journal of Orthopaedic and Sports Physical Therapy.
[American Physical Therapy Association. Orthopaedic and Sports Physical Therapy sections].
Williams and Wilkins. Monthly. ISSN 0190-6011

Journal of Sports Medicine and Physical Fitness.
Edizioni Minerva Medica, Corso Bramante. Quarterly.
ISSN 0022-4707

Medicine and Science In Sports and Exercise.
[American Society of Sports Medicine].
Williams and Wilkins. Monthly. ISSN 0195-9131

PT: The Magazine of Physical Therapy.
American Physical Therapy Association. Monthly.
ISSN 1065-5077

Pediatric Physical Therapy.
[The Official Publication of the section on Pediatrics of the American Physical Therapy Association].
The Association. Quarterly. ISSN 0898-5669

Pediatrics.
American Academy of Pediatrics. Monthly. ISSN 0031-4005

Physical and Occupational Therapy in Pediatrics.
Haworth Press. Quarterly. ISSN 0194-2638

Physical Therapy.
American Physical Therapy Association. Monthly.
ISSN 0031-9023

Physiotherapy Canada.
Canadian Physiotherapy Association. Bimonthly. ISSN 0300-0508

Research Quarterly for Exercise and Sport.
American Alliance for Health, Physical Education, Recreation, and Dance. Quarterly. ISSN 0270-1367

Sports Medicine.
Futura Pub. Co. Annual. ISSN 0271-2857

Print Materials

Ada, Louise, and Colleen Canning, eds.
Key Issues in Neurological Physiotherapy
Boston, Butterworth-Heinemann, 1990. 295p. ISBN 0750600098

This text covers neurological movement disorders, training principles, performance measurements, and a variety of related topics including literature reviews on muscle shortening and head injuries. Written for physiotherapy students, but useful for the graduate. Illustrations, photographs, summaries, and references enhance the work. Indexed. *Physiotherapy,* v. 1.

Adler, Susan S.
PNF in Practice: An Illustrated Guide
New York, Springer-Verlag, 1993. 257p. ISBN 0387526498

Practical text for physical therapists on the proprioceptive neuromuscular facilitation (PNF) treatment method. Each chapter begins with an introduction to the topic, and sections are organized by specific motions or positions for extremities. Includes halftone photographs, table of contents, and a glossary.

American College of Sports Medicine
Guidelines for Exercise Testing and Prescription
4th ed. Philadelphia, Lea and Febiger, 1991. 314p.
ISBN 0812113241

A basic text for anyone entering sports medicine or physical therapy because it provides good definitions and examples. Cardiac care is the primary focus of this text, but other special topics such as diabetes, arthritis, cancer, obesity, and pregnancy are also included. Appendices, extensive bibliographies, illustrations, and tables enhance the text.

American Occupational Therapy Association
Daily Activities After Your Hip Surgery
Rev. ed. Rockville, Md., The Association, 1990. 18p.
ISBN 0910317259

A helpful pamphlet is for the patient released from the hospital after hip surgery. The first page allows space to write in the surgical procedure; the doctor's name and phone number; the therapist's name and phone number; and the precautions for that particular patient. Different activities such as sitting, tub transfer, and using a walker are discussed. Includes explicit diagrams and illustrations.

APTA Resource Guide: Audiovisual Catalog 1992-1993
Alexandria, Va, American Physical Therapy Association, 1993. 400p.

This valuable reference tool lists audiovisual materials from eighty-five vendors and lenders. Divided into subject categories from AIDS to sports, this work includes titles of slides, audio cassettes, and VHS tapes, with ordering and lending information, prices, and brief abstracts. This resource provides a wealth of information for practitioners, students, instructors, and librarians. Format is a three-ring binder.

Baker, Margaret, and others
Controlling Movement: A Therapeutic Approach to Early Intervention
Gaithersburg, Md., Aspen Pub., 1991. 326p. ISBN 0834201925

Each page of this loose-leaf notebook describes a different activity that provides therapeutic assistance to children from birth to two years. It may be used also with school-aged children. The text lists goals for each activity and offers step-by-step instructions with illustrations. Individual educational plans and objective measurements are offered along with discussion of professional documentation and third party reimbursement. An excellent source for pediatric physical therapists and students.

Basmajian, John V., and Rich Nyberg, eds.
Rational Manual Therapies
Baltimore, Williams and Wilkins, 1993. 484p. ISBN 0683004204

This text is written for all who practice therapeutic manipulation and soft tissue mobilization. Twenty-one contributors present scientific information covering topics on exercise and training, ergonomics, types of application, and assessment. A history of the field guides the reader from general (unschooled) to specific topics. Chapter references, illustrations, photographs, observations and tables, and a thorough index enhance this work.

Bergman, Thomas
On Our Own Terms: Children Living with Physical Disabilities
North American ed. Milwaukee, Wisc., Gareth Stevens Children's Books, 1989. 48p. ISBN 155532942X

Varför blunda! was the original title of this international children's book, which describes therapeutic activities at the Caroline Hospital in Stockholm. Children with congenital handicaps are photographed and introduced with a description of their dysfunction and treatment strategies. Words like "cerebral" and "palsy" are simplistically but clearly defined, and children's most often asked questions are answered. An inspirational book for prospective pediatric therapists and an educational text for children.

Blossom, Bonnie, and Fran Ford
Physical Therapy in Public Schools: A Related Service
Roswell, Ga., Rehabilitation Publications and Therapies, 1991. 182p. ISBN 0963029401

This is the first volume of a planned two-volume set titled *Physical Therapy in Public Schools*. It incorporates both the teacher and parents as part of the therapeutic team and addresses differences between clinical and educational settings. Scenarios developed by the authors introduce readers to clients and their disorders. Includes an appendix with relevant forms, and a limited bibliography.

Bobath, Berta
Adult Hemiplegia: Evaluation and Treatment
3rd ed. Oxford, England, Heinemann Medical Books, 1990. 190p. ISBN 0433000988

Describes a treatment for patients with lesions of the upper-motor neurone, namely the Bobath treatment. Photographs and illustrations describe the treatment, which dictates routine activities in lieu of exercises. Designed for course instructors and practicing therapists. Includes index, references, a reading list, and a variety of evaluation tools.

Brimer, Mark A.
Health Care Management in Physical Therapy
Springfield, Ill., Charles C. Thomas, 1990. 315p. ISBN 0398056420

An introductory management text, this book targets physical therapists in their first supervisory position. Material covered is applicable to all health care settings. Like most management texts, this work includes budgeting, decision making, personnel appraisal, and communication skills. Also covered is recordkeeping in the context of a health care environment. A lengthy glossary, helpful index, and illustrations assist the reader.

Bullock, Margaret I., ed.
Ergonomics: The Physiotherapist in the Workplace
New York, Churchill Livingstone, 1990. 332p. ISBN 0443036128

This text presents an international exchange of ideas on a current topic. Discussions range from workplace analysis and building design to prevention of musculoskeletal disorders. A general ergonomic checklist is included as well as photographs, illustrations, tables and charts, and chapter references. An annotated bibliography of relevant journals and a list of ergonomic societies foster further research. Indexed. *International Perspectives in Physical Therapy,* v. 6.

Cailliet, Rene
Hand Pain and Impairment
4th ed. Philadelphia, F. A. Davis, 1994. 311p. ISBN 0803616198

This new edition discusses painful hand conditions and the evaluation and treatment procedures for painful disabilities. The audience for this text is physical and occupational therapists and the nonspecialist physician. A table of contents, illustrations, and index are provided.

Campbell, Suzann K., ed.
Pediatric Neurologic Physical Therapy
2nd ed. New York, Churchill Livingstone, 1991. 459p. ISBN 0443087644

Each chapter begins by defining a specific neurological condition such as Down's syndrome, then provides the disease pathology and evaluation and assessment information, and ends with treatment recommendations. Case histories and theoretical and research-based data support the text. Discusses the educational setting and early family intervention. Chapters include photographs, illustrations, tables, and references; information is accessible via the index. *Clinics in Physical Therapy* series.

Collins, Thomas S., and Amy A. Collins, eds.
Medicolegal Issues for Physical Therapists
Dubuque, Iowa, Shepherd, 1992. 86p. ISBN 0962844039

This textbook describes legal terms and past legal cases. Review questions follow each chapter to reinforce learning, and case studies or "vignettes" are numerous. This book is filled with basic legal advice, although it should not be used as a substitute for legal counsel. A good discussion of legal ethics as well as several informative appendices make this a valuable resource for practitioners.

Connolly, Barbara H., and Patricia C. Montgomery, eds.
Therapeutic Exercise in Developmental Disabilities
2nd ed. Hixson, Tenn., Chattanooga Group, 1993. 220p.
ISBN 1879971011

The authors maintain a balance of theory, assessment and practical treatment approaches to therapeutic exercise. Common pediatric dysfunctions are presented as case studies that provide opportunities for problem solving. Chapter end references are included for further study.

Davies, Patricia M.
Right in the Middle: Selective Trunk Activity in the Treatment of Adult Hemiplegia
New York, Springer-Verlag, 1990. 277p. ISBN 038751242X

Davies builds upon her earlier book *Steps to Follow* (Spinger-Verlag, 1985). Activities are broken down by physical position (lying, sitting, standing up from sitting, etc.) and are listed with page numbers for easy access. Includes a subject index, references, and photographs with narrative steps.

Davis, Carol M.
Patient Practitioner Interaction: An Experiential Manual for Developing the Art of Health Care
2nd ed. Thorofare, N.J., SLACK, 1994. 209p. ISBN 1556422326

This material is presented in a style that is easy to read and understand. It provides a well-rounded overview of health care and services. An extensive index is included for easy access. References, tables, and chapter study guides enhance the learning process.

Donatelli, Robert, and Michael J. Wooden, eds.
Orthopaedic Physical Therapy
2nd ed. New York, Churchill Livingstone, 1993. 778p.
ISBN 0443088357

Intended for physical therapists at all levels, this volume is a collection of articles by several qualified practitioners. It is divided into four parts: fundamental principles, upper and lower quarters of the body, and special considerations. Discussion of therapeutic techniques on specific areas include the elbow, foot, ankle, hip, wrist, and hand. A well-rounded source, this text is filled with photographs and illustrations, as well as an extensive bibliography and index.

Donatelli, Robert, ed.
Physical Therapy of the Shoulder
2nd ed. New York, Churchill Livingstone, 1991. 391p.
ISBN 0443087318

One of twenty-three titles in a series titled *Clinics in Physical Therapy,* this source provides a thorough explanation of shoulder anatomy followed by a discussion of a variety of shoulder ailments and injuries. Donatelli edits this compilation of articles, which were written by qualified physical therapists and orthopedic physicians. Terminology is defined, and chapter summaries are provided to assist the reader. An extensive index is provided, as well as photographs, case studies, tables, and references.

Echternach, John L., ed.
Physical Therapy of the Hip
New York, Churchill Livingstone, 1990. 221p. ISBN 0443086508

Ten contributors write on topics such as anatomy and evaluation of the hip, posture and gait, medical and surgical options, rehabilitation, and sports injuries. Includes photographs, line drawings, and illustrations. Chapter summaries or conclusions are beneficial, and chapter references promote further review of the topic. Exercises are offered, and a sample standard evaluation is applied in the work. *Clinics in Physical Therapy* series.

Eckersley, Pamela M., ed.
Elements of Paediatric Physiotherapy
New York, Churchill Livingstone, 1993. 513p. ISBN 0443038945

Discussions include development and assessment of the child, and genetic disorders in children. Written for practicing physiotherapists, this text presents theoretical and practical approaches to treating the "whole child" with the aid of parents and other health care workers. Illustrated with drawings and photographs, the text includes chapter contents, summaries, references, tables and charts, and reading lists.

Eddy, Lynne
Physical Therapy Pharmacology
St. Louis, Mo., Mosby-Yearbook, 1992. 192p. ISBN 0815130767

A textbook for PT students at all levels that addresses drug interaction and physical therapy treatment. This overview of pharmacology emphasizes drugs affecting cardiovascular, renal, nervous, respiratory, endocrine, and gastrointestinal systems. A working knowledge of physiology and pathophysiology is assumed. Over-the-counter drugs are included; antibiotics are excluded.

Edeling, Joy
Manual Therapy for Chronic Headache
Boston, Butterworths, 1988. 164p. ISBN 0407005609

Edeling states that headaches have a cervical spine component because they may reoccur when "abnormally restricted cervical movements are stretched or palpated." Headache treatment is discussed in relation to prior assessment and patient headache symptoms. Assessment questions and tips for choosing the correct therapies are provided. Cautions and contraindications, illustrated techniques, and case studies assist the reader. Includes an index, chapter references, and a reading list.

Engle, Robert P., ed.
Knee Ligament Rehabilitation
New York, Churchill Livingstone, 1991. 228p. ISBN 0443087539

A compilation by several writers, this text begins with basic anatomy and moves on to more complex abnormal considerations of "problem knee" rehabilitation, such as soft tissue and surgical treatment. Provides references, summaries, illustrations, and photographs.

Fox, Edward L., Richard W. Bowers, and Merle L. Foss
The Physiological Basis for Exercise and Sport
5th ed. Madison, Wisc., Brown and Benchmark, 1993. 710p.
ISBN 0697126269

Addresses topics such as drugs and ergogenic aids, pulmonary symbols and norms, tests of anaerobic and aerobic power, energy sources, and exercising in heat and cold. Each chapter includes goals and objectives, review questions, references, and additional reading lists for further study. Illustrations, a glossary, and an index complement the text.

Fraser, Beverly A., Robert N. Hensinger, and Judith A. Phelps
Physical Management of Multiple Handicaps: A Professional's Guide
2nd ed. Baltimore, Paul H. Brookes, 1990. 337p.
ISBN 1557660476

The five sections of this book cover everything from managing daily activities to adaptive equipment. Hand-drawn models illustrate normal motion, and descriptions of reflexes and abnormal oral motor components make this source beneficial to any caregiver of a multiple-handicapped person. Further readings are provided, as well as appendices which provide lists of resources for clothing, devices, and information. Photographs, line drawings, and an index assist the reader.

Goodman, Catherine Cavallaro, and Teresa E. Kelly Snyder
Differential Diagnosis in Physical Therapy: Musculoskeletal and Systemic Conditions
2nd ed. Philadelphia, Saunders, 1995. 424p. ISBN 0721652670

A working book for students and practitioners, particularly those in independent practice. Patient interviews, recognizing systemic disease, and timely medical referrals are addressed. Chapters review anatomy and physiology, provide overviews and summaries, and present case studies. A list of symptoms for each disease, a checklist of symptoms requiring medical referral, and special questions to ask are helpful. Includes glossary, illustrations, tables, references, and bibliographies.

Grant, Ruth, ed.
Physical Therapy of the Cervical and Thoracic Spine
New York, Churchill Livingstone, 1988. 338p. ISBN 0443085072

Pain patterns, examination and assessment, and clinical decision making are discussed in relation to cervical and thoracic functions. Charts, scientific data, case studies, and treatment histories document article findings. Prescription and evaluation of self-treatment are presented, as are preventative steps to reduce strain in the workplace. Photographs, illustrations, and text address head and neck pain and provide references for further study. Index included. *Clinics in Physical Therapy* series, v. 17.

Grisogono, Vivian, ed.
Sports Injuries
New York, Churchill Livingstone, 1989. 252p. ISBN 0443031754

Featuring an international exchange of ideas from different countries (Canada, Australia, India, United States, and Finland), this work discusses competition in sports and sport physiotherapy for experienced clinicians. The last three sections show specific scientific treatments, e.g., ultrasound, cryotherapy, and muscle measurement. Includes an annotated bibliography, index, references, photographs, tables, charts, and scientific data. *International Perspectives in Physical Therapy* series.

Hansson, Tore, Catherine A. Christensen Minor, and Deborah L. Wagnon Taylor
Physical Therapy in Craniomandibular Disorders
Chicago, Quintessence Pub. Co., 1992. 80p. ISBN 0867151927

This very specialized text consists mostly of pictures, with narrative limited chiefly to brief definitions. References are minimal since the book is intended to be a guide for dentists and physical therapists. Topics include osteoarthrosis, other craniomandibular disorders, and postoperative pain. Chapters include photographs, relevant clinical forms, and conclusions.

Hayes, Karen W.
Manual for Physical Agents
4th ed. Norwalk, Conn., Appleton and Lange, 1993. 169p.
ISBN 0838561438

Intended for physical therapy students, Hayes presents step-by-step instructions for physical therapy techniques. Illustrations and extensive reading lists also make this a valuable resource for practitioners. The learning process is supported by the same successive format for each technique. Advantages, disadvantages, precautions, and frequency are just a few of the topics addressed for seventeen different techniques. No index is provided; however, three helpful appendices are included.

Hecox, Bernadette, Tsega Andemicael Mehreteab, and Joseph Weisberg, eds.
Physical Agents: A Comprehensive Text for Physical Therapists
Norwalk, Conn., Appleton and Lange, 1994. 473p.
ISBN 0838580408

Information regarding physical agents and the body's response to them is presented at a level easily understood by the entry-level student. Thermal agents, hydrotherapy, electrotherapy, and ultraviolet, laser, and other physical agents are discussed. Clinical problems and laboratory experiments are provided for review and consideration. Illustrations and tables help explain the treatments presented. Photographs, a thorough index, and extensive references are also included.

Hedman, Glenn, ed.
Rehabilitation Technology
New York, Haworth Press, 1990. 173p. ISBN 1560240334

Begins with an explanation of technology (services) versus engineering (devices), and emphasizes collaboration between the two fields. Seating systems, facilitative toys, and prosthetics are a few of the devices introduced. Especially helpful is the annotated bibliography, "Funding Assistive Technology Services." Chapter references, summaries, photographs, and resource lists add to this valuable text. Also published as *Physical and Occupational Therapy in Pediatrics Series*, v. 10, no. 2, 1990.

Hillegass, Ellen A., and H. Steven Sadowsky, eds.
Essentials of Cardiopulmonary Physical Therapy
Philadelphia, Saunders, 1993. 801p. ISBN 0721636098

Basic information for entry-level physical therapists, based on principles of the Cardiopulmonary Section of the American Physical Therapy Association. The five sections of this text cover anatomy, diagnostic tests, pharmacology assessment, and much more. New terms are highlighted and defined in a glossary, and case studies provide suggested treatments. A wonderful feature of this text are the "bullets" with one- or two-line summaries closing each chapter. Photographs are included as well as an index.

Hunt, Gary C., ed.
Physical Therapy of the Foot and Ankle
New York, Churchill Livingstone, 1988. 338p.
ISBN 044308467X

Covers basic foot science, common deviations, pathologic categories, examination, assessment tools, and techniques. Discusses athletes' concerns, foot self-inspection, exercises, proper skin care steps, orthotic principles, and surgical options. Illustrations, photographs, and charts and tables are included for clarification.

An index, blank examination worksheets, and tips on how to observe gait are also provided. *Clinics in Physical Therapy* series, v. 15.

Irwin, Scott, and Jan Stephen Tecklin, eds.
Cardiopulmonary Physical Therapy
2nd ed. St. Louis, Mo., Mosby, 1990. 585p. ISBN 0801629071

This textbook, intended for educators, students, and clinicians, reviews basic cardiac and respiratory system dysfunctions including discussion of heart transplants and AIDS. Patient assessment, treatment procedures, and a review of related literature are discussed by thirty-three qualified contributors. To assist the reader, each chapter includes illustrations, tables and charts, references, summaries, and a chapter outline. Anatomical descriptions are provided as needed. Indexed.

Kahn, Joseph
Principles and Practice of Electrotherapy
3rd ed. New York, Churchill Livingstone, 1994. 194p. ISBN 0443089191

An update of the 1991 edition, this student/clinician manual includes expanded coverage of microampere electrical stimulation, the author's personal commentary, practical considerations, and future professional trends. Information regarding equipment purchases has also been included. Halftone illustrations.

Kendall, Florence Peterson, Elizabeth Kendall McCreary, and Patricia Geise Provance
Muscles, Testing and Function: With Posture and Pain
4th ed. Baltimore, Williams and Wilkins, 1993. 451p. ISBN 0683045768

A valuable textbook or resource for clinicians who are interested in standards of excellence in testing and treatment. Incorporates the title *Posture and Pain* by Henry Otis Kendall. Each chapter begins with its own table of contents and is written in a clear and readable style. Text includes bibliography, suggested readings, glossary, and index. Illustrated.

Kovich, Karen M., and Diane E. Bermann
Head Injury: A Guide to Functional Outcomes in Occupational Therapy
Rockville, Md., Aspen Publishers, 1988. 236p. ISBN 0871897628

Standards established by the Quality Assurance Committee at the Rehabilitation Institute of Chicago form the foundation of this five-part book on medical aspects, goal setting, treatment, daily living skills, and a myriad of other topics. This spiral-bound text includes photographs, tables and charts, reading and resource lists, and an index. *Rehabilitation Institute of Chicago Procedure Manual* series.

Lewis, Carole Bernstein
Geriatric Physical Therapy: A Clinical Approach
Norwalk, Conn., Appleton and Lange, 1994. ISBN 0838588751

This comprehensive text has three major divisions. Part One includes demographics, theories, psychological aspects, and evaluative tools, along with nutritional and pharmacological issues. Part Two discusses clinical concepts such as patient evaluation and assessment, care, and treatments. Part Three deals with administrative issues ranging from decision making to death. Includes a list of organizations, databases, and extensive references. Indexed.

Lippert, Lynn
Clinical Kinesiology for the Physical Therapist Assistants
2nd ed. Philadelphia, F. A. Davis, 1994. ISBN 080365483

A new edition of a textbook specifically designed for the PT assistant. The study of movement is addressed in relation to various systems, including skeletal, articulator, and nervous. Chapters on muscles, extremities, and normal gait are also provided. A bibliography and index are included as well as answers to the chapter study questions. Illustrated with line drawings.

Littell, Elizabeth H.
Basic Neuroscience for the Health Professions
Thorofare, N.J., Slack, 1990. 293p. ISBN 1556420536

This introductory course textbook provides an illustrated foundation for understanding the nervous system. Review exercises at the end of each chapter aid the physical therapy student, as does the list of abbreviations that precedes the first chapter. Definitions are provided for italicized words. Illustrations, tables, an extensive index, and topical reading lists provide the reader with additional tools. Includes a bibliography and index.

Mackenzie, Colin F., ed.
Chest Physiotherapy in the Intensive Care Unit
2nd ed. Baltimore, Md., Williams and Wilkins, 1989. 387p. ISBN 0683053299

This work begins with a history and literature review, then discusses topics such as clinical indications, percussion and vibration, and postural drainage. Appendices provide valuable statistical information, abbreviations and symbols, treatments, evaluation summaries, and much more. Chapter summaries aid readability, and chapter references provide additional reading opportunities. A good resource for clinicians and students. Illustrated. Indexed.

Manheim, Carol J., and Diane K. Lavett
Myofascial Release Manual
Thorofare, N.J., Slack, 1989. 126p. ISBN 1556421087

This paperback book is a simple introduction to myofascial release. There are no chapter divisions; however, the book supplies an index, bibliography, and relevant photographs.

Manual for Functional Training
3rd ed. Philadelphia, F. A. Davis, 1992. 351p. ISBN 0803667590

Intended for students and practitioners, this text would be useful for any caretaker. Readers are introduced to techniques and devices, including adaptability in home design. Functional activities are included with step-by-step instructions and line drawings. Photographs, bibliography, assessment forms, and an index are included.

McAtee, Robert E.
Facilitated Stretching
Champaign, Ill., Human Kinetics, 1993. 108p. ISBN 0873224205

Photographs, illustrations, step-by-step procedures, and a concisely written narrative allow the student to clearly see and understand stretching procedures. Special notes and cautions are provided to warn the physical therapist of potential problems. Includes glossary, index, and bibliographical references.

McGarvey, Charles L., ed.
Physical Therapy for the Cancer Patient
New York, Churchill Livingstone, 1990. 188p. ISBN 0443086672

This book is intended for practicing physical therapists, assistants, and students. Basic principles of oncology, rehabilitation concepts, assessment, and treatments well as psychosocial factors are discussed in relation to physical therapy patients. Special features include hospice protocol and instructions for breast cancer patients. Other cancer types and sites are covered: lung, leukemia and lymphoma, head and neck, and sarcoma. Photographs, anatomical illustrations, charts, tables, and references are included in each chapter. Indexed. *Clinics in Physical Therapy* series.

Montgomery, Patricia C., and Barbara H. Connolly, eds.
Motor Control and Physical Therapy: Theoretical Framework, and Practical Applications
Hixson, Tenn., Chattanooga Group, 1991. 161p.
ISBN 1879971003

Motor control from every angle is discussed including learning, production of movement, and many other facets in infants through elderly adults. Chapter goals and objectives, introductions and summaries, illustrations, graphs, and references reflect the textbook format. Case studies assess goals, outcomes, and treatment recommendations, and provide summaries and illustrations of how a patient's chart might be maintained. A good historical perspective of the field, past treatments, and theories is provided. Indexed.

Moran, Michael L., and Mark A. Brimer
Computer Principles for Physical and Occupational Therapists
Tucson, Ariz., Therapy Skill Builders, 1994. 213p.
ISBN 0884500330

This book focuses on the increasing importance of computers in health care and aims to provide basic computer information for occupational and physical therapists who are novices in computer fundamentals. Though probably too simplistic for those who use computers regularly, this work contains useful sections on computer use in therapy practices and use of on-line services.

Nelson, Roger M., and Dean P. Currier
Clinical Electrotherapy
2nd ed. Norwalk, Conn., Appleton and Lange, 1991. 422p.
ISBN 0838513344

This reference source for students and practitioners includes the standardized and updated terminology used in electrotherapy as identified by the American Physical Therapy Association. Editors and contributors present theory as well as a practical overview of the subject. Index and appendices included.

Norkin, Cynthia C.
Joint Structure and Function: A Comprehensive Analysis
2nd ed. Philadelphia, F. A. Davis, 1992. 512p.
ISBN 0803665776

Targeted to physical therapy students, this new edition includes expanded coverage of the function and dysfunction of the human musculoskeletal system. New research and theories in muscle physiology and tissue composition are included. Emphasis is on joint structure and the application of normal and abnormal force.

Norris, Christopher M.
Sports Injuries: Diagnosis and Management for Physiotherapists
Boston, Butterworth-Heinemann, 1993. 327p. ISBN 0750601566

Norris takes a holistic approach in this two-part book. The first part is more general, discussing injury and healing; assessment and modalities include manual therapy. Part Two discusses specific injuries, e.g., hip, knee, face, shoulder, and elbow. Anatomy is defined as needed, with a practitioner's guide and manual. Illustrations, index, and references are included.

Nosse, Larry J., and Deborah G. Friberg
Management Principles for Physical Therapists
Baltimore, Williams and Wilkins, 1992. 307p. ISBN 0683065769

A textbook intended to teach entry-level physical therapists the fundamentals of management and practice. Fiscal and operations management, facilities planning, and legal concepts are only a few of the topics addressed. Line drawings clarify chapter content as do chapter outlines, summaries, case studies, and references. Case study responses, a glossary, and an index are provided.

O'Connor, Linda J., and Rebecca J. Gourley
Obstetric and Gynecologic Care in Physical Therapy
Thorofare, N.J., Slack, 1990. 363p. ISBN 1556421397

The authors present historical and current trends in obstetric and gynecological care. Understanding of topics is aided by the efficient layout. Chapters contain self-assessments to reinforce material and provide references for further study. Additional learning aids, such as photographs, line drawings, tables, and relevant forms are included. Resource lists for products and information, an index, glossary, answers to assessment reviews, and topical reading lists are also provided.

O'Sullivan, Susan B., and Thomas J. Schmitz
Physical Rehabilitation: Assessment and Treatment
3rd ed. Philadelphia, F. A. Davis, 1994. 748p. ISBN 0803666993

This textbook introduces perspectives on adult rehabilitation management, decision making, psychological adjustment, and influence of values on patient care. Specific assessment procedures and treatment strategies are presented for a multitude of conditions including artery disease, multiple sclerosis, and head injuries. A very well written and understandable work for the physical therapy student.

Peringian, Lynda
Physical and Occupational Therapists' Job Search Handbook: Your Complete Job Search Strategy, How to Hire, How to Be Hired
Birmingham, Miss., Therapy Careers Press, 1989. 182p.
ISBN 0962277304

Job search checklists, salary surveys, and examples of how to log the search process make this an essential source for the job-seeking therapist. Bibliographic citations provide additional readings, and "taking stock of yourself" questions and a career quiz help the reader focus on issues to consider. Topics include networking, employment agencies, résumés, and cover letters. Indexed.

Perrin, David H.
Isokinetic Exercise and Assessment
Champaign, Ill., Human Kinetics Publishers, 1993. 212p.
ISBN 0873224647

Perrin presents isokinetic exercise and assessment as they relate to muscles from the shoulder to the trunk. Includes definitions of terminology, introduction of instruments, and information that assist a patient's return to daily activity. Photographs, tables, charts, and references are provided.

Perry, Jacquelin
Gait Analysis: Normal and Pathological Function
Thorofare, N.J., Slack, Inc., 1992. 524p. ISBN 1556421923

A valuable reference source for PT students and practitioners who specialize in normal and abnormal walking and movement patterns. Divided into four sections, the book addresses fundamentals, normal and pathological gaits, and gait analysis systems. The work is extensive and provides a glossary, index, and listing of commonly used abbreviations and acronyms.

Pollock, Michael L., and Jack H. Wilmore
Exercise in Health and Disease: Evaluation and Prescription for Prevention and Rehabilitation
2nd ed. Philadelphia, Saunders, 1990. 741p. ISBN 0721629482

The three sections of this text include information on cardiovascular function, medical screening and evaluation, and special considerations for prescribing exercise. An additional chapter on exercise for the healthy is included. Appendices provide fitness standards, sample forms, and information on rehabilitation programs. Illustrations, tables, charts, and scientific statistical data support the text. Extensive chapter references and a cookbook format make this a user-friendly text. Indexed.

Porterfield, James A., and Carl DeRosa
Mechanical Low Back Pain: Perspectives in Functional Anatomy
Philadelphia, Saunders, 1991. 209p. ISBN 0721672973

Porterfield and DeRosa address the impact that significant advances in physical therapy have had on low back pain treatment. They detail functions and dysfunctions of the low back while introducing evaluation and treatment techniques. Their active approach to rehabilitation is directed toward restoring function. Each chapter provides references. Table of contents and index.

Powers, Scott K., and Edward T. Howley
Exercise Physiology: Theory and Application to Fitness and Performance
2nd ed. Madison, Wisc., Brown and Benchmark, 1994. 1 v. (various pagings). ISBN 0697126579

A well-organized, classic textbook with outlines, objectives, and key terms preceding each chapter. Charts, subheadings, and boldfaced terms are defined throughout. Summary statements, study questions, suggested readings, and references are also included. The text's three sections discuss the physiology of exercise, health and fitness, and performance. Topics range from bioenergetics to ergogenic aids. Appendices provide numerous helpful charts.

Prentice, William E., ed.
Therapeutic Modalities in Sports Medicine
3rd ed. St. Louis, Mo., Mosby, 1994. 409p. ISBN 0801679222

Contributors to this book have earned degrees in the field, ranging from athletic trainer to PhD. Illustrates the different modalities in sports medicine and allows staff to elect treatments that are most appropriate. The text includes objectives at the beginning of each chapter, along with tables and figures. Chapter summaries, illustrations, glossaries, references, and an appendix of manufacturers where equipment may be obtained are also provided.

Richardson, Janice K., and Z. Annette Iglarsh
Clinical Orthopaedic Physical Therapy
Philadelphia, Saunders, 1994. 712p. ISBN 0721632572

Features case studies that provide opportunities for students to evaluate a variety of physical conditions, formulate appropriate diagnoses, and consider possible treatments. Tables and charts throughout the chapters assist the student, while references and bibliographies provide additional readings. Illustrated. Indexed.

Rimmer, James Hunter
Fitness and Rehabilitation Programs for Special Populations
Madison, Wisc., Brown and Benchmark, 1994. 326p.
ISBN 0697116190

Discusses fundamentals of exercise physiology for the elderly, obese, mentally retarded, and pregnant, and patients suffering from a variety of disorders. In many chapters exercise guidelines and references for additional reading are supplied. Tables, illustrations, an index, and definitions assist the reader.

Ritter, Merrill A.
Your Injury: A Common Sense Guide to Sports Injuries
Dubuque, Iowa, Brown and Benchmark, 1991. 188p.
ISBN 0697148246

This spiral-bound paperback is a quick reference to sports injuries and recommended treatments. The guide has numerous diagrams and photographs which allow the student to see clearly the area being discussed. Provides understandable explanations of symptoms for each injury, recommendations and warnings, and suggestions for doing exercises properly. This source would be very useful in the realm of physical therapy; however, its value for the home should not be overlooked.

Rocabado Seaton, Mariano, and Z. Annette Iglarsh
Musculoskeletal Approach to Maxillofacial Pain
Philadelphia, Lippincott, 1991. 244p. ISBN 0397548508

Intended for practitioners of dentistry, physical therapy, and medicine, this text discusses abnormal conditions, evaluations, and therapy results. The book's four sections cover growth and development, anatomy, etiology, and disorders and treatment of the maxillofacial region. Each section includes evaluation forms, references, and bibliographies. Photographs and diagrams clarify techniques. Marketing strategies are included. Indexed.

Rothstein, Jules M., Serge H. Roy, and Steven L. Wolf
The Rehabilitation Specialist's Handbook
Philadelphia, F. A. Davis, 1991. 1019p. ISBN 0803676298

A desktop source for all practicing clinicians, this handbook is replete with organized data. Includes temperature conversions, common equivalents, metric prefixes, English to metric conversions, volume and capacity to cubic meters conversions, plus additional information. Well-drawn illustrations, tables, and charts facilitate quick reference and retrieval of information.

Sanders, Barbara, ed.
Sports Physical Therapy
Norwalk, Conn., Appleton and Lange, 1990. 535p.
ISBN 083858652X

This wonderfully written and easy-to-read book involves thirty contributors. The articles address the needs of practicing sports and physical therapists, coaches, and athletic trainers. Topics include the role of the physical therapist, rehabilitation concepts, emergency care on the field, drugs and drug testing, nutrition, flexibility for sports, training and conditioning, protective equipment, and injuries. Includes chapter introductions, illustrations, bibliographies, and index.

Scott, Ronald W.
Health Care Malpractice: A Primer on Legal Issues for Professionals
Thorofare, N.J., Slack, 1990. 160p. ISBN 1556420730

Scott, a physical therapist practitioner who has legal experience, discusses changes in the profession and in home health care, including malpractice issues. Topics cover statute of limitations, negligence, liabilities, and contracts. Tables, chapter summaries, index, and glossary.

Scully, Rosemary M., and others, eds.
Physical Therapy
Philadelphia, Lippincott, 1989. 1319p. ISBN 0397507984

A tremendous undertaking by seventy-six contributors, this comprehensive reference tool is beneficial to the physical therapy student as well as the practicing clinician. The discussion of a variety of physical therapy issues and concerns is extensive.

Shurr, Donald G., and Thomas M. Cook
Prosthetics and Orthotics
Norwalk, Conn., Appleton and Lange, 1990. 225p.
ISBN 0838579779

Developed for entry-level physical therapy students, this textbook will assist any professional needing information on prosthetics or orthotics devices. Although not intended to be an encyclopedia of P&O devices, the overview addresses fundamental concepts for selection and use of such devices. This text assumes basic knowledge of anatomy, kinesiology, and, to a lesser degree, pathology.

Somers, Martha F.
Spinal Cord Injury: Functional Rehabilitation
Norwalk, Conn., Appleton and Lange, 1992. 365p.
ISBN 083858649X

A thorough overview of spinal cord injuries, issues related to that disability, and the therapist's role. Somers provides an understanding of spinal cord injuries and therapeutic strategies involved in a successful rehabilitation program. The text promotes independence and self-esteem for the injured patient and provides extensive charts and further readings for the student.

Stewart, Darlene L., and Susan H. Abeln, eds.
Documenting Functional Outcomes in Physical Therapy
St. Louis, Mo., Mosby, 1993. 289p. ISBN 0801663598

This text discusses documenting patient care from intake to release and discusses its importance in decision making, legal protection, and reimbursement. Although this text is slanted toward the United States health care system, it is still a must for every practitioner. Sample forms are provided, and review questions and answers conclude each chapter. Readability is complemented by good illustrations, an index, a glossary, and appendices.

Tecklin, Jan Stephen, ed.
Pediatric Physical Therapy
Philadelphia, Lippincott, 1989. 354p. ISBN 0397508220

An entry-level textbook containing current information on specific disability groups and describing skills to evaluate and treat those disabilities. Presents types of practices (such as developmental testing, school and sport physical therapy, and neonate therapy), and introduces adaptive equipment. Disorders covered include cerebral palsy, spina bifida, pulmonary disorders, juvenile rheumatoid arthritis, and mental retardation.

Trombly, Catherine A., ed.
Occupational Therapy for Physical Dysfunction
3rd ed. Baltimore, Williams and Wilkins, 1989. 629p.
ISBN 0683083899

Trombly compiles current literature on the evaluation of physically challenged adults, and presents the material in foundational layers that build on previous knowledge. Study questions throughout the text assist the reader in the learning process. When practitioners have the information necessary to analyze patient behavior, deduce problems, and establish goals which translate into therapeutic principles, they may serve as professionals rather than technicians. This text is an attempt to provide that information. (Fourth edition scheduled for publication in 1995.)

Turnbull, George I., ed.
Physical Therapy Management of Parkinson's Disease
New York, Churchill Livingstone, 1992. 208p. ISBN 0443087563

Facts about Parkinson's disease are presented along with therapeutic treatment strategies, relaxation techniques, exercises, evaluation processes, and a discussion of the impact upon the family. Akinesia, bradykinesia, disinhibition, and other impairments related to the disease are presented and concisely explained. Scientific studies are charted with explanations; illustrations and chapter references stimulate the learning process. *Clinics in Physical Therapy* series.

Umphred, Darcy Ann
Neurological Rehabilitation
2nd ed. St. Louis, Mo., Mosby, 1990. 895p. ISBN 0801652928

This comprehensive clinical textbook is divided into three areas: (1) an overview of basic neuroanatomy, control, development, and psychosocial aspects; (2) common neurological disabilities and appropriate treatment strategies/techniques; and (3) recent advances in treatment and rehabilitation. Forty-two contributors provide the student with diverse problem-solving strategies that may be incorporated in plans to solve individual patient needs. A study guide and glossary are included.

Webber, B. A., and Jennifer A. Pryor, eds.
Physiotherapy for Respiratory and Cardiac Problems
New York, Churchill Livingstone, 1993. 461p. ISBN 0443044716

Patients' problems, special populations, and therapy techniques are discussed with conclusions or summaries ending each chapter. All chapters include photographs, tables, key points, and further reading lists, some quite extensive. Includes definitions, normal values and abbreviations, and an index.

Zadai, Cynthia Coffin, ed.
Pulmonary Management in Physical Therapy
New York, Churchill Livingstone, 1992. 234p. ISBN 0443087415

This text provides anatomical descriptions and an introduction to the cardiopulmonary organ as a ventilatory pump. Organs are then discussed in relationship to their response to exercise and impairments. Evaluation and assessment strategies are explained in good detail. Illustrations, references, and an index are supplied. *Clinics in Physical Therapy* series.

Nonprint Materials

Heat and Massage for the Lower Back
Fargo, N.D., St. Luke's Meritcare; distributed by Health Sciences Consortium, 1989. 1 videocassette (9 min.) and 1 guide. VHS

Designed for physical and occupational therapists, this video introduces and explains therapeutic methods for patients who are experiencing muscle spasms and pain. Also illustrates techniques to increase circulation and improve flexibility in preparation for exercise.

Heat, Massage, and Exercise for the Neck and Upper Back.
Fargo, N.D., St. Luke's Meritcare; distributed by Health Sciences Consortium, 1989. 1 videocassette (14 min.) and 1 guide. VHS

This video examines components of thermotherapy, massage, and exercise therapy for the relief of pain. Focus is on therapy to increase a patient's likelihood of returning to normal daily activities.

Heat, Massage and Quad Setting Exercise for the Knee.
Fargo, N.D., St. Luke's Meritcare; distributed by Health Sciences Consortium, 1989. 1 videocassette (9 min.) and 1 guide. VHS

Demonstrates exercise, heat, and massage techniques that reduce muscle spasm, swelling, and pain for patients who have suffered knee injuries. Techniques assist patients further by minimizing muscle spasms and increasing circulation. Intended audience is physical and occupational therapists.

Pain Control Videorecording: A Joint Production of Dartmouth/ Hitchcock Medical Center and WNHT-TV21.
Princeton, N.J., Films for the Humanities and Sciences, 1990. 1 videocassette (26 min.)

Types of pain treatment, including acupuncture, acupressure, and electromyography are presented along with pharmaceutical options such as injections, infusions, topical sprays, inhalants, and other analgesics. Pain prevention techniques are also covered.

Respiratory Therapy Assisting

Sue Hollander
Library of Health Sciences, University of Illinois
Rockford, Illinois

and

Mary Rose Amidjaya
Rock Valley College
Rockford, Illinois

Introduction

Students in a respiratory therapy program are taught the skills needed to become respiratory technicians, certified respiratory therapy technicians, or registered respiratory therapists. Respiratory therapists treat all types of patients who have breathing disorders. Generally working in a hospital setting, respiratory therapists test the capacity of the lungs, treat breathing disorders by administering aerosolized medications, and give chest physiotherapy. They may also draw blood and analyze it to determine the amounts of oxygen and carbon dioxide in the patient's system. In addition, therapists often work with patients who require the use of ventilators to help them breathe, and adjust and troubleshoot any equipment problems.

Training for respiratory therapists may be offered by community colleges, medical schools, hospitals, vocational- technical institutes, and colleges and universities. Programs may result in a one-year technician's certificate, a two-year degree, or a four-year degree.

Both licensing and certification are used in determining the competence of a therapist to begin practice. Thirty- three states license respiratory care personnel. In addition, the National Board for Respiratory Care certifies individuals on two levels: Certified Respiratory Care Technician (CRTT) and Registered Respiratory Therapist (RRT). Regardless of the type of program a person has completed, a student must first pass an examination to become a CRTT. A separate examination is offered to those CRTTs with the appropriate experience and education to become RRTs.

This bibliography provides appropriate materials to support a curriculum leading to the eligibility to take the CRTT exam.

Accreditation and Certification

National Board for Respiratory Care
8310 Nieman Road
Lenexa, Kansas 66214
(913) 599-4200
   This Board certifies respiratory care technicians and registered respiratory technicians.

Committee on Allied Health Education and Accreditation
515 N. State St.
Chicago, Ill. 60610
(312) 464-4660

Selected Journals

AARCTimes.
Daedalus Enterprises, American Association for Respiratory Care. Monthly. ISSN 0893-8520

American Review of Respiratory Disease.
American Lung Association. Monthly. ISSN 0003-0805

Anesthesia and Analgesia.
Williams and Wilkins. Monthly. ISSN 0003-2999

Anesthesiology.
American Society of Anesthesiologists. Monthly. ISSN 0003-3022

Chest.
American College of Chest Physicians. Monthly. ISSN 0012-3692

Critical Care Medicine.
Williams and Wilkins. Monthly. ISSN 0090-3493

Heart and Lung.
Mosby. Bimonthly. ISSN 0147-9563

JAMA: The Journal of the American Medical Association.
American Medical Association. Weekly. ISSN 0098-7484

New England Journal of Medicine.
Massachusetts Medical Society. Weekly. ISSN 0028-4793

Respiratory Care.
American Association for Respiratory Care. Monthly. ISSN 0020-1324

Print Materials

American Association of Cardiovascular and Pulmonary Rehabilitation
Guidelines for Pulmonary Rehabilitation Programs
Edited by Gerilynn Connors and Lona Hilling.
Champaign, Ill., Human Kinetics Press, 1993. 139p.
ISBN 0873224027

This document was written to help set standards and establish nationally recognized guidelines for pulmonary rehabilitation programs. These guidelines assist in the establishment of specific patient goals; provide for both individual patient and program evaluation; aid in the determination of essential components of assessment, training, and follow-up; and focus on making pulmonary rehabilitation more accessible. Covers new program development, enhancement, and reimbursement.

Barnes, Thomas A., ed.
Core Textbook of Respiratory Care Practice
2nd ed. St. Louis, Mo., Mosby, 1994. 781p. ISBN 0801665507

Though written for respiratory therapy students, this text would be of interest to anyone involved in the delivery of respiratory care. A collaboration of twenty-five recognized experts in the field, this core text strives "to provide vital and essential information so that the full potential and limitations of various modalities of care and equipment can be understood" (preface). All chapters and appendices from the first edition have been updated and revised. Contains high quality illustrations, closeup views of equipment and pneumatic circuits, case studies, and chapter references. Indexed.

Barnes, Thomas A.
Respiratory Care Principles: A Programmed Guide to Entry Level Practice
3rd ed. Philadelphia, Davis, 1991. 515p. ISBN 0803606621

This edition has been redesigned and expanded to assist the student in mastering respiratory care principles needed for entry-level practice. Content is organized into sixteen chapters and includes pretests and posttests, learning exercises, and extensive diagrams and tables to facilitate learning. The author provides a 200-question exam to test the reader's understanding of the basic principles of respiratory care and to identify areas of weakness. This is a helpful resource for students reviewing for the National Board for Respiratory Care (NBRC), Certified Respiratory Care Technician (CRTT) Self-Assessment Exam, or the NBRC Entry-Level CRTT Exam. Chapter bibliographies; indexed.

Barton, Richard O.
Pulmonary Rehabilitation Homecare: From Paper to Practice
Old Town, Me., Health Educator Publications, 1989. 177p.
ISBN 0932887023

A useful reference for all institutions organizing or expanding a pulmonary rehabilitation/home care program. Part One covers organization, staffing, public relations, record keeping, home care equipment, and patient relations. Part Two discusses program curriculum and evaluation, discharge planning, and follow-up. Ideas presented in this book are intended to direct goal setting and plans of action.

Burton, George C., John E. Hodgkin, and Jeffrey J. Ward, eds.
Respiratory Care: A Guide to Clinical Practice
3rd ed. Philadelphia, Lippincott, 1991. 1043p. ISBN 039750909X

Dozens of experts in the field address topics in respiratory care. Content is organized into three parts: service and education in the modern hospital; the rational scientific basis of respiratory therapy techniques; and respiratory care in critical illness. This revised edition covers home care, sleep disorders, breathing, hyperbaric oxygen therapy, and fiber optic bronchoscopy. Materials are grouped to integrate well into classroom teaching, student review, and respiratory care practice. Extensive chapter references and appendices. Indexed.

Chatburn, Robert L., and Marvin D. Lough
Handbook of Respiratory Care
2nd ed. Chicago, Year Book Medical Publishers, 1990. 467p.
ISBN 0815155840

This handbook provides respiratory care students and practitioners with a ready reference to concise, accurate data on physical and radiographic assessment, pulmonary function testing, physiologic monitoring, gas therapy, mechanical ventilation, and general clinical laboratory data. This edition has been expanded and reorganized into three major sections: patient assessment, therapeutics, and general science. Appendices include medical abbreviations and symbols as well as translations of commonly used words. Subject and drug indexes.

Cherniak, Reuben, ed.
Current Therapy of Respiratory Disease
3rd ed. Philadelphia, Decker, 1989. 372p. ISBN 1556640625

This edition represents a continuing effort to keep readers apprised of significant changes in the treatment of respiratory disease. New contributors offer differing points of view or reinforce recommendations posed by previous contributors. Each offers an approach to therapy of a specific disorder and provides a succinct pathophysiology of the disease which forms the basis of therapy. Several new chapters have been added and include information on new imaging techniques and disorders that compound respiratory disease such as smoking and depression. Includes suggested readings and index.

Des Jardins, Terry R.
Clinical Manifestations of Respiratory Disease
2nd ed. Chicago, Year Book Medical Publishers, 1990. 385p. ISBN 0815124325

This text looks at diseases commonly encountered in respiratory care and for each disease discusses the following: (1) how the disease anatomically alters the lungs; (2) what causes the disease; (3) cardiopulmonary clinical manifestations associated with the disease; and (4) general management. Self-assessment questions are included with each chapter. Includes references, glossary, appendices, and index. Third edition scheduled for late 1994.

Dupuis, Yvon G.
Ventilators: Theory and Clinical Application
2nd ed. St. Louis, Mo., Mosby-Yearbook, 1992. 494p. ISBN 0801614627

The author succinctly presents the theory, functional specifications, and clinical applications of many types of pneumatically operated and microprocessor-based critical care ventilators. Revisions to this edition include chapters on the work of breathing during ventilatory support, mandatory minute ventilation, airway pressure release ventilation, pressure support ventilation, and flow-by mode. Chapters include self-evaluation quizzes, references, detailed diagrams, and illustrations. Complex equations are limited to the appendix. Indexed.

Eubanks, David H., and Roger C. Bone
Comprehensive Respiratory Care: A Learning System
2nd ed. St. Louis, Mo., Mosby, 1990. 1028p. ISBN 0801629322

The authors have organized and written a highly detailed text that is filled with vital facts for all respiratory therapists and technicians. This work includes learning objectives, detailed anatomic drawings, and physiologic illustrations and tables. Text is divided into modules which relate specific learning objectives to content. Knowledge is presented sequentially and covers entry-level to advanced practice concepts. Language is defined as it is used. Background science is presented immediately before, or in conjunction with, respiratory care concepts. New modules cover gerontology, critical care cardiorespiratory monitoring, and third generation microprocessor computers. Indexed.

Haas, Francois, and Kenneth Axen, eds.
Pulmonary Therapy and Rehabilitation: Principles and Practice
2nd ed. Baltimore, Williams and Wilkins, 1991. 395p. ISBN 0683038877

As with the previous edition, this edition is grounded in physiology. In addition to basic pulmonary physiology, new chapters are included to cover exercise physiology. Also included is a discussion on pediatric pulmonary rehabilitation. This text is written on a level that can be easily understood by all levels of respiratory care practitioners. Contributors were encouraged to include personal observations, philosophical thoughts, and anecdotes. Chapter references and index.

Hanowell, Leland H., and Forrest L. Junod, eds.
Pulmonary Care of the Surgical Patient
Mt. Kisco, N.Y., Futura, 1994. 400p. ISBN 0879935685

This text offers an interdisciplinary perspective on pulmonary care. Contributors from the specialties of pulmonary medicine, surgery, anesthesiology, critical care, and general medicine emphasize the importance of support and assistance between disciplines to provide the best possible perioperative assessment and management of surgical candidates who may be at risk for pulmonary complications. Extensive chapter references; indexed.

Heath, Galen G., and John M. Gallagher
Respiratory Therapy Examination Review: 600 Multiple Choice Questions with Explanatory Answers
New York, Medical Examination Pub. Co., 1991. 200p. ISBN 044401571X

This basic review is intended for students preparing for the entry-level certifying examination administered by the National Board for Respiratory Care (NBRC). Content covers respiratory procedures, equipment, and practice. Multiple-choice questions follow the same format as that used by the NBRC. Explanatory answers are referenced to a variety of information sources to allow for further study as needed.

Howder, Cynthia L.
Cardiopulmonary Pharmacology: A Handbook for Respiratory Practitioners and Other Allied Health Personnel
Baltimore, Williams and Wilkins, 1992. 315p. ISBN 0683041754

The most commonly used cardiopulmonary and critical care drugs are covered in this text. Section One introduces the reader to drug therapy. Section Two discusses individual pharmacological agents and criteria for their use, administration, actions and effects, contraindications, and adverse effects. Appendices include a comprehensive assessment test which will assist readers in preparing for national exams. Bibliography and index included.

Howder, Cynthia L.
Respiratory Care: Know the Facts
Philadelphia, Lippincott, 1989. 279p. ISBN 0397509405

The purpose of this book is to present a comprehensive quick reference guide to the clinician. Materials are arranged alphabetically in outline style to allow quick reference to facts not committed to memory. Covers all facts, figures, and equations necessary for total respiratory care. A comprehensive review section allows students to assess levels of retention and evaluate areas of weakness in preparing for the Certified Respiratory Therapy Technician (CRTT) and the Registered Respiratory Therapist (RRT) examinations. Includes bibliography.

Kacmarek, Robert M., Dean Hess, and James K. Stoller, eds.
Monitoring in Respiratory Care
St. Louis, Mo., Mosby, 1993. 841p. ISBN 0815149638

This comprehensive reference will prove useful to all clinicians who provide respiratory care in both critical and noncritical care areas, regardless of their level of expertise. Text is divided into four sections: (1) overview and engineering principles; (2) invasive monitoring techniques; (3) noninvasive monitoring techniques; and (4) integration of monitoring technology into various clinical settings. Individual contributors are experts not only in the technical aspects of monitoring, but also in teaching implications and practical implementation. Numerous illustrations, tables, and references. Indexed.

Kacmarek, Robert M., Craig W. Mack, and Steven Dimas
The Essentials of Respiratory Care
3rd ed. St. Louis, Mo., Mosby, 1990. 681p. ISBN 0815149565

The rapidly changing amount of knowledge in the respiratory care field has caused the authors to add one-third more material to this edition. Pediatric and newborn care is the area most expanded, with five chapters now devoted to it. Written in an outline format, this volume strives to be concise. Access to primary sources is offered by extensive bibliographies at the end of each chapter. Includes illustrations, charts, graphs. Indexed.

Kacmarek, Robert M., and James K. Stoller, eds.
Current Respiratory Care
Philadelphia, Decker, 1988. 338p. ISBN 1556640498

Organized as a collection of consultations with recognized experts in the field of respiratory care, this text provides useful bedside clinical information as well as personal perspectives and individual approaches to problem solving. Part One covers technical and therapeutic applications with emphasis placed on individual, technical, pharmacological, and clinical approaches to patient care. Part Two addresses approaches to patient management and discusses specific clinical conditions. Extensive chapter references are replaced by shorter lists of suggested key readings. Indexed.

Kirby, Robert R., Michael J. Banner, and John B. Downs, eds.
Clinical Applications of Ventilatory Support
New York, Churchill Livingstone, 1990. 546p. ISBN 0443086133

Update of *Mechanical Ventilation* (Churchill Livingstone, 1985). Although emphasis remains on the mechanical support of ventilation, the overall scope of this text is broader. All chapters have been revised extensively to reflect recent changes in ventilator technology and techniques. Includes new chapters on anesthesia ventilators and the outcome of treatment for acute and chronic respiratory failure. Extensive use is made of illustrations, figures, and tables. Chapter references and index.

Koff, Patricia Beck, Donald Eitzman, and Josef Neu, eds.
Neonatal and Pediatric Respiratory Care
2nd ed. St. Louis, Mo., Mosby, 1993. 525p. ISBN 0801665183

This comprehensive guide has been completely updated to include the latest research in neonatal and pediatric respiratory care. Retaining the philosophy and the format of the first edition, content is arranged according to general introductory concepts: disease states, equipment and therapy, and outcomes. Chapters on equipment and therapy have been expanded significantly to reflect important advances in the field since publication of the first edition. Chapter references; indexed.

Krider, Terrance M., Rick Meyer, and William A. Syvertsen
Master Guide for Passing the Respiratory Care Credentialing Exams
2nd ed. Claremont, Calif., Educational Resource Consortium, 1989 457p. ISBN 0931263018

This edition reflects the 1987 changes to the Entry-level CRTT and Written Registry Exams. The focus is on three main areas: (1) data (reviews patient assessment and methods of collecting and evaluating clinical data to develop effective respiratory care plans); (2) equipment (covers operational, functional, cleaning, and sterilization procedures); and (3) therapy (reviews application of therapeutic procedures, monitoring patient response, and maintaining patient records). Each major section includes references for more comprehensive discussion of major topics.

Levitsky, Michael G., Jimmy M. Cairo, and Stanley M. Hall
Introduction to Respiratory Care
Philadelphia, Saunders, 1990. 589p. ISBN 0721610900

This text is intended for entry-level students in certification programs as well as for advanced-practitioner students in registry programs. Authors present a solid background in the basic cardiopulmonary sciences before introducing clinical respiratory care areas such as assessment of cardiopulmonary disease and therapeutic intervention. Chapter outlines, learning objectives, descriptive diagrams, and useful tables all serve as effective tools for learning and retrieving information quickly. Chapter bibliographies. Indexed.

Lucas, Jeffrey
Home Respiratory Care
Norwalk, Conn., Appleton and Lange, 1988. 266p.
ISBN 0838538452

This is the first source to offer a foundation and reference book for those interested in serving the needs of the respiratory care patient in the home. Authors take a pragmatic, step-by-step, hospital-to-home approach, and discuss existing treatments modified for home care, as well as new techniques created for home care. Early chapters cover the physiologic basis for long-term oxygen therapy; home-care oxygen therapy equipment; selecting optimal oxygen systems; and traveling with oxygen. Later chapters cover psychological aspects of ventilator dependence; discharge planning for the respiratory patient; and legal issues surrounding home care. Indexed.

Malley, William J.
Clinical Blood Gases: Application and Noninvasive Alternatives
Philadelphia, Saunders, 1990. 379p. ISBN 072165861X

A comprehensive review of clinical blood gases progresses logically through six units: blood gas techniques; basic physiology; interpretation of blood gases; clinical oxygenation; clinical acid base; and noninvasive techniques and case studies. There are numerous illustrations, tables, and examples. Case studies integrate the entire scope of concepts presented throughout the text. Written primarily for the respiratory care practitioner, it is also useful for other health care professionals as well. References and index.

Marini, John J.. and Charis Roussos, eds.
Ventilatory Failure
New York, Springer-Verlag, 1991. 456p. ISBN 0387542973

Twenty-five internationally recognized authorities present state-of-the-art reviews of the latest research and practical applications in ventilatory failure. Text provides an overview of the most recent developments in the pathophysiology and therapy of ventilatory failure, and describes new modalities of partial and complete ventilatory assistance and new knowledge regarding ventilatory control and fatigue during stressful breathing situations. Contains over 200 figures and tables. Chapter references and subject index.

May, Donald F.
Rehabilitation and Continuity of Care in Pulmonary Disease
St. Louis, Mo., Mosby, 1991. 240p. ISBN 0801656796

Written for health care practitioners involved in the long-term care and rehabilitation of patients with a chronic pulmonary disorder, this text emphasizes the importance of patient participation and a team approach to long-term care. Patient care plans are discussed along with patient and family education in rehabilitation, respiratory care in the home, and the role of durable medical equipment for home care. Appendices include patient-family interview procedures, and forms for use in home care. Chapter bibliographies and index.

McPherson, Steven P., and Charles B. Spearman
Respiratory Therapy Equipment
4th ed. St. Louis, Mo., Mosby, 1990. 467p. ISBN 0801633419

This book updates previous editions as a reference source for respiratory care practitioners. The fourth edition was reorganized to make the learning process more sequential. The introductory chapter reviews principles of physics necessary to the understanding of equipment function. Early chapters review simpler devices used in respiratory care. Principles outlined in these chapters provide a basis for understanding more sophisticated devices described in subsequent chapters. New chapters cover transport and home care ventilators. All chapters have been updated to reflect new equipment. Factors of clinical significance highlighted for easy review. Contains more than 800 illustrations; indexed. Fifth edition scheduled for late 1994.

Oakes, Dana F.
Neonatal/Pediatric Respiratory Care: A Critical Care Pocket Guide
Old Town, ME, Health Educator Publications, 1989. 1 v. (various pagings) ISBN 0932887015

This quick reference represents a compilation and summarization of current knowledge necessary to ensure sound clinical judgment in recognizing, preventing, and treating respiratory disorders in neonatal and pediatric patients. Ideal for students in training and practitioners who do not work with children on a regular basis. A basic knowledge of neonatal and pediatric respiratory care is assumed.

Perel, Azriel V., and M. Christine Stock, eds.
Handbook of Mechanical Ventilatory Support
Baltimore, Williams and Wilkins, 1992. 308p. ISBN 0683068563

Written for all groups of health care professionals involved in critical care, this book addresses the technical and physiologic aspects of respiratory care. Leaders in the field provide solid, practical information needed in the clinical use and application of ventilators so that practitioners may avoid common pitfalls and make appropriate decisions resulting in the best and safest ventilatory assistance available. Chapter references; glossary; indexed.

Persing, Gary
Entry-Level Respiratory Care Review: Study Guide and Workbook
Philadelphia, Saunders, 1992. 241p. ISBN 072164533X

This comprehensive combination study guide and workbook is intended to prepare the student for the Entry Level Certification Exam offered by the National Board for Respiratory Care (NBRC). Each chapter in the study guide is preceded by a pretest to evaluate current knowledge of the subject; key concepts are highlighted. The workbook offers additional questions based on materials presented in the study guide. Final comprehensive posttest with answers parallels the NBRC Entry Level Exam.

Peters, Charles J.
Clinical Simulations: The Right Way! The Complete Guide for Respiratory Care Providers
National City, Calif., California College for the Health Sciences, 1990. 370p. ISBN 0931657202

This helpful guide provides an explanation of the clinical simulation format of the National Board for Respiratory Care Ecam, including scoring criteria, glossary of terms, and commonly used "distractors" that can lead one astray. Includes eleven clinical simulation problems with explanatory answers. Intended for students preparing to take the NBRC exam, this book is also useful as an assessment tool.

Peters, Charles J.
Mathematics, the Gas Laws and the Respiratory Practitioner
San Diego, California College for Health Sciences, 1988. 33p. ISBN 0933195257

In his booklet, Peters provides a ready reference for the gas laws of respiratory care. The author presents problems illustrating these laws as they might appear on the NBRC exam. Problems are designed to test basic understanding of the rationale or principle behind gas laws. Includes rote memory exercises designed to eliminate much mathematical calculation.

Pilbeam, Susan P.
Mechanical Ventilation: Physiological and Clinical Applications
2nd ed, St. Louis, Mo., Mosby, 1992. 649p. ISBN 0801663601

Mechanical ventilation in the adult patient is the primary focus of this text. Presents basic concepts of ventilation as simply as possible, without oversimplifying scientific truths. Each chapter includes objectives, study questions, and clear diagrams and tables. This edition has been updated and expanded to include thorough coverage of new methods of ventilation; noninvasive gas monitoring and measurement of physiologic dead space; shunt hemodynamic monitoring; neonatal resuscitation and ventilation; and home mechanical ventilation. Appendices include data about mechanical ventilators currently in use. Indexed.

Rau, Joseph L.
Respiratory Care Pharmacology
4th ed. St. Louis, Mo., Mosby, 1994. 492p. ISBN 0801671841

A well-organized, practical source of fundamental information on pharmacological concepts and drug groups commonly encountered in respiratory care. All chapters have been updated to include current agents. Information ranges from simple and factual, to theoretical and complex, to provide useful information to students as well as to all levels of respiratory care practitioners. Appendices include drug calculations and preparations. Extensive chapter bibliographies. Indexed.

Ruppel, Gregg
Manual of Pulmonary Function Testing
6th ed. St. Louis, Mo., Mosby, 1994. 507p. ISBN 0801677890

Intended for students and practitioners, this title continues to provide explanations of commonly used pulmonary functions tests, techniques used, and the pathophysiology and significance of each test for pulmonary disease. The chapter on equipment and computers has been updated to cover new technologies. Quality assurance issues have been expanded to include recommendations of the American Thoracic Society and the Centers for Disease Control. Case studies also have been expanded. Self-assessment questions are new to this edition. Chapter bibliographies are arranged according to topic. Indexed.

Scanlon, Craig L., Charles B. Spearman, and Richard L. Sheldon, eds.
Egan's Fundamentals of Respiratory Care
5th ed. St. Louis, Mo., Mosby, 1990. 1037p. ISBN 0801647371

A classic in the field of respiratory care, this edition has been largely rewritten and expanded to reflect dramatic changes in the field. Continues strong emphasis on theory and principles necessary to sound and safe practice. Underlying concepts are emphasized over details that might be learned from other texts. The text is organized to facilitate learning. Reading level has been adjusted downward. The number of explanatory illustrations has tripled. Learning objectives, key words, and chapter summaries are included. Separate workbook and study guide are available. Indexed. Sixth edition scheduled for late 1994.

Shapiro, Barry A., and others
Clinical Application of Respiratory Care
4th ed. St. Louis, Mo., Mosby, 1991. 525p. ISBN 0815176078

Authors present a serviceable text for health care professionals of varying backgrounds. Content is divided into seven sections. Sections One through Four, intended for students and trainees, cover clinically relevant, basic information concerning acute respiratory care. Section Five covers respiratory monitoring and clinical assessment of critically ill patients. Section Six covers positive airway pressure therapy and presents new technologies and concepts introduced in the 1980s. Section Seven presents clinical applications of respiratory care. Extensive references. Indexed.

Shapiro, Barry A., William T. Peruzzi, and Rozanna Kozelowski-Templin
Clinical Application of Blood Gases
5th ed. St. Louis, Mo., Mosby, 1994. 427p. ISBN 0801678390

Like its predecessors, this fifth edition focuses on aiding and promoting appropriate interpretations and clinical assessments of blood gas, pH, and related measurements. To accommodate major changes in the field, the text has been extensively revised and updated to include new materials on assessment and care of the critically ill. Text is divided into six sections and materials are presented sequentially. Case studies refer to materials in chapters. Companion workbooks available. Extensive references and index.

Shoup, Cynthia A., and Terrance J. Gilmore
Laboratory Exercises in Respiratory Care
3rd ed. St. Louis, Mo., Mosby, 1988. 418p. ISBN 0801643287

As in previous editions, this manual contains exercises designed to increase comprehension of mechanical function and principles of respiratory care equipment. Each chapter includes a goal and objectives, necessary equipment, recommended readings, and timed exercises. The National Board for Respiratory Care and Joint Review Committee for Respiratory Therapy Education were consulted to identify areas of laboratory and clinical competencies expected of respiratory care practitioners. Extensive use of illustrations.

Sills, James R.
Respiratory Care Certification Guide
St. Louis, Mo., Mosby, 1991. 412p. ISBN 0801662001

Details skills and professional attitudes required of respiratory therapy technicians. Content reflects the basic restructuring of the entry-level exam matrix. Arranged in fourteen sections, the text progresses logically from patient assessment, through various therapies, to mechanical ventilation, pulmonary rehabilitation, and home care. Each section includes self-assessment questions with answers. Extensive use of illustrations aids student understanding and retention. Indexed.

Smith, Gary A., ed.
Respiratory Care: Evolution of a Profession
Lenexa, Kans., Applied Measurement Professionals, 1989. 233p. ISBN 0962475416

The text presents important events in the development and maturation of the respiratory care profession. Discussion of clinical practice, the educational system, the credentialing system, the professional association, government regulations, and the contributions of industry are included. Appendices include a detailed chronology of respiratory care milestones. Intended for students, educators, and practitioners.

Staub, Norman C.
Basic Respiratory Physiology
New York, Churchill Livingstone, 1991. 242p. ISBN 0443087555

Directed towards beginning professional students in nursing, medicine, or respiratory therapy, this basic introductory text is intended to provide a thorough, yet concise, explanation of mechanics, ventilation and perfusion, blood gas transport, and control of breathing. Upon completion of this text students will understand normal oxygen and carbon dioxide exchange and transport. Chapters include introductory summaries, drill questions and problems, extensive illustrations, and further readings. Figures are simplified to stress features important to the beginning student. Indexed.

Surkin, Howard B., and Anna Weigand Parkman
The Respiratory Care Workbook
Philadelphia, Davis, 1990. 372p. ISBN 0803682298

This well-illustrated study guide is intended to clarify and reinforce understanding of fundamental and advanced concepts in respiratory care as presented in major textbooks in the field. The first unit reviews basic concepts in a logical and progressive sequence. The second unit covers advanced concepts including pharmacology, neonatal and pediatric respiratory care, pulmonary rehabilitation, and home care. Each chapter is arranged with an outline, study objectives, and theoretical concepts presented in a question-and-answer format. All chapters conclude with a NBRC credentialing-style quiz.

Tobin, Martin J., ed.
Respiratory Monitoring
New York, Churchill Livingstone, 1991. 258p. ISBN 0443088314

This practical text was written to keep respiratory care workers abreast of the latest issues and advances in respiratory monitoring. Reviews of the state-of-the-art are written by experts in their fields and cover such topics as arterial blood gas monitoring, control of breathing and respiratory muscle function, computerization and quality control of monitoring techniques, use of monitoring in clinical decision making, and cost-effectiveness of respiratory monitoring. Extensive chapter references and index.

Wanger, Jack
Pulmonary Function Testing: A Practical Approach
Baltimore, Williams and Wilkins, 1992. 227p. ISBN 0683086073

A how-to manual intended to assist students as well as respiratory therapists, nurses, and others who are already involved in measuring pulmonary function. Focusing on the most commonly used tests, each chapter is a self-contained unit covering brief historical information, pertinent background information, and relevant physiology as well as instrumentation, techniques, calculations used, and quality control controversies. Self-assessment questions are included with each chapter. Indexed.

West, John B.
Respiratory Physiology—The Essentials
4th ed. Baltimore, Williams and Wilkins, 1990. 185p.
ISBN 0683089420

This revised edition updates information on pulmonary circulation, metabolism, ventilation-perfusion relationships, blood-gas transport, mechanics and the control of ventilation. Supplemental audio tapes with slides are available separately. Includes questions and answers. References and index.

Whitaker, Kent B.
Comprehensive Perinatal and Pediatric Respiratory Care
Albany, N.Y., Delmar Publishers, 1992. 688p. ISBN 0827338503

Provides in-depth coverage of respiratory care equipment and therapeutic modalities used in neonatal and pediatric care. Five units cover fetal development; care of the neonatal and pediatric patient; causes and care of illness in perinatal and pediatric patients; management of ventilation and oxygenation; and transport. Each chapter includes instructional objectives, a bibliography, a self-evaluation test, and a key word list with corresponding glossary, to aid in understanding difficult words and concepts. An examination review book designed to prepare students for the NBRC Perinatal/Pediatric Specialty Exam accompanies text. Corresponding clinical lab manual available. Indexed.

Whitaker, Kent B.
Perinatal and Pediatric Respiratory Care: Clinical Lab Manual and Competencies
Albany, N.Y., Delmar Publishers, 1992. 264p. ISBN 0827338554

Accompanies Whitaker's *Comprehensive Perinatal and Pediatric Respiratory Care* (Delmar Publishers, 1992). Designed to be taken by the student into the clinical setting, this manual focuses on neonatal skills. Section One serves as a reference and review to reinforce understanding of the clinical procedures presented in the corresponding text. Section Two is dedicated to basic skills performed by the respiratory care practitioner in the clinical setting. Laboratory exercises are included. Appendices.

White, Gary C.
Basic Clinical Lab Competencies for Respiratory Care: An Integrated Approach
2nd ed. Albany, Delmar Publishers, 1993. 468p.
ISBN 0827351186

This is the only text available that integrates theoretical knowledge with laboratory skills. This combination has proven effective in facilitating retention of information. Includes both entry-level and advanced skills and procedures. Updated sections include universal precautions, phlebotomy, arterial blood gas sampling, decontamination, and equipment processing. Ventilator chapters are revised to incorporate new modes of therapy. Also included are new chapters on pharmacology, electrocardiography, noninvasive monitoring, and newborn mechanical ventilation. Includes self-evaluation posttests to assess knowledge. Indexed.

White, Gary C.
Equipment Theory for Respiratory Care
Albany, Delmar Publishers, 1992. 505p. ISBN 0827338600

This comprehensive text presents a representative sampling of equipment used in respiratory care. The author emphasizes physical and engineering principles employed in the operation of each piece of equipment. Illustrations and computer-aided design drawings are used throughout the text to facilitate learning. Assembly and troubleshooting guides are provided to assist students in preparing equipment for clinical use. Objectives are keyed to main headings in chapters. Includes review questions and self-assessment quiz. Simplified language helps the student understand complex concepts. Indexed.

Wilkins, Robert L., and James R. Dexter, eds.
Respiratory Disease: Principles of Patient Care
Philadelphia, Davis, 1993. 414p. ISBN 0803693265

This volume is organized around the care of twenty-two of the most commonly encountered respiratory conditions. The objective is to give the practitioner or student a quick reference to the underlying physical condition resulting in the signs and symptoms of each condition, the information that must be gathered for a thorough assessment of the patient, and the appropriate treatment of each condition. This unique and useful format includes case studies as well as chapter references and study questions. Indexed.

Wilkins, Robert L., Richard L. Sheldon and Susan Jones Krider
Clinical Assessment in Respiratory Care
2nd ed. St. Louis, Mo., Mosby, 1990. 319p. ISBN 0801653290

The focus of this book is on assessment techniques for health professionals treating cardiopulmonary disease. It is divided into two sections: fundamentals of respiratory assessment, and advanced assessment techniques. New to this edition are chapters on the basics of interpreting electrocardiograms and sleep-related breathing disorders. Information on patient assessment in the intensive care unit has been greatly expanded. Extensive chapter references. Indexed.

Wilkins, Robert L., John E. Hodgkin, and Brad Lopez
Lung Sounds: A Practical Guide
St. Louis, Mo., Mosby, 1988. 116p. ISBN 0801655323

This text and accompanying audiocassette were developed to help students and clinicians better understand lung sounds. Introductory chapters review pulmonary anatomy and physiology and explain how auscultation fits into patient evaluation. Later chapters review terminology used to describe lung sounds and describe clinical application of lung sounds. Case studies are presented along with auscultatory findings. Adult and neonatal lung sounds are recorded.

Witek, Theodore J., and E. Neil Schachter
Pharmacology and Therapeutics in Respiratory Care
Philadelphia, Saunders, 1994. 465p. ISBN 0721634834

The authors present a state-of-the-art reference for respiratory therapists and other health care providers involved in the care of respiratory patients. Provides comprehensive, yet practical information essential for building a foundation to understand the vast amount of information emerging in respiratory pharmacology. Divided into three parts: scientific basis for therapy; major classes of respiratory drugs; and new and emerging applications of respiratory pharmacology. Experts in the field served as consulting editors. Extensive diagrams, tables, and chapter references. Indexed.

Nonprint Materials

Chatburn, Robert L.
Theory and Application of Neonatal Ventilation
Dallas, American Association for Respiratory Care, 1988. 1 videocassette (60 min.) VHS

Respiratory therapist Chatburn discusses the physiology and chemistry of infant respiration. He presents mathematical models for various changes in ventilator procedure. Typical ventilators and humidification systems are diagrammed. A flowchart and computer algorithm for ventilating infants are shown.

Chest Tubes; Patient and System Management
Garden Grove, Calif., Medcom, 1988. 1 videocassette (27 min.) VHS

Management of patients and equipment before, during, and after instituting closed chest drainage are the major topics discussed. Measures to promote the evacuation of air or fluid are shown. This video instructs the practitioner in stripping the tubing, changing the dressing and drainage system, and troubleshooting any complications. Part of the Airway Management Series.

Endotracheal Intubation
Garden Grove, Calif., Medcom, 1988. 1 videocassette (28 min.) VHS

Assessing the need for airway assistance and the type of device to use are the focus of this video. Pertinent anatomy and physiology are reviewed. The viewer will see a demonstration of the procedure for endotracheal intubation as well as emergency oral and nasal intubation. Potential complications of these critical procedures are also discussed. Part of the Airway Management Series.

Mechanical Ventilation
Springhouse, Pa., Springhouse Corporation, 1991. 1 videocassette (45 min.) VHS, ISBN 0874343739

This videotape explains several topics of interest to a therapist or nurse caring for a patient on a mechanical ventilator. Includes the following topics: how mechanical ventilators work; assessing a patient who is on a ventilator; easing patient discomfort; dealing with high-and low-pressure alarms; and troubleshooting common problems in ventilator use. Clear examples are given for each topic. A twenty-four page booklet provides additional information. Includes a short bibliography and a video index.

Smoking Cessation Intervention Techniques for the Respiratory Care Practitioner
Dallas, American Association for Respiratory Care, 199-?, 1 videocassette (58 min.) VHS

Dr. Kathleen Smalky of Southwest Medical Center in Dallas addresses respiratory care professionals about smoking cessation techniques. This videotape focuses on how smoking causes illness and death; why people smoke; current smoking cessation programs; and what respiratory therapists can do to help their patients during smoking cessation. Specific written resources are recommended.

Suctioning: Nasotracheal, Oropharyngeal and Endotracheal Techniques
Garden Grove, Calif., Medcom, 1988. 1 videocassette (34 min.) VHS

The potential hazards of accumulated respiratory secretions are explained through an animated look at pertinent anatomy. The tape demonstrates three types of suctioning and illustrates potential complications for the patient (and options for treatment if complications occur). The effectiveness of suctioning is shown, as well as how the practitioner should document actions taken. Part of the *Airway Management Series*.

Tracheostomy Care, Tube Change, and Artificial Airway Cuff Management
Garden Grove, Calif., Medcom, 1988. 1 videocassette (24 min.) VHS

The precise movements needed to set up equipment and complete the care of a patient with a tracheostomy are shown in detail on this videotape. The steps illustrated are cleansing the stoma, changing the dressing, and maintaining proper cuff pressure. A team approach to tracheostomy care is highlighted. Part of the *Airway Management Series*.

Weaning and Extubation
Garden Grove, Calif., Medcom, 1988. 1 videocassette (25 min.) VHS

This tape covers the entire process of weaning and extubation. Particular emphasis as to which studies should be performed before weaning, and which signs the practitioner should monitor when observing the weaning process. The extubation process is demonstrated, including precautionary measures to take should the patient require emergency reintubation. Part of the Airway *Management Series*.

Building and Construction Trades

Cabinetmaking and Woodworking 121
Patty Miller

Construction, Remodeling, and Repair 127
Barbara Edwards

Electrical Technology 141
Dan Haley

Heating, Ventilation, and Air Conditioning 150
Steven Self

Masonry 155
Steven Self

Plumbing Technology 159
Joanne Kim and Tori Beyer

Cabinetmaking and Woodworking

Patty Miller
New Hampshire Technical College
Laconia, New Hampshire

Introduction

Those entering the many fields within cabinetmaking and woodworking may choose from a variety of jobs where raw wood is transformed into a product: woodworker, wood machinist, assembler, cabinetmaker, or finisher. They may be production workers in industries such as furniture, cabinetry, and other fabricated wood product manufacturing. Precision work may be done in architectural woodworking, furniture, or other specialty items. The woodworker operates machines that cut, shape, assemble, and finish, working from prints, instructions, or software.

Machining processes include layout, setup, and operating machines such as saws, planers, sanders, lathes, jointers, and routers. Components are shaped and fit, some involving hand tools such as planes, chisels, files, or sandpaper. Assemblers produce subassemblies, using fasteners and adhesives, then sand and finish the piece. Custom woodworkers can be cabinetmakers, pattern and model makers, wood machinists, or wood finishers, and may work on a custom basis. A combination of tasks may be carried out, including cutting, shaping, surface preparation, assembly, and finishing. Computer numerical control machines are used increasingly; machines exist that take in a piece of wood and turn out a finished paneled cabinet front.

Working conditions may be dusty, noisy, and include noxious air quality, but in all cases, substantial safety procedures must be followed. Considerable physical activity is involved.

Although most woodworkers learn through on-the-job training, an increasing number are finding that college programs can prepare them for positions in production, supervision, engineering, or management. Workers find they will have enhanced employment and advancement opportunities with training in mathematics and computer applications. Mechanical ability, manual dexterity, and the ability to pay attention to detail are important. The experienced cabinetmaker may become an inspector or supervisor, or may choose to go into a specialty field such as producing unique items or doing special orders.

High school completion is a must; most college programs are at the certificate or associate degree level, and there are some specialized doctoral programs.

Employment growth will be modest due to environmental considerations, automated processes, and uses of materials such as metal, plastic, and fiberglass. However, replacement workers will be in demand, and there is particular need for workers in the repair and renovation of residential and commercial property.

The focus of educational programs is not on the basics but on new skill development that integrates production, design, manufacture, and marketing. Part of the modest rise in worker demand may be attributed to increased use of computerized equipment, and workers using such equipment are highly skilled, with strong backgrounds in mathematics.

This bibliography has been created for libraries at the technical and community college level, and is intended for librarians, as well as those working in the field.

Selected Associations

The following are representative of associations that provide information on cabinetmaking and woodworking education, industry trends, and continuing education:

American Association of Woodturners
667 Harriet Ave.
Shoreview, Minn. 55126

International Woodworkers of America, U.S. AFL-CIO
25 Cornell
Gladstone, Ore. 97027

Kitchen Cabinet Manufacturers Association
1899 Preston White Dr.
Reston, Va. 22091-4326

United Brotherhood of Carpenters and Joiners of America
101 Constitution Ave. NW
Washington, D.C. 20001

Selected Journals

American Woodworker.
Rodale Press. Seven times a year. ISSN 8750-9318

Better Homes and Gardens Wood.
Better Homes and Gardens. Bimonthly. ISSN 0743-894X

Fine Woodworking.
Taunton Press. Bimonthly. ISSN 0361-3453

Wood Digest.
PTN Publishing Co. Monthly. ISSN 1045-7348

Woodshop News. Northeast.
Soundings Publications. Monthly. ISSN 0894-5403

Woodsmith.
Woodsmith Publishing. Bimonthly. ISSN 0164-4114

Woodworker's Journal.
Madrigal Publishing. Bimonthly. ISSN 0199-1892

Woodworking information for the handyperson is also found in the following popular periodicals:

Home Mechanix.
Times Mirror. Ten times a year. ISSN 8755-0423

Popular Mechanics.
Hearst Corp. Monthly. ISSN 0032-4558

Popular Science.
Times Mirror. Monthly. ISSN 0161-7370

Print Materials

Blackburn, Graham
The Illustrated Encyclopedia of Woodworking Handtools, Instruments, and Devices: Containing a Full Description of the Tools Used by Carpenters, Joiners, and Cabinetmakers
Rev. ed. Chester, Conn, Globe Pequot Press, 1992. 147p.
ISBN 0871061686

An alphabetical guide to tools from the eighteenth century to today. Includes cross references, explanations, and drawings.

Blandford, Percy W.
Percy Blandford's Favorite Woodworking Projects
Blue Ridge Summit, Pa., TAB Books, 1992. 780p.
ISBN 0830621482

Almost 200 projects in this book, which are divided into areas such as weekend projects, tables, children's furniture, outdoor and other furniture, toys and puzzles, tools and workshops, and lathework. Presents a wide variety of projects designed for those with workshop experience.

Broun, Jeremy
Electric Woodwork: Power Tool Woodworking
Lawes, East Sussex, England, Guild of Master Craftsman Publications, 1993. 280p. ISBN 0946819262

Broun introduces readers to various power tools such as the drill, router, bandsaw, lathe, glue gun, and power sander. He then moves on to projects, including a bench, tool rack, and stool. Includes good photographs and diagrams. Emphasis is on safety.

Cabinetmaking
Alexandria, Va., Time-Life Books, 1992. 144p. ISBN 0809499045

A thorough introduction to cabinetmaking, including carcass construction, frame-and-panel construction, drawers, doors, and legs. Clearly written and amply illustrated with drawings, the author presents solid instruction on the topic. Contains glossary and index. A volume in the Time-Life *Art of Woodworking* series.

Cabinets and Bookcases
Alexandria, Va., Time-Life Books, 1993. 144p. ISBN 0809499452

Projects are prefaced by a chapter about basic skills and information (lumber, preparation, sanding, joinery). Projects include a bookcase, armoire, blanket chest, and highboy. Provides detailed instructions, accompanied by many drawings and photographs. Includes glossary. A volume in the Time-Life *Art of Woodworking* series.

Carpentry: Tools, Shelves, Walls, Doors
Minnetonka, Minn., Cy DeCosse, 1989. 128p. ISBN 0865737045

Written for the homeowner, this work contains a helpful section on basic carpentry tools, fasteners, and the home workshop. Later chapters are devoted to the projects. Text is accompanied by numerous photographs. *Black and Decker Home Improvement Library* series.

Classic Woodworking Woods: And How to Use Them
Des Moines, Iowa, Meredith Books, 1993. 96p. ISBN 0696024691

The focus of this work is on judging, selecting, and drying wood to use in woodworking projects. Discusses different types of wood such as aspen and yellow birch. Contains pictures of the tree, uses of the wood, cost, and availability. *Better Homes and Gardens Wood* series.

Cliffe, Roger W.
Radial Arm Saw Basics
New York, Sterling Pub. Co., 1991. 128p. ISBN 0806972181
 A good starter book that provides information on the radial arm saws: different blades, safety, basic procedures, and adjustments. Contains a brief section about work preparation.

Cliffe, Roger W., and Michael J. Holtz
Shaper Handbook
New York, Sterling Pub. Co., 1990. 255p. ISBN 0806967986
 The authors discuss the purchase, maintenance, safety, troubleshooting, controls, and attachments of a shaper. Various operations are detailed, from basic through advanced, and are accompanied by photographs on each page. Discusses edges and joints as well as specialized operations such as balusters. Detailed, clear, and authoritative.

Conover, Ernie
The Lathe Book: A Complete Guide for the Wood Craftsman
Newtown, Conn., Taunton Press, 1993. 197p. ISBN 1561580570
 Focuses on the lathe as it relates to other essential woodworking tools. Details techniques such as turning square to round, cutting coves and beads, planing, and maintenance of the lathe. Includes glossary, suppliers list, and safety section. A *Fine Woodworking* book.

Duginske, Gene, and Mark Duginske
Band Saw Basics
New York, Sterling Pub. Co., 1990. 127p. ISBN 0806972106
 With easy-to-follow directions, photographs, and diagrams, the authors provide invaluable advice on all the intricacies of using a band saw, from basic uses to scroll sawing, circular work, straight cuts, curves, and decorations. A good introductory book with clear presentation.

Duginske, Mark
Band Saw Handbook
New York, Sterling Pub. Co., 1989. 318p. ISBN 0806963980
 The versatility of the band saw makes it suitable for many jobs. Duginske covers aligning, tracking, adjusting, and tensioning, before going on to methods of cutting, making jigs, tips, and troubleshooting. Includes diagrams, charts, and photographs.

Effner, Jim
Chisels on a Wheel: A Comprehensive Reference to Modern Woodworking Tools and Materials
Ann Arbor, Mich., Prakken Publications, 1992. 200p. ISBN 0911168826
 New materials such as panels, constructed woods, fibers, chips, and particles have resulted in the creation of new cutting tools. Effner covers materials and tools in detail: saw blades, drills, routers, moveable knives, and special tools. This book extends beyond the home workshop.

Engler, Nick
Advanced Routing: Techniques for Better Woodworking
Emmaus, Pa., Rodale Press, 1993. 123p. ISBN 0875965784
 Engler discusses joinery, patterns, lathe turning, and inlays. His instructions are clear and detailed, and are accompanied by diagrams and black-and-white photographs on all pages. The author's alternative way to do things makes this work one that should expand any woodworker's repertoire. *Workshop Companion* series.

Engler, Nick
Making Built-In Cabinets: Techniques for Better Woodworking
Emmaus, Pa., Rodale Press, 1992. 124p. ISBN 0875961398
 This work covers the planning, building, and installation of all types of cabinets, including kitchen cupboards, bathroom vanities, bookshelves, and tool cabinets. Plans are provided for a cabinetmaking jig and for a complete set of built-in kitchen cabinets, a counter, and a freestanding cabinet. Illustrated with numerous drawings and photographs. *Workshop Companion* series.

Engler, Nick
Making Desks and Bookcases: Techniques for Better Woodworking
Emmaus, Pa., Rodale Press, 1993. 124p. ISBN 0875965814
 A discussion of the many types and styles of desks and bookcases precedes detailed instructions for construction. There are chapters on the bookcase, shelves and doors, and drawers and pigeonholes. Several projects round out this helpful book: a country letter desk, adjustable bookshelves, and a hanging display case. Very useful and well illustrated. *Workshop Companion* series.

Engler, Nick
Using the Band Saw: Techniques for Better Woodworking
Emmaus, Pa., Rodale Press, 1992. 124p. ISBN 0875961401
 A specialized book about a very versatile tool. Covers miters, crosscuts, patterns, compound curves, and resawing. Selection, tuning, blades, and cutting speeds, as well as special shop tips, jigs, and charts preface the chapter containing several projects. Clearly written, well illustrated. *Workshop Companion* series.

Engler, Nick
Using the Table Saw: Techniques for Better Woodworking
Emmaus, Pa., Rodale Press, 1992. 124p. ISBN 0875961274
 A basic, but thorough, introduction to table saws. Covers selection, alignment, and instructions for use including crosscuts, rips, miters, dadoing, molding, and special techniques such as paneling, tapering, and kerf bending. Several projects show how the table saw can be used. Includes many photographs, diagrams, and hints. *Workshop Companion* series.

Engler, Nick
Workbenches and Shop Furniture: Techniques for Better Woodworking
Emmaus, Pa., Rodale Press, 1993. 124p. ISBN 0875965792
 The author discusses ways of designing a workshop to suit the owner's preferences and needs. He covers workbenches and their accessories (vices, stops, hold-downs), tool stands, tool cabinets, and shop storage. This book is clearly written, with numerous diagrams and photographs to assist the woodworker in creating an ideal workshop. *Workshop Companion* series.

Fine Woodworking Magazine
Bench Tools
Newtown, Conn., Taunton Press, 1990. 127p. ISBN 0942391845
 A volume of reprint articles in the *Best of Fine Woodworking* series. Emphasis is on traditional tools and methods. The author

presents different kinds of chisels, planes, scrapers, saws, and trimmers, and methods of grinding and sharpening. Includes many photographs and drawings.

Fine Woodworking Magazine
Fine Woodworking Design Book Six: 266 Photographs of the Best Work in Wood
Newtown, Conn., Taunton Press, 1992. 185p. ISBN 1561580171

A beautiful coffee-table book featuring woodworking, picturing cabinets, chairs, boxes, desks, beds, sculpture, turnings, and musical instruments. Fine craftsmanship is evident in the wide range of contemporary, unusual designs that are covered. Form, detail, edges, surface, hardware, and finish innovation will inspire the woodworker.

Fine Woodworking Magazine
Fine Woodworking on More Proven Shop Tips
New York, Taunton Press, 1990. 87p. ISBN 0942391438

A second compilation from the "Methods of Work" column in *Fine Woodworking* magazine. Three woodworking shop mainstays are featured: table saw, lathe, and router. Construction tips and aids present a wealth of very helpful information.

Fine Woodworking Magazine
Finishes and Finishing Techniques
Newtown, Conn., Taunton Press, 1991. 127p. ISBN 1561580031

Experts talk about such topics as finishing, compounds, safety, dyes and stains, and the best tools and methods to use. These articles reflect decades of combined experience by the writers, further shown in the many photographs, drawings, and charts. *Best of Fine Woodworking* series.

Fine Woodworking Magazine
Lathes and Turning Techniques
Newtown, Conn., Taunton Press, 1991. 127p. ISBN 156158021X

Demonstrates how chair parts, table legs, bowls, candlesticks, and a croquet set can be made with one machine and some chisels. Tips for beginners and advanced turners are included, along with information on woodworking schools and seminars available in Vermont, Ireland, and Germany. A volume of reprint articles from the *Best of Fine Woodworking* series.

Fine Woodworking Magazine
Modern Furniture Projects
Newtown, Conn., Taunton Press, 1991. 127p. ISBN 0942391918

This work's emphasis is on function, durability, simple lines, and minimal ornamentation. Projects include bookcases, credenzas, beds, and music stands. Contains much discussion and detail; it is not for beginners. Beautiful photographs. A volume of reprint articles from the *Best of Fine Woodworking* series.

Fine Woodworking Magazine
Power Saws and Planers
Newtown, Conn., Taunton Press, 1990. 128p. ISBN 0942391837

The key to fine woodworking lies with setting and tuning the machines; this book tells how to do it skillfully. Provides information on blades, fences, safety, scrolls, and jointers. Includes photographs, charts, and diagrams. A volume of reprint articles from the *Best of Fine Woodworking* series.

Fine Woodworking Magazine
The Small Woodshop
Newtown, Conn., Taunton Press, 1993. 127p. ISBN 1561580619

This work focuses on making the most of a work space. It includes information on outfitting a first shop, fundamental tools, safety, wiring, and solvent disposal. Illustrations are well drawn, well photographed, and well presented. A volume of reprint articles from the *Best of Fine Woodworking* series.

Fine Woodworking Magazine
Small Woodworking Projects
Newtown, Conn., Taunton Press, 1992. 127p. ISBN 156158018X

A volume of reprint articles in the *Best of Fine Woodworking* series. Presents a wide variety of projects to make including a fire engine, a whistle, a bracelet, wooden shoes, pool cues, and jigsaw puzzles. Not for the beginner, but these projects could be an inspiration. Beautifully photographed.

Flexner, Bob
Understanding Wood Finishing: How to Select and Apply the Right Finish
Emmaus, Pa., Rodale Press, 1994. 310p. ISBN 0875965660

Years of study in the chemistry of wood finishing leads to this comprehensive, clearly presented book that provides detailed information about wood, tools, finishes, repairs, strippers, and wood care. Includes tips, cautions, and myths about wood finishing. Contains numerous photographs.

Geary, Don
Woodworking Projects for the Great Outdoors: Forty-two Complete Step-By-Step Projects for Campers, Hunters, Boaters, Anglers, Gardeners, Bicyclists, Walkers, and Photographers
Pownal, Vt., Storey Communications, 1990. 137p.
ISBN 0882666177

Discusses tools, materials, techniques, adhesives, and finishes in its introductory chapter and then moves on to specific projects. Contains drawings, materials lists, and clear instructions for the home shop.

Gutierrez, Al, ed.
Woodworking Projects Yearbook: Favorite Projects and Techniques
New York, Meredith Press, 1993- . Annual. 1 v. (various pagings)

Organized into three sections: indoor projects, outdoor projects, and special tools. This book incorporates drawings, cutting lists, tips, skill level, and instructions for a variety of projects, from a child's toolbox to a garden shed. Well presented, with many photographs.

Hand Tools
Alexandria, Va., Time-Life Books, 1993. 144p. ISBN 0809499258

This wide-ranging introductory book provides information on how to measure, mark, saw, chisel, bore, smooth, shape, strike, fasten, and clamp. The many drawings are clear and helpful. Includes glossary. Time-Life *Art of Woodworking* series.

Handbook of Joinery
Alexandria, Va., Time-Life Books, 1993. 144p.
ISBN 080949941X

Clearly organized, the introductory chapter leads to five sections on different types of joints. Many clear pictures and drawings make the material easy to follow and understand. Includes a chapter on Japanese joinery and a glossary. Time-Life *Art of Woodworking* series.

Home Workshop
Alexandria, Va., Time-Life Books, 1993. 144p. ISBN 0809499207

This basic book is a good starting point for those setting up a home workshop, with chapters on safety, layout, the workbench, accessories, storage, and work surfaces. Includes numerous drawings and photographs. Time-Life *Art of Woodworking* series.

Hylton, Bill, and Fred Matlack
Woodworking with the Router: Professional Router Techniques and Jigs Any Woodworker Can Use
Emmaus, Pa., Rodale Press, 1993. 344p. ISBN 0875965776

This thorough and detailed manual goes beyond the basics in discussions of routers, bits, templates, surfaces, and joints. It provides ideas on alternative ways to do things, and numerous handy hints, time-savers, and cross-references to other sections for related information. Includes an excellent table of contents and clear drawings.

Leach, Noel Johnson
Modern Wood Finishing Techniques
Fresno, Calif., Linden Pub. Co., 1993. 235p. ISBN 0941936244

The complete range of finishing tasks is covered, e.g., surface preparation, abrasives, coloring wood, and various finishes and processes available, both traditional and new methods. Safety checklists are provided with each chapter, along with common questions and answers. A list of suppliers completes this wide-ranging book.

Nelson, John A.
The Weekend Woodworker: 101 Easy-to-Build Projects
Emmaus, Pa., Rodale Press, 1990. 292p. ISBN 0878578943

The focus of this book is on uncomplicated projects that require only basic joinery and common power and hand tools. The author's intended audience is people who want to work with wood but do not have much time. Materials lists, views (front, top, right views, section), and equipment suppliers are provided.

Projects for the Home Craftsman: Cabinets and Chests, Tables and Chairs, Kitchen Projects, Accents, Outdoor Projects, Toys
Emmaus, Pa., Rodale Press, 1993. 264p. ISBN 0875965938

Describes projects for a wide range of skills and of varying complexity, progressing from a single shelf to an involved corner cupboard. Forty projects are detailed suitable for the novice or the advanced woodworker. Photographs, diagrams, materials lists, step- by-step instructions, and shop tips combine to make this a handy book. *Weekend Woodworker* annual series (1993).

Routing and Shaping
Alexandria, Va., Time-Life Books, 1993. 144p. ISBN 0809499371

Like the other books in the *Art of Woodworking* series, the basics are presented, followed by specifics: edges, grooves, joinery, and shaping. Easy-to-follow drawings and photographs help illustrate the various techniques.

Saberin, Gloria
Southwestern Country Classics: Early American Woodworking Projects
Blue Ridge Summit, Pa., TAB Books, 1993. 158p.
ISBN 0830641955

The term Early American is not restricted to pine, maple, and cherry wood, or to the styles of the original thirteen colonies. This highly illustrated book includes Native American and Spanish-based motifs. Twenty projects with materials lists and clear instructions are provided. Some experience is assumed.

Salaman, R. A.
Dictionary of Woodworking Tools, c. 1700-1970, and Tools of Allied Trades
Rev. ed., revised by Philip Walker. Newtown, Conn., Taunton Press, 1990. 546p. ISBN 0942391519

Hand tools became increasing differentiated to meet new demands around the turn of the eighteenth century. Approximately 2,000 hand tools are defined, illustrated, and explained. This work provides the historical framework for modern tools. Includes many cross-references.

Schiff, David, and Kenneth S. Burton
The Woodworkers Guide to Making and Using Jigs, Fixtures and Setups: How to Get the Most From Every Tool in Your Shop
Emmaus, Pa., Rodale Press, 1992. 318p. ISBN 0875961371

Designed to help get the job done with basic tools, this work encourages the woodworker to consider a wider spectrum of uses for tools such as the jig. Hundreds of applications are demonstrated through photographs and drawings. Clearly organized with a good table of contents.

Stankus, Bill
Setting Up Your Own Woodworking Shop
New York, Sterling Pub. Co., 1993. 224p. ISBN 0806983140

Different space requirements, skill levels, and interests are factors in deciding what kind of workshop to build. The beginning and advanced woodworker will be able to conceptualize and design plans, and build a workshop based on the clear information presented here. Stankus provides information on tools, machines, dust, and safety. Includes many photographs, charts, and diagrams.

Taylor, V. J.
The Wood Worker's Dictionary
Pownal, Vt., Storey Communications, 1990. 260p.
ISBN 0882666460

This dictionary of more than 3,500 terms will serve as an excellent single source on this subject. Comprehensive descriptions include many illustrations of cabinetmaking, carpentry, joinery, and furniture repair. Separate sections on craftspersons and furniture styles round out this comprehensive source.

Three Hundred and Forty-Seven Woodworking Patterns: A Bound Set of Woodworking Patterns
Peachtree City, Ga., FC&A, 1993. 160p. ISBN 0915099268

Includes designs for accessories, decorations, toys, games, garden ornaments, clocks, lamps, plaques, door signs, jewelry, frames, racks, hangers, and whirligigs. Instructions are very brief and assume prior experience. Index.

Tolpin, Jim
Working at Woodworking: How to Organize Your Shop and Your Business
Newtown, Conn., Taunton Press, 1990. 147p. ISBN 0942391675

Written for those who want to open a small shop, Tolpin discusses organization and equipment needs, preparation for jobs, manufacturing, and general business principles. Tolpin's expertise is evident throughout. Includes a list of suppliers, reading materials, and a production flow chart.

Underhill, Roy
The Woodwright's Eclectic Workshop
Chapel Hill, N.C., Univ. of North Carolina Press, 1991. 238p. ISBN 0807820032

Roy Underhill, of PBS and Williamsburg fame, describes traditional woodworking methods. He illustrates the use of axes, mallets, chisels, and planes in the making of furniture, toys, and instruments. The coverage of nonpower tools provides a needed balance to most other books. Numerous black-and-white photographs supplement the text.

Wagner, Willis H., and Clois E. Kicklighter
Modern Woodworking Workbook: Tools, Materials, and Processes
South Holland, Ill., Goodheart-Willcox, 1991. 638p. ISBN 0870068709

Intended for high school and college programs, this work covers wood and its selection, hand and power tools, safety, woodworking procedures, and residential construction. Includes furniture, cabinetmaking, wood finishing, laminating (wood and plastic), shaping, and patternmaking. Photographs and drawings appear on every page. An instructor guide and answer key are also available.

Washer, Doug
Coated Abrasives Reference Manual: Woodworking and Furniture Industries
Hickory, N.C., Klingspor Abrasives, 1991. 64p. ISBN 0962971707

Coated abrasives are essential to woodworking at all stages, from rough planing to final finish. This book discusses different types of sanding, product recommendations, safety, storage, and usage. Also provided are troubleshooting and glossary sections.

Wearing, Robert
Woodworker's Essential Shop Aids and Jigs: Original Devices You Can Make
New York, Sterling Pub. Co., 1992. 160p. ISBN 0806985844

In this thorough presentation, Wearing discusses the maximum use of a jig for a variety of tasks such as holding, clamping, marking, routing, drilling, veneering, and lathework. More than 400 illustrations aid understanding.

Wood: Basic Woodworking Tips and Techniques
Des Moines, Iowa, Meredith Books, 1992. 96p. ISBN 0696019515

Provides a general introduction to woodworking including the following topics: guidance on how to get the most out of tools, project fundamentals, and woodworking techniques (such as mortise and tenon and miter joints.) Supplies standards for seating, tables, bookcases, shelves, cabinets, screws, and tools. Numerous drawings and photographs.

Woodworking Machines
Alexandria, Va., Time-Life Books, 1992. 144p. ISBN 0809499002

Presents basic information and diagrams on the drill press, radial arm saw, table saw, band saw, and jointer. Includes clear descriptions, many drawings, and a glossary. This solid introduction to the topic is part of the *Art of Woodworking* series.

Nonprint Materials

Kitchen Cabinet Construction
Media, Pa., Bergwall Video Productions, 1989. 4 videocassettes (77 min.) ISBN 0943008956

A clear, knowledgeable, and thorough presentation that illustrates all the steps in cabinet building, including many close-up views. Includes objectives, pretests, posttests and plans.

✦ Software

Although not previewed for this bibliography, there are many software packages available for cabinetmaking. For example, *Cabinetware, Inc.* covers casework and cabinetry, from estimating, design, and cutting to final detailing. Refer to current issues of *Wood Digest* for other titles.

CabnetWare, Inc.
35355 Country Rd. 31
Davis, Calif. 95616

Other software packages to consider are those available from Pattern Systems International, Cabinet Vision, and DrawPower 3rd.

✦ Supply Catalogs

Catalogs supply a rich source of materials for ideas and projects:

Constantine
2050 Eastchester Road
Bronx, N.Y. 10461-2297

Craft Supplies USA
P.O. Box 50300
Provo, Utah 84605-0300

Full Circle: A Woodturners Marketplace
The Mill
172 West St.
Litchfield, Conn. 06759-1179

The Sanding Catalogue
P.O. Box 3737
Hickory, N.C. 28603-9928

Woodcraft
P.O. Box 1686
Parkersburg, W. Va. 26102-1686

The Woodworker's Friend
3149 Major Street
Fort Worth, Tex. 76112-9945

Construction, Remodeling, and Repair

Barbara Edwards
Ranken Technical College
St. Louis, Missouri

Introduction

The field of construction, repair, and remodeling is a major segment of the building industry, a nationwide network of trades that employs millions of people. Those choosing a career in construction are trained in such areas as site preparation, laying foundations, floor systems, roofing, framing, stairways, or restoration. Although mechanization and power tools have created numerous changes in this field, today's professionals must be skilled in many different types of tasks and must stay current with rapid technical changes in materials and techniques. Though there is still a place for expert craftspersons who perform detailed finish work, most trained professionals are versatile and knowledgeable about the entire construction field and often work closely with other trades in the industry, such as stonemasons, electricians, plumbers, millwrights, and glaziers, to name a few.

There are four major types of construction: residential, commercial, industrial, and civil. Residential construction generally refers to the building of homes and small structures, while commercial construction deals with larger projects, such as apartment buildings, schools, churches, and office buildings. Most published materials fall into these two categories, especially residential construction, the area where most students begin their training. Industrial construction deals with the building of complex structures such as power plants and steel mills, while civil construction refers to the building of bridges and highways. These fields require special techniques and specific training in areas such as the use of heavy equipment. Although these areas are beyond the scope of this bibliography, some materials dealing with general civil and industrial construction techniques have been included.

Postsecondary educational programs in construction, remodeling, and repair are generally offered as carpentry curricula in junior colleges and vocational-technical schools. These usually lead to a two-year associate degree. In addition, apprenticeship programs offer training to students who attend classes and receive on-the-job training. Apprentices work under the supervision of journeymen, and the programs are strictly supervised locally by a Joint Apprenticeship and Training Program Committee. They must also conform to national standards established by the Bureau of Apprenticeship and Training, a division of the United States Department of Labor. The Associated General Contractors of America also works with educational programs and is the accrediting and certifying body for the trade. Trade associations are numerous, but the major ones include the National Association of Home Builders and the Building Officials and Code Administrators International.

This bibliography is intended for use by librarians; students and professionals, such as carpenters and contractors; as well as the large number of dedicated and skilled laypersons who, while not employed in the trade, build, remodel, and repair everything from decks and kitchen cabinets to an entire house from the foundation up. Many materials in this bibliography aim for comprehensive coverage, while others provide information and instruction only on specific topics such as roofing, foundations, or kitchen remodeling. Although there is some overlap, two main categories of materials emerge: (1) carpentry and construction, and (2) remodeling, repair, and restoration. Three other subdivisions classify items into reference materials, historical items, and books on construction management.

Accreditation and Certification

Associated General Contractors of America
1957 East St., NW
Washington, D.C. 20006

Selected Associations

The following associations provide information on construction and remodeling education, industry trends, and continuing education:

American Building Contractors Association
12123-A Woodruff Ave.
Downey, Calif. 90241

American Institute of Timber Construction, Inc.
11818 SE Mill Plain Blvd.
Vancouver, Wash. 98684-5092

American Society for Testing and Materials
1916 Race St.
Philadelphia, Pa. 19103-1187

Associated Builders and Contractors
729 15th St., NW
Washington, D.C. 20005

Building Officials and Code Administrators International
4051 W. Flossmoor Rd.
Country Club Hills, Ill. 60478-5795

Building Research Board
2101 Constitution Ave., NW
Washington, D.C. 20418

Building Systems Council
15th and M Sts., NW
Washington, D.C. 20005

Construction Industry Employers Association
P.O. Box 4189
Buffalo, N.Y. 14217

Construction Industry Manufacturers Association
111 E. Wisconsin Ave., Suite 940
Milwaukee, Wisc. 53202-4879

Construction Specifications Institute
601 Madison St.
Alexandria, Va. 22314-1791

Council of American Building Officials
5203 Leesburg Pike, Ste 708
Falls Church, Va. 22041

General Building Contractors Association
36 S. 18th St.
P.O. Box 15959
Philadelphia, Pa. 19103

National Association of Home Builders
1201 15th St., NW
Washington, D.C. 20005

National Association of Women in Construction
327 S. Adams St.
Fort Worth, Tex. 76104

National Constructors Association
1730 M St. NW, Suite 900
Washington, D.C. 20036

National Forest Products Association
1250 Connecticut Ave., NW, Suite 200
Washington, D.C. 20036

National Roofing Contractors Association
10255 W. Higgins Rd., Suite 600
Rosemont, Ill. 60018-5607

United Brotherhood of Carpenters and Joiners of America
101 Constitution Ave., NW
Washington, D.C. 20001

Western Wood Products Association
Yeon Building
522 SW 5th Ave.
Portland, Ore. 97204-2122

Selected Journals

Aberdeen's Concrete Construction.
Aberdeen Group. Monthly. ISSN 1051-5526

Builder.
Remodeling. Monthly. ISSN 0744-1193

Constructor.
[National Association of Home Builders.] Remodeling.
Eleven times a year. ISSN 0162-6191

Custom Builder.
Willows Pub. Group. Bimonthly. ISSN 0895-2493

Fine Homebuilding.
Taunton Press. Seven times a year (includes special issue in February). ISSN 0273-1398

Journal of Construction Engineering and Management.
Construction Engineering and Management. Quarterly.
ISSN 0733-9364

Wood and Wood Products.
Vance Pub. Co. Thirteen times a year (includes annual reference).
ISSN 0043-7662

Print Materials

✦ Reference

Ambrose, James E.
Building Structures
New York, Wiley, 1988. 843p. ISBN 0471830941
 This massive compendium presents an in-depth investigation of structures by breaking them into their separate components, such as beams, trusses, etc. The author, an expert in the field, discusses wood, steel, and concrete structures. An essential reference for students and professionals.

Ambrose, James E.
Construction Revisited: An Illustrated Guide to Construction Details of the Early Twentieth Century
New York, Wiley, 1993. 250p. ISBN 0471591300
 Containing little text, this work is composed of detailed and precisely executed drawings exploring construction techniques used prior to 1950. The emphasis is on historical information with regard to contemporary applications. Three main sections deal with construction materials, systems, and components.

Ball, John E.
The Carpenters and Builders Library series.
6th ed. Revised and edited by John Leeke. New York, Macmillan, 1991. 4 v. ISBN 0025064517 (vol. 1); ISBN 0025064525 (vol. 2); ISBN 0025064533 (vol. 3); ISBN 0025064541 (vol. 4).
 This Audel series has set standards for the carpentry profession since 1965. This comprehensive set covers all the fundamentals for students. Volume One covers builders math, plans, and specifications; Volume Two addresses layout, foundation, and framing work; Volume Three includes tools, steel square and joinery; and the final volume discusses millwork, powertools, and painting. Includes numerous tables, charts, and illustrations.

The BOCA National Building Code
Country Club Hills, Ill., Building Officials and Code Administrators International, 1986- . 1 v. (various pagings) ISSN 0897-0068
 Reissued every three years, this essential reference furnishes standards and regulations for the construction industry and provides the minimum safety requirements for government contractors. Changes in the latest edition are marked for quick reference as a convenience for users. Appendices contain a listing of cited standards and metric equivalents.

Branson, Gary D.
The Complete Guide to Lumber Yards and Home Centers: A Consumer's Guide to Choosing and Using Building Materials and Tools
White Hall, Va., Betterway Publications, 1991. 174p.
ISBN 1558702091
 Arranged according to the types of materials available in lumber yards, this compilation provides a general survey of products. Subjects addressed include tools, lumber, fasteners and connectors, sheet metals, glues and adhesives, roofing, carpet and floor covering, and paints and stains. Materials are identified by brand names when applicable, and black-and-white photographs assist in identifying specific items.

Ching, Frank
Building Construction Illustrated
2nd ed. New York, Van Nostrand Reinhold, 1991. 1 v. (various pagings). ISBN 0442234988
 A classic guide for students that presents construction basics, primarily through illustrations. The text is comprised of explanatory material for the detailed drawings and follows a logical arrangement of construction order, beginning with the building site and concluding with finish work. Appendix materials include metric conversions and a list of trade associations.

Ehrlich, Jeffrey, and Marc Mannheimer
The Carpenter's Manifesto
Rev. and updated ed. New York, Holt, 1990. 331p.
ISBN 0805012990
 This practical handbook summarizes the basic principles and proven methods carpenters and students need to know. Explanations feature not only the best and most efficient techniques, but discuss why these methods are used. Page layout places diagrams and text adjoining each other for quick reference.

Forest Products Laboratory (U.S.)
Handbook of Wood and Wood-Based Materials for Engineers, Architects, and Builders
New York, Hemisphere Pub. Corp., 1989. 1 v. (various pagings).
ISBN 0891161244
 A comprehensive reference written by professionals from the U.S. Department of Agriculture. Subjects include wood, commercial lumber, bonding wood, plywood, finishing of wood, modified woods, and paper-based laminates. Characteristics, structure, and properties are discussed for each topic. Black-and-white photographs are included.

Grow, Lawrence, comp.
The Seventh Old House Catalogue
New York, Sterling Pub. Co., 1991. 224p. ISBN 0806974362
 The alphabetical arrangement includes over 2,250 entries of products and services used in home renovation and restoration. Entries contain a general description of the product, followed by the names and addresses of sources. A list of more than 500 distributors and suppliers is also given. Cross-references assist readers in finding items quickly.

Hornbostel, Caleb
Construction Materials: Types, Uses, and Applications
2nd ed. New York, Wiley, 1991. 1023p. ISBN 0471851450
 This definitive resource covers nearly every type of material used in construction and architecture. The second edition reflects new materials recently developed, as well as those no longer in general use. The 1,500 entries, alphabetically arranged, include information on the history and manufacture of materials, physical and chemical properties, and usage applications. Includes numerous tables, charts, standards, and illustrations.

Jacobs, David H.
Home Improvement Tools and Equipment
Blue Ridge Summit, Pa., TAB Books, 1994. 156p.
ISBN 0830644202
 Provides descriptions of common tools and their applications. Tools are divided by type, and categories include saws and saw blades, drills and drill bits, routers, trimmers and shapers, hammers,

nails, and measuring devices. Features information on handling tools safely, and equipment is depicted in black-and-white photographs.

Meers, Gary D.
The Carpenter's Toolbox Manual
New York, Arco, 1989. 333p. ISBN 0131152963

This convenient pocket-sized reference covers all the basics for students and professionals. Discusses hand and power tools, basic math, blueprints, wood and construction materials, and standard carpentry procedures such as stairs, roofs, and doors. Appendices contain job safety information, codes, and numerous tables. Includes glossary and index.

Phillips, Steven J.
Old-House Dictionary: An Illustrated Guide to American Domestic Architecture, 1600-1940
Lakewood, Colo., American Source Books, 1989. 239p. ISBN 0962133361

Concise explanations of terms are accompanied by crisp line drawings. The result is a useful reference which is limited to the identification of parts of buildings. Includes an extensive list of cross-references and a bibliography.

Reader's Digest Book of Skills and Tools
Pleasantville, N.Y., Reader's Digest Association, 1993. 360p. ISBN 0895774690

A practical handbook that contains a multitude of color drawings and photographs and is easy-to-use and understand. In addition to tools, it covers materials such as metals, paints, flooring, ceramics, glass, and plastics. Information on tools includes techniques for use, with step-by-step instructions.

Smit, Kornelis, and Howard Chandler, eds.
Means Illustrated Construction Dictionary
New unabridged ed. Kingston, Mass., R. S. Means, 1991. 691p. ISBN 087629218X

Encyclopedic in scope and lavishly illustrated, this revised edition contains more than 17,000 terms, phrases, acronyms, and abbreviations. Easy-to-understand language makes this book particularly suited for students, as well as professionals.

Toenjes, Leonard P.
Building Trades Dictionary
Homewood, Ill., American Technical Publishers, 1989. 314p. ISBN 0826904033

Written by an expert in the construction field, this alphabetically arranged resource contains clear definitions frequently accompanied by illustrations. Introductory material describes in detail how to use the dictionary, and appendices contain mathematical tables and formulas. Also includes an extensive list of trade organizations.

Traister, John E.
Illustrated Dictionary for Building Construction
Liburn, Ga., Fairmont Press, 1993. 465p. ISBN 0881731730

Defines more than 1,000 terms in one alphabetical list. Explanations are simple, concise, and generally in nontechnical language. An outstanding source for students and professionals.

Walker, Aidan, ed.
The Encyclopedia of Wood: A Tree-By-Tree Guide to the World's Most Versatile Resource
New York, Facts On File, 1989. 192p. ISBN 0816021597

Identifies 150 of the world's most prevalent kinds of wood, which are described according to properties, appearance, uses, and location. Accompanying illustrations depict the color and grain for each entry. Location maps are also included. Several chapters discuss the history of wood usage and wood around the world. Features glossy color photographs.

Wing, Charles
The Visual Handbook of Building and Remodeling: The Only Guide to Choosing the Right Materials and Systems for Every Part of Your Home
Expanded professional ed. Emmaus, Pa., Rodale Press, 1990. 498p. ISBN 087857901X

Features thousands of illustrations, tables, charts, and graphs in a thoroughly researched volume filled with practical details and facts. Arrangement is by subject, and explanations include concise technical language. Includes glossary and index.

✦ History of Construction

Brown, S. Azby
The Genius of Japanese Carpentry: An Account of a Temple's Construction
New York, Kodansha International, 1989. 156p. ISBN 0870118978

Focuses on the relationship of modern craftspeople to the ancient carpenters of Japan who constructed the Yakushiji Buddhist monastery over 1,000 years ago. Stresses the pivotal role of the master carpenter through the step-by-step reconstruction of one of the temples. Follows the original methods of construction. Tradition, ceremony, and techniques emerge as highlights. Includes bibliography and index.

The Builders: Marvels of Engineering
Washington, D.C., National Geographic Society, 1992. 288p. ISBN 0870448366

Familiar structures around the world, from ancient to modern times, receive an in-depth analysis of the planning and methods that went into their construction. Addressed are human reasons for constructing buildings that reach up into the sky or tunnel down into the earth. Also discusses structures built for protection and those built to honor a higher spirit. Visual layout features lavish illustrations and color photographs comprising over half the volume.

Clarke, Somers, and Reginald Engelbach
Ancient Egyptian Construction and Architecture
New York, Dover, 1990. 242p. ISBN 0486264858

An excellent historical study of the problems confronting the ancient Egyptians in the construction of the pyramids. Provides a detailed analysis of construction methods and building processes of that era. Over 250 photographs and illustrations accompany the text.

Coaldrake, William Howard
The Way of the Carpenter: Tools and Japanese Architecture
New York, Weatherhill, 1990. 204p. ISBN 0834802317

Explores the role of the master carpenter throughout Japanese history in the construction of traditional structures such as temples and tea houses. Discusses the rituals and traditions involved, methods of design and construction, guilds and apprenticeship programs, and materials used. Also contains detailed descriptions of tools used by carpenters.

Fitchen, John
Building Construction Before Mechanization
Cambridge, Mass., MIT Press, 1989. 326p. ISBN 026256047X

An extensive account of construction methods from ancient times to the 1800s. Explores the techniques used for building cathedrals, aqueducts, and the pyramids, as well as bridges, tents, military devices, and igloos. Also considers materials used, such as clay or stone. An excellent historical survey of construction principles for students and professionals.

Hawkes, Nigel
Structures: The Way Things are Built
Marshall ed. New York, Macmillan, 1990. 240p.
ISBN 0025491059

Features explanations of building techniques for forty-three memorable architectural and engineering accomplishments. Structures discussed include the Statue of Liberty, the Vatican, Mt. Rushmore, Epcot Center, and the Great Wall of China. Serves as an informative historical survey and a source of inspiration to those involved in the construction industry.

Levy, Matthys, and Mario George Salvadori
Why Buildings Fall Down: How Structures Fail
New York, Norton, 1992. 334p. ISBN 0393033562

A thorough, scholarly study of structural failures throughout history, beginning with the Egyptian pyramids and ending in the twentieth century with the Hyatt Regency walkways in Kansas City. The authors explore various reasons for these failures and relate them to actual building failures. Provides a good understanding of the underlying principles of construction and their relationship to the laws of physics.

Rose, Walter
The Village Carpenter
New York, New Amsterdam Books, 1988. 146p.
ISBN 0941533182

Reminiscences of a family of master carpenters in Buckinghamshire, England in the Victorian era. Rose discusses tools, methods, some of the projects worked on, and village life. What emerges is an image of carpentry as a skilled craft in a rural community. Photographs and drawings contribute to its authenticity.

Salvadori, Mario George
Why Buildings Stand Up: The Strength of Architecture
New York, Norton, 1990. 232p. ISBN 0393306763

Presents an overview of construction methods throughout history. Readers gain a clearer understanding of building principles, technology, and architectural theory through the exploration of such diverse structures as bridges, skyscrapers, cathedrals, domes, and towers. An extremely thorough analysis written by an established author and professor at Columbia University.

✦ Construction Management

Alfeld, Louis Edward
Construction Productivity: On Site Measurement and Management
New York, McGraw-Hill, 1988. 234p. ISBN 0070010277

Applies the principles of performance-based management to the construction industry. Managers learn how to evaluate on-site performance and use the results to improve productivity and maintain construction schedules. Presented in straightforward language, with numerous examples. Features a step-by-step approach for contractors, supervisors, and project managers.

Clough, Richard Hudson, and Glenn A. Sears
Construction Project Management
3rd ed. New York, Wiley, 1991. 296p. ISBN 0471546089

Designed for professional contractors, this guide offers a practical approach to the effective management of construction fieldwork. Presents an overview of the construction industry, but the focus is on techniques for managing cost, time, resources, and finances of a construction project. The third edition emphasizes the technical changes that have taken place over recent years, especially the application of data generated by computers.

Dagostino, Frank R.
Estimating in Building Construction
3rd ed. Englewood Cliffs, N.J., Prentice Hall, 1989. 414p.
ISBN 0132897377

A well-organized volume by a respected authority that covers all aspects of professional estimating. Construction materials are discussed with regard to estimating quantity and cost. Also covered are bidding procedures, contract documents, and insurance. Diagrams and line drawings are clear and well labeled.

Day, David A.
Construction Equipment Guide
2nd ed. New York, Wiley, 1991. 437p. ISBN 0471888400

Equipment is arranged according to the type of job it accomplishes, such as loaders and haulers, lifting and hoisting equipment, and graders and compactors. Early chapters present a general overview, followed by extensive technical information for each category. Equipment is analyzed according to uses, conditions, and features which affect job productivity. Useful for professionals working with heavy construction equipment.

Frisby, Thomas N.
Basics for Builders: How to Survive and Prosper in Construction—Checklists for Success
Kingston, Mass., R. S. Means, 1990. 289p. ISBN 0876293429

Specifically aimed at those who manage any aspect of the construction industry. This book addresses changes which the construction process has undergone in recent years. Presents a comprehensive analysis of methods used in marketing, management, bidding and estimating, contract interpretation, job setup, and scheduling. Sixty-five checklists are a valuable feature for professionals.

Wallace, Thomas J., and Walter J. Rossiter, eds.
Roofing Research and Standards Development: 2nd Volume
Philadelphia, American Society for Testing and Materials, 1990. 228p. ISBN 0803113935

A compilation of fourteen papers presented at a 1990 symposium on roofing research. Deals with three subject areas: bituminous roofing systems, elastomeric and thermoplastic roofing systems, and analytical investigations of roofing systems. Highly technical and designed for roofing specialists.

✦ Carpentry and Construction

Ambrose, James E.
Building Construction: Site and Below-Grade Systems
New York, Van Nostrand Reinhold, 1991. 156p.
ISBN 0442002939

The scope is limited to the construction of basement walls, foundations, floors, retaining structures, and site development. The author relates design problems to construction sites, and information is selective rather than comprehensive. Language is geared to students.

Ambrose, James E.
Simplified Design of Building Foundations
2nd ed. New York, Wiley, 1988. 237p. ISBN 0471858986

This edition contains new materials on soil modification, pole foundations, braced excavations, waterfront foundations, and slope utilization. Designed for students and professionals who need simple yet specific information relating to design and construction problems without extensive training in structural engineering.

Anthenat, Kathy Smith
American Tree Houses and Play Houses: Childhood Retreats From Yesteryear—Play Houses and Tree Houses of Today—and Six "Build-it-Yourself" Play House Plans
White Hall, Va., Betterway Publications, 1991. 175p.
ISBN 1558702040

Contains plans for six construction projects, including step-by-step instructions with optional features, photographs, illustrations, and a list of required materials. A large section of the book is devoted to playhouses, both simple and elaborate, with background descriptions and ownership history.

Armpriester, Kate, and B. A. Bremer
The Homeowner's Guide to Carpentry and Cabinetry
New York, Sterling Pub. Co., 1990. 246p. ISBN 080697298X

Includes nontechnical language and precise clarification of terms and methods. Explanations of techniques aim for understanding and are accompanied by excellent drawings with two-color contrast. There is also an emphasis on proper use of tools and safety.

Bauer, Jim, and Scott Millard
Deck Planner: Twenty-five Outstanding Decks You Can Build
Tucson, Ariz., Home Planners, 1990. 112p. ISBN 0918894840

Features plans, construction drawings, and specific instructions for twenty-five decks, each adapted to a different house style. Also covers general deck-building techniques and features color photographs of completed decks. Blueprints may be ordered.

Benson, Tedd
The Timber-Frame Home: Design, Construction, Finishing
Newtown, Conn., Taunton Press, 1988. 225p. ISBN 0918804817

The nationally known author and television personality explains the craft of timber framing within the context of modern living. Covers not only the fundamentals of framing, but presents explicit instructions for adding plumbing and wiring, as well as discussing solar energy. Gives many tips for choosing options and is accompanied by color photographs and drawings.

Birchard, John
Make Your Own Handcrafted Doors and Windows
New York, Sterling Pub. Co., 1988. 288p. ISBN 0806965444

A comprehensive guide that presents practical information on making doors and windows in a home-based shop. Simplified techniques are used to explain the construction, installation, maintenance, and finishing work. Included are eight pages of color photographs of the finished products, and a special chapter on working with irregularly shaped doors and windows.

Blackburn, Graham
Floors, Walls, and Ceilings
Mt. Vernon, N.Y., Consumers Union, 1989. 231p.
ISBN 0890432457

Presents detailed background information on various types of floors, walls, and ceilings. Contains basic techniques for construction, and instructions for maintenance, repair, and refinishing. Coverage extends to materials and tools used, which are highlighted by illustrations. Terms are clearly defined, and the appendix contains a tool glossary and a product list rated by *Consumer Reports*.

Branson, Gary D.
The Complete Guide to Log and Cedar Homes: All About Buying, Building, Decorating, and Furnishing Log, Cedar, and Post and Beam Homes
Cincinnati, Betterway Books, 1993. 160p. ISBN 1558702768

Discusses the advantages of constructing log and cedar homes and the modern design changes which make them feasible for urban settings, rather than limiting them to vacation cabins. Contains instructions for construction, a section on floor plans, and a list of relevant addresses. Illustrated with numerous color and black-and-white photographs.

Brown, Robert Wade
Design and Repair of Residential and Light Commercial Foundations
New York, McGraw-Hill, 1990. 241p. ISBN 0070081921

A unique reference containing considerable technical detail on a specialized subject. Topics include site preparation, foundation design, soil stabilization, foundation failures, and repair procedures. An excellent glossary accompanies the text. Designed to increase the expertise of professionals and students with some prior knowledge.

Brumbaugh, James E.
Complete Siding Handbook: Installation, Maintenance, Repair
2nd ed. New York, Macmillan, 1992. 472p. ISBN 0025178814

Provides extensive information on all types of siding, as well as peripheral aspects such as wall preparation, finishes, gutters, and downspouts. Detailed line drawings illustrate descriptions, and appendix material contains lists of siding manufacturers and professional associations. Intended for those in the construction trade.

Bryant, Ernie
Garages: Complete Step-By-Step Building Plans
Blue Ridge Summit, Pa., TAB Books, 1990. 177p.
ISBN 0830692142

TAB publishes home improvement manuals known for good organization, simple language, and detailed instructions. This one encompasses all aspects of building a garage and includes a brief inventory of required tools, as well as five complete working drawings for different types of garages. Extensive use of illustrations.

Building Doors, Windows, and Skylights
Newtown, Conn., Taunton Press, 1989. 160p. ISBN 0942391144

This visually appealing volume, part of a series for professionals or experienced laypeople, features outstanding color drawings and photographs. The forty-four articles were originally published in *Fine Homebuilding*, and cover a diverse set of building projects, e.g., curbless skylights, greenhouse shutters, screen porch windows, and batten doors.

Collings, George
Circular Work in Carpentry and Joinery
5th ed. Fresno, Calif., Linden Pub. Co., 1992. 126p.
ISBN 0941936228

Originally published in London in 1886, this fifth edition reprint remains faithful to the text of the 1911 edition. Annotations bridge the gap for today's readers, while the text provides invaluable information on complex skills unknown to many of today's carpenters. Covers both single and double curvature and is accompanied by numerous illustrations. An important reference for a little-known area of carpentry.

Cooper, Jim
Log Homes Made Easy: Contracting and Building Your Own Log Home
Harrisburg, Pa., Stackpole Books, 1993. 245p. ISBN 0811724220

An excellent general guide for those who wish to direct and understand the construction of a log home. Topics include log home kits, general contracting, the construction process, and maintenance. Material featured in the appendix includes a cost estimate checklist and a sample week-by-week schedule.

DeCristoforo, R. J.
Housebuilding: A Do-It-Yourself Guide
San Bernardino, Calif., Borgo Press, 1990. 644p.
ISBN 0809575116

This reprint of a 1977 *Popular Science* book remains an important guide for teaching the fundamentals of house construction to serious do-it-yourselfers or students. Extremely thorough, it features over 1,000 simplified drawings and step-by-step instructions on all aspects of construction, tools, scheduling, and materials. The appendix contains charts of metric equivalents and weights and measures.

Dietz, Albert G. H.
Dwelling House Construction
5th ed. Cambridge, Mass., MIT Press, 1991. 440p.
ISBN 0262041081

The fifth edition reflects technical changes in both building construction and materials. Sections on concrete, steel, and masonry have been included, information on roofing and wood foundations have been updated, and chapters on traditional practices have been reduced. Contains technical information with detailed drawings and diagrams. A classic source for students and professionals by a respected professor at the Massachusetts Institute of Technology.

Donaldson, Barry, ed.
Exterior Wall Systems: Glass and Concrete Technology, Design, and Construction
Philadelphia, American Society for Testing and Materials, 1991. 205p. ISBN 0803114249

Offers advanced technical papers on exterior walls presented at two symposia. Topics include design, testing and analysis, structural sealant glazing, stone selection, and precast composite concrete wall systems. Intended to expand the expertise of specialists in the field of constructing exterior wall systems. Accompanied by black-and-white photographs, drawings, and an index.

Feirer, John Louis, and Gilbert R. Hutchings
Guide to Residential Carpentry
2nd ed. New York, Collier Books, 1991. 512p. ISBN 0020004915

Designed for students and apprentices, this standard guide provides detailed coverage of carpentry fundamentals written in simple English. Chapters contain numerous visuals in the form of illustrations, photographs, tables, and charts. Logical sequencing of information introduces tools and materials first, then progresses to foundations, framing, exteriors, interiors, and remodeling. Written by a respected instructor, writer, and expert in the field.

Fine Homebuilding on Finish Carpentry
Newtown, Conn., Taunton Press, 1993. 128p. ISBN 156158052X

Twenty-nine articles written by professionals are notable for their attention to detail and clear, concise writing. The authors impart tips, tested methods, and problem-solving techniques on topics such as crown molding, French doors, baseboards, curved casings, and screen doors. Page layout and color graphics are planned for maximum visual appeal.

Frame Carpentry
Newtown, Conn., Taunton Press, 1988. 160p. ISBN 094239111X

Contains thirty-four articles from *Fine Homebuilding* magazine that deal with building methods for foundations, roofing, walls, and floors. Detailed and informative articles written by experts address specific solutions to construction problems. Color photographs and detailed drawings are highlights.

Goad, Karen
Drywall Installation and Finishing
Albany, N.Y., Delmar Publishers, 1993. 182p. ISBN 0827356056

In concise and simple language Goad helps students understand professional techniques of hanging and finishing drywall. Includes essential materials on tools, taping, sanding, texturing, and repairs. Abundant drawings clarify explanations.

Hiro, John E.
Millwork Handbook
New York, Sterling Pub. Co., 1993. 284p. ISBN 0806986980

Addresses the subject of doors and windows, which often present problems in construction and renovation. Includes techniques for finishing, installation, maintenance, and restoration. Best features are the comparative drawings and an excellent glossary of technical terms.

Hop, Frederick Uhlen
Modular House Design: The Key to Complete Construction Efficiency
Englewood Cliffs, N.J., Prentice Hall, 1988. 193p.
ISBN 0135994241

Deals with both design and construction of modular homes. The author identifies materials and how these interrelate. Specifically addressed are skills of designing a house plan and stairway construction. Effective use is made of headings, subheadings, and summaries to create a useful teaching manual.

Huth, Mark W.
Construction Technology
2nd ed. Albany, N.Y., Delmar Publishers, 1989. 401p.
ISBN 0827329628

This text for students focuses on the overall interrelationship of all facets of the construction industry and analyzes how each unit fits into our social and economic structure. Good overall organization includes objectives, review questions, lists of key terms, and activities in each chapter.

Jacobs, David H.
Workshops and Outbuildings
Blue Ridge Summit, Pa., TAB Books, 1994. 144p.
ISBN 0070324026

To be used as a guide for beginners interested in building simple outdoor structures. Explanatory material is clear, succinct, and well organized. Readers gain a fundamental understanding of how to perform basic carpentry tasks, as well as an understanding of the rationale behind these tasks.

Johnson, W. E.
Roofers Handbook
Carlsbad, Calif., Craftsman Book Co., 1989. 189p.
ISBN 0910460175

In its seventh printing, this manual is still the definitive guide for roofing. Covers standard roofing procedures for professionals as well as chapters on tools and equipment, plus tips on operating a roofing company. The appendix contains valuable information on materials and specifications.

Kidder, Tracy
House
New York, Avon Books, 1990. 341p. ISBN 0380711141

Kidder presents a dramatized account of the creation and building process of a home by focusing on the actions and relationships of the people involved in its planning and construction. Presents a unique and highly readable narrative of interest to professionals, students, and laypeople. Not intended as an instruction manual.

Kilpatrick, John A.
Understanding House Construction
2nd ed. Washington, D.C., Home Builder Press, National Association of Home Builders, 1993. 120p. ISBN 0867183829

An introduction to home construction that provides basic information to homeowners and laypeople in an informal manner. Covers all aspects of the building process, e.g., plumbing, heating, electricity, and finish work. Includes a glossary and a bibliography.

Koel, Leonard
Carpentry
2nd ed. Homewood, Ill., American Technical Publishers, 1991. 721p. ISBN 0826907326

A comprehensive textbook that is well organized and clearly written. This introduction to the trade provides the necessary fundamentals for students. Includes sections on materials, tools, safety, and energy conservation, as well as construction techniques.

Locke, Jim
The Well-Built House
Rev. ed. Boston, Houghton Mifflin, 1992. 302p.
ISBN 0395629519

Writing in an informal narrative style, Locke imparts a wealth of detail and information for those contemplating building their own home or addition. The focus is on the importance of preliminary planning and paperwork, and the book contains tips and ideas to assist laypeople in avoiding the pitfalls of construction.

Love, T. W.
Construction Manual: Finish Carpentry
Carlsbad, Calif., Craftsman Book Co., 1989. 188p.
ISBN 0910460086

Originally issued in 1974, this 1989 reprint is still widely used in the profession. It focuses on solving specific problems in finish carpentry, and chapters are filled with sound practical advice rather than construction theory. It is written in language students can understand and is accompanied by many illustrations.

Love, T. W.
Construction Manual: Rough Carpentry
Carlsbad, Calif., Craftsman Book Co., 1991. 285p.
ISBN 0910460183

Companion to Love's *Finish Carpentry*, this classic volume contains basic instruction for students. Also provides information on lumber and offers many tips for saving time, materials, and labor. Extensively labeled drawings illustrate every page.

Maguire, Byron W.
Cabinetmaking: From Design to Finish
Carlsbad, Calif., Craftsman Book Co., 1990. 403p.
ISBN 0934041628

Addresses the process of designing, building, and finishing different styles of cabinets. Also focuses on the methodology for creating the joints used in the construction process. Technical language and intricate drawings provide extensive coverage.

Maguire, Byron W.
Carpentry in Commercial Construction
2nd ed. Carlsbad, Calif., Craftsman Book Co., 1988. 266p.
ISBN 0934041334

This standard manual concentrates on specific skills carpenters need when constructing larger buildings such as offices, townhouses, or stores. Maguire, a carpenter and instructor in the building trades for over forty-five years, also authors the companion volume *Carpentry for Residential Construction* (Craftsman Book Co., 1987). Includes review questions, charts, and illustrations.

Maguire, Byron W.
Outdoor Building Projects
Englewood Cliffs, N.J., Prentice Hall, 1991. 308p.
ISBN 0136433545

Features both simple and complex outdoor projects arranged by subject, with specifications and materials lists included for each. Contents include patios and decks, porches, gazebos, trellises, fencing, and terracing. Projects are designed to act as models and can be adapted to fit specific topography and circumstances. Instructions are methodical and thorough, and the use of objectives is useful for students. Detailed line drawings, index, and bibliography.

Manning, Richard
A Good House: Building a Life on the Land
New York, Grove Press, 1993. 238p. ISBN 0802115039

Explores the author's personal experience with building an earth-sheltered home. Anecdotes and experiences are interspersed with technical details, construction information, and procedures to form a delightful narrative. The emphasis is on the importance of environmental awareness in construction.

Martin, Ray, and Lee Rankin
Building Garden Furniture: More Than Thirty Beautiful Outdoor Projects
New York, Sterling Pub. Co., 1993. 160p. ISBN 0806983744

Finished projects are displayed in colorful garden settings in this collection arranged according to function (i.e., garden tables and chairs are in a section on dining out.) Also covered are garden benches, swings, and planters, plus useful information on softwood, lumber sizing, and grades. Designed for homeowners and laypeople.

McClintock, Michael
Alternative Housebuilding
New York, Sterling Pub. Co., 1989. 367p. ISBN 0806969954

Provides the reader with sound alternatives to traditional building construction. The author focuses on incorporating the construction project into its natural environment and stresses the importance of craftsmanship. Provides detailed descriptions of earth-sheltered houses, log houses, timber-framed homes, pole houses, and others. Instructions include ideas for cutting costs, site selection, and energy conserving adaptations. Useful and informative, with hundreds of photographs and drawings.

Miller, Charles, ed.
Tips and Techniques for Builders
Newtown, Conn., Taunton Press, 1988. 256p. ISBN 0942391098

An excellent sourcebook by *Fine Homebuilding,* filled with information and practical insights. Loosely arranged into chapters according to type of tool or construction skill, with each entry covering one-half page or less. Accurate, shaded drawings accompany entries.

Miller, Rex
Carpentry and Construction
2nd ed. Blue Ridge Summit, Pa., TAB Books, 1991. 547p.
ISBN 0830686789

An updated version of a core text. Comprehensive, featuring large type and detailed drawings. Includes glossary and index.

Oberschulte, William, ed.
Wood-Frame House Construction
Bowker. Carlsbad, Calif., Craftsman Book Co., 1992. 336p.
ISBN 0934041741

Recently updated, this book has been a core source for basic construction principles since 1976. Arranged in logical sequence, chapters contain standard procedures for construction from excavation to finish work. Useful information on cost estimating is included, as well as maintenance and repair tips. Illustrations clarify text.

Ramsey, Dan
Doors, Windows, and Skylights
2nd ed. Blue Ridge Summit, Pa., TAB Books, 1990. 230p.
ISBN 0830682481

Provides complete information on numerous aspects of the subject, including historical details and decorating tips. The author covers building, installing, finishing, repairing, and remodeling all types of windows and skylights. Contains practical information for both laypeople and experts, in everyday language. Index and glossary.

Ramsey, Dan
Fences, Decks, and Other Backyard Projects
3rd ed. Blue Ridge Summit, Pa., TAB Books, 1992. 278p.
ISBN 0830634932

A popular book in its third edition, this survey focuses on picket, board, rail, chain link, masonry, and livestock fences. Also contains historical information and chapters about gates, fence maintenance, and fence landscaping.

Ramsey, Dan
Hardwood Floors
2nd ed. Blue Ridge Summit, Pa., TAB Books, 1991. 176p.
ISBN 0830675299

The resurgent popularity of wood floors as a decorating feature makes this book an important resource for those working in construction or remodeling. Ramsey describes floors made of plank, block, parquet, and tongue and groove, and their repair, maintenance, and installation. Instructive, with excellent diagrams and drawings.

Roy, Greg
The Complete Guide to Residential Deck Construction: From the Simplest to the Most Sophisticated
White Hall, Va., Betterway Publications, 1992. 175p.
ISBN 1558702318

Offers in-depth information on building a deck, from the initial needs evaluation to designing, building, and maintenance. A section is devoted to explaining uses and pros and cons of tools, materials, and woods. Color photographs illustrate completed decks, but the most useful feature is the section of checklists covering all aspects of deck construction.

Roy, Robert L.
Complete Book of Cordwood Masonry Housebuilding
New York, Sterling Pub. Co., 1992. 255p. ISBN 0806985909

Provides an overview of cordwood masonry construction, including its history. The author, an expert in alternative construction methods, built several of his own homes, and subsequently opened the Earthwood Building School. Offers a practical emphasis on the planning stages with the goal of low-cost construction, self-reliance, and harmonious living in a natural setting.

Sabbagh, Karl
Skyscraper
New York, Viking, 1990. 388p. ISBN 0670832294

Examines the methodology in building the 800-foot tower of the Worldwide Plaza Complex in New York, from its initial inception through all phases of construction. The author follows the processes of architects, engineers, and construction workers and incorporates these into a readable chronology. Contains color photographs.

Savage, Craig
Trim Carpentry Techniques: Installing Doors, Windows, Base, and Crown
Newtown, Conn., Taunton Press, 1989. 185p. ISBN 0942391084

Focuses on practical advice for installing doors, windows, baseboards, and moldings. Savage also discusses techniques for solving common problems and stresses the importance of precise trim work as part of the overall construction job. Very detailed and geared to professionals.

Schuttner, Scott
Basic Stairbuilding
Newtown, Conn, Taunton Press, 1990. 121p. ISBN 0942391446

Teaches the basic principles of constructing three simple staircases: an open-riser ladder stair, a straight case, and an L-shaped stair. Features step-by-step instructions with drawings labeled in detail. Can be used by students or laypeople.

Schuttner, Scott
Building and Designing Decks
Newtown, Conn., Taunton Press, 1993. 153p. ISBN 1561580287

For the do-it-yourself carpenter, this guide provides complete instructions written in simple, nontechnical language. Liberal use of diagrams; color photographs of completed decks.

Sedan, Paul S.
The Factory-Crafted House: New Visions of Affordable Home Design
Old Saybrook, Conn., Globe Pequot Press, 1992. 118p.
ISBN 1564400611

Explores the reasons for considering construction of factory-made homes and describes the types available. Provides useful and well-researched information such as sources for finding subcontractors, mobile home builders, owner-builder schools, and cost estimation.

Self, Charles R.
Joinery: Methods of Fastening Wood
Pownal, Vt., Storey Communications, 1991. 213p.
ISBN 088266641X

Deals with six major methods of joining and fastening wood used by both carpenters and woodworkers. For each method the author explains the types of joints, when to use them, and the tools needed. Visuals consist of photographs and a few diagrams.

Sherwood, Gerald E., and Robert C. Stroh, eds.
Wood Frame House Construction
2nd ed. Albany, N.Y., Delmar Publishers, 1992. 306p.
ISBN 0827347391

A sourcebook for carpenters and students since 1955, this recently updated edition incorporates the latest technical changes in the industry. Information is presented in logical construction sequence and is written in straightforward language that often refers to detailed accompanying drawings. Glossary and index.

Small Houses
Newtown, Conn., Taunton Press, 1992. 160p. ISBN 1561580465

The homes presented in this collection of articles are unusual, economical, energy efficient, and attractive. They range from those professionally designed and built to those crafted by laypeople. Highlights are floor plans and detailed, labeled cutaway drawings. The articles were written by a variety of professionals and craftspeople.

Sobon, Jack A.
Build a Classic Timber-Framed House: Planning and Design, Traditional Materials, Affordable Methods
Pownal, Vt., Storey Communications, 1994. 202p.
ISBN 0882668420

Points out the significant differences between traditional timber framing and regular house construction. Step-by-step procedures are given for building a timber-framed house with the goal of creating an affordable, efficient home. Informative and practical.

Spence, William Perkins
Residential Framing: A Homebuilder's Construction Guide
New York, Sterling Pub. Co., 1993. 319p. ISBN 0806985941

A comprehensive survey for those constructing a home or acting as contractor. Topics include site preparation, materials, tools, framing, roofing, and finish work. Features useful appendix material including standard lumber sizes in the U.S. and Canada. Illustrated with numerous drawings.

Syvanen, Bob
Carpentry and Exterior Finish: Some Tricks of the Trade
3rd ed. Old Saybrook, Conn., Globe Pequot Press, 1993. 112p.
ISBN 1564400786

A companion volume to the author's *Carpentry and Interior Finish,* this book adopts a commonsense approach to lumber, sills, joists, roofing, bridging, decking, siding, and exterior walls. This work is written in a witty, straightforward style.

Syvanen, Bob
Carpentry and Interior Finish
2nd ed. Old Saybrook, Conn., Globe Pequot Press, 1993. 120p.
ISBN 1564402517

Written by a professional who presents tricks of the trade and proven methods to assist not only novices, but also experienced professionals. Addresses all aspects of interior work, with elementary instructions and lavish illustrations complementing the text.

Taylor, Stephen
Building Thoreau's Cabin
Wainscott, N.Y., Pushcart Press, 1991. 244p. ISBN 091636674X

An unusual guide based on the premise that learning how to construct a small building is a useful and practical skill. Taylor also conveys that it can be a spiritual and empowering experience. Novices are guided in a leisurely way through topics such as lumberyards, foundations, rough carpentry, and trim work. The focus in on practical know-how.

Timber-Frame Houses
Newtown, Conn., Taunton Press, 1992. 160p. ISBN 1561580473

The articles in this volume characterize the craft of timber framing with both old and new construction techniques. Explores traditional house styles and tools as well as new designs, styles, and building methods. Although written for professionals or experts, the articles are clear and easy to read. Color visuals are outstanding. One of a series by *Fine Homebuilding*.

Todd, Ken
Carpentry Layout
Carlsbad, Calif., Craftsman Book Co., 1988. 220p.
ISBN 0934041326

A useful manual for carpenters that demonstrates how to perform the proper mathematical calculations for stairs, rafters, foundations, walls, and joists. Simple language, drawings, diagrams, and numerous examples assure that concepts can be easily understood. Practice exercises are included for each chapter.

Wagner, John D., ed.
Advanced Framing: Techniques, Troubleshooting, and Structural Design
Richmond, Vt., Journal of Light Construction, 1992. 281p.
ISBN 0963226800

This collection of articles from *The Journal of Light Construction* includes topics on framing errors, truss installation, sizing joists and beams, demolition, curved walls, and seismic bracing. The emphasis is on commonsense solutions.

Watkins, A. M.
The Complete Guide to Factory-Made Houses
Newly rev. ed. Chicago, Ill., Longman Financial Services Publishing, 1988. 182p. ISBN 0884627098

Designed for laypeople, this work provides information on manufactured dwellings such as dome houses, A-frames, and mobile homes. Provides tips on finding high quality materials and cutting costs. Floor plans and finished photographs appear throughout the volume, and a directory of home manufacturers is included.

Williams, Elizabeth, and Robert Leonard Williams
Finish Carpentry Illustrated
2nd ed. Blue Ridge Summit, Pa., TAB Books, 1993. 200p.
ISBN 0830644105

Addressing the subject in clear and simple language, this practical overview and guide is intended for beginning students or home carpenters with little experience. Illustrations are color highlighted, and the page layout includes wide margins containing chapter subdivisions and room for notes.

Williams, Elizabeth, and Robert Leonard Williams
Rough Carpentry Illustrated
Blue Ridge Summit, Pa., TAB Books, 1990. 166p.
ISBN 0830634355

A companion to Williams' work on finish carpentry. The authors attempt to cover the main points of house framing for beginners, avoiding too much technical detail. Provides a step-by-step approach to ceilings, closets, walls, floors, stairways, garages, porches, and roof supports.

Wylde, Margaret A., Adrian Baron-Robbins, and Sam Clark
Building for a Lifetime: The Design and Construction of Fully Accessible Homes
Newtown, Conn., Taunton Press, 1994. 295p. ISBN 1561580368

Focuses on the concept of constructing homes to accommodate the aged or people with disabilities. Offers information on both design and construction of functional housing, as well as ideas for remodeling existing buildings to provide greater accessibility and foster independence. An important and necessary guide which meets specific needs in housing.

✦ Remodeling, Repair, and Restoration

Adler, Bill
The Home Remodelers' Combat Manual
New York, Harper Perennial, 1991. 247p. ISBN 0060552794

Adler intersperses facts, anecdotes, and humor with real wisdom and tips for homeowners in this useful compilation. Provides sufficient information to enable remodelers to work successfully with building professionals in the management of a renovation project. Chatty, conversational writing style keeps the reader's interest.

Allen, Sam
Remodeling and Repairing Kitchen Cabinets
New York, Sterling Pub. Co., 1988. 128p. ISBN 080696720X

Explains simple projects, from updating hardware and door fronts to more complex structural changes or complete cabinet replacement projects. Instructions are straightforward and are accompanied by black-and-white photographs.

Bryant, Ernie
Making Space: Remodeling for More Living Area
Blue Ridge Summit, Pa., TAB Books, 1992. 225p.
ISBN 0830639322

A do-it-yourself manual for amateurs featuring lucid and elementary instructions with large, clear drawings. Summarizes carpentry tools and discusses both home additions and remodeling projects. Topics include garage conversion, dormer construction, porches, decks, and attic conversions.

Cobb, Hubbard H., and Betsy Cobb
Your Barn House
New York, Holt, 1991. 239p. ISBN 080501151X

Includes historical information on American barns and case histories of actual converted barn houses. Discusses pros and cons of different construction methods and advises readers on what to expect and what to avoid. A logical and practical guide with useful floor plan sketches. Color and black-and-white photographs.

Decker, Phillip J., and T. Newell Decker
Renovating Brick Houses: For Yourself or for Investment
Pownal, Vt., Storey Communications, 1990. 247p.
ISBN 0882665936

Through their experiences as contractors and renovators, the authors have put together a manual that is both complete and detailed. Combines basic instructions for carpentry skills and also addresses important factors like inspecting old houses, estimating cost, and dealing with subcontractors. Extensive bibliographic references, a glossary, and an index.

Frechette, Leon A.
Bathroom Remodeling
Blue Ridge Summit, Pa., TAB Books, 1993. 272p.
ISBN 0830644792

Comprehensive in scope, this handbook covers every aspect of the remodeling process. Assists homeowners in analyzing their needs and helps them decide which jobs they can handle themselves. Includes excellent drawings and diagrams, but black- and-white photographs could be crisper.

Hamilton, Gene, and Katie Hamilton
Fix It Fast, Fix It Right: Hundreds of Quick and Easy Home Improvement Projects
Emmaus, Pa., Rodale Press, 1991. 312p. ISBN 0878578595

Presents the tested techniques and commonsense ideas gained by the authors over a period of twenty-three years and thirteen houses. Includes coverage of everything from paint and wallpaper to floors, windows, lighting, bathrooms, and kitchens. Best features are tips on saving money and factors to consider when determining whether or not to handle a project personally.

Hufnagel, James A.
The Stanley Complete Step-by-Step Book of Home Repair and Improvement
New York, Simon and Schuster, 1993. 478p. ISBN 0671744429

This sourcebook features more than 2,000 full-color drawings and many cross-references on home improvement. Highlights safety concerns. This is a very up-to-date, authoritative, and visually appealing manual.

Jackson, Albert, and David Day
The Complete Home Restoration Manual: An Authoritative, Do-It-Yourself Guide to Restoring and Maintaining the Older House
New York, Simon and Schuster, 1992. 256p. ISBN 0671737988

Strengths of this book include the numerous full-color photographs and illustrations, and the wealth of historical information on both house styles and interiors. Also features information on subjects not often covered, such as stucco, paving, wall tiles, ornamental glass, and chimneys.

Kitchens and Bathrooms
2nd ed. Alexandria, Va., Time-Life Books, 1989. 128p.
ISBN 080947350X

Organization and multicolored drawings are excellent, as in most Time-Life Books. Also featured in this beginner's guide are step-by-step instructions and an attractive graphic layout. Part of the *Home Repair and Improvement Series*.

Lawrence, Mike, and Derek Bradford, eds.
The Complete Home Renovation Manual
New York, Smithmark Publishers, 1993. 224p. ISBN 083171588X

Addresses all aspects of both exterior and interior renovation, while emphasizing the process of planning and analyzing the home's structure. Contains an attractive layout and design with full-color photographs and illustrations. Written in language that nonprofessionals can understand. Index.

Litchfield, Michael W.
Renovation: A Complete Guide
2nd ed. Englewood Cliffs, N.J., Prentice Hall, 1991. 566p.
ISBN 0131593366

This revised edition contains updated information from hundreds of professionals and craftspersons. Provides thorough information on planning, tools and building materials, rough carpentry, and finish work for both exterior and interior construction. Also treats plumbing, heating and cooling, and electricity, and features a section on energy conservation. Detailed information can be used by professionals and laypersons. Contains glossary, bibliography, and index.

Nash, George
Renovating Old Houses
Newtown, Conn., Taunton Press, 1992. 343p. ISBN 0942391659

Highlights of this survey on renovation include chapter subheadings and detailed labeling of drawings to encourage understanding. The author emphasizes the importance of initial evaluation of the home's potential, and teaches readers to use common sense when setting priorities for renovation. Aims at achieving a balance between modernization and maintenance of the original essence of the house.

New Complete Do-It-Yourself Manual
Pleasantville, N.Y., Reader's Digest Association, 1991. 528p.
ISBN 0895773783

This Reader's Digest reference for nonprofessionals uses logically sequenced chapters to cover both repair and remodeling work. Chapter subdivisions are listed in the table of contents to permit quick location of information. Colored and shaded drawings are often boxed for emphasis. Attractive layout.

New York Landmarks Conservancy
Repairing Old and Historic Windows: A Manual for Architects and Homeowners
Washington, D.C., Preservation Press, 1992. 207p.
ISBN 0891331859

A historical overview of how windows are used in American houses is followed by detailed descriptions on evaluating, repairing, maintaining, and replacing windows. Also contains useful information on standards and guidelines. Glossary, bibliography, and index.

Orme, Alan Dan
Reviving Old Houses: Over Five Hundred Low-Cost Tips and Techniques
Pownal, Vt., Storey Communications, 1989. 168p.
ISBN 0882665820

Focuses on the particular problems faced when remodeling an older home while trying to preserve its historic features and keep costs down. The author addresses the subject through extensive personal experience and has compiled tested tips and techniques. Written with clarity and perception.

Owen, David
The Walls Around Us: The Thinking Person's Guide to How a House Works
New York, Vintage Books, 1992. 308p. ISBN 0679741445

Owen resolves some of the complexities of home maintenance and repair for homeowners who possess few skills or prior knowledge. Readers are introduced to basic construction techniques, as well as a multitude of tips and proven methods. Tools and materials are discussed throughout the volume, which is written in a narrative style that is witty and entertaining.

Philbin, Tom
Knock It Down, Break It Up: The Definitive Guide to In-Home Demolition
New York, Avon Books, 1993. 201p. ISBN 038076850X

This unique volume deals with the demolition and preparation work often necessary before beginning a renovation or remodeling project. Discusses the easiest and most efficient ways of tearing out windows or floors, removing tile, taking down walls, removing old decks or stairs, and tearing down siding, gutters, and masonry. Philbin, a nationally known how-to expert, also deals with tools and tool rental, safety, and debris disposal.

Pollan, Stephen M., and Mark Levine
The Big Fix-Up: How to Renovate Your Home Without Losing Your Shirt
New York, Simon and Schuster, 1992. 239p. ISBN 0671760416

The authors address the general consumer with this basic guidebook in managing a home renovation project. Not a manual of construction techniques, but a survey of the process. Includes the advantages and disadvantages of remodeling, and suggestions on working with subcontractors. Guidance is also provided for scheduling, contracts, financing, and selecting designs.

Rand, Ellen, Florence Perchuk, and the editors of Consumer Reports Books
The Complete Book of Kitchen Design
Yonkers, N.Y., Consumer Reports Books, 1991. 216p. ISBN 0890434743
The contents are arranged into two logical sections: planning and design; and products and installation. Part One emphasizes de-signs that accommodate today's lifestyles; Part Two focuses on evaluating the latest appliances and products in flooring, countertops, and cabinets. Stress is on energy efficiency and cost- effectiveness.

Remodeling
Newtown, Conn., Taunton Press, 1993. 128p. ISBN 1561580627

Part of the *Fine Homebuilding* series designed for experts and professionals, this volume contains articles on unusual projects for kitchens, attics, porches, alcoves, and additions. Includes full-color photographs and drawings. Well-written and concise style.

Shivers, Natalie W.
Walls and Molding: How to Care for Old and Historic Wood and Plaster
Washington, D.C., Preservation Press, 1990. 198p. ISBN 0891331557

Offers a survey of materials and procedures for restoring and preserving interior surfaces of older buildings. Instructs readers on how to retain the character of the interior structure without allowing it to deteriorate or change drastically. Discusses architectural styles, materials, and methods. This thorough study is accompanied by black-and-white photographs and an index.

Simmons, H. Leslie
Repairing and Extending Doors and Windows: Metal, Wood, Entrances, Store Fronts, Curtain Walls, Glazing
New York, Van Nostrand Reinhold, 1991. 406p.
ISBN 0442206186

Highly technical and written for professionals, this guide contains information on preexisting wood and metal doors and windows. Simmons identifies the various types of possible damage, suggests causes, and describes materials and methods for repair. The appendix contains a list of data sources. Includes extensive bibliography, glossary, and index.

Stephen, George
New Life for Old Houses
Washington, D.C., Preservation Press, 1989. 257p.
ISBN 0891331492

This revised edition of an older classic (*Remodeling Old Houses Without Destroying Their Character*, Knopf, 1972) explores the special skills and treatment required for rehabilitating and renovating older houses. Among topics covered are basic architectural styles; protection of original details; restoration of windows, doors and porches; and interiors. Appendix covers information on standards and energy conservation. Glossary and index.

Thomas, Steve, and Philip Langdon
This Old House Kitchens: A Guide to Design and Renovation
Boston, Little, Brown, 1992. 273p. ISBN 0316841064

By the experts of the popular television series *This Old House*, this in-depth guide to kitchens describes the complete process of designing a new kitchen or renovating an existing one. Emphasis is on planning and design, as well as demolition. Many imaginative ideas, good attention to detail, and attractive color photographs enhance this work.

Time-Life Books Complete Home Improvement and Renovation Manual
New York, Prentice Hall, 1991. 479p. ISBN 0139218831

Comprehensive in scope, this manual covers every aspect of home renovation, including tools and techniques. In-depth coverage, eye-pleasing layout, and three-color drawings combine to produce a distinctive and useful volume. Index.

Wilson, Steven, ed.
Popular Mechanics Home Answer Book
New York, Hearst Books, 1991. 224p. ISBN 0688108547

Covers more than sixty everyday home-repair problems in six subject areas: plumbing; heating and ventilation; electricity; walls, floors, stairs, doors, and windows; safety and tools; and exterior work. Chapters are subdivided into individual projects (e.g., how to install vinyl gutters, how to repair a porch). Easy to interpret drawings accompany instructions.

Nonprint Materials

Kiley, Martin
National Construction Estimator
42nd ed. Carlsbad, Calif., Craftsman Book Co., 1993. 588p. and 1 computer disk. ISBN 093404187X

This software program is designed for student use and is simple enough for beginners. Designed to be used in conjunction with the book, it contains a tutorial and bases estimates on only labor and materials. The index searches a database of 30,000 items. Not suitable for more complex professional projects but ideal for classroom situations and small projects.

Paxton, Albert S.
National Repair and Remodeling Estimator
17th ed. Carlsbad, Calif., Craftsman Book Co., 1994. 408p. and 1 computer disk. ISBN 093404189X

A concise set of instructions encompassing only thirteen pages is a key feature of this easy-to-install program. Prior versions are automatically erased during installation without deleting data. Permits users to enter descriptions as text lines, as well as data, and displays the completed estimate in its entirety. The keyword search function is helpful in quickly accessing information on materials.

Electrical Technology

Dan Haley
Pasadena City College
Pasadena, California

Introduction

Electricity is already a vital part of our economy. However, given the rapid expansion of the use of computers, new developments in data transmission methods, and the emergence of alternative transportation vehicles, electricity is becoming an even more important factor in everyday life. Workers in the electrical field are mostly electricians, but a large number also work in power generation and transmission. These include power plant operators, load dispatchers, substation operators, power-line installers and repairers, and cable splicers. Electricians fall into two categories, construction and maintenance.

It is possible to obtain entry-level employment in the field with as little as a high school diploma, but postsecondary coursework or a formal apprenticeship is very desirable. Formal apprenticeship programs are sometimes sponsored by local or national labor unions. Generally they last for five years and include classroom instruction and on-the-job training. In many locations, a license is required to work as an electrician. Licensing is usually based on written examinations. Electricians are required to follow the *National Electrical Code* (NEC) and state and local codes in their work. In addition, since the materials and requirements of electrical work are constantly changing, electricians usually take periodic refresher courses. Although formal classwork is not required, there is a definite need for it, and there is also a need for access to information resources for reference.

The primary reference source for all electrical work is the *National Electrical Code*, also referred to as NEC, NE Code, or sometimes "The Code." It is published by the National Fire Protection Association and is the basis for most state and local building codes dealing with electricity. The code is updated every three years. Most texts and nonprint sources will refer to the NEC, so it is important to ensure that they are referring to the most recent update. An enormous and detailed work, it is used for building codes, and thus is written as a legal document. It is often difficult to interpret, and has generated many other books attempting to explain it. Most either concentrate on specific areas or give basic overviews of the main sections. The explanations either give interpretations with relevant examples, or rewrite the most used sections to a level understandable to a beginner.

In general, books on the electrical trade are competently written, but illustrations usually lack quality, similar to books in other construction fields. Just as electricians are often required to work in awkward positions to get to hidden installations, so too, photographs are often taken at odd angles with less than adequate lighting. Even when a good in-field photograph is achieved, it usually translates to a grainy, gray picture in a book. Drawings are often reduced in size, making them harder to decipher, and subtle shadings are lost in black-and-white reproductions. In a field where so much depends on colored wires, it is difficult to justify black-and-white illustrations at all. More colors are being used in illustrations, and some publishers are using drawings that are actually larger-than-life. Of special note is the *Black and Decker Home Improvement Series* published by Cy DeCosse. While the text is limited to the homeowner or amateur level, these books have excellent electrical illustrations. Instead of trying to photograph at a construction worksite, they recreated a worksite inside a photography studio to obtain correct lighting and adequate camera angles. It is hoped that other publishers will follow this lead.

Items in this bibliography have been coded *General Interest, Unskilled level, Skilled level,* and *Professional level* to aid librarians in the selection of appropriate materials for the curriculum that is offered.

Selected Associations

There is no single nationwide group responsible for accrediting educational programs in the electrical technology arena. Certification of technicians is usually done at the state or local level. However, there are several associations that can provide information on education, industry trends, and continuing education:

American National Standards Institute
1430 Broadway
New York, N.Y. 10018

Independent Electrical Contractors, Inc.
P.O. Box 10379
Alexandria, Va. 22310-0379

Institute of Electrical and Electronic Engineers
345 E. 47th St.
New York, N.Y. 10017

International Association of Electrical Inspectors (IAEI)
901 Waterfall Way
Richardson, Tex. 75080

International Brotherhood of Electrical Workers (IBEW)
1125 15th Street, NW
Washington, D.C. 20005

National Electrical Contractors Association (NECA)
3 Bethesda Metro Center, Suite 1100
Bethesda, Md. 20814

National Electrical Manufacturers Association
2101 L. Street NW, Suite 300
Washington, D.C. 20037

National Fire Protection Association (NFPA)
1 Batterymarch Park
P.O. Box 9101
Quincy, Mass. 02269-9101

Selected Journals

CEE News [Contractors Electrical Equipment News].
Intertec Publishing. Monthly (except twice in September). ISSN 1045-2710

Electrical Code Watch.
Intertec Publishing. Monthly. ISSN 1057-0241

Electrical Construction and Maintenance [ECM].
Intertec Publishing. Monthly. ISSN 0013-4260

Electrical Contractor.
National Electrical Contractors Association. Monthly. ISSN 0033-5118

IAEI News.
International Association of Electrical Inspectors. Bimonthly. ISSN 0020-5974

NFPA Journal.
National Fire Protection Association. Bimonthly. ISSN 1054-8793

Print Materials

Advanced Home Wiring
Minnetonka, Minn., Cy DeCosse, 1992. 128p. ISBN 0865737185
Somewhat brief in coverage, but color photograph illustrations are excellent. Includes sections on planning a wiring project and evaluating loads, and describes various materials and techniques, wiring requirements for room additions and kitchen remodels, and installation of outdoor wiring. Indexed. *(Skilled level)*

Aglow, Stanley H.
Schematic Wiring: A Step-By-Step Guide
Troy, Mich., Business News Pub., 1991. 243p. ISBN 0912524669
An instructional manual that demonstrates how to draw electrical schematics and interpret manufacturers' schematics. Includes symbols and legends. Intended for beginning students, the sections include sample schematics for basic devices such as fan relays, transformers, and thermostats, and more advanced electronic controls such as solid state relays, ignition controls, and test instruments. Some photographic illustrations. No index. *(Skilled level)*

American National Standards Institute
National Electrical Safety Code
1993 ed. New York, IEEE, 1992. 257p. ISBN 1559372109
Different from the *National Electrical Code*, this book covers safety during installation, operation, or maintenance of electric supply and communication lines and equipment. The text is legal in tone as intended. This book does not pretend to be an instruction manual, but points out generally accepted practices designed to prevent injury. Broad experience is assumed of the reader and there are few illustrations. The index is comprehensive. Anyone working in the electrical field will need to have access to a copy. *(Professional level)*

Andreas, John C.
Energy Efficient Electric Motors: Selection and Application
2nd ed. New York, Marcel Dekker, 1992. 272p.
ISBN 0824785967
While primary focus is on motor design, the data included are of importance to selection and applications of electric motors. Covers electric power costs, power factor considerations, and general application problems such as voltage unbalance and overmotoring. Adjustable speed drive systems receive special emphasis. The last chapter includes life cycle expectations and cost analysis. Illustrated and indexed. *(Professional level)*

AVO Multi-Amp Institute and John Cadick
Cables and Wiring
Albany, N.Y., Delmar Publishers, 1993. 196p. ISBN 0827354606
With the development of a multitude of cables designed for varying environmental, design, and functional specifications, cables

and wiring are becoming a specialized subfield. The first part of the book is a basic instruction for installation, splicing termination, and testing of electrical cable with a separate section covering fault location. The second part is a reference guide to over twenty-five types of cable. Black-and-white illustrations. Indexed. *(Professional level)*

Basic Wiring and Electrical Repairs
Minnetonka, Minn. Cy DeCosse, 1990. 128p. ISBN 0865737142

Aimed at homeowners attempting basic repairs or additions to residential wiring. The color photographs are excellent and would be useful to any level reader. Covers wiring basics, wall switches, receptacles, lighting fixtures, doorbells, and thermostats. Includes an "Inspectors Notebook" section that amply illustrates potential problems or hazards. *(Skilled level)*

Benfield, Jack
Benfield Conduit Bending Manual
2nd ed. Overland Park, Kans., EC&M, 1992. 109p. ISBN 0872885100

Updated edition of a basic manual on conduit bending. Much more detailed than the usual chapter coverage in a standard manual, this work covers a variety of applications such as saddle bends, back-to-backs, and rolling offsets. A number of different bending tools are explained including rigid conduit, IMC, and combination benders. The text is intended for beginners and is easy to understand. The illustrations are helpful. No index is provided, but the length and form of presentation probably make one unnecessary. *(Professional level)*

Berutti, Al, and R. M. (Ray) Waggoner, eds.
Practical Guide to Quality Power for Sensitive Electronic Equipment
Overland Park, Kans., Intertec Electrical Group, 1993. 100p. ISBN 0872885127

This slim, but information-packed volume covers an increasingly important aspect of electrical installations. A similar Intertec volume by Robert J. Lawrie, *Practical Guide to Power Distribution Systems for Computers* (1993), could be considered a companion piece. Good illustrations and clear text enhance a discussion of sensitive equipment and a review of FIPS *(Federal Information Processing Standards)* Pub. 94, treating quality power, grounding, surge and transient suppression, nonlinear loads, harmonics and how to defeat them, and coordination guidelines for sensitive loads. An index would enhance the usefulness of this work. *(Professional level)*

Bierals, Gregory P.
Applying the 1993 National Electrical Code
Lilburn, Ga., Fairmont Press, 1994. 228p. ISBN 0881731676

One of many books offering explanations and interpretations of the NEC, this one is fairly well written and gives numerous examples of actual situations where interpretation and application of the code are very difficult. The narrative tone and numerous anecdotal examples make it more readable than most. A high level of expertise is expected of the reader. The last section covers a number of revisions suggested for the 1996 edition of NEC. Illustrated. A more extensive index would be helpful. *(Professional level)*

Bierals, Gregory P.
National Electrical Code Illustrated Changes Deskbook
Lilburn, Ga., Fairmont Press, 1993. 105p. ISBN 0881731714

Provides not only explanations of changes made to the 1993 NEC but also the background that prompted the change. Intended for those already familiar with the code. The explanations for any one change refer to numerous other sections of the code. A slim volume, it is meant to be read from cover to cover, as there is neither an index nor a table of contents. The information is arranged numerically by the most relevant code section. Some illustrations. *(Professional level)*

Brodeur, Paul
The Great Power-Line Cover-Up: How the Utilities and the Government Are Trying to Hide the Cancer Hazard Posed by Electromagnetic Fields
Boston, Little, Brown, 1993. 326p. ISBN 0316109096

Researched account of high voltage power line areas with unusually high rates of cancer. Provides data on possible physiological effects of exposure to electromagnetic fields. Indexed. *(General Interest)*

Colvin, Thomas S.
Electrical Wiring: Residential, Utility, Service Areas
6th ed. Winterville, Ga., American Association for Vocational Instructional Materials, 1993. 187p. ISBN 089606302X

This book is intended as a text, but the color illustrations make it a useful reference. Intended for apprentice level, but can be used by amateurs. Covers wiring installation, connecting circuits, reading wiring plans, metallic conduits, cost estimations, and special circumstances for rural areas. Indexed. *(Skilled level)*

Earley, Mark W., ed.
National Electrical Code Handbook: Based on the 1993 Edition of the National Electrical Code
6th ed. Quincy, Mass., National Fire Protection Association, 1992. 1120p. ISBN 0877653844

Differs from most aids to the NEC in that it is published by the NFPA and contains the entire code published verbatim. In comparison to the standard NEC, this book's larger format and print are easier to read and use, although the work is less portable. The explanatory text is more an expansion and clarification of the code rather than an instructional manual. Often it only cross-references to other sections. Includes several helpful supplements. The cost is comparable to other reference books of similar size; libraries may want to consider the purchase of this instead of the standard NEC. The index is completely different from the one in NEC. *(Professional level)*

Farm Buildings Wiring Handbook
2nd ed. Ames, Iowa, Midwest Plan Service, 1992. 63p. ISBN 0893730858

This slim volume contains a wealth of information on electrical power and wiring for agricultural buildings. More of a reference manual than an instruction book, the text is still clear and readable. The illustrations and tables are numerous and well done. Covers such specialized applications as wiring for dry, damp, and dusty buildings, standby power, branch circuits, motor circuits, stray voltage, alarm systems, and lightning protection. Indexed. *(Skilled level)*

Gibilisco, Stan
Teach Yourself Electricity and Electronics
Blue Ridge Summit, Pa., TAB Books, 1993. 662p.
ISBN 0830641335

An introductory text for basic electricity and electronics. There are almost no references to actual manufacturers' equipment or materials. However, the text is clearly written and can be used for reference. Illustrated and indexed. *(Unskilled level)*

Greenwald, E. K., ed.
Electrical Hazards and Accidents: Their Cause and Prevention
New York, Van Nostrand Reinhold, 1991. 200p.
ISBN 0442237995

Provides information on correct wiring materials and procedures as they specifically relate to preventing electrical accidents. Includes general information on grounding and specific information on wiring in hazardous areas, such as those prone to fire (e.g., storage areas for flammable liquids). Separate chapters cover electrical fires, lighting hazard, static electricity prevention, and the physiological effects of electricity including death and injury. Sparse illustrations. *(Skilled level)*

Guide to the 1993 National Electrical Code
Revised by Paul Rosenberg. New York, Macmillan, 1993. 609p.
ISBN 0020777612

An explanatory text that follows the NEC section by section and provides explanations for the code that can be easily understood. This book tries to distinguish itself from the sometimes "legalese" content of NEC with aids (such as explaining that the word "shall" in NEC means that you "must" do it that way). For the most part it succeeds, but the reader still has to have more than an average familiarity with the field. The author advises that this book should only be used in conjunction with the NEC, so lack of the latter precludes buying this one. Illustrated and indexed. *(Professional level)*

Gustafson, Robert J.
Fundamentals of Electricity for Agriculture
2nd ed. St. Joseph, Mich., American Society of Agricultural Engineers, 1988. 411p. ISBN 0916150941

Provides basics of power generation, wiring, and use, specifically as they apply to farmsteads. Contains information on pertinent special problems such as lightning, farm sensors alarms, and stray voltage. Also covers isolated/self-reliant areas such as standby power units and solar and wind energy generation. Illustrated and indexed. *(Skilled level)*

Hartwell, Frederic P.
Illustrated Changes in the 1993 National Electrical Code
Overland Park, Kans., EC&M, 1992. 329p. ISBN 0872885119

A very interesting and down-to-earth explanatory text covering 1993 changes to NEC. The author provides helpful information such as when a code change is primarily meant for manufacturers as opposed to installers. The explanations are clear and to the point, especially when the author points out the new materials or products that permit or require changes. The actual text of NEC is on blue backgrounds and is arranged by the code section numbers. Contains illustrations. Lacks index. *(Professional level)*

Hartwell, Frederic P.
Understanding NE Code Rules on Grounding and Bonding
Overland Park, Kans., EC&M, 1994. 174p. ISBN 0872885437

Hartwell, Frederic P.
Understanding NE Code Rules on Medium Voltage Power Systems
Overland Park, Kans., EC&M, 1994. 202p. ISBN 0872885461

Hartwell, Frederic P.
Understanding NE Code Rules on Transformers
Overland Park, Kans., EC&M, 1993. 98p. ISBN 0872885410

This series pulls together sections of the *National Electrical Code* that are relevant to the particular application of transformers and presents the code in a readable manner. Each volume is a handy guide to the code for anyone involved in these areas, but access to the entire code is still recommended. In addition, these volumes should not be confused with straight instructional manuals for the applications. Considerable expertise is expected. Not indexed. *(Professional level) (Skilled level)*

Hiatt, Richard S., ed.
Agricultural Wiring Handbook
10th ed. Columbia, Mo., National Food and Energy Council, 1993. 92p.

Not an instruction manual, but rather a planning guide for electrical systems and installations on farms. Similar in format to NEC, but the text is much less oriented toward legal interpretations, and more toward practical advice. Gives specific guidelines for dairy, poultry, and livestock structures, feed preparation centers, and crop storage and handling buildings. Separate section for water supply requirements. Also covers details on electrical distribution including services and service loads, metering, grounding, standby service, three-phase service, and wiring for motors. Illustrated, numerous tables, and indexed. *(Skilled level)*

Hines, Paul R.
Basic Alternating Current Control Diagrams
5th ed. Chicago, Barks Pubs., 1990. 83p. ISBN 0943876001

Pocket-sized reference manual containing schematic drawings of thirty-five basic wiring diagrams, a voltage table, and various electrical formulas. Because of minimal explanations and a lack of legends, this work is for experienced workers only. *(Skilled level)*

Holzman, Harvey N.
Modern Residential Wiring
South Holland, Ill., Goodheart-Willcox, 1993. 288p.
ISBN 1566370019

Designed for entry-level instruction, this text provides brief overview of basic electricity followed by individual chapters covering professional-level installation procedures. Frequent references to the 1993 NEC. Covers wiring, service entrances, and circuits but also includes chapters on farm wiring, mobile home wiring, remodeling, swimming pool wiring, and troubleshooting. Numerous illustrations and tables in the appendix. Glossary and index. *(Unskilled level)*

Kiehne, H. A., ed.
Battery Technology Handbook
New York, Marcel Dekker, 1989. 519p. ISBN 0824781805

Provides a survey of technical levels of commercial batteries and an overview of emerging applications. Extensive coverage of current technology including theory, design, and applications. Special emphasis is provided for electric vehicles. Applications range from large-scale (e.g., telecommunications) to tiny (electronic watches). Numerous charts and illustrations. Indexed. *(Professional level) (Skilled level)*

Lawrie, Robert J., ed.
Practical Guide to Power Distribution Systems for Computers
Overland Park, Kans., Intertec Electrical Group, 1993. 90p.
ISBN 0872885283

This volume and a similar Intertec publication, *Practical Guide to Quality Power for Sensitive Electronic Equipment*, (1993), cover an area in this field that will become increasing important. The text is clear and straightforward but requires some prior experience of the reader. Topics covered include supplying uninterruptible power, distribution system design considerations, power disturbances, grounding, transfer switches, surge protection, and power conditioning. One part covers relevant sections of NEC, and another discusses an actual case history. The illustrations are numerous and above average in quality. No index. *(Skilled level)*

Loyd, Richard E.
Electrical Raceways and Other Wiring Methods
Albany, N.Y., Delmar Publishers, 1993. 223p. ISBN 0827354932

Raceways are enclosed channels or conduits designed to hold electrical wires or cables. This instructional text on the use of raceways in electrical installation covers selection factors such as safety, capacity, and flexibility; installation requirements such as expansion properties and induction heating; and accompanying fixtures such as boxes and fittings. A special section covers installation in hazardous locations like gasoline service stations. Applicable codes and standards are presented. Illustrated and indexed. *(Skilled level)*

Maloney, Timothy J.
Electricity: Fundamental Concepts and Applications
Albany, N.Y., Delmar Publishers, 1992. 472p. ISBN 0827346751

A text that covers the basics of electricity, but also skims the surface of electrical systems and installation. Designed for the reader who begins with absolutely no knowledge of electricity, it still manages to cover specialized areas such as the color coding system for resistors. Textbook arrangement with simple illustrations, examples, and review questions. Includes glossary. Indexed. *(Unskilled level)*

Means Electrical Change Order Cost Data
Kingston, Mass., R. S. Means, Annual. (various pagings)
ISSN 1044-2812

Similar data and layout to *Means Electrical Cost Data*, but costs are adjusted for change order variations such as smaller quantities usually purchased. Costs are estimated for both preinstallation and postinstallation work. A special section covers causes, procedures, and extra cost factors involving change orders. Major cities' cost indexes, and abbreviations section are included. A companion to the *Means Electrical Cost Data*; purchase of both is recommended. Indexed. *(Professional level)*

Means Electrical Cost Data
Kingston, Mass., R. S. Means, Annual. (various pagings)
ISSN 0748-7002

Provides detailed breakdowns of cost estimates for electrical work. Entries include description of work, estimated crew, daily output and worker-hours required. Costs are broken down by material, labor, and equipment. Bare cost totals and totals indicating overhead and profit are indicated. The information is broken down by unit price and by assembled functional elements. Indexes for most major cities are carried in a separate section. The abbreviations section is helpful. Indexed. Updated annually. *(Professional level)*

Miller, Rex, and Mark Richard Miller
Small Electric Motors: Use, Selection, Operation, Repair and Maintenance
New York, Macmillan, 1992. 436p. ISBN 0025849751

A very good manual that jumps quickly from electric motor basics to practical information on selection, operation, repair, and maintenance. Detailed coverage of DC, universal, split-phase, capacitor-start, shaded-pole, three-phase, stepper, synchronous, and special types of motors. Contains glossary and a number of useful appendices. Illustrated and indexed. *(Professional level)*

Mix, Floyd M.
House Wiring Simplified
South Holland, Ill., Goodheart-Willcox, 1991. 176p.
ISBN 0870068695

Basic text that can be easily understood by amateurs, but goes beyond the usual homeowners guide. Covers simple tasks like receptacles and switches but also addresses essentials needed for professional work such as installation of wiring boxes. Numerous illustrations are uniformly excellent. Of special note are the load demand and conductor tables. Includes glossary and index. *(Unskilled level)*

Mullin, Ray C., and Robert L. Smith
Electrical Wiring, Commercial
8th ed. Albany, N.Y., Delmar Publishers, 1993. 318p.
ISBN 0827350937

Basically an explanatory text of the NEC for commercial buildings. Arrangement is by learning units with review questions at the end of each. Units include plans and specifications, branch circuits and feeders, appliance circuits, lighting, panelboards, services, overcurrent protection, emergency power systems, and cooling systems. The examples given include a bakery, a beauty salon, and an insurance office. Written for someone learning, but still requires a journeyman level of expertise. Numerous illustrations. Includes an index for NEC code references. Indexed. *(Skilled level)*

Mullin, Ray C.
Electrical Wiring, Residential
11th ed. Albany, N.Y., Delmar Publishers, 1993. 504p.
ISBN 0827350953

Textbook-style manual, but still good for reference and self-instruction. Covers entry-level information. Refers to 1993 NEC and gives examples of installations that meet the code and ones that do not. Illustrations are average, but some tables provide good guidance. Separate chapters for various residential rooms and major appliances. Of note are the sections on remote control systems, smoke detector wiring, swimming pool wiring, and "smart house" units. Includes code index and foldout sample floor plans. *(Skilled level)*

Multilingual Dictionary of Electricity, Electronics, and Telecommunications
Geneva, Switzerland, International Electrotechnical Commission, 1992. 936p. ISBN 1559371854

Provides definitions in English, French, Russian, German, Spanish, Italian, Dutch, Polish, and Swedish. Arranged alphabetically, but each entry is hierarchical in nature, thus filtration, inherent filtration, and total filtration are all under "filtration." Cross-references are provided. Translation is from English to other languages only. Indexes from other languages are available in a five-volume set available from the IEC. Definitions are brief and there are no illustrations. *(General Interest)*

Nailen, Richard L.
Managing Controls
Chicago, Barks Publications, 1993. 1 v. (various pagings)
ISBN 0943876060

Basic coverage of an increasingly important aspect of electrical applications. Sections include mechanical and electrical motor starters, control relays, automatic transfer switches, and motor circuit protection. One chapter entitled "When Off Doesn't Mean Off," contains experience-based safety information not easily found elsewhere. Diagrams are adequate, but black-and-white illustrations do not do justice to the text. Indexed. *(Skilled level)*

Nailen, Richard L.
Managing Motors
Chicago, Barks Publications, 1991. 1 v. (various pagings)
ISBN 0943876044

Covers the basics of electrical motor engineering and maintenance. Includes sections on efficiency (with some surprising notations), starting conditions, environmental problems, types of insulation and windings, bearings and lubrication, and optional accessories. Also has a large section on motor installations and problems likely to be encountered. Adequately illustrated and indexed. *(Skilled level)*

National Electrical Code: 1993
Quincy, Mass., National Fire Protection Association, 1992. 917p. ISBN 0877653836

An enormous compendium of specifications primarily intended to prevent hazards such as fire or explosion. Covering virtually every possible electrical installation or application, it is used as a basis for many local electrical building codes. Clearly written, but assumes extensive experience and training of the reader. Also covers communications systems, e.g., radio and television. Some illustrations and tables included. A required reference for anyone working in the electrical field. Extensively indexed. *(Professional level)*

National Electrical Estimator
Carlsbad, Calif., Craftsman Book Co., Annual. (various pagings)

Cost estimates for electrical work arranged by the material to be installed. Entries include materials cost, labor cost indicated by craft level and number of hours, and total costs. Special sections contain an estimating checklist and an explanation of diagram symbols. A computer disk and instruction manual are included for a program entitled "Estimate Writer." Indexed. *(Professional level)*

Nunn, Richard V.
Basic Wiring.
Rev. ed. Upper Saddle River, N.J., Creative Homeowner Press, 1994. 160p. ISBN 1880029367

Aimed at amateur level, this instructional manual is notable for its color illustrations. Text is clear and easily understandable. Covers basics such as plugs, switches, and small appliances, but also addresses wiring installation, rewiring, and outdoor wiring. Includes glossary and index. *(Unskilled level)*

O'Riley, Ronald P.
Electrical Grounding: Bringing Grounding Back to Earth
3rd ed. Albany, N.Y., Delmar Publishers, 1993. 278p.
ISBN 0827352484

Covers basic reasons for grounding; system faults; grounding conductors; jumpers; equipment, system, and circuit grounding; and enclosure bonding and grounding. This book is arranged like an instructional text, but the sheer depth of coverage makes it more like a reference manual. Numerous illustrations and tables. Requires extensive knowledge of the reader. Indexed. *(Professional level) (Skilled level)*

Palmquist, Roland E.
Electrical Course for Apprentices and Journeymen
3rd ed., revised by Joseph A. Tedesco. New York, Macmillan, 1988. 471p. ISBN 0025945505

Designed as a textbook, this is still valuable as a reference. The emphasis is on basic electricity, but practical application information is included, e.g., insulation testing, polyphase circuits, instrument transformers, single and polyphase motors, synchronous motors, grounding, rectifiers, and bend specifications for conduits. Illustrated and indexed. *(Skilled level)*

Palmquist, Roland E.
House Wiring
7th ed., revised by Paul Rosenberg. New York, Macmillan, 1991. 237p. ISBN 0025946927

A nice applications manual that combines basic instruction with explanations of the relevant sections of the NEC. Pertinent definitions are supplied at the beginning of each chapter. Topics include load calculations, service installations, a variety of cable applications, tubing and conduits, dwelling circuits, boxes and fittings, and receptacles. Special sections cover farm buildings, electric heating, and mobile homes. Illustrated and indexed. *(Skilled level)*

Pansini, Anthony J., and Kenneth D. Smalling
Undergrounding Electric Lines
2nd ed. Lilburn, Ga., Fairmont Press, 1993. 215p.
ISBN 0881731625

Underground cabling is becoming increasingly prevalent in residential wiring and is already common in commercial facilities. For most localities it is considered more reliable, although it is more expensive and failures are more difficult to locate. This volume serves as an introduction to design, installation, and maintenance of such systems. Black-and-white illustrations. *(Professional level)*

Rosenberg, Paul
Installation Requirements of the 1993 National Electrical Code
New York, Macmillan, 1993. 263p. ISBN 0020777604

A compromise book between the legally oriented NEC and a basic instruction manual in electrical installation. It does not provide the depth of NEC, nor is it as easy to follow as a basic manual; however, it is unique in pulling together elements of the code into sections that cover specific applications, such as grounding, conduit wiring, and motors. This is a very useful book, but it does require some previous experience, and access to the NEC. Illustrated and indexed. *(Professional level)*

Sauselein, Theodore B.
Power Technology
Troy, Mich., Business News Pub., 1994. 138p. ISBN 0912524855

A good beginning text covering the basics of power generation, delivery and use. It lacks the detail necessary for a reference manual, but does include some very forward-looking sections covering new developments in metering methods, conservation, thermal storage, and cogeneration. The author also provides some very interesting inside information such as the way some electric utilities calculate rates for their own purposes. Includes glossary. Lacks index. *(General Interest)*

Seale, Arthur C.
Electrical Wiring
Carmel, Ind., Sams, 1990. 274p. ISBN 0672226952

Intended for beginners, but skips usual homeowner repair jobs and concentrates on professional-level work. The text expects some familiarity with the field, but is straightforward enough to be understood by an amateur. Illustrations are black-and-white, but are laid out so the reader can differentiate individual wires. Glossary and index. *(Unskilled level)*

Small AC Generator Service Manual
3rd ed. Overland Park, Kans., Intertec Pub. Co., 1991. 536p. ISBN 0872884678

Technical manual for "electrical powerplants under eight kw that generate sixty hertz alternating current and are driven by air-cooled engines." (pref.) Similar to the Chilton auto repair compilations, this book supplies the technical specifications needed for repair and maintenance of generators from over thirty different manufacturers. A twenty-page section of engine fundamentals is provided for basic instruction. Some prior experience is expected. Generally only a few pages are devoted to each make and model, so this could be used as a reference work. Numerous illustrations. No index is provided, but probably none is needed. *(Professional level)*

Smead, David, and Ruth Ishihara
Living on 12 Volts with Ample Power
Seattle, Wash., RIDES Pub. Co., 1992. 344p. ISBN 0945415028

A companion piece to *Wiring 12 Volts for Ample Power* (RIDES Pub. Co., 1991), this book covers the basics of alternative power in practical, understandable language. The author dispels the notion that alternative power is free, or even cheap, and provides in-depth, experience-based information on solar and wind-driven electricity. Topics include selection and maintenance of batteries, alternators, wind and tow generators, and solar panels. Overall emphasis is on the conservation of electrical power, and the author provides much-needed practical information on how that is achieved. Indexed. *(Professional level)*

Smead, David, and Ruth Ishihara
Wiring 12 Volts for Ample Power
Seattle, Wash., RIDES Pub. Co., 1991. 240p. ISBN 0945415036

With its companion piece *Living on 12 Volts with Ample Power* (RIDES Pub. Co., 1992), this volume provides in-depth, practical information for battery-based electrical systems. Emphasis is primarily on boats and RVs. Sections cover choice of battery types, charging sources, alternators, instrumentation, wiring, tools, and schematics. A special section covers testing and troubleshooting. Contains few illustrations, but the text is down-to-earth and readable. Indexed. *(Professional level)*

Smith, Robert L., and Stephen L. Herman
Electrical Wiring, Industrial
8th ed. Albany, N.Y., Delmar Publishing, 1993. 222p. ISBN 0827353251

An interpretation and explanatory manual of the NEC for industrial buildings, which is arranged in learning units. Review questions follow each unit. Chapters include unit substations, feeder bus systems, panelboards, trolley busways, wire table calculations, signal systems, motors and controllers, lighting, air conditioning, programmable controllers, and special equipment such as precipitation units, synchronous condensers, and capacitors. A special section covers unique fixtures for hazardous locations. An NEC code index is included. Illustrated and indexed. *(Professional level)*

Stallcup, James G.
Illustrated Changes of the 1993 NEC
Homewood, Ill., American Technical Publishers, 1992. 218p. ISBN 0826915299

One of the few books that is more illustration than text, and the result is a good companion to the NEC. Arranged by the code numbers, each entry lists the heading, type of change, a brief summary of the change, and an illustration. The illustrations are concentrated, but easy to read. Of special interest is a reference table for each entry indicating the corresponding standard from the Underwriters Laboratories, the National Fire Protection Association, OSHA, the National Electrical Manufacturers Association, the NEC Technical Committee Report, and the NEC Technical Committee Documentation. An index would have been helpful. *(Professional level)*

Standard for Health Care Facilities
1993 ed. Quincy, Mass., National Fire Protection Association, 1993. 229p.

Cites current electrical standards of the American National Standards Institute and the National Fire Protection Association. Written from a legal standpoint and requiring considerable prior experience, this volume is necessary for anyone performing electrical work on a hospital or medical facility. Serves as an adjunct to the NEC and covers special considerations necessary in a health care building. Topics include electrical systems, gas and vacuum systems, laboratories, and the use of high frequency electricity. Few illustrations. Indexed. *(Professional level)*

Stauffer, H. Brooke, and Ray C. Mullin
Smart House Wiring
Albany, N.Y., Delmar Publishers, 1993. 173p. ISBN 0827354894

Instructional manual for electrical system installation for a "smart house," a house that integrates power, control, telephone, and coaxial wiring to allow for computer automation of all electrical and communication appliances. Such systems require materials purchased from "smart house" manufacturers. Text assumes basic electrical knowledge and abilities, and covers only the specialized skills and tools necessary for this type of system. Treats specialized symbols, cables and outlets, system layout, cable installation and termination, and computerized service center. Separate chapters cover bedroom, bathroom, entry, kitchen, living room, and garage installations. Divided into learning units with review questions at the end of each unit. Indexed. *(Skilled level)*

Thumann, Albert, ed.
Lighting Efficiency Applications
2nd ed. Lilburn, Ga., Fairmont Press, 1992. 354p.
ISBN 0881731374

Overview of the design and installation of a lighting system with emphasis on energy efficiency. Includes unit selection criteria, control equipment, reflectors, ballast selection, and incorporation of natural daylight. Special sections on fluorescent lighting and use of computer software. Last section covers current research. Case studies and illustrations. Index needs expansion. *(Professional level)*

Tomal, Daniel R., and Neal S. Widmer
Electronic Troubleshooting
Blue Ridge Summit, Pa., TAB Books, 1993. 381p.
ISBN 0830643559

The emphasis here is on electronics, but several sections cover troubleshooting electrical installations. The topics include motors and generators, residential and industrial wiring, and microprocessor-based electrical systems. Small appliances such as radios, televisions, and tape players are also covered. Black-and-white photographs; appendices list troubleshooting guides. Indexed. *(Professional level)*

Traister, John E., and Paul Rosenberg
Construction Electrical Contracting
2nd ed. New York, Wiley, 1989. 300p. ISBN 0471630144

Obviously aimed at the contractor, this volume still has lots of information needed for professional work. Would be helpful for an apprentice if only to explain what is behind the decisions of the contractor. Covers basics such as pulling permits, scheduling, management advice, and installation techniques. Also has separate chapters on branch circuits, service wiring, lighting fixtures, electrical drawings and specifications, and cost estimations. Illustrated and indexed. *(Professional level) (Skilled level)*

Traister, John E.
Illustrated Dictionary for Electrical Workers
Albany, N.Y., Delmar Publishers, 1991. 223p. ISBN 0827347634

Definitions for over 3,000 electrical terms. Definitions assume some familiarity with electrical nomenclature but are definitely more explanatory than the usual glossary. Includes some black-and-white illustrations. *(Skilled level)*

Traister, John E.
Illustrated Guide to the National Electrical Code, 1993
Carlsbad, Calif., Craftsman Book Co., 1992. 268p.
ISBN 0934041806

An explanation of the NEC written for apprentice-level workers. Should be used with the NEC but is also a good text for basic principles of electric installation. Although it provides a very good explanation of fuses and overcurrent protection, it also supplies information on terms, e.g., vestibule. An extensive glossary and symbols section are included. Illustrated and Indexed. *(Skilled level)*

Traister, John E.
Residential Electrical Design Revised
Carlsbad, Calif., Craftsman Book Co., 1994. 254p.
ISBN 0934041954

Generalized overview of the planning and installation of residential electrical systems and appliances. Topics include overview of the NEC, electrical drawings, wiring for services, wiring methods, and branch-circuit layout. Half of the volume is devoted to lighting plans and heating and air conditioning. This book is aimed at the professional level, but the text is clear enough for nonprofessionals. An appendix covers work specifications. Black-and-white illustrations. Indexed. *(Professional level) (Skilled level)*

Understanding NE Code Rules On Hazardous (Classified) Locations
Overland Park, Kans., EC&M, Intertec Electrical Group, 1993. 160p. ISBN 0872885135

Understanding Regulations on–OSHA Electrical Design Safety Standards: Including the Official Regulations
Overland Park, Kans., EC&M, Intertec Electrical Group, 1991. 164p. ISBN 0872884589

Understanding Regulations on–OSHA Electrical Safety Rules for Construction Sites: Including the Official Regulations
Overland Park, Kans., EC&M, Intertec Electrical Group, 1992. 154p. ISBN 0872884600

Understanding Regulations on–OSHA Electrical Work Rules With Complete Lockout/tagout Procedures: Including the Official Regulations
Overland Park, Kans., EC&M, Intertec Electrical Group, 1991. 139p. ISBN 0872884597

This series presents a handy guide to the Occupational Safety and Health Administration regulations regarding the electrical field indicated in each title. These books do not attempt to cover all the OSHA regulations. *Electrical Design* covers Section 1910.302 through 1910.309 and Subpart S. *Safety Rules for Construction Sites* covers 1926.402 through 1926.408, 1926.416-1926.417, 1926.431-1926.432, 1926.441, and Subpart K. *Electrical Work Rules* covers 1910.331-1910.335, 1910.399, and Subparts J and S. All in all, a good attempt at presenting complex legal rules in an understandable manner. Prior experience and access to the regulations are assumed. Illustrated. Indexes would have enhanced use. *(Professional level)*

Whitaker, Jerry C.
AC Power Systems
Boca Raton, Fla., CRC Press, 1991. 327p. ISBN 084937412X

This book covers virtually all electric power sources available in the United States, including power system operation, protecting equipment loads, grounding, standby power, and safety. The text does not gloss over practical considerations; it provides a lot of "down-to-earth" advice and guidance. Aimed at systems design and power supply maintenance, it has applications for anyone working with electricity. Illustrated and indexed. *(Professional level) (Skilled level)*

Winslow, Taylor F.
Electrical Blueprint Reading
Carlsbad, Calif., Craftsman Book Co., 1991. 204p.
ISBN 0934041644

Basic instruction manual on how to read electrical sections of blueprints, schematics, site plans, wiring diagrams, electrical schedules, electrical symbols, and electrical specifications. Meant for entry level, but could also be used for reference. Typeface varies from section to section. Includes numerous illustrations, glossary, and symbols legend. Indexed. *(Skilled level)*

Wiring
Upper Saddle River, N.J., Creative Homeowner Press, 1992. 80p.
ISBN 1880029138

This instruction manual aimed at beginners covers only the very basics, but does it well. Contains numerous illustrations that are larger-than-life, which makes them more easily understood. Use of some color also helpful. The text is clear and does not assume prior knowledge. Indexed. *(Unskilled level)*

Wood, Robert W.
Home Electrical Wiring Made Easy
2nd ed. Blue Ridge Summit, Pa., TAB Books, 1993. 190p.
ISBN 0830641882

Though an instruction manual for beginners, this work goes a little beyond most. Covers ambitious home projects like upgrading a service, adding a circuit, and installing a garage door. The author also includes chapters on telephone wiring, troubleshooting techniques, and a section demonstrating possible costs for materials. Illustrated and indexed. *(Unskilled level)*

Heating, Ventilation, and Air Conditioning

Steven Self
Austin Community College
Austin, Texas

Introduction

Keeping physically comfortable has long been a deciding factor in how we design our surroundings. The temperature of an environment will influence the types of clothing worn, the foods eaten, and the shelters that are built. Although the design of a structure can have a great impact on personal comfort, the mechanical means of changing temperature are important also. Heating, ventilation, and air conditioning (HVAC) refers to the mechanical processes developed for controlling temperatures within a specific space.

HVAC (or HVAC and R as it is referred to when including principles of refrigeration) involves more than just heating and cooling air. Humidity control, efficient air distribution, contaminant removal, sound level monitoring, and air pressure control are all areas of concern in HVAC development and maintenance.

It is possible to divide HVAC into two general areas of application. The comfort applications of HVAC refer to heating and cooling used to provide people with comfortable and healthful surroundings. All areas of human activity benefit from the comfort applications of HVAC, including businesses, public buildings, homes, and public transportation. Process applications provide environmental control to support research, storage, and manufacturing of materials. Examples of HVAC process applications include environmental control of clean rooms for high-tech industries, cryogenics, food storage, greenhouses, and pharmaceutical development.

Many recent developments in HVAC technology have centered around changes in refrigerant use. Scientists have learned that chlorofluorocarbons (CFCs), chemicals used in air conditioning and refrigeration, damage atmospheric ozone. This realization has led to legislation regulating CFC use, the development of new coolants and equipment, and new maintenance standards.

Another area undergoing rapid change is HVAC controls. Controls are mechanical and electronic devices that monitor and make changes in environments subject to HVAC systems. HVAC designers are adapting innovations from the realm of microelectronics to develop ever-more sophisticated sensing equipment and digital controls.

HVAC programs in community colleges employ a combination of theory and application in teaching students the principles of heating, air-conditioning, and ventilation. Courses are offered on such topics as principles of heat transfer, electrical circuits, troubleshooting and maintenance, soldering, and repair. Often, special courses consider both residential and commercial HVAC applications. Mechanical codes, technical mathematics, and technical writing are frequently included in the course of study. As in many vocational and technical programs, labs and other hands-on exercises are essential in providing experience and ensuring proficiency.

Selected Associations

There is no single nationwide group responsible for accrediting educational programs in HVAC. Certification of HVAC mechanics and technicians is usually done at the state or local level. However, there are several national HVAC associations that can provide information on HVAC education, industry trends, and continuing education:

Air Conditioning Contractors of America
1513 16th St. NW
Washington, D.C. 20036

Air-Conditioning and Refrigeration Institute
4301 N. Fairfax Dr., Suite 425
Arlington, Va. 22203

American Society of Heating, Refrigerating, and Air-Conditioning Engineers (ASHRAE)
1791 Tullie Circle NE
Atlanta, Ga. 30329

National Association of Plumbing-Heating-Cooling Contractors
180 S. Washington St.
P. O. Box 6808
Falls Church, Va. 22040

Refrigerating Engineers and Technicians Association
c/o Smith-Bucklin Associates
401 N. Michigan Ave.
Chicago, Ill. 60611

Selected Journals

Air Conditioning, Heating and Refrigeration News.
Business News Pub. Co. Weekly. ISSN 0002-2276

ASHRAE Journal.
American Society of Heating, Refrigerating and Air-Conditioning Engineers. Monthly. ISSN 0001-2491

Heating, Piping, and Air Conditioning.
American Society of Heating and Ventilating Engineers/American Society of Heating and Air-Conditioning Engineers. Monthly. ISSN 0017-940X

Print Materials

✦ Reference

1993 ASHRAE Handbook: Fundamentals
Inch-pound ed. Atlanta, ASHRAE, 1993. 1 v. (various pagings)
ISBN 0910110964

This volume is the first of four annuals produced on a four-year rotation by ASHRAE, the primary professional association for HVAC. This is the authoritative resource for all standards in design and practice according to ASHRAE. Included at the end of this volume is a table of codes and standards from ASHRAE and other groups. The Fundamentals volume contains sections on HVAC theory, general engineering data, basic materials, load and energy calculations, and duct and pipe sizing. The final part of the book has a chapter on environmental health issues.

1992 ASHRAE Handbook: Heating, Ventilating, and Air-Conditioning Systems and Equipment
Atlanta, ASHRAE, 1992. 1 v. (various pagings)
ISBN 0910110867

The nuts-and-bolts volume of this series offers information on the types of machines used in HVAC. Examples include air conditioning and heating systems, air-handling equipment, heating equipment, and the general components of each of these types of systems. A single equipment section covers stand-alone devices such as room air conditioners and dehumidifiers. References and bibliographies appear at the end of many chapters.

1991 ASHRAE Handbook: Heating, Ventilating, and Air-Conditioning Applications
Inch-pound ed. Atlanta, ASHRAE, 1991. 1 v. (various pagings)
ISBN 0910110808

This volume of the series covers comfort air conditioning and heating, industrial air conditioning and ventilation, energy sources (including treatments of geothermal and solar energy), building operation and maintenance. A final section of the book covers general applications and unique situations such as snow melting and smoke control. Most sections conclude with references and a bibliography.

ASHRAE Handbook: Refrigeration Systems and Applications
Inch-pound ed. Atlanta, ASHRAE, 1990, 1 v. (various pagings),
ISBN 0910110697

This volume of the series concentrates on the process applications of refrigeration. The first section covers the variety of refrigeration systems used. Food refrigeration chapters describe standards and practices for cold storage of everything from meats to candies. Another section covers the complexities of transporting cold-stored materials. Finally, there are treatments of the industrial applications of refrigeration, such as the maintenance of ice rinks, mass production of ice, and cryogenics and other low-temperature applications.

Means Mechanical Cost Data, 1994
17th annual ed. Kingston, Mass., R. S. Means, 1993. 456p.
ISBN 0876293259

A standard in cost estimating for larger construction projects. Prices for HVAC and controls installation are presented in a tabular format using materials and labor prices based on a national average.

Readers can choose between a "quick start" or more detailed instruction section to begin their search. Determination of HVAC cost estimates uses either the unit price (CSI) or assemblies type system. The second half includes charts referenced to the preceding section by item, crew listings, a historical cost index, and a city cost index.

Oberg, Erik, and others
Machinery's Handbook: A Reference Book for the Mechanical Engineer, Designer, Manufacturing Engineer, Draftsman, Toolmaker, and Machinist
24th ed. New York, Industrial Press, 1992. 2543p.
ISBN 083112492X

While not specific to the topic at hand, this title offers up-to-date information on the latest specifications and standards for the mechanical field. Early sections of the book highlight basic math and physics. A useful section on strength of materials follows. This chart and diagram-filled work provides information on a wide range of items pertinent to HVAC, including bolts, nuts, screws, pipe threads, belts, air compression, and much more. This book, updated every four years or so, should be available in any general technical library.

Wang, Shan K.
Handbook of Air Conditioning and Refrigeration
New York, McGraw-Hill, 1993. 1 v. (various pagings)
ISBN 0070681384

This may be the most comprehensive, up-to-date one-volume reference tool in this field. It covers everything from engineering considerations in AC/R systems design to their practical operations. It serves as a summary of the ASHRAE handbooks in air conditioning and refrigeration. The book's organization groups chapters under broad headings. The final portion contains sample forms for system inventories, design flow charts, and maintenance procedure manuals. References and additional reading materials are listed at the end of each chapter.

✦ General

Coffin, Michael J.
Direct Digital Control for Building HVAC Systems
New York, Van Nostrand Reinhold, 1992. 220p.
ISBN 0442237979

An easy-to-understand survey of the use of computers to control HVAC systems. The book covers the basics of controls and computer control systems, and the interaction of digital controls and traditional devices. There is a thorough treatment of design and application of digital controls, selection of a system, and economic analysis of systems; includes an interesting discussion of the lack of standards in this new area of HVAC.

Goetsch, David L., Deborah M. Goetsch, and Raymond L. Rickman
Mathematics for the Heating, Ventilating, and Cooling Trades
Englewood Cliffs, N.J., Prentice Hall, 1988. 204p.
ISBN 0135625211

A resource for prospective or practicing HVAC technicians who need to enhance their math skills. The book starts with basic mathematics and rapidly moves on to algebra, geometry, and trigonometry. The second half of the book applies these skills to real-world problems that HVAC students encounter. Skills tests appear throughout, with answers provided at the back of the book.

Grimm, Nils R., and Robert C. Rosaler, eds.
Handbook of HVAC Design
New York, McGraw-Hill, 1990. 1 v. (various pagings)
ISBN 0070248419

This reference serves as a start-to-finish guide for creating HVAC systems. Each chapter, written by HVAC experts from various companies, covers a different aspect of HVAC. Some deal with equipment like boilers, centrifugal chillers, valves, and cooling towers. Solar space heating, noise control, and door heating are examples of some processes covered. Diagrams, schematic drawings, graphs, and sample forms are on almost every page. The chapters range from general overviews to specific technical treatments. Some chapters conclude with references and bibliographies.

Haines, Roger W.
Roger Haines on HVAC Controls
Blue Ridge Summit, Pa., TAB Books, 1991. 214p.
ISBN 0830676252

This book is a reprint of Haines' column on controls that appeared in *Heating, Piping, and Air Conditioning* magazine from 1980 to 1987. The articles appear in five large sections covering philosophy, theory, systems and subsystems, control devices, and computer-based control systems. Many columns were written in response to actual questions posed by readers. There are many illustrations. Not a reference tool, but a good source for the fundamentals of HVAC controls.

Kittle, James L.
Home Heating and Air Conditioning Systems
Blue Ridge Summit, Pa., TAB Books, 1990. 230p.
ISBN 0830692576

Kittle wrote this book for homeowners and handypersons interested in installing and servicing both heating and cooling systems. Its main focus, however, is on heating. Many drawings illustrate proper maintenance and connection procedures. Several manufacturers have provided wiring diagrams and cutaway drawings of their equipment, features that contribute to the book's usefulness. Following a discussion of safety issues, proper tools, fault detection, and selection of the right system, Kittle gets to the specifics of installing different classes of heating systems. Contains a troubleshooting guide and a glossary.

Krigger, John
Your Home Cooling Energy Guide
Helena, Mont., Saturn Resource Management, 1992. 79p.
ISBN 1880120046

A brief book that covers a wide range of cooling options for houses. While a large part of the book describes various mechanical means of cooling, it also acknowledges natural solutions, such as tree-planting and using proper ventilation. Written for the layperson, it contains many useful charts, graphs, and illustrations and is a good basic introduction to methods of home cooling.

Langley, Billy C.
Refrigerant Management: The Recovery, Recycling, and Reclaiming of CFCs
Albany, N.Y., Delmar Publishers, 1994. 155p. ISBN 0827355904

This slim textbook opens with chapters on atmospheric ozone depletion, theories as to its causes, and possible solutions. There are brief discussions of the Clean Air Act of 1990 and the levels of EPA certification. Chapter arrangement is by specific area of certification and every chapter stands alone. Each begins with a statement of objectives and ends with review questions. This is a good study source for students seeking EPA certification for installation or maintenance of air conditioning or refrigeration equipment.

Langley, Billy C.
Heating Systems Troubleshooting Handbook
Englewood Cliffs, N.J., Prentice Hall, 1988. 283p.
ISBN 0835928055

The first chapter is a series of tables divided by heat source. Choosing a source, the reader then skims for the problem. There is a list of possible causes and corrective actions listed next to each problem. A reference number for more details refers the reader to subsequent chapters on components, start-up procedures, and standard service. An interesting feature of this book is a chapter on analyzing problems through flame observation.

Levenhagen, John I., and Donald H. Spethmann
HVAC Controls and Systems
New York, McGraw-Hill, 1993. 334p. ISBN 0070375097

The focus of this overview is commercial systems, but residential and industrial applications are mentioned. Chapters cover all aspects of the machinery of HVAC controls including thermostats, valves, air handlers, dampers, and terminal units. The book highlights electronic and direct digital controls and all types of distribution systems.

Parmley, Robert O., ed.
HVAC Field Manual
New York, McGraw-Hill, 1988. 1 v. (various pagings)
ISBN 0070485240

One of the McGraw-Hill series of pocket-sized field manuals. This title is similar to Tenenbaum's, listed below, but it covers heating systems as well. The inclusion of various types of practical information makes this an inexpensive reference alternative for HVAC work. Math, material properties, heating systems and sources, snow and ice problems, air distribution and conditioning, climatic conditions, and more appear in easy-to-understand charts and graphs.

Sauselein, Theodore B.
Stationary Engineering for Boiler Operations
Troy, Mich., Business News Publishing, 1990. 202p.
ISBN 0912524537

Created by a vocational teacher, this book is a general guide to commercial and light industrial boiler operation. Sections detail combustion, steam traps, water treatment, pumps, turbines, bearings, and more. Definitions appear at the bottom of the page on which the words appear, rather than at the back of the book. Boxes within the text announce safety warnings. Other outstanding visual elements include clear illustrations, boldface type words, and outlines.

Steingress, Frederick M., and Harold J. Frost
High Pressure Boilers
2nd ed., Homewood, Ill., American Technical Publishers, 1993. 265p. ISBN 0826944159

Another title from American Technical Publishers useful for preparing for licensing examinations. Chapters cover boiler operation and theory, operators' duties and qualifications, and review tests consisting of true-false, multiple-choice, and identification questions. At the end of each chapter is a list of key words; corresponding definitions appear in the glossary at the back of the book. Diagrams throughout the book use arrowhead symbols coded to the types and physical states of elements flowing through a system. Pages are perforated.

Swenson, S. Don
Troubleshooting and Servicing Air Conditioning Equipment
Troy, Mich., Business News Publishing, 1990. 383p.
ISBN 0912524561

Designed for use as textbook and as technical reference for practitioners, this book covers all aspects of AC equipment repair—from condensers and compressors to refrigerants. The appendix contains diagnostic tables that direct the service technician from the original complaint to troubleshooting specific parts and systems that need repair.

Tenenbaum, David
Air Conditioning and Refrigeration Toolbox Manual
New York, Arco, 1990. 426p. ISBN 0137702647

This pocket-sized volume is a handy reference covering all the basics of HVAC. By chapter, the book groups topics such as heat theory, components, ozone, and recycling. Simple diagrams and charts provide clarification without being too detailed. The back of the book contains addresses of associations, a glossary, and mathematical tables.

Wilson, R. Dean
Boiler Operator's Workbook
Homewood, Ill., American Technical Publishers, 1991. 282p.
ISBN 0826944914

This workbook, which may also function as a practice tool for licensing examinations, consists entirely of questions. An answer, sometimes quite lengthy, follows each question. The book groups chapters into broad topics such as boiler design and steam systems. Mathematical questions are answered, showing all the steps taken toward the solution. Diagrams and photographs are directly referenced in questions. At the end of each chapter are "trade test" questions. Pages are perforated.

✦ Careers/Licensing

Dundas, James L.
3000 Questions and Answers for HVAC/R Licensing Examinations
Troy, Mich., Business News Pub. Co., 1989. 322p.
ISBN 0912524510

This book, produced by the major periodical publisher in the area of HVAC, is exactly as titled. It divides evenly into two sections—air conditioning/refrigeration and heating—with fifteen hundred sample questions listed in each. The questions are multiple-choice and appear to be in random order. The result is a generic flash-card-style study guide not geared toward any specific exam. Answers are provided at the end of the book.

Guiling, Stanley Douglas
Careers in Power Engineering and Boiler Operation: Your Guide to a Secure Future
Pontiac, Mich., PowerPlant Press, 1993. 813p. ISBN 1881861201

This book provides comprehensive job-seeking information. The first part of the book discusses the background of the fields and includes tips on job hunting. A list of representative wage rates throughout the United States follows. The bulk of the book lists information on licensing requirements by state, city, and province for the United States and Canada. This includes addresses of licensing agencies, topics covered on examinations, and fees. A list of schools offering power engineering and boiler operation concludes the book.

Nonprint Materials

General Training Air Conditioning–Fundamentals
Syracuse, N.Y., Carrier Corp., 1991. 10 videocassettes (463 min.) VHS, 656 slides, 10 sound cassettes (467 min.), 10 workbooks

An instructional series designed for classroom or individual use that covers all aspects of air conditioning and refrigeration installation, maintenance, and repair. Units may be purchased separately. A unit consists of a slide/tape set and a videotape. The video consists of images from the slides with some animation added. Unit topics include temperature/pressure, the refrigeration cycle, air conditioning systems, compressors, condensers, evaporators, metering devices, electrical and refrigeration controls, and accessories.

Masonry

Steven Self
Austin Community College
Austin, Texas

Introduction

Masonry is the art of building with brick or stone, and binding these materials with mortar. Building with bricks and stone must rank as one of humankind's earliest skills. Ancient edifices still exist—including the Great Wall of China and the pyramids of Egypt and the Americas—which bear witness to our relationship with stone and bricks. Temples, bridges, and monuments worldwide reflect the durability and beauty of masonry, and it has remained a fundamental element of building construction up to the present.

Although masons in the strictest sense are associated with brick and stone work, the net will be cast a bit wider for the purposes of this bibliography to include working with cement and concrete, and other related skills, such as tile-setting. This will allow for comprehensive coverage of similar skills that may not be covered elsewhere in this collection. Since masonry training is often a component of a broader building trades program, including these areas reflects more accurately the context in which many two-year colleges teach construction skills.

Education as a mason still carries some overtones of the medieval guild system. Even with vocational education in the field, many students of masonry still start their careers with a three-year stint as an apprentice. Apprenticeships are usually set up with individual contractors or through programs arranged between contractors and unions. After becoming journeymen, masons are free to work for any contractor they please.

Stonemasons and bricklayers interact with their materials on many levels. For more artistic applications, masons may cut the stone they will work with, or select individual stones for a particular shape, color, or composition. Bricklayers may use their skills for facade work or in designing intricate pathways. Some stonemasons specialize in working with specific types of stone, such as marble or granite.

In the last century, improvements in cement and concrete production have increased rapidly. Portland cement, made from gypsum added to burned limestone and clay, has proven to be a particularly important development. Knowledge of the principles involved in mixing and pouring concrete is essential in the field of building construction. Materials in this bibliography will describe a wide range of concrete-related projects from backyard walks to foundations and highway construction.

Items included in this list represent a wide range of skill and reading levels to meet the varied needs of vocational college students interested in masonry. The bibliography includes materials suitable for the weekend do-it-yourselfer and the skilled mason. A few items are appropriate for people at the engineering level. Not every publication will be suitable for all collections, but the items on this list should help answer questions, provide ideas for projects, and supplement classroom instruction.

Selected Associations

Certification of masons usually occurs on the local level or they may have their training tracked by contractors or unions through apprenticeship programs. Following is a list of several national masonry associations that can provide information on masonry education, industry trends, and continuing education.

Brick Institute of America
11490 Commerce Park Dr.
Reston, Va. 22091

American Concrete Institute
P.O. Box 19150
Detroit, Mich. 48219

International Masonry Institute
823 15th St. NW
Washington, D.C. 20005

National Stone Association
1415 Elliot Pl. NW
Washington, D.C. 20007

National Concrete Masonry Association
2302 Horse Pen Rd.
Herndon, Va. 22071-3406

Selected Journals

Aberdeen's Concrete Construction.
Aberdeen Group. Monthly. ISSN 1051-5526

Fine Homebuilding.
Taunton Press. Bimonthly. ISSN 0273-1398

The Masonry Society Journal.
The Society. Semiannual. ISSN 0741-1294

Print Materials

✦ Reference

1991 Masonry Codes and Specifications: With 1992 Supplements
Los Angeles, Masonry Promotion Groups of Southern California, 1992. 549p. ISBN 0940116219

A good one-volume reference work for standards and codes relating to masonry. Within this volume are those sections of the *Uniform Building Code* (UBC, 1991) and UBC standards, the American Society for Testing and Materials (ASTM) Standards, and the California building code sections that pertain to masonry. Sections list selected quality control standards that are cross-referenced to UBC and ASTM standards. This volume is updated every three years to reflect UBC changes. (New edition to be published in late 1994, ISBN 0940116294)

Cement and Concrete Terminology
Detroit, American Concrete Institute, 1990. 68p.

The American Concrete Institute promotes this volume as "the authoritative glossary" on this topic. From "Abram's law" to "zero-slump concrete," this volume covers hundreds of concrete terms. Entries run the gamut from the simple ("coat" and "swelling") to the more complex ("Kern area" and "ettringite"). Many terms contain "see" and "see also" references. Bold letters identify words used in definitions noted elsewhere in the dictionary.

Means Concrete and Masonry Cost Data
Kingston, Mass., R. S. Means, Annual. 1 v. (various pagings)
ISSN 0739-8298

A standard in cost estimating for larger construction projects over $20,000. All aspects of masonry construction are presented in a tabular format using materials and labor prices based on a national average. Charts find masonry cost estimates by either the unit price (CSI) or assemblies type systems. The second half includes reference charts to the preceding section by item, crew listings, a historical cost index, and a city cost index.

Tenenbaum, David
The Mason's Toolbox Manual
New York, Arco, 1990. 360p. ISBN 0135588758

A compact source designed for use on the job by both the professional and the do-it-yourselfer. The first half covers tools of the trade and the basics of constructing fireplaces and walls. The second half covers essential skills in more detail. A section at the end has tips on estimating, mathematics, measurement, and safety. Appendices provide a brief glossary and a list of codes and standards.

✦ Concrete

ACI Manual of Concrete Inspection
8th ed. Detroit, American Concrete Institute, 1992. 220p.
ISBN 068562675X

Quality assurance is a rapidly growing field. This ACI work looks at the application of quality assurance standards to concrete and concrete constructions. The book begins with a short explanation of the role of statistics in quality assurance, then outlines points of inspection for all areas of cement production. Chapters cover special concreting environments and unique varieties of concrete. The volume ends with sample reports, standards for inspection, and an inspection checklist.

Cowan, Henry J.
Design of Reinforced Concrete Structures
2nd ed. Englewood Cliffs, N.J., Prentice Hall, 1989. 304p.
ISBN 0132014432

A highly technical overview of reinforced concrete in a textbook format. The book begins with chapters on the history of concrete and its reinforcement with steel bars. It goes into great detail on the types of steel and concrete combinations in use today, paying particular attention to analysis of loads, stress, and torsion. Final chapters discuss the design of reinforced columns and retaining walls. Appendices include a glossary, tables of reinforcing bar areas, and answers to questions in the text.

Koel, Leonard
Concrete Formwork
Homewood, Ill., American Technical Publishers, 1988. 282p.
ISBN 0826907040

    Koel provides a thorough examination of formwork from residential applications like curbs and driveways to heavy construction projects such as casting columns and girders. Chapters cover prestressed and precast concrete. Later chapters outline math fundamentals, formwork estimating, and exercises in blueprint reading. The appendices briefly highlight American Concrete Institute (ACI) and OSHA regulations.

Newman, Morton
Structural Details for Concrete Construction
New York, McGraw-Hill, 1988. 234p. ISBN 0070463603

    This volume consists of the concrete-related detail drawings from *Newman's Standard Structural Details for Building Construction* (McGraw-Hill, 1968). Newman designed this book to serve as a graphic communication tool between those who design buildings and those who construct them. Every other page contains structural detail drawings frequently used in construction. The drawings have a grid for notes and sketching. Drawings are grouped by general applications.

Panarese, William C., Steven H. Kosmatka, and Frank Alfred Randall
Concrete Masonry Handbook for Architects, Engineers, Builders
5th ed. Skokie, Ill., Portland Cement Association, 1991. 247p.
ISBN 0893120936

    This book has been a standard reference since the 1950s. This edition is probably the best one-volume work on the topic of concrete masonry. Covers the basics of concrete masonry, mortar and grout, design of concrete masonry walls, hot and cold weather construction, applied finishes, maintenance of concrete masonry, and special applications. Appendices contain a bibliography, a list of ASTM specifications and methods, and detail diagrams.

White, George R.
Concrete Technology
3rd ed. Albany, N.Y., Delmar Publishers, 1991. 144p.
ISBN 0827336357

    A short textbook that covers the basics of working with concrete, from its chemistry to hot and cold weather pouring. Properties of concrete and selection of mixtures receive special attention. The appendix provides forms for using the unit-weight method for concrete mix design. A brief yet thorough introduction to working with concrete.

✦ Brick, Stone, and Tile

Bridge, John P.
Ceramic Tile Setting
Blue Ridge Summit, Pa., TAB Books, 1992. 226p.
ISBN 0830625720

    There are few recently published books that impart the fundamentals of tile setting. Although written generally toward the weekend handyperson and quite conversational in presentation, this book serves as a good source for the skilled setter. The beginning of the book concentrates on selecting the right tools and types of tile. Many photographs and figures illustrate sections on floor, wall, and countertop tile setting.

Cramb, Ian
The Art of the Stonemason
White Hall, Va., Betterway Publications, 1992. 174p.
ISBN 1558702253

    Cramb, a Scottish-born master stonemason, uses examples from over fifty years of experience to illustrate traditional random rubble stonemasonry. Explanations of hand-cutting and setting stones are an interesting feature. Cramb's career has focused on restoration, which is the emphasis of this book. Photographs and excellent line drawings by Cramb illustrate such topics as wall, arch, bridge, and tower construction, and the tools of the stonemason. An excellent source for amateurs and armchair stonemasons.

Fine Homebuilding on Foundations and Masonry
Newtown, Conn., Taunton Press, 1990. 127p. ISBN 0942391551

    Illustrated with diagrams and some color photographs. Comprised of twenty-nine articles from issues of *Fine Homebuilding* between 1981 and 1988. Many levels of description can be found, from the basics of mixing and pouring concrete to more detailed articles on laying hillside foundations and installing radiant floor heating. A good volume for either accomplished weekend masons or the serious student seeking more complicated undertakings.

Heddy, Edward J., and Pete Peterson
The Complete Guide to Decorative Landscaping with Brick and Masonry
White Hall, Va., Betterway Publications, 1990. 159p.
ISBN 1558701451

    Heddy, a professional mason since World War II, presents a do-it-yourself offering in a casual, conversational style. The first quarter of the book covers basic tools and techniques of working with bricks and mortar. The second part consists of progressively more complicated projects, beginning with brick walls. Each project begins with a list of tools and materials needed and ends with a brief summary of all the steps involved. A chapter on basement waterproofing may prove useful in some parts of the country.

Kicklighter, Clois E.
Modern Masonry: Brick, Block, Stone
South Holland, Ill., Goodheart-Willcox, 1991. 256p.
ISBN 0870068733

    This textbook, which may also serve as a refresher for more experienced masons, is an introduction to the trade. The beginning chapter of the book describes the qualities of concrete and clay-based masonry units. Other topics covered include tools, safety, form construction, masonry mathematics, blueprint reading, and masonry as a career.

Lawrence, Mike, ed.
Backyard Brickwork: How to Build Walls, Paths, Patios, and Barbecues
Pownal, Vt., Garden Way Publishing, 1989. 90p.
ISBN 0882665626

    A colorful, illustrated how-to book aimed at the weekend mason. The beginning section briefly covers basic tools and planning hints. Later sections outline brick home improvement projects like walkways and fish ponds. Browsers will appreciate its many photographs, which can provide some good starting points for masonry students looking for project ideas.

London, Mark
Masonry: How to Care for Old and Historic Brick and Stone
Washington, D.C., Preservation Press, 1988. 208p.
ISBN 0891331255

A fine introduction to the maintenance and repair of existing masonry. Photographs and line drawings illustrate nearly every page. The book, produced by the National Trust for Historic Preservation, seeks to impart information about masonry repair with an eye toward maintaining the historic integrity of masonry structures. Insets highlight helpful lists for preservation projects such as masonry cleaning products, problem checklists for inspection, and performing mortar analysis.

Lynch, Gerard C. J.
Gauged Brickwork: A Technical Manual
Brookfield, Vt., Gower, 1990. 115p. ISBN 0566090570

This item is a historically minded book that uses examples of brickwork in nineteenth-century England to discuss techniques of gauged brickwork. Gauged brickwork is brick used as ornamentation, as in arches and molding. Fine photographs and drawings illustrate the techniques that craftspeople have used to create elaborate niches, cornices, etc. Includes a short glossary, bibliography, and index.

Newman, Morton
Structural Details for Masonry Construction
New York, McGraw-Hill, 1988. 137p. ISBN 0070463611

This volume consists of the masonry-related detail drawings from *Newman's Standard Structural Details for Building Construction* (McGraw-Hill, 1968). Detail drawings are grouped into sections on hollow block, brick, and retaining wall constructions.

Plumridge, Andrew, and Wim Meulenkamp
Brickwork: Architecture and Design
New York, Abrams, 1993. 224p. ISBN 0810931230

A glossy, beautifully photographed book that serves as a guide to the use of brick throughout history. The beginning outlines the history of brick in architecture and its decorative uses. Later, the book offers chapters on brick manufacturing and structural techniques. Line drawings illustrate various patterns for creating brick walls and vaults. Includes a glossary and bibliography.

Portland Cement Association
The Homeowner's Guide to Building with Concrete, Brick, and Stone
Emmaus, Pa., Rodale Press, 1988. 231p. ISBN 0878577955

Geared toward the beginner, this book provides instruction in household masonry projects such as walks, patios, and barbecues. The first half of the book covers projects involving concrete only, while latter projects incorporate brick and stonework. One unique chapter features plans for concrete benches and flower boxes made from concrete molds. Tables throughout give tips on standards, coloration, and proportions for mixing. Includes index.

Schwenke, Karl, and Sue Schwenke
Build Your Own Stone House: Using the Easy Slipform Method
2nd ed., rev. and updated. Pownal, Vt., Storey Communications, 1991. 164p. ISBN 0882666398

The Schwenkes built a stone house in Vermont in the early 1970s and wrote this popular work about their experiences. Their back-to-basics philosophy and conversational style permeates their description of this technical, yet straightforward, process. Many explanatory diagramss, anecdotes, and quotations make this book useful for both experienced masons and laypersons. Contains an excellent bibliography of older sources on building with stone.

Self, Charles R.
Bricklaying: A Homeowner's Illustrated Guide
Blue Ridge Summit, Pa., TAB Books, 1992. 162p.
ISBN 0830639179

A good amateur's guide for around-the-house projects. Although geared toward the basics, this book covers areas that many similar titles do not, such as reinforcing brick masonry and corbeling. Well illustrated with photographs, line drawings, graphs, and charts. An appendix provides a list of brick and tile manufacturers in the United States with a code of the types of products they provide.

Plumbing Technology

Joanne Kim
and
Tori Beyer
Pasadena City College
Pasadena, California

Introduction

Plumbing offers an essential key to healthful living. Without clean drinking water and a good sewage system, our lives would be devastated by health problems and disease.

Plumbing involves the supply pipes that channel water or gas into a building, and that remove waste materials from toilets and sinks to ensure sanitary conditions. Acquiring plumbing skills involves learning installation, fitting, repairing, and maintenance of pipes and fixtures for liquids and gases in and around a building. In many career guidance books, plumbing and pipe fitting are combined into one category, but workers generally specialize in one or the other. Plumbers normally install and repair the water, waste disposal, drainage, and gas systems in homes and commercial or industrial buildings. They also install plumbing fixtures such as bathtubs, showers, sinks, and toilets, and appliances such as dishwashers and water heaters. This bibliography focuses on books that teach plumbing trade skills.

Although there are no uniform national licensing requirements, most communities require licenses. Licensed plumbers are expected to fulfill their jobs in accordance with the required plumbing codes legislated by states or cities. One can learn the plumbing trade through formal schooling or apprenticeship. High schools, vocational and trade schools, and community colleges offer courses in drafting, blueprint reading, mathematics, and local plumbing codes and regulations. Apprenticeship consists of four to five years of on-the-job training with at least 144 hours annually of related classroom instruction. Plumbers and apprentices need physical strength as well as stamina due to working conditions that require stooping or working in a small space and lifting heavy materials.

Plumbers require the skills and abilities to master the following:
- skillfully use the hand tools or machines needed for plumbing;
- read blueprints and drawings of the items to be made or repaired;
- measure, cut, or work on materials or objects with great precision;
- use arithmetic and shop geometry to figure amounts of materials needed, dimensions to be followed, and cost of materials;
- visualize what the finished product will look like;
- accept responsibilities for the accuracy of the work as it is produced.

As the world population increases, the job of the plumber in modern society will become more important to the welfare of the world community. Supply of clean drinking water and safe disposal and treatment of waste generated by many people will be the critical tasks of all communities. Jobs for plumbers are distributed across the country in about the same proportion as the general population. The job growth rate for plumbers is expected to increase at the same rate as the population, and will increasingly involve more complex technologies applied to modern living.

Items in this bibliography have been coded *Unskilled level, Skilled level,* and *Professional level* to aid librarians in the selection of appropriate materials for the curriculum that is offered.

Selected Associations

There is no single nationwide group responsible for accrediting educational programs in plumbing. Certification of plumbers is usually done at the state or local level. However, there are several national associations that can provide information on education, industry trends, and continuing education:

American Society of Plumbing Engineers (ASPE)
3617 Thousand Oaks Blvd., No. 210
Westlake, Calif. 91362

American Society of Sanitary Engineering (ASSE)
P.O. Box 40362
Bay Village, Ohio 44140

International Association of Plumbing and Mechanical Officials (IAPMO)
20001 Walnut Dr., S.
Walnut, Calif. 91789

Mechanical Contractors Association of America (MCAA)
1385 Piccard Dr.
Rockville, Md. 20850-4329

National Association of Plumbing-Heating-Cooling Contractors (NAPHCC)
180 S Washington St.
P.O. Box 6808
Falls Church, Va. 22040

United Association of Journeymen and Apprentices of the Plumbing and Pipe Fitting Industry of the United States and Canada
P.O. Box 37800
Washington, D.C. 20013

Selected Journals

Contractor.
Cahners Pub. Co. Monthly. ISSN 0010-7891

Official.
International Association of Plumbing and Mechanical Officials. Bimonthly. ISSN 0192-5784

Plumbing Engineer.
American Society of Plumbing Engineers. Ten times a year. ISSN 0192-1711

Plumbing-Heating-Cooling Business.
National Association of Plumbing-Heating-Cooling Contractors. Monthly.

UA Journal.
United Association of Journeymen and Apprentices of the Plumbing and Pipe Fitting Industry of the United States and Canada. Monthly. ISSN 0095-7763

Print Materials

Aglow, Stanley H.
Blueprint Reading Made Easy
Troy, Mich., Business News Pub. Co., 1992. 133p.
ISBN 0912524715

A fundamental guide to reading blueprints for all kinds of building contractors, this book provides information on interpreting blueprints. Emphasis is on understanding the relationship among blueprint types, such as electrical, mechanical, and elevation. Included are chapters on symbols, scales, and lines, with four sample assignments for practice. Contains review questions and answers for each chapter, a glossary, and index. *(Skilled level) (Professional level)*

Annis, William H.
Basic Plumbing Skills
Athens, Ga., American Association for Vocational Instructional Materials, 1989. 68p. ISBN 0896062503

A well-illustrated text that gives more attention to the fundamental aspects of plumbing than to the technical ones. The focus is on topics such as identifying equipment and tools, what kinds of pipe are used, and safety practices in plumbing. The most technical information contained in this text is how to cut and join different kinds of pipe, and how to thread unthreaded pipe. Each chapter includes study questions for review, and an index is provided. *(Unskilled level)*

Blankenbaker, E. Keith
Modern Plumbing
South Holland, Ill., Goodheart-Willcox, 1992. 363p.
ISBN 087006939X

A richly illustrated text of basic plumbing techniques, equipment, and tools. Beyond technical instruction, it includes information on how to organize plumbing jobs, and what kinds of career opportunities are available for plumbing tradespeople. A section called "Useful Information" contains lists and tables of the calculations necessary for plumbing jobs, such as length conversions, blueprint symbols, and flow capacity of pipe. A student workbook and an instructor's guide are also available. *(Skilled level) (Unskilled level)*

Duncan, Justin
Plumbing Technology
Troy, Mich., Business News Pub. Co., 1994. 370p.
ISBN 0912524766

This comprehensive text covers all aspects of plumbing, from the practical to the theoretical. Included are such topics as estimating demand, the properties of water, energy conservation, and codes and standards. Appendices contain a list of symbols, tables of water friction in various kinds and sizes of pipe, plumbing terminology, and a sample plumbing project with specifications. Includes references and an index. *(Skilled level)(Unskilled level)*

Galeno, Joseph J., and Sheldon T. Greene
Means Plumbing Estimating
Kingston, Mass., R. S. Means, 1991. 324p. ISBN 0876292120
    A complete guide to all aspects of plumbing estimation, from labor costs to material markup, this illustrated handbook begins with an explanation of the plumbing system for novice estimators. It walks the reader through the entire estimating process using a sample project. Included are sample forms for material and cost estimates, job specifications, and architectural blueprints for the project. Appendices include tables of wage rate information and average installation hours for projects. *(Professional level)*

Guest, J. Russell, Bartholomew D'Arcangelo, and Benedict D'Arcangelo
Blueprint Reading for Plumbers: Residential and Commercial
5th ed. Albany, N.Y., Delmar Publishers, 1989. 156p. ISBN 0827334591
    A manual for students of plumbing that gives instruction on interpreting blueprints and planning for installation of an indicated system. This basic guide is highly illustrated and plainly written. Actual blueprints for residential and commercial buildings are included with the book. Each chapter contains review questions or drawing assignments, and an index is provided. An instructor's guide is available. *(Skilled level) (Unskilled level)*

Home Plumbing Projects and Repairs
Minnetonka, Minn., Cy DeCosse, 1990. 128p. ISBN 086573710X
    A how-to guide that students of the plumbing trade will find extremely useful. Numerous color photographs make this an excellent visual review of plumbing tools, equipment, and procedures. The first chapter contains a photographic layout of all the necessary tools, from screwdrivers and wrenches to drain augers and miter boxes. Each demonstrated repair is accompanied by detailed photographs, with elements clearly labeled. This is an excellent resource for ESL students. *(Unskilled level)*

International Association of Plumbing and Mechanical Officials
Cumulative Analysis of Uniform Plumbing Code Changes
Walnut, Calif., The Association, Published triennially. 1 v. (various pagings)
    This guide lists and explains all changes made to the Uniform Plumbing Code since its last publication. When information has been removed, the deleted code segment is provided first, with its replacement text following; added code segments are also noted. An analysis of the addition or deletion is then given. Expected publication date is 1995. *(Professional level)*

International Association of Plumbing and Mechanical Officials
Dwelling Requirements of the Uniform Plumbing Code
Walnut, Calif. The Association, Published triennially, 1 v. (various pagings)
    Contains the latest standards and installation requirements for one- and two-family dwellings. Includes guidelines for drainage systems, vents, traps, joints, sewers, and fixtures. Expected publication date is 1995. *(Professional level)*

International Association of Plumbing and Mechanical Officials
Uniform Plumbing Code
Walnut, Calif., The Association, Published triennially. 1 v. (various pagings) ISSN 0733-2335
    A standard for industry guidelines on equipment, procedures, and safety, the Uniform Plumbing Code (UPC) includes requirements for all plumbing materials and installations, including water heaters, fuel gas piping, and sewers. Appendices contain specifications for more unusual plumbing systems such as rainwater, mobile home park, and graywater systems. The UPC is published every three years in an effort to maintain standardization and to include the latest in plumbing technology. The most recent edition of UPC at the time of this publication was 1994. *(Professional level)*

International Association of Plumbing and Mechanical Officials
Uniform Plumbing Code Illustrated Training Manual
Walnut, Calif., The Association, Published triennially. 1 v. (various pagings)
    A supplementary manual that illustrates the rules set forth in the Uniform Plumbing Code (UPC). The manual is divided into chapters that correspond to the chapters in the UPC. Each chapter contains detailed diagrams and tables illustrating sections of the UPC. Expected publication date is 1995. *(Professional level)*

International Association of Plumbing and Mechanical Officials
Uniform Plumbing Code Interpretations Manual
Walnut, Calif., The Association, Published annually. 1 v. (various pagings)
    Provides clarification of the Uniform Plumbing Code (UPC) guidelines. This manual is divided into chapters that correspond to the chapters in the UPC. Each chapter contains questions about specifics of the UPC which are then answered with an explanatory code interpretation. Expected publication date is 1995. *(Professional level)*

International Association of Plumbing and Mechanical Officials
Uniform Solar Energy Code
Walnut, Calif., The Association, Published triennially. 1 v. (various pagings)
    Contains guidelines for the installation of tanks, collectors, insulation, and other solar equipment. Includes tables of material standards and thermal conductivity of various insulation, and a list of important definitions. Next expected publication date is 1995. *(Professional level)*

International Association of Plumbing and Mechanical Officials
Uniform Swimming Pool, Spa, and Hot Tub Code
Walnut, Calif., The Association, Published triennially. 1 v. (various pagings) ISSN 1058-3114
    The latest standards and installation requirements for the swimming pool, hot tub, and spa industries are provided, along with definitions, guidelines for material quality, and regulations for constructing, installing, and locating pools, hot tubs, and spas. Next expected publication date is 1995. *(Professional level)*

Jackson, Albert, and David Day
Popular Mechanics Home How-To: Plumbing and Heating
New York, Hearst Books, 1989. 80p. ISBN 068810407X
    Although the text of this home repair guide is too fundamental for advanced students of plumbing, it contains an illustrated glossary of tools and procedures that would rival resources written for plumbers. The glossary entries are simply written, yet they include such diverse topics as sawing a groove and tinning a soldering iron. La-

beled illustrations throughout the text and glossary complement the book's simple nature. An excellent review source for beginning students. *(Professional level)*

Massey, Howard C.
Basic Plumbing with Illustrations
Carlsbad, Calif., Craftsman Book Co., 1994. 384p.
ISBN 0934041997

Completely revised, this edition is up-to-date with the changes in the plumbing codes. It is a journeyman's and apprentice's guide to installation of plumbing, piping, and fixtures in residential and light commercial buildings. Selection of the correct materials, laying out the job, and professional quality work are covered. Includes information on selecting essential tools and materials, making repairs, maintaining plumbing systems, installing fixtures, and adding to existing systems. Also provides study questions and answers. *(Skilled level) (Unskilled level)*

Means Plumbing Cost Data
Kingston, Mass., R.S. Means, Published annually. 1 v. (various pagings) ISSN 1042-3850

A detailed industry standard for estimating costs on plumbing materials and labor, every annual edition lists current prices and average labor costs for thousands of materials and assemblies. Each material price is added to an estimated installation cost for a total cost on each item. Also included is a reference section containing tables of average costs for such things as overhead, work crews, and project costs for different cities. *(Professional level)*

National Plumbing and HVAC Estimator
Carlsbad, Calif., Craftsman Book Co., Published annually.
1 v. (various pagings)

Specific guidelines are given for estimating labor and material costs for all plumbing and heating, ventilation, and air conditioning (HVAC) projects. Contains tables of equipment costs, plus cost of installation time for each item. Includes a 5 1/4" disk called "Plumbing and HVAC Estimate Writer," used to organize and process the reader's personal information. Samples of important forms are provided, such as change estimate, subcontract, and billing breakdown. A thorough index is included. *(Professional level)*

Pocket Reference to Plumbing and Pipe Fitting Calculations
2nd ed., Albany, N.Y., Delmar Publishers, 1994. 99p.
ISBN 0827356994

This pocket-sized handbook contains lists, tables, formulas, illustrations, and other data useful in the plumbing trade. Examples of included information are pipe fitting angles, temperature conversion, water pressure, and dimensions of pipe and fittings. Includes an index and a section for the reader's notes. *(Skilled level)(Professional level)*

Ripka, L. V.
Plumbing
Homewood, Ill., American Technical Publishers, 1994. 384p.
ISBN 0826906125

A richly illustrated, up-to-date, and comprehensive text for beginning students of plumbing. Covers job safety, plumbing materials, tools, practices, testing, and inspections using the new mandates of the *Energy Policy Act of 1992*. Also includes a glossary with illustrations. *(Skilled level)*

Smith, Lee
Plumbing Technology: Design and Installation
2nd ed., Albany, N.Y., Delmar Publishers, 1994. 502p.
ISBN 0827355238

Students of the plumbing trade will find this illustrated book useful as a source for basic information. This second edition includes sections on waste systems, threaded pipe, and cold and hot water systems. A section on basic knowledge covers such topics as safety, blueprint reading, and math for plumbers. Each chapter contains questions for review, and an appendix provides tables of water pressure, pipe data, and conversions. *(Skilled level)*

Woodson, R. Dodge
Home Plumbing Illustrated
Blue Ridge Summit, Pa., TAB Books, 1993. 270p.
ISBN 0830639861

The author brings nearly fifteen years of experience as a licensed plumber to his writing. Although intended for the skilled amateur home plumber, this book contains illustrations and projects that students of the plumbing trade would find useful. Instruction in professional techniques of planning, installation, and code compliance are included for all kinds of residential plumbing projects. Each chapter provides detailed photographs, and a large section of the book covers troubleshooting. *(Skilled level) (Unskilled level)*

Woodson, R. Dodge
National Plumbing Codes Handbook
New York, McGraw-Hill, 1993. 285p. ISBN 0070717699

A handbook for plumbers that supplements official plumbing codes. This book clarifies national and local codes through straightforward language and numerous illustrations. Areas covered include potable water, drainage and vent systems, and fixtures and cleanouts. Provides three practice exams (with answers) and study suggestions for passing licensing exams. *(Professional level)*

Woodson, R. Dodge
The Plumber's Troubleshooting Guide
New York, McGraw-Hill, 1994. 385p. ISBN 007071777X

The author outlines his ten-step troubleshooting system, and explains why troubleshooting is an essential skill for plumbers to master. Next follow chapters on troubleshooting in specific areas of plumbing including bathroom fixtures, spas and whirlpools, wells and septic systems, water treatment equipment, and specialty fixtures like dental equipment and drinking fountains. Detailed illustrations and examples of sample problems supplement the text; includes index. *(Skilled level)*

Woodson, R. Dodge
The Plumbing Apprentice Handbook
New York, McGraw-Hill, 1994. 366p. ISBN 0070717729

Everything an apprentice plumber needs to know is covered in this handbook. Included are the fundamentals of safety, tools, blueprint reading, and math for plumbers. Technical topics include vent, drainage, and water distribution systems, with chapters devoted to plumbing repairs and maintenance. The author focuses on the special interests of apprentices, with chapters on field experience and the apprenticeship process. *(Unskilled level)(Skilled level)*

Woodson, R. Dodge
Plumbing Contractor: Start and Run a Money-Making Business
Blue Ridge Summit, Pa., TAB Books, 1994. 278p.
ISBN 0830643230

A guide to the business aspects of being an independent plumbing contractor. Topics include budgeting, employee management, cost estimation, computers and software, and advertising. The book also contains an appendix of the federal tax forms necessary for running a business. Includes a glossary and a thorough index. *(Professional level)*

Nonprint Materials

Making Basic Plumbing Repairs
Seattle, Wa., Cinema Associates; distributed by Morris Video, 1990.
1 videocassette (30 min.) VHS format

Shows necessary plumbing tools and step-by-step instructions for home plumbing repairs. Demonstrates replacing O-rings, worn seals for kitchen and bathroom faucets, and simple remedies for fixing clogged drains. Container title: *Be Your Own Plumber. (Unskilled level)*

Business

Accounting, Bookkeeping, and Income Tax Preparation 165
Melinda Townsel

Computers and Data Processing 173
Rebecca Kroll

Cosmetology 185
Barbara A. Heiffner

Culinary Arts/Food Service Management 190
Anne Marsh Fields

Hospitality Management 201
Julie Pinnell

Office Technologies 208
Natalie Diamond

Real Estate 213
Antoinette Byers

Small Business and Entrepreneurship 222
Barbara Alper

Travel and Tourism 226
Trudy Cleveland

Accounting, Bookkeeping, and Income Tax Preparation

Melinda Townsel
Austin Community College
Austin, Texas

Introduction

Accounting and bookkeeping courses are offered at most community colleges, but the curriculum varies from one institution to another. Some institutions only offer basic bookkeeping and accounting courses, while others offer advanced accounting classes. Students completing two-year programs have a variety of job options. Positions as bookkeepers or accounting and auditing clerks offer students the flexibility to work part- or full-time. Responsibilites vary not only by job, but also by the size of the organization and the experience of the person in the position. For example, in a small organization, accounting clerks handle numerous financial transactions: preparation and analysis of general ledger accounts and financial statements, and preparation of reports and banking transactions. In larger organizations, students assume more specialized roles, from entry-level positions (posting details of accounts, totaling accounts, computing interest, and handling accounts payable and receivable) to more skilled positions (e.g., review of invoices and statements and reconciliation of more complex accounts).

This bibliography contains annotations of publications that have universal appeal to all community college students. The titles have been selected for general collections that support accounting studies. Also, many of the titles were selected because of price, availability, author, and publisher reputation. Excluded were works commonly found in large public libraries and four-year institutions, and standard professional sources published by the American Institute of Certified Public Accountants (AICPA), the American Accounting Association, and the National Association of State Board of Accountancy. Such resources will be discussed briefly in this introduction.

Accounting standards are rules developed by accountants to govern their practice. As students' skills and education increase, they will need to stay abreast of the changes in accounting principles and standards. One way to keep current is to become familiar with trusted resources like the *FASB Accounting Standards: Original Pronouncements* (Irwin, 1993), *FASB Accounting Standards: Current Text* (Irwin, 1993), and Martin A. Miller's *Miller HBJ Comprehensive GAAP Guide* (Harcourt Brace Jovanovich, 1991-93).

Students of accounting may need to consult industry ratio information. Robert Morris Associates publishes the *RMA Annual Statement Studies* (Robert Morris Associates, 1977), which is probably the most popular tool for industry standards information. In addition, students enrolled in advanced accounting classes might consider Leo Troy's *Almanac of Business and Industrial Financial Ratios* (Prentice Hall), and Dun and Bradstreet's *Industry Norms and Key Business Ratios* (Dun and Bradstreet, 1983).

The accounting profession is experiencing new trends. The field of accounting is becoming more complex, and the role of accountants more valued. Accountants are also experiencing new demands of professional responsibility as they face social and professional challenges. A growing reliance on technology has significantly affected the accounting profession as well. In most corporations, accounting procedures are computerized. For more information about trends in accounting, students should read AICPA's annual publication, *Accounting Trends and Techniques* (American Institute of Certified Public Accountants, 1962-).

Accounting, Bookkeeping, and Income Tax Preparation — Business

For career information, students should consult *Careers in Public Accounting* (Emerson Co., 1993), *Emerson's Directory of Leading U.S. Accounting Firms* (Emerson Co., 1994), and *Who Audits America* (Data Financial Press, 1991). There are three major types of accountants and auditors: Certified Public Accountant (CPA), Certified Management Accountant (CMA), and Certified Internal Auditor (CIA).

Most accountants are required to have at least a bachelor's degree in accounting or a related field. Certified Public Accountants (CPAs) must have a bachelor's degree in accounting or a related field. CPAs must also pass the CPA exam, which is given twice a year by the AICPA. Qualifications for CPA vary by state. To be awarded a CPA certificate, a candidate must also possess at least two years of work experience. Students can utilize review books such as Patrick Delaney's *CPA Examination Review* (Wiley, 1994). The first volume covers financial accounting and reporting for business enterprises. Volume Two addresses taxation, and managerial, governmental, and not-for-profit accounting and reporting. Volume Three deals with business law and professional responsibilities. The final volume addresses auditing.

A Certified Management Accountant (CMA) may hold an accredited baccalaureate degree in any area, or substitute previous work experience. CMAs are also required to pass a standardized exam and have previous work experience. The Institute of Internal Auditors confers the designation of Certified Internal Auditor (CIA). All CIA candidates must complete two years of internal auditing work experience, earn a bachelor's degree, and pass a four-part CIA examination. All positions require continuing accounting experience requirements.

Selected Associations

American Accounting Association
5717 Bessie Drive
Sarasota, Fl. 34233-2399

American Association of Hispanic CPAs
1414 Metropolitan Avenue
Bronx, N.Y. 10462

American Institute of Certified Public Accountants
1211 Avenue of the America
New York, N.Y. 10036

American Society of Tax Professionals
P.O. Box 1024
Sioux Falls, S. D. 57101

American Woman's Society of Certified Public Accountants
401 N Michigan Ave.
Chicago, Ill 60611

Asian American Certified Public Accountants
580 California Street, 16th Floor
San Francisco, Calif. 94104

Institute of Certified Management Accountants
10 Paragon Drive
Montvale, N.J. 07645

Institute of Internal Auditors
249 Maitland Ave.
Altamonte Springs, Fla. 32701-4201

National Association of Accountants
10 Paragon Dr.
Montvale, N.J. 07645

National Association of Black Accountants
220 I Street NE, Suite 150
Washington, D.C. 20012

National Association of Public Accountants
1010 N Fairfax St.
Alexandria, Va. 22314

Selected Journals

Accounting Review.
American Accounting Association. Quarterly. ISSN 0001-4826

Journal of Accountancy.
American Institute of Certified Public Accountants. Monthly. ISSN 0021-8448

Journal of Accounting Research.
University of Chicago, Graduate School of Business, Institute of Professional Accounting. Biannually. ISSN 0021-8456

Journal of Taxation.
Warren Gorham Lamont. Monthly. ISSN 0022-4863

Management Accounting.
Institute of Management Accountants. Monthly. ISSN 0025-1690

Print Materials

Accounting and Auditing Careers Passbook: Test Preparations Study Guide: Questions and Answers
Syosset, N.Y., National Learning Corp., 1992. 1 v. (various pagings). ISBN 0837334284

Consists of examination questions, answer sheets, and answers. Includes sections for accounting problems and written materials. Concluding sections contribute fundamental information about financial statements. Also includes a list of terms. Designed as a preparation guide for civil service examinations.

Allen, David Grayson
Accounting for Success: A History of Price Waterhouse in America, 1890-1990
Boston, Harvard Business School Press, 1993. 373p. ISBN 087584328X

Samuel Price, Williams Hopkins Holyland, and Edwin Waterhouse managed this British-based company from 1850-1901. This work provides a history of the company's stages of development. Includes appendix, notes, and an index.

Anthony, Robert Newton
Review of Essentials of Accounting
4th ed. Reading, Mass., Addison-Wesley, 1988. 158p. ISBN 0201059053

This supplement to the author's *Essentials of Accounting* (Addison-Wesley, 1988) covers identical information but without the frames, exercises, tests, and answers. The work is arranged in ten parts, with a glossary and a term index included. The "Key Points to Remember" feature concluding each section and the list of basic accounting concepts add to the book's usefulness.

Anthony, Robert Newton
Essentials of Accounting
4th ed. Reading, Mass., Addison-Wesley, 1988. 193p. ISBN 0201000172

Accounting language should reflect the society and business in which it functions, according to this work, designed as a self-instructional guide. It is organized in a column-frame approach: the left column contains the exercise or question; the right column provides the correct answer. Students are directed to cover the correct answers and to disclose them only after they have answered. A separate and appended booklet contains exhibits, posttests, and answers. The glossary also serves as an index.

Berton, Lee, and Jonathan B. Schiff
The Wall Street Journal on Accounting
Homewood, Ill., Dow Jones-Irwin, 1990. 471p. ISBN 155623225X

Presents material about the new world of accounting, the impact of financial accounting standards boards, taxes, government accounting, industry-specific reporting, and humor in accounting.

Boockholdt, James L.
Accounting Information Systems: Transaction Processing and Controls
3rd ed. Homewood, Ill., Irwin, 1993. 861p. ISBN 0256108412

Covers all aspects of accounting systems. Chapters are grouped into five parts: (1) accounting and systems concepts; (2) developing accounting systems; (3) technology of accounting systems; (4) controls; and (5) processing accounting transactions. Indexed.

Brimson, James A.
Activity Accounting: An Activity-Based Costing Approach
New York, Wiley, 1991. 214p. ISBN 0471539856

Activity-based accounting, the process of cost management that surveys the cost and performance of an organization's activities, is analyzed in this work, with the author clearly arguing in favor of its effective use. New concepts and terms appear in boldface and are defined in a glossary. In addition, models and charts are used to help present this accounting approach to the reader. Indexed.

Carmichael, D. R., Steven B. Lilien, and Martin Mellman, eds.
Accountant's Handbook
7th ed. New York, Wiley, 1991. 1 v. (various pagings). ISBN 0471619795

This revised and updated edition features contributions from fifty-eight leading authorities on accounting and tax practices. As expected, the handbook includes new chapters and covers the latest material on emerging issues and official pronouncements. Also covers financial accounting topics, concepts, and standards. Includes a list of acronyms, a bibliography, and a subject index.

Collett, Iris Weil
Accosting the Golden Spire
Sun Lakes, Ariz., Thomas Horton and Daughters, 1988. 205p. ISBN 091387843X

Collett has written numerous books that teach accounting basics in an entertaining manner. In this book, Lenny Cramer, a forensic accountant from the Wharton School of Finance, goes on an international jaunt to Rangoon, Burma, meets an attractive jewelry shop owner, and becomes the hero of a plot filled with intrigue, mystery, and murder. This is a representative title of many published by Thomas Horton and Daughters.

Computers in Accounting: Buyers Guide
Boston, Warren, Gorham and Lamont, 1991. 150p. ISBN 068545987X

A directory written in response to the growing interest in accounting software. The publication is a sourcebook of vendors and products, divided into three main parts: (1) software; (2) hardware; and (3) alphabetical listings. The introduction provides instructions for using the guide, fax inquiry forms, and toll-free numbers for updated vendor information.

Cottell, Philip G., and Terry M. Perlin
Accounting Ethics: A Practical Guide for Professionals
New York, Quorum Books, 1990. 171p. ISBN 089930401X

In a jargon-free style, the authors emphasize the ethics of accounting in accounting mentoring, social responsibility, and law. The case studies provided in each chapter represent real-life situations, and students should be encouraged to work through the case studies. This work is particularly useful for students serving accounting internships.

Cypert, Samuel A.
Following the Money: The Inside Story of Accounting's First Mega-Merger
New York, AMACOM, 1991. 279p. ISBN 0814450024

Cypert narrates the story of the formation of the world's largest professional services firm. The author offers perspectives from both sides of the merger between Peat Marwick International and Klynveld, Main, Goerdeler (KMG), identifying key players, conflicts, and the cultural and geographic concerns involved in this undertaking. Of particular interest is the epilogue and the chronology in chapter thirteen entitled "The Great Feeding Frenzy of 1989." Indexed.

Davidson, Sidney
Accounting: The Language of Business
8th ed. Sun Lakes, Ariz., Thomas Horton and Daughters, 1990. 146p. ISBN 0913878472

Horton and Daughters publications tend to be innovative, reliable, and applicable to community college accounting studies. This work should be considered for any core collection in bookkeeping and accounting resources. The main text is complemented by a glossary, a copy of an annual report, and a balance sheet of a company approaching bankruptcy.

Fay, Jack R.
Accounting Certification, Educational and Reciprocity Requirements: An International Guide
Westport, Conn., Quorum Books, 1992. 301p. ISBN 0899306403

Students can study the professional requirements for the Americas, Asia, the Pacific Basin, Europe, Africa, and the Middle and Near East in this international guide. In addition to reciprocity and basic certification requirements, professional activities and responsibilities, continuing education, accounting organizations and their publications, and ethics are detailed. Contains appendices, a bibliography, and an index.

Fields, Louis W.
Bookkeeping Made Simple
Rev. ed. New York, Doubleday, 1990. 144p. ISBN 0385238827

This resource, a combination guide and workbook, covers the basic techniques in a way that makes bookkeeping and accounting seem easy. The clear and easily understood examples are demonstrated on work sheets, and each chapter begins with a list of new terms. Corresponding exercises and solutions complete the guide.

Garner, C. William
Accounting and Budgeting in Public and Nonprofit Organizations: A Manager's Guide
San Francisco, Calif., Jossey-Bass, 1991. 252p. ISBN 1555423361

This book was written for managers with minimal background in accounting. Despite the obvious appeal to professionals, students could also use this source for advanced study. Beginning with the history of accounting and budgeting, Garner then discusses everything from the difference between bookkeeping and accounting to inventory controls. Includes chapter-based terms and definitions, references, a bibliography, and an index.

Gill, James O.
Understanding Financial Statements: A Guide for Non-Financial Readers
Los Altos, Calif., Crisp Publications, 1990. 104p. ISBN 1560520221

With a very clever approach and introduction to deciphering financial statements, this primer teaches the basics of analyzing balance sheets and profit and loss statements. Best suited for small business accounting classes. Students will enjoy Gill's effective use of cartoons and artwork. The clearly written text and eye-catching illustrations make this an inviting resource.

Half, Robert
Robert Half's Success Guide for Accountants
2nd ed. Englewood Cliffs, N.J., Prentice Hall, 675p. ISBN 0137815351

Accounting students on the career path should not overlook this vocational guidance guide that includes tips on personal marketing, effective management and communication, and job search strategies. The tone is positive and supportive.

Harrison, David, and John W. Yu
Spreadsheet Style Manual
Homewood, Ill., Dow Jones-Irwin, 1990. 149p. ISBN 1556232675

An all-purpose reference source for creating, editing, maintaining, and revising spreadsheets. Of the fourteen chapters, students might want to read chapter fourteen first since it is a summary of the tips covered in the rest of the book. Indexed.

Ijiri, Yuji
Momentum Accounting and Triple-Entry Bookkeeping: Exploring the Dynamic Structure of Accounting Measurements
Sarasota, Fla., American Accounting Association, 1989. 151p. ISBN 086539072X

This is number thirty-one in the *Studies in Accounting Research* series published by the American Accounting Association (AAA). Chapters include the dynamic structure of accounting measurements; accounting records and database fields; double entry bookkeeping; momentum accounting; triple entry bookkeeping; wealth utilization; managerial goals and performance measurements; objectives and postulates of momentum accounting; implementation of momentum accounting; momentum measurement and statistical forecasting; and summary of findings. Concludes with a list of references and an index.

J. K. Lasser's Your Income Tax
New York, Prentice Hall, 1936- . Published annually.

A popular how-to guide that explains income tax preparation, step by step. The 1994 edition consists of forty-eight chapters divided into eight parts. Part Eight includes a glossary and an index.

Kanatsu, Takashi
TQC for Accounting: A New Role in Companywide Improvement
Cambridge, Mass., Productivity Press, 1990. 226p. ISBN 0915299739

TQM (Total Quality Management) is the management method of choice today, and TQC (Total Quality Control) is a derivation of TQM for marketing and accounting. Although this work provides accounting students with information on current managerial theory and trends, previous knowledge of accounting, management, and marketing theory history is required. A good supplement for a special topics course.

Lehman, Cheryl R.
Accounting's Changing Roles in Social Conflict
New York, Markus Wiener Publishing, 1992. 174p.
ISBN 555876030X

Accounting and social responsibility is the issue addressed in this volume. Useful as a text or for supplemental reading for courses covering special topics and issues, including controversial subjects. The discussions of socio-historical theories are intellectually appealing. Includes an extensive bibliography.

Lerner, Joel J.
Schaum's Outline of Theory and Problems of Bookkeeping and Accounting
3rd ed. New York, McGraw-Hill, 1994. 400p. ISBN 0070375933

More than 500 problems are presented in this easy-to-follow guide, published as part of *Schaum's Outline Series*. This series has become a reliable source for libraries needing supplemental information for coursework.

Lipkin, Lawrence
Accountant's Handbook of Formulas and Tables
3rd ed. Englewood Cliffs, N.J., Prentice Hall, 1988. 627p.
ISBN 0130029572

Compiled for professional accountants, most of this material is advanced. Nonetheless, students can use the index to locate specific formulas and tables for their coursework. Formulas and tables are provided for simple and compound interest, annuities, computing and statistics, inventory, depreciation, finance, capitalized values, price level adjustments, marketing, cost and production, ratio analysis, determining sale and purchases, and pension plans.

Listman, Robert J.
Marketing Accounting Services
Homewood, Ill., Dow Jones-Irwin, 1988. 196p. ISBN 0870948970

This book is a simple look at marketing strategies used by professionals to promote accounting services. Understanding the vital aspects and duties of marketing public accounting is crucial; therefore, the advice in this book is presented in step-by-step instructions. Special features include a valuable appendix of secondary data sources, a list of business reference titles and business reference bibliographies, and a list of government documents, which are often considered primary sources. Indexed.

Livingstone, John Leslie, ed.
Portable MBA in Finance and Accounting
New York, Wiley, 1992. 524p. ISBN 0471532266

In addition to the author's intended audience of professionals, this resource could also serve as a ready reference guide for students. Livingstone provides nonbusiness professionals with an intellectual understanding of accounting at the graduate level. Students, on the other hand, are given examples of "how it should be done" as provided by Livingstone and his team of professors from top business schools. Techniques that might be most appreciated by community college students are strategies for activity-based accounting, budgetary control, and designing a business plan. Because the book is written for use by professionals in other fields, business jargon is rarely used. Includes glossary.

Louderback, Joseph G.
Survey of Accounting
St. Paul, Minn., West Pub. Co., 1993. 788p. ISBN 0314010416

The nonbusiness major will appreciate this publication's coverage financial and managerial accounting, auditing, and accounting for acquisitions. In addition, tips for analyzing financial statements and a chapter of vocabulary are provided.

Marshall, David H.
Survey of Accounting
2nd ed. Homewood, Ill., Richard D. Irwin, 1993. 615p.
ISBN 0256113017

The author of this textbook hopes to provide students with the "big picture" in accounting as he sees it. This innovative book supports the objectives of the Accounting Education Change Commission for undergraduate classes, and it is also is intended to meet the special information needs of nonbusiness majors. Includes two appendices.

McNair, Carol Jean
World-Class Accounting and Finance
Homewood, Ill., Business One Irwin, 1993. 356p.
ISBN 155623550X

Explains information storage and retrieval processes for accounting firms. Special emphasis is given to the customer's perspective. McNair includes a glossary that defines terms such as backflushing, double entry accounting, avoidable and allowable costs, and burden rate. Indexed.

Meyers, Larry F.
Tax Tips Strategies: Income Tax Hints Every Taxpayer Should Know
Boca Raton, Fla., Meyers Publishing, 1994. 116p.
ISBN 0963916602

This easy-to-read guide contains humorous illustrations of interest to small business owners, real estate investors, seniors and retirees, parents, and anyone being audited. Features material on Clinton's tax update, tax breaks for homeowners, inherited property and taxes, moving expenses, advantages of a two-car family, and the best time to schedule an audit.

Milliron, Robert R.
How to Do Your Own Accounting for a Small Business
4th ed. Wilmington, Del., Enterprise Publishing, 1988. 159p.
ISBN 091386434X

This work, in its fourth edition, is written for small business owners unfamiliar with the basics of bookkeeping. The book's step-by-step instructions of bookkeeping practices is especially useful for students. Chapter coverage is extensive, with more than adequate attention given to double entry bookkeeping, posting to general ledgers, and accounting for growing a business. Contains two appendices and an index.

Miranti, Paul J.
Accountancy Comes of Age: The Development of an American Profession, 1886-1940
Chapel Hill, N.C., Univ. of North Carolina Press, 1990. 275p.
ISBN 0807818933

Fifty-four years of accounting are covered in this volume, which includes the dawn of professional accountancy, 1880-1900; accountants in an age of progress, 1900-1916; mobilization, prosperity, and autonomy, 1916-1929; and economic, crisis, and countervailing associationalism, 1929-1940. Also discusses the struggle that accountants have faced to establish their professional status and receive social acceptance. The text also explores the battles between merging associations, and diverse opinion over the need for professional training and education. Includes notes, a bibliography, and an index.

O'Connor, Walter F.
Accounting and Taxation
New York, Barron's, 1990. 230p. ISBN 0812041542

Although aimed at business managers and owners, this step-by-step guide is useful for students taking intermediate to advanced accounting classes as well. The issues of accounting and accounting information, budgeting, company ownership, cost accounting, and performance measurement are addressed in this ready reference source. Indexed.

Pinson, Linda, and Jerry Jinnett
Keeping the Books: Basic Recordkeeping and Accounting for Small Business
2nd ed. Dover, N.H., Upstart Pub. Co., 1993. 200p.
ISBN 0936894474

A resource guide for students enrolled in a small business or managerial accounting class. The appendix information is extremely useful, containing a glossary of accounting terms, a helpful bibliography to small business resources, and a supplemental section of blank forms and worksheets.

Ragan, Robert C.
Step-by-Step Bookkeeping
Rev. ed. New York, Sterling Pub. Co., 1992. 134p.
ISBN 0806986905

Emphasizes the importance of keeping well-organized records while introducing basic bookkeeping and accounting theory. Especially valuable is the advice given for handling refunds, coupons, cash receipts, and customer check cashing.

Riahi-Belkaoui, Ahmed
The Coming Crisis in Accounting
New York, Quorum Books, 1989. 184p. ISBN 089930379X

The contemporary accounting profession and the challenges facing it are profiled in this text. Students are given an introspective look at accountants, including such issues as professional status and peer review. Further discussion focuses on the regulation of CPA firms, fraud control, and the impact of technology on accounting.

RMA Annual Statement Studies
Philadelphia, Robert Morris Associates, 1977- . Published annually. ISSN 0080-3340

This annual publication lists industry ratio information by four-digit Standard Industrial Classification (SIC) codes.

Rosenthal, Lawrence
Exploring Careers in Accounting
Rev. ed. New York, Rosen Pub. Group, 1993. 153p.
ISBN 0823915018

More than a source for career information, this is a complete advisory guide. Students are advised of advantages and disadvantages of accounting careers, various job opportunities, career planning and preparation, and new technologies. Includes four appendices and lists of definitions, accounting associations, and degree codes as used in the *Peterson's* college guides.

Sangster, Alan
Workbook of Accounting Standards
2nd ed. London, Pitman Publishing, 1993. 399p.
ISBN 0273601040

A practical source for the community college student with a bachelor's degree who is preparing for the CPA exam. Includes twenty-two chapters covering current accounting standards as set forth in Statements of Standard Accounting Practice (SSAP) and Financial Reporting Standards (FRS). Most chapters include summaries of the accounting standards and contain material about the interpretations of the standards. The index is followed by a section of solutions to the practice questions.

Sax, Nancy, and H. Steve Dashefsky
Peachtree Complete II: Accounting Made Easy
Blue Ridge Summit, Pa., Windcrest Books, 1990. 250p.
ISBN 0830633057

Since accounting in many companies is done primarily by computers, basic accounting courses for beginning students now include a segment on computerized accounting. This work helps guide the student through the program and software of the Peachtree company, producer of one of the most popular accounting systems available.

Sax, Nancy, and H. Steve Dashefsky
Peachtree Complete III
Blue Ridge Summit, Pa., Windcrest Books, 1990. 284p.
ISBN 0830635645

This update to *Peachtree Complete II* (Windcrest Books, 1990), contains slightly revised versions of the first eleven chapters of the previous work. In addition, an added chapter reviews the new purchase order module. The novice will appreciate this introduction to personal computing.

Shim, Jae K., and Joel G. Siegel
Encyclopedic Dictionary of Accounting and Finance
Englewood Cliffs, N.J., Prentice Hall, 1989. 504p.
ISBN 0132758016

A dictionary of accounting, finance, investing, and banking. This reference book offers answers and suggestions for dealing with everyday problems in a financial environment. More than 500 major accounting and finance topics are presented along with tables, charts, ratios, formulas, and applications. Appendix includes coverage of future and present values. Some entries are cross-referenced.

Siegel, Joel G., and Jae K. Shim
Thinking Finance: Everything Managers Need to Know about Finance and Accounting
New York, Harper Business, 1990. 231p. ISBN 0887304427

This title would be a good supplement to a financial accounting textbook. Although intended for managers, students will be provided with a fundamental understanding of finance and financial decision making as well. Covers cost-volume-profit analysis, financial forecasting, managing inventories, rate of return equations, and loans. This book would be especially helpful for evaluating financial cases and scenarios. Includes an index.

Siegel, Joel G., and Jae K. Shim
Accounting Handbook
New York, Barron's, 1990. 836p. ISBN 0812061764

Anyone involved in accounting will find this handbook invaluable. Students at all levels will appreciate its overview of the most important topics in the field. Chapter coverage includes financial statements, financial reporting, management and cost accounting, financial statement analysis, preparation and planning for individual income taxes, accounting terms, microcomputer applications, quantitative methods, auditing, personal finance, and accounting for government and nonprofit organizations.

Simini, Joseph Peter
Accounting Made Simple
Rev. ed. New York, Made Simple Books, 1988. 162p. ISBN 0385232802

Terms are explained as they are introduced in this inexpensive guide that is designed to supplement a textbook and to correspond with the accounting curriculum at the college level. Cost accounting gets special consideration in chapters eleven and twelve. The manual is indexed, but does not include a glossary.

Slater, Jeffrey
Simplifying Accounting Language: Don't Lose Your Balance!
3rd ed., rev. printing. Dubuque, Iowa, Kendall/Hunt, 1988. 273p. ISBN 0840345674

This resource was written by a community college accounting instructor for students in Accounting I and II, as well as for students enrolled in advanced level accounting courses who need to review basic accounting theory. Slater's theme is "accounting can be fun." Cartoons, puzzles, and other graphic illustrations help support the author's presentation.

Smith, L. Murphy, Robert H. Strawser, and Casper E. Wiggins
Readings and Problems in Accounting Information Systems
Homewood, Ill., Irwin, 1991. 438p. ISBN 0256070407

The authors, college professors, survey the problems associated with computers in accounting. Specific issues addressed are computer systems and information literacy, communication software, expert systems, and auditing in the future. Those discussions along with reprints of articles taken from the *Journal of Accounting, The CPA Journal,* and the *Internal Auditor,* among others, force the student to consider the impact that technology has had on accounting.

Spacek, Leonard
The Growth of Arthur Andersen and Company, 1928-1973: An Oral History
New York, Garland Publishing, 1989. 349p. ISBN 0824061470

Leonard Spacek, past chairman of Arthur Andersen and Co., presents an oral account of his forty-five year tenure with the company. This interview offers good insight into the development of one of the largest international public accounting firms in the world. Indexed. Part of the *Accounting History and Thought* series.

Spurling, David
Discover Bookkeeping and Accounts
London, Pitman Publishing, 1988. 353p. ISBN 0273028707

Contains practical information on accounting and bookkeeping, featuring a key to syllabus topics covered by accrediting examination boards, and examples of exams, questions, and exercises.

The handbook might be of special interest to those in international accounting because of its British influence. (All figures are expressed in pounds).

Stevens, Mark
The Big Six: The Selling Out of America's Top Accounting Firms
New York, Simon and Schuster, 1991. 271p. ISBN 0671695495

The "Big Eight" was replaced by the "Big Six" and consists of Arthur Andersen, KPMG Peat Marwick, Ernst and Young, Coopers and Lybrand, Price Waterhouse, and Deloitte and Touche. Stevens, an authority on the history of accounting firms, relates the events associated with the transitions involving these public accounting firms in the 1980s. Includes an index.

Strobel, Caroline D., ed.
Accounting for Business Combinations and Restructurings
New York, Executive Enterprises Publications Co., 1991. 221p. ISBN 155840452X

Individually authored chapters are contained in this guide to accounting for corporate mergers. Students will especially appreciate the accounts of negotiations and transitions made during the 1980s merger madness. The book lacks an index, but the detailed discussion of the history and methods used during this business decade more than make up for this shortcoming.

Wallace, Wanda A.
Accounting Research Methods: Do the Facts Speak for Themselves?
Homewood, Ill., Irwin, 1991. 105p. ISBN 0256100470

Students learn to identify and understand the relationship between information gathering and accounting in this text. Wallace makes the argument that developing research questions that are well thought out is just as vital as determining the validity of research and analysis. A model of the perfect researcher is provided, as well as helpful research tips, usually given in the form of caveats. Subject-related quotes for each chapter, a glossary entitled "jargonese," notes, and a keyword index enhance the text.

Weil, Roman L., Clyde P. Stickney, and Sidney Davidson
Vest-Pocket CPA
Englewood Cliffs, N.J., Prentice Hall, 1988. 578p. ISBN 0139422935

A practical reference to financial accounting, managerial accounting, auditing, taxation, financial planning, financial statement analysis, and quantitative analysis. An excellent guide for any student who has ever struggled with analyzing financial statements or needed tips to unraveling case studies.

Nonprint Materials

A.P.E. Surfshop Accounting Videos
Charleston, W. Va., Cambridge Business Education, 1992.
4 videocassettes (105 min.) VHS

Explains small business operations for maintaining integrated accounting systems, inventory, daily transactions, bank reconciliation, general ledger entries, and other accounting and bookkeeping procedures.

Accounting, Bookkeeping, and Income Tax Preparation

Accounting and Office Systems
Charleston, W. Va., Cambridge Career Products, 1993.
1 videocassette (20 min.) VHS
   Designed to help students develop communications skills for careers in accounting. The *Communication Skills Video Series* is drawn from the Carl Perkins Applied Technology Act. Also available in 3.5" or 5.25" disk.

Basic Accounting Video
Charleston, W. Va., Cambridge Business Education, 1990.
1 videocassette (164 min.) VHS
   A program that covers basic accounting and also gives good explanations of terminology used. Topics covered include the nature of transactions, books of entry, financial statements, and computerized accounting.

Basic System: Accounting and Bookkeeping for Small Businesses: Based on Microsoft Excel for the Apple Macintosh
Charleston, W. Va., Cambridge Career Products, 1988.
1 videocassette (112 min.) VHS
   Demonstrates the appropriate steps for setting up a workable accounting system for a small business.

Finance for Nonfinancial Professionals
Boulder, Colo., CareerTrack Publications, 1992. 3 videocassettes (240 mins.) VHS
   This three-video series comes with a workbook. Section titles include "Speaking Accounting," "Reading Financials," and "Building Budgets."

Hey—We're Being Audited!
Action Productions for the Internal Revenue Service, 1993.
1 videocassette (28 min.) VHS
   This IRS provides instructional videos to help filers with their tax problems. This video presents a typical American family as they prepare for a tax audit.

Hey, We're in Business!
Washington, D.C., Internal Revenue Service, 1993. 1 videocassette (28 min.) VHS
   This video is a step-by-step guide to tax and financial management for a small business.

Introduction to Computerized Accounting
Charleston, W.V., Cambridge Business Education, 1993. Computer disks and manual
   Students apply basic accounting principles as they work through this simulated business experience. Intended for IBMs or IBM PC compatibles.

¿Por Que Los Impuestos?
[Washington, D.C.], Department of the Treasury. Internal Revenue Service, 1993.
1 videocassette (28 min.) VHS
   This IRS videocassette presents a reporter who is researching taxation and taxpayers rights. The focus of this video is on education of the Hispanic community regarding tax issues.

¿Por Que Nosotros, Los Garcia?
[Washington, D.C.], Department of the Treasury. Internal Revenue Service, 1993.
1 videocassette (28 min.) VHS
   This IRS videocassettse provides instructional videos to help filers with their tax problems. This video presents the Garcia family as they prepare for a tax audit.

Processing with DacEasy 4.4
Pryor, Okla., ViaGrafix, 1994. 1 videocassette (76 min.) VHS
   Introduction to a computerized accounting program.

Quicken 7.0 Quick and Easy
Pryor, Okla., ViaGrafix, 1994. 1 videocassette (82 min.) VHS
   A video guide to using Quicken 7.0.

Quicken for Windows, Quick and Easy
Pryor, Okla., ViaGraphix, 1992. 1 videocassette (64 min.) VHS
   This video explains accessing Quicken via Windows software.

Red Flags: What Every Manager Should Know about Internal Crime
Austin, Tex., National Association of Certified Fraud Examiners, 1991. 1 videocassette (35 min.) VHS
   This video emphasizes the importance of internal control.

Using Computerized Accounting
Dallas, Tex., M-USA Business Systems, 1988. 1 videocassette (30 min.) VHS
   This video is an introduction to a computerized accounting system.

Why You Should File
[Washington, D.C.], Department of the Treasury. Internal Revenue Service, 1993.
1 videocassette (28 min.) VHS
   This video covers completing Form 8300, and other reporting requirements for small businesses conducting cash transactions over $10,000.

Computers and Data Processing

Rebecca Kroll
Lorain County Community College
Elyria, Ohio

Introduction

Like all technologically driven fields, the area of computers and data processing in business is in a constant state of flux, with upgrades, improvements, and innovations coming fast and furiously. Greater computer power now fits into smaller spaces than ever before, while the range of activities possible at a workstation or on a hand-held device increases geometrically over time.

Given this rapid rate of change, this bibliography stresses the importance of building and maintaining a strong journal collection in all relevant areas of computers and data processing, as well as keeping a core collection of both introductory and slightly more advanced materials covering hardware, software, programming, and applications. With the exception of a few very popular programs, emphasis is on publications that provide background information and additional titles that take a tutorial approach; it is assumed that the actual software and the accompanying manuals will be covered in classes. Since much of the most recent material in the field is now published electronically, a core collection should also include information on networking and specifically on the Internet. This applies whether the library is located on a campus already fully networked, or is serving a population with individual dial-up access.

A basic philosophy guides selections made for this collection: The library should provide basic reference works, alternative explanations, how-to books for patrons needing this learning approach, background reading materials to support more advanced work, and a comprehensive journal collection. The collection should have available the most recent information for those who wish to remain current in the field, while still providing a solid foundation for students entering the field. It is recommended that a library select one or more series of titles from reputable publishers who can be relied on to produce up-to-date coverage of hardware or software in standard formats. The titles and editions listed in this bibliography were the most recent ones available at the time of review; they should be updated on a regular basis.

The areas covered in this chapter on computers and data processing emphasize business applications and include keyboarding, word processing, simple programming, spreadsheets, computer graphics, the Internet, and local area networks. Many of the print sources may be purchased with or without accompanying software, according to the collection development policy of the library and the computer facilities available for students.

Selected Associations

Association for Computer Operations Management
742 E. Chapman Ave.
Orange, Calif. 92666

Association of Computer Professionals
230 Park Ave. Suite 460
New York, N.Y. 10169

Association for Computing Machinery
1515 Broadway
New York, N.Y. 10036

Selected Journals

Byte.
McGraw-Hill. Monthly. ISSN 0360-5280

Cadence.
Miller Freeman. Monthly. ISSN 0887-9141

Communications of the ACM.
Association for Computing Machinery. Monthly. ISSN 0001-0782

Compute.
General Media Pub. Co. Monthly. ISSN 0194-357X

Computer Language.
Miller Freeman. Monthly. ISSN 0749-2839

Computerworld.
Computerworld. Weekly. ISSN 0010-4841

Data Based Advisor.
Data Based Solutions. Monthly. ISSN 0740-5200

Datamation.
Cahners Pub. Co. Semimonthly. ISSN 0011-6963

Home Office Computing.
U.S. Scholastic. Monthly. ISSN 0899-7373

Internet World.
Meckler Pub. Co. Six times a year. ISSN 1064-3923

MacUser.
Ziff-Davis. Monthly. ISSN 0884-0997

MacWorld.
PCW Communications. Monthly. ISSN 0741-8647

PC/Computing.
Ziff-Davis. Monthly. ISSN 0899-1847

PC Magazine.
Ziff-Davis. Twenty-two times a year. ISSN 0888-8507

PC World.
PCW Communications. Monthly. ISSN 0737-8939

UNIX Review.
Miller Freeman. Monthly. ISSN 0742-3136

UNIX/World.
McGraw-Hill. Monthly. ISSN 0739-5922

Print Materials

✦ Reference

Data Sources: The Comprehensive Guide to the Information Processing Industry: Equipment, Software, Services, Companies, and People
New York, Ziff-Davis Pub. Co., Published biennially. various pagings. ISSN 0744-1673

Issued in two volumes: *Hardware/Data Communications* and *Software.* Covers data processing and data communications hardware, software, and companies. Main arrangement is by broad subject category, with detailed indexing by product and company name.

Directory of Electronic Journals, Newsletters and Academic Discussion Lists
Washington, D.C., Association of Research Libraries, Office of Scientific and Academic Publishing. Published annually.
ISSN 1057-1337

Includes a listing of journals published on CD-ROM, on disk, or over the Internet, a list of bulletin boards, and lists of discussion groups of interest to academics. The listings include Internet and BITNET addresses as appropriate, a brief statement of coverage, and a contact person. Also invaluable are the pages containing the National Library of Medicine Recommended Formats for Bibliographic Citation. Includes bibliography and index.

Hordenski, Michael F., comp.
Illustrated Dictionary of Microcomputers
3rd ed. Blue Ridge Summit, Pa., TAB Books, 1990. 430p.
ISBN 0830633685

Although titled an illustrated dictionary, illustrations are included only when absolutely necessary, but the definitions are clear and to the point. This is a handy resource for quick reference, or for help in translating a difficult computer question at the reference desk.

McDaniel, George, ed.
IBM Dictionary of Computing
New York, McGraw-Hill, 1994. 758p. ISBN 0070314896

Contains 18,000 entries covering all aspects of computing, with particular attention to highly technical areas and acronyms. Sparsely illustrated, but extremely comprehensive in coverage.

Ralston, Anthony, and Edwin D. Reilly, eds.
Encyclopedia of Computer Science
3rd ed. New York, Van Nostrand Reinhold, 1993. 1558p.
ISBN 0442276796

Covers all aspects of computing: theory, applications, history, hardware, and software. The articles provide fascinating reading, with numerous illustrations for clarification. Appendices include a timeline of computer development. Indexed.

Raymond, Eric S., comp.
New Hacker's Dictionary
Cambridge, Mass., MIT Press, 1991. 433p. ISBN 0262680696

From "ack" to "zorch," this revision of the 1983 edition provides humorous but genuinely useful translations of the jargon of the computer hacker: "one who programs enthusiastically (even obsessively)." Includes a brief annotated bibliography.

Rosenberg, Jerry Martin
Business Dictionary of Computers
New York, Wiley, 1993. 403p. ISBN 0471585750

Covers virtually all areas of computers, defining formal terms, acronyms, abbreviations, and some slang. Useful for more than just computer terminology for business.

Sochats, Ken
Networking and Communications Desk Reference
Carmel, Ind., Sams Publishing, 1992. 406p. ISBN 0672300966

Predominantly a glossary of networking and communications terms and acronyms intended for both students and professionals. The many appendices form a small trove of useful bits of information ranging from ASCII character codes to the Morse code table to an outline of American telecommunications history.

Software Digest Ratings Report
Conshohoken, Pa. National Software Testing Laboratories. Published monthly. (Various pagings.) ISSN 0893-6455

Each loose-leaf issue is devoted to a specific topic such as network management systems or executive word processors for Windows Includes comparative rating scales as well as an overview of each program.

Software Reviews on File
New York, Facts on File. Published monthly. (various pagings.) ISSN 8755-7169

This loose-leaf review service contains publication information, system requirements, and publishers' descriptions of software, along with abstracts of reviews from almost one hundred journals. Programs are arranged by broad application and indexed by name and publisher.

Spenser, Donald D.
Computer Dictionary
4th ed. Ormond Beach, Fla., Camelot Pub. Co., 1993. 459p.
ISBN 0892182393

Provides brief and basic definitions of words related to computers, including a few names of people who have made major contributions to computing. Includes serious terminology, not jargon.

✦ General

Aitken, Peter G.
The First Book of Lotus 1-2-3 for Windows
Carmel, Ind., Sams Publishing, 1991. 352p. ISBN 0672273640

One of a series of books published by Sams covering hardware and software, this title follows the *First Book* format by assuming no prior knowledge on the part of the reader. Useful for a beginner who has never used a spreadsheet or worked in a Windows environment, but equally useful for someone with experience who wants a quick overview and some useful shortcuts. Illustrations and index included.

Amon, Elenore M., Annette N. Heatherington, and Douglas D. Henderson
Intelligent LAN Management with Novell NetWare
Englewood Cliffs, N.J., Prentice Hall, 1991. 369p.
ISBN 0134722590

Designed to fill the gap between basic treatments and extremely technical manuals, this book starts by explaining basic concepts of local area networks, then advances to discussions of topologies, hardware, connectivity, and security. Although much of the coverage is brand name specific, the general explanations are useful in any LAN management context. Appendices, glossary, and index included.

Arick, Martin
The TCP/IP Companion: A Guide for the Common User
New York, Wiley, 1993. 240p. ISBN 0471556319

This book helps to demystify networking and network protocols for the inexperienced user by providing explanations of how networks operate. It also gives specific details of networking applications such as FTP (file transfer protocol), Telnet, and e-mail. Appendices, glossary, illustrations, and index included.

Ashley, Ruth, and Judi N. Fernandez
DOS
New York, Wiley, 1993. 378p. ISBN 0471590746

Many nontraditional students at the community college level prefer to teach themselves rather than trying to learn in a classroom setting. This title provides a do-it-yourself approach to learning DOS, both Versions 5 and 6, with both mouse and keyboard instructions. The language is clear and direct, and frequent warnings help the novice avoid mistakes. Includes glossary, illustrations, and index. *Self-Teaching Guide* series.

Badgett, Tom, and Corey Sandler
Business Software Companion
New York, Wiley, 1993. 412p. ISBN 0471569402

A valuable overview of the basic types of business software, this book covers word processors, spreadsheets, databases, communications software, and desktop publishing. The format includes a general description of each category, then comments on specific programs. Includes a short but excellent introduction to Windows. Includes illustrations and index.

Badgett, Tom, and Corey Sandler
WordPerfect 6 for Windows Solutions
New York, Wiley, 1994. 374p. ISBN 0471303291

Although this book starts with basic editing commands, it quickly progresses to more complicated functions, including macros, sound clips, custom templates, and printing to preprinted forms. This gold mine of information on "how to do" and "how to undo" will be in demand by staff as well as students who use WordPerfect. Includes illustrations and index.

Baran, Nicholas
Windows from the Keyboard
New York, Wiley, 1993. 302p. ISBN 0471590932

Questions on running Windows without a mouse arise from users who violently dislike the mouse, or who find "point and click" more difficult than keyboarding. Baran covers how to use keystrokes for Windows and in six Windows applications. The volume's greatest value lies in the quick reference charts at the end of the book that illustrate the keystrokes required to execute desired operations. Includes appendix, illustrations, and index.

Bernstock, Peter N.
The Handbook for Microcomputer Technicians
Boston, QED Pub. Group, 1993. 451p. ISBN 0894354248

It is unlikely that anyone interested in reading a book with this title would need an explanation of what a computer is, but Bernstock leads off with a cursory explanation. Once this is accomplished, he launches into detailed information helpful to a personal computer user with a yen to know more about the internal workings of the machine, or a student contemplating computer repair as a job path.

Bilbo, Mark K.
Macs for Beginners
Carmel, Ind., Que Corp., 1992. 433p. ISBN 1565290615

A great starter book for new users, this title takes the reader from choosing a Mac through setting it up, installing software, learning the mouse and the desktop, working with disks, and operations. Contains excellent explanations of computer drives, memory, and other basics. Also good for the experienced PC user new to the Mac. Appendices, glossary, illustrations, and index included.

Blissmer, Robert H.
Introducing Computers: Concepts, Systems and Applications
New York, Wiley, 1988/1989- . 1 v. (various pagings)

This annually updated title is one to reach for when a patron who knows nothing about computers comes to the library with preclass jitters. It covers key concepts, assuming no prior familiarity with computers, and provides clear illustrations and straightforward language. Glossary; index.

Campbell, Mary V.
Microsoft Word for Windows: The Complete Reference
Berkeley, Calif., Osborne McGraw-Hill, 1994. 858p.
ISBN 0078819482

This book starts with an introduction to Windows concepts before going on to cover Microsoft Word in detail, with special attention to Release 6. The early chapters teach creation and editing of documents; the balance of the text is a detailed explanation of commands in alphabetical order. Appendices include installation, button bars, and more. Illustrations and index included.

Chaiken, Craig
Blueprint of a LAN
Redwood City, Calif., M&T Books, 1989. 337p.
ISBN 1558510524

Although bulk of this book is dedicated to one specific local area network (LAN) software, the general introduction to computers and generic networks alone makes it worthy of inclusion in the library collection. Includes appendices, glossary, and index. Accompanying software can be ordered separately.

Christen, Jenna, and others
Using 1-2-3 Release 3.4
Special ed. Carmel, Ind., Que Corp., 1993. 1152p.
ISBN 1565290046

Follows the Que format for software self-help books, resulting in a detailed, easy-to-follow compendium of information, useful for reviewing the upgrades from previous versions or for learning Lotus 1 2 3 for the first time with this version. This edition is helpful for answering quick questions as well as for learning shortcuts and improving overall output quality. Detailed table of contents, appendices, illustrations, and index provided.

Cohen, Frederick B.
A Short Course on Computer Viruses
2nd ed. New York, Wiley, 1994. 250p. and computer disk
ISBN 0471007692

This book is written for anyone who needs or wants to learn about computer viruses: what they are, how they work, what they do, and how to circumvent them. It gives an overview of the problem and discusses ways of defending computer systems against invasion. Includes floppy disk and appendices.

Dern, Daniel P.
The Internet Guide for New Users
New York, McGraw-Hill, 1994. 570p. ISBN 0070165106

This title explains the basics of the Internet, including e-mail and FTP. A good introduction to UNIX concepts is presented also, but like many other sources, it assumes the presence of a network administrator. Appendices, glossary, illustrations, and index.

Dickinson, Bradley W.
Systems: Analysis, Design, and Computation
Englewood Cliffs, N.J., Prentice Hall, 1991. 289p.
ISBN 0133380475

Beginning with an overview of algebra, matrices, and calculus, Dickinson goes on to cover linear systems, continuous-time systems, nonlinear systems, and optimization, with the goal of providing an introduction to the skills needed for computer-aided systems analysis and design. Includes illustrations, index, and references for each chapter.

Dix, Mark, and Paul Riley
Discovering AutoCAD Release 12
Englewood Cliffs, N.J., Prentice Hall, 1994. 477p.
ISBN 013042904X

Extensive illustrations, a detailed table of contents, and step-by-step instructions for suggested learning exercises are combined in this well-organized, self-teaching guide. Intended to be used in concert with the *AutoCAD Reference Manual* (Autodesk, 1992). Includes appendices and index.

Eddy, Sandra, and John E. Schnyder
The First Book of Word for Windows 6
Indianapolis, Ind., Alpha Books, 1993. 379p. ISBN 1567613527

Like other *First Book* titles, this one gives simple instructions and examples for using Word for Windows software couched in terms basic enough for a word processing beginner, but with tips that more experienced users might find helpful. Good to have for quick consultations or to circulate to patrons who want more than the help screens. Appendix, illustrations, and index included.

Eliason, Alan L.
Online Business Computer Applications
3rd ed. New York, Macmillan, 1991. 534p. ISBN 0023324813

Starting with basic concepts of business information systems and computer programs, Eliason gives an overview of systems design, then reviews specific applications such as accounts payable, payroll, and purchasing, and the utility of information systems from the business point of view. Includes a glossary, illustrations, and an index.

Estrada, Susan
Connecting to the Internet: A Buyer's Guide
Sebastapol, Calif., O'Reilly and Associates, 1993. 170p. ISBN 1565920619

After the initial explanation of the Internet, Estrada concentrates on how to connect to the network. Covers choosing a network provider, dial-up versus a dedicated connection, and a list of public dial-up access points. Includes a brief bibliography, glossary, illustrations, and index.

Feldman, Phil, and Tom Rugg
Using BASIC
2nd ed. Carmel, Ind., Que Corp., 1993. 929p. ISBN 1565291409

Useful as a second opinion for students struggling with programming concepts in class, or as a stand-alone tutorial for the individual learner, this title covers QBASIC, BASICA, GW-BASIC, and QuickBasic as well as Visual Basic. The extensive table of contents and index make it a useful quick reference tool. Appendices, illustrations, and index included.

Gibbs, Mark
Absolute Beginner's Guide to Networking
Carmel, Ind., Sams Publishing, 1993. 468p. ISBN 0672303264

Examines and illustrates network components, topologies, software, and services, with practical examples as well as theoretical presentation. Appendices, glossary, illustrations, and index.

Gilster, Paul
The Internet Navigator
New York, Wiley, 1993. 470p. ISBN 0471597821

In addition to a history and overview of the Internet, Gilster provides one of the clearest and easiest to follow sets of instructions for new Internet users. Only one section is devoted specifically to dial-up connections; the rest of the book applies equally well to any connectivity mode. Appendix, bibliography, and index included.

Grabowski, Ralph
The AutoCAD Technical Reference
Albany, N.Y., Delmar Publishers, 1992. 381p. ISBN 0827348207

Extensive table of contents and indexing make this book useful for quick reference and problem solving or for in-depth coverage of AutoCad. Although it definitely goes beyond the basics, it serves as background reading for the perplexed or the curious. Includes appendices, illustrations, and index.

Grabowski, Ralph
The Successful CAD Manager's Handbook
Albany, N.Y., Delmar Publishers, 1994. 290p. ISBN 0827352336

This discussion of computer-aided design (CAD) conventions, standards, and software is a useful addition to a collection because it covers many different packages rather than explaining a single system. Appendices, illustrations, and index included.

Grace, Rich
Word for Windows 6: Quick Reference
Indianapolis, Ind., Que Corp., 1993. 231p. ISBN 1565294688

Apart from a quick introduction highlighting new features in Version 6 and skimming briefly over Word for Windows functions, this is purely a quick look-up tool for use by people with some word processing experience. It should be invaluable in the library to answer questions from students trying to complete classroom lessons. Indexed.

Grupe, Fritz H.
Microcomputer Applications: Using WordPerfect 5.0, Lotus 1-2-3, and dBASE III Plus
Dubuque, Iowa, Wm. C. Brown, 1990. 399p. ISBN 0697113183

This title takes a different approach by introducing basic microcomputer concepts, then word processing, spreadsheet, and database management options to give the reader an overview of the power of personal computers. Contains no index, but rather a detailed table of contents. Illustrated.

Guild, Stephen
Word for Windows 6: Self-Teaching Guide
New York, Wiley, 1994. 446p. ISBN 0471304670

A combination tutorial and guide to shortcuts, this title, like other Wiley self-teaching guides, is aimed at the computer student who learns better independently rather than in a classroom situation. Leads step-by-step through basic and then more advanced operations, teaching by practical example. Includes illustrations and index.

Hahn, Harley, and Rick Stout
The Internet Complete Reference
Berkeley, Calif., Osborne McGraw-Hill, 1994. 818p. ISBN 0078819806

A very good source of information about the Internet and ways to take advantage of it, including explanations and examples for e-mail, FTP, Telnet, Usenet, WAIS, and World Wide Web. The catalog of Internet resources gives a useful overview of types of material available. Appendices include a list of Usenet discussion groups. Indexed.

Hahn, Harley
The Internet Yellow Pages
Berkeley, Calif., Osborne McGraw-Hill, 1994. 447p. ISBN 0078820235

A hilarious and genuinely useful guide to the Internet. The main yellow pages arrangement is what makes this book unique: it covers everything by topic, rather than sorting by method of approach. Interspersed with the brief listings that show how to reach a subject by anonymous FTP or Usenet are excerpts from actual Internet files indicating what you can expect when you get to the file. Includes a list of Usenet newsgroups. Indexed.

Hanson, Larry
Everything You Wanted to Know About the Mac
2nd ed. Indianapolis, Ind., Hayden Books, 1993. 1232p.
ISBN 1568300581

If only one Macintosh book could be included in the collection, this should be it. A compilation of information and advice from experienced Mac users and well-known Mac writers, this book includes everything from history to hardware to Hypercard. Includes an extensive table of contents for quick browsing, a glossary, and an index.

Hayward, Tom
Adventures in Virtual Reality
Carmel, Ind., Que Corp., 1993. 258p. and computer disk
ISBN 1565292081

Introduces the concept of virtual reality in simple language, giving historical background, and explaining the complexities involved in managing perspective and feedback. Covers some current and near-future applications of virtual reality, and provides a sampler of 3-D programs on the disk included with the text. Indexed.

Held, Gilbert
Understanding Data Communications
4th ed. Carmel, Ind., Sams Publishing, 1994. 435p.
ISBN 0672305011

Covers computer communications systems in general, with discussions of terminals, modems, data transmission, and network design. Could be used to accompany a course or as background reading. Glossary, illustrations, and index included.

Heldman, Robert K.
Global Telecommunications: Layered Networks' Layered Services
New York, McGraw-Hill, 1992. 390p. ISBN 0070280304

Presents the complexities of worldwide telecommunications and networking, with quick overviews of specific products and lengthier discussions of possible future applications. Includes appendix, glossary, illustrations, and index.

Hershey, Gerald L., and Donna L. McAlister Kizzier
Planning and Implementing End-User Information Systems: Office and End-User Systems Management
Cincinnati, Ohio, South-Western Pub. Co., 1992. 807p. and computer disk ISBN 0538702192

Designed as an introduction to guidelines, techniques, and principles for planning and implementing end user and office information systems, this book uses a textbook format to give a wide-ranging overview suitable for beginners. Includes disk, glossary, illustrations, index, and references for each chapter.

Hixson, Amanda C.
Advanced Excel for the PC
Berkeley, Calif., Osborne McGraw-Hill, 1988. 479p.
ISBN 0078812739

This is not a book for beginners, but for spreadsheet users having basic familiarity with spreadsheets. Hixson, who has written a computer-based training program and several books, walks the reader through increasingly complex operations to increase productivity. Appendices, illustrations, and index included.

Huws, Ursula, Werner B. Korte, and Simon Robinson
Telework: Towards the Elusive Office
New York, Wiley, 1990. 276p. ISBN 0471922846

Telework, or telecommuting, the ability to perform one's job from a remote location using the power of telecommunications, is increasing in both popularity and practicality as computing power grows and costs shrink. This title offers a global overview of telecommuting, examining trends and identifying its impact on companies of various sizes. It is written from an analytical rather than an advocacy viewpoint. Includes appendices, bibliography, tables, and index.

Jamsa, Kris A.
DOS: The Complete Reference
4th ed. Berkeley, Calif., Osborne McGraw-Hill, 1993. 1121p.
ISBN 0078819040

Osborne's *Complete Reference* titles cover many different levels and aspects in one volume, and Jamsa's books in particular are virtually guaranteed to contain everything you need to know. Starting with an overview of a personal computer, Jamsa leads the novice gradually through the workings of DOS, with quick access for more experienced users who just need to brush up on specific commands. Includes sections on memory management, disk errors, and viruses. Appendices, illustrations, and index included.

Jamsa, Kris A.
Instant Multimedia for Windows 3.1
New York, Wiley, 1993. 390p. and 2 computer disks
ISBN 0471589721

Prolific computer author Jamsa guides the novice user through a brisk introduction to multimedia possibilities in Windows, using the disks to demonstrate. More a demonstration and sampler than a teaching tool, this title nevertheless provides a good introduction to multimedia products and software. Includes illustrations and index.

Jamsa, Kris A.
Microsoft C: Secrets, Shortcuts and Solutions
Redmond, Wash., Microsoft Press, 1989. 713p. ISBN 1556152035

Offers something for all levels, from rank beginner to expert, with well-delineated programming examples. Appendices include ASCII character set and a glossary. Indexed.

Jamsa, Kris A.
Simply DOS
2nd ed. Berkeley, Calif., Osborne McGraw-Hill, 1993. 215p.
ISBN 0078819148

This is the resource for the patron who walks into the library lamenting, "I don't know anything about computers and I'm taking a course." It explains the various parts of a computer and what they do, what software is and how to use it, and some of the basic DOS commands. This work should increase substantially the comfort level of the beginner. Covers up to DOS 6.0. Includes illustrations and index.

Jordan, Larry E., and Bruce Churchill
Communications and Networking for the IBM PC and Compatibles
4th ed. New York, Brady Publishing, 1992. 693p.
ISBN 0131577697

This fourth edition of a classic text covers the fundamentals of data communications, networking, and hardware and software in straightforward language with clear illustrations. Includes appendices and a detailed index.

Kent, Dorothy
AutoCAD Reference Guide: Everything You Wanted to Know About AutoCAD—Fast!
2nd ed. Gresham, Ore., New Riders Publishing, 1991. 304p.
ISBN 0934035024

As the title indicates, this is not a tutorial or a learning tool, but a reference manual for users with at least a minimum of experience. It explains commands, walks the reader through examples (clarified in many cases with screen diagrams), and gives cross-references to related commands. Covers AutoCAD Releases 10 and 11. Appendix, illustrations and index included.

Kochan, Stephen G., and Patrick H. Wood
Exploring the UNIX System
3rd ed. Carmel, Ind., Sams, 1992. 466p. ISBN 0672485168

Billing itself as "the complete beginner's guide to UNIX," this book aims to explain the UNIX operating system and offers enough commands and options to get the new user started. Includes appendices, illustrations, and index.

Kochan, Stephen G.
Programming in C
Rev. ed. Indianapolis, Ind., Hayden Books, 1988. 445p.
ISBN 067248420X

A good beginner book with a hands-on, teach-by-example approach that results in the reader being able to perform basic programming immediately. Meant as a teaching tool, not as a quick reference. Includes extensive appendices (including one on ANSI C) and an index.

Kochmer, Jonathan
The Internet Passport: NorthWestNet's Guide to Our World Online
4th ed. Bellevue, Wash., NorthWestNet, 1993. 515p.
ISBN 0963528106

One of the best organized and most comprehensive guides to the Internet, this title may deter users looking for quick results, but will be invaluable to librarians, teaching faculty, and students alike if they are willing to invest the time to follow directions. Appendices, index, glossary, and reading lists are provided.

Kraynak, Joe
The First Book of MS-DOS 6
Carmel, Ind., Alpha Books, 1993. 345p. ISBN 1567611249

Following the same format as other *First Book* titles, this one assumes absolutely no knowledge on the part of the reader. It begins with a basic introduction to computers, then continues with easily understood discussions of directories, file management, and DOS commands. Includes a section on programs for laptops. Could also be used by someone just looking for a quick review of the upgrades in DOS 6. Illustrations and index.

Kraynak, Joe
The First Book of Personal Computing
4th ed. Carmel, Ind., Alpha Books, 1993. 345p.
ISBN 1567612849

Designed as a tutorial for totally inexperienced users, this book explains the parts of a personal computer (IBM-PC and Macintosh) and the basic peripherals, walks the reader through turning on the computer, starting programs, exiting safely, and shutting down the system. Also gives an overview of word processing and database software. Appendices, glossary, illustrations, and index included.

Krol, Ed
The Whole Internet User's Guide and Catalog
2nd ed. Sebastopol, Calif., O'Reilly and Associates, 1994. 543p.
ISBN 1565920635

One of the earliest and best Internet guides, Krol's book provides a wonderful start to exploring the Internet, whether you have backup system help on campus or have to learn (and teach students) on your own. Here is an explanation of the Internet, a guide to e-mail, FTP and Telnet, and a partial catalog of Internet resources, all in usable format. Includes appendices, illustrations, and index.

Kroenke, David, and Richard Hatch
Business Information Systems: An Introduction
5th ed. New York, Mitchell McGraw-Hill, 1993. 516p.
ISBN 0070358710

This book provides an introduction to information systems from the viewpoint of the end user rather than the programmer. It covers a wide range of topics, from desktop applications to distributed processing systems. It even includes a useful "History of Information Systems" module, which presents information on the history and development of computers in a way that will thrill students doing term papers on that topic. Glossary, illustrations, and index included.

Kurzweil, Ray
The Age of Intelligent Machines
Cambridge, Mass., MIT Press, 1992. 565p. ISBN 0262610795

Beginning with the history of artificial intelligence and the development of the "thinking machine," and going on to future trends and possibilities, Kurzweil has compiled a cross between a reference work and a coffee-table book that gives insight into the reasons why computers were originally developed, and describes the transition from basic calculating machines to knowledge processors. Profusely illustrated, with an extensive bibliography and a chronology, this book is better suited to browsing than quick reference, but contains an enormous amount of information. Includes glossary and index.

Leigh, Ronald W.
AutoCAD: A Concise Guide To Commands And Features For Release 12
3rd ed. Chapel Hill, N.C., Ventana Press, 1993. 447p.
ISBN 1566040086

Intended as both a learning tool and a reference book, this work is designed to help new users master AutoCAD by progressing from the simple to the complex, covering terms, concepts, commands, and procedures. Specific commands can be looked up through either the very detailed table of contents or the index.

Louw, Eric, and Neil M. Duffy
Managing Computer Viruses
New York, Oxford Univ. Press, 1992. 171p. ISBN 0198539746

No coverage of computers in business is complete without information on computer viruses. This book has something for everyone, from senior systems people to business students who may some day encounter a virus "line" on their computer screens. Covers both prevention and recovery methods. Appendices, footnotes, and index provided.

Lu, Cary
The Apple Macintosh Book
4th ed. Redmond, Wash., Microsoft Press, 1992. 513p.
ISBN 1556152787

Possibly the best single book on the Macintosh, this covers not just Mac hardware and software, but also networking with IBM PCs and compatibles, Macs on local area networks, and multimedia applications. Includes appendices, glossary, and index.

Matthews, Martin
Excel for Windows: The Complete Reference
2nd ed. Berkeley, Calif., McGraw-Hill, 1994. 1302p.
ISBN 007881975X

This title begins with a standard introduction to *Excel* and what it can do in *Windows* that is thorough enough to stand alone as a self-teaching tool. It follows with a detailed reference section listing *Excel* commands and features and explains how to apply them correctly. One of the more useful appendices is a series of reproductions of toolbars, explaining their button functions and how to use them. Includes appendices, illustrations, and index.

Mayes, Kathleen L.
WordPerfect 5.1 for Windows
Danvers, Mass., Boyd and Fraser Pub. Co., 1993. 656p.
ISBN 0878359664

Although designed for classroom teaching, this can also be used as a self-teaching workbook for individuals needing to master quickly basic and intermediate levels of *WordPerfect 5.1 for Windows*. Uses a combination of explanations and sample exercises to ensure active learning. Appendices and index included.

Miller, Kate
The First Book of WordPerfect 6
Carmel, Ind., Alpha Books, 1993. 359p. ISBN 1567610226

Suitable for either quick reference or cover-to-cover self-study, this and other *First Book* series titles are recommended because they provide a dependable, basic approach to learning software, presented in a predictable format. Stresses the importance of keeping up-to-date-with new versions as they are released questions. The same author has done a *Windows* version of this title. Includes appendix, illustrations, and index.

Mueller, Scott
Upgrading and Repairing PCs
3rd ed. Carmel, Ind., Que Corp., 1993. 1254p. and computer disk
ISBN 156529467X

Presents both basic and advanced information on the internals of PCs, available add-ons, and upgrades. The clear explanations of concepts from memory boards to ROM BIOS are very helpful for the concerned PC owner or the confused student struggling with topics taught in class. Accompying workbook and disk are useful as teaching tools as well as for self-study. Includes appendices, illustrations, and index.

Murray, Katherine
Introduction to Personal Computers
4th ed. Carmel, Ind., Que Corp., 1993. 507p. ISBN 1565292758

Another good title to recommend to nervous students who "don't know anything about computers," and want information about personal computers. Starting with the basic "What is a computer?" question, this book covers the internals, externals, and peripherals of IBM, Macintosh, and Amiga personal computers, as well as software and applications. Glossary, illustrations, and index included.

Nance, Barry
Introduction to Networking
2nd ed. Carmel, Ind., Que Corp., 1993. 422p. and computer disk
ISBN 1565292979

Introduces concepts, components, and communications and provides a good general introduction for beginners who are just starting to learn about networking, or who must work with a network and feel the need for background information. Includes software disk, appendices, illustrations, and index.

Norton, Peter
Outside the IBM PC and PS/2: Access to New Technology
New York, Brady Publishing, 1992. 500p. ISBN 0136435866

Norton addresses some of the options for expansion of the conventional personal computer: graphics cards, memory expansion, modems, monitors, printers, and networking. Specific product information will go out of date quickly, but the overviews of communications, graphics, networking, and security options are worth having. Indexed.

Norton, Peter
Peter Norton's Inside the PC
5th ed. New York, Brady Publishing, 1993. 610p.
ISBN 1566860970

Written by one of the gurus of personal computing, this book explains what is going on inside a personal computer, and how to go beyond the obvious to get the most out of your equipment. Aside from being fascinating reading for those who are less than experts, this book contains an extensive index that aids in quickly locating information on terms and acronyms covered here in more depth than in a standard computer dictionary. Appendices, illustrations, and index included.

Nunemacher, Greg
LAN Primer: The Definitive Guide to Networking Fundamentals
2nd ed. San Mateo, Calif., M&T Books, 1992. 345p.
ISBN 155851287X

Excellent basic coverage of generic network components, specific operating systems and architectures, networking software, and interconnectivity. Includes glossary, illustrations, and index.

O'Day, Kate, and others
Understanding MS-DOS
2nd ed. Carmel, Ind., Sams, 1989. 361p. ISBN 0672272989

This self-teaching starter book assumes no prior exposure to

MS-DOS. It highlights key topics, including basic commands, directories, file management, DOS shell, and system configuration. Appendices, illustrations, and an extensive index are included.

O'Hara, Shelley
Easy DOS
3rd ed. Indianapolis, Ind., Que Corp., 1993. 226p.
ISBN 1565296400

Absolute beginner-level text based on MS-DOS 6.2, but usable with any version of MS-DOS. The author uses a nonthreatening teaching approach: simple explanations, basic examples, and perhaps best of all, quick tips on how to "undo" mistakes and avoid future mishaps. Includes glossary, illustrations, and index.

Orman, Levent V.
Elements of Information Systems: Components and Architecture
New York, Macmillan, 1991. 261p. ISBN 002389475X

A clear and concise introduction to the components of an information system, aimed at readers who already understand some basic programming concepts, and who will be involved in running or using information systems. Includes a brief bibliography, illustrations, and an index.

Pandit, Milind S.
How Computers Really Work
Berkeley, Calif., Osborne McGraw-Hill, 1993. 383p.
ISBN 0078819369

Starting with transistors and chips, this book explains the parts of a computer: what they are, what they do, and how they do it. The book does not bog down in jargon or focus on specific brands. Try this work on the next patron who walks in and remarks pathetically, "I don't know anything about computers!" Includes glossary, illustrations, and index.

Parker, Dana, and Bob Starrett
Technology Edge: A Guide to CD-ROM
Carmel, Ind., New Riders Publishing, 1992. 426p.
ISBN 1562050907

Explains CD-ROM technology, the types of disks, and how to install a CD-ROM drive, and introduces CD-ROM multimedia. Lists hardware manufacturers, software sources, and consulting services for CD-ROM projects. Includes appendices, illustrations, and index.

Pelton, Gordon E.
Voice Processing
New York, McGraw-Hill, 1993. 396p. ISBN 007049309X

Concentrating on voice processing hardware and software, this book covers the basic concepts of voice processing and the past history of attempts to develop successful systems, and gives an overview of the features of existing systems. Includes appendices, bibliography, illustrations, and index.

Perry, Greg M.
Absolute Beginner's Guide to C
Carmel, Ind., Sams, 1993. 309p. ISBN 0672303418

This very informal approach to teaching programming guarantees that an absolute beginner can learn to program in C. It is as nonthreatening as possible, and should be a hit with reluctant computer users. Appendices, illustrations, and index included.

Perry, Greg M.
Absolute Beginner's Guide to Programming
Carmel, Ind., Sams Publishing, 1993. 448p. ISBN 0672302691

The first half of this book is an introduction to computers and a history of programming languages up to the present. The second half is an introduction to programming concepts using QBASIC to demonstrate the logical thought processes needed to create and run a successful program. Appendices, glossary, illustrations, and index provided.

Rheingold, Howard
Virtual Reality
New York, Simon and Schuster, 1991. 415p. ISBN 0671778978

Presents the history, status, and some of the potential future uses for virtual reality, emphasizing the serious, the scientific, and the scholarly without in any way diminishing the fantastic nature of progress being made. Index and references included.

Ribar, L. John
The Programming Primer: A Guide to Programming Fundamentals
Berkeley, Calif., Osborne McGraw-Hill, 1994. 281p.
ISBN 0078819997

Presents the logic of programming in general on a very introductory level, starting with the basics of computer operations. Good reading for anyone interested in understanding the programming function without needing to master a specific language, or for someone contemplating taking programming who needs to know what is involved. Includes appendices, bibliography, glossary, illustrations, and index.

Rose, Carla
Mac Online! Making the Connection
Blue Ridge Summit, Pa., Windcrest/McGraw-Hill, 1993. 428p.
ISBN 0830642544

Explains what is needed for dial-up access with a Macintosh, and how to use it for everything from e-mail to online database searching and faxing. Using reproductions of actual screens to aid comprehension, this work is written at a level suitable for beginners, but provides tips for the more experienced. Includes glossary and index.

Rose, Jay, and Carla Rose
The First Book of the Mac
2nd ed. Carmel, Ind., Alpha Books, 1992. 446p.
ISBN 0672274183

Clearly written and profusely illustrated, this work covers basic Mac activities (graphics, spreadsheets, and desktop publishing). Good for beginners, or for experienced PC users who need to switch platforms. Glossary, illustrations, and index included.

Rosenborg, Victoria, and others
A Guide to Multimedia
Carmel, Ind., New Riders Publishing, 1993. 592p. and computer disk ISBN 1562050826

This book and disk set is intended for fairly intensive individual work rather than quick reference, although the introductory chapter on multimedia gives a useful overview. Includes appendices, brief glossary, illustrations, and index. (A *Technology Edge* title.)

Russel, Charlie
Murphy's Laws of DOS
2nd ed. San Francisco, Calif., SYBEX, 1993. 356p.
ISBN 0782114245

Despite the title, this book is filled with genuinely useful information, from the basic to the obscure. Good-to-know shortcuts appear on nearly every page. This work would be useful to those who need to learn more, but hate to read manuals. The second edition includes DOS version 6.2. Glossary, illustrations, and extensive index included.

Sanders, Rosanne Bryce
Administrative Procedures in the Electronic Office
Englewood Cliffs, N.J., Prentice Hall, 1991. 436p.
ISBN 0130194727

Presents basic explanations of word processing, telecommunications, records management, information retrieval, and meeting organization skills in an electronic setting. Most chapters include brief bibliographies and glossaries. Intended for introductory level only. Indexed.

Schatt, Stanley
Understanding Local Area Networks
4th ed. Indianapolis, Ind., Sams Publishing, 1993. 325p.
ISBN 0672303825

A beginning-level presentation of both the theory and technology of local area networks (LANs). Explains basic LAN components and cabling, and discusses LAN software for both PC compatibles and the Macintosh. Includes bibliography, glossary, and index.

Schildt, Herbert
Mostly Windows with Just Enough DOS
Berkeley, Calif., Osborne McGraw-Hill, 1994. 421p.
ISBN 0078819768

The title of this work, though interesting, does the book a disservice because it does not adequately depict the breadth of coverage. This book presents all the fundamentals a complete novice would need to start up and run a personal computer for the first time, yet also provides an overview of *Windows* and DOS and how to integrate their use. Organized for quick access, this work is great for the busy reader who wants all the information in one place. Includes illustrations and index.

Schulman, Mark
Introduction to UNIX
Carmel, Ind., Que Corp., 1992. 458p. ISBN 0880227451

Whether dealing with students who need to learn how UNIX works, or with frustrated e-mail users who need help with UNIX commands and editing, this book will provide a good start. It is neither a tutorial nor a textbook, but gives an overview of the UNIX operating system, a summary of UNIX text processing options, and a list of UNIX commands. Appendices, glossary, and index also included.

Schutzer, Daniel
Business Decisions with Computers: New Trends on Technology
New York, Van Nostrand Reinhold, 1991. 230p.
ISBN 0442318790

An introductory-level discussion of computers and business for students with little or no experience with either. Includes a brief overview of basic concepts such as financial statements and ratios, along with explanations of databases, networks, and systems. Intended for beginners only. Includes bibliography, illustrations, and index.

Sharp, Craig W.
AutoCAD Quick Reference
3rd ed. Carmel, Ind., Que Corp., 1992. 167p. ISBN 1565290240

This title lacks the diagrams and examples included in Dorothy Kent's *AutoCAD Reference Guide* (New Riders, 1991), but would still be useful to experienced users who just need to look something up quickly. It consists of a simple alphabetical listing of commands. For each command there is a brief definition, a display of proper syntax, and an explanation of how to use the command. Indexed.

Sheldon, Thomas
Windows 3.1: The Complete Reference
2nd ed. Berkeley, Calif., Osborne McGraw-Hill, 1993. 826p.
ISBN 0078818893

Sheldon, who has written many books and articles on computer technology and software, favors a hands-on approach that allows beginners to start using the software immediately, so that they can be productive while learning the more advanced topics. This work provides something for everyone, from novice to experienced user looking for shortcuts. Includes appendices, illustrations, and index.

Simpson, Alan, and Paul Lichtman
The First Book of Lotus 1-2-3, Release 2.4
3rd ed. Revised by Jennifer Flynn. Carmel, Ind., Sams Publishing, 1992. 452p. ISBN 0672274140

Follows the Sams *First Book* format which is designed to familiarize the beginner quickly with the essentials needed to start using the software. Liberal use of graphics recreating screen displays make the text easy to follow. Includes appendix, illustrations, and index.

Solvberge, Arne, and Chen-Ho Kung
Information Systems Engineering: An Introduction
New York, Springer-Verlag, 1993. 540p. ISBN 0387563105

Provides a basic introduction to information systems: their development, evolution, and maintenance. Recommended because it gives a good overview without requiring advanced knowledge of mathematics or programming. Includes bibliography, illustrations, and index.

Stallings, William
The Business Guide to Local Area Networks
Carmel, Ind., Sams Publishing, 1990. 365p. ISBN 0672227282

Stallings takes a business approach to local area networks, discussing market forces and business issues as well as configurations and operating systems. Also covers security and LAN management. Includes illustrations and index.

Stephenson, Peter
Introduction to Personal Computers: Self-Teaching Guide
New York, Wiley, 1991. 255p. ISBN 047154714X

Beginner-level orientation to the concepts and physical components of personal computers such as simple commands, files and directories, and basic applications, including telecommunications.

Meant less for quick reference than for easy reading and study, this title will familiarize the total novice with the basic concepts needed to get started. Includes illustrations and index.

Tanenbaum, Andrew S.
Computer Networks
2nd ed. Englewood Cliffs, N.J., Prentice Hall, 1989. 658p. ISBN 013162959X

The coverage of computer networks and protocols in this book is on a fairly sophisticated level, making it more useful as a reference manual. For the truly dedicated, annotated reading lists are provided for each section, and are tied to the extensive bibliography. Also includes appendices and an index.

Tennant, Roy, John Ober, and Anne Grodzius Lipow
Crossing the Internet Threshold: An Instructional Handbook
2nd ed. Berkeley, Calif., Library Solutions Press, 1994. 134p. ISBN 1882208072

Designed as a self-teaching tool and an instructional aid for librarians and information professionals, this book contains explanations, hands-on practice exercises, and even sample pages to be used for overheads in a training presentation. Highly recommended for anyone who teaches others how to use the Internet. Appendix and index included.

Tolhurst, William A., Mary Ann Pike, and Keith A. Blanton
Using the Internet
Special ed. Indianapolis, Ind., Que Corp., 1994. 1188p. and computer disk ISBN 1565293533

Somewhat more academic in coverage than most Internet books, this one teaches not only the basics of use, but also the background behind many Internet phenomena. It also includes a list of Usenet newsgroups, an index to RFCs (requests for comments), and an explanation of basic UNIX commands. Appendices, floppy disk, glossary, illustrations, and index are included.

Townsend, James J.
Introduction to Databases
Carmel, Ind., Que Corp., 1992. 359p. ISBN 0880228407

A generic introduction to databases in general, both flat file and relational, this title explains the theory of how to plan, organize, and report data, regardless of which software is chosen. By concentrating on the logic behind the use of databases, Townsend covers the concepts without tying them to a single product. There is also a good section on project analysis and conducting user interviews, which touches on skills many students will need later. Includes bibliography, glossary, illustrations, and the index.

Using 1-2-3 Release 4 for Windows
Special ed. Carmel, Ind., Que Corp., 1993. 1105p. ISBN 1565290054

Presents a clear, comprehensive, and well-organized guide to *Lotus 1-2-3 for Windows*, to be used as a self-teaching tool, a problem solver, or a reference for experts looking for shortcuts and improvements. Includes pullout charts for quick reference, appendices, illustrations, and index.

Van Buren, Chris
The First Book of Excel 5.0 for Windows
Indianapolis, Ind., Alpha Books, 1993. 322p. ISBN 1567613209

Valuable as a reliable backup for finding quick answers to how-to questions, or a self-paced tutorial for even the most inexperienced user. Interspersed throughout with screen representations to help clarify the text, the book goes from the basics to cosmetic improvements and shortcuts. Includes appendix, illustrations, and index.

Van Wyk, Christopher J.
Data Structures and C Programs
Reading, Mass., Addison-Wesley, 1990. 387p. ISBN 0201539853

The first part of this book presents fundamental ideas such as algorithms, data types, data structures, and complexity, which form a base for the ensuing material on searching, sorting, and problem-solving techniques. The coverage ends with advanced material on graphs and graph algorithms. Diagrams, index, and references provided.

Walkowski, Debbie
The First Book of Microsoft Works for the PC
Carmel, Ind., Sams Publishing, 1991. 286p. ISBN 0672273608

Another ideal starter book. It assumes no prior knowledge and could be used successfully by someone inexperienced with personal computers, or by an experienced user looking for a fast and efficient introduction to *Works*. Profusely illustrated, with a detailed index and convenient access to shortcuts on the inside covers.

Watkins, Paul R., and Lance B. Eliot, eds.
Expert Systems in Business and Finance: Issues and Applications
New York, Wiley, 1993. 367p. ISBN 0471919675

Goes beyond textbook coverage, using case studies to demonstrate applications of expert systems, including modeling of human judgment, validation issues, security and legal issues, and resource planning. Includes bibliography, illustrations, and index.

Wolfe, David
The BBS Construction Kit: All the Software and Expert Advice You Need to Start Your Own BBS Today!
New York, Wiley, 1994. 373p. and computer disk
ISBN 0471007978

No computer-related collection would be complete without a reference on the BBS (bulletin board) phenomenon, which is one of today's most popular computer applications. Written by a successful BBS operator, this title covers hardware, software, modems, legal ramifications, and virus protection, among other pertinent topics. This work is designed for the interested computer user, or the student who wants a really good independent study project. Includes disk, illustrations, and index.

Wolverton, Van
Running MS-DOS: Covers Version 6.0
6th ed. Redmond, Wash., Microsoft Press, 1993. 613p. ISBN 1556155425

The latest edition of this straightforward introduction to MS-DOS and personal computers assumes no prior knowledge on the part of the reader. Starting with disk drives and operating systems, it progresses using concise explanations and easy-to-follow examples. Peripherals to file management, batch files, and configuration commands are also addressed. Includes an MS-DOS command listing and a glossary.

Wyatt, Allen
Using Assembly Language
3rd ed. Carmel, Ind., Que Corp., 1992. 1126p. ISBN 0880228849

This book is intended for computer users with at least a basic grasp of programming, and who are interested in using Assembly Language for subroutines. Recommended because it starts from the very beginning, teaching fundamentals before going into more sophisticated programming. Includes an overview of Borland and Microsoft Assembly Language products, appendices, glossary, and index.

Wyatt, Allen
WordPerfect 6 for Windows: The Complete Reference
Berkeley, Calif., Osborne McGraw-Hill, 1994. 807p.
ISBN 0078819261

This work is for those who prefer to keep a printed resource at the workstation rather than rely on help screens. Starting with the basics, it then leads the reader to more expert levels. Assuming that guidance may be needed even for installing programs, Wyatt discusses installation and basic operations before going on to more complex maneuvers. Aappendices, illustrations, and index included.

Wyatt, Allen, Jim Sheldon, and Steven Namaroff
WordPerfect 6: The Complete Reference
Berkeley, Calif., Osborne McGraw-Hill, 1993. 981p.
ISBN 0078819016

This book could be useful to someone using a word processor for the first time or switching to a new program or version. It can be used as a step-by-step manual or as a quick reference tool. Having the latest edition of this and similar Osborne titles available will provide backup help for puzzled students and nervous beginners. Includes appendix, illustrations, and an index.

Cosmetology

Barbara A. Heiffner
Nicolet Area Technical College
Rhinelander, Wisconsin

Introduction

Cosmetology and barbering programs provide training for a broad range of services related to grooming and personal appearance. Students have the opportunity to develop proficiency in the following: latest hair styles, scalp and hair conditioning, hair color, permanent waves, facials, makeup, nail sculpture, and manicures, as well as simple shampoos and haircuts.

These services are often performed by cosmetologists, who are also called beauty operators, hairstylists, hair designers, beauticians, cosmeticians, or beauty specialists. In small salons, cosmetologists may perform all basic services, while in larger salons cosmetologists may specialize in one or more fields. Specialized workers include manicurists, makeup artists, electrologists, and estheticians.

A growing number of barbers work in unisex salons and cut and style women's hair. Some states require a cosmetologist's license as well as a barber's license to permanent wave or color hair. In other states, barbers are licensed to perform all the duties of cosmetologists, except skin care and nail treatment.

Educational requirements vary from state to state. Both public and private vocational schools offer barber and cosmetology training that lasts from six to fifteen months. The training program usually includes classroom study, demonstrations, and practice on mannequins and clients.

All states require barbers and cosmetologists to be licensed, but state requirements vary. Generally, applicants must be at least sixteen years old, pass a physical examination, and have graduated from a state-approved cosmetology or barbering program. They must pass written, practical, and sometimes oral tests. Some, but not all, states have reciprocity agreements for licensed cosmetologists and barbers.

Accreditation and Certification

Most states have a state cosmetology board that sets requirements for schools, salons, and individual cosmetologists. Contact the state board of barber examiners or the state board of cosmetology for information. Information about cosmetology and state licensing requirements is available from:

National Cosmetology Association
3510 Olive Street
St. Louis, Mo. 63103

A list of licensed training schools and licensing requirements is also available from:

National Accrediting Commission of Cosmetology Arts and Sciences
901 N. Stuart St., No. 900
Arlington, Va. 22203

Association of Accredited Cosmetology Schools
5201 Leesburg Pike
Falls Church, Va. 22041

Selected Journals

American Salon.
National Hairdressers and Cosmetologists Association. Monthly. ISSN 0741-5737

Beauty Education.
Milady Pub. Co. Monthly. ISSN 1052-4169

Modern Salon.
Vance Pub. Co. Monthly. ISSN 0148-4001

Nails.
Peter Grimes Productions. Monthly. ISSN 0896-193X

Print Materials

Ayala, Emma
Styling Competition: A Guide to Winning Technique
Bronx, N.Y., Milady Pub. Co., 1989. 75p. ISBN 0873503597

Competition opportunities, benefits, and attitudes are briefly explored before introducing the categories and types of competition. There are many illustrations, examples, and tips for the student who is interested in competition hairstyling.

Baran, R., R. P. R. Dawber, and G. M. Levene
Color Atlas of the Hair, Scalp, and Nails
St. Louis, Mo., Mosby, 1991. 192p. ISBN 0815104170

Numerous color photographs illustrate the majority of common hair, scalp, and nail disorders. Attention is given to normal appearances, variations of normal with aging, genetic disorders, and the effects of environment, occupations, and cosmetics. This book will be of value to cosmeticians and other professionals who deal with aspects of hair and nail care. Index.

Bergant, Kathleen Ann
Communication Skills for Cosmetologists
Albany, N.Y., Milady Pub. Co., 1993. 188p. ISBN 1562530879

Emphasizes the importance of communication skills for success in the field of cosmetology. Nonverbal communication, good listening skills, telephone techniques, résumé and cover letter preparation, skills for a job interview, and dealing with difficult people are some of the topics this book addresses.

Bigan, Tammy
Nail Art and Design
Albany, N.Y., Milady Pub. Co., 1994. 90p. ISBN 1562531182

Provides detailed descriptions and instructions for many different types of nail designs. Organized in three parts: supplies and beginning steps; creating nail designs; and promoting and selling nail art designs. Includes lists of supplies and suggestions on where to get them. Many full color photographs illustrate the designs. Index.

Bigan, Tammy
Technails: Extensions, Wraps, and Nail Art
Albany, N.Y., Milady Pub. Co., 1992. 109p. ISBN 0873503821

Techniques are given for a variety of nail care services. In addition to manicures, this work provides information about nail extensions and different supplies and procedures used to reinforce or strengthen nails. Includes tips for working with clients and information on starting a business.

Charest-Papagno, Noella
Daralism: Phases of Hair Care for the Bedridden and Disabled
Hollywood, Fla., JJ Publishing, 1994. 1 v. (various pagings) ISBN 096046106X

Daralism is a method of hair care for bedridden or disabled persons that is more efficient and pleasurable than traditional methods. This book explains not only the Daralism procedures, but it also discusses the interrelationship of hair care and self-esteem.

Charest-Papagno, Noella
Desairology: Hairstyling of the Deceased
Hollywood, Fla., JJ Publishing, 1991. 105p. ISBN 0960461051

The first chapter discusses funeral homes and what a desairologist can expect in working conditions. Other chapters present specific hairdressing skills, techniques, and guidelines. The final section attempts to answer questions and dispel some of the misgivings a hairdresser may have about working with a decedent.

Chesky, Sheldon R., Isabel Cristina, and Richard B. Rosenberg
Playing It Safe: Milady's Guide to Decontamination, Sterilization, and Personal Protection
Albany, N.Y., Milady Pub. Co., 1994. 130p. ISBN 1562531794

This resource explains the microbiology involved in contamination and details measures salon professionals should take to protect themselves and their clients. Includes compliance measures for federal agencies and state licensing boards, questions and answers regarding AIDS, and information on procedures for decontaminating materials. Glossary and bibliography.

Colletti, Anthony B.
Competency in Cosmetology: A Professional Text
3rd ed. New York, Keystone Publications, 1990. 607p. ISBN 0912126701

This book is written for students with entry-level skills. Chapters may be explored in any order and are designed to be used as a competency-based, self-contained unit of study. Includes general and task indexes, several charts, and reference tables.

Cosmetologist
Syosset, N.Y., National Learning Corp., 1991. 1 v. (various pagings) ISBN 0837322510

The first half of the book is devoted to practice examination questions and answers. The second part contains hairdressing and cosmetology licensing and practice regulations for the state of New York. The final section is a lengthy glossary of cosmetology terms.

Cosmetologists State Board Exam Review: Typical State Board Examination Questions
Albany, N.Y., Milady Pub. Co., 1990. 124p. ISBN 0873503651

This exam review contains the types of questions most frequently used by state boards of cosmetology and by national testing agencies. The entire book is composed of representative multiple-choice questions covering all aspects of cosmetology.

Cozens, Bronwyn
Milady's Color Crazy: The Complete Color Guide for Cosmetologists
Albany, N.Y., Milady Pub. Co., 1992. 161p. ISBN 1562530348

This handbook for cosmetologists who want to develop their skills as hair colorists covers topics such as becoming a hair colorist, simple color theory, color design, and information on updating skills. It is written in nontechnical language and contains illustrations, information, and exercises for all experience levels. Glossary and index.

Cozens, Bronwyn
Science in Your Salon
Albany, N.Y., Milady Pub. Co., 1992. 214p. ISBN 1562530356

Part One addresses the nature of hair; Part Two explains how hair is affected by the services offered in the salon. Provides cosmetologists and their clients with answers about salon products and services. Includes experiments, activities, discussion topics, glossary, and index.

Cutting, P., R. Ross, and R. Hill
Hairdressing: Theory, Science, and Practice
London, Pitman Publishing, 1991. 410p. ISBN 027303457X

The first section of this London-based publication, which deals with the salon, includes some topics not applicable in the United States. However, a number of sections on hair and skin, action and reactions, and more advanced work would be useful to hairdressers anywhere. Many photographs and illustrations; explanations and procedures are listed in step-by-step format. Glossary and index.

Dalton, John W.
The Professional Cosmetologist
4th ed. St. Paul, Minn., West Pub. Co., 1992. 741p.
ISBN 0314730427

The revised and updated fourth edition covers current topics such as prevention of skin cancer, and AIDS in the workplace; there is a special section on ecology. This is a comprehensive text with glossary and subject indexes. Instructor's manual also available.

Dalton, John W.
State Board Review Questions to Accompany the Professional Cosmetologist
4th ed. St. Paul, Minn., West Pub. Co., 1992. 235p.
ISBN 0314007865

Includes over seventy-five practice tests which follow testing guidelines set by the Educational Testing Service, the National Interstate Council of State Boards of Cosmetology, individual state boards, and state departments of education for technical and vocational training.

Edgerton, Leslie
Managing Your Business: Milady's Guide to the Salon
Albany, N.Y., Milady Pub. Co., 1994. 212p. ISBN 1562530844

This blueprint for opening and running a successful salon addresses formulating a business plan, financing, hiring and training personnel, pricing strategy, purchasing, and other essential salon operations. Appendices list agencies that assist small business owners, as well as references of interest to salon managers. Index.

Edgerton, Leslie
You and Your Clients: Human Relations for Cosmetology
Albany, N.Y., Milady Pub. Co., 1992. 161p. ISBN 1562530585

Stressing the importance of proper communication between the client and stylist, this work suggests concrete methods for improving communication skills. Topics include consultations, overselling, and the issues related to encouraging clients to use professional products.

Gearhart, Susan Wood
Opportunities in Beauty Culture Careers
Lincolnwood, Ill., VGM Career Horizons, 1989. 139p.
ISBN 0844265195

Topics include an overview of the field, the employment outlook, possibilities for career advancement, educational requirements, and salary opportunities. Appendices include a bibliography, standards for accreditation, curricula available in accredited schools, and addresses of state boards of cosmetology.

Harper, Victoria
Professional by Choice: Milady's Career Development Guide
Albany, N.Y., Milady Pub. Co., 1994. 143p. ISBN 1562531484

Focuses on career options and planning, résumé writing, interviewing skills, and useful tips on job performance and attitude. The book also addresses the importance of professionalism in dealing with clients and associates. Index.

Healy, Mary
Regents/Prentice Hall Textbook of Cosmetology
3rd ed. Englewood Cliffs, N.J., Regents/Prentice Hall, 1993. 543p.
ISBN 0136900097

A comprehensive text covering skills required for a successful career in cosmetology. In addition to chapters on actual techniques, other chapters treat anatomy and physiology, bacteriology and infection control, chemistry for cosmetologists, electricity, and emergency procedures. Each chapter includes a list of goals, a summary, questions, and a list of projects. All illustrations and photographs are in color. Glossary and index.

Jayne, Charlotte
Hair Additions: The Fourth Dimension
Tarrytown, N.Y., Milady Pub. Co., 1991. 259p.
ISBN 0873503902

Covers areas important to all forms of hair additions: extensions, add-ons, and replacements, with an emphasis on hair extension services. In addition to explanations of the techniques, several case studies detail methods and step-by-step procedures.

Johnson, Barbara L.
Careers in Beauty Culture
New York, Rosen Pub. Group, 1989. 101p. ISBN 0823910024

Examples and success stories from cosmetologists are used to convey the different aspects of the field of beauty culture. Topics include desirable personality traits, educational requirements, advantages and drawbacks of a career in beauty culture, how to find a job, and self-employment. A bibliography and a dictionary of terms are included.

Marcie, Dorothy Anne, and Janet Mittelsteadt
You Can Double Your Income and Build Clientele as a Stylist
St. Paul, Minn., MarSuccess Books, 1989. 111p.
ISBN 0962362840

The main focus of this book is effective communication techniques and their importance to hairdressers and salon operators. Suggestions for staff meetings, worksheets, and numerous examples are presented in a nontechnical manner.

Milady's Art and Science of Nail Technology
Rev. ed. Albany, N.Y., Milady Pub. Co., 1992. 224p.
ISBN 1562531174

This revision of *The Art and Science of Manicuring* (Milady, 1986) contains substantially new material. All illustrations are in full color and show the latest technology in the application of artificial nails and the creation of original designs on nails. Glossary and index.

Milady's Standard Textbook of Cosmetology
Rev. ed. Albany, N.Y., Milady Pub. Co., 1995. 634p.
ISBN 1562532006

Many colorful photographs and illustrations accompany the text. Covers all aspects of cosmetology, from a brief exploration of career opportunities to styling techniques, theory of massage, skin disorders, and opening and operating a salon. Includes glossary and index.

Padgett, Mark
A Contemporary Approach to Permanent Waving: A Study of Creative and Technical Elements in Perming Today
Albany, N.Y., Milady Pub. Co., 1994. 142p. ISBN 1562531018

Using the latest products and techniques, Padgett teaches how to assess clients' needs and enhance the natural qualities of hair, and offers post-perm regimens for maximum client satisfaction and loyalty. Part One addresses the fundamentals of perming; Part Two discusses creative perming. Glossary and index.

Palladino, Leo, and June Hunt
The Nail File
London, Macmillan, 1992. 154p. ISBN 0333525841

This is a comprehensive and up-to-date guide on all aspects of the manicure. It deals with established manicure techniques as well as the latest technology in nail extensions, false nails, and nail repair. The second part of the book explains the physiology of nails and nail disorders and diseases, and provides a question- and- answer review.

Patterson, Marva
Braiding
Albany, N.Y., Milady Pub. Co., 1992. 120p. ISBN 0873503864

Explains and illustrates how hair braid extensions can be applied in both visible and invisible braids for males and females with different textures of hair. Techniques are shown for treating chemically damaged hair, adding additional hair to achieve a particular style, providing for clients' comfort during braiding, and practicing to acquire good finger dexterity.

Place, Stan
The Art and Science of Professional Makeup
Bronx, N.Y., Milady Pub. Co., 1989. 298p. ISBN 0873503619

Six parts comprise this book: career options; cosmetics and the skin; understanding light and color; anatomy and makeup techniques; special makeup techniques; and business management skills. Includes diagrams, illustrations, glossary, and index.

Poignard, Renee
Waxing Made Easy: A Step-By-Step Guide
Albany, N.Y., Milady Pub. Co., 1994. 108p. ISBN 1562531719

This guide covers all aspects of the depilatory waxing process, from materials and procedures to marketing waxing services to clients. Although the manual is designed for those new to epilation, experienced epilators may find new methods and remedies for existing problems.

Rudman, Jack
New Rudman's Questions and Answers on the OCE Occupational Competency Examination in Cosmetology
Syosset, N.Y., National Learning Corp., 1992. 1 v. (various pagings)

Occupational competency examinations are intended for experienced, skilled workers who need objective evidence of competency to secure teacher certification or to obtain academic credit. The examination sections contain hundreds of sample questions and answers. The second section contains a programmed course of instruction in hairdressing and cosmetology.

Scali-Sheahan, Maura T.
Milady's Standard Textbook of Professional Barber-Styling
Rev. ed. Albany, N.Y., Milady Pub. Co., 1993. 640p.
ISBN 1562531042

This edition contains material that has been completely revised and brought up-to-date with respect to both the theory and the practice of barber-styling. Covers all aspects from history through shop management. Includes glossary and index.

Simmons, John V.
Science and the Beauty Business, Vol. 1: The Science of Cosmetics
Basingstoke, England, Macmillan Education, 1989. 247p.
ISBN 0333438450

A rather technical examination of the scientific principles of the beauty industry including the theoretical and practical aspects of the formulation of cosmetics and toiletries; skin and skin care products; hair and hair care products; perfumery, product safety, and the packaging of products. Index.

Simmons, John V.
Science and the Beauty Business, Vol. 2: The Beauty Salon and Its Equipment
London, Macmillan Education Ltd., 1989. 235p.
ISBN 0333438418

A rather technical examination of electricity and electrical equipment used in a salon; mechanical massage and electrotherapy equipment; heat rays and ultraviolet therapy; hydrotherapy and the mechanical principles in movement and exercise; and finally, salon working conditions and services associated with heating, ventilation, and lighting. Indexed.

Sollock, Tom
Corrective Haircoloring: A Hands-On Approach
Albany, N.Y., Milady Pub. Co., 1993. 106p. ISBN 1562530836

This book provides an overview of the current developments and education necessary to become an expert corrective colorist. Explanations and directions for experiments are given for basic color theory, the level system, hydrogen peroxide uses, hair color analysis, corrective color solutions, and lighteners and bleach. Includes glossary.

Taylor, Pamela
Milady's Makeup Techniques
Albany, N.Y., Milady Pub. Co., 1994. 164p. ISBN 1562531425

The clear text and step-by-step photographs reveal the world of professional makeup using a variety of live models of different ethnic types. The book is designed as an educational tool and provides detailed and technical information needed for makeup artistry.

Ventura, Judy
Milady's Salon Receptionist Training
Albany, N.Y., Milady Pub. Co., 1993. 97p. ISBN 1562530445

Designed to provide readers with proper reception procedures and techniques. It is especially tailored for students who wish to become beauty salon receptionists, but it is useful for those desiring to become receptionists in any service-related business.

Wiggins, Joanne L.
Milady's Guide to Owning and Operating a Nail Salon
Albany, N.Y., Milady Pub. Co., 1994. 151p. ISBN 1562532014

A concise, authoritative explanation of the principles and practices of salon ownership based on the author's personal experience. Step-by-step guidelines are included, as well as lists of associations, state boards, and offices that offer additional business assistance. Index.

Nonprint Materials

The A to Z's of Clipper Cutting: Everything You Have Ever Wanted to Know About Clipper Cutting
East Lansing, Mich., Jim Jones Enterprises, 1993. 1 videocassette (60 min.) VHS

Jim Jones demonstrates the fundamentals and techniques of clipper cutting. Includes clipper-over-comb techniques and complete instructions for "the fade" haircut.

Basic Clipper Techniques, Vol. 1
Racine, Wisc., Andis Co., 1989. 1 videocassette (28 min.) VHS

Demonstrates the proper use and care of clippers and trimmers and ways to utilize these tools in creating the latest haircuts.

Braiding Beautiful: Basic and Advanced Techniques
Hialeah, Fla., Good Life Products, Inc., 1990. 1 videocassette (30 min.) VHS

Teacher and hair artist Ines Yimoc shows how to divide the hair and position the hands to do the fishtail, cobra, cornrow, and invisible French braids.

Braiding Made Easy
Jupiter, Fla., Helping Hand Productions, 1989. 1 videocassette (30 min.) VHS

In step-by-step instructions, Jamie Rines Jones demonstrates three easy braiding techniques used in eleven beautiful braided hairstyles. Includes chignon twist, two-strand twist, and fishtail.

Chemistry: A Teaching Aid for the Cosmetology Student
St. Louis, Mo., Success Enterprises, Inc., 1988. 1 videocassette (140 min.) VHS

Chuck Williams lectures on the aspects of chemistry that pertain to cosmetology. He explains why a basic understanding of chemistry is important to cosmetologists and the way in which chemical changes and reactions can occur in such chemical compounds as waving lotions, conditioners, hair coloring solutions, and hair relaxers. The entire video is composed of the lecture; no demonstrations or teaching aids other than a chalkboard are used.

Haircutting for the Now Nineties, A Family Affair
East Lansing, Mich., Jim Jones Enterprises, 1991. 1 videocassette (60 min.) VHS

Jim Jones shows step-by-step details of clipper cuts for all ages, plus techniques with the razor and blending shears.

Haircutting with Clippers: Basic Techniques
Hialeah, Fla., Good Life Products, Inc., 1990. 1 videocassette (30 min.) VHS ISBN 0944460194

Step-by-step techniques to cut beards and mustaches and men's hair to create the bob and modern flattop hairstyles using clipper-over-skin, clipper-over-fingers, and clipper-over-comb techniques.

Men's Hairstyling and Beard Design: How to Double Your Income and Have Fun Doing It
East Lansing, Mich., Jim Jones Enterprises, 1990. 1 videocassette. (60 min.) VHS

Jim Jones uses seven models to demonstrate the flattop haircut, natural tapered neckline, razor cut on long hair, haircut done entirely with blending shears, buzz cut, laser-lines, and three beard styles.

Sculptured Nails
St. Louis, Mo., Success Enterprises, 1989. 1 videocassette (30 min.) VHS

Donna Louis demonstrates the techniques for applying sculptured nails as well as methods for maintenance and fills for the nails.

Sterilization and Sanitation: A Teaching Aid for the Cosmetology Student
St. Louis, Mo., Success Enterprises, Inc., 1989. 1 videocassette (52 min.) VHS

Chuck Williams lectures on the importance of sterilization and sanitation in helping to prevent the spread of disease in salons and homes.

Culinary Arts/Food Service Management

Anne Marsh Fields
Columbus State Community College
Columbus, Ohio

Introduction

Because the demand for chefs, restaurant managers, and other culinary and food service workers is expected to grow by approximately 30% in the 1990s, programs in culinary arts and food service at vocational and technical colleges also should expect high demand. Such programs prepare their students to be cooks, bakers, and manager trainees for restaurants, hotels, clubs, catering companies, food service corporations, as well as for institutions such as schools and health care facilities. Many graduates of these programs aspire to own their own restaurants or other food-related operations. Recent entrants into the field have had to demonstrate increasing knowledge of sanitation, nutrition, international cuisine, cost controls, and new sources of supply for popular foods such as seafood.

The national accrediting body for culinary arts and food service management programs is the American Culinary Federation Educational Institute (ACFEI) Accrediting Commission. The ACFEI also offers the National Apprenticeship Training Program for Cooks. Graduates of this program are certified as Certified Cooks by the ACFEI and as Journeyman Chefs by the U.S. Department of Labor. ACFEI certification is an important credential for anyone aspiring to the position of chef in a major restaurant, as are sanitation certification, on-the-job experience, and, increasingly, computer literacy. Students who graduate from two-year community college programs can earn an associate degree along with ACFEI certification if they have the appropriate work experience. A number of other organizations also provide support, educational programs, and information services to those involved in culinary arts and food service. These include, but are not limited to, the Educational Institute of the American Hotel and Motel Association (EIAHMA); the Council on Hotel, Restaurant and Institutional Education (CHRIE); the Educational Foundation of the National Restaurant Association (EFNRA), which offers the Management Development Diploma and a national certificate in Applied Foodservice Sanitation, and its state and local organizations; the International Association of Cooking Professionals (IACP); the Society for Foodservice Management; the American School Food Service Association; and the Career College Association. Certificate programs may be local, state, or national, depending on the jurisdiction.

The ACFEI currently has no core list of library resources, but they do require accredited programs to include certain knowledge and competencies in their curricula around which a collection could be organized: basic baking; beverage management; business and math skills; dining room service; food preparation; garde-manger; human relations management; introduction to the hospitality industry; menu planning; nutrition; purchasing and receiving; and sanitation and safety. The ACFEI Accrediting Commission does have a list of texts currently used by accredited programs (available from the Director of Accreditation). Both the EFNRA and the EIAHMA are excellent sources of fundamental books and audiovisual materials, many of which are published by Van Nostrand Reinhold, Wiley, or Delmar. Many EIAHMA and EFNRA texts are required for various certificates and diplomas granted by these associations.

This section limits itself to culinary arts and food service, generally excluding general nutrition and dietetics and non-restaurant related hospitality management topics covered elsewhere in this book. Hospitality management works published by or used in courses offered by the EFNRA or EIAHMA for culinary arts or food service students are included. Coverage of cookbooks is also tightly limited. For works published prior to 1988, users will be well-served by Judith M. Nixon's excellent bibliography, *Hotel and Restaurant Industries: An Information Sourcebook* (Oryx, 1988).

Business Culinary Arts/Food Service Management

Agencies Offering Educational Programs and Certificates

American Culinary Federation Educational Institute
10 San Bartola Rd.
P.O. Box 3466
St. Augustine, Fla. 32085-3466

Educational Foundation of the National Restaurant Association
250 S. Wacker Dr., No. 1400
Chicago, Ill. 60606
Management Development Diploma, including certificate in Applied Foodservice Sanitation.

American Hotel and Motel Association
Educational Institute
1201 New York Ave. NW, Suite 600
Washington, D.C. 20005

Accrediting Agencies and Professional Associations

American Culinary Federation Educational Institute Accrediting Commission
10 San Bartola Rd.
P.O. Box 3466
St. Augustine, Fla. 32085-3466
National accrediting body for postsecondary culinary and food service management programs up to the associate degree level.

American Hotel and Motel Association (state and local organizations)
1201 New York Ave. NW, Suite 600
Washington, D.C. 20005

American School Food Service Association
1600 Duke St., 7th Floor
Alexandria, Va. 22314

Career College Association
750 1st St., NE, Suite 900
Washington, D.C. 20002
Accredits postsecondary institutions offering career-specific education programs.

Council on Hotel, Restaurant and Institutional Education
1200 17th St. NW
Washington, D.C. 20036-3097
International association of hospitality and tourism educators and professionals.

International Association of Culinary Professionals
304 W. Liberty St., Suite 201
Louisville, Ky. 40202

National Restaurant Association (and state and local organizations)
1200 17th St. NW
Washington, D.C. 20036-3097

Society for Foodservice Management
304 W. Liberty St., Suite 201
Louisville, Ky. 40202

Selected Journals

Art Culinaire.
Culinaire, Inc. Quarterly. ISSN 0892-1024

Catering Today.
Pro-Tech Publishing and Communications. Quarterly. ISSN 0884-4984

Food and Wine.
[International Review of Food and Wine Associates]. American Express Pub. Corp. Monthly. ISSN 0279-6740

Food Arts.
M. Shanken Publications. Monthly. ISSN 1042-9123

Food Management.
Advanstar Communications. Monthly. ISSN 0091-018X

Food Technology.
Institute of Food Technologists. Monthly. ISSN 0015-6639

Gourmet.
Condé Nast Publications. Monthly. ISSN 0017-2553

National Culinary Review: Official Magazine of the American Culinary Federation.
The Federation. Monthly. ISSN 0747-7716

Nation's Restaurant News: The Weekly Newspaper of the Food Service Industry.
Lebhar Friedman. Weekly. ISSN 0028-0518

Restaurant Hospitality.
Penton Publishing. Monthly. ISSN 0147-9989

Restaurants and Institutions.
Cahners Pub. Co. Semimonthly. ISSN 0273-5520

Restaurants USA: The Monthly Magazine of the National Restaurant Association.
The Association. Eleven times a year. ISSN 0890-5584

School Food Service Journal.
American School Food Service Association. Eleven times a year. ISSN 0160-6271

The Wine Spectator.
Shanken Communications. Semimonthly. ISSN 0193-497X

Print Materials

Amendola, Joseph
The Bakers' Manual
4th ed. New York, Van Nostrand Reinhold, 1993. 276p.
ISBN 0442009976

With many diagrams, recipes, and charts, Amendola covers yeast-made products, cakes and tortes, cookies, pies, puddings and sauces, icings, pastries, and other desserts. In the appendix, he covers more advanced techniques, such as pulled and blown sugar. Includes a baker's dictionary and index.

Amendola, Joseph, and Donald E. Lundberg
Understanding Baking
2nd ed. New York, Van Nostrand Reinhold, 1992. 299p.
ISBN 0442009674

Emphasizes the technical side of baking, the chemistry and physics of the overall process rather than the manual skills involved. Questions and answers are scattered throughout, along with chapter reviews, in order to help students apply the principles learned. Recommended readings, glossary, and index.

Anderson, Kenneth, and Lois E. Anderson
The International Menu Speller
New York, Wiley, 1993. 202p. ISBN 0471584355

Based on *The International Dictionary of Food and Nutrition*, this word list is designed to aid in writing menus and other written food descriptions. Lists more than 7,500 words from forty languages, including hyphenations, accents, and country of origin. No definitions.

Axler, Bruce H., and Carol A. Litrides
Food and Beverage Service
New York, Wiley, 1990. 222p. ISBN 0471621765

The first in the projected three-volume *Wiley Professional Restaurateur Guide* series, this volume provides step-by-step instructions and recipes for use by bussers, servers, and captains. Topics range from loading a tray to making a milkshake, to tableside preparation and cooking. Appendices, glossary, and index.

Birchfield, John C.
Design and Layout of Foodservice Facilities
New York, Van Nostrand Reinhold, 1988. 250p.
ISBN 0442210426

This logically organized text begins with concept development, then moves through architectural and interior design and on to equipment selection and layout. Also discusses agency and industry standards, and inspection and acceptance procedures. Index.

Bowes, Anna De Planter
Bowes and Church's Food Values of Portions Commonly Used
Revised by Jean A. T. Pennington
16th ed. Philadelphia, Lippincott, 1994. 483p. ISBN 0397550871

Primarily composed of tables covering approximately 8,500 different foods. This edition reflects current heightened awareness of fats in the diet and also includes newer food groups, such as frozen meals for children.

Byers, Brenda A., Carol W. Shanklin, and Linda C. Hoover
Food Service Manual for Health Care Institutions
Rev. ed. Chicago, Ill., American Hospital Publishing, 1994. 574p.
ISBN 155648114

Introduces basic hospital food service functions, then details daily operations. Sample chapter topics include management of human resources, food purchasing, and facility planning. Chapters include bibliographies, including some computer references, forms and charts, letters, job descriptions, and tables. Index.

Chalmers, Irena, comp.
The Food Professional's Guide: The James Beard Foundation Directory of People, Products, and Services
New York, American Showcase, 1990. 325p. ISBN 0471524603

A nonevaluative listing of representative people, products, and services in the food industry. Covers everything from cooking schools to food festivals and conventions.

Chesser, Jerald W.
The Art and Science of Culinary Preparation: A Culinarian's Manual
St. Augustine, Fla., Educational Institute of the American Culinary Federation, 1992. 586p. ISBN 0963102311

Differing from other approaches to the subject, this reference text teaches the basics of food preparation in various categories, then moves on to culinary theory in order to encourage student creativity. Includes recipes and cooking suggestions, tables, diagrams, bibliography, glossary, and index.

Cichy, Ronald F.
Sanitation Management
2nd ed. East Lansing, Mich., Educational Institute of the American Hotel and Motel Association, 1993. 425p. ISBN 0866120548

Designed to function as the basis for a full-length course on sanitation, this fundamental text covers all basic control points of sanitation, as well as safety, cleaning, self-inspection, programs, and hiring and training procedures. Uses an ongoing case study approach. Includes sample forms and checklists, tables, photographs, appendices, a glossary, and an index.

Coltman, Michael M.
Hospitality Management Accounting
5th ed. New York, Van Nostrand Reinhold, 1994. 536p.
ISBN 0442016557

This potentially self-instructional text emphasizes the application of principles, with special attention paid to computers, yield management, and internal control. Each chapter features discussion questions, problems, and case studies. Includes glossary and index.

Cournoyer, Norman G., Anthony G. Marshall, and Karen L. Morris
Hotel, Restaurant, and Travel Law: A Preventive Approach
4th ed. Albany, N.Y., Delmar Publishers, 1993. 644p.
ISBN 0827352891

Covers legal fundamentals for the hospitality industry, negligence, guest and other patron relations, and special topics such as alcohol. This new edition also devotes attention to the Americans with Disabilities Act, sexual harassment, and civil rights. Includes appendices, glossary, and a subject and case index.

Culinary Institute of America
The New Professional Chef
Edited by Linda Glick Conway. 5th ed. New York, Van Nostrand Reinhold, 1991. 869p. ISBN 0442008074

Designed to be used as both text and reference, this updated classic is organized to lead students progressively through various categories of food preparation, such as yeast doughs and vegetable cooking. Also covers equipment, sanitation, nutrition, purchasing, and *mise en place*. Equal emphasis on techniques and recipes. Color photographs, appendices, glossary, bibliography, and index.

Dahmer, Sondra J., and Kurt Kahl
The Waiter and Waitress Training Manual
3rd ed. New York, Van Nostrand Reinhold, 1988. 158p. ISBN 0442204841

This new edition features information on third party liability, special patron types such as children and the handicapped, electronic service, CPR techniques, and bar service. Includes case problems and index.

Dittmer, Paul, and Gerald G. Griffin
Dimensions of the Hospitality Industry: An Introduction
New York, Van Nostrand Reinhold, 1993. 550p.
ISBN 0442007701

The restaurant portion of this reference text can stand alone, complementing a general introduction to the hospitality industry for restaurant manager trainees. It also addresses practical business applications of the principles presented. Includes bibliography and index.

Dittmer, Paul, and Gerald G. Griffin
Principles of Food, Beverage, and Labor Cost Controls for Hotels and Restaurants
5th ed. New York, Van Nostrand Reinhold, 1994. 562p.
ISBN 0442016018

Emphasizing the relationships among costs, volume, and profits, this edition provides updated coverage of labor controls, computers, minimum wage laws, and ways to increase sales. It examines the receipt, storage, and production or dispensing aspects for each of the three areas covered. Index.

Donovan, Mary Deirdre, ed.
The Professional Chef's Techniques of Healthy Cooking
New York, Van Nostrand Reinhold, 1993. 614p.
ISBN 0442011261

Designed to help the chef use techniques already at hand to create healthful, yet tasty foods. Using the seven guidelines for nutritional cooking compiled by the U.S.D.A., the U.S. Department of Health and Human Services, the American Heart Association, and the Senate Select Committee on Health, it not only supplies 300 recipes but suggests methods for identifying, modifying, and creating other healthful recipes. Includes appendices, glossaries of cooking and nutrition terms, bibliography, and index.

Dore, Ian
Fish and Shellfish Quality Assessment: A Guide for Retailers and Restaurateurs
New York, Van Nostrand Reinhold, 1991. 112p.
ISBN 0442002068

Uses photography (some color), drawings, and diagrams to illustrate differences between fresh and not-so-fresh fish and shellfish, as well as frozen seafood and seafood products. Warns of substitutions to avoid and describes how to check for short weight. Includes tables, appendices, resource lists, and index.

Drummond, Karen Eich
Human Resource Management for the Hospitality Industry
New York, Van Nostrand Reinhold, 1990. 343p.
ISBN 0442318596

This reference text offers comprehensive coverage of topics ranging from job descriptions to working with unions, discipline, and diversity. Ample tables, drawings, study questions, and chapter references are provided. Appendices include a safety self-inspection form and a list of sources of recognition, awards, and other help. Index.

Drummond, Karen Eich
Nutrition for the Foodservice Professional
2nd ed. New York, Van Nostrand Reinhold, 1994. 576p.
ISBN 0442013701

This edition addresses menu planning for various age groups and lifestyles and expands coverage of marketing, nutrition planning, and computerized nutrient analysis. Incorporates recommendations of *The Surgeon General's Report on Nutrition and Health*, the 10th edition of *Recommended Dietary Allowances*, the 3rd edition of *Dietary Guidelines for Americans*, and the USDA's *Food Pyramid*. Appendices, glossary, and index.

Educational Foundation of the National Restaurant Association
Applied Foodservice Sanitation: A Foundation Textbook
4th ed. New York, Wiley, 1992. 360p. ISBN 0471542210

Coursebook for the NRA's "SERVSAFE" certification program, this work details the Hazard Analysis Critical Control Point (HACCP) system for food service operations. Logically organized by steps in the food handling process, it covers FDA standards, training employees, working with vendors and health inspectors, and other related topics. Sample cases, charts, diagrams, and forms, as well as numerous appendices.

Educational Foundation of the National Restaurant Association
HAACP Reference Book
Washington, D.C., The Association, 1993. 198p.

Designed for those who are planning to begin a Hazard Analysis Critical Control Point (HAACP) food safety system, or who wish to make an existing one more effective, this step-by-step guide helps managers of all types of restaurants and food service institutions identify and correct food safety hazards. Sample flow charts and recipes for common items take the reader through receiving, storing, preparing, holding, and serving. Includes a glossary, list of professional and trade organizations, bibliography, and index.

Emery, William H.
The Chef's Guide to Practical Restaurant Cookery
New York, Van Nostrand Reinhold, 1988. 248p.
ISBN 0442221738

Not so much a cookbook as a guidebook for quantity cooking for the cook who is already familiar with basic techniques. Aids in the determination of quantity food preparation, such as quantities of butter, parsley, seasoning, onions, and mushrooms required to produce ten ounces of duxelles. Provides concise instructions. Index.

Ettlinger, Steve
The Kitchenware Book
New York, Macmillan, 1992. 466p. ISBN 0025363026
 Although designed for the home cook, this work should prove useful to the beginning professional cook as well. Line drawings clearly depict over 1,000 preparation, cooking, garnishing, and serving tools. For each, alternate names, description, uses, and tips for use and purchase are given. Index.

Forrest, Lewis C.
Training for the Hospitality Industry
2nd ed. East Lansing, Mich., Educational Institute of the American Hotel and Motel Association, 1990. 370p. ISBN 0866120440
 Based on a four-step training method that managers can apply to all functions, this book moves from job analysis and development, through hiring and training, to counseling and performance reviews. Includes sources of help, case studies, sample forms and checklists, and an index.

Frank, Robyn C., ed.
Directory of Food and Nutrition Information for Professionals and Consumers
2nd ed. Phoenix, Ariz., Oryx, 1992. 332p. ISBN 0897746899
 Part One lists organizations, academic programs, software, and databases, along with a term and keyword index. Part Two is a comprehensive annotated bibliography of key reference materials, journals, hotlines, government agencies, etc., including a subject index. Appendix includes the Food Pyramid and Recommended Dietary Allowances.

Gebhardt, Susan E.
Nutritive Value of Foods
Rev. ed. Washington, D.C., U.S. Dept. of Agriculture, 1991. 72p. ISBN 0160345421
 Tables include equivalents by weight and volume, nutritive value of foods by category, yields of cooked meats, RDAs, food sources of additional nutrients, and a list of foods in which the fat content provides 30% to 85% of calories. Index.

Gisslen, Wayne
Advanced Professional Cooking
New York, Wiley, 1992. 645p. ISBN 0471836834
 Emphasizes contemporary North American cuisine. Offering more elaborate or difficult techniques and recipes than a basic text, each chapter covers a different topic such as soups or poultry and includes techniques, recipes, variations, and study questions. Appendix, bibliography, glossary, and index.

Gisslen, Wayne
Professional Baking
2nd ed. New York, Wiley, 1994. 377p. ISBN 0471595098
 Emphasizing methods used in smaller baking operations, this book can be used as either a text or a manual. Approximately 400 recipes with additional variations provide practice. Many line drawings and photographs. Includes numerous appendices, a bibliography, glossary, and index.

Gisslen, Wayne
Professional Cooking
2nd ed. New York, Wiley, 1989. 808p. ISBN 0471838497
 Gisslen uses 900 recipes to promote understanding and skills adaptable to any type and size of food service operation. Each chapter moves from the general to the specific, with quantities and ingredients adjustable to various teaching situations. "The Recipe and Its Use" chapter is particularly helpful. Includes both a subject and recipe table of contents, as well as many appendices, a bibliography, a glossary, and an index.

Haines, Robert G.
Math Principles for Food Service Operations
2nd ed. Albany, N.Y., Delmar Publishers, 1988. 283p. ISBN 0827331312
 Although a textbook, this work covers math in a basic manner not covered by other food service books, and relates it specifically to necessary skills such as calculating recipe yields, converting recipes, tabulating sales checks, and inventory. Amply illustrated with sample forms and accompanied by chapter tests. Glossary, index.

Herbst, Sharon Tyler
Food Lover's Companion: Comprehensive Definitions of Over 3,000 Food, Wine, and Culinary Terms
New York, Barron's, 1990. 582p. ISBN 0812041569
 An authoritative encyclopedic dictionary for both amateur and professional. Includes definitions, etymology, pronunciation guide, numerous charts of equivalencies and adjustments, appendices, a bibliography, as well as numerous cross-references.

Hodges, Carol A.
Culinary Nutrition for Food Professionals
2nd ed. New York, Van Nostrand Reinhold, 1994. 306p. ISBN 0442017634
 Written by a registered dietitian with culinary credentials and experience, this work helps the food service professional factor into cooking and menu planning the increasing public awareness of nutrition as part of a healthy lifestyle. It is recommended by the American Culinary Federation for meeting certification requirements. Each member of the food service staff, from chef to waitstaff, is addressed. The author uses little scientific jargon and many helpful analogies. The work covers scientific aspects of nutrition, the influences of various lifestyles and government policies on food, then moves on to specific applications of food service. Includes chapter references, appendices and an index.

Johnson, Richard C., and Diane L. Bridge
The Legal Problem Solver for Foodservice Operators
6th ed. Washington, D.C., National Restaurant Association, 1993. 110p. ISBN 0929663004
 Nontechnical, question-and-answer coverage in loose-leaf format of federal laws and regulations affecting food service operations. Sample topics include meal credits, uniforms, OSHA rules, and overtime. Includes chart of federal and state minimum wage laws. Indexed.

Jones, Julie Miller
Food Safety
St. Paul, Minn., Eagan Press, 1992. 453p. ISBN 0962440736

Comprehensive coverage of all aspects of food safety such as bacteria, food processing, and pesticides. Each topic is analyzed particularly on the basis of risk and benefit to the consumer, as well as food producers, processors and handlers. Includes a glossary, index, and chapter references.

Kahn, Mahmood A., Michael D. Olsen, and Turgut Var, eds.
VNR's Encyclopedia of Hospitality and Tourism
New York, Van Nostrand Reinhold, 1993. 1008p.
ISBN 0442003463

The twenty-five articles in Part One cover food service management, including both preparation and service, with signed articles ranging in length from five to nineteen pages. Each uses tables, figures, and case studies where appropriate, and concludes with a bibliography. Sample topics include kitchen planning, the cook-chill food production process, and computers in the food service industry. Author and subject indexes.

Kaplan, Dorlene V., ed.
The Guide to Cooking Schools, 1994
6th ed. Coral Gables, Fla., ShawGuides, 1993. 326p.
ISBN 0945834187

Nonevaluative guide to 664 cooking courses, apprenticeships, tours, and culinary arts schools worldwide, for both the amateur and professional. Lists culinary and accrediting organizations and their endorsed schools and programs. Several indexes.

Kasavana, Michael L., and John J. Cahill
Managing Computers in the Hospitality Industry
2nd ed. East Lansing, Mich., Educational Institute of the American Hotel and Motel Association, 1992. 331p. ISBN 0866120645

Provides help with selecting and implementing computer systems, as well as examining the management of information systems. Also covers the essentials of hardware and software and food and beverage applications in both service and management functions. Special features include mini-case studies, sample forms, checklists, a glossary, and an index.

Katsigris, Costas
The Bar and Beverage Book: Basics of Profitable Management
2nd ed. New York, Wiley, 1991. 485p. ISBN 047184294X

Applies management theory to bar management, covering such topics as purchase, mixology, cost controls, and marketing. Bibliography and index.

Kavanaugh, Raphael R., and Jack D. Ninemeier
Supervision in the Hospitality Industry
2nd ed. East Lansing, Mich., Educational Institute of the American Hotel and Motel Association, 1991. 312p. ISBN 0866120521

A text for entry-level supervisors that covers basic principles and their applications, with some attention given to future trends in management as a career. Includes many forms, agendas, and checklists, as well as a glossary and an index.

Kazarian, Edward A.
Foodservice Facilities Planning
3rd ed. New York, Van Nostrand Reinhold, 1989. 412p.
ISBN 0442205880

This updated edition addresses the effects of increased labor and energy costs, regulatory factors, and heightened consumer interest in sanitation and quality. It covers all areas of the food service facility—kitchen, dining rooms, and restrooms—from the first through the last steps in the planning process. Includes a chapter on planning a prospectus, chapter problems, bibliographies, several helpful appendices, a glossary, and an index.

Ketterer, Manfred
How to Manage a Successful Catering Business
2nd ed. New York, Van Nostrand Reinhold, 1991. 327p.
ISBN 0442006756

This edition includes new chapters on the banquet bar, computerization, and off-premise catering. Index.

Khan, Mahmood A.
Concepts of Foodservice Operations and Management
2nd ed. New York, Van Nostrand Reinhold, 1991. 370p.
ISBN 0442003803

Covers all aspects of restaurants, cafeterias, and institutional food service operations with an emphasis on practicality. Provides updated information on facilities, equipment, and computerization. Appendices, bibliography, and index are provided.

King, Carol A.
Professional Dining Room Management
2nd ed. New York, Van Nostrand Reinhold, 1988. 214p.
ISBN 0442247060

Emphasizes efficiency and courtesy as they contribute to profitable business and satisfied customers and staff. Contains checklists and instructions for training staff in specific tasks such as changing tablecloths, as well as consideration of legal issues. Concludes with bibliography, sources of training materials, and index.

Knight, John Barton, and Lendal H. Kotschevar
Quantity Food Production, Planning, and Management
2nd ed. New York, Van Nostrand Reinhold, 1988. 445p.
ISBN 0442240163

Topics covered include food planning, food preparation and management, management of people and property, profits, and promotion. Includes numerous appendices, bibliographies, and index.

Kotschevar, Lendal Henry
Standards, Principles, and Techniques in Quantity Food Production
4th ed. New York, Van Nostrand Reinhold, 1988. 505p.
ISBN 0442256620

Contents begin with general topics such as planning and health and safety, then move to specific categories of food. This edition updates information on equipment and computerization. Includes bibliography and index.

Kotschevar, Lendal Henry, and Marcel R. Escoffier
Management by Menu
3rd ed. Chicago, Educational Foundation of the National Restaurant Association, 1994. 387p. ISBN 0915452731

Using the menu as the key to food service management, this book covers pricing, cost control, bar and beverage operations, nutrition, purchasing, production, and service. This edition updates coverage of computers and desktop publishing. Includes bibliography and index.

Culinary Arts/Food Service Management — Business

Kotschevar, Lendal Henry, and Mary L. Tanke
Managing Bar and Beverage Operations
East Lansing, Mich., Educational Institute of the American Hotel and Motel Association, 1991. 505p. ISBN 0866120599

Stressing responsible service, this work covers all aspects of the subject including selecting and training employees and establishing control systems. Includes non-alcoholic drinks and wines. Special features include forms and checklists, bibliography, glossary, and index.

Kotschevar, Lendal Henry, and Richard Donnelly
Quantity Food Purchasing
4th ed. New York, Macmillan, 1994. 553p. ISBN 0023662301

Emphasizes knowledge of products with regard to product identification and selection, with expanded coverage in this edition of the uses of computers for accountability and purchasing preparation. Includes many tables, charts, and appendices, as well as references and an index.

LaGreca, Gen
Training Foodservice Employees: A Guide to Profitable Training Techniques
New York, Van Nostrand Reinhold, 1988. 243p.
ISBN 0442259824

Adaptable to any kind of food service operation, this book suggests detailed steps for developing both back- and front- of-the-house employee training programs. Sample topics include orientation, job analysis, and teaching skills. Liquor liability is not covered. Includes many useful appendices and an index.

Lattin, Gerald W.
The Lodging and Food Service Industry
3rd ed. East Lansing, Mich., Educational Institute of the American Hotel and Motel Association, 1993. 423p. ISBN 0866120718

This introductory text details the complex nature of the food service business, but not at the expense of highlighting the many career opportunities available on an increasingly global scale. Particularly valuable for its inclusion of spot interviews with industry leaders. Includes list of recommended journals, glossary, and index.

Laventhol and Horwath
Uniform System of Accounts for Restaurants: Adopted and Recommended by the National Restaurant Association
6th rev. ed. Washington, D.C., National Restaurant Association, 1990. 163p. ISBN 0914528149

The National Restaurant Association's (NRA's) "best-seller" since 1958, this guide is primarily designed for operators of smaller restaurants, although it also suggests ways in which larger restaurants can use the *Uniform System*. It is based on statements contributed by NRA members. Includes sample financial statements and covers record-keeping for small restaurants and the use of computerized tools. Appendix lists comprehensive sample charts of accounts.

Lieberman, Judy Serra
The Complete Off-Premise Caterer
New York, Van Nostrand Reinhold, 1991. 487p.
ISBN 0442318588

Designed primarily for the prospective or established catering business owner who must be able to manage, sell, create, cook, and supervise with style. It provides an overview of the business, deals with sites and styles of events, and includes recipes. Appendices, a recipe index, and a subject index.

Lipinski, Robert A., and Kathleen A. Lipinski
The Complete Beverage Dictionary
New York, Van Nostrand Reinhold, 1992. 425p.
ISBN 0442239874

Extensive coverage of both alcoholic and nonalcoholic terms including word origins, cross-references, and some historical detail. Complements *Knight's Foodservice Dictionary* (Van Nostrand Reinhold, 1987) and Elizabeth Riely's *The Chef's Companion* (Van Nostrand Reinhold, 1986).

Lipinski, Robert A., and Kathleen A. Lipinski
Professional Guide to Alcoholic Beverages
New York, Van Nostrand Reinhold, 1989. 548p.
ISBN 0442258372

Offers global coverage of wines and other alcoholic beverages. Covers purchase, storage, cost control, service, and staff training. Numerous appendices, a glossary, and an index.

Lundberg, Donald E.
The Management of People in Hotels and Restaurants
5th ed. Dubuque, Iowa, Wm. C. Brown, 1992. 266p.
ISBN 0697084167

Topics include the theory of different management styles, personnel practices and procedures, and union relations. Special attention is paid to ethnically diverse staffs and those with high turnover. Includes an appendix on the use of tests in selecting employees, glossary, and an index.

McDowell, M. C., and H. W. Crawford
Math Workbook for Foodservice/Lodging
3rd ed. New York, Van Nostrand Reinhold, 1988. 265p.
ISBN 0442218729

Although similar to Haines' *Math Principles for the Food Service Occupations*, this text is designed to be self-instructional. The first half is devoted to basic math principles; the second applies those principles to business situations. Includes many self-tests, practice questions, charts, and an index.

Marvin, Bill
Restaurant Basics: Why Guests Don't Come Back—and What You Can Do About It
New York, Wiley, 1992. 225p. ISBN 0471551740

This work is organized by subjects including menus, dealing with elderly clientele, and table service. Marvin emphasizes the full-service restaurant and how the lack of attention to the smallest details can hinder customer satisfaction. Includes list of suggested books and tapes.

Meyer, Sylvia, Edy Schmid, and Christel Spuhler
Professional Table Service
New York, Van Nostrand Reinhold, 1990. 464p.
ISBN 0442239823

Provides detailed treatment of all aspects of table service, including preparation, bar and beverage service, breakfast, and tableside cooking. Includes a wine dictionary, glossaries of culinary and service terms, and an index.

Miller, Jack E., and David K. Hayes
Basic Food and Beverage Cost Control
New York, Wiley, 1994. 388p. ISBN 0471579181

Reviews basic math, and introduces concepts of purchasing, accounting, and cost controls. Includes sample forms, a bibliography, and an index.

Miller, Jack E., Mary Porter, and Karen Eich Drummond
Supervision in the Hospitality Industry
2nd ed. New York, Wiley, 1992. 383p. ISBN 0471549045

For beginning first-line managers, this edition is updated with legal aspects, new tables and figures, and information about recruitment, motivation, and evaluation. It also acknowledges the special demands of this industry, such as long hours and high turnover, by using case studies and discussion questions throughout. Glossary of key words, a bibliography, and an index.

Mills, Irving J.
Tabletop Presentations: A Guide for the Foodservice Professional
New York, Van Nostrand Reinhold, 1989. 178p.
ISBN 0442264720

Rather than focusing on service and tabletop layout, this work addresses the function and aesthetics of dinnerware, flatware, glassware, linens, and accessories. It considers the tabletop as an environment for customers and food. Emphasizes value and discusses relationships with suppliers. Provides interviews with food service professionals. Recommended readings list and index.

Miner, Tom
The Business Chef
New York, Van Nostrand Reinhold, 1989. 352p.
ISBN 0442207638

Applicable to all kinds of food service operations and to both experienced and novice chefs, this book offers advice on many topics, all aimed at generating a successful business. Topics range from quality control and menu pricing, to suggested recipes for trendy dishes, to a chapter on career advancement. Appendices; index.

Morrison, Alastair M.
Hospitality and Travel Marketing
Albany, N.Y., Delmar Publishers, 1989. 532p. ISBN 0827329385

Uses an overall hospitality marketing model and actual case examples to relate food service marketing functions and techniques to one another. Includes bibliographies and index.

Murfitt, Janice
First Steps in Cake Decorating
New York, Sterling Pub. Co., 1993. 159p. ISBN 0806903945

Using many color photographs, the author leads the reader from simple to more complex techniques. Sample topics cover food colorings, instant decorations, chocolate, sugar and flour pastes, and marzipan. Appendices include basic cake and frosting recipes and design templates. Index.

National Association of Meat Purveyors
The Meat Buyers Guide
2nd ed. Reston, Va., The Association, 1992. 199p.
ISBN 1878154001

This thoroughly illustrated guide reflects the USDA's latest Institutional Meat Purchase Specifications (IMPS) for trim, cutting terms, and nomenclature, with many color photographs and diagrams. Individual sections deal with different kinds of meat, each having its own index to product names. Includes nutritional information and detail about fat content.

National Restaurant Association
National Restaurant Association's Current Issues Reports
Washington, D.C., The Association, 1985- . various pagings

Individual titles in this series survey various topics, e.g., *Foodservice Manager 2000*, which projects the least and most likely developments in that career by the year 2000. *Waitstaff Compensation* and *Nutrition Awareness and the Foodservice Industry* are two other examples of titles covered by this series.

National Restaurant Association, Research and Information Service Dept.
A Guide to Preparing a Restaurant Business Plan
Washington, D.C., The Association, 1992. 139p.

Describes components of a thorough restaurant business plan such as balance sheets and advertising campaign plans, then provides an example of each. Includes bibliography, checklists, and an appendix of suggested supporting documentation.

National Restaurant Association, Research and Information Services Dept.
Foodservice Numbers: A Statistical Digest for the Foodservice Industry
Washington, D.C., The Association, 1992. 118p.

Clearly organized charts and notes document food service sales and purchases from 1970 through 1990, as well as customer attitudes and behavior, wage rates and benefits.

Ninemeier, Jack D.
Management of Food and Beverage Operations
2nd ed. East Lansing, Mich., Educational Institute of the American Hotel and Motel Association, 1990. 370p. ISBN 0866120572

With discussion questions, sample forms and checklists, cartoons, and exhibits, this basic reference text helps aspiring managers understand the best ways to coordinate food service operations with guest needs. It begins with fundamentals of management, marketing, and nutrition, then proceeds to menu development, production, service, sanitation and safety, facilities and equipment, and financial management. Also examines hardware and software for automation. Bibliography; index.

Ninemeier, Jack D.
Planning and Control for Food and Beverage Operations
3rd ed. East Lansing, Mich., Educational Institute of the American Hotel and Motel Association, 1991. 399p. ISBN 0866120556

Student text and manager reference covering cost-volume-profit analysis and menu pricing, as well as computer applications useful in controlling costs and maximizing sales. Each chapter includes key terms, endnotes, discussion questions, and problems; many examples of worksheets, formulae, and checklists. Glossary and index.

Nixon, Judith M.
Hotel and Restaurant Industries: An Information Sourcebook
Phoenix, Ariz., Oryx, 1988. 240p. ISBN 0897743768

Annotates more than 1,000 books and journals aimed at college-level audiences, emphasizing materials published in the

1980s as well as classic works. Begins with a suggested core collection, then moves on to more specialized subject sections, like history and franchising. Author, title, and subject indexes.

Ostroff, Harriet, and Tom Nichols
Specialty Cookbooks: A Subject Guide
New York, Garland Publishing, 1992. 659p. ISBN 0824069471

Nonevaluative bibliography of some 4,500 titles, most published between 1980 and 1990, along with brief descriptions. Sections cover specific ingredients, dishes and courses, meals, and special techniques. Author and title indexes and many useful cross-references.

The PMA Fresh Produce Reference Manual for Food Service
Newark, Del., Produce Marketing Association, 1989. unpaged

Covers seasonality, storage, handling, and specifications for fruits and vegetables. Suggests ways of dealing with vendors and of allaying consumer fears about pesticides. Large color photographs.

Pauli, Eugen
Classical Cooking the Modern Way
2nd ed. New York, Van Nostrand Reinhold, 1989. 560p. ISBN 0442272065

Based on Escoffier, yet updated with the latest techniques, this work deals not only with cooking but also with kitchen management. Theoretical topics covered include kitchen planning, foods, nutrition, the menu, and accounting. Cooking topics include principles, methods, recipes and variations, specific food categories, and national dishes. Many color and black-and-white photographs and diagrams, glossary, index.

Powers, Thomas F.
Introduction to the Hospitality Industry
2nd ed. New York, Wiley, 1992. 503p. ISBN 0471530549

Purposely excluding management techniques, this book relates the restaurant industry to current forces and issues shaping the industry. It also examines the role of service and the management of international hospitality, offering views of the future. Bibliography; index.

Reed, Lewis
Specs: The Comprehensive Foodservice Purchasing and Specification Manual
2nd ed. New York, Van Nostrand Reinhold, 1993. 712p. ISBN 0442007051

Describes in detail more than 2,750 processed and unprepared food items to help users develop purchasing policies, pricing, and accounting systems. Organized into chapters according to food category, (e.g., lemons), and then from general to specific within a product, (e.g., varieties of lemons). Several useful appendices, a bibliography, and an index.

Reich, Allen Z.
The Restaurant Operator's Manual
New York, Van Nostrand Reinhold, 1990. 395p. ISBN 0442206380

This loose-leaf reference contains information needed by restaurant managers to update and improve standard operating policies and procedures. Contains over one hundred suggested forms, as well as clearly written elements for employee handbooks. Topics include pre-opening training, managing waiting lists, and quality control. Index.

Reid, Robert D.
Hospitality Marketing Management
2nd ed. New York, Van Nostrand Reinhold, 1989. 399p. ISBN 0442278489

This edition adds chapters on hospitality group sales, hospitality services marketing and promotions, and public relations. Offers many mini-case studies, keyword and concept listings, and discussion questions. Index.

Rubash, Joyce
Master Dictionary of Food and Wine
New York, Van Nostrand Reinhold, 1990. 372p. ISBN 0442234651

Provides brief, simple definitions, pronunciation aids, and countries of origin. Some cross-references, synonyms, and alternate spellings included.

Schmidgall, Raymond S.
Hospitality Industry Managerial Accounting
2nd ed. East Lansing, Mich., Educational Institute of the American Hotel and Motel Association, 1990. 623p. ISBN 0866120580

Intended to help new managers set financial goals while maintaining assets and controlling costs. Includes *Managerial Accounting Practices 2.0* software, sample schedules, forms, statements and reports, numerous exhibits and examples, and computer forecasting information. Glossary, bibliography, and index.

Schmidt, Arno
The Chef's Book of Formulas, Yields, and Sizes
New York, Van Nostrand Reinhold, 1990. 338p. ISBN 0442318359

This work's goal is to help the chef purchase appropriate amounts of food to avoid waste. For each food, the author lists available forms, seasonality, most popular sizes and packages, suggested serving sizes, and the number of servings yielded by sizes and packages. Provides both U.S. and metric weights and measures, as well as calories. No methods given except for certain common items such as salad dressings. Alphabetically arranged. Index.

Scriven, Carl, and James Stevens
Food Equipment Facts: A Handbook for the Foodservice Industry
Rev. and updated ed. New York, Van Nostrand Reinhold, 1989. 436p. ISBN 0442318642

This extremely useful reference includes charts, brainstorming questions, and suggestions and specifications for facilities and equipment, ranging from receiving dock capacities to ovens, and from the size of shelving required for storing various sizes of canned goods and glassware to safety equipment. Many appendices; index.

Seaberg, Albin G.
Menu Design, Merchandising, and Marketing
4th ed. New York, Van Nostrand Reinhold, 1991. 212p. ISBN 0442319584

Updated to reflect changing food and drink preferences exhibited in all kinds of restaurants, this work covers not only how a menu should look, but what kinds of foods a menu should include to make a restaurant profitable. Offering examples from successful restaurants, the author covers papers, typeface, copy, drink and wine lists, among other topics, and discusses common mistakes made. Indexed.

Sherry, John E. H.
The Laws of Innkeepers: For Hotels, Motels, Restaurants, and Clubs
3rd ed. Ithaca, N.Y., Cornell Univ. Press, 1993. 924p.
ISBN 0801425085

Combination college text and reference manual for owners, managers, and general practice attorneys. This edition features new chapters on employment law, environmental law and land use, and catastrophic risk liability. Includes case citations, bibliography of law review articles, and subject index. Undergraduate study guide available.

Shugart, Grace Severance, and Mary Molt
Food for Fifty
9th ed. New York, Macmillan, 1993. 767p. ISBN 0024103411

Part One consists of food production information tables; Part Two details recipes, variations, nutritive values, and production techniques; Part Three deals with menu planning and special events. Appendices, glossary, and index.

Snider, Nancy, ed.
Frozen Food Book of Knowledge: A Foodservice Reference
10th ed. Harrisburg, Pa., National Frozen Food Association, 1992. 344p.

Detailed technical information on specifications, safety and sanitation, label designations, and nutrition, organized by category of food, e.g., fruits, meat, and hors d'oeuvres. Includes a glossary, charts, and an index to products, regulations, and industry terminology.

Sonnenschmidt, Frederic H., and John F. Nicolas
The Professional Chef's Art of Garde Manger
5th ed. New York, Van Nostrand Reinhold, 1993. 285p.
ISBN 0442011539

Although primarily devoted to cold food presentation, this updated edition now includes hot hors d'oeuvres and stresses lower-fat forcemeats and more healthful cold foods. Based on the American Culinary Federation's international criteria for buffet platters. Many color photographs emphasize elegant simplicity. Includes recipes, diagrams, menus, glossary, and index.

Stefanelli, John M.
Purchasing: Selection and Procurement for the Hospitality Industry
3rd ed. New York, Wiley, 1992. 632p. ISBN 0471844543

Begins with an overall examination of selection and procurement, then addresses principles of selection and procurement, and finally moves to specific items. A well-organized guide for the nonspecialist. Index.

Strianese, Anthony J.
Dining Room and Banquet Management
Albany, N.Y., Delmar Publishers, 1990. 356p. ISBN 0827333072

This comprehensive guide to service stresses the importance of both ideas and their application. Chapters include open-ended problem questions, problem cases, and references. Photographs illustrate both incorrect and correct service methods. Includes glossary, bibliography, and index.

Sultan, William J.
Practical Baking
5th ed. New York, Van Nostrand Reinhold, 1990. 882p.
ISBN 0442319568

Covers all categories of baked goods, including Passover and dietetic items, and provides new international bread and roll formularies. Reflects international variations in the availability of various flours. Appendices, glossary, index.

Tanke, Mary L.
Human Resource Management for the Hospitality Industry
Albany, N.Y., Delmar Publishers, 1990. 416p. ISBN 0827335903

Focuses especially on the challenges of multiculturalism, conflict resolution, benefits and compensation, and legal considerations. Bibliography.

Terrell, Margaret E., and Dorothea B. Headlund
Large Quantity Recipes
4th ed. New York, Van Nostrand Reinhold, 1989. 506p.
ISBN 0442204868

Recipes serve approximately fifty people, with ingredients given in both metric and U.S. weights and volumes. Organized by category; for example, veal or frozen pies. Professionally tested in public kitchens of various kinds at food service institutions. Tables, index.

Touissant-Samat, Maguelonne
A History of Food
Translated from the French by Anthea Bell. Cambridge, Mass., Blackwell Reference, 1993. 801p. ISBN 0631177418

The history, lore, and legend of food, cookery, and eating, enhanced with nutritional information.

VanEgmond-Pannell, Dorothy
School Foodservice Management
4th ed. New York, Van Nostrand Reinhold, 1990. 275p.
ISBN 0442319592

This edition acknowledges changes in the labor force, heightened consumer awareness, technological change, higher costs, and competition from commercial food service management corporations that typify school food service in the nineties. It also emphasizes a diet lower in sodium and fat. Sample topics include convenience food, central vs. satellite production, breakfasts, advertising, and computerization. Includes appendices, chapter bibliographies, charts and checklists, glossary, and index.

West, Bessie Brooks, and LeVelle Wood
Food Service in Institutions
6th ed. New York, Macmillan, 1988. 662p. ISBN 0024259403

A classic text and reference offering a full introduction to food service in such institutions as schools, hospitals, nursing homes, and the transportation industry. It addresses such topics as quantity production, purchasing, equipment arrangement, and management information systems. Includes supplementary chapter references, charts, photographs, diagrams, forms, and an index.

Wigger, G. Eugene
Catering to Every Whim: A Complete Guide to Catering Sales, Administration, and Operations
Englewood Cliffs, N.J., Prentice Hall, 1991. 274p.
ISBN 0131205021

Wigger looks at catering from the perspective of the catering executive, offering lists, forms, sample letters, and phone solicitation scripts. Includes chapters on kosher and wedding catering and beverage sales. Color photographs, glossaries, diagrams, and an index are provided.

Woods, Robert H.
Managing Hospitality Human Resources
East Lansing, Mich., Educational Institute of the American Hotel and Motel Association, 1992. 441p. ISBN 0866120688

Addressed to future managers, this work is highlighted by interviews with experienced managers and illustrative case studies. Includes coverage of the Americans with Disabilities Act and discussion on global trends. Photographs, sample forms and checklists, a glossary, and an index.

Yudd, Ronald A.
Successful Buffet Management
New York, Van Nostrand Reinhold, 1990. 346p.
ISBN 0442205759

Thorough advice on the full spectrum of buffet options from theme-oriented to regional to holiday. With each option comes suggested menus. Also covers general information on precosting, presentation, avoiding waste, etc. Provides checklists, lists of equipment suppliers, color photographs, a menu index, and a subject index.

Zaccerelli, Herman E.
Food Service Management by Checklist: A Handbook of Control Techniques
New York, Wiley, 1991. 244p. ISBN 0471530638

A handbook for developing checklists to use for training, supervising, resolving problems, and controlling costs. Appendix contains alternative blank checklist forms, which may be duplicated for personal use.

Hospitality Management

Julie Pinnell
Johnson County Community College
Overland Park, Kansas

Introduction

Hospitality management is a program of study that usually includes lodging, food, travel, and tourism service. This chapter deals only with resources supporting the lodging industry; other chapters of this book cover food service, travel, and tourism.

Large hotel companies are becoming more international both in ownership and outlook. Hotel development has been slowing with many of the larger chains expanding by acquiring existing hotels. Both of these trends are predicted to continue as is the increased use of technology.

Other industry trends are increased demand for spa facilities and increased awareness of the environmental impact of the industry. Success has come to managers offering specialized guest services. Guest services include offering as many amenities as possible (in-room coffeemakers, hairdryers, makeup mirrors, VCRs, health club privileges, complimentary newspapers, cocktail hours, breakfasts, etc.). Other amenities may include high levels of personalized services such as personal fitness trainers, concierge services, and secretarial services. All of these services may be presented in a hospitality marketing strategy.

The hotel business is, in essence, service to people. Human interaction is at the heart of hotel keeping, whether it is staff interacting with staff, or staff interacting with guests. Anyone considering a career in the field should consider this aspect of the job. There are many entry-level jobs in the industry that are unskilled or semi- skilled, requiring no previous experience or training (clerk, housekeeper, or porter). Skilled jobs that do require experience or special preparation include accountant, reservations agent, plumber, sales representative, etc. Academic training will enhance the chances of those seeking management-level positions, but these jobs are also available to employees who have worked their way up through the organization. Many of the texts in this bibliography recognize this by targeting not only students in hospitality management programs, but practitioners interested in self-education.

Selected Associations

Although there are no accrediting agencies specific to this area, there are many helpful organizations that can supply additional information.

American Hotel and Motel Association
1201 New York Avenue, NW, Suite 600
Washington, D.C. 20005

Club Managers Association of America
1733 King Street
Alexandria, Va. 22314

Council on Hotel, Restaurant and Institutional Education
1200 17th Street, NW
Washington, D.C. 20036-3097

Hospitality Sales and Marketing Association International
1300 L. St. NW, Suite 800
Washington, D.C. 20005

International Association of Hospitality Accountants
P.O. Box 203008
Austin, Tex. 78720-3008

Meeting Planners International
1950 Stemmons Freeway
Infomart Building, Suite 5018
Dallas, Tex. 75207-3109

National Executive Housekeepers Associaton
1001 Eastwind Drive, Suite 301
Westerville, Ohio 43081-3361

Professional Convention Management Association
100 Vestavia Office Park, Suite 220
Birmingham, Ala. 35216

Selected Journals

Canadian Hotel and Restaurant: CH&R.
MacLean Hunter Ltd. Monthly. ISSN 1182-9923

The Cornell Hotel and Restaurant Administration Quarterly.
Elsevier Science Pub. Co. Bimonthly. ISSN 0010-8804

FIU Hospitality Review.
Florida International University. Semiannual. ISSN 0739-7011

Hospitality and Tourism Educator.
Council on Hotel Restaurant and Institutional Education. Three times a year.

Hospitality Design.
Bill Communications. Monthly. ISSN 1062-9254

Hospitality Law: The Preventative-Law Information Service for the Lodging Industry.
Magna Publications. Monthly. ISSN 0889-5414

Hospitality Research Journal: The Professional Journal of the Council on Hotel, Restaurant, and Institutional Education.
Council on Hotel Restaurant and Institutional Education. Three times a year. ISSN 1060-9350

Hotel and Motel Management.
Advanstar Communications. Semimonthly (except monthly in Jan., Aug., Dec.). ISSN 0018-6082

Hotel and Resort Industry.
Coastal Communications. Monthly. ISSN 0149-3639

Hotel and Travel Index.
ABC International. Quarterly. ISSN 0162-9972

Hotels: The International Magazine of the Hotel and Hotel Restaurant Industry.
Cahners Pub. Co. Monthly. ISSN 1047-2975

International Journal of Hospitality Management.
Pergamon Press. Quarterly. ISSN 0278-4319

Lodging.
American Hotel and Motel Association. Monthly (except August). ISSN 0360-9235

Lodging Hospitality.
Penton Pub. Co. Thirteen times a year. ISSN 0148-0766

Meetings and Conventions.
Cahners Pub. Co. Monthly (thirteen issues per year; two in March). ISSN 0025-8652

OAG Business Travel Planner. (North American ed.)
Official Airline Guide. Annual. ISSN 1053-0002

Progress in Tourism and Recreation and Hospitality.
Pinter Publishers. Annual. ISSN 0952-5424

Sales and Marketing Management.
Bill Communications. Monthly (except in Feb., Apr., Aug., and Nov.). ISSN 0163-7517

Service Edge.
Lakewood Publications. Monthly. ISSN 1053-1734

✦ Indexes

Hospitality Index: An Index for the Hotel, Foodservice, and Travel Industries.
Washington, D.C., American Hotel and Motel Association. Quarterly.
   Gives complete bibliographic citations to articles in hotel, food service, and travel industry journals.

Lodging and Restaurant Index.
West Lafayette, Ind., Purdue Univ. Quarterly. ISSN 0894-5128.
   Indexes fifty hotel, food service, and travel journals. Annual cumulations.

Print Materials

Bardi, James A.
Hotel Front Office Management
New York, Van Nostrand Reinhold, 1990. 364p.
ISBN 0442318618
   Emphasizing human resource management principles for effective recruitment, training, development, and motivation, the author describes the application of management concepts to the front office. Indexed.

Berger, Florence, and Dennis H. Ferguson
INNovation: Creativity Techniques for Hospitality Managers
New York, Wiley, 1990. 116p. ISBN 0471527742
   Based on the idea that success depends on the ability to stand out and excel, this book concentrates on exercises for developing new and better products and services. Indexed.

Boardman, R. D.
Hotel and Catering Accounts
Oxford, England, Butterworth-Heinemann, 1991. 175p.
ISBN 0750600861
   Written with hotel and catering students in mind, this textbook begins with an introduction to accounting and double entry bookkeeping and works its way through all needed aspects of hotel and

catering accounting. Additional chapters cover graphic presentation, variance analysis, interfirm comparisons, and uniform accounting.

Borsenik, Frank D., and Alan T. Stutts
The Management of Maintenance and Engineering Systems in the Hospitality Industry
3rd ed. New York, Wiley, 1992. 594p. ISBN 0471542229
　　The purpose of this textbook is to provide a basic understanding of mechanical-electrical systems and their energy consumption and constraints. Includes references and index.

Bryson, McDowell, and Adele Ziminski
The Concierge: Key to Hospitality
New York, Wiley, 1992. 242p. ISBN 0471528935
　　Covers everything from a list of supplies to be kept on hand (shirt studs, bow ties, clear nail polish, birthday candles, etc.) and the "black book" of information for referrals and concierge wizardry, to how to interview for that important first job. The final chap-ter is a collection of stories illustrating the incredible requests that have been fielded and guest problems that have been solved. Indexed.

Coltman, Michael M.
Cost Control for the Hospitality Industry
2nd ed. New York, Van Nostrand Reinhold, 1989. 371p. ISBN 0442205910
　　A good basic text covering all aspects of cost control for hotels and restaurants. Two chapters are devoted to information systems. Indexed.

Coltman, Michael M.
Financial Control for Your Hotel
New York, Van Nostrand Reinhold, 1992. 280p. ISBN 04420073611
　　Written for finance executives and hotel general managers, these cost control techniques are promoted as necessary to successful hotel management. Strategies for long-term financial growth planning are explored. Includes glossary and index.

Coltman, Michael M.
Hospitality Management Accounting
5th ed. New York, Van Nostrand Reinhold, 1994. 536p. ISBN 0442016557
　　Intended for students taking accounting courses that are industry-related. Includes glossary and index.

Cournoyer, Norman G., Anthony G. Marshall, and Karen L. Morris
Hotel, Restaurant and Travel Law: A Preventative Approach
4th ed. Albany, N.Y., Delmar Publishers, 1993. 644p. ISBN 0827352891
　　The focus of this volume is protecting businesses from accidents, personal interactions, and incidents that could lead to lawsuits. The aim is to educate hospitality personnel as to what laws exist and to help train staff to recognize potential pitfalls and guard against them. Discusses the fundamentals of hospitality law, negligence, relationships with guests and other patrons, and special topics such as alcohol liability, travel agent relations, employee issues, licensing, and regulations. Includes cases, glossary, and index.

Craig, Stephen R.
Housekeeping Management in the Hospitality Industry
Elmsford, N.Y., National Publishers of the Black Hills, 1989. 114p. ISBN 093592065X
　　A concise manual for the housekeeping manager. Covers hiring, training, and managing housekeeping staff, and management of work flow and supplies. Specific chapters cover systems and procedures, carpet care, guest relations, periodic cleaning and preventive maintenance, cleaning common areas, and laundries. Also illustrates the basics of being a good manager: communication skills, quality assurance, planning, and budgeting. Includes glossary.

Cunningham, Stephen
Data Analysis in Hotel and Catering Management
Boston, Butterworth-Heinemann, 1991. 204p. ISBN 0750601116
　　Gives a general introduction to methods of quantitative analysis and the scope of their applications in the hotel and catering industry. References; index.

Davies, Thomas D., and Kim A. Beasley
Design for Hospitality: Planning for Accessible Hotels and Motels
New York, Nichols Publishing, 1988. 187p. ISBN 0893973181
　　Developed by the Paralyzed Veterans of America with assistance from the American Hotel and Motel Association to promote barrier-free design, this volume uses many illustrations, including scale drawings and photographs. ANSI standards are given when available. Bibliography and index.

Davies, Mary E., and others
So...You Want to Be an Innkeeper: The Complete Guide to Operating a Successful Bed and Breakfast Inn
San Francisco, Chronicle Books, 1990. 209p. ISBN 0877017212
　　Covers evaluation of self and site, building a business plan, and marketing a bed-and-breakfast inn. The most interesting feature of this publication is the inset stories of innkeepers' real life solutions and innovations. Bibliography and index.

Diekelman, Donald
Survival Spanish for the Hospitality Industry: Employer-Employee Relations
New York, Van Nostrand Reinhold, 1991. 160p. ISBN 0442007019
　　Designed as a manual for a course to prepare hotel and restaurant management students for interaction with employees whose primary language is Spanish.

Drummond, Karen Eich
Human Resource Management for the Hospitality Industry
New York, Van Nostrand Reinhold, 1990. 343p. ISBN 0442318596
　　Explores management of the complex human resource issues of recruitment, performance management, review, employee discipline, and diversity in the workplace. Delivers helpful tips on working with unions, resolving conflicts, and building a team.

Dykstra, John Jefferson
Infection Control for Lodging and Food Service Establishments
New York, Wiley, 1990. 178p. ISBN 0471623172
　　Created as a tool to educate hospitality personnel to the poten-

tial impact of infectious diseases and the need for infection control programs. Describes a simple, cost-effective program. Bibliography and index.

Fellows, Jane, and Richard Fellows
Buildings for Hospitality: Principles of Care and Design for Accommodation Managers
London, England, Pitman Publishing, 1990. 224.
ISBN 0273030744

Provides a fundamental introduction to "premises and plant" or the preparation of buildings for hospitality. Indexed.

Forrest, Lewis C.
Training for the Hospitality Industry
2nd ed. East Lansing, Mich., Educational Institute of the American Hotel and Motel Association, 1990. 370p. ISBN 0866120440

Intended for managers, this book covers every facet of training for new or established operations. Special chapters are devoted to ongoing training, management training, staff development, and pre-opening training. Indexed.

Gee, Chuck Y.
Resort Development and Management
2nd ed. East Lansing, Mich., Educational Institute of the American Hotel and Motel Association, 1988. 531p. ISBN 0866120432

Focusing on the uniqueness of resort operation and development, this book will appeal to practitioners looking for new management ideas, developers interested in self-education, and students hoping to specialize. Each chapter ends with a list of suggested readings. Indexed.

Gray, William S., and Salvatore C. Liguori
Hotel and Motel Management and Operations
3rd ed. Englewood Cliffs, N.J., Regents/Prentice Hall, 1994. 346p. ISBN 013095795X

While other books focus on marketing and human resources, this work looks at hotel keeping as a major industry and concentrates on the actions needed to create success and profit. The authors cover all steps in the development of a successful hotel or motel, from site development to opening phases. Not aimed at students in formal educational settings but to all individuals interested in management within the industry. Includes casino and spa management.

Groome, James J.
How to Make the Perfect Hotel Deal: A Working Guide for Meeting Planners
Princeton, N.J., Groome Marketing Associates, 1992. 1 v. (loose-leaf). ISBN 0963274007

Although the intended audience is meeting planners, this loose-leaf volume will be of interest to hotel industry management for insight into the expectations of meeting planners. Groome advocates fairness on both sides of the negotiation process and rules out deceptive or ruthless practices. In one chapter he outlines the elements on which a hotel is usually willing to negotiate.

Harris, Peter
Profit Planning
Boston, Butterworth-Heinemann, 1992. 204p. ISBN 0750602236

Aimed at helping hotel and restaurant managers understand how to obtain financial information and read and use financial statements and ratios. Other areas of insight include monitoring progress, conducting pricing, budgeting, and forecasting. Computer spreadsheets are discussed. Indexed.

Henkin, Shepard
Opportunities in Hotel and Motel Careers
Lincolnwood, Ill., VGM Career Horizons, 1992. 152p.
ISBN 0844281700

Dozens of positions from both "front of house" (public relations, reservations, etc.) to "back of house" (engineering, housekeeping, etc.) are described in terms of wages, prospects, and working conditions. Includes lists of colleges and schools offering training.

Hotch, Ripley, and Carl A. Glassman
How to Start and Run Your Own Bed and Breakfast Inn
Harrisburg, Pa., Stackpole Books, 1992. 182p. ISBN 0811724417

Covers the basics of the inn business: marketing, business plans, etc. Perhaps the most unique features of this book are the final two chapters entitled "Why You Don't Want to Be in This Alone" and "Taking Care Of Yourself." The first explores the advantages of association membership and the latter is a frank discussion of protecting one's physical, mental, and social health. Indexed.

Hughes, Howard L.
Economics for Hotel and Catering Students
2nd ed. Leckhampton, Cheltenham, England, Stanley Thornes Publishers, 1990. 248p. ISBN 0748703802

Of interest to managers and students, this textbook is designed to cover a full course on economics relevant to hotel and catering activities. References and index.

Hoyle, Leonard H., David C. Dorf, and Thomas J. A. Jones
Managing Conventions and Group Business
East Lansing, Mich., Educational Institute of the American Hotel and Motel Association, 1989. 349p. ISBN 0866120424

Examines various markets: associations, corporations, sports, theater, tour groups, trade unions, arts, etc. Tips on organizing, marketing, communications, and meeting technologies. Glossary and index.

Iverson, Kathleen M.
Introduction to Hospitality Management
New York, Van Nostrand Reinhold, 1989. 138p.
ISBN 0442239009

Beginning with a history of hospitality, this text proceeds with a discussion of how to choose a career path, descriptions of various fields within the industry, and continues with a discussion of how to provide quality service and manage employees. An overview of trends in hospitality management completes the volume. Glossary, references, index, and study guide.

Kasavana, Michael L., and Richard M. Brooks
Managing Front Office Operations
3rd ed. East Lansing, Mich., Educational Institute of the American Hotel and Motel Association, 1991. 435p. ISBN 0866120610

Arranged in three parts: overview of hotel operation that places the front office in context; responsibilities and tasks, and management tools. References, glossary, and index.

Kavanaugh, Raphael R., and Jack D. Ninemeier
Supervision in the Hospitality Industry
2nd ed. East Lansing, Mich., Educational Institute of the American Hotel and Motel Association, 1991. 312p. ISBN 0866120521

Designed to help hospitality managers deal with the pressures from the employees they supervise and from higher management. Glossary and index.

Khan, Mahmood A., Michael D. Olsen, and Turgut Var, eds.
VNR's Encyclopedia of Hospitality and Tourism
New York, Van Nostrand Reinhold, 1993. 1008p.
ISBN 0442003463

Divided into three parts, this reference work covers food service, lodging, and tourism. Tables, charts, and case studies are used to illustrate text that offers practical advice on current issues and methodologies. Contributors are prominent leaders in their fields, and all sections were screened by an editorial board.

Kohr, Robert L.
Accident Prevention for Hotels, Motels, and Restaurants
New York, Van Nostrand Reinhold, 1991. 314p.
ISBN 0442239556

The author indicates in the preface that he has attempted to make this a how-to book for creating an environment that is safe for guests and employees. Security issues are included here with the emphasis on loss control. Bibliography and index.

Lanier, Pamela
All Suite Hotel Guide
Rev. 7th ed. Oakland, Calif., Lanier Pub. International, 1993. 293p. ISBN 0898155800

A listing of 1,200 worldwide hotels. Detailed entries.

Lanier, Pamela
Elegant Small Hotels: A Connoisseur's Guide
Rev. 8th ed. Oakland, Calif., Lanier Pub. International, 1993. 209p. ISBN 0898155819

A directory of hotels with fewer than 200 rooms, outstanding decor, homey atmosphere and a full range of amenities: concierge, turn-down service, fresh flowers, secretarial services, etc.

Lattin, Gerald W.
The Lodging and Food Service Industry
3rd ed. East Lansing, Mich., Educational Institute of the American Hotel and Motel Association, 1993. 423p. ISBN 0866120718

An introduction to the lodging and food service industry, their interrelationships, and career opportunities. Glossary and index.

Lewis, Robert C.
Cases in Hospitality Marketing and Management
New York, Wiley, 1989. 349p. ISBN 0471508985

This book is divided into six major divisions: concepts of marketing, environmental scanning and competitive analysis, segmentation and positioning, marketing mix, research and marketing information, and strategy and the marketing plan. Each section begins with a "mini-case" followed by a more lengthy case study. Each case illustrates a mix of major marketing tools. The final chapter is a comprehensive case study that serves to showcase all the concepts previously presented.

Lewis, Robert C., and Richard E. Chambers
Marketing Leadership in Hospitality: Foundations and Practices
New York, Van Nostrand Reinhold, 1989. 699p.
ISBN 0442205317

Operating on the idea that marketing is the link between a sophisticated hospitality customer and the wide variety of hotels, restaurants, and airlines available, the authors explore marketing theory and its applications. Glossary and index.

Lundberg, Donald E.
The Hotel and Restaurant Business
6th ed. New York, Van Nostrand Reinhold, 1994. 304p.
ISBN 0442012462

This introductory textbook explains hotel and restaurant finance within the framework of the industry trend toward internationalization. Chapters on human resources, fast food, and franchising are new in this edition. Selective bibliography and index.

Lundberg, Donald E.
The Management of People in Hotels and Restaurants
5th ed. Dubuque, Iowa, William C. Brown, 1992. 266p.
ISBN 0697084167

Centered on what many hospitality managers consider their biggest problem area, this book discusses management theory applied to hospitality, personnel practices and procedures, and union relations. The final section examines a case study of a highly successful businessman who buys a hotel, but is inexperienced in the unique aspects of hotel operation. An appendix covers using tests to select employees. Glossary and index.

Makens, James C.
The Hotel Sales and Marketing Planbook
Winston-Salem, N.C., Marion-Clarence Pub. House, 1990. 255p.

A sourcebook and guide for hotel marketing and sales managers. Covers current sales concepts and writing marketing plans.

Martin, Donald J., and Donald E. Lundberg
Human Relations for the Hospitality Industry
New York, Van Nostrand Reinhold, 1991. 299p.
ISBN 0442006764

Applying transactional analysis as a model of human interaction, Martin and Lundberg cover courtesy, guest relations, special industry problems, and telephone and nonverbal communication. Index.

Martin, Robert J., and Thomas J. A. Jones
Professional Management of Housekeeping Operations
2nd ed. New York, Wiley, 1992. 458p. ISBN 0471547794

A textbook intended for managers of hotels and hospitals. In this edition, general management principles are applied to specific application settings. Covers scenarios such as bringing a new organization into being and directing and controlling the ongoing operation. Special topics such as environmental services, laundry, and computer management systems are explored. The chapter on safety and security includes OSHA and Hazardous Communication Standards (HAZCOM) regulation requirements. Glossary and index.

Medlik, S.
The Business of Hotels
2nd ed. Oxford, England, Butterworth-Heinemann, 1989. 188p.
ISBN 0750605340

Medlik wrote this book to fill a void in the literature on the hotel as a business. Written as a comprehensive outline, it includes suggested readings at the end of each chapter and ends with a bibliography of the sources used in the preparation of the text. Indexed.

Medlik, S.
Dictionary of Travel, Tourism, and Hospitality
Boston, Butterworth-Heinemann, 1993. 360p. ISBN 0750609532

This reference work provides current and complete definitions of terms in the hospitality industry, with a special section on abbreviations in the tourism field. Definitely shows a British emphasis with a large section listing tourism organizations in the United Kingdom and the international community. Includes a bibliography of sources used in compiling the dictionary.

Mill, Robert Christie
Managing for Productivity in the Hospitality Industry
New York, Van Nostrand Reinhold, 1989. 226p.
ISBN 0442264518

Explores personnel management topics including organizational culture and climate, positive reinforcement, hiring, designing the workplace, and work flow. Bibliographies and index.

Miller, Jack E., Mary Porter, and Karen Eich Drummond
Supervision in the Hospitality Industry
2nd ed. New York, Wiley, 1992. 383p. ISBN 0471549045

Focusing on the special characteristics of supervision in the industry, this book is intended for the first-line hospitality manager. All the basics are here, including a chapter on delegation. Glossary, bibliography, and index.

Morehouse, Ward
The Waldorf-Astoria: America's Gilded Dream
New York, M. Evans, 1991. 260p. ISBN 0871316633

Filled with great "gossipy" stories of the two Waldorf-Astorias: the hotel that stood at the site of the Empire State Building and the current hotel on Park Avenue. Stories range from those of parties that featured Ringling's flying trapeze or party hostesses riding elephants to profiles of well-known personalities such as the Duke and Duchess of Windsor.

Nebel, Eddystone C.
Managing Hotels Effectively: Lessons from Outstanding General Managers
New York, Van Nostrand Reinhold, 1991. 436p.
ISBN 0442238142

The basis of this book is interviews of ten successful hotel general managers. The author describes the characteristics of the hotel business and the characteristics of successful general managers. Indexed.

Negotiating International Hotel Chain Management Agreements: A Primer for Hotel Owners in Developing Countries
New York, United Nations, 1990. 60p. ISBN 9211043379

Deals with the basic issues of negotiating agreements following the format of a typical hotel chain management agreement.

Nykiel, Ronald A.
Marketing in the Hospitality Industry
2nd ed. New York, Van Nostrand Reinhold, 1989. 305p.
ISBN 0442266979

Focus is on sending the correct marketing message to the multifaceted consumer. Includes a bibliography, glossary, and index.

Paige, Grace, and Jane Paige
Hotel/Motel Front Desk Personnel
New York, Van Nostrand Reinhold, 1989. 219p.
ISBN 0442204914

Front office procedures from A to Z! Written in a concise, easy-to-understand manner. The intended audience is students in two- or four-year programs or the practitioner.

Powers, Thomas F.
Introduction to the Hospitality Industry
2nd ed. New York, Wiley, 1992. 503p. ISBN 0471530549

Hospitality is presented as a unified, interrelated industry. Special attention is given to the topics of problem-solving and industry trends. Indexed.

Powers, Thomas F.
Marketing Hospitality
New York, Wiley, 1990. 434p. ISBN 0471638463

An introduction to marketing concepts and the value of marketing in the hospitality industry. Indexed.

Reid, Robert D.
Hospitality Marketing Management
2nd ed. New York, Van Nostrand Reinhold, 1989. 399p.
ISBN 0442278489

Reid goes beyond the usual introductory text on marketing basics and includes group sales, menu design, and pricing. Bibliography and index.

Renner, Peter Franz
Basic Hotel Front Office Procedures
3rd ed. New York, Van Nostrand Reinhold, 1994. 216p.
ISBN 0442016115

Step-by-step descriptions of technical front office procedures make this a unique contribution to the hospitality literature. The last chapter introduces transactional analysis as a framework for examining front desk communication. Glossary and index.

Sawin, Philip V., and others, comps.
A Literature Guide to the Hospitality Industry
New York, Greenwood Press, 1990. 99p. ISBN 0313267219

A review of the literature divided into the following major sections: dictionaries and encyclopedias, handbooks and manuals, periodicals, indexes, databases, bibliographies, association publications, directories, and statistics. Entries are annotated. Indexed.

Schaetzing, Edgar E.
Fachwörterbuch für Hotellerie and Gastronomie, Reisebüro and Reiseveranstalter: Deutsch-Englisch, English-German
3 Aufl. Frankfurt am Main, Germany, Deutscher Fachverlag, 1989. 342p. ISBN 3871502952

A basic German-English, English-German dictionary for the hospitality industry.

Schmidgall, Raymond S.
Hospitality Industry Managerial Accounting
2nd ed. East Lansing, Mich., Educational Institute of the American Hotel and Motel Association, 1990. 623p. ISBN 0866120580

Intended for practicing managers in the industry as well as students, this text assumes a knowledge of basic accounting concepts and procedures. As with all Educational Institute of the American Hotel and Motel Association texts, there are many illustrations and inset boxes showcasing industry leaders' insights. Computer basics are included in an appendix. Glossary and index.

Schneider, Madelin, and Georgina Tucker
The Professional Housekeeper
3rd ed. New York, Van Nostrand Reinhold, 1989. 503p. ISBN 0442205120

A very comprehensive treatment, from housekeeping techniques and products to management concerns. Glossary, bibliography, and index.

Schulz, Marjorie Rittenberg
Hospitality and Recreation
New York, Watts, 1990. 96p. ISBN 0531109739

Vocational guidance for the industry. Includes bibliography and index.

Sherry, John E. H.
The Laws of Innkeepers: For Hotels, Motels, Restaurants, and Clubs
3rd ed. Ithaca, N.Y., Cornell Univ. Press, 1993. 924p. ISBN 0801425085

The objectives of this book are threefold: (1) a reference manual for hotel, restaurant, and club owners, operators, and corporate executives on site; (2) an aid for attorneys in the general practice of law; and (3) a textbook for use in schools of hospitality management. This edition has several new chapters covering employment law, environmental law, catastrophic risk liability, and land use. Includes bibliography and index.

Shock, Patti J., and John M. Stefanelli
Hotel Catering: A Handbook for Sales and Operations
New York, Wiley, 1992. 400p. ISBN 0471544183

This book begins with an overview of the subject; specialized chapters follow on working with other hotel departments, function room selection and setup, and production and service planning. Includes glossary, bibliography, and index.

Wood, Roy C.
Working in Hotels and Catering
New York, Routledge, 1992. 184p. ISBN 041504782X

Reviews special issues in hotel and catering management. Examines industry employment trends, pay, education, and workforce composition in the United Kingdom. Bibliography and index.

Zaid, Barry
Wish You Were Here: A Tour of America's Great Hotels During the Golden Age of the Picture Post Card
New York, Crown Publishers, 1990. 95p. ISBN 0517580098

A delightful collection of picture postcards created from the 1930s to the 1950s showcasing North America hotels. Most of the legends from the message side of the cards are reproduced next to beautiful representations of the postcards. A fascinating history of the development of hotel amenities and design. The majority of the postcard views are exteriors, but there are sections entitled "The Grand Foyer" and "Dinner, Drinks, and Dancing."

Nonprint Materials

Hospitality Industry
Chatworth, Calif., AIMS Media, 1988. 1 videocassette (40 min.) VHS, Beta

Part of the "Career Awareness" series intended for hotel staff training.

Irish Castles
Sandy, Utah, Century Productions, 1991, 1 videocassette (50 min.) VHS

Looks at historic Irish castles converted into hotels. For general audiences and hotel staffs.

Office Technologies

Natalie Diamond
Formerly affiliated with Indiana Vocational Technical College
Evansville, Indiana

Introduction

Office Technology is one of the most rapidly changing fields in the world of education. It seems almost impossible to maintain currency with all of the software, new techniques, and updated methods of communication that face the office worker of today. When one includes requirements such as knowledge of computer networks, awareness of repetitive stress injury and how to avoid it, and informed decision making about hardware purchases, the subject is indeed overwhelming. Many of the books that relate to this topic are almost obsolete before they are published due to the rapid changes that are constantly being made.

Office Technology has changed dramatically from the time when the equipment in the typical office consisted of a typewriter, an adding machine, and a telephone. The individual who works in the office of today (and tomorrow) must be constantly attuned to the latest developments in office technology while demonstrating proficiency in traditional office procedures. Accuracy, courtesy, and intelligence are still needed in the modern office; these traits, however, must now be enhanced with a comprehension of new technologies.

Owing to the dynamics of the field, this bibliography focuses primarily on the electronic aspects of office technology. Citations are representative of the types of materials that can augment the Office Technology curriculum in two-year programs.

Accreditation

National Association of Executive Secretaries
900 S. Washington St., No. G-13
Falls Church, Va. 22046

National Business Educators Association
1914 Association Drive
Reston, Va. 22091-1596

Professional Secretaries International
10502 NW Ambassador Drive
P.O. Box 20404
Kansas City, Mo. 64195-0404

Selected Journals

Inc.: The Magazine for Growing Companies.
Inc. Pub. Co. Thirteen times a year. ISSN 0162-8968

Managing Office Technology.
Penton Pub. Co. Monthly. ISSN 1070-4051

Office Systems: The Magazine for Small and Medium Offices.
Office Systems Management. Monthly. ISSN 8750-3441

Office Technology Management.
Business Technology Communications. Monthly.
ISSN 1061-4656

PC Magazine: The Independent Guide to IBM-Standard Personal Computing.
Ziff-Davis. Monthly. ISSN 0888-8507

PC World.
PCW Communications. Monthly. ISSN 0737-8939

The Secretary.
Professional Secretaries International. Nine times a year.
ISSN 0037-0622

Windows Magazine.
CMP Publications. Monthly. ISSN 1060-1066

WordPerfect: The Magazine.
WordPerfect Pub. Co. Monthly. ISSN 1042-5152

Business   Office Technologies

Print Materials

Barrett, Charles Francis, Grady Kimbrell, and Pattie Odgers
Modern Office Procedures
Minneapolis/St. Paul, West Pub. Co., 1993. 440p.
ISBN 0314018999
    Although focusing on traditional office procedures such as filing and telephone etiquette, this work covers the many different ways to process information when utilizing computers and electronic communication systems. Includes helpful office tips.

Blyth, W. John, and Mary M. Blyth
Telecommunications: Concepts, Development, and Management
2nd ed. Mission Hills, Calif., Glencoe/McGraw-Hill, 1990. 362p.
ISBN 0026808412
    This book provides an overview of telecommunications including voice, data, message, and image communications. The intent of the authors is to offer a balanced overview of the benefits and problems encountered when using each method of communication. Included in each section is a bibliography of reference books, periodicals, reports, and newsletters for the reader who seeks additional information.

Burns, Patrick
Windows, Word, and Excel Office Companion
2nd ed. Chapel Hill, N.C., Ventana Press, 1994. 653p.
ISBN 1566040833
    Specifically written for the person who wants quick answers to questions about *Windows, Word,* or *Excel,* and who does not have time to read all of the user documentation that accompanies these programs. The book is divided into three distinct sections, one for each of the software programs discussed. Many easy-to-understand graphics are provided, and specific informational tips are highlighted, making them easy to locate.

Campbell, Mary
The Best 1001 WordPerfect Tips Ever
Berkeley, Calif., Osborne-McGraw-Hill, 1994. 1 v. (unpaged) and computer disk ISBN 0078818192
    Information is arranged by "Tip number" in a very descriptive table of contents. Even though the book is not paginated, it is easy to use. Covers *WordPerfect* for DOS versions 5.1 and 6.0. Disk features clip art, macros, and templates.

Cohen, Allan R., and David L. Bradford
Influence Without Authority
New York, Wiley, 1991. 319p. ISBN 0471548944
    Written by two professors of management who are also training consultants to major corporations, this work is based on the fact that one of the best ways to increase organizational effectiveness is to find new ways of leadership for all employees. The authors stress that managers must constantly be concerned with the interests of their colleagues (whether they are peers, subordinates, or superiors) because cooperation among colleagues is the key to effectiveness. Includes information on topics such as how to build effective relationships, understanding your allies, and understanding your own power.

Covey, Stephen R., A. Roger Merrill, and Rebecca R. Merrill
First Things First: To Live, to Love, to Learn, to Leave a Legacy
New York, Simon and Schuster, 1994. 360p. ISBN 0671864416
    Explores reasons why people find it hard to really put "first things first." Section One discusses the conflicts faced when individuals try to balance commitments, appointments, and schedules with vision, values, and mission. The authors advise on ways to balance one's life for the effective management of time. Several appendices are included and an index arranged according to specific issues and problems discussed in the book. An abridgment on audiocassette is also available (Simon and Schuster Audio, 1994.)

Currid, Cheryl
The Electronic Invasion: Survival Guide for the Brave New World of Business Communications
Carmel, Ind., Brady Publishing, 1993. 256p. ISBN 1566860857
    This book covers the gamut of communications options (e-mail, local area networks, fax machines, voice mail, video conferences, and mobile communications) that today's office workers encounter. It offers a clear table of contents as well as a detailed glossary and index.

Day, David E.
The Trouble with Fonts
Carmel, Ind., Alpha Books, 1993. 299p. ISBN 1567610307
    This well-written book is not about problems with typography as its title might indicate. Instead, the author, a commercial phototypesetter, has chosen to write a fascinating and informative overview of fonts. Includes many excellent and descriptive illustrations of a wide variety of fonts.

De Vries, Mary A.
Complete Secretary's Handbook
7th ed. Englewood Cliffs, N.J., Prentice Hall, 1993. 707p.
ISBN 0131596748
    This edition of a well-known, reliable, and basic book for secretaries reflects the electronic office of today. Covers topics such as computer ergonomics, magnetic storage media, and CD-ROM technology. Also discusses conventional procedures for offices that are not fully automated.

Foster, Jennifer
WordPerfect for Windows: Wiley Command Reference
New York, Wiley, 1992. 270p. ISBN 0471549029
    A comprehensive source book which enables the reader to quickly locate the commands, procedures, and applications necessary to effectively use *WordPerfect for Windows*. Each chapter covers a specific function on the menu bar. All commands and features are cross-referenced in the index, and the individual chapters each have "notes" and "see also" references that guide the reader to other parts of the book. An appendix provides an overview of the *Windows* concept and discusses how *Windows* and *WordPerfect* work together. Includes glossary.

Goodman, Danny, and Katherine Murray
Fear WordPerfect No More
New York, Brady Publishing, 1993. 296p. ISBN 1566860962
    The book's authors pride themselves on what the book does *not* have: "no very technical explanations, no programming codes, no

references to procedures rarely used." This volume simply presents the essential basics of *WordPerfect* for the user who wants to learn the basics with minimal technical information.

Green, William B.
Introduction to Electronic Document Management Systems
Boston, Academic Press, 1993. 205p. ISBN 0122981804

Divided into two sections, this book offers a basic introduction to the electronic management of documents. The management systems discussed are those built around digital images of documents, and those found in many storage and retrieval systems. The first section of the book is an overview of individual components of these systems, such as image scanners. The book's second section provides information on advanced details such as image compression and bar code technology.

Jaderstrom, Susan, and Leonard Kruk
Professional Secretaries International Complete Office Handbook: The Secretary's Guide to Today's Electronic Office
New York, Random House, 1992. 573p. ISBN 067940080X

This book is a standard for the office professional and enables the reader to review the traditional areas of expertise, and varied topics concerning the electronic office environment. Includes topics such as electronic communications, office publishing, and office computer software and hardware.

Jennings, Lucy Mae, Sharon Burton, and Nelda Shelton
Procedures for the Automated Office
3rd ed. Englewood Cliffs, N.J., Regents/Prentice Hall, 1994. 313p. ISBN 0131387103

Included in this book is a section on planning an office that will meet the requirements of today's technologies as well as those that will exist in the near future. A major section of the book is devoted to ergonomic information on lighting, acoustics, and positioning of equipment. Several chapters focus on the best ways to use technology to process information.

Kallaus, Norman Francis, and B. Lewis Keeling
Administrative Office Management
10th ed. Cincinnati, Ohio, South-Western Pub. Co., 1991. 755p. ISBN 0538701951

This latest edition of a well-respected standard text broadens the concept of word processing to that of text management. Includes in-depth discussions of office furniture and equipment, and discussion of the best use of space for today's sophisticated equipment. Glossaries follow each of the twenty-three chapters.

Klein, Ruth
Where Did the Time Go? The Working Woman's Guide to Creative Time Management
Rocklin, Calif., Prima Publishing, 1993. 250p. ISBN 1559582227

This book is a down-to-earth, how-to manual to help the working woman learn to balance priorities. The author stresses identification of goals, and then makes recommendations on the management of time so that achievable goals can be reached in a nonstressful manner. Based on the results of surveys conducted by the author, she has divided working women into three groups: the homemaker, the traditional woman, and the achieving woman. Klein discusses how the women in each group might best learn to manage their time most effectively.

Krumm, Rob
Power Shortcuts: Word 2.0 for Windows
New York, MIS:Press, 1992. 559p. and computer disk
ISBN 155828236X

With a very detailed and well-planned table of contents, the book takes users of *Word for Windows* through the transition from word processing to document production. The author provides excellent summaries at the end of each chapter.

Marler, Jerilyn
Unlocking WordPerfect 6.0: Mastering Styles, Merges, Macros, and Tables
New York, MIS:Press, 1993. 464p. and computer disk
ISBN 1558282513

The author has written this book for the reader who is very familiar with *WordPerfect 6.0* and who wants to concentrate on its advanced features. Many detailed exercises are provided so that these features can be easily mastered and used with confidence by the individual interested in enhancing basic skills.

Meilach, Dona Z.
Word Processing for Business Users
Hauppauge, N.Y., Barron's, 1993. 186p. and computer disk
ISBN 0812014669

This book provides the basic information which will enable any business office to be set up as effectively as possible with state-of-the-art technology. It is really a research tool for the world of word processing. A very useful appendix provides a resource list (complete with addresses and telephone numbers) for all products mentioned in the text.

Merriam-Webster's Secretarial Handbook
3rd ed. Springfield, Mass., Merriam-Webster, 1993. 590p.
ISBN 0877792364

Covers the essentials for office professionals, including records management, business English, telephones, telecommunication, and office equipment. Other topics are varied and information is presented in depth. As an example, the chapter on travel arrangements and international business includes basic information on setting up travel plans, and also includes a detailed table naming all the countries of the world. Listed in this table are the languages spoken, currency used, time differences, major cities and geographical areas, and area codes. The chapter on accounting and bookkeeping systems includes detailed information on data processing with computers and electronic calculators, in addition to basic bookkeeping rules.

Microsoft
Word 6.0 for Windows Resource Kit
Redmond, Wash., Microsoft Press, 1994. 637p. and 2 computer disks ISBN 1556157207

Divided into twenty-one easy-to-read chapters on *Word 6.0 for Windows*. This book also includes three valuable appendices, including a directory of the names and telephone numbers to contact within the Microsoft Corporation when questions or problems arise. Includes software on training, sample files, and a *Word 6.0 for Windows* converter.

Minceberg, Mella
WordPerfect 6 Made Easy
Berkeley, Calif., Osborne McGraw-Hill, 1993. 712p.
ISBN 0078818958
  This book is very well organized. The index, in particular, is well done, which makes information very easy to locate. A helpful feature of the book is a summary section of the *WordPerfect 6.0* commands.

Nelson, Stephen L.
Field Guide to Microsoft Word 6 for Windows
Redmond, Wash., Microsoft Press, 1994. 172p. ISBN 1557155778
A well-illustrated, quick reference book that provides a very visual path to needed information. Includes a section on troubleshooting that offers solutions to many of the most common problems faced by the new or casual user of *Word*.

Sabin, William A.
The Gregg Reference Manual
7th ed. Lake Forest, Ill., Glencoe, 1992. 502p. ISBN 0028199200
  This reference manual has been updated to include information on how to edit and proofread at the computer, and discusses the impact of electronic office equipment on the formatting of letters and other business documents. The book retains a number of chapters covering the essential rules of grammar, spelling, and capitalization, rules that remain the foundation upon which all office communication rests. Welcome additions to the book are two glossaries; one provides definitions of computer terms, and the other provides brief definitions of grammatical terms. Examples of various types of business documents are included.

Sanders, Rosanne Bryce
Administrative Procedures for the Electronic Office
Englewood Cliffs, N.J., Prentice Hall, 1991. 436p.
ISBN 0130194727
  The author provides an easy-to-read book which is divided into six main sections. Each of the sections, in turn, is subdivided into appropriate chapters. Communications management and records management are discussed from both the manual and electronic viewpoints. Many illustrations and graphics are included.

Silver, Susan
Organized to Be the Best! New Timesaving Ways to Simplify and Improve How You Work
2nd ed. Los Angeles, Adams-Hall Publishing, 1991. 434p.
ISBN 0944708226
  The author is a management consultant whose theory of time management is that individuals must learn specific skills to take control of their own lives. As an initial guide to saving time, she suggests that only three of the chapters in the book are essential. She also suggests that individuals read only those chapters that apply to their own life situations. Each chapter starts with a quick scan summary of the material. Also covers management of paper files. An outstanding guide for those who seek to lead more organized lives personally and professionally.

Wallace, Patricia E.
Records Management: Integrated Information Systems
3rd ed. Englewood Cliffs, N.J., Prentice Hall, 1992. 542p.
ISBN 0137699360
  Written to offer the reader an understanding of the scope and complexity of managing records, this update includes a special section on electronic records storage and retrieval. Technology chapters discuss imaging systems and bar code scanners. Also includes a comprehensive listing of records management publications and records management software.

Nonprint Materials

Changing Workplace
Princeton, N.J., Films for the Humanities, 1988. 1 videocassette (20 min.) VHS, and 1 workbook
  The workplace changes in response to changes in society and technology, and the office reflects the larger environment in which it exists. This video recounts the history of the professional secretary and provides discussion about the future of office professionals. Part Ten of *Hub of the Wheel Series: The Office Professional*.

I'm Tired of a Messy Desk
Vancouver, Wash., Career Development Software, 1990.
1 videocassette (25 min.) VHS, and 1 workbook (12p.)
  Recommends strategies for organization of work projects and activities including the REACT method: read, earmark for filing, choose to delegate, trash.

Managing Your Time
Princeton, N.J., Films for the Humanities, 1988. 1 videocassette (20 min.) VHS
  Because office support personnel often work for several people, their time management responsibilities and problems are complicated. This video highlights the importance of time management and provides suggestions for a proactive time management plan. Part Five of *Hub of the Wheel Series: The Office Professional*.

Office Automation
Shawnee Mission, Kans., RMI Media Productions, distributor, 1988. 1 videocassette (30 min.) VHS
  Discusses office automation technologies within the framework of the workplace environment and traditional office tasks. Examines voice mail, electronic mail, word processing, and data processing. A production of the Corporation for Community College Television.

Projecting a Professional Image
Princeton, N.J., Films for the Humanities, 1988. 1 videocassette (20 min.) VHS
  Covers the importance of a professional image in the workplace. Part One of *Hub of the Wheel Series: The Office Professional*.

Sexual Harassment on the Job
Princeton, N.J., Films for the Humanities, 1988. 1 videocassette (28 min.) VHS

Phil Donahue discusses sexual harassment in the workplace with Karen Sauvigne and Susan Meyer, directors of the Working Women United Institute.

Additional Resource

Insight Media
121 W. 85th St., New York 10024

This distributor is a good source for videocassettes on computers in the office, specific applications (such as *DOS, Lotus, WordPerfect*), business communications, interviewing, and job success.

Real Estate

Antoinette Byers
M. M. Bennett Library—Clearwater Campus
St. Petersburg Junior College
Clearwater, Florida

Introduction

The Real Estate Programs in community colleges prepare students for certification in real estate. The curriculum for many of these vocational programs has been developed cooperatively with advisory committees of realtors who continually evaluate the courses in light of specific community needs.

According to the *Occupational Outlook Handbook*, career opportunities in this field are expected to grow at an average rate until the year 2005. This industry is heavily dependent on the current climate of economic trends and fluctuations. Positions within the real estate field are numerous: agent, auctioneer, appraiser, broker, counselor, developer, subdivider, educator, mortgage loan officer, planner, or manager. The majority of students enter the real estate field as real estate agents. Agents are independent sales representatives who work for a licensed broker, usually on a contract basis. Earnings range from commission-based to salaried income.

Students need fundamental background in a variety of areas, and the curriculum is developed to meet those needs. Courses prepare students in areas such as the economics of the housing market, community and neighborhood analyses, zoning and tax laws, and mortgage financing.

All real estate agents and brokers must be licensed by the state in which they practice. States require that agents be at least eighteen years of age, have completed high school, and pass a written test. The test includes questions regarding real estate practices and transactions and various aspects of real estate law. Many states also require formal classroom training before a license is issued and continuing education requirements for renewal.

This selective bibliography has been prepared to meet primarily the needs of the real estate students and those involved in continuing education. In addition, a section on popular titles has been added to assist students in other disciplines with their real estate and investment concerns.

Selected Associations

Real Estate agents, brokers, and appraisers are certified to practice by state agencies. The following associations may provide additional information on standards, certifications, and professional issues:

American Institute of Real Estate Appraisers
430 North Michigan Avenue
Chicago, Ill. 60611

American Society of Appraisers
P. O. Box 17265
Washington, D.C. 20041

Appraisal Institute
225 N. Michigan Ave., Suite 724
Chicago, Ill. 60601-7601

National Association of Realtors
430 North Michigan Avenue
Chicago, Ill. 60611

National Auctioneers Association
8880 Ballentine Street
Overland, Kans. 66214

Society of Industrial and Office Realtors
777 Fourteenth Street, NW
Washington, D.C. 20005-3279

Women's Council of Realtors
430 North Michigan Avenue
Chicago, Ill. 60611

Selected Journals

The Appraisal Journal.
Appraisal Institute. Quarterly. ISSN 0003-7087

Appraiser News.
Appraisal Institute. Semimonthly. ISSN 1054-5999

Commercial Investment Real Estate Journal.
Realtors National Marketing Institute. Quarterly. ISSN 0887-4778

Existing Home Sales.
National Association Realtors. Monthly. ISSN 0161-5882

Journal of Real Estate Taxation.
Warren, Gorham, and Lamont. Quarterly. ISSN 0093-5107

National Real Estate Investor.
Communication Channels. Monthly. ISSN 0027-9994

New Homes Magazines.
MDM Publications. Six times a year. ISSN 0192-4893

Real Estate Finance Journal.
Warren, Gorham, and Lamont. Quarterly. ISSN 0898-0209

Print Materials

✦ Reference

Estes, Jack C., and Dennis R. Kelley
McGraw-Hill's Compound Interest and Annuity Tables
2nd ed. Edited by David Conti. New York, McGraw-Hill, 1992. 247p. ISBN 0070196869

This best-selling book provides compound interest and annuity tables for monthly, quarterly, semiannual, yearly, and daily payments with interest rates ranging from 5 to 16 percent. Various mortgage periods are provided. Includes a glossary of real estate terms.

Estes, Jack C., and Dennis R. Kelley
McGraw-Hill's Interest Amortization Tables
2nd ed. Edted by David Conti. New York, McGraw-Hill, 1993. 311p. ISBN 0070196966

Catering to the needs of home buyers and borrowers of the 1990s, this easy-to-use source aids in the calculation of mortgage loan payments. Payment categories include adjustable-rate mortgages, and balloon, biweekly, and jumbo loans.

Friedman, Jack P., Jack C. Harris, and Bruce J. Lindeman
Dictionary of Real Estate Terms
3rd ed. Hauppauge, N.Y., Barron's, 1993. 389p.
ISBN 0812014340

This pocket guide offers up-to-date explanations of real estate vocabulary. The text is accompanied by diagrams, charts, mathematical formulas, and a mortgage payment table. The appendix includes an explanation of the *Internal Revenue Code* (West Pub. Co., 1954-). All users of this book will benefit, regardless of their level of expertise.

Miles, Martin J.
The Vest-Pocket Real Estate Advisor
Englewood Cliffs, N.J., Prentice Hall, 1989. 534p.
ISBN 0139450645

This portable guide to real estate presents guidelines, formulas, tables, and techniques in a quick reference. Includes index.

Reilly, John W.
The Language of Real Estate
Chicago, Real Estate Education Co., 1993. 456p.
ISBN 0793105838

This quick reference defines terms, topics, and abbreviations of the real estate, appraisal, and finance environments. Includes a subject classification of terms, 350 abbreviations and their meanings, and the text of the Realtor's Code of Ethics, and Standards of Practice of the National Association of Realtors. Includes an example of a closing problem and a settlement worksheet and statement.

Thomsett, Michael C.
The Illustrated Real Estate Dictionary
Homewood, Ill., Dow Jones-Irwin, 1988. 220p. ISBN 1556231652

Provides definitions of over 1,000 key terms used in the real estate profession. Definitions are augmented by numerous graphs, charts, and numerical calculations. Includes a real estate checklist, amortization tables, and a table of abbreviations.

Tosh, Dennis S., Jr., and Nicholas O. Ordway
Handbook of Real Estate Terms
Rev. ed. Englewood Cliffs, N.J., Prentice Hall, 1992. 522p.
ISBN 0133760707

This dictionary of real estate terms provides numerous abbreviations and their meanings, and a directory of organizations and agencies (including real estate commissions in the U.S. and Canada). Includes sample documents such as settlement statements and appraisal reports.

Wiener, Eric
Mortgage Payment Handbook: Monthly Payment Tables and Annual Amortization Schedules for Fixed-Rate Mortgages
New York, Barnes and Noble Books, 1992. 199p.
ISBN 0880299703

This pocket guide is useful for calculating mortgage payments, finding loan sources, and providing explanations of financial terms. Especially useful for those planning financing or refinancing of real estate.

✦ Appraisals

The Appraisal of Real Estate
10th ed. Chicago, Appraisal Institute, 1992. 768p.
ISBN 092215404X

This textbook on appraisal provides background information on the appraisal process: valuation, money markets, data collection, and analysis. Covers information on depreciation and comparison approaches including the income capitalization approach. Provides information on writing the appraisal report. Appendices treat professional issues, e.g., codes of ethics, legislation related to appraisal work, a mathematics review, and financial tables. Includes a bibliography and index.

Betts, Richard M., and Silas J. Ely
Basic Real Estate Appraisal
3rd ed. Englewood Cliffs, N.J., Prentice Hall, 1994. 496p.
ISBN 0130759872

This a comprehensive outline of appraisal theory and practice is valuable for community college students. Discusses skills and practices of appraisal, including formal appraisal processes, analysis techniques for neighborhoods and communities, and economic principles applicable to real estate. Includes information on property inspections, various approaches to comparisons, property valuation techniques, construction facets, and reporting practices. Vocational information, answers to chapter reviews, a glossary, and index are also provided.

O'Connell, Daniel J.
The Appraisal of Apartment Buildings
New York, Wiley, 1989. 184p. ISBN 0471509558

Provides a fundamental guide to the preparation of an apartment appraisal. Includes topics such as property inspection, income estimation, expense statements, operating expenses, and various approaches to appraisals. Appendices cover information on tax considerations, calculator procedures, and a sample appraisal report.

✦ Basic Skills and Practices

Cummings, Jack
The McGraw-Hill Thirty-Six Hour Real Estate Investing Course
New York, McGraw-Hill, 1993. 358p. ISBN 0070150486

An independent study source for real estate that includes user goals, key terms, an explanation of real estate concepts, case studies, and an open-book final examination. Fosters negotiating skills, profit calculations, financing skills, and a strong understanding of the real estate field. Key terms at the beginning of each chapter provide opportunities to increase vocabulary skills.

Galaty, Fillmore W., Wellington J. Allaway, and Robert C. Kyle
Modern Real Estate Practice
13th ed. Chicago, Real Estate Education Co., 1993. 433p.
ISBN 0793107040

Part One of this text provides a general introduction to real estate and property principles, including legal issues, brokerage agency information, listing agreements, and ownership issues such as financing. Part Two includes topics such as appraisals, land use, property development, and ethical issues. The students will appreciate the outline of essential concepts and the sample examination. (This course and study guide are also available on cassette.)

Harris, Jack C., and Jack P. Friedman
Barron's Real Estate Handbook
3rd. ed. Happauge, N.Y., Barron's, 1993. 810p.
ISBN 0812063309

This third edition serves as a complete reference to real estate. Provides information on home buyers' and home sellers' guides; real estate closings; investment analysis; real estate regulations; and mortgage loan tables. Includes a glossary and a bibliography.

Harwood, Bruce M., and Charles J. Jacobus
Real Estate Principles
6th ed. Englewood Cliffs, N.J., Prentice Hall, 1993. 695p.
ISBN 0137659342

Harwood provides a study guide for the real estate examination, as well as an overview and history of the field. Includes basic principles and practices of real estate. Appendices provide construction terms with illustrations, a review of real estate math, and conversion tables. Each chapter has study questions with answers.

Maloney, Roy T.
Real Estate Quick and Easy: Concepts of Real Estate Clearly Illustrated
Updated 10th ed. San Francisco, Dropzone Press, 1990. 384p.
ISBN 0913257036

Provides a graphical approach to real estate including graphs, charts, and illustrations. Students who prefer to study independently will find that this workbook offers a very descriptive and original approach to real estate fundamentals. Using graphs, charts, and illutrations, the author breaks down the basic concepts of real estate into six broad areas. This is also of interest to the potential home owner or beginning student.

McKenzie, Dennis J., and Richard M. Betts
Essentials of Real Estate Economics
3rd ed. Englewood Cliffs, N. J., Prentice Hall, 1992. 336p.
ISBN 0132877236

This text is geared to beginning students who have little or no background in economics. Provides basic economic background for real estate analysis: the government's role in the economy, an analysis of credit and money supply, community growth patterns, and real estate markets (residential, commercial, industrial, rural, and recreational). Discusses the influences of taxation on real estate. Includes answers to chapter reviews and an index.

Modern Real Estate Practice Study Guide
13th ed. Chicago, Real Estate Education Co., 1994. 230p.
ISBN 0793107288

This guide is designed for students who prefer to study independently. Chapter summaries, key terms, illustrations, and self-examinations and practical exercises aid understanding.

Ventolo, William L.
Mastering Real Estate Mathematics: A Self-Instructional Text
5th ed. Chicago, Real Estate Education Co., 1989. 377p.
ISBN 0884628132

Designed as a self-instruction manual or review, this work covers basic arithmetic, percentages, fractions, and decimals. Basic real estate mathematical transactions are explained: calculation of commissions, pricing, interest, and taxes. Includes chapter-based reviews, an examination, and an answer key.

Warda, Mark
How to Negotiate Real Estate Contracts: For Buyers and Sellers: With Forms
Clearwater, Fla., Sphinx Publishing, 1993. 124p.
ISBN 091382559X

Warda's contract books are designed to clearly explain the various types of clauses found in real estate contracts. Includes guidance on preparing a contract, when to consult an attorney, the art of negotiating, and federal and local laws. Covers various contract clauses such as risk of loss, lien affidavits, termites, and survival of contract. Presents options to exercise after signing the contract: backing out, rescuing the deal, closing, and after closing. Five different contracts are provided as examples. Indexed.

✦ Financing

Friedman, Jack P., and Jack C. Harris
Keys to Mortgage Financing and Refinancing
2nd ed. Hauppauge, N.Y., Barron's, 1993. 139p.
ISBN 0812014367

The primary focus of this work is on obtaining mortgage financing or refinancing during a fluctuating economy. This handy guide shares important guidelines as well as necessary cautions.

Mettling, Stephen R., and Gerald R. Cortesi, eds.
Modern Residential Financing Methods: Tools of the Trade
2nd ed. Chicago, Real Estate Education Co., 1990. 189p.
ISBN 088462885X

Real estate has undergone such a major transformation during the past ten years that an analysis of mortgage financing methods is essential. This well-organized workbook caters to both the real estate techniques of the 1990s and the vital issues that emerged in the 1980s. Includes sample forms, definitions, formulas, and descriptive diagrams that aid understanding.

Sirmans, C. F., Jr.
Real Estate Finance
2nd ed. New York, McGraw-Hill, 1989. 550p. ISBN 007057698X

Covers fundamentals of real estate finance including background on the financing decision and various financing options: varying-rate and wrap-around mortgages, construction financing, and land-development and participation mortgages.

✦ Sales Strategies and Techniques

There are numerous books recounting personal success strategies in real estate. A few titles have been included as representative examples.

Caughman, Joyce L.
Real Estate Prospecting: Strategies for Farming Your Markets
2nd ed. Chicago, Real Estate Education Co., 1994. unpaged.
ISBN 0793109450

Caughman presents a step-by-step approach to increasing sales through the tools of technology, a personal marketing program, time management, farming plans, and effective direct mail plans.

Ferry, Mike
How to Develop a Six-Figure Income in Real Estate: Super Star Selling the Mike Ferry Way
Chicago, Real Estate Education Co., 1993. 198p.
ISBN 0793104904

Ferry's "Super Star Success Secrets" provide insight into his strategies and techniques for listing and selling homes. Includes index.

Hopkins, Tom
How to Master the Art of Listing and Selling Real Estate
Englewood Cliffs, N.J., Prentice Hall, 1991. 402p.
ISBN 0134022564

Hopkins, a real estate sales trainer, presents his successful strategies and techniques. Provides tips on canvassing neighborhoods, addressing the needs of "For Sale by Owner" clients, uncertain buyers, price negotiation, and techniques for successful open houses. Hopkins includes a daily activity chart to use when tracking success. An extension of Hopkin's best seller, *How to Master The Art of Selling (*Warner Books, 1982*)*.

Kennedy, Danielle
Double Your Income in Real Estate Sales
New York, Wiley, 1993. 272p. ISBN 0471579734

Presents techniques and strategies for increasing sales performance by providing increased service. Includes marketing tips, problem solving and negotiation skills, and Kennedy's "Leadership Management Formula." Includes bibliography and index.

Kennedy, Danielle, and Warren Jamison
How to List and Sell Real Estate in the 90s
Englewood Cliffs, N.J., Prentice Hall, 1990. 481p.
ISBN 0134022491

Focus is not only on the mechanics of real estate, but also on the marketing strategies to increase earning potential. Includes neighborhood and community analyses.

Lyons, Gail G., ed.
Real Estate Sales Handbook
10th ed. Chicago, Real Estate Education Co., 1994. 274p.
ISBN 0793109477

Lyons provides an overview of the various approaches to real estate sales, including the use of technology such as computers and faxes to stimulate sales. Discusses getting started, various issues of professionalism, sales and listing qualities, time management, and negotiating sales. The focus of this work is on expanding service. Includes glossary and index.

Pivar, William H.
Power Real Estate Selling
2nd ed. Chicago, Real Estate Education Co., 1989. 250p.
ISBN 0884621529

Pivar presents his techniques and strategies to sell more properties. Using examples that employ role playing and dialogues, the reader can focus on the buyer's emotional and financial needs. Tips are provided to help the agent learn to work with buyers to resolve particular issues of concern. Includes glossary and index. Also available on cassette.

Pivar, William H., and Corinne E. Pivar
Power Real Estate Letters: A Professional's Resource for Success
Newly rev. Chicago, Real Estate Education Co., 1994. 346p. and 1 computer disk. ISBN 0793111153

Pivar provides 279 adaptable letters for self-promotion in real estate. Includes examples for listing solicitation, responses to inquiries, service letters for listings, press releases, and investment buyer letters, to name a few.

Reynolds, Monica
Multiply Your Success with Real Estate Assistants
Chicago, Real Estate Education Co., 1994. 247p.
ISBN 0793107768

Presents strategies for increasing success in the real estate market through real estate assistants. Hiring guidelines are included, along with interview questions and rating sheets. Provides advice on goal setting, telephone effectiveness training, and office procedures and policies. Lists 500 jobs to delegate.

Schwarz, Barb
How to List Residential Real Estate Successfully
Englewood Cliffs, N. J., Prentice Hall, 1991. 227p.
ISBN 0133965163

Helps novice and experienced salespersons enhance their techniques for greater listing and selling success. Provides a step-by-step approach to self-promotion and promotion of listings.

Tappan, William T.
Real Estate Exchange and Acquisition Techniques
2nd ed. Englewood Cliffs, N.J., Prentice Hall, 1989. 307p.
ISBN 0137626428

Focuses on the advantages and techniques of real estate exchanges. Presents information on opportunities, specific tax benefits, legal aspects, negotiation skills for acquisition, techniques for financing, and closing transactions. Includes a section on acquiring real estate without cash. Appendices include related Internal Revenue Code sections, sample forms, and IRS forms for reporting an exchange. A bibliography and index are provided.

✦ Vocational Guidance

Cross, Carla
How About a Career in Real Estate?
Chicago, Real Estate Education Co., 1993. 177p.
ISBN 0793107458

This career exploration manual describes the day-to-day activities of a real estate professional, including potential earnings, program description, licensing requirements, and evaluation of companies for employment. The appendix includes survey results of the expectations of new agents. Includes bibliography and index.

Cross, Carla
Real Estate Agent's Business Planning Guide
Chicago, Real Estate Education Co., 1994. 256p.
ISBN 0793109558

This graphical approach to the real estate business is a pleasant antidote to the numerous text-oriented volumes. Cross uses charts, tables, and checklists to stimulate creativity and innovation in business. Includes information on customer surveys, low-cost promotional tactics, sample business plans, marketing guidelines, and a timeline to align spending, income, and profits. Tips for effective newsletters, brochures, and direct mailings are also provided.

Edwards, Kenneth W.
Your Successful Real Estate Career
New and updated ed. New York, American Management Association, 1993. 176p.
ISBN 0814477852

In this discussion of various aspects of a career in real estate, Edwards presents selected professional traits that are assets in the field, surveys job-related activities, and considers long-term job opportunities. Includes a real estate directory, an appendix, a bibliography, and an index.

Edwards, Kenneth W.
Your Successful Real Estate Career
New and updated ed. New York, American Management Association, 1993. 176p. ISBN 0814477852

This career launch manual provides information on how to pass the licensing exam, how to choose a company to work for, and tips on listing and selling. Case studies of individuals in real estate careers provide useful insights on opportunities in the profession. Includes bibliography and index.

Evans, Mariwyn
Opportunities in Real Estate Careers
Lincolnwood, Ill., VGM Career Horizons, 1988. 149p.
ISBN 0844264768

Provides an overview of the field, employment outlook, educational requirements, salary information, and sources for addition information. Includes bibliography.

Evans, Mariwyn
Real Estate: A VGM Career Planner
Lincolnwood, Ill., VGM Career Horizons, 1989. 128p.
ISBN 0844286761

Presents a concise overview of the field, the employment outlook, guidelines and tips for career advancement, educational requirements, and salary information. *VGM Career Planner Series.*

Janik, Carolyn, and Ruth Rejnis
Real Estate Careers: Twenty-Five Growing Opportunities
New York, Wiley, 1994. 214p. ISBN 0471592048

This career guide goes beyond the traditional information on agent and broker careers. Includes insights on twenty-five professional positions in real estate.

✦ Popular Titles for Homeowners

Albrecht, Donna G.
Buying a Home When You're Single
New York, Wiley, 1994. 193p. ISBN 0471024996

This book caters to one of the fastest growing real estate customer groups today--the single person. Albrecht covers how to prequalify and determine what is affordable, and presents information on the types of loans available. Also contains suggestions for selecting a property and tips for negotiating. Glossary included.

Allen, Robert G.
Nothing Down for the 90s: How to Buy Real Estate with Little or No Money Down
New rev. ed. New York, Simon and Schuster, 1990. 362p.
ISBN 0671-72558-0

Robert Allen, a real estate investment leader, is a self-made multimillionaire in real estate. In this work he illustrates the tools and skills needed to become a successful real estate investor and explains the signs of a profit-making successful investment. Includes formulas, strategies, diagrams, and index.

Bell, W. Frazier
How to Get the Best Home Loan
New York, Wiley, 1992. 232p. ISBN 0471558508.

Presents mortgage information including explanations of the different loan types, the secondary markets, appraisals, different types of real estate available, and types of lenders. Forms are included in an appendix.

Bly, Amy Sprecher, and Robert W. Bly
How to Sell Your House, Condo, Co-op
Yonkers, N.Y., Consumer Reports Books, 1993. 242p.
ISBN 0890436711

Examining strategies for selling a house, condo, or co-op, this step-by-step guide discloses important information such as setting a price, developing rapport with a realtor or agent, tips for a depressed or slow real estate market, price negotiation, and review of contracts.

Boroson, Warren
Save Thousands on Your Mortgage: The Best Investment You Can Make
New York, Collier Books/Macmillan, 1990. 245p.
ISBN 0020283458

Boroson's work demonstrates how thousands of dollars can be saved over the life-span of a ten- to thirty-year mortgage period by applying a monthly payment formula. Includes mortgage tables.

Brill, Jack A., and Alan Reder
Investing from the Heart: A Guide to Socially Responsible Investments and Money Management
New York, Crown Publishers, 1992. 414p. ISBN 0517584956

Addresses the need for both money management and socially responsible investments (SRI), i.e., investments that better the condition of the world. Presents information on the "Investing from the Heart Action Plan" and how to build a basic financial groundwork. Includes a list of organizations designed to help investors make socially responsible investments, and a money management and SRI glossary. Indexed.

Burton, Kermit
The Non-Lawyers A-B Trust Kit
Special book ed. Tucson, Ariz., Alpha Publications of America, 1992. 1v. (various pagings). ISBN 0937434469

This do-it-yourself text provides legal guidance on the assignment and transfer instruments for deeds, and information on format and descriptions of forms for all fifty states. Burton provides examples of trust document forms and actual forms that can be removed from the book. Includes a toll-free telephone number for assistance with any questions.

Dallow, Theodore J.
How to Buy Foreclosed Real Estate for a Fraction of Its Value
Holbrook, Mass., Bob Adams, 1992. 152p. ISBN 155850026X

This source, designed to answer questions regarding the purchase of foreclosed real estate, is useful for the individual willing to invest not only money, but also hard work and time to realize a profit. Dallow, an expert in the field, clearly explains the particular facets involved in this type of real estate.

Dworin, Lawrence
Profits in Buying and Renovating Homes
Carlsbad, Calif., Craftsman Book Co., 1990. 290p.
ISBN 0934041571

Dworin, an investor and remodeler, focuses on the money to be made in buying and renovating real estate. Provides information on the selection of fixer-uppers. Also includes a checklist for repairs.

Effros, William G.
How to Sell Your Home in Five Days
New York, Workman Pub. Co., 1993. 260p. ISBN 1563054965

Utilizing the plan that Effros prepared to sell his own house within a five-day time frame, this work provides guidance on newspaper ads, the selection and use of a broker, pricing, and legal aspects to consider. This-easy-to-read guide is written in a step-by-step format and includes a checklist for organization. The author stresses the importance of consulting with a lawyer since information may vary from state to state.

Gadow, Sandy
All About Escrow: Or How to Buy the Brooklyn Bridge and Have the Last Laugh
5th ed. El Cerrito, Calif., Express, 1992. 222p. ISBN 0932956173

Provides an overview of the most common real estate questions buyers and sellers have regarding closings. Includes explanations of real estate terminology such as escrow, title searches, and mortgage insurance.

Garton-Good, Julie
All About Mortgages: Inside Tips to Finance the Home
Chicago, Ill., Real Estate Education Co., 299p. ISBN 0793109493

Provides a practical approach to mortgage options including the author's tips.

Hardwick, Phil
Two Hours of Real Estate: One Minute at a Time
Brandon, Miss., Quail Ridge Press, 1993. 160p.
ISBN 0937552534

This quick guide to real estate addresses the basic concerns most people have when buying or selling a house: determining affordable payments, best seasons to sell, how much to pay for advertising, and interpreting contracts.

Hicks, Tyler Gregory
How to Make One Million Dollars in Real Estate in Three Years Starting with No Cash
2nd ed. Englewood Cliffs, N.J., Prentice Hall, 1989. 259p.
ISBN 0134236742

Hicks contends that it is still possible to build a fortune from real estate and shares his step-by-step formula for obtaining that goal in this book. Discusses leverage, financing, and conversions of apartment houses or office buildings to condominiums. Stresses the use of private lenders.

Irwin, Robert
The For Sale by Owner Kit: Hooking the Buyer, Closing the Deal, Saving $$ on Commissions
Chicago, Dearborn Financial Publishing, 1993. 179p.
ISBN 0793103916

Irwin, an author of more than twenty real estate books, caters to the student who desires a direct approach to real estate. This book provides background on key subjects such as how to sell a house, taxes, pricing, and closing a real estate deal. Includes forms, checklists, and sample newspaper advertisements. Indexed.

Irwin, Robert
Tips and Traps When Buying a Home
New York, McGraw-Hill, 1990. 190p. ISBN 0070321388

Irwin acknowledges that purchasing real estate is a major decision that involves careful study of when to purchase and the reasons for buying. He addresses negotiating skills, determining what is affordable, and identifying real estate to meet family needs and financing. Indexed.

Jordan, Cora
Neighbor Law: Fences, Trees, Boundaries and Noise
Berkeley, Calif., Nolo Press, 1991. ISBN 0873371585

Laws relating to common disagreements between neighbors are discussed in this work. Written in a step-by-step format, it provides guidance for ending property disputes on agreeable terms.

Kiplinger's Buying and Selling a Home
4th ed. Washington, D.C., Kiplinger Books, 1993. 420p.
ISBN 0938721267

Kiplinger responds to changes in the real estate market with this broad discussion of a variety of real estate topics. Provides information on adjustable-rate mortgages, changing interest rates, and shifting economic conditions. Includes charts, work sheets (for such needs as refinancing calculations and home equity loans), and a glossary of terms. Also discusses alternatives to selling a house.

Lank, Edity
The Homebuyer's Kit
3rd ed. Champaign, Ill., Real Estate Education Co., 1994. 194p.
ISBN 0793111145

A useful guide to basic real estate practices for the first time buyer. Includes information on contract negotiation and financing, as well as working with a real estate agent.

Lerch, Gregory D.
How to Sell Your Home When Homes Aren't Selling
White Hall, Va., Betterway Publications, 1991. 254p.
ISBN 155870182-6

The needs of clients who want to sell their own homes are addressed in this work. Covers topics such as market identification, advertising, attracting prospective buyers, and negotiating the deal. Includes graphs, charts, and appendices. Indexed.

Levy, Mitchell A.
Home Ownership: The American Myth
2nd ed. Cupertino, Calif., Myth Breakers, 1993. 160p.
ISBN 0963330217

Levy debates the value of home ownership by addressing the benefits, cost, savings, realities, and rent versus purchase options. Presents an analysis of twenty-five metropolitan cities based on the median home price and the average monthly rent.

Lieberman, Dan, and Paul Hoffman
Getting the Most for Your Home in a Down Market
Holbrook, Mass., Bob Adams, 1991. 345p. ISBN 1558500359

Lieberman presents strategies for the nineties, with an understanding of a real estate market that is often depressed or competitive. Includes information on how and when to sell, common mistakes to avoid, and confronting the realities of the market. Also provides marketing tips.

Lohmar, Ceil
For Sale by Owner: A No-Nonsense Guide to Getting the Best Price for Your House
Chicago, Probus Pub. Co., 1990. 100p. ISBN 1557381615

This seven-step workbook guides homeowners in preparing to sell their own homes. Selection of a mortgage lender is reviewed, as well as pricing and the closing of the real estate deal. Includes glossary and index.

Lumley, James E. A.
How to Get a Mortgage in Twenty-Four Hours
3rd ed. New York, Wiley, 1994. 263p. ISBN 0471599379

In question-and-answer format, Lumley outlines the key issues and roles of various departments involved in a real estate transaction, providing a unique perspective of the process. New information is also presented on obtaining low-interest loans and pre-qualification before purchasing a home. Includes sample letters.

Miller, Peter G.
The Common-Sense Mortgage: How to Cut the Cost of Home Ownership by $100,000 or More
1992 ed. New York, HarperPerennial, 1992. 298p.
ISBN 0062731092

Miller presents strategies to cut the cost of traditional mortgages. In addition, he provides an overview of the lending system and various mortgage options.

Mullin, Dale H.
Your Home: A Home Buying, Selling, Building, Remodeling Guide
Nashville, Tenn., Hampshire House Pub. Co., 1991. 175p.
ISBN 0963192108

Provides suggestions on what to look for when purchasing real estate. Includes information on the advantages of buying a house, various types of financing available, designing house plans, choosing a location, and house inspections.

Walman, David
How to Negotiate a Lease on Your Terms: Guerrilla Tactics for Renting an Apartment, House, or Trailer Home
Deer Park, N.Y., The Forms Man, 1991. 159p. ISBN 1879191002

Includes useful information and techniques for both landlords and tenants. Provides budget forms, checklists, sample forms and clauses, and information on breaking a lease without being penalized. Mobile home parks are also discussed.

✦ Popular Titles on Real Estate Management

Goodwin, Daniel, and Richard Rusdorf
The Landlord's Handbook: A Complete Guide to Managing Small Residential Properties
Chicago, Longman Financial Services Publications, 1989. 237p. ISBN 0884624056

Explores the skills needed for managing residential properties for income. Discusses the five basic functions of property management: applications, leases, rental agreements, rent collection, and accounting. Includes sample letters, notices, and forms.

Newman, Jeanne D.
The Tenant's Leasing Handbook
Chicago, Dearborn Financial Publishing, 1992. 216p. ISBN 0793103177

Written by a real estate leasing specialist, this book answers questions pertaining to leasing. Focusing on the prospective tenant, information is provided on lease negotiation, leasing terms, conditions, and lease proposals. Reviews retail establishments, restaurants, shopping centers, and medical offices, as well as construction and remodeling costs. Indexes and appendices are provided.

Robinson, Leigh
Landlording: A Handymanual for Scrupulous Landlords and Landladies Who Do It Themselves
7th ed. El Cerrito, Calif., ExPress, 1994. 430p. ISBN 0932956165

Offers how-to information, timesaving tips, examples, and explanations of general and legal forms, bookkeeping, and other practical landlord issues. In addition, Robinson has provided a comprehensive introduction and analysis of some of the best tried-and-true management methods.

Stevens, Lawrence L.
Landlording as a Second Income: The Survival Handbook
Lanham, Md., Scarborough House, 1994. 215p. ISBN 0812840240

Stevens covers the essentials of landlording and its advantages and disadvantages. Discusses the possibilities associated with second incomes.

✦ Popular Titles on Real Estate Investment

Band, Richard E.
Contrary Investing for the 90s: How to Profit by Going Against the Crowd
New York, St. Martin's Press, 1989. 260p. ISBN 0312038046

Band's book reflects the trends of the nineties regarding the concepts of investing in stock or real estate. He advises the reader to investigate overlooked opportunities and to apply the art of thinking for oneself. Alternative avenues to take during periods of inflation are suggested and a profile analysis test is included.

Barr, Gary K.
J. K. Lasser's Real Estate Investment Guide
New York, J. K. Lasser Institute, 1989. 330p. ISBN 0135108845

Lasser's guide presents information on tax credits, types of financing, contracts, and the characteristics of real estate investment. The focus of this work is on developing a financial life cycle through real estate investments.

Burgauer, James
Do-It-Yourself Investment Analysis: Practical Guide to Life Cycle, Fundamental, and Technical Analysis
Chicago, International Pub. Corp., 1990. 165p. ISBN 0942641248

This do-it-yourself guide is presented in a convenient-sized format ideal for quick reference. Provides an introduction to investment analysis and explains the fundamentals and risks involved in investment. Sample formulas and graphs complement the text. Includes a list of sources for additional information and a glossary.

Cushman, Robert Frank, Michael L. Evans, and Arnold S. Levy
The Professionals' Guide to Commercial Property Development
Homewood, Ill., Dow Jones-Irwin, 1988. 354p. ISBN 0870949683

Although tax laws and the economy of the real estate market change regularly, this book still provides the reader with the basics. Includes charts for analyzing capital investments, information on selection of architects and contractors, and considerations for financing the project. Targeted audience is property developers and business owners who are seeking to develop or construct commercial properties. Indexed.

Sirota, David
Essentials of Real Estate Investment
4th ed. Chicago, Real Estate Education Co., 1991. 259p. ISBN 0793101042

This workbook discusses fundamentals: ownership interests, financing, income taxes and real estate investments, feasibility studies, financial analysis, and land investment. Includes areas outside of the residential properties such as office buildings, strip stores and shopping centers, and industrial properties. Also addresses alternate investment opportunities such as franchises, mineral rights, air rights, real estate securities, and mortgage investment conduits. Includes an answer key, glossary, and topic index. Fifth edition was scheduled for late 1994. Sirota also authored a companion volume: *Essentials of Real Estate Finance*, Prentice Hall, 1994.

Nonprint Materials

Agency: Choices, Challenges, and Opportunities
Chicago, National Association of Realtors, 1993. 1 videocassette (18 min.) VHS and 3 booklets.

This four-hour study course introduces the role of the agency, including fiduciary duties and agency relationships.

Are You Buying a Home? A Professional Guide to Home Inspection
St. Claire, Mich., dkb Productions, 1988. 1 videocassette (63 min.) VHS

Presents a complete home inspection by a licensed home builder. Includes inspection of foundations; electrical, plumbing, and heating systems; and carpentry.

Bower, John, and Linda Bower
Your House, Your Health: A Non-Toxic Building Guide
Unionville, Ill, Healthy House Institute, 1992. 1 videocassette (28 min.) 1992

Discusses the features of a "healthy," safe house, including windows, doors, heating, and air conditioning. This is helpful for the prospective home buyer or home remodeler, as well as real estate students.

From Buyer Beware to Broker Beware
Chicago, National Association of Realtors, 1988. 1 videocassette (24 min.) VHS and 3 booklets.

This three-hour study course discusses the importance of accurate presentation of property facts to potential purchasers to avoid liability for undisclosed property defects.

Schwarz, Barb
How to Prepare Your Home for Sale...so it Sells
Issaquah, Wash., Barb Inc., 1990. 1 videocassette (60 min) VHS

A helpful video for agents to give clients who are getting their homes ready for the market. Also of value to the general public.

Personal Safety for Real Estate Professionals
Indianapolis, Ind., Producers Network, 1988. 1 videocassette (18 min.) VHS

Offers personal safety tips and self-defense strategies to aid real estate agents who need to show property alone.

Safety Before Selling: A Survival Guide
Austin, Tex., Texas Association of Realtors, 1988. 1 videocassette (15 min.) VHS.

Presents a survival plan for agents and brokers who may become victims of crimes.

Welcome Home: A Consumer's Guide to Home Buying
Lisle, Ill., American Home Productions, 1993. 1 videocassette (54 min.) VHS and 1 booklet

Sally Roberts, a real estate consultant, introduces the novice homebuyer to the various procedures involved with buying a home. Provides tips, financing information, and legal aspects.

Additional Resources

There are numerous other audiocassettes and videocassettes available on techniques and strategies for success in real estate. The titles above are listed as representative examples. Many state and national associations, such as the two listed below, produce audio and videocassettes that addresss regional as well as general issues.

National Association of Realtors
430 N. Michigan Ave.
Chicago, Ill. 60611

Realtors National Marketing Institute
430 N. Michigan Ave., Suite 500
Chicago, Ill. 60611

Small Business and Entrepreneurship

Barbara Alper
Bergen Community College
Paramus, New Jersey

Introduction

Small business programs in community and junior colleges prepare students to go into business for themselves as owners or managers. Students study the role of small business in the American economy and topics of relevance to small business and entrepreneurship. This program usually covers all aspects of business, including accounting, business plans, finance, human resources, law, marketing, public relations, and site location.

Small business is predicted to be the backbone of the economy in the 1990s, and consequently government is turning to community and junior colleges to train entrepreneurs in order to create jobs and fuel the economy. In addition, many who are laid off from positions in large corporations are looking to community colleges for retraining in the methodologies of running a small business. Whether the focus is managing a franchise, buying an existing business, or starting from scratch, there are many common elements covered in the small business curriculum.

Accreditation

American Assembly of Collegiate Schools of Business (AACSB)
600 Emerson Rd., Suite 300
St. Louis, Mo. 63141-6762

International Federation for Business Education
6201 College Blvd., Suite 210
Overland Park, Kans., 66221-2422

Selected Associations

American Small Business Association
1800 N Kent Suite 910
Arlington, Va. 22209

U.S. Small Business Administration
409 3rd Street SW
Washington, D.C. 20416

Selected Journals

ASBA Today.
American Small Business Association. Bimonthly.

Entrepreneur.
Entrepreneur Group Inc. Monthly. ISSN 0163-3341

Inc.: The Magazine for Growing Companies.
Inc. Pub. Co. Thirteen times a year. ISSN 0162-8968

Journal of Small Business Management.
International Council for Small Business. Quarterly.
ISSN 0047-2778

Print Materials

Bacon, Mark S.
Do-It-Yourself Direct Marketing: Secrets for Small Business
New York, Wiley, 1992. 274p. ISBN 047153241X

This handbook for business owners, marketing managers of small businesses, or the self-employed begins with the definition and history of direct marketing. Filled with examples and checklists, and includes bibliography and sources.

Baty, Gordon B.
Entrepreneurship for the Nineties
Englewood Cliffs, N.J., Prentice Hall, 1990. 270p.
ISBN 0132822946

Discussion of the risks and rewards of owning a business. Basic, practical look at finance, marketing, and management. References at the end of each chapter.

Bell, C. Gordon, and John E. McNamara
High-Tech Ventures: The Guide for Entrepreneurial Success
Reading, Mass., Addison-Wesley, 1991. 387p. ISBN 0201563215

The authors use a diagnostic approach to analyze human applied expert systems and to produce objective questions and evaluation. Includes case studies and a good bibliography.

Berle, Gustav
Planning and Forming Your Company
New York, Wiley, 1990. 272p. ISBN 047151795X

Step-by-step guide that covers going into business, the business plan, how to buy a business, inventions, patents, copyright and trademarks, employees, taxes, and insurance. Includes a directory of useful addresses, sources of information, and a glossary.

Berle, Gustav
Small Business Information Handbook
New York, Wiley, 1990. 256p. ISBN 0471524476

Alphabetical dictionary of terms ranging from "Academy of Senior Professionals" to "Zenz Case." Definitions are brief, but useful and informative. Many definitions include further sources of information. Cross-referenced with index and appendices.

Botkin, James W.
Winning Combinations: The Coming Wave of Entrepreneurial Partnerships Between Large and Small Companies
New York, Wiley, 1992. 278p. ISBN 047153658X

International focus on partnerships between small and large companies. Includes chapters on countries such as Japan, the United Kingdom, and Sweden.

Chaganti, Rajeswararao
High Performance Management Strategies for Entrepreneurial Companies: Research Findings from over 500 Firms
New York, Quorum Books, 1991. 192p. ISBN 089930561X

Reference book for chief executive officers, business owners, and consultants that emphasizes strategic management and its link to profit. Includes an index and bibliographic references.

Cohen, William A.
The Entrepreneur and Small Business Problem Solver: An Encyclopedic Reference and Guide
2nd ed. New York, Wiley, 1990. 565p. ISBN 0471501239

An answer guide that covers sources of capital, insurance, leasing, marketing, and other aspects of starting, buying, or managing a business. Includes many forms, samples, and applications.

Day, John
Small Business in Tough Times: How to Survive and Prosper
San Diego, Calif., Pfieffer and Co., 1993. 255p.
ISBN 0893842427

Covers topics from getting there (establishing a business and fund raising) to being there (success and management). Also includes insolvency issues and a very useful chapter on where to go for help. Brief bibliography.

Diamond, Michael R., and Julie L. Williams
How to Incorporate: A Handbook for Entrepreneurs and Professionals
2nd ed. New York, Wiley, 1993. 291p. ISBN 0471585866

Useful guide to the procedures and process of incorporation.

Dyer, W. Gibb
The Entrepreneurial Experience: Confronting Career Dilemmas of the Start-Up Executive
San Francisco, Jossey-Bass, 1992. 268p. ISBN 1555424171

Reviews personal, family, and business dilemmas at various stages in a career. This psychological treatment of beginning, managing, and retiring from entrepreneurial positions is theoretical and touches on topics such as priorities, commitment, and stress.

Halloran, James W.
The Entrepreneur's Guide to Starting a Successful Business
2nd ed. New York, McGraw-Hill, 1992. 291p. ISBN 0070257981

Halloran emphasizes the entrepreneurial spirit and start-up ideas. Uses case studies and projects to analyze market, location, finance, design, advertising, and other aspects in the small business environment. Discusses import and export and home-based businesses, among other topics.

Halloran, James W.
The Right Fit: The Entrepreneur's Guide to Finding the Perfect Business
Blue Ridge Summit, Pa., Liberty House, 1989. 198p.
ISBN 0830630279

Practical self-help book filled with case studies. Includes business and marketing plans.

Hisrich, Robert D., and Michael P. Peters
Entrepreneurship: Starting, Developing, and Managing a New Enterprise
2nd ed. Homewood, Ill., Irwin, 1992. 641p. ISBN 0256086915

The focus of this work is the entrepreneurial perspective: beginning a new venture, developing a business plan, finance, and management. Each chapter begins with a profile of a business and includes a discussion of questions and terms. Includes suggestions for further reading.

How to Run a Small Business
7th ed. New York, McGraw-Hill, 1993. 328p. ISBN 0070365768

A classic from the J. K. Lasser Institute staff which covers a wide breadth of topics and many forms of businesses, from wholesale to mail order. Includes useful checklists.

Hyypia, Erik
Crafting the Successful Business Plan
Englewood Cliffs, N.J., Prentice Hall, 1992. 208p.
ISBN 0131589245

Takes the reader step-by-step through the creation of a business plan. The author develops business plans section-by-section and then details how to use plans to start and run a business profitably.

Lane, Marc J.
Legal Handbook for Small Business
Rev. ed. New York, AMACOM, 1989. 255p. ISBN 081445951X

Detailed account of how to buy and structure a business. Includes glossary, acronyms, and index.

Mancuso, Joseph
Mancuso's Small Business Resource Guide
New York, Prentice Hal, 1988. 557p. ISBN 0138130809

Arranged in dictionary format, this book spans subjects from "Accounting" to "Women Entrepreneurs." Each entry includes a brief definition, description, list of associations, services, and publications. Useful but dated (some sections originally published in 1980).

Maul, Lyle R.
Entrepreneur's Road Map to Business Success
Alexandria, Va., Saxtons River Publications, 1990. 320p.
ISBN 0929382064

Treats entrepreneurship as a state of mind. Fun and easy to use with helpful checklists, business plan outline, glossary, and a good index.

Merrill, Ronald E., and Gaylord E. Nicholas
Raising Money: Venture Funding and How to Get It
New York, AMACOM, 1990. 283p. ISBN 0814459668

This practical approach to raising capital covers topics such as writing business plans, choosing a method of finance, and making presentations. Includes bibliographies with many chapters and a glossary.

Nicholas, Ted
Small Business Course
Chicago, Enterprise Dearborn, 1994. 327p. ISBN 0793107253

This self-study course reviews general business information in a comprehensive and well-organized manner. Each chapter includes a summary and self-test accompanied by answers. Includes bibliography and index.

Pinson, Linda, and Jerry Jinnett
Anatomy of a Business Plan
2nd ed. Chicago, Ill., Enterprise Dearborn, 1993. 191p.
ISBN 0793106184

Clear, easy, basic approach to developing a business plan, filled with sample forms. Includes sources of information, glossary, blank forms, and worksheets. Also offered in computer software format.

Pinson, Linda, and Jerry Jinnett
Steps to Small Business Start-Up: Everything You Need to Know to Turn Your Idea into a Successful Business
2nd ed. Dover, N.H., Upstart Pub. Co., 1993. 242p.
ISBN 0936894504

Well-organized and comprehensive guide that covers all aspects of small business development from choosing a name to legal status and licensing.

Robert Morris Associates (RMA) Annual Statement Studies
Philadelphia, Pa., Robert Morris Associates, Annual. 1 v. (various pagings)

Industry ratio information is grouped by four-digit Standard Industrial Classification (SIC) codes.

Schwartz, Carol A., ed.
Small Business Sourcebook: The Entrepreneur's Resource
6th ed. Detroit, Gale, 1993. 2 v. ISBN 0810380765

An up-to-date annotated guide of more than 25,000 subjects. Volume One is arranged alphabetically by small business; Volume Two is arranged by topic with sections on state and federal government sources of assistance. Listings include: associations, educational programs, reference books, trade publications, trade shows, libraries, computer software, and databases.

Sherlock, Paul
Rethinking Business to Business Marketing
New York, Free Press, 1991. 188p. ISBN 0029286158

Basic introduction to marketing plans, product support, distribution channels and advertising. Advises businesses on how to position themselves and how to market to other businesses. Indexed, but lacks bibliography.

Siegel, Eric S., Brian R. Ford, and Jay M. Bornstein
The Ernst and Young Business Plan Guide
2nd ed. New York, Wiley, 1993. 194p. ISBN 0471578258

This manual explains the procedures used to develop a business plan and why the information is needed, as well as advice on how to present the business in the most effective manner. Revision of *Arthur Young Business Plan Guide* (Wiley, 1987).

Siropolis, Nicholas C.
Small Business Management: A Guide to Entrepreneurship
4th ed. Boston, Houghton Mifflin, 1990. 634p. ISBN 0395433665

Each chapter begins with a summary and includes discussion, review questions, cases, and notes. The emphasis is on social responsibility and ethics. Heavily illustrated with graphs, charts, and diagrams.

The Small Business Encyclopedia
Irvine, Calif., Entrepreneur Group, 1992. 3 v.

Loose-leaf volumes cover seventy-two topics in small business. Each chapter begins with an outline and includes forms, checklists, and sources of information.

Timmons, Jeffrey A.
Planning and Financing the New Venture
Acton, Mass., Brick House Pub. Co., 1990. 188p.
ISBN 0931790921

    Begins with the business plan and progresses through financial sources and deal negotiation. Appendices include sample business plans, investment agreements, and stock restriction agreements.

Troy, Leo
Almanac of Business and Industrial Financial Ratios
Englewood Cliffs, N.J., Prentice Hall, Annual. ISSN 0747-9107

    Contains financial and operating ratios for 160 industries. Data cover more than twenty categories arranged by four-digit Standard Industrial Classification code. Information is drawn from aggregate numbers from the IRS records. Very useful for someone preparing a business plan.

Voth, Eric R.
The New Owner: Making the Transition from Employee to Employer
Homewood, Ill., Business One Irwin, 1993. 225p.
ISBN 1556239653

    Good basic introduction to small business ownership with a focus on service businesses. Includes management, personnel, and customer service issues.

Nonprint Materials

The Entrepreneurs
Chicago, Johnson Pub. Co., 1991. 1 videocassette (35 min.) VHS

    A motivational videocassette that uses contemporary (Oprah Winfrey) and historical (Frederick Douglass) figures to introduce successful people across a variety of fields.

Franchising: How to Be in Business for Yourself, Not by Yourself
Potomac, Md., Auteur Productions, 1988. 1 videocassette (58 min.) VHS

    Beginning with the definition of a franchise, this video goes through the decision-making process of selecting the right franchise for the individual. Comprehensive and useful with sources cited for further information.

Management
Princeton, N.J., Films for the Humanities, 1989. 1 videocassette (23 min.) VHS

    Content ranges from five-year plans to strategic planning, management by objective to crisis management, Small Business Association programs, training, and supervision.

Marketing Strategy for Small Business
Sunrise, Fla., D.E.Visuals, 1989. 1 videocassette (32 min.) VHS

    Begins by explaining the need for a business plan, a marketing plan, and identified target markets prior to commencing business. Discusses implementing marketing strategies as they relate to place, product, promotion, or marketing.

Promotional Strategy for Small Business
Sunrise, Fla., D.E. Visuals, 1989. 1 videocassette (28 min.) VHS

    Small business owners discuss personal selling, advertising, sales promotion, and publicity. Provides guidelines for budgeting and the need for promotional objectives.

Travel and Tourism

Trudy Cleveland
Houston Community College
Houston, Texas

Introduction

Travel and tourism is a vocational program designed to provide the career-minded individual with basic skills to become a competent professional. Many careers fall under the broad field of travel and tourism, including reservation agent, ticket agent, travel consultant, travel agent, vacation tour guide, and cruise director. Since the field is so diverse, community colleges offer courses in the basics of the travel industry, geography, ticketing, computer training, interpersonal communications, and marketing. After completion of the program, students may earn a certificate of competency or an Associate Degree of Applied Science.

Many developments affect the travel and tourism industry which, in turn, impact course offerings by community colleges. The major recent developments are shorter, more frequent vacations (four-day weekends are replacing two- to three-week vacations), adventure travel, ecotourism, travel by the disabled, and convention or meeting travel. These trends along with required knowledge of a foreign language influence future curriculum plans.

Professional status as a travel agent is achieved through membership in the American Society of Travel Agents, which maintains stringent requirements. The Institute of Certified Travel Agents has a program for those interested in becoming a Certified Travel Counselor.

Several things should be noted in this bibliography. Nonprint materials are not listed, but major sources of travel and tourism videos are provided. Some books are part of a series. such as those by Fielding and Fodor. In these cases one title has been cited as an example. Students and faculty should also be cognizant of the *AAA Tour Guides* and the *Mobile Travel Guides*, although they are not listed in the main bibliography.

Certification

Institute of Certified Travel Agents
148 Linden St.
P.O. Box 82-56
Wellesley, Mass. 02181

Selected Associations

The following national associations are representative of the associations that provide information on travel education, industry trends, and continuing education:

American Sightseeing International
490 Post St., Suite 1701-A
San Francisco, Calif. 94102

American Society of Travel Agents
1101 King St.
Alexandria, Va. 22314

Travel Industry Association of America
2 Lafayette Center
1133 21st St. NW, Suite 800
Washington, D.C. 20036

Selected Journals

ASTA Agency Management.
American Society of Travel Agents. Monthly. ISSN 0896-4114

Condé Nast's Traveler.
Condé Nast Publications. Monthly. ISSN 0893-9683

Corporate Travel.
Miller Freeman Publications. Thirteen times a year.
ISSN 0882-8709

Cruise Travel Magazine.
Century Pub. Co. Bimonthly. ISSN 0199-5111

Travel Agent.
Fairchild Media. Weekly. ISSN 1053-9360

Travel Trade.
Travel Trade Publications. Weekly. ISSN 0041-2066

Print Materials

Berman, Eleanor
Traveling on Your Own: 250 Great Ideas for Group and Solo Vacations
New York, Clarkson N. Potter, 1990. 276p. ISBN 0517574543

An easy-to-read, comprehensive travel guide which includes comments from travelers. Topics include traveling alone, what to do with children, and information on packing. Lists addresses and phone numbers as well as other vital information. Male and female ratio of travelers is discussed.

Birnbaum, Alexandra Mayes, and Lois Spritzer, eds.
Birnbaum's Eastern Europe
New York, HarperPerennial, Annual since 1992. 1 v. (various pagings) ISSN 1056-439X

Information includes planning, budgeting, best hotels and restaurants, shopping, festivals, and events. Gives advice for special travelers (disabled, singles, seniors).

Birnbaum, Stephen, and Wendy Lefkon
Birnbaum's Disneyland
New York, Hyperion, Annual since 1992. 1 v. (various pagings)

A slim, amusing guide providing the best touring tips for the Magic Kingdom. Sample schedules are given along with helpful hints.

Boyd, Wilma
Travel Agent
New York, Arco, 1989. 276p. ISBN 0139300589

Provides information on a career as a travel agent. Topics covered include basics of U.S. and foreign travel, time zones, ticketing, world geography, and sales techniques. Includes glossary, diagrams, charts, and photographs.

Burke, James F., and Barry Paul Resnick
Marketing and Selling the Travel Product
Cincinnati, Ohio, South-Western Pub. Co., 1991. 323p ISBN 053870277X

The authors discuss principles and procedures of sales and marketing as they apply to travel. Each chapter has a list of main objectives, key terms, a summary, and discussion questions. Also includes study guides and work sheet exercises for each chapter.

Butler, Arlene Kay
Traveling with Children and Enjoying It: A Complete Guide to Family Travel By Car, Plane, and Train
Chester, Conn., Globe Pequot Press, 1991. 284p.
ISBN 0871063166

Directed to anyone traveling with children, this book discusses how to plan and what to take along. Includes tips on food, health and safety, games and songs, what to pack, and what to leave at home. Also includes a chapter on things to do when you return home. Lists toll-free numbers of major hotel and motel chains.

Butler, Brian
Europe for Free
3rd rev. ed. Memphis, Tenn., Mustang Publishing, 1994. 286p.
ISBN 0914457608

The book is alphabetically arranged by country, then by cities and towns within the country. Includes museums, attractions, and events with brief descriptions. Addresses and phone numbers are provided where available.

Clark, Jim
Cycling the U.S. Parks: Fifty Scenic Tours in America's National Parks
San Francisco, Bicycle Books, 1991, 1993 printing. 223p.
ISBN 0933201567

Arranged by geographic regions, the author, a bicycling enthusiast, provides information on planning, access, accommodations, bicycle rental, and availability of supplies. In addition to the history of each park, he provides information on the park's geology and special features. Very clear maps are provided with descriptions of the routes.

Coltman, Michael M.
Introduction to Travel and Tourism: An International Approach
New York, Van Nostrand Reinhold, 1989. 370p.
ISBN 0442206526

Overview of all aspects of travel and tourism, including a history and current trends. Discussion questions and suggested readings are included at the end of each chapter. An appendix deals with the issue of franchising.

Coltman, Michael M.
Tourism Marketing
New York, Van Nostrand Reinhold, 1989. 320p.
ISBN 0442236190

Intended for students, educators, and travel agents, this book discusses advertising promotion, marketing, transportation, food services, and accommodations. Suggested readings are provided at the end of each chapter.

Cummings, Jack
The Business Travel Survival Guide
New York, Wiley, 1991. 396p. ISBN 0471530751

The author of more than twenty books and an owner of a travel agency, Cummings provides a comprehensive text that includes discussion of critical situations, jet lag, sickness, IRS deductions, and packing. Includes information on twenty-five major cities, e.g., airports, ports, lodging, dining, climate, and transportation.

Davidoff, Philip G., Doris S. Davidoff, and J. Douglas Eyer
Tourism Geography
Elmsford, N.Y., National Publishers of the Black Hills, 1988. 502p. ISBN 093592048X

This is not intended as a travel guide, but as a text for future and present travel agents and for those people who wish to visit a country or region. The authors provide details about documentation, customs, immigration, and health information. Includes examples of tour itineraries, a glossary, maps, and black-and-white photographs.

Davidoff, Philip G., Doris S. Davidoff, and John D. Eyre
Tourism Geography Workbook
Englewood Cliffs, N.J., Prentice Hall, 1992. 156p.
ISBN 0139231528

Intended as a companion to *Tourism Geography* (National Publishers of the Black Hills, 1988). This workbook has question-and-answer forms, map exercises, crossword puzzles, and picture identification exercises. This interesting book can also be used for a self-directed study.

Delaney, John F.
Travelwise: A Guide to Safety, Security, and Convenience When You Travel
North Vancouver, B.C., Self-Counsel Press, 1989. 136p.
ISBN 0889086974

This short but informative book contains tips on how to safeguard one's house while away, as well as hints regarding pets and advice for disabled travelers. Eight appendices include information on world currencies, time zones, and clothing sizes.

DuBois, Dorothy
Traveler's Hotline Directory
San Diego, Calif., Visions Resource Publishing, 1991. 284p.
ISBN 1880581884

Arranged alphabetically within subject chapters are 12,000 toll-free numbers for travelers. Some subjects contain a short explanation or pertinent information, such as the chapter on ecotourism. This book is suitable for anyone interested in travel.

Edgell, David L., ed.
World Tourism at the Millennium: An Agenda for Industry, Government, and Education
Washington, D.C., U.S. Department of Commerce, U.S. Travel and Tourism Administration, 1993. 97p.

Looks at tourism within the context of society as a whole while informing students about the nature and complexity of tourism. Includes charts, graphs, and selected bibliography.

Farewell, Susan
How to Make a Living as a Travel Writer
New York, Shooting Star Press, 1994. 201p. ISBN 1569248907

The author, a veteran travel writer, talks about the positive and negative aspects of a travel writing career. She gives hints on writing techniques, networking, current trends, and how to get published. Two appendices list resources and writers' associations.

Fielding's Series...see Zellers, Margaret (example)

Fodor's Paris
New York, Fodor, 1974- . 1 v. (various pagings) ISSN 0149-1288

An example of one of the Fodor travel series. In addition to typical travel information, this guide also details a brief history of the city, conversion tables, and menus. Includes special sections such as "Highlights" and "Fodor's Choice," and hints for special groups of travelers, such as the disabled, children, and seniors. Includes index, maps, and city plans.

Foster, Dennis L.
First Class: An Introduction to Travel and Tourism
2nd ed. New York, Glencoe, 1994. 385p. ISBN 0028013840

A basic guide for students planning to become travel agents, with emphasis on interpersonal skills and communication. Covers all aspects of the industry. Contains black-and-white photographs, maps, a study guide, glossary, and index. There are four other books in this series dealing with travel professionals.

Frome, Michael
National Park Guide
New York, Prentice Hall Travel, Annual. 1 v. (various pagings)
ISSN 0734-7960

Recently updated and written by a park authority and environmentalist, this book lists the best seasons for certain parks, along with activities, geology, history, and points of interest. Included are color maps and photographs, addresses, and index.

Frommer series...see Greenberg, Arnold (example)

Gagnan, Patricia J., and Karen Silva
Travel Career Development
5th ed. Homewood, Ill., Irwin/Mirror Press, 1992. 309p.
ISBN 0256119775

Comprehensive guidebook and text for those considering a travel career. Illustrations, charts, glossary, appendix, and index.

Gee, Chuck Y., and others
Professional Travel Agency Management
Englewood Cliffs, N.J., Prentice Hall, 1990. 320p.
ISBN 0137255578

Basic information on management and business skills along with information on the environment, the economy, and specific areas of the travel industry. Index and glossary.

Geffen, Alice M., and Carole Berglie
Ecotours and Nature Getaways: A Guide To Environmental Vacations Around the World
New York, Clarkson Potter, 1993. 324p. ISBN 0517880687

This book is for the nature lover. Arranged by regions of the world, it gives history of the regions along with facts such as the best time to travel there, transportation, and number of people traveling in the tour. There is a geographical and trip index, recommended readings, and a list of environmental organizations.

Getz, Donald
Festivals, Special Events, and Tourism
New York, Van Nostrand Reinhold, 1991. 374p.
ISBN 0442237448

    Written for the travel agent who specializes in this type of tourism. Gives planning and marketing methods while distinguishing special event travel from regular travel. References, a glossary, charts, figures, black-and-white photographs, and case studies are part of this specialized text.

Gillard, Quentin
Travel Geography Handbook
New York, Van Nostrand Reinhold, 1991. 530p.
ISBN 0442001584

    Although lengthy, this is a basic text for the prospective travel agent. Specific destinations are listed by regions of the world. Includes cultural and physical geography. A special section lists professional associations, tourist offices, and information on weather, health, customs, and manners. Includes charts, diagrams, and maps.

Grant, Edgar
Exploring Careers in the Travel Industry
Rev. ed. New York, Rosen Pub. Group, 1989. 136p.
ISBN 0823909611

    Short but insightful look into the industry as a whole, with most attention paid to the travel agent. Provides hints, and pros and cons of various travel occupations. Includes an index, photographs, charts, and appendices.

Greenberg, Arnold, and Harriett Greenberg
Frommer's Budget Travel Guide: South America on $40 a Day
New York, Prentice Hall Travel, Annual. 1 v. (various pagings)

    One of many titles in the Frommer series, this covers such typical topics as accommodations and sightseeing, along with savvy shopping, the typical cost of things, fast facts, and interesting factual tidbits. Includes maps, index and two appendices: "Useful vocabulary" and "Metric measures."

Gregory, Aryear
The Travel Agent: Dealer in Dreams
4th ed. Englewood Cliffs, N.J., Regents/Prentice Hall, 1993. 351p.
ISBN 0139483403

    A policy and procedures manual for travel agents that can also be used as training material for beginners in the field. Includes a time chart of the world, photographs, glossary, appendix, and index.

Gunn, Clare A.
Tourism Planning: Basics, Concepts, Cases
3rd ed. Washington, D.C., Taylor and Francis, 1994. 460p.
ISBN 0844817430

    The author, who developed the tourism program at Texas A&M University, has written a very technical, detailed, and comprehensive work for those who teach tourism. Subjects addressed are the nature and scope of markets, services, and promotion. Includes charts, diagrams, tables, figures, index, and bibliography.

Harris, Dale
Help Yourself to Travel: Travel Planning Made Easy
Northridge, Calif., World View Publications, 1989. 120p.
ISBN 092967300X

    Harris provides basic information on topics such as inoculations, documents needed, and tipping. Includes photographs and a list of foreign government tourist offices in the U.S.

Harvard Student Agencies Staff
Let's Go: The Budget Guide to the USA and Canada
Edited by Mark D. Moody, New York, St. Martin's Press, Annual. 1 v. (various pagings)

    This title continues the *Let's Go* series. Includes detailed maps, advertisements, and index. Discusses the basics of travel preparation, as well as specific destinations. Travel information for the United States is divided into twelve geographic sections which are then organized alphabetically by state; the Canadian section is organized in a similar fashion.

Heath, Ernie, and Geoffrey Wall
Marketing Tourism Destinations: A Strategic Planning Approach
New York, Wiley, 1992. 226p. ISBN 0471540676

    A general text for the planning and development of tourism. Includes references, index, charts, and graphs.

Heilman, Joan Rattner
Unbelievably Good Deals and Great Adventures That You Absolutely Can't Get Unless You're Over Fifty
6th rev. ed. Chicago, Contemporary Books, 1994. 304p.
ISBN 0809236370

    The author, a veteran writer and traveler herself, provides for the fifty-plus group up-to-date information about trips, tours, discounts, insurance, and tax breaks. There is also a chapter on free college education and volunteer vacations. Includes names, addresses, and phone numbers of sources to contact. Larger than normal print. Index.

Howell, David W.
Passport: An Introduction to the Travel and Tourism Industry
2nd ed. Cincinnati, Ohio, South-Western Pub. Co., 1993. 436p.
ISBN 0538706171

    A comprehensive guide for those interested in a travel career. Discusses transportation, services, business travel, and products. Includes an index, black-and-white photographs, a glossary, and two appendices listing abbreviations and associations.

Hudman, Lloyd E., and Richard H. Jackson
Geography of Travel and Tourism
2nd ed. Albany, N.Y., Delmar Publishers, 1994. 620p.
ISBN 0827358849

    Provides thorough coverage of geography both for students of tourism and professionals in the field. Discusses specific characteristics of regions listed, as well as demographic information for countries. Sample travel itineraries are given along with bibliographies for each country or region. Includes tables, maps, glossary, index, and black-and-white photographs.

Jordon, Dorothy Ann, and Marjorie Adoff Cohen
Great Vacations with Your Kids: The Complete Guide to Family Vacations in the U.S. for Infants to Teenagers
Rev. ed., 2nd ed. New York, Dutton, 1992. 491p.
ISBN 0452269016

    This book gives very general travel advice, then lists specific destinations to visit, including the best theme parks. An index and an appendix identify more places that welcome families.

Keown, Ian
Caribbean Hideaways
New York, Prentice Hall Travel, 1993. 366p. ISBN 0671849212

This travel guide is an eclectic selection of hotels, resorts, and inns for all types of people. Details reservation and tourist information and provides a special section of helpful hints. Index and maps.

Lansky, Vicki
Trouble-Free Travel with Children: Helpful Hints for Parents on the Go
Deephaven, Minn., Book Peddlers, 1991. 148p. ISBN 916773140

Covering infancy through school-age years, this guide has helpful hints from experienced parents. Subjects include packing, sleeping away from home, entertaining, and traveling abroad. There is a list of travel agencies specializing in family travel and a list of museums for children. Index.

Lundberg, Donald E.
The Tourist Business
6th ed. New York, Van Nostrand Reinhold, 1990. 339p. ISBN 0442233760

Tracing travel and tourism historically and economically, this overview is for the student who is interested in a career as a travel agent. Discussion questions are included at the end of each chapter. Includes glossary, index, charts, figures, and black-and-white photographs.

Maltzman, Jeffrey
Jobs In Paradise: The Definitive Guide to Exotic Jobs Everywhere
Rev. ed. New York, HarperPerennial, 1993. 432p. ISBN 0062731866

A career guide for the tourism student, this work lists jobs available in the U.S., Canada, South Pacific, and the Caribbean. Gives pluses and minuses, qualities that employers desire, insider tips, and basic job hunting information. Alphabetical and geographic listings of companies.

Mancini, Marc
Selling Destinations: Geography for the Travel Professional
2nd ed. Cincinnati, Ohio, South-Western Pub. Co., 1994. 485p. ISBN 0538634502

Written for the travel professional, but also informative to the layperson, especially the "tips" section. Provides a wealth of information including transportation, cultural patterns, weather, trivia, and the client's perspective on destinations. Includes an index, maps, charts, and appendices.

McCord, Robert R.
The 479 Best Public Golf Courses in the United States, Canada, the Caribbean, and Mexico
New York, Random House, 1993. 724p. ISBN 0679739750

The author's choices for the best courses from among 15,000 are listed. Provides history of the course, length of each, fees, places to stay, directions, local attractions, and phone numbers.

McIntosh, Robert Woodrow, and Charles R. Goeldner
Tourism: Principles, Practices, Philosophies
6th ed. New York, Wiley, 1990. 534p. ISBN 047622559

Covers tourism from A to Z with information about the past, present, and future of the trade. The author talks about tourism and its products, supply and demand, and marketing. Each chapter concludes with key concepts, review questions, and selected references. Includes a glossary, two appendices, an index, and black-and-white photographs. Seventh edition is scheduled to be published in 1995.

Mieczkowski, Zbigniew
World Trends in Tourism and Recreation
New York, P. Lang, 1990. 370p. ISBN 0820411973

This scholarly reference source is intended for the serious student and professional. The objective is to bring tourism and recreation into focus with all of its ramifications and developments. Chapter topics include terminology, history, transportation, and supply and demand. Includes diagrams, tables, references, and index.

Pearce, Douglas G.
Tourist Organizations
New York, Wiley, 1992. 219p. ISBN 0582070104

Tourism organizations and their relationship to the field of travel and tourism are the focus of this work. A large portion of the book relates case studies from all over the world. Includes author and subject indexes, tables, charts, and references.

Persson, Conrad
The USA Travel Phone Book: A Quick-Help Guide to Essential Addresses and Telephone Numbers for Business and Vacation Travelers
Shawnee, Kans., Bon A Tirer Publishing, 1991. 190p. ISBN 1878446037

Alphabetical listing by state provides information on organizations that provide free materials, maps, and travel ideas for 1,700 cities and 1,600 attractions. In addition to the U.S., this work also covers Puerto Rico, Mexico, and Canada. Includes index.

Plawin, Paul
Careers for Travel Buffs and Other Restless Types
Lincolnwood, Ill., VGM Career Horizons, 1992. 130p. ISBN 0844281271

A valuable resource for those trying to decide on a career to pursue. In addition to travel and tourism, this work presents sports, sales, the military, and law enforcement as possible career choices. Names and addresses where more information can be obtained are listed at the end of each chapter.

Pond, Kathleen Lingle
The Professional Guide: Dynamics of Tour Guiding
New York, Van Nostrand Reinhold, 1993. 274p. ISBN 0442001487

Written for students of tour guiding, this work shows how it relates to travel and tourism as a whole. Includes bibliography, figures, charts, black-and-white photographs, a glossary, and appendices.

Rosenbluth, Hal F., and Diane McFerrin Peters
The Customer Comes Second: And Other Secrets of Exceptional Service
New York, Quill/Morrow, 1992. 240p. ISBN 0688132464

This work focuses on total quality management, and is written by the owner and manager of a travel service for over twenty years.

The book is not specifically about travel and tourism, although examples given that relate to the field. Includes bibliography and an extensive index.

Roth, Wendy, and Michael Tompane
Easy Access to National Parks: The Sierra Club Guide for People with Disabilities
San Francisco, Calif., Sierra Club Books, 1992. 404p.
ISBN 0871566206

One of the authors, being confined to a wheelchair, has personal as well as practical experience making safe and pleasurable trips to national parks. The authors describe the fifteen best parks for people with disabilities. There are tips on natural history, climate, terrain, and essentials. Includes a list of support groups, bibliography, glossary, index, and maps.

Rubinstein, Hilary, ed.
Europe's Wonderful Little Hotels and Inns: The Continent
New York, St. Martin's Press, Annual. 1 v. (various pagings).
ISSN 0193-4074

This annual guide is for the sophisticated traveler. Provides information about towns, hotels, inns, rates, and facilities. Any unusual qualities are noted as well as comments from readers who visited a particular establishment. Maps and exchange rates are included.

Rundback, Betty R.
Bed and Breakfast USA
New York, Plume, 1994. 737p. ISBN 0452271266

A state-by-state listing providing basic information. Also has a section on how to start a bed-and-breakfast business. Includes recipes.

Ryan, Chris
Recreational Tourism: A Social Science Perspective
New York, Routledge, 1991. 227p. ISBN 0415054249

An introduction to issues facing tourism today in relation to economics, sociology, psychology, and other social science areas. Includes figures, tables, bibliography, and indexes.

Savage, Peter
The Safe Travel Book
Rev. ed. New York, Lexington Books, 1993. 222p.
ISBN 0029277264

Written for the layperson, this book discusses crime, medical care, the ramifications of political problems, and various ways the U.S. Embassy can help Americans in a foreign country. Index included.

Schroeder, Dirk G.
Staying Healthy in Asia, Africa, and Latin America
Rev. ed. Chico, Calif., Moon Publications, 1993. 185p.
ISBN 1566910110

Very detailed medical information for travelers making weekend trips or for those staying for extended periods of time. Includes simple drawings, bibliographic references, and index. Useful for both the layperson and travel agent.

Scott, Gini Graham
The Creative Traveler: A Guidebook for All Places and All Seasons
Greensboro, N.C., Tudor Publishers, 1989. 140p.
ISBN 0936389133

Not a typical guide book, this in-depth look at countries, cities, and people is aimed at the traveler who wants to experience local culture and its people. Includes a section on how to use a camera.

Smith, Valene L., and William R. Eadington, eds.
Tourism Alternatives: Potentials and Problems in the Development of Tourism
Philadelphia, Univ. of Pennsylvania Press, 1992. 253p.
ISBN 0812213912

Experts in their fields have contributed various case studies to help illustrate concepts, management, and economic development of the tourism industry. Includes tables, figures, references, and index.

Smithsonian Guide to Historic America...see Winckler, Suzanne (example)

Stutts, Alan T.
The Travel Safety Handbook
New York, Van Nostrand Reinhold, 1990. 181p.
ISBN 0442318065

A simple text that describes factors which affect travel safety. Includes tips for protection from crime and illness. Provides selected readings at the end of each chapter.

Thompson, Douglas, and Alexander Anolik
A Personnel and Operations Manual for Travel Agencies
4th ed. San Francisco, Dendrobium Publishing, 1993. 1 v. (looseleaf) ISBN 0936831138

This manual encourages readers to create their own in-house manuals for operating travel businesses. Includes samples of forms.

Tovey, Priscilla, ed.
Smart Vacations: The Traveler's Guide to Learning Adventures Abroad
New York, St. Martin's Press, 1993. 318p. ISBN 031208823X

Consists of 200 one- to six-week vacations including study tours, outdoor adventures, archaeological digs, language studies, and others. Includes addresses, telephone numbers, and suggestions for the best time of the year to go. Sponsored by the Council on International Education Exchange.

U.S. and Worldwide Travel Accommodations Guide for $12- $24 Per Day
Fullerton, Calif., Campus Travel Service, Published every 18 months, 1 v. (various pagings). ISSN 0898-4247

Listed alphabetically by the states and countries of the world are 600 university campuses where lodging is available for one-third the cost of hotels and motels. Lists names, addresses, phone numbers, rates, dates available, and activities. Includes travel ideas for educators and additional tips about staying on campuses.

United States. National Park Service
The National Parks Camping Guide
Washington, D.C., U.S. National Park Service, 1986- . Annual

Lists camp sites alphabetically by state, then by park within state. Includes regulations, tips on planning a trip, and maps of the parks. Black-and-white photographs, charts, tables, and index.

Van Harssel, Jan
Tourism: An Exploration
3rd ed. Englewood Cliffs, N.J., Prentice Hall Career and Technology, 1994. 335p. ISBN 0139233431

A basic, comprehensive text for the student of travel and tourism. Includes black-and-white photographs, a glossary, charts, tables, graphs, and appendices. Review questions are provided at the end of each chapter.

Vladimir, Andrew, ed.
The Complete Travel Marketing Handbook: Thirty-Seven Industry Experts Share Their Secrets
Lincolnwood, Ill., NTC Business Books, 1988. 291p. ISBN 0844231479

Comprehensive collection of essays that gives advice from industry experts. Includes photographs, charts, and graphs.

Weiler, Betty, and C. M. Hall, eds.
Special-Interest Tourism
New York, Halstead Press, 1992. 214p. ISBN 1852930721

A series of essays by experts in the area of travel and tourism, which covers subjects such as education, arts and culture, sports, and outdoor pursuits. Includes plates, tables, figures, references, and index.

Whelan, Tensie, ed.
Nature Tourism: Managing for the Environment
Washington, D.C., Island Press, 1991. 223p. ISBN 155963037X

This collection of articles dealing with different aspects of ecotourism and specific destinations is intended for policy makers, conservationists, and tour operators. Includes maps, photographs, and index.

Winckler, Suzanne
Plains States
New York, Stewart, Tabori, and Chang, 1990. 461p. ISBN1556701233

This is the twelfth volume of the *Smithsonian Guide to Historic America* series. Not a travel guide, but rather a sightseeing tour book. Includes addresses and phone numbers, hours, and fees for major sites. Color photographs, maps, and index highlight the text.

Woodall's The Campground Directory for North America
Lake Forest, Ill., Woodall Pub. Co., Annual. 1 v. (various pagings)

Includes campground information for the U.S., Canada, and Mexico, arranged alphabetically by states, Canadian provinces, and then Mexico. Includes guide to seasonal sites, maps, and advertisements.

World Chamber of Commerce
Chamber of Commerce Directory
Loveland, Colo., World Chamber of Commerce, 1989- . Annual. 1 v. (various pagings) ISSN 1048-2849

Alphabetical listings by state, Canadian provinces, and foreign chambers of commerce. Gives contact person, address, phone numbers, and fax numbers.

Zellers, Margaret
Caribbean 1994: The Most In-Depth and Savvy Guide to the Islands of the Caribbean
Redondo Beach, Calif., Fielding Worldwide, 1994. 1086p. ISBN 1569520089

One in the Fielding series, written from a personal point of view. Provides hotel quick reference charts, maps, insider tips, and information on sports, life-style, politics, and medical facilities. Arranged alphabetically by country.

Zvoncheck, Juls
Cruises: Selecting, Selling, and Booking
2nd ed. New York, Prentice Hall, 1994. 235p. ISBN 0131926918

A text for the student of travel. Provides history and basic facts, as well as reservation information, menus, and floor plans. Includes review questions, black-and-white photographs, appendices, and index.

Nonprint Materials

There are numerous travel videocassettes available. The following producers will provide catalogs of currently available materials.

First Entertainment Inc.
1380 Lawrence St. #1400
Denver, Colo. 80204

International Video Network
2242 Camino Ramon
San Ramon, Calif. 94583

Preview Media
747 Front St.
San Francisco, Calif. 94111

Questar Videos
Box 11345
Chicago, Ill. 60611

Travelview International
10370 Richmond Ave.
Houston, Tex. 77042

World Video Projects, Inc.
2250 S.W. 3rd. Avenue, #205
Miami, Fla. 33129-2064

Communications/Production Technologies

Journalism 233
    Marit MacArthur

Public Relations 257
    Jane L. Crocker

Radio and Television Production Technologies 263
    Red Wassenich

Journalism

Marit MacArthur
Auraria Library
Community College of Denver
Metropolitan State College of Denver
and University of Colorado at Denver
Denver, Colorado

Introduction

Journalism is a program of studies offering varied career options. Opportunities at newspapers include reporting, writing columns and editorials, photojournalism and graphic arts, editing, and the business side, including advertising. There are also career opportunities in writing for magazines, either as an employee or free-lancer, and writing for business, trade, or professional publications. The book publishing field is also an option. Broadcast journalism and public relations, although part of the journalism field, are addressed in separate sections of this book.

The revolutionary new tools made available by computer technology are strongly influencing the field of journalism. Computerized information systems offer expanded opportunities for acquiring and analyzing information. Word processing has greatly assisted the work of writers. Desktop publishing has revolutionized the ability of individuals to make use of sophisticated printing and formatting techniques formerly available only to professional printers.

Community colleges or vocational schools offer coursework in a number of different areas of journalism, and sometimes have different degree requirements according to an emphasis option, such as advertising-public relations or news-editorial. Classes taught in journalism can include print media advertising, news reporting, editing, print media photography or photojournalism, newspaper design and production, advanced print or broadcast media reporting and writing, magazine article writing, specialized reporting areas such as public affairs or sports, and public relations. Students specializing in a news-editorial option may be required to take classes in photojournalism and newspaper production as well as writing and editing; those in a public relations option may take classes in advertising, photography, and newspaper production, as well as public relations. Whatever the specialization, journalism curricula will include a class on the news media and society. In addition to journalism classes, students are generally required to take courses in core areas, such as English, speech communications, humanities, science, social science, physical education, computer literacy, and mathematics.

Increasingly, many employers prefer to hire graduates of baccalaureate or master's journalism programs, but some community college graduates are finding writing or public relations jobs with newspapers, magazines, and other publishers or businesses. Some community college students may be sent to the college by their employers to enhance their writing and public relations skills. Students who wish to continue their education beyond a two-year program are likely to find that both general studies and journalism credits from a community college can transfer to a four-year institution. The accrediting agency in journalism is the Accrediting Council on Education in Journalism and Mass Communications. However, it does not accredit community college programs but restricts itself to professional baccalaureate and master's degree programs.

This bibliography is compiled for students in a two-year journalism program or those working in the field. In addition to materials directly supporting coursework, it includes reference materials, books on specialized areas of journalism that students might want to investigate as career options, and books on mass media and society that will interest both journalism students and those in other fields. Materials on desktop publishing will be of use to many practicing journalists, particularly those editing or publishing newsletters and business or professional publications.

Selected Associations

Newspaper Association of America
The Newspaper Center
Box 17407
Dulles International Airport
Washington, D.C. 20041

Society of Professional Journalists
16 S. Jackson
Greencastle, Ind. 46135

Student Press Law Center
1735 Eye St. NW, Suite 504
Washington, D.C. 20006

Selected Journals

American Journalism Review.
Washington Journalism Review. Ten times a year.
ISSN 1067-8654

The Bulletin of the American Society of Newspaper Editors.
The Society. Nine times a year. ISSN 0003-1178

Columbia Journalism Review.
Columbia Univ. Bimonthly. ISSN 0010-194X

Editor and Publisher.
Editor and Publisher Co. Weekly. ISSN 0013-094X

IRE Journal.
Investigative Reporters and Editors. Bimonthly. ISSN 0164-7016

Journalism Quarterly.
Univ. of South Carolina. Association for Education in Journalism and Mass Communication. Quarterly. ISSN 0196-3031

The Masthead.
National Conference of Editorial Writers. Quarterly.
ISSN 0025-5122

News Photographer.
National Press Photographers Association. Monthly.
ISSN 0199-2422

Newspaper Research Journal.
Ohio University. Scripps School of Journalism. Quarterly.
ISSN 0739-5329

Quill.
Society of Professional Journalists. Ten times a year.
ISSN 0033-6475

Report—Student Press Law Center.
Student Press Law Center. Three times a year. ISSN 0160-3825

Writer's Digest.
F & W Publications. Monthly. ISSN 0043-9525

Print Materials

✦ Advertising

Antin, Tony
Great Print Advertising: Creative Approaches, Strategies, and Tactics
New York, Wiley, 1993. 247p. ISBN 0471557137

An exceptionally clear, lively presentation of the basic principles behind good print advertising and the details needed to ensure its success. Part One covers basic principles. Part Two contains chapters covering layout, headlines, how to condense, and typography. Part Three provides a creative flow chart on how to make effective ads great. An appendix offers a checklist with a scoring system for your ads.

Book, Albert C., and C. Dennis Schick
Fundamentals of Copy and Layout
2nd ed. Lincolnwood, Ill., NTC Business Books, 1990. 263p.
ISBN 0844230308

A basic guide for the advertising beginner. Chapters cover creative philosophies, research and writing copy, layout, print media, broadcast media, and sample cases. Includes a glossary.

Ehler, R. L.
Directory of Print Media Advertising Resources: Nearly 400 Sources for Facts and Figures Needed by the Print Media Planner and Buyer
Santa Barbara, Calif., Richler and Co., 1992. 85p.
ISBN 1879299127

Lists resources by services offered. Categories include reference books, professional organizations, magazines, newspapers, buying agencies, research companies, information retrieval and tracking companies, libraries and information centers, computer software, databases, videocassette media producers, and media-related trade shows.

Floyd, Elaine, and Lee Wilson
Advertising from the Desktop: The Desktop Publisher's Guide to Designing Ads That Work
Chapel Hill, N.C., Ventana Press, 1994. 427p.
ISBN 1566040647

A guide for desktop publishers planning to promote a business or organization. Section One covers creating a marketing concept and plan, desktop visuals, typography and organizational tools, layout, and color. Section Two covers specific types of materials such as display ads, brochures, banners, signs, and order forms. Section Three includes resource lists of software and services, books and periodicals, associations and training, advertising software features, and a glossary.

Hahn, Fred E.
Do-It-Yourself Advertising: How to Produce Great Ads, Brochures, Catalogs, Direct Mail, and Much More
New York, Wiley, 1993. 246p. ISBN 0471553891

A practical, detailed, step-by-step guide. Covers newspaper and magazine advertising; flyers, brochures, and bulletins; direct mail and data-based direct marketing; catalogs; conventions, trade and consumer shows; public relations; telemarketing; audiovisual communication; and finding an advertising agency.

Klein, Erica Levy
Write Great Ads: A Step-by-Step Approach
New York, Wiley, 1990. 180p. ISBN 0471524182

A brief, chatty introduction on how to approach advertising copywriting from research to final product, with exercises to involve the reader. Suggests approaches to different formats and types of advertising. Helpful to beginners.

✦ Anthologies and Award Winners

The Best American Sports Writing, 1992
Boston, Houghton Mifflin, 1992. 348p. ISBN 0395603412

The 1992 edition contains twenty-five selections from sources as diverse as *GQ*, *Sports Illustrated*, *The Dallas Morning News*, and *The Yale Review*. The contributors have varied backgrounds and include not only those who are professional sports writers but general interest reporters, columnists, and novelists. This is an excellent source for good examples of the many varieties of sports writing.

Best Newspaper Writing
St. Petersburg, Fla., Poynter Institute for Media Studies. Annual. ISSN 0195-895X

Publishes the contributions of the winners and finalists in the American Society of Newspaper Editors Distinguished Writing Awards. Categories include nondeadline writing, deadline writing, short newswriting, editorial writing, and commentary. Short biographies are included, as are interviews with the winners. The editor has also included an "observations and questions" section analyzing the writing techniques used in each winning contribution.

The Best of Newspaper Design
Washington, D.C., Society of Newspaper Design, 1980- . ISSN 0737-2612

Presents winning entries from the Society of Newspaper Design's annual competition. The twelfth edition, published in 1991, includes 850 award-winning images from this international competition. Contains an index of winners, including both individuals and newspapers. Commercial and member editions available.

The Best of Photojournalism: The Year in Pictures
Philadelphia, Pa., Running Press Book Publishers, 1977- . Annual. ISSN 0161-4762

A compilation of photographs from the Pictures of the Year competition, presented by the National Press Photographers Association and the University of Missouri School of Journalism.

Brooks, Charles, ed.
Best Editorial Cartoons of the Year
Gretna, La., Pelican Pub. Annual. ISSN 0737-9498

The 1994 volume includes 395 cartoons with commentary grouped under eighteen subjects. Includes winners of the major awards for excellence: Pulitzer Prize, National Headliner Club Award, Fischetti Award, National Newspaper Award, and the National Society of Professional Journalists Award.

Great Newspaper Graphics
Glen Cove, N.Y., PBC International, 1990. 160p. ISBN 0866361200

Contains color illustrations of outstanding pages chosen by a panel of judges in the general groupings of news categories, feature sections, projects, magazines, and art and photos.

Meyer, Karl Ernest, ed.
Pundits, Poets, and Wits: An Omnibus of American Newspaper Columns
New York, Oxford Univ. Press, 1990. 458p. ISBN 0195060636

An anthology of work from seventy-two American columnists, including such diverse writers as Benjamin Franklin, Ambrose Bierce, Miss Manners, and Dave Barry. The editor provides a short introduction to each writer. An appendix provides capsule information on seventy-eight other columnists not included.

Rothmyer, Karen
Winning Pulitzers: The Stories Behind Some of the Best News Coverage of Our Time
New York, Columbia Univ. Press, 1991. 211p. ISBN 0231070284

Includes samples from the work of selected, diverse prize winners from the early years through the eighties. Each entry has a short introduction and a first-person account from the recipients themselves about their careers, the writing of their prize-winning stories, and what has happened since. There is also an introduction about the Pulitzer Prize Board and a commentary on the Board by former Board member Roger Wilkins.

Sloan, William David, and others, eds.
The Great Reporters: An Anthology of News Writing at its Best
Northport, Ala., Vision Press, 1992. 300p. ISBN 0963070029

Contains sample writings and biographical information on several nineteenth and early twentieth-century reporters. Among those included are Richard Harding Davis, Damon Runyan, Grantland Rice, Walter Winchell, Ernest Hemingway, Ernie Pyle, and Marguerite Higgins.

Sloan, William David, Cheryl Watts, and Joanne Sloan
Great Editorials: Masterpieces of Opinion Writing
Northport, Ala., Vision Press, 1992. 244p. ISBN 0963070002

A collection of sample writings and biographical information on editorial writers. Coverage is chronological, beginning with Benjamin Franklin in 1754, and emphasizing the late nineteenth and early twentieth centuries.

World Press Photo 1992
New York, Thames and Hudson, 1992. 121p. ISBN 0500973997

Publishes winners of the international photojournalism competition sponsored by the World Press Photo Foundation. Includes eight categories: sport news, people in the news, sports, general news, nature and the environment, the arts, science and technology, and daily life. This compilation from the thirty-fifth contest includes 150 images chosen from 17,887 submissions.

✦ Bibliographies

Belanger, Sandra E., comp.
Better Said and Clearly Written: An Annotated Guide to Business Communication Sources, Skills, and Samples
New York, Greenwood Press, 1989. 196p. ISBN 0313266417

Part One of this guide lists research sources, including reference books, periodicals, computer databases, professional associations, government agencies, and grammar hotlines. Part Two covers books on written and oral communication skills, including those on how to write documents such as business plans and proposals. Includes name, title, and subject indexes. Volume three of *Bibliographies and Indexes in Mass Media and Communications*.

Blum, Eleanor, and Frances Goins Wilhoit
Mass Media Bibliography: An Annotated Guide to Books and Journals for Research and Reference
3rd ed. Urbana, Ill., Univ. of Illinois Press, 1990. 344p.
ISBN 0252017064

An excellent guide to a core collection for research and reading, including works published through 1987. Includes sections on general communications, broadcasting media, print media, film, advertising and public relations, bibliographies, directories and handbooks, journals, and indexes to the mass communication literature. Also includes author, title, and subject indexes.

Cates, Jo A.
Journalism: A Guide to the Reference Literature
Englewood, Colo., Libraries Unlimited, 1990. 214p.
ISBN 0872877167

An extensively annotated, comprehensive guide to over 700 sources. Covers bibliographies and bibliographic guides, encyclopedias, dictionaries, indexes, abstracts and databases, biographical sources, handbooks and manuals, stylebooks, catalogs, miscellaneous sources, and core periodicals.

Sloan, W. David, comp.
American Journalism History: An Annotated Bibliography
New York, Greenwood Press, 1989. 344p. ISBN 0313263507

An annotated bibliography to 2,657 articles, books, and dissertations on American journalism history, arranged in fourteen chapters by chronological period. Includes only a subject index. *Bibliographies and Indexes in Mass Media and Communications* series, 1.

✦ Desktop Publishing and Newsletters

(See also the Desktop and Printing Technologies and the Visual Communications bibliographies in this volume.)

Adler, Elizabeth
Everyone's Guide to Successful Publications: How to Produce Powerful Brochures, Newsletters, Flyers, and Business Communications, Start to Finish
Berkeley, Calif., Peachpit Press, 1993. 400p. ISBN 156609027X

Offers an approach to the production of print materials that integrates marketing, writing, and graphic design. Sections cover understanding the medium, getting organized, getting and holding attention, writing for printed materials, designing for communication, production, and distribution. Includes worksheets, a list of resources, and a glossary.

Bivins, Thomas
Fundamentals of Successful Newsletters: Everything You Need to Write, Design, and Publish More Effective Newsletters
Lincolnwood, Ill., NTC Business Books, 1992. 208p.
ISBN 0844234834

An introduction to all aspects of producing a newsletter. Topics covered include planning, content, graphic elements, writing, layout, distribution, evaluation, basic newsletter templates, and desktop publishing. An appendix provides a stylebook for a sample newsletter.

Bivins, Thomas, and William E. Ryan
How to Produce Creative Publications: Traditional Techniques and Computer Applications
Lincolnwood, Ill., NTC Business Books, 1991. 438p.
ISBN 0844234931

An introductory text on the basics of publication. Part One covers writing, design, typography, photography, illustration, layout, and printing. For each topic, the book covers traditional approaches followed by a section on the role of computers in that area. Part Two covers how to apply this information to the four basic publication types: magazines, annual reports, brochures, and newsletters.

Brigham, Nancy, with Maria Catalfio and Dick Cluster
How to Do Leaflets, Newsletters and Newspapers
New ed. Detroit, Mich., PEP Publishers, 1991. 176p.
ISBN 096290676X

Contains succinct information on newsletter production including techniques of planning issues, gathering information, writing, editing, design, layout, typesetting, graphics, and distribution. Useful for editors at all experience levels. Discusses techniques from mimeograph machines to desktop publishing programs and professional typesetting.

Collier, David, and Bob Cotton
Basic Desktop Design and Layout
Cincinnati, Ohio, North Light Books, 1989. 160p.
ISBN 0891342850

An excellent guide to the methods of professional design. Section One describes the design process in general, with many examples. Section Two covers specific types of projects (e.g., creating logos, brochures, and newsletters). Section Three provides outstanding examples created by professionals, and the last section includes information on output and printing options. Glossary provided.

Kleper, Michael L.
The Illustrated Handbook of Desktop Publishing and Typesetting
2nd ed. Blue Ridge Summit, Pa., Windcrest, 1990. 927p.
ISBN 0830633502

A comprehensive sourcebook on typesetting and graphic reproduction. It begins with basic definitions and includes detailed information on software packages, hardware, and applications. Topics covered include basic typography and design principles, word processing, professional typesetting equipment, page description languages, networking IBM PCs and Macintoshes, optical character recognition, telecommunications, and high capacity data storage media. Includes many illustrations.

Lem, Dean Phillip
Graphics Master 4: A Workbook of Planning Aids, Reference Guides and Graphic Tools for the Design, Estimating, Preparation, and Production of Printing, Print Advertising and Desktop Publishing
4th ed. Los Angeles, D. Lem Associates, 1988. 153p.
ISBN 0914218069

A concise reference guide and tool kit that provides the technical data and working information needed by those in the design, planning, and production of printing and print advertising. Covers topics such as printing processes, graphic arts photography, printing industry trade customs, color scanners, papers, presses and inks, envelope styles and sizes, word processing and typesetting connections, and desktop publishing. Includes specimens of 1,768 typefaces, proofreaders' marks, conversion tables, and a glossary.

Mantus, Roberta
Design Guidelines for Desktop Publishing
Albany, N.Y., Delmar Publishers, 1992. 136p. and computer disk.
ISBN 0827350759

A text that teaches basic desktop design principles, with exercises and sample documents. Individual chapters cover type, special treatment of text elements, designing displays, dealing with artwork, planning the page, creating the design, executing the design, and estimating and controlling document length. An accompanying disk contains files for exercises in *WordPerfect* and ASCII formats.

Parker, Roger C.
Looking Good in Print: A Guide to Basic Design for Desktop Publishing
3rd ed. Chapel Hill, N.C., Ventana Press, 1993. 423p.
ISBN 1566040477

Provides design fundamentals to assist novices in creating effective newsletters, advertisements, sales materials, brochures, books, manuals, presentation graphics, and many other types of documents. Three sections cover elements of graphic design, makeovers that demonstrate how to enhance humdrum publications, and how to apply design principles to specific kinds of projects. Appendices cover graphics tips, clip art resources, and other desktop publishers' resources.

Parker, Roger C.
The Makeover Book: 101 Solutions for Desktop Publishing
Chapel Hill, N.C., Ventana Press, 1989. 278p.
ISBN 0940087200

Contains examples of poor design, then shows the same examples after an redesign. Covers many specific types of application such as newsletters, advertisements, brochures, business correspondence, catalogs, and charts. Emphasizes illustrations rather than explanatory text. Includes a list of the twenty-five major problem areas in desktop design and an annotated bibliography.

Williams, Robin
The PC Is Not a Typewriter: A Style Manual for Creating Professional-Level Type on Your Personal Computer
Berkeley, Calif., Peachpit Press, 1992. 92p. ISBN 0938151495

An invaluable book that introduces basic techniques of desktop publishing to those who originally learned to prepare documents on a typewriter. Covers such details as spacing between sentences, quotation marks and apostrophes, dashes, accent marks, alternatives to underlining, kerning, and leading. Appendices cover features of widely used programs such as *WordPerfect*.

✦ Editing

Baskette, Floyd K., Jack Zanville Sissors, and Brian S. Brooks
The Art of Editing
5th ed. New York, Macmillan, 1992. 518p. ISBN 0023062959

A popular text that covers copy editing skills, headline writing, pictures, graphics and design, editing for other media, and the outlook for the future. Appendices cover newspaper style, proofreading, database searches, and a glossary.

Berner, R. Thomas
The Process of Editing
Boston, Allyn and Bacon, 1991. 334p. ISBN 0205126944

A textbook on copy editing for newspapers. Topics include newspaper organization, the editor and technology, the editing function, writing, language skills, consultative editing, headlines, news evaluation, design, photographs, wire services, and how newspapers have reacted to the relative decline in their readership. Appendices include editing symbols, a condensed stylebook, a glossary, a budget for beginners, and a workbook.

Bowles, Dorothy A., Diane L. Borden, and William L. Rivers
Creative Editing for Print Media
Belmont, Calif., Wadsworth Pub. Co., 1993. 310p.
ISBN 0534190987

This text covers the role of the copy editor in the newsroom, editing details, legal and ethical concerns, headlines, editing copy from news services, pictures and infographics, typography, design, and layout. Includes an Associated Press style primer and numerous exercises.

Clark, Roy Peter, and Don Fry
Coaching Writers: The Essential Guide for Editors and Reporters
New York, St. Martin's Press, 1992. 182p. ISBN 0312068425

Assists editors in changing their traditionally adversarial relationship with reporters to a more human one in which the editor teaches, supports, and generally forms a team with the reporter. Each chapter contains a summary and a "workshop" section with exercises. Includes a reading list for coaches.

Gibson, Martin L.
Editing in the Electronic Era
3rd ed. Ames, Iowa, Iowa State Univ. Press, 1991. 312p.
ISBN 0813809649

A general textbook on copy editing, with chapters on headline writing, wire services, libel, type, layout and photographs. Not as current as other works on the recent changes in newspaper format, but especially informative on the differences between editing on paper and in electronic format.

Gilmore, Gene
Modern Newspaper Editing
4th ed Ames, Iowa, Iowa State Univ. Press, 1990. 305p.
ISBN 0813801745

Topics include copy editing, basics of printing, writing headlines, page design, the editor and journalistic writing, evaluating the news, editing wire news, picture editing, imagination in news editing, news crises, and edition changes. Also addresses issues such as the law and the copy editor, ethics, problems of policy, editorial management, newspaper research, and the future of editing. Includes a glossary and a bibliography.

Harrigan, Jane T.
The Editorial Eye
New York, St. Martin's Press, 1993. 439p. ISBN 0312041179

This textbook covers fundamentals of copy, page, and content editing. Specific topics include editing for grammar, style, accuracy, and fairness; legal and ethical issues; writing headlines; thinking visually; designing pages; working with writers as a coach; editing information and meaning; and the future for editors. Includes a glossary.

Hubbard, J.T.W.
Magazine Editing for Professionals
Rev. ed. Syracuse, N.Y., Syracuse Univ. Press, 1989. 184p. ISBN 0815624638

An entertaining introduction to the work of magazine editors. This guide describes the process of generating ideas, working with writers, editing for story organization and details, and planning issues. It also explains the business side of starting and producing a magazine. The complete planning process is illustrated with a real magazine issue.

✦ Education, Scholarship, and Career Guides

Ferguson, Donald L., and Jim Patten
Opportunities in Journalism Careers
Lincolnwood, Ill., VGM Career Horizons, 1993. 148p. ISBN 0844240141

A brief, informal summary of the nature and history of journalism, careers in newspaper work, electronic media, public relations, and supporting careers such as advertising, office support, marketing, technology, and circulation. Also discusses preparing for a journalism career and finding a job. Includes a bibliography, lists of career resources, colleges offering journalism degrees, and guidelines for journalistic conduct.

Financial Aid for Minorities in Journalism/Mass Communications
Garrett Park, Md., Garrett Park Press. Annual. Unpaged

Lists more than 300 programs and services available for minority students, ranging from scholarships to workshops and training programs. Entries are brief and arranged by organization name. Each entry provides an address where one can write for more information. Also includes a very general index and a list of books on financial aid and careers in journalism.

Guiley, Rosemary
Career Opportunities for Writers
2nd ed. New York, Facts on File, 1991. 230p. ISBN 0816024006

Describes jobs available in media, book publishing, arts and entertainment, business communication and public relations, advertising, the federal government, academic and nonprofit institutions, and free-lance services. Career profiles, position descriptions, career ladders, salaries, and employment prospects are given for each. Appendices list educational institutions and scholarships, associations and unions, periodicals, and a bibliography.

Holtz, Herman
How to Start and Run a Writing and Editing Business
New York, Wiley, 1992. 253p. ISBN 0471548324

A guide to providing free-lance writing services to business, industry, the government, and nonprofit organizations. Part One discusses various markets for one's services and how to sell to them. Part Two addresses self-publishing of specialized reports by mail. Part Three covers secrets of success in writing. Appendices include copyright, equipment and resources, writers' associations, and a recommended proposal format.

Journalist's Road To Success: A Career and Scholarship Guide
Princeton, N.J., Dow Jones Newspaper Fund, 1993- . Annual

Provides extensive information on newspaper careers and salaries, job searching, and detailed information on journalism education programs, scholarships, and professional organizations. Oriented toward four-year programs, but also of interest to community college students intending to continue their education. (Endorsed by the Association for Education in Journalism and Mass Communication and the Association of Schools of Journalism and Mass Communication.)

Morgan, Bradley J., and Joseph M. Palmisano, eds.
Magazines Career Directory: A Practical, One-Stop Guide to Getting a Job in Magazine Publishing
5th ed. Detroit, Mich., Visible Ink Press, 1993. 318p. ISBN 0810394405

Part One provides articles by active professionals on various career specialties on the editorial, art, or business sides of different kinds of publications. Part Two covers the job search process, including self-evaluation, résumés, and interviews. Part Three provides career resources, including a "job opportunities databank" that gives information on companies with frequent openings, and a listing of a wide range of other information resources.

Morgan, Bradley J., and Joseph M. Palmisano, eds.
Newspapers Career Directory: A Practical, One-Stop Guide to Getting a Job in Newspaper Publishing
4th ed. Detroit, Mich., Visible Ink Press, 1993. 344p. ISBN 0810394383

Similar format to the *Magazines Career Directory*. Includes information on opportunities for minorities, and internships. The editor also publishes a book publishing career directory.

✦ Interviewing

Biagi, Shirley
Interviews That Work: A Practical Guide for Journalists
2nd ed. Belmont, Calif., Wadsworth Pub. Co., 1992. 200p. ISBN 0534159664

A concise guide for interviewers that includes several case studies. Provides information on getting organized, finding research information, taking notes, getting interviews, conducting interviews, asking good questions, interviewing for broadcast, choosing and using quotes, and legalities and ethics.

Killenberg, George M., and Rob Anderson
Before the Story: Interviewing and Communication Skills for Journalists
New York, St. Martin's Press, 1989. 224p. ISBN 0312012381

A readable discussion emphasizing the interview as dialogue, in which there is rapport and two-way communication, rather than insensitive confrontation. Topics covered include choosing questions, listening, assessing truthfulness and deception, handling interviewees, telephone and broadcast interviews, and ethics.

Metzler, Ken
Creative Interviewing: The Writer's Guide to Gathering Information By Asking Questions
2nd ed. Englewood Cliffs, N.J., Prentice Hall, 1989. 208p. ISBN 0131897470

A comprehensive text covering the problems of interviewing, stages of the typical interview, the dynamics of conversation on which interviews are based, and listening and observation skills. Discusses ways to plan an interview, cover a newsbeat, and conduct the personality interview. Draws on the results of research in journalism and other disciplines. An appendix includes interviewing exercises, a sample interview report, and a bibliography.

Schumacher, Michael
The Writer's Complete Guide to Conducting Interviews
Cincinnati, Ohio, Writer's Digest Books, 1993. 236p. ISBN 0898795931

Topics include the interview and its uses, finding the right voices, research and preparation, lining up interviews, creative interviewing, how to use your interviewee's words, other ways to conduct and use interviews, and legalities and ethics. An appendix includes documents related to specific interviews, such as lists of questions and annotated transcripts. Also available in hardcover under the title *Creative Conversations*.

✦ Magazine, Feature, and General Nonfiction Writing

Blundell, William E.
The Art and Craft of Feature Writing: Based on the Wall Street Journal Guide
New York, New American Library, 1988. 259p. ISBN 0452261589

This readable guide emphasizes development of ideas, planning, organization, and techniques for maintaining interest. It is based on an internal manual used at *The Wall Street Journal*, where the author is a news editor for features.

Brooks, Terri
Words' Worth: A Handbook on Writing and Selling Nonfiction
New York, St. Martin's Press, 1989. 231p. ISBN 0312030053

Topics covered include the lead, transitions, verbs, description, quotes, voice, the weave, doing it, selling it, and one's rights and responsibilities in the writing business. An appendix recommends great works of nonfiction, with brief annotations.

Burgett, Gordon
The Writer's Guide to Query Letters and Cover Letters
Rocklin, Calif., Prima Pub., 1992. 261p. ISBN 1559581182

A detailed explanation of the art of writing query letters to sell an idea for an article or book, and cover letters to accompany a submission. Includes many sample letters. Explains how to use cover letters to sell the same article many times.

Cool, Lisa Collier
How To Write Irresistible Query Letters
Cincinnati, Ohio, Writer's Digest Books, 1990. 136p.
ISBN 0898793912

A guide to writing letters to summarize and sell concepts for magazine articles or nonfiction books to publishers. Topics covered include getting ideas, slanting a topic for a market, leads, tantalizing descriptions, selling oneself, research and interview shortcuts, professional details, and mistakes to avoid. Contains many sample letters.

Digregorio, Charlotte
You Can Be a Columnist: Writing and Selling Your Way to Prestige
Portland, Ore., Civetta Press, 1993. 232p. ISBN 0962331813

A practical guide for nonprofessional writers who want to share their knowledge and insights or enhance their careers in other professions. Includes chapters on creative and informational columns, style details, and getting published, with many analyzed examples. Appendices list selected columnists, directories of writers' markets, and writers' books and magazines.

Fensch, Thomas
Writing Solutions: Beginnings, Middles, and Endings
Hillsdale, N.J., L. Erlbaum Associates, 1989. 142p.
ISBN 0805804102

Presents tools for all types of writing including examples of twenty-five types of beginnings, four types of story structures, and fourteen types of endings. Provides checklists that recommend types of beginnings and endings to consider for various types of writing.

Fryxell, David
How to Write Fast (While Writing Well)
Cincinnati, Ohio, Writer's Digest Books, 1992. 198p.
ISBN 0898795141

An entertaining and useful guide to streamlining and improving all aspects of the writing process, from getting ideas, researching, interviewing, outlining, and writing drafts to editing the completed manuscript. Emphasizes organization, planning, and time management.

Graham, Betsy P.
Magazine Article Writing
2nd ed. Fort Worth, Tex., Harcourt Brace Jovanovich College Publishers, 1993. 248p. ISBN 0030750091

A comprehensive text that discusses the value of learning from others' writing, the difference between topic and thesis, and how to revise slowly and systematically. Provides information about markets, getting ideas, types of articles, querying, researching, interviewing, writer-editor relationships, and law. Includes chapter summaries, exercises, and many extensive writing samples from both professional writers and students.

Hay, Millicent V.
The Essential Feature: Writing for Magazines and Newspapers
New York, Columbia Univ. Press, 1990. 289p

Contains detailed information on subjects such as developing and selling ideas, research, interviewing, structure and writing, writing specific types of stories, packaging manuscripts, law and ethics, free-lancing, and using computers. Each chapter contains short bibliographies and suggestions for projects. Eight full-length examples of different types of features are given, with detailed marginal notes that analyze techniques. Appendices include useful punctuation rules, editors' marks, and a glossary.

Jacobi, Peter
The Magazine Article: How to Think It, Plan It, Write It
Cincinnati, Ohio, Writer's Digest Books, 1990. 247p.
ISBN 0898794501

Covers all aspects of magazine article writing, beginning with formulating ideas and proceeding to researching, structuring, and writing the article. Emphasizes the creative, personal approach. Includes many examples of good writing.

Rivers, William L., Bryce McIntyre, and Alison R. Work
Writing Opinion: Editorials
Ames, Iowa, Iowa State Univ., 1988. 114p. ISBN 0813805279

Designed to teach editorial writing skills using well-written and flawed editorials with side-by-side critiques, exercises, and tips from professional editorial writers. Emphasizes the importance of preparation and knowledge of fundamental reporting skills.

Stonecipher, Harry W.
Editorial and Persuasive Writing: Opinion Functions of the News Media
2nd ed. New York, Hastings House, 1988. 302p.
ISBN 0803893175

Discusses why persuasive writing is needed, how well it is being done, and how to do it better. Topics include choosing subjects, finding facts, avoiding libel, persuasive techniques, verbal and statistical fallacies, appeal to an audience, the question of feedback, and editorial policies.

Ward, Hiley H.
Magazine and Feature Writing
Mountain View, Calif., Mayfield Pub. Co., 1993. 367p.
ISBN 155934086X

Emphasizes writing and learning to publish from the outset. Includes chapters on trends, places for students to publish, and skills such as querying, researching, writing anecdotes, interviewing, structuring, and style. Other chapters cover specific types of articles, including how-to and investigative articles, and particular fields such as science, travel, history, religion, and writing for children. Discusses ethics, legal considerations, and getting jobs in the newspaper and magazine business.

Westfall, Patricia Tichenor
Beyond Intuition: A Guide to Writing and Editing Magazine Nonfiction
New York, Longman, 1994. 232p. ISBN 0801306749

Approaches writing of magazine nonfiction through the analytical techniques of focus, form, structure, authority, and voice.

This helps writers identify target magazines and tailor writing to specific readers. Intended to be used as a text for either magazine feature writing or an editing course.

Wilson, John M.
The Complete Guide to Magazine Article Writing
Cincinnati, Ohio, Writer's Digest Books, 1993. 294p.
ISBN 0898795478

Emphasizes finding one's particular strengths and marketing one's writing to a wide variety of sources. Covers types of articles and markets, querying, article openings, staying on track, specific writing techniques, research, interviewing, editing, packaging and submitting manuscripts, dealing with editors, managing free-lance writing as a business, and the writer's life.

Witt, Leonard, ed.
The Complete Book of Feature Writing, from Great American Feature Writers, Editors, and Teachers
Cincinnati, Ohio, Writer's Digest Books, 1991. 277p.
ISBN 0898794706

A collection of chapters by various writers with commentary by the editor. Sections cover the nature of feature writing, finding stories, basic and advanced writing techniques, particular kinds of features such as travel and fashion writing, selling free-lance stories, and finding a staff job.

Zinsser, William Knowlton
On Writing Well: An Informal Guide to Writing Nonfiction
4th ed., rev. updated and expanded. New York, HarperPerennial, 1990. 288p. ISBN 0060552727

The latest edition of one of the most widely used and respected books on nonfiction writing. Part One covers basic principles: the transaction, simplicity, clutter, style, the audience, words, and usage. Part Two covers particular forms: nonfiction as literature, the lead, the ending, the interview, writing about a place, science, technology and nature, business writing, sports, criticism, humor, and writing about oneself. The final section covers writing with a word processor, trusting one's material, and writers' decisions.

✦ Mass Communication and Society

Alexander, Alison, and Jarice Hanson, eds.
Taking Sides: Clashing Views on Controversial Issues in Mass Media and Society
2nd ed. Guilford, Conn., Dushkin Pub. Group, 1993. 352p.
ISBN 1561341207

Contains thirty-six previously published selections arguing the pros and cons of eighteen controversial media issues. General areas covered include mass media's role in society, ethics, regulation, media and politics, the media business, and the information society. Contributors are extremely varied, ranging from the president of NBC News and journalism professors, to musician Frank Zappa.

Altschull, J. Herbert
From Milton to McLuhan: The Ideas Behind American Journalism
New York, Longman, 1990. 447p. ISBN 0582285623

Traces the origins of the ideas that shape the attitudes of journalists. The author begins with English and French philosophers such as Milton and Montesquieu, and traces the development of ideas through the twentieth-century media philosopher Marshall McLuhan. Also discussed are assumptions about the role of the press in American society, the code of objectivity, and the role of investigative journalism. Includes extensive notes and suggestions for further reading.

Barbour, William, ed.
Mass Media: Opposing Viewpoints
San Diego, Calif., Greenhaven Press, 1994. 306p.
ISBN 1565101073

Contains thirty-four selections that argue opposing viewpoints on issues related to the media, with some emphasis on television. Major categories cover media bias and whether American society is accurately reflected, media regulation, media effects on politics, and whether advertising is harmful to society. A page of references to relevant periodical articles follows each section. Appendices contain questions for further discussion, organizations to contact, and a bibliography.

Barsamian, David
Stenographers to Power: Media and Propaganda
Monroe, Maine, Common Courage Press, 1992. 193p.
ISBN 0962883840

A collection of interviews with press critics from the political left including Noam Chomsky, Ben Bagdikian, Alexander Cockburn, Mark Hertsgaard, Michael Parenti, Erwin Knoll, and Jeff Cohen. Contends that the media largely repeats the official government line on foreign affairs, particularly concerning the Vietnamese and Persian Gulf Wars and the Nicaraguan Contras.

Bird, S. Elizabeth
For Enquiring Minds: A Cultural Study of Supermarket Tabloids
Knoxville, Tenn., Univ. of Tennessee Press, 1992. 234p.
ISBN 0870497286

Examines the tabloids as a source of information for the analysis of American popular culture. Discusses the kinds of stories published in several best-selling examples, emphasizing their predictability and basic conservatism, and also the reasons people read them.

Bogart, Leo
Preserving the Press: How Daily Newspapers Mobilized to Keep Their Readers
New York, Columbia Univ. Press, 1991. 327p. ISBN 0231072627

An insider account of the extremely influential Newspaper Readership Project (1977-1983), a cooperative effort by the American newspaper industry to halt the downward trend in readership and circulation. The Project conducted research and provided presentations and programs that newspapers could use to strengthen readership.

Bogart, Leo
Press and Public: Who Reads What, When, Where, and Why in American Newspapers
2nd ed. Hillsdale, N.J., L. Erlbaum, 1989. 376p.
ISBN 0805804315

Reports and summarizes the results of many studies conducted by the Newspaper Advertising Bureau between 1960 and 1988. Contains a wealth of detailed information and the insights of the author, who directed the studies.

Brasch, Walter M.
Forerunners of Revolution: Muckrakers and the American Social Conscience
Lanham, Md., Univ. Press of America, 1990. 197p.
ISBN 0819179671

A well-written, concise history of the reformist journalists who exposed corruption and greed for three decades beginning in the 1880s. Also presents information on "the modern muckrakers" who emerged during the 1960s, case overviews of investigative journalists and civil rights, and discussions of the My Lai massacre and Watergate. Includes a bibliography.

Connery, Thomas Bernard, ed.
A Sourcebook of American Literary Journalism: Representative Writers in an Emerging Genre
New York, Greenwood Press, 1992. 408p. ISBN 0313265941

Contains an overview and essays on thirty-five major representatives of literary or "new" journalism, beginning chronologically with Mark Twain. A bibliography of primary and secondary sources is given for each writer. A general bibliography is also included.

Cook, Philip S., Douglas Gomery, Lawrence Wilson Lichty, eds.
The Future of News: Television-Newspapers-Wire Services-Newsmagazines
Washington, D.C., Woodrow Wilson Center Press, 1992. 270p.
ISBN 0943875331

Contains papers and commentaries prepared for or resulting from a 1989 conference convened by the Media Studies Project of the Woodrow Wilson International Center for Scholars. Contains brief analyses of trends in the news business. There is also a bibliographical essay listing "background books" in each area covered.

Dennis, Everette E.
Of Media and People
Newbury Park, Calif., Sage Pub., 1992. 187p. ISBN 080394747X

Thought-provoking essays based on articles and speeches by the author, who is the executive director of the Freedom Forum Media Studies Center at Columbia University. Examines the relationship of the media with individuals and with institutions such as business, labor, the military, and education. The author is particularly concerned with the current trend towards merging all media, both news and entertainment, into a single electronically-based system.

Denton, Frank, and Howard Kurtz
Reinventing the Newspaper: Essays
New York, Twentieth Century Fund Press, 1993. Perspectives on the News. 119p. ISBN 0870783505

Includes essays on the newspaper's changing role and the marketplace forces that threaten it. The authors suggest remedies based on marketing and content.

Deppa, Joan
The Media and Disasters: Pan Am 103
New York Univ. Press, 1994. 346p. ISBN 0814718574

An examination of media coverage of the terrorist-caused airline explosion over Lockerbie, Scotland, in 1988. Includes a detailed description of the news gathering process and a discussion of advantages, disadvantages, and ethical issues concerning media coverage of disasters.

Emery, Michael C., and Edwin Emery
The Press in America: An Interpretive History of the Mass Media
6th ed. Englewood Cliffs, N.J., Prentice Hall, 1988. 786p. ISBN 0136990592

A straightforward, comprehensive historical text first published in the 1950s and brought up to date. Includes an extensive annotated bibliography.

Harris, Richard Jackson
A Cognitive Psychology of Mass Communication
Hillsdale, N.J., L. Erlbaum, 1989. 287p. ISBN 0805800948

An interesting and clearly written summary of research on the influence of the media on thinking and perceptions of reality. Topics covered include the psychology of mass communication, group portrayals, advertising, media and values, sports, news, politics, violence, sex, and prosocial uses of media. Includes a substantial bibliography. Very informative on its own or as a guide to further reading.

Kaniss, Phyllis C.
Making Local News
Chicago, Univ. of Chicago Press, 1991. 260p. ISBN 0226423476

Examines factors that impact coverage of local news issues in print and broadcast media, including economic interests of media owners, the professional motives and characteristics of journalists, and the media strategies and expertise of local officials. Contends that newspapers provide favorable, uncritical coverage of major urban development projects out of self-interest. Includes numerous examples and a case study of coverage of Philadelphia's new convention center.

Kennedy, William V.
The Military and the Media: Why the Press Cannot Be Trusted to Cover a War
Westport, Conn., Praeger, 1993. 167p. ISBN 0275941914

Contends that because American journalism has failed to train competent, knowledgeable specialists in military reporting, the government has had no choice but to control journalists' access to military information for security reasons, in cases such as the Persian Gulf War. The author, who has been both a military officer and journalist, feels that journalists' ignorance presents a danger to democracy and the military.

Klaidman, Stephen
Health in the Headlines: The Stories Behind the Stories
New York, Oxford Univ. Press, 1991. 249p. ISBN 0195052986

Examines the coverage of heath risks by the media through seven specific case studies. Discusses why coverage of health risks is problematical, on account of limitations of media coverage and because of the way Americans interpret and understand news.

Kurtz, Howard
Media Circus: The Trouble With America's Newspapers
New York, Times Books, 1993. 420p. ISBN 0812920228

An entertaining, anecdotal discussion of newspapers' current problems by the media reporter for the *Washington Post*. Sections cover particular stories he feels the press has botched in recent years, problems with setting boundaries for coverage and ethical standards, the interrelationship between press and the national government, and the current problems in determining what the role of newspapers should be in an information-rich society.

LaMay, Craig L, and Everette E. Dennis, eds.
Media and the Environment
Washington, D.C., Island Press, 1991. 266p. ISBN 1559631317

An important collection of essays by members of the media and three U.S. senators. Sections discuss the complexities of environmentalism as a topic for media coverage and public debate, the necessity for covering the environment "as if it mattered," economics and environmental policy making, and the importance of thinking globally. Includes an essay on classic environmental books and an annotated bibliography.

Lotz, Roy
Crime and the American Press
New York, Praeger, 1991. 170p. ISBN 0275940128

Examines popular perceptions and stereotypes about how the press covers crime and how that coverage has an effect on society. It also gives the results of research projects to prove or disprove particular theories. Places coverage of crime, courts, and prisons in historical perspective, and discusses published media criticism.

MacArthur, John R.
Second Front: Censorship and Propaganda in the Gulf War
New York, Hill and Wang, 1992. 260p. ISBN 0809085178

An anecdotal discussion of how the press was controlled and deceived by the U.S. and Kuwaiti governments during the Gulf War, with the media's apparent cooperation. Includes reports of interviews with prominent journalists such as Ben Bradlee, Katharine Graham, Dan Rather, and Peter Jennings.

Mann, Thomas E., and Gary R. Orren, eds.
Media Polls in American Politics
Washington, D.C., Brookings Institution, 1992. 172p. ISBN 0815754566

A concise, informative, and well-written study of the nature and history of media polls. Describes methodologies used for the various types of polls, sources of poll error, and their impact on the public. Concludes that polls are an important and useful tool provided they help report news, not create it.

Marzolf, Marion
Civilizing Voices: American Press Criticism, 1880-1950
New York, Longman, 1991. 233p. ISBN 0801302862

A history of media criticism in popular, professional, and scholarly literature, organized chronologically. An epilogue makes suggestions as to what we should expect from media critics. A very extensive bibliography lists critical works.

Mayer, Martin
Making News
Rev. and updated. Boston: Harvard Business School Press, 1993. 345p. ISBN 0875843719

A lively and informative account of how print and broadcast media arose in the U.S. and what they are like today. Both entertaining and thought provoking.

Miraldi, Robert
Muckraking and Objectivity: Journalism's Colliding Traditions
New York, Greenwood Press, 1990. 180p. ISBN 0313272980

An analysis of the role of the journalist in society, including the history of reformist and exposé journalism in American print and broadcast journalism. The author argues that individual journalists should be allowed to offer opinions and recommendations as well as facts on public issues, especially since the press is increasingly owned by a few big business conglomerates who are sensitive to pressure from advertisers.

Moore, Mike, ed.
Health Risks and the Press: Perspectives on Media Coverage of Risk Assessment and Health
Washington, D.C., The Media Institute in cooperation with the American Medical Association, 1989. 111p. ISBN 0937790397

Essays by three scientists and three journalists examine issues in reporting environmental health risks, including dealing with the ambiguities of scientific data and problems with reporting complex information. Includes suggestions about techniques, further reading, and organizations and programs of interest to science and health writers.

Parenti, Michael
Inventing Reality: The Politics of News Media
2nd ed. New York, St. Martin's Press, 1993. 274p.
ISBN 0312020139

Contends that the American media, controlled by the wealthy and powerful, provide a conservative view of reality skewed to support the existing social and economic elite. Written with a leftist orientation and a polemical but well-documented style, this provides a useful balance to the frequent conservative criticism of the press as excessively liberal.

Pavlik, John Vernon, and Everette E. Dennis, eds.
Demystifying Media Technology: Readings From the Freedom Forum Center
Mountain View, Calif., Mayfield Pub. Co., 1993. 194p.
ISBN 1559341459

The essays in Part One discuss the economic, social, and technical issues involved in the convergence of media technology into electronic form. Part Two discusses the state of the art in the newsroom and "next generation" technologies. Part Three covers electronic information sources, including databases, computerized public records, expert systems, facsimile newspaper transmissions, and videotex.

Picard, Robert G.
Media Portrayals of Terrorism: Functions and Meaning of News Coverage
Ames, Iowa, Iowa State Univ. Press, 1993. 147p.
ISBN 0813818427

An excellent discussion of the relationship between the media and terrorism, dispelling a number of widely held myths. Topics covered include the nature of terrorism, the setting and role of communication in terrorism, the role of government, and the content, meaning, and implications of media coverage. Suggests ways in which the media can become more sensitive to what they cover and how they cover it. Includes a substantial bibliography.

Protess, David, and others
The Journalism of Outrage: Investigative Reporting and Agenda Building in America
New York, Guilford Press, 1991. 301p. ISBN 0898623146

Reports in detail on a study of six major investigative stories, and summarizes results of a survey of more than 900 investigative reporters and editors. Also includes a historical survey of muckraking journalism and a new theory about investigative reporting. Includes a bibliography.

Puette, William
Through Jaundiced Eyes: How the Media Views Organized Labor
Ithaca, N.Y., ILR Press, 1992. 228p. ISBN 0875461840

Contends that the media consistently shows negative bias in their coverage of organized labor. Chapters discuss portrayals of unions in movies, television, newspapers, and cartoons, and case studies of coverage of local and national labor disputes. Appendices show results of a high school labor survey, and lists movies, television specials, and dramas with labor themes. Includes an extensive bibliography.

Quirt, John
The Press and the World of Money: How the News Media Cover Business and Finance, Panic and Prosperity, and the Pursuit of the American Dream
Byron, Calif., Anton/California-Courier, 1993. 364p.
ISBN 0963550403

An anecdotal but informative examination of how the American press covers business news by a former business reporter and editor for *Fortune* and *The Institutional Investor*. Covers the history of the financial press, with chapters on individual newspapers and magazines. It also includes extensive, quoted comments from writers, economists, and other prominent figures. An appendix reprints twelve examples of good business writing.

Reston, James
Deadline: A Memoir
New York, Random House, 1991. 525p. ISBN 0394585585

A political and personal memoir by one of America's most prominent newspapermen. Offers insights into many events and personalities in public affairs and in the press.

Rowan, Carl Thomas
Breaking Barriers: A Memoir
Boston, Little, Brown, 1991. 395p. ISBN 0316759775

The hard-hitting, engrossing memoir of an award-winning syndicated columnist, who is also a panelist on *Inside Washington*, and host of his own daily syndicated radio show. From a poverty-stricken childhood, he went on to become one of the first fifteen African American Navy officers, one of the first African American reporters hired by a major newspaper, the U.S. Ambassador to Finland, and the director of the U.S. Information Agency.

Sabato, Larry J.
Feeding Frenzy: How Attack Journalism Has Transformed American Politics
New York, Free Press, 1991. 306p. ISBN 0029276357

Examines the trend toward excessive, often unsubstantiated press coverage of real or suspected scandal about public figures. After putting this trend in historical perspective, the author examines thirty-six particular incidents. He also discusses cases in which potential scandals were not published, draws conclusions about the consequences of "feeding frenzies," and recommends guidelines as to what the press should and should not publish, and why. Includes bibliography.

Shoemaker, Pamela J., and Stephen D. Reese
Mediating the Message: Theories of Influence on Mass Media Content
New York, Longman, 1991. 233p. ISBN 0801303079

Summarizes research on what determines media content. Individual chapters group the results of studies into types, such as studies of influences from individual media workers, from media routines, from organizational influences, from outside the organizations, and from ideology. Each chapter is followed by a list of references. The final chapter, "Building a Theory of News Content," lists a number of hypotheses, assumptions, and propositions that would be good subjects for further research.

Singer, Eleanor
Reporting on Risk: How the Mass Media Portray Accidents, Diseases, Disasters, and Other Hazards
New York, Russell Sage Foundation, 1993. 244p.
ISBN 0871548011

Examines news stories about accidents, illnesses, natural and man-made disasters, and scientific breakthroughs. Discusses how accurate the stories are and how they may distort the public's perception of the real extent of danger from these hazards.

Soley, Lawrence C.
The News Shapers: The Sources Who Explain the News
New York, Praeger, 1992. 175p. ISBN 0275940330

Examines the political affiliations, institutional associations, and levels of expertise of the individuals chosen by the news media to interpret and explain the news. The author contends that this "expert" commentary is frequently neither objective nor expert, but significantly affects public opinion.

Squires, James D.
Read All About It! The Corporate Takeover of America's Newspapers
New York, Times Books, 1993. 244p. ISBN 0812921011

This former editor of the *Chicago Tribune* argues that the press is changing from a privately-owned, bias-free, and informative institution to a collection of corporate giants whose interest is restricted to making money through entertainment.

Stephens, Mitchell
A History of News: From the Drum to the Satellite
New York, Viking, 1988. ISBN 0140094903

An examination of news from a historical perspective that illuminates the nature, strengths, and weaknesses of news and modern-day journalism. Covers spoken news, written news, early forms of printed news, newspapers, reporting systems, and electronic news media. Although it is a highly entertaining, anecdotal, and interpretive rather than exhaustive, it is very well documented and has an extensive bibliography.

Strentz, Herbert
News Reporters and News Sources: Accomplices in Shaping and Misshaping the News
2nd ed. Ames, Iowa, Iowa State Univ. Press, 1989. 181p.
ISBN 0813818869

Examines how news is distorted, often unintentionally, by factors such as the relationship between the reporter and the source, and the reporter's or editor's concept of what is news. Provides both philosophical and practical information for the journalist or journalism student.

Surette, Ray
Media, Crime, and Criminal Justice: Images and Realties
Pacific Grove, Calif., Brooks/Cole Pub. Co., 1992. 299p.
ISBN 0534164404

Examines the media's effect on crime and justice, summarizing research results and correcting common misconceptions. Topics include the effect of media on attitudes about crime and justice, crime and justice in the entertainment and news media, the media as a cause of crime, media-based anticrime programs, the media and the courts, and the use of media technology in the justice system. Clearly written and informative. Includes an extensive bibliography.

Tebbel, John William, and Mary Ellen Zuckerman
The Magazine in America, 1741-1990
New York, Oxford Univ. Press, 1991. 433p. ISBN 0195051270

A comprehensive history that updates earlier histories of American magazines by Frank Luther Mott and Theodore Peterson. Includes individual discussions of such topics as alternative magazines, pulps and science fiction, women's magazines, the business press, and African American periodicals. Includes an extensive bibliography.

Underwood, Doug
When MBAs Rule the Newsroom: How the Marketers and Managers are Reshaping Today's Media
New York, Columbia Univ. Press, 1993. 259p. ISBN 0231080484

Explores the trend toward market-driven journalism exemplified by *USA Today,* in which information and news are replaced by entertainment factors that respond to reader surveys and what sells. Includes results of a survey of the perceived influence of marketing on newsroom management according to 429 staffers at twelve newspapers.

Weaver, David H., and G. Cleveland Wilhoit
The American Journalist: A Portrait of U.S. News People and Their Work
2nd ed. Bloomington: Indiana Univ. Press, 1991. 276p.
ISBN 0253363640

Analyzes the results of a 1982-83 telephone survey of 1,001 full-time journalists, covering their backgrounds, education, career patterns, professional values and ethics, and job conditions. This new edition adds chapters on women journalists, broadcast journalists, and what the respondents considered their best work.

Willis, William James
Journalism: State of the Art
New York, Praeger, 1990. 209p. ISBN 0275932435

Provides readable summaries of research in various areas published in *Journalism Quarterly* and *Newspaper Research Journal*. Areas covered include journalists' attitudes and backgrounds, law and ethics, predictors of readership, journalists and sources, polling and precision journalism, advertising, media effects on the public, and electronic publishing. Also discusses the feud between media researchers and practitioners. Includes a short bibliography.

Willis, William James
The Shadow World: Life Between the News Media and Reality
New York, Praeger, 1991. 260p. ISBN 0275934241

An excellent examination of the difference between reporting isolated facts accurately and providing the whole truth about issues to aid the public in acting appropriately. Covers problems with the approach taken by individual reporters and organizations, many historical examples of the media's failure to report important issues adequately, problems with specialization, secrecy, lies, and the effect of distortion on the public. Concludes with concrete suggestions on providing better coverage. Includes a bibliography.

✦ Media Law and Ethics

Black, Jay, Bob Steele, and Ralph Barney
Doing Ethics in Journalism: A Handbook with Case Studies
Greencastle, Ind., Sigma Delta Chi Foundation and the Society of Professional Journalists, 1993. 252p

The first part of this text introduces and critiques codes of ethics, discusses the role of journalism in society, and critiques ethical decisions in a Pulitzer Prize-winning story. The center section categorizes and discusses particular types of problems, with case studies and questions for further discussion. Appendices include an extensive annotated bibliography and the results of a survey of editors and news directors from 304 news organizations.

Campbell, Douglas S.
The Supreme Court and the Mass Media: Selected Cases, Summaries, and Analyses
New York, Praeger, 1990. 242p. ISBN 0275934217

Provides information on forty-five important cases grouped under the general categories of libel, privacy, and First Amendment issues. Gives background, circumstances, summaries of the Court's analysis, ruling, and significance for each case. Includes a selected bibliography.

Fishman, Stephen
The Copyright Handbook: How to Protect and Use Written Works
Berkeley, Calif., Nolo Press, 1992. 1 v. (various pagings).
ISBN 0873371305

Written in clear language by an attorney, this handbook includes a short overview of copyright law followed by a detailed discussion of its most important aspects. Designed to be useful to a range of users, from those who want to protect their own works to those who want to know all about copyright law. Includes Copyright Office forms.

Gillmor, Donald M.
Power, Publicity and the Abuse of Libel Law
New York, Oxford Univ. Press, 1992. 222p. ISBN 0195071921

The author criticizes the current abuse of libel law by those with power and prestige who use it to silence criticism. According to the author, another offender is the "self-righteous" press that idealizes free flow of information but does not supply space for replies to their attacks. The author proposes reform that would make it impossible for some categories of citizens, such as policymakers, to sue for libel.

Hausman, Carl
Crisis of Conscience: Perspectives on Journalism Ethics
New York, HarperCollins Publishers, 1992. 214p.
ISBN 0065003659

Examines contributions from various fields relevant to journalism ethics and surveys one group of practitioners. Part One covers the historical background; Part Two, the principles of accuracy and objectivity, social responsibility, fairness, and professional conduct. Part Three includes an examination of principles in conflict (e.g., privacy versus the right to know). The fourth part discusses past attempts at resolution of problems, including codes of ethics and methods of review. Appendices include annotated bibliographies in philosophy, ethics, and journalism ethics.

Kane, Peter E.
Errors, Lies, and Libel
Carbondale, Ill., Southern Illinois Univ. Press, 1992. 171p.
ISBN 0809317192

A concise, clear discussion of major recent libel cases. Discusses problems with libel law stemming from the conflict between freedom of expression and the right to protect one's name, and suggests possible reform of libel law. Includes an annotated case list and selected annotated bibliography.

Knowlton, Steven R., and Patrick Parsons, eds.
The Journalist's Moral Compass: Basic Principles
Westport, Conn., Praeger, 1994. 246p. ISBN 0275945375

A collection of writings on ethical issues in journalism, beginning with philosophers such as Milton and Hobbes and continuing to the present. Readings are grouped under three broad topics: the relationship of the press and the state, truth and objectivity, and economic issues. Also included is a bibliographic essay and the text of the ethics code of the Society of Professional Journalists.

Lambeth, Edmund B.
Committed Journalism: An Ethic for the Profession
2nd ed. Bloomington: Indiana Univ. Press, 1992. 242p.
ISBN 0253332206

Examines the moral value of journalistic work itself as well as the moral dilemmas journalists encounter in their daily activities. Sums up classical ethical theories, then suggests a framework upon which to build a practical system of journalistic ethics. Provides many real and theoretical examples in a very readable style.

Middleton, Kent R., and Bill F. Chamberlain
The Law of Public Communication
3rd ed. New York, Longman, 1994. 625p. ISBN 0801311888

A basic text covering topics such as the First Amendment, libel, privacy and personal security, intellectual property, advertising, protection of news sources, access to information, regulation of broadcasting of cable television, and the new electronic media. Appendices include a guide to finding and reading the law, the first fourteen amendments to the U.S. Constitution, and a glossary.

Powe, L. A. Scot
The Fourth Estate and the Constitution: Freedom of the Press in America
Berkeley, Univ. of California Press, 1991. 357p. ISBN 0520072901

A legal scholar's examination of the issues related to freedom of the press in America. A historical summary emphasizes the post-1964 period. Separate chapters discuss the issues of libel, prior restraints, access to sources of information, and antitrust concerns. The author then discusses two theoretical models used to justify and explain the need for this freedom: "the right to know," and the function of the "Fourth Estate" as governmental watchdog.

Rosini, Neil J.
The Practical Guide to Libel Law
New York, Praeger, 1991. 229p. ISBN 0275937828

A lawyer's clearly written guide for reporters on how to avoid legal trouble. Three sections cover how to identify potential libel, how much proof one should have to support a potentially libelous statement, and how to write a story making the best use of available legal defenses. Includes lists of dos and don'ts, quick questions with short answers, and a table of cases.

Zelezny, John D.
Communications Law: Liberties, Restraints, and the Modern Media
Belmont, Calif., Wadsworth Pub. Co., 1993. 540p. ISBN 0534134521

A basic text giving an overview of the U.S. legal system, the First Amendment, risks to public safety, damage to reputation, privacy and peace of mind, and gaining access. Also includes information on media and the justice system, creative property, corporate and government speech, commercial speech, electronic media, obscenity, and indecency. Includes excerpts from major cases, the text of the U.S. Constitution, and two codes of ethics.

✦ News Reporting and Writing

Brooks, Brian, and others
News Reporting and Writing
4th ed. New York, St. Martin's Press, 1992. 589p. and computer disk. ISBN 0312047681

This widely-used textbook is an introduction to the news: basic skills (including writing, reporting, interviewing, quotes, and attribution); basic stories (including obituaries, news releases, speeches, and meetings); beat reporting of various kinds; advanced techniques; writing news for radio and television; and rights and responsibilities. Appendices include information on wire service style. Glossary.

Cappon, Rene J.
The Associated Press Guide to News Writing
2nd ed. New York, Prentice Hall, 1991. 162p. ISBN 0130536792

An entertaining and useful guide to clear, concise, accurate, and interesting writing. Topics include news writing, leads, writing good sentences, "journalese," tone, pitfalls, how to use quotes, clichés, feature styles, and overcoming hangups about good usage. One section lists clichés to avoid.

Davidson, Margaret
A Guide for Newspaper Stringers
Hillsdale, N.J., L. Erlbaum, 1990. 211p. ISBN 0805807640

A concise overview of newspaper reporting skills for stringers, those supplementary reporters who cover local news on a free-lance basis. Covers finding news, journalistic writing, interviewing, kinds of stories frequently covered, legal and ethical concerns, and the work style of a stringer. Includes references.

Fedler, Fred
Reporting for the Print Media
5th ed. Fort Worth, Tex., Harcourt Brace Jovanovich, 1993. 817p. ISBN 0155006029

This textbook covers news writing style, use of words, quotation and attribution, interviews and polls, and improving newsgathering. Also includes information on topics such as careers, communication law, ethics, obituaries, news releases, speeches and meetings, and public affairs reporting. Includes advanced reporting exercises and software.

Gaines, William
Investigative Reporting for Print and Broadcast
Chicago, Nelson-Hall, 1994. 246p. ISBN 0830412832

A basic text emphasizing how to investigate and report findings, with many fictitious case studies, some real examples, and exercises. Includes some particular types of stories, such as personal profiles, consumer abuse, private businesses, and heath care. Appendices provide the Freedom of Information Act and federal laws for nonprofit organizations.

Hausman, Carl
The Decision-Making Process in Journalism
Chicago, Nelson-Hall, 1990. 140p. ISBN 0830412034

A concise and helpful handbook intended to train journalists to make good decisions about news. Individual chapters discuss newsworthiness, accuracy, fairness, logic, freedom from distortion, liability for libel or invasion of privacy, ethics, and possible consequences of the story. Each chapter contains a checklist to help the journalist make difficult decisions.

Kennedy, George, Daryl R. Moen, and Don Ranly
Beyond the Inverted Pyramid: Effective Writing for Newspapers, Magazines, and Specialized Publications
New York, St. Martin's Press, 1993. 276p. ISBN 031204058X

An excellent guide to the many kinds of journalistic writing, with an emphasis on engaging readers' attention through their own experiences. Sections focus on characteristics of writers and audiences, reporting skills, and writing to inform and entertain. Chapters cover particular challenges in service journalism, organization publications, opinion, and humor. Appendices include information on getting an article into print, grammar, and the lessons of general semantics.

Mencher, Melvin
News Reporting and Writing
6th ed. Madison, Wisc., Brown and Benchmark, 1994. 634p.
ISBN 0697139352

A beginning text used to teach students to become reporters. Covers topics such as vocational guidance, editing and researching stories, principles of reporting, and risks and responsibilities. Focus is on reporting accurately and writing precisely. Appendices.

Meyer, Philip
The New Precision Journalism
Bloomington, Indiana Univ. Press, 1991. 273p.
ISBN 0253337909

This new edition urges journalists to make use of the techniques of social and behavioral science research. Specific areas covered include data analysis, statistics, use of computers, field experiments, databases, and how to do an election survey.

Murray, Donald Morrison
Writing for Your Readers: Notes on the Writer's Craft from the Boston Globe
2nd ed. Old Saybrook, Conn., Globe Pequot Press, 1992. 243p.
ISBN 1564400514

This clear, concise guide provides interviews with many current writers for the prestigious *Boston Globe*. Topics covered include focus (find the tension); explore (report for surprise); rehearse (write before writing); draft (write for surprise); develop (work on what works); and clarify (represent the reader).

Parsigian, Elise K.
Mass Media Writing
Hillsdale, N.J., L. Erlbaum Associates, 1992. 355p.
ISBN 0805811303

Despite the title, this book focuses chiefly on developing critical thinking skills and gathering information using systematic methods shared with science. Useful to writers in print and broadcast news, magazines, public relations, and advertising. Topics covered include gathering background information, stating the problem, evaluating research, designing and conducting interviews, organizing and coding data, analysis, and writing copy. Includes case studies and bibliographies.

Rich, Carole
Writing and Reporting News: A Coaching Method
Belmont, Calif., Wadsworth Pub., 1994. 590p ISBN 053419074X

A comprehensive text that includes current concerns such as the trend toward briefer, more practical news stories, extensive use of technology, and multicultural sensitivity. Covers not only many techniques for general reporting and writing, but also public relations, broadcast writing, libel, and ethics, as well as techniques for reporting speeches, obituaries, crime, disasters, profiles, and features.

Schulte, Henry H., and Marcel P. Dufresne
Getting the Story: An Advanced Guide to Reporting Beats, Records, and Sources
New York, Macmillan, 1994. 459p. ISBN 002408042X

This guide covers public affairs reporting with an emphasis on research. Part One covers basic reporting skills, including newer techniques such as online database research, polling, and the use of computers. Parts Two through Four give background information and strategies for covering various levels of government, law enforcement and the courts, education, politics, business and consumerism, the work force, the environment, and science and health. The final section covers complex relationships between reporters and their sources, including manipulation, favoritism, and privacy issues.

Ullmann, John, and Jan Colbert, eds.
Reporter's Handbook: An Investigator's Guide to Documents and Techniques
2nd ed. New York, St. Martin's Press, 1991. 457p.
ISBN 0312051476

An extremely detailed and useful guide to finding information about individuals, businesses and institutions from government publications, public records, computer databases, and various organizations. Explains how to acquire information under the *Freedom of Information Act* or other legal requirements. Includes chapters on how to investigate particular categories of people (e.g., licensed professionals, politicians) and topics (e.g., health care, education).

Ward, Hiley H.
Reporting in Depth
Mountain View, Calif., Mayfield Pub. Co., 1991. 447p.
ISBN 0874848474

An advanced reporting and writing textbook that focuses on students. Includes many sample articles, by students, with explanations of how they got ideas, and researched, shaped, and wrote their articles. Particular topics covered include finding ideas, researching, interviewing, and structuring articles. Also includes chapters on reporting opportunities in specific areas, both on campus and elsewhere. A concluding chapter discusses ethics. Bibliography.

✦ Newspaper Design and Layout

Ames, Steven E.
Elements of Newspaper Design
New York, Praeger, 1989. 326p. ISBN 0275923304

Explains the technical issues in newspaper design. Sections cover the basics of the Total Page Concept, principles of design, typography, design creativity, finishing touches (color), and putting it all together. Also covers computer-initiated text and redesign. Includes bibliography.

Barnhurst, Kevin G.
Seeing the Newspaper
New York, St. Martin's Press, 1994. 222p. ISBN 0312061498

Examines the visual history and meaning of the newspaper, including influences from philosophy, science, and art. Topics covered include newspaper redesign, photography, charts, typography, and layout. Includes exercises to help the reader examine newspaper design from artistic, political, and philosophical viewpoints.

Bohle, Robert H.
Publication Design for Editors
Englewood Cliffs, N.J., Prentice Hall, 1990. 196p.
ISBN 0137375379

Approaches design as an integral part of the editing process that enhances the communication of information. Emphasizes design concepts, layout, typography, use of photographs, art, graphics, and color, with individual chapters applying these techniques to newsletters, brochures, magazines, and newspapers. Each chapter has a checklist of important points and many chapters include lists of projects.

Conover, Theodore E.
Graphic Communications Today
2nd ed. St. Paul, Minn., West Pub. Co., 1990. 517p.
ISBN 0314665706

A comprehensive and detailed text including topics such as traditional printing methods; typography; art and illustrations; color, paper and ink; layout and pasteup; desktop publishing; advertising design; magazine and newspaper design; newsletters; and the future of graphic design. Appendices include a basic communications design library, type specimens, and a glossary.

Garcia, Mario R.
Eyes on the News
St. Petersburg, Fla., Poynter Institute for Media Studies, 1991. 85p.
ISBN 0935742190

Reports the results of research conducted by the Poynter Institute on how color has an effect on newspaper reading. Includes information on other factors related to what readers process. Includes a bibliography on color in newspapers.

Finberg, Howard I., and Bruce D. Itule
Visual Editing: A Graphic Guide for Journalists
Belmont, Calif., Wadsworth Pub. Co., 1990. 270p.
ISBN 0534117368

A guide to the effective use of informational graphics, photographs, and illustrations in newspaper design and layout. Discusses the impact of visual presentations on readership. Includes many illustrations including sample front pages from forty newspapers on the same day. Also includes exercises and interviews with visual journalists.

Garcia, Mario
Contemporary Newspaper Design: A Structural Approach
3rd ed. Englewood Cliffs, N.J., Prentice Hall, 1993. 302p.
ISBN 0131748718

A lavishly illustrated guide by a well-known international consultant in newspaper design. Emphasizes the integration of writing, editing, and design to instill excitement and interest in a newspaper.

Harrower, Tim
The Newspaper Designer's Handbook
2nd ed. Dubuque, Iowa, William C. Brown Publishers, 1992. 180p. ISBN 0697133826

An excellent step-by-step guide to newspaper design in a lively, illustrated style. Major topics include fundamentals, page design, photographs and art, packaging, special effects, and infographics. Shows what works, what does not, and why. Includes exercises with answers and a glossary.

Moen, Daryl R.
Newspaper Layout and Design
2nd ed. Ames, Iowa, Iowa State Univ. Press, 1989. 263p.
ISBN 0813812275

A basic guide with more narrative than many design books. The first section covers the basic mechanics of layout, including terminology, working in modules, using graphics, packaging sections and stories, and the tabloid format. The second section on design covers the design team, general principles, use of type and color, design of sections and accessories, and the process of redesign. Includes a glossary and a bibliography.

✦ Photojournalism

Gross, Larry, John Stuart Katz, and Jay Ruby, eds.
Image Ethics: the Moral Rights of Subject in Photographs, Film, and Television
New York, Oxford Univ. Press, 1988. 382p. ISBN 0195054334

Although the essays in this volume cover far more media than print journalism, the issues they raise are applicable here. Issues covered include intrusion into private space, disclosure of true but embarrassing facts, presenting people in a false light, and use of personal photographs for profit without consent or compensation to the subject. Includes analyses of cases.

Horton, Brian
The Associated Press Photo-Journalism Stylebook
Reading, Mass., Addison-Wesley, 1990. 143p. ISBN 0201132354

Despite its title, this is not a source for technical details. The content is largely quoted information from working photographers explaining the context and approach they took in achieving successful photographs. Includes information on electronic photography.

Hoy, Frank P.
Photojournalism: The Visual Approach
2nd ed. Englewood Cliffs, N.J., Prentice Hall, 1993. 290p.
ISBN 0136655718

A manual covering all aspects of photojournalism including equipment, techniques for different purposes, history, editing, caption writing, ethics, and legal aspects. Includes a list of assignments and a bibliography.

Kobre, Kenneth
Photojournalism: The Professionals' Approach
2nd ed. Boston, Focal Press, 1991. 359p. ISBN 0240800613

A thorough, readable, and informative guide with extensive, striking illustrations. Topics covered include features, portraits, sports, picture stories, illustrations, editing, cameras and film, strobe, electronics, the law, and ethics. Includes an extensive bibliography.

LaBelle, Dave
The Great Picture Hunt: The Art and Ethics of Feature Picture Hunting
Bowling Green, Ky., Western Kentucky Univ., 1991. 93p.
ISBN 0963077082

A guide for hunting great pictures, with numerous examples to complement the text. Includes both methods and discussions of ideas. A chapter covers the ethics of publishing, captioning, and editing.

LaBelle, Dave
Lessons in Death and Life
Blackshear, Ga., Dave LaBelle, 1993. 106p. ISBN 0963077007

A very moving and personal examination of the value, pain, and ethical considerations in photographing death and grief, with many photographs by the author. Includes a bibliography.

Lacayo, Richard, and George Russel
Eyewitness: 150 Years of Photojournalism
New York, Time, 1990. 192p. ISBN 0848710223

A striking and informative history consisting largely of photographs with explanatory paragraphs. Includes introductory historical essays for each of six time periods, beginning in 1839 and continuing to the present.

Lester, Paul Martin, ed.
The Ethics of Photojournalism: A Special Report
Durham, N.C., National Press Photographers Association, 1990. 101p.

A collection of contributions on topics such as the history of photojournalism ethics, photographic invasion of privacy, the rights of those pictured, digital retouching, electronic editing, research on photojournalism ethics, and the NPPA *Code of Ethics*.

Lester, Paul Martin
Photojournalism: An Ethical Approach
Hillsdale, N.J., L. Erlbaum, 1991. 202p. ISBN 0805806717

Raises many questions intended to cause the reader to examine one's ethical perspective. Among the particular areas examined, often from a legal as well as ethical viewpoint, are victims of violence, rights to privacy, and picture manipulation. An appendix includes a review of selected literature with extended comments and a bibliography.

Lewis, Greg
Photojournalism: Content and Technique
Dubuque, Iowa, W.C. Brown, 1991. 337p. ISBN 0697042928

A comprehensive textbook that covers tools (cameras, film and exposure, lenses, darkroom, light, introduction to color); techniques (composition, news and features, sports, studio photography, the photo story, photo editing); and broader topics (ethics, legal limits, education and careers, the history of photojournalism). An epilogue covers the electronic newspaper, and an appendix covers technical processes and professional organizations. Includes a bibliography.

London, Barbara, and John Upton
Photography
5th ed. New York, HarperCollins College Publishers, 1994. 422p. ISBN 0673522237

An extremely popular photography text that covers both basic techniques and some advanced applications. Each two facing pages covers a single idea, skill, or technique. Most chapters have a "What went wrong?" section for problem-solving. Includes coverage of photojournalism and digital imaging.

McDougall, Angus, and Veita Jo Hampton
Picture Editing and Layout: A Guide to Better Visual Communication
Columbia, Mo., Viscom Press, 1990. 300p. ISBN 0962513709

A guide to choosing and using photographs that teaches through the presentation of many examples with accompanying discussion. Topics covered include the psychology of visual perception, photographic techniques for nonphotographer editors, cropping, portraits, captions, photo sequences, use of interrelated photographs, editorial illustration with created pictures, and color photography.

Ritchin, Fred
In Our Own Image: The Coming Revolution in Photography: How Computer Technology is Changing Our View of the World
New York, Aperture, 1990. 158p. ISBN 0893813982

Examines the implications of the increasing use of computers to alter photographs. Discusses the implications for photojournalism, which include the ability to enhance images but also to deceive the public. Includes many illustrative photographs.

✦ Reference Books, Stylebooks, Handbooks and Grammar References

Anderson, Laura Killen
Handbook for Proofreading
Lincolnwood, Ill., NTC Business Books, 1990. 176p.
ISBN 0844232653

A practical guide for professionals. Topics covered include who proofreads, ways to proofread, proofreading skills, using style sheets, querying writers, giving clear instructions, working with type, and understanding the printer's language. Appendices list tools, checklists, words commonly misused, and several style sheets. Includes a glossary.

Ashley, Perry J., ed.
American Newspaper Publishers
1950-1990. Detroit, Gale Research, 1993. 424p.
ISBN 0810353865

After a lengthy introductory essay, this book gives biographical entries for forty-six prominent publishers, with bibliographical references for each individual. Also included is a substantial checklist for further general reading. *Dictionary of Literary Biography*, vol. 127.

Botts, Jack
The Language of News: A Journalist's Pocket Reference
Ames, Iowa, Iowa State Univ. Press, 1994. 216p.
ISBN 081382494X

A review of grammar and punctuation, common language blunders, and problems of usage and spelling. Includes information on the news story, including problems with leads, transitions, story, and failure to humanize. Appendices include a list of problem words to memorize, a stylebook summary, and a bibliography.

Brooks, Brian S., and James L. Pinson
Working with Words: A Concise Handbook for Media Writers and Editors
2nd ed. New York, St. Martin's Press, 1993. 256p.
ISBN 0312091397

A useful, concise guide to English grammar and usage that provides information stylebooks and dictionaries don't furnish. Areas covered include basic grammatical terms and their use, punctuation, tight writing, muddled language, spelling, and a wire service style summary of capitalization, abbreviations, and numerals. Also includes a chapter on sexism, racism, and other "-isms" to avoid, and lists of words commonly misused or misspelled.

Directory of Women's Media
Washington, D.C., National Council for Research on Women, Annual

A directory of media by, for, about, and owned by women. One section lists women's periodicals, publishers, writers' groups, organizations, and library collections. Another section includes articles and documents on feminist media, e.g., a listing of "Women of Color Periodicals, 1968-1988." The final section is a directory of media women and media-concerned writers.

Downs, Robert Bingham, and Jane B. Downs
Journalists of the United States: Biographical Sketches of Print and Broadcast News Shapers from the Late Seventeenth Century to the Present
Jefferson, N.C., McFarland, 1991. 391p. ISBN 089950549X

Contains more entries in a shorter format that those in *Biographical Dictionary of American Journalism* (McKerns, 1989). Includes a brief bibliography.

Editor and Publisher International Yearbook
New York, Editor and Publisher. ISSN 0424-4923

Includes directories of daily and weekly newspapers in the U.S. and abroad, news and syndicate services, mechanical equipment suppliers, associations and trade organizations, and suppliers and services. Also includes "Who's Where," a selective telephone directory of people in the newspaper industry. Includes statistics.

Ellmore, R. Terry
NTC's Mass Media Dictionary
Lincolnwood, Ill., National Textbook Co., 1991. 668p.
ISBN 0844231851

Contains more than 20,000 entries in the areas of radio, television, cable television, film, newspapers, magazines, books, direct mail, and outdoor advertising. Includes widely used engineering terms, cross-references, and acronyms.

Felici, James
The Desktop Style Guide
New York, Bantam Books, 1991. 129p. ISBN 0553354450

A concise reference guide to typographic style. Topics covered include typographic basics, typographic style, special characters, character style, tables, and labor-saving ways to prepare an electronic manuscript. Includes a glossary.

Forbes MediaGuide 500
New York, Forbes, 1993- . Annual. ISSN 1067-4918

An important and interesting who's who and evaluation of print journalists, newspapers, and news periodicals. Includes an overview of the year and its major news stories, and rates the ten best stories and columns of the preceding year in each of five categories. Also provides an overview of the Canadian press and broadcast journalism, and discusses and rates the performance of nine newspapers and thirty-nine periodicals. The main section lists and discusses the work of more than 500 leading reporters and commentators in six categories. An index lists rated journalists by subject area of assignment.

Gale Directory of Publications and Broadcast Media: An Annual Guide to Publications and Broadcasting Stations
26th ed., Detroit, Gale, 1994. 3 v. ISBN 0810380595

This annual publication is arranged by state, with indexes. Lists 37,000 newspapers, magazines, journals, and other periodicals, as well as radio, television, and cable stations and systems. Gives advertising rates, circulation statistics, local programming, and names of personnel.

Goldstein, Norm, ed.
The Associated Press Stylebook and Libel Manual: Including Guidelines on Photo Captions, Filing the Wire, Proofreaders' Marks, Copyright. Fully updated.
Reading, Mass., Addison-Wesley, 1992. 338p. ISBN 0201567601

A standard stylebook for the newspaper industry. The main section is a detailed guide to capitalization, abbreviations, punctuation, spelling, numerals, and usage; there are separate sections on style guidelines for sports and business. Other sections cover libel, copyright, the Freedom of Information Act, photo captions, and the Associated Press. Includes a bibliography.

Gordon, Karen Elizabeth
The Deluxe Transitive Vampire: The Ultimate Handbook of Grammar for the Innocent, The Eager, and the Doomed
New York, Pantheon Books, 1993. 175p. ISBN 0679418601

The latest edition of a highly regarded, entertaining, and informative book on grammar. The author uses unusual examples and illustrations to keep the reader's attention.

Gordon, Karen Elizabeth
The New Well-Tempered Sentence: A Punctuation Handbook for the Innocent, the Eager, and the Doomed
Expanded and rev. ed. New York, Ticknor and Fields, 1993. 148p.
ISBN 0395628830

A revised version of the author's entertaining book on punctuation. It serves as a companion to *The Deluxe Transitive Vampire* (Pantheon Books, 1993).

Gunderloy, Mike, and Cari Goldberg Janice
The World of Zines: A Guide to the Independent Magazine Revolution
New York, Penguin Books, 1992. 181p. ISBN 014016720X

Provides reviews of 400 small specialized publications, generally produced by one individual. Arranged under twenty-one subject categories such as fringe culture, sports, music, and spirituality. Also provides practical information on how to get started publishing a "zine."

Johnson, Ben, and Mary Bullard-Johnson
Who's What and Where: A Directory and Reference Book on America's Minority Journalists
Columbia, Mo., Who's What and Where, 1988. 735p.
ISBN 0961441828

    Lists more than 4,000 African American, Asian American, Hispanic, and Native American journalists, with "thumbnail biographies." Also contains self-help information for minority journalists and a short history of minorities in journalism.

Kessler, Lauren, and Duncan McDonald
When Words Collide: A Media Writer's Guide to Grammar and Style
3rd ed. Belmont, Calif., Wadsworth Pub. Co., 1992. 242p.
ISBN 0534170285

    A readable, helpful, and concise guide to words, grammar, style, and punctuation that can be used either as a text or a reference. Includes chapters on spelling and avoiding racism, sexism, and other stereotypical language. An appendix contains a diagnostic review of grammar and word use.

Lutz, William
Doublespeak: From "Revenue Enhancement" to "Terminal Living": How Government, Business, Advertisers, and Others Use Language to Deceive You
New York, Harper and Row, 1989. 290p. ISBN 0060161345

    An entertaining examination of the type of language that pretends to communicate, but really obscures meaning. The author identifies four kinds of doublespeak: euphemism, jargon, gobbledegook, or "bureaucratizese"/inflated language. He then gives examples of doublespeak in everyday living, advertising, business communication, the Pentagon, government, and the nuclear industry. An index of doublespeak leads to explanations of terms in the text.

Maggio, Rosalie
The Dictionary of Bias-Free Usage: A Guide to Nondiscriminatory Language
Phoenix, Ariz., Oryx Press, 1991. 293p. ISBN 0897746538

    Lists 5,000 potentially problematic words and suggests alternatives. Entries also include definitions of kinds of bias (e.g., ageism), occupations traditionally held by one sex or the other, gender breakdowns of occupations, and discussions of "key concepts" like gender roles. Includes a section discussing general writing guidelines for avoiding bias.

Mauro, John
Statistical Deception at Work
Hillsdale, N.J., L. Erlbaum, 1992. 113p. ISBN 0805812326

    Clearly explains the basic concepts of statistics and how they can be used to create false impressions. Discusses deceptive uses of averages, percentages, modifying words, index numbers (e.g., Consumer Price Index), charts and graphs, "convenient" numbers, probabilities, large numbers, correlations, sampling, and questionnaires. Includes suggestions on how to avoid being fooled.

McKerns, Joseph P., ed.
Biographical Dictionary of American Journalism
New York, Greenwood Press, 1989. 820p. ISBN 0313238189

    Covers nearly 500 people who contributed to the development of the American media. An appendix categorizes individuals by fields such as columnists and war correspondents. Other appendices cover women journalists, minority and ethnic journalism, and Pulitzer Prizes. Entries include bibliographical references.

Miller, Casey, and Kate Swift
The Handbook of Nonsexist Writing
2nd ed. New York, Harper and Row, 1988. 180p.
ISBN 0061816027

    An interesting, entertaining, and informative discussion of the ways our language shows unconscious bias against women (and also other groups such as ethnic and racial minorities, the aged, or men). Includes many illustrative examples and suggestions for using more inclusive, less biased language. Includes a bibliography.

Monmonier, Mark S.
How to Lie With Maps
Chicago, Univ. of Chicago Press, 1991. 176p. ISBN 0226534146

    Essential information for those who use maps to illustrate text. Provides information on mapping and warns the reader about intentional or accidental misrepresentation of data. Topics include elements of the map, map generalization, blunders that mislead, maps that advertise, develop-ment maps, maps for political propaganda, defense and disinfor-mation, data maps, and the use of color in maps. Appendices explain latitude and longitude and provide a bibliography.

Nourie, Alan, and Barbara Nourie, eds.
American Mass-Market Magazines
New York, Greenwood Press, 1990. 611p. ISBN 0313252548

    This book profiles 106 magazines, both continuing and defunct, with a general circulation of more than 100,000. Gives a general description and a short history of each. Includes bibliographic information, indexing and location sources, and details of publishing such as history, editors, and circulation.

Riley, Sam G., ed.
American Magazine Journalists, 1741-1850
Detroit, Mich., Gale Research, 1989. 387p. ISBN 081034551X

    Contains fifty entries in a format similar to the editor's work covering magazine journalists of a later period. *Dictionary of Literary Biography,* 79.

Riley, Sam G., ed.
American Magazine Journalists, 1900-1960
Detroit, Mich., Gale Research, 1990. 401p. ISBN 0810345714

    Contains thirty-seven lengthy biographical articles on historically significant writers, editors, and publishers, with bibliographical references to works both by and about each. Also includes a checklist of more general references on magazine journalism of the period. *Dictionary of Literary Biography,* 91. Second series also available, *Dictionary of Literary Biography,* 137.

Riley, Sam G., ed.
Corporate Magazines of the United States
New York, Greenwood Press, 1992. 281p. ISBN 0313275696

    This guide profiles company publications, both internal and external, that are published in magazine format. Examples range from *Aramco World* to *WordPerfect, the Magazine.* Historical essays on each magazine's founding, development, editorial policies, and content are given for fifty-one titles; there are also directory entries for 273 titles. *Historical Guides to the World's Periodicals and Newspapers.*

Weiner, Richard
Webster's New World Dictionary of Media and Communications
New York, Webster's New World, 1990. 533p.
ISBN 0139697594

Contains more than 30,000 entries whose definitions can be easily understood by nonprofessionals. Covers a wide range of subject areas including journalism, newspapers, photography, printing, typography, and writing. Also includes major companies and subsidiaries in the field, as well as trademarks and trade and professional organizations. Provides contact information for the top personnel at most media organizations in the country.

The Working Press of the Nation
Chicago, National Research Bureau. Annual. 1961- . 5 v.
ISSN 0084-1323

Individual volumes cover newspapers, magazines, TV and radio, feature writers and photographers, and internal publications of U.S. companies, government agencies, clubs, and other groups. Arranged geographically or by subject areas (Black newspapers, foreign language newspapers, etc.), with indexes.

Writer's Market 1994: Where and How to Sell What You Write
Cincinnati, Ohio, Writer's Digest Books. 1993. 1001p.
ISBN 0898796075

Each publisher entry gives a description of what the firm publishes, submission requirements, and response time for submitted material. Contains a directory of book publishers, consumer magazines, and trade, technical, and professional journals. Also includes some general information on how to get published and thorough title and subject indexes. Annual.

✦ Subject and Other Specialties in Journalism

Anderson, Douglas A.
Contemporary Sports Reporting
2nd ed. Chicago, Ill., Nelson-Hall, 1994. 327p.
ISBN 0830412883

An introductory section covers trends and philosophies of sports reporting, and precontest coverage. Sections Two and Three discuss coverage of revenue-producing and nonrevenue-producing sports by individual sport. Section Four discusses coverage beyond stadiums and arenas (features, columns, ratings and all-star teams, the business of sports, etc.). The final section covers design of pages and legal considerations.

Bohle, Robert
From News to Newsprint: Producing a Student Newspaper
2nd ed. Englewood Cliffs, N.J., Prentice Hall, 1992. 259p.
ISBN 0133253252

An elementary handbook for a student newspaper. The first part covers organization, newsgathering, ethics, and specific kinds of articles; the second part covers editing, photos, design, layout, advertising, and distribution. Each chapter includes a list of suggested projects and exercises. Also included are a brief stylebook, glossary, and bibliography.

Burgett, Gordon
The Travel Writer's Guide
Rocklin, Calif., Prima Pub., 1991. 293p. ISBN 1559581158

This guide emphasizes business rather than writing. Provides information about marketing one's magazine or newspaper articles before and after traveling. Also covers getting organized, writing and sales plans, query letters, tax deductions, photographs, and interviewing. Contains a list of 300 ideas for travel articles, a bibliography, and a list of sources for travel writers.

Denniston, Lyle W.
The Reporter and the Law: Techniques of Covering the Courts
2nd ed. New York, Columbia Univ. Press, 1992. 289p.
ISBN 0231080301

A detailed introduction to how the American legal system functions, and how a legal reporter should prepare for assignments in this area. Topics include structure and process of law, legal reporting, the press as adversary, criminal cases, civil cases, prison law, special courts, regulatory law, and covering the legislature in court. Includes a glossary of legal terms and a bibliography.

Fensch, Thomas
The Sports Writing Handbook
Hillsdale, N.J., L. Erlbaum, 1988. 262p. ISBN 0805802630

A practical guide to sports writing, from interviewing and observation to structure and style, with many examples of good writing. Fensch explains and illustrates many types of article structures, leads, transitions, and endings, and discusses advance and follow-up articles, columns, sidebars, writing about women, and ethics. Appendices list specialized terms used in newspapers and magazines for thirty different sports.

Fischer, Lionel L.
The Craft of Corporate Journalism: Writing and Editing Creative Organizational Publications
Chicago, Nelson-Hall, 1992. 146p. ISBN 0830413227

Explains the role of corporate writers and publications in sup-porting corporate objectives and advocating particular actions. Provides general advice on good writing style, types of stories, knowing one's readers, and defining a publication's objectives and content.

Garfinkel, Perry
Travel Writing for Profit and Pleasure
New York, New American Library, 1988. 208p.
ISBN 0452261597

The first part of this guide discusses the art of travel writing. The second part explains how to market one's work. Includes chapters on getting ideas, matching ideas to publications, and photography. Provides a bibliography and an appendix listing travel writers' resources and tourism offices.

Garrison, Bruce, and Mark Sabljak
Sports Reporting
2nd ed. Ames, Iowa, Iowa State Univ. Press, 1993. 354p.
ISBN 0813816920

A useful text for advanced students and beginning professionals. Covers the history of sports journalism, traditional reporting strategies (sources, interviewing, observation), basic story forms (game story, features, columns, big events), advanced reporting strategies (investigative reporting, the sports business, sports and the judicial system, precision sports journalism), new technological tools, and issues and ethics.

Glasbergen, Randy
Getting Started Drawing and Selling Cartoons
Cincinnati, Ohio, North Light Books, 1993. 117p.
ISBN 0891344713

An informative, entertaining primer. Covers how to draw funny cartoon characters, how to develop one's own style and characters, how to get funny ideas, how to sketch and ink in cartoons, and where and how to sell cartoons. Includes suggestions for practice, profiles of a number of cartoonists, and a short bibliography of recommended books.

Hubbard, Benjamin Jerome, ed.
Reporting Religion: Facts and Faith
Sonoma, Calif., Polebridge Press, 1990. 196p. ISBN 0944344100.

Part One explains why coverage of organized religion in the media is essential for understanding individual and cultural issues, and discusses the problems associated with this coverage. Part Two covers use of the media by organized religion, including televangelism, the Catholic, Protestant, and Jewish presses, and religious organizations and public relations. Part Three covers ethical issues involving religion and the media.

Izard, Ralph S., and Marilyn S. Greenwald
Public Affairs Reporting: The Citizens News
2nd ed. Dubuque, Iowa, William C. Brown, 1991. 377p.
ISBN 0697086151

Emphasizes the difference between the beat reporter and the specialist, who writes from a broader perspective and with more background knowledge and latitude. Subject coverage includes cultural diversity, religion, government and politics, economics, medicine and science, business, education, courts, and law enforcement. Includes information gleaned from interviews with practitioners in the various fields, with specialized glossaries and lists of reference materials for some fields.

Killenberg, George M.
Public Affairs Reporting: Covering the News in an Information Age
New York, St. Martin's Press, 1992. 352p. ISBN 031203637X

An introductory text with an emphasis on multicultural perspectives, interpersonal competence, and ethical issues. "Beats" covered include government news, public safety (crimes to corrections), and the judicial system. Several other areas are discussed briefly in one chapter. Includes exercises.

Kluge, Pamela Hollie, ed.
The Columbia-Knight Bagehot Guide to Economics and Business Journalism
New York, Columbia Univ. Press, 1991. 400p. ISBN 0231070721

Contains essays by alumni of a Columbia University midcareer fellowship program. Part One contains basic background information on business, finance, and economics. Sample topics include how to read balance sheets and earnings statements, regulation, venture capital, and economic indicators. Part Two provides information on writing specific types of business stories, including the troubled company, acquisitions, annual meetings, consumer reporting, investment options, financial planning, taxes, and insurance. Includes a thorough glossary.

Lovell, Ronald P.
Reporting Public Affairs: Problems and Solutions
2nd ed. Prospect Heights, Ill., Waveland Press, 1993. 506p.
ISBN 0881336963

A valuable guide for those in specialized public areas of reporting, emphasizing each specialty's problems and suggested solutions. Areas covered include politics and elections; federal, state, and local government; education; investigative reporting; police; courts; science; environment; medicine and health; business and the economy. For each area there is a general discussion, sample writings, a checklist of problems and solutions, a specialized glossary, and a bibliography for further reading. Appendices cover legal problems, ethical dilemmas, and searching public records.

Titchener, Campbell B.
Reviewing the Arts
Hillsdale, N.J., L. Erlbaum, 1988. 141p. ISBN 0805802371

A brief but helpful work intended for either students or journalists beginning as arts reviewers. Includes a discussion of the differences between critics and reviewers and a provides a suggested general method for reviewing. Other chapters discuss film, television, music, drama, dance, architecture, painting, sculpture, and book reviews. Includes a chapter on ethics and a selected reading list.

Warfield, Gerald
How to Read and Understand the Financial News
2nd ed. New York, HarperPerennial, 1994. 241p.
ISBN 0062732498

A fundamental guide to the financial community including thorough explanations of important terminology. Uses examples of news stories or tables as a starting point for explanations. Topics include corporations, stock, bond, and futures markets; economic indicators; government deficits and the balance of trade; the Federal Reserve; and the money supply. Includes a glossary.

Wojciechowski, Gene
Pond Scum and Vultures: America's Sportswriters Talk About Their Glamorous Profession
New York, Macmillan, 1990. 244p. ISBN 0026308517

An informal collection of humorous anecdotes gleaned from interviews with 130 sportswriters. Chiefly useful as entertainment although it does provide some insight into what it's like to be a sportswriter. Includes a bibliography.

Zobel, Louise Purwin
The Travel Writer's Handbook: How to Write and Sell Your Own Travel Experiences
Rev. and updated 2nd ed. Chicago, Surrey Books, 1992. 310p.
ISBN 0940625474

An extremely detailed and useful guide to travel writing. Includes instruction on how to do market and subject research, how to organize travel and writing, how to write the twelve most common types of travel articles, the art of interviewing, and techniques of photography. Also covers the business side of travel writing, including how to use queries, where to obtain illustrations, what to do about freebies, and what to deduct from your income tax.

✦ Women and Minorities in Media

Beasley, Maurine Hoffman, and Sheila Silver
Taking Their Place: A Documentary History of Women and Journalism
Washington, D.C., American Univ. Press in cooperation with the Women's Institute for Freedom of the Press, 1993. 359p. ISBN 0879383098

A collection of writings by and about women in the media, beginning with the colonial era and covering such topics as suffrage newspapers, reform periodicals, war correspondents, and women's pages. The editors provide an overview chapter and an introduction to each topic. Of particular note is a collection of publishers' "language and image guidelines" that address the need to avoid sexist portrayals of women and men. Includes a bibliography.

Benedict, Helen
Virgin or Vamp: How the Press Covers Sex Crimes
New York, Oxford Univ. Press, 1992. 309p. ISBN 0195086651

Critiques press coverage of sex crimes, charging that the press shows bias based on newsroom procedures, and acceptance of false stereotypes about women, rape, and violence. Examines four prominently covered cases in great detail. Clearly written by a journalism professor who is also an expert on rape. Includes a bibliography.

Braden, Maria
She Said What? Interviews With Women Newspaper Columnists
Lexington, Ky., Univ. Press of Kentucky, 1993. 208p. ISBN 0813118190

Includes lively interviews, background information, and sample columns from Mary McGrory, Erma Bombeck, Jane Bryant Quinn, Georgie Anne Geyer, Ellen Goodman, Judith Martin (Miss Manners), Joyce Maynard, and Merlene Davis, to name a few. This book also provides an introductory chapter on the history of women as columnists, and a selected bibliography.

Dates, Jannette Lake, and William Barlow
Split Image: African Americans in the Mass Media
2nd ed. Washington, D.C., Howard Univ. Press, 1993. 574p. ISBN 0882581791

A history of African Americans in the music, film, television, news, and public relations industries. The section on the print news industry presents both the history of the Black press and of Blacks in the mainstream press. It includes a listing of African American reporters, editors, and columnists at major newspapers as well as African American prize winners.

Edwards, Julia
Women of the World: The Great Foreign Correspondents
Boston, Houghton Mifflin, 1988. 275p. ISBN 0395444861

A lively, anecdotal history of women correspondents beginning with Margaret Fuller in the nineteenth-century and continuing through Georgie Anne Geyer and Shirley Christian in the 1980s. The author was a foreign correspondent herself and draws on her own reminiscences. Includes a bibliography of consulted sources.

Faludi, Susan
Backlash: The Undeclared War Against American Women
New York, Anchor Books, 1992. 552p. ISBN 0385425074

A Pulitzer Prize-winning journalist's examination of an anti-feminist backlash in popular culture and the media. This well-documented account contrasts what the media has been saying about women's dissatisfaction with their new, wider roles and what polls show women are actually saying and doing. She emphasizes that the media are not so much intentionally hostile to women as extremely sensitive to them, and that the media reflect the prevailing political climate.

Flocke, Lynne, Dona Hayes, and Anna L. Babic
Journalism and the Aging Population: Covering the Story: Ideas for Instructors
Syracuse, N.Y., Syracuse Univ., 1990. Syracuse University Series in Gerontology Education. 81p. ISBN 0874113911

Includes materials to help journalism students become more knowledgeable about and sensitive to older people. Includes exercises, story ideas, suggested readings and resources, and transparency masters.

Johnson, Mary, and Susan R. Elkins, eds.
Reporting on Disability: Approaches and Issues, A Sourcebook
Louisville, Ky., Avocado Press, 1989. 77p

An informal guide to avoiding bias and to improving news coverage of issues affecting the disabled. Includes information on disability rights issues, how to avoid biased language, and ethical issues. Focus is on covering the issue rather than the victim or hero. Includes sample articles and a list of sources for further information.

Martindale, Carolyn, ed.
Pluralizing Journalism Education: A Multicultural Handbook
Westport, Conn., Greenwood Press, 1993. 247p. ISBN 0313285926

An informative book of resources for those interested in promoting diversity in journalism. Topics include advantages of media diversity, strategies for recruiting and retaining students and faculty of color, pluralizing the curriculum, and pluralizing the student media. Includes specific chapters on African Americans, Latinos, Asian Americans, Native Americans and women. Includes many bibliographies.

Mills, Kay
A Place in the News: From the Women's Pages to the Front Page
New York, Dodd, Mead, 1988. 378p. ISBN 03696089321

Addresses both the press coverage of women and the role of women journalists. Discusses women's influence in both areas and how it has expanded, as more women have entered the workforce and attained important roles. This is an anecdotal history which is readable and well-documented. The author is a longtime journalist and editor for the *Los Angeles Times*.

Newspapers, Diversity, and You
Princeton, N.J., Dow Jones Newspaper Fund, 1993. 48p

A guide for minorities interested in journalism. Lists grants, internships, training programs, and recruiters. Also contains information on the challenge of diversity, career and salary reports, academic preparation, and job hunting trips. Many articles were written by minority journalists.

Ricchiardi, Sherry, and Virginia Young, eds.
Women on Deadline: A Collection of America's Best
Ames, Iowa, Iowa State Univ. Press, 1991. 201p.
ISBN 0813816874

Nine outstanding American women journalists are profiled through examples of their writing and interviews. Those included are Lucy Morgan, Jacqui Banaszynski, Alice Steinbach, Bella Stumbo, Cynthia Gorney, Anna Quindlen, Molly Ivins, Christine Brennan, and Sara Terry. A concluding chapter gives paragraph profiles on twenty-five more "women worth watching."

Robertson, Nan
The Girls in the Balcony: Women, Men and the New York Times
New York, Random House, 1992. 274p. ISBN 03945842X

A lively, anecdotal account of the landmark sex discrimination class action lawsuit won by women employees of *The New York Times*. The author, a reporter for the *Times* for more than thirty years, uses this event and her own experiences to examine the history of the *Times* and the careers and roles of women journalists.

Sing, Bill, ed.
Asian Pacific Americans: A Handbook on How to Cover and Portray Our Nation's Fastest Growing Minority Group
Los Angeles, Calif., National Conference of Christians and Jews, 1989. 80p

Includes information on stereotypes, Asian Pacific community issues, Asian Pacific names, and key demographic statistics. Includes a glossary, lists of contacts and resources, and a pictorial guide to Asian Pacific images in the entertainment media.

Streitmatter, Rodger
Raising Her Voice: African-American Women Journalists Who Changed History
Lexington, Ky., Univ. Press of Kentucky, 1994. 208p.
ISBN 0813118611

Contains studies of eleven influential African American women journalists and activists from Maria W. Stewart to Charlayne Hunter-Gault.

Wilson, Clint C.
Black Journalists in Paradox: Historical Perspectives and Current Dilemmas
New York, Greenwood Press, 1991. 188p. ISBN 0313266905

A brief history of African American journalists in the United States, with many individual profiles. Nearly half the book discusses events since the Kerner Commission report, which criticized the press for failing to hire and promote African Americans. The "paradox" in the title refers to the dilemma over whether African Americans should seek employment in the mainstream media, or support the financially strapped African American press and address issues from within the standpoint of the African American community. *Contributions in Afro-American and African Studies*, no. 145.

Wolseley, Roland Edgar
The Black Press U.S.A
2nd ed. Ames, Iowa, Iowa Univ. Press, 1990. 453p.
ISBN 0813804949

Topics covered include history of the Black press, today's African American newspapers and magazines, a profile of the African American journalist, journalism education and training, publishers and their problems, auxiliaries and competitors, pros and cons of the African American press, and speculation on its future. Includes an extensive bibliography.

Nonprint Materials

Anonymous Sources
Alexandria, Va., PBS Video, 1988. 1 videocassette (58 min) VHS

Discusses policies and ethics involved in the use of anonymous sources in journalism.

The Functions of Mass Communication
Maumee, Ohio, Instructional Video, 1989. 1 videocassette (26 min) VHS

Discusses the functions of mass communications including surveillance, entertainment, cultural transmission, status conferral, enforcement of norms, narcotism, a contemporary marketplace, and the public relations arena.

Newspaper Wars
Princeton, N.J., Films for the Humanities and Sciences, 1989.
1 videocassette (21 min) VHS

Discusses the impact of television and the information explosion on the newspaper industry and examines the competition between two newspapers in Trenton, New Jersey, *The Trenton Times* and *The Trentonian*. Former editor Michael O'Neill, media analyst John Morton, and editor Watson Sims share their views on the role of the newspaper in the age of instant journalism.

Politics, Privacy, and the Press. Ethics in America
no. 10. [s.l.], Annenberg/CPB Project, 1989. 1 videocassette (60 min) VHS

A panel of prominent journalists and advocates for the ethical responsibility of the media discuss reporting on the private lives of political figures. The hypothetical case of a presidential candidate's extramarital relationships provides a basis for debate

The Public Mind
Alexandria, Va., PBS Video, 1989. 4 videocassettes (60 min. each) VHS

Examines the impact of mass culture on our democracy where basic information comes from image making, public opinion polls, public relations, and propaganda.

Racism in the News
Nashville, Tenn., Media Action Research Center, 1992.
1 videocassette (23 min) VHS

Provides an analysis of how news content impacts the public's perceptions of people of color.

When Good Journalists Do Bad Things: Truthtelling and the Public Trust
St. Petersberg, Fla., Poynter Institute for Media Studies, 1993.
1 videocassette (2 hours) VHS

Media commentator and reporter Hodding Carter moderates an examination of two cases that drew wide criticism: NBC's *Dateline* report on General Motors trucks, and *USA Today's* front-page photograph and story about the gang mood in Los Angeles. Distinguished journalists and ethicists question a panel of key players in both cases, and discuss the credibility gap between journalists and the citizens they serve.

Public Relations

Jane L. Crocker
Gloucester County College Library
Sewell, New Jersey

Introduction

In the current environment of rapid social, economic, and political change, public relations (PR) has become a major service industry. Public relations now commands respect as a specialized discipline that functions as a critical facilitator between organizations and their publics. PR specialists serve as advocates for businesses, government, and not-for-profit organizations, as creators and purveyors of changing perceptions of organizations and people, and as architects of positive and effective relationships between organizations and their publics.

PR specialists need to comprehend the vagaries of public opinion, research the marketplace, plan effective public relations programs, create clear and effective messages, promulgate these messages through all types of media to the various publics, and evaluate the effectiveness of programs. People who choose public relations as a career should have an outgoing personality, self-confidence, an understanding of human psychology, and an ability to motivate people. They should be highly competitive, yet somewhat flexible and able to function as part of a team. Creativity, initiative, good judgment, and the ability to express thoughts clearly and concisely are all critical skills for the public relations professional, as are decision-making, problem-solving, and research skills.

Positions in public relations vary considerably depending on the type of public or private sector enterprise involved and the needs of the organizations or institutions. Employment opportunities include work in community, employee, or government relations; international relations; investor/stockholder relations; media liaisons; and consumer affairs. Tasks may vary, but they usually include keeping the organization's public informed of changes, accomplishments, and future directions through press and media contacts, preparing media releases, planning programs and activities, and responding to the needs of the public.

There is no one mandated route for entry into public relations; individuals people use various options to break into the field. One career track option is to begin to prepare in high school and enter college with a major in public relations, communications, advertising, journalism, or marketing. In evaluating job candidates, some employers look for studies in electronic or print journalism, while others seek specific abilities or experience in oral and written communications.

A common public relations curriculum covers the following courses: public relations principles and techniques; public relations management and administration; writing (basic writing skills, persuasive and creative writing), preparing news releases, proposals, annual reports, scripts, speeches, and other written documents; visual communications, including desktop publishing and computer graphics; and research, particularly focusing on social science research and survey design and implementation. Courses in advertising, business administration, creative arts, journalism, political science, psychology, and sociology are valuable, as are experiences with graphics, word processing, and other related computer applications.

Professional memberships in local chapters of the Public Relations Society of America or the International Association of Business Communicators provide opportunities to exchange views with public relations specialists and establish professional contacts that may prove valuable in pursuing employment in the field.

Selected Associations

The American Society for Health Care Marketing and Public Relations
American Hospital Association
840 North Lake Shore Dr.
Chicago, Ill. 60611

International Association of Business Communicators
1 Hallidie Plaza
Suite 600
San Francisco, Calif. 94102

Public Relations Society of America
33 Irving Place
New York, N.Y. 10003-2376
A comprehensive directory of schools offering degree programs or sequence of study in public relations and a brochure on careers in the field are available from this organization for a nominal cost.

Selected Journals

Advertising Age.
Crain Communications. Weekly. ISSN 0001-8899

Adweek.
ASM Communications. Western Advertising News. Weekly. ISSN 0199-4743

Business Marketing.
Crain Communications. Monthly. ISSN 0745-5933

Communication Briefings.
Encoders. Monthly. ISSN 0730-7799

Journal of Advertising.
American Academy of Advertising, University of Houston, Department of Marketing, College of Business Administration. Quarterly. ISSN 0091-3367

Journal of Advertising Research.
Advertising Research Foundation. Bimonthly. ISSN 0021-8499

Journal of Marketing.
American Marketing Association. Quarterly. ISSN 0022-2429

Marketing News.
American Marketing Association. Biweekly. ISSN 0025-3790

Media Week.
BPI Communications. Monthly. ISSN 1055-176X

PR Reporter.
PR Pub. Co. Weekly. ISSN 0048-2609

Public Relations Journal.
Public Relations Society of America. Monthly. ISSN 0033-3670

Public Relations Quarterly.
Public Relations Quarterly. Quarterly. ISSN 0033-3700

Public Relations Review.
JAI Press. Five times a year. ISSN 0363-8111

Standard Rate & Data Service Publications:
Consumer Magazine and Agri-Media Rates and Data.
Monthly. ISSN 0746-2522

Newspaper Rates and Data.
Monthly. ISSN 0038-9544

Print Media Production Data.
Quarterly. ISSN 0038-9455

Spot Television Rates and Data.
Monthly. ISSN 0038-9552

Print Materials

Advertising's Ten Best of the Decade, 1980-1990
New York, One Club for Art and Copy, 1990. 99p.
ISBN 0929837029
   Produced by the One Club, the organization that sponsors the advertising industry's most prestigious award, this impressive collection of the very best in print, radio, and television ads reads like a photographic history of advertising excellence. Graphics and dialogue are documented; production credits are acknowledged.

Ambry, Margaret, and Cheryl Russell
The Official Guide to the American Marketplace: The Real Facts About How Well-Educated, Healthy, Family-Oriented, Rich, Productive, Demanding, and Opinionated We Are
Ithaca, N.Y., New Strategist Publications and Consulting, 1992. 477p. ISBN 0962809217
   An innovative book, this guide includes facts about the character and attitude of the American marketplace including areas such as population trends, the labor force, income, and spending habits. Includes glossary and charts.

Amos, Wally, and Camilla Denton
Man with No Name Turns Lemons into Lemonade
Lower Lake, Calif., Aslan Publishing, 1994. 148p.
ISBN 0944031579
   Wally Amos, the famous cookie man, illustrates how adversity was sweetened in a PR coup that transcended time and litigation. It is a tale of personal growth, inspiration, and determination.

Baker, Michael John
Dictionary of Marketing and Advertising
2nd ed. New York, Nichols Publishing, 1990. 271p.
ISBN 0893973718
   This dictionary of terminology is useful for any library collection that focuses on the subject of marketing, advertising, and public relations.

Beals, Melba
Expose Yourself: Using the Power of Public Relations to Promote Your Business and Yourself
San Francisco, Calif., Chronicle Books, 1990. 189p.
ISBN 0877015856

Melba Beals, in her pleasant and comfortable style of writing, presents ways to maximize opportunities for gaining publicity and outlines strategies for developing a successful campaign with powerful results. Includes worksheets, charts, and potential scenarios for students to utilize when developing effective marketing strategies.

Bergmann, Jorg R.
Discreet Indiscretions: The Social Organization of Gossip
New York, Aldine DeGruyter, 1993. 206p. ISBN 0202304671

This text studies gossip as an interesting and independent form of oral communication. Bergmann analyzes gossip as a means of social control, a mechanism of preserving social groups, and a technique of information management. Includes notes and references.

Bianco, David P., ed.
PR News Casebook: 1,000 Public Relations Case Studies
Detroit, Gale Research, 1993. 1796p. ISBN 0810389053

Presents case studies of 1,000 public relations campaigns from the period 1945-1991. Offers practical, easy-to-follow approaches to problems and situations that corporate, nonprofit, and government organizations face daily. It is arranged by subject areas such as employee relations, new products and services, creating and renewing visibility, negative perceptions, and customer relations.

Bly, Robert W.
Targeted Public Relations: How to Get Thousands of Dollars of Free Publicity for Your Product, Service, Organization, or Idea
New York, Holt, 1993. 334p. ISBN 0805019758

In a very direct and specific approach, Bly gets to the heart of PR techniques for small and medium-sized businesses. Includes an appendix of selected list of sources and resources.

Brody, E. W., and Dan Lattimore
Public Relations Writing
New York, Praeger, 1990. 267p. ISBN 0275928950

Brody focuses on writing in public relations through a discussion of roles and responsibilities in the art of persuasion and development of the message. This is a good text for developing basic writing skills as applied to the field of public relations.

Brown, Lillian
Your Public Best: The Complete Guide to Making Successful Public Appearances in the Meeting Room, on the Platform, and on TV
New York, Newmarket Press, 1989. 223p. ISBN 1557040427

This work is highly recommended for anyone making public appearances or communicating through the media. Includes practical suggestions for making effective public appearances, improving voice and speaking skills, handling the media in television and radio interviews, and ultimately communicating more directly and confidently.

Brown, Marvin
How to Buy Advertising Like the Pros—and Save 15 to 50%: A Primer for Anyone in a Small- to Medium-Sized Business or Profession
Columbus, Ohio, Valmaran Books, 1993. 200p.
ISBN 0963961802

Brown suggests how to guarantee the best value from advertising dollars spent for radio, TV, and the trade press. Gives advice on how to evaluate advertising, suggests how much to spend, and outlines budget stretchers.

Clancy, Kevin J., and Robert S. Shulman
Marketing Myths That Are Killing Business: The Cure for Death Wish Marketing
New York, McGraw-Hill, 1994. 308p. ISBN 0070111243

One hundred fifty-seven marketing myths are identified, followed by detailed explanations, facts, and statistics disproving those myths. Myths are numbered and cited in the table of contents for easy perusal and classification. Charts, graphs, and illustrations substantiate the authors' views.

Clark, Eric
The Wantmakers: Inside the World of Advertising
New York, Viking Penguin, 1990. 416p. ISBN 0140117776

Eric Clark presents a fascinating and comprehensive study of the advertising industry. Advertising is discussed through the advertising machine (researchers, rules, advertising industry); the targets (consumers, the sick, smokers, drinkers, voters); and the media (TV, radio, newspapers, magazines). Advertising is portrayed as a powerful tool that can create subtle and real change.

Cutlip, Scott M., Allen H. Center, and Glenn M. Broom
Effective Public Relations
7th ed. Englewood Cliffs, N.J., Prentice Hall, 1994. 560p.
ISBN 0132450100

The latest revision of this classic is still a standard text in the field of public relations. Covered in this edition are concepts, practitioners, contexts, and origins of PR; legal, ethical, theoretical, and media foundations; and management and practice issues.

Dilenschneider, Robert L.
Power and Influence: Mastering the Art of Persuasion
New York, Prentice Hall, 1990. 258p. ISBN 0134640411

Dilenschneider, president and CEO of one of the world's largest and most prestigious PR firms, writes about the changing business environment of the 1990s. He discusses how personal power, communication, and information management skills, along with the ability to advance personal and corporate agendas are critical to business and individual success.

Field, Shelly
Career Opportunities in Advertising and Public Relations
New York, Facts on File, 1990. 308p. ISBN 0816020809

This essential reference tool for the PR career-oriented student presents classifications of eighty positions containing career ladder information, salary expectations, employment prospects, duties, educational and experiential qualifications, personality trait prerequisites, and recommended "best" geographic locations for positions.

Includes several helpful appendices that provide information on programs, trade associations, unions, PR agencies, and advertising agencies. A glossary of acronyms and a comprehensive bibliography are also provided.

Fraser, James Howard
The American Billboard
New York, Abrams, 1991. 192p. ISBN 0810931168

Starting with a P.T. Barnum Circus billboard created in 1835, this catalog of billboard art and advertising ranges from the serious to the sublime, from war posters to high-tech creations. It is a fascinating pictorial of how the advertising message has reached the public during this 150-year period.

Goodrum, Charles A., and Helen Dalrymple
Advertising in America: The First 200 Years
New York, Abrams, 1990. 288p. ISBN 0810911876

Starting with a short history of advertising, this volume traces advertising from its origins to the present, through visual representation and the written word. It is a true record of the changes in art technique, products advertised, and messages conveyed throughout a 200-year period.

Jacobs, Marvin
Graphic Design for Desktop Dummies
Cleveland, Ohio, Ameritype and Art, 1993. 162p.
ISBN 0962970018

Provides an introduction to graphic design principles for design type, layout, logo design, symbols, stationery, newsletters, promos, ads, and business documents.

Keding, Ann, and Thomas Bivins
How to Produce Creative Advertising: Proven Techniques and Computer Applications
Lincolnwood, Ill., NTC Business Books, 1991. 181p.
ISBN 084434818

Presents specific techniques for designing ads using traditional and computer applications. The author discusses the creative process throughout the first half of the book, then demonstrates how the creative energies can be converted into dynamic products through the use of the computer. Well illustrated.

Key, Wilson Bryan
The Age of Manipulation: The Con in Confidence, the Sin in Sincere
Lanham, Md., Madison Books, 1992. 296p. ISBN 0819186538

Wilson Key continues his condemnation of media tactics that manipulate the public in this frank discussion. His self-proclaimed warning at the beginning of the book explains how readers can use the information to protect themselves from media manipulation. Readers can also use this information in preparing for profitable careers in advertising and PR. The author demonstrates his premise through dialogue and specific examples.

Lem, Dean Phillip
Graphics Master 5: A Workbook of Planning Aids, Reference Guides, and Graphic Tools for the Design, Estimating, Preparation, and Production of Technology
5th ed. Los Angeles, Dean Lem Associates, 1993. 153p.
ISBN 0914218085

A true reference manual for graphics, this text includes standard guides for printing processes, graphic arts, photography, process color printing, industry trade customs, printing papers, presses, inks, electronic composition, proofreading marks, and proportional scale conversion tables. Charts and samples include color selector guide, envelope styles and sizes, binding style examples, type size (point), line gauge and ruler, sample type fonts, proof marks, and proportional scale.

Lesly, Philip, ed.
Lesly's Handbook of Public Relations
4th ed. New York, Amacom, 1991. 874p. ISBN 0814401082

Another essential reference tool, this comprehensive resource is the ultimate handbook of PR and communications. Edited by Philip Lesly, president of Phil Lesly Company, this resource is thorough, broad based, and easy to use. Includes a list of resources, a glossary of terms, and a good bibliography of books and journals. This edition, touted as a '90s update, brings PR into the current environment.

Levine, Michael
Guerrilla PR: How You Can Wage an Effective Publicity Campaign—Without Going Broke
New York, HarperBusiness, 1993. 229p. ISBN 088730608X

Written by a top PR consultant, this book concisely explains the techniques for achieving successful exposure with limited funds. Dubbed "guerrilla PR," the author defines this concept as being constantly vigilant to the overwhelming opportunities to heighten profiles of his clients. Chapter headings reflect "war terminology" such as Basic Training, First Maneuvers, Intelligence Gathering, and the Call to Battle.

Luscher, Keith F.
Advertise! An Assessment of Fundamentals for Small Business
Columbus, Ohio, K and L Publications, 1991. 135p.
ISBN 0962597791

Through its terse disclaimer at the opening of the book, Luscher establishes the focus of this work to be small business and proclaims his nonliability for information presented. Nonetheless, this small text provides a general overview of how advertising works, with good suggestions on how to effectively implement an advertising campaign, from small space advertising and writing copy to basic black-and-white design, to the system of ad agencies.

McLuhan, Marshall, and Quentin Fiore
The Medium Is the Massage
New York, Touchstone Books, 1989. 157p. ISBN 0671689975

This classic work introduces the concept of technology as a force that reshapes and restructures patterns of social interdependence and every other aspect of personal life. McLuhan's message seems as pertinent today as it was in the 1960s when the book was originally written. It is interesting to note the impli-cations of dramatic societal change, the nature of the media through which humans communicate, and the content of communication.

Mann, Thomas E., and Gary R. Orren, eds.
Media Polls in American Politics
Washington, D.C., Brookings Institution, 1992. 172p.
ISBN 0815754566

The editors address the issues of polling and the impact of polls on American politics and democracy. This work is a product of the Brookings Institution and the Joan Shorenstein Barone Center on the Press, Politics, and Public Policy. The editors conclude that media polls can be constructive if they are used to *report* rather then *create* news. Indexed.

Moog, Carol
Are They Selling Her Lips? Advertising and Identity
New York, Morrow, 1990. 236p. ISBN 0688087043

Carol Moog looks at the psychological framework of advertising—how it creates a desire to buy and the impact of the massive advertising one is exposed to in a lifetime. Understanding these powerful and complex influences and how media controls the public is the thrust of this book. The author is a practicing psychologist who also works as a consultant to advertisers.

Morgan, Bradley J., and Joseph M. Palmisano
Advertising Career Directory: A Practical One-Stop Guide to Getting a Job in Advertising
5th ed. Detroit, Visible Ink Press, 1993. 256p. ISBN 0810394294

In addition to providing data on more than 300 U.S., Canadian, and international advertising agencies, this directory includes summary articles from advertising executives on the advertising industry and information on creative art, media, and research alternatives. Also included is a list of job opportunities and vocational guidance on finding the right job, interviewing well, and getting started on the job. Recommended for high school through graduate students.

Morgan, Bradley J., ed.
Public Relations Career Directory: A Practical One-Stop Guide to Getting a Job in Public Relations
5th ed. Detroit, Gale Research, 1993. 337p. ISBN 0810356074

This text provides information on entering the job market, salary ranges, and academic preparation for the field of public relations. Morgan outlines the educational sequence for public relations careers in compliance with the 1987 guidelines from the Commission on Public Relations Education. The curriculum is cosponsored by Public Relations Division of the Association for Education in Journalism and in Mass Communication, sections of the Public Relations Society of America. Provides suggestions for getting that first job in PR, an annotated list of career resources, and a job opportunities data bank. Indexed.

Newsom, Doug, and Bob Carrell
Public Relations Writing: Form and Style
3rd ed. Belmont, Calif., Wadsworth Pub. Co., 1991. 458p. ISBN 0534143881

Writing copy for the PR message is discussed in detail in this very focused resource. Includes topics such as conducting basic research, rules of grammar, and principles of persuasion. This easy-to-read, easy-to-comprehend book on communication is highly recommended for the beginning writer.

Newsom, Doug, Alan Scott, and Judy VanSlyke Turk
This is PR: The Realities of Public Relations
5th ed. Belmont, Calif., Wadsworth Pub. Co., 1993. 615p. ISBN 0534172628

Newsom offers a practical and academic approach to PR in this revised edition. An excellent basic text, this overview is specific and relevant to the rapidly changing public relations scene.

Rapp, Stan, and Thomas L. Collins
Beyond Maximarketing: The New Power of Caring and Daring
New York, McGraw-Hill, 1994. 319p. ISBN 0070513430

At a time when electronic technology is revolutionizing business, the authors contend that new trends are emerging in some of the most innovative and successful corporations. Keys to business success are discussed. Includes information on new media technologies and priorities, advertising and marketing realities, consumer attitudes and demands, how "maximarketing" sells products and services, and the global village concept.

Smith, Jeannette
The Publicity Kit: A Complete Guide for Entrepreneurs, Small Business, and Nonprofit Organizations
New York, Wiley, 1991. 280p. ISBN 0471545864

This book explains what publicity is and the mechanics of generating positive publicity, from the basic steps of getting the program started to the development of press kits and distribution of information. A glossary and index are included, together with several appendices that illustrate press releases forms, photo release forms, and basic brainstorming techniques. Includes directories and sources of additional information.

Wasserman, Dick
That's Our New Ad Campaign—? A Handy Guide for CEOs, Presidents, Ad Managers, Account Executives, Art Directors, Copywriters, Students, and Anyone Else Who Wants to Learn How to Create Better Ads
Lexington, Mass., Lexington Books, 1988. 188p.
ISBN 0669169749

Wasserman uses critiques of forty successful advertising campaigns and ten speculative, unsuccessful ads to demonstrate how successful ad campaigns are designed and implemented, and conversely, why unsuccessful ads fail. Examples illustrate how an ineffective or bad campaign can be extremely costly in terms of negative reactions.

Wood, Robert J.
Confessions of a PR Man
New York, New American Library, 1988. 269p.
ISBN 0453005969

This pseudobiography of the legendary Robert Wood defines PR through classic examples of faux pas and triumphs, and presents lessons to be learned from personal case histories presented. The anecdotal references reinforce how PR strategies work—or do not work.

Zimbarbo, Philip G., and Michael R. Leippe
The Psychology of Attitude Change and Social Influence
Philadelphia, Temple Univ. Press, 1991. 1 v. (various pagings).
ISBN 0877228523

This fascinating book on behavior covers the elementary factors of influence: conscious and unconscious, the legal aspects, and the health aspects. The text is well illustrated, has several appendices, and a name and subject index.

Nonprint Materials

RoAne, Susan
How to Work a Room
Los Angeles, Audio Renaissance Tapes, 1991. 1 audiocassette (90 min.) VHS

    This program focuses on quick tips and strategies for meeting, communicating, and establishing rapport with strangers. It is presented by Susan RoAne, an experienced speaker who has made numerous presentations to major corporations and associations.

Zoller, Bettye Pierce, John Arthur Watkins, and Hugh Lampman
Power Talk: Standard American English, Your Ladder to Success
Dallas, Tex., ZWL Publishing, 1994. 3 audiocassettes and booklet (159p.) (90 min.) VHS. ISBN 1884643027 (set)

    This set focuses on effective speaking skills and presents a good outline of the basics of public speaking.

Radio and Television Production Technologies

Red Wassenich
Austin Community College
Austin, Texas

Introduction

Radio and television studies in community colleges introduce students to the equipment and production skills that generally lead directly to employment—most often in corporate, governmental, or educational settings—or to advanced study in four-year colleges. This bibliography emphasizes the former to some extent while acknowledging the latter in importance. Books giving career advice specific to radio and television are included.

A survey of community college catalogs shows that the range of course offerings varies greatly; most schools offer basic and intermediate production with a variety of courses such as scriptwriting, announcing, and editing. A few colleges have extensive curriculums. The print materials below address the most frequently offered courses. Textbooks are excluded, as a rule; however, some that focus on specific aspects of production, rather than general surveys, are included. Film production is not included, although many sources listed are useful for either film or video.

Like virtually all technology-intensive areas, radio and television production are in a period of rapid change due to computerization and increased portability of most aspects of production. Books inevitably lag behind developments, so titles that seem to address more overarching concerns are emphasized. As this bibliography is aimed at the production level, material on aesthetics, and social and political aspects of electronic media, are not covered. However several entries address these issues within the context of practical and technical information.

Selected Associations

Although there are no accrediting agencies specific to this area, there are many helpful organizations that can supply additional information.

Intercollegiate Broadcasting System
Box 592
Vails Gate, N.Y. 12584-0592

National Association of Broadcasters
1771 N Street, NW
Washington, D.C. 20036

Society of Broadcast Engineers, Inc.
8445 Keystone Crossing, Suite 140
Indianapolis, Ind. 46240

Society of Motion Picture and Television Engineers
595 W. Hartsdale Avenue
White Plains, N.Y. 10607

Selected Journals

AV Video.
Montage Publishing. Monthly. ISSN 0747-1335

AVC Presentation Development and Delivery: AV, Video, and Computer.
National Business Association Directory. Monthly
ISSN 1062-2683

Videography.
PSN Publications. Monthly. ISSN 0363-1001

Print Materials

✦ General Works

Benson, K. Blair, and Jerry C. Whitaker
Television and Audio Handbook for Technicians and Engineers
New York, McGraw-Hill, 1990. 1 v. (various pagings)
ISBN 00700447871

Covers day-to-day maintenance for television and audio equipment that "does not require the background of an engineering degree." Largely adapted and simplified information from two earlier titles, *Television Engineering Handbook* (McGraw-Hill, 1986) and *Audio Engineering Handbook* (McGraw-Hill, 1988). Seventeen chapters by different authors range from basic electronics to detailed testing procedures. New technologies such as high definition television (HDTV), satellite and cable transmission, and compact disk recording are included. Separate chapter of standards and tables. Includes bibliography and index.

Broadcasting and Cable Yearbook
New Providence, N.J., Bowker, 1993- . Annual. 2 v.
ISSN 0000-1511

Formerly the *Broadcasting and Cable Marketplace* and *Broadcasting Yearbook*. Contains information on 13,000 AM and FM radio stations, 2,000 television stations, and 492 MSOs and independent owners. Volume One includes law, regulations, government agencies, radio, television, cable, satellite and other carriers, advertising and marketing services, to name a few. Volume Two consists of the radio, television and cable yellow pages.

Bryer, Richard, Peter Moller, and Michael Schoonmaker
Making Television Programs: A Professional Approach
2nd ed. Prospect Heights, Ill., Waveland Press, 1991. 349p.
ISBN 0881335894

Emphasizing the process rather than the technology, this volume covers a wide range of television production topics. From studio sets to costume to postproduction, everything is at least touched upon. Numerous examples highlight the text. Commercial, documentary, and industrial uses of television are addressed. Sample production forms are appended. A useful text for the novice. Includes appendices, bibliography, glossary, index.

Corporation for Public Broadcasting
Public Broadcasting Directory
Washington, D.C., The Corporation. Annual.

Lists national, regional, state organizations and networks, related organizations, and public radio and television stations. Includes technical and programming fact sheet. Index.

Da Silva, Raul
Making Money in Films and Video: A Freelancer's Handbook
2nd ed. Boston, Focal Press, 1992. 177p. ISBN 024080144X

Covers the business angles of starting and operating a freelance film and video operation. Production is not included. Promotional material, client relations, and dealing with various outlets, such as cable and corporate, are outlined. Useful if library patrons are advanced enough to free-lance. Includes bibliography, glossary, index.

Diamant, Lincoln, ed.
Dictionary of Broadcast Communications
3rd ed. Lincolnwood, Ill., NTC Business Books, 1991. 255p.
ISBN 0844233250

Reprint of 1989 Greenwood Press work. Six thousand short entries, rarely over fifteen words, on all telecommunications areas. Encompasses commercial, governmental, business, and educational broadcast sources, covering technology, production, marketing, media usage, and more. Cross-references. Wide breadth, little depth, accessible. Includes bibliography.

Gross, Lynne S.
New Television Technologies
3rd ed. Dubuque, Wm. C. Brown, 1990. 252p. ISBN 0697054918

The most significant recent developments affecting television such as satellites, computerization, videodiscs, and low-power television each receive a brief history and description. Details technical and social ramifications for each development. Text remains accessible in a jargon-filled area. The author is on the faculty of California State University, Fullerton. Includes bibliography, glossary, index.

Lambert, Mark
TV and Video Technology
New York, Bookwright Press, 1990. 47p. ISBN 0531183270

Written on a sixth- through eighth-grade level, this could be useful in community colleges that need material appropriate for developmentally challenged students. Technical topics such as radio waves, audio versus video taping, and the role of fiber optics in new video technology are covered in an elementary manner. Heavily illustrated. Includes bibliography, glossary, and index.

Langman, Larry, and Joseph A. Molinari
The New Video Encyclopedia
New York, Garland Pub., 1990. 312p. ISBN 0824082443

This reference work mixes entries on technology with popular topics such as HBO. Basic production areas are generally dealt with by their specific components (for instance, there are separate entries for "Lighting," "Back light," "Base light," and "Fill light"). Entries range from a short paragraph to one column. Includes short biographies of key figures. More than 1,500 entries. Good ready reference source.

Multimedia Solutions
Delran, N.J., Datapro, 1992-. 1 v. (loose-leaf)

This title is one section in the six-part *Workgroup Computing Series* reference set. Provides the latest information on computer-based multimedia services that wed video, audio, and graphics, often with a business slant. Lists products and software, giving prices and specifications. Some background needed. Appropriate for comprehensive collections.

Penney, Edmund F.
Facts on File Dictionary of Film and Broadcast Terms
New York, Facts on File, 1991. 251p. ISBN 0816019231

Provides clear definitions of a broad range of electronic media vocabulary. Entries range from ten to one hundred words and cover technical terms and jargon. Cross-references. Useful appendices contain information on twelve production firms.

Orlik, Peter B.
Critiquing Radio and Television Content
Boston, Allyn and Bacon, 1988. 344p. ISBN 0205116434

Critical theory is central here, but the integral role of production is explicitly addressed. Lighting, music, camerawork, and sound production as aspects of aesthetic evaluation, provide an angle not commonly presented. Full scripts to a *Cosby* show and a 1950s sitcom, *Ethel and Albert*, are included, as are student exercises. Appendices, references, index.

Reed, Robert M., and Maxine K. Reed
Encyclopedia of Television, Cable, and Video
New York, Van Nostrand Reinhold, 1992. 622p.
ISBN 0442006276

An enjoyable reference work with 3,100 entries, each averaging about 200 words, although some major topics receive several pages. Production and technology topics are intermixed with entries on specific television shows and biographies in alphabetical arrangement. The production angle is presented but not stressed. Includes cross-references, appendix and bibliography.

Television and Cable Factbook
Washington, D.C., Television Digest. Annual. (various pagings)
ISSN 0732-8648

Divided into three volumes: TV stations, TV and cable services, and cable systems. This publication lists all the commercial, public, and low power stations, and cable systems in the U.S., with briefer information for international stations. Major professional organizations, unions, networks, state and federal governmental regulatory agencies, and related organizations are listed. Useful for job hunters and for answering business questions. For in-depth reference collections.

Weiner, Richard
Webster's New World Dictionary of Media and Communications
New York, Webster's New World, 1990. 533p. ISBN 0139697594

Print and electronic media share coverage in this ready reference source. Short entries, usually ten to fifty words, include technical and slang terms. Solid and straightforward, its broad coverage makes it a good choice for small reference collections. Includes cross-references.

✦ Announcing and Reporting

Keith, Michael C.
Broadcast Voice Performance
Boston, Focal Press, 1989. 160p. ISBN 0240800036

The author, a community college faculty member, approaches announcing from an industry standpoint—the types of voices needed and the different formats using them. Radio announcer, newsreader, sportscaster, etc. are addressed separately. Provides sections on vocal technique and studio equipment. A useful adjunct to textbooks. Includes bibliography and index.

Shook, Frederick
Television Field Production and Reporting
New York, Longman, 1989. 386p. ISBN 0582286336

Argues that television news is a distinct field, often incorrectly forced into print journalism's mold. Emphasizes "writing" which combines words, images, and sound. Presents production and post-production basics of remote shoots. Sections on interviewing, sports reporting, on-air performance, legal and ethical aspects, and career development are included. Though this is a textbook, its focus would contribute to collections supporting television news courses. Shook is on the faculty at Colorado State University. Includes appendices, glossary, and index.

Utterback, Ann S.
Broadcast Voice Handbook: How to Polish Your On-Air Delivery
Chicago, Bonus Books, 1990. 264p. ISBN 0929387163

Very specific techniques and solutions to problems for voice. Breathing, intonation, and use of accents are addressed. Appropriate for dedicated students of broadcast announcing, but may be too detailed for collections without specific, related coursework. Index.

✦ Direction and Production

Armer, Alan A.
Directing Television and Film
2nd ed. Belmont, Calif., Wadsworth, 1990. 386p.
ISBN 0534116167

Although essentially a textbook, this work's focus is on the directing angle of production. The author is on the faculty at California State University, Northridge. Project ideas, often including sample scripts, are included in each chapter, and will help the beginning and intermediate student. Includes bibliography, glossary, and index.

Caruso, James R., and Mavis E. Arthur
Beginner's Guide to Producing TV: Complete Planning Techniques and Scripts to Shoot
Englewood Cliffs, N.J., Prentice Hall, 1990. 429p
ISBN 0139440917

A very basic guide to planning and shooting television scripts from both the producer's and director's perspectives. Eight short scripts and "101 Ideas for Videos" make this particularly useful for beginning students in search of project ideas. Highly illustrated. Includes appendix and index.

Compesi, Ronald J., and Ronald E. Sherriffs
Small Format Television Production
2nd ed. Boston, Allyn and Bacon, 1990. 500p. ISBN 0205123325

A textbook, by faculty members at San Francisco State and the University of Oregon, focuses on the use of portable TV equipment, which is a growing emphasis in many academic programs. This work is an accessible introduction and reference to camcorders and production considerations, such as lighting, audio, graphics, and postproduction. Would also serve a library's amateur video patrons. Includes sample production forms and exercises. Includes appendices, bibliography, glossary, and index.

DiZazzo, Raymond
Corporate Television: A Producer's Handbook
Boston, Focal Press, 1990. 201p. ISBN 0240800230

Looks at the entire process of putting together a video in the corporate arena. The producer's role in design, script supervision, budgeting, preproduction, production, postproduction, and client relations receive practical but not overly detailed attention. Appropriate for collections where coursework goes beyond introduction to production. Author works in corporate video. Includes appendices, bibliography, glossary, and index.

Hickman, Harold R.
Television Directing
Santa Rosa, Calif., Cole Pub., 1991. 468p. ISBN 0929635043

This textbook focuses on directing, which often gets less detailed coverage in other TV production books. Separate chapters address types of programming (e.g., news, drama, sports), from pre- to postproduction. This introductory-level book includes appendices with sample production forms and crew descriptions. Includes bibliography, glossary, and index.

Lindheim, Richard D., and Richard A. Blum
Inside Television Producing
Boston, Focal Press, 1991. 328p. ISBN 0240800192

An interesting exploration of the role of producer in commercial television. One author is a studio executive, the other is on the faculty at the University of Maryland. They compare the functions of producer in television and film. Major sections use case histories of the network series *Coach* and *Law and Order* to illustrate their points. The complete scripts to the pilots of these shows are appended. Although the practical is emphasized, some attention to the social aspects of television is given. Includes appendices, bibliography, and index.

Musberger, Robert B.
Single-Camera Video Production
Boston, Focal Press, 1993. 195p. ISBN 0240800346

A sound bite guide to the entire single-camera production process. Topics such as "physics of waves" and "editing" are covered within one page, with the facing page consisting of a related illustration or graphic. Aimed at beginners. Good for nontextbook coverage of the whole process. Most information can be used in either studio or field production. Author is on the University of Houston faculty. Includes bibliography, glossary, and index.

Nardantonio, Dennis N.
Sound Studio Production Techniques
Blue Ridge Summit, Pa., TAB Books, 1990. 293p. ISBN 0830692509

Practical guide to a wide spectrum of audio production. The author keeps the text aimed at the motivated beginner. Electronics, acoustics, recording equipment, mixing consoles, digital recording, microphones, and specific production and postproduction techniques are among the chapters. Heavily illustrated. An appendix has a review of math needed for certain areas. Includes appendix, glossary, and index.

Smith, Coleman Cecil
Mastering Television Technology: A Cure for the Common Video
Richardson, Tex., Newman-Smith Pub., 1988. 388p. ISBN 0929549007

Coverage of television technology aimed at people with some production background who need a larger view of the entire process. Includes video, audio, recorders, signals, and other video systems, such as switchers and effects generators. Five appendices have various tests and configurations for specific systems and products. Includes bibliography and index.

✦ Videoproduction Techniques

Arwady, Joseph W., and Diane M. Gayeski
Using Video: Interactive and Linear Designs
Englewood Cliffs, N.J., Educational Technology Publications, 1989. 178p. ISBN 0877781990

This book emphasizes design techniques that elicit desired audience responses. Does not stress details of production. The author presents eighteen techniques for use in continual-play, linear videos and twenty-two techniques for interactive videos, which require audience responses. Appropriate uses of irony, exaggeration, and technical methods, such as animation and simulations, are given in two- to-five-page entries. Very practical. Accessible to intermediate students although aimed more at practicing corporate video professionals. Includes bibliography and index.

Benedict, Larry, and Susan Benedict
The Video Demo Tape: How to Save Money Making a Tape That Gets You Work
Boston, Focal Press, 1992. 220p. ISBN 0240801407

For people from all walks of life who need inexpensive, quality videos for job searches or professional development. Includes sections on editing, shooting, miscellaneous technical considerations (e.g., tape formats and quality), and distribution. Short sections with "demo tips" throughout make it easy to manage. Useful for videographer and customer. Authored by two professional actors and videomakers. Glossary.

Bernard, Robert
Practical Videography: Field Systems and Troubleshooting
Boston, Focal Press, 1990. 192p. ISBN 0240800214

Details a "logic tree" methodology for dealing with technical breakdowns, useful primarily in remote shooting but also in studio settings. Separate chapters on audio, video camera, video recorders, power, lighting, and signals. Also covers preventive maintenance. Not overly technical, and appropriate for intermediate students. The author is an independent videographer and college instructor. Includes bibliography, glossary, and index.

Bishop, John Melville
Making It in Video: An Insider's Guide to the Fastest Growing Industry of the Decade
New York, McGraw-Hill, 1989. 195p. ISBN 0070054681

Divided into various video-related areas: commercial TV, public TV, cable TV, corporate and educational video, and freelance work. Describes different jobs within each area and profiles practicing professionals. Includes glossary and index.

Chesire, David F.
The Book of Video Photography: A Handbook for the Amateur Movie-Maker
New York, Knopf, 1990. 224p. ISBN 0394587448

The very high quality of the illustrations, most in color, distinguishes this volume, which targets the nonacademic audience. Provides techniques useful for increasing the professionalism of videos through camera choices, lighting, audio, editing, and simple special effects. Typical video settings, such as travel and weddings, are adaptable to introductory academic situations. Includes appendices, glossary, and index.

Richardson, Alan R., ed.
Corporate and Organizational Video
New York, McGraw-Hill, 1992. 358p. ISBN 0070523347

Designed as an intermediate textbook, but useful as an overview of current and emerging trends and technologies in non-broadcast video. Each of the twenty chapters is written by a different expert from education and business. Separate chapters on specific aspects, such as teleconferencing, computers, and video editing make it easy to use. Management and some other nonproduction matters are addressed but not stressed. Bibliographies, index.

Roth, Cliff
Low Budget Video Bible: The Essential Do-It-Yourself Guide to Making Top Notch Video on a Shoestring Budget
New York, Desktop Video Systems, 1994. 388p.
ISBN 0963521608

For the semi-professional and serious amateur rather than the traditional academic audience, but useful to students as affordable, portable equipment becomes more available. Emphasizes low- or no-cost techniques that professionalize videos—lighting, audio, and camerawork. Different types of camcorders and editing systems are covered, as well as opportunities for showing independent videos. Good balance of technical information and accessibility. Includes appendices, glossary, and index.

Schactman, Tom, and Harriet Shelare
Video Power: A Complete Guide to Writing, Planning, and Shooting Videos
New York, Holt, 1988. 102p. ISBN 080500338X

Very basic coverage of the process of making a video, written at a lower reading level, perhaps tenth through eleventh grade. In addition to simple coverage of the production angle, the author covers methods for public access broadcasting, and student competitions and contests. Useful for developmental students needing an introduction to video. Includes bibliography, glossary, and index.

Schihl, Robert J.
Single Camera Video: From Concept to Edited Master
Boston, Focal Press, 1989. 334p. ISBN 0240800397

Schihl, Robert J.
Studio Drama Processes and Procedures
Boston, Focal Press, 1992. 123p. ISBN 0240800966

Schihl, Robert J.
Talk Show and Entertainment Program Processes and Procedures
Boston, Focal Press, 1992. 101p. ISBN 0240800923

Schihl, Robert J.
Television Commercial Processes and Procedures
Boston, Focal Press, 1992. 123p. ISBN 0240800982

Schihl, Robert J.
TV Newscast Processes and Procedures
Boston, Focal Press, 1992. 148p. ISBN 024080094X

All of these books by Schihl, a faculty member from Regent University in Canada, employ the same, useful format. An initial section describes the video production topic at hand, then specific production forms for the process are printed with accompanying glossary, explanation, and advice. *Single Camera Video* is the only one not set in the studio. All the others are part of the *Multiple Camera Video Series*. Well done and practical. Excellent for both beginning and intermediate students. All volumes have bibliography, glossary, and index.

Smith, David L.
Video Communication: Structuring Content for Maximum Program Effectiveness
Belmont, Calif., Wadsworth Pub., 1991. 412p. ISBN 0534131468

Stresses theory and communications models, but ties them to production techniques. Analyzes different types of videos, such as persuasion and drama, from the perspective of production approaches and values. Separate chapter on ethics. Useful contrast to purely how-to texts. Includes appendices, bibliography, glossary, and index.

Utz, Peter
Create Excellent Video
Englewood Cliffs, N.J., Prentice Hall, 1990. 451p.
ISBN 0135471427

A chatty, nonacademic primer on the basics of video production for the amateur. Technical advice on shooting, lighting, audio, and editing for nonstudio productions. Consumer advice and ways to sell one's videos make this more useful in a public library setting, but its accessibility may increase its value for the college collection. Includes recommended readings and index.

Whittaker, Ron
Video Field Production
Mountain View, Calif., Mayfield Pub., 1989. 426p.
ISBN 0874848369

Most textbooks emphasize studio production, so the focus on fieldwork makes this worthy of library consideration. Strong on useful technical information, including lenses, luminance, and types and characteristics of video and audio equipment. Discusses effects of remote production on scripting and shooting. Includes chapters on single versus multiple camera shoots and alternate transmission formats (satellite, microwave, fiber optics). Author is on the faculty at Pepperdine. Includes bibliography, glossary, and index.

Wiese, Michael
Film and Video Financing
Studio City, Calif., M. Wiese Productions in conjunction with Focal Press, 1991. 307p. ISBN 0941188116

Target audience is independent film and video producers. Addresses specific aspects of financing such as packaging, completion bonds, loans, risk capital, and pitch techniques. Interviews and a case study of *Sex, Lies, and Videotape* emphasize the real-world angle of this book, written by an independent producer-director. Includes appendices and bibliography.

✦ Editing

Browne, Steven E.
Videotape Editing: A Postproduction Primer
2nd ed. Boston, Focal Press, 1993. 300p. ISBN 0240801512

Based on the concept that technology is radically changing but the process and goals are not, this presents a short history of video editing, followed by more detailed examinations of time codes, assemble recordings, and styles of edits. Addresses conceptual aspects of editing and probable effects of increasing computerization. For students with basic video skills background. Author is a practicing video editor. Includes appendix, bibliography, glossary, and index.

Caruso, James R., and Mavis E. Arthur
Video Editing and Post Production
Englewood Cliffs, N.J., Prentice Hall, 1992. 330p.
ISBN 0139465758

A good introductory work for an area of video that can be daunting (the glossary alone is forty-five pages). Explains how television and videotape work and moves on to the equipment and methods of postproduction. Techniques are divided into simple and advanced (e.g., fades, wipes, editing graphics). Also covers time code and audio editing. Includes sample exercises and work sheets, appendices, glossary, and index.

Ohanian, Thomas A.
Digital Nonlinear Editing: New Approaches to Editing Film and Video
Boston, Focal Press, 1993. 348p. ISBN 024080175X

A thorough explanation of computerized, interactive postproduction techniques and equipment. Comparing recent editing technologies to the difference between using a typewriter and a word processor, Ohanian, a professional editor and system designer, presents new approaches to manipulating video and film. Addresses the increasing flexibility and complexity of dealing with video, film, laser discs, digital systems, and film transfer to video. Stresses the technical over the aesthetic. For intermediate and advanced students. Includes bibliography, glossary, and index.

Schneider, Arthur
Electronic Post-Production and Videotape Editing
Boston, Focal Press, 1989. 124p. ISBN 0240517997

Brief, wide coverage that minimizes as much as possible the inevitable technical details. Gives basics of videotape and differences between on- and off-line editing, with an emphasis on time code editing. Includes short sections on film and tape transfer, proper handling of videotape, careers in editing, and future trends. Appropriate choice for beginners. Includes bibliography, glossary, and index.

Schneider, Arthur
Electronic Post-Production Terms and Concepts
Boston, Focal Press, 1990. 152p. ISBN 0240800060

Entries range from one sentence to 500 words in this reference work on video postproduction terminology. Aimed at beginners. Since most video books seem to have glossaries, this will probably only be needed in institutions with serious video production collections. Includes cross-references.

✦ Lighting and Special Effects

Caruso, James R., and Mavis E. Arthur
Video Lighting and Special Effects
Englewood Cliffs, N.J., Prentice Hall, 1991. 354p.
ISBN 0138247234

An introduction to TV studio lighting and a how-to for more than one hundred nonelectronic special effects. The lighting section does not try to go beyond the basics and has two short practice scripts. Special effects are given step-by-step and needed props are listed. Includes such effects as storms, simple miniature sets, and makeup. Index.

Millerson, Gerald
Lighting for Video
3rd ed. Oxford, England, Focal Press, 1991. 154p.
ISBN 0240513037

A respected and prolific writer on all aspects of video production, Millerson delivers practical information on a wide variety of lighting situations. Covers both studio and location shoots. Most entries consist of a single page of text with another page of illustrations on specific lighting challenges. Accessible to beginners and intermediate students. A thorough table of contents precludes the need for an index. Includes bibliography and glossary.

Mott, Robert L.
Sound Effects: Radio, TV, and Film
Boston, Focal Press, 1990. 223p. ISBN 024080029X

Artistry is emphasized over the technical in this book on audio effects, although practical considerations are covered. Discusses the rationale behind the use of effects in scripts. A brief history of sound in radio, film, and early television leads to separate, detailed chapters on their current use in each medium. Foleying gets its own chapter. This work is a solid balance of utilitarian principles, written in an accessible manner. Includes bibliography, glossary, index.

Viera, John David
Lighting for Film and Electronic Cinematography
Belmont, Calif., Wadsworth Pub., 1993. 336p. ISBN 0534128106

Written by a producer-director-college instructor, this work presents detailed information on lighting, with film receiving more emphasis than video. Five chapters are devoted to exposure, and others to the differences between interior and exterior lighting. Appendices address areas such as film stock differences and light meters. Some background is needed, although portions are accessible to beginners. Numerous, good illustrations. Includes bibliography, glossary, and index.

Wilkie, Bernard
The Technique of Special Effects in Television
2nd ed. Boston, Focal Press, 1989. 262p. ISBN 0240512847

Visual effects within the capabilities of the beginner form the bulk of this very practical handbook. One or two pages are devoted to each effect, such as cobwebs, rain, breaking glass, and some more exotic ones. Six chapters cover advanced effects involving mattes, chroma key, and stop motion. The author is a former manager of special effects at the BBC, now working as a consultant. Includes bibliography, glossary, and index.

✦ Scriptwriting

Dancyger, Ken
Broadcast Writing: Dramas, Comedies, and Documentaries
Boston, Focal Press, 1991. 102p. ISBN 0240800540

Presents radio and television scriptwriting techniques geared for story and character development, rather than technical or format concerns. There are separate sections on drama, sitcoms, soaps, and police-doctor-lawyer stories in this accessible, short book.

DiZazzo, Raymond
Corporate Scriptwriting: A Professional's Guide
Boston, Focal Press, 1992. 209p. ISBN 0240801156

If the curriculum includes intermediate or advanced scriptwriting, this is a good, practical guide to all the aspects of developing and completing a script for business. Stresses relating to clients and producers effectively, and selecting appropriate writing devices and approaches. Includes a sample business proposal and complete sample script. Author is a corporate media writer and producer. Includes a glossary and an index.

Hewitt, John
Air Words: Writing for Broadcast News
Mountain View, Calif., Mayfield Pub., 1988. 226p. ISBN 087484844X

A combination of scriptwriting and journalism. Using an exercise-filled format that verges on a workbook, this text covers how different media use words. Explains effective structures and techniques for types of news reporting. Provides a lively approach that gives many catchy short tips. The author is from San Francisco State University. Index.

Morley, John
Scriptwriting for High-Impact Videos: Imaginative Approaches to Delivering Factual Information
Belmont, Calif., Wadsworth Pub., 1992. 272p. ISBN 0534150667

Presents real world aspects of video scriptwriting, written by a faculty member of California State University, Northridge, who is also a practicing writer-producer. Assumes a basic understanding of video production and writing skills. Corporate and educational scripts are stressed. The bulk of the book covers the process of scriptwriting, from conceptualization through completion, addressing the decisions each step presents. Written in a snappy style. Appendices have sample scripts and questions to ask during script development. Includes chapters on "Realities of the Marketplace" and job searching. Bibliography, glossary, and index.

Orlik, Peter B.
Broadcast/Cable Copywriting
4th ed. Boston, Allyn and Bacon, 1990. 746p. ISBN 0205123252

A very thorough textbook from a faculty member at Central Michigan University. Particularly useful if the curriculum needs in-depth video writing support. Includes the entire writing process, with attention to incorporating production considerations into every aspect. Focuses on such projects as spot announcements, promos, and ads, with the roles different markets play in the process. Filled with sample scripts. Appendices on industry codes and regulations. Includes notes and index.

Walter, Richard
Screenwriting: The Art, Craft, and Business of Film and Television Writing
New York, New American Library, 1988. 240p.
ISBN 0452260868

Logically arranged, the author presents underlying concepts behind commercial scripts and the nuts and bolts of putting together stories—character, dialogue, etc. Includes script format and the business angle of selling stories. Written by the chair of the UCLA screenwriting department in a readable, nonacademic style. Includes an annotated bibliography.

Walters, Roger L.
Broadcast Writing: Principles and Practice
New York, Random House, 1988. 515p. ISBN 0394330897

A textbook strong on breadth of coverage. Rather than focusing on one medium, overriding principles applicable to all electronic media writing are discussed, although it gives specific ramifications of each format when appropriate. Different contexts of broadcast writing (such as educational, news, business) are presented as driving forces. Presents production techniques as a major factor in writing without going into their details. Short examples are given throughout. Coverage includes career advice and ethics. Includes bibliography, glossary, and index.

✦ Radio

Holsopple, Curtis R.
Skills for Radio Broadcasters
3rd ed. Blue Ridge Summit, Pa., TAB Books, 1988. 201p.
ISBN 0830629300

The author, an engineer in a commercial station, notes that it is "ironic that the entry-level broadcasting job usually requires a huge variety of skills" (pref.). Divided into three units—equipment, operator, and advanced skills—this title introduces the basics of radio production. Each chapter gives exercises and vocabulary. Amateurish illustrations detract from an otherwise useful book for beginners. Includes bibliography, glossary, and index.

Kaufman, Milton
Radio Operator's License Q and A Manual
11th ed. Indianapolis, Ind., Hayden Books, 1989. 553p.
ISBN 0672484447

Continuing a forty-year publication history, this very practical guide covers preparation for three tests: General Radiotelephone Operator License, Marine Radio Operator Permit, and the Radar Endorsement. A separate section covers troubleshooting. Includes sample tests and index.

Keith, Michael C.
Radio Production: Art and Science
Boston, Focal Press, 1990. 255p. ISBN 0240800176

Presents the complete process of radio production. The author argues that production is best taught as an integrated whole, combining staff descriptions, studio design, equipment, writing, mixing and editing, and formats. Descriptive emphasis rather than how-to. Good overview for introductory students. Includes many examples from actual radio stations, a bibliography, a glossary, and an index.

McLeish, Robert
The Technique of Radio Production: A Manual for Broadcasters
2nd ed. London, Focal Press, 1988. 250p. ISBN 0204512669

This survey of radio production emphasizes communications skills. Mainly divides discussion by types of productions (e.g., commercials, interviews, remotes, DJ). Covers technical matters for novices. Includes bibliography and index.

Monaco, Fred
Preparing for the FCC General Radiotelephone Operator's License Examination
New York, Merrill, 1991. 316p. ISBN 0675213134

A straightforward guide to the exam required for maintenance and testing of many communications tools. Twenty-three chapters cover basic examination content, including the Ship Radar section. Eight appendices address test-taking and practice questions. Less of a workbook format than many exam guides, so it is less likely to get defaced in the library. The author teaches at Los Angeles Trade Technical College. Appendices, index.

O'Donnell, Lewis B., Philip Benoit, and Carl Hausman
Modern Radio Production
3rd ed. Belmont, Calif., Wadsworth Pub. Co., 1993. 347p. ISBN 0534190804

A nontechnical introduction to the field of radio production and professional radio. Structured to provide a balance of techniques, theory and hands-on production skills. Stresses communication skills, rather than hardware. The third edition presents a new chapter on demographics and market segmentation. Includes updated information on digital computer technology, satellite feeds, digital audio broadcasting, station promotions, radio news, and newsroom computer systems.

Stucker, Steve
How to Get a Job in Radio
Albuquerque, N.M., Stucker Productions, 1990. 100p. ISBN 0962797030

A self-published volume by a radio personality. Low production quality, but if a collection needs a title specific to radio job hunting, the practicality may make it worthwhile. Explains different training routes for academic and trade schools and what to include in demo tapes, cover letters, and resumes.

✦ Vocational Guidance

Horwin, Michael
Careers in Film and Video Production
Boston, Focal Press, 1990. 191p. ISBN 0240800494

Describes the various jobs and responsibilities in film and TV production. This work goes beyond the usual pre- through post-production careers, highlighting related careers, such as makeup, scouting locations, and accounting. Profiles and interviews throughout add to its readability. Little on job-hunting skills. Includes appendices, bibliography, glossary, and index.

Jurek, Ken
Careers in Video: Getting Ahead in Professional Television
White Plains, N.Y., Knowledge Industry Publications, 1989. 267p. ISBN 0867291699

Corporate and nonprofit video, but not broadcast television, are the career tracks addressed here. Emphasizes job-hunting skills specific to these fields, including demo tapes, résumés, and interview advice. The book's visuals are poor, but the focus on areas where community college graduates are most likely to find employment make it worthwhile. The author is a former community college instructor and has been involved in private-sector video recruitment. Includes appendices, bibliography, and index.

London, Mel
Getting into Video: A Career Guide
New York, Ballantine Books, 1990. 310p. ISBN 0345356489

Written by a practicing professional, this book provides a solid introduction to various video jobs. The basics of video production are described as a lead-in to chapters on different career areas such as commercial, corporate, cable, and government TV. A separate chapter deals with postproduction jobs. Provides interviews with relevant personnel. Academic and private training opportunities are discussed in general terms, without mentioning specific programs. Job search strategies receive their own chapter. Includes bibliography and index.

Noronha, Shonan F. R.
Television and Video: A VGM Career Planner
Lincolnwood, Ill., VGM Career Horizons, 1989. 117p. ISBN 0844286826

A quick survey of job opportunities, from production to adjunct fields, such as marketing for television. Commercial and cable TV and institutional video are addressed. Entries cover the usual aspects of each career: educational requirements, working conditions and salaries, professional organizations. A starting point for a student, but supplemental material would probably be needed.

Rachlin, Harvey
The TV and Movie Business: An Encyclopedia of Careers, Technology, and Practices
New York, Harmony Books, 1991. 343p. ISBN 0517575787

Describes major terminology, organizations, and concepts in entries ranging from fifty words to ten pages. Technology gets relatively little attention in favor of more audience-related ideas. The text's focus on film (somewhat) over television, and its nontechnical nature may make this readable volume most useful for students exploring radio, television, and film as potential careers. Arranged alphabetically. May be suitable for the reference shelf.

Reed, Maxine K., and Robert M. Reed
Career Opportunities in Television, Cable, and Video
3rd ed. New York, Facts on File, 1990. 260p. ISBN 0816023182

An excellent, usable guide. Divided into television broadcasting (seventy jobs described) and cable, video, and media (thirty jobs described). Each job description also includes the education required; expected salary; and where the job fits into the usual organizational structure, with possible career ladders. Separate coverage given to unions, professional organizations, opportunities for minorities and women, four-year college and technical schools, financial-aid opportunities, professional workshops and internships. Includes appendices and bibliography.

Nonprint Materials

Amiga Animation Video
Peterborough, N.H., IDG Communications, 1990-1991.
2 videocassettes (138 min.) VHS

Examples of computer-generated videos done by amateurs using Amiga computers and various software. Does not have instruction, simply short examples. Useful as an introduction for students interested in this emerging technique, but not for how-to instruction.

Videography: The Successful Home Video Series
Toronto, Peter Hitchcock Productions and TV Ontario, 1992.
8 videocassettes (27 min. each) VHS, and guide (172p.)
ISBN 096962610X

This series, done for Canadian television, covers the spectrum of independent video production. Though not aimed at the academic audience, the content can be applied to them easily. Topics covered include: camcorder basics; basic video production; photography and lighting; audio techniques and the travel video; advanced camcorder; advanced editing and instructional video; the documentary; and the big event and video hi-tech. Lighthearted, professional. Has "assignments" for practice.

Criminal Justice and Law

Court Reporting 273
  Louise Treff

Law Enforcement and Criminal Justice 284
  Nancy Tenhet

Legal Secretarial 293
  Susan L. Brant

Paralegal 297
  Susan Stussy

Court Reporting

Louise Treff
Auraria Library
Community College of Denver
Metropolitan State College of Denver
and University of Colorado at Denver
Denver, Colorado

Introduction

Court reporters record verbatim the testimony given at hearings, trials, and other legal proceedings in city, county, state, and federal courts; attorneys' offices; legislatures; government agencies; as well as at conferences, conventions, or other meetings. They use stenotype machines to take shorthand notes and then transcribe them, most often using a computer-aided transcription system. A relatively new development has been the use of videotapes to record court proceedings, affecting the role of the court reporter. Another application of court reporting skills is in the field of television captioning for news and other television programs for hearing-impaired viewers.

A formal training program, usually of two years' duration, is required for court reporting. Such degree programs are offered at community colleges and private business and vocational schools; they include instruction not only in developing high-speed techniques and using stenotype machines and transcription computer programs, but in many other skills as well. Programs are utilizing new technologies in the areas of real-time writing, closed captioning, and steno interpreting for the hearing-impaired. It is essential for the court reporter to be proficient in English-language spelling and grammar, especially for words that sound alike but have different spellings and meanings. They must also be knowledgeable about legal procedures and terminology. Understanding of medical terminology is also important; court reporting curricula include courses in human anatomy and physiology. Knowledge of foreign languages is also helpful when interpreting for witnesses who are not native speakers of English.

Some states (eighteen) require court reporters to be licensed by state officials as a Certified Shorthand Reporter (CSR) or sometimes as a notary public. The National Court Reporters Association (NCRA) certifies a reporter as a Registered Professional Reporter (RPR) if the person passes a two-part examination and participates in continuing education programs. The voluntary RPR designation is recognized as a mark of distinction in the profession. Reporters are then often employed by free-lance reporting companies or directly by the courts, although courts usually hire only experienced reporters. Employment opportunities are expected to grow according to the *Encyclopedia of Careers and Vocational Guidance* (Ferguson Publishing, 1993); however, the *Occupational Outlook Handbook* (U.S. Government Printing Office, 1994) states that most openings will be replacements for those who leave the occupation, partially due to the increased use of video recordings for legal records of court proceedings.

Acknowledgement is extended to the Denver Academy of Court Reporting, the Mile Hi College in Lakewood, Colorado, and the University of Denver Law Library for the use of library materials. The use of these libraries facilitated the identification of the many types of resources needed for student and professional court reporters.

Accreditation and Certification

Accrediting Council for Independent Colleges and Schools
750 First Street NE, Suite 980
Washington, DC 20002-4241

Associated with the Career College Association, the Accrediting Council accredits independent, nonpublic colleges, career institutions, and career schools.

National Court Reporters Association
8224 Old Courthouse Rd.
Vienna, Va. 22182

Includes independent state, regional, and local associations. Also conducts research, compiles statistics, bestows awards, and offers placement service and Registered Professional Reporter Certification program. The Association also publishes or distributes many books and tapes useful to the student or practicing court reporter.

Selected Journals

Journal of Court Reporting.
National Court Reporters Association. Monthly. ISSN 1057-5847

American Bar Association Journal.
The Association. Monthly. ISSN 0747-0088

Trial.
Association of Trial Lawyers of America. Monthly.
ISSN 0041-2538

State law journals such as *Colorado Lawyer* and the *Oregon State Bar Bulletin* are also a rich resource for students in a court reporting program.

Print Materials

✦ Career Materials and Test Preparation

Barnett, Barbara
Court Reporters and Stress: How to Find the Time to Live
Vienna, Va., National Court Reporters Association, 1993. 162p.
ISBN 1881859045

An excellent book for people in any occupation, this work helps the reader understand the inner sources of stress, especially in a highly demanding profession such as court reporting. Although the majority of court reporters surveyed reported satisfaction with their work, ninety-one percent felt stressed. Offers many simple, positive actions to reduce personal stress.

Gowen, Michele C., ed.
A Guide for Legal Assistants: Roles, Responsibilities, Specializations
2nd ed. New York, Practising Law Institute, 1991. 610p.
ISBN 087224024X

Useful to the court reporter as an explanation of the various types of litigation and procedures: federal, antitrust, securities, bankruptcy, construction, pension law and employee benefits, estates and trusts, product liability, negligence, and real estate. Also includes chapters on writing, research, and computerized litigation support systems.

Knapp, Mary H.
The Complete Court Reporter's Handbook
2nd ed. Englewood Cliffs, N.J., Prentice Hall, 1991. 378p.
ISBN 0131593692

Covers the essentials of court reporting: oaths, interrogatories, depositions, court organization, juries, arraignments and sentences, exhibits, legal citations, videotaping, and computer-aided transcription. Test preparation hints. Illustrated with forms and examples. A study guide is also available.

Patterson, Nancy
Preparing for the RPR and CM Written Knowledge Test: Procedures, Medical, Legal, English, Technology
Vienna, Va., National Shorthand Reporters Association, 1990. 258p.

Compiled under the auspices of the Board of the Academy of Professional Reporters to assist test candidates to prepare for the Registered Professional Reporter or Certificate of Merit test. Consists entirely of sample test questions in English vocabulary, word usage and punctuation, legal and medical terminology, court and deposition procedures, and technology. Answer key included.

Rudman, Jack
Passbooks for Career Opportunities Series
Syosset, N.Y., National Learning Corp.

Court Hearing Reporter. (C-172-6) 1988. 1 v. (various pagings). ISBN 0837301726

Court Law Stenographer. (C-173-4) 1990. 1 v. (various pagings). ISBN 0837301734

Court Reporter. (C-174-2) 1992. 1 v. (various pagings). ISBN 0837301742

Court Reporter I. (C-967-0) 1991. 1 v. (various pagings). ISBN 0837309670

Court Reporter II. (C-968-9) 1991. 1 v. (various pagings). ISBN 0837309689

Senior Court Reporter. (C-3543-4) 1989. 1 v. (various pagings). ISBN 0837335434

Senior Shorthand Reporter. (C-724-4) 1988. 1 v. (various pagings). ISBN 0837307244

Shorthand Reporter. (C-471-4) 1991. 1 v. (various pagings). ISBN 0837307414

Examination study guides include questions and answers. Paperbound and hardbound editions available.

Your Career as a Court Reporter: Law Courts, Legislative, Government, Business, Public Hearings
Rev. ed. Chicago, Ill., Institute for Research, 1993. 24p.

Designed to introduce a prospective student to the career, this booklet describes typical duties, earnings, demands of the profession, and recommended preparation and education.

✦ Computers and Office Management

Ayres, James J.
Law Office Software: Attorney's Guide to Selection
New York, Wiley Law Publications, 1990. 425p.
ISBN 0471613851 (and Cumulative Supplement 1993, ISBN 047158228X)

With more than 500 software programs for lawyers and other legal professionals to choose from, a work such as this is invaluable. The guide advocates an analysis and review of office policies and procedures before automating to increase efficiency. Software is examined for both lawyers and support staff to handle the billing of time and fees, accounting, word processing, document assembly, litigation support (document management), utilities, multiple users, legal research, and telecommunications. Appendices summarize computer basics and terminology and provide directories of products and vendors.

Baker, Kim, and Sunny Baker
Office on the Go: Tools, Tips, and Techniques for Every Business Traveler
Englewood Cliffs, N.J., Prentice Hall, 1993. 324p.
ISBN 0136309062

Ideal for the traveling court reporter, this is a state-of-the art guidebook to equipment and software, and a sourcebook for traveling office tips. Nontechnical language is used in the evaluations of the current equipment: notebook computers, electronic organizers, modems, portable fax machines, personal digital assistants, and cellular phones. The section "Putting Mobile Office Technology to Work" offers many tips for improving productivity and time management on the road.

Brophy, Maureen, and Patsy Pressley
Paralegal Management Handbook
New York, Wiley Law Publications, 1993. 247p.
ISBN 0471587958

Contains many useful sections for the paralegal as well as the court reporter in a management position: policies, planning, communication, procedures manuals, profitability, tracking productivity, billable and nonbillable hours, hiring and termination, staff evaluation, litigation case management, and using contract personnel. Includes sample letters, forms, and reports.

De Vries, Mary Ann
Legal Secretary's Complete Handbook.
4th ed. Englewood Cliffs, N.J., Prentice Hall, 1992. 606p.
ISBN 0135298768

A comprehensive handbook on law office procedures that could be helpful to the court reporter. Topics covered include client files and documents; contracts with clients; legal and business correspondence; accounting and recordkeeping; legal research; legal instruments such as affidavits, records on appeal, briefs, and citations; preparation of legal and court papers; understanding courts and their functions; acting as a corporate secretary; and handling commercial collections. Many good step-by-step checklists and examples of letters and forms are provided.

Differding, Debra A., and Sandra Halsne
Law Office Transcription
Cincinnati, Ohio, South-Western Pub. Co., 1992. 227p.
ISBN 0538705507

A workbook designed to teach legal secretarial students how to use a transcribing machine and how to format legal correspondence and legal documents directly from a dictation tape. Includes ten legal cases researched from actual law office files. Examples of documents include a mechanic's lien, wills and probate, personal injury, and litigation. Provides samples and guides to legal and court documents, forms, and terminology. *A Manual* and *Instructor's Solutions Manual* with suggested added activities are also available.

Freer, Carolee
Computer Shorthand: Skillbuilding and Transcription
Englewood Cliffs, N.J., Prentice Hall, 1991. 375p.
ISBN 0131734105

Freer presents fifty lessons that cover skill drills, vocabulary enrichment, phrase and number reviews, letter-by-letter spelling drills, and a review of jury charge instructions. Includes information on question-and-answer testimony. (New edition to be published in the summer of 1995.)

Freer, Carolee
Computer Shorthand: Speedbuildnig and Transcription
Englewood Cliffs, N.J., Prentice Hall, 1991. 371p.
ISBN 0131734024

This text provides fifty chapters on congressional record material, multiple-voice testimony, Latin and French vocabulary words, jury charge, and special tips for court reporters in the field. (New edition to be published in the summer of 1995.)

Larson, Steven B.
Using DOS in Court Reporting
Vienna, Va., National Court Reporters Association, 1994. 122p. and computer disk ISBN 188185907X

Covers the essential DOS commands used by court reporters for MS-DOS versions 3.3, 4.0, 5.0, 6.0 and 6.2. All screen illustrations are based on MS-DOS Version 6.2. Covers directories, files, batch files, backups, error messages, and more. Addresses using Windows for access. Illustrates the parts of the microcomputer and computer-aided transcription (CAT) systems. Includes glossary and disk for completing practice activities.

Lipson, Ashley
Law Office Automation for Paralegals, Administrators, and Legal Secretaries
Englewood Cliffs, N.J., Prentice Hall, 1989. 318p.
ISBN 0135265835

Purports to offer revolutionary techniques for increasing productivity and efficiency in the court reporter's office. Whether in file folders or on a computer, client files, forms, and documents can be reorganized with a simple numerical classification system. Shows how creating master forms and using a computer can speed up document production, yet allow for customization. Also discusses selection of equipment and software for word processing, billing, docketing, and drafting documents.

Lynton, Jonathan S., Donna Masinter, and Terri Mick Lyndall
Law Office Management for Paralegals
Rochester, N.Y., Lawyers Cooperative Publishing, 1992. 355p.
ISBN 0827348657

Malpractice avoidance is the major unifying theme of this text, which covers marketing of legal services; office systems and procedures; accounting and budgeting; time management; records management; the law library; computers, telecommunication, and other office equipment; document management; ethics; and skills in communication, stress management, and career management. Appendices address legal research and filing rules, and provide a sample outline of an office manual. Illustrated with sample forms.

McDaniel, George, ed.
IBM Dictionary of Computing
New York, McGraw-Hill, 1994. 758p. ISBN 0070314888

Defines 18,000 terms and abbreviations covering virtually all of personal computing, information processing, telecommunications, office systems, and IBM-specific terms. Designed for all levels of users. Includes alternative definitions. Some illustrations.

Pulsifer, Nancy, Dana L. Graves, and Jill S. Levin
How to Survive in a Law Firm: Client Relations
New York, Wiley Law Publications, 1993. 1 v. (various pagings). ISBN 0471593672

Attorney relations, client billing, and confidentiality are the three main topics of this guide for paralegals; all are applicable to the court reporter as well. Discusses different billing options, such as hourly, contingency, and flat fee; how to determine what operations are profitable; billing for travel and overtime; and record-keeping, each with detailed examples. "Unwritten Rules" offers many practical tips for working with attorneys and clients. Also includes many time and task management tips on filing, record-keeping, document production, and communication. Also addresses confidentiality issues, phone calls, office space, interviewing prospective employees, work habits, and meetings. Includes sample policy statements and ABA rules.

Randall, Lynn M.
Litigation Organization and Management for Paralegals
New York, Wiley Law Publications, 1993. 320p.
ISBN 0471587982

Aside from pure casework, paralegals, including court reporters, often must handle various administrative, coordinating, supervisory, and management activities. This text deals with a wide variety of tasks and assignments, most of which carry a significant level of responsibility involving design, implementation, creative thinking, problem solving, and decision making. Includes guidance for working with witness files, document production and management, manual and computerized data retrieval systems, and administrative controls (calendars and time and task management).

Saari, David J.
The Court and Free-Lance Reporter Profession: Improved Management Strategies
New York, Quorum Books, 1988. 169p. ISBN 0899302343

Strategies for managing a successful court reporting business either as an individual or as a manager of other reporters. Discusses the development of the profession, organizational theory, ethics, and personnel management, finances, marketing, and technology.

Williams, Ric, and Jean Gonzalez
Advanced Court Reporting Technology: Computer Concepts
Huntington Beach, Calif., Middleton/Wasley, 1991. 1 v. (various pagings)

Textbook for students and practicing reporters on the basics of computer use in court reporting, the system now most used. Defines terminology and discusses selection and use of a computer-aided transcription (CAT) system, real-time translation and closed captioning, abstracting/indexing database programs to organize documents, videotaping, use of expert systems, and future technologies such as CD-ROM and WORM. Includes chapter review quizzes with answers. A teacher's guide, which includes transparency masters, and computer tutorial program are also available.

✦ English Language and Style Guides

Barber, Steve
Legal Writing for Paralegals
Cincinnati,. Ohio, South-Western Pub. Co., 1993. 292p.
ISBN 0538706325

Introduces the basics of technical writing, starting with research strategies: legal publications, encyclopedias, cases, digests, codes, and citations. Addresses composing legal memoranda, briefs, and litigation documents. Includes study and writing exercises with answers and an appendix of cases and briefs. The instructor's manual includes learning objectives, lecture material, teaching suggestions, and answers for the writing exercises.

Butcher, Judith
Copy-Editing: The Cambridge Handbook for Editors, Authors, and Publishers
3rd ed. New York, Cambridge Univ. Press, 1992. 471p.
ISBN 0521400740

Covers all aspects of the editorial processes involved in converting text or disk to the printed page. Includes how to read and correct proofs.

Castilaw, Diane
Court Reporting: Grammar and Punctuation
2nd ed. Cincinnati, Ohio, South-Western Pub. Co., 1993. 305p.
ISBN 0538705779

Basic grammatical rules illustrated with examples. Presents exercises, tests, and situations specific to the court reporter's job. Explains hyphenation, capitalization, homonyms, and other confusing words. Glossary of grammatical terms included.

Finkelstein, Karen, Darlene Leasure, and Tammie Shedd
Real-Time Writing: The Court Reporter's Guide to Mastering Real-Time Skills
Falls Church, Va., National Captioning Institute, 1991. 59p.

The National Captioning Institute's machine shorthand theory and writing style techniques can be used by court reporters to produce a more accurate, conflict-free record. Presents many suggestions and codes or symbols to distinguish words or phrases that are frequently confusing. Utilizes rules for prefixes, suffixes, punctuation, abbreviations, and acronyms. Includes a short drill.

Floyd, Sally, and Dot Mathias
Speedbuilding for Court Reporting
1990 ed. Melrose, Fla., Stenotype Educational Products, 1990. 2 v.
ISBN 0938643037 (vol. 1); ISBN 0938643045 (vol. 2)

This reprint of the 1981 edition consists of practice oral dictation exercises for students of court reporting. Duplicates real-life situations and speaking styles in a variety of contemporary legal areas such as entrapment, negligence, manslaughter, circumstantial evidence, and assault and battery. Includes warm-up exercises at three speed levels, a word preview list, a jury charge exercise, and testimony for each topic. Seventy-two companion practice cassettes that coordinate with the lessons are also available.

Gilman, Mary Louise
Fairly Familiar Phrases
Vienna, Va., National Court Reporters Association, 1993. 50p. ISBN 1881859037

A convenient reference guide to colloquial expressions used in the English language, including those taken from Latin, French, and other languages. Many words were chosen because they have been known to give reporters problems in the past. Includes slang, computer, drug, and legal terms, Black English, and other words of recent origin with short definitions.

Gilman, Mary Louise
One Word, Two Words, Hyphenated?
New ed. Vienna, Va., National Court Reporters Association, 1992. 83p. ISBN 1881859010

As the English language continues to evolve, with a trend away from hyphenation toward single words, this new edition offers quick answers on whether or not to hyphenate more than 8,000 common words. Includes words from foreign languages.

Gilman, Mary Louise, with the assistance of Dorothy P. Kennedy
6,000 Soundalikes, Lookalikes, and Other Words Often Confused
Vienna, Va., National Court Reporters Association, 1992. 200p. ISBN 1881859002

Spellings and definitions for words that sound or look alike, with many cross-references. Includes medical and legal terms. This revision of previous editions includes additional common words suggested by practicing court reporters and a great number of more sophisticated sound-alikes, look-alikes, and other problem words.

Glossaries for Court Reporters
Vienna, Va., National Court Reporters Association, 1994. 281p. ISBN 1881859061

A collection of more than 100 glossaries of specialized terms on many subjects, including AIDS, coal mining, finance, forensic testing, and transportation. Combines and updates the earlier volumes I and II (covering 1975-1986) with all the glossaries published in the *Journal of Court Reporting* from November 1986 to June 1994. Compiled by various authors, the definitions vary in length.

Kocar, Marcella J.
Court Reporter's Language Arts Workbook
Englewood Cliffs, N.J., Regents/Prentice Hall, 1992. 283p. ISBN 0131843915

Presents basic to complex rules of punctuation and grammar, with examples and plentiful practice exercises. Emphasizes why the rules are used, and asks the student to provide reasons for decisions made in completing the exercises, using the abbreviation guide provided.

Levi, Judith N., and Anne Graffam Walker, eds.
Language in the Judicial Process
New York, Plenum Press, 1990. 373p. ISBN 0306435519

A collection of eleven original essays on the dynamic social influence of language in the legal process. Walker's essay addresses the "Customs, Conventions, and Consequences of Court Reporting." Other essays discuss the study of language in the judicial process, bilingual court proceedings, plea bargaining, client communication, cross-examination, linguistic analysis, social conversation, language and cognition in product liability, and cigarette package warnings.

Mager, N. H., and Sylvia K. Mager
Prentice Hall Encyclopedic Dictionary of English Usage
2nd ed. Revised by John Domini. Englewood Cliffs, N.J., Prentice Hall, 1993. 427p. ISBN 0132768585

A reference for the educated professional who frequently writes, this book combines the elements of a dictionary, a style manual, and a grammar guide. Dictionary entries include new words such as "fax" and "MTV," place names, and proper names. Style and grammar guides help the user with definitions, spelling, pronunciation, measurements, abbreviations, capitalization, differentiation of words, and word usage rules. Single alphabetic arrangement of topics and words, with cross-references to other entries as needed.

Making the Record
Vienna, Va., National Court Reporters Association, 1993. 13p.

Provides numerous tips, with examples of attorneys' speech and actions, to help the court reporter create an accurate record. Includes examples of names, words, numbers, measurements, exhibits, quotations, and objections.

Ray, Mary Barnard, and Jill J. Ramsfield
Legal Writing—Getting It Right and Getting It Written
2nd ed. St. Paul, Minn., West Pub. Co., 1993. 361p. ISBN 0314022554

Discusses writing in general, including writing blocks, pressure, priorities, organization, language, and revision. The dictionary format section deals with technical areas such as punctuation, paragraphs, and citations. Addresses client letters, memos, briefs, oral arguments, and legal research.

Reporters Reference Manual
Vienna, Va., National Court Reporters Association, 1991.
1 v. (loose-leaf)

A training guide for evaluating and improving skills in language, office procedures, court structure and procedures, computer knowledge, transcription, and proofreading. Includes separate tests with answer keys and practice exercises.

Stitt-Gohdes, Wanda
Business English: Quick and Easy
Eden Prairie, Minn., Paradigm Publishing, 1992. 260p. ISBN 1561183660

Covers the basics of business English: nouns, verbs, punctuation, capitalization, abbreviations, and constructing sentences. Includes practice exercises.

Weiss, Irwin
Grammar for Court Reporters
Vienna, Va., National Court Reporters Association, 1993. 142p. ISBN 1881859029

Covers the essentials of grammar for court reporters: punctuation, adjectives, adverbs, verbs, pronouns, hyphens, apostrophes, quotation marks, capitalization, and avoiding run-on sentences and fragments. Each section includes short quizzes with answers and practice exercises. Updates and revises the 1978 edition to reflect changes in court reporting and society.

✦ Legal Practice and Procedures

Berk-Seligson, Susan
The Bilingual Courtroom: Court Interpreters in the Judicial Process
Chicago, Univ. of Chicago Press, 1990. 299p. ISBN 0226043711

The *Court Interpreters Act of 1978* requires the use of foreign language interpreters in federal courts and has stimulated parallel action in state and municipal courts. This book shows how the courtroom is transformed by interpretation and how that influences judicial proceedings. Only what is said aloud in English goes into the court record; thus, even minor changes to the original intent of the testimony as a result of inaccurate interpretation can affect the official record taken by the court reporter and hence later appeals.

Cooper, David G., and Michael J. Gibson
An Introduction to Paralegal Studies
Cincinnati, Ohio, South-Western Pub. Co., 1994. 351p.
ISBN 0538707291

This is a practical guide to the basic skills of investigation, legal research, and legal analysis. Provides a step-by-step introduction to the litigation process, including a basic legal vocabulary. Also addresses legal ethics.

The Deposition Handbook
Vienna, Va., National Court Reporters Association, 1989. 75p.

Provides an indispensable handbook to taking testimony outside the courtroom from a witness who will not appear in court. Designed for both the new and experienced reporter, it gives step-by-step procedures for stipulations, oaths, exhibits, off-the-record discussions, reporting through an interpreter, and many other situations the reporter may encounter. Sample forms are appended.

Edwards, Linda L., and J. Stanley Edwards
Tort Law for Legal Assistants
St. Paul, Minn., West Pub. Co., 1992. 445p. ISBN 0314934472

Designed as a textbook for legal assistants to explain civil wrongs for which the victim may receive compensation for damages. Outlines the steps and procedures involved in a typical case: reading, analyzing, and briefing. Specifics of negligence, defamation, malpractice, liability, insurance, and other situations are discussed in detail and are illustrated with case examples. Also covers interviewing clients and witnesses. Special terminology is defined in both the chapters and in the glossary. Instructor's manual is available.

Federal Rules of Appellate Procedure: December 1, 1993
Washington, D.C., U.S. Gov. Printing Office, 1993. 42p.
ISBN 0160432383

Federal Rules of Civil Procedure: With Forms: December 1, 1993
Washington, D.C., U.S. Gov. Printing Office, 1993. 128p.
ISBN 016043257X

Federal Rules of Criminal Procedure: December 1, 1993
Washington, D.C., U.S. Gov. Printing Office, 1993. 55p.
ISBN 0160432375

Federal Rules of Evidence: December 1, 1993
Washington, D.C., U.S. Gov. Printing Office, 1993. 25p.
ISBN 0160432367

All federal rules have been promulgated by the U.S. Supreme Court or public law and have been amended by Acts of Congress and the U.S. Supreme Court. Updated editions of these rules are published each year. For the convenience of the user, any rule that has been amended is listed, along with has a reference to the date the amendment was promulgated and the date the amendment became effective.

Hatch, Scott A., and Lisa Zimmer Hatch
Paralegal Procedures and Practices
St. Paul, Minn., West Pub. Co., 1993. 114p. ISBN 0314013482

Supplements other textbooks with workbook format exercises to assist the student in learning the U.S. court system, the litigation process, filing documents with the court, trial procedures, evidence, legal research, and time management. Answers to exercises are available in the *Instructor's Manual*.

Kerley, Peggy N., Paul A. Sukys, and Joanne Banker Hames
Civil Litigation for the Paralegal
Albany, N.Y., Lawyer's Cooperative, 1992. 442p.
ISBN 0827347715

A textbook introducing civil litigation procedures: pleadings, motions, depositions, interrogatories, physical and mental examinations, documents, requests for admissions, settlements, dismissals, and trial techniques. Sample documents are included. Unique computerized application sections in each chapter show how paralegal tasks can be simplified through computerized procedures and information, usually featuring a specific product. Sidebars define terms. Selected state-specific pocket parts are also available, presenting procedural variations in the courts of those states.

Rules of Practice Before the United States Court of Appeals for the Federal Circuit: Federal Rules of Appellate Procedure (With Amendments Through December 1, 1993); Federal Circuit Rules, Practice Notes, and Guide for Pro Se Petitioners and Appellants
Washington, D.C., United States Court of Appeals for the Federal Circuit, 1993. 170p.

The text of the federal rules of appellate procedure are distinguished from those of the federal circuit by bold-faced type. In addition to the text, there is an appendix of forms.

✦ Legal System

Abadinsky, Howard
Law and Justice: An Introduction to the American Legal System
2nd ed. Chicago, Nelson-Hall, 1991. 403p. ISBN 083041228X

Designed for courses on the law and judicial process. Topics discussed include a definition of law (statutory, common, case, administrative); the history and organization of law and court systems; lawyers and judges; plea bargaining and negotiation (most frequently used); criminal and civil procedures; and alternative methods.

Grilliot, Harold J., and Frank A. Schubert
Introduction to Law and the Legal System
5th ed. Boston, Houghton Mifflin, 1992. 684p. ISBN 0395472830

Designed for a one-term basic law course at the graduate or undergraduate level. Uses the case study method to introduce basic legal terminology. Introduces the court system and judicial procedures from beginning to end for civil and criminal proceedings. Includes case questions, suggested readings, and a glossary of terms. The *Study Guide* gives an overview of each chapter and more review questions. The *Instructor's Manual* contains chapter objectives, a summary of cases, and an answer key.

Kiely, Terrence F.
Modern Tort Liability: Recovery in the '90s
New York, Wiley Law Publications, 1990. 501p.
ISBN 0471614890

A theoretical and practical guide to contemporary tort issues. Discusses sexual harassment, child abuse, emotional distress, witness protection, clergy malpractice, product liability, and children against parents. Cites many specific cases.

Pyle, Ransford Comstock
Foundations of Law for Paralegals: Cases, Commentary, and Ethics
Albany, N.Y., Delmar Publishing, 1992. 444p. ISBN 0827345720

Comprehensive introductory textbook for legal assistants that provides a general overview of American law and the legal system. Includes chapters on ethics; legislation as related to judicial action; the court system; civil and criminal procedures; torts, personal injury, and compensation; contracts and commercial law; property; and administrative law. Illustrated with numerous excerpts from real court cases. Includes an appendix on how to read a case, exercises in legal analysis, and a glossary of terms. *Instructor's Guide* also available.

Taylor, Lawrence
Drunk Driving Defense
3rd ed. Boston, Little, Brown, 1991. 927p. ISBN 0316833940

The most common offense in courts today gets a thorough treatment here from pretrial through post-trial procedures: arraignment, jury selection, evidence, illustrative trial examinations, plea bargaining, and sentencing. Sample checklists, forms, pleadings, and motions are plentiful throughout the text. Offers the material needed by counsel to prosecute or defend a case and assists the court reporter in preparing the final transcript of the court proceedings by providing a greater understanding of the basic procedures.

Wenke, Robert A.
The Art of Selecting a Jury
2nd ed. Springfield, Ill., Charles C. Thomas, 1989. 141p.
ISBN 0398055254

A guide to the jury selection process, whether by court or by counsel. Although aimed at the lawyer, this book will benefit the court reporter through a familiarization with questioning techniques, tactics, and evaluation of jurors. Suggests typical questions for common and special situations.

Workshops for Legal Assistants: Litigation, Legal Research, and Writing, Environmental Law and Toxic Tort Litigation
New York, Practising Law Institute, 1991. 472p.
ISBN 0685499421

Many useful sections for court reporters: conflict of interest, litigation document management by personal computer, drafting document requests, summarizing depositions, time sheets, legal research and writing, hazardous substances, real estate transactions, and an overview of federal environmental law.

◆ Reference—General Legal

Baker, Brian L., and Patrick J. Petit, eds.
Encyclopedia of Legal Information Sources: A Bibliographic Guide to Approximately 29,000 Citations for Publications, Organizations, and Other Sources of Information on 480 Law-Related Subjects
2nd ed. Detroit, Gale Research, 1993. 1083p. ISBN 0810374390

Arranged by topic such as abortion, bail, drug testing, emigration and immigration, psychology, and sexual harassment, this comprehensive guide provides addresses or basic publishing information for loose-leaf services, handbooks, periodicals, bibliographies, directories, associations, professional societies, on-line databases, research centers, statistics sources, audiovisuals, and computer-assisted legal software. Sources for individual states are listed by state.

Bieber's Dictionary of Legal Abbreviations: Reference Guide for Attorneys, Legal Secretaries, Paralegals, and Law Students
Prince's 4th ed. By Mary Miles Prince. Buffalo, N.Y., William S. Hein, 1993. 791p. ISBN 0899418473

Defines more than 4,000 letter abbreviations used in and for legal encyclopedias, dictionaries, reporters, loose-leaf services, treatises, reference books, citators, and other popular material found in a medium-sized law library. It is not, however, a guide for legal citation forms. In determining proper citation form, one should consult *The Bluebook* (Harvard Law Review Association, 1991-) or *Bieber's Dictionary of Legal Citations* (W.S. Hein, 1992).

Black, Henry Campbell
Black's Law Dictionary: Definitions of the Terms and Phrases of American and English Jurisprudence, Ancient and Modern
6th ed. By the publishers editorial staff, coauthors, Joseph R. Nolan, and others. St. Paul, Minn., West Pub. Co., 1990. 1657p. ISBN 0314771654

The standard authority since 1891, *Black's* is more than basic definitions of legal words and terms. It includes multiple definitions for different areas of the law, examples of usage, and citations to pertinent court decisions that have superseded, modified, or supplemented the law. It also references major uniform or model acts and rules.

The Bluebook: A Uniform System of Citation
15th ed. Cambridge, Mass., Harvard Law Review Association, 1991- . Irregular. ISSN 1062-9971

Compiled by the editors of the *Columbia Law Review*, the *Harvard Law Review*, the *University of Pennsylvania Law Review*, and the *Yale Law Journal*. This is the authority for citing practically any type of court case, statutes, regulations, administrative and executive materials, and secondary materials such as books or periodicals in court documents and legal memoranda. It sets forth the basic scheme and rules for legal citations, with examples for each rule. A new section for practitioners details typeface conventions, citation sentences and clauses, parallel citations, short forms, special citation forms, capitalization, and abbreviations. Quick reference guides to common citation forms and footnotes are printed inside the front and back covers.

Brunner, Thomas W., Julie P. Hamre, and Joan McCaffrey Wegrzyn
Legal Assistant's Handbook
2nd ed. Washington, D.C., Bureau of National Affairs, 1988. 232p. ISBN 0871795906

Based on an in-house manual at BNA, this work is a basic introduction to the court systems, their facilities, and personnel; legal concepts such as torts, probate, corporate and securities, antitrust, employment, environment, and criminal law; and typical court proceedings for civil and criminal cases. It discusses the legal assistant's role, such as preparation of documents, information retrieval, digesting depositions, legal research, rules and examples for citation of cases and other documents, basic rules on style and form, ethics for paralegals, and managing other legal assistants. Includes glossary and examples of digests of depositions.

Cohen, Morris L., Robert C. Berring, and Kent C. Olson
Finding the Law: An Abridged Edition of How to Find the Law
9th ed. St. Paul, Minn., West Pub. Co., 1989. 570p. ISBN 0314545875

This abridged version retains the full text and illustrations of those chapters of the larger work that focuses on American law. A chapter on U.S. practice in international law discusses treaties, case law, statutes, regulations, restatements, periodicals, treatises, and foreign relations documentation, all excerpted from the fuller text. For the more specialized coverage of international and foreign law (including English), readers are referred to the full edition.

Cohen, Morris L., Robert C. Berring, and Kent C. Olson
How to Find the Law
9th ed. St. Paul, Minn., West Pub. Co., 1989. 716p. ISBN 0314533184

This standard work, known for its comprehensive treatment of legal research, integrates computer-based research methods into this ninth edition. It discusses both the essentials of present bibliographic procedures and possible future developments. It provides in-depth discussion of research strategies, court reports, regional reporters, case finding and verification, updating, statutes, constitutional law, legislative history, administrative and executive publications, court rules and practice, loose-leaf services, legal periodicals, encyclopedias, restatements, texts, international and foreign law (including English law), and nonlegal research sources. The text is liberally illustrated with examples from the various publications, and are essential to understanding their use.

Cohen, Morris L., and Kent C. Olson
Legal Research in a Nutshell
5th ed. St. Paul, Minn., West Pub. Co., 1992. 370p. ISBN 0314007830

Combines discussion of the research process with a description of the bibliographic tools needed to learn the essentials of legal research. Includes a separate section on computerized legal research, but shows the symbiotic relation to essential print resources. The focus is American legal research, but this edition has expanded coverage of international and comparative law. The "Nutshell" edition introduces the basic tools and methodology in legal literature: court reports, case finding, statutes, legislative history, administrative law, loose-leaf services, and secondary materials, with examples from printed and on-line sources. Although indexed, this is not intended as a comprehensive reference source as are the authors' *How to Find the Law* (West Pub. Co., 1989) and its abridged edition, *Finding the Law* (West Pub. Co., 1989).

Dworsky, Alan L.
User's Guide to the Bluebook
Rev. for the 15th ed. of the *Bluebook*. Littleton, Colo., Fred B. Rothman, 1991. 54p. ISBN 0837705584

Especially good for the beginner trying to apply the rules in the *Bluebook*, this guide is a how-to manual primarily aimed at explaining step-by-step the rules required for everyday legal writing. It gives examples as they apply to a memorandum or brief. This guide is not meant to stand alone; it is intended to be used with the *Bluebook* to assist in mastering the rules.

Eskew, Mike
Legal Procedures and Terminology for Court Reporters and Paralegals
Englewood Cliffs, N.J., Regents/Prentice Hall, 1993. 206p ISBN 0135292158

Provides a broad introduction to criminal and civil law and the most commonly used legal terms. Includes information on contracts, legal research, torts, real estate, wills, corporations, equity, domestic relations, trials, and the origin of law. Quizzes with answers, a glossary for quick reference, and an index of all legal terms used are included. Instructor's manual is available with additional quizzes and tests.

Gifis, Steven H.
Dictionary of Legal Terms: A Simplified Guide to the Language of the Law
2nd ed. Hauppauge, N.Y., Barron's, 1993. 517p. ISBN 0812014111

Geared for the layperson and redrafted from the *Law Dictionary* (Barron's, 1991), this work provides a ready, accessible, and useful source for understanding the law and law-related processes and concepts. Terms were included that may not be thought of as technically "legal," but have special meaning and arise in legal contexts. Hundreds of definitions were added from securities, finance, and taxation. Entries include a pronunciation guide, alternate expressions in brackets, cross-references to related terms, and examples of use. Appendices contain a "Consumer's Guide to Law and Lawyers" and basic information on common legal procedures such as divorce, wills, real estate, small claims, bankruptcy, and injury compensation.

Kavass, Igor I., and Mary Miles Prince
World Dictionary of Legal Abbreviations
Buffalo, N.Y., William S. Hein, 1991- . 1 v. (loose-leaf) ISBN 0899417817

For the user of law and literature outside the U.S., this dictionary defines 8,400 entries in French, Italian, Portuguese, and Spanish, indicating the country of origin.

Kunz, Christina L., and others
The Process of Legal Research: Successful Strategies
3rd ed. Boston, Little, Brown, 1992. 603p. ISBN 0316507202

Teaches legal research by enabling the student to explore the sources in a self-paced method, using extensive problem exercises.

Covers searching strategies for CD-ROM and on-line sources as well as print. Employs a step-by-step approach with plentiful illustrations to dictionaries, encyclopedias, case reporters, statutes, administrative materials, legislative materials, rules, and model and uniform laws. A generous number of problem sets are included at the back of the book for each chapter.

Lawyer's Desk Book
9th ed. Englewood Cliffs, N.J., Prentice Hall, 1989. 764p.
ISBN 0135267730

A concise introduction to common areas of law: bankruptcy, compensation and fringe benefits, corporate law, estates, intellectual property, real estate, social security programs, and taxes. Updated with supplements.

Lynton, Jonathan S., and Terri Mick Lyndal
Legal Ethics and Professional Responsibility
Albany, N.Y., Lawyers Cooperative Publishing, 1993. 392p.
ISBN 0827355041

Written for the legal paraprofessional, this book deals with the critical issues of confidentiality, computer security, conflict of interest, the unauthorized practice of law, advertising and solicitation, and client funds. Illustrated with hypothetical and real cases. It includes the text and analysis of the relevant rules and regulations promulgated by the American Bar Association, the National Association of Legal Assistants, and the National Federation of Paralegal Associates.

Oran, Daniel
Oran's Dictionary of the Law
2nd ed. St. Paul, Minn., West Pub. Co., 1991. 500p.
ISBN 0314846905

Helps the reader understand and use the technical vocabulary used in American law, while aiding in the recognition of words that are often vague in meaning. Typical technical and general jargon is also presented in the appendix. The substantial appendix also includes legal research methodology and tools.

Redden, Kenneth R., and Gerry W. Beyer
Modern Dictionary for the Legal Profession
Buffalo, N.Y., William S. Hein, 1993. 802p. ISBN 0899418295

Different from other law dictionaries in that it emphasizes popular, technical, and slang terms in various fields. Presents more than 8,000 modern terms from banking, advertising, sports, religion, physics, sociology, literature, and many others. Also includes 250 computer terms and 200 Spanish legal terms. Examples of new terminology include "kiddie tax," "hostile takeover," "Seven Sisters," and "voodoo economics."

Roderick, Wanda Walker
Legal Studies, to Wit: Basic Terminology and Transcription
3rd ed. Cincinnati, Ohio, South-Western Pub. Co., 1990. 474p.
ISBN 0538701048

Intended to impart knowledge and understanding of approximately 800 terms commonly used in the legal profession. The workbook format includes pronunciation guide, typing practice, and quizzes at the end of each of the thirty lessons. Lessons cover general legal terms, litigation, civil action, criminal law, probate, real property, contracts, domestic relations, commercial paper, bankruptcy, partnerships, and corporations. Thirty practice cassette tapes (ISBN 0538701056) for dictation and eight evaluation cassettes to be administered by the instructor are also available.

A Student's Guide to Am Jur 2nd, ALR, and USCS
Rochester, N.Y., Lawyers Cooperative Publishing, 1990. 47p.

The purpose of this guide is to acquaint the user with the publisher's major sets: *American Jurisprudence 2nd*, *American Law Reports*, and the *United States Code Service*. The text explains what each tool is and gives instructions for use providing numerous examples. Quite comprehensive and direct.

✦ Reference Materials—Business

Davidson, Daniel V., and others
Business Law: Principles and Cases
4th ed. Belmont, Calif., Wadsworth Publishing, 1993. 1199p.
ISBN 0534932808

Aids in the comprehension of business law topics by using plain language, relevant and realistic examples, and pertinent cases. Includes discussion of crimes and business, civil lawsuits, the attorney-client relationship, an accountant's legal liability, contracts, bankruptcy, and business partnerships. Discussion questions and case problems end each chapter. Appendix includes index, glossary, and *Uniform Commercial Code*.

Goldman, Arnold J., and William D. Sigismond
Business Law, Principles and Practices
3rd ed. Boston, Houghton Mifflin, 1992. 664p.
ISBN 0395472806

Intended for postsecondary students, this text addresses the foundations of law, and topics such as, torts, crime, litigation, contracts, commercial paper, property, bailments, insurance, consumers, and creditors. Summary of concepts, questions, and cases for discussion end each chapter. Legal research principles are summarized in a two-page appendix.

✦ Reference Materials—Medical and Scientific

Campbell, Robert Jean
Psychiatric Dictionary
6th ed. New York, Oxford Univ. Press, 1989. 811p.
ISBN 0195052935

Since the fifth edition was published in 1981, new terms have been added resulting from new research in areas such as agoraphobia and panic disorder; sleep, speech, and eating disorders; depression; neuroscience; and psychopharmacology. The nomenclature of the revised third edition of the *Diagnostic and Statistical Manual of Mental Disorders* (American Psychiatric Association, 1987) has also been incorporated.

Chabner, Davi-Ellen
The Language of Medicine: A Write-In Text Explaining Medical Terms
4th ed. Philadelphia, Saunders, 1991. 768p. ISBN 0721632440

Assuming no previous knowledge of biology, this work introduces anatomy, physiology, pathology, clinical procedures, and lab tests. It includes abbreviations related to each body system. Practical applications sections contain short examples of medical language in an actual written context, such as case reports, operative and diagnostic lists, autopsy records, and lab and X-ray reports. Illustrated with diagrams, charts, illustrations, and tables. Supplemented with pronunciation lists and exercises with answers.

Delong, Marilyn Fuller
Medical Acronyms and Abbreviations
2nd ed. Oradell, N.J., Medical Economics Books, 1989. 355p. ISBN 0874895413

Defines more than 6,000 medical abbreviations and acronyms in all medical and surgical specialties, as well as nursing, administration, quality assurance, dietetics, pharmacology, and laboratory medicine. This expanded edition adds new material on dentistry and hospital finance.

Diehl, Marcy Otis, and Marilyn Takahashi Fordney
Medical Typing and Transcribing: Techniques and Procedures
3rd ed. Philadelphia, Saunders, 1991. 485p. ISBN 0721634796

An introduction to medical terms and practices for the student of court reporting, legal secretarial studies, or medical transcription. Includes new medical diseases and procedures. Each chapter was reviewed by three experts to ensure accuracy. Appendixes include answers to practice tests and performance evaluation sheets. Instructor's manual available.

Dupayrat, Jacques
Dictionary of Biomedical Acronyms and Abbreviations
2nd ed. New York, Wiley, 1990. 162p. ISBN 0471926493

More than 7,000 definitions of the 4,000 most common acronyms in medicine, biology, and biochemistry. Multiple definitions exist for numerous acronyms.

Fordney, Marilyn Takahashi, and Marcy Otis Diehl
Medical Transcription Guide: Do's and Don'ts
Philadelphia, Saunders, 1990. 535p. ISBN 0721637981

A how-to guide of forms and records, including examples and practice tests. Introduces terms, medical charts, patient history, physical records, and psychiatric reports. Addresses punctuation and common medical homonyms.

Griffiths, Joan
Handbook for the Hospital Medical Secretary
Cambridge, Mass., Blackwell Scientific Publications, 1993. 237p. ISBN 0632035846

Describes the different hospital departments, something not usually found in other medical terminology books. Also explains medical basics and terminology, e.g., body systems, root words, prefixes and suffixes, medical investigations, drugs, surgery, and specialties. Includes sample medical and surgical summaries and X-ray reports. Through not a dictionary or encyclopedia of terminology, an appendix lists terms and their definitions.

Hamilton, Betty, and Barbara Guidos
MASA: Medical Acronyms, Symbols, and Abbreviations
2nd ed. New York, Neal-Schuman, 1988. 277p. ISBN 1555700128

A comprehensive dictionary of more than 32,000 entries (an increase of 12,000 over the 1st ed.) of "medical shorthand" in all major specialties and subspecialties of medicine and related fields. Includes contemporary terms, and obsolete terms that may be found in older medical records or literature.

Lawrence, Eleanor
Henderson's Dictionary of Biological Terms
10th ed. New York, Wiley, 1989. 637p. ISBN 0470214465

Updated to include new vocabulary in cell biology, molecular genetics, and recombinant DNA technology. Abbreviations (bot., zool., and mycol.) indicate the field of usage. Common abbreviations and acronyms appear at the front of each letter section; suffixes and prefixes are in the body of the dictionary. Appendices provide outlines of the plant and animal kingdoms and present the structural chemical formulae of some important biological compounds.

Leonard, Peggy C.
Building a Medical Vocabulary
3rd ed. Philadelphia, Saunders, 1993. 558p. ISBN 0721646905

Step-by-step method for learning medical terminology using a fill-in-the-blank workbook style, crossword puzzles, and review exercises. Illustrated. Includes index and glossary of more than 2,000 terms.

Merck Index: An Encyclopedia of Chemicals, Drugs, and Biologicals
11th ed., Centennial ed. Rathway, N.J., Merck Research Laboratories, 1989. 1 v. (various pagings). ISBN 091191028X

Provides concise descriptions of the preparation and general properties of compounds, and correlates their trivial, generic, and chemical names with structures, trademarks, and company affiliations. Also offers succinct information on the use, principal pharmacological action, and toxicity of these substances.

Merck Manual of Diagnosis and Therapy
Rahway, N.J., Merck Research Laboratories, 1950- . ISSN 0076-6526

The classic, thorough guide to most kinds of physical and mental disorders: their causes, symptoms, signs, diagnosis, prevention, and treatment, as well as common clinical procedures and laboratory tests. The descriptions are comprehensive, yet concise.

Physicians' Desk Reference: PDR
Montvale, N.J., Medical Economics Co., 1974- . ISSN 0093-4461

An annual publication, the *PDR* is the only compendium of official, FDA-approved drug labeling. The *Code of Federal Regulations* mandates that the wording for such information as dosages, warnings, hazards, side effects, and precautions must be in the same language as the FDA-approved labeling for the products. The latest available information is given for 3,000 specific pharmaceutical products, prescription and nonprescription. Also issued in CD-ROM format.

Rice, Jane
SPELLRIGHT: A Medical Wordbook
Norwalk, Conn., Appleton and Lange, 1991. 772p. ISBN 0838562906

More than 90,000 medical words, laboratory terms, drugs, and abbreviations are listed in alphabetical order in this handbook. Drugs are listed by their trade name with the generic name in parentheses. In addition, more than 6,000 abbreviations are defined in an appendix. This is a sourcebook to rely on for spelling accuracy, but it is a spelling dictionary only; there are no pronunciation guidelines or definitions. Easy-to-read typeface.

Solomon, Eldra Pearl
Introduction to Human Anatomy and Physiology
Philadelphia, Saunders, 1992. 314p. ISBN 0721639666

Uses familiar examples together with tables, color illustrations, new terms in boldface, a glossary, summaries, review questions, and posttests. Study guide for students that highlights concepts and provides test exercises with answers. A *Test Bank* that provides two tests for each chapter is also available for instructors.

Sormunen, Carolee
Terminology for Allied Health Professionals
2nd ed. Cincinnati, Ohio, South-Western Pub. Co., 1990. 563p. ISBN 053870070X

Intended to build a basic medical vocabulary through an introduction to anatomy, medical histories, and physical examinations. Workbook format with illustrations and exercises. Appendix includes abbreviations and drug names.

Stanfield, Peggy, and Y. H. Hui
Medical Terminology: Principles and Practices (With Self-Instructional Modules)
Boston, Jones and Bartlett Publishers, 1989. 541p. ISBN 0867204095

Makes terminology easier to learn by breaking down words into their roots, prefixes, and suffixes, with definitions of those common parts and common examples. Sample clinical histories are also included with related practice exercises in a programmed instruction format. Answer keys are provided.

Tennenhouse, Dan J.
Attorney's Medical Deskbook
3rd ed. Deerfield, Ill., Clark Boardman Callaghan, 1993. 3 v.

Assists in better and faster utilization of medical records, medical literature, and medical expert witnesses for case evaluation and presentation by providing explanations of medical procedures and terminology and samples of numerous forms. Includes a typical medical history questionnaire; information on diagnostic, lab, and special tests; a section on autopsies, poisoning, and alcoholism; anatomical terminology and diagrams; diseases and medical disorders often litigated; and drug uses and effects.

✦ Videotaping Court Proceedings

Hewitt, William E.
Videotaped Trial Records: Evaluation and Guide
Williamsburg, Va., National Center for State Courts, 1990. 175p. ISBN 0896560996

First used on a regular basis in Kentucky in the 1980s, videotapes of trials were the object of a formal study in six courts in the states of Kentucky, Michigan, North Carolina, and Washington. The results of the study are reported here, and include survey responses from attorneys and judges in the appendix, along with the names and addresses of respondents. Comments from the study are incorporated into the chapters on evaluation criteria for deciding whether to adopt such a system. The use of court reporting with computer-aided transcription (CAT) is discussed within the report as a competing or complementary technology. Implementation of a video system is discussed, and a list of vendors and a bibliography are included.

Video Court Reporting: A Bibliography of Recent Information Sources
Williamsburg, Va., National Center for State Courts, 1991. 3p.

A bibliography of recent National Center for State Courts publications, journal articles, and studies about operatorless video court reporting systems. Assistance with obtaining the publications is available through the NCSC.

Video Court Reporting: A Primer for Trial and Appellate Court Judges
Williamsburg, Va., National Center for State Courts, 1991. 15p.

An introduction to the technology of video court reporting. Question-and-answer describes the basics, advantages and disadvantages of video reporting, the use of appeals, attorney comments, and the outlook for the future. Bibliography included.

Video Court Reporting: A Review of Recording Systems and Vendors
Williamsburg, Va., National Center for State Courts, 1991. 19p.

Results of a workshop discussing the advantages of using video, and compares different equipment, vendors, and features. Also reviews service and maintenance records and the outlook for the future. Bibliography included.

Video Court Reporting: Tips for Court Managers: Summary Observations by Participants in the Institute for Court Management Workshop on Videorecording, Louisville, KY, October 22-24, 1990
Williamsburg, Va, National Center for State Courts, 1991. 13 leaves

Discusses making a record with videotape (little controversy) versus using videotape as the record on appeal (much controversy). Includes points to consider in the selection of a system, installation and tuning, management and procedures for recording, logging tapes, reviewing, transcribing, use of the record for appeals, storage and access, and dealing with attorney resistance.

Nonprint Materials

There are numerous audiocassettes and videocassettes available for court reporting students. Topics covered include dictation, speed development, learning medical and business terminology, dealing with stress, and vocational guidance. These nonprint materials are used actively in the classroom; thus they are not available for interlibrary loan. The following publishers will provide catalogs and descriptions of their media.

National Court Reporters Association
8224 Old Courthouse Rd.
Vienna, Va. 22182

Stenograph Corporation
1500 Bishop Court
Mt. Prospect, Ill. 60056

Law Enforcement and Criminal Justice

Nancy Tenhet
Hinds Community College
Raymond Campus
Raymond, Mississippi

Introduction

The law enforcement or criminal justice programs in the two-year college serve two general purposes. They offer basic introductory courses leading to a two-year degree and immediate employment and they also provide courses that serve as an introductory curriculum for students transferring to a four-year college or university. The program must prepare students for their immediate goals, while exposing them to the realities of the profession and to pertinent resources in their field.

Law enforcement personnel protect the life and property of the citizens of the United States. Their duties are varied, but generally revolve around four main responsibilities: preserving the peace, preventing crime, enforcing the laws, and making arrests. Applicants for law enforcement work undergo strenuous entrance requirements. They must pass health and physical requirements, written and oral tests, and sometimes psychological tests. Their personal history is investigated and their integrity and honesty scrutinized. Often college courses and specialized training are required.

Police officers may be employed by city, county, state, or federal agencies. They may work for a large department where the duties are specialized and mainly involve one area, such as criminal investigation, fingerprint identification, traffic control, or canine corps work. Other employment opportunities may be in a small town or rural area, where the duties are varied and the officer must be knowledgeable in all areas, including administration and training.

Many police departments encourage their employees to seek college degrees or additional schooling. Often two-year colleges are chosen because of their proximity and flexible scheduling. These students often have the basic knowledge but may need to update their skills or broaden their backgrounds for advancement.

This bibliography is for students of law enforcement and police science programs on a two-year level, as well as those already employed in the field. Most of the items chosen for this bibliography cover a variety of the basic areas of police work such as protection, weapons, and report writing. Other books focus on popular topics such as stress, use of force, ethics, and DNA testing. Omitted were items on specific areas such as juvenile justice, gangs, and organized crime. The focus of this bibliography is on materials that reflect the range of job positions and responsibilities.

Selected Associations

Accreditation is generally done at the statewide or local levels. However, these are selected organizations to contact for more information.

American Correctional Association
8025 Laurel Lakes Court
Laurel, Md. 20707-5075

American Federation of Police
3801 Biscayne Blvd.
Miami, Fla. 33137-3732

International Association of Chiefs of Police
515 N. Washington Street
Alexandria, Va. 22314-2357

National Association of Chiefs of Police
3801 Biscayne Blvd.
Miami, Fla. 33137-3732

National Police Officers Association of America
P.O. Box 22129
Louisville, Ky. 40252-0129

Selected Journals

FBI Law Enforcement Bulletin.
U.S.G.P.O. Monthly. ISSN 0014-5688

Journal of Criminal Law and Criminology.
Northwestern University School of Law. Quarterly.
ISSN 0091-4169

Law and Order.
Law and Order. Monthly. ISSN 0023-9194

The Police Chief.
International Association of Chiefs of Police. Monthly.
ISSN 0032-2571

Print Materials

✦ Reference

Bailey, William G., ed.
Encyclopedia of Police Science
New York, Garland, 1989. 718p. ISBN 0824066278
    Provides introductory information on over 140 topics such as arrest, deadly force, fingerprinting, police ethics, and serial murder. Includes some articles on persons and organizations. The attributed articles range from two to ten pages with brief bibliographies. Concludes with a bibliography on police science, a bibliography of bibliographies, an index of legal cases, and a general index. Second edition due to be published in 1995 (ISBN 0815313314).

Bentley, William K., and James M. Corbett
Prison Slang: Words and Expressions Depicting Life Behind Bars
Jefferson, N.C., McFarland, 1992. 120p. ISBN 0899506461
    Contains definitions of approximately 1,000 words and expressions arranged alphabetically by categories. Covers areas of prison life such as drugs and alcohol, sexuality, violence, and security. Index of terms included.

Criminal Justice Abstracts
Monsey, N.Y., Willow Tree Press, Quarterly. (various pagings)
ISSN 0146-9177
    Includes abstracts of books, journals, dissertations, and reports published worldwide. Some issues contain a review or bibliography on a current topic. Includes subject, geographic, and author indexes. Cumulates annually. Available in print, on CD-ROM or on-line through WESTLAW.

Criminal Justice Periodicals Index
Ann Arbor, Mich., Indexing Services, University Microfilms.
Published three times a year. ISSN 0145-5818
    The basic index for criminal justice periodicals, this resource indexes over 100 English language journals, newsletters, and law reports that represent broad coverage of criminal justice. Includes both author and subject indexes in an easy-to-use format. Cumulates annually. Available in paper format or online on DIALOG file 171.

Encyclopedia of Criminology
Sue Titus Reid, editor in chief
New York, Macmillan, 1995. 4 v. ISBN 0028840003
    A four-volume encyclopedia consisting of more than 700 entries on legal, corrections, police and social issues. Contributors include professors of law, criminology and sociology, as well as practitioners, such as chiefs of police, prison wardens, and parole officers. Publication date: Fall 1995.

Fay, John J.
The Police Dictionary and Encyclopedia
Springfield, Ill., Charles C. Thomas, 1988. 370p.
ISBN 0398054940
    Nearly 5,000 terms, phrases, and concepts in law enforcement are briefly defined. Includes abbreviations, slang terms, drug names, as well as summaries of pertinent Supreme Court cases. Appendixes include definitions of felonies, lists of felony sentences, capital offenses, and execution methods. Includes bibliography.

Ferdico, John N.
Ferdico's Criminal Law and Justice Dictionary
St. Paul, Minn., West Pub. Co., 1992. 476p. ISBN 0314933107
    This easy-to-use dictionary contains concise definitions and numerous cross-references. Included are abbreviations and foreign terms. Court cases are cited in some definitions.

Kurian, George Thomas
World Encyclopedia of Police Forces and Penal Systems
New York, Facts on File, 1989. 582p. ISBN 0816010196

Covers the national law enforcement and penal systems of 183 countries and is divided into two sections, by the size of the countries. The countries are listed alphabetically within each section. Basic information for each country's police force includes history, structure and organization, education and training. An overview of the various penal systems is also given. Contains charts of crime statistics and police organizational charts. Four appendices include information about Interpol, a world police directory, a brief bibliography, and a chart giving comparative statistics on police protection. Index.

Nash, Jay Robert, ed.
Encyclopedia of World Crime: Criminal Justice, Criminology, and Law Enforcement
Wilmette, Ill., CrimeBooks, 1989-1990. 6 v. ISBN 0923582002 (set)

More than 50,000 biographies, of mostly American criminals, are contained in this six-volume set. Three and one-half volumes contain the biographies. The other volumes contain chronologies of specific crimes, definitions, a prison directory, acronyms, landmark federal and state court decisions, important U.S. crime legislation, the U.S. crime commissions, and FBI offices listed by city. Volume Six contains subject and name indexes. Photographs accompany the text.

National Police Chiefs and Sheriffs Information Bureau
National Directory of Law Enforcement Administrators
1992-93 ed. Stevens Point, Wisc., The Bureau, 1992. 637p. ISBN 1880245035

This annual directory lists all types of law enforcement agencies including municipal, county, campus, highway patrol, state correctional, airport and harbor, and federal. Divided by type of agency, it is arranged alphabetically by state. For each agency, basic information is given: name of administrator, address, phone, and FAX number.

O'Block, Robert L.
Criminal Justice Research Sources
3rd ed. Cincinnati, Ohio, Anderson Pub. Co., 1992. 189p. ISBN 0870846655

Each chapter in this comprehensive bibliography is on a broad type of source, such as books, indexing sources, journals, bibliographies, etc. Most items are annotated. The chapters on government documents and statistical data are particularly informative.

Rush, George E.
The Dictionary of Criminal Justice
4th ed. Guilford, Conn., Dushkin Pub. Group, 1994. 374p. ISBN 0879678984

Contains over 3,600 definitions, charts and illustrations. Includes summaries of over 800 Supreme Court cases important to criminal justice (organized by topic), a bibliography of sources, an alphabetical listing of the Supreme Court cases listed, and appendices listing criminal justice doctoral programs, forensic agencies, and juried or refereed journals.

Sourcebook of Criminal Justice Statistics, 1992
Albany, N.Y., Hindelang Criminal Justice Research Center, 1993. 794p. ISBN 0160424402

This annual work contains in one volume national data compiled from a variety of sources. To be included, data must be national in scope and methodologically sound. It provides the characteristics of criminal justice systems, public opinion toward crimes, nature and distribution of known offenses, number of arrests, court statistics, and correctional statistics. Appendices and index.

Terrill, Richard J.
World Criminal Justice Systems: A Survey
2nd ed. Cincinnati, Ohio, Anderson Pub. Co., 1992. 410p. ISBN 0870848364

Gives an overview of the criminal justice systems in five countries: England, France, Sweden, Japan, and the Soviet Union. (Information is prior to the breakup of the Union of Soviet Socialist Republics in September of 1991.) Critical issues in some of the countries are discussed. Appendix has comparative statistics for four of the countries from 1976 to 1985. Bibliography and index.

United States. Federal Bureau of Investigation
Uniform Crime Reports for the United States
Washington, D.C., The Bureau, Annual. 1 v. (various pagings) ISSN 0082-7592

The Uniform Crime Reports (UCR) give a nationwide view of crime, based on statistics contributed by over 16,000 local and state law enforcement agencies. Statistics for major types of crimes such as rape, murder, and theft are given for local areas. Also gives information on number of arrests and law enforcement personnel. Appendices define terminology and explain the methodology. Annual. Good color charts included.

Vanden Heuvel, Nichole, comp.
Directory of Criminal Justice Information Sources
Washington, D.C., U.S. Dept. of Justice, Office of Justice Programs, National Institute of Justice, Published biennially. (various pagings)

To encourage the exchange of information among those in criminal justice, the National Institute of Justice produces this listing of organizations offering such informational services as computerized literature searches and reference services. Included alphabetically are 176 organizations. A summary is given for each of the informational services offered, the contact person, phone number and address, and publications produced. Indexed by subject.

✦ Administration

Ferrari, Mario J.
Supervision Handbook
Binghampton, N.Y., Gould Publications, 1990. 159p.

An easy-to-read manual especially appropriate for a beginning or future supervisor. Fifteen chapters cover areas such as leadership traits, discipline, cooperation, on-the-job instruction, grievances, public speaking, public relations, safety, and management. Includes index.

Guyot, Dorothy
Policing As Though People Matter
Philadelphia, Temple Univ. Press, 1991. 357p. ISBN 0877227667

By analyzing the transformation of the police department in Troy, New York, under the leadership of George W. O'Connor, the author illustrates what makes a good police department. The standards for judging the quality of police service are explained and ways to improve services are illustrated. The appendix contains questions for each chapter to aid in assessing and improving the quality of police services. Includes bibliography and general and author indexes.

International Association of Chiefs of Police
Managing the Small Law Enforcement Agency
Arlington, Va., The Association, 1990. 185p. ISBN 0840379315

Designed as a practical reference, this work proposes solutions to management issues in the small department. Covers all aspects of the department, from goals and objectives, policies and procedures, ethics, conduct of officers, screening and selection of applicants, and training, to fiscal management, labor relations, and planning. Review questions follow the chapters. Indexed. (Available from Kendall/Hall.)

✦ Courtroom Testimony and Criminal Law

Holtz, Larry E.
Contemporary Criminal Procedure, 1993-1994
Longwood, Fla., Gould Publications, 1993. 674p.
ISBN 0875263658

Presents pertinent judicial decisions from the U.S. Supreme Court and the U.S. Circuit Courts of Appeal. Gives the case name and its citation, date decided, and an extended explanation of the decision. Covers the many aspects of arrest, search, seizure, confessions, eyewitness identification, Sixth Amendment rights, drugs, and liability. Looseleaf format, updated periodically. Good for identifying and understanding relevant court cases.

Reynolds, D. W.
The Truth, the Whole Truth, and Nothing But—: A Police Officer's Guide to Testifying in Court
Springfield, Ill., Charles C. Thomas, 1990. 78p.
ISBN 0398056560

The author states that the goal of this book is to enlighten, expose, educate, and equip. That is exactly what he does. Reynolds, who has extensive experience in courtroom testimony, exposes the tricks and tactics of defense attorneys. He explains what is happening in court and how to avoid the pitfalls of testifying. Tips and examples illustrate the main points.

Rutledge, Devallis
Courtroom Survival: The Officer's Guide to Better Testimony
Placerville, Calif., Copperhouse Pub. Co., 1993. 188p.
ISBN 0942728157

Numerous examples illustrate the practical tips that fill this easy-to-read book. Written in an informal style, it covers all the basics from preparations before court, courthouse conduct, and what to expect, to how to testify and handle cross-examinations.

✦ Criminal Investigation and Evidence

Abrams, Stan
The Complete Polygraph Handbook
Lexington, Mass., Lexington Books, 1989. 270p.
ISBN 0669153451

Having nearly twenty years experience administering polygraphs and researching them, the author shares his vast knowledge of the topic in this comprehensive work. It covers the history of the polygraph, other lie detectors, and their psychophysiologic basis, the instrument, test procedures, techniques, validity and reliability of the test, and legal status of the polygraph and polygraphist as expert witness.

Billings, Paul R., ed.
DNA on Trial: Genetic Identification and Criminal Justice
Plainview, N.Y., Cold Spring Harbor Laboratory Press, 1992. 154p. ISBN 0879693797

Compilation of articles on DNA and the issues associated with it. Covers legal issues, use in court, reliability, impact on civil liberties, and DNA data banking. Articles contain references. Index.

Collins, Clarence Gerald
Fingerprint Science: How to Roll, Classify, File, and Use Fingerprints
Placerville, Calif., Copperhouse Pub. Co., 1994. 178p.
ISBN 0942728181

Designed for those with little or no fingerprinting experience, the basics of fingerprinting are covered. Also included are major case prints, preparation for court testimony, and fingerprinting the dead. Two appendices have basics for finding and preserving fingerprints at the crime scene, and an overview of advanced fingerprint technology.

Cowger, James F.
Friction Ridge Skin: Comparison and Identification of Fingerprints
Boca Raton, Fla., CRC Press, 1993. 219p. ISBN 084939502X

Very detailed information on taking, classifying, photographing, and comparing fingerprints. Also discusses palm and sole prints, evidence prints, basis of print comparisons, and testifying concerning conclusions. Originally part of the *Elsevier Series in Practical Aspects of Criminal and Forensic Investigations*. Includes references and index.

Douglas, John E., and others, eds.
The Crime Classification Manual
New York, Lexington Books, 1992. 374p. ISBN 0669246387

Developed by criminal justice professionals, the purposes of the manual are to standardize terminology, to facilitate communication, and to educate concerning types of crimes. The first part of the manual lists major crime categories, and for each category, provides defining characteristics, victimology, crime scene indicators, forensics, and illustrated cases. The second part discusses modus operandi, crime scene photography, etc. Useful appendices. Indexed.

Fisher, Barry A. J.
Techniques of Crime Scene Investigation
5th ed. New York, Elsevier, 1992. 526p. ISBN 0444016368

    Head of one of the largest municipal crime labs in the United States, the author bases this resource text and investigation guide on his experience and research. Covers all aspects from first officer at the crime scene and establishing identity, to the various types of evidence such as blood, body fluids, and impressions. Also covers special types of investigations. Numerous illustrations. Contains an appendix with a listing of equipment needed for crime scene investigations. (Available from CRC Press.)

Handbook of Forensic Science
Washington, D.C., U.S. Dept. of Justice. Federal Bureau of Investigation, 1992. 119p.

    A guide to legally accepted and practical procedures for collecting, preserving, and handling physical evidence. Gives advantages and disadvantages of some collection techniques, and covers such types of evidence as fingerprints, blood, DNA, body fluids, shoe prints, and tire treads. Discusses services of the FBI laboratory, the importance of initial crime scene search, and shipment of evidence. Includes sample FBI laboratory reports.

Holtz, Larry E.
Criminal Evidence for Law Enforcement Officers
Longwood, Fla., Gould Publications, 1994. 326p.
ISBN 0875263933

    Similar to the author's *Contemporary Criminal Procedure* (Gould Publications, 1993), this book is a guide to understanding the basic principles governing criminal evidence. Presents pertinent federal judicial decisions with the case name and citation, date decided, and a summary of the decision. Covers such aspects of evidence as forms and preservation, and discusses witnesses, proof, and privileges. Appendix contains federal rules of evidence.

National Research Council (U.S.). Committee on DNA Technology in Forensic Science
DNA Technology in Forensic Science
Washington, D.C., National Academy Press, 1992. 185p.
ISBN 0309045878

    Addresses quality and reliability in DNA typing, problems of standardization, DNA data banks and privacy, use of DNA typing in the courtroom, and the economic and ethical aspects of DNA typing. Offers recommendations for many of the emerging questions involving DNA. Includes glossary, index, listing of the participants in the study, and biographies of the committee members.

Zulawski, David E., and Douglas E. Wicklander
Practical Aspects of Interview and Interrogation
New York, Elsevier, 1992. 337p. ISBN 0444016333

    The authors, who are experts in this field, present the characteristics of a good interview and interrogation. They cover everything from preparation and strategy to obtaining the confession. Legal aspects and interpretation of verbal and nonverbal behavior are included. Photographs illustrate nonverbal behavior and example interrogation questions are provided. Part of the *Elsevier Series in Practical Aspects of Criminal and Forensic Investigations*. Index. (Avilable from CRC Press.)

✦ Education and Training

Alpert, Geoffrey P., and Lorie A. Fridell
Police Vehicles and Firearms: Instruments of Deadly Force
Prospect Heights, Ill., Waveland Press, 1992. 167p.
ISBN 0881336130

    Discusses deadly force, use of force and the law, research on deadly force and firearms, policy and training for firearms use, and research and policy on pursuit driving. Relevant court cases are cited and examples of departmental policies are given. References and index.

Bintliff, Russell L.
Training Manual for Law Enforcement Officers
Englewood Cliffs, N.J., Prentice Hall, 1990. 326p.
ISBN 0139268901

    This easy-to-understand manual has good, practical ideas illustrated with charts and examples. It covers everything from planning and organizing to conducting actual training. Discusses training evaluation, performance appraisals, making and using training plans and records, and writing in-service training manuals. Indexed.

Cane, Andries C.
Basic Arrest and Prisoner Control Tactics: Practical Techniques, Fast, Simple, Effective: The Cane Method
Springfield, Ill., Charles C. Thomas, 1989. 224p.
ISBN 0398056021

    The author has taught the techniques presented here to police officers for over twenty years. The four basic techniques are presented with photographs to help illustrate methods. Covered are offense, defense, body parts that can inflict pain, stances, and combinations of the four basic techniques.

Durkin, Kenneth Wayne
Internship Program Policies and Procedures
Dubuque, Iowa, Kendall/Hunt Pub. Co., 1989. 82p.
ISBN 0840351690

    Developed for the students completing the requirements for a degree in Law Enforcement Administration at Western Illinois University, this book contains information helpful to anyone considering an internship. It explains internships and the responsibilities of those involved. Contains numerous forms and samples, examples of cover letters and résumés, and tips on interviewing.

Flesch, R. A.
Basic Knife Defense for Criminal Justice
Longwood, Fla., Gould Publications, 1993. 44 leaves
ISBN 0875264220

    Illustrates with numerous photographs the basic techniques for surviving different types of knife attacks. Discusses the use of force and being prepared for an emergency or attack. Includes a glossary and references.

Garner, Gerald W.
High-Risk Patrol: Reducing the Danger to You
Springfield, Ill., Charles C. Thomas, 1990. 212p.
ISBN 0398056722

    High-risk patrol is described and preparations suggested for risk reduction, including mental attitude and physical skills and training. Potential situations are covered. Most chapters have a "Risk-Reduction Checklist." Indexed.

Harr, J. Scott, and Kären M. Hess
Seeking Employment in Law Enforcement, Private Security, and Related Fields
St. Paul, Minn., West Pub. Co., 1992. 244p. ISBN 0314934448

Covers law enforcement and private security jobs and discusses such related topics as physical fitness and testing, résumés, the application process, presenting yourself, and interviews. Appendices contain examples of résumés and cover letters, sources of job information, and Equal Employment Opportunity Guidelines. Indexed.

Knight, Les, and others
Self-Defense for Criminal Justice
2nd ed. Lakeland, Fla., ISC Division of Wellness, 1988. 1 v. (various pagings)

Presents a self-defense program planned to minimize injury to both the officer and the subject, which can be learned in a minimum amount of training time. Includes the techniques developed by the Florida Criminal Justice Standards and Training Commission's Defense Tactics Curriculum Revision Task Force. Sections included are fitness testing, stretching, and performance under stress. Well illustrated with over 300 photographs.

Marshall, Evan P., and Edwin J. Sanow
Handgun Stopping Power: The Definitive Study
Boulder, Colo., Paladin Press, 1992. 224p. ISBN 0873646533

Based on years of data collected on street shootings and the results of test firings, the authors produce an accurate prediction of the stopping power of specific handgun loads and some specialty rounds. Several theories are discussed concerning the lack of correlation between predictions of stopping power and results. Summaries of actual shootings are grouped by caliber, and followed by charts of brand and bullet type, shootings, and number and percentage of one-shot stops.

Reintzell, John F.
The Police Officer's Guide to Survival, Health, and Fitness
Springfield, Ill., Charles C. Thomas, 1990. 138p.
ISBN 0398057117

The author, a veteran of law enforcement and a certified instructor, has given lectures on aerobic fitness, stress, and nutrition to police officers. Practical information is presented to inform and as a guide to better health. Sections cover such topics as sleep, stress, body and health, nutrition, alcohol, alienation, and a holistic life-style. Includes chapter notes, selected references, and index.

Stern, Ron
Law Enforcement Employment Guide
2nd ed. Mt. Shasta, Calif., Lawman Press, 1990. 136p.
ISBN 0944711049

Arranged by state with federal agencies last, this guide covers only about eighty agencies in thirty-three states and the District of Columbia, but it provides relevant data on a variety of law enforcement agencies. For each agency it gives the name and address, entrance requirements, salary and benefits, retirement, application and testing process, career ladder, number of officers, and anticipated openings.

Warner, John W.
Federal Jobs in Law Enforcement
New York, Prentice Hall, 1992. 262p. ISBN 0136399150

Provides insight into law enforcement careers with more than one hundred federal agencies. For most departments it gives the mission, history, functions and activities, training, qualifications, and an address and phone number. Concludes with information about the federal law enforcement training facilities.

✦ Psychology and Ethics

Barker, Thomas, and David L. Carter
Police Deviance
3rd ed. Cincinnati, Ohio, Anderson Pub. Co., 1994. 444p.
ISBN 0870847147

Describes and analyzes police misconduct including off-duty behavior, vice and corruption, drug use, lies and perjury, sexual misconduct, deadly force, prejudice, and abuse of authority. Ways of managing police deviance, disciplinary procedures, investigations of complaints, and civil and criminal liabilities of police officers are discussed. Notes or references follow each chapter. Some chapters are reprints.

Blau, Theodore H.
Psychological Services for Law Enforcement
New York, Wiley, 1994. 454p. ISBN 0471559504

Divided into four sections which cover the history of policing; psychological assessment methods and procedures; intervention techniques and applications (including critical incident counseling, stress reduction, family problems, drug and alcohol abuse); and operational assistance in hostage negotiation and psychological profiling. Numerous examples are given. Eleven appendices provide practical information, samples, and programs.

Cohen, Howard, and Michael Feldberg
Power and Restraint: The Moral Dimension of Police Work
New York, Praeger, 1991. 166p. ISBN 0275938565

This book is based on a series of summer workshops in humanities for police personnel first presented in the 1980s. It is divided into two sections. Part One develops a system of ethical standards by which to measure responsible police behavior: fair access to services, power not abuse, safety and security of citizens, cooperation with other officials, and objectivity. Part Two presents cases illustrating how police officers can develop appropriate moral responses to difficult situations. Includes bibliography and index.

Conroy, Dennis L., and Kären M. Hess
Officers at Risk: How to Identify and Cope with Stress
Placerville, Calif., Custom Pub. Co., 1992. 271p.
ISBN 0942728483

Gives an overview of officers and stress, the effects of stress on the individual, and the implications of stress for the officer as well as for family and supervisors. Identifies factors causing stress and discusses controlling negative stress, preventing burnout, and establishing programs to reduce burnout and stress. Includes glossary and index.

Gellman, Stuart
Cops: The Men and Women Behind the Badge
Rev. ed. Tucson, Ariz., Horizon Press, 1993. 288p.
ISBN 0962762849

    Documents the personal and professional experiences of nine police recruits through police training academy and five years as police officers. By "telling it like it is" and presenting the problems and successes, as well as the views and feelings of these officers, this book offers valuable insight for the prospective police officer of actual life at the academy or on the street.

Hennessy, Stephen M.
Thinking Cop, Feeling Cop: A Study in Police Personalities
Scottsdale, Ariz., Leadership Incorporated of Scottsdale, 1992. 118p. ISBN 0963077201

    Discusses using the Myers-Briggs Type Indicator (MBTI) with police personnel as a team building tool, to help explain behaviors, to aid in understanding why some job situations are uncomfortable, and to understand personal strengths and weaknesses. The MBTI also offers insights as to why police personnel look at crime differently than do other professionals. Numerous comments from interviews with thirty-three police executives.

Jones, Clarence E.
After the Smoke Clears: Surviving the Police Shooting; An Analysis of the Post Officer-Involved Shooting Trauma
Springfield, Ill., Charles C. Thomas, 1989. 104p.
ISBN 0398055270

    Overview of Post Traumatic Stress Disorder (PTSD) or, as the author calls it, Post Officer-Involved Shooting Trauma (POST). Gives the immediate and long-range effects on the officer and his family and what departments are doing and should do about POST. Presents model policy and training suggestions. Includes brief bibliography and glossary.

Russell, Harold E., and Allan Beigel
Understanding Human Behavior for Effective Police Work
3rd ed. New York, Basic Books, 1990. 448p. ISBN 0465088597

    Coverage includes normal and abnormal behavior, mental illness, assessing and managing abnormal behavior in the field, behavioral aspects of crisis situations (disasters, riots, hostage situations), stresses in police work, and the mental health professional and police work. Refers to DSM-III R (Diagnostic and Statistical Manual) of the American Psychiatric Association. Chapter bibliographies and index.

Thompson, George J., and Jerry B. Jenkins
Verbal Judo: The Gentle Art of Persuasion
New York, Morrow, 1993. 222p. ISBN 0688122639

    Written by a former English professor with experience in law enforcement and a tenth-degree black belt in karate, this book explains coping techniques for confrontations. It shows how to resolve volatile situations without resorting to physical methods. Very readable, it uses numerous examples to illustrate the communication skills it teaches.

✦ Report Writing

Holtz, Larry E.
Investigative and Operational Report Writing
2nd ed. Binghampton, N.Y., Gould Publications, 1990. 36p.
ISBN 0875263720

    This compact book explains in ten steps ways to improve and professionalize official reports. It emphasizes documenting all actions and observations, maintaining perspective, outlining for coherence, being accurate and clear, and previewing and reviewing.

Miller, S. Dennis
How to Write a Police Report
Albany, N.Y., Delmar Publishers, 1993. 148p. ISBN 0827347286

    Well organized with ample illustrations, this work covers the basics of writing reports, stressing their importance and suggesting ways to write and organize them. Demonstrates how to write clearly and effectively. Includes chapters on diagrams, photo display identification, police field notebooks, and testimony in court. Appendices and index included.

Rutledge, Devallis
The New Police Report Manual
Placerville, Calif., Copperhouse Pub. Co., 1993. 172p.
ISBN 0942728122

    This informative, easy-to-read manual presents ways to simplify police writing while making it more effective, briefer, less time-consuming, and easier to support in court.

✦ Security

Bintliff, Russell L.
The Complete Manual of Corporate and Industrial Security
Englewood Cliffs, N.J., Prentice Hall, 1992. 612p.
ISBN 0131596411

    Covers results-oriented security management, including budgets, the foundations of security, technology and intrusion detection, protection for specialized areas such as headquarters, and security officer career development. Numerous types of security measures are discussed with the advantages and disadvantages of each given. Many useful illustrations and checklists are included. Appendices include checklists for security inspections and information on managing bomb threats. Indexed.

Fennelly, Lawrence J., ed.
Handbook of Loss Prevention and Crime Prevention
2nd ed. Boston, Butterworths, 1989. 721p. ISBN 040990144X

    Written by over forty crime prevention professionals, this handbook is divided into four main parts. "Methods" covers environmental design, security surveys, and emergency planning. "Operations and equipment" covers locks, lighting, alarms, closed-circuit television, and physical barriers. "Applications" discusses the security needs of specific situations, such as retail trade, banks, and computers. "Management" covers such broad concepts as planning and evaluation, budgeting, and crime analysis. Many chapters have checklists or forms and references. Indexed.

✦ Series

ACJS/Anderson Monograph Series
Cincinnati, Ohio, Anderson Pub. Co.

This series, published jointly by the Academy of Criminal Justice Sciences and Anderson Publishing Company, includes such titles as *The Death Penalty in America, What Works in Policing?*, and *Violent Crime and Gun Control*. Most volumes are a collection of articles presenting a variety of research, analysis, and thoughts on different aspects of a specific topic.

Annual Editions: Criminal Justice
Guilford, Conn., Duskin Pub. Group

Annual Editions arranges articles from numerous sources by broad topics. Includes a topic guide, statistics, glossary, and index. Good source for current articles on a topic.

Information Plus: The Information Series on Current Topics
Wylie, Tex., Information Plus

Contains over twenty titles that are updated every two years and cover a variety of aspects of various topics in a very readable format. Includes background, attitudes, statistics, pros and cons of the issue, and causes. Numerous charts and tables. Titles include *Capital Punishment, Crime, Domestic Violence, Gambling*, and *Gun Control*.

Opposing Viewpoints Series
San Diego, Calif., Greenhaven Press

Titles in this series concerning criminal justice include: *America's Prisons, Crime and Criminals, Criminal Justice, The Death Penalty*, and *Violence in America*. Each title contains articles from a variety of sources, including position papers from individuals and organizations. Approximately five issues are presented in each volume.

Social Issues Resources Series
Boca Raton, Fla., The Series

Social Issues Resources series (SIRS) consists of loose-leaf notebooks covering thirty-four broad subject areas. Each year, twenty articles that represent a variety of opinions, are added to each subject. Two titles are *Crime* and *Corrections*.

Taking Sides: Clashing Views on Controversial Issues in Crime and Criminology
Guilford, Conn., Dushkin Pub. Group

Presented in a debate style, several controversial issues are discussed with contrasting points of view. Each issue begins with an introduction and ends with a postscript that reviews the arguments and suggests additional readings. Indexed. Representative titles are *Drugs and Society* and *Legal Issues*.

Nonprint Resources

National Criminal Justice Reference Service (NCJRS)

The National Criminal Justice Reference Service is part of the National Institute of Justice (NIJ), the research branch of the U.S. Department of Justice. The NCJRS disseminates information from research, evaluations, and special programs. It maintains a collection of over 120,000 documents. The numerous services of NCJRS include:

- Interlibrary Loan: Most items in its collection are available through interlibrary loan for a small fee.
- Topical Searches and Topical Bibliographies: These are pre-packaged listings of abstracts on a single topic from the NCJRS database of documents in their collection.
- Custom Searches: The NCJRS staff will perform a custom search of their collection for a topic not already available.
- *National Institute of Justice Journal*: A bimonthly journal that includes abstracts of significant additions to the NCJRS collection. The journal is free.
- *NIJ Publications Catalog*: A free catalog of items published by the National Institute of Justice and available through NCJRS. Most items are free.
- Free distribution of many of the publications of the other areas of NIJ, including the Justice Statistics Clearinghouse and the Bureau of Justice Assistance Clearinghouse. Contact NCJRS or the clearinghouse for free catalogs of their publications.
- NCJRS database is available for searching on DIALOG file 21.

For additional information, contact NCJRS (1-800-851-3420) or write, NCJRS, Box 6000, Rockville, Md. 20850.

State Justice Sourcebook of Statistics and Research: A Cooperative Project of the State Statistical Analysis Centers, the Bureau of Justice Statistics, and the Justice Research and Statistics Association
Washington, D.C. U.S. Department of Justice. Bureau of Justice Statistics, 1992. 4 microfiche.

Consolidates information on crime, criminal justice systems, and special programs and research of each state, the District of Columbia, and Puerto Rico. Arranged by state, it profiles the state criminal justice systems (law enforcement, courts, corrections, statutory provisions) and the State Statistical Analysis Centers.

Technology Assessment Program Information Center

Part of the National Institute of Justice (NIJ), and publisher of materials based on research and product evaluations. Some of their current titles include: *Police Body Armor: Consumer Product List, Oreosin Capsicum: Pepper Spray As a Force Alternative*, and *Tire Testing for Police Vehicles*. For a free catalog of their publications write: Technology Assessment Program Information Center, Box 6000, Rockville, Md. 20850.

✦ Internet

This global computer network is rich in federal and private information, including data on law enforcement. It is impossible to enumerate all of the resources that are available through Internet, but the following suggestions may aid in research.

Bureau of Justice Statistics documents:
- Available full text through gopher uacs2.albany.edu/united nations justice network.

U.N. reports on crime and criminal justice in 123 countries:
- Available through this same gopher under U.N. criminal justice country profiles.

The Library of Congress gopher MARVEL.LOC.GOV:

A good source of information on numerous areas of law enforcement. Federal information is arranged by branch of government. It includes information on Bureau of Justice statistical documents and the National Institute of Justice, which has its own gopher.

Regular full text articles from periodicals can be searched on-line for law enforcement information. One source is the electronic newsstand on GOPHER.INTERNET.COM. Through the device GOPHER.LIB.UMICH.EDU both full text press releases and the U.N. Justice Information Network can be searched by subject.

✦ Satellite Broadcast Resources

Two free sources of video tapes covering all aspects of law enforcement are Peace Officer Standards and Training (POST), and the Law Enforcement Satellite Training Network (LESTN). Both broadcast regularly by satellite, are intended for all levels of law enforcement personnel, may be recorded for future training use by the downlink sites, and are available to all law enforcement agencies.

California's Commission on Peace Officer Standards and Training uses satellite broadcast for monthly telecourse training presentations and interactive teleconferences. These are two-hour presentations. POST also has broadcasts of training films produced by various agencies (local, state, etc.) and selected by a POST review board to ensure quality. Tapes range in time from three to thirty minutes and usually cover practical tips on various topics of interest.

To get more information on POST, a schedule of broadcasts, or to locate the nearest downlink, contact POST, 1601 Alhambra Blvd., Sacramento, Calif. 95816.

The Law Enforcement Satellite Training Network is a multi-agency broadcast training network sponsored by the FBI. For additional information on LESTN or to receive their program announcements, contact the Teleconference Program Manager, FBI Academy, Quantico, Va. 22135.

Legal Secretarial

Susan L. Brant
Nicolet Area Technical College
Rhinelander, Wisconsin

Introduction

The Legal Secretary program provides students the opportunity to become acquainted with legal procedures and court structure, and to become familiar with legal terminology and all aspects of municipal, district, state, and federal court proceedings. Students master skills and acquire knowledge for employment in a law office, a government office, or a legal department in industry. Students need to be able to communicate well in English, both written and oral. Proficiency in spelling, recordkeeping, and time management, as well as other business office skills, are crucial. Other necessary skills include transcription of court documents, correspondence, word processing, human relations, and interpersonal communications. A strong code of personal and professional ethics is essential.

Community and junior colleges, as well as a number of business schools, offer a variety of programs which support the legal profession. There are programs for training legal secretaries, legal assistants, paralegals, and legal receptionists. This bibliography concentrates on legal secretarial and assistant materials and covers a few basic reference materials used by all secretaries. Graduates can become certified after completing specific courses and passing the Certified Professional Legal Secretary examination. For other materials on general office technologies, please refer to the Office Technologies bibliography in the Business section of this book.

Certification and Licensure

National Association of Legal Assistants (NALA)
2250 E. 73rd St., Suite 550
Tulsa, Okla. 74136
    The CLA Certification program is administered by NALA. A two-day comprehensive examination is given which bestows a measure of professional recognition for those who achieve significant competence in the field.

National Association of Legal Secretaries (NALS)
1516 S. Boston, Suite 200
Tulsa, Okla. 74119
    The Certified Professional Legal Secretary (PLS) and the Accredited Legal Secretary (ALS) are the only national certification programs for legal secretaries. They set recognized standards of measurement for knowledge and skills. Three years of experience in the field are required before secretaries can sit for the PLS examination. The (ALS) examination is designed for those just entering the field as legal support staff. Three areas of legal secretarial practice and procedures are tested during the six-hour exam. To take the ALS exam and accredited secretarial course of study, students must take the NALS Legal Training Course, or possess at least one year of experience as a secretary.

Selected Associations

National Association of Legal Assistants (NALA)
1516 S. Boston, Suite 200
Tulsa, Okla. 74119

National Association of Legal Secretaries (NALS)
2250 E. 73rd St., Suite 550
Tulsa, Okla. 74136

Selected Journals

The Docket.
National Association of Legal Secretaries. Bimonthly.
ISSN 0895-1659

Law Technology Product News.
New York Law Pub. Co. Monthly. ISSN 1071-9121

The Legal Assistant Today.
James Publishing. Bimonthly. ISSN 1051-3663

The Secretary.
Professional Secretaries International. Nine times a year.
ISSN 0037-0622

Print Materials

Bieber's Dictionary of Legal Abbreviations: Reference Guide for Attorneys, Legal Secretaries, Paralegals, and Law Students
Prince's 4th ed. by Mary Miles Prince. Buffalo, N.Y., William S. Hein, 1993. 791p. ISBN 0899418473
   Defines more than 4,000 abbreviations used in legal encyclopedias, dictionaries, reporters, loose-leaf services, treatises, reference books, citators, and other popular material found in a medium-sized law library. It is not, however, a guide for legal citation forms. In determining proper citation form, one should always consult the *Bluebook* or *Bieber's Dictionary of Legal Citations*.

Bieber's Dictionary of Legal Citations: Reference Guide for Attorneys, Legal Secretaries, Paralegals, and Law Students
Prince's 4th ed. Buffalo, N.Y., William S. Hein, 1992. 372p.
ISBN 0899418244
   Prince designed her dictionary as a companion to the fifteenth edition of *The Bluebook* (Harvard Law Review Association, 1991-). Paralegals will find helpful models here that illustrate the complex *Bluebook* rules.

Black, Henry Campbell, and others
Black's Law Dictionary: Definitions of the Terms and Phrases of American and English Jurisprudence, Ancient and Modern
6th ed. St. Paul, Minn., West Pub. Co., 1990. 1657p.
ISBN 0314771654
   Due to the growth of laws and regulations, every legal assistant and secretarial program should have this easy-to-use legal dictionary, which is organized alphabetically for easy access to current legal terms.

The Bluebook: A Uniform System of Citation
15th ed. Cambridge, Mass., Harvard Law Review Association, 1991. 343p. ISSN 1062-9971
   Compiled by the editors of the *Columbia Law Review*, the *Harvard Law Review*, the *University of Pennsylvania Law Review*, and the *Yale Law Journal*. *The Bluebook* is the authority for citing practically any type of court case, statutes, regulations, administrative and executive materials, and secondary materials such as books or periodicals in court documents and legal memoranda. It sets forth the basic scheme and rules for legal citations with examples for each rule. A new section for practitioners details typeface conventions, citation sentences and clauses, parallel citations, short forms, special citation forms, capitalization, and abbreviations. Quick reference guides to common citation forms and footnotes are printed inside the flyleaves.

Certified PLS Resource Manual: Study Guide, Mock Examination, and Recommendations for the Certified Professional Legal Secretary Examination
Rev. ed. Tulsa, Okla., National Association of Legal Secretaries, 1989. 1 v. (loose-leaf)
   A study manual prepared to assist legal secretaries in the preparation for the certification exam for Professional Legal Secretaries (PLS). Provides information about the PLS program, a study guide and a mock examination. This certification is a recognized standard for assessing the qualifications of a legal secretary. Also includes discussions on ethics, legal accounting and terminology, procedures and skills, as well as reference materials.

Crocker, Dee
Litigation Handbook for the Lawyer's Assistant
St. Paul, Minn., West Pub. Co., 1992. 304p. ISBN 0314009396
   This manual presents guidelines for the legal assistant to assist the lawyer in litigating cases or resolving legal disputes between parties in a court of law. The background work necessary to prepare for litigation is covered in the fifteen chapters of this handbook. Checklists are provided, basic legal research sources are discussed, and legal forms are presented (including pleadings, summons, removal to U.S. District Court, pretrial motions, settlements, discovery, subpoenas, trials, post-trial motions and appeals). Includes various appendices on rules, use of citations, and timetables. A table of statutes and rules and a useful index complete this handbook.

Eason, Susan, and others
Essentials of Accounting
7th ed. Cincinnati Ohio, South-Western Pub. Co., 1991. 335p.
ISBN 0538806842
   This text presents a system of accounting that can be used in any business office, including law offices. Special attention is paid to accounting problems and situations encountered by attorneys and other professionals. Presents fundamentals such as the accounting cycle, income statement, and balance sheet. Includes appendices, exercises, and index.

Finn, Molly M.
Crucial Skills for Legal Secretaries—and Receptionists
Boston, Massachusetts Continuing Legal Education, 1993. 72p.
   A workbook containing tips, exercises, articles, and references published as an aid in maintaining professional competence. Topics include dealing with angry or volatile clients, bosses, or coworkers; managing an office crisis; and self-improvement of skills. Includes a bibliography and reference materials.

Koerselman, Virginia
CLA Review Manual: A Practical Guide to CLA Exam Preparation
Minneapolis, Minn., West Pub. Co., 1993. 937p.
ISBN 0314013490

A comprehensive manual to prepare the legal assistant for the CLA certification examination. Topics covered include legal terms, communication skills, ethical issues, judgment and analytical ability, legal research, human relations and interviewing, all types of law ranging from criminal to bankruptcy, contract and probate, litigation and real estate. Includes self-tests at the end of each section, answer keys, a comprehensive bibliography, and an index.

Morton, Joyce
Legal Secretarial Procedures
3rd ed. Englewood Cliffs, N.J., Prentice Hall, 1993. 385p.
ISBN 0135297362

Morton captures the essentials of law office management including duties, office systems and technologies, a brief history of law, and how courts are structured. Provides information on legal research and lawsuits, family law, wills and probate, corporation law, real estate and criminal law, and job seeking. Includes a bibliography of materials, forms, and practice sets on various legal matters, as well as an index.

National Association of Legal Assistants
NALA Manual for Legal Assistants
2nd ed. St. Paul, Minn., West Pub. Co., 1992. 397p.
ISBN 0314807802

This revised manual is intended as a quick reference guide for the legal assistant and contains a collection of techniques and procedures. There are many successful examples of solutions to actual assignments accomplished by working legal assistants. Sections cover communications, information on the American legal system, ethics, legal research, interviewing and investigations, litigation skills, and document discovery cases. Includes a bibliography, a glossary, and several helpful appendices.

National Association of Legal Secretaries
The Career Legal Secretary
Rev. ed. St. Paul, Minn., West Pub. Co., 1993 reprint, 1987. 770p.
ISBN 031432237X

A training aid for legal secretaries or for those just entering the field. (Not intended to be used as a legal authority.) Coverage includes chapters on accounting, ethics, the structure of a law office, preparing documents, information on the courts and administrative agencies, legal issues including litigation, torts, real estate, family, criminal, bankruptcy estate planning, and contract law. Includes a brief bibliography of sources, a useful glossary of common legal terms, and an index.

National Association of Legal Secretaries
Manual for the Lawyer's Assistant
2nd ed. St. Paul, Minn., West Pub. Co., 1988. 1118p.
ISBN 0314411623

This broad-based reference aid for legal support staff can also be used in advanced courses beyond the postsecondary level. Includes duties and responsibilities relating to work in a law office, research and legal writing, and information relating to agencies, courts, and legal procedures. Addresses specialized litigation, torts, civil and criminal procedure, all areas of general legal practice, and examples of more specialized practice areas (such as environmental, water and oil and gas law, patents, immigration, and labor law). Provides a bibliography, samples of legal forms, a glossary, and an index. Teacher's manual also available.

Nevine, Kathleen B.
Probate Handbook for the Lawyer's Assistant
St. Paul, Minn., West Pub. Co., 1993. 403p. ISBN 0314023518

This overview of the law of probate or estate administration can assist the lawyer's assistant but is not intended to be used as a legal text on the subject. Discussion centers around the *Uniform Probate Code*, wills, intestacy, notices, decedent's benefits, valuation of assets and their safekeeping, debts and liabilities of decedent, expenses and liquidation and distribution of assets, aspects of record keeping, and tax returns. Includes a short bibliography of other resources, a table of statutes, and an index.

Renstrom, Peter G.
The American Law Dictionary
Santa Barbara, Calif., ABC-Clio, 1991. 308p. ISBN 0874362261

Renstrom focuses on terms, phrases and concepts common to the American judicial system. Entries are arranged alphabetically, and each is assigned its own number within subject-specific chapters. Cross-references and an index are also provided.

Rotunda, Ronald D.
Professional Responsibility
3rd ed. St. Paul, Minn., West Pub. Co., 1992. 492p.
ISBN 031492146X

This text should be used by legal secretaries as a reference to define and understand the issues of legal professional responsibility. Addresses ethical issues such as rules of professional conduct, overviews of the lawyers obligation to the client, obligations as a member of the firm, obligations regarding advertising and solicitation, not misusing the office of government, and pro bono activities. Includes many appendices, a glossary, a table of cases, and an index.

Sabin, William A.
The Gregg Reference Manual
7th ed. Lake Forest, Ill., Glencoe, 1992. 502p. ISBN 0028199200

A basic secretarial reference tool for anyone who writes, transcribes or types. Presents the basic rules that apply to writing, and definitive answers to problems encountered when writing. Includes a glossary of word processing terms. Each section of the volume deals with a specific issue. A useful index is also provided.

Salser, Carl Walter, C. Theo. Yerian, and Mark R. Salser
Handbook for Beginning Legal Assistants and Receptionists
Portland, Ore., National Book Company, 1989. 314p.
ISBN 0894202596

A general introduction to the legal system for legal secretaries and legal assistants. The seven chapters cover frequently used vocabulary, management of a law office, various courtroom procedures, examples of legal opinions and testimony, a selection of business law concepts, and concluding with a chapter on producing legal documents, along with samples of completed documents.

Wasserman, Paul, Brian L. Baker, and Patrick J. Petit
Encyclopedia of Legal Information Sources: A Bibliographic Guide to Approximately 29,000 Citations for Publications, Organizations, and Other Sources of Information on 480 Law-Related Subjects
2nd ed. Detroit, Mich., Gale Research, 1993. 1083p.
ISBN 0810374390

    This easily accessible reference tool provides access to books, periodicals, databases and organizations that can aid in answering legal questions. It is arranged by broad subject headings, covering over 450 topics chosen for their timeliness and importance to the field of law. Each section includes statutes, handbooks, textbooks, loose-leaf services, digests, law reviews, and associations pertinent to the topic covered. The third edition is scheduled for publication in 1995.

Nonprint Materials

Insight Media
121 W. 85th St., New York, N.Y. 10024
    This distributor is a good source for videocassettes on the legal office, computers in the office, and specific applications (such as *DOS, Lotus, WordPerfect*), business communications, interviewing, and job success.

Wisconsin Bar Association
402 W. Wilson St., Madison, Wisc. 53707
    The state bar issues videocassettes on topics such as law office confidentiality, legal ethics, and professionalism in the office, including telephone strategies, time management, and client service.

Paralegal

Susan Stussy
Texas Graduate School of International Management
Corpus Christi, Texas
(formerly Barton County Community College,
Great Bend, Kansas)

Introduction

The paralegal provides highly skilled support to attorneys in all aspects of legal practice. Currently, paralegals hold approximately 95,000 jobs in private law firms, various levels of federal, state, and local governments, legal service projects, and industry. According to the *Occupational Outlook Handbook* (1994-1995), paralegal employment is expected to increase faster than the average for all occupations through 2005.

At present there are several routes to paralegal practice. Some legal secretaries move up from the clerical level and become paralegals. Others enter from formal paralegal training programs in two-year community college programs, four-year programs in colleges and universities, and other programs in law schools, business schools, and private, independent schools.

As the responsibilities of paralegals increase, there is growing concern about establishing a clearer definition of the responsibilities of a paralegal and refining of certification standards. Always a consideration is the danger of unauthorized practice of the law. Highly skilled paralegals will inevitably be tempted to cross the poorly defined boundary between paralegal and attorney, and this could subject them to serious legal problems.

Acknowledgment is extended to Carol Crocker of the Washburn University law library staff, who assisted the author by obtaining numerous publications through interlibrary loan.

Accreditation and Certification

Many programs do not require or seek American Bar Association approval; however, 177 schools held ABA approval in 1993.

American Bar Association
750 North Shore Dr.
Chicago, Ill. 60611

Paralegals need not be certified, but voluntary certification is offered through the National Association of Legal Assistants. Paralegals who meet basic requirements of education and experience are eligible to take a two-day examination. An advanced paralegal competency exam is also offered by the National Federation of Paralegal Associations.

National Association of Legal Assistants, Inc.
1516 S. Boston, Suite 200
Tulsa, Okla. 74119

Selected Associations

American Association for Paralegal Education
P.O. Box 40244
Overland Park, Kans. 66204

Legal Assistant Management Association
638 Prospect Ave.
Hartford, Conn. 06105-4298

National Association of Legal Assistants, Inc.
1516 S Boston, Suite 200
Tulsa, Okla. 74119

National Federation of Paralegal Associations
P.O. Box 33108
Kansas, City, Mo. 64114-0108

Selected Journals

The Docket.
National Association of Legal Secretaries. Bimonthly.
ISSN 0895-1659

Law Technology Product News.
New York Law Pub. Co. Monthly. ISSN 1071-9121

The Legal Assistant Today.
James Publishing. Bimonthly. ISSN 1051-3663

Print Materials

✦ Education and Career Guidance

American Bar Association. Standing Committee on Legal Assistants
Guidelines and Procedures for Obtaining ABA Approval of Legal Assistant Education Programs: As Amended 1992
Chicago, American Bar Association, Standing Committee on Legal Assistants, 1992. 40p. ISBN 0897077711

This slim pamphlet should be in every junior college library supporting paralegal studies, whether or not the institution has plans to seek ABA accreditation. It offers an authoritative source for institutional curriculum planning and is comprehensive in scope. Includes a section on library resources.

Bernardo, Barbara
Paralegal: An Insider's Guide to the Fastest Growing Occupation of the 1990s
2nd ed. Princeton, N.J., Peterson's Guides, 1993. 214p.
ISBN 1560792426

Bernardo provides a good overview of the paralegal profession's origins, development, and projected future from the point of view of an enthusiastic practitioner. She discusses the varied ways of entering the profession and the diverse employment opportunities. She strengthens her book with an excellent bibliography and index.

Brooks, Teddar S., and Wesley P. Hackett
Hiring Legal Staff: Determining Cost and Value
Chicago, American Bar Association, 1990. 45p. and 2 computer disks. ISBN 1560792426

The authors, who are both attorneys and solo practitioners, have written a brief but important work that should be useful to attorneys and paralegals. This work illustrates the complete financial impact of employment relationships. A computer spreadsheet template to calculate the approximate cost of hiring full- and part-time employees is provided in addition to sample job descriptions and job applications.

Brunner, Thomas W., Julie P. Hamre, and Joan McCaffrey Wegrzyn
The Legal Assistant's Handbook
2nd ed. Washington, D.C., Bureau of National Affairs, 1988.
232p. ISBN 0871795906

Provides an overview of the work of legal assistants. This work is notably strong on cite checking, ethics, and general research methods. In addition to the text, there are eight well-planned appendices illustrating the types of work done by paralegals. Samples of key types of legal documentation are also provided. Includes glossary and index.

Estrin, Chere B.
Paralegal Career Guide
New York, Wiley Law Publications, 1992. 298p.
ISBN 047158794X

This work provides an excellent guide for paralegal career planning. In a rapidly changing field, it is notable that supplements are provided to update this monograph. Estrin's target audience is the career-oriented paralegal who wishes to maximize the opportunities for professional growth in an occupation still subsidiary to the organized bar. She describes how to negotiate an appropriate salary and the mechanics of changing positions. In addition, her coverage is broadened by treatment of free-lance and temporary paralegals, as well as paralegal entrepreneurs. Includes appendices, bibliography, and index.

Garrett, Vena
Introduction to Legal Assisting
New York, Glencoe, 365p. 1993 ISBN 0028002776

Garrett provides a good introduction to paralegal studies, and summarizes the role of legal assistants in the American legal system and how they interact with law firm clients. She provides an excellent overview of different specializations and a guide to career development. Law office procedures are also addressed. Includes good illustrations, an excellent glossary, and an index.

Goodrich, David Lee
The Basics of Paralegal Studies
Englewood Cliffs, N.J., Prentice Hall, 1991. 360p.
ISBN 0136504825

This thoughtful overview of paralegal work defines the field and the different types of employment opportunities available. It provides an introduction to the legal system, legal terminology, and legal analysis and writing skills. Criminal law and civil discovery and specialties such as decedents' estates, domestic relations, and real estate law receive fuller, chapter-length treatment.

Gowen, Michele C., ed.
A Guide for Legal Assistants: Roles, Responsibilities, Specializations
2nd ed. New York, Practising Law Institute, 1991. 610p.
ISBN 087224024X

The collected essays in this volume provide a good survey of paralegal work in general, and of recognized paralegal specialties. Each chapter treats a different topic and is followed by appendices directly related to the topic covered. Indexed.

Hermann, R. L., L. P. Sutherland, and J. Sobajian
The Paralegal's Guide to U.S. Government Jobs: How to Land a Job in Seventy Law-Related Careers
6th ed. Washington, D.C., Federal Reports, 1993. 76p.
ISBN 0929728203

This is an invaluable guide for paralegals desiring the security of federal employment. It explains the intricacies of the employment application process and the mysteries of the SF-l71 federal job application form. In addition, other law-related civil service titles appropriate to paralegal applicants are described. Includes a list of agencies that employ paralegals or other legal paraprofessionals.

Kisiel, Marie
How to Find a Job As a Paralegal: A Step-by-Step Job Search Guide
2nd ed. St. Paul, Minn., West Pub. Co., 1992. 137p.
ISBN 0314920536

This slim volume is helpful to the beginner, yet it also addresses paralegals who have been working in the field and are considering free-lancing or moving on to other careers. The appendices are practical, and the index is well done.

Koerselman, Virginia
CLA Review Manual: A Practical Guide to CLA Exam Preparation
Minneapolis, Minn., West Pub. Co., 1993. 937p.
ISBN 0314013490

The first chapter of this work describes the Certified Legal Assistant (CLA) exam and the justification behind it. This manual provides a comprehensive overview of all areas of the law, with experienced paralegals as the target audience. Each chapter contains a self-test and an answer key. Includes a helpful bibliography and an index.

Legal Assistant I
Syosset, N.Y., National Learning Corp., 1991. 1 v. (various pagings). ISBN 0837329884

For the civil service job candidate, this book provides a practical guide to test preparation and test-taking techniques. Includes sample multiple-choice questions and keys to the correct answers, a practice answer sheet, and a glossary of legal terms designed to meet practical needs. This book should be used in conjunction with related titles by the same publisher. *Passbooks for Career Opportunities* series.

Legal Assistant Trainee
Syosset, N.Y., National Learning Corp., 1989. 1 v. (various pagings). ISBN 0837329795

This is a practical guide for the civil service job candidate seeking employment as a legal assistant. Includes advice on test-taking techniques, sample multiple-choice questions with keys to the correct answers, plus a practice answer sheet. The questions are in all major areas of law, and they probe the candidate's ability to read carefully and react to legal issues. Readers should also see related titles by the same publisher. *Passbooks for Career Opportunities* series.

Legal Assistants: Are You Ready for the 90s
Raleigh, N.C., North Carolina Academy of Trial Lawyers, 1991. 1 v. (various pagings).

Though this book contains a collection of presentations offered to North Carolina paralegals by the trial lawyers of that state, the information is applicable to all legal assistants. Two presentations may be of particular interest: Russ McCarter's "Legal Assistants: Managing Stress" and Harvey L. Kennedy's "Civil Law: Sexual Harassment."

Legal Clerk
Syosset, N.Y., National Learning Corp., 1988. 1 v. (various pagings). ISBN 0837333946

This text is a focused guide for the civil service test taker who desires employment as a legal clerk. In addition to commentary, there are ten sample tests with keys to the correct answers and a practice answer sheet. *Passbooks for Career Opportunities* series.

Magryta, Leslie
Introduction to Paralegal Studies: A Skills Approach
Homewood, Ill., Irwin Mirror Press, 1993. 354p.
ISBN 025612390X

Magryta, an attorney, wrote this text for the beginning paralegal student. She uses a step-by-step skills approach to introduce basic paralegal tasks. The text contains numerous examples and documents designed to illustrate legal procedures. Each chapter contains a list of objectives and concludes with a discussion and review questions. Glossary and index are provided.

Nemeth, Charles P.
The Paralegal Resource Manual
Cincinnati, Ohio, Anderson Pub. Co., 1989. 476p.
ISBN 087084606X

Nemeth provides a broad overview of the paralegal profession and an introduction to major types of law for paralegals. A unique facet of Nemeth's work is chapter-length coverage of the *Uniform Commercial Code* (West Pub. Co., 1972-) for the paralegal audience. In addition, there is a strong concentration on proper pleading formats. In comparison to other similar works, Nemeth's monograph is notable for the detailed bibliographic footnotes in each chapter and meaningful charts, forms, and illustrations. Each chapter ends with problems for consideration, with answers to these problems at the end of the text.

Paralegal Aide
Syosset, N.Y., National Learning Corp., 1992. 1 v. (various pagings). ISBN 0837322456

This work is one in a series oriented toward explaining test taking and interview strategies for the potential civil service employee. It includes sample tests, an answer sheet, and a glossary of legal terms. Like other books in this series, it is a useful tool to help students assess their skills and prepare themselves for the job market. *Passbooks for Career Opportunities* series.

Pyle, Ransford Comstock
Foundations of Law for Paralegals: Cases, Commentary, and Ethics
Albany, N.Y., Lawyers Cooperative Publishing and Delmar Publishers, 1992. 444p. ISBN 0827345720

Pyle writes an excellent introduction to law directed at a paralegal audience. Topics covered include the role of paralegals, ethics, sources of the law, court structure, criminal law, civil law, administrative law and procedure, as well as law and computers. The appendices cover helpful topics such as "How to Read a Case." Bibliography and index included.

Statsky, William P.
Essentials of Paralegalism
2nd ed. St. Paul, Minn., West Pub. Co., 1993. 543p.
ISBN 0314010831

Statsky, a respected authority in the field, provides an excellent overview of paralegal work. Libraries may prefer this work to his longer *Introduction to Paralegalism* (West Pub. Co., 1992). The book contains an excellent glossary and index and relevant appendices.

Statsky, William P.
Introduction to Paralegalism: Perspectives, Problems, and Skills
4th ed. St. Paul, Minn., West Pub. Co., 1992. 877p.
ISBN 0314925740

Provides a comprehensive overview of the knowledge, practice, and skills characteristic of paralegals. Statsky's writing is easy for beginners in the field to follow, with each chapter including an outline, chapter summary, and list of key terms. Includes bibliography, glossary, and index.

Statsky, William P.
Paralegal Employment: Facts and Strategies for the 1990s
St. Paul, Minn., West Pub. Co., 1988. 188p. ISBN 0314593543

Statsky provides a practical sourcebook for the would-be paralegal and the paralegal seeking career growth in a second job. He describes the different categories of paralegal employment and paralegal specialties and includes guidance for organizing a job hunt. Four useful appendices cover federal job information centers, proposed legislation to increase paralegal employment in government, paralegal associations, and how to start a free-lance paralegal business. Includes index.

Vernon, Julia, and Francis Reagan, eds.
Improving Access to Justice: The Future of Paralegal Professionals: Proceedings of a Conference Held 19-20 February 1990
Canberra, Australia, Australian Institute of Criminology, 1991. 163p. ISBN 0642157146

Although this work is clearly an optional purchase that can be recommended only to well-funded libraries, it illustrates how Australia is attempting to deal with questions of legal access and the possible role that paralegals might play in increasing this access. This work is helpful for comparison studies.

Wagner, Andrea
How to Land Your First Paralegal Job
Santa Monica, Calif., Estrin Publishing, 1992. 337p.
ISBN 0963011219

Wagner provides detailed practical information on entering the profession and organizing an effective job search campaign. Contains helpful appendices listing buzzwords, legal assistant and paralegal schools, and legal assistant and paralegal associations. Includes suggested reading lists and index.

Warner, Ralph
The Independent Paralegal's Handbook
2nd ed., edited by Karen Chambers. Berkeley, Calif., Nolo Press, 1993. 1 v. (various pagings). ISBN 0873371194

The author is a controversial attorney who favors independent paralegal practice and wants to break the bar's monopoly control of legal services. Warner believes that in the near future paralegals should handle divorce, name change, adoption, bankruptcy, small business incorporation, child support collection, and landlord-tenant forms. Paralegals should also read this book, but they should be aware of the inherent dangers in the unauthorized practice of law.

✦ Reference

Albrecht, Steve
The Paralegal's Desk Reference
New York, Prentice Hall, 1993. 209p. ISBN 0671847155

This short volume offers basic information on the role of the paralegal in case investigation, legal writing and research, and reading police reports. One of the most helpful sections discusses the use of police reports and working with investigators. Includes a list of paralegal training programs.

American Bench: Judges of the Nation
7th ed. Sacramento, Calif., Foster Long, 1993. 2526p.
ISBN 0931398390

This comprehensive guide to the American judiciary is published biennially. It gives biographical data on more than 17,500 judges in the federal and state courts and includes information on the courts they serve. A glossary of legal terms used in the volume is provided. Readers should note that the cutoff date for this particular volume was April 30, 1993.

American Jurisprudence Desk Book: Historical and Legal Documents, Facts, Tables, Charts, and Statistics of Special Interest to Lawyers
Prepared by the Editorial Staff of the Publishers. 2nd ed. Rochester, N.Y., Lawyers Cooperative Pub. Co., 1992. 1290p.

This is an invaluable ready reference tool. Subject matter is divided into four sections: federal, international, national, and state. Includes research and practice guide, an April 1994 supplement, and an index.

Black, Henry Campbell
Black's Law Dictionary: Definitions of the Terms and Phrases of American and English Jurisprudence, Ancient and Modern
Abridged 6th ed.; edited by the Publisher's Editorial Staff. St. Paul, Minn., West Pub. Co., 1991. 1136p. ISBN 0314885366

This desk dictionary will serve most paralegal needs. It has been extensively updated to reflect changes in legal terminology since the publication of the abridged fifth edition in 1983. This edition has increased coverage of financial terminology and also has incorporated the terminology used by the *Uniform Commercial Code, Restatements of the Law, Model Penal Code*, and the *Federal Rules*.

The Bluebook: A Uniform System of Citation
Cambridge, Mass. Harvard Law Review Association, 1991- . 1 v. (various pagings). ISSN 1062-9971

This work is vital to all paralegal collections because a thorough familiarity with correct citation forms is essential. Other notable features of this edition are the treatment of citations from computer sources and international law sources. The quick reference section is particularly helpful.

Cohen, Morris L., and Kent C. Olson
Legal Research in a Nutshell
5th ed. St. Paul, Minn., West Pub. Co., 1992. 370p.
ISBN 0314007830

The authors, both law librarians, have provided a brief but comprehensive introductory guide to legal research and legal bibliography. In addition, Cohen includes four excellent appendices: state research guides, contents of the national research system, current status of major official state reports, and major loose-leaf services. Part of West's *Nutshell Series*. Includes subject and title indexes.

Corbin, John
Find the Law in the Library: A Guide to Legal Research
Chicago, American Library Association, 1989. 327p.
ISBN 0838905021

Corbin provides an introduction to legal research that will effectively meet the needs of his target audience of librarians and library school students. He uses sample cases efficiently to show librarians how they can help patrons research family, community, and business concerns. The appendices include a legal abbreviations table, state uniform commercial code references, and a glossary.

Corpus Juris Secundum: A Contemporary Statement of American Law As Derived from Reported Cases and Legislation
St. Paul, Minn., West Pub. Co., 1936- .

In addition to annual pocket parts, new volumes are published on a regular basis to keep this encyclopedic guide to American law up-to-date. This set presents American case precedent as it develops. While it reflects change, it also addresses the older principles of law.

Doyle, Francis R.
Searching the Law: The States; A Selective Bibliography of State Practice Materials in the Fifty States
Dobbs Ferry, N.Y., Transnational Publishers, 1989. 525p.
ISBN 0941320472

This bibliography provides an authoritative guide to the literature of the fifty states. Arrangement is first by state and then by subject. The author notes that the availability of legal literature varies dramatically from state to state. Updated by supplements.

Encyclopedia of Legal Information Sources: A Bibliographic Guide to Approximately 29,000 Citations for Publications, Organizations, and Other Sources of Information on 480 Law-Related Subjects
2nd ed. Detroit, Gale Research, 1993. 1083p. ISBN 08100374390

This is a comprehensive guide to legal bibliography and computer-assisted resources that is useful for collection building. Selected phone numbers are included for associations and research centers.

Eskew, Mike
Legal Procedures and Terminology for Court Reporters and Paralegals
Englewood Cliffs, N.J., Regents/Prentice Hall, 1993. 206p.
ISBN 0135292158

The key audience for this work is the court reporter; however, the legal terminology provided is suited for paralegals as well.

Following an introduction, each chapter covers the terminology of a different area of law. Includes an excellent glossary and index.

Index to Legal Periodicals
New York, H. W. Wilson, Published irregularly. 1908- . (various pagings). ISSN 0019-4077

This authoritative index to legal literature is available in CD-ROM format as well as in print. It indexes articles contained in legal periodicals from the United States, Canada, Great Britain, Ireland, Australia, and New Zealand. Yearbooks, proceedings of annual institutes, and annual reviews are also indexed.

Johnson-Maloney, Nancy
Westlaw Reference Manual
5th ed. St. Paul, Minn., West Pub. Co., 1993. 443p.

This manual concentrates on the terms and connectors method of using WESTLAW. Novice researchers, who are more likely to feel comfortable with EZ-ACCESS and WIN searching, may not benefit as much from this work as more experienced researchers. WESTLAW citation services receive comprehensive treatment, and the citation sections are especially helpful to paralegals checking briefs. Includes sections devoted to billing and passwords.

The Lawyer's Almanac, 1994: A Complete Reference to Vital Facts and Figures about the Legal Profession
14th ed. Englewood Cliffs, N.J., Prentice Hall Law and Business, 1981/82- . ISSN 0277-9544

This is the best ready reference guide to the legal profession as a whole. It includes exhaustive information about the nation's 500 largest firms and the American Bar Association. There is also a wealth of information about the federal and state judiciary. Includes a helpful section devoted to commonly used abbreviations.

Leiter, Richard A., ed.
National Survey of State Laws
Detroit, Gale Research, 1993. 427p. ISBN 081038406X

This reference source is a valuable aid for finding basic information on the laws of all fifty states. Key topics are enhanced by well-constructed graphs including abortion, adoption, the right to die and durable power of attorney, child custody, marriage and annulment, compulsory education, gun control, drug laws, and interest rates.

Martindale-Hubbell International Law Directory
Summit, N.J., Martindale-Hubbell, Annual. 27 v.

This 126-year-old set is the most comprehensive guide to practicing attorneys in the United States. It is also available in CD-ROM format. In addition to the basic sixteen-volume set, volume seventeen covers both corporate law departments and law schools. Four unnumbered volumes provide a two-volume alphabetical index and a two-volume areas of practice index. There are also three unnumbered volumes devoted to international law and three unnumbered digest volumes.

National Association of Legal Assistants Manual for Legal Assistants
2nd ed. St. Paul, Minn., West Pub. Co., 1992. 397p.
ISBN 0314807802

Despite the fact that most of the information this volume contains is duplicated elsewhere, it can be strongly recommended. This manual reflects the quality of the professional association responsible for its preparation. There is an excellent glossary of basic legal terminology and a good index. The appendices include several court cases demonstrating increased court recognition and acceptance of the role of paralegals, including paralegal costs in normal billing for legal services.

Olson, Kent C., and Robert C. Berring
Practical Approaches to Legal Research
New York, Haworth Press, 1988. 150p. ISBN 0866568530

This excellent introduction to legal research was published as a monographic supplement to *Legal Reference Services Quarterly* (Haworth Press, 1981-). Both authors are librarians with law degrees. Lawyers, librarians, and paralegals could all benefit by using this brief yet comprehensive survey of case law, statutes, legal periodicals, and other forms of legal literature on both the federal and state levels.

Prince, Mary Miles
Bieber's Dictionary of Legal Citations: Reference Guide for Attorneys, Legal Secretaries, Paralegals, and Law Students
Prince's 4th ed. Buffalo, N.Y., William S. Hein, 1992. 372p. ISBN 0899418244

Prince designed her dictionary as a companion to the fifteenth edition of *The Bluebook* (Harvard Law Review Association, 1991- .). When quoting from law journals or cases, paralegals can find helpful models here that illustrate the often complex *Bluebook* rules. Although she includes some foreign sources, the author notes that these often do not provide enough information for proper legal citation.

Shapiro, Fred R., comp.
The Oxford Dictionary of American Legal Quotations
New York, Oxford Univ. Press, 1993. 582p. ISBN 0195058593

This comprehensive work contains short, thoughtful quotations arranged by topic. The sources include historical figures such as John Marshall, Oliver Wendell Holmes, Learned Hand, and Benjamin Cardozo as well as contemporary figures. Includes author and keyword indexes, and an appendix with the text of the *Constitution of the United States*.

Shepard's Acts and Cases by Popular Names, Federal and State
4th ed. Colorado Springs, Colo., Shepard's/McGraw-Hill, 1992. 3 v. ISSN 0080-9233

This three-volume set provides a useful ready reference when combined with its softbound supplement. Useful abbreviations for acts and cases appear at the beginning of Volume One. Alphabetically lists federal and state acts and cases with references to their popular names.

Shepard's United States Citations. Cases: A Compilation of Citations to United States Supreme Court Cases
6th ed. Colorado Springs, Colo., Shepard's/McGraw-Hill, 1988- . Updated by supplements.

This reference set assists users in determining whether United States Supreme Court decisions are still "good law." Paralegal students should become familiar with this set along with state and regional citators by the same publisher. Also valuable are *Shepard's Restatement of the Law Citations* (Shepard's/McGraw-Hill, 1994) and *Shepard's Law Review Citations* (Shepard's/McGraw-Hill, 1986).

United States Code, Containing the General and Permanent Laws of the United States
1988 ed. Washington, D.C., U.S. Government Printing Office, 1989- . Updated by supplements.

This is the standard source for federal statutes. First published in 1926, this work arranges the laws by subject into forty-eight titles. Revised editions are published every six years. Bound cumulative supplements are issued in the intervening years.

Wasserman, Steven R., Jacqueline Wasserman O'Brien, Bonnie Shaw Pfaff
Law and Legal Information Directory
7th ed. Detroit, Gale Research, 1993. 2 v. ISBN 0810376245

The authors provide an excellent source for quick information on all topics of concern to lawyers and paralegals. Includes topics such as federal and state courts, bar associations, law schools, regulatory agencies, sources of legal aid information, sources of speakers on law-related topics, and national and international organizations.

West's Legal Desk Reference
Special deluxe ed. St. Paul, Minn. West Pub. Co., 1991. 1564p. ISBN 031448146X

This is a comprehensive desk reference useful to the paralegal. There is a unique "Biography of Litigation" section as well as a legal dictionary and a guide to research arranged by subject. Other key features are the text of the *Constitution of the United States* and "Quick Cites" sections devoted to selected *United States Code* statutes and Supreme Court cases.

West's Reporter Series
St. Paul, Minn., West Pub Co. (Regional Editions.)

This series provides excellent coverage of American case law. *Regional Reporters* cover various areas of the country, for example, *The Pacific Reporter* covers a multistate region including Arizona, California, Colorado, Hawaii, Idaho, Kansas, Montana, Nevada, New Mexico, Oklahoma, Oregon, Utah, Washington, and Wyoming.

Who's Who in American Law
New Providence, N.J., Marquis Who's Who, 1977/78- . Annual. ISSN 0162-7880

This basic, easy-to-use reference source for biographical information includes a table of abbreviations, a fields of practice index, and a professional index.

Wren, Christopher G.
Using Computers in Legal Research: A Guide to LEXIS and WESTLAW
Madison, Wisc., Adams and Ambrose Publishing, 1994. 771p. ISBN 0916951219

If only one book can be acquired on the computer in legal research, it should be this one, which provides a comprehensive introduction to the principles of effective on-line searching as well as more than eighty pages of detailed analysis of LEXIS and WESTLAW.

✦ Litigation Support, Office Management, and Office Relations

Brophy, Maureen, and Patsy Pressley
Paralegal Management Handbook
New York, Wiley Law Publications, 1993. 247p.
ISBN 0471587958

This handbook provides guidance for the supervision of paralegals in corporations and law firms. Both general management skills and skills specific to the legal setting are considered. In addition, there are appendices with related useful information, a short bibliography, and an index.

Bruno, Carole A.
Paralegal's Litigation Handbook
2nd ed. Minneapolis, Minn., West Pub. Co., 1993. 416p.
ISBN 0314011773

This volume provides a comprehensive survey of the litigation process from the initial filing of a case to appeals. In addition, it discusses computerized litigation support and general research skills needed by the litigation paralegal. Numerous forms and exhibits complement the text. Includes appendices, glossary, and index.

Eyres, Patricia S.
Smart Litigating with Computers: The Complete Step-By-Step Guide to Computer Applications for the Nineties and Beyond
Los Angeles, Estrin Publishing, 1992. 404p. ISBN 0963011227

The author is an attorney experienced in litigation who has become a computer consultant to all types of legal practitioners. While her target audience is attorneys, her work is also useful to paralegals, who may do a large part of the work involved in law office automation. Her work is clearly written and enhanced by helpful figures. Contains a good glossary and index.

Fawcett, Marcy Davis
Computerized Litigation Management for Paralegals
Santa Ana, Calif., James Pub. Group, 1990. 1 v. (loose-leaf)
ISBN 0938065483

Computer literacy is a must for paralegals, and Fawcett shows where to begin. After considering computer basics important to paralegals, Fawcett covers litigation support as well as docket control, conflict of interest, and file management. She also considers remote or portable computer applications. Indexed.

Feeney, Kerri W.
The Paralegal's Role at Trial
New York, Wiley Law Publications, 1993. 235p.
ISBN 0471593435

Since most paralegal literature emphasizes the paralegal's role in preparation for trial, Feeney has performed an important service in writing this monograph. Periodic updating supplements are available and enhance the book's value to the paralegal member of the trial team. Contains an appendix covering selected federal rules and a good index.

Hutson, Beverly K.
Paralegal Trial Handbook
Santa Ana, Calif., James Pub. Group, 1991. 1 v. (various pagings).
ISBN 0938065564

Provides practical information for the litigation paralegal who needs to develop skills for effective support to attorneys. The book stresses organizational skills and provides an extensive selection of checklists that lead to good trial preparation. The chapter on visual aids and demonstrative evidence is particularly well done. Includes a fine index.

Kerley, Peggy N., Paul Sukys, and Joanne Banker Hames
Civil Litigation for the Paralegal
Albany, N.Y., Lawyers Cooperative Pub. Co. and Delmar Publishers, 1992. 442p. ISBN 0827347715

Kerley explains the basics of American civil trials for paralegals and paralegal students in well-written and easy-to-understand terms. The final chapter covers post-trial practice. The appendices are helpful, and an excellent glossary and index are included. Illustrations and figures supplement the text.

Lipson, Ashley S.
Law Office Automation for Paralegals, Administrators, and Legal Secretaries
Englewood Cliffs, N.J., Prentice Hall, 1989. 318p.
ISBN 0135265835

Lipson provides a comprehensive, illustrated introduction to the use of computers in law offices. For the paralegal assigned to organize a firm's files, this work gives an excellent overview of how to organize different forms of legal information. Includes a bibliography and index.

Lynton, Jonathan S., Donna Masinter, and Teri Mick Lyndall
Law Office Management for Paralegals
Albany, N.Y., Lawyers Cooperative Pub. Co. and Delmar Publishers, 1992. 355p. ISBN 0827348657

This monograph targets law firm office managers. The coverage is broad enough to address all of the major concerns of office managers, including topics such as office accounting, records management, and time management.

Nemeth, Charles P.
Litigation, Pleadings, and Arbitration
Cincinnati, Ohio, Anderson Pub. Co., 1990. 508p.

Nemeth provides a competent introduction to litigation and pleadings for paralegals, with an entire chapter devoted to pleadings exercises. In addition to his treatment of litigation and pleadings, Nemeth also considers arbitration as a means of settling legal disputes. Practical forms are included throughout the text.

Randall, Lynn M.
Litigation Organization and Management for Paralegals
New York, Wiley, 1993. 320p. ISBN 0471587982

This good overview of what paralegals can and should do in litigation support contains a detailed description of how to create, organize, and update documentary records needed for litigation support. Although there is no bibliography, there is a good index. The publisher offers essential supplements to update this volume.

Roper, Brent D.
Computers in the Law: Concepts and Applications
St. Paul, Minn., West Pub. Co., 1992. 481p. ISBN 0314933743

Roper's overview of computers in the law office treats office application software as well as legal-specific software. Includes information on computers and the law, computer hardware, and computer software. Specific discussions of ABA/NET as well as LEXIS and WESTLAW are provided. Case studies focusing on small, medium, and large firms as well as a corporate legal department help to expand the usefulness of the book. The supplemental information adds to the quality of this work.

Roselle, Laurie P., and Laurie Beth Zimet, eds.
Workshop for Legal Assistants: Basic Litigation, Legal Research, and Writing
New York, Practising Law Institute, 1993- . 744p.

This volume contains materials used in continuing education courses between April and June 1993. Twenty-five of the thirty-one chapters cover basic litigation, while the remaining chapters discuss legal research and writing. All the contributions are well written and helpful to practicing paralegals. A list of faculty concludes the volume. Both the pretrial and the trial role of the paralegal are covered, with the importance of ethics and professional responsibility receiving ample attention as the subject of the first five chapters. *Litigation and Administrative Practice Series.*

Roselle, Laurie P., Laurie Beth Zimet, and Norman A. Polizzi, eds.
Workshops for Legal Assistants: Litigation, Legal Research and Writing, Environmental Law and Toxic Tort Litigation
New York, Practising Law Institute, 1992. 560p.

This collection of essays contains continuing education materials used in course instruction from April to June 1992. Eighteen of the twenty-two essays deal with litigation in general or the developing specialty of toxic tort litigation. The four concluding essays deal with legal research and writing. *Litigation and Administrative Practice Series.*

Signey, Phillip J.
Litigation Paralegal.
Edited by Debra Jurczynski. Santa Ana, Calif., James Pub. Group, 1989- . ISBN 0938065475 1 v. (loose-leaf)

Provides practical information to support all aspects of the paralegal's role in litigation from the precomplaint stage to the point of post-trial action. Useful forms and checklists are provided throughout the text. In addition, the fine index increases access to points of particular concern.

✦ Paralegal Practice—Ethics

Orlik, Deborah K.
Ethics for the Legal Assistant
Encino, Calif., Marlen Hill Publishing, 1994. 355p.
ISBN 0963327623

Orlik addresses issues of professional responsibility related to interpersonal relations, misrepresentation, calendaring systems, and time records. She offers guidance on topics such as the unauthorized practice of law, preserving confidences, negligence issues, and regulation of the profession. Appendices include the *ABA Model Rules, ABA Model Code, NFPA Model Code,* and the *NALA Model Standards and Guidelines.*

Pulsifer, Nancy, Dana L. Graves, and Jill S. Levin
How to Survive in a Law Firm: Client Relations
New York, Wiley Law Publications, 1993. 1 v. (various pagings).
ISBN 04171593672

Pulsifer, Graves, and Levin provide practical survival tips to paralegals. Covers topics such as paralegal and attorney relationships, time and billing techniques, and confidentiality. Ethics is stressed in the text and in the appendix, which lists selections from the ABA's *Model Rules.* Indexed.

Statsky, William P.
Paralegal Ethics and Regulation
2nd ed. Minneapolis/St. Paul, West Pub. Co., 1993. 211p.
ISBN 0314012095

Statsky has produced an excellent work for inclusion in paralegal studies collections. He provides a comprehensive glossary of the legal terms used in this volume and a detailed index. The appendices provide needed bar association and paralegal association addresses as well as opinions on paralegal practice and a survey of paralegal appearances before federal agencies.

✦ Types of Law—Bankruptcy

Berger, Steven N., J. Daria Westland, and Thomas J. Salerno
Bankruptcy Law and Procedure: A Fundamental Guide for Law Office Professionals
2nd ed. Eau Claire, Wisc., Professional Education Systems, 1989. 1 v. (loose-leaf). ISBN 0941161617

Two attorneys, Berger and Salerno, have combined with Westland, a paralegal, to provide an excellent introduction to bankruptcy law. After an introductory chapter and two overview chapters on bankruptcy law and bankruptcy courts, both consumer and personal bankruptcy are discussed. There are also chapters on family farmer bankruptcy, discovery, and appeals in bankruptcy cases.

Coda-Wagener, Dawn, ed.
Workshop for Legal Assistants, Bankruptcy, 1992
New York, Practising Law Institute, 1992. 344p.

Coda-Wagener collected these authoritative essays for workshops held between April and June 1992. They provide a good introduction to bankruptcy practice for paralegals. In addition, objections to claims and other bankruptcy litigation issues receive consideration. Useful appendices include a reproduction of *Wolcotts All Purpose Bankruptcy Litigation Kit.* A volume in the *Commercial Law and Practice Course Handbook Series,* no. 619.

Pearson, John K., Charles M. Tatelbaum, and Herbert Katz
Bankruptcy Rules and Forms Handbook
New York, Wiley Law Publications, 1992. 2 v. (loose-leaf).
ISBN 0471559091

This reference set is invaluable for lawyers and paralegals engaged in bankruptcy practice. The forms are organized around the Federal Rules of Bankruptcy Procedure, although readers are warned that local courts may deviate from these rules. Features a detailed list of its contents. Supplements are issued on a regular basis. *Bankruptcy Practice Library* series.

Williamson, Darcy D.
Bankruptcy and Collections: The Paralegal Perspective
2nd ed. Kansas City, Mo., National Federation of Paralegal Associations, 1992. 404p.

Williamson provides a comprehensive and well-written introduction to bankruptcy law and procedure for paralegals.

Examples of bankruptcy problems are effectively interspersed throughout the text, and extensive references to the appropriate *U.S. Code* sections are cited. Each chapter is followed by a thoughtful summary and pertinent tables.

✦ Types of Law—Civil Procedure

Greene, Christine Bergren
Deposition Manual for Paralegals
2nd ed. New York, Wiley Law Publications, 1993. 388p.
ISBN 0471581275

Greene provides an excellent summary of the deposition process including information on depositions, notices and subpoenas, the preparation of a client for deposition, and motions for depositions. She also covers preparation for both written and videotaped depositions, file management, summarizing depositions, and using the computer to analyze depositions. Includes appendices on the *Federal Rules of Civil Procedure* and an index. Supplements are issued periodically.

McClellan, Dorien Smith
Paralegal Drafting Guide
New York, Wiley Law Publications, 1993. 258p.
ISBN 0471587974

McClellan provides a well-written introduction to the preparation of pleadings and discovery documents. She is a paralegal and stresses that she writes for paralegals, not attorneys. The text is based on the *The Federal Rules of Civil Procedure*. A detailed, subject-specific index makes the book easy to use, and periodic supplements are available. McClellan reminds paralegals to consider state rules in drafting documents.

Medina, Pat
Paralegal Discovery: Procedures and Forms
Santa Ana, Calif., James Pub. Group, 1989. 1 v. (loose-leaf).
ISBN 0938065459

Medina, a paralegal who has taught legal assistants, has compiled a practical guide to discovery aimed at the intermediate- to advanced-level legal assistant. She includes both the practical mechanics of discovery and a sophisticated discussion of discovery rationale. Indexed.

Nemeth, Charles P.
Evidence Handbook for Paralegals
New York, Wiley Law Publications, 1993. 258p.
ISBN 047158746X

Nemeth provides an excellent introduction to the principles of evidence for paralegals. Although the author demonstrates considerable scholarship in his writing, he keeps his focus on the practical needs of the paralegal, rather than on theory. The text is enhanced by good footnotes, pertinent illustrations, and a comprehensive index. Updated by supplements.

Zalewski, Dianne Dupre, and Joyce Samson
Paralegal Discovery
Edited by Stuart D. Ogilvie. Santa Ana, Calif., James Pub. Group, 1991. 1 v. (various pagings). ISBN 0938065246

This monograph is a practical tool for the paralegal involved in the discovery process. The authors cover case management, depositions, request for production of documents, interrogatories, and requests for admissions. Other topics include computerization of documents, legal research, systems management, and time management. An invaluable work for the paralegal needing a guide through the maze of formalities that are involved in the discovery process.

✦ Types of Law—Contract

Baldwin, Diane M., and Frances B. Whiteside
Introduction to Contracts for Paralegals
Dallas, Tex., Pearson Publications, 1993. 132p. ISBN 0929563115

Baldwin, an attorney, and Whiteside, a paralegal, provide a short and readable introduction to contract law for paralegals. Contract law basics, the formation of contracts, the discharge of contract obligations, contract remedies, sales agreements, and commercial paper receive chapter-length consideration. Appendices provide examples of basic contract types. Includes brief bibliography and index.

✦ Types of Law—Criminal

Hall, Daniel
Criminal Law and Procedure
Albany, N.Y., Lawyers Cooperative Pub. Co. and Delmar Publishers, 1992. 514p. ISBN 0827345593

Provides an introduction to criminal law for paralegals, assuming no previous background in the area. Different types of crime, trial procedure, sentencing, and the appeal process are described. Each chapter includes review questions, problems, and notes. Pertinent appendices contain the *Constitution of the United States* and excerpts from the *Model Penal Code*. Includes glossary and index.

✦ Types of Law—Family

Statsky, William P.
Family Law
3rd ed. St. Paul, Minn., West Pub. Co., 1991. 670p.
ISBN 0314718516

Statsky provides a comprehensive introduction to family law for paralegals. Each chapter contains a list of key terminology and assignments to test mastery of the law covered, as well as "exhibits" designed to illustrate important topics. Appendices provide good documentary support to the text. Includes an extensive and easy-to-use index.

✦ Types of Law—General Business

Harrington, Linda, and Neal S. Millard, eds.
Legal Assistants, 1989
New York, Practising Law Institute, 1989. 1088p.

This work contains a selection of essays on topics of interest to paralegals. Following an introduction devoted to paralegal regulation, the essays consider estate administration, real estate, bankruptcy, and the role of the paralegal in preparing for and participating in litigation. The estate planning and bankruptcy sections include helpful appendices. *Commercial Law and Practice Course Handbook Series*, no. 496.

In 1990, this annual title split into the following works:
- *Workshops for Legal Assistants: Bankruptcy*
- *Workshops for Legal Assistants: Employee Benefits*
- *Workshops for Legal Assistants: Litigation, Legal Research and Writing, Environmental Law, and Toxic Tort Litigation*

Harrington, Linda, and Neal S. Millard, eds.
Legal Assistants, 1990
New York, Practising Law Institute, 2 v.

This work consists of collected essays by experts on the role of paralegals in two specialized areas of law: bankruptcy proceedings, and real estate administration and probate. Helpful forms illustrate bankruptcy and estate practice. *Commercial Law and Practice Course Handbook Series*, no. 537.

Schneeman, Angela
The Law of Corporations, Partnerships, and Sole Proprietorships
Albany, N.Y., Lawyers Cooperative Pub. Co. and Delmar Publishers, 1993. 608p. ISBN 0827352549

The author is an experienced paralegal, who now works on a free-lance basis and writes a column for *Legal Assistant Today*. Her work targets the paralegal student who plans to work in corporate law, but it is also useful for general readers who need to know the legal basics behind corporate formation, functioning, and dissolution.

Wheeler, Patricia R., ed.
Legal Assistants, 1988
New York, Practising Law Institute and Delmar Publishers, 1988. 1080p.

Following an introductory essay by the editor on the role of the legal assistant and paralegal regulation, the collected essays in this volume cover litigation, bankruptcy, real estate, estate planning, administration, and probate. They were prepared for legal assistant workshops held in May and June of 1988. A list of faculty concludes the volume. *Commercial Law and Practice Course Handbook Series*, no. 456.

Zalewski, Dianne Dupre, and Cheryl D. Evans
Construction Litigation for Paralegals: Claims Preparation and Management
New York, Wiley Law Publications, 1993. 444p. ISBN 0471589586

The authors provide a good introduction for paralegals on the use of computers to manage information in a highly specialized area of law. Covers fact gathering, as well as information on legislation and the construction industry. Practical figures are interspersed throughout the text. Includes an excellent glossary and index.

✦ Types of Law—Real Property

Kearns, Michael P.
The Law of Real Property
Albany, N.Y., Lawyers Cooperative Pub. Co. and Delmar Publishers, 1994. 460p. ISBN 0827348789

Both commercial and residential real estate are covered in this practical monograph. The text is written clearly and each new term is explained when it first appears. Contains numerous illustrations and figures, and a discussion and review section. Includes a brief table of cases in the field, a glossary, and an index.

Novak, Theodore J., Brian William Blaesser, and Thomas F. Geselbracht
Condemnation of Property: Practice and Strategies for Winning Just Compensation
New York, Wiley, 1994. 421p. ISBN 0471574031

With property issues becoming increasingly important, this comprehensive monograph assumes a crucial place in library collections. Includes constitutional challenges to eminent domain powers, an impressive table of cases, and a good index. Although there are good running footnotes, there is no bibliography.

✦ Types of Law—Tax

Raabe, William A., Gerald E. Whittenberg, and John C. Bost
West's Federal Tax Research
2nd ed. St. Paul, Minn., West Pub. Co., 1991. 648p. ISBN 0314787348

Raabe and Whittenburg are CPAs and Bost is an attorney. Together these authors have provided a comprehensive and excellent guide to this complex area of law. Recent changes in tax law are not included.

Smith, Robert Sellers
West's Tax Law Dictionary: Definitions of Terms, Words, and Phrases Used in Modern American Tax Law
1994 ed. St. Paul, Minn., West Pub. Co., 1994. 1162p. ISBN 031404034X

This volume is a good ready reference source for tax terminology. It also includes four helpful appendices which contains sources for free federal tax forms and publications, a listing of federal tax return and related forms, an index to forms, federal tax instructions, a listing of free federal tax publications, and an index to publications.

✦ Tyes of Law—Torts

Appleby, Kristyn S., and Joanne Tarver
Paralegal Medical Records Review
New York, Wiley Law Publications, 1991. 1 v. (loose-leaf) ISBN 0471587559

This excellent manual will prove invaluable to paralegals evaluating medical records for settlement and trial. It contains comprehensive coverage on the use of consultants or experts by case type, including product liability, toxic exposure, workers' compensation, personal injury, and medical malpractice cases. Confidentiality receives chapter-length treatment. Includes annual supplements.

Eldridge, Burgess C., and Sharon Atwood
Personal Injury Paralegal
Revision 4. New York, Wiley Law Publications, 1991. 1 v. (loose-leaf) ISBN 0938065327

This is a practical guide for paralegals who work with attorneys representing the plaintiff in personal injury and workers' compensation claims. Emphasis is on files, logs, and worksheets, as well as the paralegal's role in investigation. The authors consider in some detail the paralegal's role in accepting new clients, pretrial investigation, trial, and settlement.

Walden, Joyce
Personal Injury Paralegal Forms and Procedures
New York, Wiley Law Publications, 1993. 395p.
ISBN 047158799

This volume takes the personal injury case from start to finish, and explains all aspects of the paralegal's role in such cases. The forms are excellent and provide helpful guidance for information gathering. Includes a comprehensive index to enhance the book's usefulness both to the student and the practicing paralegal. In addition, Wiley provides an updating service that supplements the book on a regular basis.

✦ Types of Law—Wills and Trusts

Nemeth, Charles P.
Estate Planning and Administration for Paralegals
New York, Wiley, 1993. 550p. ISBN 0471587451

Nemeth provides an excellent introduction to the field of estates and trusts for paralegals. His thorough coverage includes the implications of intestacy and the tax implications of estate planning. Estate administration and probate also receive chapter-length treatment. Wiley issues periodic supplements to keep the book up-to-date.

Nonprint Resources

✦ Videocassettes

Legal Research Made Easy: A Roadmap Through the Law Library Maze
Berkeley, Calif., Nolo Press, 1990. 1 videocassette (145 min.) VHS

This video provides an excellent introduction to use of the law library. The narrator, Robert C. Berring, is Professor of Law and Director of the Law Library at Boalt Hall School of Law, University of California, Berkeley. For instructional purposes, the tape can be divided into three parts. Berring provides a useful six-step road map to law library research. He also provides "patterns" to aid novice researchers.

✦ Additional Resources

American Bar Association
750 North Shore Dr.
Chicago, Ill. 60611

National Association of Legal Assistants, Inc.
1516 S Boston, Suite 200
Tulsa, Okla. 74119

The ABA and NALA offer many videocassettes for professionals and paraprofessionals on topics related to law and legal practice.

Insight Media
121 W 85th St., New York, N.Y. 10024

This distributor is a good source for videocassettes on legal office topics, computers in the office, specific applications (such as *DOS, Lotus, WordPerfect*), business communications, interviewing, and job success.

Wisconsin Bar Association
402 W. Wilson St., Madison, Wisc. 53707
The state bar issues videocassettes on topics such as law office confidentiality, legal ethics, and professionalism in the office, including telephone strategies, time management, and client service.

✦ Online Resources

LEXIS/NEXIS
Dayton, Ohio, Mead Data Central, Inc.

WESTLAW
St. Paul, Minn., West Pub. Co.

These online databases provide access to the full text of judicial cases from federal and state courts in the United States. Also included are statutory and reulatory materials. Data are organized by the judicial system, and searchers may explore the various levels of case law. Specialized law may also be explored. Both systems provide information directly from the courts.

Differences in these two online services lie in the coverage of additional resources and in delivery options, organization of information, retrieval methods, and front-ends for users.

Education

Early Childhood Education 309
  Peggy Holleman
Library Technology 328
  Diane J. Turner

Early Childhood Education

Peggy Holleman
Pima Community College
Tucson, Arizona

Introduction

Early Childhood Education certificate and associate degree programs prepare students for employment as teacher aides or assistants in elementary and preschools, or as directors of day-care, preschool, or after-school centers. Paraprofessionals perform a wide variety of instructional, technical, or clerical duties, including supervising students in classrooms or on playgrounds and field trips, setting up audiovisual equipment, preparing instructional materials, tutoring, reading to students, assisting with microcomputers, and preparing exhibits and bulletin boards. They may also assist with bilingual, special education, art, music, or health and recreation programs. Employment opportunities for teacher aides are expected to grow at a faster rate than average, with increased attention to special education; the increased number of working mothers is contributing to this employment growth. However, employment in this field is dependent upon the financial status of local school districts. Those entering this field should have good interpersonal skills and enjoy working with young children.

According to the National Association for the Education of Young Children (NAEYC), "early childhood" covers the period from birth to age eight. ECE programs generally include courses dealing with child growth and development (through adolescence); educational theories and philosophies; teaching techniques; classroom management; effective parenting; administration and staffing of preschools, day-care centers, and after-school programs; special education; children's literature; and instructional methods for language arts, social studies, math, science, music, and fine arts. ECE programs may also be designed to enable students to transfer to upper-division programs in child development and family relations, and elementary, secondary, special, or early childhood education. Such students must work with advisers at the four-year institutions to determine lower-division curricular needs. Teachers and day-care and preschool staff frequently take ECE classes to upgrade their knowledge and skills.

Major trends in early childhood education are (1) increased parental choice; (2) "back to basics" versus "child-centered" philosophies; (3) an "ecological" outlook of child development, with attention to political, social, cultural, physical, and moral aspects; (4) nonsexist and multicultural curricula; (5) integration of the disabled; and (6) intergenerational approaches. Other factors generating popular interest in early childhood education are the growing numbers of working mothers and teenage and/or single parents; the desire of corporate America to improve the skills of the work force; the success of Head Start and Follow Through programs; national and international attention on prenatal care, child abuse and neglect, and children's rights; and efforts to restructure the public schools.

In 1990, Congress passed the Head Start Reauthorization Bill and the Child Care and Development Block Grant. Each state was required to develop a plan to use its share of the CCDBG funds by September 30, 1994. Each Head Start classroom in a center-based program must be assigned one teacher who has the following qualifications: (1) a Child Development Associate (CDA) credential appropriate to the age of the children being served; or, (2) a state-awarded certificate for preschool teachers that meets or exceeds the requirements for a CDA credential; or, (3) an associate, baccalaureate, or advanced degree in ECE; or, (4) a degree in a field related to ECE with experience in teaching preschool children and a state-awarded certificate to teach in a preschool program. NAEYC issued guidelines for associate degree ECE programs in 1985, criteria for quality child care centers in 1986, and in 1989, policies essential for achieving developmentally appropriate early childhood programs.

Selected Associations

Child Care Action Campaign
330 7th Ave., 18th Floor
New York, N.Y. 10001

Child Care Employee Project
6536 Telegraph Ave., Suite 201A
Oakland, Calif. 94609

National Association for the Education of Young Children
1834 Connecticut Ave., NW
Washington, D.C. 20009

Selected Journals

Booklist.
American Library Association. Twenty-two times a year. ISSN 0006-7385

Child and Youth Care Forum.
Human Sciences Press. Bimonthly. ISSN 1053-1890

Child Care Information Exchange.
Exchange Press. Six times a year. ISSN 0164-8527

Child Development.
Univ. of Chicago Press. Bimonthly. ISSN 0009-3920

Childhood Education.
Association for Childhood Education. Five times a year. ISSN 0009-4056

Children Today.
U.S.G.P.O. Bimonthly. ISSN 0361-4336

Children's Literature Association Quarterly.
The Association. Quarterly. ISSN 0885-0429

Day Care and Early Education.
Plenum Press. Quarterly. ISSN 0092-4199

Developmental Psychology.
American Psychological Association. Bimonthly. ISSN 0012-1649

ERIC/EECE Newsletter.
ERIC. Biannual. [also available on Internet: gopher to *ericps.ed.uiuc.edu* At menu, select ERIC/EECE Publications and Services.]

Exceptional Children.
Council for Exceptional Children. Six times a year. ISSN 0014-4029

Five Owls.
Five Owls, Inc. Five times a year. ISSN 0892-6735

Growing Child.
Dunn & Hargitt. Monthly. ISSN 0193-8037

Horn Book Magazine.
Horn Book, Inc. Bimonthly. ISSN 0018-5078

Horn Book Guide to Children's and Young Adult Books.
Horn Book, Inc. Semiannually. ISSN 1044-405X

Instructor.
Scholastic. Nine times a year. ISSN 1049-5851

International Journal of Early Childhood.
Early Childhood. Semiannually. ISSN 0020-7187

Journal of Child and Youth Care.
Univ. of Calgary Press. Quarterly. ISSN 0840-982X

Journal of Computing in Childhood Education.
Association for the Advancement of Computing in Education. Quarterly. ISSN 1043-1055

Journal of Learning Disabilities.
Pro-Ed. Ten times a year. ISSN 0022-2194

Language Arts.
National Council of Teachers of English. Eight times a year. ISSN 0360-9170

Learning.
Springhouse Corp. Eight times a year. ISSN 0090-3167

New Directions for Child Development.
Jossey-Bass. Quarterly. ISSN 0195-2269

Parent and Preschooler.
Preschool Publications. Monthly. ISSN 0887-0365

Parents Magazine.
Education Today Co. Monthly. ISSN 0195-0967

Reading Teacher.
International Reading Association. Eight times a year. ISSN 0034-0561

Scholastic Pre-K Today.
Scholastic, Inc. Monthly. ISSN 0888-3009

Sesame Street Parents' Guide.
Children's Television Workshop. [Included as part of Sesame Street Magazine]. Monthly. ISSN 0049-0253

Teaching Exceptional Children.
Special Education Instructional Materials Network. Quarterly. ISSN 0040-0599

Teaching K-8.
Early Years, Inc. Eight times a year. ISSN 0891-4508

Totline.
Warren Publishing House. Six times a year. ISSN 0734-4473

Young Children.
National Association for the Education of Young Children. Bimonthly. ISSN 0044-0728

Zero to Three.
National Center for Clinical Infant Programs. Six times a year. ISSN 0736-8038

Print Materials

All You Wanted to Know About Day Care Centers But Didn't Know Who to Ask
Cape Coral, Fla., Syndicated Capital Pub. Co., 1989. 295p.

In loose-leaf format, this work covers all aspects of operating a day-care center: hiring, setting policies, public relations, food preparation, working with parents, disciplining and guiding children, liability insurance, and program planning.

Allen, Judy, Earldene McNeill, and Velma Schmidt
Cultural Awareness for Children
Menlo Park, Calif., Addison-Wesley, 1992. 254p.
ISBN 0201287315

An excellent program for multicultural education, which can be adapted to the abilities of children from four to ten years. Units cover diverse cultures: African and African American, American Indian, Chinese and Chinese American, Japanese and Japanese American, Korean and Korean American, Mexican and Mexican American, Thai, and Southeast Asian. Each of these units includes stories, songs, book lists, recipes, clothing, food, art, dress, animals, plants, folktales, puppets, music, dance, games, and celebrations.

Althouse, Rosemary
Investigating Science with Young Children
New York, Teachers College Press, 1988. 200p.
ISBN 0807729124

Describes how to teach science by an active learning, "process approach" for children three-to five-years old. Includes eighty-five activities in which children work with water, colors, pets, sand, foods, seeds, boxes, wood, and other materials. Children learn the processes of observing, classifying, measuring, predicting, making inferences, communicating, using numbers, using space-time relationships, and drawing conclusions.

Barad, Dianne, and Wyonne Hegland
Language Lessons for the Curriculum: Social Studies and Science/Level K
East Moline, Ill., LinguiSystems, 1991. 149p.

Like others in this series formerly titled *Collaborate! Celebrate!*, this work was prepared by two speech-language pathologists for language impaired students. It is one of three books at kindergarten level. The two others deal with language arts and math. Activities in this work are related to ten themes: environment, community, animals, nutrition, holidays, other cultures, seasons, home, space, anatomy, and nutrition.

Barchers, Suzanne I.
Teaching Language Arts: An Integrated Approach
Minneapolis, Minn., West Pub. Co., 1994. 462p.
ISBN 0314025030

Gives an overview of theories of language acquisition and provides excellent examples of ways to combine listening and speaking, reading and writing, and literature and viewing. Also covers evaluation, classroom management, and creation of thematic units. Appendix includes a list of multicultural literature, magazines, and newspapers for children. Emphasis is on creating literate school and home environments.

Barlin, Anne Lief, and Nurit Kalev
Goodnight Toes!: Bedtime Stories, Lullabies, and Movement Games
Pennington, N.J., Princeton Book Co., 1993. 122p. and audiocassette. ISBN 0871271907

An illustrated book of activities and games for preschoolers that utilizes music, movement, and simple toys to help children think, exercise, and relax. Music cassette is available to use with the dance games.

Barnett, David W., and Karen T. Carey
Designing Interventions for Preschool Learning and Behavior Problems
San Francisco, Calif., Jossey-Bass, 1992. 469p. ISBN 1555424090

This book presents case studies intended to help educators and psychologists develop interventions for preschoolers with learning and behavior problems. Sections include intervention design, basic interventions, family/home interventions, and school-based interventions. Book discusses problem situations rather than disabled children. Authors allow for cross-cultural variables.

Barron, Marlene
I Learn to Read and Write the Way I Learn to Talk: A Very First Book about Whole Language
Katonah, N.Y., Richard C. Owen Publishers, 1990. 27p.
ISBN 1878450077

Explains the Montessori whole language approach to language arts education.

Barton, Bob, and David Booth
Stories in the Classroom: Storytelling, Reading Aloud, and Roleplaying with Children
Portsmouth, N.H., Heinemann, 1990. 194p. ISBN 0435085271

Based on the authors' thirty years of experience using stories with children in schools, this book outlines the value of stories in school programs, and explains ways in which educators can use oral literature to stimulate students to tell and create their own stories.

Bauch, Jerold P., ed.
Early Childhood Education in the Schools
Washington, D.C., National Education Association, 1988. 352p.
ISBN 0810614642

Ten sections include articles on historical perspectives, policy decisions, issues, trends, curriculum, evaluation, technology, parent and community involvement, behavior and discipline, and preschool programs.

Baumgardner, Jeannette Mahan
Sixty Art Projects for Children: Painting, Clay, Puppets, Prints, Masks, and More
New York, Clarkson Potter, 1993. 112p. ISBN 0517880083

Activities cover drawing, painting, paper, clay, plaster, puppets, masks, prints, fabrics, yarns, baskets, wood, straws, and cardboard. Does not indicate for which ages projects are appropriate.

Beardsley, Lyda
Good Day Bad Day: The Child's Experience of Child Care
New York, Teachers College Press, 1990. 158p.
ISBN 0807730408

Intended for policy makers, advocates, researchers, administrators, teachers, and teacher trainers interested in quality programs. This book describes two centers (one bad, and one good) to illustrate aspects of quality. Includes information on adult-child ratios, group sizes, and health and safety measures.

Beaty, Janice J.
Observing Development of the Young Child
3rd ed. New York, Merrill, 1994. 389p. ISBN 0023077417

A guide to observing and recording the emotional, social, motor, cognitive, language, and creative development of children, aged two to six. Includes a Child Skills Checklist and suggested activities for children who have not demonstrated specific skills. Each chapter presents current research and an observation of a child, followed by interpretation of the data.

Bee, Helen L.
Lifespan Development
New York, HarperCollins, 1994. 583p. ISBN 0065009819

Describes physical, cognitive, social, and personality development for various stages of life, including preschool years, ages six to twelve, adolescence, young and middle adulthood, and old age, ending with dying, death, and bereavement. The appealing format features goals, critical thinking questions, insets with cross-cultural or cross-ethnic research, and links between child and adult development.

Berndt, Thomas J.
Child Development
Fort Worth, Tex., Harcourt Brace Jovanovich, 1992. 776p.
ISBN 0030227127

Emphasizes heavy reliance on research and theories for practical applications, incorporating cross-cultural perspectives. Each chapter begins with a vignette illustrating child development in other countries. Includes chapters on genetic influences; motor, perceptual, cognitive, social, and moral development; and sex roles. Chapters open with goals, objectives, and outlines. Includes highlights, lists of key terms and people, and practice exams.

Biller, Henry B.
Fathers and Families: Paternal Factors in Child Development
Westport, Conn., Auburn House, 1993. 325p. ISBN 0865692084

This book outlines ways in which paternal involvement, either in co-parenting or single parenting, affects child development at home and in school.

Bittinger, Gayle
Exploring Sand and the Desert
Everett, Wash., Warren Pub. House, 1993. 93p. ISBN 0911019588

Presents sand play in the curriculum to develop children's basic math and science skills, language, and fine and gross motor skills.

Brazelton, T. Berry
Toddlers and Parents: A Declaration of Independence
Rev. ed. New York, Delacorte Press/Seymour Lawrence, 1989. 249p. ISBN 0385297874

Describes parenting during the stage when children are trying to establish autonomy. Chapters cover sibling rivalry; single parents; withdrawn, demanding, and unusually active children; daycare experiences; and learning inner control and self-awareness.

Brazelton, T. Berry
Families: Crisis and Caring
Reading, Mass., Addison-Wesley, 1989. 251p. ISBN 0201092646

From the author's television series, *What Every Baby Knows,* these five case histories describe how five families handled various crises, including cross-cultural adoption, serious illness, death, divorce, and stepparenting. Case studies include office visit dialogues, discussions of issues, and follow-ups.

Brazelton, T. Berry
Touchpoints: Your Child's Emotional and Behavioral Development
Reading, Mass., Addison-Wesley, 1992. 479p. ISBN 0201093804

The "touchpoints" of the title refers to periods when very young children (aged birth through three years) seem to regress emotionally and behaviorally. The book explains how parents can deal with these periods and other problems, such as bedwetting, imaginary friends, separation anxiety, etc.

Bronner, Simon J., ed.
American Children's Folklore
Little Rock, Ark., August House, 1988. 281p. ISBN 0874830680

A fascinating collection of communications unique to children such as tongue twisters, insults, secret languages, rhymes of play, autograph inscriptions, song parodies, riddles, jokes, tales, legends, beliefs, and street games.

Brooks, Jane B.
Process of Parenting
3rd ed. Mountain View, Calif., Mayfield Pub. Co., 1991. 552p.
ISBN 1559340134

Primary focus is on parenting as a joyful process to be shared with teachers and day-care workers. Covering conception through adolescence, this work includes material for working, single, and stepparents, and those with special needs children. Presents a problem-solving approach that utilizes strategies from many different experts such as researchers, professionals, theorists, and parents.

Buckleitner, Warren
High/Scope Buyer's Guide to Children's Software 1993: Annual Survey of Computer Programs for Children Aged 3 to 7
Ypsilanti, Mich., High/Scope Press, Annual. 1 v. (various pagings). ISSN 1060-9504

Includes annotated reviews of 515 commercial software titles, many mouse-controlled, and some on CD-ROM. Lists criteria, producers, and topics, computer brands, producers, and software no longer available. Includes a glossary of terms.

Burchard, Arliss
No-Fail Art Projects: 100 Success-Oriented Lessons for the Primary Grades
West Nyack, N.Y., Parker Pub. Co., 1990. 206p.
ISBN 0136224814

Each illustrated project lists objectives, supplies and materials, directions and patterns.

Campbell, Robin
Reading Real Books
Philadelphia, Open Univ. Press, 1992. 90p. ISBN 0335157947

British author recommends a whole language approach to reading that employs meaningful children's books. Urges caregivers to post charts in the classroom that children will read, such as attendance charts, messages, list of birthdays, finished work, poems, and daily weather. Also recommended are reading and writing activities tied to themes.

Campbell, Susan B.
Behavior Problems in Preschool Children: Clinical and Developmental Issues
New York, Guilford Press, 1990. 270p. ISBN 0898623952

The author outlines many factors that may contribute to behavioral problems, such as genetic, biological, environmental, and developmental factors. She specifically discusses family, sibling, and peer relationships and concludes with treatment approaches, follow-up, and outcomes. Also covers social policy implications of research findings. Includes four clinical case studies.

Catron, Carol E., and Jan Allen
Early Childhood Curriculum
New York, Merrill, 1993. 479p. ISBN 0023202653

A comprehensive guide for planning a holistic, creative play curriculum for children from birth to age five. Incorporates concerns regarding developmental assessment, cross-cultural perspectives, and special needs students. The reader is asked to solve a problem at the beginning of each chapter; issues are identified and the solution is provided at the end. Chapters Ten through Fifteen include teaching activities. Identifies key terms.

Chapin, June R., and Rosemary G. Messick
Elementary Social Studies: A Practical Guide
New York, Longman, 1989. 224p. ISBN 0801300436

Includes separate chapters on primary and middle grade social studies. Includes activities for learning about local and global citizenship, other cultures, time, space, and microcomputers, and for incorporating communications skills into social studies units.

Charlesworth, Rosalind
Understanding Child Development
3rd ed. Albany, N.Y., Delmar Publishers, 1992. 703p.
ISBN 0827348916

Provides coverage of developmental and learning theories; prenatal, infancy, and toddler stages; the motor, affective, and cognitive development stages of preschool and primary years; and sections on play, technology, special needs children, and the holistic approach.

Coleman, Margaret Cecil
Behavior Disorders: Theory and Practice
2nd ed. Boston, Allyn and Bacon, 1992. 368p. ISBN 0205132065

Includes a historical perspective and definition of behavior disorders. Also defines biophysical, psychodynamic, behavioral and ecological models of disturbance, and describes internalizing, externalizing, and severe behavior disorders. Explains assessment and instruction and a systems approach to education. A new chapter deals with behavior disorders in adolescents and another with trends and the future.

Collins, H. Thomas, and Fred R. Czarra
Global Primer: Skills for a Changing World
Rev. ed. Denver, Colo., University of Denver. Center for Teaching International Relations, 1991. 169p. ISBN 0943804604

Includes tested supplemental classroom activities for K-8 to teach the concept of global interdependence. Discusses students' homes and communities in the context of other continents and nations. Includes maps and globe skills.

Cooper, Kenneth H.
Kid Fitness: A Complete Shape-Up Program from Birth through High School
New York, Bantam Books, 1991. 367p. ISBN 055307332X

An exercise and nutrition guide for young children. Part of the book is devoted to wholesome menus and recipes.

Cullum, Carolyn N.
Storytime Sourcebook: A Compendium of Ideas and Resources for Storytellers
New York, Neal-Schuman Publishers, 1990. 177p.
ISBN 1555700675

This reference book for planning story hours is arranged alphabetically by topics common to the preschool curriculum. For each topic, the author has included a list of books, filmstrips, films, a finger play, crafts, physical activities, and a song. The appendix provides lists of filmstrips, films, videos, picture book titles, publishers, and an extensive bibliography.

Davidson, Neil, ed.
Cooperative Learning in Mathematics: A Handbook for Teachers
Menlo Park, Calif., Addison-Wesley, 1990. 409p.
ISBN 0201232995

Though aimed at those working with elementary, middle school, or secondary math students, this book also has value for preschool personnel. It explains cooperative education in small groups, a teaching technique that research has proven effective in boosting academic achievement, self-esteem, and intergroup relations, including cross-racial and cross-cultural interactions, and social acceptance of the mainstreamed.

Decker, Celia Anita, and John R. Decker
Planning and Administering Early Childhood Programs
5th ed. New York, Merrill, 1992. 580p. ISBN 0023279656

Emphasizes the importance of choosing a specific program rationale and discusses development of the whole child. Helps teachers meet literacy demands and anticipate appropriate developmental processes. Includes sections on staffing, equipment, and finance, planning activities, and meeting nutritional and health and safety needs. Provides a list of tests and a parent's handbook.

Donavin, Denise Perry, ed.
Best of the Best for Children: Books, Magazines, Videos, Audio, Software, Toys, Travel
New York, Random House, 1992. 366p. ISBN 0679404503

An excellent annotated list of books, magazines, videos and audiotapes, stories, folklore, computer software, toys, and travel destinations divided by age groups and accompanied by "connections"— additional material for follow-up. Includes many award winners.

Dutton, Wilbur H., and Ann Dutton
Mathematics Children Use and Understand: Preschool through Third Grade
Mountain View, Calif., Mayfield Pub. Co., 1991. 330p. ISBN 0874849683

The authors provide a mathematics activity chapter for each age level (four-to eight-years old), basing the sequence of activities upon Piaget's theories. The authors incorporate children's literature, features of the environment, and recipes. Other chapters deal with computers, calculators, classroom organization, and materials.

Dworetzky, John
Introduction to Child Development
5th ed. Minneapolis, Minn., West Pub. Co., 1993. 606p. ISBN 0314011358

Designed in SQ3R format (survey, question, read, recite, and review). Includes an overview of history and research, inheritance, the birth and infancy period, learning, language, cognition, social processes, and adolescence. Inviting format includes learning checks and applications.

Early Childhood Education and the Elementary School Principal: Standards for Quality Programs for Young Children
Alexandria, Va., National Association of Elementary School Principals, 1990. 60p.

Lists standards for quality programs based on current research, including an integrated curriculum, developmentally appropriate operations, and parent and community involvement.

Essa, Eva
Practical Guide to Solving Preschool Behavior Problems
2nd ed. Albany, N.Y., Delmar Publishers, 1990. 434p. ISBN 0827339658

Emphasis is on behavior modification techniques for solving aggressive, antisocial, disruptive, emotional, and dependent behaviors. Author also discusses nonparticipation, shyness, and various types of eating problems.

Field, Tiffany
Infancy
Cambridge, Mass., Harvard Univ. Press, 1990. 170p. ISBN 0674452623

Covers the social-emotional, perceptual, cognitive, and motor development of the infant up to age two through day-care experience. Focus is on prelanguage development.

Finn, Patrick J.
Helping Children Learn to Read
Rev. ed. New York, Longman, 1990. 372p. ISBN 0801303214

The author recommends a literature-based reading program, a language experience approach to beginning reading, and a process approach to writing. He describes two teaching methods: bottom-up empirical and top-down holistic. He also discusses two learning methods: individual and social. He recommends the top-down holistic, social pattern.

Flavell, John H., Patricia H. Miller, and Scott A. Miller
Cognitive Development
3rd ed. Englewood Cliffs, N.J., Prentice Hall, 1993. 408p. ISBN 0131400398

Describes five views of the nature and development of human cognition: (1) Piaget's assimilation-accommodation, (2) information processing, (3) neo-Piagetian, (4) contextual, and (5) biological-maturational. Covers cognitive development through infancy, early childhood, middle childhood, and adolescence. Concluding chapters deal with social cognition, memory, and language.

Foster, David R., and James L. Overholt
Indoor Action Games for Elementary Children: Active Games and Academic Activities for Fun and Fitness
West Nyack, N.Y., Parker Pub. Co., 1989. 225p. ISBN 0134591240

This is a collection of games or activities that can be conducted in limited space with a limited amount of equipment. For each, authors indicate appropriate grade level, purposes, equipment, and directions.

Frank, Marjorie
Kids' Stuff Book of Math for the Primary Grades
Nashville, Tenn., Incentive Publications, 1988. 240p. ISBN 0865300402

Contains supplemental math activities for primary teachers— some for teachers, others to be used by students. Includes an appendix with useful tables, terms, and definitions.

Fredericks, Anthony D.
Integrated Curriculum: Books for Reluctant Readers, Grades 2-5
Englewood, Colo., Teacher Ideas Press, 1992. 187p. ISBN 0872879941

Includes descriptions of specific books and the methods for utilizing these books for critical thinking. Geared to second through fifth grades, this work includes the following categories: reading/language arts, science/health, art, math, music, social studies, and physical education. Includes additional useful book lists. Author has done similar books such as *Involving Parents through Children's Literature* (Teacher Ideas Press, 1993) for preschool through kindergarten; Grades 1-2 (1992); Grades 3-4 (1993); and Grades 5-6 (1993).

Fredericks, Anthony D., Anita Meyer Meinbach, and Liz Rothlein
Thematic Units: An Integrated Approach to Teaching Science and Social Studies
New York, HarperCollins, 1993. 347p. ISBN 0065008928

Written by reading teachers, this book describes how to integrate children's literature into elementary science and social studies programs, providing fourteen thematic units for first through third grades. Also included are fourteen units for fourth through sixth grades.

Freeman, Yvonne S., and David E. Freeman
Whole Language for Second Language Learners
Portsmouth, N.H., Heinemann Educational Books, 1992. 257p. ISBN 0435087231

This work provides a good introduction to whole language instruction in strategic, authentic contexts for those who speak English as a second language.

Friedes, Harriet
Pre-School Resource Guide: Educating and Entertaining Children Aged Two through Five
New York, Plenum Press, 1993. 247p. ISBN 030644464X

Extremely useful for parents, teachers, caregivers, and librarians, this book gives selection criteria, annotated listings, and sources for acquiring books, magazines, toys, games, software, and audio and videocassettes. Also provides lists of book clubs, professional associations, and free or inexpensive materials.

Friedmann, Barb, and Cheri Brooks, eds.
On Base! The Step-by-Step Self-Esteem Program for Children from Birth to Eighteen
Kansas City, Mo., Westport, 1990. 201p. ISBN 0933701438

Professionally evaluated at the University of Kansas, BASE (Behavioral Alternatives through Self Esteem) is a series of short, user-friendly exercises which can be used at fifteen different age levels by parents, day-care providers, babysitters, teachers, relatives, or friends.

Fry, Edward Bernard
How to Teach Reading: For Teachers, Parents, Tutors
Laguna Beach, Calif., Laguna Beach Educational Books, 1992. 144p. ISBN 0876730233

Originally developed for training Peace Corps volunteers at Rutgers, this manual contains six steps for teaching reading: determining reading ability, selecting the right reading materials, reading aloud with comprehension, teaching vocabulary, developing phonics skills, and writing, speaking, and listening.

Fry, Prem S.
Fostering Children's Cognitive Competence through Mediated Learning Experiences: Frontiers and Futures
Springfield, Ill., Charles C. Thomas, 1992. 338p. ISBN 0398057761

Presents a conceptual treatment of research on cognitive development and provides broad strategies for promoting higher-order thinking skills to develop creative, self-directed learners. Helps teachers recognize appropriate age levels for fostering special skills.

Gabbard, Carl, Betty LeBlanc, and Susan Lowry
Physical Education for Children: Building the Foundation
2nd ed., Englewood Cliffs, N.J., Prentice Hall, 1994. 514p. ISBN 0136667287

This text for elementary physical education covers current trends and chapters on curriculum and planning, class management and discipline, children with special needs, and program activities for nonlocomotor, locomotor, and manipulative skill themes, movement awareness, physical fitness, games, rhythm/ dance, and gymnastics.

Garbarino, James, Frances M. Stott, and Faculty of the Erikson Institute
What Children Can Tell Us: Eliciting, Interpreting, and Evaluating Critical Information from Children
San Francisco, Calif., Jossey-Bass, 1992. 373p. ISBN 1555424651

Contributors emphasize that communication with children, particularly on significant issues, is facilitated by experience with them and knowledge about child development, particularly cognitive and language development. Chapters discuss cultural issues and adult biases in eliciting information, communicating through play and storytelling, guidelines for interviewing, and using tests. Later chapters deal with communication in special settings.

Garrod, Andrew, ed.
Approaches to Moral Development: New Research and Emerging Themes
New York, Teachers College Press, 1993. 244p. ISBN 0807732478

This collection examines the beginnings of moral development in children in the context of social concerns. Discusses gender differences that underscore the need for a whole child approach to education.

Garvey, Catherine
Play
Enlarged ed. Cambridge, Mass., Harvard Univ. Press, 1990. 184p. ISBN 0674673646

Includes current research on play including its use for building social competence. Discusses multicultural concerns, social pretending, the role of family members in play, gender identities and play styles, and the relationship between play and friendship, popularity, or peer status.

Genishi, Celia, ed.
Ways of Assessing Children and Curriculum: Stories of Early Childhood Practice
New York, Teachers College Press, 1992. 220p. ISBN 0807731862

Provides case studies in which teachers revealed their experiences with children and parents in a teacher training institution. They used developmentally appropriate practices and integration of experiences across subjects.

Gerard, Patty Carmichael, and Marian Cohn
Teaching Your Child Basic Body Confidence
Boston, Houghton Mifflin, 1988. 217p. ISBN 0395442532

Written by a world-class gymnast who has combined her own training techniques with the Body Color Theory principles of progressive skill learning, this book provides directions for developmentally appropriate activities for children from birth through six years. Illustrated.

Gillespie, John Thomas, and Corinne J. Naden, eds.
Best Books for Children: Preschool through Grade 6
5th ed. New York, Bowker, 1994. 1002p. ISBN 083523455X

Citations for more than 15,000 titles include author, illustrator, title, subject, price, and grade-level indexes. Materials are recommended for recreational reading and the typical school curriculum. Fiction titles are arranged by interest areas, and nonfiction by curricular topics.

Ginsburg, Herbert, and Sylvia Opper Brandt
Piaget's Theory of Intellectual Development
3rd ed. Englewood Cliffs, N.J., Prentice Hall, 1988. 264p. ISBN 013675158X

A brief introduction to Piaget's basic theories, suitable for undergraduates. Covers the periods of infancy, ages two through eleven, and adolescents. Concludes with a chapter on learning and development that incorporates his last works and presents the overall implications of his theories for education.

Goffin, Stacie G.
Curriculum Models and Early Childhood Education: Appraising the Relationship
New York, Merrill, 1994. 252p. ISBN 0675211549

Describes and compares five early childhood curriculum models: Montessori, developmental interaction, direct instruction, Kamii-DeVries, and High/Scope. Discusses the historical and contemporary forces which shaped them and the dilemma the profession faces in adopting a single model. Provides a good overview of current research.

Golomb, Claire
Child's Creation of a Pictorial World
Berkeley, Calif., Univ. of California Press, 1992. 353p. ISBN 0520070844

A scholarly treatment of children's art, from first graphic forms through differentiation, spatial dimensions, color, expression, and composition. Filled with children's black-and-white pencil drawings, the book discusses gifted children, art as a diagnostic tool, the child as art critic, and cultural variables.

Gonzalez-Mena, Janet
Multicultural Issues in Child Care
Mountain View, Calif., Mayfield Pub. Co., 1993. 91p. ISBN 1559342056

A practical approach to cultural differences in important aspects of child care. Discusses nonverbal communication among adults, and subjects such as preferences regarding a child's toilet training, sleeping, stimulation in the play environment, internalized controls, and punishment. Author gives examples of differing practices and ways caregivers can be sensitive to such differences and solve conflicts which arise.

Gonzalez-Mena, Janet, and Dianne Widmeyer Eyer
Infants, Toddlers, and Caregivers
Mountain View, Calif., Mayfield Pub. Co., 1989. 247p. ISBN 0874848792

Uses a ten-step Hungarian philosophy advanced by Magda Gerber and Tom Forrest for the care of children from birth through three years. According to this philosophy, education consists of facilitating problem solving. Provides clear examples of daily activities, ways to foster the development of the whole child, and ways to establish a quality program.

Gordon, Thomas
Discipline That Works: Promoting Self-Discipline in Children
New York, Plume, 1991. 258p. ISBN 0452266432

Gordon, developer of Parent Effectiveness Training (P.E.T.) and Teacher Effectiveness Training (T.E.T.), provides a number of alternatives to discipline and control, including participatory management, the Corsini Four-R system (C4R), and a six-step problem-solving process, all of which lead to greater self-esteem in children.

Gottfried, Adele Eskeles, and Allen W. Gottfried, eds.
Redefining Families: Implications for Children's Development
New York, Plenum Press, 1994. 237p. ISBN 030644559X

Discusses research on alternatives to the traditional nuclear family, such as those headed by single fathers, single mothers, lesbian mothers, gay fathers, or custodial grandparents. Although some research is tentative, most indicates that children do equally well in any family configuration that provides key factors such as love, encouragement, respect, and empathy.

Gould, Toni S.
Get Ready to Read: A Practical Guide for Teaching Young Children at Home and in School
New York, Walker, 1991. 166p. ISBN 0802773613

Revised edition of *Reading: The Right Start* (Educators Pub. Service, 1988). Describes a structural reading approach that combines phonics and a modified linguistic approach using linguistically regular forms of words to promote reading readiness. Describes and illustrates games and activities that teach reading, writing, and spelling.

Hammett, Carol Totsky
Movement Activities for Early Childhood
Champaign, Ill., Human Kinetics Books, 1992. 133p. ISBN 0873223527

This book covers developmentally appropriate activities to help children develop locomotor, ball-handling, gymnastic, and rhythmic skills. Author combines skill themes (actions) with movement concepts. Most activities appear to require relatively inexpensive supplies or equipment, which are described in the appendix.

Handbook for Public Playground Safety
Washington, D.C., Consumer Product Safety Commission, 1991. 31p.

This free publication is to be used in conjunction with the ASTM standard F1487-93 *Specifications for Playground Equipment for Public Use* (ASTM, 1993).

Harms, Thelma, Debby Cryer, and Richard M. Clifford
Infant/Toddler Environment Rating Scale
New York, Teachers College Press, 1990. 39p. ISBN 0807730106

Contains thirty-five items for rating the quality of child care centers for children up to thirty months of age. The definition of quality is consistent with the criteria listed by NAEYC and the requirements of the Child Development Associate credential. The seven-point evaluation scale has been tested for both reliability and validity.

Harrison, Annette
Easy-to-Tell Stories for Young Children
Jonesboro, Tenn., National Storytelling Press, 1992. 86p. ISBN 1879991128

Consists of directions for telling twelve simple stories to young children.

Hauser-Cram, Penny, and others
Early Education in the Public Schools: Lessons from a Comprehensive Birth-to-Kindergarten Program
San Francisco, Calif., Jossey-Bass, 1991. 261p. ISBN 1555423280

Part One demonstrates need for and benefits of ECE programs; Part Two describes the Brookline Early Education Project of the early 1970s; and Part Three outlines issues for policy makers, including the need to expand care throughout the day and year. Includes list of health and developmental measures for various ages.

Heltshe, Mary Ann, and Audrey Burie Kirchner
Multicultural Explorations: Joyous Journeys with Books
Englewood, Colo., Teacher Ideas Press, 1991. 276p.
ISBN 0872878481

Provides classroom-tested, integrated units on six different geographical locations: Hawaii, Australia, Japan, Italy, Kenya, and Brazil. The lesson plans for each include such aspects as flowers, food, animals, celebrations, music, dress, folklore, games, recipes, famous people, sites, and folk art. Can be adapted for different ages.

Herberholz, Barbara J., and Lee Hanson
Early Childhood Art
4th ed. Dubuque, Iowa, Wm. C. Brown, 1990. 292p.
ISBN 0697058638

Deals with art education from preschool through third grade in four categories: aesthetic perception, creative expression, aesthetic judgment, and art heritage. One chapter explains how to challenge the artistically talented and disabled. Others describe and illustrate activities involving prints, face painting, weaving, masks, collages, murals, paper sculptures, clay and dough modeling, puppets, and multicultural and holiday crafts.

Heward, William L., and Michael D. Orlansky
Exceptional Children: An Introductory Survey of Special Education
4th ed. New York, Merrill, 1992. 1 v. (various pagings).
ISBN 0675222001

This text emphasizes looking at individuals first and disabilities second. The inviting format includes focus questions, essays, interventions, key terms, chapter summaries, additional resources, and organizational lists. Provides a good overview of current legislation, school practices, ways of working with parents, early intervention, and adult alternatives.

Hinitz, Blythe Simone Farb
Teaching Social Studies to the Young Child: A Research and Resource Guide
New York, Garland Publishing, 1992. 164p. ISBN 0824044398

The first section gives an overview of teaching strategies and resources for geography, economics, and history for preschool children. The second covers these topics for grades K-3. The last section discusses current and future issues. Evaluates many resources, including the contents of publishers' series.

Holland, Kathleen E., Rachael A. Hungerford, and Shirley B. Ernst, eds.
Journeying: Children Responding to Literature
Portsmouth, N.H., Heinemann, 1993. 326p. ISBN 0435087584

Recommends literature that elicits verbal, artistic, dramatic, and written responses from children. Includes picture books, poetry, informational books, folk and fairy tales, science fiction, fantasy, historical and realistic fiction, and biography of interest to elementary and middle school classes.

Huck, Charlotte S., Susan Hepler, and Janet Hickman
Children's Literature in the Elementary School
5th ed. Fort Worth, Tex., Harcourt Brace Jovanovich, 1993. 775p.
ISBN 0030475287

Another classic text on children's literature, this new edition includes children's work from new literature-based school programs, an increase in the literature from other cultures in each genre chapter, criteria for each genre, trends in informational books, exercises in critical thinking, book talks, and readers' theater. Appealing format. Many suggestions for incorporating books into the curriculum.

Humphrey, James Harry
Elementary School Child Health: For Parents and Teachers
Springfield, Ill., Charles C. Thomas, 1993. 149p.
ISBN 0398058687

This book provides information for parents and teachers on health education for children ages five through twelve. It also includes material on children's emotional and social health, nutrition, stress, and relaxation.

Iatridis, Mary D.
Teaching Science to Children
2nd ed. New York, Garland Publishing, 1993. 199p.
ISBN 0815300905

Author reviews books on methods and activities for teaching science in elementary schools. A special chapter by Miriam Marecek reviews children's literature science books. The last chapter deals with science instruction for the special child. Emphasizes hands-on, developmentally appropriate activities.

Jalongo, Mary Renck
Early Childhood Language Arts
Boston, Allyn and Bacon, 1992. 364p. ISBN 0205132812

Covers topics of language acquisition such as literacy, listening, speaking, storytelling, children's books, writing, drawing, and assessment. Defines key concepts and gives practical examples readers can use in a child-centered language arts classroom. Appendix contains a glossary and listings of stories, poetry, wordless books, computer resources, and language tests.

Jansma, Paul, and Ronald W. French
Special Physical Education: Physical Activity, Sports, and Recreation
Englewood Cliffs, N.J., Prentice Hall, 1994. 596p.
ISBN 0138270562

Although this text is intended for physical education majors, it offers an extensive overview of disabling conditions and ways to mainstream students. Provides useful information for anyone in education. Describes how to adapt activities for various age groups.

Jones, Carroll J.
Social and Emotional Development of Exceptional Students: Handicapped and Gifted
Springfield, Ill., Charles C. Thomas, 1992. 201p.
ISBN 0398057818

The author examines the social, emotional, and self-concept theories of those who are mentally or physically impaired, behaviorally disordered, learning disabled, or gifted.

Joseph, Joanne M.
Resilient Child: Preparing Today's Youth for Tomorrow's World
New York, Insight Books, 1994. 355p. ISBN 0306446464

Though aimed at parents, this book is also useful for caregivers who want to develop self-esteem in young children. Describes how to evaluate individual temperaments and how to use stories and communication to teach children how to think constructively and make good decisions. Lists stories by grade level.

Justice, Jennifer, ed.
Ghost and I: Scary Stories for Participatory Telling
Cambridge, Mass., Yellow Moon Press, 1992. 126p.
ISBN 0938756370

Justice has collected ghost stories from famous storytellers for this volume. The first section (for ages five to eight) features brave girls, humor, and happy endings. In the second section (for ages nine to eleven), clever tricks defeat the ghosts. The stories in the third section for ages twelve and up feature the mysteries of life and death. The editor adds directions for participation and includes some songs.

Karen, Robert
Becoming Attached: Unfolding the Mystery of the Infant-Mother Bond and Its Impact on Later Life
New York, Warner Books, 1994. 500p. ISBN 0446516341

This work provides an overview of research on the infant-mother bond which follows a nature versus nurture approach and examines secure and anxious attachments in infants.

Kaye, Peggy
Games for Learning: Ten Minutes a Day to Help Your Child Do Well In School—From Kindergarten to Third Grade
New York, Farrar, Straus, Giroux, 1991. 251p. ISBN 0374522863

Collection of eighty-four clever games to use with young children that will help them with school subjects. Games enhance thinking, reading, mathematics, senses, writing, categorizing, and social studies concepts. Recommended ages included.

Kohl, Mary Ann F.
Mudworks: Creative Clay, Dough, and Modeling Experiences
Bellingham, Wash., Bright Ring Publishing, 1989. 150p.
ISBN 0935607021

Recipes and commercial sources for play dough, bread dough, plaster of paris, papier-mâché, edible dough art, sandworks, and other modeling mixtures. Each page contains a recipe and activity and is coded to indicate appropriate age level, whether cooking is needed before use, if the material is to be baked or air-dried, whether it is edible or not, and whether adult supervision is required.

Konner, Melvin
Childhood
Boston, Little, Brown, 1991. 451p. ISBN 0316501840

Written by an anthropologist to accompany a nine-part video series by WNET/Boston, this book examines the nature vs. nurture problem from birth through adolescence. Many parts reflect a nostalgia for childhood; others show great concern regarding serious problems facing children in the U.S.

Kostelnik, Marjorie J., and others
Guiding Children's Social Development
2nd ed. Albany, N.Y., Delmar Publishers, 1993. 547p.
ISBN 0827358598

Using an ecological perspective, the authors cover social development from birth through age twelve, including verbal and nonverbal communication, emotions, play, self-discipline, friendships, and prosocial behavior. Chapters open with student objectives and an overview of theory and research skills relevant to the subject. They end with summaries, discussion topics, and field assignments. Instructor's manual available.

Laughlin, Mildred, and Patricia Payne Kardaleff
Literature-Based Social Studies: Children's Books and Activities to Enrich the K-5 Curriculum
Phoenix, Ariz., Oryx Press, 1991. 148p. ISBN 0897746058

Outlines units around the report of the Curriculum Task Force of the National Commission on Social Studies in the Schools: *Charting a Course: Social Studies for the 21st Century* (The Commission, 1989). The sequence is traditional, starting with the family, moving to the community, then dealing with the nation, but it includes a number of multicultural topics such as Native Americans, world neighbors, immigrant groups, and folklore.

Lauter-Klatell, Nancy, comp.
Readings in Child Development
Mountain View, Calif., Mayfield Pub. Co., 1991. 178p.
ISBN 187484942X

Writings in this work were selected to accommodate a holistic approach to development. Includes major issues and writers in the field.

Leach, Penelope
Your Baby and Child: From Birth to Age Five
New ed., rev. and expanded. New York, Knopf, 1989. 553p.
ISBN 0394579518

A classic parenting manual, with a very appealing format. The first three chapters deal with the first year of life; Section Four focuses on the toddler; and Section Five addresses the needs of the preschool child. The appendix offers practical advice on topics such as first aid, safety, infectious diseases, and playthings.

Lehr, Susan S.
The Child's Developing Sense of Theme: Responses to Literature
New York, Teachers College Press, Columbia Univ., 1991. 203p.
ISBN 0807731064

Young children tend to interpret stories from their own experiences. The author provides examples of eliciting themes from picture books or folktales through class discussions, art, or writing assignments. She describes groups comparing various versions of folktales.

Lief, Nina R., and Rebecca M. Thomas
The Third Year of Life
New York, Walker, 1991. 205p. ISBN 0802711553
   Each chapter deals with developmental highlights and related issues and is followed by discussions with parents. Covers such topics as play, personality, discipline, changing family structure, preparing for nursery school, anger, and social behavior. Part of *The New York Medical College's Early Childhood Development Center's Parenting Series.*

Lima, Carolyn W., and John A. Lima
A to Zoo: Subject Access to Children's Picture Books
3rd ed. New York, Bowker, 1989. 989p. ISBN 0835225992
   Includes 12,000 picture book titles listed under 700 subjects. Entries are not annotated. Includes subject, title, and illustration indexes, a bibliographical guide, and a chapter describing the genesis of the English-language picture book.

Livo, Norma J., and Sandra A. Rietz
Storytelling Folklore Sourcebook
Englewood, Colo., Libraries Unlimited, 1991. 384p.
ISBN 0872876012
   This is a fascinating collection of many types of folklore and references to additional sources of folklore. It includes sections on cultural heroes, story artifacts, customs, divination, magic and superstition, games, numbers, motifs, and humor.

Maffei, Anthony C., and Theresa M. Hauck
Purposeful Play with Your Preschooler: A Learning-Based Activity Book
New York, Plenum Press, 1992. 291p. ISBN 0306443252
   Discusses the intellectual development of children from birth through age five, covering language, thinking, and concepts of number, space, and time. Chapters describe play activities for home, park, car, store, or anywhere. The activities develop gross motor and self-help skills, and skills in mathematics, writing, science, and language. Appendices include lists of equipment and supplies, literature, recordings, videos, and software.

Mallory, Bruce L., and Rebecca Staples New, eds.
Diversity and Developmentally Appropriate Practices: Challenges for Early Childhood Education
New York, Teachers College Press, Columbia Univ., 1994. 295p. ISBN 0807733008
   This collection of essays challenge the "appropriate and inappropriate practices" outlined in NAEYC's 1987 guidelines, particularly in reference to the cultural, developmental, class, and gender differences in children. Presents perceived differences between the values, beliefs, and goals reflected in a typical program and those held by the parents and children served by the program.

Maurer, Daphne, and Charles Maurer
World of the Newborn
New York, Basic Books, 1988. 293p. ISBN 0465092306
   Based on extensive research, this book presents the latest information on how babies experience the world through the senses and physical interactions.

McLane, Joan Brooks, and Gillian Dowley McNamee
Early Literacy
Cambridge, Mass., Harvard Univ. Press, 1990. 161p.
ISBN 0674221648
   Based on the authors' years of observing young children, this work describes various ways in which youngsters have been motivated to try to read and write in their homes, neighborhoods, and schools.

McMillan, Dana, and the Learning Exchange, Inc.
Teacher Tips for Action Time
Palo Alto, Calif., Monday Morning Books, 1990. 79p.
ISBN 0878279092
   This work includes indoor and outdoor activities for children ages three to seven that require minimal materials and preparation. They feature pantomime, role-playing, dramatic play, beanbags, balloons, bubbles, dance, marches, games, and problem-solving. Also includes one-on-one activities for parents to use.

McMillan, Dana, and the Learning Exchange, Inc.
Teacher Tips for Circle Time
Palo Alto, Calif., Monday Morning Books, 1990. 64p.
ISBN 1878279084
   The activities described in this book for children ages three to seven promote listening, motor skills, discrimination, problem-solving, patterning, and music appreciation.

McMillan, Dana, and the Learning Exchange, Inc.
Teacher Tips for Quiet Time
Palo Alto, Calif., Monday Morning Books, 1990. 64p.
ISBN 1878279106
   Activities described in this book for children ages three to seven feature labeling, matching, categorizing, measuring, sequencing, and letters and sounds.

Mertzlufft, Bonnie, Brenda Morton, and Virginia Woolf
Beginning Bulletin Boards: Basic Skills
Palo Alto, Calif., Monday Morning Books, 1990. 128p.
ISBN 187827905X
   Provides patterns for four bulletin boards that involve active learning activities for kindergarten and first-grade children.

Miller, Teresa, comp.
Joining In: An Anthology of Audience Participation Stories and How to Tell Them
Cambridge, Mass., Yellow Moon Press, 1988. 125p.
ISBN 0938756214
   Each story in this anthology is introduced by the contributor. The text appears on the left-hand side, while directions for the audience's response or teller's actions appear on the right. Stories originated in a variety of cultures and countries.

Mills, Heidi, Timothy O'Keefe, and Diane Stephens
Looking Closely: Exploring the Role of Phonics in One Whole Language Classroom
Urbana, Ill., National Council of Teachers of English, 1992. 68p.
ISBN 0814130313
   Utilizing children's literature, the authors describe a whole language approach to teaching reading which incorporates phonics, syntax, and semantics.

Mohr, Carolyn, Dorothy Nixon, and Shirley Vickers
Thinking Activities for Books Children Love: A Whole Language Approach
Englewood, Colo., Teacher Ideas Press, 1988. 206p.
ISBN 0872876977

For fifteen titles popular with third to sixth graders, the authors have compiled lists of vocabulary, questions testing knowledge and comprehension, and applied activities, related to art, drama, creative writing, music, social studies, and science. This work can be used to promote critical thinking, problem-solving, cooperative teams, and independent study, or to motivate gifted students. Teachers can apply these models to other books.

Montgomery, Paula Kay
Approaches to Literature through Subject
Phoenix, Ariz., Oryx Press, 1993. 243p. ISBN 0897747747

For use by teachers and librarians, this book explains how to diagnose students' interests in order to match students with books about people, places, things, or events. Primarily for upper-elementary grade students. Although book lists are not annotated, sample activities are included to motivate student interest.

Moore, Matthew S., and Linda Levitan
For Hearing People Only: Answers to Some of the Most Commonly Asked Questions about the Deaf Community, Its Culture, and the 'Deaf Reality'
2nd ed. Rochester, N.Y., Deaf Life Press, 1993. 334p.
ISBN 0963401610

This book would be useful for anyone working with hearing impaired students. It provides factual information from the perspective of the hearing-impaired. Includes topics such as American Sign Language, deaf children's learning problems, closed captioning, and the experience of parenting by the deaf.

Moore, Robin
Awakening the Hidden Storyteller: How to Build a Storytelling Tradition in Your Family
Boston, Shambhala, 1991. 153p. ISBN 0877735999

Presents step-by-step instructions that readers can use to learn to tell stories or to teach others to tell stories, especially in a family context. Includes good lists of resources.

Morgan, John D., ed.
Young People and Death
Philadelphia, Pa., Charles Press, 1991. 202p. ISBN 0914783491

Provides advice and guidelines for caregivers to help children while they integrate a loss or the threat of a loss. Chapters cover the child's understanding of death, the bereaved child and family, and the role of the school.

Morrison, George S.
Early Childhood Education Today
5th ed. New York, Macmillan, 1991. 549p. ISBN 0675213428

Provides good historical overview of early childhood education with separate chapters on Montessori and Piaget. Chapters discriminate between early child care, preschool, kindergarten, and primary years. Presents excellent coverage of current issues and future trends. Appendix includes NAEYC's "Code of Ethical Conduct," a sample Individualized Family Service Plan, and lists of journals and relevant associations.

Moses, Kathy
It's a Child's World
Minot, N.D., North American Heritage Press, 1989. 168p.
ISBN 0942323068

Chapters outline useful information on aspects of running day-care centers, including routines, safety, pets, and discipline. Provides a list of ways to recognize children with learning disabilities.

Munger, Richard L.
Changing Children's Behavior Quickly
Landham, Md., Madison Books, 1993. 213p. ISBN 1568330014

A book for parents and professionals which takes some of the most successful techniques in child management and presents them clearly and concisely so they can be applied quickly. Chapters deal with rules, time-outs for young children, taming the hyperactive child, helping to control bed-wetting, and addressing the needs of a child with divorced parents.

Newkirk, Thomas, and Patricia McClure
Listening In: Children Talk About Books (and Other Things)
Portsmouth, N.H., Heinemann, 1992. 158p. ISBN 0435087134

The authors recorded children's book discussions in a combined first- and second-grade classroom at Mast Way School in Lee, New Hampshire. This work recreates the dynamics and daily operations of the class where the activity center is central and the teacher assesses development of children through their book discussions.

Paley, Vivian Gussin
The Boy Who Would Be a Helicopter
Cambridge, Mass., Harvard Univ. Press, 1990. 163p.
ISBN 0674080300

Case study of a master teacher using storytelling techniques in a preschool to foster social integration, alternative behaviors, moral education, listening skills, and imaginative problem solving. Focuses on play therapy as it transforms one isolated child.

Paley, Vivian Gussin
You Can't Say You Can't Play
Cambridge, Mass., Harvard Univ. Press, 1992. 134p.
ISBN 0674965892

A teacher who uses storytelling techniques introduces the concepts of friendship and feelings of rejection to her kindergarten class. She adds a magical character to the storytelling in an attempt to deal with childhood relationships.

Pangrazi, Robert P., and Victor P. Dauer
Dynamic Physical Education for Elementary School Children
10th ed. New York, Macmillan Pub. Co., 1992. 723p.
ISBN 0023278218

This text presents physical fitness and developmentally appropriate activities in order of difficulty. An accompanying text titled *Lesson Plans for Dynamic Physical Education for Elementary School Children* (Macmillan, 1992) organizes the activities into lesson plans for an academic year.

Pellowski, Anne
Hidden Stories in Plants: Unusual and Easy-to-Tell Stories from Around the World Together with Creative Things to Do While Telling Them
New York, Macmillan, 1990. 93p. ISBN 0027706117

Pellowski groups these stories around the uses the storyteller makes of the flowers or plants: ornaments, disguises, playthings, dolls, or musical instruments. She provides the sources of these and similar stories, and lists holidays and other times to celebrate plants. The author uses natural objects found in woods, fields, parks, gardens, or kitchens.

Pellowski, Anne
World of Storytelling
Expanded and rev. ed. Bronx, N.Y., H.W. Wilson, 1990. 311p. ISBN 0824207882

For teachers, librarians, storytellers, or those interested in incorporating multicultural elements into the classroom, this book is a treasure. Pellowski distinguishes between different types of storytelling—bardic, religious, folk, theatrical, library, camp, and therapeutic—and describes stories told in these different modes and in different cultures. She tells how to incorporate musical instruments, pictures, and objects and provides training materials. Includes lists of festivals, books, articles, and collections.

Perdue, Peggy K.
Small Wonders: Hands-On Science Activities for Young Children
Glenview, Ill., Scott, Foresman, 1989. 66p. ISBN 0673381986

For both preschool and primary school teachers, this book contains science lab projects which utilize everyday materials. For each project the author indicates materials and preparation required, a focusing activity, procedure for the lab, and evaluation and extension activities.

Raines, Shirley C., and Robert J. Canady
Whole Language Kindergarten
New York, Teachers College Press, 1990. 272p.
ISBN 0807730491

Authors describe the whole language kindergarten as one in which children are immersed in spoken and written language with emphases on listening, speaking, reading, and writing, i.e., meaningful communication. Instructors build on the skills the children bring with them. Discusses play, writing, reading, science, art, music, housekeeping, and blocks in the kindergarten classroom.

Ramirez, Gonzalo, and Jan L. Ramirez
Multiethnic Children's Literature
Albany, N.Y., Delmar Publishers, 1994. 158p. ISBN 0827354339

Includes annotated bibliographies of children's books for and about Latinos/Hispanics, African Americans, Asian Americans, and Native Americans. Includes valuable activities for using multiethnic stories in the classroom.

Rickert, Colleen M., and Mary Gold
Teaching Reading in the Social Studies: A Global Approach for Grades K-5
Rev. ed., by Diane R. Simmons. Denver, Colo., Univ. of Denver. Center for Teaching International Relations, 1991. 83p.
ISBN 0943804744

This work shows how to teach reading skills in the social studies curriculum. Themes discussed include change, environment, the world, other cultures, and travel.

Roemer, Joan, as told to Barbara Leslie Austin
Two to Four from 9 to 5: The Adventures of a Daycare Provider
New York, Harper and Row, 1989. 294p. ISBN 0060160853

Chronicles everyday adventures of the director of a day-care program in Oakland, California.

Roopnarine, Jaipaul L., James E. Johnson, and Frank H. Hooper, eds.
Children's Play in Diverse Cultures
Albany, N.Y., State Univ. of New York Press, 1994. 234p. ISBN 0791417530

The current emphasis on multicultural approaches to early childhood education makes this book especially useful through its coverage of the way society and schools value play in East India, Taiwan, Japan, Polynesia, Puerto Rico, Italy, and Africa.

Ruff, Thomas P.
Teaching Social Studies in Grades K-8: Information, Ideas, and Resources for Classroom Teachers
Boston, Allyn and Bacon, 1994. 232p. ISBN 0205146066

Designed to complement social studies texts, this excellent book provides a multitude of recommended activities, books, primary resources, and pertinent information regarding geography, cultural diversity, history of the U.S., music, folklore, art, literacy, ecology, and the calendar.

Russell, David L.
Literature for Children: A Short Introduction
New York, Longman Pub. Group, 1991. 178p. ISBN 0801306736

Provides a concise overview of the history of children's literature and its place in child development. Defines and illustrates the various genres of imaginative and folk literature, provides chapters on biographical and informational books, and ends with listings of children's book awards.

Rust, Frances O'Connell
Changing Teaching, Changing Schools: Bringing Early Childhood Practice into Public Education: Case Studies from the Kindergarten
New York, Teachers College Press, Columbia Univ., 1993. 125p.
ISBN 080773289

The author recommends transforming elementary education by infusing the early childhood experience-based instructional model into the traditional approaches.

Sanders, Danielle M.
Teaching Deaf Children: Techniques and Methods
Boston, Little, Brown, 1988. 158p. ISBN 0316770159

With many years of experience teaching deaf children, the author outlines factors that influence the total development of deaf children and how to apply these factors in preparing materials and selecting teaching techniques.

Santrock, John W.
Child Development
6th ed. Madison, Wisc., Brown and Benchmark, 1994. 605p. ISBN 0697145123

Color insets enhance reader interest and help to highlight concepts, questions, and examples. This excellent text has a global viewpoint and deals with current issues such as teen pregnancy, diversity, STDs, working and divorced parents, and homelessness. Focuses on learner-centered education and emotional, physical, moral, cognitive, and social factors. Includes a good review of Piaget.

Saracho, Olivia N., ed.
Cognitive Style and Early Education
New York, Gordon and Breach, 1990. 231p. ISBN 2881247458

The emphasis of this book is on identifying and matching teaching and learning styles to enhance learning for low-income and cross-ethnic groups and to provide a greater variety of learning experiences for children in ECE programs.

Scales, Barbara, and others, eds.
Play and the Social Context of Development in Early Care and Education
New York, Teachers College Press, Columbia Univ., 1991. 275p. ISBN 080773067X

These articles, based on papers presented given at a symposium, highlight the need for improved communication between researchers and practitioners. Focus is on developmentally appropriate practices in ECE.

Schiller, Pamela Byrne, and Patricia Dyke
Managing Quality Child Care Centers: A Comprehensive Manual for Administrators
New York, Teachers College Press, Columbia Univ., 1990. 133p. ISBN 0807729760

Addresses managing a child care center including common problems and suggested solutions for finances, personnel, public relations, safety, and developmental issues.

Schirrmacher, Robert
Art and Creative Development for Young Children
Albany, N.Y., Delmar Publishers, 1988. 312p. ISBN 0827330332

Written for ECE practitioners by a community college ECE instructor, this work provides a discussion on designing art programs, children's creativity and artistic development, elements of art, establishing an art center, and art activities. Art recipes are also provided along with lists of resources for art materials and information.

Sciarra, Dorothy June, and Anne G. Dorsey
Developing and Administering a Child Care Center
2nd ed. Albany, N.Y., Delmar Publishers, 1990. 418p. ISBN 0827336667

A how-to book that discusses preplanning for a child care center including assessment of community need and establishment of the program. Also contains information on licensing, certifying, handling finances, planning, equipping and staffing a facility, selecting children, maintaining health and safety, and managing staff. Discusses working with parents, staff, volunteers, and the community served.

Seefeldt, Carol
Social Studies for the Preschool-Primary Child
4th ed. New York, Merrill, 1993. 312p. ISBN 0024084514

Based on recommendations of *The National Commission on Social Studies in the Schools in Charting a Course* (The Commission, 1989), this work moves away from previous emphases on home and neighborhood, social living, and holidays. Chapters cover lesson planning, resources, and key concepts; and appropriate ECE activities for history, geography, social studies, and cross-cultural education. Includes discussions on skills, attitudes, and values.

Seefeldt, Carol, ed.
Early Childhood Curriculum: A Review of Current Research
2nd ed. New York, Teachers College Press, Columbia Univ., 1992. 339p. ISBN 080773196X

A critical look at what ECE staff are teaching and what children are learning in light of current research on the integrated curriculum, teaching strategies, play, language, reading, mathematics, social studies, movement development, music, and art. Describes integrated Italian programs more fully explained in *The Hundred Languages of Children: The Reggio Emilia Approach to Early Childhood Education* by Carolyn P. Edwards and Lella Gandini (Ablex, 1993).

Shea, Thomas M., and Anne M. Bauer
Learners with Disabilities: A Social Systems Perspective of Special Education
Madison, Wisc., Brown and Benchmark, 1994. 493p. ISBN 0697153703

This text uses a social systems or ecological perspective. Each chapter begins with learning objectives, outlines key words and phrases, and gives clear descriptions, illustrations, and summaries. Includes issues, trends, and a glossary.

Shell, Ellen Ruppel
A Child's Place: A Year in the Life of a Day Care Center
Boston, Little, Brown, 1992. 264p. ISBN 0316783765

An in-depth profile of a multicultural day-care center in Cambridge, Massachusetts. Focuses on the teacher and three families who are involved in it. The journalist/author supports minimum national standards for such centers.

Shelov, Steven P., and others
Caring for Your Baby and Young Child: Birth to Age Five
New York, Bantam Books, 1991. 676p. ISBN 0553071866

Covers growth and development and basic care of the young child, as well as safety factors, part-time care, medical emergencies, health, behavior, family problems, and immunizations.

Sierra, Judy, and Robert Kaminiski
Multicultural Folktales: Stories to Tell Young Children
Selected translations by Adela Artola Allen, Phoenix, Ariz., Oryx Press, 1991. 126p. ISBN 0897746880

Authors provide directions for telling stories with flannel boards and puppets and include a section on folktales for preschool children and another for ages five to seven. The folktales are from many different countries. Includes patterns for flannel characters and instructions for hand puppets.

Sierra, Judy, and Robert Kaminiski
Twice Upon a Time: Stories to Tell, Retell, Act Out, and Write About
New York, H. W. Wilson, 1989. 232p. ISBN 0824207750

Authors present stories consistently successful with children ages eight to twelve. The opening chapters outline proven techniques for storytelling, creative dramatics, and writing. Each story is accompanied by tips for effective storytelling.

Slavin, Robert E., Nancy L. Karweit, and Barbara A. Wasik, eds.
Preventing Early School Failure: Research, Policy, and Practice
Boston, Allyn and Bacon, 1994. 237p. ISBN 0205139914

Reviews research on various options to keep at-risk students from failing. Includes suggested interventions for infants to three year olds, preschool programs for four-year-olds, an expanded Head Start Program, extra-year kindergarten programs, nongraded primaries, transitional first grades, and tutoring.

Smith, Charles A.
From Wonder to Wisdom: Using Stories to Help Children Grow
New York, New American Library, 1989. 304p. ISBN 0453006973

The author recommends stories on eight self-development themes, such as setting goals, confronting challenges, solving problems, and dealing with loss and grief. Smith reviews picture books for children ages two to eight.

Smutny, Joan F., Kathleen Veenker, and Stephen Veenker
Your Gifted Child: How to Recognize and Develop the Special Talents in Your Child from Birth to Age Seven
New York, Ballantine Books, 1991. 212p. ISBN 0345368304

Useful for both parents and ECE staff, this book distinguishes between bright and gifted children and suggests ways to stimulate creativity and enhance special skills. Appendix includes useful skills checklists for children from birth to kindergarten.

Spock, Benjamin, and Michael B. Rothenberg
Dr. Spock's Baby and Child Care
6th ed. New York, Pocket Books, 1992. 832p. ISBN 0671759671

This classic text covers parenting from birth through adolescence with special sections on health problems, first aid, and family relations.

Spodek, Bernard, and Olivia N. Saracho, eds.
Language and Literacy in Early Childhood Education
New York, Teachers College Press, Columbia Univ., 1993. 210p. ISBN 080773280X

The editors of this fourth volume in the *Yearbook in Early Childhood Education* series present material that reflects major changes in developmental theory, cross-cultural school enrollments, and the teaching of reading. Topics addressed are language skills instruction, multicultural children's literature, the classroom environment, and the role of parents and teachers in literacy development.

Spodek, Bernard, and Olivia N. Saracho
Dealing with Individual Differences in the Early Childhood Classroom
New York, Longman, 1994. 402p. ISBN 0801304512

Focusing on regular classrooms that include gifted, at-risk, and disabled children, this book deals with identification and assessment of student needs, working with parents, organizing the classroom, fostering social learning, educational play, creativity, teaching regular classroom subjects, and professional development.

Spodek, Bernard, and Olivia N. Saracho, eds.
Early Childhood Teacher Preparation
New York, Teachers College Press, Columbia Univ., 1990. 240p. ISBN 0807730424

Chapters cover the historical foundations of ECE teacher preparation, describe current training programs at many levels, and place teacher preparation in a cross-cultural perspective. Predicts that in the future, care and education may be combined into "educare," which will have an impact on vocational and community colleges and their ECE personnel training programs. *Yearbook in Early Childhood Education* series, 1.

Standard Consumer Safety Playground Specification for Playground Equipment for Public Use, ASTM F1487-93
Philadelphia, Pa., ASTM, 1994. unpaged.

An essential standard to consult when designing or evaluating playground equipment. Use in conjunction with the *Handbook for Public Playground Safety* (Consumer Product Safety Commission, 1991).

Starr, Richard
Woodworking with Your Kids
New rev. ed. Newtown, Conn., Taunton Press, 1990. 205p. ISBN 0942391616

Written by a public school teacher, this book contains four chapters of woodworking projects appropriate for children aged five to ten. It features step-by-step directions and appealing photographs of children making a toy man, airplane, car, and box.

Stegelin, Dolores, ed.
Early Childhood Education: Policy Issues for the 1990s
Norwood, N.J., Ablex Pub. Corp., 1992. 227p. ISBN 0893917974

A compilation of articles for ECE teachers and students as well as legislators that highlight public policy issues relating to children and families: standards of accountability, early intervention for the disadvantaged and developmentally delayed and handicapped, multicultural influences, and developmentally appropriate practices. Describes the Perry Preschool Program Study.

Stephens, Diane, ed.
What Matters? A Primer for Teaching Reading
Portsmouth, N.H., Heinemann, 1990. 76p. ISBN 0435085247

This book discusses the teacher's need to understand what readers bring to the text, to capitalize on the fact that reading is strategic, and to use specific questions and responses to elicit learning.

Stern, Daniel N.
Diary of a Baby
New York, Basic Books, 1990. 165p. ISBN 0465016421

Case study of the psychological development of a child from birth to age four.

Stewart, Bernice, and Julie S. Vargas
Teaching Behavior to Infants and Toddlers: A Manual for Caregivers and Parents
Springfield, Ill., Charles C. Thomas, 1990. 256p.
ISBN 0398056382

This manual emphasizes positive reinforcement to shape appropriate infant behavior. Discusses streamlining routine caregiving tasks within mandated standards common to the U.S., Canada, and Australia.

Stewig, John W., and Sam Leaton Sebesta, eds.
Using Literature in the Elementary Classroom
Rev. and enl. ed. Urbana, Ill., National Council of Teachers of English, 1989. 132p. ISBN 0814156185
Written in response to the whole language approach to instruction, this work includes articles on the use and nature of picture books, and using children's literature across the curriculum to advance reading, writing, and drama.

Stone, Janet I.
Hands-on Math: Manipulative Math for Young Children
Glenview, Ill., Scott, Foresman, 1990. 148p. ISBN 0673384632

Stone describes cross-curricular math projects for children aged three to six. The last section includes playsheets required for some of the projects.

Sunal, Cynthia S.
Early Childhood Social Studies
Columbus, Ohio, Merrill Pub. Co., 1990. 251p. ISBN 0675209609

Describes three models for teaching young children social studies: social exploration, experimentation, and concept attainment. Includes chapters on multiculturalism; psychology and values; history; geography; consumerism; and aging, the elderly, and death.

Sutherland, Zena, and May Hill Arbuthnot
Children and Books
8th ed. New York, HarperCollins, 1991. 768p. ISBN 0673463576

This updated classic in the field provides an extensive overview of the history of children's literature, ways to use literature with children, and chapters describing the various genres, giving outstanding examples and extensive bibliographies for each. Also focuses on major authors and artists and lists criteria for judging each genre. Discusses censorship, research, and book awards.

Taylor, Barbara J.
Science Everywhere: Opportunities for Very Young Children
Fort Worth, Tex., Harcourt Brace Jovanovich, 1993. 289p.
ISBN 0030541948

This work is devoted to principles and theories of teaching science to young children. Discusses methods of teaching science through discovery learning, keeping safety considerations in mind. Includes activities and resource materials on a variety of science topics.

Thomas, Glyn V., and Angele M. J. Silk
Introduction to the Psychology of Children's Drawings
New York Univ. Press, 1990. 177p. ISBN 0814781845

A fascinating book, though most of the research on the psychological theories reviewed seems inconclusive or unscientific. The authors examine the significance of various aspects of drawings, especially those done by children and exceptional or emotionally disturbed individuals. They also speculate on whether drawing may assist children's thinking and problem-solving skills.

Thomas, James L.
Play, Learn, and Grow: An Annotated Guide to the Best Books and Materials for Very Young Children
New Providence, N.J., Bowker, 1992. 439p. ISBN 0835230198

Includes information about using books (in print and other formats) with infants, toddlers, and preschoolers. Annotated entries are cross-indexed by age and subject categories. Includes criteria for selection and information for purchase.

Thompson, Charles L., and Linda B. Rudolph
Counseling Children
3rd ed. Pacific Grove, Calif., Brooks/Cole Pub. Co., 1992. 576p.
ISBN 0534171966

For use by teachers and parents as well as counselors, this book presents practical, up-to-date methods for helping children with particular developmental, social, or behavioral problems, including exceptional students and those who have problems dealing with divorce, death, abuse, homelessness, alcoholism, or AIDS. Describes nine counseling theories (and the techniques associated with each) and cross-cultural counseling.

Tobin, Joseph Jay, David Y. H. Wu, and Dana H. Davidson
Preschool in Three Cultures: Japan, China, and the United States
New Haven, Conn., Yale Univ. Press, 1989. 238p.
ISBN 0300042353

Useful for those wanting a multicultural approach to ECE, this book provides a comparative perspective on the values and goals of parents and schools in these three cultures.

Toman, Walter
Family Constellation: Its Effects on Personality and Social Behavior
4th ed. New York, Springer Pub. Co., 1993. 307p.
ISBN 0826104967

Describes the effects on children of various social and family dynamics, including sibling roles and positions, types of parents, friendships, and parent-child relations. Includes case studies.

Trawick-Smith, Jeffrey W.
Interactions in the Classroom: Facilitating Play in the Early Years
New York, Merrill, 1994. 383p. ISBN 0024125113

Defines play as "any activity self-chosen, open-ended, spontaneous, and enjoyable" which helps children learn in areas such as literacy, mathematics, science, arts, and motor development. Describes strategies for facilitating play at appropriate times with a consideration of cultural differences.

Trelease, Jim
New Read-Aloud Handbook
New York, Penguin Books, 1989. 290p. ISBN 0140468811

This well-known book explains the best ways to read to young children at home or at school, outlines the effects of television on children, and provides lists of proven, recommended books—predictable, wordless, reference, picture, short novels, novels, poetry, and anthologies. Many are annotated.

Turecki, Stanley, and Leslie Tonner
Difficult Child
Rev. ed. New York, Bantam Books, 1989. 258p.
ISBN 0553344463
 Written by a family and child psychiatrist, this book helps readers assess a difficult child, then regain authority and establish appropriate scheduling and discipline.

Vander Woude, Judith, and Kimberly Montero
Language Lessons for the Curriculum: Level 1
East Moline, Ill., LinguiSystems, 1991. 157p. ISBN 1559991631
 One of a series of books written by school speech-language pathologists to incorporate language arts into the curriculum. Each has fifty thematic lessons aimed at a particular interest level.

Vasquez-Nutall, Ena, Ivonne Romero, and Joanne Kalesnik, eds.
Assessing and Screening Preschoolers: Psychological and Educational Dimensions
Boston, Allyn and Bacon, 1992. 481p. ISBN 0205132804
 This text provides an overview and an ecological model to guide the process of assessing preschoolers. Surveys normal and abnormal development of preschoolers and describes the categories detailed in P.L. 94-142 and P.L. 99-451 for children up to age five. Also deals with ethical and technical issues, including evaluating special populations, e.g., the culturally different, newborns, the disabled and gifted; and concludes with recording and implementing the results of assessments or screening activities.

Vasta, Ross, ed.
Six Theories of Child Development: Revised Formulations and Current Issues
Philadelphia, Jessica Kingsley Publishers, 1992. 285p.
ISBN 1853021377
 Originally published as six volumes of the *Annals of Child Development* (JAI Press, 1989), this work covers the six most important theories now used by researchers in the field: social cognitive, behavior analysis, Piagetian, information-processing, ecological, and ethnological.

Vasta, Ross, Marshall M. Haith, and Scott A. Miller
Child Psychology: The Modern Science
New York, Wiley, 1992. 665p. ISBN 0471887544
 This research-oriented study focuses on current issues, integrates family and school influences on child development, and takes a balanced theoretical approach. Part One covers history, theory, and research methods; Part Two addresses biological and physical development; Part Three looks at sensory and perceptual development; and Part Four discusses social, emotional, gender, and moral development. Information is presented on an appealing format with highlighted text. Can accompany the Annenberg *Seasons of Life* video series.

Warren, Jean
1-2-3 Math: Pre-Math Opportunities for Working with Young Children
Everett, Wash., Warren Pub. House, 1992. 157p.
ISBN 0911019529
 The author describes activities to develop pre-math skills and includes a section listing ways to incorporate these into language arts, music, art, science, and playtime.

Warren, Jean
1-2-3 Reading and Writing: Pre-Reading and Pre-Writing Opportunities for Working with Young Children
Everett, Wash., Warren Pub. House, 1992. 157p.
ISBN 0911019472
 Contains activities that prepare children for reading and writing such as developing eye-hand coordination; visual, auditory, and sensory discrimination; using various tools and surfaces; and recognizing letters and numbers. The second half includes pre-reading and writing activities.

Weis, Lois, and others, eds.
Critical Perspectives on Early Childhood Education
Albany, N.Y., State Univ. of New York Press, 1991. 282p.
ISBN 0791406970
 This collection of essays focuses on teachers, schools, and the social and economic contexts of early childhood education.

Wilson, George, and Joyce Moss
Books for Children to Read Alone: A Guide for Parents and Librarians
New York, Bowker, 1988. 184p. ISBN 0835223469
 Includes an annotated list of books on subjects that appeal to children pre-kindergarten through third grade. Young children actually evaluated the books and pictures. Includes readability, subject, author, and title indexes and a list of popular books in series.

Wodrich, David L., and Sally A. Kush
Children's Psychological Testing: A Guide for Nonpsychologists
2nd ed. Baltimore, Md., Paul H. Brookes, 1990. 288p.
ISBN 1557660468
 Authors attempt to make readers informed consumers of psychological tests, explaining how to make a referral, how to use infant scales to diagnose early developmental delays, and how to use preschool tests that diagnose mental retardation and learning disabilities. Also covers tests to measure intelligence, ability, academic achievement, and emotional problems. Deals with contemporary problems and issues regarding testing.

Zabel, Mary Kay
Teaching Young Children with Behavioral Disorders
Reston, Va., Council for Exceptional Children, 1991. 23p.
ISBN 0865862001
 A book in the series produced by the Council for Exceptional Children, *Working with Behavioral Problems*. This work gives a historical overview of research on the topic, provisions of laws or prevailing school policies, and current strategies for behavior modification and social skills development.

Zubrowski, Bernie
Balloons: Building and Experimenting with Inflatable Toys
New York, Morrow Junior Books, 1990. 79p. ISBN 0688083250
 A title in the *Boston Children's Museum Activity Books* series, which could be used for teaching elementary science. Other books deal with ballpoint pens, bubbles, clocks, baking, drinking straws, water pumps, milk carton blocks, raceways, tops, and wheels.

Nonprint Resources

Beatrix Potter: Artist, Storyteller and Countrywoman
Weston, Conn., Weston Woods, 1993. 1 videocassette (55 min.) VHS

Based on the biography by Judy Taylor and narrated by Lynn Redgrave, this videocassette presents Potter's life, her books, her animals, and the English countryside around her homes. Winner of the CINE Golden Eagle Award.

The Caldecott Video Library, Volume Four: Stories from the Black Tradition
Weston, Conn., Weston Woods, 1992. 1 videocassette (46 min.) VHS

Includes four Caldecott winners: *A Story, A Story, Mufaro's Beautiful Daughters, The Village of Round and Square Houses*, and *Why Mosquitoes Buzz in People's Ears*.

Child Development: One to Three
Bloomington, Ill., Meridian Education Corp., 1990. 1 videocassette (17 min.) VHS

Covers primary developmental features of children ages one to three in regard to motor and language abilities and initial socialization skills. Each feature described is then highlighted by interviews with parents who describe their experiences.

Considerations of Discipline
Bloomington, Ill., Meridian Education Corp., 1990. 1 videocassette (13 min.) VHS

This video covers three types of discipline—emotional, rational, and developmental—and explains the superiority of the latter. Outlines various ways parents can positively reinforce age-appropriate behavior through consistency and cooperation, and provides examples in home settings.

Developing Child
Boston, WGBH Educational Foundation, 1989. 1 videocassette (27 min.) VHS

Host Philip Zimbardo focuses on the heredity versus environment debate in child development. Demonstrates tests of object permanence and symbolic understanding and emphasizes that inherited temperaments can be modified for individuals to realize their potentials. *Discovering Psychology Series*, program 5.

Discipline: Appropriate Guidance for Young Children
Washington, D.C., National Association for the Education of Young Children, 1988. 1 videocassette (28 min.) VHS

Describes developmentally appropriate methods of handling various discipline problems for healthy social and emotional development.

Discipline: What Lily Learned
Evanston, Ill., Altschul Group Corp., 1993. 1 videocassette (10 min.). VHS

This parenting tape describes techniques that a young mother learns at a workshop on toddler discipline: achieving parental self-discipline, giving limited choices and time-outs, being firm and consistent, modeling behavior, letting toddlers experience the consequences of their actions, explaining restrictions, steering children to other activities, and rewarding good behavior.

Elkind, David
Piaget's Developmental Theory: An Overview
Davis, Calif., Davidson Films, 1989. 1 videocassette (27 min.) VHS

Dr. David Elkind, once a student of Piaget's, narrates this overview of the four stages of Piaget's work in developmental psychology. Discusses genetic epistemology, egocentric and sociocentric stages, the conservation concept, and assimilation and accomodation. Illustrates his four stages of child development: sensorimotor, preoperational, concrete operational, and formal operational.

Eric Carle: Picture Writer
New York, Searchlight Films; dist. by Philomel Books/Scholastic, Inc., 1993. 1 videocassette (27 min.) VHS ISBN 0590469177

Visits the studio of this well-known children's author and illustrator who reads from his books and creates a "hungry caterpillar" collage while the music of Mozart plays in the background. The director, Rawn Fulton, received the Carnegie Award from the Association for Library Service to Children.

Granpa
Weston, Conn., Weston Woods, 1991. 1 videocassette (27 min.) VHS

Based on the picture book by John Burningham and narrated by Peter Ustinov, this video could be used in the social studies curriculum. Death, aging, and love in families are some of the topics addressed.

Language Development
Boston, Mass., WGBH Educational Foundation, 1989. 1 videocassette (27 min.) VHS

Host Philip Zimbardo outlines the various contributions of both heredity and environment to the sequence of language acquisition in young children. Discusses the necessity for Chomsky's "language acquisition device," social interaction, and the ability of the child to symbolize, manipulate tools, and perceive underlying syntax rules. *Discovering Psychology Series*, program 6.

Living the Life of a Pre-Schooler: A Day in the Life
Vancouver, Wash., The School Co., 1990. 1 videocassette (30 min.) VHS

Kathy Bobula outlines five important ways that preschool children can grow through play and pretend: using symbolic functions in dramatic play, expanding egocentric thinking, developing self-identity, sharing, and obeying rules. *The Parenting and Human Development Video Series*.

Lovaas, O. Ivar
Behavioral Treatment of Autistic Children
Huntington Station, N.Y., Focus International, 1988. 1 videocassette (43 min.) VHS

Describes the intensive behavioral treatment that two groups of similar autistic children (aged thirty-two to thirty-four months) received at UCLA under Dr. Lovaas and his staff. Follows the progress of a few of the subjects as adolescents, some of whom continued to receive such treatment and others who did not.

Maurice Sendak Library
Weston, Conn., Weston Woods, 1992. 1 videocassette (35 min.). VHS

Includes *Where the Wild Things Are, In the Night Kitchen, The Nutshell Library*, and a brief talk by the author.

My Child Has a Disability
Seattle, University of Washington Press, 1992. 1 videocassette (30 min.) VHS ISBN 0023592346

Kent Gerlach asks the mother of a fourteen-year-old with Down's syndrome to chronicle the sequence of emotions and reactions that parents of disabled children experience. She uses an outline from Robert Perske's *Hope for the Families* (Abingdon, 1981).

Parenting Special Children
Niles, Ill., United Learning Corp., 1993. 2 videocassettes (30 min. each) VHS ISBN 1560073926

This videocassette is one of a series of ten parenting tapes called *Practical Parenting*, hosted by Dick Van Patten and featuring parenting expert Dr. Bill Wagonseller. This video outlines ways that parents of special children can cope, including setting up individual family service plans and individualized education programs for preschool early intervention. Another video in this series, *The Art of Communication* (1987), provides examples of effective parent-child communication in real life situations.

Preschool Power 4!
Bethesda, Md., Concept Associates, 1993. 1 videocassette (31 min.) VHS ISBN 1563190184

One of a series of videotapes featuring preschool-aged children engaged in creative learning and play activities. Includes appealing songs and music.

Training Tape for Living with F.A.S. and F.A.E.: The Early Years, Birth to 12
Evanston, Ill., Altschul Group Corp., 1992. 1 videocassette (32 min.) VHS

This video describes the characteristics of fetal alcohol syndrome such as distinctive facial features, low IQ and birth weight, small stature, and brain damage. Also discusses the fetal alcohol effects: lack of attention skills, behavioral problems, and difficulty learning and socializing. Practical examples are given for adapting school programs and home environments to accommodate these children.

Why Do These Kids Love School?
Santa Monica, Calif., Pyramid Film and Video, 1990. 1 videocassette (60 min.) VHS

This documentary focuses on one independent and eight public schools that are innovative and student-centered in distinctive ways. All focus on human values, communication, relevance, involvement, cooperation, mutual respect, and leadership. Principals empower teachers to develop integrated curricula and supportive environments and to maintain high standards. Students become enthusiastic, self-confident, independent learners.

With a Little Help from My Friends
Toronto, Ontario, Centre for Integrated Education and Community; distributed by Ontario Association for Community Living; licensed by Merrill/Macmillan, 1992. 1 videocassette (20 min.) VHS ISBN 0023387785

Shows MAPS (McGill Action Planning System) session involving a parent, peers, school principal, counselor, and a teacher of a disabled student being mainstreamed. A facilitator and recorder elicit from the group the following about the individual: MAPS history, dreams, nightmares, identity, strengths, needs, and plan of action. The plan reveals how all participate in creating an ideal school situation for the student.

World of Abnormal Psychology, Program 11: Behavior Disorders of Childhood
South Burlington, Vt., Annenberg/CPB Collection, 1992. 1 videocassette (60 min.) VHS

One of thirteen videocassettes in the Annenberg/CPB Collection. Describes symptoms and causes of attention-deficit hyperactivity disorder, conduct disorder, separation anxiety, and autism, and outlines current treatments.

Years 3-5: What Lily Learned
Evanston, Ill., Altschul Group Corp., 1991. 1 videocassette (15 min.) VHS

Illustrates the various things a young mother learns in a parenting support group regarding discipline, potty training, play, sharing, motor skills, nutrition, health, temperament, bedtime, and separation.

✦ Internet

The following listservs discussion groups are available on the Internet:

- ECENET-L@UIUCVMD.BITNET

- ECEOL-L@MAINE.BITNET

Send mail to askeric@ericir.syr.edu [for e-mail inquiries]

Gopher to eric.syr.edu (port #70) or access the National Gopher System through: gopher.micro.umn.edu. Then move through the directories to AskERIC or Telnet to: eric.syr.edu, login with word *gopher*. Access National Gopher System and move through the directories to AskERIC.

For more information Call ERIC Information and Technology Clearinghouse 1-800-464-9107.

Library Technology

Diane J. Turner
Auraria Library
Denver, Colorado

Introduction

Position descriptions in the library field are changing rapidly. The knowledge required for a well-qualified library technician or entry-level librarian is becoming increasingly complex. Today's technician is doing work that only a librarian was permitted to do yesterday. Librarians are assuming managerial and administrative tasks or finding themselves working with advancing technologies. Low budgets and high demands have combined to create new challenges for librarian and technician alike. Every area, from collection development to public services, is being influenced and transformed by technology. As a library grows and responds to innovative change, library personnel are asked to take on unfamiliar and sometimes very different responsibilities; they may request or be asked to cross-train, or assume a new position in another department.

Several states offer library technical assistant programs, but the majority of library staff employed throughout the United States learn through on-the-job training. Library technicians work in all types of libraries, and in a wide variety of areas: circulation, reserves, acquisitions, bibliographic control, periodicals, media, instruction, or reference. Because of the changing and increasing responsibilities they assume and the new situations they face, library technicians need to know about sources they can access for answers and solutions to questions and problems.

This bibliography has been created as a guide to current information for students in certification programs for library technicians in community colleges, and as a resource for new librarians, technicians, and staff who have moved into new and unfamiliar areas of their libraries. Rather than presenting a list of standard reference works, the focus of this bibliography is on background materials and core lists that provide a springboard for further study. Although traditional sources are the backbone of any library and information about them remains vital, review sources and sources covering emerging electronic information are emphasized in this bibliography.

This is not an exhaustive list by any account. Librarians and library technicians must remain alert to all nature of resources and trends. Staying current will aid in the performance of specialized jobs, keep the library collection geared to the needs of the community it serves, and help maintain quality service.

Selected Associations

American Library Association
50 E. Huron St.
Chicago, Ill. 60611

Council on Library-Media Technical-Assistants
Library/Media Technology, Dept. SC 126
2900 Community College Ave.
Cuyahoga Community College
Cleveland, Ohio 44115

Selected Journals

American Libraries.
American Library Association. Eleven times a year.
ISSN 0002-9769

Choice.
Association of College and Research Libraries. Eleven times a year. ISSN 0009-4978

Collection Building.
Neal-Schuman. Quarterly. ISSN 0160-4953

Collection Management.
Haworth Press. Quarterly. ISSN 0146-2679

College and Research Libraries.
American Library Association. Eleven times a year.
ISSN 0010-0870

College and Research Libraries News.
American Library Association. Bimonthly. ISSN 0099-0086

Community and Junior College Libraries.
Haworth Press. Semiannually. ISSN 0276-3915

Emergency Librarian.
Dyad Services. Bimonthly. ISSN 0315-8888

Information Technology and Libraries.
American Library Association. Quarterly. ISSN 0730-9295

Journal of Academic Librarianship.
Mountainside Publishing. Bimonthly. ISSN 0099-1333

Journal of Interlibrary Loan and Information Supply.
Haworth Press. Quarterly. ISSN 1042-4458

Library Acquisitions: Practice and Theory.
Pergamon Press. Quarterly. ISSN 0364-6408

Library Administration and Management.
American Library Association. Quarterly. ISSN 0888-4463

Library Hotline.
Cahners Pub. Co. Fifty times a year. ISSN 0740-736X

Library Journal.
Cahners Pub. Co. Twenty times a year. ISSN 0363-0277

Library Mosaics.
Yenor, Inc. Bimonthly. ISSN 1054-9676

Library Personnel News.
ALA Office for Library Personnel Resources. Six times a year.
ISSN 0891-2742

Library Resources and Technical Services.
American Library Association. Quarterly. ISSN 0024-2527

Publishers Weekly.
Cahners Pub. Co. Weekly. ISSN 0000-0019

Reference Librarian.
Haworth Press. Semiannually. ISSN 0276-3877

Reference Services Review.
Pierian Press. Quarterly. ISSN 0090-7324

Research Strategies: RS.
Mountainside Publishing. Quarterly. ISSN 0734-3310

RQ.
American Library Association. Quarterly. ISSN 0033-7072

Serials Librarian.
Haworth Press. Quarterly. ISSN 0361-526X

Serials Review.
Haworth Press. Quarterly. ISSN 0098-7913

Special Libraries.
Special Library Association. Quarterly. ISSN 0038-6723

Technicalities.
Westport Publishers. Monthly. ISSN 0272-0884

Wilson Library Bulletin.
H. W. Wilson. Monthly. ISSN 0043-5651

Print Materials

✦ Academic Libraries

Baldwin, David A.
Supervising Student Employees in Academic Libraries
Englewood, Colo., Libraries Unlimited, 1991. 194p.
ISBN 0872878694

Guide to the employment of student assistants. Includes information on federal student financial aid, hiring, orientation and training, basic supervisory skills, resolving problems, and many other issues related to the employment of students. Includes bibliography and index.

George, Susan C., comp.
Emergency Planning and Management in College Libraries
Chicago, College Library Information Packet Committee, College Libraries Section, Association of College and Research Libraries, 1994. 142p. ISBN 0838977103

The importance of taking the time to plan for emergencies, such as fire, flood, typhoon, or earthquakes, is underscored in this informational packet.

Katz, William A., ed.
Community College Reference Services: A Working Guide for and by Librarians
Metuchen, N. J., Scarecrow Press, 1992. 356p. ISBN 0810826151 (v. 1)

A vital source for technicians planning to work in the reference area at the community college library. Although the guidelines and ideas presented are for providing quality reference services in a community college setting, the useful ideas can be tailored to fit almost any library.

Lipow, Anne Grodzins, ed.
Rethinking Reference in Academic Libraries: The Proceedings and Process of Library Solutions Institute No. 2, University of California, Berkeley, March 12-14, 1993, Duke University, June 4-6, 1993
Berkeley, Calif., Library Solutions Press, 1993. 242p.
ISBN 1882208021

Proceedings of a Library Solutions Institute, which discusses the changes required in reference services as a result of technological innovations and budget cutbacks. Includes bibliography and appendices.

McCabe, Gerard B., ed.
The Smaller Academic Library: A Management Handbook
New York, Greenwood Press, 1988. 380p. ISBN 0313250278

A reference handbook for directors, librarians, and staff members that discusses topics such as user programs and services, personnel management, budgets, collection management, and technical services. Suggests ideas and looks at trends in academic libraries. Includes bibliographic essay and index.

✦ Aquisitions and Collection Development

Gorman, G. E., and Brian R. Howes
Collection Development for Libraries
New York, Bowker-Saur, 1989. 432p. ISBN 0408301007

This excellent source covers the formulation of collection development policies, various methods of collection evaluation and user studies, types of selection sources, and the ongoing weeding of materials necessary to keep the collection useful. Includes index and a select bibliography.

Magrill, Rose Mary, and John Boyd Corbin
Acquisitions Management and Collection Development in Libraries
2nd ed. Chicago, American Library Association, 1989. 285p.
ISBN 0838905137

A revised and updated textbook which will introduce students, technicians, and beginning librarians to the management of acquisitions programs in libraries. Cost-effective collection development is of primary importance to all libraries; this book provides a basic understanding of both the acquisition and collection development process.

Sapp, Gregg, ed.
Access Services in Libraries: New Solutions for Collection Management
New York, Haworth Press, 1992. 245p. ISBN 1560244178

Reprinted from volume 17, number 1/2 of the journal *Collection Management*, this work is a compilation of essays on access services. It provides a variety of opinions and ideas on the importance of circulation, the broad definition of what access service really is, and the challenges being faced. Includes bibliographical references to aid further research.

Schmidt, Karen A., ed.
Understanding the Business of Library Acquisitions
Chicago, American Library Association, 1990. 322p.
ISBN 0838905366

Treating all aspects of acquisitions work, this book is a must for technicians involved in this technical service area. Issues addressed are the domestic and foreign publishing industry, vendors, domestic and foreign out-of-print and secondhand markets, nonprint publications, and methods of accounting and business practices. Includes index.

Spiller, David
Book Selection: Principles and Practice
5th ed. London, England, Library Association Publishing, 1991. 213p. ISBN 0851574645

Covers the fundamentals and background of book selection. This standard work guides the user in the selection and maintenance of resources that are appropriate and useful for a specific library. References included.

✦ Archives Management

Bellardo, Lewis J., and Lynn Lady Bellardo, comps.
A Glossary for Archivists, Manuscript Curators, and Records Managers
Chicago, Society of American Archivists, 1992. 45p.
ISBN 0931828791

A basic tool for records managers, manuscript curators, and archivists, the glossary provides a standard vocabulary and reflects changes in practice and terminology for professionals in this field. This source gives the beginning student a sampling of the terms used for archival purposes. Additional reading section included. *Archival Fundamental Series.*

Dearstyne, Bruce W.
The Archival Enterprise: Modern Archival Principles, Practices, and Management Techniques
Chicago, American Library Association, 1993. 295p.
ISBN 0838906028

Introducing archival theory, principles, and techniques, this source emphasizes the importance of maintaining records of our past to enhance our society in the future. Deciding what to preserve and how to preserve it is one of the primary missions of the archivist. Excellent overview for students in any library certification program. Includes bibliography and index.

DePew, John N.
A Library, Media, and Archival Preservation Handbook
Santa Barbara, Calif., ABC-CLIO, 1991. 441p. ISBN 0874365430

This handbook, designed as an introduction to the preservation of library materials, explains procedures that are used by archivists. Its coverage of conservation and restoration techniques will help students understand more about the changing and expanding field of preservation. Includes index.

DePew, John N., with C. Lee Jones
A Library, Media, and Archival Preservation Glossary
Santa Barbara, Calif., ABC-CLIO, 1992. 192p. ISBN 0874365767

Designed to bring together the terminology related to the preservation of archival, library, and media center materials, this glossary reflects the spectrum of archival and library preservation practices. This is the companion volume to *A Library, Media, and Archival Preservation Handbook* (ABC-CLIO, 1991).

✦ Bibliographic Instruction

Brottman, May, and Mary Loe, eds.
The LIRT Library Handbook
Englewood, Colo., Libraries Unlimited, 1990. 125p.
ISBN 0872876640

This handbook offers practical advice on how to develop or improve library instruction programs in a variety of library settings. Prepared by experienced bibliographic instruction librarians, it is of value to students and practitioners alike. Assessment of needs, the program's audience, budget, and other practical concerns are covered. Includes bibliography and index.

Hardesty, Larry L., Jamie Hastreiter, and David Henderson, eds.
Bibliographic Instruction in Practice: A Tribute to the Legacy of Evan Ira Farber
Ann Arbor, Mich., Pierian Press, 1993. 157p. ISBN 0876503288

This easy-to-read book of current models of library instruction was based on the Earlham College-Eckerd College Bibliographic Instruction Conference held at Eckerd College in 1992. Includes a selective annotated bibliography of readings on bibliographic instruction.

Shirato, Linda, ed.
What Is Good Instruction Now? Library Instruction for the 90s: Papers and Session Materials Presented at the Twentieth National LOEX Library Instruction Conference Held at Eastern Michigan University, 8 to 9 May 1992, and Related Resource Materials Gathered by the LOEX Clearinghouse
Ann Arbor, Mich., Pierian Press, 1993. 178p. ISBN 087650327X

This source provides a wealth of ideas on library instruction for the '90s and beyond. Includes a useful bibliography covering library instruction in all types of libraries, and for all types of patrons.

✦ Children's Services

Benne, Mae
Principles of Children's Services in Public Libraries
Chicago, American Library Association, 1991. 332p.
ISBN 0838905552

This resource will aid librarians or technicians who assume responsibility for children's services in any public library.

Children's Books in Print
New Providence, N. J., Bowker/Reed Reference Publishing,
1969- . ISSN 0069-3480

This is a valuable set for anyone selecting and using children's books in public, school, or academic libraries.

Connor, Jane Gardner
Children's Library Services Handbook
Phoenix, Ariz., Oryx Press, 1990. 128p. ISBN 0897744896

This easy-to-read handbook covers the fundamentals of library service to children emphasizing the practical side of providing children's services.

✦ Computers and Automation

Boss, Richard W.
The Library Manager's Guide to Automation
3rd ed. Boston, G. K. Hall, 1990. 202p. ISBN 0816119422

Hardware basics, software, data communications, bibliographic and networking standards, planning and implementation, costs of automation, and trends for the future are covered in this edition. With technology changing at such a rapid pace, automation is one of the most significant factors in libraries today. This book introduces valuable ideas to help readers learn about, and keep up with, advancing technologies. Includes glossary and bibliography.

Cibbarelli, Pamela, ed.
Directory of Library Automation Software, Systems, and Services.
1993 ed. Medford, N. J., Learned Information, 1993. 370p.
ISBN 0938734652

Provides information on 240 software packages and automation systems and services specific to libraries. Includes a library automation consultants' directory, a section on CD-ROM distributors, and an alphabetical index.

✦ Distance Education

Harry, Keith, John Magnus, and Desmond Keegan, eds.
Distance Education: New Perspectives
New York, Routledge, 1993. 348p. ISBN 0415089417

Addresses the education of people from a distance using video technology and other electronic means, a practice that has been gaining momentum for years. Index.

Willis, Barry Donald
Distance Education: A Practical Guide
Englewood Cliffs, N. J., Educational Technology Publications,
1993. 138p. ISBN 0877782555

Covers the development of useful distance education courses. Compact and understandable, it provides information on making intelligent decisions concerning the best type of video or electronic education for various clientele.

Willis, Barry Donald, ed.
Distance Education: Strategies and Tools
Englewood Cliffs, N. J., Educational Technology Publications,
1994. 350p. ISBN 0877782687

This source on the expanding field of distance education provides information on audiovisual education, computer-assisted instruction, and a variety of other useful distance education tools. With the virtual library around the corner, this is a useful book. Includes bibliographical references and index.

✦ Electronic Reference and Resources

Abbott, Tony, ed.
Internet World's on Internet: An International Guide to Electronic Journals, Newsletter, Texts, Discussion Lists, and Other Resources on the Internet
Westport, Conn., Mecklermedia, 1994- . ISSN 1066-9973

Billed as "the most comprehensive guide available," this book lists more than 750 Internet resources. Covers electronic mail, anonymous FTP, archie, standard file formats, mailing lists, and other information of use to new and experienced Internet surfers.

Dern, Daniel P.
The Internet Guide for New Users
New York, McGraw-Hill, 1994. 570p. ISBN 0070165106

How to join and use the massive information highway is covered in this basic guide for beginning users. Written for a UNIX audience, this is still a prime source for anyone interested in the Internet. Includes appendices, bibliography, and index.

Engle, Mary, and others
Internet Connections: A Librarian's Guide to Dial-Up Access and Use
Chicago, American Library Association, 1993. 166p. ISBN 0838976778

This overview of the Internet includes history, access, and network tools. A beginner's guide, it is one of the many sources that will help users get onto the "information highway." Includes an itemized list of service providers.

Goldmann, Nahum
Online Information Hunting
Blue Ridge Summit, Pa., Windcrest/McGraw-Hill, 1992. 236p. ISBN 0830639454

A practical guide to on-line searching, highlighting useful databases and the subject expert searching techniques. This source will give the beginner a good background of the search process.

Hahn, Harley
The Internet Yellow Pages
Berkeley, Calif., Osborne McGraw-Hill, 1994. 447p. ISBN 00778820235

A directory that guides users to resources on the Internet. As computer networks are continually expanding, finding the appropriate place to go for information is becoming increasingly important. This source answers that need.

Hahn, Harley, and Rick Stout
The Internet Complete Reference
Berkeley, Osborne McGraw-Hill, 1994. 818p. ISBN 0078819806

Provides a wealth of information on the Internet in an easy-to-understand format. Includes a catalog of Internet resources and a list of Usenet discussion groups.

Jaffe, Lee David
Introducing the Internet: A Trainer's Workshop
Berkeley, Calif., Library Solutions Press, 1994. 92p. and computer disk. ISBN 1882208056

This supplement to Roy Tennant's *Crossing the Internet Threshold* (Library Solutions Press, 1993) aids training and instruction in the use of the Internet. A 3 1/2" computer disk contains files of presentation slides that can be displayed using Windows on DOS-based computers.

Jacsó, Péter
CD-ROM Software, Dataware, and Hardware: Evaluation, Selection, and Installation
Englewood, Colo., Libraries Unlimited, 1992. 256p. ISBN 0872879070

Covers various types of automation in libraries and explains how to make intelligent selections of products for particular needs. Provides information on optical disks, CD-ROMs, and other electronic forms of information.

Kehoe, Brendan P.
Zen and the Art of the Internet: A Beginner's Guide
3rd ed. Englewood Cliffs, N. J., Prentice Hall, 1994. 193p. ISBN 0131214926

A clear, logical book describing the Internet and its various uses. Packed with information, this slim source is easy to read and understand.

Krol, Ed
The Whole Internet User's Guide and Catalog
2nd ed. Sebastopol, Calif., O'Reilly and Associates, 1994. 544p. ISBN 1565920635

One of the best books on the topic, this guide takes readers on an Internet tour. Written with a sense of humor and packed with tips, it is appropriate for anyone who wants to use Internet resources. Contains a resource catalog arranged alphabetically.

Lambert, Steve, and Walt Howe
Internet Basics: Your Online Access to the Global Electronic Superhighway
New York, Random House Electronic Publishing, 1993. 495p. ISBN 0679750231

Another guide on how to access and surf the electronic superhighway. The step-by-step guidance teaches how to use Telnet, Archie, Gopher, and WAIS, among others. Also covers e-mail and transferring files with FTP.

LaQuey, Tracy L., and Jeanne C. Ryder
The Internet Companion: A Beginner's Guide to Global Networking
Reading, Mass., Addison-Wesley, 1993. 196p. ISBN 0201622246

Showing people how to use the Internet in the real world is the purpose of this guide, which gives basic concepts and user applications. The easy-to-read style, low cost, and manageable size make this book less intimidating than many other recent texts on the subject.

Levine, John R., and Carol Baroudi
The Internet for Dummies
San Mateo, Calif., IDG Books, 1993. 355p. ISBN 1568840241

Another winner in the *Computer Book Series for Beginners*, this guide is an easy and unintimidating way to learn about the Internet. Covers popular services and commands, hints for moving around the Internet, and helpful shortcuts.

MacDonald, Linda Brew, and others
Teaching Technologies in Libraries: A Practical Guide
Boston, G. K. Hall, 1991. 275p. ISBN 0816119066

This book offers practical suggestions and guidelines for teaching electronic resources to patrons by using the new technologies (i.e., CAI, videotapes, help screens, etc.). Contains good ideas to help implement instruction programs.

Saunders, Laverna M., ed.
The Virtual Library: Visions and Realities
Westport, Conn., Meckler, 1993. 165p. ISBN 0887368603

This work is based on the seventh annual Computers in Libraries Conference. The book consists of essays covering the practical and philosophical issues of the virtual library, from librarians as mediators to the potential role of the virtual library as a part of the virtual university.

Shirato, Linda, ed.
Working with Faculty in the New Electronic Library: Papers and Session Materials Presented at the Nineteenth National LOEX Library Instruction Conference Held at Eastern Michigan University, 10 to 11 May 1991, and Related Resource Materials Gathered by the LOEX Clearinghouse
Ann Arbor, Mich., Pierian Press, 1992. 189p. ISBN 087650294X

Part of the *Library Orientation Series.* Includes the four main papers presented, nine instructive sessions, three poster sessions, and an annotated bibliography.

Strangelove, Michael, Diane Kovacs, and Ann Okerson, eds.
Directory of Electronic Journals, Newsletters, and Academic Discussion Lists
Washington, D.C., Association of Research Libraries, 1993- .
ISSN 1057-1337

This directory has increasing value as librarians and their patrons explore the information highway. The discussion lists for paraprofessionals will be of great value to library technicians learning about libraries both in the United States and elsewhere.

Tennant, Roy
Crossing the Internet Threshold: An Instructional Handbook
Berkeley, Calif., Library Solutions Press, 1993. 134p.
ISBN 1882208013

Although designed to teach students, faculty, or other patrons about the Internet, this handbook is essential for anyone providing instruction. Includes lectures, sample overheads, and exercises for e-mail. An excellent tool for learning and teaching.

Widman, Rudy, and Jimmie Annen Nourse
Electronic Access to Information: A Guidebook for High-Technology Library Instruction
Dubuque, Iowa, Kendall/Hunt Pub. Co., 1991. 142p.
ISBN 0840367074

Instruction in electronic sources is escalating in libraries nationwide. This text, for instruction in twelve different CD-ROM products, provides a basic introduction to getting users started with electronic information retrieval. It could be tailored to fit the needs of varied instruction programs.

✦ General Works

ALA Handbook of Organization and Membership Directory
Chicago, American Library Association. Annual. ISSN 0273-4605

This source can help library personnel network with others on committees or in libraries that have similar interests. Published by the American Library Association on an annual basis, it is a good resource for keeping current with activities in this organization.

American Library Association. Office for Intellectual Freedom
Intellectual Freedom Manual
4th ed. Chicago, American Library Association, 1992. 283p.
ISBN 0838934129

This manual covers the basics of handling complaints, lobbying legislators, developing confidentiality reports, and the selection of materials. Valuable for any library, but of particular value to public libraries.

American Library Directory
New Providence, N. J., Bowker, 1923- . Annual. ISSN 0038-6723

Annual which Lists public, academic, government, and special libraries in the United States and Canada. Documents changes in library budgets, personnel, expenditures, and automated capabilities. Includes directory of library schools and training courses.

Baker, Sharon L., and F. Wilfrid Lancaster
The Measurement and Evaluation of Library Services
2nd ed. Arlington, Va., Information Resources Press, 1991. 411p.
ISBN 0878150617

This second revised edition covers detailed information on techniques for evaluating services in public, school, and special libraries. Concentrating heavily on the evaluation of collections and materials availability, as well as on public services such as reference, this resource provides the library technician with useful information on how to evaluate a library and the services that it provides.

Blitzer, Roy J.
Office Smarts: 222 Tips for Success in the Workplace
Old Saybrook, Conn., Globe Pequot Press, 1994. 1 v. (various pagings). ISBN 1564403866

Provides helpful hints for how to get along in a world run by office politics and other unseen or unacknowledged forces unique to a particular workplace. Although not tailored specifically to libraries, those who work there will find much of the information useful.

Boone, Morell D., Sandra G. Yee, and Rita Bullard
Training Student Library Assistants
Chicago, American Library Association, 1991. 110p.
ISBN 08389905617

This manual incorporates a "critical-thinking" emphasis on training programs for student assistants. In addition to interview and orientation checklists, the authors focus on a full training cycle that incorporates all its aspects: planning, preparation, development of materials, administration of training, evaluation of the students, and program evaluation. Includes bibliography.

Buckland, Michael Keeble
Redesigning Library Services: A Manifesto
Chicago, American Library Association, 1992. 82p.
ISBN 0838905900

This compact, thought-provoking work discusses how services need to change in an environment of advancing technology and decreasing budgets. Provides a framework for understanding future needs in planning library services.

Caputo, Janette S.
Stress and Burnout in Library Service
Phoenix, Ariz., Oryx Press, 1991. 172p. ISBN 0897746023

Explains the causes of stress and burnout and why they are becoming more of a problem in libraries today. Working with increasing numbers of people, the need to learn rapidly expanding technology, and dealing with ever-present budget cuts, are just a few of the issues highlighted.

Chernik, Barbara E.
Introduction to Library Services
Englewood, Colo., Libraries Unlimited, 1992. 230p.
ISBN 0872879313

Replaces and largely updates *Introduction to Library Services for Library Technicians* (Libraries Unlimited, 1982). Covers personnel resources, organization, types of libraries, library networks, facilities, and auxiliary services. A concept of what constitutes good library service is also included.

DuBrin, Andrew J.
Winning Office Politics: DuBrin's Guide for the 90s
Englewood Cliffs, N. J., Prentice Hall, 1990. 337p.
ISBN 0139649581

This update of the 1978 edition is essential for anyone working with other people. Intended as a manual for job success, this resource offers tips on how to survive office politics and get ahead. No matter what job, or at what level, this guide will provide practical advice for practicing ethical politics.

Eberhart, George M., comp.
Whole Library Handbook: Current Data, Professional Advice, and Curiosa about Libraries and Library Services
Chicago, American Library Association, 1991. 490p.
ISBN 0838905730

One of the most beneficial and practical sources for anyone working in libraries. Provides a variety of useful and factual information on such issues as support staff resources, pay equity, censorship, and the *Library Bill of Rights*, to name a few.

Fisher, James C., and Kathleen M. Cole
Leadership and Management of Volunteer Programs: A Guide for Volunteer Administrators
San Francisco, Jossey-Bass, 1993. 208p. ISBN 1555425313

Written for administrators in nonprofit organizations such as libraries, as well as for those in government and business, this book addresses the needs of anyone working with and supervising volunteers. Includes references and index.

Fortson, Judith
Disaster Planning and Recovery: A How-To-Do-It Manual for Librarians and Archivists
New York, Neal-Schuman, 1992. 181p. ISBN 1555700594

Safety measures in libraries, conservation and restoration of archival and other library materials, and library planning for disaster relief are some of the topics addressed in this handbook. Includes select bibliography and appendices.

Giesecke, Joan, ed.
Practical Help for New Supervisors
Chicago, American Library Association, 1992. 69p.
ISBN 0838934080

A guide to help both new and experienced supervisors acquire better supervisory skills. Provides commonsense ideas and tips on managing employees. Bibliography included.

Haass, Richard
The Power to Persuade: [How to Be Effective in Government, the Public Sector, or Any Other Unruly Organization]
Boston, Houghton Mifflin, 1994. 260p. ISBN 0395675855

This valuable resource presents ideas and methods for dealing with a variety of situations that arise when working with the public or in a government bureaucracy.

Henn, Harry G.
Henn on Copyright Law: A Practitioner's Guide
3rd ed. New York, Practicing Law Institute, 1991. 690p.
ISBN 0872240207

This guide covers copyright issues in a logical, readable format. From fair use reproduction for classroom use to satellite cable programming, this is a valuable resource for any library collection.

Karp, Rashelle S.
Volunteers in Libraries
Chicago, American Library Association, 1993. 11p.
ISBN 0838957560

Part of the *Small Libraries Publications* series, this publication explains the importance of volunteers in libraries. Recruiting, interviewing, hiring, training, evaluating, and recognizing the contributions of volunteers are covered in an easy-to-read, logical way.

Kinney, Lisa
Lobby for Your Library: Know What Works
Chicago, American Library Association, 1992. 189p.
ISBN 0838934102

This readable book is an introduction to the wonderful world of political lobbying for anyone who seeks funding for a library. Explains how declining budgets and bleak economic situations throughout the country make it a necessity for library staff to lobby for funds to maintain full services for their users.

Orr, Lisa, ed.
Censorship: Opposing Viewpoints
San Diego, Calif., Greenhaven Press, 1990. 238p.
ISBN 0899084540

Part of the *Opposing Viewpoints Series*, this text covers both sides of the censorship issue. Chapter Four, on school and library censorship, provides an overview of the issues currently affecting libraries and schools. Includes a list of organizations and a bibliography.

Peterson, Lorna
Library Paraprofessionals: A Bibliography
Monticello, Ill., Vance Bibliographies, 1991. 6p.
ISBN 0792007441

A brief bibliography on paraprofessionals (or support staff) that will provide the incoming technician with background concerning issues impacting support staff in libraries.

Roney, Raymond G., ed.
Directory of Institutions Offering Programs for the Training of Library Media Technicians
7th ed. Cleveland, Ohio, Council on Library Media Technicians, 1992. 61p.

Provides descriptions of LTA programs and information on degrees offered. Includes the number of students, purpose of program, courses available, and faculty teaching the courses. Index included.

Sheal, Peter R.
The Staff Development Handbook: An Action Guide for Managers and Supervisors
East Brunswick, N. J., Nichols Pub., 1992. 234p.
ISBN 0893973807

A guide for those responsible for supervising others. Includes tips on improving motivation and performance, conducting on-the-job training, holding productive meetings, and coaching. Gives practical advice on staff development. Also provides a useful appendix on developing visual aids for training.

Steele, Victoria, and Stephen D. Elder
Becoming a Fundraiser: The Principles and Practice of Library Development
Chicago, American Library Association, 1992. 139p.
ISBN 0838905897

Covers the basics of fundraising and provides step-by-step guidelines on the solicitation of gifts for the library.

Stueart, Robert D., and Barbara B. Moran
Library and Information Center Management
4th ed. Englewood, Colo., Libraries Unlimited, 1993. 402p.
ISBN 1563081350

Focuses on the management functions common to all libraries, such as staffing, directing, planning, and organizing. Useful appendices included.

Stueart, Robert D., and Maureen Sullivan
Performance Analysis and Appraisal: A How-to-Do-It Manual for Librarians
New York, Neal-Schuman, 1991. 174p. ISBN 1555700616

This useful handbook covers the complexities of the performance appraisal process, including rating library employees and writing job descriptions. Includes bibliography and index.

Tools of the Profession
Washington, D.C., Special Libraries Association, 1988. 129p.
ISBN 0871113384

Presents more than a thousand bibliographic entries in a wide variety of subject areas. Covers the basic library tools used in disciplines from advertising to transportation.

Watts, Tim J.
Confidentiality in the Use of Library Materials: A Bibliography
Monticello, Ill., Vance Bibliographies, 1989. 26p.
ISBN 0792000552

This bibliography lists works which apply to the rights of library patrons to use library materials without having their privacy invaded. Covers topics such as policies on confidentiality and technology's impact on privacy.

Wildman, Iris J., and Rhonda Carlson
Researching Copyright Renewal: A Guide to Information and Procedure
Littleton, Colo., F. B. Rothman, 1989. 85p. ISBN 0837713528

Used to help investigate the status of copyright renewals on pamphlets and books, this work briefly covers copyright law and refers users to other sources that can help with more in-depth copyright questions.

✦ Government Documents

Ekhaml, Leticia T., and Alice J. Wittig
U.S. Government Publications for the School Library Media Center
2nd ed. Englewood, Colo., Libraries Unlimited, 1991. 172p.
ISBN 0872878228

Written with any school library media center in mind, this easy-to-use source offers a bibliography of current materials that would be useful in a school library. It provides the user with ideas on obtaining and organizing government information and suggestions on how to integrate government publications into the curriculum. A good starting point for the beginning library technician.

Guide to U.S. Government Publications
McLean, Va., Documents Index, 1973- ISSN 0092-3168

This is an annotated guide to series, periodicals, and important reference publications published by a variety of U.S. government agencies. Provides a complete listing of Superintendent of Documents classification numbers. Agency and title index included.

Pokorny, Elizabeth J., and Suzanne M. Miller
U.S. Government Documents: A Practical Guide for Non-Professionals in Academic and Public Libraries
Englewood, Colo., Libraries Unlimited, 1989. 143p.
ISBN 0872875075

A guide to U.S. government documents for library assistants and paraprofessionals who have no experience with them. Written in a readable style, the authors present an overview of publishers and their distribution of government documents. Acquisition tools, ordering, and processing procedures are also included.

Robinson, Judith Schiek
Tapping the Government Grapevine: The User-Friendly Guide to U.S. Government Information Sources
2nd ed. Phoenix, Ariz., Oryx Press, 1993. 227p.
ISBN 0897747127

This expanded, updated second edition includes a wealth of information on how to track down government information. Giving practical advice and tips, the author helps the inexperienced user make sense out of the massive amounts of information generated by the government. Includes an index.

Sears, Jean L., and Marilyn K. Moody
Using Government Information Sources: Print and Electronic
2nd ed. Phoenix, Ariz., Oryx Press, 1994. 539p.
ISBN 0897746708

This outstanding work provides information on specific government topics, and suggests how to find information on topics not covered. Also addresses electronic publishing in government publications.

✦ Interlibrary Loan

Dearie, Tammy Nickelson, and Virginia Steel, comps.
Interlibrary Loan Trends: Making Access a Reality
Washington, D.C., Association of Research Libraries, Office of Management Services, 1992. 184p.

Another in the valuable Association of Research Libraries *SPEC kit* series, this work covers circulation, interlibrary loan, the importance of quick access, and the increasing cooperation among American libraries which makes rapid access possible. *SPEC kit*, no. 184.

Morris, Leslie R., and Sandra Chass Morris
Interlibrary Loan Policies Directory
4th ed. New York, Neal-Schuman, 1991. 785p.
ISBN 155570090X

Provides information on more than 1,500 lenders in the U.S., Canada, and Puerto Rico. Includes topics such as average turn-around time, billing procedures, and lending requirements. Includes indexes.

✦ Media Services

Gale Directory of Publications and Broadcast Media
Detroit, Mich., Gale Research Co. 1990- . Published annually.
ISSN 1048-7972

Formerly the *Ayer Directory of Publications*, this work has been a definitive source since 1869. Covers the entire media industry for the U.S. and Canada, including newspapers, magazines, journals, radio, television, and cable stations and systems.

Vlcek, Charles W., and Raymond V. Wiman
Managing Media Services: Theory and Practice
Englewood, Colo., Libraries Unlimited, 1989. 426p.
ISBN 0872877159

Providing an overview of managing media services, this book gives background on media centers and their basic function, that of facilitating the integration of communication resources with instructional learning situations. Introduces some basics on designing and managing a facility, and provides a view into the future of media technology. Appendixes and index included.

✦ Public Libraries

Arozena, Steven, ed.
Best Books for Public Libraries: The 10,000 Top Fiction and Nonfiction Titles
New Providence, N. J., Bowker, 1992. 840p. ISBN 0835230732

This concise source is a guide to critically acclaimed books suitable for general readers. Useful for library staff looking to expand their collections in specific subject areas, or just to peruse what is available. Does not include books of a scholarly nature, reference titles, or children's books. Indexes included.

Dewey, Barbara I., and Sheila D. Creth
Team Power: Making Library Meetings Work
Chicago, American Library Association, 1993. 123p.
ISBN 0838906168

Suggests how to improve meetings to enhance the decision-making process in libraries.

Jones, Patrick
Connecting Young Adults and Libraries: A How-To-Do-It Manual
New York, Neal-Schuman, 1992. 278p. ISBN 1555701086

This tool is for building or restructuring young adult services to fit the needs of that user population. Sample documents, forms, and checklists that can be tailored to user needs are provided in this useful guide.

Patrick, Gay D.
Building the Reference Collection: A How-To-Do-It Manual for School and Public Librarians
New York, Neal-Schuman, 1992. 187p. ISBN 1555701051

This handy, practical work covers how to develop a basic reference collection for school and public libraries. Useful for beginning school media and public librarians.

Weingand, Darlene E.
Managing Today's Public Library: Blueprint for Change
Englewood, Colo., Libraries Unlimited, 1994. 185p.
ISBN 0872878074

Covering management principles in the administration of public libraries, this book provides insight on planning and marketing library services to fit community needs. Useful to all levels of public library personnel.

✦ Public Services—General

Bingham, Karen Havill
Building Security and Personal Safety
Washington, D.C., Office of Management Services, Association of Research Libraries, 1989. 141p.

The issue of safety is, unfortunately, becoming increasingly prominent in libraries today. This brief source provides valuable information which can help keep patrons and staff more secure.

Chaney, Michael, and Alan MacDougall, eds.
Security and Crime Prevention in Libraries
Brookfield, Vt., Ashgate Pub. Co., 1992. 307p. ISBN 1857420144

An overview of crime in libraries, this book reviews legal perspectives, presents planning and management of crime prevention strategies, covers book detection systems, and provides additional useful information.

Library Safety and Security: A Comprehensive Manual for Library Administrators and Police and Security Officers
Goshen, Ky., Campus Crime Prevention Program, 1992. 123p.

This manual covers safety and security issues increasingly faced by libraries.

Paietta, Ann Catherine
Access Services: A Handbook
Jefferson, N.C., McFarland, 1991. 208p. ISBN 0899505996

Written to help library employees understand how services in each department in the library have an impact on the library's overall mission to provide access to information.

Salter, Charles A.
On the Frontlines: Coping with the Library's Problem Patrons
Littleton, Colo., Libraries Unlimited, 1988. 170p.
ISBN 0872876586

A handy guide to coping with difficult and sometimes intractable library users. Useful for any library in providing staff training on dealing with problem patrons.

Steele, Virginia, comp.
Access Services: Organization and Management
Washington, D.C., Association of Research Libraries, Office of Management Services, 1991. 123p.

Covers reference services, circulation, and processing in libraries. Emphasizes ways to make information more accessible to readers. Bibliographical references included.

✦ Reference Services

American Reference Books Annual
Englewood, Colo., Libraries Unlimited, 1970- . Annual.
ISSN 0065-9959

An excellent resource for finding reference books to meet a collection's needs. Index included.

Balay, Robert, ed.
Guide to Reference Books: Covering Materials from 1985-1990. Supplement to the 10th ed.
Chicago, American Library Association, 1992. 613p.
ISBN 0838905889

This third supplement to a distinguished reference source covers 4,668 titles in a variety of subject areas. Titles listed were published from December 1984 to December 1990. Emphasis is on reference sources used for academic and library research. Includes index.

Berle, Gustav
Business Information Sourcebook
New York, Wiley, 1991. 374p. ISBN 0471529761

Gives easy access to the information needed to solve business questions and problems. Includes currently published business books, periodicals, directories, government publications, and electronic media.

Bowker Annual: Library and Book Trade Almanac
New Providence, N.J., Bowker, 1961- . Annual. ISSN 0068-0540

Reports statistics and updates data on libraries and publishing both in the U.S. and abroad, and provides information on legislation, funding programs, library salaries, and a wide variety of other library-related information. Subject and organization indexes included.

Chernow, Barbara A., and George A. Vallasi, eds.
The Reader's Advisor: A Layman's Guide to Literature
New York, Bowker, 1964- . Published irregularly.
ISSN 0094-5943

Volumes One through Five of this set give the student a survey of the best literature in the humanities, social sciences, and sciences. Volume Six provides a directory of publishers and separate name, title, and subject indexes.

Daniells, Lorna M.
Business Information Sources
3rd ed. Berkeley, Univ. of California Press, 1993. 725p.
ISBN 0520081803

This is a guide to the vast amount of business information sources currently available. Finding up-to-date business information is becoming increasingly complex, and this work shows users what information is accessible in print and on-line sources.

Evans, G. Edward, Antony J. Amodeo, and Thomas L. Carter
Introduction to Library Public Services
5th ed. Englewood, Colo., Libraries Unlimited, 1992. 374p.
ISBN 0872878597

Describes differences in public services by type of library. Although particularly strong on reference services, it also includes a discussion of computerization, library instruction, interlibrary loan, circulation, reserves, special collections, serials, media, government information, and security issues. Previous editions were titled *Introduction to Public Services for Library Technicians*.

Hillard, James M.
Where to Find What: A Handbook to Reference Service
3rd ed. Metuchen, N. J., Scarecrow Press, 1991. 333p.
ISBN 0810824043

This ready-reference tool provides the user with a starting point in an information search. Features include 607 subject headings covering everything from abortion to zoology, and bibliographic citations with brief annotations for resources that will help library staff find answers to commonly asked questions.

Katz, William A.
Introduction to Reference Work
6th ed. New York, McGraw-Hill, 1992. 2 v. ISBN 0070336385 (vol. 1); ISBN 0070336393 (vol. 2)

This resource is an introduction to basic sources, services, and processes. Updated for the '90s, this revision of the classic work continues to be useful for introductory studies.

Katz, William A., comp.
Reference and Information Services: A Reader for the Nineties
Metuchen, N. J., Scarecrow Press, 1991. 415p. ISBN 0810824868

This collection of essays surveys the reference and information services of the '90s. Discusses topics such as computerized reference services, the reference interview, cultural literacy, and evaluation of reference books.

Kinder, Robin, and William A. Katz, eds.
Serials and Reference Services
New York, Haworth Press, 1990. 457p. ISBN 0866568107

This collection of essays covers the importance of serials to public services (such as reference) and discusses serial records and serials classification. It contains a wealth of practical articles that will give an entry-level technician a sense of the burgeoning field of serials.

Lang, Jovian, ed.
Reference Sources for Small and Medium Sized Libraries
Chicago, Reference Sources Committee of the Reference and Adult Services Division, 1992. 317p. ISBN 0838934064

Contains almost two thousand print and electronic reference sources with standard bibliographic elements and concise annotations.

Lavin, Michael R.
Business Information: How to Find It, How to Use It
2nd ed., new and expanded ed. Phoenix, Ariz., Oryx Press, 1992. 499p. ISBN 0897745566

This revised and enlarged second edition is a comprehensive, hands-on manual that provides in-depth descriptions of the most frequently used business resources and how to use them. Includes index.

McDaniel, Julie Ann, and Judith K. Ohles
Training Paraprofessionals for Reference Service: A How-to-Do-It Manual for Librarians
New York, Neal-Schuman, 1993. 180p. ISBN 1555700845

Gives a step-by-step plan for maximizing reference services by training library paraprofessionals in the essential parts of reference work, including reference sources and the all-important reference interview. Includes bibliography.

O'Block, Robert L.
Criminal Justice Research Sources
3rd ed. Cincinnati, Ohio, Anderson Pub. Co., 1992. 189p. ISBN 0870846655

A guide to the types of materials that will assist criminal justice students or practitioners in their research. Covers general reference sources, journals, bibliographies, and other materials..

Puccio, Joseph A.
Serials Reference Work
Englewood, Colo., Libraries Unlimited, 1989. 228p. ISBN 0872877574

This guide presents serials reference techniques for coping with this growing collection. Most sources used in reference are serials and some can be frustrating and difficult to use. This work aids technicians and librarians in the understanding of serials, and in assisting patrons with questions about serial publications.

Recommended Reference Books for Small and Medium-Sized Libraries and Media Centers
Englewood, Colo., Libraries Unlimited, 1983/84- . Annual. ISSN 0277-5948

This source is an abridged version of the 1993 *American Reference Books Annual* (ARBA) designed specifically for smaller libraries and media centers. Lends consideration to libraries with a limited budget that cannot afford the more comprehensive ARBA.

Reference Books Bulletin
Chicago, American Library Association, 1983- . Annual. ISSN 8755-0962

A listing of reference sources of general interest for small-to-medium-sized public, academic, and school libraries. This is a good source for evaluating the reference collections in various disciplines.

Sader, Marion, ed.
General Reference Books for Adults: Authoritative Evaluations of Encyclopedias, Atlases, and Dictionaries
New York, R. R. Bowker, 1988. 614p. ISBN 0835223930

Provides useful information to help develop and maintain any library collection. Divided into five parts, this work presents evaluations of general reference books. Excellent for library-technician training programs. Includes index and bibliography.

Sader, Marion, ed.
Topical Reference Books: Authoritative Evaluations of Recommended Resources in Specialized Subject Areas
New Providence, N. J., R. R. Bowker, 1991. 892p. ISBN 0835230872

Reviews of reference books in specific areas serve as a selection guide. Subjects range from advertising to zoology. An excellent tool for librarians or technicians who want to build core collections.

Schlessinger, Bernard S., Rashelle S. Karp, and Virginia S. Vocelli, eds.
The Basic Business Library: Core Resources
2nd ed. Phoenix, Ariz., Oryx Press, 1989. 278p. ISBN 0897744519

A basic business resource for small to medium-sized libraries. Provides a core list of business reference sources, explains how to organize materials in a business library, and includes a bibliography of pertinent business literature. Indexed.

Taylor, Margaret, and Ronald R. Powell
Basic Reference Sources: A Self-Study Manual
4th ed. Metuchen, N.J., Scarecrow Press, 1990. 319p. ISBN 081082244X

Excellent source for training entry-level librarians or paraprofessionals on the reference sources most frequently used for answering questions at the reference desk. This manual can be used in any library for staff development and cross-training.

Wasserman, Paul, Gary McCann, and Patricia Tobin, eds.
Encyclopedia of Legal Information Sources: A Bibliographic Guide to Approximately 19,000 Citations for Publications, Organizations, and Other Sources of Information on 460 Law-Related Subjects
Detroit, Mich., Gale Research Co., 1988. 634p. ISBN 0810302454

An easy-to-use tool for locating books, periodicals, databases, and organizations that can help lawyers, paralegals, and the public find information that is related to law. Arranged alphabetically.

Wasserman, Steven R., Martin A. Smith, and Susan Mottu, eds.
Encyclopedia of Physical Sciences and Engineering Information Sources: A Bibliographic Guide to Approximately 16,000 Citations for Publications, Organizations, and Other Sources of Information on 425 Subjects Relating to the Physical Sciences and Engineering
Detroit, Mich., Gale Research Co., 1989. 736p. ISBN 0810324989

Another subject-specific reference source published by Gale, this encyclopedia can help anyone, from researcher to the general reader, to find information and resources in the disciplines of engineering or the physical sciences.

✦ School Libraries

Garrett, Linda J., and Joanne Moore
Teaching Library Skills in Middle and High School: A How-To-Do-It Manual
New York, Neal-Schuman, 1993. 253p. ISBN 1555701256
  Focuses on ideas for teaching library skills to sixth- through twelfth-graders. Provides forty-four lesson plans that reinforce library skills.

Rudin, Claire
The School Librarian's Sourcebook
New York, Bowker, 1990. 504p. ISBN 0835227111
  This bibliography lists resources covering management of a school library media center, building and maintaining a collection, and service and education of students.

✦ Special Libraries

Asantewa, Doris
Strategic Planning Basics for Special Libraries
Washington, D.C., Special Libraries Association, 1992. 57p. ISBN 0871113996
  In a workbook format, Asantewa presents specific steps for strategic planning, including assembling a team, user surveys, and use budgeting as a planning tool. The appendix contains a sample strategic plan and some helpful charts.

Christianson, Elin B., and others
Special Libraries: A Guide for Management
3rd ed. Washington, D.C., Special Libraries Association, 1991. 92p. ISBN 0871113805
  Provides answers to frequently asked questions from paraprofessionals regarding staffing and services of special libraries. Includes criteria for establishing special libraries as well as information on staffing and maintaining them.

Griffiths, José-Marie, and Donald W. King
Special Libraries: Increasing the Information Edge
Washington, D.C., Special Libraries Association, 1993. 198p. ISBN 0871114143
  Presents a study of the nature and use of corporate and organizational special libraries. This work was drawn from twenty-seven different studies over a twelve-year period. Discusses all major facets of special libraries and the delivery of service and information to their specialized clientele.

✦ Special Populations

Americans with Disabilities Act Handbook
Washington, D.C., U.S. Equal Employment Opportunities Commission, U.S. Dept. of Justice, 1992. 1 v. (various pagings). ISBN 0160381487
  A handbook to provide the public, businesses, and institutions with assistance in understanding the ADA (Americans with Disabilities Act). Intended as a basic resource document, it contains annotated regulations for Titles I, II, and III. Appendix included.

Bringing Us Together: A Selected Resource Guide to Cultural Diversity Activities in the Library Community
Chicago, American Library Association, 1992. 19p.
  This compact resource guide on services to minorities will give technicians ideas about how to provide outreach services for their libraries.

Library of Congress. National Library Service for the Blind and Physically Handicapped
Library Resources for the Blind and Physically Handicapped
Washington, D.C., Library of Congress, 1977- . Published annually. ISSN 0364-1236
  Providing information for physically challenged patrons can sometimes be problematic. This directory lists available resources and how to find them.

McNulty, Tom, and Dawn M. Suvino
Access to Information: Materials, Technologies, and Services for Print-Impaired Readers
Chicago, American Library Association, 1993. 162p. ISBN 0838976417
  This guide aids library staff in the location of materials and services available to blind or visually-impaired patrons. Understanding the problems faced by print-impaired readers helps libraries offer more effective services. Includes appendices on agencies and associations. Includes bibliography and index.

Salter, Jeffrey L., and Charles A. Salter
Literacy and the Library
Englewood, Colo., Libraries Unlimited, 1991. 212p. ISBN 0872878732
  Illuminating the role that all libraries must assume in dismantling roadblocks to literacy, this source enlightens and informs readers about problems potentially faced by students or patrons when using the library. Includes an excellent bibliography.

Trotta, Marcia
Managing Library Outreach Programs: A How-to-Do-It Manual for Librarians
New York, Neal-Schuman, 1993. 154p. ISBN 1555701213
  Making library services accessible to a broader population is the mission of this compact manual which provides useful tips and guidelines on promoting library services. Although it is written for public libraries, the ideas presented here are useful for anyone in library outreach.

Wright, Kieth C., and Judith F. Davie
Library and Information Services for Handicapped Individuals
3rd ed. Englewood, Colo., Libraries Unlimited, 1989. 242p. ISBN 0872876322
  Mirrors the changing scene of information services to the handicapped patron and aids in the evaluation of those library services provided. The chapters on services for AIDS victims, and the impact of technology on services to the handicapped, are especially informative. Each of the ten chapters includes a section on employee development activities designed to help adapt attitudes toward these patrons.

✦ Technical Services

Boyd, Alan, and Elaine Druesedow, comps.
Library of Congress Rule Interpretations for AACR2, 1988 Revision
Oberlin, Ohio, Oberlin College Library, 1989- . 1 v. (loose-leaf)

Updated quarterly, this source cumulates the revisions and interpretations of the cataloging rules as reported in the *Cataloging Service Bulletin* (Library of Congress, Processing Services, 1978-).

Cargill, Jennifer S., ed.
Library Management and Technical Services: The Changing Role of Technical Services in Library Organizations
New York, Haworth Press, 1988. 154p. ISBN 0866567798

Covers the impact of technology on technical service personnel, and the value of technical services in providing effective services to patrons.

Crawford, Walt
MARC for Library Use: Understanding Integrated USMARC
2nd ed. Boston, G. K. Hall, 1989. 359p. ISBN 0816118876

Since MARC is the most important factor in U.S. library automation, a complete understanding of MARC formats and tags is vital. This handbook is updated and revised to include changes in USMARC standards. Recommended for anyone working in bibliographic control.

Evans, G. Edward, and Sandra M. Heft
Introduction to Technical Services
6th ed. Englewood, Colo., Libraries Unlimited, 1994. 534p. ISBN 0872879399

This newly revised source covers all facets of technical services and provides a wealth of information for the beginning technician. Includes index.

Fecko, Mary Beth
Cataloging Nonbook Resources: A How-to-Do-It Manual for Librarians
New York, Neal-Schuman, 1993. 204p. ISBN 1555701248

One of the many *How-to-Do-It* manuals for libraries, this source covers major formats in nonbook cataloging, such as visual materials, sound recordings, maps, computer files, kits, and electronic resources. Meant to be used in conjunction with *AACR2*, examples are provided to aid the nonbook cataloger.

Ferl, Terry Ellen
Subject Cataloging: A How-to-Do-It Workbook
New York, Neal-Schuman, 1991. 92p. ISBN 1555700993

This workbook focuses on structured practice in the principles of subject cataloging: assigning subject headings and classifying materials. Each exercise provides information from a title page, a dust jacket, preface, and (occasionally) table of contents. Emphasis is on book materials. Answers are provided with explanations and interpretations. Includes bibliography.

Frost, Carolyn O.
Media Access and Organization: A Cataloging and Reference Sources Guide for Nonbook Materials
Englewood, Colo., Libraries Unlimited, 1989. 265p.
ISBN 0872875830

This book provides instruction on descriptive cataloging of nonbook materials. Developed to help catalogers as well as reference staff, explanations are given on organization and access to materials such as maps, sound recordings, and films.

Gorman, Michael
The Concise AACR2, 1988 Revision
Chicago, American Library Association, 1989. 161p.
ISBN 0838933629

This concise edition of the *Anglo-American Cataloguing Rules* (American Library Association, 1988) offers a solid introduction to a complicated set of rules for beginning catalogers or technicians. This edition is intended for individuals learning the rudiments of cataloging without the structure and detail of the full text.

Gorman, Michael, comp.
Technical Services Today and Tomorrow
Englewood, Colo., Libraries Unlimited, 1990. 207p.
ISBN 087287608X

Examines the present state of technical services and the changes that occur due to advancing technologies. Includes bibliographical references and index.

Hallam, Adele
Cataloging Rules for the Description of Looseleaf Publications: With Special Emphasis on Legal Materials
2nd ed. Washington, D.C., Office for Descriptive Cataloging, Library of Congress, 1989. 64p.

A guide on how to catalog loose-leaf publications. Since a great deal of legal material is published in loose-leaf format, this work puts a special emphasis on legal materials.

Haynes, Kathleen J. M.
Cataloging and Classification for the Small Library
2nd ed., rev. and enlarged, Santa Fe, N.M., New Mexico State Library, 1993. 1 v. (loose-leaf)

Provides programmed instruction on classification and cataloging for smaller libraries. Includes bibliographical references.

Hirons, Jean L., ed.
CONSER Cataloging Manual
Washington, D.C., Serial Record Division, Library of Congress, 1993. 2 v. (loose-leaf) ISBN 0844407844

Necessary addition to any cataloging department using the MARC standards serials cataloging. Provides the rational and procedures for serials work.

Holzberlein, Deanne
Cataloging Sound Recordings: A Manual with Examples
New York, Haworth Press, 1988. 300p. ISBN 0866567909

Designed to help individuals cataloging sound recordings, this manual provides examples of descriptive cataloging of different types. The useful appendices will help the novice clear up general cataloging problems.

Johnson, Peggy, ed.
Guide to Technical Services Resources
Chicago, American Library Association, 1994. 313p.
ISBN 0838906249

Providing an overview of technical services, this manual helps the technician understand acquisitions, descriptive cataloging, authority control, and many other issues in technical services. Includes bibliographical references and index.

Lavender, Kenneth, and Scott Stockton
Book Repair: A How-to-Do-It Manual for Librarians
New York, Neal-Schuman, 1992. 119p. ISBN 1555701035

The basics of book repair are presented in this straightforward guide for technicians or librarians. Gives step-by-step instructions on fundamental repair techniques, and the necessary tools, supplies, and equipment.

Leong, Carol L. H.
Serials Cataloging Handbook: An Illustrative Guide to the Use of AACR2 and LC Rule Interpretations
Chicago, American Library Association, 1989. 313p. ISBN 0838905013

An essential aid to technical services departments, this guide emphasizes the areas in serials cataloging that pose unique problems. It contains 178 cataloging examples, illustrating the use of AACR2 and LC rule interpretation.

Library of Congress. Cataloging Policy and Support Office
Free-Floating Subdivisions: An Alphabetical Index
5th ed. Washington, D.C., Cataloging Distribution Service, Library of Congress, 1993. 185p. ISBN 10521445

Useful for catalogers and other users of *Library of Congress Subject Headings*, this source gives an alphabetical list of all free-floating subdivisions. Although this work aids in the assigning and retrieving of free-floating subdivisions, it should be used with caution and in conjunction with the *Subject Cataloging Manual* (Library of Congress, Cataloging Distribution Service, 1991- .).

Library of Congress. Serial Record Division.
CONSER Editing Guide
Washington, D.C., The Division, 1994- 1 v. (loose-leaf) ISBN 0844408441

This manual describes the cataloging of serial publications using the MARC standards. Setting forth policies, procedures, and technical guidelines, this guide is useful for any institution creating machine-readable cataloging records for serials that must be consistent with CONSER (*Cooperative ONline SERials*) standards. Updated with supplements.

Library of Congress
Library of Congress Subject Headings
Washington, D.C., Cataloging Distribution Service, The Library. 1975- . Published annually. ISSN 1048-9711

This is easily the most important reference tool in academic libraries. Used in every area of the library, from cataloging to instruction, it provides the key to determining the subject headings to be used in subject cataloging, and retrieval of, library materials. Arranged alphabetically, it is updated monthly by the *L. C. Subject Headings Weekly Lists* (Library of Congress, 1984).

Library of Congress. Network Development and MARC Standards Office
Format Integration and Its Effect on the USMARC Bibliographic Format
1992 ed. Washington, D.C., Cataloging Distribution Service, Library of Congress, 1992. 92p. ISBN 0844407542

This manual describes format integration and its effects on the USMARC format. Since format integration will soon be a reality, this publication is essential for those working in bibliographic control.

Library of Congress. Network Development and MARC Standards Office
USMARC Code List for Countries
1993 ed. Washington, D.C., Cataloging Distribution Service, Library of Congress, 1993. 55p. ISBN 0844408255

A document to help catalogers designate countries by codes in the USMARC record. Contains a list of place names and their two- or three-character lower-case alphabetic codes.

Library of Congress. Network Development and MARC Standards Office
USMARC Code List for Geographic Areas
Washington, D.C., Cataloging Distribution Service, Library of Congress, 1988. 53p. ISBN 0844406074

This manual provides the cataloger with geographic area codes for the USMARC record. The codes, which consist of lowercase alphabetic letters and seven-character strings, give a hierarchical arrangement of geographic and political entities.

Library of Congress. Network Development and MARC Standards Office
USMARC Code List for Languages
1989 ed. Washington, D.C., Cataloging Distribution Service, Library of Congress, 1989. 154p. ISBN 0844406562

Vital to anyone in bibliographic control, this list specifies designation of languages in USMARC records. This revised edition contains 372 discrete codes and adds almost 200 languages which have been assigned group codes.

Library of Congress. Network Development and MARC Standards Office
USMARC Format for Authority Data: Including Guidelines for Content Designation
1993 ed. Washington, D.C., Cataloging Distribution Service, Library of Congress, 1993. 1 v. (loose-leaf) ISBN 0844408026

This standard source for those working in bibliographic control covers authority files and the use of MARC standards.

Library of Congress. Network Development and MARC Standards Office
USMARC Format for Bibliographic Data: Including Guidelines for Content Designation
Washington, D.C., Cataloging Distribution Service, Library of Congress, 1988- 2 v. (loose-leaf) ISBN 0844405957

This basic text describes the USMARC format for bibliographic data. Supplements keep the user current with any changes to the MARC format. The LC guidelines and standards for machine-readable bibliographic data are covered in depth.

Merrill-Oldham, Jan, and Paul A. Parisi
Guide to the Library Binding Institute Standard for Library Binding
Chicago, American Library Association, 1990. 62p. ISBN 0838933912

Discusses the conservation and preservation of books in understandable language. Covers the Library Binding Institute Standard and provides illustrations and useful appendices to introduce beginners to library binding.

Miller, Rosalind E., and Jane C. Terwillegar
Commonsense Cataloging: A Cataloger's Manual
4th ed., rev. New York, H. W. Wilson, 1990. 180p.
ISBN 0824207890

    Incorporates the AACR2 cataloging rules and covers changes in practices due to the impact of new technologies. Discusses typical cataloging problems and the issues in cataloging nonbook formats.

Olson, Nancy B., and Edward Swanson
The Complete Cataloging Reference Set: Collected Manuals of the Minnesota AACR2 Trainers
DeKalb, Ill., Minnesota Scholarly Press, 1988. 2 v.
ISBN 093347444X (set)

    Provides the user with a wealth of examples of descriptive cataloging using AACR2. Designed for use as a training manual, this set can help the novice cataloger apply the *Anglo-American Cataloguing Rules*.

Penn, Ira A., and others
Records Management Handbook
Brookfield, Vt., Gower Pub. Co., 1989. 249p. ISBN 0566056666

    Valuable for new and experienced staff working in records management, this handbook covers the creation, maintenance, and disposition of all types of information in records. Records discussed include any reproducible form of information, from handcopy to electronic mass storage. The vast number of records kept in libraries makes this a valuable reference source. Includes selected readings and index.

Pierce, Sydney J., ed.
Weeding and Maintenance of Reference Collections
New York, Haworth Press, 1990. 173p. ISBN 1560240016

    A collection of essays that provides ideas on evaluating a reference collection to meet the needs of clientele. Topics covered include politics and practices, methods of measuring use, and CD-ROM sources. Includes bibliographical references.

Racine, Drew, ed.
Managing Technical Services in the 90s
New York, Haworth Press, 1991. 150p. ISBN 1560241667

    This work was previously published in the *Journal of Library Administration*. Essays on managing and working with new emerging technologies in all areas of technical services give a newcomer insight into the changes in technical services resulting from technological advances.

Saye, Jerry D.
Manheimer's Cataloging and Classification: A Workbook
3rd ed., rev. and expanded, New York, Marcel Dekker, 1991.
ISBN 0824784936

    This workbook uses detailed examples and exercises to supplement a curriculum in descriptive and subject cataloging practices using *AACR2R* (ALA, 1988), *Library of Congress Subject Headings*, and the Dewey Decimal and Library of Congress classification systems. The examples focus on the most commonly encountered cataloging and classification problems.

Saye, Jerry D., and Sherry L. Vellucci
Notes in the Catalog Record Based on AACR2 and LC Rule Interpretations
Chicago, American Library Association, 1989. 581p.
ISBN 0838933483

    Notes have traditionally been left to the discretion of the cataloger. Note field instruction, however, is becoming more important in an age when many catalogs are machine readable, and accessed through keyword searching. This book gives examples of specific types of notes for more inclusive retrieval using the on-line catalog.

Slote, Stanley J.
Weeding Library Collections: Library Weeding Methods
3rd ed. Englewood, Colo., Libraries Unlimited, 1989. 283p.
ISBN 0872876330

    Describes the important function of responsible deaccession strategies for books, periodicals, and other materials. Covers current weeding practices and related library standards. Contains a new chapter on computer-assisted weeding. Includes bibliography, appendices, and index.

Training of Technical Services Staff in the Automated Environment
Washington, D.C., Association of Research Libraries, Office of Management Services, 1991. 157p.

    A guide for training new employees, or cross-training seasoned employees, in technical services. Includes bibliographical references. *SPEC kit*, no. 171.

Wellisch, Hans H.
Indexing from A to Z
Bronx, N.Y., H. W. Wilson, 1991. 461p. ISBN 0824208072

    This practical, clearly written text covers book indexing, indexing of periodicals, and indexing for nonprint materials. Useful examples of correct and incorrect indexing will help beginning indexers and experienced professionals. Cites the latest national and international standards and contains an extensive bibliography and index.

Wynar, Bohdan S.
Introduction to Cataloging and Classification
8th ed. Edited by Arlene G. Taylor. Englewood, Colo., Libraries Unlimited, 1992. 633p. ISBN 0872878112

    A guide for the inexperienced cataloger or library technician, this source reflects the dynamic changes in almost every area of bibliographic control from descriptive cataloging to authority control. Includes appendices, glossary, bibliography, and index.

Yee, Martha M., comp.
Moving Image Materials: Genre Terms
Washington, D.C., Cataloging Distribution Service, Library of Congress, 1988. 108p.

    Provides a standard list of genre terms and form headings for moving-image materials.

Nonprint Materials

✦ Special Populations

—And Access for All: ADA and Your Library
Towson, Md., ALA Video/Library Video Network, 1993.
1 videocassette (47 min.). VHS ISBN 1566410096

This video training package, consisting of a resource manual and a videotape, will enable staff to respond effectively to patrons with disabilities. Covers ADA accessibility guidelines and includes a resource list.

People First: Serving and Employing People with Disabilities
Baltimore, Md., ALA Video/Library Network, 1990.
1 videocassette (38 min.) VHS. ISBN 0838921132

Understanding mentally and physically challenged patrons is the focus of this video that helps staff become more sensitive to special needs.

✦ Public and Reference Services

Customer Service: More Than a Smile
Baltimore, Md., ALA Video/Library Video Network, 1991.
1 videocassette (13 min.) VHS. ISBN 0838921167

Designed to help promote good customer service to all, this video training package serves as a good orientation for new staff members or as a refresher for existing staff.

How to Deal with Difficult Patrons
Fort Wayne, Ind., Allen County Public Library, 1993.
1 videocassette (87 min.) VHS

Provides a three-part approach to dealing and interacting with problem patrons. Also covers security measures and good public relations.

Smith, Robin
Circulation
Fullerton, Calif., North Orange Community College District, 1989.
1 videocassette (15 min.) VHS

Instructs library technicians on paraprofessional job duties at the circulation desk. *Library Technicians At Work* series.

Valuing Diversity: The Library Reference Assistant Program as a Student Empowerment Model
Oswego, N.Y., State University of New York College at Oswego, Learning Resource Center, 1992. 1 videocassette (24 min.) VHS

Discusses the cultural diversity initiative at Oswego's Penfield Library and Office of Learning Support Services, where nontraditional and minority students work with reference librarians at the information desk.

✦ Technical Services

The Future Is Now: The Changing Face of Technical Services
Dublin, Ohio, OCLC, 1994. 2 videocassettes (158 min.) VHS

Presents the proceedings of the OCLC Symposium at the 1993 ALA Midwinter Conference, and a discussion of how technology is changing the role of technical service staff.

Smith, David, and others
Using the New AACR2: An Expert Systems Approach to Choice of Access Points
Rev. ed. London, Library Association Publishing., 1993. 97p. and computer disk ISBN 1856040860

Provides step-by-step programmed instruction on descriptive cataloging using the *Anglo-American Cataloging Rules* (ALA, 1988).

Song, Jizhong
PRISM: A Computerized Tutorial for IBM PC and Compatibles
MS-DOS version. Englewood, Colo., Libraries Unlimited, 1992.
1 computer disk and 1 pamphlet (9p.) ISBN 087287995X

This tutorial teaches the search structures of the OCLC on-line database. Its primary use is to allow self-paced instruction in most of the OCLC PRISM search commands: numeric search keys, derived search keys (title, personal name, corporate name and name/title), use of qualifiers, and title browsing. Procedures to review search results are also included. System requirements: IBM PC or compatible; 256K RAM; MS-DOS 3.1 or higher; monochrome or color monitor.

Engineering and Technology

Aircraft Technologies 345
Marcia Miller

Automotive Technology 351
Ellen Tiedrich

Aviation 357
Joyce Hopkins

Diesel Mechanics and Heavy Equipment Operations 362
Marcia Miller

Environmental Technologies 366
Kathy C. O'Gorman

Industrial and Mechanical Design (including CAD and CAM) 371
Debbie Bogenschutz

Laser/Electrooptics Technology 376
Debbie Bogenschutz

Small Engine Repair 380
Barbara A. Heiffner

Surveying 383
Susan L. Brant

Welding 388
Karen Fischer and Dona Mitoma

Aircraft Technologies

Marcia Miller
Western Nebraska Community College
Sidney, Nebraska

Introduction

The airframe and power plant (A and P) mechanic services, repairs, and rebuilds fixed-wing or rotary civilian aircraft of all sizes, from ultralight homebuilt craft to large jumbo jets. Because the general field of aircraft is so large and varied, most mechanics specialize in one category of aircraft. Mechanics can also be certified by the Federal Aviation Administration (FAA) to perform regularly scheduled FAA inspections according to FAA- prescribed standards. Airframe mechanics are qualified to work on the fuselage, wings, landing gear, and other structural components of the plane; power plant mechanics are qualified to work on the engine. Generally the student will become qualified in both areas. When students complete the program, they have a thorough knowledge of the current *Federal Aviation Regulations* (FARs). Mechanics must maintain a rigorous program of continuing education to stay abreast of the changes in regulations.

This decade has seen the virtual collapse of the light plane industry, due in large part to an increase in product liability litigation. This development has had two immediate effects: an increase in homebuilt and experimental aircraft; and, older light planes are being refurbished and reconditioned to give longer service than originally intended. Both developments have resulted in the need for skilled mechanics who can perform necessary maintenance and inspection of these aircraft.

The FAA specifies course requirements in the FAR Part 147. The certification program consists of a rigorous schedule of written, oral, and practical examinations that are taken after the completion of a prescribed number of hours of both classroom and hands-on training. A and P mechanics rarely become certified through an apprenticeship program, although certain knowledge and skills can be obtained through military experience.

This bibliography was prepared with A and P students, instructors, and practicing mechanics in mind, as well as light plane owners and private pilots who frequent community college libraries. Materials were chosen based upon their usefulness as supplements to curriculum contents. Both recreational and commercial aircraft materials are included, as well as materials of a historical nature that illustrate the development of general aviation. Representative biographical works are included only as they relate to significant aeronautical developments (e.g., works about the Wright Brothers and Burt Rutan). Examples of significant marques or types of aircraft (e.g., Boeing 747, F-117a) are also included. Military aircraft and personalities are generally not included, with the exception of the Stealth bomber.

Accreditation

Federal Aviation Administration
800 Independence Ave. SW
Washington, D.C. 20591

Selected Associations

There are many helpful organizations that can supply additional information on airframe and power plant technology.

Aviation Maintenance Foundation International
P.O. Box 2826
Redmond, Wa. 98073

Future Aviation Professionals of America
4959 Massachusetts Blvd.
Atlanta, Ga. 30337

Professional Aviation Maintenance Association
500 Northwest Plaza, Suite 1016
St. Ann, Mo. 63074-2209

Selected Journals

Aircraft Maintenance Technology.
Johnson Hill Press. Bimonthly. ISSN 1072-3145.

Aircraft Technician.
Johnson Hill Press. Six times a year. ISSN 1044-8012

Aviation Consumer.
Belvoir Publications. Semimonthly. ISSN 0147-9911

Aviation Equipment Maintenance.
Delta Communications. Monthly, and twice in October.
ISSN 0745-0214

Aviation Mechanics Bulletin.
Flight Safety Foundation. Bimonthly. ISSN 0005-2140

General Aviation News and Flyer.
General Aviation News and Flyer. Every other week.
ISSN 1052-9136

International Aviation Mechanics Journal.
Inving-Cloud Pub. Co. Monthly. ISSN 0045-1193

Kitplanes.
Fancy Publications. Monthly. ISSN 0891-1851

Light Plane Maintenance.
Belvoir Publications. Monthly. ISSN 0278-8950

Print Materials

AIM/FAR [Airman's Information Manual/Federal Aviation Regulations]
Blue Ridge Summit, Pa., TAB/McGraw-Hill, annual. (various pagings). ISBN 083064380X
  A compilation of the Federal Aviation Administration's *Airman's Information Manual* and the *Federal Aviation Regulations*. A necessary tool for both mechanics and pilots to stay abreast of federal aviation regulation changes. New edition published annually in November.

Aircraft Ignition and Electrical Power Systems
Casper, Wyo., IAP, 1992? 133p. ISBN 0891000631
  One of a series of specialized study guides for aviation maintenance personnel. Discusses battery ignition systems, magneto ignition systems, spark plugs, DC generators and controls, A/C alternators and controls, electrical motors, and overhauls. Useful information for gasoline engines, as well as aviation applications. Includes glossary, study questions and answers.

Askue, Vaughan
Flight Testing Homebuilt Aircraft
Ames, Iowa, Iowa State Univ. Press, 1992. 177p.
ISBN 0813813085
  Useful for both the homebuilt aircraft owner and the A and P mechanic, this book takes the reader step-by-step through the process of learning the limitations of the aircraft, defining and correcting problems, and determining its capability and optimum flying techniques. Includes many diagrams and an index.

The Best of AMJ Maintenance Tips: Proven Methods from Technicians in the Field
Casper, Wyo., IAP, 1989. 176p. ISBN 089100341X
  Intended to make the mechanic's job easier and more efficient, this little book is a compilation of fifteen years of helpful hints from readers of *Aviation Mechanics Journal*. Each tip is accompanied by a drawing, a photograph, or both. Topical index.

Bygate, J. E.
Aircraft Electrical Systems: Single and Twin Engine
Casper, Wyo., IAP, 1990. 128p. ISBN 0891003576
  Intended to familiarize the A and P technician with the types of electrical systems in use in light general aircraft. Wiring schematics and excerpts are included for illustration purposes. A basic understanding of DC electricity is assumed. Includes glossary, review questions and answers.

Crane, Dale, ed.
ASA Aviation Mechanic's Handbook
Seattle, Wash., ASA Publications, 1989. 1 v. (various pagings).
ISBN 0940732890
  Contains a wealth of information compiled from a variety of industry and government sources. It is divided into nine sections, with each section having its own table of contents in addition to the general one. Sections include physical and chemical; mathematics; electrical; tools and equipment; aircraft fabrication; assembly and servicing; aircraft materials; aircraft hardware; and general information.

Crouch, Tom D.
The Bishop's Boys: A Life of Wilbur and Orville Wright
New York, Norton, 1989. 606p. ISBN 0393026604
  A biography that reads more like a novel as it tells of the arduous process of inventing a lighter-than-air powered craft. Contains extensive reference notes, bibliography, and index.

Crouch, Tom D.
A Dream of Wings: Americans and the Airplane, 1875-1905
Washington, D.C., Smithsonian Institution Press, 1989. 349p.
ISBN 0874743257
  This is a reprint of the 1981 original work. Presents the lives and contributions of early designers who preceded the Wright Brothers and Kitty Hawk. Details the contributions of Chanute, Langley, Whitehead, and Herring and culminates with the Wrights. Includes extensive notes, a bibliography, and an index.

Dzik, Stanley J.
Aircraft Detail Design Manual
3rd ed. Appleton, Wisc., Aviation Publications, 1988. 99p.
ISBN 0877940115

This book resulted from the findings of FAA investigations that many homebuilt plane crashes are due to the failure of one or more poorly designed or installed parts. Detailed drawings with measurements are given for mounts, fittings, and installations that are typically used in building a recreational aircraft. Emphasis is on eliminating unnecessary materials without sacrificing strength. A separate section deals with wing design, followed by various engineering tables and aircraft hardware specifications.

Eichenberger, Jerry A.
General Aviation Law
Blue Ridge Summit, Pa., TAB Books, 1990. 222p.
ISBN 0830634312

A general, nontechnical, very readable survey. A major section deals with the Federal Aviation Administration (FAA) regulations regarding both certification for airframe and power plant mechanic and FAA enforcement procedures. Another major portion of the book details principles of negligence liability, applications of negligence law, and product liability issues. Indexed.

Engine Troubleshooting Guide
By the editors of *Light Plane Maintenance* magazine. Riverside, Conn., Belvoir Publications, 1988. 96p.

Designed to be used in the shop or hangar by the light aircraft owner or mechanic, the information is divided into six sections, headed by tab dividers, and bound in a three-ring loose-leaf notebook. Each of the sections begins with an extensive overview of the section, followed by diagnostic charts showing the problem, the solution, and additional comments. Topics include hard starting, the turbocharged engine, troubleshooting with the EGT (exhaust gas temperature) gauge, oil pressure and temperature, high oil consumption, and miscellaneous conditions (rough engine, erratic RPM, manifold pressure problems).

FAR '93 Handbook for Aviation Maintenance Technicians: An FAA Extract
Casper, Wyo., IAP, 1993. 495p. ISBN 0891004262

This book is an acceptable alternative to the massive volume of FAA circulars in either paper or microfiche formats. Updated on a continuing basis, the recent edition contains the most recent amendments to the *Federal Aviation Regulations*. Updates include additions, deletions, and selected *Special Federal Aviation Regulations* (SFARS). All revisions are noted by bracketed bold type.

Firewall Forward—Maintaining Power: Basic and Advanced Light Plane Maintenance
By the editors of *Light Plane Maintenance* magazine. Riverside, Conn., Belvoir Publications, 1988. 180p. ISBN 0961313919

Written in conversational style, as one owner and operator to another, this work broadens the student's understanding of the "bottom end" of the light plane engine. Topics include judging the engine's "health," oil leaks, cold weather operations, turbochargers, propeller governors, time between overhauls (TBO), and protecting the engine during inactivity. Especially useful for the student and mechanic is a list of twenty questions to ask the overhauler. The appendix describes how to use the exhaust gas temperature (EGT) as a troubleshooting aid. Includes photographs and index.

Foreman, Cindy
Advanced Composites: An IAP, Inc. Training Manual
Casper, Wyo., IAP, 1990. 189p. ISBN 0891003584

A comprehensive guide to composite materials, structures, and repairs intended for the airframe technician. Covers applications of composites in military and commercial aircraft, business jets, and helicopters. Includes glossary.

Foye, James
Aircraft Technical Dictionary
3rd ed. Casper, Wyo., IAP, 1992. 502p. ISBN 0891004106

Technical terms are explained in everyday language. Appendices include aviation abbreviations, Greek alphabet symbols, standard symbols and abbreviations, and chemical element symbols. Illustrated.

Fuel and Oil Systems: Basic and Advanced Light Plane Maintenance
By the editors of *Light Plane Maintenance* magazine. Riverside, Conn., Belvoir Publications, 1988. 179p.
ISBN 0961313927

Carburetion, fuel injection, fuel and oil additives, and the pros and cons of oil analysis are a few of the topics discussed in this work—topics that few other books mention. Written in the same conversational style found in the magazine, this work is well illustrated with photographs and drawings. Includes appendices and index.

Garrison, Paul
Illustrated Encyclopedia of General Aviation
2nd ed. Blue Ridge Summit, Pa., TAB Books, 1990. 462p.
ISBN 083068316X

This one-volume work provides terms and acronyms that range from one-line definitions to multipage entries. Liberally illustrated with black-and-white photographs, diagrams, charts, tables and examples. Includes names and addresses of aviation organizations, aviation publications, FAA Flight Standards District Offices (FSDO) and General Aviation District Offices (GADO), state aviation departments, airframe manufacturers, avionics, engine and equipment manufacturers.

Gunston, Bill, ed.
Chronicle of Aviation
Liberty, Mo., JL International Publishing, 1992. 984p.
ISBN 1872031307

Follows the format of other publications, most notably *The Chronicle of the Twentieth Century* (Chronicle, 1987), which reports events of the past as though they had just happened. Each year begins with a page that chronologically summarizes the major aviation events of the year. The opposite page presents a photograph of an aircraft that flew for the first time in that year or another significant event. One appendix includes an explanation of the evolution of the piston and gas turbine engines. Another lists the major airlines and includes logos, a brief history, and statistics on current fleets. Indexed.

Gunston, Bill
Giants of the Sky: The Biggest Aeroplanes of All Time
Sparkford, Nr Yeovil, Somerset, England, Patrick Stephens, 1991. 320p. ISBN 1852602589

Discusses "leviathan" aircraft divided into groups according to the decade of their appearance, pre-1920 to the present day. Addresses the original requirement for a particular oversized aircraft, the background politics, the reasons for the design choices, and why the airplane succeeded or failed. Includes tables showing comparative wing areas, wing spans, horsepower-per-pounds of thrust, and maximum weights for takeoff. Indexed.

Hall, R. J., and R. D. Campbell
Dictionary of Aviation
Chicago, St. James Press, 1991. 346p. ISBN 1558621067

The first section of this book covers the definitions of aeronautical words, terms, and phrases; the second explains acronyms and abbreviations. Cross-references are made between terms where necessary, especially to differentiate between American and European procedures. Entries of significance (e.g., radar) are boxed. Useful appendices include radio and radar frequencies used in aircraft navigation, specifications for jet fuels, and factors to convert between SI and metric systems.

Holden, Henry M.
The Douglas DC-3
Blue Ridge Summit, Pa., TAB Books, 1991. 205p.
ISBN 0830634509

Provides a history of the DC-3 and a discussion of its impact upon the entire field of air transportation. Includes a bibliography and an index.

Hubin, Wilbert N.
The Science of Flight: Pilot-Oriented Aerodynamics
Ames, Iowa, Iowa State Univ. Press, 1992. 352p.
ISBN 0813803985

Because the intended audience is private operators of both standard civil aircraft as well as experimental and ultralight craft, the author has adopted a conversational tone, even when dealing with some rather complex mathematics and technical topics. Charts, diagrams, and photographs complement the text. Includes appendices and index.

Irving, Clive
Wide-Body: The Triumph of the 747
New York, Morrow, 1993. 384p. ISBN 0688099025

A well-researched behind-the-scenes story of the creation of the 747 and the transformation of America's transportation habits. Reads like a novel. Includes illustrations, photographs, bibliography, and index.

Jakab, Peter L.
Visions of a Flying Machine: The Wright Brothers and the Process of Invention
Washington, D.C., Smithsonian Institution Press, 1990. 263p.
ISBN 0874744563

Although this book is partly biographical, the focus is on the process of invention and how the Wright brothers solved the range of technical puzzles that had baffled scientists for a century. This book complements Tom Crouch's *The Bishop's Boys* (Norton, 1989).

Jones, Robert T.
Wing Theory
Princeton, N.J., Princeton Univ. Press, 1990. 216p.
ISBN 0691085366

Covers classical wing theory, including the behavior of wings and airflow at both high and low speeds. Intended audience is physicists, aeronautical engineers, and advanced students. Includes diagrams and calculations, a bibliography, and an index.

Leary, William M., ed.
Aviation's Golden Age: Portraits from the 1920s and 1930s
Iowa City, Iowa, University of Iowa Press, 1989. 201p.
ISBN 0877452423

Chronicles the development of both civil and military aviation during the period between the World Wars. Discusses contributions to the field by people not usually connected with aviation, such as Henry Ford and Herbert Hoover.

Lombardo, David A.
Advanced Aircraft Systems
Blue Ridge Summit, Pa., TAB Books, 1993. 359p.
ISBN 0830639977

The majority of the material in this book is extracted from the three airframe and powerplant mechanic's handbooks from the FAA, but it has been rewritten to be more easily understood by the layman. As the author states in the introduction, the pilot, as well as the A and P mechanic, has become less a practitioner of a skill and more a manager of systems. Thus, this book treats each system individually. Topics include turbine engines, lubrication and cooling, propellers, electrical, hydraulic, and pneumatic systems, fuel, environmental concerns, landing gear, fire protection, and aerodynamic control.

Lombardo, David A.
Aircraft Systems: Understanding Your Airplane
Blue Ridge Summit, Pa., TAB Books, 1988. 287p.
ISBN 0830608230

Although intended for private airplane owners and pilots, the material is covered in "layman's" terms and is useful to the A and P student as well. Useful introductory information includes a discussion of aging elements; external and interior preflight inspections; preventative maintenance; and parts manufacturing processes. Also includes information on choosing the best parts source. Major sections discuss the power plant, electrical system, aircraft systems, and instrumentation. The author is an FAA Volunteer Accident Prevention Counselor.

Middleton, D. H., ed.
Composite Materials in Aircraft Structures
New York, Wiley, 1990. 394p. ISBN 0582017122

Following a brief history of the development of composite materials and technical information regarding the types of composites, the author discusses their design and applications. Testing and quality assurance are treated in separate chapters, followed by a section on repair techniques. Scholarly tone. Includes case histories, references, and index.

inton, David
Boeing 737
Blue Ridge Summit, Pa., TAB Books, 1990. 76p.
ISBN 0830686185

Historical account of both commercial and military versions of the Boeing 737. Shows details of various configurations of the aircraft, including wings, tails, and pylons, as well as the power plant. A section is devoted to major accidents and airframe failures in the 737. Sections on fleet census as well as modeling tips and a listing of available model kits are particularly interesting. Indexed.

Minton, David
The Boeing 747
Blue Ridge Summit, Pa., TAB Books, 1991. 114p.
ISBN 0830635742

This history of the Boeing 747 covers the civil, military, foreign, and domestic versions of the plane. Discusses significant historical events associated with the 747, such as terrorism and accidents. The final chapter gives tips for model construction, as well as a listing of available model kits by manufacturer. Includes a complete fleet listing featuring serial and registration numbers of every 747 produced. Photographs, drawings, and tables enhance the text. Indexed.

Neese, William A.
Aircraft Hydraulic Systems
3rd ed. Malabar, Fla., Krieger Pub. Co., 1991. 518p.
ISBN 0894645625

Well-labeled exploded drawings and black-and-white photographs illustrate all components of aircraft hydraulic and pneumatic control systems. Emphasizes correct installation techniques and use of appropriate tools for the job. Includes an extensive section on abbreviations, followed by logic, electrical, and hydraulic symbols. An excellent shop reference.

Otis, Clark E., and Peter A. Vosbury
Aircraft Gas Turbine Engines of the World and Dictionary of the Gas Turbine
Casper, Wyo., IAP, 1991. 455p. ISBN 0891003908

Over half of this book is devoted to the listing of manufacturers' data sheets for international, multinational, and American-made turbine engines. Another 150 pages comprise an easy-to-understand, jargon-free dictionary. Also included is a directory of turbine-powered aircraft manufacturers. Appendices.

Pace, Steve
F-117A Stealth Fighter
Blue Ridge Summit, Pa., TAB Books, 1992. 70p.
ISBN 0830627952

Concentrates on the technological breakthroughs achieved by this revolutionary aircraft. A major section of the book discusses the unique features of the airframe and power plant systems and structures of the F-117A. Appendix provides specifications, major suppliers, and subcontractors list. Includes a bibliography and an index.

Porter, Donald J.
Learjets: The World's Executive Aircraft
Blue Ridge Summit, Pa., TAB Books, 1990. 120p.
ISBN 0830624406

Details the inception and development of this aircraft. Photographs and diagrams illustrate the engine and fuselage changes. Nine pages are devoted to specifications and performance data of various models. An appendix gives a chronology of major events in the development of the Learjet. Indexed.

Shevell, Richard Shepherd
Fundamentals of Flight
2nd ed. Englewood Cliffs, N.J., Prentice Hall, 1989. 438p.
ISBN 0133390608

Designed for a one-semester undergraduate class in the basics of aeronautics, including flight theory and fluid mechanics fundamentals. Includes diagrams, tables, and appendices, and chapter-based problems and references.

Smith, Hubert
The Illustrated Guide to Aerodynamics
2nd ed. Blue Ridge Summit, Pa., TAB Books, 1992. 337p.
ISBN 0830639020

Explains in everyday language what enables an airplane to fly, to be stable and controllable. Shows in a straightforward manner how aerodynamic principles are applied to the design of efficient airplanes. Mathematical equations are used only when they are essential to the understanding of some important relationships. Could be used as an introductory text. Includes many diagrams, review questions and answers, sample problems, a bibliography, and an index.

Standard Aircraft Workers' Manual
15th pocket ed. Sedalia, Mo., Fletcher Aircraft Co., 1989. 1 v. (various pagings)

Target audience includes A&P students and mechanics in the field. Treats only all-metal constructions, not composites. Sections address the following topics: aircraft materials (aluminum, steel, titanium); aircraft finishes and processes; standard parts (fasteners, pulleys); flat pattern layout; and shop practices (tools, fabrication). Pocket-sized format.

Stits, Don
Synthetic Fabric Coverings: The Complete Guide to Aircraft Finishing
Casper, Wyo., IAP, 1990. 80p. ISBN 089100307X

Intended as a general source of information, not as a replacement for manufacturers' product literature. Provides general guidelines regarding fabrics, finishing, and patching techniques for homebuilt aircraft. Metals and metal finishes are also covered, as well as tools and equipment. Includes appendix and glossary.

Taylor, Richard L.
Aircraft Systems: Really Knowing Your Airplane
Riverside, Conn., Belvoir Publications, 1991. 221p.
ISBN 1879620049

Uses case anecdotes and photographs as guides to inspect and troubleshoot small civil aircraft systems. Conversational writing style. Brief index. *Command Decision Series*, 3.

Thomas, Kas
EGT Systems: Basic and Advanced Light Plane Maintenance
Greenwich, Conn., Belvoir Publications, 1989. 171p.
ISBN 0961313986

This book goes beyond the manufacturers' literature and takes a generalist's approach to exhaust gas temperature (EGT) systems.

It describes how the mechanic or pilot can use the EGT gauge to determine the fuel-air mixture, and also troubleshoot the entire combustion system of an airplane. Includes both digital and analog displays, and an appendix section detailing the graphic engine monitor (GEM) by Insight Instrument Corporation. Includes appendices, glossary, and index.

Wanttaja, Ron
Kitplane Construction
Blue Ridge Summit, Pa., TAB Books, 1991. 400p.
ISBN 0830635653

Deals with every aspect of the decision-making process and preparations involved in the selection and building of a homebuilt aircraft. Also provides sections on composite, metal monocoque, steel and aluminum tube, and wood and fabric construction techniques. Includes ground testing, FAA inspection, and flight- testing information. The appendix lists basic comparison data for a variety of kits, grouped by type of construction.

Wright, Orville
How We Invented the Airplane: An Illustrated History
New York, Dover, 1988. 87p. ISBN 0486256626

Presents the reprint of two historical writings by Orville Wright: *How We Invented the Airplane* and *After the First Flights*. Includes seventy-six black-and-white photographs, most of them taken by Wilbur or Orville Wright. Fred C. Kelly, the official biographer of the Wright brothers, contributes an article, "After Kitty Hawk: A Brief Resume" and the book concludes with an article the brothers wrote for *The Century Magazine* in 1908 called "The Wright Brothers' Aeroplane."

Yenne, Bill
The World's Worst Aircraft
New York, Dorset Press, 1990. 159p. ISBN 0880294906

Provides two-page entries on problem aircraft designs. In addition to American aircraft misfits, Italian, Russian, Japanese, and German "lemons" are described as well. Includes index.

Nonprint Materials

Basic Aerodynamics
Englewood, Colo., Jeppesen Sanderson, Inc., 1990
1 videocassette (ca. 30 min.) VHS

This video explains the fundamental concepts of flight, beginning with an analysis of the four forces of flight, and continuing with segments on maneuvering flight, airplane balance, and stability. Bernoulli's principle and airfoil design are explained and demonstrated. Presents technical information in clear language. Part of the *Flightime Video Series*.

Computerized Aviation Publications
Golden, Colo., Summit Aviation, 1990. CD-ROM

This compact disk is a convenient single source for a wide variety of FAA publications. It contains all FARs, CFRs and preambles, FAA Handbooks, ATC, 8300, 8400, and 8700. Contains information on federal law as it pertains to the Americans with Disabilities Act and the aviation industry. Also includes FAA legal opinions, more than 300 advisory circulars, and more than 3,700 illustrations. Upgraded approximately every four months, to coincide with the updates of the AIM and ATC handbooks. Annual subscription fee. Information may be searched by words, phrases, related words, and truncations. Test may be either downloaded to a DOS text file or printed. Includes context-sensitive help screens.

Automotive Technology

Ellen Tiedrich
Gloucester County College
Sewell, New Jersey

Introduction

Automotive Technology is a sophisticated field as the automobiles produced today are complex, intricate, and highly electronic. The automotive technician of today must be prepared to work on automobiles having integrated circuits that monitor and control the engine, transmission, suspension, braking, emission control, and other systems. The individual must be able to maintain, service, diagnose, and repair these cars. It is essential that individuals entering this field have good reasoning ability, a thorough knowledge of automobiles and their components, and a comprehensive understanding of how the complex systems interrelate.

Automotive mechanic training programs are offered in high schools, community colleges, and public and private vocational and technical schools, but postsecondary programs generally provide more thorough career preparation than do high school programs. Most postsecondary programs provide intensive training through a combination of classroom instruction and hands-on practice.

Community college programs normally spread the training over two years, supplementing the automotive training with instruction in English, basic mathematics, and other subjects, and an associate degree is awarded on completion. Various automobile manufacturers and their participating dealers sponsor two-year associate degree programs at many community colleges throughout the nation. Manufacturers provide service equipment and late-model cars on which students practice new skills, and ensure that the programs teach the latest automotive technology. Students in these programs typically spend alternate blocks of time attending classes full-time and working full-time in the service departments of sponsoring dealers.

The National Automotive Technicians Education Foundation (NATEF), an affiliate of the National Institute for Automotive Service Excellence (NIASE), certifies automobile mechanic training programs offered by high schools and postsecondary trade schools, technical institutes, and community colleges. Certification by the National Institute of Automotive Service Excellence is not a requirement, but it is a widely recognized standard of achievement for automobile mechanics. Mechanics are certified in one or more of eight different service areas: electrical systems; engine repair; brakes; suspension and steering; heating and air conditioning; engine performance; manual drive train and axles; and automatic transmission and transaxle. Master automotive mechanics are certified in all eight areas. Automotive body repairers can achieve certification by NIASE by passing a written examination and completing two years of experience in the trade. Completion of a recognized program can be substituted for one year of work experience.

This bibliography has been created for libraries that support students enrolled in an automotive technology program in a postsecondary school environment. Attempts were made to locate some audiovisual materials that might be used in a library environment; unfortunately, the only identified materials are those produced by the cooperating manufacturers for student use in a laboratory or classroom situation. Each film is specifically designed to relate to the automobiles that are produced by the sponsoring manufacturer. Traditionally these films are kept in the laboratory established to support the program. Only if the instructors wish to place them on reserve in the library should they be available as part of the library collection; they are too program-specific to incorporate into this bibliography.

Certification

National Institute for Automotive Service Excellence (NIASE)
13505 Dulles Technology Drive
Herndon, Va. 22071

Selected Associations

Following is a list of several national associations that can provide information on automotive education, industry trends, and continuing education:

ASSET Program/Training Department
Ford Parts and Service Division
Ford Motor Company
Room 109
3000 Schaefer Rd.
Dearborn, Mich. 48121

Information on automobile manufacturer-sponsored two-year associate degree programs in automotive service technology.

Automotive Service Association
1901 Airport Fwy., Suite 100
P.O. Box 929
Bedford, Tex. 76095-0929

Chrysler Dealer Apprenticeship Program, National C.A.P. Coordinator, SIM 423-21-06
26001 Lawrence Avenue
Center Line, Mich. 48015

General Motors Automotive Service Educational Program
National College Coordinator
General Motors Service Technology Group
30501 Van Dyke Ave.
Warren, Mich. 48090

National Automotive Technicians Education Foundation
13505 Dulles Technology Drive
Herndon, Va. 22071-3415

Provides a list of certified mechanic training and automotive body repair programs.

Selected Journals

Automotive Engineering.
Society of Automotive Engineers. Monthly. ISSN 0098-2571

Automotive News.
Crain Communications. Weekly. ISSN 0005-1551

Auto Service Today.
ATCOM, Inc. Biweekly. ISSN 1042-7414

AutoWeek.
Crain Communications. Weekly. ISSN 0192-9674

Blue Seal.
National Institute for Automotive Service Excellence. Semiannually. ISSN 0897-9421

Body Language.
Sarco Management and Publications. Monthly.

BodyShop Business.
Babcox Publications. Monthly. ISSN 0730-7241

Car and Driver.
Hachette Magazines. Monthly. ISSN 0008-6002

Hot Rod.
Petersen Pub. Co. Monthly. ISSN 0018-6031

Impact: A Journal of Safety Litigation News from the Center for Auto Safety.
The Center. Bimonthly. ISSN 0162-4989

Motor Service.
Hunter Pub. Co. Monthly. ISSN 0027-1977

Motor Trends.
Petersen Pub. Co. Monthly. ISSN 0027-2094

Print Materials

Bartz, Wilfried J., ed.
Engine Oils and Automotive Lubrication
New York, Marcel Dekker, 1993. 801p. ISBN 0824788079

This comprehensive text deals with the state-of-the-art in the field of automotive lubrication. It contains seven chapters, each with numerous sections authored by experts from the mineral oil, additive, and automotive industries as well as from research institutes. This book is well referenced and, where appropriate, appendices are presented to support ideas and theories.

Bastow, Donald
Car Suspension and Handling
3rd ed. Revised by Geoffrey P. Howard. Warrendale, Pa., Society of Automotive Engineers, 1993. 362p. ISBN 1560914041

The third edition of this well-written and comprehensive text reflects a major update. The original theory presented in the second edition is kept intact; information that applies only to later automobiles is incorporated in the text. Several completely new chapters include drive layout arrangements and computer use in suspension design. Each new chapter includes many references, excellent illustrations, diagrams, and charts.

Bono, Saverio G.
Auto Technology: Theory and Service
Albany, N.Y., Delmar Publishers, 1990. 819p. ISBN 0827338112

An excellent text that begins with a discussion of automotive career opportunities and the basics of the automobile. It addresses all facets of the car from the engine, chassis, and emissions to auxiliary systems such as heating and air conditioning. This book is comprehensive but very approachable. There is a well-constructed glossary with a comprehensive index.

Bosch, Robert
Automotive Handbook
Cambridge, Mass., Robert Bentley Publishers, 1993. 852p. ISBN 0837603307

This handbook is a pocket-sized technical reference tool whose primary purpose is to provide the automotive engineer, the mechanic, and the student with a wealth of reliable technical data as well as an insight into present day state-of-the-art automotive technology. This text was originally published in Germany. This is an excellent reference tool; it contains many illustrations and is extremely comprehensive for its size.

Brejcha, Mathias F.
Automatic Transmissions and Transaxles: Diagnostics and Service
Englewood Cliffs, N.J., Regents/Prentice Hall, 1993. 267p. ISBN 0130520101

This text is divided into two distinct sections: diagnostic procedures and service practices. and contains more than 500 line and photo illustrations. It is designed for use on its own, or as a companion to the author's *Automatic Transmissions and Transaxles: Fundamentals of Operation* (Prentice Hall, 1993). English (also known as "customary" or "U.S.") and metric units of measurement are combined throughout the text with English/ metric conversion tables provided in the appendix.

Deem, Bill R.
Electronics Math
3rd ed. Englewood Cliffs, N.J., Prentice Hall, 1990. 713p. ISBN 0132483602

As mathematics plays an indispensable role in the world of modern automotive technology, this text is recommended for individuals entering the field. Specialized math texts geared to specific curricula aid students in the mathematical calculations commonly used in that specific field. Although the first part of this book is standard in its approach, the second part deals very specifically with DC and AC circuitry, with the mathematics involved in understanding circuitry and applications to current automotive technocology.

Deroche, A. G.
The Principles of Auto Body Repairing and Repainting
5th ed. Englewood Cliffs, N.J., Prentice Hall, 1992. 82p. ISBN 0136780539

This twenty-chapter text, with its fine index, multiple appendices, and glossary, offers a good summation and explanation of all practical repair operations in use today. Chapters have clearly stated objectives and provide excellent illustrations and diagrams. The questions at the end of each chapter clearly give it a textbook character. Chapter Twenty, "Fundamentals of Writing an Estimate," is extremely well done and is of particular note.

Duffy, James E.
Auto Electricity, Electronics, Computers
South Holland, Ill., Goodheart-Willcox, 1989. 528p. ISBN 0870066943

This text offers thorough coverage of the principles and operations of electrically or electronically controlled automotive systems. There are extensive illustrations and graphics that help explain the theory and operations of the systems described. This book can serve as a text for those studying the field, and as a review for those who are working in the field.

Duffy, James E.
Modern Automotive Mechanics
South Holland, Ill., Goodheart-Willcox, 1990. 1056p. ISBN 087006777X

This text provides an easy summary of the operation and repair of all makes and models of cars. Section One introduces basic information on safety, tools, shop manuals, electricity, ASE certification, and vehicle maintenance. Each automotive system is further described in several chapters. This book can serve as a refresher course for those working in the field, and as a text for those entering the field.

DuPuy, Richard, ed.
Automotive Electrical and Electronic Systems
3rd ed. New York, HarperCollins, 1994. 2 v. ISBN 006500759X

Contains descriptive text and illustrations that enhance understanding of the material presented and aid in review. Each chapter is fully illustrated and key words are printed in boldface type. Review questions are provided for each chapter. Includes a sample test similar to those given for National Institute for Automotive Service Excellence (NIASE) certification. The shop provides detailed instructions for test procedures and overhaul and service information. Part of the *Harper Collins/Chek-Chart Automotive Series*.

Garrett, T. K.
Automotive Fuels and Fuel Systems
Warrendale, Pa., Society of Automotive Engineers, 1991. 370p. ISBN 1560911581 (Vol. 1)

The phasing out of lead additives in automotive fuels and the implementation of strict controls over evaporative and exhaust emissions have heavily impacted the constitution of fuels and the design of fuel systems. This text contains details on fuels, tanks, delivery, metering, mixing and combustions, carburetion, environmental considerations, and many other areas related to gasoline.

Gillespie, T. D.
Fundamentals of Vehicle Dynamics
Warrendale, Pa., Society of Automotive Engineers, 1992. 495p. ISBN 1560911999

This ten-chapter text, complete with bibliographical references and an excellent index, introduces the reader to the basic mechanics governing vehicle dynamic performance. After a thorough introduction to the subject, the author details acceleration, braking, suspension, steering, and tires in the chapters that follow. There are two appendices: the first addresses vehicle dynamics terminology; the second is a ride-and-vibration data manual.

Goodsell, Don
Dictionary of Automotive Engineering
Warrendale, Pa., Society of Automotive Engineers, 1989. 182p.
ISBN 0898837669

There are more than two thousand entries in this work; many of the terms are receiving their first published definition, while others have become part of everyday communication. Many of the defined terms originated with the engineer, inventor, or manufacturer who designed the item. All of the terms have become part of the vernacular of the automotive industry.

Gott, Philip G.
Automotive Air-Conditioning Refrigerant Service Guide
Warrendale, Pa., Society of Automotive Engineers, 1992. 90p.
ISBN 1560912626

This text aids automotive air-conditioning service professionals in complying with federal and state requirements. Discusses subjects such as the Clean Air Act, the environmental impact of some refrigerants, refrigerant and system properties, and the service procedures for the containment of automotive air-conditioning refrigerants. Appendices contain applicable SAE Standards and federal certification and record-keeping requirements.

Gott, Philip G.
Changing Gears: The Development of the Automotive Transmission
Warrendale, Pa., Society of Automotive Engineers, 1991. 437p.
ISBN 1560910992

This informative text begins with the history of transmission devices, explains why they are essential, and traces their development throughout the twentieth century. This book follows chronologically the evolution of these devices up to the 1990s and beyond. The appendix shows schematics of hydraulic coupling, gear trains, friction clutch, band, and one-way clutch arrangements for most of the automatic transmissions discussed in the book. Well referenced; excellent index.

Hughes, James G.
Guide to the Automobile Certification Examination
3rd ed. Englewood Cliffs, N.J., Regents/Prentice Hall, 1993. 322p.
ISBN 0133656934

This newly revised text includes material on the new section of the Automotive Service Excellence (ASE) exam. All questions appear in ASE format. Areas incorporated into this text include engine repair, automotive transmission, manual drive train and axles, suspension and steering, brakes, electrical systems, heating and air conditioning, engine performance, and the engine machinist. Contains numerous illustrations.

Jacobs, David H.
How to Repair and Restore Bodywork
Osceola, Wisc., Motorbooks International, 1991. 160p.
ISBN 0879385146

This illustrated paperback text focuses on the management of minor-to-moderate collision repairs. It gives particular attention to safety procedures during the auto-body repair process. Although not comprehensive in nature, it is a well-illustrated manual with a good index and bibliography. Addresses such issues as parts replacement versus parts repair, tools and materials selection, and damage assessment.

Layne, Ken
Automotive Engine Performance: Tune Up, Testing, and Service. Volume 1: Text
2nd ed. Englewood Cliffs, N.J., Regents/Prentice Hall, 1993. 528p. ISBN 0130597759

This set is intended for classroom use. It offers a thorough introduction to the theory and skills necessary to service engine combustion control systems. The first volume focuses on the operation of fuel, ignition, emission control, and electrical and electronic systems. Each chapter presents review questions that follow the requirements of ASE certification testing. Each of the twenty-one chapters is carefully constructed so that the student feels at ease with the material presented.

Layne, Ken
Automotive Engine Performance: Tune Up, Testing, and Service. Volume 2: Practice Manual
Englewood Cliffs, N.J., Regents/Prentice Hall, 1993. 544p.
ISBN 0130611778

This second volume provides the testing and repair instructions necessary to learn the service skills of a master mechanic. It emphasizes safety practices for servicing engines and includes many warnings and cautions to help students do repair jobs safely.

Lenz, Hans Peter
Mixture Formation in Spark-Ignition Engines
Warrendale, Pa., Society of Automotive Engineers, 1992. 400p.
ISBN 1560911883

The focal point of this volume is the mixture formation in spark-ignition engines. It includes both the theory and actual design of the mixture formation units and appropriate intake manifolds. This book was originally published in German and translated into English. Most of the symbols and abbreviations used in illustrations, equations, and charts of the German edition have been retained in the English translations. Numerous diagrams and graphs add to the value of this text.

Limpert, Rudolf
Brake Design and Safety
Warrendale, Pa., Society of Automotive Engineers, 1992. 460p.
ISBN 1560912618

This text is an ideal reference tool for those involved in brake system design and analysis. Contains all essential concepts, guidelines, and design checks required for designing safe brakes. It introduces the elements of braking performance and follows with design and analyses of friction, and mechanical, hydraulic, and air brake systems. Also included is a discussion of anti-lock brake systems and an analysis of brake failure.

Mitchell International, Inc.
Mitchell Automechanics
2nd ed. Englewood Cliffs, N.J., Prentice Hall, 1991. 788p.
ISBN 0135837820

This second edition focuses on the latest-model import and domestic cars. It is specifically aimed at helping the student understand the latest automobile technology, and it covers such concepts as body control modules, four-wheel steering, and computer controls. It uses bold colors to highlight terms and bring photographs and illustrations to life. It includes a new chapter on tires and wheels, and a short section entitled "A Day in the Life of an Automotive Technician."

Owen, K.
Automotive Fuels Handbook
Warrendale, Pa., Society of Automotive Engineers, 1990. 684p.
ISBN 156091064X

This text addresses the commonly used automotive fuels: gasoline and diesel. Discusses alternative fuels, such as methanol and ethanol, and racing fuels, which are rarely discussed. Includes an extensive glossary of terms and a list of commonly used abbreviations. There are nine appendices at the end of the text as well as a comprehensive index.

Ribbens, William B.
Understanding Automotive Electronics
4th ed. Carmel, Ind., Sams, 1992. 392p. ISBN 0672273586

This self-paced tutorial introduces the user to the most vital areas of electronics within the automotive environment. This edition explores electronic engine control and the development of new automotive subsystems. It is an essential reference tool with numerous illustrations that helps one understand how sensors and actuators operate, the elements of automotive instrumentation and diagnostics, and the latest digital engine control techniques. Each chapter has a quiz that enables the reader to evaluate comprehension of the material presented.

Rudman, Jack
New Rudman's Questions and Answers on the OCE, Occupational Competency Examination in Auto Body Repair.
Syosset, N.Y., National Learning Corp., 1992. 1 v. (various pagings). ISBN 083735705

This review book prepares applicants for the competency exams in auto body repair. Includes a question-and-answer format of factual and technical information and also provides exercises in problem solving.

Rudman, Jack
New Rudman's Questions and Answers on the OCE, Occupational Competency Examination in Auto Mechanics
Syosset, N.Y., National Learning Corporation, 1992. 1 v. (various pagings). ISBN 0837357071

This is another in the series of review books that is intended for applicants who seek to present objective competency in the area of auto mechanics. This question-and-answer text prepares candidates for the Automotive Service Excellence (ASE) certification exam. Covers factual knowledge, technical information, understanding of principles, and problem solving.

Scharff, Robert
Complete Automotive Engine Rebuilding and Parts Machining
Albany, N.Y., Delmar Publishers, 1991. 504p. ISBN 0827336128

This text on engine rebuilding includes chapters with clearly stated objectives. Provides review questions so that students can check their understanding. This is an excellent tool for those in the field and those aspiring to enter the field. Includes glossary.

Scharff, Robert, ed.
Motor Auto Body Repair
2nd ed. Albany, N.Y., Delmar Publishers, 1992. 824p.
ISBN 0827346670

All car bodies manufactured today are unibody-constructed. When working on most collision-damaged unibody vehicles, it is necessary to have a complete knowledge of the mechanical components, including operation and diagnosis of problems. An autobody specialist must know how to make adjustments to the following: strut suspension system, independent rear wheel suspension, rack-and-pinion steering, and trailing rear axles. This text also contains an excellent chapter on plastic repair techniques.

Schultz, Morton J.
Mort Schultz's Electronic Fuel Injection Repair Manual
New York, Crown Publishers, 1989. 168p. ISBN 0517572400

Almost all new cars designed today have electronic fuel injection (EFI) systems in place of carburetors. This up-to-date guide is an excellent tool for troubleshooting and repairing new EFI systems. After an introductory chapter that explains the roles of vacuum and electricity and how they can cause so many performance problems, following chapters discuss General Motors, Ford, Chrysler, and many foreign cars' electronic fuel-injection systems. This manual is an excellent tool for diagnosing, troubleshooting, and explaining problems relating to EFI.

Society of Automotive Engineers
SAE Transactions: Journal of Passenger Cars. Section 6
Warrendale, Pa., Annual Publications, Annual. (various pagings)
ISSN 0096-736X

This publication contains the best papers on passenger cars from the Society of Automotive Engineers presented in a given year; they have been judged by technical experts to be worthy of preserving in the permanent technical literature. Each of these scientific articles is comprehensive and includes an abstract at the beginning and an excellent conclusion at the end. Includes intricate diagrams, charts and illustrations to help explain the text.

Taylor, Don
Paint and Body Handbook
New York, HP Books, 1994. 144p. ISBN 1557880824

This newly revised and updated text contains over 450 illustrations and an easy-to-read format. Includes information on the following: repair of dents and rust, replacement of panels, aluminum and fiberglass repair, sectioning, lead and plastic filler work, welding of patches and panels, and shaping sheet metal. In addition, this work covers preparation of the surface for paint, equipment, and painting techniques. Well indexed, with an excellent glossary at the end.

Thiessen, F. J.
Automotive Engine Repair and Rebuilding
Englewood Cliffs, N.J., Prentice Hall, 1992. 338p.
ISBN 0130510122

Designed for a course in engine repair, this illustrated text covers all aspects of this specialty. At the end of each chapter there are review questions. This book is well indexed with excellent appendices.

Thiessen, F. J.
Automotive Principles and Service
3rd ed. Englewood Cliffs, N.J., Prentice Hall, 1989. 988p.
ISBN 013053935X

This text, divided into nine sections, contains forty chapters. Each section is distinct and topics range from basic shop routines to ventilating, heating, and air-conditioning systems. At the beginning of each chapter, the student performance objectives are given; at the conclusion of each chapter a self-check with test questions is provided so that the user can evaluate comprehension and retention of the information.

White, John H.
Automatic Transmissions
Albany, N.Y., Delmar Publishers, 1990. 408p. ISBN 0827326068

This comprehensive text on automatic transmissions prepares the student for work on all types of automobiles and light trucks. It contains information on front-wheel drive, all-wheel drive transmissions, and computer-controlled transmissions. Separate chapters are devoted to servicing the latest domestic and imported transmissions. It is arranged in textbook fashion with each of its fifteen chapters containing a self-test and a final test. It is well indexed and contains an excellent glossary.

Automotive Repair Manuals

A library should have available an assortment of automotive repair manuals that reflects the nature and demands of its general clientele. This general clientele might be individuals who graduated from an Automotive Technology program, do-it-yourself patrons, or just trade people who wish to reference some specific information. Three publishers are well regarded and recognized for the quality of motor manuals: Chilton, Mitchell, and Motor. It is suggested that all three publishers' works be reviewed and that a commitment be made to maintain a collection that best meets the needs of the library's patrons. For ease and simplicity of patron use, it is recommended that a library commit to one publisher.

Chilton Book Company
Chilton's Manuals
Radnor, Pa.

Chilton publishes several hundred automotive service and repair guides. This series includes repair guides for many specific vehicles such as GMC vans 1967-1986 and Mercedes-Benz 1959-1975. In addition, they publish some general manuals for certain classes of vehicles, such as turbocharged cars and trucks, import car parts, and labor guides. There are also manuals that address the repair and maintenance of certain systems, e.g., fuel injection and electronic engine controls. These manuals are extremely popular.

Cordura Publications, Inc.
Mitchell Manuals
San Diego, Calif.

Mitchell publishes a wide variety of repair and service manuals; however, the number of titles is more limited than Chilton's. This series of manuals contains the latest labor and parts information on all major repair operations. There are exploded illustrations with call- outs, parts/labor estimating guides, model identification charts, and quick-service labor operations tables. Many professional mechanics find this series to be the most comprehensive. These manuals lack dialogue, but are good reference tool for estimating time, selecting parts, and reviewing illustrations that display how parts or systems fit together.

Motor Publishing Company
Motors Manuals
New York, N.Y.

This series covers all makes and models of vehicles in a fashion similar to Mitchell's; it does not offer the numerous formats that Chilton's publishes. Motors' strengths are in the numerous charts and illustrations, in conjunction with comprehensive dialogue that guides individuals through a repair or service. These manuals are particularly helpful to individuals who need a refresher course to perform a particular job.

Aviation

Joyce Hopkins
Lee Davis Library
San Jacinto College Central
Pasadena, Texas

Introduction

Aviation or Aeronautical Technology programs are designed to provide students with the knowledge and skills to pass the Federal Aviation Administration's written and flight examinations, required for the licensing of pilots and other aircraft personnel. Though many airports offer private flight training, the major air carriers prefer to hire graduates of a collegiate flight program. *Federal Aviation Regulations,* Part 141 (Pilot Schools) outlines the basic requirements of the curriculum.

The Aviation Technology courses prepare students for the following career options: pilot (private, commercial, instrument, or multiengine), flight engineer, flight attendant, flight instructor, aircraft dispatcher (sometimes called flight superintendent), air traffic controller, and aviation management.

Historically, most commercial pilots were former military pilots. With the downsizing of the armed forces, the pilot candidate pool has greatly diminished. To meet future needs, some of the major air carriers contract with local community colleges to offer pilot training programs. These programs train students without previous flight experience for entry-level positions with regional air carriers. Those students who are hired receive additional training in that specific carrier's operations. The major carriers then hire from the regional carriers.

This bibliography can be used as a guide to the literature of aviation for the prospective airline employee, or by practitioners who want to upgrade their technical knowledge.

Certification

Federal Aviation Administration
800 Independence Ave., SW
Washington, D.C. 20591

Selected Associations

Further information on aviation careers and licensing may be obtained from the following organizations:

Air Traffic Control Association
2300 Clarendon Blvd., Suite 711
Arlington, Va. 22201

Air Transport Association of America
1301 Pennsylvania Ave., Suite 1100
Washington, D.C. 20004-7017

Airline Pilots Association International
1625 Massachusetts Ave., NW
Washington, D.C. 20036

Association of Flight Attendants
1625 Massachusetts Ave., NW
Washington, D.C. 20036

Flight Engineers International Association
1926 Pacific Coast Highway, No. 202
Redondo Beach, Calif. 90277

Future Aviation Professionals of America
4959 Massachusetts Blvd.
Altanta, Ga. 30337

Selected Journals

Air Line Pilot.
Air Line Pilots Association. Monthly. ISSN 0002-242X

Air Transport World.
Penton Publishing. Monthly. ISSN 0002-2543

Airport/Facility Directory.
(Seven regional eds.) National Oceanic and Atmospheric Administration, National Ocean Service. Issued every 8 weeks.

Aviation Tradescan.
Aerospace Research Group. Twelve times a year. ISSN 0899-1928

Aviation Week and Space Technology.
Aviation Week. Weekly. ISSN 0005-2175

Flying.
Hachette Magazines. Monthly. ISSN 0015-4806

NTSB [National Transportation Safety Board] Reporter.
Peter Katz Productions. Monthly. ISSN 0745-9874

Print Materials

Aftermath
Blue Ridge Summit, Pa., TAB Books, 1994. 225p.
ISBN 083064282X

True-life accounts of aviation accidents, with analysis of what happened and why—and the lessons to be learned. Based on in-flight recordings, National Transportation Safety Board (NTSB) reports, and firsthand accounts, each story first appeared in *Flying* magazine's column of the same title.

Ahrens, C. Donald
Essentials of Meteorology: An Invitation to the Atmosphere
Minneapolis/St. Paul, Minn., West. Pub. Co., 1993. 437p.
ISBN 0314012451

An introduction to weather and how weather has an effect on principles of flight. Includes multicolor maps and photographs. A fold-out cloud chart is included for ready reference.

Airman's Information Manual/Federal Aviation Regulations: AIM/FAR
Blue Ridge Summit, Pa., TAB-Aero. Annual. (various pagings).
ISSN 0886-9200

A compilation of the Federal Aviation Administration's *Airman's Information Manual* and the *Federal Aviation Regulations*. A necessary tool for both mechanics and pilots to stay abreast of federal aviation regulation changes. (Editions also available from other publishers).

Aviation Week Group Newsletters
The Aviation and Aerospace Almanac
Washington, D.C., McGraw-Hill, 1993. 1 v. (various pagings)
ISBN 0076070689

Includes names and addresses of key personnel in the aviation and aerospace industry. Contains statistical data on aerospace, airlines, airports, general aviation trends, government agencies, military and defense aviation, and the space program. International in scope.

Cadogan, Mary
Women with Wings: Female Flyers in Fact and Fiction
Chicago, Academy Chicago, 1993. 278p. ISBN 0897333853

A factual and entertaining look at women throughout aviation history, from balloonists to astronauts. Illustrated.

Calderone, Robert
The Complete Aviation/Aerospace Career Guide
Blue Ridge Summit, Pa., Aero, 1989. 254p. ISBN 0830683801

A sourcebook of job descriptions, organizations, companies, and publications, to assist the student who is considering a career in aviation. Indexed.

Christy, Joe
American Aviation: An Illustrated History
2nd ed. Rev. and updated by LeRoy Cook. Blue Ridge Summit, Pa., TAB-Aero, 1994. 506p. ISBN 0830644806

A pictorial survey of flight in America, from hot air balloons to the space shuttle. Includes brief biographies of pioneer men and women aviators. Chapter objectives and review questions aid its use as a textbook, as well as a historical reference.

Collins, Richard L.
Flying IFR
3rd ed. Charlottesville, Va., Thomasson-Grant, 1993. 232p.
ISBN 1565660439

Revised to include the latest changes in equipment and Instrument Flight Rules (IFR) flight procedures. Includes section on the risks and rewards of instrument flying in a small plane.

Davis, John P., and William B. Sanders
The Complete DUAT Book
San Diego, Calif., Sandlight Publications, 1992. 200p.
ISBN 0931145155

An introduction and user's guide to the Direct User Access Terminal (DUAT) which allows the pilot access to raw or translated FAA weather data, and to file flight plans via any computer terminal modem. Includes a key to abbreviations used in weather briefings. Index.

Dunlop, Reginald
Come Fly With Me: Your Nineties Guide to Becoming a Flight Attendant
Chicago, Maxmillian Publishing, 1993. 92p. ISBN 0963274996

Illustrated guide to the industry. Provides information on the life-style and insider secrets to getting a job in this highly competitive field.

FAR '93 Handbook for Aviation Maintenance Technicians: An FAA Extract
Casper, Wyo., IAP, 1993. 495p. ISBN 0891004262

This book is an acceptable alternative to massive quantities of FAA circulars in either paper or microfiche formats. Updated on a continuing basis, the recent edition contains the most recent amendments to the *Federal Aviation Regulations*. Updates include additions, deletions, and selected *Special Federal Aviation Regulations* (SFARS). All revisions are noted by bracketed bold type.

Gero, David
Aviation Disasters: The World's Major Civil Airliner Crashes Since 1950
Sparkford, England, Patrick Stephens Ltd., 1993. 224p. ISBN 1852603798

Decade-by-decade surveys of the major crashes in commercial aviation. Lists date and time, location, commercial operator, and type of aircraft. Also includes short narrative on each crash and resulting investigations. Includes maps, photographs, glossary, and index.

Gesell, Laurence E.
Aviation and the Law
2nd ed. Chandler, Ariz., Coast Aire Productions, 1993. 827p. ISBN 0960687483

An introduction to the American aviation law and legal research. Discusses criminal law (security of aircraft, airports and passengers), contracts, ownership, commercial law (goods and services), insurance, product and tort liability, and international law. Includes source documents, table of cases, and bibliography.

Gleim, Irvin N.
Private Pilot and Recreational Pilot FAA Written Exam
7th (1993-1995) ed. Gainesville, Fla., Gleim Publications, 1993. 346p. ISBN 0917539435

One of a series of practical test prep books for all levels of flight certification. Outline format contains basic aeronautical concepts and flight techniques. Each section is followed by sample questions, answers, and explanations. Revised biennially.

Gunston, Bill, ed.
Chronicle of Aviation
Liberty, Mo., JL International Publishing, 1992. 984p. ISBN 1872031307

More than 4,000 illustrations tell the story of flight from the earliest times through the stealth bomber. Worldwide coverage. The early years of aviation are arranged by broad time frames; the twentieth century is discussed year by year. Each year includes a chronology and lists aviation records broken. Includes articles, statistics and photographs (including new aircraft flown each year). Appendices include an evolution of the engine and current airlines of the world (name, national origin, logo, brief history, and fleet data). Detailed index.

Hall, R. J., and R. D. Campbell
Dictionary of Aviation
Chicago, St. James Press, 1991. 346p. ISBN 1558621067

Terms, jargon, abbreviations, and organizations associated with the fields of aviation, aeronautics, and aerospace technology are contained in this work, which is international in scope. Entries vary in length from brief descriptions to expanded definitions, especially when usage varies among countries. Written for the aviation professional, but suitable for the layman. Cross-referenced.

Holden, Henry M., with Lori Griffith
Ladybirds: The Untold Story of Women Pilots in America
Freedom, N.J., Black Hawk Pub. Co., 1991. 215p. ISBN 1879630117

Covers the often overlooked role of women in aviation, from its earliest days to the space shuttle missions. Format includes short biographies, anecdotes, and many photographs. A sequel, *Ladybirds II: The Continuing Story,* was published by Black Hawk Pub. Co. in 1993. Includes bibliography and index.

Illman, Paul E.
The Pilot's Handbook of Aeronautical Knowledge
Rev. and expanded ed. Blue Ridge Summit, Pa., TAB Books, 1991. 408p. ISBN 0830635173

Based on the FAA publication of the same name last issued in 1980 (Advisory Circular 61-23B), this work provides technical and operational information necessary to prepare for both private and commercial pilot certification. Includes charts and diagrams, key to acronyms and abbreviations, and index.

Illman, Paul E.
The Pilot's Radio Communications Handbook
4th ed. Blue Ridge Summit, Pa., TAB Books, 1993. 232p. ISBN 0830641408

Stresses the importance of communication between the pilot and air traffic control. Includes the essentials of visual flight rules communications under both normal flying conditions and emergency situations. Includes key to air traffic control abbreviations and index. *TAB Practical Flying Series.*

Jane's All the World's Aircraft
London, England, Sampson Low, Marston and Co., Annual. 1 v. (various pagings) ISSN 0075-3017

This is an identification guide for commercial and private aircraft, microlights, sailplanes, and lighter-than-air craft. Includes a section on engines. Provides a record of first flights of new aircraft. Illustrated with diagrams and photographs.

Lankford, Terry T.
The Pilot's Guide to Weather Reports, Forecasts, and Flight Planning
Blue Ridge Summit, Pa., TAB Books, 1990. 383p. ISBN 083066582X

Emphasizes the importance of weather in preflight planning. Stresses the use of analysis charts and air traffic control weather briefings. Summarizes steps in preparation of visual flight rules and instrument flight rules plans. Includes appendices, glossary, and index. *TAB Practical Flying Series.*

Leary, William M.
The Airline Industry
New York, Facts on File, 1992. 525p. ISBN 0816026750

Contains company profiles (passenger and freight carriers, past and present), key agencies and organizations, personalities, a review of government policy, and industry trends. Includes illustrations and index. *Encyclopedia of American Business History and Biography* series.

Mills, Thomas S., and Janet S. Archibald
The Pilot's Reference to ATC Procedures and Phraseology
4th ed. Van Nuys, Calif., Reavco Publishing, 1975- . 1 v. (various pagings) ISBN 0935695052

    Combines into one source civilian air traffic control procedures for both visual flight rules and instrument flight rules from the following government publications: *Air Traffic Control Handbook* (GPO, 1975-), *Federal Aviation Regulations* (GPO), and *Airman's Information Manual* (GPO). Includes glossary and index.

Newton, Dennis W.
Severe Weather Flying
2nd ed. Renton, Wash., Aviation Supplies and Academics, 1991. 178p. ISBN 1560270713

    Discusses nontechnical explanations of the types of adverse weather conditions a pilot may encounter. Offers suggestions on ways to avoid them or to fly through them safely. Index.

O'Hare, David, and Stanley Roscoe
Flightdeck Performance: The Human Factor
Ames, Iowa, Iowa State Univ. Press, 1992. 308p. ISBN 0813801737

    A study of the human factors most identified as causes of aviation accidents.

O'Malley, Penelope Grenoble
Takeoffs Are Optional, Landings Are Mandatory: Airline Pilots Talk About Deregulation, Safety, and the Future of Commercial Aviation
Ames, Iowa, Iowa State Univ. Press, 1993. 227p. ISBN 0813824141

    Pilots share their insights on topics such as training, experience, pilot error, the FAA, air traffic control, and factors that influence the mental and physical condition of the flight crew.

Oster, Clinton V., John S. Strong, and C. Kurt Zorn
Why Airplanes Crash: Aviation in a Changing World
New York, Oxford Univ. Press, 1992. 200p. ISBN 0195072235

    A study of airplane accidents and safety practices in the post-deregulation era. Examines not only American and Canadian scheduled air carriers, but also the unscheduled and general aviation sectors which are the current training ground for today's commercial pilots. Includes statistical charts, graphs, glossary, and index.

Padfield, R. Randall
Learning to Fly Helicopters
Blue Ridge Summit, Pa., TAB Books, 1992. 354p. ISBN 0830620923

    The principles and techniques of helicopter operation. Appendices include photographs and descriptions of common civilian helicopters in use today, flight test standards, and other information useful to those piloting a helicopter. Includes glossary and index. *TAB Practical Flying Series.*

Park, Edwards
Fighters: The World's Greatest Aces and Their Planes
Charlottesville, Va., Thomasson-Grant, 1990. 226p. ISBN 0934738653

    A wide-ranging discussion of airplanes, from World War II biplanes to today's high-tech war machines and the personalities who made them famous. Illustrated.

Perret, Geoffrey
Winged Victory: The Army Air Forces in World War II
New York, Random House, 1993. 549p. ISBN 0679404643

    One-volume history of the U.S. Army Air Force from its creation to its successful role in World War II. Based on previously unavailable oral histories and papers. Illustrated.

Reinhart, Richard O.
Fit to Fly: A Pilot's Guide to Health and Safety
Blue Ridge Summit, Pa., TAB Books, 1993. 192p. ISBN 0830620702

    A pilot's guide to mental and physical well-being before and during flight. Includes signs and symptoms of aviation-related medical problems. Includes charts, diagrams, and index.

Sandler, Stanley
Segregated Skies: All-Black Combat Squadrons of WW II
Washington, D.C., Smithsonian Institution Press, 1992. 217p. ISBN 1560981547

    The operational and combat history of the famous Tuskegee Airmen. Illustrated.

Scorer, Richard Segar, and Arjen Verkaik
Spacious Skies
Newton Abbot, England, David and Charles, 1989. 192p. ISBN 0715391399

    Though not a technical meteorological treatise, this is an interesting study of clouds and the weather they produce. Introduces the art of sky-watching. Beautifully illustrated.

Siuru, William D., and John D. Busick
Future Flight: The Next Generation of Aircraft Technology
2nd ed. Blue Ridge Summit, Pa., TAB/Aero, 1994. 194p. ISBN 0830643761

    Discusses the latest aviation technology including aircraft systems, design aerodynamics, propulsion, artificial intelligence, and manufacturing of components and the shift in development from military to civilian aviation. Illustrated.

Smith, Hubert
The Illustrated Guide to Aerodynamics
2nd ed. Blue Ridge Summit, Pa., TAB Books, 1992. 337p. ISBN 0830639012

    An introduction to the physics of flight and how it affects aircraft performance.

Smith, Myron J.
Passenger Airliners of the United States, 1926-1991: A Pictorial History
Rev. ed. Missoula, Mont., Pictorial Histories Pub. Co., 1991. 204p. ISBN 0933126727

    A history of commercial aircraft, organized by plane type. Data on each plane includes date plane was first put into service, its purpose, seating capacity, dimensions, engine type, performance capabilities, first company to use it, and subsequent users. Presented in outline format and accompanied by photographs of the planes.

Taylor, Richard L.
Aircraft Performance: The Forces Without
Greenwich, Conn., Belvoir Publications, 1991. 171p.
ISBN 1879620073

An examination of preflight and in-flight factors that affect the performance of an aircraft and how it responds to emergency situations: weight, speed and degree of ascent or descent, engine indicators, and whether it is a single or multiengine plane. Includes charts, graphs, photographs, and index. Information is based on case histories of actual situations and offers practical advice to sharpen the pilot's decision-making skills. Volume Five of the author's *Command Decisions Series*.

Thomas, Michael
Managing Pilot Stress
New York, Macmillan, 1989. 210p. ISBN 0026177609

Presents the causes of pilot stress and burnout and offers practical stress reduction techniques. References and index.

United States. Federal Aviation Administration
FAA Statistical Handbook of Aviation
Washington, D.C., FAA, Annual. (various pagings)
ISSN 0566-9618

Statistical data pertaining to the FAA, National Airspace System, airports, airport activity, U.S. Civil Air Carrier Fleet, U.S. Civil Air Carrier Operating Data, air personnel, general aviation aircraft, accidents, aeronautical production, and import/export.

Wells, Alexander T.
Commercial Aviation Safety
Blue Ridge Summit, Pa., TAB Books, 1991. 340p.
ISBN 0830621946

An introduction to the regulatory procedures of the FAA and the work of the National Transportation Safety Board (NTSB). Lists NTSB reports on the major crashes investigated by the Board.

World Aviation Directory
Washington, D.C., American Aviation Publications, Semiannual. (Annual edition also available). 3 v. ISSN 0043-826X
Includes company profiles, key personnel, and a buyer's guide for the international commercial aviation industry.

Nonprint Materials

Computerized Aviation Publications
Golden, Colo., Summit Aviation, 1990. CD-ROM

This compact disk is a convenient single source for a wide variety of FAA publications. It contains all FARs, CFRs, and preambles, FAA Handbooks, ATC, 8300, 8400, and 8700. Contains information on federal law as it pertains to the Americans with Disabilities Act and the aviation industry. Also includes FAA legal opinions, over 300 advisory circulars, and more than 3,700 illustrations. Updated approximately every four months to coincide with the updates of the AIM and ATC handbooks. Annual subscription fee. Information may be searched by words, phrases, related words, and truncations. Test may be either downloaded to a DOS text file, or printed. Context-sensitive help screens.

Additional Resources

The choices of videos for all aspects of the aviation industry are numerous. Computer software for test preparation and flight simulation are also available. The following companies will provide catalogs of currently available materials:

Aviation Book Company
25133 Anza Drive
Santa Clarita, Calif. 91355

Historic Aviation
1401 Kings Wood Road
Eagan, Minn. 55122

Zenith Books
P. O. Box 1
Osceola, Wisc. 54020

Diesel Mechanics and Heavy Equipment Operations

Marcia Miller
Western Nebraska Community College
Sidney, Nebraska

Introduction

The fields of Heavy Equipment Operations, Diesel Mechanics, and Trucking are so closely related that they have been combined for the purposes of this bibliography. The common thread among these areas is the diesel engine.

The diesel mechanic deals with diesel power units in heavy earthmoving or construction machinery, farm tractors and combines, long-haul trucks, stationary power units, large and small marine craft, or railroad locomotives. The mechanic must be able to operate the power unit, troubleshoot and diagnose problems with the engine, overhaul the engine when necessary, and maintain it on a day-to-day basis. The diesel mechanic learns about all the major systems of a diesel engine, such as fuel injection, hydraulics, and transmission. Additional areas of study may include such things as welding, electronics and electronic engine controls and sensors, safe shop habits and skills, and customer relations. Once the mechanic enters the workplace, the employer may require additional training at the training facility operated by the original equipment manufacturer.

The Heavy Equipment Operations Program student learns to operate a wide variety of heavy construction or mining equipment. Additional subjects studied in this program may include such areas as welding, general surveying, and staking and grading. A technician who also learns to maintain equipment, as well as operate it, is virtually assured of year-round, as opposed to seasonal, employment.

Community colleges, vocational and technical colleges, and proprietary schools may offer a certificate, diploma, or degree program in Diesel Mechanics or Heavy Equipment Operations. These programs generally are acceptable substitutes for apprentice programs in the field. Colleges may offer a certificate in truck driving as well. It is more likely, however, that the goal of the truck driver is to earn the national Commercial Driver's License (CDL) issued by each state.

As with other mechanical fields, the practitioner in any of the fields that deal with the diesel engine must engage in regular continuing education. New regulations that emanate from the Environmental Protection Agency (EPA) are of concern to the diesel mechanic as they relate to such topics as emissions standards and noise. The various state departments of transportation regulations impact the trucking industry as well.

This bibliography will be useful for students, instructors, and working professionals in the related areas of diesel mechanics, heavy equipment operations, and trucking. An effort was made to avoid information specific to one make or model of engine, equipment, or truck. Because of the lack of general information, some texts were included. The periodicals included were selected in part because they regularly contain new product announcements. Many include reader response cards that can be used to obtain in-depth information regarding various products or equipment. These periodicals regularly contain information that is transferable from one setting to another (i.e., a piston is a piston, whether it is in a small diesel engine in a car or in a large diesel engine in a locomotive).

Appreciation is extended to Jack L'Heureux, a faculty member of the Diesel Mechanics Program at the Western Nebraska Community College, for his advice and assistance in the selection of items for this bibliography.

Certification

National Institute for Automotive Service Excellence (ASE)
13505 Dulles Technology Drive
Herndon, Va. 22071

While certification is not mandatory, a heavy-duty truck mechanic may be certified by the National Institute for Automotive Service Excellence, which recognizes six different areas of competence. A mechanic who becomes certified in all six areas is recognized as a Master Heavy-Duty Truck Technician.

There is no certification or accreditation agency for Heavy Equipment Operations. However, several unions in the construction industries have proficiency standards for workers, and some states license crane operators. The U.S. Department of Labor, Bureau of Apprenticeship and Training maintains standards to which local unions must conform.

Selected Associations

American Trucking Associations, Inc. Maintenance Council
2200 Mill Road
Alexandria, Va. 22314-4677

Automotive Service Industry Association
444 N. Michigan Ave.
Chicago, Ill. 60611-3975

National Automotive Technicians Education Foundation
13505 Dulles Technology Dr.
Herndon, Va. 22071-3415

Selected Journals

DES (Diesel Equipment Superintendent).
Business Journals. Monthly. ISSN 0884-6324

Equipment Today.
Johnson Hill Press. Monthly. ISSN 0891-141X

Fleet Equipment.
Maple Publishing. Monthly. ISSN 0747-2544

Motor Age Mechanics Newsletter.
Chilton. Irregular. ISSN 0278-9418.

Motor Magazine.
Hearst Corp. Monthly. ISSN 0027-1748

Overdrive.
Overdrive. Monthly. ISSN 0030-7394

Owner Operator.
Chilton. Bimonthly. ISSN 0475-2112

Print Materials

Brady, Robert N.
Heavy-Duty Truck Fuel Systems: Operation, Service, and Maintenance
Englewood Cliffs, N.J., Prentice Hall, 1991. 557p.
ISBN 0133856755
　　This text covers engines in class 6, 7, and 8 truck models and discusses various types of fuel injection systems, routine service procedures, troubleshooting, and tune-up procedures to comply with EPA exhaust smoke emission standards. Intended for students, apprentices, and experienced mechanics. Liberally illustrated with drawings and photographs. Includes review questions, glossary, and index.

Brady, Robert N.
Heavy-Duty Truck Power Trains: Transmissions, Drive Lines, and Axles
Englewood Cliffs, N.J., Prentice Hall, 1989. 514p.
ISBN 0133858324
　　Topics include transmissions and transfer cases, gearing, clutches, drive shafts, U-joints, CV joints, rear axles, and differentials. Includes many photographs, exploded diagrams, charts, and index. Helpful text for both students and professional mechanics.

Brady, Robert N.
Heavy-Duty Truck Suspension, Steering, and Braking Systems
Englewood Cliffs, N.J., Prentice Hall, 1989. 503p.
ISBN 0133858243
　　In addition to the title topics, this text also discusses wheels and hubs, alignment, rims and tires, chassis, frames, trailers, and fifth wheels. Intended for both the student and the practicing mechanic. Includes illustrations, diagrams, charts, tables, index, and spec sheets for trucks, engines, and drive trains.

Capachi, Nick
Excavation and Grading Handbook
Revised 2nd ed. Carlsbad, Calif., Craftsman Book Co., 1987, 1991 reprint with revisions. 380p. ISBN 0934041296
　　Includes information on understanding survey stakes and using laser levels. Other topics discussed include excavation and compaction for building sites, highways, drainage channels, or trenching; installation of water, sewer, and drainpipe; laying asphalt pavement; and handling rural roads, commercial sites, and subdivisions. Very readable, nontechnical language. Includes glossary, abbreviations key, and index.

Day, David A., and Neal B. H. Benjamin
Construction Equipment Guide
2nd ed. New York, Wiley, 1991. 437p. ISBN 0471888400
　　Stresses design features and equipment operations conditions that apply to all heavy equipment rather than to particular machines. The authors emphasize performance on the job. Chapters conclude with applicable industry standards for equipment. This work is part of the *Wiley Series of Practical Construction Guides.*

Dempsey, Paul
How to Repair Diesel Engines
2nd ed. Blue Ridge Summit, Pa., TAB Books, 1991. 359p. ISBN 0830681671

The target audience is mechanics, owners, and operators of diesel engines. Dempsey combines practical how-to information with theory to produce understanding as well as knowledge. He takes a generic approach—the information is not specific to one make or model. Also includes an overview of specialized topics such as injection pump rebuilding and engine balancing. Includes index.

Fitchen, John
Building Construction Before Mechanization
Cambridge, Mass., MIT Press, 1989. 400p. ISBN 0262061023

Intended for laypeople, architectural historians, and students of construction, this text covers hand-crafted building methods from prehistory to approximately the nineteenth century when "people-power" was replaced by machine power. Structures and building methods discussed include buildings, bridges, defensive structures, funeral structures, quarries, tunnels, water supply structures, and roofs. Illustrated with black-and-white photographs, diagrams, and cutaway drawings. Includes extensive notes, bibliography, and index.

Goering, Carroll E.
Engine and Tractor Power
2nd ed. St. Joseph, Mich., American Society of Agricultural Engineers, 1989. 404p. ISBN 0929355024

Although this is a text for students in colleges of agriculture, it is useful for the owner/operator and the mechanic as well. The focus is on farm tractors, but many of the concepts apply to other self- propelled, diesel-powered vehicles as well. The volume addresses engine components, various systems, power trains, weight transfer, traction, maintenance, and safe and efficient operation. Some math is included and a knowledge of algebra is assumed. Includes review questions, problems, bibliography, standard hydraulic circuit symbols, and index.

Highway Users Federation for Safety and Mobility, and Robert M. Calvin
Truck Driver's Guide to CDL: Commercial Driver License
New York, ARCO; distributed by Prentice Hall, 1990. 644p. ISBN 0131522582

This is a study guide that is useful for anyone wanting to take the CDL exam. It covers all aspects of truck driving, maintenance, and procedures necessary for over-the-road (OTR) drivers. Each chapter ends with a glossary of keywords and their definitions. Includes practice test questions, numerous diagrams, and charts. Extensive appendices provide supplementary information. Includes index.

Mike Byrnes and Associates
Bumper to Bumper: The Complete Guide to Tractor-Trailer Operations
Tempe, Ariz., Mike Byrnes and Associates, 1988. 558p. ISBN 096216870X

This study guide covers everything the student needs to know about over-the-road trucks before beginning a career in the trucking industry. Topics are treated in complete units so the reader can utilize only the components needed at the time. There is a large section dealing with the various mechanical systems of trucks and trailers. Includes information on inspection, cargo types, loading and unloading, hazardous materials, and more. Each chapter ends with a self-quiz. Index included.

Ober, Gary J.
Operating Techniques for the Tractor-Loader-Backhoe
3rd ed. Northridge, Calif., Talus Resources, 1992. 181p. ISBN 0911785000

Material is presented sequentially so that each procedure illustrated builds upon the previous one(s). Emphasis is on finding intelligent, safe, and workable solutions to the tasks that the tractor-loader-backhoe will perform. Part One deals with the basics of control, safety, and maintenance; Part Two explains specific techniques. Includes glossary and index.

Shapiro, Howard I., Jay P. Shapiro, and Lawrence K. Shapiro
Cranes and Derricks
2nd ed. New York, McGraw-Hill, 1991. 465p. ISBN 0070564221

Emphasizing adequate preplanning of crane operation, this publication discusses and illustrates various configurations, loads and forces, stability, and mobile, tower, or derrick installations. It contains an entire chapter on safety and liability issues, and provides references to specific industry codes and standards. Includes conversion tables, glossary, and index.

Williams, Michael
Tractors Since 1889
Alexandria Bay, N.Y., Diamond Farm Books, 1991. 136p. ISBN 0852362234

This book chronicles the evolution of farm tractors from steam engines to internal combustion engines. Of particular interest is information on semi-diesel and diesel engine technology. Includes many color and black-and-white photographs. Indexed.

Nonprint Materials

Loader-Backhoe Safety: Part I—Operator Safety
Northridge, Calif., Talus Resources/Distributed by Vocational Marketing Services, 1991. 1 videocassette (20 min.) VHS

The aim of this video is to ensure that the operator protects both property and equipment while operating the loader-backhoe. Some topics included are condition and stability of the soil, reducing the risk of rolling over, weight distribution, and field markings. Still shots are interspersed with live-action video.

Loader-Backhoe Safety: Part II—Worker Safety
Northridge, Calif., Talus Resources/Distributed by Vocational Marketing Services, 1991. 1 videocassette (20 min.) VHS

This video demonstrates and illustrates safety considerations for workers on the ground in the vicinity of the backhoe-loader. Describes shoring and sloping procedures to avoid cave-ins of excavations and trenches. Cautions the worker to "treat the machine as if it's out to get you."

Safe Crane Operation and Practices
Lexington, Ky., Hammond Productions, 1989. 1 videocassette (110 min.) VHS

With the exception of a brief "infomercial" at the beginning, this video does not promote the use of specific equipment. Segments are of varying lengths: crane operating safety (40 min.); crane setup (30 min.); hand signals (10 min.); and lift capacity charts (30 min.). A booklet containing the script of the video is included in the container.

Trenchers: Stay Alert—Stay Alive
Northridge, Calif., Talus Resources/Distributed by Vocational Marketing Services, 1992. 1 videocassette (13 min.) VHS

The emphasis of this video is safety awareness and job preplanning. It takes the viewer through a prework inspection and demonstrates the correct way to load the trencher for transport. Also stresses the importance of protective gear for the operator.

Other Sources of Information

Midland Press
3218 E 35th Street Court
Davenport, Iowa 52807

This press will reproduce copies of manuals for more than 10,000 bulldozers, tractors, backhoe loaders, wheel loaders, truck loaders, motor graders, pipelayers, and skidders. Engine manuals also available. Price varies from $25 to $90.

John Deere DSC
Dept. S-P 1
P.O. Box 186
Moline, Ill. 61266-0186

John Deere provides operation maintenance manuals, service instruction booklets, and parts books reprinted from originals, ranging in price from $5 to $100.

Caterpillar Antiques Club
10816 Monitor-McKee Rd. NE
Woodburn, Ore. 97071
(503) 634-2496 (Dave Smith)

Environmental Technologies

Kathy C. O'Gorman
Cincinnati State Technical and Community College
Cincinnati, Ohio

Introduction

Environmental Technology programs in two-year colleges vary throughout the country. Most programs are designed to provide students with the training and knowledge to become environmental technicians. Environmental technicians work for cities, counties, states, or companies and deal with environmental regulations and cleanup.

Environmental issues affect the operations of the transportation industry, manufacturers who use or supply chemicals, petroleum producers, defense and energy industries, construction companies, and the health care field. Within the realm of government agencies, environmental issues are a continuing concern.

A typical curriculum features core courses designed to prepare students for the collection of soil and water samples, monitoring of treatments, management of cleanup activities, preparation of reports or recommendations concerning solid and hazardous waste management, and the performance of laboratory testing. The curriculum orients students to work with regulations, laws and permits, air pollution control, solid waste management, hazardous waste management, water and wastewater technologies, treatment technologies, and environmental chemistry. Other areas of study include computer-aided drafting, surveying measurements, and civil and environmental topics.

Information sources in this bibliography range from general topics such as the quality of natural resources, the Environmental Protection Agency, and pollution prevention methods, to specific topics such as waste management and treatment, global warming, recycling, pollution prevention, hazardous waste management, and engineering aspects of environmental cleanup. The scope is wide and pertinent to a collection supporting this curriculum.

The Environmental Protection Agency (EPA) is a prolific publisher of books, videotapes, information databases, manuals, and handbooks in this field. This current bibliography presents representative titles to illustrate the scope of EPA materials. There are two sources of note from the EPA: *Access EPA* and the *Core List for an Environmental Reference Collection*. These are the pathfinders to the resources of the EPA, in essence, guidebooks to the myriad information sources and services available. Many currently available EPA documents are free of charge or are available for a nominal fee from the Government Printing Office (GPO) or the National Technical Information Service (NTIS). Government Institutes and Noyes are two publishers who republish noteworthy sources from the EPA.

Accreditation

Each area of environmental technician training has different requirements. Currently there are no national groups or organizations that provide certification in any single area.

Accrediting Board for Engineering And Technology. Technical Accreditation Commission
345 E. 47th St.
New York 10017

Selected Journals

Biocycle.
JG Press. Monthly. ISSN 0276-5055

E: The Environmental Magazine.
Earth Action Network. Six times a year. ISSN 1046-8021

EPA Journal.
U.S.G.P.O. Bimonthly. ISSN 0145-1189

Environment Abstracts.
Bowker. Monthly, except bimonthly in May/June and Nov./Dec.
ISSN 0093-3287

Environmental Science and Technology.
American Chemical Society. Monthly, except semimonthly in September. ISSN 0013-936X

Journal of Environmental Education.
Heldref Publications. Quarterly. ISSN 0095-8964

Journal of Environmental Engineering.
American Society of Civil Engineers. Six times a year.
ISSN 0733-9372

Pollution Engineering.
Cahners Pub. Co. Twenty-one times a year. ISSN 0032-3640

Wastetech News.
Waste Tech News. Twenty-four times a year. ISSN 1040-8916

Print Materials

✦ Reference

United States. Environmental Protection Agency
Access EPA
Washington, D.C., The Agency, Annual. 1 (various pagings)
   This guide to EPA resources and materials includes public information tools, major EPA dockets, clearinghouse and hotline numbers, major EPA databases and access information, library and information services, state environmental libraries, and EPA scientific models. (EP 1.8/13:AC 2)

United States. Environmental Protection Agency
Core List for an Environmental Reference Collection
Washington, D.C., EPA, 1993. 35p.
   Lists seventeen areas of environmental references including commercial and governmental publishers and distributors. This is a good acquisitions checklist to use when ordering or updating a collection in this area. (EPA 220-R93-007)

✦ General

Air and Waste Management Association
Air Pollution Engineering Manual
Edited by Anthony J. Buonicore and Wayne T. Davis.
New York, Van Nostrand Reinhold, 1992. 918p.
ISBN 0442008430
   Covers the latest technology in this rapidly changing field. Discusses all types of emissions including gaseous, particulate, chemical, including those from food and agriculture, petroleum, and mineral and metal processing. The intended audience is teachers, students, technicians, and engineers.

Andrews, Lori, P., and others
Worker Protection During Hazardous Waste Remediation
New York, Van Nostrand Reinhold, 1990. 391p.
ISBN 0442238991
   Focuses on worker safety when dealing with hazardous waste. Includes topics such as rights and responsibilities of workers, medical surveillance, hazard recognition, toxicology, engineering controls, safe work practices, and the safe use of field equipment, respirators, clothing, and accessories.

Arbuckle, J. Gorden, and others
Environmental Law Handbook
Rockville, Md., Government Institutes, 198?- . v.
ISSN 0147-7714
   Environmental attorneys lend their expertise to this series. For example, the 1993 volume includes information on the National Environmental Policy Act, the Resource Conservation and Recovery Act, and the Toxic Substances Control Act. Legislation is explained in everyday terminology. Two companion volumes *Environmental Regulatory Glossary* (Government Institutes, 1993) and *Environmental Statutes* (Government Institutes, 1994) provide additional information in the complex area of environmental law and regulation.

Bellandi, Robert, ed.
Hazardous Waste Site Remediation: The Engineer's Perspective
New York, Van Nostrand Reinhold, 1988. 422p.
ISBN 0442272103
   Written under the auspices of a firm that specializes in engineering the cleanup of hazardous waste sites, this work speaks to college students in a technical program. Each chapter has been written by a staff person at O'Brien and Gere Engineers, a firm that is directly involved in cleanup processes and designs.

Blackman, William C.
Basic Hazardous Waste Management
Boca Raton, Fla., Lewis Publishers, 1993. 399p.
ISBN 0873717929
   Includes information on a wide variety of environmental waste topics, such as minimization, reuse, recycling, permits, compliance and enforcement, assessment techniques for site remediation, technologies, practices, and regulations. Addresses medical, biomedical, infectious, and radioactive waste management. Includes a glossary.

Cahill, Lawrence B., ed.
Environmental Audits
6th ed. Rockville, Md., Government Institutes, 1989. 1 v. (various pagings). ISBN 0865877769
   This is a practical guide for businesses seeking to avoid environmental liability. Written in a handbook format, it provides step-by-step methods for beginning and managing an audit program including planning, conducting, evaluating, and implementing changes. Legal issues are also discussed.

Carberry, Judith B.
Environmental Systems and Engineering
Philadelphia, Saunders College Publishing, 1990. 263p.
ISBN 0030296579
   This source grew out of class notes for a junior-level class in environmental engineering. It is an excellent supplement to similar

courses or parts of curriculum on this topic. Chapters covered include environmental systems, pollutant characteristics, sludge treatment and disposal, and primary and secondary water treatment.

Christensen, Thomas, H., and others
Sanitary Landfilling: Process, Technology, and Environmental Impact
San Diego, Academic Press, 1989. 592p. ISBN 0121742555

An international collaborative effort, this work presents a concise discussion on sanitary landfills. Topics include the degradation process, biogas, leachate, lining and drainage, environmental impacts, and design.

Conner, Jesse R.
Chemical Fixation and Solidification of Hazardous Wastes
New York, Van Nostrand Reinhold, 1990. 692p.
ISBN 0442205112

Addresses chemical fixation and solidification technology for the technician as well as the chemistry student. A comprehensive source for all aspects of this technology, written for use at all levels.

Energy and Pollution Control Opportunities to the Year 2000
Liburn, Ga., Fairmont Press, 1994. 824p. ISBN 0881731897

This source is valuable for determining various career options in environmental technology that are available to students. Chapters on environmental management, indoor air quality, global warming, and energy management inform job seekers of the trends in these areas.

Freeman, Harry, ed.
Hazardous Waste Minimization
New York, McGraw-Hill, 1990. 343p. ISBN 0070220433

Written by an EPA employee in the Risk Reduction Engineering Laboratory, this source covers waste minimization in industry, public sector activities, and case studies of successful waste minimization. Addresses the potential for waste reduction and various techniques and technologies.

Freeman, Harry, ed.
Standard Handbook of Hazardous Waste Treatment and Disposal
New York, McGraw-Hill, 1989. 1 v. (various pagings).
ISBN 0070220425

Comprehensive in its approach to hazardous waste, this handbook covers legislation for hazardous waste treatment and disposal, alternative waste minimization and recycling methods, and process description, including selection of appropriate treatments, expectation of results, and implementation of each procedure. A standard work in this area of environmental technology.

Gilbert, Charles E., and Edward J. Calabrese, eds.
Regulating Drinking Water Quality
Boca Raton, Fla., Lewis Publishers, 1992. 328p.
ISBN 0873715950

The Safe Drinking Water Act of 1974 and the 1986 amendments are this book's foundation. Chapters address regulations pertaining to lead in drinking water, radon in water, and microbes and disinfection by-products in water. Also included are discussions on risk estimation and the setting of standards.

Greenberg, Arnold E., and others, eds.
Standard Methods for the Examination of Water and Wastewater
18th ed. Washington, D.C., American Public Health Association, 1992. ISBN 0875532071 1 v. (various pagings)

This classic text for examining water and wastewater is essential for any library collection in environmental technologies. Provides a thorough presentation of examination techniques for physical and aggregate properties, metals, inorganic nonmetals, individual organic compounds, and radioactive and biological properties. This work is prepared and published jointly by the American Public Health Association, the American Water Works Association, and the Water Environment Federation.

Hauser, Barbara A.
Practical Hydraulics Handbook
Chelsea, Mich., Lewis Publishers, 1991. 347p. ISBN 0873715489

Written for the person learning about water and wastewater treatment facilities, this text stresses how hydraulics are an essential part of successful day-to-day operations in this field. Mathematical calculations are kept simple and practical applications are described in a very readable manner.

Holmes, Gwendolyn
Handbook of Environmental Management and Technology
New York, Wiley, 1993. 651p. ISBN 047158584X

A basic text that provides an overview to air pollution, water pollution, and solid and hazardous waste management issues. Additional chapters address noise pollution, architectural considerations, and management of the environment.

Karnofsky, Brian, ed.
Hazardous Waste Management Compliance Handbook
New York, Van Nostrand Reinhold, 1992. 454p.
ISBN 0442011067

Presents detailed, practical procedures for compliance with the regulations under the Resource Conservation and Recovery Act and its amendments pertaining to hazardous waste. Presents laws, definitions, emergency response procedures, and requirements for hazardous waste transportation, disposal, storage, and treatment.

Kaufman, Donald G.
Biosphere 2000: Protecting Our Environment
New York, HarperCollins College Publishers, 1993. 1 v.
(various pagings). ISBN 0060435763

Designed for an introductory environmental science course, this textbook offers a broad view of the environment and related issues. Provides good supplementary material for the two-year program.

Kimball, Debi
Recycling in America: A Reference Handbook
Santa Barbara, Calif., ABC-CLIO, 1992. 254p.
ISBN 0874366631

This reference work covers various aspects of recycling including history, economics, recycling organizations, state laws and regulations, and reference materials.

Manahan, Stanley E.
Hazardous Waste Chemistry, Toxicology, and Treatment
Chelsea, Mich., Lewis Publishers, 1990. 378p. ISBN 0873712099

Presents information on the chemical aspects of hazardous waste: regulation, treatment, remediation, biological effects, chemical phenomena, transport, source reduction, and research. Some chapters are very technical and are written for the professional.

Marowski, Daniel G., ed.
Environmental Viewpoints: Selected Essays and Excerpts on Issues in Environmental Protection
Detroit, Gale Research, 1992. 554p. ISBN 0810388448

The scope of this series is to provide information about important environmental issues. This first volume is a compilation of articles and selections from books on topics such as acid rain, desertification, nuclear energy, ozone depletion, wetland ecology, and the world's water supply. The series' goal is to keep current with environmental concerns and changes. *Gale Environmental Library*, v. 1.

Miller, Joseph Arthur, and others
Island Press Bibliography of Environmental Literature
Washington, D.C., Island Press, 1993. 396p. ISBN 1559631899

An annotated bibliography covering aspects of the natural and human environments. Includes books and journals in each subject area, a quick-reference index, and author-title and subject indexes.

Montgomery, John H.
Groundwater Chemicals Field Guide
Chelsea, Mich., Lewis Publishers, 1991. 271p. ISBN 0873715543

This easy-to-use guide to groundwater chemicals is organized by substance name. It is a resource for practitioners including lawyers, chemical engineers and cleanup contractors.

National Association of Safety and Health Professionals
Handbook of Emergency Response and Toxic Chemical Releases
Matawan, N.J., SciTech Publishers, 1992. 315p.
ISBN 0925760579

Presents a practical approach to dealing with emergencies. Discusses release inventory reports, corrective action technologies, and computer systems for chemical emergency planning. Also provided is an employer's guide to community right-to-know reporting.

National Institute for Occupational Safety and Health
NIOSH Bookshelf
Rev. ed. Cincinnati, Ohio, U.S. Dept. of Health and Human Services, Public Health Service, Centers for Disease Control, National Institute for Occupational Safety and Health, Division of Standards Development and Technology Transfer, 1992. 41p.

This document provides a listing of basic reference publications of NIOSH. Sources address occupational safety and health recommendations for chemical, physical, and other hazards in the workplace. Includes bibliographies.

National Institute for Occupational Safety and Health
NIOSH Pocket Guide to Chemical Hazards
Washington, D.C., The Institute, 1990. 245p.

This guidebook describes 400 hazardous chemicals or substances that may be encountered in the workplace. Recommends appropriate actions and provides data on health hazards, exposure signs and symptoms, emergency treatment, exposure limits, personal protection, respirator recommendations, and cleanup precautions.

Novotny, Vladimir, and others, eds.
Karl Imhoff's Handbook of Urban Drainage and Wastewater Disposal
New York, Wiley, 1989. 390p. ISBN 0471810371

This book has its roots in Germany, but a team of authors updated it to include U.S. technology and practice. It provides a simple and clear guide to designing urban drainage and treatment systems, and includes information on computer-generated designs.

Noyes, Robert, ed.
Pollution Prevention Technology Handbook
Park Ridge, N.J., Noyes Publications, 1993. 683p.
ISBN 0815513119

Presents a logical outline of technical information relating to current and potential pollution prevention and waste minimization techniques in thirty-six industries. The following information is provided for each industry: description of the manufacturing process, types of waste generated, and specific pollution prevention and waste minimization methods.

Nyer, Evan K.
Groundwater Treatment Technology
2nd ed. New York, Van Nostrand Reinhold, 1992. 306p.
ISBN 0442005628

A broad analysis of various technologies used in the treatment of contaminated groundwater. Chapters address definitions of the treatment systems, the lifecycle design, treatments for organic and inorganic contaminants, and actual field applications of design methods.

Pennell, Allison A., ed.
Business and the Environment: A Resource Guide
Washington, D.C., Island Press, 1992. 364p. ISBN 1559631597

A useful source for bringing together lists of books, journals, videos, case studies, and networks that illustrate how business and the environment are linked. Areas addressed include accounting and finance, government and society, management, marketing, production and operations management, and strategic management. Indexes provide multiple access points.

Shields, J., ed.
Air Emissions, Baselines, and Environmental Auditing
New York, Van Nostrand Reinhold, 1993. 280p.
ISBN 0442010923

The *Clean Air Act Amendments of 1990* demanded an increase in environmental auditing. This book focuses on the legal requirements, principles, and practices, and provides specific situations and case studies. Each chapter presents theory, practice, problems, caveats, and practical tips.

Sullivan, Thomas F. P., ed.
Directory of Environmental Information Sources
4th ed. Rockville, Md., Government Institutes, 1992. 262p.
ISBN 0865873267

A "must have" for libraries, this reference work is filled with valuable information and ways to maximize the use of the resources and information provided. Chapters cover federal and state government resources; professional, scientific, and trade organizations; reference books; newsletters; magazines; periodicals; and databases.

Treatment Technologies
2nd ed. Rockville, Md., Government Institutes, 1991. 1 v. (various pagings). ISBN 0865872635

This EPA manual explains twenty-four methods for treating hazardous wastes including thermal, chemical, and physical treatments. The format allows readers to compare the benefits and drawbacks of individual technologies.

Upp, E.L.
Fluid Flow Measurement: A Practical Guide to Accurate Flow Measurement
Houston, Tex., Gulf Pub. Co., 1993. 178p. ISBN 0884150178

A well-organized source that discusses various aspects of fluid evaluation including standards, "good" and "bad" measurements, measurement techniques for various gases and liquids, and the devices used to take those measurements.

Urbonas, Ben
Stormwater: Best Management Practices and Detention for Water Quality, Drainage, and CSO Management
Englewood Cliffs, N.J., Prentice Hall, 1993. 449p.
ISBN 0138474923

Revised edition of *Stormwater Detention* (Prentice Hall, 1989). Discusses types of storage facilities, flow regulation, storage, estimation of storage volumes, and stormwater quality enhancements. Focus is on the needs of the practicing engineer or stormwater manager.

Wastewater Engineering: Treatment, Disposal, and Reuse
3rd ed. New York, McGraw-Hill, 1991. 1334p.
ISBN 0070416907

This edition incorporates new technologies and regulations that have come into effect since the previous edition was published in 1979. Comprehensive in topics, chapters include wastewater characteristics, the design of facilities for physical, biological, and chemical treatment of wastewater, small wastewater treatment systems, and waste reclamation.

Wigglesworth, David T., ed.
Pollution Prevention: A Practical Guide for State and Local Government
Boca Raton, Fla., Lewis Publishers, 1993. 200p.
ISBN 087371654X

An extremely well-organized guide that is also practical. Discusses the pollution prevention framework, integration, fostering partnerships, and finding resources that aid pollution prevention. This work is aimed at the practitioner in the state and local government arenas.

Wild, Alan
Soils and the Environment: An Introduction
New York, Cambridge Univ. Press, 1993. 287p.
ISBN 0521432804

Examines soils of all varieties, offering detailed descriptions of properties, various processes they undergo, and the relationship of soils to the environment. Written with the student and layperson in mind, this source could be used to supplement courses dealing with water and hazardous materials.

World Wildlife Fund
Getting at the Source: Strategies for Reducing Municipal Solid Waste
Washington, D.C., Island Press, 1991. 142p. ISBN 1559631627

This work presents the results of the final report of the Strategies for Source Reduction Steering Committee for the World Wildlife Fund. It is a practical guide to the solid waste production problem, and provides a step-by-step approach to implementation of changes.

Nonprint Resources

✦ Online Databases

There are several information databases available from the EPA. One notable source for chemical information is the EPA's *Integrated Risk Information Information Database*. This work provides assessments of data quality with respect to carcinogenesis and other toxic effects. The EPA National Online Library System is also accessible through Internet. For current Internet address information, refer to the current issue of *EPA Access*.

Industrial and Mechanical Design
(including CAD and CAM)

Debbie Bogenschutz
Cincinnati State Technical Community College
Cincinnati, Ohio

Introduction

Industrial and mechanical design technicians are the intermediaries between the invention and the manufacture of a product. Working from verbal instructions, rough sketches, notes, and the calculations of engineers, they prepare scale drawings of the object to be made as seen from all sides and angles. They also calculate details such as the strength and amount of material needed, and estimate manufacturing costs. They strive to improve design to maximize ease and comfort of use, desirability of appearance, and economy of production. Industrial and mechanical design technicians work with architects, and mechanical, electrical, or civil engineers, usually specializing in one area.

Although the traditional tools of the industrial and mechanical design technician have been paper and pencil, the computer has dynamically changed this field. Computer-aided design or computer-aided drafting (CAD) and computer-aided manufacturing (CAM) are revolutionizing industrial and mechanical design. CAD technicians integrate their knowledge of traditional drafting and design with their computer knowledge and skills to develop new and improved product design. This design is done in considerably less time, with more opportunities for product testing built into the process. CAM technicians, on the other hand, set up the manufacturing processes to run at optimum efficiency, and operate the computer-controlled manufacturing apparatus.

Jobs for the technician without computer skills are scarce, while the job outlook for CAD/CAM technicians is mixed. Although jobs are plentiful now, some futurists foresee a leveling off, since one person can produce work so much more efficiently with a computer. Others predict that the expanded capabilities of the designer and drafter will raise expectations of end users and will actually result in expanded job opportunities, especially in 3-D visualization and simulation, and in linking databases and other programs to CAD.

Community colleges, technical institutes, and vocational schools offer courses and degree programs in industrial and mechanical design/CAD/CAM technology. Degree programs are generally completed in two years. Some apprenticeship programs combine classroom instruction with on-the-job training that takes three or four years to complete. Other technicians learn their skills in the armed forces.

Accreditation

Accrediting Board for Engineering and Technology
345 E. 47th St.
New York, N.Y. 10017

Selected Associations

American Design and Drafting Association
P.O. Box 799
Rockville, Md. 20848-0799

American Society for Engineering Education
11 Dupont Cir., Suite 200
Washington, D.C. 20036

Computer-Aided Manufacturing, International
1250 E. Copeland Rd. No. 500
Arlington, Tex. 76011

Industrial Designers of America
1142 E. Walker Rd.
Great Falls, Va. 22066

International Federation of Professional and Technical Engineering
8701 Georgia Ave., Suite 701
Silver Spring, Md. 20910

Selected Journals

Cadalyst.
Advanstar Communications. Monthly. ISSN 0820-5450

Cadence.
Miller Freeman. Monthly. ISSN 0887-9141

Computer Graphics World.
PennWell Pub. Co. Monthly. ISSN 0271-4159

Design News.
Cahners Pub. Co. Twenty-four times a year. ISSN 0011-9407

Electronic Design.
Penton Pub. Co. Twenty-six times a year. ISSN 0013-4872

Machine Design.
Penton Pub. Co. Twenty-eight times a year. ISSN 0024-9114

Print Materials

Adams, Lee
High-Performance CAD Graphics in C
Blue Ridge Summit, Pa., Windcrest, 1989. 524p.
ISBN 083069059X
   A very basic introduction for the PC user with no prior CAD experience. A nicely illustrated book that walks the user through the process. Includes glossary and index.

Bedworth, David D., Mark Richard Henderson, and Philip M. Wolfe
Computer-Integrated Design and Manufacturing
New York, McGraw-Hill, 1991. 653p. ISBN 0070042047
   Not as basic as some texts, this work addresses the application of CAD/CAM to discrete-item manufacturing, particularly the manufacture of automobiles, refrigerators, electronic circuit boards, and computers. Follows a logical progression from design through implementation with line drawings illustrating the text. Indexed.

Bethune, James D., and Bonnie A. Kee
Modern Drafting: An Introduction to CAD
Englewood Cliffs, N.J., Prentice Hall, 1989. 415p.
ISBN 0135910587
   A generic approach to computer-aided drafting, illustrated throughout with line drawings. Uses a step-by-step approach to introduce and reinforce the concepts. Indexed.

Bethune, James D.
Modern Electronic and Electrical Drafting with Computers
Englewood Cliffs, N.J., Prentice Hall, 1990. 288p.
ISBN 0135933102
   Generic introduction to CAD, with extensive coverage of electronic symbols, schematics, and circuit diagrams. Includes sample problems and index.

Cox, John, Peter Hartley, and Doug Walton
Keyguide to Information Sources in CAD/CAM
New York, Mansell, 1988. 257p. ISBN 072011974X
   An annotated bibliography especially useful for its essays defining aspects of the field. Includes a glossary and an index.

Dimarogonas, Andrew D.
Computer Aided Machine Design
New York, Prentice Hall, 1988. 729p. ISBN 0131664972
   A nicely illustrated introductory text using real life situations. Indexed.

Foster, Robert J., Hugh F. Rogers, and Richard F. Devon
Graphical Communication Principles: A Prelude to CAD
New York, McGraw-Hill, 1991. 333p. ISBN 0070216436
   Provides the basic knowledge of graphic interpretation needed before attempting CAD. Nicely illustrated and includes problems and index.

French, Thomas Ewing, and others
Mechanical Drawing: CAD-Communications
11th ed. New York, Gregg Division, McGraw-Hill, 1990. 616p.
ISBN 0070223378
   A standard text for mechanical drawing since 1919, this edition updates the material for the computer age. Lavishly illustrated, with text complementing the pictures. Most of the book is geared to generic topics, but special attention is given to VersaCAD, CADKEY, and FastCAD. Includes glossary and index.

Greco, Joseph, ed.
The Macintosh CAD/CAM Book
Glenview, Ill., Scott, Foresman, 1989. 390p. ISBN 0673384462
   A reference manual for the Macintosh user. Includes glossary and index.

Hunt, V. Daniel
Computer-Integrated Manufacturing Handbook
New York, Chapman and Hall, 1989. 322p. ISBN 0412016516
   General introduction to integrated technologies. Includes extensive appendices, glossary, and index.

Ingham, Peter
CAD Systems in Mechanical and Production Engineering
New York, Industrial Press, 1990. 177p. ISBN 0831130083
   A well-written, nicely illustrated introduction to the integration of CAD throughout various branches of engineering. Indexed.

Jacobs, Stephen Paul
The CAD Design Studio: 3D Modeling as a Fundamental Design Skill
New York, McGraw-Hill, 1991. 120p. ISBN 0070322279
   Nicely illustrated introduction to the principles and concepts of design. Places 3-D CAD within the full spectrum of design techniques. Illustrated, with bibliographies and index.

Juvinall, Robert C., and Kurt M. Marshek
Fundamentals of Machine Component Design
2nd ed. New York, Wiley, 1991. 804p. ISBN 0471622818
   Designed as a text for beginning students, but also serves as a reference for practitioners. Part One discusses fundamentals; Part Two focuses on applications. Includes sample problems, illustrations, appendices, bibliographies, and index.

Kerlow, Isaac Victor, and Judson Rosebush
Computer Graphics for Designers and Artists
2nd ed. New York, Van Nostrand Reinhold, 1994. 306p. ISBN 0442014309
   Although not strictly written for industrial and mechanical designers, this lavishly illustrated text would be an excellent source of enrichment for students in the field. Includes bibliography and index.

Kirkpatrick, James M.
The AutoCAD Book: Drawing, Modeling, and Applications
2nd ed. New York, Macmillan, 1993. 558p. ISBN 0675222885
   Provides hands-on exercises to allow users practical experience with AutoCAD. The exercises were designed by classroom teachers hoping to give the same information to independent learners. Amply illustrated with glossary and index.

Kusiak, Andrew, ed.
Concurrent Engineering: Automation, Tools, and Techniques
New York, Wiley, 1993. 589p. ISBN 0471554928
   Emphasizes the lifecycle approach to the design process, which evaluates not only the primary functionality but also producibility, assembly, testing, serviceability and even recyclability. Takes basic CAD a step further. Illustrated, with bibliographies and index.

Kusiak, Andrew
Intelligent Design and Manufacturing
New York, Wiley, 1992. 753p. ISBN 0471534730
   A more advanced text emphasizing the concurrent design process and its ability to shorten the timeline for design. Illustrated, with bibliographies and index.

Machover, Carl
The C4 Handbook: CAD, CAM, CAE, CIM
Blue Ridge Summit, Pa., TAB Professional and Reference Books, 1989. 438p. ISBN 083069398X
   An excellent introduction to the tools of the trade and their applications, with discussion of selecting and implementing a system. Nicely illustrated. Includes a glossary, bibliography, and index.

Omura, George
Learn CAD Now
Redmond, Wash., Microsoft Press, 1991. 439p. and computer disk ISBN 1556152817
   Tutorial introduction to EasyCAD, with computer disk for practice exercises. Includes appendices and index.

Pike, Jayna
An Introduction to Computer Graphics Concepts: From Pixels to Pictures
Reading, Mass., Addison-Wesley, 1991. 209p. ISBN 020156789x
   A nicely illustrated, nontechnical text. Includes glossary, bibliography, and index.

Rodriguez, Walter
Computer-Aided Engineering Design Graphics
2nd ed. Edited by B. J. Clark and John M. Morriss. New York, McGraw-Hill, 1989. 550p. ISBN 0070533962
   A generic introduction, with numerous examples and exercises. Includes illustrations and index.

Rodriguez, Walter
The Modeling of Design Ideas: Graphics and Visualization Techniques for Engineers
New York, McGraw-Hill, 1992. 272p. ISBN 0070533946
   A generic introduction to the principles and techniques of two- and three-dimensional modeling. Includes glossary and index.

Rooney, Joe, and Philip Steadman
Principles of Computer-Aided Design
North and South American English-language ed. Englewood Cliffs, N.J., Prentice Hall, 1988. 341p. ISBN 0137093462
   Essays on conceptual and mathematical foundations, geometric modeling, applications, and current research. Heavy math emphasis. Includes illustrations and bibliographies.

Snyder, John M.
Generative Models for Computer Graphics and CAD: Symbolic Shape Design Using Interval Analysis
Boston, Academic Press, 1992. 311p. ISBN 0126540403
   Introduces generic modeling principles and applications in a highly illustrated, math-intensive text. Includes bibliography and index.

Stark, John
Managing CAD/CAM: Implementation, Organization, and Integration
New York, McGraw-Hill, 1988. 189p. ISBN 0070608768
   Provides a basic introduction to CAD/CAM technology directed to the manager rather than the engineer. Indexed.

Taylor, Dean
Computer-Aided Design
Reading, Mass., Addison-Wesley, 1992. 492p. ISBN 020116891X
   A more scientific introduction to computer-aided design, beginning with a thorough definition of design itself. Includes appendices and index.

Tinder, Richard F.
Digital Engineering Design: A Modern Approach
Englewood Cliffs, N.J., Prentice Hall, 1991. 685p.
ISBN 013211707X
　　Practical introduction to engineering design, including examples, exercises, and problems. Includes annotated references, glossary, and index.

Turbide, David A.
Computers in Manufacturing
New York, Industrial Press, 1991. 207p. ISBN 0831130334
　　Introduces the major aspects of computer applications in manufacturing, including integration of various manufacturing processes. Includes glossary and index.

Ullman, David G.
The Mechanical Design Process
New York, McGraw-Hill, 1992. 337p. ISBN 0070657394
　　An excellent introduction to mechanical design that examines the design process from a practical standpoint through a balanced combination of text and illustration. Appendices discuss materials and their properties, normal probability, and safety. Includes bibliographies and index.

Voisinet, Donald D.
Mechanical Design Using CADD
New York, McGraw-Hill, 1989. 120p. ISBN 0070675678
　　Introduces mechanical design on CADD, assuming knowledge of CAD, algebra, trigonometry, and mechanical drafting concepts. Extensive appendices are provided.

Zeid, Ibrahim
CAD/CAM Theory and Practice
New York, McGraw-Hill, 1991. 1052p. ISBN 0070728577
　　An encyclopedic introduction to CAD/CAM. Non-software-specific material useful to beginning students as well as professionals. Includes bibliographies and index.

✦ AutoCAD

Brittain, James L., and George O. Head and A. Ted Schaefer
The AutoCAD Productivity Book for Releases 10, 11, 12
Chapel Hill, N.C., Ventana, 1993. 369p. ISBN 1566040264
　　Subtitled *The Nonprogrammers Guide to Customizing AutoCAD*, the emphasis here is on creating macros, basic customization, and increasing productivity in AutoCAD. Includes clear exercises for developing skills. Indexed.

Fuller, James Edward
Using AutoCAD: Release 11 with AME, AutoLISP, and Customizing
5th ed. Albany, N.Y., Delmar Publishers, 1992. 727p.
ISBN 0827353448
　　Practical tutorial in the use of AutoCAD. Includes an index and helpful appendices.

Gesner, Rusty
Inside AutoCAD: The Complete AutoCAD Guide
6th ed. New Riders Publishers, 1990. 1 v. (various pagings).
ISBN 0934035555
　　Complete guide suitable for the beginner through the experienced user of AutoCAD for 2-D, 3-D, and customized applications. Includes index.

Goetsch, David L., and Raymond L. Rickman
Computer-Aided Drafting with AutoCAD
Columbus, Ohio, Merrill, 1990. 396p. ISBN 0675209153
　　For novice users through professionals, this work begins with a generic introduction to CAD, then goes into the specifics. Illustrated.

Grabowski, Ralph
The AutoCAD Technical Reference
Albany, N.Y., Delmar Publishers, 1992. 381p. ISBN 0827348207
　　Advanced practical applications of AutoCAD. Includes appendices and index.

Guenther, Jeff, and Ed Ocoboc
AutoCAD: Methods and Macros
2nd ed. Blue Ridge Summit, Pa., TAB Professional and Reference Books, 1991. 474p. ISBN 0830675442
　　Provides both an introductory tutorial to AutoCAD, and encyclopedic coverage of its concepts and commands. Includes illustrations, problems, and index.

Knight, Robert L., and William R. Valaski
AutoCAD Quick Reference
2nd ed. Carmel, Ind., Que Corp., 1991. 153p. ISBN 0880226226
　　As the title says, this is a quick reference handbook providing easy answers to problems that the AutoCAD user might encounter.

Merickel, Mark
AutoCAD: Drafting and 3D Design
Carmel, Ind., New Riders Publishers, 1991. 632p.
ISBN 1562050346
　　Practical applications in the use of AutoCAD for students as well as engineers, designers, and technical drafters. Includes a good index and helpful appendices.

Shumaker, Terence M., and David A. Madsen
AutoCAD and its Applications: Release 12 for Windows
South Holland, Ill., Goodheart-Willcox, 1994. 1 v. (various pagings). ISBN 1566370205
　　A combination workbook and text providing an introduction to the skills needed to master AutoCAD. Fully illustrated with detailed explanations for all sorts of processes. Includes exercises, chapter tests, and professional "tips." Comprehensively indexed.

Tumility, Thomas
AutoCAD for Electronics: A Tutorial
Englewood Cliffs, N.J., Prentice Hall, 1991. 267p.
ISBN 0130510955
　　A very basic introduction with computer exercises. Includes useful appendices and index.

✦ CADKey

Duff, Jon M.
CADKey Light: Computer Aided Design and Drafting for Engineers and Technologists
Englewood Cliffs, N.J., Prentice Hall, 1991. 395p. and computer disk ISBN 0131173839

   An introduction to the low-end, more affordable subset of CADKey. Excellent illustrations, problems, and exercises provided. Includes a glossary, bibliography, and index.

Goss, Larry D.
Fundamentals of CAD with CADKey for Engineering Graphics
New York, Macmillan, 1990. 178p. ISBN 0023452919

   An easy-to-use reference for CADKey, with glossary and exercises.

Reichard, David C.
Exploring CADKey 3
Englewood Cliffs, N.J., Prentice Hall, 1990. 336p. and computer disk. ISBN 0132961121

   A hands-on introduction to CADKey with practical exercises. Indexed.

✦ Generic CADD

Newton, Randall S.
Inside Generic CADD
Carmel, Ind., New Riders Publishers, 1991. 516p.
ISBN 156205001X

   Practical applications in the use of Generic CADD for students as well as engineers, designers, and technical drafters. Includes a good index and helpful appendices.

✦ VersaCAD

Buehrens, Carol
VersaCAD: A Practical Approach to Computer-Aided Design
Blue Ridge Summit, Pa., TAB Books, 1988. 305p.
ISBN 0830603034

   Nicely illustrated, step-by-step tutorial. Includes exercises.

Foster, Steven R.
Introduction to VersaCAD Macintosh
Englewood Cliffs, N.J., Prentice Hall, 1992. 489p. and computer disk. ISBN 0134890639

   Nicely illustrated applications-based introduction. Includes a computer disk with exercises.

Morrissey, Maria, and Jack Mitchell
Versacad Corporation's Training Guide 2D and 3D Tutorials
Albany, N.Y., Delmar Publishers, 1989. 1 v. (various pagings) and computer disk. ISBN 0827335628

   Tutorial text with beginner-level through advanced-level exercises. Illustrated.

Laser/Electrooptics Technology

Debbie Bogenschutz
Cincinnati State Technical and Community College
Cincinnati, Ohio

Introduction

Laser/electrooptics technicians design, test, install, and repair lasers and electrooptical equipment. These technicians work with engineers and scientists in the private sector or in governmental positions. Their duties vary greatly, depending upon the laser applications with which they work. Lasers have many applications, including surgery, surveying and measurement, industrial testing, and manufacturing. The work of the technician varies not only with the application, but with the type of laser used. Principal types are gas or solid state. Some specific tasks of technicans include troubleshooting and repairing systems that use lasers; aligning optical systems; performing calculations, gathering data, and preparing reports; operating and calibrating a variety of scientific equipment; and directing the fabrication and assembly of components for lasers and electrooptic devices and services.

Employment opportunities for laser electrooptics technicians are expected to continue to increase. However, these technicians should expect to acquire new skills and knowledge on a continuing basis.

Laser electrooptics technicians graduate from two-year degree programs at community colleges and technical institutes. Students should come to college with strong preparation in science and mathematics. College programs include intensive scientific and technical study, and require many hours of laboratory work.

There is no specific certification or licensing requirements for laser electrooptics technicians; many positions may still require security clearance, while clearances in other commercial and industrial sectors have been dropped. Materials in this bibliography range from a beginning, entry level to a more sophisticated level. Items have been coded with a recommended level to aid librarians in the selection of appropriate materials to meet their distinctive program needs.

Selected Associations

Laser Institute of America
12424 Research Pkwy., Suite 130
Orlando, Fla. 32826

Optical Society of America
2010 Massachusetts Ave. NW
Washington, D.C. 20036

SPIE—The International Society for Optical Engineering
P.O. Box 10
1000 20th St.
Bellingham, Wa. 98227

Selected Journals

American Machinist.
Penton Pub. Co. Monthly. ISSN 1041-7958

Applied Optics.
Optical Society of America. Thirty-six times a year.
ISSN 0003-6935

Journal of the Optical Society of America. B, Optical Physics.
The Society. Monthly. ISSN 0740-3224

Laser Focus World.
PennWell Pub. Co. Monthly. ISSN 1043-8092

Optical Engineering Report.
Society of Photo-Optical Instrumentation Engines. Monthly. ISSN 0091-3286

Photonics Spectra.
Laurin Pub. Co. Monthly. ISSN 0731-1230

Laser electrooptical information can also be found in the popular periodical collections. Some of the most useful titles include:

New Scientist.
IPC Magazines. Weekly. ISSN 0262-4079

Popular Mechanics.
Hearst Corp. Monthly. ISSN 0032-4558

Popular Science.
Times Mirror. Monthly. ISSN 0161-7370

Print Materials

Alfano, R.R., ed.
The Supercontinuum Laser Source
New York, Springer-Verlag, 1989. 458p. ISBN 0387969462
   Deals with ultrafast laser and nonlinear optics technologies, mixing theoretical and experimental knowledge. Illustrated. Includes bibliographies and index. *(Advanced level)*

Banerjee, Partha P., and Ting-Chung Poon
Principles of Applied Optics
Homewood, Ill., Irwin, 1991. 347p. ISBN 0256088608
   Provides a general, broad introduction to the principles and applications of modern optics. Indexed. *(Intermediate level)*

Bromberg, Joan Lisa
The Laser in America, 1950-1970
Cambridge, Mass., MIT Press, 1991. 310p. ISBN 0262023180
   Discusses the early years of laser technology in both military and civilian life. Contemporary laser technicians should enjoy this account of the history. *(General interest)*

Cheo, Peter K., ed.
Handbook of Solid-State Lasers
New York, Marcel Dekker, 1989. 619p. ISBN 082477857X
   Brings together recent research on a select group of solid-state lasers. A useful reference for more advanced study. Includes bibliography and index. *(Advanced level)*

Chin, S. L.
Fundamentals of Laser Optoelectronics
Teaneck, N.J., World Scientific Pub. Co., 1989. 362p. ISBN 9810200722
   Introduces the fundamental principals of laser beam control in optoelectronics. Heavily dependent on mathematics. Includes bibliographies, index, and illustrations. *(Advanced level)*

Crafer, R. C., and P. J. Oakley, eds.
Laser Processing and Manufacturing
New York, Chapman and Hall, 1993. 292p. ISBN 0412415208
   Although the laser has become "just another machine tool" in the last thirty years, its use does require specialized training. This text is highly practical and well illustrated. Includes bibliographies and index. *(Beginning level)*

Control Optics Optical Engineering Series
Baldwin Park, Calif., Control Optics Corp.
   This monographic series covers a wide variety of topics of interest to electrooptics students at the intermediate level. Representative titles include:

   Behzadizadeh, Arman. *Geometrical Optics Experiments: An Instructional Lab Manual,* 1991.
   Bernal, Jose, and others. *An Introduction to Radiometry, Photometry, and Colorimetry,* 1991.
   Cook, David, and Kevin Ear. *Holography: An Instructional Lab Manual,* 1991.
   Eichel, Margarethe Eichel, and Joseph Diep. *Experiments in Wave Optics,* 1991.
   Ng, Joseph. *Fiber Optics Laboratory Experiments and Projects: An Instructional Lab Manual,* 1989.

Das, Pankai K.
Lasers and Optical Engineering
New York, Springer-Verlag, 1991. 470p. ISBN 0387971084
   A beginning textbook on lasers and optical engineering designed for the student with only a basic background in physics. A good overall reference. *(Beginning level) (Intermediate level)*

Dawes, Christopher
Laser Welding
New York, McGraw-Hill, 1992. 258p. ISBN 0070161232
   Introduces the laser as a welding tool, focusing on practical applications. Includes glossary and index. *(Beginning level)*

Delone, N. B.
Fundamentals of Nonlinear Optics of Atomic Gases
New York, Wiley, 1988. 221p. ISBN 0471893919
   Presents the basics of nonlinear optics and describes all interrelated phenomena. Designed for use by beginning students through professionals. Includes bibliography and index. *(Beginning level) (Intermediate level)*

Gibilisco, Stan
Understanding Lasers
Blue Ridge Summit, Pa., TAB Books, 1989. 169p. ISBN 0830692754
   An elementary introduction to the laser providing a discussion of the way lasers work; various types of lasers; uses in industry, communication, and medicine; holography; and future uses of laser technology. Indexed. *(Beginning level)*

Heavens, O. S., and R. W. Ditchburn
Insight into Optics
New York, Wiley, 1991. 309p. ISBN 047192769
   The first nine chapters provide the background and theory for the discussions in subsequent chapters. Includes topics such as electro- and magneto-optics, holography, waveguides, fibres and

Wilson, J., and J. F. B. Hawkes
Optoelectronics: An Introduction
2nd ed. Englewood Cliffs, N.J., Prentice Hall, 1989. 470p. ISBN 0136384617

    Introduces optoelectronics and optical communication systems, providing basics in fundamental physics. Includes detailed information on lasers and semiconductor devices and optical radiation detection devices. *(Beginning level) (Intermediate level)*

Yariv, Amnon
Optical Electronics
4th ed. Philadelphia, Saunders, 1991. 713p. ISBN 0030474442

    A theoretical introduction to optical electronics, with examples drawn from real life. Indexed. (*Advanced level*)

Small Engine Repair

Barbara A. Heiffner
Nicolet Area Technical College
Rhinelander, Wisconsin

Introduction

Small engine repair programs provide training in the basic operation, maintenance, and repair of lawn and garden equipment, outboard motors, motorcycles, snowmobiles, and chain saws. Students learn to locate problems; replace parts; rebuild, tune, and adjust engines; perform preventive maintenance; and repair other operating parts of these machines.

Technicians, or mechanics, usually specialize in the service and repair of one type of equipment, although they may work on closely related products with dealers or manufacturers in small businesses—engine repair shops and service stations or in their own small engine repair business. Graduates of small engine repair programs may also work as factory service representatives or as engine maintenance technicians in industrial operations.

Vocational schools and two-year colleges offer courses in small engine repair with specializations in areas such as motorcycle or marine repair. Certificates or associate degrees are available in marine propulsion technology. Studies for this program include subjects such as marine ignition and fuel systems, marine propulsion systems, and electronic navigation. As electrical and fuel systems become more complex, marine mechanics will require more advanced skills.

A knowledge of basic electronics is becoming increasingly important for lawn and garden and motorcycle technicians because electronics are often part of the engine controls, instrument displays, and other components.

Selected Associations

Although there are no accrediting agencies specific to this area, there are many helpful organizations that can supply additional information.

Motor and Equipment Manufacturers Association
300 Sylvan Ave.
P.O. Box 1638
Englewood Cliffs, N.J. 07632-0638

North American Equipment Dealers Association
10877 Watson Road
St. Louis, Mo. 63127

Society of Automotive Engineers
400 Commonwealth Drive
Warrendale, Pa. 15096-0001

Print Materials

Atkinson, Henry F.
Mechanics of Small Engines
New York, Gregg Division, McGraw-Hill, 1990. 182p.
ISBN 0070025371
   Contains many illustrations and step-by-step descriptions that will aid the beginning student in small engine mechanics. Introduces two- and four-stroke engines, lubricating, carburetion, cooling, and ignition systems. Index.

Chilton Small Engine Repair: 2 Hp to 12 Hp
Radnor, Pa., Chilton Book Co., 1993. 561p. ISBN 0801983231
   Including six popular brands of small engines, this book covers routine maintenance, tune-up, fuel system, engine overhaul, and specifications for each engine type.

Chilton Small Engine Repair: 13 Hp to 20 Hp
Radnor, Pa., Chilton Book Co., 1993. 388p. ISBN 080198324X

Five popular brands of large horsepower small engines are covered, and topics include maintenance, tune-up, repair, and overhaul procedures for each engine type.

Choate, Curt, and John Harold Haynes
The Haynes Small Engine Repair Manual: The Haynes Workshop Manual for Small Engine Repair
Newbury Park, Calif., Haynes Publications, 1990. 290p.
ISBN 1850106665

This small engine repair manual covers the most popular and widely used engines from the leading manufacturers. An appendix lists engine service specifications for several brands of engines. Index.

Dempsey, Paul
Small Gas Engine Repair
2nd ed. Blue Ridge Summit, Pa., TAB Books, 1993. 230p.
ISBN 0830641416

Dempsey emphasizes troubleshooting and repair information. The many flow charts and exploded diagrams help to identify specific malfunctions. Index.

Electrical Systems: Compact Equipment
Moline, Ill., Deere and Co., 1982- . 1 v. (various pagings).

This manual explains basic electrical principles such as electrical circuits, electric motors and generators, battery and charging circuits, and coil ignition systems. Includes wiring diagrams, information on special tools, diagnostic procedures, and a glossary.

Engines: Compact Equipment
3rd ed. Moline, Ill., Deere and Co., 1992. 1 v. (various pagings).
ISBN 0866911464

A manual that explains how to test and repair engines up to 40 hp—those found on such equipment as chain saws, lawn mowers, garden tractors, and weed trimmers. Part One includes an overview of the components and operation of engines; Part Two covers engine service, repair, adjustment, diagnosis, and safety. Index.

Hoerner, Harry, Donald L. Ahrens, and W. Forrest Bear
Small Gasoline Engines, Operation, and Maintenance
Rev. ed. St. Paul, Minn., Hobar Publications, 1992. 176p.
ISBN 0913163260

This is a revised edition of the original manual and reflects the latest technical changes in small engine design and operation. Classroom and laboratory exercises can be used for instruction in repair and maintenance and for evaluation activities in an instructional program.

Miller, Rex, and Mark Richard Miller
Small Gasoline Engines: Service and Repair
3rd ed. New York, Macmillan, 1993. 1v. (various pagings).
ISBN 0025849913

A well-illustrated book that covers practical information needed to repair and service small gasoline engines. Follows standard procedures recommended by the manufacturers of engines. Includes appendices and index.

Power Trains: Compact Equipment
2nd ed. Moline, Ill., Deere and Co., 1991. 1 v. (various pagings).
ISBN 0866911367

A book designed to provide basic knowledge and skills required of a beginning compact equipment service technician. The power trains discussed are found on chain saws, lawn mowers, garden tractors, tillers, and snow blowers, and are limited to equipment up to 40 PTO horsepower. Chapters are devoted to explanations, operating procedures, maintenance, and repair. Index.

Roth, Alfred C.
Small Gas Engines
South Holland, Ill., Goodheart-Willcox Co., 1992. 352p.
ISBN 0870069195

Provides practical information about small engine construction, operation, lubrication, maintenance, troubleshooting, service, rebuilding, and repair. Includes detailed technical information about one- and two-cylinder, two- and four-cycle gasoline engines.

Rudman, Jack
New Rudman's Questions and Answers on the OCE, Occupational Competency Examination in Small Engine Repair
Syosset, N.Y., National Learning Corp., 1993. 1 v. (various pagings)

Tips for taking the exam as well as hundreds of practice questions are provided in this work. Also includes a section on the basic fundamentals of electricity.

Shoemark, Pete
Motorcycle Basics Manual
Newbury Park, Calif., Haynes Publications, 1991. 164p.
ISBN 1850100837

A book for the motorcycle owner with little understanding of how the machine works. Information is presented in a clear, basic manner with many line drawings to illustrate components and systems. Glossary and index.

Shuster, William A.
Small Engine Technology
Albany, N.Y., Delmar Publishers, 1993. 348p. ISBN 0827349270

This is a text designed to be used as a bridge between an instructor's lectures and the information furnished by the engine manufacturer. Many photographs and illustrations supplement the text, which includes data on the most common small air-cooled engines. Emphasis is on the teardown and assembly of the small Briggs and Stratton 3.5 hp engine.

Thiel, Richard
Keep Your Outboard Motor Running
Camden, Maine, International Marine, 1992. 77p.
ISBN 0877423288

This is a basic introduction to outboard motors with simple operational and maintenance guidelines. The emphasis is on proper use and preventative maintenance with very little reference to actual repair procedures. Index.

Additional Resources

The following publishers are excellent resources for specific vehicle or engine workshop practice manuals, owners manuals, repair manuals, and shop manuals:

Chilton Book Co.
201 King of Prussia Road
Radnor, Pa. 19089-0230

Clymer Publishing
P.O. Box 12901
9800 Metcalf
Overland Park, Kans. 66282-2901

Intertec Publishing Corp.
P.O. Box 12901
Overland Park, Kans. 66282-2901

Motorbooks International
729 Prospect Ave.
Osceola, Wisc. 54020

The following companies specialize in the reprint of manuals:

John Deere DSC
Dept. S-P 1
P.O. Box 186
Moline, Ill. 61266-0186

Operation maintenance manuals, service instruction booklets, and parts books reprinted from originals, ranging in price from $5 to $100. Specific to this area are manuals on hydraulics and power trains.

Midland Press
3218 E 35th Street Court
Davenport, Iowa 52807

This press will reproduce copies of manuals for more than 10,000 bulldozers, tractors, backhoe loaders, wheel loaders, truck loaders, motor graders, pipelayers, and skidders. Engine manuals also available. Price varies from $25 to $90.

Surveying

Susan L. Brant
Nicolet Area Technical College
Rhinelander, Wisconsin

Introduction

The Land Surveying Technician program at community, junior, and technical colleges prepares students for entry-level positions in land surveying careers. Courses in math, science, English, and surveying provide students the requisite knowledge and skills to succeed as land surveying technicians.

Land surveyors measure and plot the locations of land and water boundaries. Data are gathered about the features of the land and water areas. Using reference points, they measure lines, elevation contours, and distances between points. Surveyors also make legal descriptions for deeds and leases and establish and mark property lines. They are assisted by land survey technicians who operate surveying instruments and collect information.

Recent trends in surveying technology have significantly affected surveying practices. Field data collection is taught using electronic surveying instruments. Data acquisition systems, the use of GPS (Global Positioning Systems), along with increased computer utilization have all changed the methods used by surveyors in the field, as well as the work that is completed in their offices. This new technology is reflected in many of the titles chosen for this bibliography.

Most persons prepare for licensing by combining postsecondary school courses in surveying with extensive on-the-job training. Community and junior colleges offer one-, two-, and three-year programs in both surveying and surveying technology. All fifty states license surveyors. In order to become licensed, most states require some formal postsecondary education courses and five to twelve years of surveying experience.

Surveyors work for private surveying or engineering firms, state or municipal highway departments, or are self-employed. They generally work eight-hour days, five days per week. Much of the work is done outdoors, except when doing computations, writing reports, drawing maps, or testifying in court.

Certification and Licensure

National Society of Professional Surveyors
5410 Grosvenor Lane
Bethesda, Md. 20814-2122

American Congress of Surveying and Mapping
5410 Grosvenor Lane
Bethesda, Md. 20814-2122

The NSPS-ACSM certification program is registered with the U.S. Department of Labor as part of the National Apprenticeship Program. Technicians can be certified at one of three progressive levels depending on experience. The certification program is not intended to be a part of a professional track.

National Council of Examiners for Engineering and Surveying
P.O. Box 1686
Clemson, S.C. 29633

This service organization is for engineering and land surveying registration. It promotes uniform standards of registration and coordinates interstate registration of engineers.

Selected Journals

Journal of Surveying Engineering.
American Society of Civil Engineers. Four times a year.
ISSN 0733-9453

Point of Beginning-P.O.B.
POB Pub. Co. Six times a year. ISSN 0739-3865

Professional Surveyor.
Harrison Communications. Six times a year. ISSN 0278-1425

Surveying and Land Information Systems.
American Congress on Surveying and Mapping. Quarterly.
ISSN 1052-2905

Print Materials

Bell, Frank
Surveying and Setting Out Procedures
Brookfield, Vt., Ashgate Pub. Co., 1993. 276p. ISBN 1856284360

Bell's comprehensive and practical volume combines the details of the work involved in land surveying and setting out procedures for building construction. Covers the use of levels, tapes and theodolites and their latest counterparts. Presents modern instruments and techniques as well as traditional methods. This British publication uses the metric system when discussing measurements. Includes a recommended reading list and useful index.

Broadus, Jerry R.
The Surveyor and Property Law
Canton, Mich., POB Pub. Co., 1990. 1 v. (various pagings)

A seminar manual that aids in the understanding of complex legal issues in the practice of surveying. Chapters deal with topics such as easements, deed interpretations, title insurance, surveyors and the courtroom, contracts, and testifying as an expert witness. Appendices follow with article reprints, court cases, useful examples, diagrams and sample problems.

Brown, Curtis M., Walter G. Robillard, and Donald A. Wilson
Evidence and Procedure for Boundary Location
3nd ed. New York, Wiley, 1994. 512p. ISBN 0471552194

New material has been added to this edition, earlier material has been updated, and the text has been rearranged to better assist students in progressing through the subjects presented. A new section on using measurement evidence to prove the proximity of monuments and corners has been added, along with a section on the evidence of measurement. There are adequate graphics, clear photographs, and excellent line drawings throughout the text. A famous surveying treatise by Justice Cooley on the quasi-legal functions of surveyors is provided in an appendix.

Buckner, R. B.
Land Surveying Computations
Rancho Cordova, Calif., Landmark Enterprises, 1989. 108p.
ISBN 0910845506

The emphasis of this volume centers on plane surveying and the application of trigonometry and analytic geometry to typical land surveying problems. Chapters include material on basic mathematical concepts in surveying, basic traverse calculations and applications of plane surveying computations. Includes an appendix on programming HP calculators, and a list of illustrations and tables.

Buckner, R. B.
Land Survey Review Manual
2nd ed. Rancho Cordova, Calif., Landmark Enterprises, 1993. 436p. ISBN 0910845492

Consists of review material for the surveying exam. The coverage is broad and is not simply limited to problems. In addition to test-taking tips, this work includes basic surveying and mapping concepts, surveying computations which include traverse and route surveying, earthwork and coordinates, and questions on hydrographic and mine surveys. Appendices cover examination standards and scoring, and give the format used by the National Council of Examiners and Engineers.

Evett, Jack B.
Surveying
2nd ed. Englewood Cliffs, N.J., Prentice Hall, 1991. 246p.
ISBN 0138788855

Although still considered a basic surveying text, this edition has been updated to include material on hydrographics, mining, and municipal surveying. Theodolites, recording of field data, aerial photography, and state plane coordinate systems have also been added to appropriate chapters. Includes a change of emphasis from probable error to standard error, and a computer program change from FORTRAN to BASIC to allow easier access by PC users. Sample problems follow each chapter; answers to half of the questions are provided at the end of the book. Includes stadia reduction tables and an index.

Harbin, Andrew L.
Land Surveyor Reference Manual
2nd ed. Belmont, Calif., Professional Publications, 1989. 1 v. (various pagings). ISBN 0912045094

Harbin's reference manual aids students in studying for surveying examinations by covering mathematics and measurements, topographic surveying and mapping, legal issues, various aspects of field practice, astronomical observations, and the restoration of lost corners. Appendices include mathematical tables, a list of organizations, and a glossary of terms. Also includes practical problems and a useful index.

Harbin, Andrew L.
The Solutions Manual for the Land Surveyor Reference Manual
2nd ed. Belmont, Calif., Professional Publications, 1992. 74p.
ISBN 0912045426

Provides answers to the practice problems offered in the *Land Surveyor Reference Manual* (Professional Publications, 1989). They are organized according to chapters.

Herubin, Charles A.
Principles of Surveying
4th ed. Englewood Cliffs, N.J., Prentice Hall, 1991. 301p.
ISBN 0137176953

Covers surveying practices and training in the traditional use of equipment (tape, transit, and level). There is a logical progression from easier to more difficult topics. Emphasizes concepts dealing with space visualization and logical thinking. Includes a set of reference tables and problems to solve, as well as a comprehensive index.

Kavanagh, Barry F.
Surveying with Construction Applications
2nd ed. Englewood Cliffs, N.J., Prentice Hall, 1992. 464p.
ISBN 0138789509

While the second edition contains all the material from the first, it also incorporates updated information on digital levels, electronic theodolites, total stations, Geographic Information Systems, and Global Positioning Systems. The first eight chapters cover the fundamental principles with sections on topics such as distance measurement, leveling, transits, and traverse computations. Part Two presents more advanced material. Comprehensive coverage of construction surveying provides students with "real world" applications. Includes a useful field note index, a glossary of common terms, answers to selected problems, and a helpful guide to abbreviations. Metric equivalents and symbols are printed on the inside of the front and back covers.

Kavanagh, Barry F., and S. J. Glenn Bird
Surveying: Principles and Applications
3nd ed. Englewood Cliffs, N.J., Prentice Hall, 1992. 667p.
ISBN 0138789274

Presents relevant new techniques and instrumentation while also retaining topics that continue to be important to surveying curricula. Information concerning total stations, digital levels, and electronic theodolites have been updated in this edition. Also included is a new section on the historical evolution of surveying. There is a glossary offering definitions of surveying terms, answers to selected chapter problems, and an index. Numerous illustrations add to the quality of this text.

Keen, John E.
Land Surveying Law
Rev. ed. Spruce Pine, N.C., Land Surveyor's Workshops, 1990. 278p. ISBN 156569001X

Covers legal statutes, ordinances, and laws relating to the practice of land surveying. Each of the sixteen chapters is followed by a series of questions relating to the material covered. An answer key is provided at the end of the book. Some of the subjects included here deal with conveyances, lot proration, easements, eminent domain, equitable estoppel, monumentation, common law dedications, water law, and deed interpretation.

Keen, John E.
Trouble Shooting Boundary Line Problems: Questions and Answers
Spruce Pine, N.C., Land Surveyor's Workshops, 1989. 226p.
ISBN 1565690028

This study guide is designed to help students prepare for professional examinations. The format consists of a series of questions and answers relating to boundary lines. Keen recommends specific reference titles for students to consult.

Kratz, Kenneth E.
Survey Crew Manual
Canton, Mich., POB Pub. Co., 1991. 161p. ISBN 0962412414

This beginner's guide is a manual intended to assist survey field crews in their jobs. Provides a historical framework and a handbook on modern surveying practice. The manual is divided into three sections which delineate the responsibilities and instruments used by the rodman, instrumentman, and the chief of party. Includes glossary.

Landon, Robert P.
Practical Surveying for Technicians
Albany, N.Y., Delmar Publishers, 1994. 299p. ISBN 0827339410

The focus of this practical guide is on presenting the essential skills and techniques required for complete surveying projects. Covers the initial planning and field reconnaissance to the final design and map. Written in trade language. Includes a chapter on survey history, a glossary, and a bibliography.

A Manual of the Principal Instruments Used in American Engineering and Surveying
21st ed. Mendam, N.J., Astragal Press, 1993. 217p.
ISBN 1879335344

This reprint of the 1874 edition has preserved a historical record of the instruments manufactured by the firm of W. and L. E. Gurley. Engravings of many of the instruments are found throughout the book, as well as explanations of how the instrument is used. Includes price list and a table of contents at the end of the manual.

McCormac Jack C.
Surveying Fundamentals
2nd ed. Englewood Cliffs, N.J., Prentice Hall, 1991. 567p.
Includes 1 computer disk. ISBN 0138780269

Broad-based applications of both modern and traditional surveying techniques are provided for the fields of civil engineering, architecture, construction, and forestry. A computer disk is included for IBM PCs and compatibles so that tedious math problems can be easily handled. Chapters include material on determining land areas, preparing topographic maps, the computation of earthwork volume, astronomical surveys, construction and control surveys, and photogrammetry. Discusses new technologies including the Global Positioning System, total station instruments, and the hour angle method for making astronomical observations. Also includes material on professional ethics and registration. Appendices include mathematical tables and sample problems. Index.

National Council of Examiners for Engineering and Surveying
Professional Land Surveying Candidate Handbook
2nd ed., Clemson, S.C., The Council, 1992. 73p.

This handbook assists students preparing for national land surveying examinations. Discusses licensing requirements and gives a description and overview of the examination development. Land surveying task statements are presented for examination specification; a final section is devoted to sample questions.

Reilly, James P.
Improving Your Field Procedures
Canton, Mich., POB Pub. Co., 1991. 1 v. (various pagings)

This seminar publication covers four different areas in field procedures which address the quality of field measurements: keys to achieving accuracy, do it right the first time (DIRFT), measurement techniques, and evaluating the performance of the equipment being used. The appendices include ten different checklists and offer photocopies of equipment surveys.

Reilly, James P.
Practical Surveying with GPS
Canton, Mich., POB Pub. Co., 1992. 1 v. (various pagings).

The discussion in this seminar on Global Positioning Systems focuses upon the following areas: GPS as a surveying tool; GPS equipment, including hardware and software; geodesy basics for GPS; the economics of GPS; getting started with GPS; making GPS work; and deciding when to incorporate GPS into a surveying practice. Includes a helpful glossary of specific GPS terms, a bibliography, and photocopies of articles on GPS.

Robillard, Walter G., and Lane J. Bouman
Clark on Surveying and Boundaries
6th ed. Charlottesville, Va., Michie Co., 1992. 1118p. ISBN 155834022X

A revision of the classic volume *A Treatise on the Law of Surveying and Boundaries,* by Frank Emerson Clark (Michie Co., 1987). Provides the surveyor with a synopsis of laws, legal cases, and regulations relating to the practice of surveying and the subdivision of public lands. Courts of law cite Clark as an authority. Coverage includes information on eminent domain, using maps as evidence, the relationship between private and government surveys, and the conveyance of legal and equitable title to land. An excellent index refers to subjects by section number. An annual supplement updates the volume.

Roth, Alan W.
Successful Field-to-Office Automation
Canton, Mich., POB Pub. Co., 1990. 1 v. (various pagings)

This seminar manual offers the reader techniques that may increase efficiency using an automated system. Discussion centers around data collectors, total stations, computer workstations, data reduction, design, CAD software, and field coding systems. Provides discussion on what can be accomplished with field automation. Highlights three different surveying jobs as pertinent examples. Two appendices offer surveys on total stations and data collectors.

Rudman, Jack
Assistant Land Surveyor
Syosset, N.Y., National Learning Corp., 1991. 1 v. (various pagings). ISBN 0837330319

An examination study guide that covers information on land surveying, preparation of maps, legal requirements relating to property line boundaries, mathematics, and supervision. Includes tips for taking the written and oral interview tests. *Passbook for Career Examination Series.*

Rudman, Jack
City Surveyor
Syosset, N.Y., National Learning Corp., 1988. 1 v. (various pagings). ISBN 0837311888

An examination guide for city surveyors consisting of photocopied pages and hand-drawn diagrams. Includes sections on the basic fundamentals of grading, excavating, and blueprint reading, along with samples of questions and answers covering technical problems relating to surveying. *Passbook for Career Examination Series.*

Rudman, Jack
Land Surveyor
Syosset, N.Y., National Learning Corp., 1991. 1 v. (various pagings). ISBN 0837330297

Helps to prepare students planning to take the general land surveying exam by providing sample questions with answers. Several different tests are incorporated in this volume. Offers helpful suggestions to prepare the student for studying, as well as taking the test. Includes a glossary of surveying terms. *Passbook for Career Examination Series.*

Rudman, Jack
Land Surveyor Trainee
Syosset, N.Y., National Learning Corp., 1990. 1 v. (various pagings). ISBN 0837330300

This practice examination study guide examines material on land surveying, on preparing maps and interpreting them, and discusses legal issues relating to line boundaries and mathematics. Offers tips for preparing and taking the written and interview portions of the test. *Passbook for Career Examination Series.*

Van Sickle, Jan
1001 Solved Surveying Fundamentals Problems
Belmont, Calif., Professional Publications, 1993. 1 v. (various pagings). ISBN 091204554X

Designed to help the reader focus on surveying mathematics, boundary law, legal descriptions, and public land surveying systems. Includes such topics as photogrammetry, taping, surveying instruments, EDM, geodetic and control surveys, plats, and mapping. The format is similar to that used in examinations given by the National Council of Engineering Examiners (NCEE).

Wolf, Paul R., and Russell C. Brinker
Elementary Surveying
9th ed. New York, HarperCollins, 1994. 760p. ISBN 0065003993

A textbook of basic theory and materials for the field and office. A substantial revision of the earlier edition that incorporates emerging technologies such as geographic information systems (GIS), total stations, and automatic data collectors. Offers advanced and specialized topics such as satellite and inertial surveying systems, and photogrammetry.

Nonprint Materials

Espenschied, Roland F.
Differential Leveling
Winterville, Ga., American Association for Vocational Instructional Materials, 1994. 1 computer disk

This computer program explains surveying terms which the student needs to know to completely understand the concept of differential leveling procedure. Color graphics illustrate the terms. Students are taught about note-taking through the use of a hypothetical survey. *Surveying Series Computer Program.*

Espenschied, Roland F.
Land Measurement
Winterville, Ga., American Association for Vocational Instructional Materials, 1994. 1 computer disk

The many applications of taping, use of the equipment, and information about the duties of the taping crew are presented in this fifteen-minute computer program. Questions are included to reinforce the material covered by the colorful graphics and brief text. *Surveying Series Computer Program.*

Espenschied, Roland F.
Profile Leveling
Winterville, Ga., American Association for Vocational Instructional Materials, 1994. 1 computer disk

This mouse-driven computer program for surveying students should follow the other four programs. Colorful graphic illustrations are presented within the four individual lessons which range from ten to twenty minutes in length. These lessons present aspects related to profile leveling: the procedures involved, note-taking, contour map reading, and preparation of contour maps. *Surveying Series Computer Program.*

Espenschied, Roland F.
Recording Field Notes
Winterville, Ga., American Association for Vocational Instructional Materials, 1994. 1 computer disk

Presents the rationale behind the accurate recording of field notes: no erasures, the proper use of pencils for recording notes, calling out correct numbers, and not copying notes from someone else. Color graphics illustrate the subject matter and questions are included to help test the student's knowledge of the information presented. *Surveying Series Computer Program.*

Espenschied, Roland F.
Surveying Skills
Winterville, Ga., American Association for Vocational Instructional Materials, 1994. 1 computer disk

This program acquaints the student with the use of the self-reading leveling-rod. It's simple, but effective colored graphics help to illustrate the proper handling and reading of the device. Also presents the duties and titles of the leveling crew members. Test questions follow. *Surveying Series Computer Program.*

Welding

Karen Fischer
University of Minnesota, Morris
Morris, Minnesota
and
Dona Mitoma
Pasadena City College
Pasadena, California

Introduction

Welding is an ancient craft. For thousands of years there was only one type of welding, the kind the blacksmith practiced—forge welding. The twentieth century brought new ways of welding metals, especially relating to steel and nonferrous metals. For example, the advent of widely available electricity made possible the electric arc welding torch. An electric current that is run through a carbon electrode produces a visible arc when brought close to metal, and generates enough heat to weld steel.

Many improvements on this basic model make it one of the most widely used welding processes today. Currently, hundreds of different flux-coated consumable electrodes are available. In the 1940s, a nonconsumable tungsten electrode shielded with an inert gas was developed to weld aluminum, magnesium, and stainless steel. Called TIG (tungsten inert gas) or GTAW (gas tungsten arc welding), it eliminated the need for fluxes to prevent oxidation, but was limited to welding metal less than one-quarter inch thick. MIG (metal inert gas), or GMAW (gas metal arc welding), was developed with a consumable electrode that allowed the welding of thicker metals. Plasma arc welding (PAW) involves running a gas past a tungsten electrode. Proximity to the arc heats and ionizes the gas into a hot plasma jet. This plasma is also known as the *fourth* state of matter, not a solid, liquid, or gas, but a superheated material that can weld or cut virtually any metal.

Another turn-of-the-century type of electric welding is resistance welding (RW). Two pieces of metal are heated by the application of low-voltage, high-amperage electrodes, one on each side of the metals to be joined; the heated pieces are then joined by pressure. Only a small spot is heated at one time, giving the process the alternate name of spot welding. Also developed commercially in the early 1900s, oxyacetylene welding (OAW) is still one of the most widely used and versatile types of welding. When acetylene gas is burned with oxygen, the resulting flame is extremely bright and extremely hot (approximately 6,000 degrees F). It is a standard welding method taught in community colleges.

Pipe welding is considered a more advanced welding skill regardless of whether the welder is using oxyacetylene or an arc welding process. Pipe welding involves working from beneath as well as above the pipe. The welder cannot maintain the same position for the entire weld but must move around the pipe while being consistent in the quality of the weld—no easy task.

Today, welding keeps evolving. One protective coating process, called cladding, involves explosive welding. Two pieces of unlike metal, which normally have such different melting points that they cannot be joined by heating, are placed together. An explosive charge is set off against one of them. The exact placement and strength of the charge are important. The two pieces are pressed together with enough force from the explosion to join them at the molecular level. This process is used in shipbuilding when a noncorroding surface, such as magnesium or aluminum, is needed to cover steel under the waterline. Pressure vessels created for nuclear power plants also use this process.

Laser beam technology and robotic welding systems are two relatively new additions to the welding field. Although they may not be the mainstay of a community college welding program at the present, they will grow in importance in the next decade as they are widely used in industry. The library should include information on these systems.

Most welding programs in community colleges include course work on both oxyacetylene and arc welding processes. A good welding program also includes basic instruction on the properties of metals and the physics of the welding process, reading blueprints, and basic math skills. Because acetylene is explosive at pressures greater than fifteen pounds per square inch and pure oxygen supports fire, oxyacetylene welding demands strict adherence to safety standards. Arc welding has its own electrical safety concerns. In the past ten years, increased concern for worker safety and health has led to revised standards. All current textbooks should cover health and safety issues and libraries supporting these programs need to keep current on the latest safety standards.

Not all books on this list will be appropriate for all libraries. Some are more useful as references for experienced welders; some are helpful for preengineering programs; others are for community members with a background in welding. The American Welding Society catalog is a good source for welding books and specifications. It lists materials related to welding from other publishers, making it particularly convenient and comprehensive.

Paul Easlon of Central Oregon Community College and Alan Armstrong of Pasadena City College are thanked for their advice and assistance in the selection of items for this bibliography.

Selected Association

American Welding Society
550 LeJeune Rd. NW
P.O. Box 351040
Miami, Fla. 33135

Selected Journals

Lincoln Electric Stabilizer
Lincoln Electric Co. Quarterly.

Welding Design and Fabrication.
Penton Pub. Co. Monthly. ISSN 0043-2253

The Welding Innovation Quarterly.
James F. Lincoln Arc Welding Foundation. Quarterly.

Welding Journal.
American Welding Society. Monthly. ISSN 0043-2296

Print Materials

✦ Reference

American Welding Society. Committee on Brazing and Soldering
Brazing Handbook
4th ed. Miami, American Welding Society, 1991. 493p.
ISBN 0871713594
　　This manual explains the fundamental concepts of brazing, the necessary preparation of surfaces, and the brazing process. Included are separate chapters on various kinds of metals as well as ceramics and graphite. Contains a chapter on safety and health. Definitions of terms, and properties of metals and alloys being brazed are discussed. Graphs of thermal expansion data are appended.

AWS Committee on Definitions and Symbols
Standard Symbols for Welding, Brazing, and Nondestructive Examination
Miami, American Welding Society, 1993. 99p. ISBN 0871713705
　　This standard establishes a method of specifying certain welding, brazing, and nondestructive examination information by means of symbols. Detailed information and examples are provided for the construction and interpretation of these symbols. This system also clarifies examination method, frequency, and extent. ANSI approved.

American Welding Society. Structural Welding Committee
Structural Welding Code—Reinforcing Steel
4th ed. Miami, American Welding Society, 1992. 39p.
ISBN 0871713780

This code covers the requirements for welding reinforcing steel in most reinforced concrete applications, including metal inserts and connections in reinforced concrete construction. It contains a body of rules for the regulation of welding reinforcing steel and provides suitable acceptance criteria for such welds.

American Welding Society. Structural Welding Committee
Structural Welding Code—Sheet Steel
2nd ed. Miami, American Welding Society, 1989. 51p.
ISBN 087171308X

This code covers the requirements associated with welding sheet steel having a minimum specified yield point no greater than 80,000 psi (550MPa). The code requirements cover any welded joint made from structural quality low-carbon hot-rolled and cold-rolled sheet and strip steel, with or without zinc coatings (galvanized). Includes rules for the specific jurisdiction of this code, rules for the implementation of welding activities, and requirements for the qualification of welding procedure specifications, welders, and welding operators.

American Welding Society. Structural Welding Committee
Structural Welding Code—Steel
13th ed. Miami, American Welding Society, 1994. 447p.
ISBN 0871714191

New editions of this code are issued whenever significant changes warrant, generally every two years. It constitutes a body of rules for the regulation of welding any type of structure made from commonly used carbon and low-alloy constructional steels. Includes rules for specific types of constructions, such as bridges and tubular structures, as well as guidelines for inspecting and testing welds.

Campbell, Hallock C., comp.
AWS Film Directory
Miami, American Welding Society, 1989. 45p. ISBN 0871713020

A listing of more than 300 films, filmstrips, and videos produced by the American Welding Society, and many other titles on topics related to welding. Includes distributor and price information for both purchase and rental.

Geary, Don
The Welder's Bible
2nd ed. Blue Ridge Summit, Pa., TAB Books, 1993. 313p.
ISBN 0830638261

This is a do-it-yourself manual for the home oxyacetylene welder. In clear language it covers the basics of working with oxygen and acetylene, safety, home equipment and procedures for basic welding, brazing, and cutting of metals. Discusses elements of a home workshop and how to handle typical home welding projects on rain gutters, car mufflers, etc. Well illustrated. Includes glossary and index.

Houldcroft, P. T.
Which Process? An Introduction to Welding and Related Processes and a Guide to Their Selection
Cambridge, England, Abington Publishing, 1990. 93p.
ISBN 1855730081

Presents a unique scheme for selecting the most suitable welding process. The book begins at the drawing board stage, where the need for a connection is first decided. It then presents a series of diagrams and tables to show which processes are most feasible for each particular joint. The book includes clear illustrations and practical descriptions of twenty-eight joining processes, and it explains the methods of use and applications.

Welding and Fabricating Data Book.
Cleveland, Ohio, Penton Pub. Co., 1981- . 1 v. (various pagings)
ISSN 0278-7067

A welding and fabrication source guide containing engineering data and directory information. Published biennially, this databook continues the *Welding Data Book.*

Welding Handbook
8th ed. Miami, American Welding Society, 1994. 3 v.
ISBN 0871712814 (vol. 1); ISBN 0871713543 (vol. 2); ISBN not available (vol. 3)

The key reference set for the field—no library supporting a welding program should be without it. It covers standards, specifications, safety, processes, and equipment, and provides references for further information. The three-volume eighth edition will supersede the five-volume seventh edition. (Volume 3 is scheduled to be published late Spring 1995). The eighth edition, Volume 1, includes a section on welding safety available separately as *Safe Practices*.

✦ Blueprint Reading

Bennett, A. E., and Louis J. Siy
Blueprint Reading for Welders
5th ed. Albany, N.Y., Delmar Publishers, 1993. 340p.
ISBN 0827355793

This standard covers the fundamentals of blueprint drawing interpretation as well as the latest American Welding Society welding symbols and American National Standards Institute (ANSI) standards. Instructor's guide and videocassettes also available.

Walker, John R.
Welding Print Reading
South Holland, Ill., Goodheart-Willcox, 1991. 206p.
ISBN 0870068903

Provides instruction on interpreting and using the type of engineering drawings or prints found in the welding trade. This workbook begins with the basics and progresses to more specialized coverage of welding symbols and notations. Text follows recently revised standards established by the American Welding Society (AWS) and the American National Standards Institute (ANSI). Exercises at end of chapters test knowledge. Many illustrations of good quality. Instructor's guide also available.

✦ Mathematics for Welders

American Welding Society. Committee on Metric Practice
Metric Practice Guide for the Welding Industry
3rd ed. Miami, American Welding Society, 1989. 22p.
ISBN 0871712970

By law, federal agencies must now use the metric system in their procurements and other business-related activities. This standard text is intended to facilitate this recent change within the welding industry. It contains the accepted definition of the International System of Units (SI). It includes the base units, supplementary units, derived SI units, and rules for their use in Society documents and by the welding industry. Provides factors and rules for converting from U.S. inch-pound units to SI units, and recommendations to the industry for managing the transition.

Marion, Nino
Math for Welders
South Holland, Ill., Goodheart-Willcox, 1990. 224p.
ISBN 0870067834

A workbook that teaches basic mathematics skills. Covers whole numbers, common fractions, decimal fractions, measurement, percentages, and the metric system, incorporating many welding-related drills and exercises. Includes a glossary of terms and answers to odd-numbered questions.

Schell, Frank R., and Bill J. Matlock
Practical Problems in Mathematics for Welders
3rd ed. Albany, N.Y., Delmar Publishers, 1988. 225p.
ISBN 0827332947

Welding involves measuring materials, determining angles for cutting, and finding weights of materials. This textbook provides mathematical problems that relate to actual welding situations. It can be used in conjunction with a comprehensive mathematics text. Intended for basic math students from junior high through the two-year college level.

✦ Metallurgy and Design

Easterling, K. E.
Introduction to the Physical Metallurgy of Welding
2nd ed. Oxford, England, Butterworth-Heinemann, 1992. 270p.
ISBN 0750603941

Intended for either undergraduate or graduate engineering courses in metallurgy. It explains the fusion welding process in terms of its effect on the weld metals, the heated zone surrounding them, and cracking and fracturing in welds. Includes a case study of the Alexander Kielland oil platform disaster in the North Sea.

Lancaster, J. F.
Handbook of Structural Welding
New York, McGraw-Hill, 1992. 436p. ISBN 0070316848

Describes the current state of welding technology as applied to large structures. The book covers processes, metallurgical effects of fusion welding, and welds in service, structures, pipelines, and process plants, and their reliability. Includes information on how to avoid fatigue failure. An excellent resource for metallurgists, welding engineers, and designers.

Lancaster, J. F.
Metallurgy of Welding
5th ed. New York, Chapman and Hall, 1993. 389p.
ISBN 0412478102

This source covers the joining of metals, both to other metals and to nonmetals, including soldering, brazing, solid-phase welding, and fusion welding. The fusion welding of steel is treated in depth, as are nonferrous metals, glasses, and ceramics. Intended users are those attending specialist welding courses as well as undergraduate and graduate students of metallurgy.

✦ Inspection, Qualification, and Certification

American Welding Society. Qualification and Certification Committee
Standard for Accreditation of Test Facilities for AWS Certified Welder Program
Miami, American Welding Society, 1989. 6p. ISBN 087171342X

This standard work describes a program directed by the American Welding Society for a third-party accreditation of test facilities used to perform welding qualification testing.

American Welding Society. Qualification and Certification Committee
Standard for AWS Certification of Welding Inspectors
14th ed. Miami, American Welding Society, 1988. 7p.
ISBN 087171289X

Defines the requirements and program for the American Welding Society to certify welding inspectors. The certification of inspectors requires documentation of experience, satisfactory completion of an examination, and proof of visual acuity. The examination tests the inspector's knowledge of welding processes and procedures, welder qualification, destructive and nondestructive testing, terms, definitions, symbols, reports, records, safety, and responsibilities.

American Welding Society. Committee on Mechanical Testing of Welds
Standard Methods of Mechanical Testing of Welds
Miami, American Welding Society, 1992. 61p. ISBN 0871713934

Eight mechanical test methods that are applicable to welds and welded joints are described. They are recognized as the standard methods for bend, tension, nick break, hardness, and fracture toughness testing, as well as fillet weld break, shear testing, and stud weld testing. For each testing method, information is provided concerning applicable American National Standards Institute (ANSI), American Society for Testing and Materials (ASTM), and American Petroleum Institute (API) documents. The required testing, apparatus, specimen preparation, procedure to be followed, and report requirements are included.

Jackson, Henry L.
Welding Certification Questionnaire
Riverside, Calif., BJ Publications, 1992. 159p. ISBN 1881870006

This text presents questions as they would appear on a welding certification test. Answers and supportive evidence for the correct answer are referenced for easy verification. The focus of this workbook is on test questions typical for written examinations, such as "The Los Angeles City Welder and Deputy Welding Inspector, ICBO Inspector, AWS CWI or AWS CAWI." It may be useful for other regional tests as well.

Rowh, Mark
Opportunities in Welding Careers
Lincolnwood, Ill., VGM Career Horizons, 1990. 144p.
ISBN 0844285986

Explains the importance of welding and describes the many careers available in the welding industry. Also discussed are new technologies, special training programs, earnings and benefits, organizations, and certifications. Schools offering welding programs are also listed.

✦ Health and Safety

Accredited Standards Committee Z49, Safety in Welding and Cutting
Safety in Welding and Cutting: Superseding ANSI Z49.1-83
Miami, American National Standards Institute, 1988. 49p.
ISBN 0871712903

Includes American National Standards Institute standard Z49.1, covering all aspects of safety and health in the welding environment, emphasizing oxyfuel gas and electric arc welding processes. Covers protection of personnel, ventilation, fire prevention, and precautionary labeling. Includes a bibliography of standards and specifications, and a list of publishers.

American Welding Society. Committee on Fumes and Gases
A Sampling Strategy Guide for Evaluating Contaminants in the Welding Environment
Miami, American Welding Society, 1991. 12p. ISBN 0871713527

This guide provides advice on contaminants that may be present in the welding environment, and presents a strategy for collecting valid samples from the welder's breathing zone. Recommendations for fume analysis for various elements found in AWS filler metal specifications are presented in a table. An appendix includes a checklist to use when evaluating the workplace.

✦ General Texts

Althouse, Andrew Daniel, and others
Modern Welding: Complete Coverage of the Welding Field in One Easy-to-Use Volume!
South Holland, Ill., Goodheart-Willcox, 1992. 736p.
ISBN 0870069667

This comprehensive course text for secondary and postsecondary students, apprentices, and journeyman welders covers theory, fundamentals, and basic welding processes: oxyfuel gas processes, shielded metal arc welding, gas tungsten and gas metal arc welding, arc cutting, resistance welding, special processes, metal technology, and professional welding. Uses metric and conventional measurement units and American Welding Society terminology as well as conventional terms. Throughout the book, safety information is highlighted in red print within the processes being discussed; a separate safety index is at the end of the book. Includes general index and glossary of terms. Also available are a laboratory manual, instructor's guide, and test creation software. The software allows instructors to select test questions, write their own questions and answers, or modify existing questions.

Bowditch, William A., and Kevin E. Bowditch
Welding Technology Fundamentals
South Holland, Ill., Goodheart-Willcox, 1991. 367p.
ISBN 0870067516

Introduces the most widely used welding and cutting equipment and techniques. Safety hints and cautions are printed in red typeface for emphasis. Covers certification, inspection, and testing. Includes glossary. Laboratory manual and answer key also available.

Cary, Howard B.
Modern Welding Technology
2nd ed. Englewood Cliffs, N.J., Regents/Prentice Hall, 1989. 787p. ISBN 0135992834

One of the best overall texts, this work covers welding fundamentals, safety, training and certification of welders, and specific welding processes, concentrating on the arc welding processes. It also includes a discussion of arc welding power sources, computer controls of welding equipment, laser and plasma cutting, welding of plastics, automated and robotic welding, pipe, maintenance, design, cost, and quality control. Up-to-date in its information, it is one of the few texts with chapter references for further information. Third edition is scheduled for late 1994.

Davies, A. C.
The Science and Practice of Welding
10th ed. New York, Cambridge Univ. Press, 1992-1993. 2 v. (various pagings). ISBN 052143565X (vol. 1) ISBN 0521435668 (vol. 2)

A standard British text in two volumes that uses British terminology, but refers to American terms. Volume One covers the physics and chemistry of metallurgy, equilibrium diagrams, basic electrical principles, the inverter power unit, weld testing, welding symbols, and classification of aluminum, steel, duplex stainless steel, and magnesium alloys. Volume Two describes the various types of welding processes, including robotics; underwater, laser, explosive, and plastics welding; and various cutting processes and safety equipment. Has examination questions from City and Guilds of London Institute.

Jeffus, Larry F.
Welding: Principles and Applications
3rd ed. Albany, N.Y., Delmar Publishers, 1992. 764p.
ISBN 0827350481

A comprehensive text that covers the major welding processes: oxyfuel, shielded metal arc, and gas shielded welding. Also included is a section on other welding processes such as soldering, brazing, welding of plate and pipe, robotic welding, and automation. Discusses metallurgy, weldability of metals, quality control, and welding joint design. Provides experiments and practical exercises. Instructor's guide also available.

Stinchcomb, Craig
Welding Technology Today: Principles and Practices
Englewood Cliffs, N.J., Prentice Hall, 1989. 468p.
ISBN 0139244166

Provides comprehensive, state-of-the-art information simplified for the beginning student who has no prior welding background. There is no glossary; terms are explained within the text. Covers safety, joint design, and standard welding processes. Includes chapters on the classification of electrodes for shielded metal arc welding, pipe welding, oxyfuel and plasma arc cutting, cast iron welding, metallurgy, blueprints and welding symbols, welding codes, and quality control and testing.

✦ Specialized Texts

American Welding Society. Committee on Oxyfuel Gas Welding and Cutting
Operator's Manual for Oxyfuel Gas Cutting
Miami, American Welding Society, 1990. 20p ISBN 0871713160

This manual for oxygen cutters explains the equipment and supplies needed and the proper procedures to follow for oxyfuel gas cutting of iron-bearing metals. It gives the most recent safety requirements.

AWS Committee on Arc Welding and Cutting
Recommended Practices for Gas Metal Arc Welding
Miami, American Welding Society, 1989. 65p. ISBN 0871713012

This versatile welding process has a number of variations and many applications. This set of recommended practices presents basic concepts to assist the welder or engineer in selecting the best welding process for the materials to be joined. Includes definitions, fundamental principles, equipment, process requirements, joint design, spot welding, inspection and quality control, training of welders, process and welder qualifications, and safe practices.

Baird, Ronald J.
Oxyacetylene Welding: Basic Fundamentals
South Holland, Ill., Goodheart-Willcox, 1991. 104p.
ISBN 0870069136

The combination text and workbook approach allows students to work at their own pace. Fundamentals of gas welding processes are explained using both U.S. conventional and metric terms. Topics include safety, gases, metal identification, torches, welding symbols, backhand and forehand methods, and welding joints. A chapter on welding careers and a glossary of terms are included.

Beard, F. Richard
Basics of GMAW and GTAW: Gas Metal Arc Welding, Gas Tungsten Arc Welding
Athens, Ga., American Association for Vocational Instructional Materials, 1992. 87p. ISBN 0896062864

Only the manual (nonautomated or semiautomated) welding processes of gas metal arc welding and gas tungsten arc welding are discussed. Topics include fundamentals, equipment, adjustment and operation, welding mild steel, welding aluminum, and welding defects. Computer software also available.

Berge, James M.
Automating the Welding Process: Successful Implementation of Automated Welding Systems
New York, Industrial Press, 1994. 198p. ISBN 0831130512

Provides the information needed to fully understand automated welding machinery, techniques, and philosophies. The text explains how to estimate start-up costs and calculate returns on robotic and automated machinery investments. Common questions and answers about robotic welding are addressed. Each chapter is summarized with an outline.

Dawes, Christopher
Laser Welding
New York, McGraw-Hill, 1992. 258p. ISBN 0070161232

This book explains the principles of laser welding and its many aspects. It also provides examples of industrial applications. One chapter is devoted to safety and another to installing and operating a laser. A glossary of common terminology is included.

Gerken, John M.
Gas Tungsten Arc Welding
Cleveland, James F. Lincoln Arc Welding Foundation, 1991. 42p.

This book describes the Gas Tungsten Arc Welding (GTAW) process and covers the equipment needed. Both manual and automatic techniques for the GTAW process are outlined, and specific information for welding the more important metals and alloys is included. A history of the development of the process and safe practices are discussed.

Houldcroft, P.T., and Robert John
Welding and Cutting: A Guide to Fusion Welding and Associated Cutting Processes
New York, Industrial Press, 1989. 232p. ISBN 0831111844

Coverage includes thermal welding and associated cutting processes on manual metal arc and MIG processes; cored wire, submerged-arc, and electro-slag welding; and TIG and plasma arc processes. Gas welding and cutting, the use of power beams, robots, and welding productivity are discussed. A glossary, bibliography, and chapter on health and safety are included.

Walker, John R.
Arc Welding: Basic Fundamentals
South Holland, Ill., Goodheart-Willcox, 1993. 128p.
ISBN 0870060163

A write-in text consisting of short chapters with many illustrations of arc welding. Text in two colors assists readers in differentiating data. Chapters on quality control and careers in welding, a dictionary of terms, metric conversion tables (SI metric to U.S. conventional and U.S. conventional to SI metric), and service welding specifications complete the text. An instructor's guide is also available.

Welding
7th ed. Moline, Ill., Deere and Co., 1991. 150p. and slide set
ISBN 0866911383

Created as a training manual for beginners and as a reference for the journeyman welder, this manual explains gas and arc welding, cutting, brazing, soldering, surfacing, design of welds, properties of metals, and safety. Includes chart of welding symbols, a glossary of terms, and a slide set.

Nonprint Materials

Basics of GMAW and GTAW: Gas Metal Arc Welding and Gas Tungsten Arc Welding
Athens, Ga., American Association for Vocational Instructional Materials, 1992. computer software

This software package contains seventy-nine questions taken from the material in F. Richard Beard's text, *Basics of GMAW and GTAW* (American Association for Vocational Instructional Materials, 1992). Includes review, testing, printout, random selection, and password lockout. Instructors can use this software to produce tests, create review sheets, or develop independent study assignments.

Hobart Institute
Blueprint Reading for Welders and Fitters
Albany, N.Y., Delmar Publishers, 1988. 3 videocassettes (87 min.) VHS. ISBN 0827360436

A set of three videos to accompany the Bennett and Siy text *Blueprint Reading for Welders* (Delmar Publishers, 1993). Includes mathematics review, metric conversion, review of welding symbols, standard drawing lines and symbols, scale drawings, dimensional tolerance, stock allowance, and blueprint interpretation.

Welding
7th ed. Moline, Ill., Deere and Company, 1991. 147 slides

A set of slides to accompany the text *Welding* (Deere and Co., 1991).

Welding as a Career
Madison, Wisc., MATC Telecommunications, 1990.
1 videocassette (9 min.) VHS

This video was produced by Wisconsin's Madison Area Technical College to promote student interest in welding as a career. It portrays a range of welding career options and describes the benefits of a career in welding such as possessing a productive skill, interesting work, good pay, and growth opportunities.

Graphic and Apparel Arts (Applied Arts)

Desktop Publishing and Printing Technologies 395
Charles R. James

Fashion Design and Apparel Arts 411
Dianna Thor

Interior Design 428
Sue Swanson

Photography 443
Kate Hickey

Visual Communication/Graphic Arts 452
Judy Goodyear

Desktop Publishing and Printing Technologies

Charles R. James
Lexington Community College
Lexington, Kentucky

Introduction

The rapid convergence of printing technologies with digital technologies presents new challenges in education and curriculum development for community colleges and vocational/technical schools. Although printing technologies have been taught in industrial, printing, or graphic arts departments, desktop publishing (DTP) is taught in computer science, business, communications, or visual design courses. The arrival of DTP in 1985, along with sophisticated technology and computer software and hardware, has caused many of the traditional methods of design, prepress, printing, and publishing to change and merge, opening up entirely new training, educational, and career opportunities for students.

At present, DTP careers may include, but are not limited to, corporate communications specialists, writers, positions in publishing (book, newspaper, and magazine), designers, typographers, educational support staff, advertising representatives, illustrators, graphic artists, print shop managers, and operators of service bureaus and prepress businesses. Careers in printing may include specialists in typographic composition, commercial art, pressroom operations, estimating, binding, fine arts printing, and commercial printing methods such as lithography, letterpress, flexography, or gravure.

The current trend in computer and printing technologies points to an ever-increasing importance of new desktop publishing applications in the traditional industries of prepress, composition, color separation, and production of printed materials. To meet the increased need for a skilled work force having a broadened knowledge base, many community colleges and vocational/technical schools now offer a more integrated curriculum leading to degrees, diplomas, or certificates in electronic publishing and graphic arts technologies. These programs often require the development of diverse skills in design, illustration, typography, printing operations, communications, and management.

This bibliography has been created to meet the educational and informational needs of students in community colleges and vocational/technical schools who are involved in printing technologies, graphic design, desktop publishing, and related fields.

Selected Associations

Although there are no accrediting or certifying agencies specific to this area, there are many helpful organizations that can supply additional information.

American Institute of Graphic Arts (AIGA)
1059 3rd Ave.
New York, N.Y. 10021

Graphics Arts Technical Foundation (GATF)
4615 Forbes Ave.
Pittsburgh, Pa. 15213

National Association of Desktop Publishers
462 Old Boston St.
Boston, Mass. 01983

National Association of Printers and Lithographers (NAPL)
780 Palisade Ave.
Teaneck, N.J. 07666

National Association of Quick Printers
401 N. Michigan Ave.
Chicago, Ill. 60611

Society for Technical Communications
901 N. Stuart St., Suite 304
Arlington, Va. 22203

XPLOR International, the Association for Electronic Printing Professionals
2550 Via Tejon, Suite 3L
P. O. Box 1501
Palos Verdes Estates, Calif. 90274

Selected Journals

AIGA Journal of Graphic Design.
American Institute of Graphic Arts. Quarterly. ISSN 0736-5322

Aldus Magazine.
Aldus Corp. Bimonthly. ISSN 1046-0616

American Printer.
MacLean Hunter Pub. Co. Monthly. ISSN 0744-6616

Before and After.
PAGELAB. Bimonthly. ISSN 1049-0035

Business Publishing.
Hitchcock Pub. Co. Monthly. ISSN 1060-2208

Byte.
McGraw-Hill. Monthly. ISSN 0360-5280

Communication Arts.
Communication Arts. Eight times a year. ISSN 0010-3519

Computer Graphics.
Association for Computing Machinery. Quarterly. ISSN 0097-8930

Computer Graphics World.
PennWell Pub. Co. Monthly. ISSN 0271-4159

Desktop Communications.
International Desktop Communications. Bimonthly. ISSN 1050-1800

Electronic Composition and Imaging.
Youngblood Pub. Co. Bimonthly. ISSN 0838-9535

Graphic Arts Monthly.
Cahners Pub. Co. Thirteen times a year. ISSN 1047-9325

Graphic Design: USA.
[American Institute of Graphic Arts]. Watson-Guptill Publications. Monthly. ISSN 0274-7499

How: the Magazine of Ideas and Techniques in Graphic Design.
F & W Publications. Bimonthly. ISSN 0886-0483

In House Graphics.
United Communications Groups. Monthly. ISSN 0883-6973

Journal/National Association of Desktop Publishers.
The Association. Quarterly. ISSN 0897-6503 (*Desktop Publisher Forum*, ISSN 0897-4764, is issued with this journal.)

MacUser.
Ziff-Davis. Monthly. ISSN 0884-0997

MacWorld.
PCW Communications. Monthly. ISSN 0741-8647

PC Magazine.
Ziff-Davis. Every two weeks. ISSN 0888-8507

PC World.
PCW Communications. Monthly. ISSN 0737-8939

Page.
Page. Ten times a year. ISSN 1056-6023

Print.
RC Publications. Bimonthly. ISSN 0032-8510.

Publish!
PCW Communications. Monthly. ISSN 0897-6007

Ready, Set, Go! In-Depth.
MindCraft Pub. Co. Monthly. ISSN 1064-7120

Seybold Report on Desktop Publishing.
Seybold Publications. Monthly. ISSN 0889-9762

Step-By-Step Graphics.
Dynamic Graphics. Seven times a year. ISSN 0886-7682

TypeWorld.
PennWell Pub. Co. Biweekly. ISSN 0194-4851

Windows Magazine.
CMP Publications. Monthly. ISSN 1060-1066

Print Material

✦ Reference

Beach, Mark
Graphically Speaking: An Illustrated Guide to the Working Language of Design and Printing
Manzanita, Ore., Elk Ridge Pub., 1992. 322p. ISBN 094338107X

An illustrated dictionary containing more than 2,500 entries dealing with the technical language used in graphic arts. Covers a diverse field, including offset, gravure, flexography, screen, and letterpress printing, as well as terms relating to materials, films, inks, advertising, publishing, and design. Many of the definitions are made more useful by including examples of their usage. Includes a Spanish-to-English index.

Brenner, Donald J.
The Language of Computer Publishing
San Diego, Calif., Brenner Information Group, 1989. 221p.
ISBN 09295350300

A dictionary of terms, abbreviations, and acronyms associated with the merger of publishing, printing, graphic arts, and computer technologies. Contains 4,300 entries compiled from related magazines and books.

The Chicago Manual of Style
14th ed. Chicago, The Univ. of Chicago Press, 1993. 921p.
ISBN 0226103897

First published in 1906, this standard reference covers bookmaking, manuscript preparation, copyediting, rules governing text, documentation, tables, indexes, abbreviations, punctuation, graphic style, typography, production, and printing. Discusses the role of computers in production of manuscripts and printing. The glossary and bibliography have been extensively revised.

Collin, P. H.
Dictionary of Printing and Publishing
Teddington, England, Collin Publishing, 1989. 260p.
ISBN 0948549092

Covers vocabulary associated with printing, publishing, and related industries. More than 5,000 words and expressions are included, many with notes and additional encyclopedic information. A supplement contains charts and useful forms such as examples of printer quotations, print order forms, printer's job sheets, rate cards, and imposition schemes.

Dewey, Patrick R.
101 Desktop Publishing and Graphics Programs
Chicago, American Library Association, 1993. 134p.
ISBN 0838906060

Basic, handy reference for librarians profiling DTP, graphics, word processing, paint and draw, presentation, and scanning software. Also includes information on utility programs and sources for type fonts and clip art. Reviews training programs that are available in video and audio formats. A complete vendor list and useful glossary make this a valuable library resource.

Li, Xia, and Nancy B. Crane
Electronic Style: A Guide to Citing Electronic Information
Westport, Conn., Meckler Publishing, 1993. 65p.
ISBN 088736909X

The proliferation of electronic resources has necessitated the development of standard principles in citing information available in databases and other computer resources. Intended for librarians, scholars, and students, this work presents standard citation rules that have been adapted from the American Psychological Association style manual. Rules are suggested for citing government documents, legal sources, court decisions, graphic images, full text files, bibliographic databases, Internet sources, electronic journals, discussion lists, bulletin boards, e-mail, and commercial on-line documents.

Peacock, John, and Michael Barnard
The Blueprint Dictionary of Printing and Publishing
London, England, Blueprint Publishing, 1990. 280p.
ISBN 0948905476

Contains the technical terminology of electronic publishing and computer technologies as well as traditional trade language. More advanced concepts are illustrated.

Spencer, Donald D.
Illustrated Computer Graphics Dictionary
Ormond Beach, Fla., Camelot Pub. Co., 1993. 305p.
ISBN 0892181176

More than 2,100 entries and numerous illustrations define words, concepts, acronyms, and abbreviations associated with computer-assisted graphics. Emphasis is on painting and drawing applications. Graphic arts professionals, teachers, students, and computer users are the intended audience. Also includes terminology from desktop publishing, computer science, and communications.

✦ General Materials

Adler, Elizabeth
Everyone's Guide to Successful Publications: How to Produce Powerful Brochures, Newsletters, Flyers, and Business Communications, Start to Finish
Berkeley, Calif., Peachpit Press, 1993. 400p. ISBN 156609027X

Adler, a communications consultant to universities, corporations, and government agencies, presents a comprehensive instruction manual on planning, writing, designing, electronic publishing, printing, and distributing virtually all types of business publications. What sets this work apart is its integrated approach to the publishing process and the helpful sidebars that summarize important considerations.

Aldrich-Ruenzel, Nancy, ed.
Designer's Guide to Print Production
New York, Watson-Guptill, 1990. 159p. ISBN 0823013146

Very practical and lavishly illustrated reference for designers, art directors, production staff, editors, and others involved in graphic arts training and education. Many useful charts, diagrams, and tables concerning print production and standards are provided, covering all aspects of planning and design. Easy-to-follow instructions on evaluating and reproducing graphic and photographic images are also included. Written by the editorial director of *Step-By-Step Graphics* magazine.

Anderson, Scott R.
Desktop Publishing: Dollars and Sense
Hillsboro, Ore., Blue Heron Publishing, 1992. 193p.
ISBN 0936085517

Concise, practical guide to starting and operating a small, one- or two-person desktop publishing business. Based on interviews with working professionals and the author's more than twenty years experience as a designer, writer, and editor, this work focuses on defining a market, establishing a business identity, the cost of doing business, the needs of the client, contracts and agreements, copyright and legal issues, choosing hardware and software, and diversifying the business.

Ayres, Julia
Printmaking Techniques
New York, Watson-Guptill, 1993. 160p. ISBN 0823043991

Traditional fine arts printing is handsomely illustrated with detailed descriptions of processes and methods. Highlights include discussion of studio equipment and techniques not usually found in overviews, such as chine collé, photoetching, photogravure, cliché verre, and computer-generated imagery. Also indexes important information on the use of hazardous materials and nontoxic substances.

Baker, Kim, and Sunny Baker
Color Publishing on the Macintosh: From Desktop to Print Shop
New York, Random House, 1992. 479p. and computer disk
ISBN 0679739777

A single resource for beginners and advanced users pulling together practical techniques and strategies, hardware and software options, color image creation and manipulation, color separation, troubleshooting, and working with printshops and service bureaus. Includes a 3.5 inch high-density disk with fourteen programs and utilities for image manipulation, file compression/decompression, communication and management tools. Invaluable for Macintosh users who have experience in black-and-white desktop publishing and want to move to color.

Bann, David, and John Gargan
How to Check and Correct Color Proofs
Cincinnati, Ohio, North Light Books, 1990. 143p.
ISBN 0891343504

An essential work for printer, designer, and production personnel involved in proofing. Fifty of the most common problems are identified, illustrated, analyzed, and explained. Section One concerns originals for reproduction, reproduction techniques, print processes, the four-color process, methods of proofing, and color separation. The second section includes a step-by-step method for checking proofs and mark-up.

Beach, Mark
Getting It Printed
Rev. ed. Cincinnati, Ohio, North Light Books, 1993. 199p.
ISBN 0891345108

From planning to finished product, this how-to guide for designers, publishers, and graphic arts students describes working with printers and service bureaus and making the right design decisions. Subjects include typography, illustration, flat and process color, prepress, paper, printing methods, finishing, and binding. The section on choosing the right printer at the right price presents especially useful information on requesting estimates, evaluating quotations, and negotiating problems.

Beach, Mark, and Kathleen Ryan
Papers for Printing: How to Choose the Right Paper at the Right Price for All Your Design and Printing Needs
2nd ed. Portland, Ore., Coast to Coast Books, 1991. 76p.
ISBN 0943381061

Paper can make up forty percent of the cost of a printing job. This guidebook covers grades, ratings, color, surface texture, weight, size, specifying and buying, and business issues involved in selecting paper. Includes forty-four full-page samples, including bond, coated, offset, text, cover, and recycled stock, which are shown with full-color printing to help in making informed decisions. Recommended by the National Paper Trade Association as a standard guide for graphic artists.

Binns, Betty
Designing with Two Colors
New York, Watson-Guptill, 1991. 127p. ISBN 0823013340

Demonstrates the impact of the minimal use of color in publications. Four color categories—warm, cool, pure, and dirty (mixed with black)—are shown over a wide range of illustration types allowing for direct comparison. Topics include selecting colors, screening, surprinting, knocking out and trapping, halftones, duotones, and special effects. Also includes a guide to production considerations such as printing inks, paper choices, and prepress proofs.

Bivins, Thomas, and William E. Ryan
How to Produce Creative Publications: Traditional Techniques and Computer Applications
Lincolnwood, Ill., NTC Business Books, 1992. 438p.
ISBN 0844234958

"The framework of this book is composed of 'hows' held firmly together by a mortar of 'whys'" (pref.). Bivins lays out the age-old approach to writing, editing, and designing, and applies new computer technology to the methodology. This book is written as a beginner's guide to be read sequentially and in its entirety. Four basic publication types are used as examples: newsletters, magazines, annual reports, and brochures. Hardware and software considerations are presented only in a generic fashion.

Bixby, Robert, Craig Danuloff, and Deke McClelland
The Micrografx Designer Companion
Homewood, Ill., Business One Irwin, 1992. 424p.
ISBN 1556234104

This is a good companion text, guide, and reference for beginners using the popular *Windows* program for technical illustrations, maps, cartoons, and architectural designs for desktop publishing. Material is presented through structured lessons and exercises demonstrating the creation of logos, business cards, and presentations. Computer graphics, layout, color, drawing, fill patterns, gradients, and other design elements are also discussed.

Black, Alison
Typefaces for Desktop Publishing: A User Guide
London, England, Architecture Design and Technology Press, 1990. 106p. ISBN 1854548417

A very detailed, well-illustrated look into the intricacies of typography, with emphasis on making informed decisions when selecting type for various publication styles. Written for self-publishers, design consultants, corporate publishers, and student designers in general. Covers terms within the full range of currently available equipment. Particularly helpful are the chapters on legibility and typeface performance in context.

Black, Roger
Roger Black's Desktop Design Power
New York, Bantam Books, 1991. 320p. ISBN 0553347527

Computer-based production and art direction from a designer who has worked for *Rolling Stone*, *Newsweek*, and *The New York Times*. Focuses on analysis, development of critical judgment, and creating dynamic documents in an electronic environment. The first half of the work concerns design fundamentals for the nondesigner; the second half consists of case studies of publications, applying basic layout principles. Additional topics include the history of magazine design, newsletter layout, color publishing, production management, and equipment limitations.

Blatner, David
Desktop Publisher's Survival Kit
Berkeley, Calif., Peachpit Press, 1991. 172p. and computer disk.
ISBN 0938151762

Neither a how-to book on electronic publishing techniques nor a detailed software applications manual, this unique guide bridges the gap between "techies" and designers, complementing the knowledge that most users acquire after years in the publishing arena. Each chapter stands alone, covering a specific topic such as graphic file formats, code-based typesetting, and scanning. Includes a program disk of useful utilities, fonts, and clip art in Macintosh format.

Blatner, David, Keith Stimely, and Eric Taub
The QuarkXPress Book
3rd ed. Berkeley, Calif., Peachpit Press, 1993. 682p.
ISBN 1566090679

Arguably the best publication on the subject of this high-end publishing program (version 3.0-3.2 for the Macintosh). This comprehensive how-to guide complements the software documentation; it is presented for any level of user, explaining all program features in detail, as well as technical issues such as system customization. Designed to be read cover-to-cover or used for ready reference. Loaded with tips, step-by-step procedures, and shortcuts.

Blatner, David, and Stephen F. Roth
Real World Scanning and Halftones: The Definitive Guide to Scanning and Halftones from the Desktop
Berkeley, Calif., Peachpit Press, 1993. 275p. ISBN 1566090938

Essential resource for Macintosh and PC users providing detailed instruction on the techniques, tools, and concepts of image production and manipulation for desktop-published documents. Both practical and conceptual information covers topics ranging from frequency, angle, and image resolution, to file formats, scanners, photo CD, and optical character recognition (OCR) technologies. Helpful user tips are also provided for *Photoshop, PhotoStyler, Illustrator, CorelDraw, FreeHand, PageMaker, QuarkXPress,* and other major software packages.

Borman, Jami Lynne
Desktop Publishing Sourcebook: Fonts and Clip-Art for the IBM PC and Compatibles
Rocklin, Calif., Prima Publishing and Communications, 1990. 423p. ISBN 1559580313

Quite simply, what's available and how to get it. This fully illustrated, quick reference contains hundreds of samples from major software publishers, all produced in 300-dpi laser print allowing accurate evaluation of the products. Presented for the novice or experienced user, with tips on design considerations and hardware integration. Products listed are compatible with most page layout, design, and word processing packages.

Borowsky, Irvin J.
Opportunities in Printing Careers
Lincolnwood, Ill., VGM Career Horizons, 1992. 150p.
ISBN 0844281786

Borowsky, founder and chair of the largest publisher of magazines for the printing and allied graphic arts industries, identifies various careers from administration and prepress to pressroom and binding. Career entries include work description, employment outlook, advancement, educational requirements, salaries, and how to obtain more information. Advice is also given on job hunting, emphasizing pre-interview planning, résumé and cover letter writing, developing a prospect list, and what to expect in an interview.

Bradley, Julia Case, and others
Desktop Publishing Using Publish It!—IBM Version
Dubuque, Iowa, Wm. C. Brown, 1992. 335p. and computer disk
ISBN 0697133974

Tutorial format training manual for *Publish It*, a personal page-layout program. Includes practice exercises that can be used to supplement teaching or for self-paced learning. Presents software features for the design of effective publications, including imported text and graphics, and instruction on the use of scanned images. Includes numerous design tips and 550 illustrations. Available in a Macintosh version.

Bristow, Nicholas
Screen Printing: Design and Technique
London, England, Batsford, 1990. 160p. ISBN 0713466316

An introduction to silk-screening as a fine art, stressing design fundamentals and basic equipment needed. This detailed do-it-yourself guide begins with the most elementary methods and progresses to more advanced processes, covering building frames, making screens and ink, stenciling and blocking out, color registration, mixed media techniques, and print studio design. Includes a list of suppliers.

Browne, David
Teach Yourself PageMaker 5.0 for Windows
New York, MIS Press, 1993. 443p. ISBN 1558282459

Introduction and guide to the popular page-composition software designed to quickly turn the novice into a power user. From installation of the program to the "Weekend Tour" chapter and the reference section covering every command and menu, concepts are presented in functional arrangements to assist the reader in learning all of the basic techniques to begin desktop publishing. Shortcuts and notes are graphically delineated within the text for clarity.

Browne, David
Welcome to—Desktop Publishing
New York, MIS Press, 1993. 409p. ISBN 1558282955

A guide to publishing in the *Windows* environment. Section One concerns production components, basic software functions, and a discussion of *Microsoft Publisher, PageMaker,* and *QuarkXPress.* Section Two is a design primer covering layout, typography, graphics, templates, and printing. Section Three is a design companion presenting planning, layout, refinements, and printing of specific business, presentation, and long documents. Appendix of practical hardware and software recommendations is also included.

Busch, David D.
The Complete Scanner Handbook for Desktop Publishing
Macintosh ed. Homewood, Ill., Business One Irwin, 1991. 386p.
ISBN 1556233396

The definitive resource for reviews and evaluations of recent scanners and scanning software. Presents in-depth discussions on virtually every related topic from using scanners, integration with desktop publishing software, and optical character recognition developments, to tips on increasing productivity and video capture. Available in an IBM version.

Busch, David D.
The Font Problem Solver: How to Install and Use Type with Your Software and Printer
Macintosh ed. Homewood, Ill., Business One Irwin, 1993. 400p.
ISBN 1556235054

Designed for both the novice and experienced desktop publisher who does not want to get bogged down in lengthy technical discussion and theory. This is not a typography book but specifically answers questions concerning font management and getting what is on the computer screen to the printer. Topics include bitmaps, outline, screen, resident and downloadable fonts, *PostScript* versus *TrueType*, installation in System 7, and building fonts from scratch. Software utilities for editing and managing fonts are included.

Busch, David D.
The Hand Scanner Handbook
PC ed. Homewood, Ill., Business One Irwin, 1993. 271p.
ISBN 1556237189

Authoritative discussion of the broad applications of image capture for desktop publications, presentations, and image databases. This excellent reference companion and hands-on guidebook answers basic questions on portable scanning, including how the device works, installation, software, optical character recognition (OCR) applications, and a thorough comparison of twenty-eight models. A very helpful "Tips and Trends" section includes increasing productivity, improving quality, handling difficult originals, and color balancing. Also available in a Macintosh version.

Busché, Don, and Bernice Glenn
The Desktop Design Workbook
Englewood Cliffs, N.J., Regents/Prentice Hall, 1992. 330p.
ISBN 013202425X

A basic design course with review questions and worksheets divided into four easy-to-follow sections covering an introduction to the publishing process, design as communication, business and commercial applications, planning, production, and printing. The interdependence of communication and design is stressed throughout. Can be used with any page layout program.

Carroll, Marianne
Marianne Carroll's Super Desktop Documents with Microsoft Word 5.0 for the Macintosh
New York, Brady Publishing, 1992. 541p. ISBN 0139641564

A well-organized, detailed guide to the execution of business documents. Topics include creating and using letterhead, preparing lists and financial tables, formatting envelopes, generating forms and expense reports, and structuring long documents. Each chapter is illustrated with screen shots of production methods and with notes that anticipate possible questions and problems. Tips are given for creating more than a hundred different desktop documents.

Carroll, Marianne
Perfect WordPerfect 6 Documents: A Visual Step-By-Step Guide
New York, Brady Publishing, 1993. 534p. ISBN 1566860520

Task-oriented approach to efficient, successful document formatting rather than a comprehensive reference manual to the software. The production of letterhead, flyers, legal documents, and more are demonstrated with more than 750 screen shots illustrating important program features. Tips on working with multiple documents, macros, mail merge, and linking information from other programs are presented to increase the reader's efficiency and productivity. For beginners and intermediate users who understand the basics of DOS and WordPerfect.

Clark, James F.
Power Publishing with Ventura 3.2
Macintosh version. Dubuque, Iowa, Wm. C. Brown, 1993. 209p.
ISBN 0697165116

Not a comprehensive guide, this book is an applications-oriented approach to the *Ventura* software, stressing the necessity of understanding design principles in the development of publications. It introduces features of the program that can be used to enhance communication and organization of documents; supporting exercises are suggested for each new element discussed. Spiral-bound workbook format.

Clifton, Merritt
The Samisdat Method: A Do-It-Yourself Guide to Printing
4th ed. Shushan, N.Y., Samisdat, 1990. 56p. ISBN 0318500515

Opinionated, humorous, but nevertheless highly useful, this guide to offset printing is designed for those who are interested in running a small independent press. In print for more than fifteen years, it was assembled from the combined wit and wisdom of hundreds of outlaw publishers in an attempt to make the method "the gospel on all cheap forms of reproduction." Also covers mimeography, spirit duplication, xerography, letterpress, layout, and binding. Includes a list of material suppliers.

Collier, David
Collier's Rules for Desktop Design and Typography
Reading, Mass., Addison-Wesley, 1991. 135p. ISBN 0201544164

Produced by a team of internationally renowned desktop publishers, this slim yet essential reference provides extensive coverage of the mechanics of typography and design, adequately explaining the industry jargon. Organization and readability are uniquely achieved by defining specific topics or problems and then grouping related points into subsequent double-page spreads. Extensive examples and illustrations explain typefaces, kerning, punctuation, hyphenation, justification, grids, style sheets, color, graphics, and hypermedia.

Cookman, Brian
Desktop Design: Getting the Professional Look
2nd ed. New York, Blueprint Publishing, 1993. 126p.
ISBN 0948905840

A thorough beginner's guide to the computerized integration of text and graphics. Independent of hardware or software platforms, Cookman carefully communicates fundamental design principles through illustrations and assists the reader by providing checklists and key points for each topic. A wide variety of publication types and their individual characteristics are covered. The section on working with service bureaus is particularly informative.

Cosnett, Jay
PageMaker 5.0 for Macintosh: Techniques and Applications
New York, MIS Press, 1993. 503p. and computer disk.
ISBN 1558282866

Directed to readers who have some experience with the software but want to enhance their productivity in the development of professional-looking documents. Organized to be read cover-to-cover or randomly, by specific topics. For quick reference, graphic symbols are used to highlight important information, shortcuts, warnings, and new features. Accompanied by a disk containing templates, macros, and sample scripts. A Windows version of the book is also available.

Craig, James
Production for the Graphic Designer
2nd ed. New York, Watson-Guptill, 1990. 207p.
ISBN 0823044165

An overview presented in straightforward informative style, helping to bridge the gap between design and production for designers and design students. This edition expands its coverage into the rapidly changing world of computer technology in the communications industry. Subjects include typesetting, digital composition, desktop publishing, platemaking, letterpress, gravure, offset, four-color printing, quality control, inks, paper, folding, binding, and working with mechanicals.

Crawford, Walt
Desktop Publishing for Librarians
Boston, G. K. Hall, 1990. 403p. ISBN 0816119295

The general description of hardware and software is quickly becoming dated; however, document design and production sections are very useful. The fundamental principles of graphic design, layout, and typography are clearly presented. Typical library publications, e.g., signs, posters, press releases, current events brochures, bibliographies, bookplates, new title lists, newsletters, operating manuals, and annual reports, are illustrated and design criteria discussed for each. A unique resource on applied DTP.

Crosby!
The QuarkXPress for Windows Companion
Homewood, Ill., Business One Irwin, 1993. 528p.
ISBN 1556239130

Instruction, document samples, and nine tutorials guide the user through the fundamentals, with emphasis on the quick creation of documents simply and efficiently. Designed for first-time users, those familiar with other layout programs, and anyone wanting information beyond the companion software manuals. Basic principles of design and typography are also presented. Information is also supplied on obtaining free clip art and backgrounds.

Crouse, David B., with Robert J. Schneider
Web Offset Press Operating
3rd ed. Pittsburgh, Pa., Graphic Arts Technical Foundation, 1989. 282p. ISBN 0883621185

Most newspapers, books, magazines, and catalogs are printed on offset presses. This book illustrates with photographs and line drawings the three major types of presses: in-line, common-impression cylinder, and blanket-to-blanket. All operating systems are discussed, as well as maintenance and adjustment procedures and auxiliary equipment. This new edition includes chapters on in-line finishing, test images, and pressroom safety.

Davies, Jacqueline
PageMaker for the IBM PC
Redwood City, Calif., The Benjamin/Cummings Pub. Co., 1992. 174p. ISBN 0201546442

Covers the basic tools of the program for the novice. Tutorial approach to version 4.0 with self-tests and exercises. One hundred and sixty screen shots. Topics include installing the software, creating a document, text and graphics, and creating a four-page newsletter.

Davis, Sally Prince
The Graphic Artist's Guide to Marketing and Self-Promotion
Rev. and updated ed. Cincinnati, Ohio, North Light Books, 1991. 136p. ISBN 0891344160

A creative and highly informative work for anyone wanting to make more money in the graphic arts field. Highlights include the development of self-promotion packages, identifying marketable skills, contacting potential clients, and pricing strategies. Successful promotional materials and résumés are interspersed with interviews of artists and clients describing the do's and don'ts of marketing. Includes interviews with successful DTP professionals.

Destree, Tom, ed.
Guide to Hardware and Software for Desktop Publishing
Pittsburgh, Pa., Graphic Arts Technical Foundation, 1992. 1 v. (various pagings). ISBN 0883621657

Overview of components based on GATF's *Market and Product Study 2* (Graphic Arts Technical Foundation, 1991?). Provides a thorough discussion of UNIX, Macintosh, and PC systems, as well as system and application software, detailing the advantages and disadvantages of each. Peripheral devices such as modems, scanners, printers, and monitors are discussed generically. Designed to help the reader make informed purchasing decisions.

Dewsnap, Don
Desktop Publisher's Easy Type Guide: The 150 Most Important Typefaces
Rockport, Mass., Rockport Publishers, 1992. 176p.
ISBN 1564960072

A presentation of one hundred and fifty typefaces from more than twenty type families from *Adobe Systems International PostScript Font Collection*, displayed in eight-, ten-, and twelve-point sizes. Includes discussion of the specialized characteristics and uses of each and factors to consider when purchasing type collections, and provides a directory of suppliers. Additionally, a helpful cross-reference guide to similar typefaces is included to match a client's request for a substitute font selection.

Doty, David B.
Power of—FrameMaker 4.0 for Windows
New York, MIS:Press, 1993. 446p. ISBN 1558283145

Complete reference to the software for intermediate and advanced desktop publishers. Covers the unique qualities and features that make this program ideal for publishing books, technical and scientific manuals, and hypertext documents. Publication aesthetics and suggestions from design professionals are clearly indicated throughout the text.

Eckhardt, Robert C., Bob Weibel, and Ted Nace
Desktop Publishing Secrets
Berkeley, Calif., Peachpit Press, 1992. 494p. ISBN 0938151742

Three hundred tips, techniques, and solutions from five years of *Publish* magazine's "Q and A" and "Tips" sections are illustrated, expanded, and updated to provide a very useful resource for PC and Mac users. This comprehensive encyclopedia of hands-on solutions is divided into chapters covering particular subject areas, such as time-saving tips; hardware and software questions; specific publishing, word processing, and graphics software; and design products. A buying guide with prices and phone numbers is included.

Emery, Richard S.
Color on Color: How Overprinting Two Colors Creates a New Third Color
Rockport, Mass., Rockport Publishers, 1993. 157p. ISBN 1564960331

An excellent color tool for designers and students facilitating the selection of transparent inks in the two-color printing process. Section One shows examples of successful designs. Section Two presents 460 color examples in solid and screened values. Using the forty-four transparent colors printed on acetate sheets with these examples, more than 150,000 variations can be created. This is a useful tool for the novice who is learning to judge the effects of using transparent inks.

Fanson, Barbara A.
Producing a First-Class Newsletter: A Guide to Planning, Writing, Editing, Designing, Photography, Production, and Printing
North Vancouver, British Columbia, Self-Counsel Press, 1994. 171p. ISBN 0889082960

Each procedure is described in terms of traditional methodology and desktop publishing technology. Important tips and planning worksheets make this a useful tool for designers, editors, writers, students, and organizations involved in promotion and information dissemination. Planning, budgeting, editorial content, style, design and layout, legal considerations, equipment, and distribution are some of the issues presented.

Felici, James
The Desktop Style Guide
New York, Bantam Books, 1991. 129p. ISBN 0553354450

Presents advice and basic rules of typographic style and page design in a concise manner, making it useful as a quick reference tool and a complete instruction manual. Guidelines for hyphenation, justification, leading, point size, kerning, tracking control, punctuation, margins, indentions, footnoting, and abbreviations are just a few of the many topics discussed. The last chapter is of particular interest for anyone involved in coding text files and preparing electronic manuscripts for publication.

Fellers, Patrick W.
Page Design with QuarkXPress 3.2 for the Mac
San Francisco, Sybex, 1993. 576p. ISBN 0782110436

A sequential how-to reference for beginning and intermediate users thoroughly covering the software and its advanced capabilities. A guided tour of the program environment and step-by-step creation of a document prepares the reader for more advanced prepress procedures. Each chapter begins with a listing of featured topics and a useful description of each command and its function. A list of more than a hundred shareware add-ons with information on suppliers is a welcome feature.

Floyd, Elaine, and Lee Wilson
Advertising from the Desktop: The Desktop Publishers Guide to Designing Ads that Work
Chapel Hill, N.C., Ventana Press, 1994. 427p. ISBN 1566040647

The traditional advertising methodology of defining the audience and producing effective designs that communicate to the defined group is discussed in relation to new computer technologies. The author assumes that the reader already has some background in copywriting and concentrates on design discussion and principles that guarantee advertising success. Many examples of promotional materials are illustrated. The last section includes a unique and very informative comparative evaluation of the major software packages.

Fraase, Michael
Rapid Reference Guide to Adobe Illustrator
Homewood, Ill., Business One Irwin, 1992. 96p. ISBN 1556237421

Introduction to effective use of the high-end *PostScript* illustration software version 3.2 for Macintosh. Covers tracing, editing, shaping, and layering plus shortcuts and keyboard equivalents to program functions. Not a replacement for the software documentation, but a presentation of basic information to enable users to become productive quickly. Illustrated with screen shots, and organized for rapid reference.

Fraase, Michael
Structured Publishing from the Desktop: Frame Technology's FrameMaker
Homewood, Ill., Business One Irwin, 1992. 336p. ISBN 1556236166

Practical manual for users of version 3.0 in producing long documents such as books, catalogs, and manuals on a Macintosh. Intended for readers familiar with the software, emphasis is placed on efficient production management and the collaborative work group environment. Covers indexing, cross-referencing, hypertext, and templates, and compares the pros and cons of various word processing and publishing software. Practical examples of applications and tutorials make this a good addition to the *FrameMaker* user manuals.

Gater, D. W.
Electronic Publishing: Evaluating, Procurement, and Management
Blue Ridge Summit, Pa., TAB Books, 1989. 296p. ISBN 0830691146

Neither an introduction to desktop publishing nor a guide to software, the work concerns issues of high-end systems development and management. These include needs surveys, justifying purchases, systems selection, employee training, establishing system operations, upgrades and add-ons, maintenance costs, and staying current. Appropriate for would-be systems managers and working professionals.

Gioia, Louis V.
Learning Desktop Publishing with PageMaker 4.0
Englewood Cliffs, N.J., Prentice Hall, 1992. 350p. and computer disk. ISBN 0135215501

Beginner's manual for versions 3.X and 4.0 based on the author's experience in producing documents and classroom teaching. Covers text formatting and creating and using graphics. Hands-on activities at the end of each chapter require the reader to apply the design techniques discussed. A disk is included with files needed for each exercise. Complements the manuals that accompany the software by adequately providing a self-paced approach to each page layout tool.

Glover, Gary
Clip Art: Image Enhancement and Integration
New York, Windcrest/McGraw-Hill, 1994. 296p.
ISBN 0830642625

Intended for electronic publishers desiring to make more effective use of graphic images and resources. Includes relevant information on sources for clip art, design techniques, file formats, scanning, tracing and image manipulation, object linking and embedding, and design hardware and software. More than forty-five products are used in the development of demonstration projects to assist the reader in the appropriate selection of tools and resources.

Gosney, Michael, John Odam, and Jim Benson
The Gray Book: Designing in Black and White on Your Computer
2nd ed. Chapel Hill, N.C., Ventana Press, 1990. 263p.
ISBN 1566040736

A useful guide to the creative possibilities of desktop publishing without color. Subjects include contrast, shading, halftones, design choices, professional techniques, and a section on sample documents. New to this edition is the "Annotated Design Gallery," illustrations with the artists describing the techniques and methods used in their creations. A valuable information resource and idea book.

Gosney, Michael, Linnea Dayton, and Jennifer Ball
The Verbum Book of Digital Typography
Redwood City, Calif., M&T Books, 1991. 209p.
ISBN 1558510923

This is a comprehensive resource for the desktop publisher and illustrator covering font design and the effective use of type in communication. Not just another collection of typefaces, but an instruction manual in which a variety of working professionals provide more than one hundred examples and design solutions for applications. Each is presented in a detailed step-by-step procedure with sidebars providing supplemental information.

Goss, Tom, ed.
Print's Best Letterheads and Business Cards: Winning Designs from Print Magazine's National Competition
New York, RC Publications, 1990. 173p. ISBN 0915734664

Collected designs from *Print's Regional Design Annual* (RC Publications, 1981-), reproduced in color, promoting architects, illustrators, photographers, corporations, hotels, and restaurants. This companion volume to *Print's Best Logos and Symbols* (RC Publications, 1990-) is a highly useful visual reference and sourcebook for designers and desktop publishers.

Graphic Artists Guild Handbook: Pricing and Ethical Guidelines
7th ed. New York, Graphic Artists Guild, 1991. 235p.
ISBN 0932102077

Based on national surveys of professional organizations for graphic arts, this work is specifically designed for those involved in buying and selling creative works; however, it is also of particular interest to students, showing the complex relationships between artist and client in the communication industry. Includes discussion on pricing artwork, trade customs, contracts, legal practices, writing proposals, licensing and royalties, computer art, comparative salaries, and current laws and legislation.

Grotta, Sally Wiener, and Daniel Grotta
Digital Imaging for Visual Artists
New York, Windcrest/McGraw-Hill, 1994. 661p.
ISBN 0830644423

Singularly unique, comprehensive, and thoroughly readable guide to computer hardware and software decisions, including scanners and printers for high-quality imaging. Written primarily for professional designers and illustrators, but appropriate as a primer for intermediate-level graphic arts students and desktop publishers. Issues covered include impact on careers and creativity, ethical and legal considerations, PC versus Mac systems, publishing and imaging software, filmless cameras and film recorders, storage and data management, calculating costs, and service bureaus.

Hartmann, Thom, Brad Walrod, and Don Arnoldy, eds.
The Best of the Desktop Publishing Forum on Compuserve
Berkeley, Calif., Peachpit Press, 1993. 262p. ISBN 1566090644

This electronic forum has always been an invaluable resource for information on new products and a place to discuss software and trade design tips. This print summary of the forum, although organizationally difficult to follow, offers a lively, opinionated discussion of *QuarkXPress, CorelDraw*, 300-dpi printing, scanning, free-lancing, printing inks, and type and font management in Windows. Of greatest value are the 175 *PostScript* and *TrueType* fonts that can be downloaded from the forum's libraries.

Hengesbaugh, Mark
Typography for Desktop Publishers
Homewood, Ill., Business One Irwin, 1991. 173p.
ISBN 1556234287

A well-crafted compendium on basic document design and a useful ready reference when making design decisions concerning the use of type. The focus is on creating readable, persuasive documents that are well organized, to assist the busy reader in quickly comprehending the intended message. Sections on display ads and presentation graphics are quite useful for the novice designer.

Holmes, Alan, and others
Desktop Publishing: Design Basics
London, England, Blueprint Publishing, 1989. 120p.
ISBN 0948905425

A brief, simple, and informative guide for training personnel new to desktop design and layout. Topics include measurement systems, developing house styles, margins, grids, copyfitting, typography, spacing, rules and boxes, and general document production. Originally developed at Watford College to assist college staff and companies requiring guidance in using personal computers for publishing.

Holzgang, David A., and Lesley Strother
Welcome to—Concepts in Graphic Design: From Mystery to Mastery
New York, MIS:Press, 1993. 247p. ISBN 1558283064

    Detailed discussion of the relationship between traditional and computer-based design tools with emphasis on practical application and technique. Each chapter covers an independent subject, making this work a good topical reference source. Topics include selecting hardware, creating graphics, layout techniques, digital imaging, color reproduction, document assembly, and camera-ready production. Enhancing productivity and avoiding common problems are discussed for each subject.

Hudson, Howard Penn
Publishing Newsletters
Rev. ed. New York, Scribner's, 1988. 240p. ISBN 0684189542

    Offers sound advice and instructions for effective publication and production. Hudson, founder of the Newsletter Association, discusses targeting of audiences and markets, editing, style, design and composition, management, and production steps. This new edition incorporates information on desktop publishing. A "newsletter gallery" illustrates sixty-eight corporate and organizational examples.

Hughes, Kenneth L.
Desktop Publishing Handbook: For Users of DOS-Based PCs
Plano, Tex., Wordware Publishing, 1994. 152p.
ISBN 1556223668

    Offers a general overview of the history of computer publishing, publication types, typography, operating systems, related software applications, hardware, and an introduction to presentation graphics. Comparative reviews of major page-layout programs as well as charts listing the features offered in various hardware peripherals are very useful.

Jacobs, Marvin
Graphic Design for Desktop Dummies
Cleveland, Ohio, Ameritype and Art, 1993. 162p.
ISBN 0962970018

    The focus of this book is on design and training for the computer typist. Jacobs begins with a trendy treatment of the subject and offers compelling ideas for the novice. Section One addresses the basics: predesign planning, type, illustration, and step-by-step design procedures. Section Two covers designing logos, business stationery, newsletters, multipage documents, promotional materials, business documents, and forms.

Johnson, Lois, and Hester Stinnett
Water-based Inks: A Screenprinting Manual for Studio and Classroom
Rev. ed. Philadelphia, Philadelphia Colleges of the Arts Printmaking Workshop, 1990. 38p.

    Developed over many years of research and practical classroom application, the methodology presented is very useful for the professional printer as well as for the school or college teacher and student. Complete information is presented on all aspects of the silk-screening process, from setup, stencil methods, printing equipment, and ink characteristics, to choosing paper and reclaiming or disposing of waste materials.

Kerlow, Isaac Victor, and Judson Rosebush
Computer Graphics for Designers and Artists
2nd ed. New York, Van Nostrand Reinhold, 1994. 306p.
ISBN 0442014309

    A survey, in nontechnical terms, of the latest techniques, computer technologies, and two- and three- dimensional commputer graphics applications for graphic design students. More than 500 drawings and photographs are used to illustrate hardware, software, peripheral devices, publications, multimedia, animation, and color and black-and-white image processing. Key words are highlighted and defined within the text. DTP students will find particularly helpful the sections on basic concepts of computer graphics and two-dimensional image processing.

Kieran, Michael
Desktop Publishing in Color
New York, Bantam Books, 1991. 394p. ISBN 0553351400

    Comprehensive reference including an overview of the complexities of color theory, hardware and software issues, and design techniques. RGB, CIE, CMYK and Pantone color systems models are discussed in relation to Mac, PC, and UNIX platforms. Important topics of reproduction and color separation processes, digital halftones, trapping, color calibration, scanning, file compression, and electronic prepress considerations are presented. Product comparisons and buying tips are highlights of this easy-to-read text.

King, Jean Callan, and Tony Esposito
The Designer's Guide to PostScript Text Type
Rev. ed. New York, Van Nostrand Reinhold, 1993. 415p.
ISBN 0442014546

    Desktop publishers, designers, editors, graphic arts students, and production managers will find this to be a useful visual reference. Ninety-six text types from Americana to Weiss are displayed in 8-, 9-, 10-, 11-, and 12-point. Each is shown in complete alphabet, with Roman, Italic, bold, and bold Italic styles. Sample text is set with 1-, 2-, and 3-point leading, providing adequate examples of each type's overall characteristics as well as copyfitting calculations.

Lamar, Laura
Desktop Design: Using the Macintosh Computer as a Graphic Design and Production Tool
Los Altos, Calif., Crisp Publications, 1990. 95p.
ISBN 1560520019

    The author uses her twenty years as a graphic designer and eight years of teaching typographic design to produce this work for professionals and students who understand the design profession but are new to computers. Part One serves as an introduction to the history of desktop publishing and as a general overview of hardware and software. Part Two covers writing, drawing, and layout software for prepress work, with a detailed explanation of font and printing technologies.

Leach, Robert H., and others
The Printing Ink Manual
4th ed. Wokingham, England, Van Nostrand Reinhold, 1988.
[800]p. ISBN 0747600007

    A unique, authoritative reference to the formulation, manufacturing, and working characteristics of printing ink. New to this edition are chapters on radiation-curable systems and health, safety, and the environment. Especially valuable for technicians, students, and manufacturers of printing materials.

Lem, Dean Phillip
Graphics Master 5: A Workbook of Planning Aids, Reference Guides, and Graphic Tools for the Design, Estimating, Preparation, and Production of Typography
5th ed. Los Angeles, D. L. Associates, 1993. 153p.
ISBN 0914218085

Print production workbook and reference containing the planning aids, guides, and tables needed for estimating and producing typography, electronic prepress imaging, printing, binding, and finishing. This concise easy-to-use manual of technical data also provides, in a single source, everything needed for process color selection, and paper specification. Includes 1,768 typeface specimens and graphic arts tools such as a line and gauge ruler and a copyfitting calculator. Invaluable for publishing and production work.

Lichty, Tom
Desktop Publishing with Word for Windows Through Version 6
Chapel Hill, N.C., Ventana Press, 1994. 348p. ISBN 1566040744

Written for the software user whose supplementary task is to prepare more sophisticated documents than those allowed by traditional text word processing. Using the graphic design tools incorporated in *Word*, Lichty offers easy-to-understand design advice, illustrations and procedures for proper typesetting, creation of tables, multicolor documents, use of headers and footers, style sheets and templates, and data linking from other programs. Covers *Word* through version 6.

Luna, Paul
Understanding Type for Desktop Publishing
London, England, Blueprint Publishing, 1992. 144p.
ISBN 094890576X

Introductory-level material on the basic principles and conventions of typography. Part One concentrates on clarity and readability of text rather than catchy design. Part Two is a guide to nearly 400 typefaces grouped by style and suggested use. All examples are applicable to DTP. Included are major manufacturers' addresses and phone numbers.

Makarius, Theodore
Operation of the Offset Press
Chula Vista, Calif., Printer Shopper, 1992. 281p.

A reference guide to operating procedures written for both craftsperson and student apprentice, stressing troubleshooting through observation, anticipation, and sound reasoning. Twelve case histories document cause-and-effect printing problems.

Mansfield, Richard
Desktop Publishing with WordPerfect 6
Chapel Hill, N.C., Ventana Press, 1993. 261p. ISBN 1566040493

Techniques of effective design and the creation of professional-looking documents are discussed in relation to the graphic features included in the new version of this word processing software. Through examples, suggestions, and general guidelines, the author stresses personal creativity over the strict adherence to inflexible rules. Includes sections on new features, common mistakes, creating macros and styles, and a very useful chapter illustrating publication make-overs.

Mantus, Roberta
Design Guidelines for Desktop Publishing
Albany, N.Y., Delmar Publishers, 1992. 136p. ISBN 0827350759

There is no accounting for taste, but successful documents still depend on specific principles of design for effective communication. Type, artwork, and page makeup are presented in a straightforward manner for beginners and nonprofessionals, stressing planning, designing, identifying problems, and obtaining approval for the finished document. Each chapter closes with suggested activities that can be used to supplement training on any publishing software. PC formatted disk with exercises included.

Marshall, George R.
The Manager's Guide to Desktop Electronic Publishing
Englewood Cliffs, N.J., Prentice Hall, 1990. 267p.
ISBN 0131685848

Step-by-step guide for developing a corporate infrastructure integrating desktop publishing, data processing, office operations, and manual production methods. Very beneficial for executives who want to introduce this technology into their organizations to leverage productivity and profit. Although the sections on major page layout programs and hardware are dated, sections on analyzing needs, introducing change in the corporate structure, and financial considerations make this work a good addition to the library collection.

McGrew, P. C., and W. D. McDaniel
In-House Publishing in a Mainframe Environment
2nd ed. New York, McGraw-Hill, 1991. 377p. ISBN 0070462712

Written for corporate managers but equally informative for students, this manual presents an overview of current application practices, computer information systems, and high-end desktop publishing. Issues discussed include in-house publishing, available technologies, establishing publishing departments, managing corporate printing, and employee training. Highlighted with corporate scenarios and actual solutions.

Meggs, Philip B.
A History of Graphic Design
2nd ed. New York, Van Nostrand Reinhold, 1992. 508p.
ISBN 0442318952

Chronicling visual communications from the invention of writing to the advent of computers, Meggs stresses the importance of technological advancements and individual pioneers in the evolution of graphic design in Western civilization. Five general chronological sections, each with detailed timelines, are divided thematically into chapters tracing the history of design with its relationship to social, political, and economic factors. Particularly useful for the student interested in a conceptual and pictorial overview of the field. Heavily illustrated.

Misanchuk, Earl R.
Preparing Instructional Text: Document Design Using Desktop Publishing
Englewood Cliffs, N.J., Educational Technology Publications, 1992. 307p. ISBN 0877782415

Using three sources—educational and psychological research, publishing industry practices, and personal experiences in teaching and developing printed materials—the author describes the fundamental differences between educational text-based materials development and other forms of publishing. Noting contradictions in graphic design advice, and using empirical research, he establishes guidelines for educators and small work groups addressing the production of textbooks, class handouts, and lab manuals. Well researched and clearly presented.

Murray, Katherine
Using Microsoft Publisher 2
Indianapolis, Que Corp., 1994. 338p. ISBN 1565292847

Example-oriented approach to the production of brief publications for users with varying levels of experience. Topics include installing the software; working with text and pictures; layout and printing; and publication ideas and illustrations for business cards, stationery, calendars, flyers, newsletters, invoices, and coupons. For those interested in quick creation of documents, the section on PageWizards and templates is particularly useful. Glossary and index.

Nadler, James
The Dover Electronic Clip Art Library
PC ed. New York, Random House, 1993. 285p. ISBN 067979097

Useful collection of copyright-free clip art digitized from previous Dover publications. Four hundred compressed files in Computer Graphics Metafile (CGM) format for PCs contained on four accompanying 3.5" high-density disks cover the topics of dining, entertainment, education, health, navigation, transportation, and business. Ornaments and borders are also included. Each illustration is reproduced in the text. Additional text includes information on file conversion utilities, image cataloging programs, graphics applications, copyright issues, and clip art vendors. Available in Macintosh version with EPS formatted files.

Nelson, Roy Paul
Publication Design
5th ed. Dubuque, Iowa, Wm. C. Brown, 1991. 319p.
ISBN 0697086208

A detailed discussion of each step of the publishing process, from the role of the art director, photographer, illustrator, and writer to the production of the finished document. Traditional methods and computer technology are described in relation to magazine, newspaper, book, and corporate publication design. Suggested assignments, extensive bibliography, glossary, and dictionary of symbols make this an appropriate textbook or supplemental reading for graphic arts and DTP students.

Nolan, Michael J.
Page Maker 5: Expert Techniques for Macintosh
Carmel, Ind., Hayden Books, 1993. 185p. ISBN 1568300174

Intended to "...bridge the knowledge gap between utilizing *PageMaker* and understanding how it applies to the creation of beautiful publications" (pref.), the author begins with the premise that the reader can already manipulate the software but needs guidance in the development of graphic designs and printing skills. Numerous illustrations highlight discussion of layout principles, using type, and postproduction work such as working with service bureaus and commercial printers, and using *Aldus PressWise*.

Ots, Peter
Learn Ami Pro in a Day: For Versions 2.0 and 3.0
Plano, Tex., Wordware Publishing, 1993. 135p. and computer disk. ISBN 1556223013

As a classroom text or tutorial for self-teaching, this manual on the popular Lotus word-processing program for PCs teaches the fundamentals necessary to master the software through exercises and the creation of simple documents. The author assumes that the reader has a general working knowledge of *Windows*.

Paddock, Bruce T.
Graphics for the Desktop Publisher
New York, MIS:Press, 1993. 436p. ISBN 1558282580

Comprehensive introduction to computer graphics in all of its many manifestations written in nontechnical language for the beginner. Well-illustrated with practical advice and recommendations on hardware and software. Some of the additional topics include computer memory, storage media, monitors, printers, input devices, IBM versus MAC, publishing and graphics software, clip art, charts and graphics, scanned images, CDs, layout principles, service bureaus, print shops, and workplace design.

Padgett, JoAnn, ed.
Start Your Own Desktop Publishing Business
San Diego, Calif., Pfeiffer, 1994. 240p. ISBN 0893842451

A how-to book on starting and developing a profitable business on a tight budget. Topics include: start-up basics, equipment and supplies, developing a business plan, daily operations, marketing and promotion, finances, home versus commercial locations, saving on operating expenses, developing a bookkeeping system, pricing guidelines, time management, hiring practices and legal issues such as taxes, licenses, and permits. Practical, thorough, and easy-to-read.

Parker, Roger C.
Looking Good in Print: A Guide to Basic Design for Desktop Publishing
3rd ed. Chapel Hill, N.C., Ventana Press, 423p. 1993
ISBN 1566040477

One of the most widely read companions on effective design, and an excellent supplemental text for use with training manuals on any page-layout software. Highlights include details on designing common applications such as newsletters, sales materials, magazines, books, presentation graphics, business communications, and a chapter on the twenty-five most common pitfalls in publication design. Arguably, the most practical single source on the subject.

Parker, Roger C.
The Makeover Book: 101 Design Solutions for Desktop Publishing
Chapel Hill, N.C., Ventana Press, 1989. 278p. ISBN 0940087200

"Before and after" examples are used to demonstrate the effects of good design principles. Common applications examined include ads, newsletters, brochures, flyers, letterheads, charts, catalogs, surveys, résumés, and reports. Descriptive text is minimized to encourage visual analysis.

Parker, Roger C.
Newsletters from the Desktop: Designing Effective Publications with Your Computer
Chapel Hill, N.C., Ventana Press, 1990. 306p. ISBN 0940087405

An idea book and basic survey emphasing a design approach to document production over detailed discussion of particular hardware or software. More than 200 illustrations highlight the text. A detailed study of the production process of two sample publications and a close look at the author's ten favorite designs make this a valuable and informative resource. Includes bibliography, glossary, and resource guide.

Paulsen, Deborah
How to Use Ami Pro for Windows, Release 2 and 3
Addison, Ill., OneOnOne Computer Training, 1992. 95p., 4 audiocassettes, and computer disk. ISBN 1565620186

Easy-to-follow procedures are divided into four approximately two-hour lessons for first-time users in a format that leaves the user's eyes free to concentrate on the keyboard and screen. Special features allow basic or in-depth exploration of topics. Lessons include basic tools and functions, formatting with style, and special features, all in a format useful for self-teaching or the training of others.

Peacock, John, Charlotte Berrill, and Michael Barnard, eds.
The Print and Production Manual
5th ed., rev. and updated. New York, Chapman and Hall, 1991- .
1 v. (loose-leaf). ISBN 0948905484

A standard reference for the commercial printing industry in loose-leaf notebook format with annual updates. Recent updates include desktop publishing, monochrome scanning, current postal regulations, and industry trends. Covers a full range of topics including typography, planning procedures, platemaking and printing, inks, paper, finishing, distribution, contracts, customs and trade regulations, and standard reference tables.

Rimmer, Steve
The Graphic File Toolkit: Converting and Using Graphic Files
Reading, Mass., Addison-Wesley, 1992. 335p. and 1 computer disk (5.25") ISBN 0201608464

Lack of standardization of PC graphics formats and incompatibility of applications software can be a major barrier in desktop publishing. This work provides many of the answers covering popular bit-mapped and vector file formats, as well as presenting problems that may be encountered in *Ventura*, *WordPerfect*, *Word*, and *PageMaker*. Shareware in PC format is included for converting bitmap images, color reduction, dithering, gray-scale conversion, and image organization.

Rose, Carla, and Rita F. Lewis
The Essential PageMaker 5.0
Carmel, Ind., Alpha Books, 1993. 400p. ISBN 1567612490

Software instruction and strategies for creating effective publications are presented through examples and expert advice for both Macintosh and Windows operating systems. Tips, areas of caution, and useful projects are clearly indicated in each section. In addition to software fundamentals, coverage is provided on planning publications, copyrighting work, commercial color printing, optimizing hardware, and program installation.

Ross, John, Clare Romano, and Tim Ross
The Complete Printmaker: Techniques, Traditions, Innovations
Rev. and expanded ed. New York, Free Press, 1990. 352p.
ISBN 0029273714

For students of printing or graphic design, this survey offers a detailed and well-illustrated discussion of most major fine arts printing techniques. Provides information on materials, papers, and inks used in each process. Topics include relief methods, intaglio, collagraphs, screen printing, lithography, dimensional printing, monotypes, and computer prints. Information is also provided on book design, the business of printing, health concerns, school projects, and sources of material.

Ruggles, Philip Kent
Printing Estimating: Principles and Practices
3rd ed. Albany, N.Y., Delmar Publishers, 1991. 626p.
ISBN 0827338058

This work is designed to meet the needs of graphic design students, serve as a general reference for the printing industry, and provide a useful resource for production personnel and professional estimators. Covers in great detail methodology, pricing strategies, production time, budgeted hour cost rates, marketing, and material costs. New to this edition is the inclusion of information on computer prepress and desktop publishing.

Sanders, Linda
Forty-Seven Printing Headaches (And How to Avoid Them)
Cincinnati, Ohio, North Light Books, 1991. 136p.
ISBN 0891343660

This solution book for designers, production artists, and publishers presents information necessary for working with printers and for successful designing. Problems gathered from sixty printers and related professionals are integrated into discussions of letterpress, gravure, offset, lithography, and screen printing. Issues of planning, production, proofing, materials, and color make this one-of-a-kind resource invaluable for anyone, regardless of experience level.

Schenck, Mary, and Randi Benton
The Official New Print Shop Handbook: Ideas, Tips, and Design for Home, School, and Professional Use
New York, Bantam Books, 1990. 298p. ISBN 0553349678

Handbook of suggestions for using *Print Shop* software with easy-to-follow instructions for copying or adapting illustrated designs. Hundreds of projects for creating greeting cards, banners, signs, letterheads, business cards, and flyers are organized into categories of use such as home, school, or professional applications. Prepared in collaboration with the software publisher.

Schiff, Kenny
Opportunities in Desktop Publishing Careers
Lincolnwood, Ill., VGM Career Horizons, 1993. 147p.
ISBN 0844240656

Gives an overview of the field as a specific career or as part of a broader job description by providing information on educational requirements, employment outlook, salary, advancement opportunities, job hunting techniques, and how to apply for a position. Introduction includes a definition of DTP and a history of printing technology, followed by a discussion of opportunities in corporate and government publishing, service bureaus, educational institutions, and self-employment. Of most help are the numerous profiles and interviews with working professionals.

Shushan, Ronnie, and Don Wright
Desktop Publishing by Design: Blueprints for Page Layout Using Aldus PageMaker on IBM and Apple Macintosh Computers: Includes Hands-On Projects
3rd ed. Redmond, Wash., Microsoft Press, 1994. 485p.
ISBN 1556155662

Award-winning electronic page-layout and design instruction manual. Divided into four sections: elements of design (covers effective communication, typography principles, and composition using the grid system); *PageMaker* portfolio (well-illustrated presentation of tips and techniques used in promotional materials, periodicals, and data sheets); *PageMaker* basics (an overview of software fundamentals); and hands-on projects (the application of good design principles in a variety of publication types).

Silver, Gerald A., and Myrna Silver
Layout, Design, and Typography for the Desktop Publisher
Dubuque, Iowa, Wm. C. Brown, 1991. 321p. ISBN 0697123022

Many books are available concerning traditional graphic design issues or particular software and hardware; however, this one approaches design fundamentals with strict application to electronic publishing without concern for particular software or hardware configurations. Written for the student, professional, or home computer user, the book has hands-on projects at the end of each chapter. Includes useful sections on paper selection, printing, and reproduction processes not usually covered in similar works.

Silver, Linda, ed.
Print's Best Typography: Winning Designs from Print Magazine's National Competition
New York, RC Publishers, 1992. 189p. ISBN 0915734818

This handsomely designed reference tool provides ideas for publications and projects. A collection of 175 full-color examples from winners of regional design competitions is presented, illustrating the current typographic trends in advertising, business, and magazine publications. Highlighted are the new forms of expression in text and graphic integration made possible by emerging computer technologies.

Simone, Luisa
Microsoft Publisher by Design, Version 2: An Example-Packed Guide to Desktop Publishing Using Microsoft Publisher
Redmond, Wash., Microsoft Press, 1994. 449p. ISBN 1556155654

This guide for the popular small business and home publishing software package is a comprehensive, example-filled, how-to book including sections on optimizing program performance and troubleshooting. Simone, a contributing editor to *PC Magazine*, provides complete information on each of the commands and tools. Provides eight projects that reflect good design and publishing principles. These include logos, business forms, letterheads, ads, organizational charts, catalogs, newsletters, and brochures.

Smith, Robert Charles
Basic Graphic Design
2nd ed. Englewood Cliffs, N.J., Prentice Hall, 1993. 162p.
ISBN 0130658146

This work is particularly useful in outlining fundamental competencies for students of the visual arts, including printing and design technologies. Smith offers a straightforward introduction to skills necessary in translating visual elements into graphic communication. Skills are well defined through illustration of design applications and discussion of commercial printing methods. Also includes a historical overview of graphic design.

Sosinsky, Barrie A.
Beyond the Desktop: Tools and Technology for Computer Publishing
New York, Bantam Books, 1991. 732p. ISBN 055335244X

A comprehensive survey of computer hardware, software, and new technologies written primarily for intermediate and advanced computer users. Invaluable for anyone investigating the purchase or upgrade of DTP systems. Comparative data are also provided for related topics such as modems, scanners, optical character recognition (OCR), file formats, page composition languages, networking, imaging, and multimedia.

Stopke, Judy, and Chip Staley
An Eye for Type
3rd expanded ed. Ann Arbor, Mich., Promotional Perspectives, 1992. 153p. ISBN 0963268902

A complete and informative design seminar, clearly laid out and illustrated for the novice or accomplished desktop publisher. Concepts are presented through the use of checklists, production tips, and step-by-step procedures. Useful specimen and display pages of the twenty most popular typefaces as well as full-page, editorial format examples of compatible typestyles highlight this essential reference.

Swann, Alan
How to Understand and Use Grids
Cincinnati, Ohio, North Light Books, 1989. 144p.
ISBN 0891342842

Rules and formulas used to arrange information and composition for readability, balance, and overall unity are described in this structured approach to graphic design. Principles are illustrated using examples. Both manual and computer layout methods are discussed.

Testa, Laura J.
PageMaker 5.0 for the Mac Page Design
San Francisco, Sybex, 1993. 564p. ISBN 0782111718

Different from other books on the subject, this one is written for graphic designers skilled in traditional methods and novices at electronic page layout and design. Also useful as a guide for any new user, the book offers a hands-on guided tour of the program and suggested exercises. Good coverage of Aldus additions, add-on programs, and color design. Quick reference appendices offer software shortcuts, and tips on troubleshooting hardware/software problems.

Thomas, Marilyn H.
Desktop Publishing Using PFS, First Publisher (Version 3.0)
2nd ed. Englewood Cliffs, N.J., Prentice Hall, 1992. 225p. and computer disk. ISBN 0132046113

An easy-to-follow, spiral-bound, self-instruction workbook for the PC-based software. Lessons cover the complete functions of the program, with screen illustrations as guides to assist in the execution of the assigned tasks.

Toale, Bernard
Basic Printmaking Techniques
Worcester, Mass., Davis Publications, 1992. 145p.
ISBN 0871922371

Introduction to three fine art printing techniques: relief, monotype, and silk-screen. This beginner's guide details processes and materials including linoleum, wood-block, collagraphs, and cardboard plates, with additional information on planning a studio and papermaking. Extensively illustrated with examples by renowned artists. A short list of material suppliers is also provided.

Tschihold, Jan
The Form of the Book: Essays on the Morality of Good Design
Point Roberts, Wash., Hartley Marks, 1991. 185p.
ISBN 0881790346

Twenty-three short essays written between 1937 and 1975 on basic principles of typographical layout and book production. The collection offers a highly recommended philosophical approach for anyone attempting to communicate through print. Subjects range from the broad principles of planning the layout of books with illustrations, and house rules for typesetting, to the finest details of dashes, ellipsis, point sizes, and proper paragraph indention.

Venit, Sharyn
Using PageMaker 5 for Windows
Special ed. Carmel, Ind., Que Corp., 1993. 1044p.
ISBN 1565290534

Complete and authoritative guide to *PageMaker 5 for Windows* from the author of the original *PageMaker* classroom training materials produced for Aldus Corporation. For professional publishers or beginners, this work offers a look at every feature of the program through practical examples, sample documents, and discussion of design and production issues. This edition also includes a brief tutorial for beginners and a removable reference card listing keyboard shortcuts.

Walker, Lisa, and Steve Blount
Letterhead and Logo Design: Creating the Computer Image
Rockport, Mass., Rockport Publishers, 1990. 255p.
ISBN 0935603379

Lavishly illustrated, full-color presentation of some of the best logotype and stationery design produced in recent years. The text is limited to the identification of the designer or design firm and a brief description of paper and printing methods. Ideal sourcebook for graphic artists and desktop publishers.

Walkin, Colin
Relief Printmaking : A Manual of Techniques
Marlborough, Wiltshire England, Crowood Press, 1991. 176p.
ISBN 185223427X

A well-written introduction to and procedural description of press and nonpress printing. Written for any skill level, it is thoroughly illustrated with student and professional work. Suggestions for possible projects and a discussion of design principles, ink, paper, and equipment cover the printing methods of linocuts, paper blocks, found objects, and woodcuts.

Watkins, Christopher, Alberto Sadun, and Stephen Marenka
Modern Image Processing: Warping, Morphing, and Classical Techniques
Boston, Academic Press, 1993. 234p. and computer disk.
ISBN 012737860X

Instruction is presented covering the acquisition, enhancement, manipulation, and processing of high-resolution images obtained from scanning or electronic imaging, using examples from the fields of art, science, engineering, and medicine. Currently available hardware and software products are discussed and evaluated. Intended for the more advanced user who has some knowledge of the C programming language. A PC data disk with sample programs and images is included.

Webster and Associates (N.S.W.)
PageMaker 5 by Example
Macintosh ed. New York, M&T Books, 1993. 537p. and computer disk. ISBN 1558512926

This complete guide contains tutorials detailing all of the program's important features in nineteen modules of progressive sophistication. Extensive use of screen shots and practice exercises are used to introduce concepts and software operations. A 3.5" disk accompanies the text, providing exercise files. Appropriate for individual, classroom, or self-paced instruction.

Webster, Carrie, and Paul Webster
PageMaker 5.0 for Windows: Visual Quickstart Guide
2nd ed. Berkeley, Calif., Peachpit Press, 1993. 243p.
ISBN 1566090342

A unique visual approach relying heavily on screen shots to guide the beginning desktop publisher through the concise, no-nonsense instruction of basic software features. Designed to be used as a step-by-step guide or quick reference on selected topics. Covers installation, basic program layout, program features, table editor, and dictionary editor utilities.

White, Jan V.
Color for the Electronic Age
New York, Watson-Guptill, 1990. 207p. ISBN 0823007324

An exceptionally well-illustrated manual on using color efficiently in charts, graphs, typography, and art. Intended for experienced publishers, corporate communicators, and beginning graphic designers, this informative work, by one of the authoritative writers in the design field, is not about using color to make a publication "pretty," but rather about its functional use as a tool to help turn information into knowledge.

Williams, Robin
The PC Is Not a Typewriter: A Style Manual for Creating Professional-Level Type on Your Personal Computer
Berkeley, Calif., Peachpit Press, 1992. 92p. ISBN 0938151495

Concise and thorough compendium of guidelines used by professional typesetters governing quotation marks, apostrophes, dashes, special characters, accent marks, underlining, capitals, kerning, tabs and indents, widows and orphans, hyphenation, leading, justification, punctuation, typefaces, and general layout. An excellent guide for desktop publishing classes, with tips on *WordPerfect*, *Word*, *Word for Windows*, *PageMaker*, and *Ventura*. The Mac version of this book won the Franklin Award from the Publishers Marketing Association.

Nonprint Materials

Corel Draw! Version 2.0
Ottawa, Ontario, Canada, Corel Systems Corp., 1991.
1 videocassette (63 min.) VHS

An instructional videocassette that explains how to use this computer graphics software for the IBM and compatibles.

Paste-Up 1: Pre-Press Fundamentals for Professional Production
Peoria, Ill., Step-by-Step Video, 1988. 1 videocassette (28 min.) VHS

Introduces basic materials and tools for keylines, basic paste-up techniques, type, and lettering, plus valuable tips for helping artists do fast, clean paste-ups. Describes professional procedures that save time and money at the camera and platemaking stages.

Paste-Up 2: Practical Examples of Production Art
Peoria, Ill., Step-by-Step Video, 1988. 1 videocassette (28 min.) VHS

Focuses on complex one- and two-color paste-ups for print projects.

Paste-Up 3: Complex Examples of Production Art
Peoria, Ill., Step-by-Step Video, 1989. 1 videocassette (45 min.) VHS

Covers the fundamentals of full-color production. Examples include a multipanel direct mailer, a capabilities brochure, and a package design.

Printing Basics for Non-Printers: An Abridged Guide to Printing Fundamentals
Peoria, Ill., Step-by-Step Video, 1989. 1 videocassette (55 min.) VHS

Covers the fundamentals of printing and the printer's expertise, especially when it comes to reducing costs and improving quality. Intended for designers, production artists, and individuals interested in preprinted materials. Also helpful for those working with print production staff.

Fashion Design and Apparel Arts

Dianna Thor
Pima Community College
Tucson, Arizona

Introduction

Programs of study in fashion design/apparel arts are designed to assist individuals in acquiring the necessary background knowledge and technical skills, as well as stimulate the creativity needed to succeed in any number of related fields in this profession.

Fashion design and apparel arts encompass a broad range of vocational possibilities within the fashion industry, and a student entering a program of study is greeted with an enormous array of career opportunities. The industry is concerned with producing clothing for men, women, and children—and all types of clothing from underwear to business suits, sportswear to evening wear, as well as the accessories to attend each type. Additionally, both the most unique, expensive haute couture creations and the more readily available mass-produced, ready-to-wear clothing have their place in this field.

Career opportunities related to creating apparel may include textile design in natural or manufactured fibers, textile production, fashion design, interpreting others' designs into flat patterns for production, grading those patterns to achieve a design in an entire size range, or facilitating apparel production as a costing clerk or engineer, production assistant or manager, or quality control engineer, among other positions. Once the finished clothing product is produced, there still remains a multitude of career options that involve getting the product to the customer. Several of the many options are showroom sales or showroom buyer, positions in buying offices, and retail positions such as merchandise manager, department head, visual display specialist, fashion coordinator, store manager, or even store owner. Additionally, fashion art in the form of sketching, illustrating, photography, and advertising offers still more possibilities that directly support this cycle of design conception, production, and distribution.

A significant number of related careers exist apart from this merchandising cycle. A more specialized area students may explore is the dynamic field of costume design for television, film, and theatrical productions. As with retail apparel, production of costumes involves more than designing. It may involve historical research, costume sketching, collecting materials, drafting patterns, costume construction, and fitting and alterations. In large productions, this may entail many costume technicians and one or more costume shop managers. Another area, though admittedly more limited, is that of costume curator or preservationist in museum settings.

This bibliography identifies resources of importance to libraries supporting students enrolled in junior and community college fashion design/apparel arts programs. Many of these programs emphasize the fashion design, clothing construction, or fashion merchandising aspects of this broad field. Hence, these areas have been emphasized in this bibliogrpahy, though several other areas have also been taken into account. It has not been possible to provide equal coverage to each topic with appropriate materials from the past six years, but where the bibliography cannot be comprehensive, an attempt has been made to be representative.

Each item included has been selected for the student with little or no background in the field, or for the student who wishes to explore the field in more depth. In some areas, particularly those involving fashion designers or the history of fashion, preference has been given to materials that are visually appealing and stimulating. Many publications included are not only instructional and educational, but inspirational as well.

Selected Associations

American Apparel Education Foundation, a division of the American Apparel Manufacturers Association
2500 Wilson Blvd., Suite 301
Arlington, Va. 22201

International Association of Clothing Designers and Executives
475 Park Ave. S., 17th Floor
New York, N.Y. 10016

Selected Journals

✦ Children's Fashions

Kids Fashions Magazine.
Larkin-Pluznick-Larkin. Monthly. ISSN 0362-6660

✦ Men's Fashions

Details.
Details Pub. Corp. Monthly. ISSN 0740-4921

DNR: The Magazine.
Fairchild Publications. Monthly. ISSN 1041-1119

Esquire.
Hearst Corp. Monthly. ISSN 0194-9535

GQ.
Conde Nast Publications. Monthly. ISSN 0016-6979

✦ Women's Fashions

BBW: Big Beautiful Woman Magazine.
Larry Flynt Publications. Bimonthly. ISSN 0192-5938

Elle.
Elle Pub./Hachette Publications. Monthly. ISSN 0888-0808

Glamour.
Conde Nast Publications. Monthly. ISSN 0017-0747

Harper's Bazaar.
Hearst Magazines. Monthly. ISSN 0017-7873

Mademoiselle.
Conde Nast Publications. Monthly. ISSN 0024-9394

Seventeen.
K-lll Magazines. Monthly. ISSN 0037-301X

Vogue.
Conde Nast Publications. Monthly. ISSN 0042-8000

Women's Wear Daily.
Fairchild Fashion and Merchandising Group. Daily. ISSN 0043-7581

✦ Sewing and Needlework

Threads Magazine.
Taunton Press. Bimonthly. ISSN 0082-7370

✦ Textiles

International Textiles.
Benjamin Dent, Ltd. Monthly. ISSN 0020-8914

Print Materials

✦ Reference

Calasibetta, Charlotte Mankey
Fairchild's Dictionary of Fashion
2nd ed. New York, Fairchild Publications, 1988. 749p.
ISBN 0870056352

Completely revising the 1975 edition, this new work succinctly presents over 15,000 historical and contemporary terms from the world of fashion, accompanied by line drawings and full-color illustrations. Features an appendix with brief designer biographies, portraits, and photographs of their designs.

Harris, Jennifer, ed.
Textiles, 5,000 Years: An International History and Illustrated Survey
New York, Abrams, 1993. 320p. ISBN 0810938758

Compiled by twenty-four experts from museums and other institutions, this work claims to be the first comprehensive survey of textile art and production from prehistory to the present. Exquisitely illustrated, almost entirely in color, the text begins with a survey and explanation of production techniques. This is primarily a survey of world textiles, with consideration of the role that textiles have played in daily economic and social life in all cultures.

Jerde, Judith
Encyclopedia of Textiles
New York, Facts on File, 1992. 260p. ISBN 0816021058

Presents readily understandable entries on a wide range of textiles, their history, methods of production, and care, as well as information on prominent individuals in the history of textiles. Jerde holds a master's degree in the history and design of clothing and textiles and is currently a consultant in those fields. This work is cross-referenced, with bibliography and subject index.

Meller, Susan, and Joost Elffers
Textile Designs: Two Hundred Years of European and American Patterns for Printed Fabrics Organized by Motif, Style, Color, Layout, and Period
New York, Abrams, 1991. 464p. ISBN 0810938537

A compilation of 1,823 color plates, each identified as to country of origin, period, fabric type, and method of printing. Plates are organized according to motif and divided into five chapters: formal, geometric, conversational, ethnic, and art movements and period styles. The plates show textiles used by the common person rather than rare museum pieces. Includes an explanation of terms, a bibliography, and an index. The introduction and list of contents are in French, German, Italian, Spanish, and Japanese. Inspirational!

Naylor, Colin, ed.
Contemporary Designers
2nd ed. Chicago, St. James Press, 1990. 641p. ISBN 0912289694

Presents over 600 currently active or contemporary designers, many who are involved in apparel arts through design of textile, fashion, or costume for stage and screen. Entries consist of a personal and professional biography, a list of design works, a bibliography, a statement from the fashion designer, occasional photographs, and an evaluative essay by a design critic or historian.

Shaeffer, Claire B.
Fabric Sewing Guide
Radnor, Pa., Chilton Book Co., 1989. 531p. ISBN 0801978025

An easy reference guide offering complete, practical information for the selection, wear, care, and sewing of all fabrics from classic to novelty. Fabrics are organized by fiber content, fabric structure, and surface characteristics so the reader may identify the fabric in hand. Covers topics such as natural and synthetic fibers, linings and interfacing, equipment and supplies, and sewing techniques. Reader-friendly text and instructions are accompanied by illustrations. Includes a dictionary, full-color photographs, a glossary, a bibliography, and an index.

Steele, Valerie
Women of Fashion: Twentieth-Century Designers
New York, Rizzoli International Publications, 1991. 224p. ISBN 0847813940

Though still in the minority, female fashion designers are gaining prominence in this traditionally male-dominated field. From Jeanne Paquin and Coco Chanel to Rei Kawakubo and Donna Karan, Steele presents contemporary designers from Paris, Milan, New York, London, and Tokyo, and examines international innovations and trends. Contains lively text and a multitude of black-and-white and color photographs. Bibliography and index included.

Stegemeyer, Anne
Who's Who in Fashion
2nd ed. New York, Fairchild Publications, 1988. 243p. ISBN 0870055747

Presents over 200 brief biographies of international fashion designers of the twentieth century. Includes great names from the past, current designers with established track records, and newcomers showing promise for the future. Includes selective portraits and examples of creations. Includes an index.

✦ Clothing Construction Techniques

Amaden-Crawford, Connie
The Art of Fashion Draping
New York, Fairchild Publications, 1989. 307p. ISBN 0870056344

Provides step-by-step instructions and illustrations presenting basic methods and principles involved in draping for bodices, dartless and princess shapes, skirts, pants, bustiers, dresses, and other garments. Covers basic designs and their variations. Includes discussion of the dress form, elements of fabric, and instructions for transferring the finished drape to the final pattern. Includes an arm pattern to be sewn, stuffed, and attached to a dress form to check the drape of sleeves.

Bensussen, Rusty
Shortcuts to a Perfect Sewing Pattern
New York, Sterling Pub. Co., 1989. 159p. ISBN 0806968222

Easy-to-follow instructions showing how to turn a series of accurate, personal measurements into a custom-made bodice sloper, skirt sloper, and pants sloper. Illustrates how these slopers may be used over and over to correct commercial patterns for custom use, or to create new patterns for one's own fashion designs. Discusses flattering styles for various figure types. Includes a metric conversion chart and an index.

Betzina, Sandra
Fear of Sewing: Survival Skills for the "Non-Sewer," Primer for the "Would-be-Sewer"
San Francisco, Sandra Betzina, 1991. 59p. ISBN 0961561475

This work can be used independently or as a companion to Betzina's video by the same name (Practicality Press, 1993). Presents in simple language beginning sewing projects such as hemming or tapering jeans, shortening skirts or trousers, sewing on a button, installing zippers, or doing gathers, pleats, and darts. Also ventures into some alterations and includes some household projects such as making pillows and curtains. A gentle introduction to sewing, accompanied by helpful line drawings.

Betzina, Sandra
More Power Sewing: Masters Techniques for the Twenty-First Century
Rev. ed. New York, Practicality Press, 1993. 271p. ISBN 1880630141

This book covers many topics discussed in the author's videos. Appropriate for the experienced sewer looking for greater challenges or creative ideas for one-of-a-kind garments. Covers jacket, pants, and skirt construction details, offers tips on working with fake fur, leather, and suede, or choosing styles for a particular figure, and on maximizing one's wardrobe. Includes line drawings, photographs, a source list, and an index. The hidden spiral binding enables the reader to lay the book flat.

Betzina, Sandra
Power Sewing: New Ways to Make Fine Clothes Fast
Rev. ed. New York, Practicality Press, 1993. 271p. ISBN 1880630133

In a personable style, the author presents essential information and little-known tips for virtually any sewing project. Techniques from both the ready-to-wear industry and couture houses are included. Can be used independently or with the video by the same name. Line drawings and color photographs; indexed.

Boswell, Suzanne
Menswear: Suiting the Customer
Englewood Cliffs, N.J., Prentice Hall, 1993. 216p.
ISBN 0135714230

Though presented in textbook format, this book is not strictly for the classroom. Provides an overview of how the menswear industry operates, presenting issues of relevance to the customer, retailer, designer, and manufacturer. Though focusing on tailored clothing, it also treats formal wear and sportswear and explains the evaluation of garment style, fit, and quality. Presents good coverage of this aspect of the fashion industry all too often neglected. Includes numerous line drawings, some photographs, a bibliography, glossary, and index.

Brackelsberg, Phyllis, and Ruth Marshall, eds.
Unit Method of Clothing Construction
7th ed. Ames, Iowa, Iowa State Univ. Press, 1990. 280p.
ISBN 0813817110

Provides the basic information needed by the beginning sewer, from selecting the pattern to completing the final hem in women's, men's, and children's garments. Discusses pattern alterations and components of fit, and provides construction alternatives depending on the fabric used and the skill level of the sewer. A highly readable text with illustrations and shaded line drawings.

Carr, Roberta
Couture: The Fine Art of Sewing
Portland, Ore., Palmer/Pletsch, 1993. 208p. ISBN 093527828lX

Provides numerous couture techniques for clothing construction and embellishment in step-by-step diagrams accompanied by discussion. Tips are provided throughout the text to help resolve various sewing problems. Includes a brief history of couture.

Coffin, David Page
Shirtmaking: Developing Skills for Fine Sewing
Newtown, Conn., Taunton Press, 1993. 173p. ISBN 1561580155

By concentrating on just one garment type, this work presents all the necessary information and detail to create shirts of superb fit (for men or women) in a variety of styles and fabrics. Discusses fabrics, tools needed, "the classic shirt," fit, developing a pattern, sewing techniques, variations on a theme, and more. Written in an extremely readable style by a shirtmaker of fifteen years. Illustrated with line drawings and some color photographs.

Cooklin, Gerry
Pattern Grading for Children's Clothes: The Technology of Sizing
Boston, BSP Professional Books, 1991. 304p. ISBN 063202612X

Presents a pattern grading system developed through the analysis of children's sizing systems and applicable to most size charts for children's clothes. Covers specific sizing requirements for each age and size group. The applications of these are demonstrated in twenty-eight basic and styled grades. Step-by-step, illustrated instructions clearly describe the process, which is a practical and systematic approach for both students and professionals. Also covers mechanical and computerized gradings. Children's size charts included.

Cooklin, Gerry
Pattern Grading for Men's Clothes: The Technology of Sizing
Boston, Blackwell Scientific Publications, 1992. 382p.
ISBN 0632033053

Designed to help the advanced student and professional with the practical principles and applications of pattern grading for a variety of garment types available to today's male. Examines international sizing systems, shows methods of calculating grading increments, and demonstrates working methods for manual grading as well as explaining computerized grading. Applications of the grading system are demonstrated across a range of men's garments with fully illustrated instructions. Includes size charts.

Cooklin, Gerry
Pattern Grading for Women's Clothes: The Technology of Sizing
Boston, BSP Professional Books, 1990. 382p. ISBN 0632022957

Written by an acknowledged expert, this work presents a basic pattern grading system that allows for creating a range of sizes from one pattern while maintaining design form and proportions. Provides over forty-eight examples of basic and styled grades, with step-by-step instructions and clear illustrations. Includes background on the development of the system and discussion of computerized pattern grading. Includes size charts, a brief glossary, and a bibliography. Designed for use by the more advanced student.

Creative Sewing Ideas
Minnetonka, Minn., Cy DeCosse, 1990. 127p. ISBN 0865732582

An inspiring collection of possibilities for being creative with fabric, construction details, or embellishments of a garment. Explains techniques for twisted silk, acid-washed silk, felted wool, discharge dyeing, screen painting, and fabric painting. Also covers creative seaming techniques, piping, pockets, buttonholes, ribbonwork, and beadwork. A few projects for getting started are included. The step-by-step instructions are accompanied by clear, full-color photographs. Indexed.

Jaffe, Hilde, and Rosa Rosa
Childrenswear Design
2nd ed. New York, Fairchild Publications, 1990. 303p.
ISBN 0870057065

A professor at the Fashion Institute of Technology presents her three-part work: Part One provides background information on the childrenswear industry, preferred fabrics, size ranges, and clothing requirements from infants to juniors; Part Two discusses creative patternmaking using both draping and flat pattern methods; Part Three treats special problems in designing childrenswear. Includes measurement charts, a suggested reading list, and an index. Illustrated with photographs and line drawings.

Jaffe, Hilde, and Nurie Relis
Draping for Fashion Design
2nd ed. Englewood Cliffs, N.J., Regents/Prentice Hall, 1993. 226p. ISBN 0131058428

This is an expanded and revised version of a 1973 edition with an added section dealing with tailored garments. The authors attempt to transcend the changing fashion cycle by reducing design to basic principles that can be adapted to meet the needs of current trends. Intended to be used as a basic instructional text, it includes fundamental and more complex draping techniques, with nearly every step accompanied by line illustrations. Includes selected bibliography, and index.

Komives, Margaret
The Ins and Outs of Interfacings: A Complete Guide to Their Selection and Use
Mequon, Wisc., Maggi K. Enterprises, 1992. 56p.
ISBN 0878017020

Written by a sewing teacher of thirty years and current technical college instructor, this guide provides all the information needed to choose from the vast array of interfacings available on the market, based on the particular fabric used, location on the garment, and desired finished effect. Covers woven, nonwoven, knit, fusible, and sew-in interfacings. Preshrinking, application, testing, and care are also discussed. Contains photocopies of interfacing structure for easy identification, a selection table, a glossary, and an index.

Kopp, Ernestine, and others
Design Apparel Through the Flat Pattern
6th ed. New York, Fairchild Fashion and Merchandising Group, 1992. 517p. ISBN 0870057375

Designed as a companion text to *How to Draft Basic Patterns* (Fairchild Fashion and Merchandising Group, 1991) this work thoroughly presents, with step-by-step illustrated instructions, the utilization of three basic slopers—master, block, or foundation patterns—for waist, sleeve, and skirt. Style variations of these three basic patterns are also presented as well as other garment details. Miniature (one-quarter inch scale) sloper patterns are included. Designed to assist the beginner as well as those already working at a professional level.

Kopp, Ernestine, and others
How to Draft Basic Patterns
4th ed. New York, Fairchild Fashion and Merchandising Group, 1991. 135p. ISBN 0870057472

Discusses slopers as the master patterns for the structured designs on which all other designs are based. Provides the principles and instructions for drafting slopers using standard or individual measurements. Covers terminology, tools and materials needed, how to take measurements, and creating the actual slopers. Line drawings clearly illustrate every step. Includes a plastic neckline curving pattern, a metric conversion table, and standard measurement tables.

Ladbury, Ann
The Sewing Book: A Complete Practical Guide
Stamford, Conn., Longmeadow Press, 1992. 216p.
ISBN 0681416408

A guide for the beginning sewer written in a clear style and augmented by explanatory step-by-step, two-color illustrations. Contents include choosing and using sewing machines; hand sewing; glossary of textiles; notions; patterns; altering and fitting; garment construction; traditional tailoring techniques for coats, jackets, and trousers; sewing for the home; and decorative sewing such as patchwork, quilting, and appliqué. Indexed.

Rasband, Judith
Fabulous Fit
New York, Fairchild Publications, 1994. 160p. ISBN 0870057391

Rasband presents her unique "seam method of alteration" in a work that relies more on clear illustrations than text for instruction. She also identifies eight basic body types and forty-six body variations that can cause fitting problems, then shows how to alter any pattern to accommodate each variation. Also offers tips on how to adapt or alter store-bought clothes for a perfect fit. Limited index.

Reader's Digest Complete Guide to Sewing
Westmount, Quebec, Canada Reader's Digest Association, 1995.
ISBN 0888502478

The new edition of this 1976 publication provides a comprehensive encyclopedia for sewing, organized by technical areas. Includes extremely clear diagrams and illustrations and is color-coded for ease of use. It is especially valuable for the student who needs a self-instruction book or a quick reference.

Sewing Pants That Fit
Minnetonka, Minn., Cy DeCosse, 1989. 128p. ISBN 0865732515

A well-written, colorfully illustrated guide in the "Singer Sewing Reference Library." Provides a thorough treatment of the topic from selecting the pattern and fabric, to constructing the pants and varying the design. Includes a particularly good discussion of what consititutes "good fit" and how to achieve it through accurate measurements, proper figure analysis, pattern adjustments, and final alterations that are made after machine-basting the pants. Addresses typical figure problems and variations. Indexed.

Shaeffer, Claire B.
Couture Sewing Techniques
Newtown, Conn., Taunton Press, 1993. 217p. ISBN 0942391888

Presents haute couture from the inside handfinishing to the completed product. Details of construction techniques are illustrated in clear diagrams and colored photographs of original couture garments. A background on couture, and its business and history, is also part of this revealing book. Includes glossary, bibliography, sources of supplies, and an index.

Singer Sewing Step-By-Step
Minnetonka, Minn., Cy DeCosse, 1990. 320p. ISBN 0865732574

This hefty volume is designed to be a help and inspiration to the new, experienced, or returning sewer. Vivid color photographs provide a closeup look at each step of instruction. Begins with the basics—machine, equipment, notions, fabric, interfacing, layout, cutting, and marking—and moves through to techniques for pattern adjustment, garment construction, tailored garments, children's clothing, activewear, and home decorations. Text is accompanied by charts. Great as a reference source or constant companion.

Vogue and Butterick's Designer Sewing Techniques
New York, Simon and Schuster, 1994. 126p. ISBN 0671888781

Profiles a wide selection of well-known fashion designers and analyzes their specific construction and design techniques. With photographs and diagrams, students learn how they too can accomplish similar techniques.

The Vogue/Butterick Step-by-Step Guide to Sewing Techniques
New York, Prentice Hall Press, 1989. 415p. ISBN 0139441255

Designed as a reference guide for quick access to more than five hundred of the most essential and most frequently used sewing procedures and techniques as they appear in Vogue and Butterick patterns. Divided into forty-seven broad categories such as buttonholes, collars, darts, facings, linings, pleats, pockets, and zippers, each section presents various methods of construction, all in a detailed, illustrated, step-by-step format. A useful guide whether one is designing original fashions or sewing with commercial patterns. Indexed.

Zamkoff, Bernard, and Jeanne Price
Creative Fashion Skills for Fashion Design
New York, Fairchild Publications, 1990. 207p. ISBN 0870056824

Promotes flat pattern design as a creative, speedy, and accurate method of producing fashion. Concentrates on sleeve/bodice combinations such as dolman or semi-set-in sleeves, and collar/bodice combinations, including shawl or notched collars. Discussion of each design includes the theory behind the design, step-by-step instructions for creating the basic draft of the design, then variations on that basic design. A well-written text that includes line drawings illustrate the methods discussed.

Zangrillo, Frances Leto
Fashion Design for the Plus-Size
New York, Fairchild Publications, 1990. 191p. ISBN 0870056778

Profiles the four body types of the plus-size figure and offers information on planning, designing, and marketing lines for them, as well as for interpreting existing fashions. Presents the basic principles for draping and drafting various garments. Also explains how to create slopers, and demonstrates their use in the process of developing five different fashion designs. Includes charts and a diagram covering pattern grading, body proportions, and body changes that occur from one size to another. The clear text is augmented by illustrations and photographs. Includes index.

✦ Computers in the Fashion Industry

Miller, Phyllis Bell
AutoCAD for the Apparel Industry
Albany, N.Y., Delmar Publishers, 1994. 540p. ISBN 0827352247

This combination textbook and reference work explains applying AutoCAD software to fashion illustration, fabric surface design, drafting, patternmaking, alterations, and grading in nontechnical language. Assumes the reader has existing knowledge of these areas and builds on that. Covers the basics of the program and more complex design projects, making it useful for both novice and experienced operators. Includes troubleshooting guide, glossary of computer terms and AutoCAD commands, bibliography, suggested references, line drawings, tables, and index.

Steinhaus, Nancy Hoover, and Isabelle M. Lott
I CAD—Can You? AutoCAD for Apparel Design and Pattern Development
Grand Rapids, Mich., PW Publications, 1991. 276p. ISBN 0963052004

Recognizing the new directions in the apparel industry, and realizing that even entry-level employees will be expected to have apparel-related computer experience, the authors have collaborated in this volume to produce a series of lessons, exercises, problems, and questions for review to give the beginner hands-on experience. Based on AutoCAD by Autodesk, Inc. software and MS-DOS. Written clearly and with humor. Includes appendices and command list.

Taylor, Patrick J.
Computers in the Fashion Industry
London, Heinemann Professional, 1990. 180p. ISBN 0434919160

Gives an overview of the fashion industry and how computer technology has been established in all areas. Covers computer-aided design, pattern design systems, grading and lay planning with CAM, computer bulk cutting, computer-aided production machinery, and management and production control. Includes a list of hardware and software companies, flow charts, black-and-white photographs, line drawings, and an index.

✦ Costume Design

Cunningham, Rebecca
The Magic Garment: Principles of Costume Design
New York, Longman, 1989. 395p. ISBN 0801300623

Provides basic information for the student interested in theatrical costume design. Focuses on the necessary creative and practical skills of research, conceptualization, interpretation, and organization. Discusses the elements and principles of design, aspects to be considered in developing costumes, and techniques for rendering sketches and choosing fabrics. Provides very little on the actual construction of costumes. Includes three appendices: historical research sources, a historical costume outline, and a play synopsis. Bibliography and index are provided.

Ingham, Rosemary, and Liz Covey
The Costume Designer's Handbook: A Complete Guide for Amateur and Professional Costume Designers
2nd ed., rev. and updated. Portsmouth, N.H., Heinemann Educational Books, 1992. 286p. ISBN 0435086073

A detailed yet approachable discussion of the process of costume design and the role and responsibilities of the designer from first reading of the play to opening night. Includes an outline for a designer's playscript analysis and information on developing a costume plot; research, preliminary and final sketching techniques; and working with the costume shop. Also discusses the job market and the job hunt. Reference section includes a list of painters for design research, an extensive annotated bibliography, a supplies shopping guide, and an index.

Ingham, Rosemary, and Liz Covey
The Costume Technician's Handbook: A Complete Guide for Amateur and Professional Costume Technicians
Rev. ed. Portsmouth, N.H., Heinemann Educational Books, 1992. 458p. ISBN 0435086103

A comprehensive yet highly readable text that picks up where *The Costume Designer's Handbook* (Heinemann Educational Books, 1992) leaves off, treating the actual construction of costumes. Covers the work of costume shops, equipment required, health and safety practices in the shop, fabrics, pattern development, costume construction, fitting and alterations, fabric dyeing and painting, hair and hats, and costume accessories and properties. Includes an extensive annotated bibliography, a supplies shopping guide, and an index.

Motley (pseudonym)
Designing and Making Stage Costumes
Rev. ed. Edited by Michael Mullin. New York, Theatre Arts Books/ Routledge, 1992. 141p. ISBN 087830021X

Written by "Motley" (the pseudonym for three costume designers for stage and screen who were active from 1932 to 1976), this work touches on the work of the costume designer, costume sketching, the use of color, interpreting character through design, the meaning and making of stage and costume properties, facts about fabrics, cutting and fitting period costumes, and the influence of history and architecture. Illustrated predominantly with color costume sketches from actual productions. Indexed.

Pecktal, Lynn
Costume Design: Techniques of Modern Masters
New York, Back Stage Books, 1993. 256p. ISBN 082308311X

A collection of eighteen interviews with some of the most successful and exciting contemporary costume designers worldwide. Beautifully illustrated with color sketches from actual productions that show the work of these eighteen designers as well as others. Text summarizing each interviewee's career precedes the interviews covering such topics as beginnings, inspirations, and techniques. Design credits for Broadway plays, theater, dance, and opera productions are listed for each. This is an inspirational, informative look at careers in costume design. Includes index.

✦ Ethnic Costumes

Adler, Peter, and Nicholas Barnard
African Majesty: The Textile Art of the Ashanti and Ewe
London, Thames and Hudson, 1992. 192p. ISBN 0500236399

Provides a cultural history of these striking textiles that are produced by the narrow-strip weaving process in West Africa. Illustrations and photographs present these textiles worn as garments. Includes extensive notes with the colored plates, a bibliography, a glossary, and an index.

Altman, Patricia B., and Caroline D. West
Threads of Identity: Maya Costumes of the 1960s in Highland Guatemala
Los Angeles, Fowler Museum of Cultural History, Univ. of California, 1992. 190p. ISBN 0930741234

This historical analysis of the Maya provides the foundation for an understanding of the evolution of this regional costume up to the 1960s. Prepared for an exhibition, the color plates show display garments on museum mannequins. Supporting diagrams and illustrations clearly show the cut and wear of the traditional garments. Maps and many black-and-white photographs complement the text. Includes appendices and bibliographical references.

Gittinger, Mattiebelle, ed.
To Speak with Cloth: Studies in Indonesian Textiles
Los Angeles, Museum of Cultural History, Univ. of California, 1989. 256p. ISBN 0930741188

This anthropological approach to the textiles and garments of Indonesia covers specific regions and areas with the textiles and apparel for each. Maps and diagrams help clarify the differences among the areas discussed. Includes glossary and bibliography.

Gluckman, Dale Carolyn, and Sharon Sadako Takeda
When Art Became Fashion: Kosode in Edo-Period Japan
Los Angeles, Los Angeles County Museum of Art, 1992. 351p. ISBN 0875871631

Published in conjunction with the museum exhibition held November 1992-February 1993, this work provides numerous color photographs of mounted kimonos. A discussion accompanies the illustrations and provides the historical, social, and political framework for these garments. The primary focus of this work is to illustrate the textile design inherent in the wide variety of garment styles shown.

Hitchcock, Michael
Indonesian Textiles
New York, Icon Editions, 1991. 192p. ISBN 0064302172

Color and black-and-white photographs illustrate the textiles native to Indonesia, accompanied by diagrams of their production. A wide range of techniques is used to create the surface design styles and to produce these textiles. The emphasis is placed more on the textile itself, rather than on its use and application as apparel. Includes a list of museums, a bibliography, and an index.

Marshall, John
Make Your Own Japanese Clothes: Patterns and Ideas for Modern Wear
New York, Kodansha International, 1988. 130p.
ISBN 087011865X

Clear line drawings and diagrams carefully take the reader through the entire process of clothing construction—Japanese style. Traditional Japanese garments are defined and illustrated, from pattern cut through final finishing stitches. Additional chapters cover care and storage, metric conversion charts, sources of supplies, tools, and textiles. Includes bibliography and index.

Rajab, Jehan S.
Palestinian Costume
New York, Kegan Paul, 1989. 160p. ISBN 0710302835

Palestinian garments are traditionally characterized by the use of textiles and embroidery. The author focuses on the costumes of three groups—townsfolk, villagers, and the Bedu—and examines the tribal significance and technical aspects of their colorful designs. Includes a glossary, bibliography, and an index.

Sayer, Chloe
Mexican Patterns: A Design Source Book
New York, Portland House, 1990. 1 v. (various pagings).
ISBN 0517014912

This visual reference provides a closeup view of Mexican textile embellishment. Illustrations include apparel details such as blouse strips, capes, and borders of both men's and women's garments. Illustrated with full color plates. Some text accompanies the illustrations.

✦ Fashion and Wardrobe Coordination

Allen, Jeanne
Dressing with Color: The Designer's Guide to Over 1,000 Color Combinations
San Francisco, Chronicle Books, 1992. 124p. ISBN 0811800946

Illustrated with colorfully clothed silhouettes, this work takes forty-nine basic colors and shows how they work, or do not work, with other colors treated in groups labeled monotones, brights, pastels, deeps, and naturals. The author, a clothing designer, also addresses styles and accessories as she explains why certain colors work together. An invaluable sourcebook covering an essential element of fashion largely ignored elsewhere. Glossary.

Gross, Kim Johnson, Jeff Stone, and Christina Worthington
Chic Simple: Clothes
New York, Alfred A. Knopf, 1993. 192p. ISBN 0679421696

This text of "affordable style" presents a philosophy of paring one's wardrobe down to the essential, timeless, quality basics that can be mixed, matched, and accessorized to suit every occasion. Addresses the elements of style and versatility in an entertaining and evocative manner. Includes a worldwide retail guide and thoughts on fashion from Karan, Armani, Chanel, Beene, and Zoran.

Kidwell, Claudia Brush, and Valerie Steele, eds.
Men and Women: Dressing the Part
Washington, D.C., Smithsonian Institution Press, 1989. 188p. ISBN 0874745500

Explores the relationship between movements in fashion and evolving concepts about masculinity and femininity. Discusses clothing's role in expressing one's identity as men or women and how it conveys what one thinks about oneself, and how one wants to be seen by others. Covers topics such as children's clothes, work clothes, sportswear, and the realtionship between clothing and sexuality. Black-and-white and color photographs, painting reproductions, and drawings illustrate the concepts presented. Indexed.

Martin, Richard, and Harold Koda
Jocks and Nerds: Men's Style in the Twentieth Century
New York, Rizzoli, 1989. 223p. ISBN 0847810453

Presents an exquisite visual history of menswear in this century. With a unique approach, fashion is discussed in light of twelve basic social roles adopted by men: worker, cowboy, rebel, military man, sportsman, hunter, businessman, Joe College, man-about-town, dandy, jock, and nerd. Includes bibliography.

Moss, Miriam
Street Fashion
New York, Crestwood House, 1991. 32p. ISBN 0896866114

This brief and easy-to-read work concentrates on fashion for young people as a distinct group. Moss looks at various styles worn by ordinary people on the street and investigates what influences fashion choices. Presents street fashion by looking at the people who design, manufacture, and sell the clothes: Claire McCardell, Vivienne Westwood, and Katharine Hamnett, to name a few. Includes colorful photographs and a brief index.

Ogawa, Yoko
Color in Fashion: A Guide to Coordinating Fashion Colors
Rockport, Mass., Rockport Publishers, 1990. 159p. ISBN 0935603387

Similar in format to Allen's *Dressing with Color* (Chronicle Books, 1992), but more extensive. Fundamentals of color theory such as the color wheel, hue, tone, mass, and contrast are presented, then 107 colors are treated in twenty-eight groups. Ogawa provides numerous examples of color combining; a theme color is chosen and then enhanced with secondary colors. Presents casual, career, and high fashion looks across the seasons. Provides simple, practical instructions in a concise text.

Parsons, Alyce, and Diane Parente
Universal Style: Dress for Who You Are and What You Want
Ross, Calif., Parente and Parsons, 1991. 150p. ISBN 0962740500

Presents a system of assessment for determining an appropriate style or styles for any woman and her particular lifestyle. Covers seven basic styles: sporty, traditional, elegant, feminine, alluring, creative, and dramatic. Discussing the elements of each style and how to put them together. Written by two professional image consultants, this helpful guide is useful for students planning to go into the image consulting and retailing end of the fashion industry. Includes line drawings.

Rouse, Elizabeth
Understanding Fashion
London, BSP Professional Books, 1989. 256p. ISBN 0632018917

One of the few texts to explore clothing from a sociological perspective with the students of fashion design and marketing in mind. Examines the social and cultural significance of clothing and fashion, and assesses the influence that social changes have had on styles of dress along with the development of the fashion industry, providing specific historical examples from the Victorian era through the twentieth century. Includes illustrations and index.

Stamper, Anita A., Sue Humphries Sharp, and Linda B. Donnell
Evaluating Apparel Quality
2nd ed. New York, Fairchild Books, 1991. 336p. ISBN 0870057154

Provides an in-depth look at the construction components that affect the appearance, function, and cost of apparel items— components such as fabric, stitching, darts, seams, pockets, closures, decorative detail, and neckline, sleeve, waistline, and hem treatments. Black-and-white photographs and line drawings illustrate and clarify the concepts. Bibliography and index included.

Villarosa, Riccardo, and Giuliano Angeli
The Elegant Man: How to Construct the Ideal Wardrobe
New York, Random House, 1993. 192p. ISBN 0679421017

Focuses on men's suits and briefly addresses sportswear and formal attire. Provides the information necessary to identify and evaluate natural fiber fabrics as well as the essential elements of a quality suit—shape, style, and cut. Superbly detailed photographs of suit fabrics accompany the descriptions and photographs of fine suits. Also discusses maintenance and care of a wardrobe and touches on the vocabulary of textile manufacturing. Includes bibliography and index.

✦ Fashion Design Sketching and Illustration

Abling, Bina
Advanced Fashion Sketchbook
New York, Fairchild Fashion Group, 1991. 133p.
ISBN 0870056794

Picks up where *Fashion Sketchbook* (Fairchild Publications, 1988) leaves off, focusing on style and rendering techniques as they relate to fashion illustration and design detail. Discusses media and supplies, reviews the human figure, addresses layout, and devotes one chapter each to line, marker, and wash. The idea of personal style in illustration is demonstrated with inspired fashion drawings contributed by different artists in the field.

Abling, Bina
Fashion Sketchbook
New York, Fairchild Publications, 1988. 113p. ISBN 0870055623

Presents the beginner with the essentials needed to sketch and clothe the human figure. After proportion, form, structure and balance are established, each body part is treated individually. Gesture and motion, as well as elements of clothing construction, are also addressed individually and with some detail. Illustrated with clean, simple line-drawn examples. Abling is a graduate of Parsons and a teacher at Parsons and the Fashion Institute of Technology.

Baker, Georgia O'Daniel
A Handbook of Costume Drawing: A Guide to Drawing the Period Figure for Costume Design Students
Boston, Focal Press, 1992. 163p. ISBN 0240801121

Written for the student pursuing theatrical costume design, this handbook presents drawings to aid in creating images that accurately reflect a period and its mood. Details of the human figure, how clothes fit, and the materials and techniques needed for costume sketching are addressed. Sixty-four full-page examples of men's and women's dress are provided along with detailed sketches of hairstyles, footwear, undergarments, and accessories. Covers major periods from ancient Egypt through the 1960s. Bibliography and index included. Baker is a costume designer and professor.

Barnes, Colin
The Complete Guide to Fashion Illustration
Cincinnati, Ohio, North Light Books, 1988. 160p.
ISBN 0891342508

Demonstrating more through contemporary, dramatic fashion illustrations rather than through actual step-by-step instructions, the author presents how to render color, texture, and pattern. Barnes, Britain's leading fashion illustrator, discusses the materials and techniques to use to obtain the results that industry demands. A brief history of fashion illustration is included. Best suited to the more advanced student. Indexed.

Gersten, Rita
Fashion Art for the Fashion Industry
New York, Fairchild Publications, 1989. 117p. ISBN 087005676X

Combining limited written instruction with an array of black-and-white fashion drawings, Gersten presents the fashion industry from the point of view of the illustrator, and shows the role of fashion art from idea to end result. Includes discussion of research and marketing sketching, fashion forecasting, textile houses, fashion designing, workboards, spec drawings, presentations, and more. Gersten is a professor of fashion design at the Fashion Institute of Technology in New York City.

Ireland, Patrick John
Introduction to Fashion Design
Petaluma, Calif., Unicorn Books for Craftsmen, 1992. 95p.
ISBN 0962558621

Ireland is a free-lance fashion designer, illustrator, and lecturer. In this work he presents a step-by-step guide to the techniques of fashion drawing. Covers topics such as figure proportions, details of clothing, reproducing patterns and textures, and using different media for desired effect. Storyboards and tips for the effective presentation and display of work are also discussed. This is most appropriate for students with some previous experience. Numerous photographs and drawings, a suggested books list, and an index.

Packer, William
Fashion Drawings in Vogue: Carl Erickson
London, England, Michael Joseph, 1989. 128p. ISBN 0863501982

Presents the work of Carl Erickson, fashion artist and illustrator, as represented in *Vogue* magazine from the mid-1920s to the mid-1950s. Includes a brief professional and personal biography, followed by more than one hundred black-and-white or color illustrations in chronological order. Accompanied by text.

Packer, William
Fashion Drawings in Vogue: René Bouët-Williaumez
London, Michael Joseph, 1989. 128p. ISBN 0863501974

Intended as a companion volume to *Fashion Drawings in Vogue: Carl Erickson* (Michael Joseph, 1989), this volume presents the work of Erickson's rival fashion illustrator, Bouët-Williaumez. Includes a brief personal and professional biography, followed by black-and-white or color illustrations presented in loosely chronological order with accompanying explanatory text.

Parish, Justine Limpus
Drawing the Fashion Body
South Pasadena, Calif., Parish Enterprises, 1994. 117p.

Parish, a fashion illustrator, art director, and designer, created and served as director of the Fashion Department at the Academy of Art in San Francisco. With this text, she takes the novice fashion illustrator through entry-level skill exercises and provides sound advice for problems such as proportion, perspective, and the structure of the human body.

Robinson, Julian
The Fine Art of Fashion: An Illustrated History
New York, Bartley and Jensen, 1989. 208p. ISBN 1862563020

Emphasizes fashion as portrayed in fashion drawings in popular and exclusive French magazines from the eighteenth century to the present. Magnificent illustrations are presented by such fashion artists as Desrais, Watteau, Barbier, and Lepape. Focus is on fashion that captures the spirit of the fashions, the mood of the times, and demonstrates each artist's individual style. The author has extensive experience as a professional designer and design consultant with famous designers and fashion magazines.

Takamura, Zeshu
The Use of Markers in Fashion Illustrations
Tokyo, Graphicsha, 1991. 139p. ISBN 4766106458

Takamura uses a stepped approach to instruction, beginning with an explanation of figure proportion. Students are guided through the various techniques and processes of drawing with markers. Includes information on production sketches and textile rendering. Clear, close-up photographs complement the text.

Thames, Bill
Drawing Fashion
New York, McGraw-Hill, 1985. 182p. ISBN 0070637229

Thames breaks down the procedure of fashion drawing into component steps that can be easily understood by the novice illustrator. Includes instruction on marker techniques; drawing for men, women, and children; and drawing for specific fabric and garment types.

✦ History of Fashion and Costume

Bailey, Adrian
The Passion for Fashion
Limpsfield, Surrey, Dragon's World, 1988. 190p.
ISBN 1850280630

Bailey, a journalist, photographer, illustrator, and fashion artist, traces the influence of fashion on society, and society on fashions from 1670 to 1955 in France, England, and America. He focuses on women's fashion with brief mention of men's clothing. Maintains a nice balance between text, which is readable and entertaining, and illustrations that are either fashion plates of the day or staged photographs of costumes against period backgrounds. Includes a selected bibliography and index.

Beard, Tyler
100 Years of Western Wear
Salt Lake City, Utah, Gibbs-Smith Publisher, 1993. 159p.
ISBN 0879055812

An overview of the history of western wear, covering 1890 to 1990, focusing predominantly on the "golden era," 1930 to 1970. From simple bandannas to elaborate rhinestone suits, from the wild prairies to Nashville and Hollywood, this book covers it all with stunning color photographs and entertaining, informative text. Included is information on Nudie Cohn, fashion designer to western stars, and illustrations of western influence in collections such as those of Ralph Lauren. Includes a source guide to retail and custom western clothing.

Carter, Alison J.
Underwear: The Fashion History
New York, Drama Book Publishers, 1992. 161p.
ISBN 0896761207

Discusses women's underwear generally from 1490, and in some depth from 1840 to the present. Explores functionality and fashionability, the social context, the history of the industry, technical innovations, and the key role advertising has played. Provides an impressive array of illustrations and photographs, movie stills, reproduced catalog pages, advertisements, and famous paintings. Includes a glossary of terms and fabric names, lists of underwear and corsetry companies, trade names, and an index.

Chenoune, Farid
A History of Men's Fashion
Paris, France, Flammarion, 1993. 336p. ISBN 2080135368

Translated from the French, this work covers 1760 to the present, concentrating on the fashions emerging from London, Paris, Rome, New York, and more recently, Tokyo and Milan. Though primarily addressing business attire, there is a brief look at topics such as avant-garde artists, underworld fashion, ethnic minorities, and gay men. If the text at times seems difficult, the treasury of illustrations more than make up for it. Bibliography and index.

Coleman, Elizabeth A.
The Opulent Era: Fashions of Worth, Doucet, and Pingat
New York, Thames and Hudson, 1989. 208p. ISBN 0500014760

This catalog accompanied an exhibition held at the Brooklyn Museum, December 1989 - February 1990. Presents a wide range of primary resources on these three founders of the couture system. Includes photographs of existing garments on museum mannequins, supported by black-and-white illustrations. Addresses the philosophy and sociopolitical world surrounding these designers and presents a comprehensive view of high fashion during the late 1800s. Includes bibliographical references.

De La Haye, Amy
Fashion Source Book
Secaucus, N.J., Wellfleet Press, 1988. 191p. ISBN 1555213391

From haute couture to street fashion, this text charts the history and contributions of the leading couturiers, as well as the styles available to all the social classes. A one-page summary of the trends of each decade through the 1980s is followed by illustrative drawings and photographs. Indexed.

De Marly, Diana
Christian Dior
New York, Holmes and Meier, 1990. 96p. ISBN 0841912785

A concise assessment of the work of Christian Dior. Traces how he came to couture, the "new look" for which he became famous, his influence on fashion in the 1940s and 1950s, his ideals, those who influenced him (Worth and Molyneux), and the activities of the Maison Dior after his death in 1957. The numerous black-and-white photographs capture his style. Appendices include his film and stage designs and list museum collections of his creations found worldwide. A chronology and an index are provided.

Demornex, Jacqueline
Madeleine Vionnet
New York, Rizzoli, 1991. 305p. ISBN 0847813878

Sumptuously showcases in hundreds of full-page photographs the work of Vionnet, a designer best known for her contribution of the bias-cut dress. The brief text presents her early life, her arrival in and influence on the fashion world in the 1920s and 1930s, her preference for supple, free-moving clothes, and her belief that women should discard constricting and deforming garments. Her philosophy and style were in keeping with the fashion climate of the time and are well documented in the photographs presented here.

Engelmeier, Regine, and Peter W. Engelmeier, eds.
Fashion in Film
New York, Neues Pub. Co., 1990. 245p. ISBN 379131100X

Presents a collection of 372 magnificent black-and-white movie stills and stock photographs, many never before published, documenting costume designs in the movies of the past seven decades. Essays by specialists provide an insider's viewpoint. Includes a full list of all feature films to have won an Oscar for costume design, along with biographical information on the designer, and a representative still from the movie. Indexed by actors, costume designers, directors, and film titles.

Finlayson, Iain
Denim: An American Legend
New York, Simon and Schuster, 1990. 126p. ISBN 0671723685

Traces the history of denim jeans from the gold mines of California, through the rebellious youth of the postwar era, to the incredible variety of present-day fashions. Provides a discussion of marketing hype and selling denim through sex. Alternates pages of text with full-page photographic portraits and design ads. Although this work has a narrow focus, the topic is one that will be of interest to students.

Gold, Annalee
Ninety Years of Fashion
New York, Fairchild Fashion Group, 1991. 137p.
ISBN 0870056808

This simplified version of fashion history focuses on women's clothing since 1900, accompanied by succinct text and basic line-drawn figures. The author takes various garments such as dresses, coats, suits, sportswear, beachwear, and tennis fashions, and traces the history of each separately. Includes accessories. Indexed.

Hall, Lee
Common Threads: A Parade of American Clothing
Boston, Little, Brown, 1992. 324p. ISBN 0821219006

From 1492 through the twentieth century, this work follows the clothing styles worn by the "common" resident of the American colonies. Presented in three parts—1492-1800, the nineteenth century, and the twentieth century—the book illustrates the clothing trends taken from a wide range of fresh primary sources. Actual garments and photographs comprise the major portion of this very real look into American apparel. Includes bibliography and index.

Hunt, Marsha
The Way We Wore: Styles of the 1930s and '40s and Our World Since Then
Fallbrook, Calif., Fallbrook Publishing, 1993. 429p.
ISBN 1882747003

Presents a treasury of over 500 fashion layouts, production stills, and publicity shots from the private collection of Marsha Hunt, fashion model and film star. Provides a comprehensive look at what American women of "moderate to fortunate means" actually wore during the thirties and forties. Designers represented include Howard Shoup, Dolly Tree, Irene, Edith Head, Orry-Kelly, Edward Stevenson, and Travis Banton. This is a useful reference for students of costume or fashion design. Includes apparel index.

Kennedy, Shirley
Pucci: A Renaissance in Fashion
New York, Abbeville Press, 1991. 214p. ISBN 1558590579

Kennedy, adjunct professor in the fashion department at New York University, presents the Pucci phenomenon from the 1950s to the present, concentrating on the first twenty years. Information is gleaned from interviews with fashion experts and Pucci himself.

Full-color pictures taken from fashion magazines and individual collections of his wildly colorful and patterned designs highlight the text. Includes bibliography and index.

Jouve, Marie-André
Balenciaga
Text by Jacqueline Demornex. New York, Rizzoli, 1989. 376p. ISBN 0847810798

With 600 artistic and high-quality photographs and supporting text, Balenciaga archivist Jouve and journalist Demornex present the life, career, and creations of one of the leading couturiers of the twentieth century, from the opening of his first couture house in 1919 at age twenty-four, to his retirement in 1971. Includes fashion sketches by Balenciaga, a chronology, and lists of theater productions and films for which he created costumes.

Leese, Elizabeth
Costume Design in the Movies: An Illustrated Guide to the Work of 157 Great Designers
New York, Dover Publications, 1991. 171p. ISBN 048626548X

Generously illustrated with predominantly black-and-white movie stills, this work provides biographical and career data for costume designers who worked on films between 1909 and 1987. Includes an index by film title (over 6,400 titles listed), and lists of Oscar and British Academy of Film and Television Arts (BAFTA) nominations for best costume. Of value if only for the photographs.

Lobenthal, Joel
Radical Rags: Fashions of the Sixties
New York, Abbeville Press, 1990. 256p. ISBN 0896599302

Researched for ten years and with many previously unpublished photographs capturing the styles and spirit of the decade, this engaging text evokes the cultural context within which a radical fashion industry was created. Features the pivotal designers of the time, as well as the boutiques, models, celebrities, social movements, and subcultures. Focuses on London, Paris, New York, and the West Coast. Includes a good index and a bibliography.

Mackrell, Alice
Paul Poiret
New York, Holmes and Meier, 1990. 96p. ISBN 0841912793

This is a concise presentation of this innovative and influential French designer, the first couturier to successfully relate fashion to other arts. Poiret created his own line of fragrances, cosmetics, decorative arts, and painted fabrics, and imbued fashion illustration with new life. The mostly black-and-white photographs or fashion illustrations by Georges Lepape and Paul Iribe capture his directoire, Orientalism, and post-World War I fashions. A glossary, bibliography, chronology, list of museum collections, and brief index are included.

Martin, Richard, and Harold Koda
Flair: Fashion Collected by Tina Chow
New York, Rizzoli, 1992. 160p. ISBN 0847814955

Published on the occasion of an exhibition at the Fashion Institute of Technology in New York City, this work presents photographs of clothing produced by nineteen of the great twentieth century designers, all collected by Tina Chow, model and jewelry designer. Alaia, Balenciaga, Lagerfeld, Miyake, and Vionnet are represented, among others. Captions and limited text discuss the innovations and designs for which each designer is best known and which demonstrate their influence on the field. Includes bibliography and index.

Martin, Richard, and Harold Koda
Giorgio Armani: Images of Man
New York, Rizzoli, 1990. 224p. ISBN 0847812987

After a nine-page introduction and discussion of Armani's style, the rest of this work presents his Spring/Summer and Fall/Winter collections from 1975 to 1990. Artistically rendered full-page photographs effectively present his tailored yet relaxed style, the unconstructed jackets, muted colors, and drapeable fabrics for which he is famous and which have dramatically altered men's clothing in this century.

Martin, Richard, and Harold Koda
Splash! A History of Swimwear
New York, Rizzoli, 1990. 140p. ISBN 0847811867

This oversized volume presents the history of swimwear "as a study in style" through spectacular color and black-and-white photographs. Features designers such as Patou, Schiaparelli, Gernreich, and McCardell. Changing mores are followed from the modest cover-ups of the early century to the revealing postwar bikini and thong. Inspirational for fashion design students with a special interest in swimwear. The text can be difficult, but is very informative. Includes glossary and chronology.

McDowell, Colin
Hats: Status, Style, and Glamour
New York, Rizzoli, 1992. 224p. ISBN 0847815722

This is a delightful historical look at hats as fashion statements or symbols of power and honor. Covers the hat trade; hats on stage, screen, and in sports; hat etiquette; hats signifying social status; occupational headgear; the heyday of the hat; and the fall and eventual rise again of the hat. Presents information on the great early and contemporary milliners, and discusses the couturier as milliner: Chanel, Balenciaga, Schiaparelli, and Dior, among others. Fantastically illustrated. Chronology and index included.

McDowell, Colin
Shoes: Fashion and Fantasy
New York, Rizzoli, 1989. 224p. ISBN 0847811123

Much like his book on hats, this work is an extensive and enthralling look at shoes from ancient times through the twentieth century with the emergence of the great shoe designers such as Perugia, Ferragamo, Vivier, and Clergerie, and the birth of the shoe as a fashion accessory. Contents includes the shoe and the shoemaker; the cult of the shoe; form and function; show-business and recreational footwear; the designer; and the development of style. Includes an impressive array of illustrations. Indexed.

Milbank, Caroline Rennolds
New York Fashion: The Evolution of American Style
New York, Abrams, 1989. 303p. ISBN 0810913887

After a brief look at New York fashion in the nineteenth century, this work focuses on the twentieth century and devotes a chapter to each decade through the 1980s. Milbank looks at the fashion industry in terms of the social conditions within which it existed, then briefly sums up the contributions of the most prominent designers in each decade. Covers high fashion to sportswear. Sumptuously illustrated with photographs and fashion drawings. Includes bibliography and index.

Moffitt, Peggy
The Rudi Gernreich Book
New York, Rizzoli, 1991. 224p. ISBN 084781422X

Documents the career of one of America's most original and controversial designers. Comprised largely of full-page, previously unpublished photographs of Peggy Moffitt, Gernreich's favorite model, the book focuses on his designs of the 1960s and 1970s. With a biographical essay by Mary Lou Luther and photographs by William Claxton, this book provides an accurate view of his work seldom seen in the fashion magazines.

Mulvagh, Jane
Vogue History of Twentieth Century Fashion
New York, Viking, 1988. 410p. ISBN 0670801720

Illustrated generously with black-and-white photographs taken from the pages of *Vogue* magazine, this work presents trendsetting developments and less well-known creations year-by-year from 1909 to 1986. The accompanying text is insightful and informative and provides a balanced view of national and international trends. Useful as both a quick reference or for more in-depth reading.

Murray, Maggie Pexton
Changing Styles in Fashion: Who, What, Why
New York, Fairchild Publications, 1989. 252p. ISBN 0870055852

Focus of this work is on the apparel world: what motivates it, how it functions, how fashion evolves, and why. Traces clothing from ancient times to the present, demonstrating through fashion photographs how trends from other eras have been adapted to our own. Includes brief essays on French and American designers and discussions of the French and American fashion industries in the twentieth century. Includes black-and-white photographs, a bibliography, and an index.

Olian, JoAnne, ed.
Authentic French Fashion of the Twenties: 413 Costume Designs from L'Art et la Mode
New York, Dover Publications, 1990. 138p. ISBN 0486261875

Consists of 138 pages of attractive line drawings from the pages of one of the most famous French fashion journals, featuring the great French couturiers. An impressive collection from one of haute couture's most influential decades, this work covers outfits for sporting pursuits along with high fashion gowns for soirées. Includes a historical introduction by Olian, curator of the costume collection at the museum of the City of New York. Illustrations are copyright-free.

Peacock, John
Twentieth-Century Fashion: The Complete Sourcebook
London, Thames and Hudson, 1993. 240p. ISBN 0500015643

Provides a survey of fashions of the twentieth century in 1,100 detailed, original, full-color drawings by the author. The couture wear, underwear, leisure wear, day wear, evening wear, bridal wear, and accessories of each decade are presented along with descriptions of the designs. An annotated chart traces important developments and notes the principal changes in such areas as shape, length, color, fabric, and accessories for each five-year period. Includes extremely brief biographies of the couturiers and designers represented.

Pelle, Marie Paule
Valentino: Thirty Years of Magic
Text by Patrick Mauries. New York, Abbeville Press, 1990. 330p. ISBN 1558592377

This lavish work consists predominantly of full-color, full-page photographs of Valentino's creations from 1960 to 1990. The 250 selected atelier sketches representing Spring and Fall collections help to make this a magnificent presentation of this designer's career. Includes brief biographical text and lengthy quotations by Valentino.

Perl, Lila
From Top Hats to Baseball Caps, from Bustles to Blue Jeans: Why We Dress the Way We Do
New York, Clarion Books, 1990. 118p. ISBN 0899198724

This brief history of fashion written for younger or introductory-level readers focuses mainly on the development of trousers and changes in skirts, shoes, and hats. Draws examples from the Egyptians, Greeks, Romans, as well as the nineteenth and twentieth centuries. Discusses fashion as a reflection of history and as an expression of individual lifestyles. Provides black-and-white line drawings and reproductions of catalog pages. Includes an index and a brief bibliography.

Sudjic, Deyan
Rei Kawakubo and Commes des Garçons
New York, Rizzoli, 1990. 160p. ISBN 0847811964

A look at the twenty years since Kawakubo created Commes des Garçons, and the deep impression she has made on the fashion world since her 1981 Paris debut. Text and photographs go behind the scenes at Paris collections, Japanese textile mills, and cutting rooms. The reader also gets a look at her creations and the specially designed minimalist retail areas in Japan and worldwide fashion capitals where they are sold. Chronology and bibliography are included.

Yarwood, Doreen
Fashion in the Western World, 1500-1990
New York, Drama Book Publishers, 1992. 176p.
ISBN 0896761185

An author of several books on fashion history, Yarwood presents an authoritative though brief survey of fashion trends from the Renaissance through the twentieth century, with an emphasis on the last 200 years. Traces men's, women's, and children's dress and accessories in their social, cultural, and technological context. This work is generously illustrated with hundreds of black-and-white line drawings and color painting reproductions representative of the different eras. Glossary, bibliography, and index included.

✦ Merchandising

Bell, Judith A.
Silent Selling: The Complete Guide to Fashion Merchandise Presentation
Cincinnati, Ohio, ST Publications, 1988. 136p. ISBN 0911380779

Written by a retailer of seventeen years, this guide presents topics such as creating displays that promote sales, outfit coordination, store layout, fixtures, lighting, accessory presentation, mannequins, and props. Includes line drawn illustrations. Suggests exercises involving the observation of other store displays. Indexed.

Diamond, Ellen
Fashion Retailing
New York, Delmar Publishers, 1993. 431p. ISBN 0827356218

Explores traditional principles and practices as well as the latest innovative concepts, offering practical and theoretical information for fashion retailing. Covers the retailing environment, management and control functions, merchandising, and communication with clientele. Successful present-day companies are spotlighted as examples. An appendix provides a brief description of positions in retailing and tips for job hunting. Includes numerous illustrations, a glossary, and an index.

Jernigan, Marian H., and Cynthia R. Easterling
Fashion Merchandising and Marketing
New York, Macmillan, 1990. 581p. ISBN 0023313501

Designed to provide a basic working knowledge of the terminology, activities, and interrelationships of the fashion business. Industry profiles illustrate key points. Sparsely illustrated with black-and-white photographs and line drawings. Includes end-of-chapter summaries of key points, key words and concepts, and discussion questions and topics. Glossary, bibliography, and index provided.

Stone, Elaine
Fashion Merchandising: An Introduction
5th ed. New York, Gregg Division, McGraw-Hill, 1990. 454p. ISBN 0070617449

Covers all the general areas of the apparel industry; terminology; roles of designers, manufacturers, retailers; merchandising and marketing activities and trends; textiles; domestic and foreign markets; women's, men's, and children's apparel and accessories; career possibilities, and more. Provides brief discussions of fashion careers. Illustrated with black-and-white photographs. Includes review questions, suggested projects, a glossary, and an index.

✦ Textiles

Colchester, Chloë
The New Textiles: Trends and Traditions
New York, Rizzoli, 1991. 192p. ISBN 0847814181

Focuses on the work of individual, independent textile manufacturers worldwide, and their role in textile design, hence fashion design, in the 1980s. Traces the role that the textile industry played in the shift from simple, printed fabrics to more complex, woven fabrics and prints in both fashions and furnishings. Not a how-to book, but valuable for its ability to inspire creativity. Celebrates textiles as both an artistic and service enterprise. Includes brief biographies, hundreds of color photographs, a bibliography, a glossary, and an index.

Jerstorp, Karin, and Eva Köhlmark
The Textile Design Book: Understanding and Creating Patterns Using Texture, Shape, and Color
Asheville, N.C., Lark Books, 1988. 160p. ISBN 0713631309

Written by two textile artists and instructors, this work presents inspiring techniques and exercises for designing patterns in textiles, often using patterns in nature as a starting point. Featured is an in-depth look at color theory, providing the necessary knowledge for effective use of color in design. Illustrations are both pleasing and instructional. Indexed.

Joseph, Marjory L.
Joseph's Introductory Textile Science
6th ed. Edited by Peyton B. Hudson, Anne Calvert Clapp, and Darlene Knees. Fort Worth, Tex., Harcourt, Brace, Jovanovich College Publishers, 1993. 417p. ISBN 0030507235

Designed to provide a functional knowledge of the field or a solid foundation for advanced study, this text emphasizes the principles of textile science and the processes of natural and synthetic textile manufacturing. The five major sections covering fibers, yarns, fabrics, colors and finishes, and fabric end use, are divided into thirty chapters with study questions, activities, and suggested readings in each. Photographs and line drawings are found throughout the text. Includes a good glossary and an index.

Parker, Julie
All about Cotton: A Fabric Dictionary and Swatchbook
Seattle, Wash., Rain City Publishers, 1991. 120p.
ISBN 0963761218

With 2" x 4" swatches of forty-two different cotton fabrics and two pages of text devoted to each, this is a "hands-on" dictionary. Text for each fabric includes a description, manufacturing technique, related items or terms, similar fabrics, main characteristics, average price per yard, suggested fit, styles, methods of care, and sewing difficulty rating. Some history and other interesting tidbits are also included. Mail order sources, a glossary of terms, bibliography, and index are provided.

Parker, Julie
All about Silk: A Fabric Dictionary and Swatchbook
Rev. ed. Seattle, Wash., Rain City Publishers, 1992. 92p.
ISBN 096376120X

Presents and describes thirty-two different silk fabrics complete with swatches. Information provided for each includes recommended garments, styles, and fit; sewing characteristics (what to expect when working with the fabric), and difficulty rating; fabric behavior after construction and how to care for fabric; what to expect to pay per yard, and other pertinent information. Also lists mail-order companies, fabric clubs, and swatching services. Includes glossary, bibliography, and index.

Tuckman, Diane, and Jan Janas
The Complete Book of Silk Painting
Cincinnati, Ohio, North Light Books, 1992. 122p.
ISBN 0891344225

Written by two teachers who are experts on the topic, this work presents the technical information needed to master the basics, and includes a large "gallery" of works by talented silk painters to inspire creativity. The easy-to-follow text explains what materials and supplies are needed, basic procedures, and specific techniques for unusual effects. Colorfully illustrated with examples of painted silk garments and fashion accessories. Indexed.

✦ Vocational Guidance

Daria, Irene
The Fashion Cycle: A Behind-the-Scenes Look at a Year with Bill Blass, Liz Claiborne, Donna Karan, Arnold Scaasi, and Adrienne Vittadini
New York, Simon and Schuster, 1990. 240p. ISBN 0671667297

The title says it all: engaging, informative, intimate reading for students interested in the ins and outs of the fashion industry. Written in diary format, this work follows five designers as they develop their creations from idea to final product and ad campaign. Daria, reporter for *WWD* and writer for *Harper's Bazaar*, effectively captures the frenetic pace of a designer's life.

Dolber, Roslyn
Opportunities in Fashion Careers
Lincolnwood, Ill., VGM Career Horizons, 1993. 149p.
ISBN 0844240230

Written by the director of placement at the Fashion Institute of Technology in New York City, this book provides a brief overview of five areas in the fashion industry: apparel design, apparel production, textiles, fashion merchandising, and manufacturing. Dolber surveys the career opportunities in each, including the educational and personality requirements for these careers, résumé writing and interviewing. Written in an upbeat, familiar, easy-to-read style.

Tate, Sharon Lee
Inside Fashion Design
3rd ed. New York, Harper and Row, 1989. 432p.
ISBN 0060466782

Illustrated throughout with black-and-white photographs and line drawings, and presented in a textbook format, this work offers a complete and detailed look at the designer's role in the fashion industry. Examines the creation and manufacturing of men's, women's, and children's fashions. Presents the vocabulary, talents, and skills basic to all fashion careers. Indexed.

Wolfe, Mary Gorgen
Fashion! A Study of Clothing Design and Selection, Textiles, the Apparel Industries, and Careers
South Holland, Ill., Goodheart-Willcox Co., 1989. 475p.
ISBN 0870067273

Presented in simple, direct language and with hundreds of color photographs, this textbook may serve as an introductory guide to students considering a career in the apparel industry. Addresses fashion terms, garment style, textile, pattern, and apparel production, use of computers in the industry, and elements of design. Dedicates six chapters to a discussion of careers in the apparel industry. An interesting, visually attractive presentation. Includes comprehensive glossary and index. Wolfe is an educator and apparel consultant.

Nonprint Materials

✦ Career Opportunities

Careers in Clothing
Bloomington, Ill., Meridian Education Corp., 1991. 1 videocassette (15 min.) VHS

A brief overview of the textile, apparel, and retail industries and the various career opportunities in each, with an emphasis on retail. Scenes from the workplace are combined with interviews with a clothing manufacturer, retail store owner, buyer, display artist, salesperson, and marketing director. Opportunities outside retail are touched upon, e.g., as fashion photographer, theatrical costume designer, costume shop manager, museum curator, and teacher. Discussion questions are included on accompanying brochure.

Fashion Merchandising: Concepts and Careers
Bloomington, Ill., Meridian Education Corp., 1994. 1 videocassette (20 min.) VHS

Presents a realistic view of this career possibility by pointing out the long hours, hard work, and extensive preparation behind the perceived excitement and glamour. Through interviews and narration, this work explains the principles behind merchandising from design concept to retail display, and looks at various career opportunities such as manager, visual display specialist, salesperson, quality control specialist, fashion buyer, and advertising, with examples drawn from actual department stores. Emphasis is on the need to combine experience with education. A good overview and discussion questions are contained on the accompanying pamphlet.

Introduction to Fashion Merchandising
Charleston, W. Va., Cambridge Career Products, 1991.
1 videocassette (57 min.) VHS and 1 workbook

Set in the heart of New York's garment district, this program presents the wholesaling and retailing aspects of fashion merchandising through narration and interviews with fashion designers, showroom sales representatives, wholesale buyers, and retail professionals. Gives an honest look at possible career choices, playing down the glamour while addressing the demands and challenges of the business and the personality traits needed to succeed. Focuses on designer clothing and a level of merchandising above that found in the department store.

✦ Textiles

Fibers: Manufactured and Natural
Bloomington, Ill., Meridian Education Corp., 1991. 1 videocassette (9 min.) VHS and 1 pamphlet

Briefly discusses natural and synthetic fibers, touching on the manufacturing processes, some terminology, characteristics of the various fibers, and the advantages and disadvantages of each. The main emphasis of this tape is on the role of these fibers in today's society. Discussion questions for before and after viewing are included on the accompanying pamphlet.

Textiles
Sunrise, Fla., D.E. Visuals, 1990. 1 videocassette (18 min.) VHS and 1 pamphlet

Presents the fundamental concepts of textiles and related terminology. Covers topics such as natural and manufactured fibers, their sources, generic and brand names, spun and filament yarns, knit and woven fabric structures, and the characteristics of each. Accompanying pamphlet includes discussion questions and a quiz.

✦ Clothing Construction

Power Sewing Volume 1: Fear of Sewing
New York, Multi Media Communications, 1993. 1 videocassette (82 min.) VHS

With a lively and spontaneous manner, Sandra Betzina presents in this first video some basic sewing survival skills. Includes what to look for when choosing a sewing machine, threading and operating the machine, hemming skirts and pants, tapering pants, sewing on a button, and making a scarf (from first cut to finished product.)

Power Sewing Volume 2: Pattern Sizing and Alteration
New York, Practicality Press, 1993. 1 videocassette (63 min.) VHS ISBN 1880630044

Presents simple yet effective techniques for achieving a custom fit from commercial patterns. Provides information on transferring measurements to the pattern, making adjustments, and detailed information on altering a pattern. Brief tips on cutting and marking are provided. This and other *Power Sewing* videos are designed to provide basic information for the beginner, and other helpful information for the more advanced student.

Power Sewing Volume 3: Fitting Solutions
New York, Practicality Press, 1993. 1 videocassette (63 min.) VHS ISBN 1880630052

Enables viewers to overcome two obstacles: taking measurements correctly and recognizing figure/fitting problems. Live models illustrate common fitting problems; demonstrations lead into a demonstration of pattern alteration and a final modeling of the now well-fitting garments. Also offers style tips to minimize figure problems that do not require pattern alteration.

Power Sewing Volume 4: Foolproof Pants Fitting
New York, Practicality Press, 1993. 1 videocassette (75 min.) VHS ISBN 1880630060

Shares professional tips for sewing pants that fit perfectly and flatter even an imperfect figure. Tips include information on transferring measurements to a commercial pattern and making the necessary adjustments to correct fitting problems. Also provides information on inserting a pant stay for tummy control, eliminating a baggy seat, constructing an expandable designer waistband, eliminating pockets and pleats that pop open, and setting snaps the easy way.

Power Sewing Volume 5: Construction Difficulties
New York, Practicality Press, 1993. 1 videocassette (120 min.) VHS ISBN 1880630079

Offers professional tips to help minimize construction problems. Areas covered are pattern and fabric compatibility, fly front zippers, reverse curved seams, square corner seams, yoke seams, mitering, needle selection, sleeve ease, shoulder pads, sleeve plackets, walking ease, perfect lapels, lining jackets, and changing grainline to flatter a figure. Both beginning and experienced sewers will benefit.

Power Sewing Volume 6: Easy Linings
New York, Practicality Press, 1993. 1 videocassette (75 min.) VHS ISBN 1880630087

Demonstrates techniques used in the industry to dramatically cut the amount of time spent on hand sewing. Jackets and vests are covered in depth; pants are covered briefly. Correct fabric lining choices are discussed, and finishing touches for high-quality appearance are demonstrated. This video is most suitable for students who have already covered other lining methods in the classroom.

Power Sewing Volume 7: Hassle-Free Designer Jackets
New York, Practicality Press, 1993. 1 videocassette (62 min.) VHS ISBN 1880630095

Demonstrating on a variety of styles of jackets and coats, this program shares construction tips to help guarantee professional-looking results. Topics included are interfacings, finishing techniques for seams, welt pockets, and sewing and pressing difficult fabrics, such as gabardine, cashmere, and velvet.

Power Sewing Volume 12: Handwoven and Quilting
New York, Practicality Press, 1993. 1 videocassette (105 min.) VHS ISBN 1880630176

Offers many tips for working with these two fabric types which, by their very nature, can present difficulties. More professional-looking results can be expected when using these techniques involving interfacings, taped and finished seams, finishing edges without facings, button loops or faced button holes, Chanel-type finishes and trims, and more. All examples are demonstrated on handwoven fabrics.

Sewing ABCs
Charleston, W. Va., Cambridge Educational, 1988. 1 videocassette (120 min.) VHS and 1 guidebook

For the student with little or no sewing background, this video presents basic principles such as taking measurements, selecting a pattern, and using the information on the pattern envelope and instruction guide. Demonstrations are given for elastic waistbands, side pockets, hems, facings, darts, seam finishes, zippers, pleats, and fasteners. Provides helpful camera close-ups of sewing details. A reproducible student guidebook accompanies the video.

Shirtmaking Techniques
Newtown, Conn., Taunton Press, 1993. 1 videocassette (45 min.) VHS ISBN 1561580422

In this high-quality video, David Page Coffin presents the techniques necessary to produce professional results on shirt jackets, blouses, sport shirts, dress shirts, and work shirts. Such details as rolled hems, flat-felled seams, precisely sewn plackets, cuffs, and collars are explained in an easygoing manner, and further clarified with illustrative computer-generated graphics and precise camera work. Can be used separately or in conjunction with the book (Taunton Press, 1993).

✦ Merchandising

Display Lighting
Sunrise, Fla., D.E. Visuals, 1991. 1 videocassette (20 min.) VHS and 1 pamphlet

Examines lighting as an integral part of any visual merchandising display. Discusses the psychology behind lighting, the lighting fixtures and light sources available to the display technician, and the characteristics of those sources such as incandescent, fluorescent, and HID metal halide. Covers special lighting needs and situations, and the use of lighting to create ambiance and desired effects. Discussion questions and a quiz are included in accompanying pamphlet.

Fashion Display Skills
Sunrise, Fla., D.E. Visuals, 1991. 1 videocassette (19 min.) VHS

Discusses fashion display as a nonverbal form of salesmanship, and stresses the need to combine creativity with basic hands-on techniques. Demonstrates how to dismantle, dress, and reassemble a mannequin, and how to dress a shirt or coat form. Presents the possibility of using less expensive nontraditional body forms, particularly when wanting to create a special effect. Covers using floor and wall-mounted waterfall racks to their best effect for suggestion selling.

Pinning and Flying Display
Sunrise, Fla., D.E. Visuals, 1991. 1 videocassette (13 min.) VHS

Focuses on two specific techniques of store visual merchandising. With examples drawn mainly from specialty stores, it is shown how these techniques can be used to the best effect in small spaces. Demonstrates topics such as directing the line of sight, focusing attention, suggesting motion, positioning and arrangement of merchandise, and suggestion selling. Discussion questions and a quiz are included in accompanying pamphlet.

Visual Merchandising
Sunrise, Fla., D.E. Visuals, 1992. 1 videocassette (20 min.) VHS

With examples drawn from well-known department and specialty stores, this video emphasizes the need for effective visual merchandising in order to be competitive in retail. Demonstrates attention-getting devices such as repetition, contrast, motion, and themes. Discusses the psychology behind displays, and the concepts of eye movement and point of emphasis. Also illustrates arrangements such as pyramid, step, zig-zag, vertical, and horizontal, as well as the use of props, special lighting, and color. Accompanying pamphlet includes discussion questions and a quiz.

Wholesale in Fashion Merchandising
Charleston, W. Va., Cambridge Educational, 1994. 1 videocassette (30 min.) VHS

Through interviews with fabric designers, fashion designers, and showroom sales representatives, this program discusses the process of design from ideas to samples and then to the showroom. Included are considerations of price, making final decisions on the line, incorporating those decisions into production, and issues of how the product ultimately reaches the store. Focuses on the wholesaling of designer clothing at a level above that of department stores.

Additional Resources

The following companies publish excellent resources and videocassettes for the fashion student:

Fairchild Books and Visuals
7 West 34th Street
New York N.Y. 10001

Taunton Press
63 South Main Street
Newtown, Conn. 06470-5506

Interior Design

Sue Swanson
Lasell College
Newton, Massachusetts

Introduction

The National Council for Interior Design Certification defines the professional interior designer as "qualified by education, experience and examination to enhance the function and quality of interior spaces." To achieve this, the designer must be familiar with architecture, relevant safety standards, and the psychological and physical impact an environment has on its inhabitants.

Early in the twentieth century, interior decorators provided professional assistance in the selection of interior furnishings. However, in the last half of the century, the profession has widened its scope to address areas previously considered the province of architects, engineers, and contractors. Although interior designers do not need a complete technical understanding of structures and materials, they do need to understand how these elements work to enhance or detract from the aesthetic and functional success of an environment.

Increasingly, interior designers are problem solvers, working with clients to create or redesign spaces in order to improve quality of life, increase productivity, or protect the health, safety, and welfare of the inhabitants. Interior designers need good oral and written communication skills to identify problems and goals; they need visual presentation skills to communicate possible solutions to clients and to subcontractors; and they need business and computational skills to prepare accurate bids and specifications so that projects are carried out correctly, on time, and within budget.

Interior designers may work independently, with colleagues in an interior design firm, or in large organizations, such as architectural firms, retail businesses, or manufacturing firms. Some may specialize in residential interiors, others in various commercial, public, or institutional spaces. Still others design the interiors of public transportation vehicles such as planes, trains, and buses.

Many interior designers are graduates of two-year associate degree programs, although three-year certificates and baccalaureate degrees are available. Accreditation and certification are offered at all three levels from the Foundation for Interior Design Education Research (FIDER).

The National Council for Interior Design Qualification administers the National Council for Interior Design Certification examination to practitioners with six years of combined education and work experience. Seventeen states and the District of Columbia now require successful completion of this examination for professional registration within their jurisdictions. Information on professional registration can be obtained from the National Legislative Council for Interior Design.

This bibliography attempts to reflect the evolution of the profession from residential decorator to interior designer. Titles from architecture, engineering, and construction have been included only when they offer information specifically useful to the interior designer. Information on universal accessibility and the special needs of contract interiors (commercial, public, institutional) is included. The traditional fields of wall and floor coverings, window treatments, furniture, and accessories are also covered.

The journals listed are generally considered the core journals for the field of interior design; however, they are by no means the only ones a library needs. Additional titles in residential and contract design should be included in most collections, as well as trade publications and consumer-oriented decorating magazines. Interior design faculty and, if possible, local designers should be consulted for recommendations about specific titles.

An index worth considering, particularly for libraries that also support programs in architecture, landscape design, or construction, is *Search: The Periodical Index for Architecture, Interior Design, Housing and Construction* (ISSN 1043-0946). Published quarterly with an annual cumulation by Search Publishing Inc., *Search* indexes the thirty titles most likely to be used by practitioners. The "Picture File" provides access to illustrations; indexes list articles under specific projects, individual designers, and design teams, as well as subjects.

Some of the interior design oriented titles indexed by *Search* include *Abitare, Architectural Digest, Architectural Lighting, Architectural Record, Contract Design, Domus, Hospitality Design, House and Garden, Home, Interior Design, Interiors,* and *Metropolitan Home*. Copies of all the articles indexed are also available for sale.

Accreditation and Certification

Foundation for Interior Design Education Research
60 Monroe Center NW
Grand Rapids, Mich. 49503

National Council for Interior Design Qualification
50 Main Street, 5th Floor
White Plains, N.Y. 10606

Selected Associations

American Society of Interior Designers (ASID)
608 Massachusetts Ave. NE
Washington, D.C. 20002

Foundation for Interior Design Education Research
60 Monroe Center NW
Grand Rapids, Mich. 49503

Institute of Business Designers (IBD)
341 Merchandise Mart
Chicago, Ill. 60654

Institute of Store Planners
25 N Broadway
Tarrytown, N.Y. 10591

Interior Design Educators Council
14252 Culver Drive, Suite A-311
Irvine, Calif. 92714

International Society of Interior Designers
1933 S. Broadway, Suite 138
Los Angeles, Calif. 90007

National Council for Interior Design Qualification
50 Main Street, 5th Floor
White Plains, N.Y. 10606

Selected Journals

Architectural Digest.
Architectural Digest. Monthly. ISSN 0003-8520

Interior Design.
Cahners Pub. Co. Monthly. ISSN 0020-5508

Interiors.
Billboard Publications. Monthly. ISSN 0164-8470

Interior design and decoration information is also found in popular periodical collections. Some of the most useful and popular titles include:

Colonial Homes.
Hearst Corp. Bimonthly. ISSN 0195-1416

Home.
Hachette Magazines. Ten issues a year. ISSN 0278-2839

House Beautiful.
Hearst Corp. Monthly. ISSN 0018-6422

Metropolitan Home.
Hachette Magazines. Monthly. ISSN 0273-2858

Print Materials

Abercrombie, Stanley
A Philosophy of Interior Design
New York, Harper and Row, 1990. 180p. ISBN 006430194X

    An eminent designer invites readers to begin looking at interior design philosophically, not merely as problem-solving challenges, but with a more abstract understanding of the role interiors and their designs play in human society. While many students may not be interested in such a theoretical approach, Abercrombie provides an important conceptual framework for practical and applied course work. An extensive cross-disciplinary bibliography is included.

American Society of Architectural Perspectivists
Architecture in Perspective: A Five Year Retrospective of Award- Winning Illustration
New York, Van Nostrand Reinhold, 1992. 208p.
ISBN 0442007000

    Showcases the 1986-1990 award winners of juried competitions in architectural illustration and rendering. While none of the illustrations are of interiors, interior design students will find inspiration in these examples of the best current work in the field.

Americans with Disabilities Act Handbook
Indianapolis, Ind., JIST Works, 1993. 1 v. (various pagings).
ISBN 1563700808

    This is a paperbound reprint of the Equal Employment Opportunity Commission (EEOC) three-ring binder available from the Government Printing Office (GPO). Includes the full text of the 1990 Americans with Disabilities Act and the federal "Accessibility Guidelines."

Arends, Mark W.
Interior Presentation Sketching for Architects and Designers
New York, Van Nostrand Reinhold, 1990. 121p.
ISBN 0442206437

    This paperback is an excellent guide to the use of sketching and rendering in the development and presentation of interiors. It includes suggestions for both monochromatic and color sketching, with a long appendix of materials needed for color rendering. Includes an index and a short annotated bibliography.

Aves, Melanie, and John C. Aves
Interior Designers' Showcase of Color
Washington, D.C., American Institute of Architects Press, 1994. 165p. ISBN 1558351124

    The use of color in contemporary interiors is illustrated by over 250 color photographs and more than 90 color palettes. Professional interior designers give tips on how to use colors to create moods and special effects. Includes bibliography and an index of designers.

Aves, Pirrie B., and others
Best: From the Interior Design Magazine Hall of Fame
Grand Rapids, Mich., Vitae Publishing, 1992. 238p.
ISBN 0823062481

    Both an idea book for students and a career guide, this glossy volume has brief biographies and examples of the work of sixty-one of the Hall of Fame inductees. The Hall of Fame was started as a fund-raiser for the Foundation for Interior Design Education and Research, the accrediting body for interior design programs. It is an international organization, with about one-third of the members from the U.S.

Ballast, David Kent
Interior Construction and Detailing for Designers and Architects
Belmont, Calif., Professional Publications, 1994. 397p.
ISBN 0912045671

    This profusely illustrated, comprehensive handbook explains how and why components are put together and provides detailed information about materials, finishes, systems, building codes, and design standards. Features unique to barrier-free design are introduced in the final chapter. Includes an extensive bibliography and indexes.

Ballast, David Kent
Interior Design Reference Manual
Belmont, Calif., Professional Publications, 1992. 326p.
ISBN 0912045418

    Ballast, a licensed architect, certified interior designer, educator, and author provides a manual for the National Council for Interior Design Certification exam. The introduction discusses the examination itself, including eligibility, content, format, and tips. All ten subject areas included in the exam are reviewed, providing a useful resource for students and practitioners alike. Includes references and index.

Barr, Vilma, and Charles E. Broudy
Designing To Sell: A Complete Guide To Retail Store Planning and Design
2nd ed. New York, McGraw-Hill, 1990. 202p. ISBN 0070038880

    The authors begin with information on how to develop a retail design program, including questions to ask the merchant, a checklist of considerations, and presentation strategies. The program outline is extended in subsequent chapters on basic layout, exterior and interior design, colors, materials and finishes, lighting, signs and graphics, and displaying merchandise. Heating, ventilating, and air-conditioning systems are also considered. Case studies in innovative retail design illustrate most chapters. Appendix includes "Retail Graphic Standards." Also includes a bibliography, a glossary, and an index.

Bayer, Patricia
Art Deco Architecture: Design, Decoration, and Detail from the Twenties and Thirties
New York, Abrams, 1992. 224p. ISBN 0810919230

    Reviews exterior representations of art deco, noting the historical context and major artistic features of residential and public buildings, originals and revivals. Includes bibliography and indexes.

Bayer, Patricia
Art Deco Interiors: Decoration and Design Classics of the Twenties and Thirties
Boston, Little, Brown, 1990. 224p. IBN 0821218131

    An historical survey of art deco and modern interiors featuring showcase exhibitions, private homes, and public places in Europe and the U.S. Concludes with a look at the art deco revival of the past twenty years. Bibliography and index included.

Birren, Faber
Light, Color and Environment: Presenting a Wealth of Data on the Biological and Psychological Effects of Color, with Detailed Recommendations for Practical Color Use, Special Attention to Computer Facilities, and a Historic Review of Period Styles
2nd rev. ed. West Chester, Pa., Schiffer Publishing, 1988. 136p. ISBN 0887401317

This book offers students a clearly organized introduction to light and color theory without being too technical. Most of the illustrations are black and white. Bibliography reflects the first edition.

Brandt, Peter B.
Office Design
New York, Whitney Library of Design, 1992. 176p.
ISBN 0823033430

This manual covers the complete process of office design from initial marketing to final move-in. The chapter on furnishings is particularly useful to students in contract design. Includes bibliography and index.

Branson, Gary D.
The Complete Guide to Barrier-Free Housing: Convenient Living for the Elderly and the Physically Handicapped
White Hall, Va., Betterway Publications, 1991. 176p.
ISBN 1558701885

Although this was written before the American with Disabilities Act (ADA) Guidelines were developed, it has valuable information on designing interior spaces to enhance access and reduce maintenance. Unlike the Cynthia Leibrock volume *Beautiful Barrier-Free* (1993), this book does not include furnishings. The index of manufacturers is easy to access and the extensive list of sources (agencies, publications, and product catalogs) is very helpful.

Bullivant, Lucy
International Interior Design
New York, Abbeville Press, 1991. 255p. ISBN 1558592350

Seventy innovative and exciting public interiors from all over the world are illustrated with color photographs and plans. Includes brief biographies of the designers. Indexes.

Burchell, S. C.
A History of Furniture: Celebrating Baker Furniture—100 Years of Fine Reproductions
New York, Abrams, 1991. 176p. ISBN 0810931079

Half the book is a survey of furniture from antiquity to modern times, illustrated by museum pieces, artwork, and Baker reproductions. The chapters on the Baker Furniture Company present the manufacturer's point of view. Includes bibliography and index.

Burden, Ernest E.
Design Presentation: Techniques for Marketing and Project Proposals
2nd ed. New York, McGraw-Hill, 1992. 248p. ISBN 0070089388

Written by a well-known author on architectural drawing and presentation, this volume includes detailed information on preparing and presenting visual aids for design projects. A technical mastery of basic drawing concepts is assumed. This second edition covers computers, video cameras, and desktop publishing techniques. Includes case studies, competition winners, and sample presentations.

Burden, Ernest E.
Entourage: A Tracing File for Architecture and Interior Design Drawing
2nd ed. New York, McGraw-Hill, 1991. 280p. ISBN 0070089337

This useful book includes human figures, trees, and other items in different sizes, poses, and styles of presentation. Warning: Pages are perforated for incorporation into a practitioner's binder; beware that the library copy doesn't likewise disappear!

Burstein, David, and Frank Stasiowski
Project Management for the Design Professional: A Handbook for Architects, Engineers, and Interior Designers
Rev. ed. New York, Whitney Library of Design, 1991. 174p.
ISBN 0823044130

Written to keep projects on time and within budget, this step-by-step manual is probably more useful to practitioners than students, but it has a wealth of information for students with a bent for business.

Bush-Brown, Albert, and Dianne Davis
Hospitable Design for Healthcare and Senior Communities
New York, Van Nostrand Reinhold, 1992. 263p.
ISBN 0442239599

Over sixty contributors have each written and illustrated essays addressing the challenges in architectural and interior design of health care and senior communities. Both the physical and psychological needs of the residents are discussed, as are the many codes and regulations governing such facilities. Includes glossary, bibliography, and index.

Calloway, Stephen, and Elizabeth C. Cromley, eds.
Elements of Style: A Practical Encyclopedia of Interior Architectural Details from 1485 to the Present
New York, Simon and Schuster, 1991. 544p. ISBN 0671739816

This wonderful reference is a chronological arrangement by design period from the Tudors to "beyond modern." Each period is introduced and then subdivided into architectural categories such as doors, walls, fireplaces, staircases, built-in furniture, kitchen stoves, and lighting. Color-coded tabs for each subheading facilitate following the development of a single feature throughout history. Both British and U.S. terms are given, with British usage first. Despite its British slant, this is a required purchase for all reference collections serving art and design history. A glossary, bibliography, and index are included.

Calloway, Stephen, and Stephen Jones
Style Traditions: Recreating Period Interiors
New York, Rizzoli, 1990. 223p. ISBN 084781131X

Originally published in Great Britain in 1989 as *Traditional Style*, this book surveys interiors from the Tudor period to the "New Ruralism" and "Alternative Lifestyles" of the late twentieth century. British in focus, it nonetheless includes an extensive directory of manufacturers and suppliers on both continents. Especially useful to students interested in renovation and historical preservation is the inclusion of both authentic period interiors and their modern interpretations.

Chase, Linda, and Laura Cerwinske
In the Romantic Style: Creating Intimacy, Fantasy, and Charm in the Contemporary Home
New York, Thames and Hudson, 1990. 160p. ISBN 0500235929

Arranged by type of furnishing (surfaces, beds, curtains, etc.), rather than by room, this idea book may help students working on residential decoration projects. Includes an extensive resource directory.

Ching, Frank
Drawing, A Creative Process
New York, Van Nostrand Reinhold, 1990. 206p.
ISBN 0442318189

Written by a well-respected authority on architectural drawing, this text begins with an overview of the purpose and process of drawing, then discusses individual features, and concludes with suggestions on envisioning and speculative (i.e., imaginative) drawing. Some readers may find the hand-lettered text distracting. Includes bibliography and index.

Ching, Frank, and Cassandra Adams
Building Construction Illustrated
2nd ed. New York, Van Nostrand Reinhold, 1991. 1 v. (various pagings). ISBN 0442234988

Introduces the reader to "basic principles of how buildings are built, examining the relationships between major systems and available materials" (preface). The extensive appendix includes sections on "Planning for the Handicapped" and "Acoustics and Sound Control," as well as a bibliography and index. Ching's hand-lettered text may distract some readers.

Colchester, Chlöe
The New Textiles: Trends and Traditions
New York, Rizzoli, 1991. 192p. ISBN 0847814181

This book is somewhat unusual in that it discusses textiles used to create crafts and artwork, as well as those developed for commercial use in fashions and furnishings. Construction techniques are described. Appendices include a biographical dictionary of designers and directories of suppliers. Includes a bibliography, a glossary, and an index.

Curtains, Draperies, and Shades
Menlo Park, Calif. Sunset, 1993. 128p. ISBN 037601735X

This practical volume reviews decorating basics and instructions for measuring and making a variety of window treatments. Like other works in the Sunset's *Home Improvement Books* series, this is a useful source for those wanting information on the actual construction designs. Indexed.

De Chiara, Joseph, Julius Panero, and Martin Zelnik, eds.
Time-Saver Standards for Interior Design and Space Planning
New York, McGraw-Hill, 1991. 1160p. ISBN 0070162999

An essential reference for all interior design programs, this volume provides illustrative details of all kinds as well as formulas and other reference data to expedite planning and presentation of contract and residential interior designs. It even includes formulas for library shelving by subject category!

Designing and Planning Bedrooms
Upper Saddle River, N.J., Creative Homeowner Press, 1992. 96p. ISBN 0932944949

This colorful and inexpensive book provides both design ideas and practical information on lighting, furniture, and bedding, as well as storage ideas for different types of bedrooms. Especially interesting is the chapter, "Three Stages to a Perfect Bedroom," where furnishings and accessories are prioritized so that even modest budgets can afford a comfortable and attractive room.

Dizik, A. Allen
Concise Encyclopedia of Interior Design
2nd ed. New York, Van Nostrand Reinhold, 1988. 220p.
ISBN 0442221096

Written for students and consumers, this revision of a 1976 reference is now itself in need of updating; however, what is included is still useful. Definitions of technical design terms are included as well as longer essays on topics such as color and furniture arrangement. Prominent designers and architects are covered, either individually or in discussions of styles and movements.

Dorf, Martin E.
Restaurants That Work: Case Studies of the Best in the Industry
New York, Whitney Library of Design, 1992. 223p.
ISBN 0823045404

While this book's primary audience is the restaurateur rather then the designer, students working on contract interior projects for the hospitality industry will find much that is useful here, including lighting considerations and standards for the physically disabled. Each case study includes floor plans and colored photographs. Extensive bibliography and index.

Drpic, Ivo D.
Architectural Delineation: Professional Shortcuts
New York, Van Nostrand Reinhold, 1988. 198p.
ISBN 0442221053

Drpic begins with a thorough discussion of the equipment needed, continues with a series of exercises and step-by-step demonstrations of sketch preparation for architectural drawing, and concludes with a portfolio of finished work. Although all the illustrations are of exteriors, the process described here is equally useful for the interior design student.

Drpic, Ivo D.
Sketching and Rendering Interior Spaces
New York, Whitney Library of Design, 1988. 176p.
ISBN 0823048543

Cover subtitle: *Practical Techniques for Professional Results*. In this volume, Drpic focuses on interior space, beginning with the basic techniques and materials, continuing with twenty-one demonstrations of the process, and concluding with a portfolio of the final presentations of offices, homes, health care facilities, and malls. The hand-lettered text makes browsing difficult, but an index is included.

Falcone, Joseph D.
Architectural Drawing and Design: Principles and Practices
Englewood Cliffs, N.J., Prentice Hall, 1990. 324p.
ISBN 0130441325

Although directed toward architectural presentation, the information on equipment, techniques, and construction details is also useful to the interior designer. Includes glossary and index.

Farren, Carol E.
Planning and Managing Interior Projects
Kingston, Mass., R. S. Means, 1988. 314p. ISBN 0876290977

Focuses on commercial (contract design) rather than residential projects. Provides practical information on relating to clients, financial management, and purchasing and moving materials. Includes sample forms such as agreements and lists.

Fiell, Charlotte, and Peter Fiell
Modern Furniture Classics Since 1945
Washington, D.C., American Institute of Architects Press, 1991. 192p. ISBN 1558350403

This survey is divided into decades, with individual pieces briefly described and placed in their historical context. Appendices include information on furniture collecting as well as brief biographies of the designers and a directory of dealers, galleries, and museums featuring the furniture. Endnotes, bibliography, and index.

Fitch, Rodney, and Lance Knobel
Retail Design
New York, Whitney Library of Design, 1990. 256p.
ISBN 0823045501

This helpful volume begins with an introduction to the history of retail design. Part One deals with practical considerations in the design process, including materials and finishes, circulation and traffic, lighting, and graphics. Part Two includes case studies of various successful designs from the U.S., Europe, and Japan. The book concludes with a discussion of future trends. Includes glossary, bibliography, and index.

Fleming, John, and Hugh Honour
The Penguin Dictionary of Decorative Arts
New ed. New York, Viking, 1989. 935p. ISBN 0670820474

Although the scope of this reference book is broader than interior design, many entries are of interest to the field, from classic architectural details to contemporary furniture designers. Entries vary from a few lines to over a page in length and include people, companies, and individual items. There are cross-references to other articles where appropriate, and some of the longer articles include bibliographies. Some colored photographs, but most illustrations are black-and-white drawings and photographs.

Floors, Stairs, and Carpets
Alexandria, Va., Time-Life Books, 1990. 144p.
ISBN 0809462362

Like others in the *Fix-it-Yourself* series, this guide provides clear, no-nonsense, well-illustrated instructions on the repair of floors, stairs, and carpeting. Safety measures, tools, and techniques are presented within each section. This series is a useful addition to collections on renovation and rehabilitation of interiors. Includes index.

Foa, Linda
Furniture for the Workplace
New York, Architecture and Interior Design Library, 1992. 240p. ISBN 0866361758

Written for the contract interior designer, this book includes information on over 200 products arranged in seven chapters including furniture systems, office seating, computer support systems, stacking chairs, multi-purpose table systems, filing and storage systems, and accessories and lighting. Appendices include chronology, directory, and glossary. Indexed.

Garrett, Wendell D.
Victorian America: Classical Romanticism to Gilded Opulence
New York, Rizzoli, 1993. 300p. ISBN 0847817474

The editor of *Antiques* magazine has compiled a lavishly illustrated book of fifty Victorian interiors from many states. Each chapter discusses a particular region. Captions describe the histories of the antique furniture and furnishings.

Gere, Charlotte
Nineteenth-Century Decoration: The Art of the Interior
New York, Abrams, 1989. 408p. ISBN 0810913828

This rather expensive volume presents a detailed look at the design and decoration of domestic interiors in Europe and America. A long introductory essay discusses the period's notable aesthetic and technological developments and includes information on influential architects and designers. This is followed by a survey of decoration according to the type of structure (townhouse, country cottage, conservatory), and another following the development of decoration in a more chronological fashion. A rare treat for history buffs is the appendix, "The Working Library of a Nineteenth-Century Architect and Designer." Virtually all of the many illustrations are period piece—authentic photographs, paintings, and drawings of the time.

Gore, Alan, and Ann Gore
English Interiors: An Illustrated History
New York, Thames and Hudson, 1991. 192p. ISBN 0500235937

This profusely illustrated volume covers English interiors from the Norman period to the mid-twentieth century (1066-1966). The development of each element of a room is traced and the merging of various elements into particular styles is discussed. Illustrations include period art as well as photographs of preserved historic interiors. Includes glossary, chronology, bibliography, and indexes.

Gorman, J. R., and others
Plaster and Drywall Systems Manual
3nd ed. Los Angeles, BNI Books, 1988. 415p. ISBN 007032199X

A very practical manual with reference standards and specifications for all aspects of plaster and drywall installation, this book also contains some pleasant surprises for the artist. In a delightful essay, "The History and Romance of Lath and Plaster," wall surfaces are surveyed from earliest times. A color insert includes useful illustrations of different plaster (stucco) finishes available today. Perhaps not a "first purchase," this volume does a nice job of bridging the gap between art and technology.

Green, William R.
The Retail Store: Design and Construction
2nd ed. New York, Van Nostrand Reinhold, 1991. 185p.
ISBN 0442001606

Written for the retailer, this step-by-step guide provides a useful introduction for students of commercial design as well. This edition includes information on consumer psychology and new technologies. Includes glossary, references, and index.

Guild, Robin, with Vernon Gibberd and others
The Victorian House Book
New York, Rizzoli, 1992. 320p. ISBN 084781095X

This book is a systematic survey of Victorian architecture and interior design for the prospective renovator. While the British terminology may put off some students, the text is helpful and the captions both descriptive and analytical. A final chapter deals with sympathetic conversions and provides guidelines for energy conservation and adapting Victorian art and technology to twentieth-century standards of safety, comfort, and efficiency. A directory of suppliers, bibliography, and index are all included.

Guild, Tricia, and Elizabeth Wilhide
Tricia Guild on Color: Decoration, Furnishing, Display
New York, Rizzoli, 1993. 191p. ISBN 0847816435

Introductory chapters discuss color theory and the scope and potential of color in interior decoration. The use of seven different colors is then illustrated with examples from many cultures. Final chapters include color palettes with wallpapers and fabrics identified by manufacturer's stock number.

Habegger, Jerryll, and Joseph H. Osman
Sourcebook of Modern Furniture
New York, Van Nostrand Reinhold, 1989. 469p.
ISBN 0442232764

Divided into seventeen chapters by type of furniture, entries in this book are very brief, including identifying name or number of the piece, date, designer, manufacturer, materials, and dimensions. An introductory essay places the furniture in context, beginning with the Corbusier dining chair of Thonet and concluding with an Italian tea-for-two side table. The directory of distributors is probably out of date now, but could be used as a starting point. Ample indexing makes access to specific entries quite easy.

Hall, Dinah
Ethnic Interiors: Decorating with Natural Materials
New York, Rizzoli, 1992. 180p. ISBN 0847815757

One of several Rizzoli publications on regional styles, this book surveys folk art from fourteen cultures and four continents. The use of natural materials is assumed, rather than explicitly developed. The directory lists designers and suppliers specializing in folk arts. Indexed.

Hall, William R.
Contract Interior Finishes: A Handbook of Materials, Products, and Applications
New York, Whitney Library of Design, 1993. 240p.
ISBN 0832009335

Covers the technical aspects of finish materials and products for interior systems, evaluating different materials and components for quality, style, durability, and price. Includes finish evaluation, and scheduling and specification charts and tables. Includes a directory of professional organizations and an index.

Halse, Albert O.
Architectural Rendering: The Techniques of Contemporary Presentation
3rd ed. Edited by Spencer L. George, and Helen A. Halse. New York, McGraw-Hill, 1988. 233p. ISBN 0070256292

Although the bibliography still reflects the first edition (c1960), the discussion of rendering techniques has been updated to include felt-tip pens, Mylar pencils, and a brief introduction to computer-aided graphics. Theoretical bases are discussed and extensive practical assistance offered, including specific brand-name recommendations for implements and surfaces.

Hammer, Nelson
Interior Landscape Design
New York, McGraw-Hill, 1991. 288p. ISBN 0070258619

This helpful book provides the basics on designing the documentation of interior landscape projects with detailed information on specifications, lighting, irrigation, and cost estimates. General considerations are well illustrated with case studies of five different projects, including one private residence.

Hampton, Mark
Legendary Decorators of the Twentieth Century
New York, Doubleday, 1992. 293p. ISBN 0385263619

Little is available about the history of interior design as a profession, which makes this breezy collected biography worth a second look. The careers of twenty-two of the most prominent interior designers are included, illustrated by the author's watercolor renderings of their work.

Harris, Cyril M., ed.
Dictionary of Architecture and Construction
2nd ed. New York, McGraw-Hill, 1993. 924p. ISBN 0070268886

About 10 percent of the 22,500 terms are new to this second edition. Definitions are short (two to six lines) and about 10 percent of the terms are illustrated by line drawings in the page margins. Subject coverage is broad, from historical and artistic terms to current technology, tools, and techniques. Although interior design is not the focus of this reference, all interior design students will need some acquaintance with this terminology.

Helsel, Marjorie Borradaile
The Interior Designer's Drapery, Bedspread, and Canopy Sketchfile
New York, Whitney Library of Design, 1990. 1 v. (unpaged)
ISBN 0823025462

This useful manual includes over 350 sketches of "designs for draperies, canopies and bedspreads that can be used as a designer's working tool, catalog or idea book." The focus of the book is on draperies, arranged by style (historical, formal, and casual), with separate sections for related fabric treatments of windows, beds, and tables. An index aids access to specific treatments.

Hiro, John E.
Millwork Handbook
New York, Sterling Publishing, 1993. 284p. ISBN 0806986980

Written for the consumer, this book covers the selection, installation, and repair of millwork. Available options for windows, doors, and moldings are evaluated for suitability in various situations. Includes glossary and index.

Hogg, Min, and Wendy Harrop
Interiors
New York, Clarkson Potter, 1988. 256p. ISBN 0517571064

Eight decorative styles are featured in essays from the British magazine, *The World of Interiors*, of which Hogg was the editor. Included are such unusual styles as "cluttered" and "shabby chic." Index.

Home Decorating Institute (Minnetonka, Minn.)
Creative Window Treatments: Forty-Five Styles Shown Step-By-Step
Minnetonka, Minn., Cy DeCosse, 1992. 128p.
ISBN 086573352X

In addition to basic and designer fabric treatments, accessories and nonfabric alternatives are also featured in this practical manual. Students interested in construction as well as design will find the step-by-step instructions clear and easy to follow. Part of the series *Arts and Crafts for Home Decorating*.

Hope, Augustine, and Margaret Walch
The Color Compendium
New York, Van Nostrand Reinhold, 1990. 360p.
ISBN 0442318456

This reference book is really a handbook and an encyclopedia. Included are thirty long essays on topics such as "Architecture: Interior Color" and "Language and Color." Interspersed alphabetically among these essays are many short entries. A color insert features seven more specialty articles, including "Sources of Historic Color" and "Systems of Color." Includes an extensive bibliography and an index.

Hull, Alastair, and Jose Luczyc-Wyhowska
Kilim: The Complete Guide: History, Pattern, Technique, Identification
London, Thames and Hudson, 1993. 352p. ISBN 0500015651

This coffee-table book is actually a treasure trove of information for the interior designer. Covering the history and cultural context, pattern identification, and construction techniques of many ethnic kilim, it is an excellent guide to a popular decorating item. The authors include fascinating information on the cultures from which the kilim come. A chapter on the care of the carpets is useful, as are the directories, glossary, bibliography, and index.

International Furniture Design for the 90s
New York, Library of Applied Design, PBC International, 1991. 238p. ISBN 0866361367

Although there is no text to accompany this catalog of functional art, the furniture and office and home furnishings are clearly illustrated. Captions give product name, designer, and a brief description of the materials used. Indexed.

Jackman, Dianne R., and Mary K. Dixon
The Guide to Textiles for Interior Designers
2nd ed. Winnipeg, Canada, Peguis, 1990. 195p.
ISBN 0920541925

Written as a textbook, this title is also a valuable reference tool. The authors systematically describe the inherent characteristics, chemical reactions, and environmental sensitivities of most natural and factory-made fibers. Aesthetic, functional, economic, and safety considerations are discussed within the context of selecting appropriate textiles for specific interiors. Much of the information is summarized in the appendix, "Textile Selection Charts." The bibliography reflects the first edition (1983). Glossary and index.

Jankowski, Wanda
Designing with Light: Residential Interiors
New York, Library of Applied Design, PBC International, 1991. 235p. ISBN 0866361421

A variety of ideas and techniques are illustrated within a range of prices. The needs of both remodeling and new construction are addressed. A brief catalog of new products and short interviews with designers are included. Indexed.

Jones, Chester
Colefax and Fowler: The Best in English Interior Decoration
Boston, Mass., Little, Brown, 1989. 224p. ISBN 0821217461

This is a biography of the professional career of English decorator John Fowler, whose work in the mid-twentieth century profoundly influenced the renovation of British National Trust properties and the subsequent propagation of the English country decor to North America. Final chapters describe the distinctive style developed by the firm of Colefax and Fowler, as well as its role in the contemporary design milieu. Includes glossary and index.

Jones, Frederic H.
A Concise Dictionary of Interior Design
Los Altos, Calif., Crisp Publications, 1990. 215p.
ISBN 1560520671

Professing to cover both architecture and design in the introduction, Jones really concentrates on interior design. Definitions vary in length from one-liners (size sheet) to over a page (glass). Included are many French terms used in architecture and decoration, as well as the terminology for the newly emerging electronic technologies. This source is weak on illustrations, but provides a useful reference nonetheless.

Kaufman, Donald, and Taffy Dahl
Color: Natural Palettes for Painted Rooms
New York, Clarkson Potter, 1992. 224p. ISBN 0517576600

What differentiates this volume from other showcase and coffee-table books is its effective use of a single residence. This enables the designer to show how color works in specific settings and how the use of color allows rooms to effectively relate to each other. One note of caution: color reproductions are not always true to the textual descriptions. Appendices include technical information on mixing paints and using colors.

Kleeman, Walter
Interior Design of the Electronic Office: The Comfort and Productivity Payoff
New York, Van Nostrand Reinhold, 1991. 294p.
ISBN 0442006136

It's the bottom line that counts. Case studies and examples are used to show that proper design strategy is an investment that pays off. The history of office design, current trends, and the special needs of the electronic office are discussed. Includes a long list of references, suggested readings, and an index.

Knackstedt, Mary V., and Laura J. Haney
The Interior Design Business Handbook
2nd ed. New York, Van Nostrand Reinhold, 1992. 352p.
ISBN 0442011288

This handbook includes more than fifty sample forms and letters covering scores of real life scenarios in a design practice, including startup, marketing, managing finances, dealing with contractors, and professional development and networking. A bibliography and index are included.

Larkin, David, and Bruce Brooks Pfeiffer, eds.
Frank Lloyd Wright: The Masterworks
New York, Rizzoli, 1993. 311p. ISBN 0847817156

Exquisite photographs illustrate an extensive discussion of Wright's architecture and furnishings as expressions of his philosophy. Also included are many original line drawings of elevations and details. One of the more useful of the many new works on Wright.

Larsen, Jack Lenor
Material Wealth: Living with Luxurious Fabrics
New York, Abbeville Press, 1989. 240p. ISBN 1558590072

Written by an American fabric designer, this attractive work surveys trends in international fabric design. Current design practices in specific applications are placed in a historical and artistic context. The photography effectively captures both the color and texture of these luxurious fabrics.

Larson, George A., and Jay Pridmore
Chicago Architecture and Design
New York, Abrams, 1993. 256p. ISBN 0810931923

In this beautifully illustrated example of regional architecture and interior design, Larson and Pridmore place the buildings in their political and social context.

Laseau, Paul
Architectural Drawing: Options for Design
New York, Design Press, 1991. 197p. ISBN 083068008X

An important author on the presentation of design, Laseau here discusses the role of drawings in design and thinking, rather than specific tools and techniques. The examples are primarily architectural but the principles certainly apply to interior designers as well. Bibliography; index.

Laseau, Paul
Graphic Thinking for Architects and Designers
2nd ed. New York, Van Nostrand Reinhold, 1989. 243p.
ISBN 0442258445

Talking primarily to architects, Laseau shows how visual presentations can enhance understanding between client and designer, thereby improving the end result. Although computer applications are mentioned, the author is ambivalent toward their use. Bibliography reflects the 1982 edition. Index.

Leach, Sid DelMar
Photographic Perspective Drawing Techniques
New York, McGraw-Hill, 1990. 191p. ISBN 0070368147

Somewhat iconoclastic in his approach, Leach uses photographs to create visually correct scale drawings of interior spaces. His approach assumes a prior understanding of perspective drawing techniques.

Leibrock, Cynthia, with Susan Behar
Beautiful Barrier-Free: A Visual Guide to Accessibility
New York, Van Nostrand Reinhold, 1993. 192p.
ISBN 0442008821

This is a comprehensive directory to all types of adaptive technology and equipment, from kitchen and bathroom fixtures to the smallest accessories. Illustrated.

Linley, David
Classical Furniture
New York, Abrams, 1993. 192p. ISBN 0810931885

Written by a renowned British exponent of finely crafted furniture, this book discusses the history and aesthetic value of classical furniture. Includes a pictorial directory of eighteenth-century furniture. Index.

Linsley, Leslie
Nantucket Style
New York, Rizzoli, 1990. 228p. ISBN 0847811654

This is a good example of regional decor. Twenty-three houses are featured, each with four to eight pages of photographs. Includes sources and index.

Madden, Chris Casson
Rooms with a View: Two Decades of Outstanding American Interior Design from the Kips Bay Decorator Show Houses
Glen Cove, N.Y., PBC International, 1992. 191p.
ISBN 0866361901

Arranged by room, each chapter surveys several entries from this premier showcase event. The text briefly describes the design work done and why each designer chose a particular theme. Indexed.

Malnar, Joy Monice
Interior Dimension: A Theoretical Approach to Enclosed Space
New York, Van Nostrand Reinhold, 1991. 365p.
ISBN 0442237391

This book is unabashedly theoretical and might therefore put off some students, but this theory is attractively presented and appropriately illustrated. The author strives to show that without this deeper understanding of the nature of interior space, the practical, aesthetic, psychological, and ethical problems of interior design cannot be satisfactorily answered.

McGarry, Richard M.
Tracing File for Interior and Architectural Rendering
New York, Van Nostrand Reinhold, 1988. 279p.
ISBN 0442205309

This is a file of people and foliage, useful in detailed renderings for final presentation. It is valuable for students whose drafting skills may be better than their artistic ones. Also included is a checklist of points to consider when composing or evaluating a rendering. Indexed.

McGrath, Norman
Photographing Buildings Inside and Out
Rev. and expanded 2nd ed. New York, Whitney Library of Design, 1993. 208p. ISBN 0823040178

In addition to artistic considerations of composition and lighting, the author addresses legal concerns and marketing. The sites have been carefully selected for design quality, making this an idea book as well as a manual.

Miller, Judith, and Martin Miller
Period Finishes and Effects
New York, Rizzoli, 1992. 178p. ISBN 0847815692

A step-by-step guide to interior decoration techniques, this survey provides both a history lesson and detailed instructions on reproducing each effect. General chapters on equipment and the historical use of color frame a series of chapters on specific paint techniques such as limewashing, trompe l'oeil, and stenciling. Includes a directory of suppliers and an index.

Nakamura, Sadao
The Color Source Book for Graphic Designers
Kyoto, Japan, Mitsumura Suiko Shoin, 1990. 110p.
ISBN 4838101104

Over 700 different color schemes are extracted from the published work of internationally known designers and then carefully described using the Munsell notation system. An appendix explains the Munsell notation system, and the index provides access to the work of individual designers.

National Council for Interior Design Qualification
Examination Guide
7th ed. rev. White Plains, N.Y., National Council for Interior Design Qualification, 1994. 60p.

This loose-leaf binder is intended to provide access to the information needed to prepare for the NCIDQ examination, but not a study guide. Although two-year college students rarely have the six years of education and experience needed to sit for the exam, the guide can help students focus their attention on what is perceived by practitioners to be the core areas of knowledge in the field of interior design. Librarians will appreciate the "Reference List" of books, codes, and standards. (Available only from the publisher.)

Naylor, Colin, ed.
Contemporary Designers
2nd ed. Chicago, St. James Press, 1990. 641p
ISBN 0912289694

One in a series of art-oriented directories, this volume features over 800 designers in architecture, interior design, display, textiles, product development, and other fields. Information provided includes brief personal and professional biographies, lists of exhibitions, collections and major works, bibliographies of primary and secondary sources, and personal statements from many of the designers who are still living. No subject access. The third edition is scheduled for 1995.

Nielson, Karla J.
Window Treatments
New York, Van Nostrand Reinhold, 1990. 430p.
ISBN 0442268092

Written by an interior design educator and practitioner for fellow practitioners and students, this resource covers both theoretical background and practical applications. Information on calculations, specifications, and installation is given for different types of treatments. Includes directories, glossaries, topical bibliography, and index.

Oles, Paul Stevenson
Drawing the Future: A Decade of Architecture in Perspective Drawings
New York, Van Nostrand Reinhold, 1988. 161p.
ISBN 0442270038

Although virtually all the illustrations are of exteriors, this overview of postmodern architecture offers interior design students excellent examples of rendering technique, accompanied by helpful explanations. Indexed.

Pegler, Martin M., ed.
Home Furnishings Merchandising and Store Design
New York, Retail Reporting Corp., 1990. 222p.
ISBN 0934590362

Arranged by type of store and including examples from Great Britain and North America, this book is one of several designed to promote various designers and provide examples of current store design. Each entry includes several illustrations and a text that describes the store and its design. Indexed.

Peloquin, Albert A.
Barrier-Free Residential Design
New York, McGraw-Hill, 1994. 239p. ISBN 007049326X

Although some chapters are more applicable to architects than interior designers, there is plenty here to assist the latter in creating fully accessible living spaces. Each chapter surveys the barriers associated with a particular kind of room and presents ways to remove them. Clearly arranged, comprehensive, and well-illustrated, this book is a valuable addition to this important field.

Phillips, Alan
The Best in Lobby Design, Hotels, and Offices
New York, Watson-Guptill, 1991. 224p. ISBN 0823061361

One in a series of pictorial directories of commercial projects published by Watson-Guptill in recent years, this idea book is international in scope, covering architectural and interior design projects. This series presents ideas for students and provides an illustrated professional directory for prospective clients. Arranged by style, each entry has clear colored photographs with brief descriptive captions. The introductory material places the portfolios in a broader context. Indexes to the specific projects and designers are included. Others in the series include *The Best in Office Interior Design* and *The Best in Point-of-Purchase Design*.

Phillips, Barty
Fabrics and Wallpapers: Sources, Design, and Inspiration
Boston, Little, Brown, 1991. 195p. ISBN 0821218719

This is not a comprehensive study, but a pleasant introduction to the textures and patterns of historical and contemporary design. Photographs aim more for artistic effect than accurate representation. The directory of manufacturers and other appendices are useful.

Pile, John F.
Dictionary of Twentieth Century Design
New York, Facts on File, 1990. 312p. ISBN 0816018111

Industrial and graphic design subjects as well as architectural and interior design and furnishings are all represented here. Covers techniques, objects, design movements and important written works, with many entries for people and associations. A paperback edition is now available (De Capo Press, 1994).

Pile, John F.
Furniture, Modern and Postmodern: Design and Technology
2nd ed. New York, Wiley, 1990. 312p. ISBN 0471854387

Pile, professor of design at the Pratt Institute and a renowned author in the field, has revised his 1979 book to reflect the trends, products, and designers of the postmodern world. Appendices include information on standard dimensions, design piracy and protection, and safety considerations. Bibliography; index.

Pile, John F.
Interior Design
New York, Abrams, 1988. 541p. ISBN 0810911213

This classic text is too full of good information to be left off library shelves. It covers theoretical, historical, aesthetic, and practical aspects of interior design. A list of graphic symbols is only one of several useful appendices. Includes a glossary, extensive bibliography, and an index.

Piotrowski, Christine M.
Professional Practice for Interior Designers
New York, Van Nostrand Reinhold, 1989. 336p.
ISBN 0442275196

Somewhat dated now, this handbook provides both vocational guidance and professional understanding of the field. Still valuable are the copies of codes of ethics from the ASID and IBD. Includes extensive bibliography. Piotrowski's more recent *Interior Design Management* (Van Nostrand Reinhold, 1992), may provide the needed update.

Quiller, Stephen
Color Choices
New York, Watson-Guptill, 1989. 144p. ISBN 0823006964

Early chapters on the color wheel and different types of color schemes are helpful to novice interior designers. Later chapters deal with the needs of the fine artist. Bibliography; index.

Ramsay, Linda M.
Secrets of Success for Today's Interior Designers and Decorators
Oceanside, Calif., Touch of Design, 1992. 307p.
ISBN 096299183X

This paperback manual attempts to give novice and experienced designers practical advice on how to "easily sell the job, plan it correctly and keep the customer coming back for repeat sales" (cover). In addition to sections on marketing and customer relations, this work includes an extensive guide to selecting and measuring materials.

Ramsay, Linda M.
Start Your Own Interior Design Business and Keep It Growing: Your Guide to Business Success
Oceanside, Calif., Touch of Design, 1994. 367p.
ISBN 0962991805

Written for interior designers who are interested in starting their own business, this paperback text would be very useful to entrepreneurial-minded students as well. Starting with profiles of successful and unsuccessful members of the business community, it covers virtually all issues a new entrepreneur need consider, from location and capital to marketing, bidding, and reimbursement. The appendix contains useful lists of professional organizations and trade publications.

Randall, Charles T.
The Encyclopedia of Window Fashions: 700 Decorating Ideas for Windows, Bedding, and Accessories
3rd ed. Orange, Calif., Randall International, 1992. 146p.
ISBN 0962473634

This is a very useful idea book for students doing presentation boards. It is practical, concise, and well illustrated. Prefatory pages include a glossary, charts, and measurement guides.

Raschko, Bettyann Boetticher
Housing Interiors for the Disabled and Elderly
New York, Van Nostrand Reinhold, 1991. 360p.
ISBN 0442009836

A room-by-room consideration of layout, access, and furnishings is provided in the context of current research, anthropometric considerations, and classifications of disabilities. Information on assistive devices and technical standards is included, as well as a glossary, bibliography, and index.

Rees, Yvonne
Window Style
New York, Van Nostrand Reinhold, 1990. 160p.
ISBN 0442302959

This is not a how-to manual, but rather an idea book of exterior and interior window treatments. It is a good addition to an interior design collection because of its broad international coverage, historical context, and enthusiastic artistic perspective. This is the work of a person who loves windows!

Rettinger, Michael
Handbook of Architectural Acoustics and Noise Control: A Manual for Architects and Engineers
Blue Ridge Summit, Pa., TAB Professional and Reference Books, 1988. 247p. ISBN 0830626867

Discusses the science of sound and acoustics and examines ways to control noise. Many charts, tables, and graphs illustrate the text. It may be too technical for most interior designers, but the subject is of increasing importance in commercial and public projects, and the topic is covered in the NCIDQ examination.

Reznikoff, S. C.
Specifications for Commercial Interiors
Rev. ed. New York, Whitney Library of Design, 1989. 319p.
ISBN 0823048934

   This title includes much more illustrative material than Carol Farren's *Planning and Managing Interior Projects* (R. S. Means, 1988). Includes checklists and sample specifications as well as bibliographic notes and glossaries. Focus is more on the legal aspects of contract design (avoiding liability, complying with current fire regulations) than on designer/client relations. Both aspects are needed to support courses in contract design.

Riccardi-Cubitt, Monique
The Art of the Cabinet: Including a Chronological Guide to Styles
New York, Thames and Hudson, 1992. 224p. ISBN 0500236429

   This introductory survey covers cabinets from Egyptian times into the twentieth century. Discusses the cabinet's function as furniture, as cultural symbols, and as art. Appendices include an illustrated guide to styles and brief biographies of many of the cabinetmakers, craftspeople, and designers. Includes glossary, end notes, bibliography, and index.

Rieman, Timothy D., and Jean M. Burks
The Complete Book of Shaker Furniture
New York, Abrams, 1993. 400p. ISBN 0810938413

   Carefully researched, with many illustrations and an extensive bibliography, this work places Shaker furniture in its historical and cultural context with period photographs and a thoughtful background essay on Shaker life and beliefs, design theory, and technology. This work presents useful criteria for the identification of Shaker furniture, especially in regard to time and place of origin. Includes glossaries and an index.

Riggs, J. Rosemary
Materials and Components of Interior Design
3rd ed. Englewood Cliffs, N.J., Prentice Hall, 1992. 192p.
ISBN 0135713242

   Concentrating on the nuts-and-bolts information needed for accurate specifications, Riggs covers paints, floors, walls, cabinets, kitchens, bathrooms, and other components. New to this edition is a chapter on carpeting. Chapter bibliographies focus on trade publications. Indexed.

Rossbach, Sarah
Interior Design with Feng Shui
New York, Arkana, 1991. 178p. ISBN 0140193529

   The Chinese philosophy of universal harmony and balance is applied here to interior design. Offers an unusual perspective on problem solving. This book is a good choice for libraries interested in diversifying their collections in this area.

Rupp, William E., and Arnold Friedmann
Construction Materials for Interior Design: Principles of Structure and Properties of Materials
New York, Whitney Library of Design, 1989. 176p.
ISBN 0823009300

   Covers finishing materials from stone and brick to fabric and paint. Each chapter includes information on physical properties, construction techniques, and the use of the material in interiors. Organized by material type, rather than by system. Indexed.

Russell, Beverly
Architecture and Design, 1970-1990: New Ideas in America
New York, Abrams, 1989. 143p. ISBN 0810918900

   Each of four mainstream movements are placed in their cultural and social context. Included are historical recall, cybernetic influence, deconstructionism, and process design. Indexed.

Russell, Beverly
Women of Design: Contemporary American Interiors
New York, Rizzoli, 1992. 224p. ISBN 0847816141

   Presents professional profiles of thirty-two successful women in the world of commercial interior design. Arranged thematically by style, each entry includes a two-page biography, a black-and-white portrait, and a few color photographs of their work. These brief glimpses make the reader wish for more information!

Sampson, Carol A.
Estimating for Interior Designers
New York, Whitney Library of Design, 1991. 176p.
ISBN 0823016005

   This valuable manual shows how to accurately estimate costs for paint, wallcoverings and other furnishings. Blank work order forms are included, as well as a bibliography and index.

Showcase of Interior Design
Southern ed. Grand Rapids, Mich., Vitae Publishing, 1993. 235p.
ISBN 0962459658

   One in a series of *Showcase* editions representing various regions and specialties. Others include the Pacific edition and the International Commercial edition. Textual matter is limited to a directory listing and brief quotations from each of the designers. Originally conceived as an extended advertisement to promote the use of professional interior designers, care is taken to select well-recognized practitioners. For students, these volumes become picture books of current practice as well as sources for potential job leads.

Sloan, Annie, and K. Gwynn
The Complete Book of Decorative Paint Techniques
New York, Random House, 1990. 223p. ISBN 0517022656

   Assuming that the reader has a substantial understanding of painting in general, Sloan presents abbreviated instructions for a vast repertoire of paint techniques. This is of particular value to hands-on designers who want to expand their technical skills.

Sloan, Annie, and Kate Gwynn
Color in Decoration
Boston, Little, Brown, 1990. 191p. ISBN 0316798452

   An overview of how colors vary with different lighting, textures, and neighboring designs is followed by an extensive series of studies in color as used in historic and contemporary interiors from Europe, South Asia, and North America. The concluding section provides practical advice on paints and pigments, including recipes for color mixing. Bibliography; index.

Smith, Nancy A.
Old Furniture: Understanding the Craftsman's Art
2nd rev. ed. New York, Dover, 1991. 186p. ISBN 0486263398

   This inexpensive volume contains a fairly thorough examination of furniture-making techniques, followed by wear, repairs, and fakes. Illustrated by line drawings and black-and-white photographs, it is a practical addition to any collection of materials on the design and construction of furniture.

Spencer, Dorothy
Total Design: Objects by Architects
San Francisco, Chronicle Books, 1991. 187p. ISBN 0877016658

One hundred years of design history are covered in these heavily illustrated chapters. The textual matter of the earlier chapters is particularly useful and the photographs are excellent throughout, providing a good visual summary of the evolution of modern design. Indexed.

Staebler, Wendy W.
Architectural Detailing in Contract Interiors
New York, Whitney Library of Design, 1988. 256p.
ISBN 082300242X

Sixty-three different types of detail, from archways to library reference desks and magazine racks to waterwalls, are illustrated with up to three examples of each. Plans, sections, and elevations are included as needed. Design teams and their addresses are provided.

Steffy, Gary R.
Architectural Lighting Design
New York, Van Nostrand Reinhold, 1990. 202p.
ISBN 0442207611

This textbook on lighting covers both conceptual design and practical application. Various lighting options and the docu- mentation needed for specifications are included. Since not every program in interior design offers a full course on lighting, this may be a valuable library resource to supplement coverage in other courses.

Stem, Seth
Designing Furniture from Concept to Shop Drawing: A Practical Guide
Edited by Laura Tringali. Newtown, Conn., Taunton Press, 1989. 215p. ISBN 0942391020

This concise guide by a professor of furniture design shows how furniture evolves from conception through model building to construction. Illustrations are mostly of chairs, but principles could apply to any piece. Appendices include practical equipment and techniques for the craftsperson and a dimensional glossary. Includes bibliography and index.

Stores of the Year
New York, Retail Reporting Bureau, 1980- . ISSN 0192-8732

This biennial survey of current retail merchandising design covers a variety of stores from North America and Europe. The text describes the design and function of the various interiors as visual examples of merchandising goals. Indexed.

Sweet's Contract Interiors
New York, Sweet's Group, McGraw-Hill, 1993. 2 v. (various pagings)

Sweet's Information Services provides annual compilations of manufacturers' catalogs for the construction industry. The two-volume set on contract interiors includes catalogs for doors and windows and finishes (floor and wall coverings) among other products. Other titles of interest to interior designers include Sweet's *Homebuilding and Remodeling, Accessible Building Products,* and *Sweet's Light Source.* Indexed.

Thompson, Jo Ann Asher, ed.
ASID, American Society of Interior Designers Professional Practice Manual
New York, Whitney Library of Design, 1992. 224p.
ISBN 082300371X

Covers professional practice (philosophy, theory, education, ethics, and business management) and practical applications (forms useful at all steps of a design project, including historic preservation and specialized commercial projects). Includes notes, bibliography, and index.

Thornton, Peter
The Italian Renaissance Interior, 1400-1600
New York, Abrams, 1991. 407p. ISBN 0810934590

Thornton, an eminent museum curator in England, places interiors and furnishings in historical context, considering social, political, and economic conditions. The illustrations are largely reproductions of period art, with detailed explanatory notes and diagrams explaining the interiors.

Tilley, Alvin R., comp.
The Measure of Man and Woman: Human Factors in Design
New York, Whitney Library of Design, 1993. 96p.
ISBN 0823030318

This revised edition of the 1959 title *The Measure of Man* is an indispensable source of anthropometric data for virtually all human beings. Especially useful are the chapters "Residential Space Considerations," "The Elderly," and "Differently Abled People."

Torrice, Antonio F.
In My Room: Designing for and with Children
New York, Fawcett Columbine, 1989. 167p. ISBN 0345354303

This unusual book uses current research in child development to help create a spatial environment that enhances the physical and psychological growth of normal and special needs children. The author invites children to participate in the design process. Includes glossary, resource list, and index.

Weinhold, Virginia Beamer
Interior Finish Materials for Health Care Facilities: A Reference Guide for All Installations Where Durable Surfaces Are Needed
Springfield, Ill., Charles C. Thomas, 1988. 358p.
ISBN 0398053979

This accessible and authoritative reference helps specifiers effectively use materials in health care facilities. Contract design students will find the consideration of practical, aesthetic, and psychological effects of materials useful, even though the specific materials investigated may now be dated.

Weisman, Leslie
Discrimination by Design: A Feminist Critique of the Man-Made Environment
Urbana, Ill., Univ. of Illinois Press, 1992. 190p.
ISBN 0252018494

Developing the themes that "the appropriation of space is a political act" (Introd.) and that gender and race also have spatial components, Weisman shows how allotment of space is fundamentally related to status and power. This is an intriguing essay with important implications for the interior designer.

White, Antony, and Bruce Robertson
Architecture and Ornament: A Visual Guide
New York, Design Press, 1990. 111p. ISBN 0830633529

This visual dictionary, together with its companion volume, *Furniture and Furnishings*, helps students identify objects and/or terms. Cross-references between plates and a glossary are provided. Illustrations are simple line drawings and definitions are short. Also includes timelines of styles and architects, and maps of major architectural sites.

Who's Who in Interior Design
International ed. Laguna Beach, Calif., Barons Who's Who, 1988- .
1 v. (various pagings) ISSN 0897-5914

Over three thousand design professionals are listed in this biennial publication, including authors, editors, educators, and practitioners. Selection is based on "achievement, occupational stature, and professional excellence" (preface) and coverage is international. The geographic and specialty indexes make this reference particularly useful.

Wilkoff, Wm. L., and Laura W. Abed
Practicing Universal Design: An Interpretation of the ADA
New York, Van Nostrand Reinhold, 1994. 210p.
ISBN 0442013760

Part One discusses the needs of different populations and the role of the designer in creating an inviting public place. Part Two identifies problems found in paths of travel and general function areas (such as banks, health clubs, and restaurants). Solutions to these problems of contract design are found in the appendix, along with the text of the Americans with Disabilities Act. Includes references and index.

Wood, Ernest
Historic Homes of America
New York, Smithmark, 1992. 176p. ISBN 0831744766

With very little text, this book relies on colored photographs to show the historical and geographical variety of American architecture. Captions are more historical than descriptive, but many views of interiors are included. Indexed.

Wrey, Lady Caroline
The Complete Book of Curtains and Draperies
Woodstock, N.Y., Overlook Press, 1991. 135p. ISBN 0879514302

This practical instruction manual begins with chapters on planning projects, equipment, and measurement. Step-by-step directions are then given for draperies, valances, tiebacks, and rosettes. Students who need an understanding of how these items are constructed will appreciate this book. Appendices include directory of suppliers, glossary, and index.

Wilhide, Elizabeth
William Morris: Decor and Design
New York, Abrams, 1991. 192p. ISBN 0810936232

This book begins with a brief biography of Morris, continues with a discussion of his philosophy of design, and concludes with practical applications of his designs in contemporary settings. Appendices include an extensive glossary of patterns, a directory of suppliers, and a selected bibliography. Indexed.

Wylde, Margaret A., Adrian Baron-Robbins, and Sam Clark
Building for a Lifetime: The Design and Construction of Fully Accessible Homes
Newtown, Conn., Taunton Press, 1994. 295p. ISBN 1561580368

Includes both theoretical information and specific details on making housing accessible. Useful to the interior designer are the sections on interiors and remodeling/retrofitting. Includes an extensive bibliography and an index.

Yeager, Jan
Textiles for Residential and Commercial Interiors
New York, Harper and Row, 1988. 434p. ISBN 0060473185

Before dealing with individual applications, Yeager discusses fundamentals such as selection and evaluation of textiles, fibers, weaving techniques, color, labeling, and fire retardancy. Written as a textbook, this volume is a valuable resource for the library as well, especially where specific courses on textiles are not offered. Includes appendices, glossary, bibliography, and index.

Nonprint Materials

American House: A Guide to Architectural Styles
Lake Zurich, Ill., Learning Seed, 1992. 1 videocassette (23 min.) VHS

This video introduces the viewer to basic American architectural styles and descriptive vocabulary. Neither technical nor comprehensive, it is a pleasant supplement to more formal instruction. Includes teacher's guide.

Beyer, Jinny
Color Confidence!
Bethesda, Md., Concept Videos, 1991. 1 videocassette (42 min.) VHS. ISBN 1563190125

Using an innovative alternative to the usual approach, Beyer advocates color schemes where any two colors can go together when blended with intermediaries on the color wheel. Variety in texture, pattern, and intensity are illustrated to add interest.

Certification: The Measure of a Professional
New York, National Council for Interior Design Qualification, 1990. 1 videocassette (13 min.) VHS

Useful for programs where students are encouraged to prepare for the NCIDQ examination, this video discusses the change in the profession from interior decorator to interior designer. The purpose, nature, and content of the exam are introduced.

Color in Everyday Life
Lake Zurich, Ill., Learning Seed, 1992. 1 videocassette (25 min.) VHS

This video is an introduction to the psychology and language of color, presented in a clear and straightforward manner. The emphasis is on interior design, but other fields are also illustrated. Includes teacher's guide.

Interior Lighting: Bringing Rooms to Life
Lake Zurich, Ill., Learning Seed, 1990. 1 videocassette (22 min.) VHS

Illustrates the basics of interior lighting in a humorous, live-action format. The influences on perception of angle, quality, and intensity of light are discussed. Various functions and types of lighting are illustrated. Includes a teacher's guide.

Living by Design
New York, National Council for Interior Design Qualification, 1991. 1 videocassette (8 min.) VHS

This polished career video distinguishes between the traditional interior decorator and the more demanding career of interior designer. The purpose and content of the NCIDQ exam are very briefly described.

New American Home
Los Angeles, Wood Knapp Video, 1990. 1 videocassette (35 min.) VHS

This is the first *Home Magazine* showcase home to be videotaped. Covering architectural features, furnishings, and accessories, it is a nonprint idea book for residential interiors.

Schrank, Jeffrey
Styles of American Furniture
Lake Zurich, Ill., Learning Seed, 1988. 55 color slides

The evolution of American furniture design and construction is illustrated with reproductions from the Metropolitan Museum of Art and the Winterthur Museum, among other places. Includes teacher's guide.

Understanding Color
Bloomington, Ill. Meridian Education Corp., 1991. 1 videocassette (10 min.) VHS

This is a very short yet clever summary of the color wheel and coordinating color schemes, which briefly touches on color's psychological impact.

Photography

Kate Hickey
and
Robert Johnston
Pennsylvania College of Technology
Williamsport, Pennsylvania

Introduction

Photography is both the science and the art of capturing images on film. According to the *Occupational Outlook Handbook* (1994-95 ed.), photographers and photographic process workers hold more than 118,000 jobs in the United States. Most have reached their positions through individual initiative and on-the-job training, but many have completed formal coursework at one of the hundreds of colleges offering such programs.

Many two-year college catalogs list photography as a major; even more offer courses as part of a communications, media technology, or arts program. Specialized curricula such as photo equipment technology and underwater photography are offered at selected schools. Cinematography may or may not be included, depending on the program.

Photographers often are self-employed and usually specialize in commercial work, portraiture, or photojournalism. Specialties may include architectural photography, scientific illustration, or pet portraiture. Artists often use the camera to create images that stand alone as fine art. Photographic process workers, employed in both large photofinishing laboratories and small studios, develop, process, and retouch film and prints.

Photography is unique as both an art and a skill; and success in this vocation often requires, in addition, good business sense. The books chosen for this bibliography represent all aspects of the subject, but emphasis has been placed on technical and applied skills needed to get started in a career opportunity. Because of the vast array of titles in this area, most of the artistic selections focus on collections rather than individual artists. Effort has been made to present the novice student with as wide a variety of examples of photographic excellence as possible.

Theoretical works are listed in the "Analysis" section and explore recent trends. In the past, photographs usually were studied either as works of art or as factual documents. Today the field has expanded to include popular culture and its relationship to photography—considering photographs as descriptors not only of events but also of the attitudes and culture that produced them. With current technologies it has become increasingly difficult to trust photographs to be "factual." The merging of traditional graphic arts with computer wizardry blurs the line of reality, producing thought-provoking imagery. Several books in this section consider the possibilities and the dangers of these technologies.

Selected Associations

The *Encyclopedia of Associations* (28th ed., Gale, 1994) lists more than twenty professional photography associations, ranging from the Industrial Photographers of New Jersey (60 members) to the Associated Photographers International with more than 30,000 members. Major organizations of interest to community college faculty and students include:

Advertising Photographers of America
27 W. 20th St.
New York, N.Y. 10011

American Society of Magazine Photographers
419 Park Ave. S., No. 1407
New York, N.Y. 10016

National Press Photographers Association
3200 Croasdaile Drive, Suite 306
Durham, N.C. 27705

Professional Photographers of America
1090 Executive Way
Des Plaines, Ill. 60018

Professional Women Photographers
c/o Photographics Unlimited
17 W. 17th St., No. 14
New York, N.Y. 10011

Society of Photo-Technologists
6535 S. Dayton, Suite 2000
Englewood, Colo. 80111

Selected Journals

American Photo.
Hachette Magazines. Bimonthly. ISSN 1046-8986

Aperture.
Aperture Foundation. Quarterly. ISSN 0003-6420

Camera and Darkroom.
Camera and Darkroom Photography. Monthly. ISSN 1056-8484

Creative Camera.
Cornerhouse Publications. Bimonthly. ISSN 0011-0876

Darkroom and Creative Camera Techniques.
Preston Publications. Bimonthly. ISSN 0195-3850

Petersen's Photographic.
Petersen Pub. Co Monthly. ISSN 0199-4913

Photo Design.
Billboard Publications. Monthly. ISSN 0888-5680

Photo Electronic Imaging.
Professional Photographers of America. Monthly. ISSN 0146-0153

Popular Photography.
Hachette Magazines. Monthly. ISSN 0032-4582

Step-By-Step Graphics.
Dynamic Graphics. Six times a year. ISSN 0886-7682

Print Materials

✦ Reference

Cason, Jeff
The Photo Gallery and Workshop Handbook
New York, Images Press, 1991. 196p. ISBN 0929667085
   This unusual book for the photographer or photography lover includes lists of U.S. and international galleries and museums that exhibit photographs. Also included are tips for presenting the portfolio. The second section lists workshops, seminars, and photographic tours around the world. Indexed.

DuBoff, Leonard D.
The Photographer's Business and Legal Handbook
New York, Images Press, 1989. 116p. ISBN 0929667026
   A clearly written and readable guide covering most of the generally applicable legal questions that a photographer should consider. In addition to covering copyright, model releases, privacy, contracts, taxes (including deductions), and insurance, the book also provides sample forms, guidelines, and a directory of photographic organizations.

McBroom, Michael
McBroom's Camera Bluebook: A Complete, Up-To-Date Price and Buyers Guide to Cameras, Lenses, and Accessories—Fully Illustrated
1994 ed. Amherst, N.Y., Amherst Media, 1993. 239p. ISBN 0936262214
   A comprehensive buyer's guide to new and used photographic equipment. Begins with a concise but clear essay on buying equipment and continues with detailed descriptions of all major cameras and lens lines available in 35mm, medium, and large formats. Current prices are listed for new and used equipment.

Naylor, Colin, ed.
Contemporary Photographers
2nd ed. Chicago, St. James Press, 1988. 1145p. ISBN 0912289791
   Detailed biographies of 750 of the best and most prominent contemporary photographers, selected by a jury of advisors. Entries include biographical data, lists of individual exhibitions, selected group exhibitions, publications, the photographer's own statement on his or her work, and a signed critical essay.

Shaw, Susan, and Monona Rossol
Overexposure: Health Hazards in Photography
2nd ed. New York, Allworth Press, 1991. 320p.
ISBN 0960711864

This is probably the most comprehensive work on health and safety hazards involved in photography. The text clearly and frankly discusses the physiological effects of photographic chemicals and processes, breaking down the processes into six different sections. The authors also include information on criteria for setting up a safe darkroom and studio, and safe disposal of hazardous materials. Includes a detailed bibliography and index.

Stroebel, Leslie, and Richard Zakia, eds.
The Focal Encyclopedia of Photography
3rd ed. Boston, Focal Press, 1993. 914p. ISBN 0240800591

Entries range from concise definitions to comprehensive signed articles on thousands of concepts, terms, processes, and people in photography and related disciplines. Emphasis is on still photography, but also includes motion picture, video, and electronic still photography. Appendices include extensive tabular data.

Willins, Michael, ed.
Photographer's Market: Where and How to Sell Your Photographs
Cincinnati, Ohio, Writer's Digest Books, 1993. 615p.
ISBN 0898796083

An essential handbook for those wanting to make money with their photographs. In addition to essays and how-to articles on selling photographs, this guide offers an annotated list of ad agencies, studios, book publishers, galleries, magazine publishers, and stock photo agencies. Published annually.

✦ Technique and Equipment

Adams, Ansel
Examples, the Making of Forty Photographs
Boston, Little, Brown, 1989. 177p. ISBN 082121750X

Adams answers the question, "How did you make this photograph?" for forty of his favorite and best known prints. Each photograph is accompanied by a narrative describing the conditions, technical details and aesthetic considerations involved with its creation. Includes a glossary.

Brown, Alan, Joe Braun, and Tim Grondin
Lighting Secrets for the Professional Photographer
Cincinnati, Ohio, Writer's Digest Books, 1990. 134p.
ISBN 0898794129

Each of the chapters briefly describes a particular lighting problem such as highlights or shadows, followed by professionally produced examples with specific explanations and diagrams of the problem and solution.

Gibbons, Bob, and Peter Wilson
Night and Low-Light Photography: A Complete Guide
New York, Sterling Pub. Co., 1989. 192p. ISBN 0713721278

Initial chapters on equipment and film are followed by detailed discussions on photographing people, animals, plants, building, and the sky—all in low-light situations. Includes index.

Gleason, Roger
Seeing for Yourself: Techniques and Projects for Beginning Photographers
Chicago, Chicago Review Press, 1992. 187p. ISBN 155652160X

Aimed at the young adult but suitable for all ages. This work guides the reader through a well-planned and clearly explained set of assignments using pinhole cameras, 35mm equipment, and darkroom equipment and processes. Safety is stressed. Assignments and projects are provided and each chapter includes a bibliography.

Harvey, Liz, ed.
Shoot!: Everything You Ever Wanted to Know About 35mm Photography
New York, Amphoto, 1993. 256p. ISBN 0817458697

A hands-on introduction to the fundamentals of using 35mm cameras and film. Provides clear explanations and excellent visual examples of specific techniques in composition, exposure, lighting, and printing. It concludes with a brief appendix on selling the finished product.

Hedgecoe, John
John Hedgecoe's Complete Guide to Photography
New York, Sterling Pub. Co., 1991. 224p. ISBN 0806984260

Divided into three main sections (basic techniques, projects, and broadening your scope), this book provides clear instruction using diagrams and photographs. Each two-page spread deals with a specific concept or project that can be used independently or in sequence.

Hicks, Roger
Successful Black and White Photography: A Practical Handbook
Newton Abbot, England, David and Charles, 1992. 192p.
ISBN 0715398253

This work assumes a certain level of practical experience in handling cameras and darkroom equipment. The author explains in detail the variables in cameras, lenses, film, and darkroom processes, and how they have an effect on the creative process in black-and-white photography. The last half of the book offers suggestions for applying the creative techniques in specific types of photography such as portraiture, landscapes, documentary, and travel.

Hunter, Fil, and Paul Fuqua
Light: Science and Magic: An Introduction to Photographic Lighting
Boston, Focal Press, 1990. 308p. ISBN 0240517962

Provides the beginning photographer a clear and comprehensive explanation of lighting theory. Through each of the several classic lighting problems, the text offers detailed explanations, clear diagrams, and photographic results for a variety of solutions.

Langford, Michael John
Michael Langford's 35mm Handbook
3rd ed., ed. by Judith More
New York, Knopf, 1993. 224p. ISBN 067974634X

A well-designed, clearly written and illustrated guide to 35mm photography. Individual concepts are considered under the seven major headings of cameras, film, solving picture problems, special project, flash, accessories, and special effects.

Lovell, Ronald P.
Handbook of Photography
3rd ed. Albany, N.Y., Delmar Publishers, 1993. 337p.
ISBN 0827352794

Designed for the beginner, this handbook covers the basics of camera operation and technique, film development, and printmaking. Includes a chapter on ethics and law. Contains bibliography, glossary, and index.

Lovell, Ross
Matters of Light and Depth: Creating Memorable Images for Video, Film, and Stills Through Lighting
Philadelphia, Broad Street Books, 1992. 224p. ISBN 1879174030

An in-depth yet practical study of the uses of lighting in photography. Covers theory and practice using concise explanations, clear diagrams, and photographic examples. Includes an excellent glossary of lighting terms.

Marvullo, Joe
Color Vision
New York, Amphoto, 1989. 144p. ISBN 0817436758

More than one hundred color plates illustrate Marvullo's treatise on how to see color and to interpret it on film. Combines color theory of hue and saturation with practical techniques. Brief index.

Meehan, Joseph
The Comprehensive Book of Photographic Lenses
New York, Amphoto, 1991. 144p. ISBN 0817436979

Describes how lenses work and their impact on the creative photographic process. In addition to discussing the relationship of composition and focal length in general, the author specifically examines the particular problems and applications of standard, wide-angle, telephoto, and special lenses. Also includes appendices on view camera movements and on lens testing.

Paduano, Joseph
The Art of Infrared Photography: A Comprehensive Guide to the Use of Black and White Infrared Film
Rev. ed. Amherst, N.Y., Amherst Media, 1990. 76p.
ISBN 0936262036

This book is aimed at the experienced photographer who knows of infrared film but not how it works. It concentrates on the unique aspects of focus, filtering, and exposure. It then briefly discusses the requirements for developing, printing and toning. The final two-thirds of the book is devoted to examples of infrared photography.

Schaefer, John Paul
Basic Techniques of Photography: An Ansel Adams Guide
Boston, Little, Brown, 1992. 389p. ISBN 0821218018

This text provides the beginning photographer with the technical knowledge needed to produce creative photographs in the style of Ansel Adams. It offers clear explanations of camera and lens and filter systems, discusses visualization, and describes black-and-white film exposure, processing, and printmaking. The last chapters cover color photography in less detail. Quotations from Adams and examples of his work are used throughout the text to demonstrate his techniques.

Tomosy, Thomas
Camera Maintenance and Repair.
Ed. by Michael McBroom Amherst, N.Y., Amherst Media, 1993. 172p. ISBN 0936262095

A most clear and comprehensive work covering the process of cleaning, maintaining, and repairing cameras and photographic accessories. Includes step-by-step procedures, tool lists (including test equipment you can make), and parts suppliers. The second section provides specific instructions for dismantling more than two dozen popular cameras.

Zuckerman, Jim
Outstanding Special Effects Photography on a Limited Budget
Cincinnati, Ohio, Writer's Digest Books, 1993. 135p.
ISBN 0898795532

Provides very clear explanations of the creation of photographic special effects. Addresses both the creative aspect of formulating ideas, as well as the technical elements involved with the actual creation. Chapters include a discussion of low cost techniques and equipment, as well as detailed instruction on the specific techniques of multiple exposure, sandwiching, and masking.

✦ Applied Photography

American Photography Showcase
New York, American Showcase, 1982- . Annual ISSN 0278-8314

This annual showcases outstanding examples of commercial photography, grouped regionally by studio or artist. Extensive "grey pages" provide directories of representatives, photographers, stock services, and production services.

Brown, Nancy
Photographing People for Stock
New York, Amphoto, 1993. 144p. ISBN 0817455000

The author, a commercial photographer in New York City for seventeen years, explains the nature of the stock photo market, and presents detailed technical and stylistic suggestions for success. Notes and index.

Davis, Harold
Photographer's Publishing Handbook
New York, Images Press, 1991. 183p. ISBN 0929667077

A comprehensive manual that guides photographers in compiling, printing, publishing, and marketing their work. Includes interviews, lists of resources, a glossary, and sample forms.

Evans, Arthur G.
Photo Business Careers
Redondo Beach, Calif., Photo Data Research, 1992. 159p.
ISBN 0962650854

Provides a comprehensive and readable discussion of career options available in photography. In addition to a basic summary of each option, essays by professionals in the field offer more personal insights. Also includes guidelines on setting up a business.

Fitzharris, Tim
The Audubon Society Guide to Nature Photography
Boston, Little, Brown, 1990. 167p. ISBN 0316284491

A clear step-by-step approach, combined with color photographs and drawings, introduces the reader to all aspects of nature photography. Chapters cover the camera and accessories, film selection, composition, and techniques for specific situations.

The Idealizing Vision: The Art of Fashion Photography
New York, Aperture Foundation, 1991. 127p. ISBN 0893814628

Profusely illustrated, this work considers contemporary fashion photography and its relationship to journalism and popular culture. Varied essays and interviews explore the power of the fashion photograph.

Kelvin, George V.
Illustrating for Science: [A Problem-Solving Approach to Rendering Subjects in Biology, Chemistry, Physics, Astronomy, Space Technology, Medicine, Geology, and Architecture]
New York, Watson-Guptill, 1992. 192p. ISBN 0823025403

Designed for the artist entering the expanding field of scientific illustration, this book features many techniques dealing specifically with photography. Index.

Kieffer, John
The Photographer's Assistant
New York, Allworth Press, 1992. 208p. ISBN 0960711880

Suggests strategies for obtaining and keeping a job as a photographer's assistant, with tips on making the most of the opportunity. Chapters cover basic techniques and feature "summary of responsibilities" checklists. Professional photographers discuss their assistants in candid interviews. Includes brief appendices and index.

Kopelow, Gerry
How to Photograph Buildings and Interiors
New York, Princeton Architectural Press, 1993. 127p. ISBN 0910413703

A basic primer on architectural photography for readers wishing to execute, commission, or simply enjoy fine images. Covers basic technical aspects, specific applications, and business considerations. Glossary and index.

Norton, Boyd
The Art of Outdoor Photography: Techniques for the Advanced Amateur and Professional
Stillwater, Minn., Voyageur Press, 1993. 152p. ISBN 0896581594

Discussing both equipment and technique, photographer Boyd Norton combines solid information with personal anecdotes. Included are chapters on wildlife and underwater photography. Indexed.

Orenstein, Vik
Creative Techniques for Photographing Children
Cincinnati, Ohio, Writer's Digest Books, 1993. 139p. ISBN 0898795435

Written for the parent or amateur but suitable for the professional photographer as well, this manual covers both technical and artistic aspects of achieving natural, lively portraits of children. Brief index.

Photographers at Work: A Smithsonian Series
Washington, D.C., Smithsonian Institution Press, 1990- .

This outstanding series by the Smithsonian provides great insight into the creative process of contemporary photographers. Each title contains a section of interview questions and the photographer's response, a series of photographs, a signed critical essay about the photographer, and a page of technical information on the photographer's camera(s), darkroom, equipment, and materials. Titles include:

> *Creating a Sense of Place: Photographs by Joel Meyerowitz.* 1990. 63p. ISBN 1560980044
> *Dancers: Photographs by Annie Liebovitz.* 1992. 54p. ISBN 156098208X
> *Everyday Things: Photographs by Neil Winokur.* 1994. 63p. ISBN 1560984139
> *Family Pictures: Photographs by Nicholas Nixon.* 1991. 59p. ISBN 1560980494
> *Friends and Relations: Photographs by Tina Barney.* 1991. 59p. ISBN 1560980486
> *Hotel Room with a View: Photographs by Bruce Weber.* 1992. 61p. ISBN 1560981474
> *Maria: Photographs by Lee Friedlander.* 1992. 63p. ISBN 1560982071
> *Minor League: Photographs by Andrea Modica.* 1993. 64p. ISBN 156098290X
> *On Assignment: Photographs by Jay Maisel.* 1990. 63p. ISBN 1560980028
> *Panoramas of the Far East: Photographs by Lois Conner.* 1993. 60p. ISBN 1560983310
> *Partial to Home: Photographs by Birney Imes.* 1994. 63p. ISBN 1560984120
> *Photo Essay: Photographs by Mary Ellen Mark.* 1990. ISBN 1560980036
> *Pure Invention—The Tabletop Still Life: Photographs by Jan Groover.* 1990. 63p. ISBN 1560980052
> *Travels in the American West: Photographs by Len Jenshel.* 1992. 61p. ISBN 1560981482
> *The Wild West: Photographs by David Levinthal.* 1993. 63p. ISBN 1560982918

Satterwhite, Joy, and Al Satterwhite
Lights! Camera! Advertising!
New York, Amphoto, 1991. 144p. ISBN 0817442065

A frank and detailed discussion of photographic problems in major advertising campaigns. Issues covered include shooting on location, set construction, and aerial photography. Each chapter uses a major ad campaign for demonstration purposes.

Saunders, Dave
Professional Advertising Photography
London, Merehurst Press, 1988. 160p. ISBN 1853910139

Begins with the overall advertising process and includes topics such as concept development, psychology, and campaign process, with an eye toward photographic considerations. The second half deals with topics such as image manipulation, use of animals, and taboo themes.

✦ Retrospective Collections

Adams, Robert
To Make It Home: Photographs of the American West
New York, Aperture Foundation, 1989. 175p. ISBN 0893813516

Adams photographs the open spaces of the American West, juxtaposed with man's impact upon this landscape. For over two decades, his images have captured both the beauty and the ugliness of this combination. Includes illustrative quotations, an essay by the artist, and a detailed chronology.

Aperture Masters of Photography
New York, Aperture, 1987- .
This series has taken the world's most historically significant photographers and provided a sampling of their greatest works. With a minimum of text the publishers have gathered a comprehensive collection of the world's greatest images. Titles include:
    Alfred Stieglitz, 1989. 95p. no. 6 ISBN 0893813087
    Edward Weston, 1988. 95p. no. 7 ISBN 0893813044
    Walker Evans, 1993. 96p. no. 10 ISBN 0893815519
    Andre Kertesz, 1993. 95p. no. 11 ISBN 0893813621

Bunnell, Peter C.
Minor White: The Eye That Shapes
Princeton, N.J., Art Museum, Princeton University, 1989. 289p. ISBN 0943012090
This work is a catalog for the first retrospective exhibition of American photographer and teacher Minor White. In addition to nearly two hundred photographs, the catalog includes a biographical chronology, examples of White's unpublished writings, a chronological survey of images, and a bibliography of his published writings.

Celant, Germano
Ugo Mulas
New York, Rizzoli, 1990. 1 v. (unpaged). ISBN 0847812723
A catalog to accompany a major retrospective exhibition in Milan, this volume includes essays, a biography, a bibliography, and an exhibition list celebrating the remarkable body of work of Italian photographer Ugo Mulas (1928-1973).

Chiarenza, Carl
Chiarenza: Landscapes of the Mind
Boston, Mass., Godine, 1988. 159p. ISBN 0879237244
An in-depth look at Chiarenza's brooding abstract images is accompanied by a record of his career up to the time of this publication. The black-and-white photographs challenge the viewer and blur the boundaries between reality and imagination.

Collins, Douglas
Photographed by Bachrach: 125 Years of American Portraiture
New York, Rizzoli, 1992. 192p. ISBN 084781615X
Bachrach, the oldest photography studio in the U.S., has been creating portraits of illustrious Americans since the 1860s. The book's photographs illustrate the "intelligent flattery" of the Bachrach style, while the text details the unique success of a family business. Brief index.

Ewing, William A.
Flora Photographica: Masterpieces of Flower Photography, 1835 to the Present
New York, Simon and Schuster, 1991. 224p. ISBN 067174447X
Noted historian and curator Ewing has gathered more than 180 works by over 100 photographers, covering a century and a half of creative imagery. The technical variety is matched by the aesthetic range; both are enhanced by high quality reproduction. Includes notes and index.

Fulton, Marianne
Mary Ellen Mark, 25 Years
Boston, Little, Brown, 1991. 192p. ISBN 0821218379
Mark excels in the documentary photoessay and is known for her depiction of people on the fringes of society. Many of her most striking images are presented in this volume, which is the first complete retrospective of her work.

Galassi, Peter
Pleasures and Terrors of Domestic Comfort
New York, Museum of Modern Art, 1991. 127p. ISBN 0810960974
Sixty-two photographers illustrate families at home and suggest that "everyday life" is often far from ordinary. The photographs were gathered for a 1991 major exhibition at the Museum of Modern Art, and are accompanied by an extensive introduction and brief biographical information on the photographers represented.

Gleick, James
Nature's Chaos
New York, Viking, 1990. 125p. ISBN 0670835323
Color photographs by Eliot Porter illustrate the patterns, relations, and interactions present in nature's disorder and wildness. Gleick, the noted author of *Chaos*, comments on the images in an extensive essay.

Greenough, Sarah, and others
On the Art of Fixing a Shadow: One Hundred and Fifty Years of Photography
Washington, D.C., National Gallery of Art, 1989. 510p. ISBN 0821217577
This work is the catalog of a traveling exhibition celebrating the development of the art of photography from 1839 to 1989. Organized by the National Gallery of Art and the Art Institute of Chicago, the catalog includes more than 400 photographs representing the work of 221 photographers.

Hambourg, Maria Morris, and Christopher Phillips
The New Vision: Photography Between the World Wars: Ford Motor Company Collection at the Metropolitan Museum of Art, New York
New York, Metropolitan Museum of Art, 1989. 318p. ISBN 0870995502
This catalog of the Ford Motor Company Collection includes photographs by Man Ray, Evans, Stieglitz, Abbott, and others. The authors contribute historical essays discussing the photographs' relationships to technology and the rapidly changing modern world. Includes extensive notes and index.

Heyman, Therese Thau, ed.
Seeing Straight: The f.64 Revolution in Photography
Oakland, Calif., Oakland Museum, 1992. 158p. ISBN 029597219X
Published to accompany a traveling exhibit, this work is an account of the photographers who formed Group f.64 in Oakland during the 1930s. In addition to excerpts from the photographers' own writings and the exhibit photographs, there are also three essays analyzing the group's work.

Karsh: The Art of the Portrait
Ottawa, National Gallery of Canada, 1989. 176p.
ISBN 088884591X

This work is a companion publication to an exhibit of Canadian portrait photographer Yousef Karsh. In addition to nearly 100 plates of Karsh's best portraits, the book includes essays on "The Art of the Portrait," "The Psychological Portrait," and "The Portraitist at Work."

Lorenz, Richard
Imogen Cunningham: Ideas without End: A Life in Photographs
San Francisco, Chronicle Books, 1993. 180p. ISBN 0811803902

Cunningham was one of the first successful female photographers and was an early advocate of the medium as an art form. This volume is the first major retrospective of her work.

McKenna, Rosalie Thorne
A Life in Photography
New York, Knopf, 1991. 279p. ISBN 0394573943

Part autobiography, part photographic album, this work highlights Rollie McKenna's career over four decades. An internationally celebrated portraitist, McKenna has been honored for her photographs of noted poets and authors.

Mann, Sally
At Twelve: Portraits of Young Women
New York, Aperture, 1988. 53p. ISBN 0893813303

An outstanding example of the contemporary single-artist photography collection. Mann's girl/woman scenes, photographed in her Virginia hometown, are both delightful and disturbing.

Master Photographs: Master Photographs from PFA Exhibitions, 1959-67
New York, International Center of Photography, 1988. 168p.
ISBN 0933642121

Selections of illustrative, documentary, and expressive photographs, accompanied by thoughtful essays, attempt to answer the question, "What makes a great photograph?" Includes appendices and index.

Monk, Lorraine
Photographs That Changed the World
New York, Doubleday, 1989. 51p. ISBN 0385261950

A collection of fifty-one of the most significant photographs in history, beginning with Joseph Niepce's first "permanent" photograph of Paris rooftops, taken in 1826, up to William Anders' "Earthrise" taken from lunar orbit in 1968. Each photograph has accompanying text.

Mora, Gilles, and John T. Hill
Walker Evans: The Hungry Eye
New York, Abrams, 1993. 366p. ISBN 0810932598

This work is the definitive retrospective of Evans' career. The authors trace the photographer's work through five periods beginning with his early abstractions and continuing with his architectural work, travels to Cuba and Tahiti, work for the Farm Security Administration, and finally his work for Time, Inc. Each section includes a brief commentary on the period.

Mutter, Scott
Surrational Images: Photomontages
Urbana, Ill., University of Illinois Press, 1992. 1 v. (unpaged).
ISBN 0252019350

Mutter is a modern master of photomontage, the combining of two or more negatives to create thought-provoking images. Thirty-five selected photographs are accompanied by commentary.

National Museum of American Art (U.S.)
Between Home and Heaven: Contemporary American Landscape Photography from the Consolidated Natural Gas Company Foundation Collection of the National Museum of American Art, Smithsonian Institution
Washington, D.C., National Museum of American Art, 1992. 176p.
ISBN 0826313647

This 1992 exhibit catalogue presents an unusual collection of contemporary landscapes, enlivened both by modern photographic techniques and by society's increased understanding of ecological interdependence.

Photography at the Bauhaus
Cambridge, Mass., MIT Press, 1990. 362p. ISBN 0262061260

More than 400 illustrations, with extensive text, create a comprehensive reference work that analyzes the legacy of Bauhaus photography—as art, as reportage, and as documentary. Includes index, biographies and a selected bibliography.

Plachy, Sylvia
Sylvia Plachy's Unguided Tour
New York, Aperture, 1990. 1 v. (unpaged) with 45 rpm sound recording ISBN 0893814318

This unique compilation of *Village Voice* photojournalist Plachy's work is accompanied by an original music composition by singer and songwriter Tom Waits.

Sobieszek, Robert A.
The Art of Persuasion: A History of Advertising Photography
New York, Abrams, 1988. 208p. ISBN 0810914697

An exhibition catalogue that illustrates and analyzes over a century of advertising photography, touting primarily American products. Large black-and-white and color prints are arranged chronologically and illustrate the increasingly blurred distinction between commercial and "fine art" photography.

Sullivan, Constance, ed.
Women Photographers
New York, Abrams, 1990. 263p. ISBN 0810939509

Presents the work of seventy-three female artists from the mid-nineteenth century to the present. The 200 photographs include vintage, black-and-white, and color prints, many never before published. An essay by art historian Eugenia Parry Janis accompanies the plates.

Szarkowski, John
Photography Until Now
New York, Museum of Modern Art, 1989. 343p.
ISBN 0870705733

Reviewing the 150 years of photography, this history emphasizes the interrelationships between technical innovation and artistic expression. More than 160 plates and numerous smaller photographs illustrate the text. Includes extensive notes and index.

Willis-Thomas, Deborah
VanDerZee: Photographer, 1886-1983
New York, Abrams, 1993. 192p. ISBN 0810939231

Extensive plates and thoughtful text illuminate the life and art of James VanDerZee, a prominent Harlem photographer from 1916 to the 1940s. After declining into obscurity and poverty, his work was rediscovered in the '70s. His career continued with a series of contemporary portraits, which are presented in this volume. Includes notes and bibliography.

Willumson, Glenn Gardner
W. Eugene Smith and the Photographic Essay
New York, Cambridge Univ. Press, 1992. 351p. ISBN 0521414644

Author Willumson examines the work of Post-World War II photojournalist W. Eugene Smith using four works produced for *Life* magazine. Each photo-essay is analyzed for concept, narrative and aesthetic reading, political and ideological reading, and public reception. The photo-essay concept is also examined. Many of the photographer's notes are included in appendices.

Wood, John, ed.
America and the Daguerreotype
Iowa City, Iowa, Univ. of Iowa Press, 1991. 273p. ISBN 0877453349

Eight scholarly essays accompany hundreds of previously unpublished daguerreotype images, primarily from the nineteenth century. Coverage goes beyond traditional portraiture to feature active images of America at work and play. Includes notes, list of contributors, bibliography, and index.

✦ Analysis

Barrett, Terry Michael
Criticizing Photographs: An Introduction to Understanding Images
Mountain View, Calif., Mayfield Pub. Co., 1990. 180p. ISBN 0874849063

Examines the process of photographic criticism using the four major activities of describing, interpreting, evaluating, and theorizing. Uses works by Richard Avedon, Henri Cartier-Bresson, and Robert Mapplethorpe for analysis. Appendices include guidelines for "Talking About Photographs" and "Writing About Photographs."

Carlebach, Michael L.
The Origins of Photojournalism in America
Washington, D.C., Smithsonian Institution Press, 1992. 194p. ISBN 1560981598

A fascinating study of early attempts to combine photographs and words to inform and influence the reading public in nineteenth-century America. Contains glossary, extensive notes, a selected bibliography, and index.

Curtis, James
Mind's Eye, Mind's Truth: FSA Photography Reconsidered
Philadelphia, Temple Univ. Press, 1989. 139p. ISBN 087722627X

In the 1930s, the Farm Security Administration dispatched staff photographers to document rural conditions throughout America. The resulting photographs awakened the public to conditions of drought, poverty, and disease. Curtis analyzes the work of five of the most famous of these photojournalists: Roy Stryder, Walker Evans, Dorothea Lange, Arthur Rothstein, and Russell Lee. Their cultural and political values, along with those of their intended audience, influenced and subtly manipulated their persuasive images. Extensive notes and an index are provided.

Goldberg, Vicki
The Power of Photography: How Photographs Changed Our Lives
New York, Abbeville Press, 1991. 279p. ISBN 1558590390

Goldberg examines more than one hundred famous photographs and their interactions with history, popular culture, and the media. In-depth essays reveal these images as both truth and fiction, with the power to influence minds and events. Extensive notes and index.

Mitchell, William J.
The Reconfigured Eye: Visual Truth in the Post-Photographic Era
Cambridge, Mass., MIT Press, 1992. 273p. ISBN 0262132869

With an analysis and technical discussion of digital image technology, this book asserts that traditional film-based photography is "dead" and discusses the legal, social, and journalistic implications. Includes notes, extensive bibliography, and index.

Ritchin, Fred
In Our Own Image: The Coming Revolution in Photography: How Computer Technology Is Changing Our View of the World
New York, Aperture, 1990. 158p. ISBN 0893813982

Computer-based imaging technology increasingly "enhances" the photographs we view, bringing into question the objectivity of the image. Ritchin considers the ways the public must understand, interpret, and ultimately control this new medium. Includes notes and index.

Trachtenberg, Alan
Reading American Photographs: Images as History: Mathew Brady to Walker Evans
New York, Hill and Wang, 1989. 326p. ISBN 0809080370

The author "reads" photographs as cultural events beginning with the Civil War images of Brady and others and continuing to the documentaries of Evans. Trachtenberg suggests ways in which these artists both shaped and were influenced by their worlds. Includes notes and index.

Nonprint Materials

American Image: 150 Years of Photography
Rochester, N.Y., Kodak, 1988. 1 videocassette (60 min.) VHS

Narrated by Hal Holbrook, this program traces the development of photography from the daguerreotype to film packs and roll film, from black-and-white to color. Shows how technological change led to changes in the use of photography. Also demonstrates the role that photography has played in history, social change, and art, through the works of Matthew Brady, William Henry Jackson, Alfred Stieglitz, Dorothea Lange, and Ansel Adams.

Exploring the Mysteries of Film
Wilmington, Del., Dupont Imaging Systems Dept., 1988.
1 videocassette (16 min.) VHS

Aimed at the student, this program provides a brief history of film. It discusses various topics on photographic film: how it works, its physical structure, characteristics, uses, and potential problems. Also included is a companion booklet and troubleshooting guide.

High Heels and Ground Glass: Pioneering Women Photographers
New York, Filmmaker's Library, 1990. 1 videocassette (29 min.) VHS

Introduces the viewer to the careers of five outstanding women photographers: *Harper's* photographer Louise Dahl-Wolfe; photojournalist Gisele Freund; Hollywood photographer Maurine Loomis; teacher/consultant Lisette Model; and Japanese artist/photographer Eiko Yamazawa. Through interviews and examples the viewer learns about their backgrounds, what they looked for in a photograph, and what factors influenced their growth as photographers.

Portrait of Imogen
New York, New Day Films, 1988. 1 videocassette (28 min.) VHS

Meg Partridge presents an insightful biography of Imogen Cunningham through the representation of more than 250 examples of her work. Partridge enhances the image with the text of Imogen Cunningham's own comments on her works and her life. The program shows how Cunningham has influenced, and been influenced by, American photography.

Visual Communication/Graphic Arts

Judy Goodyear
Carroll Community College
Westminster, Maryland

Introduction

If you have tried to find your favorite cereal in the supermarket lately, you have some idea of the dazzling array of competing products available on the market. As more products and services vie for consumers' attention, the value of graphic design as a force in marketing is increasingly appreciated. Concurrently, technology has changed the graphic design process in the creation and production stages. Desktop publishing has enabled the uninitiated to enter the realm of graphic design without having the corresponding visual expertise. These factors have contributed to a greater need and demand for information about visual communication, reflected in the overwhelming quantity of material published since 1987.

For the purposes of this bibliography, and in accordance with current thinking in the discipline, the entire field of what formerly was called "commercial art" or "graphic art" may be more appropriately referred to as "visual communication." Although the term could be construed as encompassing many disciplines (such as psychology, semiotics, and human physiology), for the purposes of this bibliography, its focus is material for which a need might be expected in an art-oriented curriculum in a community college.

The bibliography is organized into seven primary areas, and is then subdivided by its related components:

- advertising design, the use of text and images to sell a product or service, environmental design;
- graphic design, concerned with the two-dimensional design process;
- publication design;
- computer graphics, creation of computer-generated images;
- illustration, creating an image that enhances our understanding of ideas;
- printing and desktop publishing, and related areas;
- vocational resources

Criteria for books chosen for this bibliography were that they should have reasonably appropriate subject matter, acceptable quality of manufacture, a practical rather than a theoretical approach, a publication date of 1988 or later, and inclusion in *Books in Print* at the time the bibliography was being compiled. Although material published in North America has been given preference, innovative graphic design being done in Japan and Great Britain warranted inclusion. Translations, reprints, and paperbound editions of books whose original publication date was earlier than 1988 were not included. Items were included that were fun, whimsical, or challenging; but that may be peripheral if money is not available.

Computer graphics was included as a separate category, as its increasing importance to the field makes buying and keeping a current collection imperative. Much of this material was not specifically oriented to design, but may be of interest to design students. Materials on specific computer graphics software and manuals have not been included. Reviews of specific software are readily available in sources such as *The Datapro Software Finder* (Datapro Research Group, 1991-), *The Software Encyclopedia* (Bowker, 1985/1986-), and numerous other periodicals. Disks and CD-ROMS that accompanied texts were not installed, nor evaluated separately.

It is important to mention what is not included in the bibliography. The visual communications field overlaps with other subjects, particularly in the areas of advertising, photography, multimedia, printing, desktop publishing, data processing, communications, and fine arts. Material that overlapped other subjects has either been included minimally or not at all. Appropriate areas of this bibliography and other sources should also be consulted for possible purchases. (For example, a directory of advertising agencies or a book of the best ads of the century would be useful to a visual communications student, but they are not included here.) Within the fine arts, material on drawing the human figure and anatomy for the artist has been excluded but would be used by visual communications students. Graphic artists may use all of the traditional studio media that other artists use to create their art, but the quantity of newly published materials prohibited its inclusion. Any visual communication collection should have up-to-date material on the tools, techniques, and media used in the creation of studio art., especially airbrush and markers. No clip art, formulaic, or rule-oriented materials have been listed on the basis that they stifle creativity, but students may demand such material and librarians may feel that they are needed in the collection. Dover Press and North Light Books are well-known sources of clip art and there are other sources of copyright-free images, both still and video, readily available.

Placing of many audiovisual materials in the bibliography proved to be impractical for several reasons. There seem to be very few releases in the visual communications field and even fewer in the review media. Many of those that are available do not seem to be especially suitable for classroom use. Potential entries were locally unavailable and could not be obtained through interlibrary loan. Previewing or renting any material in this field is highly recommended before making any purchases, to ensure appropriateness to the classroom setting.

Acknowledgment is extended to Lew Fifield, chair of the Visual Communications Department at the Maryland Institute College of Art, for generously sharing his wide-ranging knowledge and his invaluable advice. In addition, I would like to acknowledge H. Joanne Harrar, Director of Libraries at the University of Maryland, College Park, for providing access to that wonderful resource, and the library staffs at the University of Maryland, College Park, Enoch Pratt Free Public Library, Maryland Institute College of Art, and Carroll Community College for their kind and patient help.

Accreditation

National Association of Schools of Art and Design
11250 Roger Bacon Dr., Suite 21
Reston, Va. 22090

Selected Associations

Currently there are no certifying or licensing agencies for graphic designers. However, there are numerous professional organizations that are influential in setting standards and can provide assistance. The following are representative of the many associations that provide information on education and industry trends in the visual communications fields.

American Center for Design
233 East Ontario, Suite 500
Chicago, Ill. 60611

American Institute of Graphic Arts
1059 Third Avenue
New York, N.Y. 10021

Art Directors Club
250 Park Avenue South
New York, N.Y. 10003

Cartoonists Guild
30 East 20th Street
New York, N.Y. 10003

IDEA-International Design by Electronics Association
c/o Sparkman and Associates
1120 Connecticut Avenue, NW, Suite 270
Washington, D.C. 20036

National Computer Graphics Association
2722 Merrilee Drive, Suite 200
Fairfax, Va. 22031

The One Club for Art and Copy, Inc.
3 West 18th Street
New York, N.Y. 10011

Society for Environmental Graphic Design
One Story Street
Cambridge, Mass. 02138

Society of Illustrators
128 East 63rd Street
New York, N.Y. 10021

Society of Publication Designers
60 East 42nd Street
New York, N.Y. 10165

Type Directors Club of New York
60 East 42nd Street, Suite 1130
New York, N.Y. 10165

Selected Journals

Art Direction.
Advertising Trade Publications. Monthly. ISSN 0004-3109

Communication Arts.
Coyne and Blanchard. Eight times per year. ISSN 0010-3519

Computer Artist.
PennWell Publications. Six times per year. ISSN 1063-312X

Graphis.
Armstutz and Herdeg. Bimonthly. ISSN 0017-3452

How: the Magazine of Ideas and Techniques in Graphic Design.
F and W Publishing. Bimonthly. ISSN 0886-0483

Print.
RC Publications. Six times per year. ISSN 0032-8510

Step-By-Step Graphics.
Dynamic Graphics. Six times per year. ISSN 0886-7682

U & LC.
International Typeface Corp. Quarterly. ISSN 0362-6245

Print Materials

✦ Advertising Design—General

Angeli, Primo
Designs for Marketing: Number One
Rockport, Mass., Rockport Publishers, 1988. 143p.
ISBN 0935603654

A short explanation of the design philosophy of this well-known author and designer is followed by case studies of the design process with twelve clients. Focus is on package design. Additional samples of finished designs are provided. Contains excellent color reproductions.

Beebe, Jack M.
International Video Graphic Design
New York, Library of Applied Design, PBC International, 1991. 187p. ISBN 0866361332

Presents color reproductions of video graphic images used for openings, closings, logos, and IDs, to name a few. Provides brief information on the company or designers and a short explanation of how the images were created.

Business Cards: Dynamic Graphic Design
Glen Cove, N.Y., Graphic Details, 1993. 186p. ISBN 086636188X

Provides well-selected color examples of business cards, arranged by categories such as two-sided, die-cut and embossed, photography and typography, and logos. Lists firms, client, and designer. Includes an index of firms.

Keding, Ann, and Thomas Bivins
How to Produce Creative Advertising: Proven Techniques and Computer Applications
Lincolnwood, Ill., NTC Business Books, 1991. 181p.
ISBN 0844234818

Part One gives general advice on creative problem solving and solutions including market information and strategy. Focuses on print media but some information on broadcast advertising is also included. Includes chapters on self-promotion and portfolios. Part Two deals with the use of the computer as a tool in typography, photography, illustration, layout, and printing. The appendix provides advice on setting up a desktop publishing system. Glossary; index; black-and-white illustrations.

Konikow, Robert B.
Sales Promotion Design
Glen Cove, N.Y., PBC International, 1990. 253p.
ISBN 086636112X

Presents large color photographs of winners of the annual competition sponsored by the Council for Sales Promotion Agencies. Designs are divided by marketing objective (e.g., trial—those intended to attract first-time buyers). The concept of each section is introduced by someone in the field; then case studies of the winners are presented. Includes indexes by agency, client, and awards.

Nelson, Roy Paul
The Design of Advertising
7th ed. Madison, Wisc., Brown and Benchmark, 1994. 413p.
ISBN 0697129330

Nelson's basic introduction provides concise coverage of almost every aspect of advertising: newspaper, magazine, broadcast, direct mail, and posters. Includes information on design principles, layout and production, copy, type, and artwork. A good bibliography is organized by chapters with most items dating from the 1980s. Includes a glossary and a comprehensive index. This work can be used as a textbook and has a "suggested assignments" section.

Silver, Linda, ed.
Print's Best Letterheads and Business Cards 3: Winning Designs from Print Magazine's National Competition
New York, RC Publications, 1994. 180p. ISBN 0915734842

These large color examples of stationery systems were originally published in *Print's* regional design annual. Includes information on the client, design firm, and designer for each. Indexed by clients, firms, and directors/designers.

Squibb, Sharon, with David Squibb
Studio Techniques for Advertising Agencies and Graphic Designers
New York, Watson-Guptill, 1991. 144p. ISBN 0823049396

This is a good, readable introduction to studio techniques that addresses equipment and supplies, typology and typesetting, reproduction methods, preparing art for mechanicals, pasteup and mechanicals, presentation methods, practical studio hints, and the basics of computer use. Includes a helpful glossary and index.

Velarde, Giles
Designing Exhibitions
New York, Whitney Library of Design, 1989. 188p.
ISBN 082301326X

Velarde provides a comprehensive treatment of all types of show spaces, including trade shows, museum and gallery shows, science center exhibitions, and traveling exhibits. Presents an overview of types of exhibits and the skills needed by the designer. Chapters are devoted to design, management, production, completion, maintenance, and evaluation. Includes numerous black- and-white photographs. Examples are mainly British, but the concepts may apply to any country. Includes glossary, bibliography, and index.

✦ Advertising Design—Packaging

Biondo, Charles
Packaging Design 4: PDC Gold Awards: Ten Years of Excellence in Packaging
New York, PBC International, 1988. 250p. ISBN 0866360654

Presents the best designs in several categories such as food, health and beauty, and hardware. Each category has a brief introduction. Each entry lists the product category, award and year, design firm, client, and creators. Includes an index of directors, designers, and clients and firms. Clear, color photographs accompany the text.

Holland, D. K., ed.
Great Package Design
Rockport, Mass., Rockport/Allworth Press, 1992. 191p.
ISBN 1564960013

Interviews with designers, manufacturers, and retailers are interspersed with examples by categories, such as food and beverages, toys, games and pastimes, home and office products. Lists firm, location, creators and, often, brief comments for each design. Includes large color photographs, and an index to design firms and manufacturers.

International Brand Packaging Awards
Rockport, Mass., Rockport Publishers, 1993. 160p.
ISBN 1564960595

Shows one hundred sixty winning designs, largely judged on the basis of how well they address marketing issues. A short preface and introduction summarizing the importance of good package design and trends in the field are followed by large color photographs of the designs. Designer (or firm) and location are listed for each design. Includes a firms index and directory.

Pfeifer, Ken
Compact Disc Packaging and Graphics: The Best Promotional and Retail Packaging
Rockport, Mass., Rockport Publishers, 1992. 186p.
ISBN 156496003X

Presents large color photographs of promotional, retail, and innovative CD packaging. Each example lists CD, artist, label, design participants, and sometimes a brief comment. Specialized, but inspirational.

Sonsino, Steven
Packaging Design: Graphics, Materials, Technology
New York, Van Nostrand Reinhold, 1990. 176p.
ISBN 0442303033

Geared to the design professional and student, this work reviews the design process and uses case studies and large color photographs to emphasize points. Includes sections on computer-aided design, flexible packaging (such as plastic, films, and foils), and other containers such as glass, metal, and board. Final section deals with labeling, barcodes, and legal requirements for the U.S. and U.K. Includes a chronology of packaging, a glossary, bibliography, and index.

✦ Advertising Design—Posters/Billboards/Signage

Ades, Dawn
The Twentieth-Century Poster: Design of the Avant-garde
2nd ed. New York, Abbeville Press, [1990?]. 227p.
ISBN 1558591303

This exhibition catalog from a show at Walker Art Center presents posters mainly from the collection of Merrill C. Berman. Essay topics range from Russian film posters to general history; an interview with Berman is also included. Provides short biographies of notable artists such as Arp, Will Bradley, Chermayeff, Chwast, Delanay, Duchamp, Greiman, Hoech, Lissitzky, Rand, Man Ray, Rodchenko, Schlemmer, Tzara, Van de Velde, and Vignelli. The majority of works are from the 20's. The quality of the plates (mostly color) is good. Indexed.

Fraser, James Howard
The American Billboard: One Hundred Years
New York, Abrams, 1991. 192p. ISBN 0810931168

This handsome book with many full-page color illustrations presents the billboard in a historical and cultural context. An essay precedes the chapter on each decade. Discusses the circumstances surrounding the creation of various billboards and mentions the artists, designers, companies, (and others) involved. The epilogue summarizes information on the artists, technological changes, and trends for the future. A short bibliographic essay highlights collections and other resources. Indexed. The author is the chief librarian at Fairleigh Dickenson University, which houses the archives of the Outdoor Advertising Association of America.

Graphic Design: USA (Firm)
Sign Design: Environmental Graphics
New York, PBC Graphic Details, 1992. 186p. ISBN 0866361782

Presents color photographs of well-designed signs submitted from around the U.S. and selected by a panel of well-known designers for this book. Signs are grouped by categories such as restaurants or medical facilities. Each project spread is preceded by an inset listing client, designers, firm, fabricator, photographer, materials, and a short explanation of the design concept and how it was created. Includes an index of projects, designers, and design firms.

Successful Sign Design: Number 2.
By the editors of *Signs of the Times* magazine. New York, Retail Reporting Corp., 1992. 237p. ISBN 0934590496

Presents large color photographs of signs selected from submissions to the annual contests sponsored by the *Signs of the*

Times, the sign industry trade journal. Categories include ground signs, entry monuments, mounted signs, wall and fascia signs, projecting signs, signage systems, and specialty signs. Designer, fabricator, client, and description of materials and technique are listed for each sign. Most examples are from the United States. Earlier (1989) edition still in print.

Yew, Wei, ed.
Gotcha!: The Art of the Billboard
Edmonton, Alberta, Quon Editions, 1990. 195p.
ISBN 0969443218

Consists chiefly of large color photographs of various billboards, listing art director, design agency, writer, client, and other creative personnel involved in the creation. Includes a name index.

✦ Advertising Design—Trademarks and Logotypes

Carter, David E., ed.
Logos of Major World Corporations
New York, Art Direction Book Co., 1989. 156p.
ISBN 0881080713

Color reproductions of logos. Company index. Companion volume, *Logos of America's Largest Companies* (Art Direction Book Co., 1988).

Goss, Tom, ed.
Print's Best Logos and Symbols: Winning Designs from Print Magazine's National Competition
New York, RC Publications, 1990. 189p. ISBN 0915734656

Presents large color reproductions of over two hundred logos, symbols, and marks published in past editions of *Print's Regional Design Annual* (RC Publications, 1981-). Client, firm, and designer are provided for each entry. Includes an index of clients, firms, art directors/designers.

Kuwayama, Yasaburo
International Logotypes
Rockport, Mass., Rockport Publishers, 1990. 1 v. (various pagings). ISBN 0935603417 (vol. 1)

First published in Japan as *Logotypes of the World* (Kashiwa Shobo, 1984). Includes over fifteen hundred trademarks and logotypes produced between 1970 and 1983 in thirty-four countries. Arranged by categories indicating their distinguishing features, such as letters with shadows, or letters formed by dots. For each, the business, art director, designer, client, year and place designed, and color are listed, when known. Includes an artist and business category index and black-and-white reproductions. No selection criteria listed.

Kuwayama, Yasaburo
International Logotypes
Rockport, Mass., Rockport Publishers, 1991. 206p.
ISBN 0935603670 (vol. 2)

First published in Japan as *Trademarks Collection Europe 1* (Kashiwa Shobo, 1988). Illustrates over two thousand logotypes and trademarks in use in Europe in 1987. Arranged according to categories like "overlap and contact" or "gradation." For each entry, name of organization, business type, location, designer, year of design, and color are indicated, when available. Includes black-and-white reproductions, an industry index, and very brief artist/designer index (only a few entries have an artist or designer listed.) No selection criteria given.

Murphy, John M., and Michael Rowe
How to Design Trademarks and Logos
Cincinnati, Ohio, North Light Books, 1988. 144p.
ISBN 0891342435

This practical, readable guide discusses design aesthetics and the design process. Provides case studies with preliminary designs and illustrates applications of design and design updates. Concludes with advice on developing new brand names, and legal ramifications. Attractive layout and reproductions. Includes a short bibliography. Indexed.

Young, Doyald
Logotypes and Letterforms: Handlettered Logotypes and Typographic Considerations
New York, Design Press, 1993. 301p. ISBN 083063956X

These numerous, well-reproduced designs have been drawn mainly by the author for a variety of clients. Designs are grouped by categories and preceded by introductory essays. Listed for each design are the client, creators, use, and a detailed explanation of the concept and process. Presentation sketches often accompany the design. Includes a type classification scheme, glossary, and bibliography, as well as indexes on typeface, type design, and client.

✦ Graphic Design—General

Beach, Mark
Graphically Speaking
Manzanita, Ore., Elk Ridge, 1992. 322p. ISBN 094338107X

This dictionary covers the fields of advertising, computers, marketing, photography, postal service, public relations, and publishing as they relate to graphic design and printing. Provides clear definitions which reflect both traditional and changing meanings. Illustrated. Cross-references are italicized. Spanish index.

Graphic Design in America: A Visual Language History
Minneapolis, Minn., Walker Art Center, 1989. 264p.
ISBN 0810910365

This exhibition catalog emphasizes graphic design history. Essays on a wide range of topics, such as protest posters and changing technology, are interspersed with short biographies and interviews with fifteen well-known contemporary graphic artists (such as Ivan Chermayeff, Milton Glaser, April Greiman, Leo Lionni, and Paul Rand). Thirty short biographies are in a separate section. Contains selected plates from the exhibition, focusing on the environment, mass media, and government and commerce. The color plates and numerous illustrations in the text are of high quality. Includes a chronology of graphic design from 1829 to 1989 and a bibliography. Indexed.

Grundberg, Andy
Brodovitch
New York, Documents of American Design, 1989. 159p.
ISBN 0810907240

    This is a biography that focuses on the work of the highly influential designer and teacher, Alexey Brodovitch, who was art director for *Harper's Bazaar* from 1934 to 1958. Includes excellent color and black-and-white photographs and chronologies of Brodovitch's life and graphic design. *Masters of American Design* series.

Hanks, Kurt, and Larry Belliston
Rapid Viz: A New Method for the Rapid Visualization of Ideas
Los Altos, Calif., Crisp Publications, 1990. 149p.
ISBN 1560520558

    This text for students and practitioners in architecture, landscape architecture, engineering and design presents a rapid method of drawing for the purposes of expressing ideas and communicating, rather than of creating a work of fine art. Hanks provides a series of principles and imaginative exercises accompanied by black-and-white illustrations.

Heller, Steven, and Seymour Chwast
Graphic Style: From Victorian to Post-Modern
New York, Abrams, 1988. 239p. ISBN 0810910330

    Color and black-and-white illustrations of the graphic art of each period are preceded by a readable synopsis of the style's characteristics and historical and technological contexts. Each style includes a country-by-country survey. Prominent designers are mentioned and all work identified. Includes bibliography, index of designers, and a chronology of styles.

Labuz, Ronald
Contemporary Graphic Design
New York, Van Nostrand Reinhold, 1991. 156p.
ISBN 0442318871

    The purpose of this book is to "document the significant styles and stylists between 1975 and 1990." Provides an overview, followed by chapters introducing the styles and their proponents. International in scope. Covers computer graphics and typographic innovations. Includes references by chapters, illustrations, and index.

Livingston, Alan, and Isabella Livingston
The Thames and Hudson Encyclopedia of Graphic Design and Designers
New York, Thames and Hudson, 1992. 215p. ISBN 0500202591

    This compendium of information on the leading figures in graphic design since 1840 also presents information on artistic movements and technical advances. Includes more than 700 entries with small black-and-white illustrations. A short subject index with cross-references in the text of articles, and a chronology on evolution of graphic design are provided.

Meggs, Philip B.
A History of Graphic Design
2nd ed. New York, Van Nostrand Reinhold, 1992. 508p.
ISBN 0442318952

    This sweeping history of graphic design in five parts begins in Part One with the invention of writing and the alphabet. The second part deals with the origins of printing and typography in Europe. Part Three addresses the industrial revolution and includes Victorian graphics, the Arts and Crafts Movement, and Art Nouveau. Final sections concentrate on the growth and development of modern graphic design in the twentieth century. This readable and informative study details movements, ideas, and personalities. Black-and-white illustrations on nearly every page and short section of color plates. Includes chapter bibliographies. Indexed.

Naylor, Colin, ed.
Contemporary Designers
2nd ed. Chicago, Ill., St. James Press, 1990. 641p.
ISBN 0912289694

    Follows the same general format as other titles in the *Contemporary Arts Series*—a staccato listing of facts, list of exhibitions, works and publications, followed by a signed article that analyzes the designer's contributions. Over 600 entries cover the most prominent designers in the following design fields: architectural, interior, graphic, display, textile, fashion, product, industrial, stage and film. Contemporary refers to "those who are currently active as well as many who have died since 1970, but whose reputations remain essentially contemporary." Oriented to Western design; and seems to emphasize European designers. Includes black-and-white illustrations and cross-references in text.

Pile, John F.
Dictionary of Twentieth-Century Design
New York, DaCapo Press, 1994. 312p. ISBN 0306805693

    The focus of this encyclopedic dictionary is on product design. Includes graphic design, not illustration. Readable descriptions highlight major points for about 1,000 entries. Includes black-and-white photographs and cross-references. Good index.

Remington, R. Roger, and Barbara J. Hodik
Nine Pioneers in American Graphic Design
Cambridge, Mass., MIT Press, 1989. 179p. ISBN 0262181339

    An attractive book with interesting biographies, design philosophies, techniques, and samples of the work of Agha, Brodovitch, Coiner, Golden, Beall, Burtin, Lustig, Sutnar and Thompson in color and black-and-white. Works include posters, advertising, magazines, book jackets, business graphics, and signage. Includes chronology, notes, and bibliography for each designer. Indexed.

Supon Design Group
International Women in Design
New York, Madison Square Press, 1993. 198p. ISBN 094260430X

    Features the work of twenty-three contemporary female graphic designers mainly from Europe and North America. Short biographies, quotations and analysis of the work, with several pages of attractive color examples of the work itself, give a good sense of each designer.

Thomson, Ellen Mazur, comp.
American Graphic Design: A Guide to the Literature
Westport, Conn., Greenwood Press, 1992. 282p.
ISBN 0313287287

    This excellent and thorough annotated bibliography by Thomson, a librarian and designer, provides a comprehensive current and retrospective review of literature of American graphic design and related disciplines. Selection criteria for reference were very comprehensive and include material like the *Encyclopedia of World Art*. Criteria for other areas include frequency of citation, uniqueness of approach, quality of reproductions, and currency. Includes information on theory and practice; education and vocational, professional practice; standards and law; production and layout (including desktop and computer graphics); typography and calligraphy; picture resources; illustration; cartoon and caricature; photography and editorial illustration; and applications. Excluded are "book art," cartography, and stamp design. Includes a final section on annuals and periodicals, and an appendix of associations and organizations. Indexed.

✦ Graphic Design—Process

Arntson, Amy E.
Graphic Design Basics
2nd ed. New York, Holt, Rinehart and Winston, 1993. 228p.
ISBN 0030554837

This broad treatment of graphic design defines and briefly explores the design process and design careers. Covers history, composition, perception, type, layout, camera-ready art, mechanicals, color, advertising, photography, illustration, and computer graphics. Short exercises and projects at the end of each chapter enable its use as a text. The mainly black-and-white illustrations are well chosen. The appendix has information on tools and equipment. Includes a glossary, a bibliography, and an index.

Berry, Susan, and Judy Martin, eds.
Designing with Color: How the Language of Color Works and How to Manipulate it in Your Graphic Designs
Cincinnati, Ohio, North Light Books, 1991. 143p.
ISBN 0891344047

Discusses briefly the functions of color and color associations and gives basic information on hue, tone, and saturation. The balance of the text is arranged in nine sections by opposites, like cool and warm, passive and active, feminine and masculine. Included within each section are color associations, combinations of hue, tone and saturation, shapes and edges, size and proportion, and pattern and texture. Each pair is presented on opposing pages and each page has a summary of the main ideas, often in terms of a marketing concept. Provides large color examples of packaging, promotional materials, book covers, and posters. All but the color association pages have an inset of color variations at the bottom of each page. Indexed.

Binns, Betty, with Sue Heinemann
Designing with Two Colors
New York, Watson-Guptill, 1991. 127p. ISBN 0823013340

Intended to be used as guide to designing with two colors (one color plus black). Begins with overview of two-color printing techniques, preparing art for printing, basic color theory, and production considerations. The balance of the book is devoted to sample color illustrations used with warm or cool, high- or low-intensity colors.

Block, Jonathan, and Gisele Atterberry
Design Essentials: A Handbook
Englewood Cliffs, N.J., Prentice Hall, 1989. 160p.
ISBN 0132001551

In glossary format, the authors present an unorthodox textbook and handbook for two-dimensional design. Contains clear, useful definitions covering visual elements, design principles, media, materials and critical ideas, styles and art, and movements. Black-and-white reproductions amplify the definitions. Provides a brief review of the design process with fifteen design exercises, twenty questions to assist in analyzing a work of art, and critique strategies and questions. Includes an artist index and cross-references.

Eiseman, Leatrice, and Lawrence Herbert
The Pantone Book of Color—Over 1000 Color Standards: Color Basics and Guidelines for Design, Fashion, Furnishings, and More
New York, Abrams, 1990. 160p. ISBN 0810937115

This standard reference to the four-color mixing process used in printing and computer graphics presents color samples with name or number and printing ink formula. This is part of a whole color system that is also available on software for computer applications. The introductory text explains the system, discusses color in general, color terms, color mixing principles, the color wheel, and color combinations. An index of color names is provided.

Gill, Bob
Graphic Design Made Difficult
New York, Van Nostrand Reinhold, 1992. 160p.
ISBN 0442010982

Presents color photographs of the innovative solutions to design problems. Includes large color photographs with short explanations and a client index.

Goldstein, Nathan
Design and Composition
Englewood Cliffs, N.J., Prentice Hall, 1989. 278p.
ISBN 0131999109

A comprehensive treatment which includes discussion of the principles of composition: balance, emphasis, simplicity, hierarchy, and unity followed by chapters on the design elements of line, shape, value, volume, space, texture, and color. Final sections deal with composition topics such as the interaction of elements, compositional structures, and an analysis of the composition of specific art works. Filled with reproductions (mainly black-and-white) of famous works of art which enhance the text. Words in boldface within the text are defined in the glossary. Indexed.

Graphic Design, New York: The Work of Thirty-Nine Great Design Firms from the City that Put Graphic Design on the Map
Organized by D. K. Holland, Michael Bierut, William Drenttel; written by Steven Heller, and others. Rockport, Mass., Rockport/Allworth Editions, 1992. 336p. ISBN 093560362X

Presents color photograph examples of the work of thirty-nine prominent New York design firms interspersed with essays by well-known designers. A brief analysis of each firm's work is included. Indexed.

Lambert, Patricia
Controlling Color: A Practical Introduction for Designers and Artists
New York, Design Press, 1991. 92p. ISBN 0830635599

This readable book of color basics discusses the physical properties of light, mixing theory, hue, value, intensity, temperature, synchronicity, and contrasts. Includes sections on mixing technique and studio projects by chapter. Indexed. Illustrated with color.

Lauer, David A.
Design Basics
3rd ed. Ft. Worth, Tex., Holt, Rinehart and Winston, 1990. 274p. ISBN 0030304229

Lauer provides concise, readable explanations of the elements of two-dimensional design. Text is accompanied by excellent examples (mostly paintings) and illustrations (mainly black-and-white). Includes work by women artists and examples from non-Western cultures. Thorough, logical, and current. Includes a bibliography by subject. Indexed.

Martinez, Benjamin, and Jacqueline Block
Visual Forces: An Introduction to Design
2nd ed. Englewood Cliffs, N.J., Prentice Hall, 1995. 228p. ISBN 0139482903

This well-organized and very complete analysis of general design elements uses clear language accompanied by numerous examples and illustrations from various cultures, time periods, and media to illustrate its points. In addition to examples from the fine arts, this work includes some graphic arts illustrations, predominantly in black-and-white photographs (except for chapters on color). The chapter on color components covers hue, value, intensity; and temperature. Each chapter has an explanatory overview, a discussion of elements, and cross references. Includes a bibliography and index.

Meggs, Philip B.
Type and Image: The Language of Graphic Design
New York, Van Nostrand Reinhold, 1989. 200p. ISBN 0442258461

Discusses the nature of graphic design and its elements. Chapters include information on combining type and pictures, composition of graphic space, "graphic resonance," and a review of the design process with interesting case studies. Illustrated with numerous color illustrations. Includes bibliographical references and index.

Porter, Tom, and Sue Goodman
Designer Primer: For Architects, Graphic Designers, and Artists
New York, Scribner, 1988. 144p. ISBN 0684184575

This introduction to drawing and drafting covers the basics of visual perception and spatial illusion. Includes information on drawing tools, media and techniques, diagrams, projection drawing, orthographics, and multiview drawing.

Rosentswieg, Gerry, ed.
Graphic Design: Los Angeles
New York, Madison Square Press, 1988. 299p. ISBN 0823048918

Presents color photographic examples of the work of thirty-five prominent Los Angeles design firms. Includes concise introductory comments on L.A. design and a brief analysis of each firm's work. Indexed.

Russell, Dale
Colorworks
Cincinnati, Ohio, North Light Books, 1990. 5 v.
ISBN 0891343334 (vol.1: *The Red Book*); ISBN 0891343342 (vol. 2: *The Blue Book*); ISBN 0891343350 (vol. 3: *The Yellow Book*); ISBN 0891343407 (vol. 4: *The Pastels Book*); ISBN 0891343415 (vol. 5: *The Black-and-White Book*)

Based on the four-color system of yellow, magenta, cyan, and black, these books display color combinations for designers. Each book includes a general introduction followed by an explanation of how to use the set. A discussion follows of the featured color in that volume with regard to optical illusion, proportion, texture, psychology, marketing and cultural connotations, and the use of the color in recent history. The main section of each book focuses on four types of color combinations of the twenty-five shades or tints of the featured color. Some of the colors also have additional sample design applications, e.g., such as logos, brochures.

Shibukawa, Ikuyoshi, and Yumi Takahashi
Designer's Guide to Color 5
San Francisco, Chronicle Books, 1991. 128p. ISBN 0877018715

This fifth volume in the *Designer's Guide to Color* series contains over 1,000 color combinations classified by image or impression such as elegance, sporty, or romantic. Classifications are listed in the table of contents followed by words related to the concept. Each section begins with a discussion of the colors followed by the combinations. Each color in the combination has a separate swatch with density and printing designation. Includes a discussion of color combination techniques. Color charts (at end) show the difference of color printed on high-quality coated and uncoated paper.

Smith, Robert Charles
Basic Graphic Design
2nd ed. Englewood Cliffs, N.J., Prentice Hall, 1993. 162p. ISBN 0130658146

This historical overview of graphic design also provides descriptions of different sorts of design jobs and the design process, including fundamentals of layout and production. Includes black-and-white illustrations, a glossary, a bibliography, and an index.

Swann, Alan
Designed Right!
Cincinnati, Ohio, North Light Books, 1990. 144p. ISBN 0891343326

An examination of the use of design elements in different graphic styles. Section One is a brief review of design elements as they were used in some major styles of the past. Section Two looks at design elements like shape, size and format, type, and color. Style as a concept (traditional, retrospective, classic, young, and mass market) is featured in the next section. The final section addresses gearing style to target a particular market. Includes good visuals and a brief glossary. Indexed.

Swann, Alan
Graphic Design School
New York, Van Nostrand Reinhold, 1991. 192p.
ISBN 0442304234

This attractive book presents the basic ideas, principles, techniques, and areas of specialization in graphic design through a combination of brief explanations and exercises. Text is augmented with the liberal use of color reproductions of student solutions. Indexed.

Thornton, Richard S.
The Graphic Spirit of Japan
New York, Van Nostrand Reinhold, 1991. 240p.
ISBN 0442303769

Surveys graphic design in Japan from its early origins through the 1980s with a focus on 1950-1970. Provides a synopsis of trends for each period and reviews major events and designers. Includes lavish color illustrations of the designers' work accompanied by explanations and translations. A glossary of Japanese terms and a bibliography are provided. Indexed.

VanDyke, Scott
From Line to Design: Design Graphics Communication
3rd ed. New York, Van Nostrand Reinhold, 1990. 198p.
ISBN 0442001134

The goal of the book is to integrate graphics into the design development process for environmental design and landscape architecture, using graphics to communicate design ideas to different users. Covers conceptual graphics, representative graphics, construction graphics, and computer-aided drafting and design. Text is accompanied by black-and-white examples. Indexed.

White, Jan V.
Color for the Electronic Age
New York, Watson-Guptill, 1990. 207p. ISBN 0823007324

Although aimed at the desktop publisher, this work would be suitable for beginning design students as well. The premise is that color is more than pretty decoration and should be used as a functional communication tool. Includes chapters on color basics (terminology, perception, characteristics, and connotations), color on the screen (slides and overheads), charts and graphs, type, adding color to black-and-white illustrations, and color photography. The appendix includes information on color specification with information on the Pantone, Munsell, Natural Color, and CIE Notation systems. Glossary, bibliography, and index are provided.

Wilde, Judith, and Richard Wilde
Visual Literacy: A Conceptual Approach to Graphic Problem Solving
New York, Watson-Guptill, 1991. 191p. ISBN 0823056198

These creative graphic design problems and assignments are intended to teach conceptual thinking. A problem is stated, followed by analysis, notes, and sample student solutions, sometimes with comments. This text with its attractive layout and mostly color reproductions could serve as a resource for instructors and as inspiration for students. Indexed.

Wong, Wucius
Principles of Two-Dimensional Form
New York, Van Nostrand Reinhold, 1988. 98p. ISBN 0442291809

A straightforward, elementary guide to design possibilities using form (a shape with volume and thickness). Well illustrated.

Zelanski, Paul, and Mary Pat Fisher
Color
North and South American ed. Englewood Cliffs, N.J., Prentice Hall, 1989. 144p. ISBN 0131512595

This thorough and readable work covers basic information on perception, psychology, color as a compositional element, and color theories. Chapters on color mixing include systems, printing, photography, and light mixtures. Presents brief sections on video and computer graphics. Illustrated with large color photographs. Includes glossary and index.

✦ Publication Design

Ames, Steven E.
Elements of Newspaper Design
New York, Praeger, 1989. 326p. ISBN 0275923304

Espouses the "total page concept" which emphasizes the need for a meshing of typography, photography, and illustration with text and content. Discusses the concept, design principles, typography, creative design, computerization, and redesigning of a paper. Reviews the design style guides of four large newspapers. Includes a bibliography and index.

Harrower, Tim
The Newspaper Designer's Handbook
2nd ed. Dubuque, Iowa, Wm. C. Brown, 1992. 180p.
ISBN 0697133826

The eye-catching layout creates a readable introduction for beginners, giving a brief history of the subject and future trends. The chapter on fundamentals covers vocabulary, tools, principles, and techniques. Additional chapters provide information on story and page design, illustrations, packaging, special effects, and info-graphics. Exercises are included at the end of some chapters, with solutions in an appendix. Glossary. Indexed. Contains numerous illustrations.

✦ Computer Graphics and Multimedia

Badgett, Tom, and Corey Sandler
Creating Multimedia on Your PC
New York, John Wiley, 1994. 469p. ISBN 0471589284

A guide to producing multimedia presentations on IBM compatibles. Includes a CD-ROM of IBM's *Ultimedia* series software with demonstrations of their applications. The text provides an overview of hardware and software applications as well as information on specific packages such as *Storyboard Live*, *Action!* and *IconAuthor*. Appendices contain a reference guide to software including *Ultimedia*, information on data compression technologies, digital audio, video, and file formats, and extensions for various packages. Includes glossary and index. Paperbound.

Burger, Jeff
Desktop Multimedia Bible
Reading, Mass., Addison-Wesley, 1993. 635p. ISBN 0201581124
   This comprehensive overview is useful for people from all backgrounds, including graphic artists. Places multimedia in context for beginners, but also serves as an aid to those needing in-depth information. Provides a basic introduction to the science, technology, and hardware, followed by specific sections on computer graphics, audio, video, and media integration. Within each of these sections, chapters review the technology, the tools, and actual production. Contains some black-and-white illustrations. Appendices contain a lengthy bibliography, a list of professional organizations, and a comprehensive software and product directory. Indexed.

Campbell, Mary V.
Friendly Multimedia
New York, Random House Electronic Publishers, 1994. 210p. ISBN 0679753303
   This paperback might not withstand much use, but at its low price, it could easily be replaced. Designed with the uninitiated in mind, this work provides explanations of multimedia: audio, graphics, video, animation, sound (MIDI), hardware, software in general, and some specific packages. Covers skills needed and techniques for product development including specific applications such as education, business, and entertainment. Limited use of black-and-white illustrations. Indexed.

Carter, David E.
How to Design Logos on Your Computer: For Macintosh and PC Users
New York, Art Direction Book Co., 1993. 240p. ISBN 0881081175
   Provides an introduction to multimedia for the beginning computer artist. Includes fundamentals of hardware and software, design, samples, selling and presentation, type, line, shape, color, logo design, and legal aspects of design. Provides a step-by-step logo design project and a lengthy section of examples, with descriptions of how each design was created. Black-and-white illustrations.

Collin, Simon
The Way Multimedia Works
Redmond, Wash., Microsoft Press, 1994. 128p. ISBN 1556156510
   This popular approach to the concept of multimedia design includes information on hardware setup, sound, graphics, video, animation, and using *Windows*. Presents an overview of authoring packages and provides steps to the creation of a simple presentation using *Microsoft Viewer* or *HSC InterActive*. The reference section addresses common questions, software packages, and copyright issues. Illustrated. Includes glossary and index.

Fetterman, Roger L., and Satish K. Gupta
Mainstream Multimedia: Applying Multimedia in Business
New York, Van Nostrand Reinhold, 1993. 278p. ISBN 0442011814
   Designed for the business person who needs an overview of the technology and potential of multimedia. Covers basic terminology, how it evolved, the market, applications, and future applications. Potentially of interest to graphic artists for its context. Includes glossary and index.

Fisher, Scott
Multimedia Authoring: Building and Developing Documents
Cambridge, Mass., AP Professional, 1994. 286p. and computer disk ISBN 0122575601
   Provides an overview of authoring for anyone who wants to create multimedia documents. Explores the human factor and ways to structure information, with a focus on educational and professional applications rather than on entertainment multimedia. Includes a survey of multimedia, nonlinear thinking, applications, music, and storyboards. Projects and exercises have been integrated with the text. Contains a disk with Macintosh *HyperCard* stacks and Unix shell scripts. Appendices contain information about the accompanying disk. Includes a glossary and an index.

Frater, Harald, and Dirk Paulissen
Multimedia Mania: Experience the Excitement of Multimedia Computing
Grand Rapids, Mich., Abacus, 1993. 513p. and CD-ROM ISBN 1557551669
   This clear presentation provides an overview of current and future applications. Includes a discussion of the fundamental aspects of multimedia: audio, CD-ROM, electronic image processing, and printer output. Also addresses hardware issues, *Windows* and sound recording, MIDI, authoring systems, video, virtual reality, and cyberspace. Appendices include a glossary and a products locator. Indexed.

Gertler, Nat
Multimedia Illustrated
Indianapolis, Ind., Que, 1994. 176p. ISBN 1565299361
   Novices will enjoy this attractive, easy-to-read book with many large color illustrations of multimedia. Provides an explanation (and CD-ROM illustrations) of how multimedia works as well as illustrations of various types of multimedia systems. Discusses the current state and future of multimedia. Includes glossary and index. Paperbound.

Glassner, Andrew S.
Three-D Computer Graphics: A User's Guide for Artists and Designers
2nd ed. New York, Design Press, 1989. 213p. ISBN 0830610030
   This reader-friendly introduction to basic concepts and techniques includes information on geometry and color hardware, surfaces, materials and textures, lighting and shading, fractals, curved surfaces, modeling and polygonal models, rendering, and computer-assisted animation. Indexed.

Greiman, April
Hybrid Imagery: The Fusion of Technology and Graphic Design
New York, Watson-Guptill, 1990. 158p. ISBN 0823025187
   This beautifully printed survey of the graphic design of the well-known artist April Greiman focuses on computer-generated imagery and electronic collaging. The technique and background for each design is clearly explained by the artist.

Grotta, Sally Wiener, and Daniel Grotta
Digital Imaging for Visual Artists
New York, Windcrest, 1994. 661p. ISBN 0830644423
   This readable text provides an overview of digital imaging and its impact on the profession. Discusses copyright and the ethics of using manipulated images, software, hardware, color, prepress, and

practical considerations such as cost. Includes tips for novices. Illustrated with black-and-white illustrations. Includes a lengthy glossary and index.

Gurewich, Ori, and Nathan Gurewich
Easy Multimedia: Sound and Video for the PC Crowd
New York, Windcrest/McGraw-Hill, 1994. 276p. and computer disk. ISBN 0070252572

The authors present the fundamentals of multimedia creation using the accompanying disk of DOS and *Windows* programs. Provides novices with step-by-step directions for creating a slide show using WAV files, MIDI files, animation, and *Video for Windows* and for working with resources on CD-ROM drives. Includes two chapters on programming for readers with some previous experience. Indexed. Paperbound.

Haskin, David
The Complete Idiot's Guide to Multimedia
Indianapolis, Ind., Alpha Books, 1994. 357p. and CD-ROM. ISBN 1567615058

The humorous approach and well-designed layout of this guide painlessly provides novices with the fundamentals of multimedia: what multimedia is, why people like it, hardware, installation, and steps to creating presentations. Inset boxes list important facts and vocabulary. The accompanying CD-ROM provides demonstrations of programs and games. Includes glossary and index. Paperbound.

Holsinger, Erik
How Multimedia Works
Emeryville, Calif., Ziff-Davis, 1994. 198p. and CD-ROM. ISBN 1562762087

A graphical approach to multimedia technologies and production. Illustrates the inner workings of hardware and software with schematic drawings that accompany the text. Explains how multimedia is created from start to finish using techniques of video and audio production such as graphics, morphing, three-dimensional work, and animation. The glossary in the text is also accompanied by illustrations. Indexed.

Jerram, Peter, and Michael Gosney
Multimedia Power Tools
New York, Random House Electronic Publishing, 1993. 640p. and CD-ROM. ISBN 0679791183

A survey of multimedia's tools and technologies including sound, animation, and video. Contains interactive tutorials and offers digital resources for use in publications. Covers hardware and software, authoring tools and techniques, project organization, and information on working across platforms. The CD-ROM includes project case studies, 200 Mb of "power tools," including video clips, animations, and sound clips. Appendices include product information, the text of the *Copyright Act of 1976,* and product copyright information. Glossary and index are provided.

Labuz, Ronald
The Computer in Graphic Design: From Technology to Style
New York, Van Nostrand Reinhold, 1989. 212p.
ISBN 0442009712

Focuses on biographies and analyses of the work of important graphic designers who use computers. Categorizes work according to how visible the use of the computer is in the finished design. Includes a section on type designers. Excellent color reproductions and bibliography.

Lindstrom, Robert L.
Business Week Guide to Multimedia Presentations
Berkeley, Calif., Osborne McGraw-Hill, 1994. 456p. and CD-ROM. ISBN 007882057X

The focus of this work is a business approach to multimedia. Discusses technologies and where to begin, the capabilities of multimedia, tips and techniques for design, the use of sound and motion, and the overall impact of multimedia on an audience. Stresses the value of interactive presentations in marketing and the powers of persuasion. Includes a CD-ROM with copyright-free media clips, sound effects, and animations that can be incorporated into presentations. Indexed.

Luther, Arch C.
Authoring Interactive Multimedia
Boston, AP Professional, 1994. 298p. and CD-ROM.
ISBN 0124604307

This introduction to authoring (or the process of creating multimedia) discusses the overall process of design with examples of specific software packages. Includes information on platforms and operating systems; interfaces; languages; use of still images, graphics, and text; audio, video, and animation; and integrating objects from other applications. Luther also discusses the selection of an authoring tool and current trends. Appendices provide an authoring tool selector, a list for sources of hardware and software, and information about the accompanying CD-ROM (which includes an IBM *Ultimedia* series multimedia sampler). Includes a glossary, a bibliography, and an index.

McClelland, Deke
Painting on the PC: A Non-artist's Drawing Guide to PC Paint, Dr. Halo, Publisher's Paintbrush, and Many Others
Homewood, Ill., Dow Jones-Irwin, 1989. 342p. ISBN 1556232667

An introductory guide to computer graphics with simple, basic information on elementary drawing and graphic arts concepts, like appropriateness, purpose, scale, proportion, depth, volume, and color. Types of computer tools and hints for using them are discussed with some sample projects. Final chapters cover using clip art and scanning images. Includes thorough reviews of specific software, which include hardware requirements.

McCormick, John
Create Your Own Multimedia System
New York, Windcrest/McGraw-Hill, 1995. 398p. and CD-ROM.
ISBN 0070460345

McCormick's target audiences are professionals such as graphic artists and musicians with no computer skills, as well as skilled computer users with limited multimedia experience. Provides basic information on building multimedia with a variety of software and hardware options. Includes a chapter on electronic publishing. Appendices contain information on CD-ROM publishing, various types of boards, scanners and pointing devices, MIDI software, authoring tools, and standard file formats. In addition, a bibliography, glossary, and index are provided. The accompanying disk contains various freeware, shareware, and *Hypertext.*

Mitchell, William J.
The Reconfigured Eye: Visual Truth in the Post-photographic Era
Cambridge, Mass., MIT Press, 1992. 273p. ISBN 0262132869

The author's intent is to provide an analysis of the digital imaging revolution. A brief history and comparison between traditional photography and digital imaging is followed by detailed information on techniques. Chapters include electronic tools, digital brushstrokes, virtual cameras, synthetic shading, and computer collage. Emphasis is on the aesthetic possibilities and the ethical considerations involved in image generation and manipulation.

Norman, Richard B.
Electronic Color: The Art of Color Applied to Graphic Computing
New York, Van Nostrand Reinhold, 1990. 186p ISBN 0442235399

Feels like a textbook with exercises and references at the end of each chapter, but is geared to architects, designers, and students. Discusses color as a design element, color theory, and color systems, then considers the specifics of electronic color. Chapters on color interaction, using color to create the illusion of space, three-dimensionality, and color psychology. Includes bibliography, index, and good color reproductions.

Olsen, Gary
Getting Started in Computer Graphics
Cincinnati, Ohio, North Light Books, 1989. 144p.
ISBN 089134330X

Advice for artists and graphic designers who are computer novices. Designed to be used with any software. Covers hardware, software, drawing and painting, and working with service bureaus. Includes color examples of the work of various artists with explanation of technique on nearly every page. Sample projects with preliminary artwork shows the process. Indexed.

Pike, Jayna
An Introduction to Computer Graphics Concepts: From Pixels to Pictures
Reading, Mass., Addison-Wesley, 1991. 209p. ISBN 020156789X

A book of basic information that discusses how computer graphics are used, followed by information on hardware, software, displays, and technique (surface rendering, scaling, rotating and two- and three-dimensional images, image processing, and visualization). Applications in various fields such as medical imaging are mentioned briefly. Includes lengthy glossary, bibliography, and index.

Rathbone, Andy
Multimedia and CD-ROMs for Dummies
San Mateo, Calif., IDG Books, 1994. 336p. ISBN 15668840896; ISBN 1568842252 (with CD-ROM)

This simple introduction for novices includes hardware, setup, basic instructions, and custom setups. Includes suggestions and tips, a glossary, and index. The accompanying CD-ROM disk addresses some fundamentals, and contains video, graphic, and sound clips. The appendix provides installation instructions and a guide to the CD-ROM.

Rogondino, Michael, and Pat Rogondino
Computer Color: 10,000 Computer-generated Process Colors
San Francisco, Chronicle Books, 1990. 108p. ISBN 0877017395

Directed to anyone making electronic artwork who wants to use computer-generated film output for color separations based on the four-color process. Gives percentages for each color which can be used with Adobe Illustrator 88, Corel Draw, Aldus FreeHand, PageMaker, and Ventura among others. Clearly explains the process. Color swatches make up bulk of book.

Smedinghoff, Thomas J.
The Software Publishers Association Legal Guide to Multimedia
Reading, Mass., Addison-Wesley, 1994. 633p. and computer disk
ISBN 0201409313

This is an excellent guide to all aspects of law which apply to multimedia including copyright, trademarks, unfair competition, right of publicity, right of privacy, defamation, trade secret law, and patent law. Part One discusses the substance of the law in those areas. The second part deals with the impact of the law on the development of multimedia products. Part Three reviews protecting intellectual property rights. Part Four covers the law involved with marketing and distribution of a product. A final section contains checklists and sample agreements. Appendices list unions, professional associations, groups that will assist with gaining clearance for material used, and a glossary. Indexed. Paperbound. A disk with templates for sample agreements and planning checklists is included.

Thorell, L. G., and W. J. Smith
Using Computer Color Effectively: An Illustrated Reference
Englewood Cliffs, N.J., Prentice Hall, 1990. 258p.
ISBN 0139398783

Written primarily for system designers but also intended for creators of graphics and other artists. The same principles could apply to someone designing a system or the graphics. This work covers technical topics, for the most part in nontechnical language comprehensible to the layperson, and focuses on maximizing practical applications. Topics include the physical properties of light, color perception and how it relates to the computer, color applications, displays, hard copy, image quality, and color specification. Book is well organized. Excellent preface and introductions to chapters. Good layout and illustrations (most in color). Each chapter has a summary of useful facts and references. Includes comprehensive glossary and index.

Vaughan, Tay
Multimedia: Making It Work
2nd ed. Berkeley, Calif., Osborne McGraw-Hill, 1994. 560p. and CD-ROM disk ISBN 0078820359

The first four chapters introduce the concept of multimedia including hardware, software, platforms, and basic media (text, sound, images, animation, and video). Chapter Five focuses on assembling and delivering a project: cost analysis, design, production, delivery, and packaging. Includes vocational information for a variety of multimedia positions and a discussion of training opportunities. Includes a CD-ROM which covers limited working versions of all Macromedia products, and hands-on exercises appropriate for Macintosh and *Windows*. Appendix includes information on the CD-ROM and a list of font manufacturers.

Wodaski, Ron
Absolute Beginner's Guide to Multimedia
Indianapolis, Ind., Sams Publishing, 1994. 390p. and 2 CD-ROMs.
ISBN 0672305240

Novices will enjoy this appealing, readable overview of multimedia, accompanied by attractive color illustrations. Includes a brief discussion of common packages, start-up ideas, and tips for getting started. The accompanying CD-ROMs include PC compatible software and sound, video and photo clips. Indexed.

Wodaski, Ron
Multimedia Madness
2nd ed. Indianapolis, Ind., Sams Publishing, 1994. 1100p.
ISBN 0672304139

A reference guide to multimedia products including hardware, software, and tools that interface with multimedia programs. Provides basic background information on multimedia including sound, video, animation, and evaluations of specific software and hardware products. Also addresses the use of add-ons such as MIDI boards. Includes a shopper's guide and two CD-ROMs: the *Nautilus Demo Treasure Chest*, and 1.3 gigabytes of graphic, sound and video clips. Indexed.

Wolfgram, Douglas E.
Creating Multimedia Presentations
Indianapolis, Ind., Que, 1994. 276p. ISBN 1565296672

Explores methods for improving communication skills through the use of multimedia. Introduces background information on the interactive presentation industry, and provides a step-by-step guide through the design and production process, detailed explanations on how to organize staff and resources to produce successful presentations, and exercises for animation. Includes a list of resources for multimedia and directory layout information for CD-ROMs. Paperbound.

✦ Illustration—General

Gikow, Jacqueline
Graphic Illustration in Black-and-White
New York, Design Press, 1991. 143p. ISBN 0830690077

Gikow's helpful guide provides elementary information on the illustration process, working from sources such as photographs, silhouettes, line techniques, wet media, and computer illustration. Explains the various markets for illustration and presents an overview of communication, style, and business issues. Includes suggested exercises and portfolio projects at end of chapters, a brief bibliography, an index of illustrators, and a general index. Illustrations enhance the text.

Gordon, Elliott, and Barbara Gordon
How to Sell Your Photographs and Illustrations
New York, Allworth Press, 1990. 128p. ISBN 0927629054

Geared to the beginning or professional free-lancer, this work provides general advice about free-lancing as well as practical information on potential markets, portfolios, promotion, pricing, business practices, and agents. Includes lists of professional organizations, directories, periodicals, and sample business forms. Authors are well known in the graphic design field, and are partners in a firm that represents photographers and illustrators. Indexed.

Heller, Steven, and Karen Pomeroy
Designing with Illustration
New York, Van Nostrand Reinhold, 1990. 224p.
ISBN 0442232772

The premise of this book is that illustration is still a valid design technique. Consists of interviews with prominent illustrators, designers, and art directors who use illustration in their designs. Interviews discuss the designers' backgrounds, designing with illustration, and aspects of specific works that are featured in handsome color illustrations on nearly every page. Indexed.

Meehan, Joseph
Photography for Graphic Designers
New York, Watson-Guptill, 1993. 128p. ISBN 0823040127

Meehan, a professional photographer, presents a clear, attractive introduction to basic photographic language and techniques. The audience is designers who frequently need to communicate with photographers. The goal is to enhance communication, improve evaluation and use of photographs, and enable designers to do simple photography themselves. Provides information on cropping and retouching as well as electronic photography. Indexed.

Weinberg, Robert
A Biographical Dictionary of Science Fiction and Fantasy Artists
New York, Greenwood Press, 1988. 346p. ISBN 0313243492

A dictionary of more than 250 artists selected for their importance and influence, amount of work done in the field, and historical significance. Brief introductory history is followed by artists listed alphabetically by last name with cross-references from other names. Entries give birth and death dates, nationality, and a list of works published. Biographies were written by many of the artists themselves. Appendices contain narrative information on collections and collectors, art awards, a bibliography, and biographical and general indexes.

✦ Illustration—Books and Magazines

Bicknell, Treld, and Felicity Trotman, eds.
How to Write and Illustrate Children's Books and Get Them Published!
Cincinnati, Ohio, North Light Books, 1988. 144p.
ISBN 0891342648

Includes chapters on picture books in general, illustration, techniques, and production. Types of picture books are discussed, as well as hints on interpreting the text, presentation, portfolios, self-promotion, job specifications, and fees. Techniques chapter includes planning, layout, roughs, style, media, and the "life history of a book." A short glossary of print terminology, information on the protection of artwork, brief legal advice, and sample working methods of two artists are also included. Appendix lists recommended books, periodicals, and organizations. Bibliography and index of books. General index.

Cummins, Julie, ed.
Children's Book Illustration and Design
New York, Library of Applied Design, PBC International, 1992. 240p. ISBN 0866361472

Short biographies and color samples of the work of more than eighty well-known contemporary illustrators, such as Anno, Carle, de Paola, Galdone, Keats, Lionni, Marshall, Tafuri, Zelinsky, and Zemach. In most cases the artists discuss the rationale and technique behind the illustrations. Author, illustrator, and title indexes. Cummins is coordinator of Children's Services for the New York Public Library.

Heller, Steven, and Anne Fink, comps.
Covers and Jackets!: What the Best Dressed Books and Magazines Are Wearing
Glen Cove, N.Y., Library of Applied Design, 1993. 191p. ISBN 0866361952

Divided into book and magazine sections, then by categories like fiction or travel. Brief overviews and history precede each section. Contemporary (most from 1990s). Large, color examples list artists and designers, art directors, and publishers. Indexed. Heller is a well-known designer and author of several books on graphic design.

Marantz, Sylvia S., and Kenneth A. Marantz
Artists of the Page: Interviews with Children's Book Illustrators
Jefferson, N.C., McFarland, 1992. 255p. ISBN 0899507018

Interviews with thirty-one illustrators of children's books, mainly British and American. Some are well-known authors. Includes John Burmingham, Helen Oxenbury, and Paul Zelinsky. Illustrators respond to questions about their backgrounds, how they began, training, and working methods. Interviews are dated and fairly recent or have short updates. Gives a sense of the personalities of those interviewed. Includes photographs of the illustrators and selective lists of their publication. Bibliography. Indexed.

Rollock, Barbara
Black Authors and Illustrators of Children's Books: A Biographical Dictionary
2nd ed. New York, Garland, 1992. 234p. ISBN 082407078X

Provides short, elementary biographies of one hundred-fifty African American authors and illustrators whose work has been published in the United States. Included are those whose work had a historical impact, even if their work is not about African American themes or experiences. Excluded are those who have only had one work published for children. Includes selective bibliographies at end of each sketch. A section of photographs has some portraits and dust jackets. Appendices list awards and honor books, publishers' series, publishers, bookstores, and distributors. Indexed.

✦ Illustration—Cartoon, Caricature, Comics, and Animation

Benton, Mike
The Comic Book in America: An Illustrated History
Dallas, Tex., Taylor Pub. Co., 1989. 207p. ISBN 0878336591

This attractive book with color illustrations traces evolution of the comic book from prehistory, then from 1933 to 1989. Includes chapters on publishers and different genres like jungle, teen, and romance. Indexed.

Benton, Mike
Horror Comics: The Illustrated History
Dallas, Tex., Taylor Pub. Co., 1991. 144p. ISBN 0878336613

Traces the history of horror comics from its origins, then continues by decade from the '50s to the '90s. Discusses the controversy surrounding the genre and reprints the Code of the Comics from the Magazine Association of America, Inc. Provides information on collecting. A guide and checklist provides title, publisher, date, code information, and short descriptions. Bibliography and index.

Benton, Mike
The Illustrated History—Crime Comics
Dallas, Tex., Taylor Pub. Co., 1993. 166p. ISBN 0878338144

A narrative history of crime comic books from the 1930s to the 1990s. Includes large color illustrations, a guide, and checklist with names of publishers, artists, and writers, dates of titles, and brief descriptions. Bibliography. Indexed.

Benton, Mike
Science Fiction Comics: The Illustrated History
Dallas, Tex., Taylor Pub. Co., 1992. 150p. ISBN 087833789X

A decade-by-decade history of science fiction comics from the beginning to the present day. Includes comics from television and movies. The final chapter is a guide and checklist with title, publisher, dates of first and last issues, significant artists, and brief descriptions. Includes numerous color reproductions, many of which are comic book covers. Bibliography and index.

Benton, Mike
Superhero Comics: The Illustrated History
Dallas, Tex., Taylor Pub. Co., 1991. 224p. ISBN 0878337466

Documents the comeback of superhero comics in the late '50s and '60s. Provides career summaries of artists with their art credits. Provides information on over one hundred comic book titles with publishers, dates, major characters, significant issues, artists, and a short analysis. Includes many color reproductions, a chronology (1953-1970), bibliography, and index.

The Best Comics of the Decade
Selected by the Editors of the Comics Journal. Seattle, Wash., Fantagraphics Books, 1990. 2 v. ISBN 0560970367 (vol. 1); ISBN 0560970383 (vol. 2)

A short introduction followed by the comics dating from 1980 to 1990. Comics are contemporary and usually from the alternative press, which may not appeal to every taste.

Callahan, Bob, ed.
The New Comics Anthology
New York, Collier Books, 1991. 287p. ISBN 0020093616

Provides an introductory discussion of contemporary comics that are characterized by nihilism, a violent "doomed" world view, and a punk sensibility. A list of artists is provided with brief summaries of their work. Mainly black-and-white illustrations.

Cawley, John, and Jim Korkis
How to Create Animation
Las Vegas, Pioneer Books, 1990. 191p. ISBN 1556982852

Focus is on traditional cel animation (cartoons) or hand-drawn animation, but includes brief coverage of clay, computer, puppet, silhouette or shadow, and stop-motion animation. Lists the steps needed to produce animation and the jobs related to each step. This work largely consists of interviews with twenty successful animators. Topics include how they got started, advice they could give to beginners, a description of a typical work day, and what qualities sought when evaluating animation. Includes a list of resources. Black-and-white illustrations.

Gerberg, Mort
Cartooning: The Art and the Business
New York, Morrow, 1989. 272p. ISBN 1557100179

Gerberg presents personal advice and reminiscences, geared toward beginners who want to break into the business. Focus of the first half is on the magazine gag cartoon. The balance of the book deals with other areas such as syndicated comic strips, editorial cartoons, television, animation, comic books, advertising, and greeting cards. Appendices contain a list of magazines that publish cartoons with the name of the cartoon editor, standard rate, and type of cartoons published, as well as names and addresses of other buyers of cartoons. Includes a list of resources, a bibliography, and index.

Goulart, Ron
The Comic Book Reader's Companion: An A-to-Z Guide to Everyone's Favorite Art Form
New York, HarperPerennial, 1993. 191p. ISBN 0062731173

Provides informative entries on comic books and their characters by the author of two other books on comics. Information on comic book creators is presented within the text of the articles. Includes black-and-white illustrations.

Goulart, Ron, ed.
Encyclopedia of American Comics
New York, Facts on File, 1990. 408p. ISBN 0816018529

Substantive signed articles about comic strips and their creators provide a good history and the flavor for various comics. Actual strips are reproduced in the text. Includes index.

Halas, John
The Contemporary Animator
Boston, Focal Press, 1990. 128p. ISBN 0240512804

This overview, directed to anyone considering a career in animation, reviews the necessary skills, education, and potential employment opportunities. Explains general production methods, then specific production areas, such as celluloid, cutouts, collage, stop-motion cameras, puppets, clay, and paper. Includes a chapter on computer animation and a comparison of traditional and computer animation. Covers the basic mechanics of sound tracks and musical composition. Includes separate glossaries of traditional and computer animation terms. Bibliography and index.

Heller, Steven, and Gail Anderson
The Savage Mirror: The Art of Contemporary Caricature
New York, Watson-Guptill, 1992. 160p. ISBN 0823046443

This history and survey of portrait caricaturists and their work provides an interesting analysis. Includes good, mostly color, reproductions and an appealing layout. Main focus is post-World War II and contemporary artists. Indexed.

Lenburg, Jeff
The Encyclopedia of Animated Cartoons
New York, Facts on File, 1991. 466p. ISBN 0816022526

A general history of animated cartoons is followed by information on every U.S. television cartoon production "exhibited theatrically" or broadcast on television from 1911 through 1990. Includes silent, theatrical, full-length animated, animated television specials, television series, and award winners. Each entry lists series history, voice credits, year, and a history or summary. Contains awards and honors list, short bibliography, and index.

Locke, Lafe
Film Animation Techniques: A Beginner's Guide and Handbook
White Hall, Va., Betterway Publications, 1992. 160p.
ISBN 1558702369

This unintimidating text provides basic explanation of how animation works, how it developed, and advice on choosing a story and characters. Covers elementary animation (without film or camera) and stop-motion, cutouts, cels, backgrounds, foregrounds, special effects, titles, and sound. Provides brief mention of computer animation. Simple black-and-white drawings. Includes glossary, bibliography, and index.

Marschall, Richard
America's Great Comic-Strip Artists
New York, Abbeville Press, 1989. 295p. ISBN 0896599175

Introduces sixteen comic artists from the Cartoonists Hall of Fame who are representative of the history of the comics. Covers the general history of comic strip art, then describes and analyzes the work of the individual cartoonists. Includes biographies and large reproductions of the artists' works, many in color. Cartoonists include Outcault (Buster Brown), Winsor McCay (Little Nemo), Harold Gray (Little Orphan Annie), Milton Caniff (Terry and the Pirates), Chester Gould (Dick Tracy), Al Capp (Li'l Abner) and Walt Kelly (Pogo). The most contemporary artist included is Charles Schulz. Indexed.

O'Sullivan, Judith
The Great American Comic Strip: One Hundred Years of Cartoon Art
Boston, Little, Brown, 1990. 200p. ISBN 0821217542

A survey history of the comics from the 1890s to the 1980s, with analysis of the strips and the personalities. Includes mostly black-and-white illustrations on nearly every page. Provides a useful section of who's who in the comics containing about one hundred entries for well-known and lesser-known individuals, with a list of references for each. Notes and index.

Roncarelli, Robi
The Computer Animation Dictionary: Including Related Terms Used in Computer Graphics, Film and Video, Production, and Desktop Publishing
New York, Springer-Verlag, 1989. 124p. ISBN 0387970223

Presents clear, concise definitions of approximately fourteen hundred words. Cross-referenced.

Scott, Randall W.
Comic Books and Strips: An Information Sourcebook
Phoenix, Ariz., Oryx Press, 1988. 152p. ISBN 089774389X

This annotated bibliography covers the literature of comic books and strips excluding cartoons and animation. Includes suggestions for a core collection. Also provides lists of books about comics and those that reprint comics. Includes information on periodicals, journals, and library collections. Includes an author, title and subject index. Most entries are from the 1970s and 1980s.

Sennett, Ted
The Art of Hanna-Barbera: Fifty Years of Creativity
New York, Viking Studio Books, 1989. 270p. ISBN 067082781

This large, handsome book interweaves the biographies of William Hanna and Joseph Barbera with a chronological history of their work. Additional features include a chronology of their animated series which lists network, year, principal characters, and voices. Also includes a chronology of their live-action features and series in order of release. Includes color illustrations, a Flintstone glossary, a short bibliography, and an index.

Solomon, Charles
Enchanted Drawings: The History of Animation
New York, Knopf, 1989. 322p. ISBN 0394546849

Covers history of animation, including the personalities, the characters, television, and films from 1914 to 1989. Includes two chapters on Disney and more than fourteen hundred large reproductions—over half in color. Bibliography and index.

Thomas, Bob
Disney's Art of Animation: From Mickey Mouse to Beauty and the Beast
New York, Hyperion, 1991. 208p. ISBN 1562829971

Provides a short history of animation and biographical information integrated with the story of some of the characters. Includes a history of the company, its films and techniques. Details the making of *Beauty and the Beast*, giving a sense of the excitement and complexity of animation. Includes fine, mainly color, illustrations and a short glossary.

Van Hise, James
How to Draw Art for Comic Books: Lessons from the Masters: Corben, Elder, Foster, Kane, Kurtzman, Raymond, Spiegleman, Sprang, Williamson
Las Vegas, Pioneer Books, 1989. 146p. ISBN 1556982542

Discusses the life and work of ten artists, mostly contemporary. Includes quotations and helpful advice on drawing for comic books, black-and-white illustrations, and information on the Joe Kubert School of Cartoon and Graphic Art.

✦ Illustration—Fashion

Abling, Bina
Advance Fashion Sketchbook
New York, Fairchild Fashion Group, 1991. 133p.
ISBN 0870056794

Assumes some knowledge of figure drawing but does have a review chapter on drawing the fashion figure. Covers the rudiments of layout and discusses drawing equipment and techniques, such as using markers, washes, and making line drawings. Covers specifics such as conveying fabric textures. The final section contains brief interviews with fashion illustrators and examples of their work.

Allen, Anne, and Julian Seaman
Fashion Drawing: Basic Principles
London, B.T. Batsford, 1993. 109p. ISBN 0713470968

Presents simple concepts and straightforward, practical advice. Covers anatomy, proportion, composition, light, clothing shapes, accessories, drawing from photographs, media, textures, and storyboards.

✦ Illustration—Lettering/Calligraphy

Emery, Richard
The Creative Stroke: Communicating with Brush and Pen in Graphic Design
Rockport, Mass., Rockport Publishers, 1992. 191p.
ISBN 0935603611

In this antidote to computer imaging, Emery provides striking examples of the use of freehand art, primarily lettering, in contemporary design. Organized into the following categories: media advertising and direct mail, packaging, covers and posters, corporate identity, and gallery. Includes an index of contributors and large attractive color reproductions.

Fink, Joanne, and Judy Kastin
Lettering Arts
Glen Cove, N.Y., PBC International, 1993. 192p.
ISBN 0866362258

This brief introduction to the history of calligraphy includes large color photographs of complete alphabets, commercial lettering, calligraphic art, and a mixture of work originally presented in *Calligraphic Review*. Each section has its own short introduction. Work is identified by artist, title, date, medium, client, and other pertinent information. Includes a bibliography, a list of calligraphic societies, and an artists index.

Folsom, Rose
The Calligrapher's Dictionary
New York, Thames and Hudson, 1990. 144p. ISBN 0500014892

This clear and concise illustrated dictionary provides comprehensive coverage from the arcane to the prosaic, from ancient to contemporary. Includes information on type design, materials and pigments, parts of letters, and manuscript collections. Boldface in text indicates separate entry. Appendices contain a chart on parts of the letter, a list of libraries with manuscript collections, and calligraphic societies and publications. Includes bibliography.

Halliday, Peter, ed.
Calligraphy Masterclass
Woodstock, N.Y., Overlook Press, 1990. 128p. ISBN 0879514000

Fifteen contemporary calligraphers (including the editor) discuss examples of their work. Presents excellent illustrations of the work at various stages, with explanations of the concept and technique. Also provides brief biographies of the artists.

Harvey, Michael
Calligraphy in the Graphic Arts
London, Bodley Head, 1988. 64p. ISBN 0370311396

The preface places calligraphy in its contemporary context in graphic arts. Includes a short section on tools, techniques, and strokes. The rest of the book introduces various typefaces and uses them as the jumping-off point to explore creative typographic possibilities. Introduction by famous type designer, Hermann Zapf.

Neuenschwander, Brody
Letterwork: Creative Letterforms in Graphic Design
London, Phaidon Press, 1993. 160p. ISBN 0714828017

This survey of contemporary hand and computer-generated lettering advocates the use of creative lettering and suggests a reformation of the graphic arts curriculum. Examples are classified into sections where lettering is used for emotional impact, as elements in composition, and combined with typography. Includes well-chosen, attractive color photographs of artwork with brief commentary about technique, function, artist, and client. Provides a section on designing letters with unusual techniques and materials, such as cloth, film, found letters, letters manipulated by fax, and letters projected on a model. Includes international directories of designers, lettering artists, and design and lettering organizations. Bibliography and index.

Sutherland, Martha
Lettering for Architects and Designers
2nd ed. New York, Van Nostrand Reinhold, 1989. 139p. ISBN 0442282141

Very basic, unintimidating book on hand-lettering. Provides an explanation of how to form the letters with pencils, pens, and markers. Includes tips on techniques and tools with examples of lettering by the author and some well-known architects.

Wilson, Diana Hardy
The Encyclopedia of Calligraphy Techniques
Philadelphia, Running Press, 1990. 191p. ISBN 0894718509

The first part is an encyclopedia of different topics and techniques in alphabetical order. The second part groups sample work by categories, such as formal, experimental, and three-dimensional calligraphy. Each category consists of a short introduction followed by examples which are discussed briefly. Includes photographs of equipment. The arrangement and attractive layout and photographs are conducive to browsing. Indexed.

✦ Illustration—Scientific and Technical

Council of Biology Editors. Scientific Illustration Committee
Illustrating Science: Standards for Publication
Bethesda, Md., Council of Biology Editors, 1988. 296p. ISBN 0914340050

The purpose of this work is to develop specific standards and guidelines for illustrated scientific material and to serve as a style manual, but it has a wealth of practical information designed to help the illustrator meet quality standards. Provides information on prepress, preparation of artwork, graphs and maps, computer graphics, camera-ready copy, halftone printing, color, and legal and ethical considerations, such as copyright. Includes a good glossary, annotated bibliography by subject, and index.

Hodges, Elaine R. S., and others, eds.
The Guild Handbook of Scientific Illustration
New York, Van Nostrand Reinhold, 1989. 575p. ISBN 0442236816

This thorough and authoritative coverage of scientific illustration includes basic information every illustrator should know: techniques, tools and tips, preservation of artwork, light on forms, and rendering techniques—some unique to scientific illustration. Deals with illustration in particular fields, such as fossils, extinct vertebrates, fish, plants, human, and medical. The section on business aspects covers copyright, contracts, and free-lancing. Appendix lists art supply and equipment sources. Black-and-white illustrations. Indexed.

Kelvin, George V.
Illustrating for Science
New York, Watson-Guptill, 1992. 192p. ISBN 0823025403

Intended "as a style guide, a reference source for tools, materials, and techniques and a handbook of shortcuts, secrets, and tricks of the trade." Considers the qualifications and role of the scientific illustrator, and makes some basic design recommendations. Includes information on preparing illustration for print, typical presentations (such as cutaway views and cross sections), and working with models and photography. Separate chapters are devoted to specific fields such as astronomy. Appendix has a summary of basic business information, such as calculating overhead and estimating a job. Provides samples of high-quality illustrations done by various artists, often with photographs of earlier sketches and explanatory comments. Indexed.

Martin, Judy
High Tech Illustration
Cincinnati, Ohio, North Light Books, 1989. 160p. ISBN 0891343113

This introduction contains a helpful overview of the field, including information on materials and equipment, drawing methods, line drawing, diagrams and schematics, color, airbrushing, exploded views, cutaways, and ghosting. Excellent illustrations, chiefly in color. Glossary and index. Most addresses in the directory are located in England.

Wood, Phyllis
Scientific Illustration: A Guide to Biological, Zoological and Medical Rendering Techniques, Design, Printing, and Display
2nd ed. New York, Van Nostrand Reinhold, 1994. 158p. ISBN 0442013167

Intended for the artist or scientist who wants to produce art for publication, display, or projection (slides or television). Premise is that scientific illustration should be as well-designed and innovative as any artwork and should communicate clearly and quickly. Presents basic information in a simple format covering the essentials of drawing, such as perspective, light and shading, and tone and color. Answers questions on topics such as whether or not a specimen should be treated in an individualistic or representative manner, how to deal with live and preserved specimens, how to design diagrams and exhibits, and how to do artwork that will be printed for publication. Includes a short guide to careers and business practices, a chapter on computer graphics, and a brief bibliography. Indexed. Paperbound.

Zweifel, Frances W.
A Handbook of Biological Illustration
2nd ed. Chicago, Univ. of Chicago Press, 1988. 137p.
ISBN 0226997006

Directed to both the biologist and the artist who need to create illustration. Covers topics such as printing processes, materials and tools, preliminary and finished drawings (includes tips for drawing live animals), graphs and maps, lettering, working with photographs (but not creating photographs), and packing illustrations. Includes list of resources and index.

✦ Printing and Desktop Publishing

Aldrich-Ruenzel, Nancy, ed.
Designer's Guide to Print Production
New York, Watson-Guptill, 1990. 159p. ISBN 0823013146

This comprehensive, attractive, and useful resource provides information on preparing copy for typesetting, copyfitting and markup. Art and photography reproduction advice covers camera formats and film sizes, buying photography, cropping, and halftone reproduction. The color reproduction section discusses color separation, color electronics, prepress systems, dot etching, and gray component replacement. The printing and paper section provides information on printing trade customs and troubleshooting. Good glossary, index. This clear, readable layout is enhanced with helpful charts, diagrams, and illustrations. Some of the material was originally published by *Step-by-Step Graphics* magazine.

Blackwell, Lewis
Twentieth-Century Type
New York, Rizzoli, 1992. 256p. ISBN 084781596X

An introduction is followed by a decade-by-decade survey of typographic history from 1890 into the 1990s. This easy-to-use format and attractive layout are enhanced with excellent color illustrations. The table of contents summarizes major concepts and technological innovations, people, and typefaces of each decade. Includes supplementary material on analysis of characters and measurements, type description, and classification. Glossary, bibliography, and index.

Brady, Philip
Using Type Right
Cincinnati, Ohio, North Light Books, 1988. 119p.
ISBN 0891342559

Emphasizes the importance of design in communication. Reviews basic design guidelines, technical and special considerations, such as kerning, swash letters, tables, and roman numerals. Includes hints on working with typesetters and pasteup artists, doing reproductions, cutting expenses, and self-development. Practical, readable style. Lengthy illustrated glossary. Indexed.

Brown, Alex
In Print: Text and Type
New York, Watson-Guptill, 1989. 191p. ISBN 0823025446

Geared to the graphic artist, this work has basic information on print production and types of printing. Provides clear explanations of the parts of a letter, the classification and recognition of type, type measurement, and specification. Includes sections on the preparation of text, typesetting, and desktop publishing. Bibliography and index. Includes an index to typefaces.

Carter, Rob
American Typography Today
New York, Van Nostrand Reinhold, 1989. 159p.
ISBN 0442221061

Weaves critical analysis and some biography into the profiles of twenty-four influential twentieth-century type designers. Includes attractive photographs of the designers' works, many in color, and portraits of the designers. Bold type in the profiles refers the reader to a "Major Resources" section that functions as a brief encyclopedia of designers and movements influencing those profiled. Includes a chronology of American pioneers, a bibliography, and an index.

Carter, Rob, Ben Day, and Philip B. Meggs
Typographic Design: Form and Communication
2nd ed. New York, Van Nostrand Reinhold, 1993. 278p.
ISBN 0442007590

Begins with an illustrated timeline of events significant to typography and moves on to describe parts of the letter, type classification and families, type measurement, communication, grids, legibility, design elements such as space and color, and concise review of printing methods and technology. Provides chapters on student solutions to design problems and case studies of actual design projects. The final chapter has type specimens for major type classifications. Includes an in-depth glossary, character count table, guide to copyfitting, design strategy outline, and proofreader's marks. Also provided are a chronology of typeface designs, bibliography, general index, and type specimens index.

Conover, Theodore E.
Graphic Communications Today
2nd ed. St. Paul, Minn., West Pub. Co., 1990. 517p.
ISBN 0314665706

Directed to the student or professional in communications, this work addresses the basics of new technology as well as traditional printing methods. Discusses topics such as typography, handling art and illustrations, paper, ink, design, and color production. Also includes topics on desktop publishing and design fundamentals. Includes good examples, some in color. Appendices contain a bibliography and type specimens. Includes glossary and index.

Cook, Alton, and Robert Fleury
Type and Color: A Handbook of Creative Combinations
Rockport, Mass., Rockport Publishers, 1989. 157p.
ISBN 0935603190

Part One contains illustrated essays by successful designers, sections on design theories and ideas, and new examples of color use in typography. The second part is a color selector that shows the interaction of type and color. It consists of six hundred color bands with their four-color process formulas, to be used in conjunction with ten transparencies contained in a pocket on the inside cover. Each transparency contains two colors in four different typefaces in three sizes.

Craig, James
Basic Typography: A Design Manual
New York, Watson-Guptill, 1990. 189p. ISBN 0823004511

This practical guide is enhanced by a nice, readable layout. The top of each page defines or explains the concept that is illustrated visually below. Covers basic terminology, design options, and copyfitting. Appendix contains information on the point system, the metric system, and punctuation. Includes an illustrated glossary, bibliography, and index. Paperbound. Formerly titled *Phototypesetting: A Design Manual* (Watson-Guptill, 1978).

Craig, James
Designing with Type: A Basic Course in Typography
Edited by Susan Meyer. Rev. 3rd ed. New York, Watson-Guptill, 1992. 175p. ISBN 0823013057

This comprehensive manual covers origins of the alphabet, basics of letters and type, printing and publishing techniques, type measurement, type families, display types, tools and techniques, comping, type design, copyfitting, type arrangements, and grids. Also includes information on preparing copy and examples of design projects. Glossary, bibliography, and index are provided. Good readability.

Cullinane, Robert
The Complete Book of Comprehensives
New York, Van Nostrand Reinhold, 1990. 190p.
ISBN 0442217420

Intended as a text but valuable as a resource for practical, detailed information on the overall process of creating the completed designer's version of the work (comprehensives). Covers equipment, tools and materials, tips, and techniques such as transferring and scaling artwork. Suggests how to present comprehensives. Includes step-by-step samples of the comprehensive development process applied to particular cases and color examples of student solutions. A helpful glossary; bibliography; index.

Dewey, Patrick R.
One Hundred and One Desktop Publishing and Graphics Programs
Chicago, American Library Association, 1993. 134p.
ISBN 0838906060

Intended for librarians but useful for anyone interested in the topic. Categories of programs include desktop, specialty graphics, word processing programs with graphics, painting and drawing, presentation graphics, utility, typefaces and fonts, clip art, and scanning. Each program lists vendor, cost, hardware requirements, additional software needed, and short, helpful descriptions. Includes vendor list, current bibliography by subject, and software index. *101 Micro Series.*

Dodt, Lorette C.
Graphic Arts Production
Homewood, Ill., American Technical Publishers, 1990. 302p.
ISBN 0826926843

This textbook, with clear well-illustrated directions, covers how to do the basics of production. Includes very specific, practical information on tools, comprehensives, ruling, type, tint screens, camera copy, color and registration, design specifications like paper coatings, designing with photographs, desktop publishing, designing forms, and postal regulations. Sections of activities, projects, and review questions are provided for each chapter. The textbook format does not intrude on its usefulness as a resource. Black-and-white illustrations (mostly line drawings). Includes glossary and index.

Elam, Kimberly
Expressive Typography: The Word as Image
New York, Van Nostrand Reinhold, 1990. 165p.
ISBN 0442233566

Explores the integration of image and typography within its historical context. Includes handcrafted methods of creating typography such as handwriting, illustration, modeling, and collage. Intended as a source of visual inspiration and features attractive reproductions of creative solutions on nearly every page. Indexed.

Gottschall, Edward M.
Typographic Communications Today
New York, International Typeface Co., 1989. 249p.
ISBN 0262071142

Book is a "critical review of twentieth century typographic design" by noted author Gottschall. Places recent technological developments in a historical context. Begins with a chapter on various approaches such as clarity and order, then discusses art and design movements. Final chapters review recent typography, give examples of typefaces since 1969, survey design in various countries, and summarize design ideas of well-known designers such as Chermayeff and Rand. Includes more than nine hundred illustrations (over half in color). Bibliography and index.

Haley, Allan
Typographic Milestones
New York, Van Nostrand Reinhold, 1992. 146p.
ISBN 0442236425

Profiles the lives of eighteen type designers from Gutenberg to the present day. Originally published in the journal *U and LC*, they are detailed and entertaining portraits that reveal their personalities through quotations and anecdotes.

Hinrichs, Kit, and Delphine Hirasuna
Type Wise
Cincinnati, Ohio, North Light Books, 1990. 160p.
ISBN 0891343563

The first third of the book consists of an interview with Hinrichs, noted designer and partner in the renowned firm Pentagram Design. He discusses the creative aspects and importance of type design. Questions and responses provide inspiration and guidance. The balance of the book consists of attractive color photographs of sample designs from the firm's projects. Indexed.

The International Typebook
New York, Van Nostrand Reinhold, 1990. 383p.
ISBN 0442305036

Presents an overview of the history of typography placing typefaces in their proper context, then describes nineteen representative ones from different time periods in detail. Presents for each type: history, distinguishing characteristics, full-page upper- and lowercase alphabets, numerals and punctuation marks, different size alphabets, and samples of its italic and bold print. Sample paragraphs with different type sizes and spacing follow, then sample print applications such as letterheads and brochures. Forty-five other typefaces are shown in less detail. No glossary or index.

Labuz, Ronald
Typography and Typesetting: Type Design and Manipulation Using Today's Technology
New York, Van Nostrand Reinhold, 1988. 272p.
ISBN 0442259662

This practical orientation gives an overview of typesetting, describes different specialities within the industry, and recounts the history of type. Includes specific information on legibility, layout, specifying type, copyfitting, typesetting, and typesetting equipment. Provides explanations of technique and samples of typographic applications in publication, book design, advertising, and job work. Individual chapters discuss the role of the type house, computer basics, and the use of computers in industry. Includes extensive and concise glossary, bibliography by chapters, and index.

Lem, Dean Phillip
Graphics Master 4: A Workbook of Planning Aids, Reference Guides and Graphic Tools for the Design, Estimating, Preparation, and Production of Printing, Print Advertising, and Desktop Publishing
4th ed. Los Angeles: D. Lem Associates, 1988. 153p.
ISBN 0914218069.

This concise reference guide and tool kit provides the technical data and working information needed by those in the design, planning, and production of printing and print advertising. Covers such topics as printing processes, process color, graphic arts photography, printing industry trade customs, electronic color scanners, printing papers, presses and inks, envelope styles and sizes, typography, typesetters, copyfitting, word processing and typesetting connections, type processing command codes, and desktop publishing. Includes specimens of 1,768 typefaces, proofreaders' marks, conversion tables, and a glossary.

Parker, Roger C.
Looking Good in Print: A Guide to Basic Design for Desktop Publishing
3rd ed. Chapel Hill, N.C., Ventana, 1993. 423p.
ISBN 1566040477

Although geared to the desktop publisher with little or no design background, this work has basic information on graphic design that could be helpful to beginning design students. Subjects discussed include type, using graphics, and common design pitfalls. Chapters cover business graphics such as newsletters, newspapers, advertisements, promotional materials, books, presentation graphics, and business communications (such as letterheads, business cards, and résumés). Appendices have information on photo and clip art resources, tips and techniques, and a bibliography of resources that includes books, periodicals, seminars, and training. Indexed.

Perfect, Christopher
The Complete Typographer: A Manual for Designing with Type
Englewood Cliffs, N.J., Prentice Hall, 1992. 223p.
ISBN 0130456675

Focuses on seven typeface categories. For each group there is an introduction to the main stylistic features, history, an attractive chart of alphabet specimens, sample texts, and samples of printing styles, such as italic and bold. The section ends with display of other typefaces in the category and sample applications. Also contains sections on the history of typography and working with type, which gives basic information on measuring, structuring and choosing type, enlivening text, display typography, type, color, and desktop publishing. Good illustrations; bibliography; and index.

Poyner, Rick, and Edward Booth-Clibborn
Typography Now: The Next Wave
London, Internos, 1991. 223p. ISBN 0904866904

An introductory essay, "Type and Deconstruction in the Digital Era," is followed by large color examples of experimental typography from around the world. Whenever applicable, designs are identified by title, function, company and country of origin, date, and designer. Includes designer index.

Rogondino, Michael
Computer Type: A Designer's Guide to Computer-generated Type
San Francisco, Chronicle Books, 1991. 464p. ISBN 0877018022

Presents forty-eight examples of computer-generated type in upper and lowercase, in various sizes with sample paragraphs. Gives an overview of the history of type, type classifications, and computer type. Basic information is provided on measurement, copyfitting, proofreader's marks, and "hidden" or special characters in Macintosh, Windows, Ventura, and IBM PC environments. Glossary.

Rosen, Ben
Digital Type Specimens: The Designer's Computer Type Book
New York, Van Nostrand Reinhold, 1991. 534p.
ISBN 0442235011

Includes over 2,500 digital type specimens selected for the quality of design and general availability. Ten type families are shown in italics, light through bold, condensed through extended, and various point sizes. Provides text samples and historical discussions. Includes supplementary, display, and one-line specimens from three vendors with diminishing variety for each. Specimens are arranged in alphabetical order within each section.

Rosen, Ben
Type and Typography: The Designer's Type Book: Hot Metal Type
2nd ed. New York, Van Nostrand Reinhold, 1989. 414p.
ISBN 0442235038

Rosen's practical workbook for the graphic designer may be used as a reference to typefaces available from 1950 to 1970. It is these fonts on which the new computer-driven phototypesetting styles, variations, and developments are based. Includes a short section on the history and technical aspects of type. Type families are shown in a range of sizes, in both Roman and italic, with a variety of weights and size modifications. Provides samples of text, and a section of supplementary faces (with less detail on each one), which includes such categories as "romantic" and "exotic" styles. Indexed.

Schlemmer, Richard M.
Handbook of Advertising Art Production
4th ed. Englewood Cliffs, N.J., Prentice Hall, 1990. 312p.
ISBN 0133808823

Begins with descriptions of various types of employment for the advertising artist, followed by a historical survey of how images have been produced in the past. Chapters with basic explanations of each of several printing processes like gravure, silk-screen, and flexographic are enhanced with simple diagrams. Presents overview of computer graphics, reprography, and image-processing systems. Other sections deal with the preparation of illustrations and mechanicals, and production for the various types of printing. Final chapter on paper explains how it is made, varieties and classifications, and selection criteria. Includes definitions in each chapter, short bibliography, and index.

Typo-Graphic Design
New York, Madison Square Press, 1992- . ISBN 0942604237 (vol.1)

Displays the "best" editorial, promotional, corporate, and advertising designs from 1980 to 1990, selected by a panel of judges prominent in graphic design. Categories are logotypes, stationery, editorial packaging posters, and miscellaneous. Includes large, well-reproduced color illustrations, an index of design firms, designers, letterers, and clients. A directory of design firms is also provided.

Wallis, L. W., ed.
Modern Encyclopedia of Typefaces, 1960-90
New York, Van Nostrand Reinhold, 1990. 192p. ISBN 0442308094

Presents examples of 345 typeface families released to the market after 1960. Most are in eighteen point and four weights—Roman, italic, bold, and bold italic—and include upper- and lowercase, numerals, and punctuation. Arrangement is in alphabetical order by type name. For each specimen, the designer, date, original manufacturer or design agency, systems on which the typeface is available, any alternate names, weights and versions available, and occasional notes are given. Includes short biographies of over fifty designers. Includes a bibliography; a chronological list of typefaces; indexes for designer, manufacturer, and design agencies; and alternative typeface names.

West, Suzanne
Comping Techniques: Visualizing and Presenting Graphic Design Ideas
New York, Watson-Guptill, 1991. 160p. ISBN 0823008673

Provides practical information on how to create a comprehensive (a polished representation of the graphic design idea that illustrates what it will look like when it is printed). In addition to techniques, tools, and materials for creating the comp, this book emphasizes an understanding of prepress and printing methods and the design process. The use of a computer as a tool is well integrated with the appropriate subjects. Appendices contain sample type specimens, and companion charts for comping common printing techniques using paper or a computer screen. A floppy disk for the Macintosh with comping files is included with instructions.

White, Alex
Type in Use: Effective Typography for Electronic Publishing
New York, Design Press, 1992. 191p. ISBN 0830637966

White's guide to effective type design emphasizes ideas and the intelligent use of type. Includes chapters on text, headlines, subheads, breakouts, captions, department headings, covers, contents, "bylines and bios," folios, and foot lines. Gives basic concepts, then pages of examples (black-and-white illustrations) with short explanatory captions. A glossary, bibliography, and index are provided.

White, Jan V.
Graphic Design for the Electronic Age
New York, Watson-Guptill, 1988. 211p. ISBN 082302122X

Provides commonsense advice for nondesigners or novice designers. Covers type, type options (such as line length, capitals, serifs, and backgrounds), and type information (including pitch, kerning, punctuation, foreign languages, and proofreader's marks). Includes a lengthy section on the elements that make up a page, followed by details of publication construction. Appendices have information on paper sizes, binding, type measurement and conversions, a glossary, "a communication timeline," and an index.

✦ Vocational Resources

Conner, Floyd, and others
The Artist's Friendly Legal Guide
Rev. and updated ed. Cincinnati, Ohio, North Light Books, 1991. 142p. ISBN 0891343652

This user-friendly guide covers copyright (including computer-generated work), ethics and business practices, contracts, record keeping, and taxes. Inset boxes with answers to commonly asked questions, and easy reference checklists for complicated issues (like contracts), are nice features. Includes sample forms and addresses for Volunteer Lawyers for the Arts, listed by state. Indexed.

Craig, James
Graphic Design Career Guide
2nd rev. ed. New York, Watson-Guptill, 1992. 159p. ISBN 0823021637

Written by a practicing design director who is also a teacher and author of books on graphic design. Nearly half of the book is devoted to clear explanations of career areas and includes the personality characteristics and skills needed by people in that field. Part Two covers portfolios, résumés, and interviewing. Part Three gives excellent advice on finding a job, what to expect on the job, and free-lancing. Successful designers tell how they got their first assignment. The final section deals with the study of graphic design including information on schools, course descriptions, and getting the most out of school. Black-and-white illustrations enhance the text. Appendices list major ad agencies, book and magazine publishers, and art schools. Includes bibliography, a list of resources by topic, glossary, and index.

Craig, James, and William Bevington
Working with Graphic Designers
New York, Watson-Guptill, 1989. 159p. ISBN 0823058670

Although intended for those who work with the designer, it would actually serve very well as an introduction to what graphic designers do. Includes information on fees, schedules, approvals. Also considers typography, with samples of common typefaces, printing processes, and desktop publishing. Sample completed project and a good glossary are provided. Indexed.

Fleishman, Michael
Starting Your Own Small Graphic Design Studio
Cincinnati, Ohio, North Light Books, 1993. 122p. ISBN 0891344667

Good, practical advice in a readable format. Topics include

self-evaluation, business plans, setting up a studio, financial plans, pricing, contests, project and money management, and promotion. Includes checklists and designer case studies.

Goldfarb, Roz
Careers by Design: A Headhunter's Secrets for Success and Survival in Graphic Design
New York, Allworth Press, 1993. 223p. ISBN 1880559056

Discusses the role of graphic designers and the various fields of graphic design including computer technology. Gives advice on the job search, choosing a job, and preparing a portfolio. Covers hiring practices, salaries, changing jobs, and design education. A list of professional societies is included. Indexed.

Gordon, Barbara
Opportunities in Commercial Art and Graphic Design Careers
Lincolnwood, Ill., VGM Career Horizons, 1992. 148p. ISBN 0844240052

This basic, practical guide presents information on how to begin a career, including choosing an art school. Focuses on what jobs involve, job requirements, and helping the reader see if there is a career match. Includes clear, concise definitions and explanations of jobs.

Graphic Artists Guild (U.S.)
Graphic Artists Guild Handbook: Pricing and Ethical Guidelines
7th ed. New York, Graphic Artists Guild, 1991. 235p. ISBN 0932102077

The Guild's indispensable guide gives insight into what is entailed in various graphic arts fields, as well as information on standard business practices, customs, and fees. General chapters provide detailed, practical reviews of copyright, fair practices, contests, legal rights, pricing, negotiation, contracts, and computer-generated art. Includes sample forms, information on the Guild and its benefits, and a good glossary. Indexed.

Metzdorf, Martha
The Ultimate Portfolio
Cincinnati, Ohio, North Light Books, 1991. 144p. ISBN 0891343709

Metzdorf presents specific, practical advice on starting a portfolio. Includes color examples of successful designers' portfolios geared to different types of jobs, a detailed section with examples on presentation of work and interviews with employers discussing what qualities they look for when hiring designers and illustrators. Traces the changing portfolio of one designer over the span of his career. A list of sources for more information is appended.

Neff, Jack
Designer's Guide to Making Money with Your Desktop Computer
Cincinnati, Ohio, North Light Books, 1992. 115p. ISBN 089134439X

In a practical, easy-to-read layout, Neff presents the fundamentals of desktop design including "setting up shop" (hardware and software), service bureaus, and working at home. Other chapters discuss free-lancing and marketing, finding clients, and business aspects such as business plans, taxes, contracts, and using the computer for business management. Useful insets have comparisons, checklists, and sample forms. Includes black-and-white illustrations and a list of organizations and vendors. Indexed.

Poggenpohl, Sharon Helmer, ed.
Graphic Design: A Career Guide and Education Directory
New York, AIGA Press, 1993. 158p. ISBN 1884081002

This essential publication published by the American Institute of Graphic Arts answers a series of basic questions for beginners: What is graphic design? What do graphic designers need to know? What is design school like? How do you select one? How do schools differ? Includes suggestions on how to find a job, portraits of graphic designers in different areas, and a section on future trends. Appendix lists AIGA chapters, design organizations, basic design periodicals, and a list of four-year schools with design programs.

Sher, Paula
The Graphic Design Portfolio: How to Make a Good One
New York, Watson-Guptill, 1992. 160p. ISBN 0823021629

This work, written by a prominent designer as an outgrowth of an upper-level portfolio class, presents a series of assignments in the design areas of newspaper, retail, book jacket, poster, record package, magazine, and promotional material. Each assignment is explained thoroughly in terms of relevance and design considerations, then student solutions are presented and briefly analyzed. Provides practical information on putting together and presenting the portfolio. Fine, large black-and-white and color illustrations make it a source for student inspiration and perhaps for instructor assignments for lower-level classes. Indexed.

Nonprint Materials

Behind the Scenes: The Advertising Process at Work
New York, Advertising Educational Foundation, 1990.
1 videocassette (30 min.) VHS

Although not limited to the graphic design aspect, this is a professional effort which gives a sense of what it is like to be part of a major ad campaign including the personal dynamics, excitement, creative process, disappointments, and client relationship. Featured are campaigns for Johnson & Johnson contact lenses, Coca-Cola, and Jell-O.

Graphic Design One: Demonstrating the Values of Good Design.
Northfield, Ill., Goldsholl Learning Videos, 1989. 1 videocassette (45 min.) VHS ISBN 0939437635

Presents a project-oriented approach to graphic design. Uses multicolor ideas from sketch to final production to illustrate current trends and contemporary styles.

Graphic Design Two: Demonstrating the Values of Good Design.
Northfield, Ill., Goldsholl Learning Videos, 1989. 1 videocassette (45 min.) VHS ISBN 0939437643

Creative designers guide the viewer through project planning to comprehensive presentations. Includes information on project conception, type and color selection, project modification, thumbnail sketches, and preparation for final presentations.

Sciences

Agricultural Sciences and Agronomy **475**
Mary E. Coffin, Sandra J. Donavan, Crystal Havely Stratton, Debora Thomas, and Kelly Willmarth

Fire Services **491**
Terri Propes

Forestry **497**
Charlotte Cooper

Funeral Services **510**
Harolyn Cumlet and Lenora Lockett

Landscape Horticulture Technologies **517**
Connie Barber

Veterinary Technology **523**
Dr. Gail Staines

Agricultural Sciences and Agronomy

Mary E. Coffin
Sandra J. Donovan
Crystal Havely Stratton
Debora Thomas
Kelly Willmarth

Laramie County Community College
Cheyenne, Wyoming

Introduction

Agriculture is one of the most dynamic subjects taught in community colleges today, and materials in this bibliography reflect that. Since the Great Depression, the farming landscape in America has changed dramatically and rapidly. The agricultural industry is undergoing intense reexamination of its policies and methods of operating.

Today's students of agriculture will be faced with a changing and expanding curriculum. On many college campuses, the agricultural department is also reexamining and redefining its mission. The curriculum is expanding to include new developments in biotechnology. With growing concern about the environment and limited resources, topics such as alternative farming methods, sustainable agriculture, range management, groundwater management, and new farm chemicals will be prominent in the study of agriculture. A global economy has already impacted the agricultural community, and students will need to have an understanding of international politics and economics.

On-line computer services are changing how people will access information. Many of the agricultural statistics and business forms are now on computer databases. A broad knowledge base that is a composite of many disciplines is being taught in addition to the traditional agricultural curriculum. One of the newest technologies applicable to agriculture is Geographic Information Systems (GIS). General information on this topic has only recently become readily available through print and electronic sources. Future publications can be expected to be more specific regarding applications of GIS to agriculture.

Since the focus of this bibliography is on material to be used in the community college, it lists sources which cover subjects in broad terms. For example, titles were selected which dealt with several farm chemicals or alternative farming methods rather than one that covered one or two items in greater detail. Since many agricultural departments are changing their curricula, materials in agriculture education were added. Although an agricultural science such as food science is an area of study, such a curriculum in a two-year course of study would be more likely covered in a survey of several topics.

Given the vast changes the agricultural industry has undergone, it is surprising that subjects such as farm machinery are represented by relatively few new titles. Some titles were too advanced or had gone out of print. Journal titles, in particular, often seemed to cease publication after a few years. Indexing of agricultural journals was also difficult to locate. Many of the journals in this bibliography simply were not indexed; others were indexed in advanced, expensive indexes that community colleges would not typically purchase. The focus of many of the selections was a practical, hands-on approach; other materials consisted of broader models, theories, and procedures.

Acknowledgment is extended to Linda Hollingsworth, Interlibrary Loan Technician, for her patience and perserverance in processing numerous interlibrary loan requests.

Agricultural Sciences and Agronomy

Selected Associations

Although there are no accrediting agencies specific to this area, there are many helpful organizations that can supply additional information.

Cooperative Extension Services
These agencies are funded through a partnership of the county government and land-grant universities. Agents in each office have training and expertise in general agriculture as well as knowledge of local agricultural regions. They disseminate research conducted by state and federal networks, providing consultations and educational classes which cover a wide variety of technical and scientific interests for individual groups. Professional staffs are located in 3,150 counties of the United States. They are affiliated with state universities and land grant colleges.

National Agricultural Library
10301 Baltimore Boulevard
Beltsville, Maryland 20705-2351

Society for Range Management
1839 York Street
Denver, Colo. 80206

U.S. Department of Agriculture
Fourteenth Street and Independence Avenue, SW
Washington, D.C. 20250

Selected Journals

Ag Consultant.
Meister Pub. Co. Nine times a year. ISSN 0894-7155

Ag/Innovator.
Agricultural Information Management. Monthly. ISSN 1074-1186

The Agricultural Education Magazine.
Agricultural Education Magazine. Monthly. ISSN 0732-4677

Agricultural Outlook.
U.S. Dept. of Agriculture. Economic Research Service. Monthly (except combined Jan./Feb. issue). ISSN 0099-1066

Agricultural Research.
U.S. G.P.O. Monthly. ISSN 0002-161X

American Journal of Alternative Agriculture.
Institute for Alternative Agriculture. Quarterly. ISSN 0889-1893

Application Technology.
Meister Pub. Co. Bimonthly. ISSN 1077-9205

Bio/Technology/Diversity Week.
Institute for Agriculture and Trade Policy. Biweekly.

Farm Chemicals.
Meister Pub. Co. Monthly. ISSN 0092-0053.

Farmer's Digest.
Farmer's Digest. Monthly (bimonthly except June/July and Aug./Sept.). ISSN 0046-3337

Farm Journal.
Farm Journal. Thirteen times a year. ISSN 0014-8008

Farm Smart : [The Journal of Computerized Farm Management].
FBS Systems. Bimonthly.

FarmFutures.
Miller Pub. Co. Monthly (except Feb., Mar., and Apr.). ISSN 0091-1305

Hot Line Farm Equipment Guide.
Hot Line, Inc. Monthly. ISSN 1047-725X

Implement and Tractor.
Intertec Pub. Co. Monthly (except two issues in Mar., Oct.). ISSN 0019-2953

Journal of Natural Resources and Life Sciences Education.
American Society of Agronomy. Semiannually. ISSN 0094-2391

Journal of Soil and Water Conservation.
The Journal. Bimonthly. ISSN 0022-4561

NACTA Journal.
National Association of Colleges and Teachers of Agriculture. Quarterly. ISSN 0149-4910

NBIAP News Report. [National Biological Impact Assessment Program]
Virginia Tech. Monthly.

The New Farm.
Rodale Press. Seven times a year. ISSN 0163-0369

No-Till Farmer.
No-Till Farmer. Monthly. ISSN 0091-9993

Pest Control Technology.
GIE Pub. Co. Monthly. ISSN 0730-7608

Rangelands.
Society for Range Management. Bimonthly. ISSN 0190-0528

Rural Heritage.
Rural Heritage. Quarterly. ISSN 0889-2970

Successful Farming.
Meredith Corp. Thirteen times a year. ISSN 0039-4432

Top Producer.
Farm Journal. Monthly. ISSN 1056-0831

Print Materials

✦ Reference

The Agriculture Fact Book
Washington, D.C., U.S. Department of Agriculture, 1994.

After a lapse of a few years, this handbook is once again scheduled for publication in late 1994. Prepublication indications are that the text will be similar to past editions and reflect the structure and trends in modern agriculture and with up-to-date statistical information. Presents a very useful resource.

Arntzen, Charles J., and Ellen M. Ritter, eds.
Encyclopedia of Agricultural Science
San Diego, Calif., Academic Press, 1994. 4 v. ISBN 0122266706

This comprehensive reference work is intended for undergraduates. Among the topics presented are animal, plant, range, and soil science; agricultural economics; pest management; water resources; food processing, storage, and distribution; and rural sociology.

Hanson, A. A., ed.
Practical Handbook of Agricultural Science
Boca Raton, Fla., CRC Press, 1990. 534p. ISBN 0849337062

Tables of information (422 in all) are profuse in this handbook developed from a variety of both private and government information sources for the nonspecialist. A multitude of topics are covered: climate, soil, plants, tillage, pests, crops, and animals. Includes an extensive glossary and index and general agriculture statistical tables.

Herren, Ray V., and Roy L. Donahue
The Agriculture Dictionary
Albany, N.Y., Delmar Publishers, 1991. 553p. ISBN 0827340958

This agricultural dictionary spans a variety of areas, such as economics, agronomy, horticulture, animal science, and agricultural mechanics. This work presents more than 15,000 definitions and over 700 illustrations. Includes a table of conversion factors for weights and measures.

National Agricultural Library (U.S.)
Guide to Services of the National Agricultural Library
Beltsville, Md., The Library, 1991. 22p.

This concise publication summarizes the services offered by the National Agricultural Library. Distant users may access the collection through the database AGRICOLA and keep up-to-date on agricultural products and services through an electronic bulletin board.

Sine, Charlotte, ed.
Farm Chemicals Handbook
Willoughby, Ohio, Meister Pub. Co. Annual. ISSN 0430-0750

This is the most comprehensive reference available on pesticides and fertilizers. The 1993 edition emphasizes environmental and ecological information and provides additional advice on biological controls. Contains separate pesticide, biocontrols, and fertilizer dictionaries; regulatory files; safety and application guide; and an alphabetical list of agricultural chemical manufacturers and suppliers. To effectively use this exceptional source, use the "Sine" index which lists all products, services, terms, and regulations; alternate names of products are cross-referenced.

United States. Bureau of the Census
Census of Agriculture
Washington, D.C., U.S. Government Printing Office. Irregular.

Since the first agriculture census in 1840, this publication has been a leading source of statistical data on county, state, and national levels. Currently published on a five-year cycle in years ending in 2 and 7. Four volumes include the geographic area series (state and county data may be purchased separately for each state and Puerto Rico); subject reports (an agricultural atlas maps US agriculture); related surveys; and a census of horticultural specialties.

United States. Dept. of Agriculture
Agricultural Statistics
Washington, D.C., U.S. Government Printing Office. Annual.
ISSN 0082-9714

This basic source is available in paper or on CD-ROM, and its contents may be retrieved through gophers on the Internet. Statistics involving main crops, animals, dairy and poultry products, farm resources, price support programs, and consumption are detailed by states and years. Some world production and trade statistics are also included, both from U.S. and foreign sources.

United States. Dept. of Agriculture
List of Available Publications of the United States Department of Agriculture
Washington, D.C., U.S. Government Printing Office, 1993. 49p.
ISSN 0887-9540

Provides a bibliography of print and nonprint materials that may be acquired from the Department of Agriculture, the National Agricultural Library, the National Technical Information Service, and the Consumer Information Center. Included are pamphlets, research reports, bibliographies, handbooks, informative bulletins, periodicals, computer and audiovisual software. Published irregularly.

✦ General Works

Bartlett, Peggy F.
American Dreams, Rural Realities: Family Farms in Crisis
Chapel Hill, N.C., Univ. of North Carolina Press, 1993. 305p.
ISBN 0807820679

Bartlett's study of Dodge County, Georgia, provides valuable documentation for understanding the role of government in agriculture, the interaction of farmers with environment, the tensions between agrarian values and the industrial world, and the patterns of women's involvement with farms drawn from their experiences during the 1980 farm crisis. Her competent field research provides insight into the social and economic factors of agriculture.

Burton, L. DeVere
Agriscience and Technology
Albany, N.Y., Delmar Publishers, 1992. 326p. ISBN 0827340168

This easy-to-read textbook format provides an overview of the blending of science, technology, and agriculture. Partial contents include biotechnology, computer-aided management, environmental technology, food and fiber, and energy and power technology.

Buttel, Frederick H., Olaf F. Larson, and Gilbert W. Gillespie, Jr.
The Sociology of Agriculture
New York, Greenwood Press, 1990. 263p. ISBN 0313264449

Social aspects in the general field of agriculture are analyzed by these accomplished writers who show how the emphasis and study of agriculture have changed in this century. Three sections chronicle different developments such as village-centered agriculture, the social-psychological approaches influenced by the 1970s, and recent studies with a variety of emphases including gender, politics, environment, and economics. Highly informative, especially for faculty, providing an overview of the changing study of agriculture.

Carroll, C. Ronald, John H. Vandermeer, and Peter Rosset, eds.
Agroecology
New York, McGraw-Hill, 1990. 641p. ISBN 007052923X

Thirty-two contributors examine the multidisciplinary aspects of agroecology. The initial overview of agroecosystems is followed by writings on the ecological background, management considerations, and research in agricultural ecology. Bibliographic references are supplied for each chapter. This is a basic resource on agroecology for the advanced student.

Cochrane, Willard Wesley
The Development of American Agriculture: A Historical Analysis
2nd ed. Minneapolis, Univ. of Minnesota Press, 1993. 500p. ISBN 0816622825

Cochrane's broad background of fifty-five years in agriculture makes him eminently qualified to write this easily read, basic book. Half of his analysis is a chronological history, with additional writings on various forces of change such as mechanization, physical and social infrastructure, international influences, environmental policy, and the role of government. His concluding ideas will help students to focus on considering a future in agriculture.

Damerow, Gail
Fences for Pasture and Garden
Pownal, Vt., Storey Communications, 1992. 154p. ISBN 0882667548

Whether the purpose is to keep livestock in or predators out, to block drifting snow, or to provide a trellis for plants, this comprehensive guide offers practical advice for planning, selecting, and installing fences. Supplementary illustrations and tables enhance the text.

Erickson, Duane E., Royce A. Hinton, and Ronald D. Szoke
Microcomputers on the Farm: Getting Started
2nd ed. Ames, Iowa, Iowa State Univ. Press, 1990. 99p. ISBN 0813811570

Provides a good, simple overview of the essentials of selecting, purchasing, and using a PC in agricultural operations. Includes checklists for software and hardware evaluation. Outlines computer networks available through a modem, and details examples of how actual farm families are currently using PCs in their farming operation.

Ferleger, Lou, ed.
Agriculture and National Development: Views on the Nineteenth Century
Ames, Iowa, Iowa State Univ. Press, 1990. 363p. ISBN 0813803144

Twelve well-known teachers, historians, and economists have written substantial and informative pieces on change in agriculture during the nineteenth century, an important time in American agriculture. The essays are grouped under regional headings. The North/Midwest section analyzes economics and trends, such as development of the independent landowning farmer. The history of the South deals with post-Civil War issues. Two informative chapters on women and immigrants are included under "Special Topics."

Friedberger, Mark
Farm Families and Change in Twentieth-Century America
Lexington, Ky., Univ. of Kentucky Press, 1988. 282p. ISBN 0813116368

Friedberger's analysis of family farms uses a case study approach focusing on the Iowa Corn Belt and the southern central valley of California. Narrowing his study to two time periods—the Depression of the 1920s and the farm crises of the 1980s—he discusses land ownership, inheritance, credit, family, and community, especially in regard to the medium-sized farm. The well-written introduction and chapter conclusions will inform younger students who did not experience these unique periods.

Guither, Harold D., and Harold G. Halcrow
The American Farm Crisis: An Annotated Bibliography with Analytical Introductions
Ann Arbor, Mich., Pierian Press, 1988. 164p. ISBN 0876502400

The expertise of the two authors, plus the organizational treatment in this bibliography, make it an exceptional resource. Eight chapters with many subsections divide this large topic into sociological, scientific, and economic perspectives. They are introduced by a concise paragraph in a sidebar with lengthy essays, providing a broad base for students' understanding of issues. The 800 well-written annotations cite many government documents and journals, as well as monographs.

Haney, Wava G., and Jane B. Knowles, eds.
Women and Farming: Changing Roles, Changing Structures
Boulder, Colo., Westview Press, 1988. 390p. ISBN 081337605X

A variety of topics are covered in brief chapters by professors of history, sociology, anthropology, and agricultural economics. Regional aspects of rural women's lives are addressed, along with such specialized topics as Native American experiences, effects of technology, and political or economic change. The historical emphasis focuses on this century.

Hart, John Fraser
The Land That Feeds Us
New York, Norton, 1991. 398p. ISBN 0393029549

Hart has written a narrative history combining research with interviews with present-day farmers in the eastern U.S. Twenty-three brief chapters, supplemented with maps and charts, bring agriculture to life. Students will be exposed to the political, technological, and social forces which have an impact on farming.

Heilman, Grant
Farm
New York, Abbeville Press, 1988. 287p. ISBN 0896598896

Two hundred dramatic, creative photographs depict farming in regions of the U.S. This well-written text covers about a third of each page along with concise paragraph-length captions. The format includes six sections covering general aspects of agriculture: crops, animals, vegetables, fruits, nuts, and alternative methods.

Koch, John N., and Jean A. Gilbertson
Agriculture: Illustrated Search Strategy and Sources
Ann Arbor, Mich., Pierian Press, 1992. 152p. ISBN 0876503032

Presents a guide to library resources in the field of agriculture useful to both faculty and students. The breadth of coverage ranges from explanations of a card catalog, to *Biological Abstracts,* to computer resources. Although written primarily for upper-level undergraduates, the book is a valuable resource for community college students who want an overview of the literature as well as a guide to effective research strategies. Appendix includes a bibliography of reference sources.

Logsdon, Gene
At Nature's Pace: Farming and the American Dream
New York, Pantheon, 1994. 208p. ISBN 00679427414

Logsdon's philosophy in support of sustainable farming is based on his own research on a small homestead in Ohio as well as his knowledge of other farmers and extensive reading. These essays have been previously published in popular magazines and cover his comparisons of large-scale farming with Amish methods, his critical review of state university agricultural education, and interest in small garden farms. His views are well respected and will be thought provoking to students.

Logsdon, Gene
The Contrary Farmer
Post Mills, Vermont, Chelsea Green Publishing, 1993. 238p. ISBN 0930031679

Logsdon writes for the small, noncommercial farmer. He defines himself as "contrary," i.e., he does not subscribe to some of the current theories of production. Instead, he provides alternative but supportive, how-to essays which also have a philosophical base. He explores topics such as agricultural economics and mechanics, water sources, animals, and tree farming, and then provides tips in areas such as selection of various crops.

Martin, Philip L., and David A. Martin
The Endless Quest: Helping America's Farm Workers
Boulder, Colo., Westview Press, 1994. 258p. ISBN 0813317681

For agriculture students who may work with migrant farmhands, this study will provide helpful background information. Seeking to correct misconceptions or stereotypes about migrant workers, the authors first examine agricultural employment. They believe it is essential to reform the farm labor market, rather than continue to develop special programs, and to coordinate existing programs with their accurate assessment. Their projections for the future take possible changes from NAFTA into consideration. Appendix of tables and assistance programs will be useful.

McConnell, Primrose
Primrose McConnell's The Agricultural Notebook
18th ed., edited by R. J. Halley and R. J. Soffe. Boston, Butterworths, 1988. 689p. ISBN 0408030607

Published in Great Britain, this comprehensive handbook has become a standard work for college students. It is divided into four main sections: crop production, animal production, farm equipment, and farm management. Numerous tables are included.

Nabhan, Gary Paul
Enduring Seeds: Native American Agriculture and Wild Plant Conservation
San Francisco, North Point Press, 1989. 225p. ISBN 0865473439

Nabhan provides a different perspective on agriculture with his background as historian and naturalist. Although his expertise is in Southwestern ecology, he uses common examples of plants grown throughout the U.S., such as sunflower, maize, wild rice, and gourds to illustrate his ideas of biodiversity, cultural traditions, and local adaption of crops. His discussion of seed banks and preservation methods for propagation of plants provides relevant information for agriculture students.

National Research Council (U.S.). Committee on the Role of Alternative Farming Methods in Modern Production Agriculture
Alternative Agriculture
Washington, D.C., National Academy Press, 1989. 448p. ISBN 0309039878

This volume explains various practices using alternative agricultural methods and utilizes findings based on extensive research. Economic and environmental consequences of traditional practices are explained along with the economic potential of alternative systems. Case studies from farms in nine states demonstrate varying degrees of alternative and traditional practices. Helpful graphs, charts, and photographs accompany the readable text.

Parker, Edwin B., and others
Rural America in the Information Age: Telecommunication Policy for Rural Development
Lanham, Md., Aspen Institute and University Press of America, 1989. 170p. ISBN 0819174939

An outgrowth of a conference sponsored by the Aspen Institute and funded by the Ford Foundation, this report examines the role of telecommunications in rural economic development. As the year 2000 approaches, rural residents are still struggling to upgrade telephone services that urban dwellers have long enjoyed. Goals and recommendations to achieve those goals are outlined for state and federal government leaders. A very readable report.

Pittman, Nancy P., ed.
From The Land
Washington, D.C., Island Press, 1988. 478p. ISBN 0933280661

Literary writings on agriculture and conservation presented in this anthology were originally published in *The Land* magazine from 1941-1954. Articles and essays are grouped under two main headings, "Farming and the Soil" and "Water," and a section on poetry is also included. Early writings of prominent authors are represented, including Stegner, Muir, Stuart, Bromfield, Leopold, and Carson. An excellent choice when students need a literary collection on agriculture.

Rhodes, Richard
Farm: A Year in the Life of an American Farmer
New York, Simon and Schuster, 1989. 336p. ISBN 0671636472

A full year's cycle at a present-day Missouri farm is chronicled to show the human side of agriculture. Rhodes's portrayal is full of realism and emotion. He depicts the work life and values of typical small operators who cope with government programs and survive by upholding traditional family values which are supported by their communities.

Rosenblatt, Paul C.
Farming Is in Our Blood: Farm Families in Economic Crisis
Ames, Iowa, Iowa State Univ. Press, 1990. 187p.
ISBN 0813802385

Rosenblatt interviewed farm families who faced financial difficulties in 1986. Parts of the interviews are quoted with conclusions drawn on how they handled the stresses and subsequent decisions they had made within the year-and-a-half period of study. A helpful, easily read book which provides a reality check to students on the monetary difficulties of small farming operations.

Salamon, Sonya
Prairie Patrimony: Family, Farming, and Community in the Midwest
Chapel Hill, N.C., Univ. of North Carolina Press, 1992. 297p.
ISBN 0807820458

Based on extensive interviews in rural Illinois, Salamon's study analyzes the relationship between ethnicity and agriculture. Her research contrasts communities and families of German descent and those she terms "Yankee," with regard to farming and family business decision making. She projects her findings to future practices, suggesting that farm families of Germanic origin will easily adapt to sustainable farming systems because of their unique values. Her findings have relevance for the Midwest's agrarian sociology, which is influenced by this ethnicity.

Schwab, James
Raising Less Corn and More Hell: Midwestern Farmers Speak Out
Urbana, Ill., Univ. of Illinois Press, 1988. 301p.
ISBN 0252013980

Interviews with farm men and women and rural businessmen in the Middle West portray their views of family farming in the 1980s. Thirty-eight oral histories record responses to specific questions involving the politics and economic developments of that era. Different points of view concerning life on a farm are contrasted among business and environmental interests and the farmers themselves. This is a valuable resource for analyzing how people felt about this critical time in American agriculture.

Solbrig, Otto T., and Dorothy J. Solbrig
So Shall You Reap: Farming and Crops in Human Affairs
Washington, D.C., 1994. 284p. ISBN 1559633085

The Solbrigs explore how the evolution of farming and agriculture has paralleled societal change. The complexities and challenges of agriculture today have transformed farming; modern agricultural practices have also threatened the environment. By understanding the history of agricultural practices, students can better anticipate future needs and concerns.

Strange, Marty
Family Farming: A New Economic Vision
Lincoln, Neb., Univ. of Nebraska Press, 1988. 311p.
ISBN 0803241569

Strange, as co-director of the Nebraska Center for Rural Affairs, provides readers with the cultural and historical context of family farming issues. He discusses financial questions of agriculture, including causes and effects of federal farm policies and tax laws. He also raises the technological and industrial problems which need to be addressed. A readable book on a topic most undergraduates will find relevant.

Tarrant, John, ed.
Farming and Food
New York, Oxford Univ. Press, 1991. 256p. ISBN 0195209176

In this exceptional volume, experts summarize how agriculture is practiced in the largest agricultural-producing countries and geographic areas. Color photographs, drawings, graphs, charts, and maps support and amplify the text. An outstanding edition to support students' understanding of agriculture beyond their native region.

United States. Dept. of Agriculture
The Yearbook of Agriculture
Washington, D.C., U.S. Government Printing Office. Published annually until 1992. 1 v. (various pagings). ISSN 0886-7690

Each edition of this publication focused on a different agricultural topic. Recent titles included *New Crops, New Uses, New Markets* (1992); *Agriculture and the Environment* (1991); *Americans in Agriculture: Portraits of Diversity* (1990); *Farm Management, How to Achieve Your Farm Business Goals* (1989); and *Marketing U.S. Agriculture* (1988). Coverage of a wide variety of topics of current interest made this a basic resource for academic libraries. This annual publication was discontinued in 1992.

Vasey, Daniel E.
An Ecological History of Agriculture: 10,000 B.C.-A.D. 10,000
Ames, Iowa State Univ. Press, 1992. 363p. ISBN 0813809096

Vasey believes that the history of agriculture must include the aspect of ecology, i.e., organisms and their environment and human ecology. This study of agricultural systems includes physical, biological, and cultural aspects. Chapters detail the preindustrial, pastoral, and industrial periods and also deal with topics like dry and humid lands. Sections on agriculture for development and the future complete this study.

Wagner, John D.
Building a Multi-Use Barn for Garage, Animals, Workshops, Studio
Charlotte, Vermont, Williamson Pub. Co., 1994. 221p.

A comprehensive, step-by-step guide to the construction of a barn from initial design to painting the exterior. The basics of laying the foundation, framing, roofing, siding, wiring, plumbing, and heating are all detailed, with options for construction of a structurally sound building. Plentiful illustrations and photographs supplement the text. The appendix provides an excellent list of resources, including trade associations.

✦ Agricultural Machinery

Cooper, Elmer L.
Agricultural Mechanics: Fundamentals and Applications
2nd ed. Albany, N.Y., Delmar Publishers, 1992. 666p.
ISBN 0827334699

Cooper's text is one of the few recently published texts that delves into the world of agricultural mechanics. Well illustrated and clearly written, topics include arc welding, drainage and irrigation technology, small engine maintenance, propane and oxyacetylene equipment, and safety. Although in textbook format, this book meets the need for recent materials on these subjects.

Goering, Carroll E.
Engine and Tractor Power
3rd ed. St. Joseph, Mich., 1992. 539p. ISBN 092935530X

Intended for college agriculture students, this text covers construction, operation, and maintenance of farm tractors and engines. Keywords are italicized and defined without interrupting the flow of the text. It will be easily understood by students without mechanical or technical backgrounds. Illustrated.

Isern, Thomas D.
Bull Threshers and Bindlestiffs: Harvesting and Threshing on the North American Plains
Lawrence, Kans., Univ. Press of Kansas, 1990. 248p.
ISBN 0700604685

Historical photographs enhance this overview of grain farming on the Great Plains before the arrival of combines. These fascinating accounts of harvest hands (the bindlestiffs) and the machinery (the bull threshers) are enlightening for farmers of the nineties. The evolution of technological innovations parallels the social interactions of plains families. Well documented.

Leffingwell, Randy
The American Farm Tractor: A History of the Classic Tractor
Osceola, Wisc., Motorbooks International, 1991. 191p.
ISBN 0879385324

A history of the classic American farm tractor begins with the portable steam engine, the gas-powered engine, and the diesel. Coverage is comprehensive through 1960 of both the famous and the so-called "orphan makes." Color photographs, design drawings, and vintage advertisements enrich the text.

Pripps, Robert N.
How to Restore Your Farm Tractor
Osceola, Wisc., Motorbooks International, 1992. 176p.
ISBN 0879385936

In his introduction, Pripps explains the difference in approaches to restoration of a work tractor versus a show tractor. A comprehensive guide follows from engine rebuilding to body work and painting. Well illustrated with photographs and diagrams. Appendices include a list of sources for repair manuals, a bibliography of specialized tractor books, an extensive list of parts sources, and tips on safe disposal of parts and chemicals.

Pripps, Robert N.
Threshers
Stillwater, Minn., Motorbooks International, 1992. 128p.
ISBN 0879386177

This well-illustrated volume in Motorbooks' *Farm Tractor Color History* series traces the history of harvesting tools from biblical times to the twentieth-century, self-propelled combine. Descriptive history of the thresher and the "threshing time" camaraderie offers students a new perspective.

✦ Biotechnology

Baumgardt, Bill R., and Marshall A. Martin, eds.
Agricultural Biotechnology: Issues and Choices: Information for Decision Makers
West Lafayette, Ind., Purdue Univ. Agricultural Experiment Station, 1991. 181p. ISBN 0931682282

Chapter summaries and conclusions with attractive graphics and sidebars all make this a visually inviting and easy-to-use resource for the beginning researcher. Biotechnology as a topic is realistically placed within the context of agriculture in understandable language. A concluding section on how it will influence the future emphasizes issues and choices.

Gendel, Steven M., and others, eds.
Agricultural Bioethics: Implications of Agricultural Biotechnology
Ames, Iowa, Iowa State Univ. Press, 1990. 357p.
ISBN 081380129X

Essays by thirty-four experts provide varied points of view from researchers, professors in several disciplines, and editors of newsletters on the topic. Six sections, including safety and regulatory issues, cover the impact on science and industry, public perceptions, economics, and social and ethical considerations. The breadth of subtopics is impressive—from the interaction of biotechnology and the military, to bovine growth hormones.

Hobbelink, Henk
Biotechnology and the Future of World Agriculture: The Fourth Resource
Atlantic Highlands, N.J., Zed Books, 1991. 159p.
ISBN 0862328365

Hobbelink believes that genetic resources are the foundation of all living things; therefore, next to soil, water, and air, they are the fourth most important national resource. He places this topic in the context of world production by supplying much information on how biotechnology impacts farmers in Third World countries, and provides examples for students to consider. He also cites favorable uses of biotechnology with examples such as development of plants which have wider climatic tolerances or adaptability.

Juma, Calestous
The Gene Hunters: Biotechnology and the Scramble for Seeds
Princeton, N.J., Princeton Univ. Press, 1989. 288p.
ISBN 0691042586

In this well-researched book, agricultural-related innovations in biotechnology industries are explained in relation to history, biology, and economics. Juma views innovations in three major phases and emphasizes important ethical aspects in these developments. His analysis of intellectual property issues and policy considerations are placed in a global context. A case study of Kenyan agriculture and the role of germ plasm development is also included.

Kloppenburg, Jack Ralph
First the Seed: The Political Economy of Plant Biotechnology, 1492-2000
New York, Cambridge Univ. Press, 1988. 349p.
ISBN 0521326915

The history of the parallel development of plant breeding and the seed industry is given extensive treatment by Kloppenburg. He makes strong recommendations involving Third World countries and their rights to claim important contributions of germ plasm. He questions the sole use of plant breeding in the hands of proprietary private companies, and urges development of public plant breeding programs which would incorporate the new biotechnologies. Provides extensive notes and bibliography with numerous tables.

✦ Crops

Barnes, Robert F., and James B. Beard, eds.
Glossary of Crop Science Terms
Madison, Wisc., Crop Science Society of America, 1992. 88p.
ISBN 0891185356

This glossary provides a one-stop reference for approximately 1,800 terms used in the various disciplines of crop science. Terms are divided into three sections: general terms that are common to all facets of crop science, cell biology and molecular genetics, and field crops by nomenclature with scientific and common names. Usage of terms within the definition, cross-references, and synonyms are provided where appropriate.

Brooker, Donald B., Fred W. Bakker-Arkema, and Carl W. Hall
Drying and Storage of Grains and Oilseeds
New York, Van Nostrand Reinhold, 1992. 450p.
ISBN 0442205155

The drying of staple cereals (corn, rice and wheat) and oilseeds (soybeans and canola) is explained with accompanying graphs, diagrams, and charts, based on extensive research. Advantages and disadvantages of various methods of grain drying are explained. Although parts of this book are technical in terms of engineering data, its content will be valuable for agricultural students.

Fageria, N. K., V. C. Baligar, and Charles Allan Jones
Growth and Mineral Nutrition of Field Crops
New York, Marcel Dekker, 1991. 476p. ISBN 0824783867

Information in this volume is presented in a straightforward manner so that students can locate it easily. Addresses factors impacting the production of crops, such as climate, soil, and nutrients. Eleven separate chapters dwell on the basic crops grown in various (U.S.) geographic regions. Summaries at the conclusion of each chapter, many charts, graphs, and line drawings promote learning.

Fowler, Cary, and Pat Mooney
Shattering: Food, Politics, and the Loss of Genetic Diversity
Tucson, Ariz., Univ. of Arizona Press, 1990. 278p.
ISBN 0816511543

Two authoritative authors analyze why it is important to maintain a diverse genetic basis for food crop breeding. Ten chapters are grouped under "Legacy of Diversity," including a history of why this topic is important, and "Genetic Technology and Politics," including the role of biotechnology. Students will gain a wealth of information about genetic conservation as it pertains to agriculture. Extensive notes provide a basis for further reading.

Fussell, Betty Harper
The Story of Corn
New York, Knopf, 1992. 356p. ISBN 0394578058

Fussell provides a detailed background based on history, art, anthropology, and general science to show the part that corn has played in the U.S. and how it has been viewed by various American cultures. The historic photographs and illustrations will also be of interest.

Hill, Lowell D.
Grain Grades and Standards: Historical Issues Shaping the Future
Urbana, Ill., Univ. of Illinois Press, 1990. 423p.
ISBN 025201670X

Students who need background information on grain marketing, grading, and standards will be well served by this analysis. Hill outlines issues from both the nineteenth and twentieth centuries. He provides extensive background on the 1916 Grain Standards Act and subsequent legislative acts and violations that became well-known scandals. Tables in the appendix are concise and complete.

Langer, R. H. M., and G. D. Hill
Agricultural Plants
2nd ed. New York, Cambridge Univ. Press, 1991. 387p.
ISBN 0521405459

Students' questions regarding crop and pasture species of the world are addressed in this volume. With an emphasis on temperate regions, but including some subtropical areas, fourteen plant families are presented with separate chapters of varying length, including a discussion of prominent species, their cultivation and use. Line drawings and a bibliography are helpful supplements.

MacKenzie, James J., and Mohamed T. El-Ashry, eds.
Air Pollution's Toll on Forests and Crops
New Haven, Yale Univ. Press, 1989. 376p. ISBN 0300045697

Discussion of this important topic is based on two years of global scientific research by the World Resources Institute. Three chapters directly assess crop losses. Policy analysis is considered in terms of multiple pollutants. Policy recommendations are made within a framework of short- and long-term national energy planning. Evidence of damage is summarized in readable conclusions with numerous tables, graphs, and charts.

Wallace, Henry Agard, and William L. Brown
Corn and Its Early Fathers
Rev. ed. Ames, Iowa, Iowa State Univ. Press, 1988. 141p.
ISBN 0813800129

This chronological history of corn in America consists of brief chapters stressing the importance Indian corn has played in agriculture and how it has been hybridized by many individuals and institutions. The contributions of scientists involved in genetics from Cotton Mather and Darwin through James Logan, William Beal, and several others are featured in chapters. This work discusses the effect that other species of "forgotten corn" may have on seed banks of the future. *Henry A. Wallace Series on Agricultural History and Rural Studies.*

✦ Economic Policy

Browne, William P., and others, eds.
Sacred Cows and Hot Potatoes: Agrarian Myths in Agricultural Policy
Boulder, Colo., Westview Press, 1992. 151p. ISBN 0813385571

Challenging powerful myths such as "agrarian values are simple," "rural America is synonymous with farming," and that "an 'average' farm family exists," the authors explain how these misconceptions have adversely affected U.S. agricultural policy. The work advocates reform of agricultural policies and recommends that new policies demonstrate beyond some reasonable doubt that the public well-being is served.

Cochrane, Willard Wesley, and C. Ford Runge
Reforming Farm Policy: Toward a National Agenda
Ames, Iowa, Iowa State Univ. Press, 1992. 279p.
ISBN 0813804485

Advocates the need for policy reforms in food, farm, and rural sectors. After formulating a set of beliefs and values, the authors appraise the shortcomings of the current farm policy. A desirable farm policy is presented along with a discussion of the possible economic and social consequences of this proposed policy. Includes discussion about the specific political support a coalition would need to implement such proposed reform.

Demissie, Ejigou
Small Scale Agriculture in America: Race, Economics, and the Future
Boulder, Colo., Westview Press, 1990. 135p. ISBN 0813378230

There has been a substantial decline in the number of U.S. small-scale farms. Dr. Demissie examines the causes and effects of this decline and suggests policies which directly aid the unique needs of the small farmer. He particularly focuses on the additional obstacles African-American farm operators face.

Hallberg, M. C.
Policy for American Agriculture: Choices and Consequences
Ames, Iowa, Iowa State Univ. Press, 1992. 374p.
ISBN 0813813689

Specifically targeted for the undergraduate student who has a background in basic economic principles, this work provides a brief history of farm policy and exposure to the policy process. Examines the benefits and costs of price and income support for agriculture. Concludes by analyzing various policy-making options, such as protectionism in international trade, income support through marketing control, and price supports with no supply control.

Hamlin, Christopher, and Philip T. Shepard
Deep Disagreement in U.S. Agriculture: Making Sense of Policy Conflict
Boulder, Colo., Westview Press, 1993. 319p. ISBN 0813387035

Rather than outlining specific policies, this book examines the underlying assumptions and belief systems that have an effect on the formation of agricultural policy. The authors have divided agricultural policy into three ideologies: conventional productivism advocates a free market; ecological progressivism supports society control; and radical humanism advances self-determination in addition to strong social responsibility. The writings of Wendell Berry, Charles Walters Jr., and the Battelle report are cited as examples of these three ideologies.

Knutson, Ronald D., J. B. Penn, and William T. Boehm
Agricultural and Food Policy
2nd ed. Englewood Cliffs, N.J., Prentice Hall, 1990. 437p.
ISBN 0130187895

This undergraduate text discusses the process of policy formation, macroeconomics of agriculture, influences of interest groups, and effects of the international environment. It also examines the changing nature of farm and food problems and the consequent revision of farm policies. Attempts to give students the tools to evaluate policies and to make educated decisions.

Pasour, E. C.
Agriculture and the State: Market Processes and Bureaucracy
New York, Holmes and Meier, 1990. 258p. ISBN 0841912726

Presents a description of the policy-making process in agriculture and outlines the ineffectiveness of U.S. farm programs. Discusses the market system versus central direction. Part of the *Independent Studies in Political Economy* series.

✦ Farm and Ranch Business Management

Fleisher, Beverly
Agricultural Risk Management
Boulder, Colo., Lynne Rienner Publishers, 1990. 149p.
ISBN 1555871690

Aimed at the nonspecialist, this book provides an overview of the issues involved with agricultural risk management. Organized into three major areas, the book explores risks common to agricultural firms, examines risk management tools, and considers the effect of government policies. Each chapter concludes with a selected reading list.

Luening, R. A., Richard M. Klemme, and William P. Mortenson
The Farm Management Handbook
7th ed. Danville, Ill., Interstate Publishers, 1991. 607p.
ISBN 0813428726

Presents basic economic principles of farm management integrated with the science of production. Includes information on how to calculate capital item ownership costs and capital recovery charges. Includes appendices with farm business agreements and examples of various farm records.

Makeham, J. P., and L. R. Malcolm
The Farming Game Now
New York, Cambridge Univ. Press, 1993. 399p.
ISBN 0521404525

The focus of this work is on managing a farm in the twenty-first century. Covers techniques such as analyzing farm potential, capital, returns, costs, profit, and budgets. Much discussion is included about the farm as a separate entity and where it fits into the whole of a developed economy. This model is developed in the context of the Australian economy. Glossary of terms follows the text.

✦ Finance

Ferguson, Roy C.
Managing for Profit in Commercial Agriculture
Englewood Cliffs, N.J., Prentice Hall, 1990. 220p.
ISBN 0135518709

Outlines the "Ferguson System of Financial Analysis" which pioneered the broad-based application of financial ratios and index performance analyses in commercial agriculture. Although some of the chapters would be studied in an advanced course, the beginning chapters on business forecasting and control, accounting in agriculture, financial documentation, and fourteen signals of financial health provide a good introduction to lower-division students. Addresses the fact that many agricultural students are inadequately versed in business and financial management procedures.

Harl, Neil E.
The Farm Debt Crisis of the 1980s
Ames, Iowa, Iowa State Univ. Press, 1990. 305p.
ISBN 0813811880

Harl, the Charles F. Curtiss Distinguished Professor in Agriculture and Professor of Economics at Iowa State University, analyzes the government policies and precipitating factors that led to the farm debt crisis. Harl advocates federal government intervention during a major agricultural crisis. He believes that agricultural crises are inevitable, and outlines twelve important lessons learned from the debt crisis of 1980 which can be studied and used to respond more responsibly to the next crisis. Part of the *Henry A. Wallace Series on Agricultural History and Rural Studies*.

✦ International Agriculture

Bonanno, Alessandro, ed.
Agrarian Policies and Agricultural Systems
Boulder, Colo., Westview Press, 1990. 331p. ISBN 0813377307

The purpose of this book is to inform North Americans about the European Economic Community countries. Gives a comparative analysis of agricultural policies and systems. The history and present organization and trends of the European Community comprise the largest part of the book.

Goodman, David, and Michael Redclift, eds.
The International Farm Crisis
New York, St. Martin's Press, 1989. 296p. ISBN 031202682X

The farm crisis is not unique to the United States. Contributors from Great Britain, the United States, France, Spain, and Australia explore the nature of the crisis in various countries and the international interrelationships.

Harris, Jonathan M.
World Agriculture and the Environment
New York, Garland Publishing, 1990. 227p. ISBN 082400468X

Harris analyzes world agricultural production and then elaborates on various environmental influences such as erosion, fertilizer use, irrigation, and pesticides. He provides models with basic scenarios and shows their implications as policy. Some of the data may be more in-depth than students require, but the overall organization of the book would not be overwhelming for beginning researchers.

Helmuth, John W., and Don F. Hadwiger, eds.
International Agricultural Trade and Market Development Policy in the 1990s
Westport, Conn., Greenwood Press, 1993. 217p.
ISBN 0313286140

Social scientists specializing in trade and market development have contributed essays which analyze the changing world economy and political environment. Helmuth evaluates the effectiveness of past and present Agency for International Development (AID) programs developed by the United States. Discusses the impact these programs and policies will have on determining the United States position in this new global market. Debates the pros and cons of a market-oriented strategy of agricultural development versus comprehensive governmental market control.

Johnson, D. Gale
World Agriculture in Disarray
2nd ed. New York, St. Martin's Press, 1991. 365p.
ISBN 0312057997

Updating his earlier edition, Johnson lays out the problems of national governments dealing with agriculture. He urges national governments to understand the changes which have occurred in agriculture and advocates reduction of price supports, replacing them with payments for low-income farm families. Chapters include a discussion of the European Community and its involvement in agriculture. This work would be particularly useful for students who have an understanding of basic economics.

Rodale, Robert
Save Three Lives: A Plan for Famine Prevention
San Francisco, Sierra Club Books, 1991. 253p. ISBN 0871566214

Prevention, not relief, is the key to defeating famine according to Rodale, who maintains that indigenous cultures are actually harmed by well-intended relief efforts and should return to traditional farming techniques. Rodale is committed to a goal of sustainable agriculture worldwide. Bibliography.

✦ Professional Issues

National Research Council (U.S.). Board on Agriculture
Agriculture and the Undergraduate: Proceedings
Washington, D.C., National Academy Press, 1992. 280p.
ISBN 0309046823

Undergraduate education in agriculture is undergoing profound changes to meet the needs of farmers of the '90s. These are proceedings of a 1991 conference that addresses a rethinking of the purpose of agricultural undergraduate education. Curriculum issues, cultural diversity, the environment, scientific literacy, and excellence in teaching, are just a few of the issues that are critical to the revitalization of undergraduate education in the field of agriculture.

✦ Rangeland Management

Hage, Wayne
Storm over Rangelands: Private Rights in Federal Lands
3rd ed. Bellevue, Wash., Free Enterprise Press, 1994. 265p.
ISBN 0939571153

Traces the history of the rangeland controversy citing appropriate legislative acts and court cases. The work's central thesis is that valid private property rights exist in federal lands and that the anti-grazing movement is out to destroy those property rights. This is not intended to be a balanced portrayal of rangeland issues and is an excellent tool to spark "who owns the range?" debates. The table of cases is extremely useful in documenting judicial history. Bibliography and index.

Heitschmidt, Rodney K., and Jerry W. Stuth, eds.
Grazing Management: An Ecological Perspective
Portland, Ore., Timber Press, 1991. 259p. ISBN 0881921904

This is an excellent supplement to more basic rangeland and grazing works. Chapters written by thirteen expert contributors provide a more detailed viewpoint of many ecological and environmental issues crucial to grazing and range management. Of particular importance are two chapters devoted to social and economic influences on grazing management and decision-making and planning paradigms. Useful tables and illustrations are included. Extensive bibliography and index are especially valuable.

Holechek, Jerry, Rex D. Pieper, and Carlton H. Herbel
Range Management: Principles and Practices
Englewood Cliffs, N.J., Prentice Hall, 1989. 501p.
ISBN 0137527918

This comprehensive source integrates theoretical concepts, scientific research, and managerial aspects of grazing and range management, especially regarding multiple uses of the land. Partial contents: rangeland physical characteristics and types, range ecology and plant pathology, range management history, and inventory and monitoring. Five chapters are devoted to rangeland grazing topics including wildlife management issues. Excellent illustrations, tables, graphs, and maps are used effectively throughout the text. Extensive bibliographies follow each chapter.

James, Lynn F., and others, eds.
Noxious Range Weeds
Boulder, Colo., Westview Press, 1991. 446p. ISBN 0813383951

This is not an identification book, but a compilation of recent research from seventy-one experts on the management of range weeds. Important issues, such as global change and vegetative dynamics, biological and chemical control, plant and seed dynamics, naturalization, range ecology, economics of control, and other range management strategies, are discussed. Contains useful information on the control of specific species. Bibliographic references are included at end of each chapter. The index includes common and scientific name of range weeds and plants.

National Research Council (U.S.). Committee on Rangeland Classification
Rangeland Health: New Methods to Classify, Inventory, and Monitor Rangelands
Washington, D.C., National Academy Press, 1994. 180p.
ISBN 0309048796

Current range evaluation criteria and monitoring methods are critically analyzed and recommendations for improvement are given. Throughout the text, a variety of case studies are presented. The development of current theory and practice of range assessment is illustrated by a timeline dating from 1500 to 1990. A comparative table of definitions of assessment terms used by different agencies is very useful. Excellent tables and figures present information and data in a clear and concise manner.

Stubbendieck, James L., Stephan L. Hatch, and Charles H. Butterfield
North American Range Plants
4th ed. Lincoln, Neb., Univ. of Nebraska Press, 1992. 493p.
ISBN 0803242182

This is a primary identification reference for over 200 species. Line drawings include enlarged features of specific plant parts. Each entry includes scientific and common names; life span; habitat; growth characteristics; floral, fruit, and vegetative characteristics; season of growth; forage value estimates; and distribution maps. Noxious plants also include type of animal affected, toxic agent, and consequential effects. The glossary, list of authorities, selected bibliography, and comprehensive index increase the value of this guide.

United States. Bureau of Land Management
Rangeland Reform '94: Draft Environmental Impact Statement
Washington, D.C., U.S. Department of the Interior, Bureau of Land Management, 1994. 1 v. (ca. 550p.)

A draft of proposed changes in rangeland policies and regulations. After extensive public participation, a summary of proposed rangeland management and fee alternatives are outlined. Since the BLM does not manage rangelands in the east, the outlined policies are limited to fifteen Western states. The grazing fee policies described apply to seventeen Western states. This environmental impact statement is also available in an executive summary.

Vallentine, John F.
Grazing Management
San Diego, Calif., Academic Press, 1990. 533p.
ISBN 0127100008

This comprehensive reference manual addresses the grazing of domestic stock and wild game on native and seeded rangelands as well as temporary pastures such as crop aftermath. Partial contents: grazing effects on plants and soil, spatial patterns, grazing activities and behavior, herbivore nutrition, grazing capacity inventory, grazing intensity, and grazing systems. Presents an impressive bibliography of over 1,000 entries. Includes an index of plants with common and scientific names, a glossary, and a subject index. Includes good illustrations, photographs, and tables.

Vallentine, John F.
Range Development and Improvements
3rd ed. San Diego, Calif., Academic Press, 1989. 524p.
ISBN 0127100032

Comprehensive coverage makes this an invaluable reference. The benefits of range improvements and guidelines for planning and implementation are outlined, as well as political, legislative, social, biological, and educational issues. The value of this reference is greatly enhanced by an extensive table of plant characteristics, soil adaptation, forage usability, regional adaptation, growth season, water needs, and grazing tolerance. Scientific findings are cited and used to illustrate practices effectively. Includes excellent tables and illustrations. An impressive list of over 1,200 references provides an easy means to identify original research studies and additional information.

✦ Soil and Water

Baden, John B., and Stephen B. Lovejoy, eds.
Agriculture and Water Quality: International Perspectives
Boulder, Colo., Lynn Rienner, 1990. 224p. ISBN 1555871836

Authorities from many regions discuss problems which must be addressed as a result of pollution of water supplies by agriculture. Consideration is given to the realities of farming as a business as well as the economics of environmental necessities. Policy applications involving the National Agroenvironmental Incentives Program and California's Proposition 65 are analyzed. Two contributors from Australia and Sweden provide information for comparison with foreign situations. This book may also be useful to students in environmental studies.

Great Plains Agricultural Council. Water Quality Task Force
Agriculture and Water Quality in the Great Plains: Status and Recommendations
College Station, Tex., Texas Agricultural Experiment Station, 1992. 24p.

This study is relevant because farmers are confronted with regulations dealing with the impact of agriculture on water quality. Fourteen experts from agricultural universities and governmental agencies outline recommendations on eight important issues. Although focusing on the ten Great Plains states, this brief and very readable report will be useful for basic research on the topic.

Hillel, Daniel
Out of the Earth: Civilization and the Life of the Soil
New York, Free Press, 1991. 321p. ISBN 0029150604

Hillel presents an historical perspective on how civilizations have interacted with soil. His treatment, illustrative rather than encyclopedic, is supported with past abuses and the appropriate lessons to be learned. Human ecology, mankind's evolution, subsequent interaction with soil, and the agricultural revolution are all interestingly described. Global problems of the present are analyzed. Extensive notes and the selected bibliography fully support additional study on the topics of both soil and water.

Magdoff, Fred
Building Soils for Better Crops: Organic Matter Management
Lincoln, Neb., Univ. of Nebraska Press, 1992. 176p. ISBN 0803231601

Magdoff's work provides a good introduction to the management of organic soil matter for farmers who support sustainable agriculture. He explains the basics, using simple illustrations, and covers ways to decrease soil erosion by a variety of methods including manures, cover crops, crop rotation, reduced tillage, and composts. The author uses an overview of soil chemistry using simple graphs and charts to explain influences which have an effect on organic matter levels.

National Research Council (U.S.). Committee on Long-Range Soil and Water Conservation
Soil and Water Quality: An Agenda for Agriculture
Washington, D.C., National Academy Press, 1993. 516p. ISBN 0309049334

Faculty at major academic institutions and chief government agricultural agencies have collaborated to produce this extensive report. These experts propose a program of preventing soil degradation and water pollution while sustaining a profitable agricultural sector through a systems approach. Current policies are analyzed from a broader base. Includes chapters on pesticides, nitrogen, phosphorus, sediments, trace elements, and manure.

Opie, John
Ogallala: Water for a Dry Land
Lincoln, Neb., Univ. of Nebraska Press, 1993. 412p. ISBN 0803235577

Opie concentrates on this aquifer as an example of the agricultural use of water because of its physical location in the center of the agriculture industry and its importance in producing feed for America's beef. He describes how there are conflicting demands in formulating a sustainable agricultural policy when it involves extensive irrigation in a multistate environment. His historical approach within an environmental context will provide students with food for thought. Contains extensive notes. *Our Sustainable Future*, v. 1.

Postel, Sandra
Water for Agriculture: Facing the Limits
Washington, D.C., Worldwatch Institute, 1989. 54p. ISBN 0916468941

Postel addresses the problems of securing water to meet the world's need to grow food within a sustainable agricultural framework. She presents relevant data about irrigation in the U.S. and other countries using data from government and private sources. Taking into account the difficulties of global warming and other political problems, she suggests strategies for the decade. Eighty-five references support the easy-to-read text.

✦ Sustainable Agriculture

Bender, Jim
Future Harvest: Pesticide-free Farming
Lincoln, Neb., Univ. of Nebraska Press, 1994. 159p. ISBN 080321233X

Bender uses his own 640-acre working farm as a case study to encourage farmers to develop systems which will reduce and finally eliminate the use of pesticides. Chapters emphasize the need for goals and a gradual conversion to alternative methods, with consideration of weed management and livestock. He explains the benefits, which include convenience, soil conservation, cashflow and productivity. He also debates the philosophical reasons for trying organic farming, with discussions of technology, the farm chemical industry, and EPA regulations. *Our Sustainable Future*, v. 5.

Callaway, M. Brett, and Charles A. Francis, eds.
Crop Improvement for Sustainable Agriculture
Lincoln, Neb., Univ. of Nebraska Press, 1993. 261p.
ISBN 0803214626

The editors focus on the adaption of crop plants to systems and conditions, rather than the modification of the environment to fit current crops. In addition to background information and the history of crop improvement, they provide additional information on crop breeding objectives, and methods for increasing plants' resistance to insects and tolerance to weeds. The role of seed companies, biotechnology, and intercropping are also covered by noted researchers. *Our Sustainable Future*, v. 4.

Granatstein, David
Reshaping the Bottom Line: On-farm strategies for a Sustainable Agriculture
Lewiston, Minn., Land Stewardship Project, 1988. 63p.

Intended as an introduction to the topic, this brief book is a concise summary of principles. Tables of data from government sources support the chapters on nitrogen, soil fertility, weeds, insects, and pastures. The chapter explaining alternative crops provides examples suited to the Midwest farming climate.

Netting, Robert
Smallholders, Householders: Farm Families and the Ecology of Intensive, Sustainable Agriculture
Stanford, Calif., Stanford Univ. Press, 1993. 389p.
ISBN 0804720614

Netting applies a British term, "smallholders", to the ownership and operation of a cultivated piece of land to supplement a principal income. Discussed within the framework of American and international cultures, he sees this type of agriculture as fulfilling a need for growing food in the world. He contrasts smallholders with large operations in the U.S. since the First World War and compares characteristics common to various cultures. Useful in classes when a worldwide perspective of uses of small plots of land is needed.

Soule, Judith D., and Jon K. Piper
Farming in Nature's Image: An Ecological Approach to Agriculture
Washington, D.C., Island Press, 1992. 286P. ISBN 0933280890

The authors examine the roots of the ecological crises of modern ecosystems in order to learn to achieve sustainable agricultural production. The requisite shift in perspective among farmers is also addressed by these Land Institute researchers. This is a readable approach to solving problems caused by industrialized agriculture.

✦ Plant Pathology and Pest Control

Ashton, Floyd M., and Thomas J. Monaco
Weed Science: Principles and Practices
3rd ed. New York, Wiley, 1991. 466p. ISBN 0471600849

This excellent text covers basic weed control principles, herbicides, and practices in specific crops, pastures and rangelands, and horticultural plants. Partial contents: weed biology and ecology, herbicide registration and environmental impact, herbicides, plants and soil. The herbicide section includes over 120 herbicides in twenty-three chemical classes. General herbicide properties, uses, soil influence, effects, herbicide combinations, and rate and method of application are given. Includes excellent tables, figures, and photographs, bibliographical references, and an outstanding index.

Lucas, George Blauchard, C. Lee Campbell, and Leon T. Lucas
Introduction to Plant Diseases: Identification and Management
2nd ed. New York, Van Nostrand Reinhold, 1992. 364p.
ISBN 0442005784

An excellent introduction for students with limited background in biological sciences. Included is a history of plant pathology, basic management strategies including Integrated Pest Management (IPM), and an excellent overview of chemical options, with emphasis on pesticide safety and environmental protection. This new edition features new information on identification of diseases and pathogens; expanded coverage of biological control methods; and the application of biotechnology. The majority of the book is organized by pathogen types, effectively using specific diseases and management techniques as examples.

Maloy, Otis C.
Plant Disease Control: Principles and Practice
New York, Wiley, 1993. 346p. ISBN 0471573175

This work presents an overview of practical and theoretical concepts of disease management. Specific chapters include: disease development and forecasting, plant disease losses, strategies for exclusion and eradication, fumigation, fungicide development, characteristic, and use, biological control, resistance and Integrated Pest Management. Examples are used to illustrate management concepts. Appendices include scientific names and authorities for pathogens and host plants. Selected references are provided at ends of chapters. Includes useful tables and figures.

Metcalf, Robert Lee, and Robert A. Metcalf
Destructive and Useful Insects: Their Habits and Control
5th ed. New York, McGraw-Hill, 1993. 1 v. (various pagings).
ISBN 0070416923

Provides an excellent overview of the morphology, physiology, metamorphosis, and classification of insects. Various control measures and techniques are discussed, including Integrated Pest Management, biological control by natural enemies, and a general survey of chemical insecticides. [For information on specific insecticides consult a more current source such as *Farm Chemicals Handbook* (issued annually).] Specific insect pests are grouped by host plants and crops. Excellent line drawings or photographs illustrate insects and afflicted plants and make this a useful identification manual. Extensive index and bibliographical references are provided.

Nyvall, Robert F.
Field Crop Diseases Handbook
2nd ed. New York, Van Nostrand Reinhold, 1989. 817p.
ISBN 0442267223

An outstanding identification manual containing approximately 1,200 diseases of twenty-five plants found worldwide. Organization is first by crop, then by disease-causal organism (e.g., bacterial, fungal, nematodes), and then alphabetically by disease common name. Each entry includes cause, distribution, symptoms, and control measures. Includes an extensive index that provides access via common and scientific names of diseases and crops. Bibliographies include journal articles and research reports on specific diseases and treatments. Extensive glossary included.

Pedigo, Larry P.
Entomology and Pest Management
New York, Macmillan, 1989. 646p. ISBN 0023933100

Excellent review of general and applied entomology. Pest management theories and practices are discussed in terms of ecological and environmental factors, profitability, and sustainability. Various examples are used to illustrate concepts. "Diagnostic boxes" containing detailed information on life history, appearance, importance, and distribution are featured for specific species or species groups. Appendices include keys to insect orders, a list of common and scientific names of insects, and a list of popular insecticides by common and trade names. Includes a glossary and excellent illustrations and photographs.

Schumann, Gail L.
Plant Diseases: Their Biology and Social Impact
St. Paul, Minn., APS Press, American Phytopathological Society, 1991. 397p. ISBN 0890541167

Intended to be used as a source in general education curricula to emphasize the link between agricultural food production and social and human welfare issues. Specific issues covered include political legislation governing use of pesticides, biotechnology, environmental protection, and land preservation for agricultural use. Specific chapters are devoted to pathogens and quarantines, genes and genetic engineering, pesticides, environmental diseases, and problems. No previous biological or agricultural background is necessary to benefit from this work. Includes a good glossary and index.

Southern Weed Science Society (U.S.). Weed Identification Committee
Weed Identification Guide
Champaign, Ill., Southern Weed Science Society, 1985- . 6 v. (loose-leaf)

This is an outstanding source for approximately 300 United States weed species. Color photographs of various stages of development, and line drawings of specific features indiscernible in photographs, make this an excellent source for seedling, juvenile, and mature specimen identification. Common and scientific names are listed and indexed separately. Provides detailed descriptions including plant habitat and life history, vegetative and reproductive characteristics, unique identifying features, and distribution maps.

Whitson, Tom D., and others
Weeds of the West
Laramie, Wyo., The Western Society of Weed Science in cooperation with the Western United States Land Grant Universities Cooperative Extension Services, 1991. 630p. ISBN 0941570134

An excellent identification book of approximately 364 species located primarily in the West; some midwestern species are also included. Full-page color photographs of the adult specimens are featured, as well as other color photographs of immature plants, seeds, blooms, and fruit. The narrative includes basic description, habitat, origin, and effects of noxious plants. Includes a basic glossary, key to weed families, and a bibliography of state and regional weed guides. Index includes both scientific and common names.

Nonprint Resources

Agriculture: America's Most Crucial Industry
Washington, D.C., U.S. Dept. of Agriculture. Radio and Television Division; Distributed by National Audiovisual Center, 1990.
1 videocassette (15 min.). VHS, Beta, or U-matic

Presents a basic overview of the influence of agriculture in the nation's economy. Discusses how agriculture is a major source of employment in many industries such as food processing, textiles, transportation, and timber, and how agriculture supports other industries, such as chemicals and machinery manufacturing. Program lacks depth, but would be useful for an introduction on the first day of class, especially for students without agricultural background.

America the Bountiful
San Luis Obispo, Calif., California Polytechnic State Univ. Foundation; Distributed by Visual Education Productions, 1992.
6 videocassettes (23 min. each) VHS, and learning guide (39p.)

An outstanding series on American agricultural history from 8000 BCE to the present. Specific topics include Native American agriculture, the rise of tobacco and cotton plantations, slavery's role in agriculture, westward migration, the industrial revolution, the dust bowl era, the Great Depression and the New Deal, food production and global politics, and the "green revolution" and biotechnology. The learning guide includes a chronology, study questions, and bibliographical resources.

Biology and Agriculture Courseware: An Interactive Multimedia Kit
Rutherford, N.J., Tech Pubs Hal Inc./Technical Writing Education Ltd., 1993. 10 v.

This exceptional multimedia CD-ROM software promotes interactive and individualized learning. Students can modify different elements in a simulation to determine different outcomes and effects. Excellent graphics and easy user interface with online help. Separate volumes include the water and plant relationship; the soil and water relationship; plant needs; database and glossary; introduction to fertilization, fertilizers, fertilization; water management and irrigation systems. An instruction manual and workbook accompany each volume. Available in PC and Macintosh versions.

Herbicide Effect on Plants
Corvallis, Ore., Oregon State Univ., Dept. of Crop and Soil Science, 1988. 1 videocassette (45 min.) VHS

Using time-lapse photography, the physiological effects of specific herbicides on particular plants are shown and explained in twenty-two segments. Narration explains how the herbicide impacts on plant development and describes the symptoms of a treated plant. Many segments also demonstrate the concept of herbicide selectivity by showing a tolerant plant and targeted plant. Although the instructor may need or want to use additional information, this is an excellent means to effectively demonstrate plant reactions to herbicides.

Integrated Pest Management in Agriculture
San Luis Obispo, Calif., San Luis Video Productions, 1990.
1 videocassette (29 min.) VHS

An exceptional overview of IPM principles and management strategies. Specific examples in field crop, orchard, and vegetable production are used to illustrate basic concepts. Specifically, pest identification, effective monitoring systems, biological, cultural, chemical, and physical control measures are explained and illustrated.

"It's Not Gonna Happen to Me": [Tractor Accidents]
San Luis Obispo, Calif., Vocational Education Productions, 1990.
1 videocassette (25 min.) VHS

Emphasizes the importance of adhering to basic tractor safety practices. Preoperation safety, tractor transportation safety, operation safety, why accidents happen, and emergency responses are discussed. Computerized graphics and animation simulate safety hazards.

Marketing in a Global Economy
St. Paul, Minn., Farm Credit Services; Distributor: Agribank (St. Paul, Minn.), 1990. 1 videocassette (23 min.) VHS

Excellent overview and summary of the internationalization of the American agricultural economy. Succinctly summarizes the evolution of U.S. agricultural policy and trends. Emphasizes the need for U.S. farmers to become more attuned to other cultures, and discusses factors which will greatly influence the U.S. agricultural economy, such as free trade negotiations and various world markets. Includes facilitator's guide with exercises and answer key. The video has places to pause for discussion and to work on exercises. This is designed to reinforce important concepts.

The New Range Wars
Coproduced by the National Audubon Society, Turner Broadcasting, and WETA, Alexandria, Va., PBS Video, 1991. 1 videocassette (59 min.) VHS

Discusses the controversial issue of grazing on public lands. The ecological effects of grazing are emphasized to support the argument of many environmentalists that grazing is destroying the environment. Biologists, range specialists, environmentalists, and ranchers present a variety of opinions. Although alternative range management strategies such as holistic, rotation, intensive, and moderate are discussed, the overriding sentiment of this program is antigrazing. Use in conjunction with *Western Ranching: Culture in Crisis* (Summit Films, 1992) to present opposing viewpoints on the grazing issue.

Pesticides in Agriculture
San Luis Obispo, Calif., Vocational Education Productions, 1990.
1 videocassette (30 min.) VHS

Discusses recent trends of agricultural pesticide use in the context of economic, health, and environmental concerns. Specific topics include: definition of a pesticide, current developments of narrow spectrum chemicals with lower toxicity levels, integrated pest management principles, biological controls, and the use of biotechnology to produce genetically tolerant plant species.

Sustainable Agriculture
San Luis Obispo, Calif., Visual Education Production, 1991.
1 videocassette (23 min.) VHS

Examines major concepts and strategies of sustainable agriculture. Discusses misconceptions of the term and clearly defines other principles, including the efficient use of natural resources, economic issues, and environmentally sound practices which are both socially responsible and promote biodiversity. Specific strategies addressed include detailed monitoring of the crop cycle, soil fertility and nutrient cycling, conservation tillage techniques, crop rotations, Integrated Pest Management, promoting biological diversity, forecasting, and decision-making principles.

Sustaining America's Agriculture: High-Tech and Horse Sense
League City, Tex., National Association of Conservation Districts, 1992. 1 videocassette (29 min.) VHS

Highlights strategies used by farmers throughout the U.S. to limit soil erosion, groundwater contamination, nutrient runoff, and the destruction of wildlife habitat, while still maintaining or enhancing productivity and profitability. Biological pest control, crop rotation, strip intercropping, reduced tillage, precision irrigation, application of fertilizers, wetland preservation and wildlife corridors, and probiotics are discussed. New technologies and farm implements are being developed to meet the needs of sustainable agriculture.

Western Ranching: Culture in Crisis
[Denver?], Summit Films, 1992. 1 videocassette (57 min.) VHS

Discusses the economic, cultural, environmental, and political issues of grazing, especially on federal lands. A variety of experts testify, including ranchers, environmentalists, and range specialists. Grazing fees are discussed, but it is emphasized that sound land use is the fundamental issue. Basic concepts of grazing, range improvements, and sustainable strategies are discussed. Overall, the perspective is prograzing and can be effectively used with *The New Range Wars* (PBS Video, 1991) to present pro and con arguments of the controversial grazing issue.

✦ Electronic Resources

Ag Ed Network
West Bend, Wisc., Stewart-Peterson Group, Inc., Distributed by DTN (Data Transmission Network), Omaha, Neb.

Delivered via satellite technology, twenty-four hours a day, seven days a week, the *Ag Ed Network* provides the latest in agricultural news and information, curriculum support, weather maps, and market reports of interest to both faculty and students. Subscribers may access library materials of Future Farmers of America (FFA). The *Ag Ed Network* was launched in 1984; DTN delivery began in 1994.

AGRICOLA
Beltsville, Maryland, National Agricultural Library, 1970- .
Agricultural On-line Access (AGRICOLA) is the computerized bibliographic database for the National Agricultural Library and cooperating institutions. This is an excellent source for government documents, print and nonprint, in the field of agriculture. AGRICOLA may be searched online through DIALOG, Files 10 and 110, or on CD-ROM through SilverPlatter.

ARI Network Services, Inc.
330 East Kilbourn Avenue Suite 200
Milwaukee, Wisc. 53202-3166

   Provides dial-up access to *AP Alert,* an online agricultural information service. Includes access to agricultural and financial news summaries, futures price quotes, industry trends, trading data analysis, international commodity exchanges, cash commodity exchanges, and crop and livestock reports. Data is updated continually. Fees are usage-based.

Drew, Wilfred
Not Just Cows: A Guide to Internet/Bitnet Resources in Agriculture and Related Sciences
Release 2.0, 1993. 86 leaves

   Drew has compiled an extensive list of agricultural resources available through Internet or Bitnet. Includes access to libraries with comprehensive collections in agriculture. Includes electronic bulletin boards, gophers, and listservers among the many resources.

To locate this resource on the Internet:

Anonymous FTP:ftp.sura.net
Path:/pub/nic/agricultural.list
Also available as computer file via gopher at host=SNYMORVB.cs.snymor.edu ; port=70 ; path=Ogopher_root1 ; directory=librarydocs ; filename=not just cows.guide

Fire Services

Terri Propes
Lee Davis Library
San Jacinto College Central
Pasadena, Texas

Introduction

Fire Services studies prepare students for careers in a variety of areas such as structural fire fighting, fire inspection, fire and arson investigation, hazardous materials, and fire protection instruction. In addition, many fire departments also provide emergency medical services. Students prepare for this vocation through a combination of classroom study and field training. There currently is no national certification program for fire fighters, although a movement is now underway to formulate a program. Fire fighters are certified by individual state agencies or local county commissions. The National Fire Protection Association publishes *NFPA 1001 Fire Fighter Professional Qualifications,* guidelines that are presently accepted as the performance standard.

Selected Associations

Certification of fire fighters is usually done at the state or local level. However, there are several associations that can provide information on education, industry trends, and continuing education:

International Association of Fire Chiefs
1329 18th Street, NW
Washington, D.C. 20036-6516

International Association of Fire Fighters
1750 New York Ave., NW
Washington, D.C. 20006-5395

International Fire Service Training Association
Fire Protection Publications
Oklahoma State University
Stillwater, Okla. 74078-0118

National Association of Fire Investigators
20 E. Jackson, Stuite 1000
Chicago, Ill. 60604

National Fire Protection Association
1 Batterymarch Park
P.O. Box 9101
Quincy, Mass. 02269-9101

Selected Journals

American Fire Journal.
The Journal. Monthly. ISSN 0739-3709

Fire Chief.
Fire Chief. Monthly. ISSN 0015-2552

Fire Engineering.
PennWell Pub. Co. Monthly. ISSN 0015-2587

Fire Safety Journal.
Elsevier. Eight times a year. ISSN 0379-7112

Fire Technology.
National Fire Protection Association. Quarterly.
ISSN 0015-2684

Firefighter's News.
Lifesaving Communications. Eight times a year. ISSN 1061-4818

Firehouse.
PTN Pub. Co. Monthly. ISSN 0145-4064

Industrial Fire Safety.
PennWell Pub. Co. Six times a year.

Industrial Fire World.
Industrial Fire World. Bimonthly. ISSN 0749-890X

NFPA Journal.
National Fire Protection Association. Bimonthly. ISSN 1054-8793

National Fire Codes.
National Fire Protection Association. Annual. ISSN 0077-4545

Print Materials

Aircraft Rescue and Firefighting
3rd ed. Stillwater, Okla., Fire Protection Publications, 1992. 247p. ISBN 0879390999

General information on techniques of aircraft fire fighting are incorporated with additional sources to meet local standards. Identifies types of military and civilian aircraft, and discusses theory, methods, and the specialized equipment and procedures for aircraft fire and rescue operation.

Asken, Michael J.
Psycheresponse: Psychological Skills for Optimal Performance by Emergency Responders
Englewood Cliffs, N.Y., Regents/Prentice Hall, 1993. 172p. ISBN 0893038393

Discusses skills that are designed to help emergency responders improve performance and cope with job-related stress. Includes bibliography.

Bachtler, J. R.
Fire Instructor's Training Guide
2nd ed. New York, Fire Engineering, 1989. 283p. ISBN 0878149120

Covers the learning environment for fire instruction including concepts, course development, lesson planning, methods of instruction, training aids, testing and evaluation, and use of computers. Includes index.

Baker, Charles J.
Firefighter's Handbook of Hazardous Materials
5th ed. Indianapolis, Ind., Maltese Enterprises, 1990. 321p. ISBN 0962705209

A pocket reference of hazardous materials including information on toxicity, temperature stability, and extinguishing materials. Information is arranged in tabular form for quick access on a fire scene.

Bochnak, Peter M.
Fire Loss Control: A Management Guide
2nd ed. New York, Marcel Dekker, 1991. 326p. ISBN 0824784138

Discusses fire protection practices and property loss control principles for managers, architects, and students with little or no background in fire services.

Brannigan, Francis L.
Building Construction for the Fire Service
3rd ed. Quincy, Mass., National Fire Protection Association, 1992. 667p. ISBN 087765381X

Describes the construction of a building from a fire fighter's point of view. Traces how fire spreads through various structures and how to anticipate the dangers of a collapsed building during a fire.

Bryan, John L.
Fire Suppression and Detection Systems
3rd ed. New York, Macmillan, 1993. 595p. ISBN 0023159901

Theoretical and practical coverage of extinguishers, extinguishing agents, and suppression and detection systems. Excludes sprinkler and standpipe systems.

Bukowski, Richard, and Robert J. O'Laughlin
Fire Alarm Signaling Systems
2nd ed. Quincy, Mass., National Fire Protection Association, 1994. 430p. ISBN 0877653399

Complete overview of fire alarm signaling systems and their installation and maintenance. Reflects changes mandated by the Americans with Disabilities Act and recent updates of the *National Fire Alarm Code* (National Fire Protection Association, 1993) and the *Life Safety Code* (National Fire Protection Association, 1967-).

Carson, Wayne G., and Richard L. Klinker
Fire Protection Systems: Inspection, Test, and Maintenance Manual
2nd ed. Quincy, Mass., National Fire Protection Association, 1992. 280p. ISBN 0877653879

Handbook on the maintenance of fire protection systems. Covers water-based and alternative chemical systems as well as portable extinguishers. Provides information on NFPA and OSHA requirements and equipment manufacturers.

Carter, Harry R., and Erwin Rausch
Management in the Fire Service
2nd ed. Quincy, Mass., National Fire Protection Association, 1989. 391p. ISBN 0877653577

Presents effective management principles for fire departments including goal setting, personnel management, motivation, managing by objectives, and evaluation of programs.

Cassidy, Kevin A.
Fire Safety and Loss Prevention
Boston, Butterworth-Heinemann, 1992. 197p. ISBN 0750690399

Addresses the balance between fire safety and building security. Includes information on the need for cooperation between fire departments and building management, especially in regard to state codes. Includes bibliography, glossary, and index.

Clark, William E.
Firefighting Principles and Practices
2nd ed. New York, PennWell Pub. Co., 1991. 473p. ISBN 0878149201

A comprehensive text on fire fighting principles and practices. Discusses the characteristics of fire, fire fighting tactics, techniques and strategy, fireground safety, control and coordination, ladder and engine company operations, and prefire planning. Includes appendices and index.

Coleman, Ronny J., and John A. Granito, eds.
Managing Fire Services
2nd ed. Washington, D.C., International City Management Association, 1988. 506p. ISBN 0873260783

Includes a brief history of fire, followed by sections on the planning and evaluation of fire services, managing resources and programs, EMS and rescue services, training and personnel management, fire prevention, code administration, and the legal aspects of fire department management.

Coleman, Ronny J.
Opportunities in Fire Protection Services
Lincolnwood, Ill., VGM Career Horizons, 1990. 145p.
ISBN 0844286230

Introduction to various careers in fire protection services including descriptions of jobs, training required, and advice on getting started. Appendices list organizations to contact for further information. Includes bibliography.

Cote, Arthur E., ed.
Fire Protection Handbook
Quincy, Mass., National Fire Protection Association, 1962- .
1 v. (various pagings) ISSN 0734-5984

Employs a systems approach to fire prevention, design of buildings, detection of fire suppression, confinement of fire, and evacuation of occupants. Appendices included.

Ditzel, Paul C.
Fireboats: A Complete History of the Development of Fireboats in America
New Albany, Ind., Fire Bluff House Publishers, 1989. 125p.
ISBN 0925165010

A history of fireboats and their use in America. This is one of the few volumes that presents essentials of waterfront and shipboard techniques. Includes index and bibliography.

Downey, Ray
The Rescue Company
Saddle Brook, N.J., Fire Engineering Books and Videos, 1992.
328p. ISBN 091221225X

Present an overview of the operations of a rescue company from planning through managing the incident. Actual rescue situations illustrate techniques and execution of rescue maneuvers.

Dunn, Vincent
Safety and Survival on the Fireground
Saddle Brook, N.J., Fire Engineering Books and Videos, 1992.
377p. ISBN 0912212233

A survey of fireground dangers and safe fire fighting techniques and practices. Descriptions of various dangers are followed by recommended safe procedures for dealing with each situation. Includes index.

Eckman, William F.
Fire Department Water Supply Handbook
Saddle Brook, N.J., Fire Engineering Books and Videos, 1994.
440p. ISBN 0912212357

A comprehensive text dealing with the rural, suburban, and urban water supply. Provides information on planning, risk analysis, testing apparatus, hydrants, fill sites, developing specifications, standard operating procedures, and the water supply officer.

Essentials of Fire Fighting
3rd ed. Stillwater, Okla., Fire Protection Publications, 1992. 590p.
ISBN 0879391014

Provides an introduction to the basics of fire fighting. Chapters correspond to the chapters of the *NFPA 1001 Fire Fighter Professional Qualifications*.

Fire, Frank L., Nancy K. Grant, and David H. Hoover
SARA Title III: Intent and Implementation of Hazardous Materials Regulations
New York, Van Nostrand Reinhold, 1990. 279p.
ISBN 0442238037

Intended for fire departments and industrial fire brigades, this work addresses community differences in implementing hazardous materials regulations.

Fire Department Aerial Operations
Stillwater, Okla., Fire Protection Publications, 1991. 386p.
ISBN 0879390921

Covers operation, maintenance, testing, and evaluation of aerial apparatus such as aerial ladders, platforms, and water towers. Also includes information on vehicle operation, stabilization, and positioning. Methods of evaluating equipment for purchase and information on equipment manufacturers conclude this volume.

Fire Department Company Officer
2nd ed. Stillwater, Okla., Fire Protection Publications, 1989. 278p.
ISBN 0879390840

Surveys the training and operations of a fire department company officer. Includes company management, department inspections, company training, goal implementation, safety procedures, preincident planning, and problem solving, along with evaluations of existing officer training programs.

Fire Department Occupational Safety
2nd ed. Stillwater, Okla., Fire Protection Publications, 1991. 366p.
ISBN 0879390972

Addresses equipment and tool maintenance, fitness and health requirements, protective equipment, training and hazard responses, and how to set up and implement a safety program. Based on the *NFPA 1500, Standard on Fire Department Occupational Safety and Health Program* (NFPA, 1987).

Fire Department Pumping Apparatus
7th ed. Stillwater, Okla., Fire Protection Publications, 1989. 374p.
ISBN 0879390786

Presents general principles of pumping apparatus operations including theory, vehicle operation, maintenance, and selection of new equipment. Includes appendices and glossary.

Fire Protection Guide on Hazardous Materials
10th ed. Quincy, Mass., National Fire Protection Association, 1991. 1 v. (various pagings). ISBN 0877653666

A compilation of information from four National Fire Protection Association publications on hazardous chemicals, the properties of flammable liquids, gases, and volatile solids, and hazardous chemical reactions. Includes an identification system for the fire hazards of these materials, cross-references to common names, and topic finder.

Fire Service Instructor
5th ed. Stillwater, Okla., Fire Protection Publications, 1990. 326p. ISBN 0879390875

Provides an overview of classroom training for fire departments. Covers instructor qualifications, lesson preparation, development of teaching aids, testing and evaluation, and training of classroom instructors.

Fire Stream Practices
7th ed. Stillwater, Okla., Fire Protection Publications, 1989. 464p. ISBN 0879390832

Presents principles of hydraulics pertaining to fire hoses and nozzles, as well as methods of obtaining optimum equipment performance. Includes material on the selection and purchase of fire fighting equipment and evaluates specific brands.

Fornell, David P.
Fire Stream Management Handbook
Saddle Brook, N.J., Fire Engineering Books and Videos, 1991. 448p. ISBN 0878149279

Provides a clear explanation of fire stream management that integrates water movement and extinguishment into the overall fire fighting operation.

Frank, James A., and Jerrold B. Smith
Rope Rescue Manual
2nd ed. Santa Barbara, Calif., CMC Rescue, 1992. 177p. ISBN 0961833718

A comprehensive text on rope rescue operations including information on knots, equipment, safety procedures, and case histories.

Friedman, Raymond
Principles of Fire Protection Chemistry
2nd ed. Quincy, Mass., National Fire Protection Association, 1989. 254p. ISBN 0877653631

A review of chemistry principles followed by specific information on fire protection chemistry and its processes.

Haessler, Walter M.
Fire: Fundamentals and Control
New York, Marcel Dekker, 1989. 248p. ISBN 0824780248

Discusses all aspects of combustion and flammability of materials including fire extinguishment and environmental factors contributing to combustion. Indexed.

HazMat for First Responders
Stillwater, Okla., Fire Protection Publications, 1988. 375p. ISBN 087939076X

A discussion of the various types of hazardous materials and how they interact with other environmental factors such as weather. Emphasizes the importance of preplanning, documentation of the incident number, and evaluation of performance. Comprehensive appendices include aids in material identification, hazard assessment, and sources of technical assistance.

Holt, Francis X.
Emergency Communications Management
Saddle Brook, N.J., Fire Engineering Books and Video, 1991. 198p. ISBN 087814918X

Focuses on the eight most important areas of emergency communications: the communications center, dispatchers, scheduling, managing personnel, operations, training, computer usage, and planning.

Hoover, Stephen R.
Fire Protection for Industry
New York, Van Nostrand Reinhold, 1991. 367p.
ISBN 0442239297

Discusses the industrial use of water sprinkler systems (including automatic systems) and their water supply, alternate methods of fire suppression, and special fire protection concerns of industrial plants. Includes glossary, bibliography, and index.

Hose Practices
7th ed. Stillwater, Okla., Fire Protection Publications, 1988. 245p. ISBN 0879390751

Complete overview of the use, maintenance, and construction of the fire hose. Discusses techniques for storage, handling, and load capacities of various types of hoses.

Kennedy, Patrick M., and John Kennedy
Explosion Investigation and Analysis: Kennedy on Explosions
Chicago, Investigations Institute, 1990. 451p. ISBN 0960787623

An introduction to the investigation of explosions including the basics of investigation, fuels, explosion ignition, determining origins, and arson. Concludes with a tabular listing of material data, formulas, and conversion factors.

Kidd, J. Steven, and John D. Czajkowski
Vehicle Extrication: A Training Manual
Saddle Brook, N.J., Fire Engineering Books and Videos, 1991. 253p. ISBN 0878149155

This is a complete reference work on vehicle extrication. Discusses car construction, methods of extrication, evaluation of the condition of a trapped person, safe operation of tools, and incident control. Includes chapter summaries and extensive illustrations.

Kirk, Paul Leland
Kirk's Fire Investigation
3rd ed. Englewood Cliffs, N.J., Prentice Hall, 1990. 416p.
ISBN 0893037257

Covers the mechanics of fire and fire investigations including the methods of identifying various types of fire and their causes, information on complying with laws governing collection of evidence, and the presentation of evidence in court. Appendices, glossary, and index are included.

Lathrop, James K.
Life Safety Code Handbook
5th ed. Quincy, Mass., National Fire Protection Association, 1991. 1038p. ISBN 0877653798

Based on the 1991 edition of the *Life Safety Code* (National Fire Protection Association, 1967-). Provides background information on the development of the code, code requirements, and the rationale for inclusion. Includes practical tips on code implementation. Appendices and index provided.

Mahoney, Gene
Fire Suppression Practices and Procedures
Englewood Cliffs, N.J., Brady, 1992. 286p. ISBN 0893032158

An introductory text to fire suppression including topics such as fire chemistry and behavior, fireground planning and coordination, engine company and truck company operations, salvage, basic fire fighting tactics, fire in buildings, and mobile equipment. Includes review questions and index.

Meyer, Eugene
Chemistry of Hazardous Materials
2nd ed. Englewood Cliffs, N.J., Prentice Hall, 1989. 509p. ISBN 089303133X

Includes a review of basic chemistry followed by hazardous materials regulations with illustrative examples. Discusses corrosive and water-reactive materials, toxic chemicals, organic compounds, plastics and textiles, chemical explosives, and radioactive substances. Indexed.

Mitchell, Jeffrey T., and Grady P. Bray
Emergency Services Stress: Guidelines for Preserving the Health and Careers of Emergency Services Personnel
Englewood Cliffs, N.J., Prentice Hall, 1990. 183p. ISBN 0893036870

Explains the types of stress experienced by emergency response personnel, the danger signals of stress, and various types of stress management techniques found most effective for emergency responders.

Moore, Wayne D.
National Fire Alarm Code Handbook
Quincy, Mass., National Fire Protection Association, 1994. 1 v. (various pagings). ISBN 0877653925

Based on the *NFPA 72 National Fire Alarm Code* (NFPA, 1993), this handbook contains the complete text of the code with explanatory material. Covers all aspects of installation, maintenance, and use of fire alarm systems.

NFPA Inspection Manual
Quincy, Mass., National Fire Protection Association, 1950-. 1 v. (various pagings). Published irregularly

Covers identification of fire hazards and methods of correcting problems with special emphasis on safety. Absorbed *Conducting Fire Inspections* (NFPA, 1989).

Norman, John
Fire Officer's Handbook of Tactics
Saddle Brook, N.J., Fire Engineering, 1991. 541p. ISBN 0878149228

A hands-on approach to various tactics to be employed in fighting fires and rescuing victims. Presentation of general tactics leads to more specific fire situations and their resolutions. Index included.

Principles of Extrication
Stillwater, Okla., Fire Protection Publications, 1990. 365p. ISBN 0879390867

Presents methods of extricating victims from various types of vehicles and machinery. Covers situation assessments, tools, rescue vehicles and protective equipment. Includes various types of extrication such as from automobiles, trains, and industrial and agricultural equipment.

Roblee, Charles L., Allen J. McKechnie, and William Lundy
Investigation of Fires
2nd ed. Englewood Cliffs, N.J., Prentice Hall, 1988. 222p. ISBN 0893036420

Covers the basic principles of investigation for fire fighters, fire inspectors, and law enforcement personnel. Discusses legal aspects relating to arson, the search of a fire scene, the handling of physical evidence, safety procedures, and documentation. Bibliography and index included.

Rossotti, Hazel
Fire
New York, Oxford Univ. Press, 1993. 288p. ISBN 0198557221

Provides a history of fire fighting, descriptions of the destruction caused by fire, and discussions of fire drawn from biology, theology, and other sociological aspects of fire and fire prevention. Includes bibliography and index.

Schroll, R. Craig
Industrial Fire Protection Handbook
Lancaster, Pa., Technomic Pub. Co, 1992. 305p. ISBN 0877628874

Presents the principles of fire protection for industrial and commercial properties. Includes discussion of planning, life safety, methods of fire suppression including portable extinguishers, installed protection systems (such as automatic sprinkler systems, foam, halon systems), and organization of emergency teams and fire brigades. Appendices contain relevant OSHA regulations and NFPA standards. Indexed.

Self-Contained Breathing Apparatus
2nd ed. Stillwater, Okla., Fire Protection Publications, 1991. 360p. ISBN 087939093X

This edition has been completely revised to adhere to new guidelines in the *NFPA 1404 Standard for a Fire Department Self-Contained Breathing Apparatus Program* (NFPA, 1989). Covers training in the use of self-contained breathing apparatus, requirements for use, types of apparatus, methods of utilization, and maintenance of equipment. Includes glossary, appendices, and review questions at the end of each chapter.

SFPE Handbook of Fire Protection Engineering
Quincy, Mass., National Fire Protection Association, 1988. 1 v. (various pagings). ISBN 0877653534

This reference work covers basic fire science, hazard analysis calculations, design calculations, and fire risk calculation. Produced in conjunction with the Society of Fire Protection Engineers and the National Fire Protection Association. Index included.

Solomon, Robert E., ed.
Automatic Sprinkler Systems Handbook
6th ed. Quincy, Mass., National Fire Protection Association, 1994. 790p. ISBN 0877653976

All aspects of automatic sprinkler systems are presented in this comprehensive handbook. Includes the complete text of the *NFPA 13 Standard for the Installation of Sprinkler Systems* (NFPA, 1994).

Spahn, Edwin J.
Fire Service Radio Communications
New York, Fire Engineering, 1989. 437p. ISBN 0878149090

A detailed text on fire service radio communications including information on system design, explanations of electronics, radio signals, and evaluation of hardware.

Teele, Bruce W., ed.
NFPA 1500 Handbook: Based on the 1992 Edition of NFPA 1500, Standard on Fire Department Occupational Safety and Health Program
Quincy, Mass., National Fire Protection Association, 1993. 538p. ISBN 0877653887

Based on the 1992 edition of the *NFPA 1500 Standard on Fire Department Occupational Safety and Health Program* (NFPA, 1992). Includes explanatory chapters on administration, training and education, protective clothing, emergency operations, facility safety, and medical and fitness requirements.

Uniform Fire Code Standards
Whittier, Calif., International Conference of Building Officials, Published irregularly. ISSN 0896-9744

A companion to the *Uniform Fire Code* (International Conference of Building Officials, 1991). Contains National Fire Protection Association standards and *Uniform Building Code* (International Conference of Building Officials, 1994) standards referenced by the *Uniform Fire Code*.

Water Supplies for Fire Protection
4th ed. Stillwater, Okla., Fire Protection Publications, 1988. 268p. ISBN 0879390735

Provides an introduction to waterworks systems and the rural water supply. Discusses water distribution systems including requirements for mains and hydrants, and the need to maintain accurate records. Presents calculations for determining water supply for fire flow and the capacity of available water sources, along with a discussion of alternative methods of water delivery.

York, Kenneth J.
Hazardous Materials/Waste Handling For the Emergency Responder
New York, Fire Engineering, 1989. 379p. ISBN 0878149104

A guide to hazardous materials response with case studies of actual incidents. Includes information on the identification of hazards, surveying the scene, exposure protection, and mitigation.

Nonprint Materials

There are numerous videocassettes available. The following producers will provide catalogs of currently available materials.

Fire Engineering Books and Videos
250 Fifth Ave.
New York, N.Y. 10001

National Fire Protection Association
1 Batterymarch Park
P. O. Box 9101
Quincy, Mass. 02269-9101

Forestry

Charlotte Cooper
Finger Lakes Community College
Canandaigua, New York

Introduction

Forestry Science is a program of study for students who want to learn about the management of forested areas and the lands that surround them. Generally, forestry management is broken down into five subcategories: timber, hydrology and watershed management, wildlife, recreation, and range management.

Students complete a one- or two-year program in forestry, forest technology, or natural resources conservation at a community college, technical institute, vocational, or ranger school. Many students enter into the workforce at the end of these studies. Others prefer to continue their studies, pursuing a baccalaureate or more advanced degree.

Graduates with a certificate or associates degree are eligible for many technical-level positions in areas such as forest management, wildlife and watershed management, soil conservation, law enforcement, research, recreation management, range management, timber management, timber harvesting, wood science or wood products, forest fire management (preparedness, detection, or control), and vegetation/fuels management. Professions in the field of forestry combine a knowledge of nature and the natural world with modern technology.

There are a wide variety of employment options for graduates seeking employment. The federal government has many departments and agencies that employ forestry science graduates. Among these are the Forest Service, the Soil Conservation Service, the Department of Agriculture's Extension Service, the Bureau of Land Management, the National Parks Service, the Fish and Wildlife Service, the Bureau of Indian Affairs, the Army Corps of Engineers, the Department of Defense, the Bureau of Reclamation, the Bureau of Mines, the Department of Transportation, the Environmental Protection Agency, and the Geological Survey. Individual states have forestry agencies and parks systems, as well as fish and wildlife departments. Counties often employ individuals to manage their forests, parks, and watersheds, and to work in their soil and water conservation offices. Cities sometimes employ graduates in the area of urban forestry.

Private industry offers many opportunities as well. Timber and forest product companies, paper and pulp manufacturers, lumber companies, and timber framing companies often hire persons with forest science degrees. Graduates may also be self-employed, working as consultants, loggers, sawmill operators, tree farmers, nursery operators, tree surgeons, or landscapers.

In the 1960s, societal awareness of environmental issues and problems increased. This led to a growing interest in forestry science careers. As concern for the environment continues to increase, so will job opportunities for "green" collar workers. Citizens are beginning to question the policies and priorities used to determine how our forests are used and managed. This increased awareness should result in more resources and more responsible management of the world's natural resources.

Accreditation

The Society of American Foresters
5400 Grosvenor Lane
Bethesda, Md. 20814

Selected Associations

American Forestry Association
1516 P. St., NW
Washington, D.C. 20005

The Society of American Foresters
5400 Grosvenor Lane
Bethesda, Md. 20814

National Wildlife Federation
1400 16th St. NW
Washington, D.C. 20814

Wildlife Society
5410 Grosvenor Lane
Bethesda, Md. 20814

Selected Journals

American Forests.
American Forestry Association. Monthly. ISSN 0002-8541

American Midland Naturalist.
University of Notre Dame. Quarterly. ISSN 0003-0031

American Naturalist.
American Society of Naturalists. Monthly. ISSN 0003-0147

Audubon.
National Audubon Society. Bimonthly. ISSN 0097-7136

The Auk.
American Ornithologists' Union. Quarterly. ISSN 0004-8038

The Canadian Field-Naturalist.
Ottawa Field-Naturalists' Club. Quarterly. ISSN 0008-3550

Canadian Journal of Zoology.
National Research Council of Canada. Monthly. ISSN 0008-4301

The Condor: A Journal of Avian Biology.
The Cooper Ornithological Society. Quarterly. ISSN 0010-5422

Defenders.
Defenders of Wildlife. Quarterly. ISSN 0162-6337

Earth Journal
Buzzworm, Inc. Bimonthly. ISSN 0898-2996

Ecology.
Ecological Society of America. Eight times a year. ISSN 0012-9658

E: The Environmental Magazine.
Earth Action Network. Bimonthly. ISSN 1046-8021

Environment.
Heldref Publications. Ten times a year. ISSN 0013-9157

Forest Science.
Society of American Foresters. Quarterly. ISSN 0015-749X

International Wildlife.
National Wildlife Federation. Bimonthly. ISSN 0020-9112

Journal of Forestry.
Society of American Foresters. Monthly. ISSN 0022-1201

Journal of Mammology.
American Society of Mammalogists. Quarterly. ISSN 0022-2372

Journal of Range Management.
Society for Range Management. Bimonthly. ISSN 0022-409X

Journal of Soil and Water Conservation.
Soil and Water Conservation Society. Monthly. ISSN 0022-4561

Journal of Wildland Fire Management.
Wildlife Society. Quarterly. ISSN 0022-541X

Journal of Wildlife Management.
The Wildlife Society. Quarterly. ISSN 0022-541X

Living Bird.
Cornell Laboratory of Ornithology. Quarterly. ISSN 1059-521X

National Parks.
National Parks and Conservation Association. Bimonthly. ISSN 0276-8186

National Wildlife.
National Wildlife Federation. Bimonthly. ISSN 0028-0402

Nature Study.
American Nature Study Society. Frequency varies. ISSN 0028-0860

Northwest Science.
Washington State University Press. Cooper Publications. Quarterly. ISSN 0029-344X

Parks and Recreation.
National Recreation and Parks Association. Quarterly. ISSN 0031-0779

Sierra.
Sierra Club. Bimonthly. ISSN 0161-7362

Weatherwise.
Helfdref Publications. Ten times a year. ISSN 0043-1672

Wilderness.
The Wilderness Society. Quarterly. ISSN 0736-6477

Wildlife Society Bulletin.
The Wildlife Society. Quarterly. ISSN 0091-7648

The Wilson Bulletin.
The Wilson Ornithological Society. Quarterly. ISSN 0043-5643

Print Materials

✦ General Topics

Barbosa, Pedro
Introduction to Forest and Shade Tree Insects
San Diego, Academic Press, 1989. 639p. ISBN 0120781468

Barbosa's comprehensive introductory work on insects and their relationship to trees discusses methods of control including Integrated Pest Management and more traditional chemical treatments. Includes numerous illustrations and references.

Burns, Russell M., and Barbara H. Honkala
Silvics of North America
Washington, D.C., U.S. Dept. of Agriculture, Forest Service, 1990. 2 v. ISBN 0160271452 (vol. 1); ISBN 0160292603 (vol. 2)

This standard reference describes the silvical characteristics of over sixty conifers and 130 hardwoods commonly found in the United States. Each entry is written by a knowledgeable specialist and includes habitat, life history, special uses, genetics, and literature cited. *Agriculture Handbook*, 654.

Clark, F. Bryan, and Jay G. Hutchinson, eds.
Central Hardwood Notes
St. Paul, Minn., U.S. Dept. of Agriculture, Forest Services, North Central Experiment Station, 1989- . 1 v. (loose-leaf)

Provides management notes for the forests found in the central section of the United States. Covers silviculture, regeneration, site quality, forest growth and yield, stand management, economics, potential damage agents and protection, wildlife habitat, recreational use, and watershed management. Includes appendices.

Feininger, Andreas
Trees
New York, Rizzoli, 1991. 168p. ISBN 0847813258

Feininger's photographic study and homage to trees addresses their importance in the world's ecosystem, tree character, structure, leaves, trunk, bark, and wood.

Harlow, William Morehouse, and others
Textbook of Dendrology: Covering the Important Forest Trees of the United States and Canada
7th ed. New York, McGraw-Hill, 1991. 501p. ISBN 0070265712

Entries in this classic text on the study of trees include scientific and common names, botanical features (leaves, cones, buds, bark, flowers, fruit, twigs), geographic distribution, economic importance, and general description. A glossary, bibliographical references, and an index are provided.

Hartzell, Hal
The Yew Tree: A Thousand Whispers; Biography of a Species
Eugene, Ore., Hulogosi, 1991. 319p. ISBN 0938493140

Presents a very readable history of the yew tree, an increasingly important species because its bark is used to make the experimental anticancer drug taxol. Provides a good discussion on the current status of the tree. Includes appendices, bibliographic references, and an index.

Hunter, Malcolm L.
Wildlife, Forests, and Forestry: Principles of Managing Forests for Biological Diversity
Englewood Cliffs, N.J., Prentice Hall, 1990. 370p.
ISBN 0139594795

An introduction to managing temperate and boreal forests with diversity emphasized, this book is for professional foresters, students, and interested laypersons. Appendices include a summary of U.S. national policies on biological diversity in forests, a metric system primer, and diversity indices. A lengthy literature cited list; subject, geographic, scientific names, and taxonomic indices.

Kricher, John C.
A Field Guide to Eastern Forests, North America
Boston, Houghton Mifflin, 1988. 368p. ISBN 0395353467

This is one representative title in the *Peterson Field Guide Series*. This series stresses the interrelationships between plants and animals found in various types of forested areas. Accompanied by color identification plates, discussions of nature by seasons, and a list of common and scientific names. Indexed.

Laarman, Jan G., and Roger A. Sedjo
Global Forests: Issues for Six Billion People
New York, McGraw-Hill, 1992. 337p. ISBN 0070357021

A worldwide look at both the ecological and economic importance of forests. All types of forests from all parts of the world are considered. Many solutions and challenges are offered. Each chapter ends with a references list. Indexed. *McGraw-Hill Series in Forest Resources*.

Lauriault, Jean
Identification Guide to the Trees of Canada
Markham, Ontario, Canada, Fitzhenry and Whiteside, 1989. 479p. ISBN 0889025649

A basic identification guide to trees found commonly in Canada and the northern United States. Brief entries include common and scientific names, geographic distribution map; black-and-white drawings of leaf, fruit, stem, and tree shape; distinctive feature description; and general information. Includes bibliography and a detailed index.

Lipkis, Andy, and Katie Lipkis
The Simple Act of Planting a Tree: A Citizen Forester's Guide to Healing Your Neighborhood, Your City, and Your World
Los Angeles, Jeremy P. Tarcher, 1990. 237p. ISBN 0874776023

An outgrowth of citizen forester trainings presented by the organization TreePeople and geared to laypersons interested in improvement of the environment. Covers how to start an organization in one's own community, planning and carrying out a project, and planting and caring for trees. Includes a sample planting project, tree care project workbooks, and resources list.

Manion, Paul D.
Tree Disease Concepts
2nd ed. Englewood Cliffs, N.J., Prentice Hall, 1991. 402p.
ISBN 0139294236

Manion's basic text introduces basic tree diseases, including both the biological and ecological effects on trees. Provides information on modes of action, disease cycles, symptoms, methods of recognition, photographic examples, and suggested controls. Includes bibliographic references and a detailed index.

Norse, Elliott A.
Ancient Forests of the Pacific Northwest
Washington, D.C., Island Press, 1990. 327p. ISBN 1559630175

Norse, a senior ecologist for the Wilderness Society, has written extensively on the old growth forests of the Northwest. Like forests all over the world, their existence is being threatened. Norse discusses the importance of forests and makes recommendations on the use of sustainable forestry to save them. Includes glossary, suggested readings list, and an index.

Oliver, Chadwick Dearing, and Bruce C. Larson
Forest Stand Dynamics
New York, McGraw-Hill, 1990. 467p. ISBN 0070477930

A study of forests, how they grow, and even more importantly, how they respond to disturbances, both naturally occurring and intentionally induced. Written to assist those working with and managing forests for harvest, preservation, or recreation. Extensive reference list and an index are included.

Pielou, E. C.
The World of the Northern Evergreens
Ithaca, N.Y., Comstock Pub. Associates, 1988. 174p.
ISBN 0801421160

A basic introduction to common conifers found above forty-five degrees north latitude. Covers groups and species; life, growth, and reproduction; pests and parasites; and relationships with the elements, mammals, birds, and companion plants. Includes illustrations, chapter references, a suggested reading list, and an index.

Robinson, Gordon
The Forest and the Trees: Guide to Excellent Forestry
Washington, D.C., Island Press, 1988. 257p. ISBN 0933280408

Robinson, chief forester for the Southern Pacific Land Company and the Sierra Club, has written a guide to forest management and policy aimed at both the professional and the layperson. After a brief history of the forestry and forest management policy practiced in the United States, he defines what he considers to be excellent forestry practices for true multiple use. Included is information to help those looking at forestry as it is practiced today by the U.S. Forest Service and private industry. A lengthy section on research and opinions is included.

Shigo, Alex L.
New Tree Biology: Facts, Photos, and Philosophies on Trees and Their Problems and Proper Care
2nd ed. Durham, N.H., Shigo and Trees Associates, 1989. 618p.
ISBN 0943563046

A heavily illustrated, detailed, easy-to-understand book on all aspects of tree biology. Includes bibliographic references.

Spurr, Stephen Hopkins, and Burton Verne Barnes
Forest Ecology
3rd ed. Malabar, Fla., Krieger Pub. Co., 1992. 687p.
ISBN 0894646591

Spurr and Barnes present a basic text on forest ecology. Includes information on trees, the forest environment, forest ecosystems, forest history, and various types of forests found worldwide. Includes documentation, an index, and conversion factor tables.

Williams, Michael
Americans and Their Forests: A Historical Geography
New York, Cambridge Univ. Press, 1989. 599p.
ISBN 0521332478

This is a comprehensive history of the relationship between Americans and their forests from pre-1600 to the late twentieth century. Extensively documented, it covers lumbering, agriculture, and forestry from both environmental and historical perspectives. A lengthy bibliography and index are included.

Young, James A., and Cheryl G. Young
Seeds of Woody Plants in North America
Rev. and enlarged ed. Portland, Ore., Dioscorides Press, 1992. 407p. ISBN 0931146216

Revised and updated version of the classic *Seeds of Woody Plants in the United States* (U.S.G.P.O., 1974). Presents 386 genera with excellent drawings and photographs. Includes growth habits; occurrences and uses; collection, extraction, and storage of seeds; pregermination treatment and germination; flowering and fruiting; and nursery and field practice where appropriate.

Young, Raymond Allen, and Ronald J. Geise, eds.
Introduction to Forest Science
2nd ed. New York, Wiley, 1990. 586p. ISBN 0471856045

A clearly written basic text on all aspects of forestry including the following: trees; soils; forest genetics and tree breeding; ecology and ecosystems; insects and diseases; timber, rangeland, watershed, forest fire, recreation, and multiple use management; silviculture; remote sensing and geographic information systems; and timber harvesting. Also provides information on forest and wildlife interactions, forest products, and forestry careers.

✦ Career Resources

Alliance For Environmental Education
Education for the Earth: A Guide to Top Environmental Studies Programs
Princeton, N.J., Peterson's Guides, 1993. 175p. ISBN 1560791640

This Peterson guide meets the growing need for information on environmental education and careers. The CareerWatch 2000 section addresses job and career opportunities in both the public and private sector, as well as in education. Programs are grouped in five major categories. Includes a glossary, indexes, and a bibliography.

Basta, Nicholas
The Environmental Career Guide: Job Opportunities with the Earth in Mind
New York, Wiley, 1991. 195p. ISBN 0471534161

Includes a history of environmentalism, as well as a discussion of the types of careers available and the various organizations looking for employees. Appendices list nonprofit environmentally related organizations, environmental publications, federal addresses, state agencies, and educational organizations. Includes references and an index.

Morgan, Bradley J., and Joseph M. Palmisano, eds.
Environmental Career Directory: A Practical, One-Stop Guide to Getting a Job Preserving the Environment
Detroit, Gale Research, 1993. 348p. ISBN 0810391538

Part One is a collection of essays written by professionals working in a variety of environmentally oriented jobs. Part Two covers the job search process including résumés, applications, cover letters, and interview questions. Part Three lists companies with entry-level jobs and internship opportunities. Part Four concludes with publications and organizations that can help in the job search process. *Career Advisor Series.* Indexed.

Sharp, Bill
The New Complete Guide to Environmental Careers
2nd ed. Washington, D.C., Island Press, 1993. 364p. ISBN 1559631783

Provides information on careers in environmental planning, education, and communication, as well as environmental protection and natural resource management. Each suggested career contains a definition, a brief history, future outlooks and trends, career opportunities and salary, education, and case studies. Indexed.

Wille, Christopher M.
Opportunities in Forestry Careers
Lincolnwood, Ill., VGM Career Horizons, 1992. 143p. ISBN 0844285714

This typical example of the *VGM Opportunities Series* presents a basic overview of the careers open to forestry graduates of all levels. Includes a brief history of the profession, job descriptions and training, places of employment, education and experience necessary, and an overview of the current job market. Includes several useful appendices.

✦ Forest Fires

Buckley, John
Hotshot
Boulder, Colo., Pruett Pub. Co., 1990. 148p. ISBN 0871088096

An action-filled look at the unsung heroes of the U.S. Forest Service who fight the most difficult forest fires. Written by a thirteen-year veteran, the book presents a very realistic picture of what it is like to fight the largest and worst fires.

Fuller, Margaret
Forest Fires: An Introduction to Wildland Fire Behavior, Management, Firefighting, and Prevention
New York, Wiley, 1991. 238p. ISBN 0471521892

In the wake of the devastating forest fires of the 1980s, Fuller wrote this book that looks at U.S. policies and practices regarding forest fires. Includes chapters on prevention, fire fighting methods, and prescribed fires. Bibliography and index are provided.

Morrison, Micah
Fire in Paradise: The Yellowstone Fires and the Politics of Environmentalism
New York, HarperCollins, 1993. 253p. ISBN 0060163038

Reporter Morrison covered the devastating 1988 fires in Yellowstone National Park. In this book, he explores the politics and controversy behind the effort to control the fire, and provides an interesting look at U.S. policy on wildfires. Includes a glossary, notes and source list, and an index.

Walstad, John D., Steven R. Radosevich, and David V. Sandberg, eds.
Natural and Prescribed Fire in Pacific Northwest Forests
Corvallis, Ore., Oregon State Univ. Press, 1990. 317p. ISBN 0870713590

An overview and summary of forest fires, focusing specifically on the Pacific Northwest, but presenting principles and ideas applicable to the field as a whole. Includes natural history and ecology of forest fires, purpose and methods of prescribed burning, and the effects of prescribed burns on wildfire occurrence, forest productivity, and nontimber resources. Also discusses policy and regulation issues, economic outcomes, and public attitudes toward prescribed fire.

✦ Parks and Rangers

Douglass, Robert W.
Forest Recreation
4th ed. Prospect Heights, Ill., Waveland Press, 1993. 373p. ISBN 0881337145

An excellent introduction to the recreational use of forests and public lands. A historical introduction is followed by chapters that discuss recreational lands and the policies governing them. Sections on planning, siting, and developing lands are included. Several chapters are devoted to individual types of recreation areas such as campgrounds, picnic areas, and trails. Literature cited lists end each chapter. Appendices and index.

Frome, Michael
Regreening the National Parks
Tucson, Univ. of Arizona Press, 1992. 289p. ISBN 0816509565

Frome calls attention to the growing crisis of overpopulation, pollution, underfunding, and violence in our nation's parks. He offers recommendations and solutions to save these national treasures. Well documented, with a lengthy notes section, a bibliography, and an index.

Hartzog, George B.
Battling for the National Parks
Mt. Kisco, N.Y., Moyer Bell, 1988. 284p. ISBN 0918825709

Hartzog, director of the U.S. National Parks Service from 1964 to 1973, recounts his years in charge of that agency. In this fascinating account, the reader learns of the inside workings of government and Hartzog's fight to preserve the national parks. Indexed.

King, R. T.
The Free Life of a Ranger: Archie Murchie in the U.S. Forest Service, 1929-1965
Reno, Nev., Univ. of Nevada Oral History Program, 1991. 409p. ISBN 1564750000

King presents a fascinating oral history of a renowned forest ranger and his work in five western national forests. This work relates not only the life of a ranger, but also the policies and challenges facing the U.S. Forest Service. Glossary and index included.

Miller, Arthur P., and Marjorie L. Miller
Park Ranger Guide to Rivers and Lakes
Harrisburg, Pa., Stackpole Books, 1991. 200p. ISBN 0811730387

This guide presents an excellent introduction to understanding and interpreting natural areas near America's freshwaters. A general discussion of rivers and lakes is followed by chapters devoted to regional treatments: the Appalachian and eastern woodlands, southern mountains and lowlands, prairies and plains, the Southwest deserts, and the mountainous West. Chapters cover common animal and plant species found in each region.

Miller, Arthur P., and Marjorie L. Miller
Park Ranger Guide to Seashores
Harrisburg, Pa., Stackpole Books, 1992. 227p. ISBN 0811730395

Presents commonly found plants and animals that inhabit the various shores of the United States. Regions included are the rocky North Atlantic coast, the back bays and marshes of the South Atlantic coast, the barrier islands and sandy shores of the Gulf Coast, the bluffs and beaches of the southern Pacific coast, and the forested coast of the Northwest. A list of parks and index are included.

Miller, Arthur P.
Park Ranger Guide to Wildlife
Harrisburg, Pa., Stackpole Books, 1990. 179p. ISBN 0811722899

Miller provides a basic introduction to wildlife found in the following areas: the Appalachians and eastern woodlands, pines and marshes, prairies and plains, western mountains, and deserts. Discusses what rangers are, what they do, and where they are found. Includes an appendix of useful addresses and an index.

Sholly, Dan R., and Steven M. Newman
Guardians of Yellowstone
New York, Morrow, 1991. 317p. ISBN 0688092136

Sholly and Newman give the reader an insider's look at challenges faced by rangers as they try to meet the needs of the thousands of visitors, and at the same time protect the wildlife and natural resources of one of the most heavily used wilderness parks in the country. Sholly is chief ranger of Yellowstone and former chief of ranger activities for the entire National Parks Service.

Sternloff, Robert E., and Roger Warren
Park and Recreation Maintenance Management
3rd ed. Scottsdale, Ariz., Publishing Horizons, 1993. 295p. ISBN 0942280628

This introduction to the field of maintenance management covers basic principles, planning and organization, computer applications, and personnel, as well as maintenance of building and structures, general outdoor maintenance, grounds, and equipment. This work is of particular value to those working in parks. Chapter bibliographies; notes; and an index.

✦ Philosophy and Politics

Endicott, Eve, ed.
Land Conservation Through Public/Private Partnerships
Washington, D.C., Island Press, 1993. 364p. ISBN 1559631775

An interesting look at the role that private organizations can play in the acquisition of federal lands. The Nature Conservancy, the American Farmland Trust, and the Trust for Public Land are highlighted as representative organizations. Federal, state, and local partnerships are addressed along with information on funding. Includes bibliography and index.

Forest Resource Management in the Twenty-First Century: Will Forestry Education Meet the Challenge? Proceedings of the October 30-November 2, 1991 National Association of Professional Forestry Schools and Colleges, Society of American Foresters, Forest Resources Education Symposium
Bethesda, Md., Society of American Foresters, 1992. 86p. ISBN 0939970503

These proceedings explore issues such as the need to attract quality students, improve teaching quality, and strengthen curricula.

Hudson, Wendy E., ed.
Landscape Linkages and Biodiversity
Washington, D.C., Island Press, 1991. 196p. ISBN 1559631082

Based on the 1990 symposium on Biodiversity Conservation sponsored by Defenders of Wildlife, this work challenges the premise that the Endangered Species Act of 1973 does enough to save our endangered ecosystem. This thought-provoking work is well documented. Indexed.

O'Toole, Randal
Reforming the Forest Service
Washington, D.C., Island Press, 1988. 247p. ISBN 0933280491

Written by a forest economist for Cascade Holistic Economic Consultants, a nonprofit forestry consulting firm, this book challenges the policies and practices of the U.S. Forest Service. O'Toole analyzes the forest management philosophies and techniques currently operating and offers new solutions certain to generate much debate. Documented and indexed.

Rothenberg, David
Is It Painful to Think? Conversations with Arne Naess
Minneapolis, Univ. of Minnesota, 1993. 204p. ISBN 0816621527

Presents conversations with the Norwegian nature philosopher who originated the concept "deep ecology." This fascinating and challenging book includes photographs and lists of selected works by and about Arne Naess. Indexed.

Sample, V. Alaric
Land Stewardship in the Next Era of Conservation
Milford, Pa., Grey Towers Press, 1991. 43p. ISBN 0938549049

This challenging and essential work is an outgrowth of a conference on land stewardship held in 1990 that celebrated the centennial of the national forests. The resulting principles help guide resource managers in future conservation efforts. Includes environmental ethics materials and references, in addition to a bibliography on land stewardship.

Tober, James A.
Wildlife and the Public Interest: Nonprofit Organizations and Federal Wildlife Policy
New York, Praeger, 1989. 220p. ISBN 0275925811

This study looks at the important role independent nonprofit groups play in helping to shape U.S. policies on wildlife, specifically endangered species. The California condor and the bobcat are two of the endangered species cited as detailed examples. Includes bibliographical references and index.

Waterman, Laura, and Guy Waterman
Wilderness Ethics: Preserving the Spirit of Wilderness
Woodstock, Vt., Countryman Press, 1993. 239p.
ISBN 0881502561

In this thought-provoking book, the Watermans explore the true meaning of wilderness and the impact that increased numbers of visitors have had on wilderness areas. Includes discussions on the effects of high use on the regulation of wilderness areas. Bibliography and index provided.

✦ Rain Forests

Collins, N. Mark, ed.
The Last Rain Forests: A World Conservation Atlas
New York, Oxford Univ. Press, 1990. 200p. ISBN 0195208366

Collins presents an excellent, illustrated atlas that defines and explains rain forests. Discusses the importance of rain forests to the global environment and outlines the stresses on the world's remaining forests. Maps, illustrations, a glossary, and an index.

Collins, N. Mark, Jeffrey Sayer, and Timothy C. Whitmore, eds.
The Conservation Atlas of Tropical Forests: Asia and the Pacific
New York, Simon and Schuster, 1991. 256p. ISBN 013179227X

This is the first in a series on the rain forests of the world produced under the auspices of the International Union for Conservation of Nature and Natural Resources. The first part discusses broad topics such as forest wildlife, peoples, and forest management and is followed by a discussion of specific forests by country. Includes many charts, graphs, color drawings, and photographs. Reference lists, a list of acronyms, a glossary, and indexes are provided. The second in this series is on Africa.

Kirk, Ruth, and Jerry Franklin
The Olympic Rain Forest: An Ecological Web
Seattle, Univ. of Washington Press, 1992. 128p.
ISBN 0295971959

Heavily illustrated with beautiful color photographs, this work introduces novices to what the Olympic forest is and how the ecosystem works in one of the last remaining old growth rain forests in North America. Includes glossary, reading list, and an index.

Newman, Arnold
Tropical Rainforest: A World Survey of Our Most Valuable and Endangered Habitat with a Blueprint for Its Survival
An Eddison-Sadd ed., New York, Facts on File, 1990. 256p.
ISBN 0816019444

In this large format, well-illustrated book on rain forests worldwide, Newman discusses rain forest ecology, the threats to their existence, and what the world will lose if they become extinct. He provides suggestions on how to save them. Includes an appendix, a lengthy bibliography, and an index.

✦ Rangeland, Watershed, Wetlands, and Wildlands Management

Biswell, Harold H.
Prescribed Burning in California Wildlands Vegetation Management
Berkeley, University of California Press, 1989. 255p.
ISBN 0520064828

Authored by one of the early proponents of the use of fire in the management of wildland environments, this book covers fire behavior and fire as a viable management tool. After historical considerations are presented, Biswell offers a guide to planning and techniques and discusses the effects of prescribed burning on resources such as soils, water, wildlife, wilderness, timber, and forage. Includes a brief list of suggested readings and an index.

Brookes, Andrew
Channelized Rivers: Perspectives for Environmental Management
New York, Wiley, 1988. 326p. ISBN 0471919799

An introduction to managing and reclaiming rivers for flood control, land drainage, navigation, or erosion control. Written by a British environmental consultant, the book contains examples that are international in scope, with numerous cases cited from the United States. Topics cover river engineering, environmental legislation, channelization's physical effects, biological impacts, downstream consequences, and recommendations. Includes an extensive reference list, and indexes.

Holechek, Jerry, Rex D. Pieper, and Carlton H. Herbel
Range Management: Principles and Practices
Englewood Cliffs, N.J., Prentice Hall, 1989. 501p.
ISBN 0137527918

The authors define and introduce the science of range management: uncultivated land that provides the necessities of life for grazing and browsing animals. Ranges may include standing forage or edible leaves and twigs from woody plants. This work provides basic background information on an area of forestry often overlooked. Well referenced.

Hunt, Constance Elizabeth, and Verne Huser
Down by the River: The Impact of Federal Water Projects and Policies on Biological Diversity
Washington, D.C., Island Press, 1988. 266p. ISBN 0933280483

This is a highly readable account of how government policies are having an impact on the remaining riparian ecosystems. Discusses management of rivers and the impact of river development. Hunt relates how federal water projects and policies are devastating riparian ecosystems and provides a plan for reversing their impact before it is too late. Includes literature cited lists at the end of each chapter, a suggested reading list, glossary, and an index.

Hunter, Chris
Better Trout Habitat: A Guide to Stream Restoration and Management
Washington, D.C., Island Press, 1991. 320p. ISBN 0933280785

This comprehensive, readable handbook on the protection, management, and restoration of trout streams provides an excellent management text useful to laypersons as well as professionals. The author is an aquatic ecologist working under the auspices of the Montana Land Reliance. Illustrated and well documented. Includes a glossary and an index.

MacKenzie, James J., and Mohamed T. El-Ashry, eds.
Air Pollution's Toll on Forests and Crops
New Haven, Conn., Yale Univ. Press, 1989. 376p.
ISBN 0300045697

This is a startling collection of essays by experts who discuss the effects of air pollution on European and North American environments. Provides policy recommendations and solutions. Each chapter includes references. Indexed.

Meade, James W.
Aquaculture Management
New York, Van Nostrand Reinhold, 1989. 175p.
ISBN 0442205708

A basic manual on all aspects of aquaculture management useful to both students and professionals. Covers practices and principles as well as ethical philosophies. Includes chapter references, appendices, and an index.

Meehan, William R., ed.
Influences of Forest and Rangeland Management on Salmonid Fishes and Their Habitats
Bethesda, Md., American Fisheries Society, 1991. 751p.
ISBN 0913235687

This collection of essays demonstrates the interrelationship between wildlife and their environments. Discusses the impact of such factors as timber harvesting, forest chemicals, road construction and maintenance, and recreation on a specific species of fish. Includes extensive reference list, glossary, and index. *American Fisheries Society Special Publications,* 19.

Mitsch, William J., and James G. Gosselink
Wetlands
2nd ed. New York, Van Nostrand Reinhold, 1993. 722p.
ISBN 0442008058

After a basic introduction to wetlands, the authors address topics such as hydrology, biogeochemistry, ecology, management, restoration, classification, and inventory. Included are coastal, inland, and riparian wetlands. This is a definitive work on all types of wetland ecosystems. References and index.

Niering, William A.
Wetlands of North America
Charlottesville, Va., Thomasson-Grant, 1991. 160p.
ISBN 0934738815

Through photography, Niering documents beautiful marshes, swamps, and bogs of the United States and Canada. As more and more acres of wetlands are lost each year to development, this presentation and discussion will be of increasing interest to both students and the general public.

Payne, Neil F.
Techniques for Wildlife Habitat Management of Wetlands
New York, McGraw-Hill, 1992. 549p. ISBN 0070489556

This comprehensive introduction defines the various types of wetlands, discusses physical, chemical, and biological vegetation management, and artificial nesting and loafing sites. Includes references, numerous appendices, and an index.

Satterlund, Donald R., and Paul W. Adams
Wildland Watershed Management
2nd ed. New York, Wiley, 1992. 436p. ISBN 0471811548

Presents the theory and practice of managing watershed areas including forest, range and alpine lands. Covers hydrology, forestry, ecology, and soil science. Includes bibliographic references, an index, and a table of units of measurement with conversion factors used in hydrology.

✦ Related Ecology and Nature Titles

Duke, James A.
Handbook of Edible Weeds
Boca Raton, Fla., CRC Press, 1992. 246p. ISBN 0849342252

This guide to common edible weeds found in the United States provides scientific and common names, description, distribution, utility, and black-and-white illustrations. Includes bibliography, plant names, and general indexes.

Ehrlich, Paul R., and Anne H. Ehrlich
Healing the Planet: Strategies for Resolving the Environmental Crisis
Reading, Mass., Addison-Wesley, 1991. 366p. ISBN 0201550466

Picking up where *The Population Bomb* (Ballatine Books, 1968) left off, the Ehrlichs warn that our overconsumption of resources and destructive technological advances are leading to an environmental crisis from which the world may never recover. They challenge the reader to help solve the problems and thereby help save the world. Includes extensive documentation and index.

Flora of North America Editorial Committee
Flora of North America: North of Mexico
New York, Oxford Univ. Press, 1993- . ISBN 0195057139 (vol.1); ISBN 0195082427 (vol.2)

Certain to become a classic work on the flora of North America, this mammoth project is being issued over a ten-year period. Entries include habitat and ranges, synonymies, descriptions, and other relevant information. Excellent black-and-white illustrations. Extensively annotated.

Foth, H. D.
Fundamentals of Soil Science
8th ed. New York, Wiley, 1990. 360p. ISBN 0471522791

This good general introduction to soil science is geared to first and second year college students. Covers the nature and properties of soils, erosion, ecology, mineralogy, chemistry, plant and soil relations, fertilizers, genesis, and taxonomy. Includes chapter-end references, several useful appendices, a glossary, and an index.

Gagne, Raymond J.
The Plant-Feeding Gall Midges of North America
Ithaca, N.Y., Comstock Associates, 1989. 356p.
ISBN 0801419182

Presents information on 732 galls and 891 acidomyiids including damage, host plants, biology, and common distribution. Illustrated with mainly black-and-white line drawings in addition to four pages of color photographs. References, glossary, and an index.

Gleason, Henry A., and Arthur Cronquist
Manual of Vascular Plants of Northeastern United States and Adjacent Canada
2nd ed. Bronx, N.Y., The New York Botanical Garden, 1991. 910p. ISBN 0893273651

This update of the standard reference on wild vascular plants focuses on helping students identify often-confused species. Includes a glossary and detailed index.

Gore, Albert
Earth In the Balance: Ecology and the Human Spirit
Boston, Houghton Mifflin, 1992. 407p. ISBN 0395578213

This thought-provoking discussion of the state of the environment was written when Gore was a senator. He outlines the ecological problems he perceives and poses possible strategies. Includes notes, bibliography, and index.

Leopold, Aldo
Aldo Leopold's Wilderness: Selected Early Writings by the Author of A Sand County Almanac
Edited by David E. Brown and Neil B. Carmony. Harrisburg, Pa., Stackpole Books, 1990. 250p. ISBN 0811718646

A collection of Leopold's writings that will be of interest to anyone, including those who have read *A Sand County Almanac* (Oxford University Press, 1949.) Included are essays on game restoration, game management, and land health. Essential for all libraries. Bibliography and brief index.

Leopold, Aldo
The River of the Mother of God and Other Essays
Edited by Susan L. Flader and J. Baird Callicott. Madison, Wisc., University of Wisconsin Press, 1991. 384p. ISBN 0299127605

This is a collection of previously unpublished or hard-to-find essays by the author of *A Sand County Almanac,* one of the most famous nature books ever written. As an employee of the Forest Service for nineteen years, Leopold was known for his interest in forest, wildlife, and soil management. Includes a list of Leopold's publications and an index.

Lodge, Thomas E.
The Everglades Handbook: Understanding the Ecosystem
Delray Beach, Fla., St. Lucie Press, 1994. 228p. ISBN 1884015069

An excellent introduction to the unique ecosystem of the Florida everglades. The first chapters give an overview of the topic. Subsequent chapters focus on the various environments including freshwater marshes, wetland tree areas, tropical hardwood hammocks, pinelands, mangrove swamps, coastal lowland vegetation, coastal estuarine and marine waters. Other chapters cover flora, fauna, invertebrates, amphibians, reptiles, mammals, birds, and freshwater, estuarine, and marine fishes. The final sections consider human's impact on this delicate ecosystem and how it can be preserved and restored. Bibliography index.

Meine, Curt
Aldo Leopold: His Life and Work
Madison, Wisc., University of Wisconsin, 1988. 638p. ISBN 0299114902

This definitive biography of Aldo Leopold, one of the early leaders in the conservation movement, is also the biography of the conservation movement in the early part of this century. Extensively documented with notes and a bibliography. Indexed.

Merideth, Robert W.
The Environmentalist's Bookshelf: A Guide to the Best Books
New York, G. K. Hall, 1993. 272p. ISBN 0816173591

An annotated bibliography of the best 500 books on nature and the environment based on responses to a survey sent to environmentalists and leading experts. The bibliography is divided into four categories: the top forty books, core books, strongly recommended titles, and other recommended works. Contains a list of questionnaire respondents and a bibliography. Indexed.

Miller, G. Tyler
Environmental Science: Sustaining the Earth
3rd ed. Belmont, Calif., Wadsworth Pub. Co., 1991. 1 v. (various pagings). ISBN 0534134580

An introductory text meant to accompany the Annenberg/CPB television course *Race to Save the Planet*. Includes many color illustrations, charts, and graphs. Appendices provide lists of recommended periodicals, environmental and resource organizations, and major conservation and environmental legislation. Chapters have reference lists. Glossary and index provided.

Regens, James L., and Robert W. Rycroft
The Acid Rain Controversy
Pittsburgh, Pa., Univ. of Pittsburg Press, 1988. 228p. ISBN 0822935821

In this review of acid rain, the authors explain what acid rain is, where it comes from, and when it began. Discusses strategies for control and mitigation of acid rain, as well as economic and political costs. Poses questions on how the problem may be addressed. Includes several appendices, a bibliography, and index. *Pitt Series in Policy and Institutional Studies.*

Smith, Robert Leo
Ecology and Field Biology
4th ed. New York, Harper and Row, 1990. 922p. ISBN 0060463317

A general up-to-date introduction to ecology and field biology, aimed at the undergraduate student. Presents an introduction to ecology, the ecosystem, population ecology, ecological communities, and comparative ecosystem ecology. Includes useful illustrations, glossary, lengthy bibliography, and index.

Thoreau, Henry David
Faith in a Seed: A Dispersion of Seeds and Other Late Natural History Writings
Edited by Bradley P. Dean. Washington, D.C., Island Press, 1993. 283p. ISBN 1559631813

This is the first publication of four of Thoreau's previously unpublished works: *The Dispersion of Seeds; Wild Fruits; Weeds and Grasses;* and *Forest Trees*. This new book reflects Thoreau's philosophy of plant species and seed dispersal. A necessity for every college library.

Treshow, Michael, and Franklin K. Anderson
Plant Stress from Air Pollution
New York, Wiley, 1989. 283p. ISBN 0471923745

An understandable, basic introduction and reference to pollution's effects on plants written for the nonspecialist. Covers the effects of sulfur dioxide, heavy metals, fluoride, smog, ozone, and acid rain. Includes appendices, an extensive references list, and an index.

✦ Surveying and Measurement

Avery, Thomas Eugene, and Harold E. Burkhart
Forest Measurements
4th ed. New York, McGraw-Hill, 1994. 408p. ISBN 0070025568

This basic discussion on the measurement of forests focuses on the inventory of tree stands and related plants and animals. Chapter-end references; many useful appendices; and an index.

Bell, John F., and J. R. Dilworth
Log Scaling and Timber Cruising
Corvallis, Ore., Oregon State University Bookstore, 1988. 396p. ISBN 0882461427

Provides detailed, step-by-step instructions for estimating the board feet and cubic feet of usable lumber in various types and shapes of logs. Includes rules and formulas for log scaling and grading, with examples showing how they are applied in different situations, as well as how measurements should be reduced to account for diseases, odd shapes, and other defects. The second half of the book covers timber cruising, including measurement techniques, sampling patterns, and data mapping, as well as field and office procedures for various sampling systems.

Wolf, Paul R., and Russell C. Brinker
Elementary Surveying
9th ed. New York, HarperCollins, 1994. 760p. ISBN 0065003993

A textbook of basic theory and materials for the field and office. A substantial revision of the earlier edition that incorporates advancing technologies such as geographic information systems (GIS), total stations, and automatic data collectors. Offers advanced and specialized topics such as satellite and inertial surveying systems, and photogrammetry.

✦ Wildlife

Baldassarre, Guy A., and Eric G. Bolen
Waterfowl Ecology and Management
New York, Wiley, 1994. 609p. ISBN 0471597708

A comprehensive discussion of the biology, ecology, and management of North American waterfowl. Covers waterfowl classification, courtship, mating, reproduction, feeding, nesting, brooding, mortality and harvest management, wetland management, and waterfowl policy and administration. Chapters end with excellent bibliographies. Detailed index.

The Birds of North America
Philadelphia, Pa., American Ornithologists' Union, 1992- . Issued bimonthly. ISSN 1061-5466

A joint effort by the American Ornithologists' Union and the Academy of Natural Sciences of Philadelphia, this project was undertaken to replace Arthur Cleveland Bent's *Life Histories* series with current, comprehensive, and authoritative profiles of North American birds. Destined to become a standard reference, each profile covers distribution, systematics, migration, habitat, food habits, sounds, behavior, breeding, demography, conservation and management, appearance, and measurements. Includes references.

Brown, Gary
The Great Bear Almanac
New York, Lyons and Burford, 1993. 325p. ISBN 1558212108

Brown, a former National Park Service ranger and bear management specialist, provides a fascinating reference work on bears. Discusses bear species of the world—their anatomy, physiology, and behavior, as well as their relationship to humans. Includes several appendices, a glossary, a bibliography, and an index.

DeGraaf, Richard M., and others
Forest and Rangeland Birds of the United States: Natural History and Habitat Use
Washington, D.C., U.S. Dept. of Agriculture, Forest Service, 1991. 625p. ISBN 016028547X

Each natural history entry includes common and scientific names, a color illustration of the species, and information on range, status, habitat and special habitat requirements, nesting, food, and references. Includes bird/cover type matrices, a lengthy literature cited list, and a bird species index. *Agricultural Handbook,* 688.

Grzimek, Bernhard, ed.
Grzimek's Encyclopedia of Mammals
New York, McGraw-Hill, 1990. 5 v. ISBN 0079095089

This is an updated section of *Grzimek's Animal Life Encyclopedia* (Van Nostrand Reinhold, 1972-75) covering mammals. Includes numerous color photographs and illustrations, references, author credentials, and indexes.

Interagency Scientific Committee to Address the Conservation of the Northern Spotted Owl
A Conservation Strategy for the Northern Spotted Owl
Portland, Ore., The Committee, 1990. 427p.

Presents the official strategy adopted to preserve the northern spotted owl. The report includes a summary, the strategy, numerous appendices with references, a glossary, and several maps inserted in the back of the report.

Lowe, David W., John R. Matthews, and Charles J. Moseley, eds.
The Official World Wildlife Fund Guide to Endangered Species of North America
Washington, D.C., Beacham Pub., 1990-1992. 3 v. ISBN 0933833172 (set)

Volume One of this comprehensive guide covers plants and mammals; Volume Two contains birds, reptiles, amphibians, fishes, mussels, crustaceans, snails, insects, and arachnids. Volume Three updates the first two with additional species added from August 1989 to December 1991. Entries include description, behavior, habitat, historic range, current distribution, conservation and recovery, and contact. Several appendices, a bibliography, and a cumulative index are provided in the third volume.

Meyer, Fred P., and Lee A. Barclay
Field Manual for the Investigation of Fish Kills
Washington, D.C., U.S. Dept. of the Interior, Fish and Wildlife Service, 1990. 120p. ISBN 0160246865

This manual was developed to aid in the investigation of fish kills. Of use to students as well as practicing professionals, this work contains photographs, references, additional readings lists, and several appendices. *U.S. Fish and Wildlife Services Resource Publication,* 177.

Moyle, Peter B.
Fish: An Enthusiast's Guide
Berkeley, Univ. of California Press, 1993. 272p.
ISBN 0520079779

Moyle, a fish biologist, has written a very readable book introducing the field of fish biology and behavior. Topics covered include ecology, conservation, trout and warmwater streams, lakes, reservoirs, ponds, estuaries, and reefs. Contains illustrations, an excellent resources chapter, and an index.

Nowak, Ronald M.
Walker's Mammals of the World
5th ed. Baltimore, Johns Hopkins Univ. Press, 1991. 2 v.
ISBN 080183970X (set)

This standard reference on mammals has been updated to include new information on more than 1100 mammals and more than 4000 species. Includes extensive bibliographical references and an excellent index. Essential for any library.

Patton, David R.
Wildlife Habitat Relationships in Forested Ecosystems
Portland, Ore., Timber Press, 1992. 392p. ISBN 0881922021

An excellent, readable discussion that considers the factors affecting both habitats and wildlife populations. Patton illustrates the relationship between wildlife management and forest and habitat ecology management. Numerous useful appendices; glossary.

Pennak, Robert W.
Fresh-Water Invertebrates of the United States: Protozoa to Mollusca
3rd ed. New York, Wiley, 1989. 628p. ISBN 0471631183

This new edition of the standard reference includes updated references with each chapter, numerous illustrations, photographs and drawings, an appendix, and an index.

Thorp, James H., and Alan P. Covich, eds.
Ecology and Classification of North American Freshwater Invertebrates
San Diego, Calif., Academic Press. 1991 911p. ISBN 0126906459

This comprehensive discussion of freshwater invertebrates begins with a general introduction followed by information on freshwater habitats. Remaining chapters cover specific groups of invertebrates. Chapters typically include an introduction and cover anatomy and physiology, ecology and evolution, current and future research problems, collecting, rearing, and identification techniques, classification; and cite literature. Includes a glossary, as well as taxonomic and subject indices.

Whitehead, G. Kenneth
The Whitehead Encyclopedia of Deer
Shrewsbury, U.K., Swan Hill Press, 1993. 597p.
ISBN 1853103624

This comprehensive, worldwide treatment of deer includes general information and trivia about deer, deer species of the world, hunting and stalking, trophies, and diseases and parasites. Includes numerous appendices, indexes, and an extensive bibliography.

Wright, R. Gerald
Wildlife Research and Management in the National Parks
Urbana, Ill., Univ. of Illinois Press, 1992. 224p.
ISBN 0252018249

Wright, a National Parks Service research scientist and professor, provides an interesting look at wildlife management history and practices in national parks in the U.S. As the use of the parks has increased, the additional responsibilities in the preservation of various species have also grown. Wright does an excellent job of demonstrating the need for research and management strategies. Includes list of common and scientific names of animals, lengthy bibliography, and an index.

✦ Wood and Lumbering

Beattie, Mollie, Charles Thompson, and Lynn Levine
Working with Your Woodland: A Landowner's Guide
Rev. ed. Hanover, N.H., Univ. Press of New England, 1993. 279p.
ISBN 0874516226

An excellent treatise on woodland management written by Beattie, the coordinator of the U.S. Fish and Wildlife Service, along with Thompson and Levine, owners of private consulting firms. Includes a New England forest history, information on woodland assessment, types of forests, management plans, techniques, products, and finances. Includes bibliography, several useful appendices, and an index.

Brown, W. H.
The Conversion and Seasoning of Wood
Fresno, Calif., Linden Pub., 1989. 222p. ISBN 0941936147

Brown provides a thorough discussion and reference on the drying of timber and lumber. Includes information on converting timber to lumber; moisture content; shrinkage and movement of wood; various dryings, seasoning, and stabilizing methods; and the care and protection of dry wood. This work is of particular value to lumber professionals, foresters, woodlot owners, wood crafters, and students.

Devall, Bill, ed.
Clearcut: The Tragedy of Industrial Forestry
San Francisco, Sierra Club, 1994. 291p. ISBN 0871564947

This large format book contains fifteen essays and over 150 photographs illustrating the devastation caused by the clearcutting approach to forestry. Essayists include Dave Foreman, Mitch Lansky, and Chris Maser. Includes excellent photographs and a bibliography.

Dietrich, William
The Final Forest: The Battle for the Last Great Trees of the Pacific Northwest
New York, Simon and Schuster, 1992. 303p. ISBN 0671729675

Dietrich, a Pulitzer Prize-winning journalist, presents a well-written, balanced account of the struggle between the logging industry and those fighting to save both the old growth forests and the spotted owl in the Olympic Peninsula of Washington. Dietrich looks specifically at how this controversy is impacting the logging community of Forks.

The Encyclopedia of Wood
Rev. ed. New York, Sterling Pub. Co., 1989. 1 v. (various pagings). ISBN 0806969946

An excellent reference on the physical and structural properties of commercially available wood in the United States. Covers lumber, fastenings, bonding wood and wood products, glued structural members, bent wood members, wood-based building materials, moisture content, fire safety, and wood finishing, protection, and preservation. Includes references and glossary.

Fritz, Edward C.
Clearcutting: A Crime Against Nature
Austin, Tex., Eakin Press, 1989. 124p. ISBN 0890156743

Aiming at the layperson, Fritz has written a passionate treatise against the practice of clearcutting. Using examples from across the country, he shows the devastation associated with the common method of timber harvesting. This thought-provoking book will add balance to any collection. Glossary and index.

Haygreen, John G., and Jim L. Bowyer
Forest Products and Wood Science: An Introduction
2nd ed. Ames, Iowa, Iowa State Univ. Press, 1989. 500p. ISBN 081381801X

For the student of wood science, wood products, and forestry, this work covers the nature and properties of wood, wood products and their manufacture, and the future of the industry. Illustrations are clear and easy to understand. References and supplemental reading lists are provided for each chapter. Includes appendices with many tables and an index.

Hoadley, R. Bruce
Identifying Wood: Accurate Results with Simple Tools
Newtown, Conn., Taunton Press, 1990. 223p. ISBN 0942391047

This straightforward introduction to the art and science of wood identification focuses on native North American species. Covers classification structure, physical and chemical properties of wood, and equipment and techniques for identification. Includes several appendices as well as a glossary and bibliography. For the lumber professional, wood enthusiast, or student.

Lansky, Mitch
Beyond the Beauty Strip: Saving What's Left of Our Forests
Gardiner, Maine, Tilbury House, 1992. 453p. ISBN 0884480941

A well-written, extensively documented look at clearcutting as it is practiced in the state of Maine. Lansky offers many possible alternatives and solutions as well as suggestions for action appropriate for both the layperson and the professional. Includes several useful appendices, a glossary, and an index.

Manning, Richard
Last Stand: Logging, Journalism, and the Case for Humility
Salt Lake City, Utah, Peregrine Smith Books, 1991. 179p. ISBN 0879053895

While writing a series on large logging operations in Montana for the *Missoulian* newspaper, reporter Manning quit his job rather than succumb to the pressure of the coverage. After his departure, the paper accepted a journalism award for that series. This book covers the events leading up to his resignation and his growing conviction that the logging industry must change its practices before it is too late. Bibliography.

Vardaman, James M.
How to Make Money Growing Trees
New York, Wiley, 1989. 296p. ISBN 0471609196

A nontechnical introduction to forest management aimed at persons owning under 4,000 acres. Vardaman covers all aspects of the business from its basic nature to sales, management costs, accounting and tax laws, loans, and the future of the industry. Useful case histories are included as well as appendices, a glossary, and an index.

Williston, Ed M.
Lumber Manufacturing: The Design and Operation of Sawmills and Planer Mills
Rev. ed. San Francisco, Miller Freeman Publications, 1988. 486p. ISBN 0879301740

An up-to-date look at the lumber industry beginning with topics such as the use and procurement of raw material, the processing of that material into rough lumber, and the manufacturing of rough materials into products (such as dry lumber, chips, and fuel). There is also a discussion of equipment used and the principal types of sawmills. Includes numerous useful tables in the appendix, end of chapter references, a glossary, and an index.

Nonprint Materials

America's Wild Turkey
Rhinebeck, N.Y., Griffen Productions, 1990. 1 videocassette (50 min.) VHS

Explores the life of the wild turkey in its natural surroundings. An excellent look at the American symbol that is being restored and managed differently in many areas.

Ancient Forests
Washington, D.C., National Geographic Society, 1992. 1 videocassette (25 min.) VHS

Describes the old growth forests that once covered much of Africa, Asia, Europe, and North and South America. Shows how development of the world has almost destroyed these vital ecosytems

Aquaculture: Farming the Waters
Athens, Ga., American Association for Vocational Instructional Materials, 1990. 1 videocassette (30 min.) VHS

Covers the cultivation of three species of fish: catfish, crayfish, and trout.

Videoguide to Birds of North America
New York, MasterVision, 1988. 5 videocassettes (approx. 60-80 min. each) VHS

This excellent series is useful for the identification of bird species and their songs. Produced in cooperation with Cornell University's Laboratory of Ornithology. Cover title: *Audubon Society's Videoguide to Birds of North America.*

Careers in the Forest Service
Portland, Ore., U.S. Department of Agriculture, Forest Service, Pacific Northwest Region, 1991. 1 videocassette (10 min.) VHS

A brief look at twenty-five possible careers with the Forest Service.

The Politics of Trees
Alexandria, Va., PBS Video, 1992. 1 videocassette (58 min.) VHS

Originally broadcast on PBS as part of the *Listening to America with Bill Moyers* series. This discussion of timber harvesting and forestry management practices includes the perspectives of loggers, environmentalists, politicians, and journalists.

Rain Forests: A Part of Your Life
Hightstown, N.J., American School Publishers, 1989.
1 videocassette (30 min.) VHS

Explores the world's threatened rain forests and how they are being destroyed as their economic value is being exploited. Part One, "The Balancing Act," covers the role these great forests play in our world's ecosystem. Part Two, "The Disappearing Act," considers the long-range consequences of their destruction.

Ruffed Grouse
Milwaukee, Wisc., Anderson Video, 1989. 1 videocassette (51 min.) VHS

Covers identification, care, and habitat management, specifically woodlot management, to increase grouse production. *Field to Feast* series.

Water Quality I: Introduction to Water Quality Management
Clemson, S.C., Clemson Univ. Cooperative Extension Service, 1990. 1 videocassette (34 min.) VHS

Presents several water management options that can be employed when dealing with ponds.

Water Quality II: Water Quality Dynamics
Clemson, S.C., Clemson Univ. Cooperative Extension, 1990.
1 videocassette (25 min.) VHS

Explores water quality characteristics and how the proper balance is essential for fish to thrive.

Water Quality III: Procedures for Water Quality Management
Clemson, S.C., Clemson Univ. Cooperative Extension Service, 1990. 1 videocassette (41 min.) VHS

Demonstrates how to test water quality and how to make necessary corrections.

Wolves
National Audubon Society Specials
Alexandria, Va., PBS Video, 1989. 1 videocassette (60 min.) VHS

Covers the projects being undertaken to reinstate the wolf species in several areas. Discusses the controversy surrounding these projects.

Funeral Services

Harolyn Cumlet
and
Lenora Lockett
Delgado Community College
New Orleans, Louisiana

Introduction

Funeral Services education programs prepare students to handle all aspects of the funeral industry. Students gain expertise in embalming and preparing the body, making arrangements with the survivors, planning services, transporting the body, arranging for burial or cremation, and follow-up services. Regulations of the funeral industry as well as Federal Trade Commission (FTC) and Occupational Safety and Health Administration (OSHA) regulations are included in all programs. Some programs may be geared to training funeral directors who are responsible for all details surrounding the funeral ceremony and burial. Other programs focus on embalmers, who prepare the remains by using chemicals to disinfect and preserve, and the use of cosmeticians to assist in the restoration of the body to a more natural appearance. Many programs focus on both.

All states, except Colorado, require a license for embalming and a license for funeral directing, with some states having a dual license system (funeral director and/or embalmer). To determine who is qualified for a license, state licensing boards now use a standard National Board Examination which must be passed with a state qualifying score in the public health and technical section and in business management and the social sciences.

Licensing laws vary from state to state, with only one state requiring four years of college, and others requiring from one to three years of college. As of 1990, the American Board of Funeral Service Education (ABFSE), the accrediting agency for funeral service schools, recognizes the associate degree as the minimum recommended educational standard for preparation for the funeral service profession, although there is speculation among educators that in the future the ABFSE may recommend a bachelor's degree as the minimum educational requirement. At present the ABFSE requires that the curriculum taken by funeral service students consist of general education course work and at least fifty semester (seventy-five quarter) credits in the funeral service concentration. One to three years of apprenticeship or resident training are also required, with the length varying from state to state.

There are approximately forty schools approved by the American Board of Funeral Service Education, with many of these schools being affiliated with a community college. The funeral service curriculum typically stresses embalming, but must involve study in chemistry, microbiology and public health, anatomy, pathology, restorative art, funeral service management and practice, business studies, mortuary law, dynamics of grief, counseling, and communication skills.

In the 1990s, more states have introduced, or are considering, mandatory continuing education requirements for license renewal. Continuing education courses cover topics such as new regulations affecting the industry and new chemical solutions and preservatives. As of 1992, fourteen states required between five and twelve continuing education hours each year for license renewal.

Many changes confront the funeral industry today. Although nearly 85 percent of all funeral homes are family operated, there is a growing trend for many independent homes to be acquired by conglomerates. More women and minorities continue to enter the profession in management positions. In recent years, consumer demand has led to the increase of funeral home aftercare services, including support groups, remembrance services, newsletters, consumer resource libraries, and community referrals. Funeral homes increasingly will become resource centers for matters relating to dying, death, and bereavement. The future also promises an increase in cremations, memorial services in which the body is not present, nontraditional services, and prearrangement and prepayment of services.

This bibliography will serve three audiences: the funeral service student, the funeral service professional, and the librarian who wishes to expand and update a basic funeral services library collection.

Accreditation

American Board of Funeral Service Education (ABFSE)
14 Crestwood Rd.
Cumberland Center, Me. 04021

Formulates and enforces rules and regulations and sets up standards concerning schools teaching mortuary science. Accredits schools and colleges of mortuary science.

Selected Associations

Conference of Funeral Service Examining Boards of the U.S.
2404 Washington Blvd., Suite 1000
Odgen, Utah 84401

National Funeral Directors Association (NFDA)
11121 W. Oklahoma Ave.
Milwaukee, Wisc. 53227-4096

National Funeral Directors and Morticians Association
1800 E. Linwood Blvd.
Kansas City, Mo. 64109

Selected Journals

American Funeral Director.
American Cemetery. Monthly. ISSN 0002-8576

The Director.
National Funeral Directors Association. Monthly.
ISSN 0199-3186

Mortuary Management.
Mortuary Management. Monthly. ISSN 0027-1268

Print Materials

Aiken, Lewis R.
Dying, Death, and Bereavement
2nd ed. Boston, Allyn and Bacon, 1991. 308p. ISBN 0205126502

An interdisciplinary survey text on thanatology. Topics include characteristics and causes of death, cultural beliefs and practices concerning death, attitudes toward death, and problems of survivors. Funerary rituals are discussed from both historical and contemporary perspectives. Appendix lists organizations and periodicals concerned with dying, death, and widowhood. Includes glossary, extensive references, and separate author and subject indexes.

Anderson, Patricia
Affairs in Order: A Complete Resource Guide to Death and Dying
New York, Macmillan, 1991. 315p. ISBN 0025019910

A practical sourcebook describing options for people facing death. Part One discusses preplanning, including making a will, plans for disposition and commemoration, and living wills. Part Two addresses imminent death; Part Three deals with the aftermath of death for survivors. Resources and support groups are described after each chapter. Includes lists of resources for survivors, source notes, a bibliography, and an index.

Conference of Funeral Service Examining Boards
National Board Examination Study Guide for Applicants for Certification in the Funeral Service Profession
Ogden, Utah, The Conference, 1993. 142p.

Contains a test section for each topic area in the curriculum recommended by the American Board of Funeral Service Education, with an answer key in the appendix. Other appendices include glossaries of technical terms in relevant subject areas, a list of recommended textbooks, funeral service examining boards, mortuary science programs, state educational requirements for funeral directors and embalmers, a funeral service task statement, and professional expectations.

Corr, Charles A., Clyde M. Nabe, and Donna M. Corr
Death and Dying, Life and Living
Pacific Grove, Calif., Brooks/Cole Pub. Co., 1994. 1 v. (various pagings) ISBN 0534211380

Designed as a text for courses in death, dying, and bereavement. Topics examined include cultural differences in death-related practices and in attitudes towards death for African Americans, Hispanic Americans, Asian Americans, and Native Americans. One chapter describes American funeral practices and memorial rituals. Questions for review and suggested readings follow each chapter. Appendices offer annotated listings of selected literature for children and adolescents. Includes illustrations, extensive references, and separate name and subject indexes.

Counts, David R., and Dorothy Ayers Counts, eds.
Coping with the Final Tragedy: Cultural Variation in Dying and Grieving
Amityville, N.Y., Baywood Pub. Co., 1991. 326p.
ISBN 0895030829

A collection of cross-cultural studies by anthropologists on death and mourning which is directed at students, hospice workers, and others who work with the dying and their families. This work examines the functions mourning rituals serve, such as permitting the bereaved to resolve their grief, and the reintegration of the bereaved into society. Includes references, brief profiles of the contributors, and an index. *Perspectives on Death and Dying* series.

Crenshaw, David A.
Bereavement: Counseling the Grieving Throughout the Life Cycle
New York, Continuum, 1990. 181p. ISBN 0826404634

Provides practical advice for caregiving professionals, including funeral directors, who work with grieving children and adults, but who have little or no clinical training. The author proposes seven tasks of mourning, examines how these tasks are impacted by developmental stages, and offers guidelines for intervening in the different stages of the life cycle. Includes an appendix of helping resources and notes.

Doka, Kenneth J., ed.
Disenfranchised Grief: Recognizing Hidden Sorrow
Lexington, Mass., Lexington Books, 1989. 347p.
ISBN 066917081X

In complex societies where funeral rites are no longer communal but familial, the right to express grief is limited to kin. The concept of "disenfranchised" grief, i.e., that of those not directly related, is explored in this collection. Two chapters are of special interest, one which addresses unsanctioned and unrecognized grief from a funeral director's perspective, and a second which focuses on postdeath rituals and the disenfranchised griever. Notes or references follow each chapter. Index.

Funeral Directing Investigator
Syosset, N.Y., National Learning Corp., 1990. 1 v. (various pagings) ISBN 0837331129

A study guide to prepare for the civil service examination for funeral directing investigator. Sample examination questions and answers cover modern funeral directing practices, state laws, rules and regulations, and investigative techniques. *Passbooks for Career Opportunities* series.

Hartnett, Johnette
The Funeral: An Endangered Tradition; Making Sense of the Final Farewell
South Burlington, Vt., Good Mourning, 1993. 65p.
ISBN 1883171989

In the introduction, the author discusses the trend of Americans to skip the traditional funeral and opt for little or no funeral service; she then considers how the abbreviated funeral does not allow the bereaved structured mourning time. The rest of the book, written in question-and-answer format, provides basic information on funeral etiquette, aftercare programs, viewing the deceased, funeral arrangements, embalming, cremation, and burial and the cemetery. Includes footnotes and references. *Good Mourning Series*, vol. 3.

Houlbrooke, Ralph A., ed.
Death, Ritual, and Bereavement
New York, Routledge, 1989. 250p. ISBN 0415011655

Derived from the Conference on Death, Ritual, and Bereavement held in Oxford, 1987, this scholarly collection of essays on the social history of death concentrates on the deathbed, funerals, burial, grief, and mourning in England between the end of the Middle Ages and the early twentieth century. Topics examined for this time period include: the influence of religious beliefs and social demands on funeral rites, grief manifestations and consolation of survivors, and changes in the methods of disposal of the dead. Includes extensive notes, bibliography, and index.

Irish, Donald P., Kathleen F. Lundquist, and Vivian Jenkins Nelsen, eds.
Ethnic Variations in Dying, Death, and Grief: Diversity in Universality
Washington, D.C., Taylor and Francis, 1993. 226p.
ISBN 1560322772

In recognition of the need to sensitize professional workers in death-related fields to multicultural differences, this collection of essays examines the great variety of attitudes towards death and dying in multiethnic societies. Coverage includes: mourning and funeral customs of African Americans, Mexican American perspectives on death, Hmong death customs, Judaic traditions, Native American death rituals, Buddhist and Islamic customs, and Quaker and Unitarian memorial services. The appendix presents ethnic population data and trends for the United States and Canada. Includes a general bibliography, a bibliography for children, and an index.

Iserson, Kenneth V.
Death to Dust: What Happens to Dead Bodies?
Tucson, Ariz., Galen Press, 1994. 709p. ISBN 1883620074

A scholarly treatise describing the usually hidden postdeath treatment of the body. Topics examined include: autopsy, anatomical dissection, organ donation, embalming techniques, the funeral industry, funeral rituals and costs, cremation, burial and exhumation, and environmental impact and costs of different methods of corpse disposal. Also includes maxims and poetry relating to death. Chapter references, illustrations, a glossary, and an index are provided. Appendices include the Federal Trade Commission's "Funeral Rule" of 1984; the *Embalmers Classification: Types of Corpses;* and the *National Selected Morticians Code of Good Funeral Practice*, 1987.

Jackson, Kenneth T., and Camilo J. Vergara
Silent Cities: The Evolution of the American Cemetery
New York, Princeton Architectural Press, 1989. 129p.
ISBN 0910413223

Through color photographs and interpretive text, the authors examine over 300 cemeteries to trace the development of the American cemetery. Considers the role of ethnicity, religion, and class on cemetery architecture, as well as the decline of the cemetery as a place of commemoration. Bibliography.

James, John W., and Frank Cherry
The Grief Recovery Handbook: A Step-by-Step Program for Moving Beyond Loss
New York, Harper and Row, 1988. 175p. ISBN 0060159391

Written by the cofounders of the Grief Recovery Institute. Designed for professionals working with the bereaved, this book stresses the importance of accepting grief as a normal reaction to loss, demonstrates how to grow through grief, and defines the five stages necessary for recovery.

Kapleau, Philip
The Wheel of Life and Death: A Practical and Spiritual Guide
New York, Doubleday, 1989. 370p. ISBN 038526058X

Presents a religious orientation to dying and emphasizes the importance of the funeral service as a rite of passage. Includes chapters on cremation and burial, and information on creating a memorial service. Supplements include a section on living wills and guidelines for consoling the bereaved. References, glossary, bibliography, index.

Kastenbaum, Robert
Death, Society, and Human Experience
4th ed. New York, Merrill, 1991. 347p. ISBN 0675211891

A comprehensive survey focusing on death issues with particular attention given to suicide, AIDS, death education and counseling, the right to die, hospice care, and bereavement. One chapter examines the funeral process from premortem preparations through memorialization and includes a section on ethnic cemeteries. Ways of improving the funeral process are also discussed. Includes chapter references and an index.

Kubasak, Michael W.
Cremation and the Funeral Director: Successfully Meeting the Challenge
Malibu, Calif., Avalon Press, 1990. 156p. ISBN 0962769207

The author encourages funeral directors to examine their attitudes toward cremation. He challenges them to confront any fears they may have of cremation reducing their profit margins. He then offers specific advice to the profession on how to be more helpful to the increasing number of families choosing cremation and nontraditional funeral services. Includes a chapter on merchandising cremation products. The appendix presents a brief historical overview of cremation, references, and resources.

Kutscher, Austin, H., and others, eds.
For the Bereaved: The Road to Recovery
3rd ed. Philadelphia, Charles Press, 1990. 282p.
ISBN 0914783327

A multidisciplinary collection of articles concerning experiences in bereavement, interpretation of these experiences, and possible solutions. Contributors examine the beneficial role of the funeral and analyze how each element of the funeral process plays a part in successful grief resolution.

Leming, Michael R., and George E. Dickinson
Understanding Dying, Death, and Bereavement
2nd ed. Fort Worth, Tex., Holt, Rinehart and Winston, 1990. 475p. ISBN 0030283779

Designed as a comprehensive text on social thanatology for students in a wide array of death-related studies. Chapters address the history of bereavement, burial practices in American culture, and the funeral as an expression of contemporary American bereavement. Each chapter includes a conclusion, summary, glossary, discussion questions, references, and suggested readings. Illustrated. Indexed.

Litten, Julian
The English Way of Death: The Common Funeral since 1450
London, Robert Hale, 1991. 254p. ISBN 0709043503

A heavily illustrated treatise tracing the development of the English funeral trade over the past 500 years. Topics include the rise of the undertaking trade, embalming techniques, winding sheets and shrouds, the coffin, funerary transport, the burial vault, and changing etiquette governing burial. Includes chapter notes and references, bibliography, and an index.

Matunde, Skobi
Crossing the Great River: A Glimpse into the Funeral Rites of Afrikan-Amerikan People
Philadelphia, Freeland Publications, [1990?]. 64p.
ISBN 0936868201

Describes various funeral practices and interments of African Americans and points out links with African funeral rituals. Gives instructions on writing a will and includes a checklist of things to be done after a death. Illustrated.

Mayer, Robert G.
Embalming: History, Theory, and Practice
Norwalk, Conn., Appleton and Lange, 1990. 475p.
ISBN 0838521851

An essential text for students and a reference for practitioners.

This joint project of the American Board of Funeral Service Education and the National Funeral Directors Association covers legal aspects of embalming, origin and history of embalming, embalming chemicals, body preparation, arterial fluid and drainage techniques, and determining types of embalming treatment. "Selected readings" address health hazards from chemicals and preservatives and handling bodies with communicable diseases. Includes a glossary of terms, chapter references, illustrations, and an index.

Metcalf, Peter, and Richard Huntington
Celebrations of Death: The Anthropology of Mortuary Ritual
2nd ed. New York, Cambridge Univ. Press, 1991. 236p.
ISBN 0521413125

A comparative study of funeral rites in various cultures with an emphasis on the fate of the corpse. The final chapter examines and interprets American ways of death, emphasizing the uniformity of American death rituals despite the cultural heterogeneity of American society. Includes critiques of the funeral industry as a political lobby and as a tightly organized group of specialists controlling every phase of the disposal of bodies. Includes illustrations, a bibliography, and an index.

Meyer, Richard E., ed.
Cemeteries and Gravemarkers: Voices of American Culture
Ann Arbor, Mich., UMI Research Press, 1989. 347p.
ISBN 0835719030

This collection of essays examines American burial grounds from the seventeenth through the twentieth centuries in terms of the history, ethnicity, and culture they reveal as artifacts. Includes numerous black-and-white photographs, illustrations, extensive notes, bibliography, and index.

Meyer, Richard E., ed.
Ethnicity and the American Cemetery
Bowling Green, Ohio, Bowling Green State Univ. Popular Press, 1993. 239p. ISBN 0879726016

Through a series of essays by scholars in a variety of disciplines, this volume explores the ways selected ethnic groups in America have made their cemeteries and funerary practices serve as a voice for their varied world views. Ethnic groups represented include Italians, Czechs, Gypsies (Scottish, Irish, and Rom), Native Americans, Jewish Americans, Hispanics, Ukrainians, Chinese, Japanese, and Polynesians. Includes illustrations, photographs, chapter references, and an annotated bibliography of additional readings.

Morgan, Ernest
Dealing Creatively with Death: A Manual of Death Education and Simple Burial
12th rev. ed., edited by Jenifer Morgan. Bayside, N.Y., Barclay House, 1990. 167p. ISBN 0935016791

A practical handbook covering subjects such as death education, living with death, bereavement, burial and cremation, the right to die, memorial societies, death ceremonies, and anatomical gifts. This edition also addresses spiraling funeral costs and offers advice on preplanned sales. Appendices correspond to chapter topics and include annotated bibliographies, directories of organizations, and samples of burial contracts, death ceremonies, and living wills. Indexed.

Morgan, John D., ed.
Young People and Death
Philadelphia, Charles Press, 1991. 202p. ISBN 0914783491

An anthology written to help caregivers working with young people facing death. The book is divided into five sections that cover a child's understanding of death, the dying child, the bereaved child, the bereaved family, and the role of the school. One chapter written by a funeral director addresses the role funeral service personnel play in the grief process and the difficulties in explaining death to children. Some chapters include references.

Norrgard, Lee E., and Jo DeMars
Final Choices: Making End-of-Life Decisions
Santa Barbara, Calif., ABC-CLIO, 1992. 258p. ISBN 0874366135

Written primarily for older adults and their children. Specific topics are the right to die, coping with death, funeral arrangements and payment, and handling an estate. Includes a directory of organizations associated with dying and funerals, an annotated bibliography of topics addressed, a sample letter of instruction, and a state-by-state listing of funeral regulatory information, right-to-die laws, and schools that will accept body donations. Indexed.

Panati, Charles
Panati's Extraordinary Endings of Practically Everything and Everybody
New York, Perennial Library, 1989. 470p. ISBN 0060159200

A hodgepodge of information and anecdotes on death and dying. Includes sections on the history of inhumation, cremation, and embalming. One chapter traces the development of cemeteries. Includes references and index.

Phipps, William E.
Cremation Concerns
Springfield, Ill., Charles C. Thomas, 1989. 105p.
ISBN 0398055327

An overview of past and present attitudes about cremation. Topics include cremation in ancient cultures, religious opposition, scientific influences of the nineteenth century, contemporary attitudes, changing Christian perspectives on cremation, and the advantages of preplanning. Includes a sample cremation planning form, subject index, and illustrations.

Price, W. D.
The Embalmer's Guide to Cardiovascular Anatomy
Woodbury, Conn., Muirfield Publications, 1990. 109p.

Written as a general cardiovascular guide for both the embalming student and the licensed practitioner, the book's focus is upon the cardiovascular system and its role in vascular embalming. Describes major vessels and how these vessels might be used by professional embalmers. Illustrated, with references and index.

Professional Training Schools (Dallas, Tex.)
Funeral Rites and Customs
Dallas, Professional Training Schools, 1991. 60p.

Recognizing the decline in traditional funeral services, this text attempts to provide funeral service practitioners some basic information on the views, practices, and procedures of various religious, military, and fraternal organizations regarding funeral service and its accompanying activities. Each chapter presents a brief history of a religion or fraternal or military organization, with an analysis of the effects of the religion or group on the contemporary funeral service. Includes illustrations, photographs, a glossary of terms, and bibliographical notes.

Quigley, Christine
Death Dictionary: Over 5,500 Clinical, Legal, Literary, and Vernacular Terms
Jefferson, N.C., McFarland, 1994. 195p. ISBN 0899508693

A collection of definitions of mostly English words, phrases, and initialisms pertaining to death. These terms have been taken from the jargon of funeral directors, medical personnel, police officers, grief counselors, and life insurance sales representatives. Includes a thesaurus and a bibliography.

Rabinowicz, Tzvi
A Guide to Life: Jewish Laws and Customs of Mourning
Northvale, N.J., Jason Aronson, 1989. 242p. ISBN 0876688334

An introductory sourcebook on the Jewish laws of mourning. Sections of particular interest include details on funeral arrangements for Jews, Judaic opposition to embalming and cremation, ritual washing of the body, coffins for Jewish funerals, Jewish cemeteries, and burial service rites. Includes a memorial prayer, extensive chapter notes, a glossary of terms, a bibliography, and an index.

Raether, Howard C.
The Funeral Director's Practice Management Handbook
Englewood Cliffs, N.J., Prentice Hall, 1989. 444p.
ISBN 0133453154

An essential reference manual. Part One covers all aspects of funeral arrangements, the funeral director as grief facilitator, the relationship of clergy to the funeral director, and professional growth. Part Two examines specific business aspects, such as accounting, compliance with laws and regulations, marketing and public relations, upgrading the facility, computerizing the funeral home, and the future of the funeral service industry. Includes illustrations and an index.

Raether, Howard C.
Funeral Service: A Historical Perspective
Milwaukee, National Funeral Directors Association, 1990. 115p.

This book focuses on post-World War II funeral services in the United States. Topics include a chronological list of developments in the United States relating to the care and burial of the dead, the value of the funeral, funeral service education, clergy and funeral director relations, the FTC funeral rule, and prearranged, prefunded funerals. Covers the history, structure, and annual conferences of the National Funeral Directors Association (NFDA).

Ross, E. Betsy
After Suicide: A Ray of Hope: A Guide for the Bereaved, the Professional Caregiver, and Anyone Whose Life Has Been Touched by Suicide, Loss or Grief
Iowa City, Iowa, Lynn Publications, 1990. 231p.
ISBN 0940179016

Ross outlines issues that survivors of those who commit suicide must face. Presents procedures and attitudes needed for grief resolution. One chapter stresses the importance of the funeral director's role in influencing the survivor's grief resolution process, and it offers suggestions to funeral directors for successful intervention. Endnotes, references, and a grief resource guide are included.

Sanders, Catherine M.
Grief: The Mourning After; Dealing with Adult Bereavement
New York, Wiley, 1989. 260p. ISBN 0471627283

Intended for professional caregivers, including funeral directors, who work with the bereaved. This book serves as a practical guide and also provides a theoretical framework for understanding the processes of bereavement. After an overview of earlier studies and theories of bereavement, the author focuses on each phase of grief and offers strategies for intervention with the bereaved. Includes references, and author and subject indexes.

Schoeneck, Therese S.
How to Form Support Groups and Services for Grieving People
Syracuse, N.Y., Hope for Bereaved, 1989. 90p.

In addition to providing guidance on how to start and maintain support groups for the bereaved, this book also offers suggestions on formats for group meetings, methods of fund raising, community education, and newsletters. Includes a section on how to help grieving people. Provides Catholic, Jewish, and Protestant perspectives on grieving. Lists national organizations that offer help to the bereaved.

Sloane, David Charles
The Last Great Necessity: Cemeteries in American History
Baltimore, Johns Hopkins Univ. Press, 1991. 293p.
ISBN 0801840686

Traces the transformation of the American cemetery from the frontier graves of the late-eighteenth century to the memorial parks of the late-twentieth century. Examines the increased commercialization and institutionalization of the cemetery as well as its diminished role as a repository of the history and memoirs of the local community. Contains illustrations, extensive notes, a bibliographic essay, and an index.

Sokoll, Gary J.
The Art of Facial Reconstruction
Yukon, Okla., Pretty Good Publishing, 1992. 80p.

Offers detailed instructions for reconstruction of the nose, mouth, chin, eyes, and ears, with accompanying photographic figures and a glossary of terms. Appendices present drawings of facial features and markings.

Southard, Samuel, comp.
Death and Dying: A Bibliographical Survey
New York, Greenwood Press, 1991. 514p. ISBN 0313264651

A scholarly, comprehensive, and multidisciplinary annotated bibliography of death-related resources. Includes references to articles, chapters, and monographs. Most works cited are by researchers or practitioners rather than by the dying or bereaved and are limited to the English language. Features a section on funeral and mourning customs and the mortuary industry. Separate author, title, and subject indexes. *Bibliographies and Indexes in Religious Studies*, no. 19.

Stueve, Thomas F.
Mortuary Law
9th ed. Cincinnati, Ohio, Cincinnati Foundation for Mortuary Education, 1988. 188p. ISBN 188303101X

A comprehensive review of the laws that apply to morticians. Topics include: disposal of the dead, rights of parties undertaking disposal, duties of the funeral director, liability for funeral expenses, prepaid and prearranged funeral contracts, disinterment, mortuaries, cemeteries, wage and hour laws, and OSHA. Review questions follow each chapter. Appendices provide various authorization forms and the Federal Trade Commission rule regarding funeral industry practices. Provides an index of cases and statutes cited.

Sublette, Kathleen, and Martin Flagg
Final Celebrations: A Guide for Personal and Family Funeral Planning
Ventura, Calif., Pathfinder Publishing of California, 1992. 134p.
ISBN 0934793433

Offers practical information about funeral and burial planning, procedures, ceremonies, and costs. Detachable forms include letters of instruction, cremation information sheets, organ donation forms, and prearrangement information. Lists organizations and reading resources. Subject index.

Szabo, John F.
Mortuary Science: A Sourcebook
Metuchen, N.J., Scarecrow Press, 1993. 196p. ISBN 0810827190

A partially annotated bibliography of English language monographic materials published between 1900 and 1991 in the field of mortuary science. Specific subject areas include funeral directing, restorative art, cemeteries, cremation, and funeral rites. Appendices list professional associations and organizations, college programs in funeral service education, a state-by-state listing of funeral service examining boards, accreditation standards of the American Board of Funeral Service Education, and mortuary science periodicals. Title index.

Watts, Tim J.
The Funeral Industry: Regulating the Disposition of the Dead
Monticello, Ill., Vance Bibliographies, 1988. 13p.
ISBN 1555908543

Prefaced by a very brief introduction on funeral service regulations, this bibliography cites twenty-four monographs (mainly government publications) and over 100 articles (from law journals or popular periodicals) from the 1970s and 1980s that pertain to regulation of the funeral industry. Covers guidelines issued by consumer groups, prepaid arrangements, and liabilities of those dealing with the deceased. *Public Administration Series—Bibliography, 02454.*

Willson, Jane Wynne
Funerals Without God: A Practical Guide to Non-Religious Funerals
Buffalo, N.Y., Prometheus Books, 1990. 59p. ISBN 0879756411

In recognition of the growing need for nonreligious ceremonies, this booklet is intended to provide a basic format for a secular humanist funeral ceremony. Includes a brief list of suggestions for further reading.

Winship, D. R.
Embalming Notes
Hackettstown, N.J., Donald R. Winship, 1990. 22p.
ISBN 0963003216

A practical reference manual of embalming procedures for the embalmer or student. Graphic color illustrations supplement the author's instructions for various tasks such as embalming, setting features, injecting arteries or veins, inserting a heart tap for drainage, arterial injection, and suturing.

Wolfelt, Alan
Death and Grief: A Guide for Clergy
Muncie, Ind., Accelerated Development, 1988. 200p.
ISBN 091520276X

Although written primarily for clergy, this text provides information on the psychological and social aspects of grief for anyone working with the dying or bereaved. One chapter examines the funeral ritual as a structure that assists the mourner through the initial stages of grief. Includes chapter references, a listing of support groups and organizations, and an index.

Wolfelt, Alan
Interpersonal Skills Training: A Handbook for Funeral Service Staffs
Muncie, Ind., Accelerated Development, 1990. 219p.
ISBN 1559590254

A how-to text addressing counseling, mourning and grief therapy, and interpersonal and communication skills for the funeral service employee, areas that the author considers to be often neglected in funeral service schools. The book includes a section on job stress and burnout in the profession and recognizes the need to care for the caregiver. Includes activity exercises, chapter references, a list of training opportunities for funeral home staffs, and an index.

Nonprint Materials

Living with OSHA
Milwaukee, Wisc., National Funeral Directors Association, 1993.
1 videocassette (38 min.) VHS

Designed to supplement employee training programs required by OSHA. Includes standards on formaldehyde exposure, hazard communication, and protection against bloodborne pathogens.

NFDA Satellite Seminar 1994 Mandatory OSHA Staff Training
Milwaukee, Wisc., National Funeral Directors Association, 1993.
3 videocassettes (4 hours) VHS

The first annual satellite seminar covering occupational safety and health in funeral homes was broadcast live throughout the United States in 1993. It is available with a study guide from NFDA.

Landscape Horticulture Technologies

Connie Barber
North Harris College
Houston, Texas

Introduction

Vocational programs in Horticulture Landscape Technologies prepare students for a variety of positions such as landscape contractor or supervisor, nursery manager or employee, plant propagator, tree pruner or trimmer, floral designer, interior plant service employee, greenhouse manager, or groundskeeper. The level of entry into positions varies with the individual's education and experience. Continued studies at a four-year college or university may lead to more specialized occupations in areas such as landscape architecture or botany.

Although the American Association of Nurserymen does not have a certification program, certification by testing is recommended by most state associations. There are many associations for specialized interests and some of these, such as pesticide handling, require specialized certification. In addition, individuals starting a business usually need a license from the state department of agriculture. Students should contact the local agricultural extension service, state department of agriculture, or nursery associations for specific requirements.

The literature in the field of horticulture is vast and written at many levels, from the highly scientific to the basic level aimed at the home gardener. Each geographic region has particular considerations regarding plant species and climate; each is represented by books on those topics. Various plant types such as roses, orchids, ferns, trees, mushrooms, and cacti are well represented in print. Garden types have volumes devoted to them, such as cottage gardens and water gardens; various types of plants such as arid varieties, annuals, and perennials; and specific techniques, such as pruning and propagating.

This bibliography is intended to provide a representative selection of titles for students of landscape horticulture, as well as general students. Some regional treatments have been included as representatives of the types of material available.

Selected Associations

American Association of Nurserymen
1250 I. St. NW, Suite 500
Washington, D.C. 20005

Associated Landscape Contractors of America
405 N. Washington St.
Falls Church, Va. 22046

National Landscape Association
1250 I St. NW, Suite 500
Washington, D.C. 20005

Professional Grounds Management Society
10402 Ridgland Rd., Suite 4
Cockeysville, Md. 21030

Selected Journals

American Horticulturist.
American Horticulture Society. Monthly. ISSN 0096-4417

American Nurseryman.
American Nurseryman. Semimonthly. ISSN 0003-0198

Grounds Management Forum
Professional Grounds Management Society. Monthly.
ISSN 0742-5511

Horticulture
[Massachusetts Horticultural Society.] Horticulture Pub. Co. Ten times a year. ISSN 0018-5329

National Gardening
National Gardening Association. Six times a year.
ISSN 1052-4096

Plants and Gardens
Brooklyn Botanic Garden. Quarterly. ISSN 0362-5850

Print Materials

✦ Reference

Bagust, Harold
Gardeners Dictionary of Horticultural Terms
New York, Sterling Pub. Co., 1992. 377p. ISBN 0304341061
   Provides brief, nontechnical definitions of over 2900 terms, accompanied by line drawings. Appendices with tables illustrate terminology such as leaf shape and grafting methods. Includes some British terms not common in the U.S.

Barrett, Thomas M., ed.
North American Horticulture: A Reference Guide
2nd ed. New York, Macmillan, 1992. 427p. ISBN 0028970012
   This work is a directory of horticultural organizations, plant societies, educational programs, public and historic gardens, scholarly organizations, botanical and horticultural libraries, conservation groups, and garden club associations. Contains contact information and, when applicable, the organization's purpose, publications, scholarship information, and a list of grants.

Brickell, Christopher, ed.
The American Horticultural Society Encyclopedia of Garden Plants
New York, Macmillan, 1989. 608p. ISBN 0025579207
   Includes descriptions of 8,000 plants for all zones in the U.S., arranged by plant type, size, color, and season of interest. Also includes brief propagation notes and a planters guide that provides a lists of plants suggested for particular situations such as hedges, shade, and exposed areas.

Brickell, Christopher, and others, ed.
The American Horticultural Society Encyclopedia of Gardening
New York, Dorling Kindersley Press, 1993. 648p.
ISBN 1564582914
   Covers sites, planting, plants, propagation, maintenance, and pruning. Provides information on all types of plants, including ornamentals and vegetables. Illustrated with color photographs.

Huxley, Anthony, ed.
Dictionary of Gardening
New York, Stockton Press, 1992. 4 v. ISBN 1561590010 (set)
   An expensive, but highly regarded revision of the 1951 edition from the Royal Horticultural Society. Arranged alphabetically, this work describes 50,000 worldwide plants including 20,000 cultivars. Botanical descriptions are accompanied by cultivation information. Essays from well-known experts discuss various horticultural and conservation topics. Good black-and-white photographs illustrate many of the plants.

Mabberley, D. J.
The Plant-Book: A Portable Dictionary of the Higher Plants Utilising Cronquist's An Integrated System of Classification of Flowering Plants (1981) and Current Botanical Literature Arranged Largely on the Principles of Editions 1-6 (1896/97-1931) of Willis's A Dictionary of the Flowering Plants and Ferns
New York, Cambridge Univ. Press, 1993. 707p.
ISBN 0521340608
   This is a 1993 reprint of the 1987 pocket-sized identification guide to all the known genera in the world.

✦ Botany

Capon, Brian
Botany for Gardeners: An Introduction and Guide
Portland, Ore., Timber Press, 1990. 220p. ISBN 0881921637
   Surveys the physical structures and life processes of angiosperms and gymnosperms. Provides a scientific approach but is intended for the layperson. Includes clear illustrations, some color photographs, a glossary, and an index.

Flora of North America: North of Mexico
New York, Oxford Univ. Press, 1993- , 14 v. (proposed).
ISBN 095057139 (vol. 1); ISBN 0195082427 (vol. 2)
   This work is intended to serve as a means of identification and a systematic conspectus of regional flora. Volume One contains introductory essays on climate, physiography, soils, plant history, phytogeography, and classification systems. Volumes Two through Thirteen will cover all of the vascular plant species of the U.S. and Canada. Volume Fourteen will provide a comprehensive index and bibliography.

✦ Gardens and Gardening

Anglade, Pierre, ed.
Larousse Gardening and Gardens
New York, Facts on File, 1990. 624p. ISBN 0816022429
   Provides historical information on famous world gardens and botanists. Includes a detailed discussion and gardening techniques for various types of gardens: ornamental, flower, rock, water, and orchards. A major portion of this book provides an alphabetical discussion of individual plants. Originally published in French and revised by British editors, this work has a European slant.

Ball, Jeff
Rodale's Landscape Problem Solver: A Plant-By-Plant Guide
Emmaus, Pa., Rodale Press, 1989. 439p. ISBN 0878578021
   Focuses on prevention of problems by proper placement of plants, and proper soil, light, and water. Lists optimum growing conditions and potential problems of many individual plant species. Includes a lengthy discussion of pests with an emphasis on organic control, diseases, soil improvements, and weed control. Sources for supplies are also provided.

Barton, Barbara J.
Gardening by Mail: A Source Book
4th ed. Boston, Houghton Mifflin, 1994. 1 v. (various pagings).
ISBN 0395680794

This directory provides mail-order resources for gardeners in the U.S. and Canada. Includes a directory of seed companies, nurseries, garden supply companies, horticultural and plant societies, magazines, libraries, and gardening books. Provides indexes by plant source, geographical area, product name, society name, and magazine title.

Boisset, Caroline
The Plant Growth Planner
New York, Prentice Hall, 1992. 192p. ISBN 0136812309

This garden planning guide includes ornamental trees, specimen shrubs, ground covers, climbers, perennials, and hedges. Includes color growth charts that aid the selection of plants by sizes. Plants are shown at one, three, six, and twelve years, and at every season. Includes information on habitat, planting care, and maintenance.

Bradley, Fern Marshall, and Barbara W. Ellis, eds.
Rodale's All-New Encyclopedia of Organic Gardening: The Indispensable Resource for Every Gardener
Emmaus, Pa., Rodale Press, 1992. 690p. ISBN 0878579990

Alphabetically arranged entries range from a few sentences to ten pages. Includes four core categories: gardening techniques, organic garden management, food crops, and ornamental plants. Illustrated, with index and cross-references.

Dale, John
The Gardener's Palette: The Complete Guide to Selecting Plants by Color
New York, Harmony Books, 1992. 192p. ISBN 0517588536

Dale presents a popular color directory of plants that includes basic information on planting, cultivation, propagation, as well as cutting and preserving blooms. Includes additional information on flowering seasons, flower size, and stem length.

Damrosch, Barbara
Garden Primer
New York, Workman Publishers, 1988. 673p. ISBN 0894803174

Aimed at the home gardener, this work covers basic information: planning, equipment, plants, care, and maintenance. Appendices include a hardiness zone map, a list of plant societies, and mail order source lists. Includes bibliography and an index.

Ellefson, Connie Lockhart, Thomas L. Stephens, and Doug Welsh
Xeriscape Gardening: Water Conservation for the American Landscape
New York, Macmillan, 1992. 323p. ISBN 0026141256

A xeriscape is defined as a selection of plants that thrive with minimal or no water. Covers basic principles and presents a regional guide with detailed specifications on plants to use. Includes botanical names, common names, and height. Appendices include recommendations for annuals, perennials, wildflowers, ornamental grasses, prairie grasses, drought-tolerant ground covers, trees and shrubs, vines, bulbs, and dry, shade, and seaside gardens. Indexed.

Harris, Richard W.
Arboriculture: Integrated Management of Landscape Trees, Shrubs, and Vines
2nd ed. Englewood Cliffs, N.J., Prentice Hall, 1992. 674p.
ISBN 0130442801

Covers landscape trees, shrubs and vines, plant structure, selection of plants and sites, water and soil management, and chemical control. Provides a chapter on preventative maintenance and repair, and diagnosis of plant problems.

Heriteau, Jacqueline
National Arboretum Book of Outstanding Garden Plants: The Authoritative Guide to Selecting and Growing the Most Beautiful, Durable and Care-Free Garden Plants in North America
New York, Simon and Schuster, 1990. 292p. ISBN 0671669575

Plants selected for inclusion are considered by the National Arboretum to be the most beautiful, durable, adaptable, disease and pest resistant, and require the least care. Rated for zones. Includes good color photographs.

Imes, Rick
Wildflowers: How to Identify Flowers in the Wild and How to Grow Them in Your Garden
Emmaus, Pa., Rodale Press, 1992. 160p. ISBN 0875961185

Provides a survey of flowers that grow in the wild, and discusses how they attract butterflies, birds, and wildlife—not only in the wild, but in a garden setting as well. Describes propagation and special considerations for rare and endangered species. A habitat and species directory of 250 flowers with color illustrations is provided, along with a guide to plant and seed sources. A bibliography and index are included.

The National Wildflower Research Center's Wildflower Handbook
Austin, Tex., Texas Monthly Press, 1989. 337p.
ISBN 0877191670

Over half of this book is a directory of wildflower information sources organizations, botanic gardens, and nurseries. The balance of the book discusses wildflower seed mixes, planting on a large scale, low-maintenance gardening, and propagation from seed. Includes numerous bibliographies, but few illustrations.

Punch, Walter T.
Keeping Eden: A History of Gardening in America
Boston, Bulfinch Press, 1992. 277p. ISBN 0821218182

This is a beautifully illustrated compendium of the history of gardening with an emphasis on the nineteenth and twentieth centuries. Essays are contributed by America's top garden writers, historians, and designers. Covers regional gardening as well as specific topics such as horticulture and the American character, and artistic gardening. Includes bibliography and index.

Rice, Laura Williams, and Robert P. Rice
Practical Horticulture
2nd ed. Englewood Cliffs, N.J., Regents/Prentice Hall, 1993. 419p. ISBN 0136788068

This work's nonregional approach makes it equally useful to residents of the United States and Canada. Provides background information on plant science, plant growth and development, anatomy, and propagation. One large section is devoted to houseplants. Includes black-and-white photographs and drawings.

Sperry, Neil
Neil Sperry's Complete Guide to Texas Gardening
Dallas, Tex., Taylor Pub. Co., 1991. 388p. ISBN 0878337997

This regional guide to home landscaping and gardening covers trees, shrubs, vines, ground covers, flowers, fruits, and vegetables. Provides detailed information on landscape planning, and discussions on irrigation systems, pest control, and structures such as patios, walks, and lighting. Includes color photographs and index.

Sunset Western Garden Book
5th ed. Menlo Park, Calif., Lane Pub. Co., 1988. 592p. ISBN 0376038535

This highly regarded, illustrated guide to western horticulture provides a concise encyclopedia for plants and gardening. Includes both botanical and popular names.

Toogood, Alan R.
The Garden Trees Handbook: A Complete Guide to Choosing, Planting, and Caring for Garden Trees
New York, Facts on File, 1990. 223p. ISBN 0816022755

Lists the characteristics of over 300 species of ornamental and fruit trees with information on hardiness, size and shape, and descriptions of flowers, fruits, and leaves. Discusses planting, care, pruning, and potential diseases. Includes common and scientific name indexes.

Welsh, Pat
Pat Welsh's Southern California Gardening: A Month-By-Month Guide
San Francisco, Chronicle Books, 1992. 341p. ISBN 0877016291

This calendar approach to gardening is of interest to Southern California gardeners, as well as those who have Mediterranean climates, subtropical climates, or greenhouses. Includes step- b y-step instructions, checklists, and special techniques. Includes a glossary, and a "Rose-Pro Calendar." Indexed.

Welch, William C.
Perennial Garden Color: For Texas and the South
Dallas, Tex., Taylor Pub. Co., 1989. 268p. ISBN 0878336281

This regional guide for the southern climate includes information on 125 perennials and more than 100 varieties of old garden roses. Discusses the use of color, shape, and texture in planning a landscape project. The author, a horticulturist at Texas A&M University, uses many photographs and examples from Texas plantings. Provides a list of mail-order plant sources, a bibliography, and an index.

Wilson, Jim
Landscaping with Container Plants
Boston, Houghton Mifflin, 1990. 212p. ISBN 0395498643

Provides suggestions for the various designs of a container landscape. Describes the best ornamental and edible plants to use, and advises on climate zones and weather restrictions. Also includes information on soilless planter mixes, fertilizers, watering techniques, insects, and diseases.

✦ Greenhouses and Nurseries

Ball, Vic, ed.
Ball Red Book
15th ed. Chicago, Greenhouse Crowing, 1991. 802p. ISBN 0962679623

Discusses the physical plant and operation of the greenhouse, including mechanization and irrigation. Also provides information on insect control, computers, xeriscaping, fertilizers, heaters, soils, and water testing. Includes an extensive section on culture by crop.

Davidson, Harold, Roy Mecklenburg, and Curtis Peterson
Nursery Management: Administration and Culture
2nd ed. Englewood Cliffs, N.J., Prentice Hall, 1988. 413p. ISBN 0136272823

Provides a brief history of nursery management and an overview of the industry. Includes such topics as management principles, selection guidelines for nursery sites, inventory control, and organization and development. Also discusses topics such as soil, nutrition, irrigation, control of weeds, insects, and diseases. Numerous appendices, a bibliography, and an index.

Jozwik, Francis X.
The Greenhouse and Nursery Handbook: A Complete Guide to Growing and Selling Ornamental Container Plants
Mills, Wyo., Andmar Press, 1992. 511p. ISBN 0916781070

Part One presents issues related to container plant production: greenhouses, equipment, nursery management, propagation, growing conditions, diseases, and pests. Part Two provides cultivation notes on specific plants including topics such as bedding plants, herbaceous perennials, trees, shrubs, roses, and flowering potted plants. Illustrated with black-and-white photographs.

Nelson, Paul V.
Greenhouse Operation and Management
4th ed. Englewood Cliffs, N.J., Prentice Hall, 1991. 612p. ISBN 0133651983

Provides a historical overview of floriculture and greenhouse management including construction, heating, cooling, root media, watering, fertilization, cropping systems, light, temperature, and insect and disease control. Also addresses organizational issues such as marketing and business management.

✦ Irrigation

Melby, Pete
Simplified Irrigation Design
Mesa, Ariz., PDA Publishers, 1988. 189p. ISBN 0914886401

Covers the basics of sprinkler and drip irrigation. Presents a stepped approach; each chapter builds on the information presented in earlier chapters. Includes information on basic hydraulics, pipe sizing, friction loss calculations, and determining water pressure. The appendix provides friction loss charts and irrigation installation specifications. Includes index.

✦ Pests and Diseases

Briggs, Shirley A.
Basic Guide to Pesticides: Their Characteristics and Hazards
Washington D. C., Taylor and Francis, 1992. 283p.
ISBN 1560322535

This comprehensive reference guide provides a tabular presentation of pesticide information. Includes definitions and information on common trade and chemical names, CAS Registry numbers, pesticides and uses, persistence factors, effects on mammals, and adverse effects on nontarget species.

Dreistadt, Steve H., Jack Kelly Clark, and Mary Louise Flint
Pests of Landscape Trees and Shrubs: An Integrated Pest Management Guide
Oakland, Calif., Univ. of California, Division of Agriculture and Natural Resources, 1994. 327p. ISBN 187990618X

Designed for landscaping professionals, pest managers, and homeowners to encourage the suppression of pest problems with minimal adverse effects on the environment, health, and nontarget organisms. Provides information on growing healthy trees and shrubs, including helpful guidance on insects, mites, snails, slugs, diseases, abiotic disorders, weeds, and nematodes. Contains a problem-solving guide for the reader's use.

Ellis, Barbara W., and Fern Marshall Bradley, eds.
The Organic Gardeners Handbook of Natural Insect and Disease Control: A Complete Problem-Solving Guide to Keeping Your Garden and Yard Healthy Without Chemicals
Emmaus, Pa., Rodale Press, 1992. 534p. ISBN 087596124X

This organic approach to gardening provides a plant-by-plant guide to problem solving and prevention. Includes information on friendly and troublesome insects, diseases, and organic controls. Color photographs and drawings, and an index.

Flint, Mary Louise
Pests of the Garden or Small Farm: A Grower's Guide to Using Less Pesticide
Oakland, Calif., Univ. of California, Division of Agriculture and Natural Resources, 1990. 273p. ISBN 0931876893

Flint's emphasis is on the design of an integrated pest management program. Includes descriptions of common insects, mites and other arthropods, snails, and slugs. Also discusses diseases and disease management, nematodes, and crop tables. Bibliography and index.

Johnson, Warren T., and Howard H. Lyon
Insects That Feed on Trees and Shrubs
2nd ed., rev. Ithaca, N.Y., Comstock Publishing Associates, 1991. 560p. ISBN 080142108X

This large-format reference manual covers major biotic pests and environmental problems associated with woody ornamental plants. The large color photographs of insects and the damage they cause are helpful for identification. Scientific information is presented in nontechnical language.

Larson, Roy A.
Introduction to Floriculture
2nd ed. San Diego, Academic Press, 1992. ISBN 0124376517

Provides detailed information on every major floriculture crop and procedure. Includes two sections: miscellaneous cut flowers and flowering potted plants. The new edition features added chapters on specialty cut flowers and geraniums.

Morgan, Donald P., ed.
Recognition and Management of Pesticide Poisonings
4th ed. Washington, D.C., U.S. Environmental Protection Agency, 1989. 207p.

Presents descriptions, short-term harmful effects, and treatments for pesticide poisoning, providing health professionals with a quick reference. Includes consensus recommendations for managing poisonings and the injuries caused by them.

Ogawa, Joseph M., and Harley English
Diseases of the Temperate Zone: Tree, Fruit, and Nut Crops
Oakland, Calif., Univ. of California, Division of Agriculture and Natural Resources, 1990. 461p. ISBN 0931876974

This encyclopedia of diseases affecting pome fruit, stone fruit, nuts, olives, figs, and minor fruit crops provides more than 200 listings with history, causes, symptoms, development, and control. Emphasis is placed on temperate climates, particularly California. Includes color illustrations and index.

Powell, Charles C., and Richard K. Lindquist
Ball Pest and Disease Manual
Geneva, Ill., Ball Publishing, 1992. 332p. ISBN 096267964

This reference guide is helpful for the student needing information on insect and disease detection and control. The authors provide strategies for determining whether to use biological, environmental, or chemical measures.

Watterson, Andrew
Pesticide Users' Health and Safety Handbook: An International Guide
New York, Van Nostrand Reinhold, 1988. 504p.
ISBN 0442234872

Designed for individuals involved in the frequent handling of pesticides, this work analyzes pesticide use, toxicity, application procedures, and safety factors. Contains numerous appendices including lists of health effects and organizations. Glossary, bibliography, and index.

Westcott, Cynthia
Westcott's Plant Disease Handbook
5th ed., revised by R. Kenneth Horst. New York, Van Nostrand Reinhold, 1990. 953p. ISBN 0442318537

Provides information on diagnostic and disease control for trees, shrubs, vines, flowers, and vegetables, and discusses chemicals and their application. Pathogens include fungi, bacteria, viruses, and nematodes. Plants are listed with diseases common to each. Illustrated with photographs and drawings.

✦ Seeds, Germination, and Propagation

Hartmann, Hudson, Dale E. Kester, and Fred T. Davis
Plant Propagation Principles and Practices
5th ed. Englewood Cliffs, N.J., Prentice Hall, 1990. 647p.
ISBN 0136810160

Describes numerous useful techniques for the propagation of plants and covers general topics such as structures, media, fertilizers, sanitation, and containers. Furnishes information on seed propagation, vegetative propagation, and properties of selected plants.

Nau, Jim
Ball Culture Guide: The Encyclopedia of Seed Germination
2nd ed. Batavia, Ill., Ball Publishing, 1993. 143p.
ISBN 1883052017

This guide provides information on lighting, general temperature, days to germination, days from sowing to transplanting, crop time, pinching information, garden height, location, staking, and specification of tender versus hardy plants.

✦ Vocational Guidance

Moore, Stanley B.
Ornamental Horticulture as a Vocation
2nd ed. Dayton Beach, Fla., Mor-Mac Pub. Co., 1988. 1 v. (various pagings). ISBN 091278019

Provides brief discussions of the occupational aspects of horticulture, and then discusses various treatments of horticultural techniques. Includes information on identification, propagation, crop production, greenhouses and nurseries, pests and diseases, and landscape designs. Based largely on bulletins from the University of Kentucky Horticulture Department.

Nonprint Materials

ABC's of Landscape Pruning
Advanced ed. Idea Bank, 1989. 1 videocassette (20 min.) VHS

The various tools used for pruning and cutting are illustrated. Provides tips on neglected or badly pruned plants and how to restore or replace them.

Backyard Composting
St. Louis, Mo., Missouri Botanical Garden, 1992. 1 videocassette (26 min.) VHS

The horticultural staff of the Center for Home Gardening share basic information about composting, including materials to use, selection of compost bins, construction of the pile, biological reactions, and potential problems.

Calibrating, Mixing, and Applying Pesticides
Tempe, Ariz., Idea Bank, 1989. 1 videocassette (15 min.) VHS

A basic introduction to pesticide safety is presented to gardeners and landscapers.

Exterior Tree and Shrub Pruning
Lafayette, Ind., Purdue Univ., 1989. 1 videocassette (28 min.) VHS

Illustrates the technique of manicuring shrubs to look like artistic creations. Includes section on limb removal.

Landscape Equipment Safety
Tempe, Ariz., Idea Bank, 1988. 4 videocassettes (110 min.) VHS

Provides basic safety guidelines for push mowers, riding mowers, power hand tools, and large equipment.

Ornamental Plant Diseases
Lafayette, Ind., Purdue Univ., 1989. 1 videocassette (10 min.) VHS

Provides information about the diseases that can affect plants grown in ornamental situations.

Midwest Wildflowers: Plants of the Tall Grass Prairie
Lafayette, Ind., Purdue Univ., 1990. 1 videocassette (53 min.) VHS

Surveys seventy-three species of flowers and five grasses native to prairie communities. Part One features spring and summer varieties; Part Two features varieties that bloom in the late summer and early fall. Supplementary materials are available.

Midwest Wildflowers: Wildflowers of Woodlands and Wetlands
Lafayette, Ind., Purdue Univ., 1990. 1 videocassette (103 min.) VHS

Presents ninety-five species of wildflowers found throughout the Midwest. Program is in two parts: one for the growing season of spring and one for fall. Supplementary material is available.

Planting and Staking Landscape Trees
Tucson, Ariz., College of Agriculture, University of Arizona, 1988. 1 videocassette (26 min.) VHS

Provides an overview of essential factors in the planting and staking of trees.

Success with Bedding Plants
Tempe, Ariz., Idea Bank, 1990. 1 videocassette (20 min.) VHS and 1 video reference guide (13 sheets)

Presents information on selection, installation, and maintenance of bedding plants.

Working with Pesticides: A Video Safety Course
Tempe, Ariz., Idea Bank, 1989. 4 videocassettes (60 min.) VHS

This chemical field handbook provides information on safe storage and disposal methods; clothing, equipment, and environmental safety; calibrating, mixing, and applying pesticides; and sprayer operation and maintenance.

Additional Resources

There are numerous horticulture, gardening, and landscape resources available on audio and videocassette. The following selected publishers have a rich collection of resources and will provide catalogs of currently available materials.

A. C. Burke and Co.
2554 Lincoln Blvd., Suite 1058
Marina Del Rey, Calif. 90201

This catalog lists "intelligent tools for your garden: videos, software, books, and accessories."

San Luis Video Publishing
P.O. Box 6715
Los Osos, Calif. 93412

Produces educational and training videocassettes of interest to the horticultural, agricultural, and design professions. Includes a wide range of works on design, use of equipment, and greenhouses.

Veterinary Technology

Dr. Gail Staines
Niagara County Community College
Sanborn, New York

Introduction

Veterinary technology is a program of study designed to educate students for a career in animal care. Veterinary technicians have numerous interesting and rewarding career paths available upon completion of this degree. Nearly 85 percent of students obtain positions in private practices. However, students may also seek employment in pharmaceutical sales, humane societies, as a livestock health manager on farms and ranches, in biomedical research or diagnostic laboratories, in zoo or wildlife medicine, or as a teacher of veterinary technology. Similar to the relationship between physicians and registered nurses, veterinary technicians work as part of an animal health care team with veterinarians.

Students who choose to become veterinary technicians should expect to perform technical duties as professional assistants to veterinarians. Such responsibilities include caring for animals in a hospital setting, conducting clinical pathological procedures, administering medications, and participating in client education. Radiology, anesthesiology, dental prophylaxis, and assisting in surgery are specialized areas of responsibility required of the veterinary technician. Technicians also routinely assist in office and hospital management.

Veterinary technology is a regulated profession. According to the North American Veterinary Technician Association (NAVTA), thirty-nine states have an agency or organization responsible for regulating the profession. Each state has separate guidelines for credentialing veterinary technicians. Some states require candidates to successfully pass the Veterinary Technician National Examination developed by Professional Examination Services and overseen by the American Veterinary Medical Association's (AVMA) Veterinary Technician Testing Committee. Other states, such as California, use their own examinations.

The first college-level program for veterinary technology was established in 1961 at the State University of New York at Delhi. Today, sixty-five accredited veterinary technology programs exist in the United States. Guidelines and accreditation of veterinary technology programs in the United States are established by the American Veterinary Medical Association's Committee on Veterinary Technician Education and Activities (CVTEA). Upon graduation from a two-year accredited program, students are eligible to become certified, registered, or licensed as a veterinary technician.

More than one hundred professional associations represent veterinary technicians in the United States and Canada at the local, state, and provincial levels. NAVTA is the international organization that represents veterinary technicians. Founded in 1981, NAVTA has several goals: to promote the professional and educational advancement of veterinary technicians; to enhance knowledge and skills through continuing education programs; to promote interests of the profession via state, provincial, and federal legislation; to maintain high ethical standards; to continue to work cooperatively with the veterinary medical profession; and to promote humane medical care for all animals.

The purpose of this bibliography is to provide librarians with recent sources that support veterinary technology program instruction in community colleges. Special acknowledgment is given to Nancy Millard, Ilona Middleton, and Val Macer of Medaille College for extending the use of their college's collection and for their expertise; acknowledgment is also extended to Carlene Decker of NAVTA and Dr. Janet Donlin of the American Veterinary Medical Association's Committee on Veterinary Technician Education and Activities for reviewing this bibliography.

Veterinary Technology   Sciences

Accreditation

American Veterinary Medical Association. Committee on Veterinary Education and Activities
940 N. Meachum Road
Schaumburg, Ill. 60196

Selected Associations

American Veterinary Medical Association
940 N. Meachum Road
Schaumburg, Ill. 60196

Animal Welfare Information Center
National Agricultural Library
10301 Baltimore Boulevard
Beltsville, Maryland 20705

North American Veterinary Technical Association
P.O. Box 224
Battle Ground, Ind. 47920

Selected Journals

Journal of the American Animal Hospital Association.
The Association. Bimonthly. ISSN 0587-2871

Journal of the American Veterinary Medical Association.
The Association. Semimonthly. ISSN 0003-1488

Journal of Zoo and Wildlife Medicine.
Journal of Zoo and Wildlife Medicine. Quarterly. ISSN 1042-7260

Lab Animal.
Nature Publishing. Ten times a year. ISSN 0093-7355

Perspectives.
Veterinary Learning Systems. Six times a year. An excellent new publication for women in veterinary medicine.

Topics in Veterinary Medicine.
Smithkline Beecham Animal Health. Three times a year.
ISSN 1064-5101

Veterinary Clinics of North America.
Saunders.
- *Food Animal.*
  Three times a year. ISSN 0749-0720
- *Equine Practice.*
  Three times a year. ISSN 0749-0739
- *Small Animal Practice.*
  Six times a year. ISSN 0195-5616

Each issue of the *Veterinary Clinics of North America* series is devoted to a current topic in veterinary medicine. Authorities in the field serve as guest editors. They are essential for veterinary technology collections.

Veterinary Technician.
Veterinary Learning Systems. Eleven times a year.
ISSN 8750-8990

Veterinary Update Clinical Abstract Service.
American Veterinary Pub. Co. Monthly.
A monthly service that abstracts articles from veterinary periodicals. Abstracts are divided into small animals and large animals (food animals and equines). This source is useful for students and faculty needing to update textbook information.

Supplementary Periodicals

Subscriptions to periodicals that focus on specific animals are important supplements to any veterinary technology program. Some examples are listed below.

Cat Fancy.
Fancy Publications. Monthly. ISSN 0892-6514

Dog Fancy.
Fancy Publications. Monthly. ISSN 0012-4834

Equus.
Fleet Street. Monthly. ISSN 0149-0672

Print Materials

✦ Reference

Ackerman, Norman
Radiology and Ultrasound of Urogenital Diseases in Dogs and Cats
Ames, Iowa, Iowa State Univ. Press, 1991. 187p.
ISBN 0813815274
   This brief yet informative source on the radiology and ultrasound of urogenital diseases provides information on normal and abnormal anatomy radiographs. Each figure has a brief description. An index of figures is also provided.

American Kennel Club
The Complete Dog Book: The Photograph, History, and Official Standard of Every Breed Admitted to the AKC Registration, and the Selection, Training, Breeding, Care, and Feeding of Pure-Bred Dogs
18th ed. New York, Howell Book House, 1992. 724p.
ISBN 0876054645
   First published in 1929, this volume updates the 1985 edition. Veterinary technology students will find this book useful for obtaining historical and confirmational information on 134 dog breeds recognized by the American Kennel Club (AKC). American Kennel Club history, descriptions of dog shows and trials, and AKC official standards are also included.

Bacha, William J.
Color Atlas of Veterinary Histology
Philadelphia, Lea and Febiger, 1990. 269p. ISBN 0812113039
    Bacha provides background information on histology including the basics of microscopy, and tissue and slide preparation. Includes color photographs and black-and-white drawings.

Beech, Jill, ed.
Equine Respiratory Disorders
Philadelphia, Lea and Febiger, 1991. 458p. ISBN 081211325X
    This text fills a niche in the literature by giving readers in-depth information on equine respiratory disorders. An index, appendices, charts, tables, radiographs, and sonographs are provided.

Bennett, Kristina, ed.
Compendium of Veterinary Products
2nd ed. Port Huron, Mich., North American Compendiums, 1993. 1152p.
    This source received rave reviews in the December 15, 1993, issue of *The Journal of the American Veterinary Medical Association*. Students will find it useful in obtaining active ingredient(s), dosage, and administration information, and precautions for animal health care products. Access to entries is through six different indexes. A brief one-page guide at the front of the *Compendium* gives students easy instructions on how to use this source.

Blowey, R. W., and A. David Weaver
Color Atlas of Diseases and Disorders of Cattle
Ames, Iowa, Iowa State Univ. Press, 1991. 223p. ISBN 0813804876
    This pictorial reference provides information on more than 360 conditions affecting cattle. The text and color photographs are from countries around the world. Blowey also addresses cattle diseases in developing countries.

Campbell, James B., and K. M. Charlton, eds.
Rabies
Boston, Kluwer Academic Pub., 1988. 431p. ISBN 0898383900
    As the editors write in the preface, "Rabies is an ancient disease and a fearsome one." This text provides students with a strong background in rabies, from the structure of the rabies virus to diagnosis, vaccines, and control of wildlife rabies. Includes index and references. An essential purchase for reference collections.

Campbell, Terry W.
Avian Hematology and Cytology
Ames, Iowa, Iowa State Univ. Press, 1988. 101p. ISBN 0813800641
    This is a useful resource for the veterinary technology student to learn about avian cells and determining changes in cell structure. Basic avian cytology, from respiratory tracts to synovial fluids and internal organs, are discussed. An overview of avian hematology is presented in the first chapter. Includes an index, bibliography, glossary, and a list of scientific names of birds.

Cochran, Phillip E.
Student's Guide to Veterinary Medical Terminology
Goleta, Calif., American Veterinary Publications, 1991. 274p. ISBN 0939674319
    Both instructors of veterinary technology and vet tech students will find this book useful for teaching and learning terminology related to veterinary medicine. Pronunciation, word construction, prefixes, and suffixes are covered. Includes a glossary of definitions and lesson plans that include exercises for practice.

Colahan, Patrick T., and others
Equine Medicine and Surgery
4th ed. Goleta, Calif., American Veterinary Publications, 1991. 2 v. ISBN 0939674270
    This work has been updated to reflect current changes in the diagnosis and treatment of equine ailments. The first section of Volume One contains information useful to veterinary office staff's initial contact with clients. The remainder of Volume One and all of Volume Two contain detailed information on diseases of the cardiovascular, respiratory, alimentary, nervous, reproductive, ocular, musculoskeletal, urinary, skin, endocrine, and hemolymphatic systems. Photographs, diagrams, and bibliographies are included.

Colville, Joann
Diagnostic Parasitology for Veterinary Technicians
Goleta, Calif., American Veterinary Publications, 1991. 266p. ISBN 0939674327
    Veterinary experts present students with commonly found internal and external parasites on dogs, cats, horses, food animals, rabbits, rodents, and birds. Colville presents a good study and reference source on parasites. Laboratory procedures for use in parasitism diagnosis are described. Includes black-and-white photographs, a glossary, and an index.

Cowell, Rick L., and Ronald D. Tyler, eds.
Cytology and Hematology of the Horse
Goleta, Calif., American Veterinary Publications, 1992. 242p. ISBN 0939674343
    This reference book was designed to facilitate the study of cytology. Topics covered are lesions; eyes; oral and nasal cavities; lower respiratory and gastrointestinal tracts; lymph nodes; pleural, peritoneal, synovial, and cerebrospinal fluids; endometrium; semen evaluation; peripheral blood smears; and bone marrow. Color plates, references, and an index are included.

Cowell, Rick L., and Ronald D. Tyler, eds.
Diagnostic Cytology of the Dog and Cat
Goleta, Calif., American Veterinary Publications, 1989. 259p. ISBN 0939674254
    Students will find this practical handbook of diagnostic cytology useful. This resource provides step-by-step instructions on aspirates, sample collection, slide preparation, smears, swabs, and washes. Color plates, charts, reference, and an index are included.

Crawford, Jane Diehl
The Preveterinary Planning Guide: Preparation, Application, and Admission Procedures to Veterinary Medical College
2nd ed. Bethesda, Md., Betz Pub. Co., 1990. 194p. ISBN 0941406202
    This invaluable planning guide will assist those students in veterinary technology who plan to become veterinarians. Crawford provides essential information on required courses, standard examinations, the application process, and the interview stage. Provides references, statistics on number of applicants and enrollment, and minority student information.

Dik, Kees J., and Ilona Gunsser
Atlas of Diagnostic Radiology of the Horse: Part 2: Diseases of the Hind Limb
Philadelphia, Saunders, 1989. 148p. ISBN 0721629598

This reference source provides students with clear radiographs of the hip, stifle, and hock of the horse. Each entry, clearly written, is in both English and German.

Dik, Kees J., and Ilona Gunsser
Atlas of Diagnostic Radiology of the Horse: Part 3: Diseases of the Head, Neck and Throat
Philadelphia, Saunders, 1990. 172p. ISBN 0721634486

Similar in format to *Part 2*, this reference book presents radiographs of the head, neck, and thorax of the horse. Veterinary technology collections should contain all three parts. *Part 3* contains radiographs of diseases of the front limb, metatarsal bones, and phalangeal area of the hind limb of the horse.

Ettinger, Stephen J., and Edward Seepo, eds.
Textbook of Veterinary Internal Medicine: Diseases of the Dog and Cat
4th ed. Philadelphia, Saunders, 1994. 2 v. ISBN 0721667953

This work provides the reader with important information on diseases of the dog and cat related to internal medicine. Veterinary specialists address various topics in feline and canine care and treatment, such as emergency medicine, ear diseases, small animal zoonoses, and small animal problems in developing countries.

Fiore, Mariano S. H. di.
Atlas of Normal Histology
6th ed., edited by Victor P. Eroschenko. Philadelphia, Lea and Febiger, 1989. 267p. ISBN 0812111265

First published three decades ago, this standard reference on histology has changed its name from *The Atlas of Human Histology;* the name change reflects the inclusion of animal tissues and organs. This is an essential reference.

Guttman, Helene N., ed.
Guidelines for the Well-Being of Rodents in Research: From a Conference Held by the Scientists Center for Animal Welfare in Research, Triangle Park, North Carolina, on December 8, 1989, with Additional Material Provided by the Authors
Bethesda, Md., Scientists Center for Animal Welfare, 1990. 105p.

Founded in 1978, The Scientists Center for Animal Welfare (SCAW) provides information on all types of research animals. Students planning to work in a laboratory setting will find this to be an important reference. Provides guidelines for animal laboratory care and answers to frequently asked questions.

Hawkey, C. M., and T. B. Dennett
Color Atlas of Comparative Veterinary Hematology: Normal and Abnormal Blood Cells in Mammals, Birds, and Reptiles
Ames, Iowa, Iowa State Univ. Press, 1989. 192p.
ISBN 0813804493

Supported by clinical data, this atlas provides comparative analyses of mammals, birds, and reptiles on hematological topics. Amphibians and fish are not included. This is a useful ready reference source for veterinary technology students. Glossary, bibliography, and species and subject indexes are included.

Howard, Jimmy L., ed.
Current Veterinary Therapy: Food Animal Practice, 3
Philadelphia, Saunders, 1993. 966p. ISBN 0721636330

A necessary resource for any library supporting a veterinary technology program, this work provides students with information regarding food animal diseases. Includes appendices, references, and index.

Jones, William E., ed.
Equine Sports Medicine
Philadelphia, Lea and Febiger, 1989. 329p. ISBN 0812111001

Equine sports medicine is one of the most rapidly growing fields in the United States. Chapters in this work include an introduction to equine sports medicine, exercise physiology, nutrition, and sports injuries. Veterinary technology students working with horses should have access to this essential text. Includes index, references, diagrams, and graphs.

Kaneko, Jiro J., ed.
Clinical Biochemistry of Domestic Animals
4th ed. San Diego, Calif., Academic Press, 1989. 932p.
ISBN 0123963044

This updated edition includes new information on DNA technology, laboratory rodents, leukocyte function, normality in clinical biochemistry, trace minerals, and vitamins. The expanded appendices cover avian, lab animal, and large animal species. Includes an index, charts, and graphs.

Kirk, Robert Warren, and John D. Bonagura, eds.
Current Veterinary Therapy XI: Small Animal Practice
Philadelphia, Saunders, 1992. 1346p. ISBN 0721632939

First published in 1964, this is a standard work in the field of small animal practice. Revisions reflect current applications in veterinary science. Chapters cover topics such as critical care, infectious diseases, and immunology. Includes appendices and an extensive index.

Lederberg, Joshua, ed.
Encyclopedia of Microbiology
San Diego, Calif., Academic Press, 1992. 4 v. ISBN 0122268911 (vol. 1); ISBN 012226892X (vol. 2); ISBN 0122268938 (vol. 3); ISBN 0122268946 (vol. 4)

Veterinary technology students should have access to this high-quality resource on microbiology. This encyclopedia provides readers with supplemental material not found in course texts. Charts, tables, diagrams, photographs, an index, and a list of contributors are included.

McEntee, Kenneth
Reproductive Pathology of Domestic Animals
San Diego, Calif., Academic Press, 1990. 401p.
ISBN 0124833756

This resource fills a gap in the literature by providing students with information on the pathology of domestic animals, specifically concerning the reproductive system. Includes references, an index, and black-and-white photographs.

Kastenbaum, Robert
The Psychology of Death
2nd ed. New York, Springer Pub. Co., 1992. 266p.
ISBN 0826119220

This particular text is recommended as a supplement to the voluminous works on the psychological aspects of death. Index and references are included.

Lose, M. Phyllis
Blessed are the Brood Mares
2nd ed. New York, Howell Book House, 1991. 271p.
ISBN 0876058489

As the cover states, this is "the classic source book on breeding and foaling." Topics range from pregnancy determination through delivery, to preparing for the next foal. This text is a must for students working with equines. Photographs, illustrations, glossary, subject index, and an index of signs and symptoms are included. Libraries owning this text should also have M. P. Lose's *Blessed are the Foals* (Howell Book House, 1987).

McCurnin, Dennis M., ed.
Clinical Textbook for Veterinary Technicians
3rd ed. Philadelphia, Saunders, 1994. 816p. ISBN 0721637922

The third edition of this standard reference has twenty-six new contributors. Highly recommended for its comprehensive treatment.

McCurnin, Dennis M., and Ellen M. Poffenbarger
Small Animal Physical Diagnosis and Clinical Procedures
Philadelphia, Saunders, 1991. 222p. ISBN 0721659314

This practical guide to complete physical examination of small animals provides step-by-step procedures for obtaining the health history, information on the physical exam as a diagnostic tool, and performing the actual exam. Clinical procedures are also outlined. Includes photographs, diagrams, appendices, references, and index.

Mench, Joy A., and Lee Krulisch, eds.
Canine Research Environment: From a Conference Held by the Scientists Center for Animal Welfare in Bethesda, Maryland on June 22, 1989
Bethesda, Md., Scientists Center for Animal Welfare, 1990. 82p.

This compilation of conference proceedings is useful for students interested in working in a laboratory setting with canines. Regulatory issues, the environment and management of the lab, and canine exercise are discussed. Includes question-and-answer sections for review.

Mench, Joy A., and Lee Krulisch, eds.
Well-Being of Nonhuman Primates in Research: From a Conference Held by the Scientists Center for Animal Welfare in Bethesda, Maryland, on June 23, 1989
Bethesda, Md., Scientists Center for Animal Welfare, 1990. 86p.

Students planning to work in a laboratory setting with nonhuman primates should read these conference proceedings. This work discusses regulatory perspectives, biological perspectives, veterinary issues, and financial and administrative positions. Includes references and a question- and- answer section for review.

Morgan, Joe P., ed.
Techniques of Veterinary Radiography
5th ed. Ames, Iowa, Iowa State Univ. Press, 1993. 482p.
ISBN 0813817277

Designed for the veterinary technician and practitioner working in a clinical setting and interested in diagnostic radiology, this work provides clear descriptions of correct radiographic positions, accompanied by large black-and-white photographs. The editor has written several books on veterinary radiography, including *Equine Radiology* (Iowa State Univ. Press, 1992) and *Radiographic Diagnosis and Control of Canine Hip Dysplasia* (Iowa State Univ. Press, 1985). Authoritative and clearly written, this source could only be improved with the inclusion of an index. Includes a glossary and references.

Pettit, Thomas H.
Hospital Administration for Veterinary Staff
Goleta, Calif., American Veterinary Publications, 1994. 180p.
ISBN 093967453X

This text recognizes the trend of veterinary technicians to assume management positions as their education and experience increase. Filling a gap in the literature, this volume covers the topics of managing emergencies, supervising employees, and educating clientele, and also addresses basic management topics such as scheduling and records management. A chapter on grief counseling is also included. Review questions, recommended readings, and an index round out this useful text.

Pratt, Paul W.
Laboratory Procedures for Veterinary Technicians
2nd ed. Goleta, Calif., American Veterinary Publications, 1992. 601p. ISBN 0939674386

This source is *the* guide for veterinary technicians responsible for performing laboratory procedures. Practical and authoritative, this handbook should be in all library collections serving veterinary technology students.

Pratt, Paul W.
Medical, Surgical and Anesthetic Nursing for Veterinary Technicians
2nd ed. Goleta, Calif., American Veterinary Publications, 1994. 627p. ISBN 0939674491

As the preface states, "Veterinary technicians have become an indispensable part of the veterinary health care team." This book is intended to aid veterinary technicians in performing various clinical procedures. Chapters cover general patient management, supportive care, anesthesia and anesthetic nursing, surgical nursing, theriogenology, and treatment techniques for dogs, cats, birds, small exotics, horses, and food animals. Includes recommended readings, black-and-white photographs, diagrams, and index.

Pratt, Paul W., ed.
Review Questions and Answers for Veterinary Technicians
Goleta, Calif., American Veterinary Publications, 1993. 409p.
ISBN 0939674440

Students preparing for veterinary technology licensing examinations may have purchased their own copy of this study guide. Not an essential purchase for libraries, but useful to have nonetheless. Contains more than 2,100 study questions.

Roush, James K.
Stifle Surgery
Philadelphia, Saunders, 1993. 920p.

This overview of current techniques in small animal stifle surgery covers advances in veterinary science. Although scholarly, two-year college students will find the articles interesting and useful in furthering their own knowledge in veterinary technology. Part of *The Veterinary Clinics of North America: Small Animal Practice*, vol. 23, no. 4.

Schaeffer, Dorcas O., Kevin M. Kleinow, and Lee Krulisch
The Care and Use of Amphibians, Reptiles and Fish in Research
Bethesda, Md., Scientists Center for Animal Welfare, 1992. 196p. ISBN 0685651568

This compilation of proceedings from a Scientists Center for Animal Welfare/Louisiana State University School of Veterinary Medicine conference fills a gap in the literature on the care and use of "lower" animals in research. Students will use this information to gain a clear understanding of caring for amphibians, reptiles, and fish in a research environment. These animals have needs that are different from mammals traditionally used in research, especially in terms of food, water, environment, and drug requirements. Includes regulations and guidelines as well as a resource list.

Sonsthagen, Teresa F.
Restraint of Domestic Animals
Goleta, Calif., American Veterinary Publications, 1991. 149p. ISBN 0939674289

Written by a practicing veterinary technician, this lucid and well-illustrated guide provides students with step-by-step techniques for restraining cats, dogs, horses, cattle, pigs, sheep, goats, birds, rodents, rabbits, and ferrets. Includes an index, glossary, and appendices giving gender names and brief physiological data of animals.

Stufflebeam, Charles E.
Genetics of Domestic Animals
Englewood Cliffs, N.J., Prentice Hall, 1989. 295p. ISBN 0133512142

This text is unique as a genetics book because it uses domestic animals as examples. Divided into three sections: cell biology and gene chemistry, qualitative genetics, and quantitative genetics. Useful as a study and practice guide.

Tannenbaum, Jerrold
Veterinary Ethics
Baltimore, Md., Williams and Wilkins, 1989. 358p. ISBN 0683081020

Tannenbaum provides the student and the practitioner with information and arguments surrounding ethical issues in the veterinary profession. This text is not as comprehensive as James Wilson's *Law and Ethics of the Veterinary Profession* (Priority Press, 1988), but it provides details on issues of veterinary ethics and religion, moral theory, and problem-solving in practice.

Taylor, David
Vet on the Wild Side
New York, St. Martin's Press, 1990. 224p. ISBN 0312055293

A fascinating look a zoo vet's life. The author of *Zoo Vet* (Unwin, 1984) and *Next Panda, Please* (Stein and Day, 1982) Dr. Taylor traces his experiences as a vet to exotic animals around the world. An excellent resource for students interested in zoology.

Weidensaul, Scott
The Birder's Miscellany: Fascinating Collection of Facts, Figures, and Folklore from the World of Birds
New York, Simon and Schuster, 1991. 135p. ISBN 0671695053

A readable guide to birds. Chapters include kinds of birds, flight and migration, nesting habits, nutrition, folklore, and other interesting facts. Photographs, bibliography, and index.

Welsh, Heidi J.
Animal Testing and Consumer Products
Washington, D.C., Investor Responsibility Research Center, 1990. 167p. ISBN 093103539

A concise look at the use of animal testing of consumer products. Chapters cover the history of the animal protection movement, federal laws and regulations, alternatives to animal testing, profiles of United States companies and animal testing, and U.S. Department of Agriculture data analysis. Includeds an appendix and bibliography.

Wilson, James F., and others
Law and Ethics of the Veterinary Profession
Yardley, Pa., Priority Press, 1990. 532p. ISBN 0962100706

This basic text on law and ethics of the veterinary profession is written by a veterinarian who is also a lawyer. Topics addressed include basics of United States law, laws establishing veterinary medicine, veterinary and animal ethics, animals and the law, settling disputes, professional liability, antitrust and advertising, contract law, and the legal use of veterinary drugs.

Nonprint Materials

✦ Videocassettes

Administering Medication
Denver, Colo., American Animal Hospital Association, 1991.
1 videocassette (26 min.) VHS and 1 guide.

Professionally produced, this video presents general information on medicating animals, including dogs, cats, small mammals, and birds. The demonstrations and voice-over are clear. Color diagrams, use of live animals, and outlines of procedures make this a worthwhile videotape. Comes with an instructional guide. Part of the *AAHA Educational Videotape Training Program*.

Animal Handling and Restraint
Denver, Colo., American Animal Hospital Association, 1991.
1 videocassette (29 min.) VHS and 1 guide.

This video demonstrates safe and effective techniques for handling and restraining animals to perform exams or give medications. A dog, cat, bird, and small mammals are used in demonstrations. Techniques for handling and restraining puppies and kittens are also included. Comes with an instructor's guide. Part of the *AAHA Educational Videotape Training Program*.

McIlwraith, C. Wayne
Diagnostic and Surgical Arthroscopy in the Horse
2nd ed. Philadelphia, Lea and Febiger, 1990. 227p.
ISBN 0812111869

An extremely valuable resource for students, teachers, and practitioners interested in arthroscopy. Updated from the first edition, this text presents large black-and-white photographs, illustrations, and clearly written text. Index and references.

Muir, William, and John A. E. Hubbell
Equine Anesthesia: Monitoring and Emergency Therapy
St. Louis, Mosby Yearbook, 1991. 515p. ISBN 0801635764

The audience for this book is veterinarians, veterinary students, and those interested in specializing in equine surgery and anesthesia. This resource addresses the specialized skills and knowledge needed for equine anesthesia, a rapidly evolving field. Includes references, index, and appendices of drug schedules, along with a list of anesthetic equipment companies.

Plunkett, Signe J.
Emergency Procedures for the Small Animal Veterinarian
Philadelphia, Saunders, 1993. 277p. ISBN 0721667813

This guide provides students and practitioners with the necessary information to quickly and proficiently handle emergency cases. Clearly presented, each entry gives easily readable information on diagnosis, prognosis, and treatment. Useful as a reference in and out of the clinic.

Robinson, N. E.
Current Therapy in Equine Medicine, 3
Philadelphia, Saunders, 1992. 847p. ISBN 0721634753

A traditional source that updates information in the field of equine medicine. Standard sections on each organ system are included. Robinson presents new information on the musculoskeletal system, medical problems, use of technology and pharmaceuticals, exotic diseases, and the expanding role of veterinarians.

Rubel, G. Alexander, and others
Atlas of Diagnostic Radiology of Exotic Pets: Small Mammals, Birds, Reptiles, and Amphibians
Philadelphia, Saunders, 1991. 224p. ISBN 0721634931

A well-written, clear pictorial reference providing students with radiographs of small mammals, birds, reptiles, and amphibians. Brief descriptions accompany each radiograph.

Sherding, Robert G., ed.
The Cat: Diseases and Clinical Management
New York, Churchill Livingstone, 1989. 2 v. ISBN 0443084610

Volume One of this comprehensive clinical reference on the cat covers clinical foundations, feline nutrition and behavior, intoxication and injury, infectious and parasitic diseases, immunology, hematology, oncology, and the cardiopulmonary system. Volume Two details the feline digestive, endocrine, neuromuscular, skeletal, urinary, and reproductive systems, as well as the skin, ears, and eyes. Includes photographs, charts, diagrams, and references.

Smith, Bradford P., ed.
Large Animal Internal Medicine
St. Louis, Mo., Mosby, 1990. 1787p. ISBN 0801650623

A necessary book for students working with cattle, goats, horses, or sheep, this reference work provides students and practitioners with detailed information on more than 130 clinical diseases, problems, and signs. Includes a detailed table of contents, an index, references, diagrams, and photographs.

Smith, W. J., D. J. Taylor, and R. H. C. Penny
Color Atlas of Diseases and Disorders of the Pig
Ames, Iowa, Iowa State Univ. Press, 1990. 192p.
ISBN 0813800692

Along the same lines as R.W. Blowey's *Color Atlas of Diseases and Disorders of Cattle* (Iowa State Univ. Press, 1991), this atlas focuses on the pig. Includes large color pictures and easy-to-read text. Useful for students, teachers, practitioners, and agricultural workers. Indexed.

Stachowitsch, Michael
The Invertebrates: An Illustrated Glossary
New York, Wiley-Liss, 1992. 676p. ISBN 0471832944

This very helpful glossary of terms addresses invertebrates. Illustrations and definitions are clear, concise, and readable. An index is provided. This is an excellent addition to any reference collection.

Stoskopf, Michael K., ed.
Fish Medicine
Philadelphia, Saunders, 1993. 882p. ISBN 0721626297

The study of fish medicine is emerging as a specialty, and it is different from fish pathology, fisheries biology, and fisheries management. Stoskopf's work is essential for students interested in studying fish. Covers topics such as immunology, genetics, surgery, and hospitalization. Index, references, and appendices are provided.

Swain, Steven F., and Ralph A. Henderson
Small Animal Wound Management
Philadelphia, Lea and Febiger, 1990. 252p. ISBN 0812112393

This practical handbook presents the basics of small animal wound management and repair. Bandaging, medications, sutures, and other techniques are illustrated and described in a straightforward, step-by-step manner. An excellent study guide or on-the-job resource.

Varner, Dickson D., and others
Diseases and Management of Breeding Stallions
Goleta, Calif., American Veterinary Publications, 1991. 349p.
ISBN 0939674335

An essential guide to the care and maintenance of breeding stallions. Diseases, behavior, anatomy, and physiology are discussed in detail. Includes photographs, index, and references.

Whitford, Ronald
D.V.M. Series
Lincoln, Neb., PetCom, 1992. 3 v.

This series will be useful to students when they become practitioners in the field, but it is also a good study series. *The DVM Survival Manual: A Protocol for Treatment*, a management tool for improving veterinary practice, includes general protocol, laboratory and pharmacy, diagnosis, estimates, and coverage of different conditions. *Client Education* (2nd ed.) contains handouts on animal care and treatment. Handouts are designed to be photocopied and given to clients when the the animal is taken home. *The Forms and Management Manual* contains forms and explanations for procedures in overseeing a veterinary practice. This series is not an essential purchase, but may be very useful for students as study guides and in understanding the management of a veterinary practice.

✦ Supplemental Sources

Attenborough, David
The Trials of Life: A Natural History of Animal Behavior
Boston, Little, Brown, 1990. 320p. ISBN 0316057517

A companion text to the television series broadcast on TNT, it is third in a series, following *Life on Earth* and *The Living Planet.* This fascinating text of animal behavior describes hunting, maturation, fighting, reproduction, and other facets of animal interaction. Includes beautiful color pictures and an index.

Baker, Edward
Small Animal Allergy: A Practical Guide
Philadelphia, Lea and Febiger, 1990. 144p. ISBN 0812112407

Veterinary technology students will find this text useful in identifying and treating animal allergies. Descriptions of allergies, clinical signs, environmental allergens, allergy skin testing, and treatments are accompanied by black-and-white and color photographs. Includes an index and a bibliography.

Beeman, G. Marvin, ed.
Examination for Purchase
Philadelphia, Saunders, 1992. 425p. ISSN 0749-0739

Examining a horse prior to purchase is a common procedure veterinarians are asked to perform. This issue of *The Veterinary Clinics of North America-Equine Practice* gives the reader an extensive overview of what is involved. Chapters discuss the history of the pre-purchase exam, recording the evaluation, legal aspects, examination of the horse's anatomy and physiology, and examining the horse at an auction. A key resource for veterinary technology students working with horses.

Bill, Robert
Pharmacology for Veterinary Technicians
Goleta, Calif., American Veterinary Publications, 1993. 301p.
ISBN 0939674505

Students and practicing veterinary technicians searching for a no-nonsense guide to pharmacology should consult this source. Chapters discuss basic principles of pharmacology, antimicrobials, disinfectants, anti-inflammatories, and antiparasitics. Information on drugs that affect the nervous, respiratory, cardiovascular, gastrointestinal, and endocrine systems is presented in a straightforward and descriptive manner. Indexed.

Boggs, Donald L., and Robert Anthony Merkel
Live Animal Carcass Evaluation and Selection Manual
3rd ed. Dubuque, Iowa, Kendall Hunt Pub. Co., 1990. 221p.
ISBN 0840353936

The purpose of this text is to provide students with information on the selection and evaluation of livestock. Includes a discussion of beef cattle, sheep, and breeding; and feeder, and market swine. Useful for students working with these types of animals. Black-and-white photographs, tables, charts, appendices, and index are included.

Drum, Sue, and H. Ellen Whiteley
Women in Veterinary Medicine: Profiles of Success
Ames, Iowa, Iowa State Univ. Press, 1991. 270p.
ISBN 0813806682

This book documents the lives of twenty women veterinarians. An overall theme is gender bias as a barrier to women seeking a place in a traditionally male-dominated profession. By including women from diverse backgrounds, age groups, and geographic settings, Drum and Whiteley illustrate various career paths that can be taken by veterinarians. Veterinary technology as a career is briefly mentioned.

Feeney, Daniel A., Thomas F. Fletcher, and Robert M. Hardy
Atlas of Correlative Imaging Anatomy of the Normal Dog: Ultrasound and Computed Tomography
Philadelphia, Saunders, 1991. 382p. ISBN 0721627447

The purpose of this source is to provide students and practitioners with an "image-atlas" of normal canine anatomy. This work has four sections covering the head and neck, spine, thorax, and abdomen and pelvis. The appendix gives readers an overview of computed tomography and diagnostic sonography. Essential for students working in this field.

Gage, Loretta, and Nancy Gage
If Wishes Were Horses: The Education of a Veterinarian
New York, St. Martin's Press, 1993. 295p. ISBN 0312088175

Loretta Gage and her sister, Nancy, provide readers with a realistic view of the road to becoming a veterinarian. Writing in an easily readable style, Loretta Gage explains her attempts to be accepted at Colorado State University's College of Veterinary Medicine, the detailed course work, arduous clinical and lab tasks, and long hours of study. Veterinary technology students with hopes of attending veterinary school will want to read this book.

Hoskins, Johnny D.
Veterinary Pediatrics: Dogs and Cats from Birth to Six Months
Philadelphia, Saunders, 1990. 556p. ISBN 0721623549

This reference source is a guide to the medical and health needs of dogs and cats, up to six months of age. Discusses preventive health maintenance as well as diagnosing and treating common diseases and ailments. Includes photographs, appendices, and an index.

Ikram, Muhammed, and Eloyes Hill
Microbiology for Veterinary Technicians
Goleta, Calif., American Veterinary Publications, 1991. 213p.
ISBN 0939674300

Brief chapters of this text cover the history and background of microbiology, mycology, and virology. This easy-to-read resource was written for veterinary technician students and practitioners and works well as a study guide. Charts, tables, diagrams, glossary, references, and an index.

Jordan, F. T. W., ed.
Poultry Diseases
3rd ed. Philadelphia, Saunders, 1990. 497p. ISBN 0702013390

As the preface clearly explains, the purpose of this text is to "provide a concise account of the more important diseases of poultry with reference to their cause, epidemiology, diagnosis. and control." Includes appendix, index, and references.

Clipper Maintenance and Blade Care
Milwaukee, Wisc., Oster, 1988. 1 videocassette (12 min.) VHS

The focus is on the basics of maintaining clippers and blades, with descriptions of blades and blade care. This video is professionally done and can be used as a general overview and refresher for students and practitioners.

IV Catheter Placement
Denver, Colo., American Animal Hospital Association, 1991. 1 videocassette (26 min.) VHS and 1 instructor's guide.

Demonstrates intraveneous catheter placement: preparing the animal, inserting the catheter, monitoring fluid administration, and maintaining the catheter. Principles of fluid therapy are described and shown. *Educational Videotape Training Program* series.

North American Veterinary Technician Association
The World of the Veterinary Technician
[s.l.], Purdue University, 1990. 1 videocassette (13 min.) VHS

Covers career opportunities in veterinary technology in the areas of small animal practice, pharmaceutical sales, equine and food animal practice, and education. Describes the duties of the veterinary technician and discusses career preparation and future employment opportunities. Provides an overview of the North American Veterinary Technician Association.

We Care...The Effective Veterinary Receptionist: First Impressions
Denver, Colo., American Animal Hospital Association, 1988. 1 videocassette (25 min.) VHS

This video explains the importance of veterinary receptionists and the first impressions they make.

We Care...The Effective Veterinary Receptionist: Special Situations
Denver, Colo., American Animal Hospital Association, 1988. 1 videocassette (30 min.) VHS and 1 guide.

This video discusses how to deal with dissatisfied clients, and how to act in emergencies and in situations involving death and euthanasia. Although somewhat gender biased (use of "she" as the receptionist), this video is nonetheless useful for receptionists and veterinary technicians who work directly with clients. Communication techniques are described and demonstrated. An instructional guide that discusses pet owner anxiety is included.

We Care...The Effective Veterinary Receptionist: Telephone Techniques
Denver, Colo., American Animal Hospital Association, 1988. 1 videocassette (34 min.) VHS

Demonstrates appropriate phone techniques and discusses the characteristics of a good receptionist. Although this video is somewhat gender biased (the use of "she" as the receptionist), it is still useful. A guide of tips and techniques accompanies this video.

✦ Electronic Databases and Resources

Focus on Veterinary Science and Medicine [database]
Philadelphia, Institute for Scientific Information, 199?- . Updated monthly, Computer disks, ISSN 1067-8964

This database fully indexes more than 200 journals covering veterinary science and medicine topics. Includes tables of contents and abstracts of journals, articles, books, and conference proceedings, and articles and chapters from other sources. This easy-to-search database is available in both Macintosh and IBM versions. System requirements: IBM-PC or compatible; 5 MB RAM. Full-text articles are available through ISI's document delivery service, *The Genuine Article*.

Network of Animal Health (NOAH)

Electronic discussions of interest to veterinarians and auxiliary personnel are available on the Network of Animal Health (NOAH). NOAH resides on CompuServe and is accessible via Telnet and gopher from the Internet. A NOAH account is a requirement. An account is available from:

AVMA Network of Animal Health
1931 N. Meacham Road, Suite 100
Schaumburg, Ill. 60173-9724

✦ Other Sources of Information

Animal Welfare Information Center
National Agricultural Library
10301 Baltimore Boulevard
Beltsville, Maryland 20705
e-mail: awic@nalusda.gov

The Animal Welfare Information Center (AWIC) was created in 1986 by the United States Department of Agriculture. According to technical information specialist Michael D. Kreger, AWIC's purpose is "to provide information on any animal welfare subject covered under Federal legislation and policy to the regulated community." Moreover, AWIC provides materials to anyone needing information on animal-related topics. Most information services are free, but some, such as on-line searching of multiple databases, carry a fee.

Social Services

Alcohol and Drug Abuse Counseling 533
Pam Kessinger

Social Work and Human Services 551
Lynn Namsick

Alcohol and Drug Abuse Counseling

Pam Kessinger
Danville Area Community College
Danville, Illinois

Introduction

The purpose of a substance abuse counselor training program is multidimensional. A certificate program or Associate in Applied Science degree (AAS) prepares students for immediate employment in entry-level positions in substance abuse treatment. A certificate or AAS is also designed for professionals who hold other degrees and need additional training (e.g., a public school teacher who may also provide intervention, or a personnel director administering an Employee Assistance Program). In addition, students may opt to gain credits for an Associate of Arts (AA) or Associate of Science (AS) degree for transfer to clinically oriented programs.

Several community colleges offer substance abuse counselor training as a specialization within a general human services or social services program. Students prepare for employment in substance abuse treatment programs in a variety of inpatient and out-patient settings, such as youth service agencies, community service agencies, residential treatment centers, correctional programs, and mental health organizations. National certification is available from the National Association of Alcoholism and Drug Abuse Counselors (NAADAC). States vary in their certification requirements, but an increasing number offer reciprocal certification agreements.

Substance abuse counseling is a field fraught with debate about the division of labor, professional roles, specialization and subspecialization, and training. Terminology is also in a state of transition. Substance abuse counseling may be called addictions, chemical dependence, drug, and alcohol (or alcoholism) counseling (or rehabilitation).

The entry-level substance abuse counseling arena is unique in that many individuals enter the field as a result of dealing with their own addiction issues. The Twelve Step programs (originally of Alcoholics Anonymous) and self-help movements actively encourage former or "recovered" substance abusers to reach out to other abusers. Being a recovering user or co-dependent is often a compelling factor in considering a career as a counselor.

Beyond progressing toward recovery and being able to communicate that experience, the substance abuse counselor must develop a broad knowledge base. Programs should include basic psychology, human development, anatomy and physiology, family dynamics, pharmacology of psychoactive drugs, and the political and legal aspects of substance abuse. Strong communication skills should be developed with courses on interpersonal communications, helping skills, and rhetoric. Specific areas of study may include casework methods, crisis intervention, group therapy, and first aid.

Several factors have resulted in the exponential growth of literature in the area of substance abuse counseling. A significant factor is the continuing prevalence of substance abuse with subsequent governmental efforts at prevention and interdiction. Secondly, the 1987 revision of the *Diagnostic and Statistical Manual of Mental Disorders* (*DSM-III-R*) outlined the treatment of addictions from a multidimensional perspective and detailed the diagnosis of addiction on a continuum from "abuse" to "dependence." The fourth edition of this work, *DSM-IV,* was published in 1994. Other factors include the increased public awareness of addiction issues andtreatment modalities, the rapid growth of self-help movements, and the increased clinical legitimacy of the concept of co-dependence. Finally, recent advances in neurochemistry offer intriguing insight into the function of psychotropic substances and neurologic reactions.

In this state of rapid development and controversy, it is impossible to provide an exhaustive bibliography. Listed below are selected titles appropriate for an undergraduate level of study, along with some professional-level material that students should be exposed to for future reference. Every attempt has been made to present various points of view, ranging from popular, self-help material to professional, therapeutic, and clinical publications.

Accreditation

Joint Commission on Accreditation of Healthcare Organizations
1 Renaissance Blvd.
Oakbrook Terrace, Ill. 60181
   Administers accreditation for treatment programs.

Commission on Accreditation of Rehabilitation Facilities
101 North Wilmot Road, Suite 500
Tucson, Ariz. 85711
   Administers accreditation for treatment programs in rehabilitation/habilitation organizations.

National Commission on Accreditation of Alcoholism and Drug Abuse Counselor Credentialing Bodies
107 Lincoln St.
Worcester, Mass. 01605
   Acts as an accrediting body for state certification boards as independent agencies.

Selected Associations

Alcohol and Drug Problems Association of North America
444 N. Capitol St. NW, Suite 706
Washington, D.C. 20005
   ADPA sponsors conferences. Publications include directories, pamphlets, and the bimonthly newsletter *ADPA Professional*.

Employee Assistance Professionals Association
2101 Wilson Blvd., Suite 500
Arlington, Va. 22201-3062
   Provides certification and sponsors conferences, education, and training. Publications include *The Exchange* (monthly), *EAP Journal of Research* (biannual), and directories.

National Association of Alcoholism and Drug Abuse Counselors
3717 Columbia Pike, Suite 300
Arlington, Va. 22204
   Provides certification for Addictions Counselor, levels I and II. Sponsors an annual national conference. Publications include *The Counselor* and *NAADAC Newsletter* (monthly), directories, conference proceedings, and professional guides.

National Association of Prevention Professionals and Advocates
1228 East Breckinridge St.
Louisville, Ky. 40204
   Sponsors an annual national conference. Provides training in prevention and grantsmanship. Publications include *NAPPA Bulletin* (quarterly) and a directory.

National Clearinghouse for Alcohol and Drug Information
P.O. Box 2345
Rockville, Md. 20852
   Under the auspices of the Health and Human Services Department, Center for Substance Abuse Prevention, NCADI provides information and publications on drug abuse problems and programs.

National Institute on Alcohol Abuse and Alcoholism (NIAAA): National Institute on Drug Abuse
5600 Fisher's Lane
Rockville, Md. 20857
   The NIAAA conducts and sponsors research on the prevention and treatment of drug abuse. Publishes research studies and conference proceedings.

Substance Abuse and Mental Health Services Administration
Center for Substance Abuse Treatment; Center for Substance Abuse Prevention
5600 Fisher's Lane
Rockville, Md. 20857
   Under the auspices of the Health and Human Services Department, this agency sponsors NCADI and administers grants that support private and public addiction prevention and treatment services. Evaluates programs and delivery systems.

Substance Abuse Librarians and Information Specialists
P.O. Box 9513
Berkeley, Calif. 94709-0513
   Open to professional representatives of research and academic institutions, libraries, information clearinghouses, hospitals, and treatment centers. An international organization for prevention advocates, information providers, and policy makers. Sponsors an annual conference. Publications include *SALIS News* (quarterly) and a directory of members and organizations.

Selected Journals

Alcohol Health and Research World.
National Institute of Alcohol Abuse and Alcoholism. Quarterly. ISSN 0090-838X

Alcoholism Treatment Quarterly.
Haworth Press. Quarterly. ISSN 0734-7324

The Counselor.
National Association of Alcoholism and Drug Abuse Counselors. Bimonthly. ISSN 1047-7314

EAP Digest [The Voice of Employee Assistance Programs].
Performance Resource Press. Bimonthly. ISSN 0273-8910

International Journal of the Addictions.
Marcel Dekker. Monthly. ISSN 0020-773X

Journal of Drug Issues.
Florida State University. Quarterly. ISSN 0022-0426

Journal of Substance Abuse Treatment.
Elsevier Science. Bimonthly. ISSN 0740-5472

Professional Counselor Magazine.
A&D Pubs. Bimonthly. ISSN 1042-7570

Print Materials

Abel, Ernst
Fetal Alcohol Syndrome
Oradell, N.J., Medical Economics Books, 1990. 221p.
ISBN 0874895782
  An overview of contemporary scientific research on fetal alcohol syndrome that covers pharmacology, diagnosis, and maternal risk factors. The opening chapter debunks common myths and documents the gradual recognition of the teratogenicity of alcohol. Appendices include research data reports on aspects such as birth characteristics, postnatal measurements, and eye, ear, and facial anomalies. Extensive references, graphs, tables, and an index.

Ascher, L. Michael, ed.
Therapeutic Paradox
New York, Guilford Press, 1989. 385p. ISBN 0898623936
  Presents eleven essays divided between literature review and clinical applications on the "therapeutic paradox," a technique used in behavior therapy. The introductory chapter describes the historical basis of this technique, and the divergent definitions and applications in current behavioral therapy. Subsequent essays cover the use of this technique in treatment of eating disorders, emotional avoidance, borderline personality disorder, and family therapy. Includes references and an index.

Ashery, Rebecca Sager, ed.
Progress and Issues in Case Management
Rockville, Md., National Institute on Drug Abuse, 1992. 401p.
ISBN 0160381908
  Contains definitions and research on case management and substance abuse, with applications of the concept to substance abuse treatment, intervention, special populations, linkages to health care providers (or other relevant agencies), and criminal justice. Program models are provided. Graphs, charts, and reading lists are given with each essay. *NIDA Research Monograph,* no. 127.

Baer, John Samuel, G. Alan Marlatt, and Robert J. McMahon, eds.
Addictive Behaviors Across the Life Span
Newbury Park, Calif., Sage Publications, 1993. 358p.
ISBN 0803950780
  An in-depth study of the multidimensional approach to addiction. Thirteen chapters comprise four parts: the etiology of alcoholism; prevention models, including community based programs; treatment programs; and policy. Features treatment issues for specific populations, ages, and genders. Extensive reference lists are included at the end of each chapter. Authors are well-recognized experts in their fields, from academic and treatment backgrounds.

Beasley, Joseph D.
Diagnosing and Managing Chemical Dependency
Dallas, Essential Medical Information Systems, 1990. 273p.
ISBN 092924012X
  This handy pocket guide for the treatment professional contains a multitude of facts and dosages. Features several tables that summarize information such as "examples of alcohol and drugs that do not mix" and "maternal drugs and neonatal consequences." All aspects of substance abuse diagnosis and treatment are summarized. Includes references at the end of each chapter and a directory of support groups and national organizations.

Beattie, Melody
Codependent's Guide to the Twelve Steps
New York, Prentice Hall, 1990. 273p. ISBN 0131400541
  A supportive, shared-experience type of book that details the meaning of each of the Twelve Steps, and how they apply to recovery from a variety of behavioral problems including substance abuse, overeating, gambling, and sex addiction. Includes directories, a book list, and a glossary.

Beck, Aaron T., and others
Cognitive Therapy of Substance Abuse
New York, Guilford Press, 1993. 354p. ISBN 0898621151
  This handbook for counselors and therapists provides a case study approach to methods of diagnosis and treatment techniques. Through an accessible, explanatory approach, the authors review concepts of DSM-III-R and current research and theory. Appendices include questionnaires on beliefs about substance abuse and craving, and a relapse prediction scale. Also appended is a form for "Daily Record of Cravings" and a "Patient's Report of Therapy Session." Includes references and index. Illustrated with graphs and charts.

Bennett, Gerald, and Donna S. Woolf, eds.
Substance Abuse: Pharmacologic, Developmental, and Clinical Perspectives
2nd ed. Albany, N.Y., Delmar Publishers, 1991. 352p. ISBN 0827342055

This is a standard text for nursing students, but applicable to any health professional who may encounter patients with substance abuse symptoms. This edition is updated with information on AIDS, children of alcoholics, eating disorders, drug screening, and addictions nursing. Appendices include a directory of a wide variety of organizations; the psychoactive substance use disorders from DSM-III-R; and a position statement of the National Nurses Society on Addictions on "the role of the nurse in alcoholism." An extensive index is provided.

Bianco, David P., ed.
Professional and Occupational Licensing Directory: A Descriptive Guide to State and Federal Licensing, Registration, and Certification Requirements
Detroit, Gale Research, 1993. 1304p. ISBN 0810388944

Contains descriptions of career fields and state-by-state licensing and certification requirements. Descriptions include directory information on the credentialing agency, duration and requirements of license and certificate, type of exam required, fees, governing statute (if applicable), and reciprocity. Substance abuse counselors are listed under "Human Service Workers." Includes indexes.

Black, Claudia
Double Duty: Dual Dynamics Within the Chemically Dependent Home
New York, Ballantine Books, 1990. 587p. ISBN 0345361520

Black explores the concept of "double duty": the experience of Adult Children of Alcoholics (ACOA) or of other drug dependent parents, who also have suffered physical or sexual abuse. In addition, she examines the "dual identity ACOA," i.e., those who are from at-risk ethnic groups, homosexual, physically disabled, have eating disorders, or are self-medicating. Detailed case studies written in a conversational style are presented in each section. Includes a bibliography and an index.

Blum, Kenneth, and James E. Payne
Alcohol and the Addictive Brain: New Hope for Alcoholics from Biogenetic Research
New York, The Free Press/MacMillan, Inc., 1991. 320p. ISBN 0029037018

The authors provide a brief historical review, summarize treatment approaches, and then present a detailed, explanatory discussion of the neurochemical factors in addiction and possible future "cures." This work is directed toward a general audience, although it is fairly complex in presentation. Includes diagrams, suggested readings, notes, and a glossary. Indexed.

Bradshaw, John
Bradshaw On—The Family: A Revolutionary Way of Self-Discovery
Pompano, Fla., Health Communications, Inc., 1988. 242p. ISBN 0932194540

Bradshaw expands on his PBS TV series and provides insight into family systems theory and the origins of dysfunctional family systems (including compulsive/addictive behavior, co-dependence, and sexual/physical abuse). He refers to the work of many others in the fields of psychology, sociology, philosophy, theology, and addictions theory, but maintains a very readable, easy to understand style. Includes brief case studies, summary charts, and diagrams. A list of recommended books is appended.

Bradshaw, John
Homecoming: Reclaiming and Championing Your Inner Child
New York, Bantam, 1990. 288p. ISBN 0553057936

Bradshaw, popular author of *Healing the Shame That Binds You* (Health Communications, 1988), again draws on many sources to teach readers to self-heal with behavioral techniques. He provides a cogent, cohesive combination of deterministic theories in a very readable, conversational style. Several charts and cartoon-like drawings illustrate this book. Exercises are provided for practicing various techniques, and a reading list is included.

Brandler, Sondra, and Camille P. Roman
Group Work: Skills and Strategies for Effective Interventions
New York, Haworth Press, 1991. 256p. ISBN 0866568905

Brandler and Roman, experienced social work professionals, outline the fundamental skills necessary for working with groups. They then provide case studies to demonstrate particular group work techniques. Includes numerous charts, a bibliography, and an index.

Breggin, Peter Roger
Toxic Psychiatry: Why Therapy, Empathy, and Love Must Replace the Drugs, Electroshock, and Biochemical Theories of the "New Psychiatry"
New York, St. Martin's Press, 1991. 464p. ISBN 0312059752

This is a challenging, clearly written, and well-reasoned polemic against defining depression and substance abuse in terms of disease. Breggin challenges the "medical" approach of psychiatrists to these problems and he condemns the "psycho-pharmaceutical complex" for victimizing consumers. The long-term effects and side effects of psychopharmaceutical drugs are presented. Includes a detailed index, a reading list, and a directory of organizations.

Burns, Elizabeth M., Arlene Thompson, and Janet Kiplinger Ciccone, eds.
An Addictions Curriculum for Nurses and Other Helping Professionals
New York, Springer Publishers, 1993. 276p. ISBN 0826181902

This first volume of a two-volume set provides comprehensive coverage and an excellent summary of all salient issues, at a level appropriate for undergraduates. This guide also provides a section for the training of faculty. Clearly outlined information and detailed diagrams and charts supplement the text. Includes a glossary, an index, and a bibliography that lists books, articles, audiovisual materials, and organizations. *Springer Series on the Teaching of Nursing*, 14.

Cahalan, Don
An Ounce of Prevention: Strategies for Solving Tobacco, Alcohol and Drug Problems
San Francisco, Jossey-Bass, 1991. 290p. ISBN 1555423485

Directed toward health care administrators and public policy makers, Calahan examines substance abuse as a health care issue from a public policy perspective. He then provides recommendations for designing prevention programs. Includes numerous statistics, references, and name and subject indexes.

Cermak, Timmen L.
A Time to Heal: The Road to Recovery for Adult Children of Alcoholics
New York, J.P. Tarcher, 1988. 227p. ISBN 0874774543

Writing with conviction and clarity, Cermak shares his personal experience and professional expertise. He discusses the need for Adult Children of Alcoholics (ACOA) to help individuals recognize how they are affected by their experience in an alcoholic family. He examines post-traumatic stress disorder and co-dependency as elements of the ACOA experience. He also outlines the value of Twelve Step programs for recovery. Included are a suggested reading list and an appendix containing "Diagnostic criteria for co-dependence."

Clark, Walter B., and Michael E. Hilton, eds.
Alcohol in America: Drinking Practices and Problems
SUNY Series in New Social Studies On Alcohol and Drugs, Albany, N.Y., State Univ. of New York Press, 1991. 380p. ISBN 0791406954

This comparative study of adults in the United States, based on surveys from 1964 through 1984, analyzes overall trends in reported attitudes, consumption, and dependency problems, as well as regional differences and specific populations (Black and Hispanic). Includes statistical tables, a references list, and an index.

Cocores, James
The 800-COCAINE Book of Drug and Alcohol Recovery
New York, Villard Books, 1990. 255p. ISBN 0394574044

Cocores' book is designed for narcotic addicts and alcoholics, with a Twelve Step recovery model. This guide, written in a conversational style, with brief case studies for illustration, is based on information derived from callers to the national 800-COCAINE hotline. Outpatient recovery center treatment programs are explained in some detail. Includes an extensive list of sources and an annotated list of recommended readings.

Cohen, Sidney
The Chemical Brain: The Neurochemistry of Addictive Disorders
Irvine, Calif., Care Institute, 1988. 132p. ISBN 0917877010

Cohen begins this illuminating book with the basic function of neurons and gradually progresses to an explanation of the neurochemistry of addiction and detoxification. He also describes various technologies available to examine brain functions. Includes a glossary and a list of additional readings.

Collins, R. Lorraine, Kenneth E. Leonard, and John S. Searles, eds.
Alcohol and the Family: Research and Clinical Perspectives
New York, Guilford Press, 1990. 386p. ISBN 0898621690

Fourteen scholarly essays review the current research on alcoholism including genetics, family processes, and family- oriented treatment. Each chapter features an in-depth literature survey and numerous illustrations.

Daley, Dennis C., Howard Moss, and Frances Campbell
Dual Disorders: Counseling Clients with Chemical Dependency and Mental Illness
2nd ed. Center City, Minn., Hazelden, 1993. 237p.
ISBN 0894864491

This practical guide for mental health and substance abuse counseling professionals focuses mainly on diagnosis and counseling approaches in dual disorders, utilizing the *Diagnostic and Statistical Manual of Mental Disorders* (DSM-III-R) and state-of-the-art research. Updated information includes new chapters on relapse, and on developing dual diagnosis treatment programs. Coverage includes personality (antisocial and borderline); bipolar, anxiety, and organic mental disorders; depression; and schizophrenia. Addresses a broad range of chemical dependencies. Appended selective resources list identifies books, organizations, and videos.

Daley, Dennis C., and Miriam S. Raskin, eds.
Treating the Chemically Dependent and Their Families
Newbury Park, Calif., Sage Publications, 1991. 220p.
ISBN 0803932979

Provides an overview of current views on addiction including theoretical background and information on assessment and relapse prevention. Other aspects of addiction are defined, such as effects on the addict and the family, intervention, and recovery. Issues are addressed by various treatment professionals and educators. Conclusions are based on research literature and an overall practical approach is emphasized. Includes tables and references. *Sage Sourcebooks for the Human Services Series*, 16.

Dorris, Michael
The Broken Cord
New York, Harper and Row, 1989. 300p. ISBN 0060160713

This is the story of how Dorris adopted Adam, a child who exhibited the perplexing symptoms of fetal alcohol syndrome. This excellent case study details the process of diagnosis and learning to cope. Extensive notes, a bibliography, and filmography are provided.

Drug, Alcohol and Other Addictions: A Directory of Treatment Centers and Prevention Programs Nationwide
2nd ed. Phoenix, Ariz., Oryx Press, 1993. 646p.
ISBN 0897746236

Describes over 12,000 programs and facilities in the United States for the prevention and treatment of alcohol, drug, and other behavioral addictions. Each entry includes treatment methods, programs available, setting (resident or outpatient), statistics, payment method accepted, and accommodations (weekend and evening hours, wheelchair accessible, child care provided).

Dulfano, Celia
Families, Alcoholism and Recovery
Rev. ed. San Francisco, Jossey-Bass, 1992. 167p.
ISBN 1555424848

With significant experience in the field of family therapy and alcoholism treatment, Dulfano presents a practical and understandable family systems approach for recovery. Case studies are provided for illustration. Includes suggested readings, references, and index. *The Jossey-Bass Social and Behavioral Science Series*.

Ettorre, E. M.
Women and Substance Use
New Brunswick, N.J., Rutgers Univ. Press, 1992. 204p.
ISBN 0813518636

Ettorre presents a thoughtful feminist critique of the underlying assumptions made in the prevailingly male dominated, middle-class view of addictions treatment. The focus is on alcohol, tranquilizers, heroin, tobacco, and food. Includes an extensive bibliography and an index.

Evans, Katie, and J. Michael Sullivan
Dual Diagnosis: Counseling the Mentally Ill Substance Abuser
New York, Guilford Press, 1990. 191p. ISBN 0898624363

This text outlines diagnosis, treatment, aftercare, and relapse prevention. Special attention is given to the specifics of treating adolescents. Although the book is written primarily for a professional audience, the authors present concepts with a clarity and simplicity easily understandable to a novice. Appendices include "Modified Stepwork," "School Behavior Checklist," "Checklist for Parents," and "Home Behavior Contract." Includes references, a glossary, and an index.

Fassel, Diane
Working Ourselves to Death: The High Cost of Workaholism, the Rewards of Recovery
San Francisco, Harper, 1990. 164p. ISBN 0062548697

This text follows up on concepts developed in Anne Scheef's *The Addictive Organization* (Harper and Row, 1988.) Fassel traces the development and process of recovery from what she defines as a "killer disease." This is a clear application of the disease model of addiction, written in a readable style.

Fingarette, Herbert
Heavy Drinking: The Myth of Alcoholism as a Disease
Berkeley, Calif., Univ. of California Press, 1988. 166p.
ISBN 0520062906

Fingarette argues that research does not support the simplistic "disease concept" of alcoholism. He calls for a broader view that acknowledges heavy drinking as a part of a drinker's way of life, and treatment options that consider the drinker's environment. Includes an extensive bibliography and an index.

Finnegan, John, and Daphne Gray
Recovery from Addiction: A Comprehensive Understanding of Substance Abuse with Nutritional Therapies for Recovering Addicts and Co-Dependents
Berkeley, Calif., Celestial Arts, 1990. 246p. ISBN 0890875995

With the caveat that readers should also seek the advice of a physician, the authors explain nutritional and medical therapies to correct biochemical disorders that underlie substance abuse disorders. Describes the use of medicinal herbs, vitamin supplements, and whole food extracts. Recipes for a nutritional diet are included. A directory of national organizations and support groups is included in the appendix. An annotated "recommended reading list" and a bibliography are provided.

Fleming, Michael F., and Kristen Lawton Barry, eds.
Addictive Disorders
St. Louis, Mo., Mosby-Yearbook, 1992. 439p. ISBN 0815133693

This practical, up-to-date guide for clinicians, presented in an outline format, summarizes pertinent information regarding the treatment of substance abuse, gambling, and eating disorders. Appendices include questionnaires to assist in screening patients for addictive behavior, a list of drug street names, and criteria for mental health disorders. References, a glossary, an index, and a resources list of organizations, books, and videos are provided.

Flynn, John C.
Cocaine: An In-Depth Look at the Facts, Science, History and Future of the World's Most Addictive Drug
New York, Carol Publishing Group, 1991. 167p.
ISBN 1559720603

The author takes a primarily historical view of the development of cocaine as an illicit substance. Statistics cited are as current as 1989. Includes numerous diagrams and tables.

Freeman, Edith M., ed.
Substance Abuse Treatment: A Family Systems Perspective
Newbury Park, Calif., Sage Publications, 1993. 316p.
ISBN 0803948891

Freeman presents essays from academic researchers and therapists that are comprehensive enough to meet the needs of graduate study, yet readable enough for a lower level of study. Covers family life-span issues, including the "family systems framework" that is broad and inclusive, allowing for variance in culture, economic level, gender, age, sexual orientation, and ethnicity. Includes case studies, tables, checklists, references, and indexes. *Sage Sourcebooks for the Human Services Series*, 25.

Galanter, Marc
Network Therapy for Alcohol and Drug Abuse: A New Approach in Practice
New York, BasicBooks, 1993. 225p. ISBN 0465000991

Galanter defines network therapy as a rehabilitation technique in which selected family members and friends provide active, ongoing support to promote attitude change. They become part of the therapist's working team. This volume examines addiction theory and psychopathology, and provides a thoroughly integrated approach to addictions treatment and relapse prevention. A very readable, innovative guide with many case studies for illustration. Includes a bibliography and an index.

Galanti, Geri-Ann
Caring for Patients from Different Cultures: Case Studies from American Hospitals
Philadelphia, Univ. of Pennsylvania Press, 1991. 138p.
ISBN 0812230655

Though focused on medical care, this intriguing book has broad application for counselors who deal with patients from cultures, ethnic backgrounds, or religions different from their own. The author takes an anthropological view of patients' reactions by discussing communication styles, folk medicine, religions, and cultural history. Extensive bibliography and an index.

Gelenberg, Alan J., Ellen L. Bassuk, and Stephen C. Schoonover, eds.
The Practitioner's Guide to Psychoactive Drugs
3rd ed. New York, Plenum Medical Book Co., 1991. 504p.
ISBN 030643461X

This is a comprehensive, encyclopedic guide for clinicians, updated with information on treating borderline personality and eating disorders. The chapter "Psychoactive Substance Use Disorders" covers opioids, sedatives, alcohol, stimulants (cocaine, amphetamines), hallucinogens, phencyclidine, inhalants, designer drugs (MDMA, MDEA), and marijuana. Pharmacology, patterns of abuse, tolerance, acute intoxication, withdrawal, and prenatal effects are noted for each. Includes tables, inset text, references in each chapter, and an index.

Glantz, Meyer, and Roy W. Pickens, eds.
Vulnerability to Drug Abuse
Washington, D. C., American Psychological Association, 1992. 533p. ISBN 1557981426

A collection of sixteen essays examines the etiology and psychology of substance abuse, evaluating risk factors as well as protective factors for individuals. Based on research presented at the "Transition from Drug Use to Abuse/Dependence" conference (1989, National Institute on Drug Abuse, and the American Psychological Association). Tables, graphs, and bibliographies are provided in each chapter. Includes author and subject indexes.

Gold, Mark S., and Andrew Edmund Slaby, eds.
Dual Diagnosis in Substance Abuse
New York, Marcel Dekker, 1991. 341p. ISBN 082478457X

This collection of essays by treatment professionals reviews current literature on diagnosing and treating dual diagnosis patients, including adolescents. Introductory essays by Andrew E. Slaby and William Coryell present clear arguments for acknowledging concurrent psychiatric disorders and substance abuse, and possible genetic links in transmission.

Goldstein, Avram
Addiction: From Biology to Drug Policy
New York, W. H. Freeman and Company, 1994. 321p. ISBN 0716723840

Provides an excellent introductory text on the biological basis of addiction, with an engaging discussion of brain chemistry. Goldstein's writing is clear and explanatory. While carefully delineating key scientific concepts he also provides insight into the process of research and discovery in neurochemistry. Drugs discussed include nicotine, alcohol (with benzodiazepines and volatile solvents), opiates, cocaine and amphetamines, cannabis, caffeine, and hallucinogens. The author generally accepts the disease model of addiction and calls for public policy to consider addiction as a public health problem, arguing for the "harm reduction model" for treatment programs. Appended is an annotated list of suggested readings, with items designated as "technical" or "nontechnical." Graphs, diagrams, and detailed index.

Goode, Erich
Drugs in American Society
4th ed. New York, McGraw-Hill, 1993. 434p. ISBN 0070239231

This sociological overview of drug use in the United States focuses on recent developments within a historical perspective. Goode clearly defines the "social construct" in the public perception of substance abuse, and debunks common fallacies. Coverage also includes detailed chapters on the effects of particular classes of drugs. A lengthy reference list, and name and subject indexes are provided.

Gilman, Alfred Goodman, and others, eds.
Goodman and Gilman's The Pharmacological Basis of Therapeutics
8th ed. New York, Pergamon Press, 1990. 1811p. ISBN 0080402968

This medical standard covers pharmacokinetics, pharmacodynamics, toxicology, and therapeutics, and provides extremely detailed information and instruction on dosage, interaction, tolerance, and cross tolerance. Jerome Jaffe authors the chapter "Drug Addiction and Drug Abuse." Includes diagrams and references throughout.

Grinspoon, Lester, and James B. Bakalar
Marihuana: The Forbidden Medicine
New Haven, Conn., Yale Univ. Press, 1993. 184p. ISBN 0300054351

Through a skillful combination of research studies and personal testimonies, the authors present a convincing case for utilizing marihuana in treating various conditions. Some of the conditions considered are cancer, glaucoma, multiple sclerosis, AIDS, and epilepsy. The authors do acknowledge possible risks with use, such as a slight chance of addiction (less than tobacco or alcohol, they claim); impaired short-term memory and coordination; and damage to the respiratory system. Includes footnotes and an index.

Hamstra, Bruce
How Therapists Diagnose: Seeing Through the Psychiatric Eye: Professional Secrets You Deserve to Know—And How They Affect You and Your Family
New York, St. Martin's Press, 1994. 334p. ISBN 0312104766

Hamstra argues that it is only fair for consumers to understand the various assessment procedures and diagnostic categories used by mental health professionals, in much the same way that consumers demand to know the meaning and implications of medical diagnostic measures. He opens with a clear warning that his text is not meant for use in self-diagnosis. Rather, his intent is to explain some of the major criteria of the DSM and Mental Status Examination. Chapters on specific diagnostic categories (such as mood, anxiety, eating, personality, cognitive, and substance related disorders) summarize the major features and additional symptoms of each disorder, and include a case example. The appendix includes an annotated directory of mental health associations, referral sources, and self-help groups. Glossary and index.

Heinemann, Allen W., ed.
Substance Abuse and Physical Disability
New York, Haworth Press, 1993. 289p. ISBN 1560242906

These essays by medical and psychiatric clinicians, substance abuse workers, and nursing professionals cover the sociological context of the disabled in relation to substance abuse; the causes and prevalence of substance abuse in this particular population; and treatment and prevention. Alcohol and prescription medication (pain control) are the primary focus. Diagrams, graphs, and an extensive index are provided.

Henricks, Lorraine
Kids Who Do, Kids Who Don't: A Parents Guide to Teens and Drugs
Summit, N.J., PIA Press, 1989. 136p. ISBN 0929162110

An introduction to the signs of addiction and treatment options for adolescents and teens, written in a supportive tone. Provides guidelines for confronting a teenaged substance abuser, with guidance for developing communication skills.

Herlihy, Barbara, and Larry B. Golden, eds.
AACD Ethical Standards Casebook
4th ed. Alexandria, Va., American Association for Counseling and Development, 1990. 233p. ISBN 1556200692

Presents the revised Ethical Standards (adopted in 1988) of the

American Association for Counseling and Development. Each standard is illustrated with several examples. Includes case studies of contemporary, complex situations; essays on emerging ethical issues are also included. Written with clarity and full appreciation of the subtleties of complicated ethical challenges. Appendices include ethics statements from various groups.

Hubbard, Robert, and others
Drug Abuse Treatment: A National Study of Effectiveness
Chapel Hill, N.C., Univ. of North Carolina Press, 1989. 213p.
ISBN 080781864X

The authors review and summarize the results of the national Treatment Outcome Prospective Study (1979-1981), and compare them to data from previous national studies. Includes statistical tables, an extensive bibliography, and an index.

Hutchings, Donald E., ed.
Prenatal Abuse of Licit and Illicit Drugs
New York, The New York Academy of Sciences, 1989. 388p.
ISBN 0897665228

This is a collection of essays and poster papers from a 1988 conference by the Behavioral Teratology Society, the National Institute on Drug Abuse (NIDA), and the New York Academy of Sciences. Covers a broad range of contributors and major drugs (licit and illicit). Issues include prenatal exposure to alcohol, smoking and passive smoke, opioids (methadone), cannabinoids, caffeine, cocaine, amphetamines, and phencyclidine. History and epidemiological issues are addressed. Several essays examine the scientific basis of developmental toxicology, focusing on animal studies. *Annals of the New York Academy of Sciences,* 562.

Inaba, Darryl S., and William E. Cohen
Uppers, Downers and All Arounders: Physical and Mental Effects of Drugs of Abuse
Ashland, Ore., Cinemed, 1990. 263p. ISBN 0926544004

This is an easy-to-read guidebook on the effects of psychoactive drugs and the process of addiction, with a brief overview of treatment options. Questions and suggestions for further study at the end of each section could be useful for discussion group design. A detailed index, a brief bibliography and filmography, and numerous illustrations supplement the text.

Inciardi, James A., Frank M. Tims, and Bennett W. Fletcher, eds.
Innovative Approaches in the Treatment of Drug Abuse: Program Models and Strategies
Westport, Conn., Greenwood Press, 1993. 242p.
ISBN 0313284229

Describes fifteen projects funded by the National Institute on Drug Abuse to test innovative treatment approaches in community programs. Program models include residential, case management, and multi-modality; contributors are from a variety of professional backgrounds including clinicians, social workers, program administrators, and academic researchers. The evaluation of particular strategies provides a clear view of the substance abuse treatment field—what works, and what the future may hold. The editors state in the introduction that this volume is the first in a series, with the second to cover program implementation issues and problems. The third volume will address treatment outcome information. Includes references and index. *Contributions in Criminology and Penology,* 39.

Inciardi, James A.
The War on Drugs II: The Continuing Epic of Heroin, Cocaine, Crack, Crime, AIDS, and Public Policy
Mountain View, Calif., Mayfield Pub. Co., 1992. 315p.
ISBN 1559340169

Written in a conversational tone, Inciardi traces the history of drug use in the U.S. from the late 1800s through the late 1980s, including sociological, geographic, and economic perspectives. Each chapter features extensive notes and source lists. Includes maps, statistical charts, and black-and-white illustrations. Appendices provide information on drug terms and concepts and the "scheduling provisions" of the Federal Controlled Substances Act.

Julien, Robert M.
A Primer of Drug Action: A Concise, Nontechnical Guide to the Actions, Uses and Side Effects of Psychoactive Drugs
6th ed. New York, W. H. Freeman, 1992. 434p.
ISBN 0716722615

In generally nontechnical language the author provides a broad overview of psychopharmacology. This edition of the standard text expands coverage of benzodiazepines, caffeine, and nicotine. It includes new material on the fetal effects of psychoactive drugs, and the neurochemical basis of alcoholism and specific mental disorders. Study questions conclude each chapter. Illustrations, graphs, and references complement the text.

Kaminer, Wendy
I'm Dysfunctional, You're Dysfunctional: The Recovery Movement and Other Self-Help Fashions
Reading, Mass., Addison-Wesley, 1992. 180p. ISBN 0201570629

Kaminer rejects the religiosity she finds in Twelve Step programs, and the bipolar view often presented in the recovery movement (i.e., "co-dependent/in denial"). The author finds self-help books fatuous and despises the spectacle of testifying to one's problems on television. She indicts the "positive thinking" movements for promoting conflict avoidance in the place of critical thinking and attention to social relations. Largely a critique, with few solutions offered.

Katz, Stan J., and Aimee E. Liu
The Codependency Conspiracy: How to Break the Recovery Habit and Take Charge of Your Life
New York, Warner Books, 1991. 233p. ISBN 0446515957

Katz and Liu carefully differentiate mutual support groups (which can offer empathy and support without proscribing behavior), from self-help groups (which can demand rigid adherence to a set of beliefs.) The authors protest the proliferation of self-help groups that offer little more than victimization by making members dependent on the group and not allowing members to really change. They then describe a recovery process that stresses individual responsibility and independence.

Kaufman, Edward, and Pauline Kaufmann, eds.
Family Therapy of Drug and Alcohol Abuse
2nd ed. Boston, Allyn and Bacon, 1992. 319p. ISBN 0205134300

Updated and revised, this collection by family treatment professionals now features essays on integrating treatment modalities, stepfamily issues, and use of a Twelve Step model. An opening chapter by Salvador Minuchin concisely and clearly outlines the family-as-system model, and the role of therapist in joining, then changing, the transactional patterns of the family. Bibliographies are provided for each chapter.

Khantzian, Edward J., Kurt S. Halliday, and William E. McAuliffe
Addiction and the Vulnerable Self: Modified Dynamic Group Therapy for Substance Abusers
New York, Guilford Press, 1990. 176p. ISBN 0898621720

This manual describes the practice of modified dynamic group therapy (MDGT) and is directed to clinicians with professional background in psychodynamic theory. Features case vignettes and provides an extensive bibliography and index. The authors' approach was developed from their work in the Harvard Cocaine Recovery Project.

Kinney, Jean, and Gwen Leaton
Loosening the Grip: A Handbook of Alcohol Information
4th ed. St. Louis, Mosby Yearbook, 1991. 411p.
ISBN 0801627699

This work is well organized and well illustrated, often with humorous (sometimes poignant) cartoon drawings. In addition to the authors, several other clinicians and treatment professionals contribute chapters. Provides a detailed but readable introduction to all aspects of alcohol dependence, evaluation, and treatment. Includes a section on professional ethics, resource lists, and a detailed index.

Kitchens, James A.
Understanding and Treating Codependence
Englewood Cliffs, N.J., Prentice Hall, 1991. 212p.
ISBN 0139334823

An academic approach to the study of adult children of dysfunctional families and co-dependence. This guidebook for therapists is presented with the caveat that this is a field of study still in the process of development. Diagnostic techniques as well as group, family, and individual therapy techniques are discussed. Provides a brief, but authoritative historical overview augmented by diagrams, brief reference lists, and an index.

Kominars, Sheppard B.
Accepting Ourselves: The Twelve-Step Journey of Recovery from Addiction for Gay Men and Lesbians
San Francisco, Harper and Row, 1989. 186p. ISBN 0062504940

Writing with the conviction of someone who has been there—through "coming out" and beginning a recovery process from addiction—Kominars presents a very supportive and informative self-help guide. He works through each of the Twelve Steps, providing examples relevant to a gay and lesbian audience. Checklists and lists of resources are included in an appendix.

Kramer, Peter D.
Listening to Prozac
New York, Viking, 1993. 409p. ISBN 0670841838

Kramer sees Prozac (fluoxetine) as more than merely a new antidepressant; he asserts that this drug is capable of actually transforming patients' views of themselves. The author does not address specific uses in treating major mental illness. An appended chapter briefly acknowledges some of the controversial issues surrounding the use of this drug (such as tendencies toward violent and suicidal behavior). Extensive notes and an index are provided.

L'Abate, Luciano, Jack E. Farrar, and Daniel A. Serritella, eds.
Handbook of Differential Treatments for Addictions
Boston, Allyn and Bacon, 1992. 337p. ISBN 0205132375

This well-organized textbook defines addiction in the very broadest sense. Coverage includes addictive behaviors such as substance abuse, domestic violence, eating disorders, co-dependency, and religious fanaticism. Chapters are written by a variety of therapists and academic researchers. Each offers definitions of fundamental concepts of assessing and treating particular addictive behaviors, coordinated with a current literature review. Each topic is presented within the context of differentiating the characteristics of each individual case and each type of addiction. Includes checklists and name and subject indexes.

Lawson, Gary W., and Ann W. Lawson, eds.
Alcoholism and Substance Abuse in Special Populations
Rockville, Md., Aspen Publishers, 1989. 370p. ISBN 0834200074

A collection of practically oriented essays by treatment workers, most with a behavioral approach. Broad coverage includes dual-diagnosis, women, elderly, disabled, Black, Hispanic, Native American, homosexual, sexually abused, and poor populations. Also included are essays on substance-abusing physicians, athletes, and military personnel. Adult children of alcoholics are also discussed. Each chapter includes bibliographies.

Lecca, Pedro J., and Thomas D. Watts
Preschoolers and Substance Abuse: Strategies for Prevention and Intervention
New York, Haworth Press, 1993. 112p. ISBN 1560242345

This slim volume reviews the current literature and programs in substance abuse, and recommends public policy and prevention methods. Includes an extensive references list and a bibliography of bibliographies. "State Prevention Contacts" and "State Authorities" directories are appended.

Leukefeld, Carl G., and Frank M. Tims, eds.
Drug Abuse Treatment in Prisons and Jails
Rockville, Md., National Institute on Drug Abuse, 1992. 299p.
ISBN 0160361559

Based on a technical review held in May, 1990, sponsored by the National Institute on Drug Abuse, these essays evaluate program models in various state and federal prison settings, including juvenile justice. The editors point out the need for substance abuse rehabilitation, pleading for more consistent support and increased funding. Authors of the essays are mental health and corrections professionals, and academic researchers. Tables and references are found throughout the book. *NIDA Research Monograph*, no. 118.

Lewis, Judith A., Robert Q. Dana, and Gregory A. Blevins
Substance Abuse Counseling: An Individualized Approach
2nd ed. Pacific Grove, Calif., Brooks/Cole Pub. Co., 1994. 287p.
ISBN 0534200532

This text may be used by counselors of all types who may at some point need to deal with substance abusing clients. In this clear, easy-to-understand work, the authors provide a comprehensive overview of the field including definitions of terms, various counseling techniques, abuse and relapse prevention, and program planning and evaluation. Includes name and subject indexes. Appendices include checklists and questionnaires.

Light, William J. Haugen
Alcoholism and Women, Genetics, and Fetal Development
Springfield, Ill., Charles C. Thomas, 1988. 170p.
ISBN 0398053995

Light's technical volume is intended for health care professionals and clinical therapists. Provides a comprehensive literature review, with an emphasis on the genetic basis of, and subsequent fetal development disorders due to, maternal alcoholism. The chapter on fetal alcohol syndrome is extremely detailed. Appended chapters include basic concepts in genetics and central nervous system development. Numerous diagrams and tables are provided.

Lin, Geraline C., and Lynda Erinoff, eds.
Anabolic Steroid Abuse
Rockville, Md., National Institute on Drug Abuse, 1990. 249p.

This collection of papers presented at a technical review held in 1989 addresses steroid topics such as history, illicit use, health risks, dependence, and mental and behavioral changes. Includes diagrams, references, and tables. *Research Monograph,* 102; *DHHS Publication* (ADM) 91-1720.

Lovern, John D.
Pathways to Reality: Erickson-Inspired Treatment Approaches to Chemical Dependency
New York, Brunner/Mazel, 1991. 228p. ISBN 0876306334

Using his professional experience, Lovern demonstrates the application of Milton H. Erickson's psychotherapeutic techniques in treating substance abuse. Group and family therapy are covered in some detail. Includes a bibliography and an index.

Lowinson, Joyce H., Pedro Ruiz, and Robert B. Millman, eds.
Substance Abuse: A Comprehensive Textbook
2nd ed. Baltimore, Williams and Wilkins, 1992. 1110p.
ISBN 068305211X

This comprehensive, authoritative text provides chapters written by well-recognized clinicians and researchers in the field of substance abuse prevention, intervention, and treatment. This edition is updated with more information on the determinants of dependence, as well as new chapters on eating disorders, anabolic steroids, HIV, and co-dependence. Needs of special populations are evaluated, as well as public policy. Graphs, charts, and extensive reference lists are provided.

Marshall, Shelly
Teenage Addicts Can Recover: Treating the Addict, Not the Age
Littleton, Colo., Gylantic Pub. Co., 1992. 128p.
ISBN 1880197022

The author speaks with experience as a therapist, a child of alcoholic parents, and a recovering alcoholic. Her approach is forthright, practical, and soundly based in substance abuse research. She supports the disease concept of alcoholism and "mainstreaming" (adolescents treated with adults), and rejects the common dismissive view of relapse in adolescent patients. Includes diagrams, checklists, and references throughout. Appended studies include "Treatment Paternalism in Chemical Dependency Counselors," "Adolescent Recovery Rate," and "Homogeneous Versus Heterogeneous Age Group Treatment of Adolescent Substance Abusers."

Maternal Drug Abuse and Drug Exposed Children: Understanding the Problem
Washington, D.C., U.S. Department of Health and Human Services, 1992. 82p.

This work was a project of the Substance Abusing Women and Their Children Subcommittee of the Department of Health and Human Service's Ad Hoc Drug Policy Group. Summarizes recent survey studies and current research and outlines current efforts in governmental research, prevention, and treatment. An appendix provides summaries of case reports on maternal drug abuse and reference lists (primarily journal articles and government documents). Material presented is stated to be in the public domain and not subject to copyright.

McGrath, Ellen
When Feeling Bad Is Good
New York, Henry Holt, 1992. 350p. ISBN 0805014748

Based on her personal and professional experience, the author relates her theory of various levels of depression, from "good" to "bad." This self-help manual is an insightful and readable text, focused primarily on issues relevant to women and the particularities of their experience in relation to depressive states. Suggests when and how to seek professional intervention. Includes diagrams, drawings, checklists, notes, and an index.

Mellody, Pia, Andrea Wells Miller, and Keith Miller
Facing Codependence: What It Is, Where It Comes From, How It Sabotages Our Lives
New York, HarperSan Francisco, 1989. 222p. ISBN 0062505890

Speaking from her own experience, the author charts the development of her understanding of the co-dependent personality and progress towards recovery. She traces the roots of co-dependence to various forms of childhood abuse and asserts that co-dependence is a disease requiring progressive treatment, such as a Twelve Step program and professional counseling.

Milkman, Harvey B., and Stanley G. Sunderwirth
Pathways to Pleasure: The Consciousness and Chemistry of Optimal Living
New York, Lexington Books, 1993. 302p. ISBN 0029212731

The authors present a positive yet provocative view of enhancing pleasure in life by pursuing a "natural high." They clearly delineate how to pursue this in order to avoid addictive behaviors. Based on current research in neurobiology. A "Personal Pleasure Inventory" is included, along with detailed diagrams and tables.

Milkman, Harvey B., and Lloyd I. Sederer, eds.
Treatment Choices for Alcoholism and Substance Abuse
Lexington, Mass., Lexington Books, 1990. 395p.
ISBN 0669200190

Essays from a 1988 conference sponsored by the Alcohol and Drug Abuse Division, Colorado Department of Health. Thirty-six chapters cover a wide range of perspectives. Contributors include academic researchers, substance abuse counselors, health care providers, psychiatric clinicians, and law enforcement personnel.

Miller, Norman S., and Mark S. Gold
Alcohol
New York, Plenum Medical Book Company, 1991. 275p.
ISBN 0306436418

A good summary of current research and psychological theory on alcoholism. References and tables support the text. Indexed. *Drugs of Abuse: A Comprehensive Series for Clinicians*, 2.

Miller, William R., and Stephen Rollnick
Motivational Interviewing: Preparing People to Change Addictive Behavior
New York, Guilford Press, 1991. 348p. ISBN 0898625661

In the first two-thirds of the book the authors contrast their "client-centered" approach to motivational interviewing with the more traditional confrontational approach. The last third of the book features case studies from seventeen clinicians in the application of motivational interviewing. A bibliography and an index are included.

Minkoff, Kenneth, and Robert E. Drake, eds.
Dual Diagnosis of Major Mental Illness and Substance Disorder
San Francisco, Jossey-Bass, 1991. 113p. ISBN 1555427944

The editors present eight essays by clinicians and researchers on design components of integrated, or "hybridized," programs for the treatment of the dual diagnosis patient. Existing, innovative programs are highlighted as examples. References are included in each chapter. *New Directions for Mental Health Services*, 50.

Monti, Peter M., and others
Treating Alcohol Dependence: A Coping Skills Training Guide
New York, Guilford Press, 1989. 240p. ISBN 0898622042

Outlines the authors' "coping skills" approach to alcoholism treatment based on the "cognitive-social learning theory" of alcoholism. Provides a step-by-step training guide for clinicians and administrators of alcoholism treatment programs. "Reminder Sheets," practice exercises, a comprehensive bibliography, and an index are included.

Muisener, Philip P.
Understanding and Treating Adolescent Substance Abuse
Thousand Oaks, Calif., Sage Publications, 1994. 252p.
ISBN 0803942753

Muisener presents a biopsychosocial model for diagnosing and treating adolescent substance abusers, a group he acknowledges as heterogeneous. His book is intended for mental health professionals and substance abuse treatment professionals. It offers both theoretical frameworks and practical information regarding treatment issues. Clear and concise, this is an excellent current source. The Twelve Steps of Alcoholics Anonymous and Narcotics Anonymous are appended. References, name, and subject indexes are provided. *Sage Sourcebooks for the Human Services Series*, 27.

Nowinski, Joseph
Hungry Hearts: On Men, Intimacy, Self-Esteem, and Addiction
New York, Lexington Books, 1993. 163p. ISBN 002923221X

Asserting that statistics show men more predisposed to addiction than are women, Nowinski dismisses purely genetic factors, focusing more on social issues. He posits that men are socialized to adopt a "positional orientation" (as opposed to "relational"), and that this has an impact on men's capacity for intimacy. He examines definitions of masculinity, changing male roles, and differences in defining intimacy for men and women, and how factors relate to vulnerability to addiction. Case studies are given for illustration. Includes notes and an index.

Nowinski, Joseph
Substance Abuse in Adolescents and Young Adults: A Guide to Treatment
New York, Norton, 1990. 246p. ISBN 0393700976

Nowinski defines substance abuse within the context of adolescent development and measures levels of drug use on a continuum from experimental to compulsive. This introduction to current diagnosis and treatment models contrasts the disease model of the "recovery movement" (including Twelve Step programs) to the psychodynamic model of mental health professionals. Covers patterns and causes of substance abuse, assessment, recovery, and relapse prevention. An index is included.

Nuckols, Cardwell C.
Cocaine: From Dependency to Recovery
2nd ed. Blue Ridge Summit, Pa., TAB Books, 1989. 208p.
ISBN 0830692037

The author describes in detail the pharmacology of cocaine and presents treatment and relapse prevention strategies for cocaine dependence. This comprehensive, accessible text is based on the author's experience as a recovered addict. Designed for clinicians and counselors, but appropriate for the layperson. The appendix includes an overview of the Controlled Substances Act and the author's "Schedule C: A Clinical Self-Assessment for Cocaine Dependency." Includes a glossary, a bibliography, and an index.

Othmer, Ekkehard, and Sieglinde C. Othmer
The Clinical Interview Using DSM-III-R
Washington, D.C., American Psychiatric Press, 1989. 507p.
ISBN 0880483156

This handbook is directed toward mental health professionals who are dealing with DSM-III-R for the first time. This guide is filled with case studies and interview examples. The glossary is very detailed and the index is extensive. The text is supplemented by numerous drawings, tables, and charts.

Peele, Stanton, ed.
Visions of Addiction: Major Contemporary Perspectives on Addiction and Alcoholism
Lexington, Mass., Lexington Books, 1988. 244p.
ISBN 0669130923

Peele presents a collection of challenging and thought- provoking readings by researchers and addiction treatment specialists. Every effort has been made to present varying points of view. In the concluding chapter, Peele presents his clear opposition to the prevailing "disease model" of addiction. Includes diagrams, tables, and bibliographical notes. Indexed.

Pratsinak, George, and Robert Alexander, eds.
Understanding Substance Abuse and Treatment
Laurel, Md., American Correctional Association, 1992. 212p.
ISBN 092931073X

This manual designed for use in Federal Bureau of Prisons facilities includes lists of recommended films, books, pamphlets, and lesson plans for education purposes and for the design of treatment programs. Offers clear, concise discussion of models of addiction, overviews of drugs, and the transmission of HIV. Also includes a directory of organizations that can provide substance abuse information. The information in this manual is not copyrighted, so it may be copied and distributed as needed.

Read, Edward M., and Dennis C. Daley
Getting High and Doing Time: What's the Connection? A Recovery Guide for Alcoholics and Drug Addicts in Trouble with the Law
Washington, D.C., St. Mary's Press, 1990. 80p. ISBN 0929310314

This handy guide for use in correctional settings outlines basic facts, interspersed with brief case histories and questionnaire forms. Includes a list of readings and a directory of self-help organizations and publishers.

Reid, William H., and Michael G. Wise
DSM-III-R Training Guide
New York, Brunner/Mazel, 1989. 269p. ISBN 0876305052

Designed for use with the American Psychiatric Association's *Diagnostic and Statistical Manual of Mental Disorders* (3rd ed., rev.), this guide covers all individual diagnostic criteria and includes an appendix of relevant ICD-9-CM V Codes and E codes. Includes brief case studies for illustration. The preface recommends using the guide with videotapes and slides available in the authors' *DSM-III-R Training Program*.

Rice-Licare, Jennifer, and Katharine Delaney-McLoughlin
Cocaine Solutions: Help for Cocaine Abusers and Their Families
New York, Haworth Press, 1990. 142p. ISBN 1560240350

This slim guidebook covers essential information on adult and adolescent cocaine addiction in a straightforward, objective style. A preface is provided by self-admitted cocaine addict Thomas Henderson, a former NFL linebacker. Lists of Twelve Step programs and public information groups are appended. Checklists and questionnaires are provided for self assessment, and brief case studies are provided for illustration. *Haworth Series In Addictions Treatment*, 4.

Rivinus, Timothy M., ed.
Children of Chemically Dependent Parents: Multiperspectives from the Cutting Edge
New York, Brunner/Mazel, 1991. 364p. ISBN 0876305958

Rivinus provides a comprehensive guide to current research, theory, and public policy for the treatment and self-help of children of substance abusers (both adult and child). The fifteen essays included in this collection are by academic researchers, treatment professionals, and therapists in private practice. Chapter Four, "Children of Chemically Dependent Parents: A Theoretical Crossroads" by Stephanie Brown, outlines the divergent methods of diagnosis and treatment in the chemical dependency field, the Adult Children of Alcoholics (ACOA) social movement, and of the traditionally based mental health professionals; this chapter should be required reading. A bibliography is provided in each chapter.

Robak, Rostyslaw
A Primer for Today's Substance Abuse Counselor
New York, Lexington Books, 1991. 152p. ISBN 0669269352

This is an overview of the concepts basic to substance abuse counseling, such as personality development, group counseling techniques, behavioral approaches, biological and disease models of addiction, and the three major modalities of treatment: Twelve Step, therapeutic, and cognitive-behavioral. The author's stated intent is to ease confusion in the field by clearly defining terms, and to provide practical advice gleaned from his professional experience. Includes a brief bibliography, author, and subject indexes.

Robertson, Bozena-Eva
Alcohol Disabilities Primer: A Guide to Physical and Psychosocial Disabilities Caused by Alcohol Use
Boca Raton, Fla., CRC Press, 1993. 193p. ISBN 0849389666

Robertson provides a very readable review of current research on the effects of alcohol use, geared to a medical and counseling professional audience. All major points are covered, including fetal effects, chronic illness, nutritional deficiencies, social effects, and effects on the family. Concluding chapter contains assessment tools, and a brief review of treatment methods. Includes tables, reference lists in each chapter, and a detailed index.

Rogers, Ronald, and Chandler Scott McMillin
Freeing Someone You Love from Alcohol and Other Drugs: A Step by Step Plan Starting Today!
Los Angeles, Body Press, 1989. 273p. ISBN 0895867648

The authors credit James Milam with originating the concept of the "chronic disease model" of alcoholism (from his *Under the Influence: A Guide to the Myths and Realities of Alcoholism*, Madrona Publishers, 1981). They use this model as a core component of their approach to intervention and treatment. Written in a straightforward, factual way, this guide is designed for use by the general public. Includes numerous illustrations, a glossary, an index, and a bibliography.

Rosellini, Gayle, and Mark Worden
Barriers to Intimacy: For People Torn by Addiction
New York, Ballantine Books, 1990. 163p. ISBN 0345367359

In a thoughtful, conversational style, the authors explain how to identify and overcome the emotional blocks that develop in relationships overshadowed by substance abuse. Habit, co-dependency, self-knowledge, trust, sex, and communication skills are some of the key points discussed. Includes references and notes.

Ross, George R.
Treating Adolescent Substance Abuse: Understanding the Fundamental Elements
Boston, Allyn and Bacon, 1994. 238p. ISBN 0205152554

Provides a detailed guide written for substance abuse workers, with step-by-step directions on how to implement specific treatment strategies. Ross identifies eight elements in the cognitive-behavioral approach to treating adolescent substance abuse: rationale, screening, care, staff, environment, strategies, family, and evaluation of treatment. He presents a Twelve Step model and acknowledges spiritual concepts as an important part of treatment modalities. Appendices include "Psychosocial Assessment" checklist and "Twelve Steps of Alcoholics Anonymous." Includes charts and diagrams, author and subject indexes, and a reference list.

Roth, Paula, ed.
Alcohol and Drugs Are Women's Issues: Volume One: A Review of the Issues
Metuchen, N.J., Women's Action Alliance/Scarecrow Press, 1991. 192p. ISBN 0810823608

Featuring 24 essays by women of various professional backgrounds, including academic, clinical, law enforcement, social work, public health, and treatment professionals. A wide range of issues are addressed, e.g., special populations, drug use in pregnancy, mass media, and AIDs in the workplace. Index.

Roth, Paula, ed.
Alcohol and Drugs Are Women's Issues: Volume Two: The Model Program Guide
Metuchen, N.J., Women's Action Alliance/Scarecrow Press, 1991. 143p. ISBN 0810823896

A very practical, step-by-step guide for setting up a treatment program and designing staff training sessions, this volume can stand alone from the first. Includes questionnaires, a "resource section" with a directory of organizations and self-help groups, an annotated bibliography, and short lists of journals, videos, and audiocassettes. The authors also provide a directory of funding sources, both governmental and private, and a state by state directory of drug-abuse prevention authorities. Index.

Schaef, Anne Wilson, and Diane Fassel
The Addictive Organization
San Francisco, Harper and Row, 1988. 232p. ISBN 0062548417

The authors apply addiction theory to the structure and (dys)functioning of organizations. They define the roles of addicted employees, adult children of addicts, the organization as addictive, and the organization as addict. This is a helpful guide with practical suggestions for improving the functioning of the organization. Some discussion of employee assistance programs is included. Written in a clear, readable style.

Schilit, Rebecca, and Edith Lisansky Gomberg
Drugs and Behavior: A Sourcebook for the Helping Professions
Newbury Park, Calif., Sage Publications, 1991. 355p.
ISBN 0803934610

Designed as a comprehensive, introductory text for undergraduate study, this work covers psychopharmacology, various theories of substance abuse etiology, modalities of different treatments, and abuse problems and treatments in special populations. This is an in-depth, yet approachable text. Drugs described in detail include alcohol, narcotics, hallucinogens, nicotine, and central nervous system depressants and stimulants. Some chapters include a short directory of organizations. A glossary and an index are provided.

Schinke, Steven Paul, Gilbert J. Botvin, and Mario A. Orlandi
Substance Abuse in Children and Adolescents: Evaluation and Intervention
Newbury Park, Calif., Sage Publications, 1991. 100p.
ISBN 0803937482

The authors summarize data from the High School Seniors Survey (University of Michigan, 1975-1986) in analyzing patterns of abuse. They also provide a comparative review of major use studies in a convenient tabular format, and a detailed evaluation of prevention programs. An extensive reference list and name and subject indexes are included. *Developmental Clinical Psychology and Psychiatry,* 22.

Schuckit, Marc Alan
Drug and Alcohol Abuse: A Clinical Guide to Diagnosis and Treatment
3rd ed. New York, Plenum Medical Book Co., 1989. 307p.
ISBN 030643041X

This guidebook for physicians and mental health professionals provides an overview of drug and alcohol abuse including the classification of drugs with street names. Chapters cover specific classes of drugs, including over-the-counter and prescription, caffeine and nicotine, glues, solvents and aerosols, and other major classes of addictive substances. An "Emergency Problem" section is organized by symptoms, with treatment guidelines. References, tables, and index are provided. *Critical Issues in Psychiatry* series.

Seeburger, Francis F.
Addiction and Responsibility: An Inquiry into the Addictive Mind
New York, Crossroad Pub. Co., 1993. 200p. ISBN 0824513657

In defining addiction in philosophical terms, Seeburger takes on a massive task, but he does it masterfully. His presentation is distinguished by remarkable clarity and insight. The discussion is centered on alcoholism (with reference to the Twelve Step approach for recovery) but he provides broad coverage of addictive behaviors. Includes notes and index.

Shernoff, Michael, ed.
Counseling Chemically Dependent People with HIV Illness
New York, Haworth Press, 1991. 172p. ISBN 1560242590

This collection of practical essays, written primarily by social workers, covers aspects of treatment and counseling of various populations such as minorities, adolescents, intravenous drug users, gays, and the incarcerated. Includes information on some of the medical aspects of HIV. Also provides information on the "harm reduction model" whereby substance abusers can reduce their risk of exposure to HIV and receive essential social services even while refusing to abstain. Some references are included with each chapter. (Also published as *Journal of Chemical Dependency Treatment,* vol. 4, no. 2, 1991.)

Skoll, Geoffrey R.
Walk the Walk and Talk the Talk: An Ethnography of a Drug Abuse Treatment Facility
Philadelphia, Temple Univ. Press, 1992. 198p. ISBN 0877229171

Skoll draws on his experiences as a counselor in a residential treatment facility over a span of two years and provides a close examination of the ideology evident in counselors' treatment strategies. He evaluates how communication is constrained to facilitate their ideology, and he studies discourse in relation to social stratification within the facility. Provides an interesting and revealing look at this type of environment. Includes references and an index.

Small, Jacquelyn
Becoming Naturally Therapeutic: A Return to the True Essence of Helping
Rev. ed. New York, Bantam Books, 1990. 162p.
ISBN 0553348000

Writing in a conversational style, Small explains the basic characteristics effective counselors model such as empathy, confrontation, warmth (versus seduction), and self-actualization. This is appropriate for undergraduate readers with application for professional therapists. Includes sample dialogues, exercises, and glossary.

Smith, Carol Cox
Recovery at Work: A Clean and Sober Career Guide
San Francisco, Harper and Row, 1990. 228p. ISBN 0062553828

This conversational guide is packed with practical tips for reentry to work after initiating the recovery process from addiction. Covers topics such as tough questions at interviews, legal rights, and bankruptcy. This book is designed as a self-help guide, but it would also be useful for Employee Assistance Program directors.

Sobell, Mark B., and Linda C. Sobell
Problem Drinkers: Guided Self-Change Treatment
Treatment Manuals for Practitioners, New York, Guilford Press. 1993 187p. ISBN 0898622123

This guidebook for treatment professionals outlines the use of motivational intervention as a primary treatment modality. The authors are well-recognized authorities in addictions research. Their first three chapters provide an insightful review of alcohol dependence research. Specific treatment techniques and procedures are then discussed, and case studies given for illustration. Charts, graphs, and forms are included. The publisher states that permission is granted to reproduce handouts and forms for "professional use with their clients." Reference list and index provided.

Sonderegger, Theo, ed.
Perinatal Substance Abuse: Research Findings and Clinical Implications
Baltimore, Md., Johns Hopkins Univ. Press, 1992. 355p. ISBN 0801842751

These findings are directed toward health care and child protective services professionals. The opening chapter outlines methodological issues (such as the use of animal and human studies) and examines the effects of common drugs of abuse. A final chapter discusses legal issues relevant to protecting the fetus and maternal rights. Contributors are clinical and academic researchers. Tables and references support the text. *Johns Hopkins Series in Environmental Toxicology.*

Sparks, Shirley N.
Children of Prenatal Substance Abuse
San Diego, Calif., Singular Pub. Group, 1993. 204p. ISBN 1565930711

This clearly written guide covers prenatal exposure to, and effects of, cocaine, alcohol, and other substances of abuse. Intervention and prevention are covered also. In addition, the author discusses educators' and public welfare professionals' response to PL 99-457 (education of handicapped children under three years of age). A directory of hotlines and agencies, and a list of videos, books, curriculum guides, and posters are included. *School-Age Children Series.*

Stafford, Peter G.
Psychedelics Encyclopedia
3rd expanded ed. Berkeley, Calif., Ronin Pub., 1992. 420p. ISBN 0914171518

This is essentially the 1977 edition, with current information and a separate index placed in the front. The opening material includes the chapter "Psychedelic Renaissance" by Dan Joy, the author's "MDMA Update," and an obituary list. The book provides a sociological, historical, biographical, psychological, and medical look at all classes of legal and illegal hallucinogens. Contains many illustrations and photographs. Bibliographical references and two indexes are included.

Stephens, Richard C.
The Street Addict Role: A Theory of Heroin Addiction
Albany, N.Y., State Univ. of New York Press, 1991. 223p. ISBN 0791406199

Stephens takes a sociocultural view of heroin addiction and defines the "street addict role," a readily identifiable (and not easy to abdicate) role in the culture of addiction. He evaluates current treatment options and provides recommendations for public policy. This book is primarily theoretical in presentation, but case studies are included for illustration. Includes tables, notes, references, and an index. *SUNY Series, the New Inequalities.*

Stout, Chris E., John L. Levitt, and Douglas L. Ruben, eds.
Handbook for Assessing and Treating Addictive Disorders
New York, Greenwood Press, 1992. 371p. ISBN 031327634X

This readable guide for therapists and physicians begins with a broad overview of current theoretical controversies and some historical background. Subsequent essays by leading academic researchers and clinicians cogently address a broad spectrum of topics, including assessment, treatment (including eating disorders), and special groups (Native Americans, elderly, brain injured, and adolescents). An appended resource guide identifies organizations and directories of programs. Also appended are state-by-state offices for credentialing. Includes a reading list and an index.

Sweet, Eileen Smith, ed.
Special Problems in Counseling the Chemically Dependent Adolescent
New York, Haworth Press, 1991. 137p. ISBN 1560241632

Six scholarly essays, provided by clinicians and counselors, describe disorders of adolescents, such as chemical addiction, mental illness, child abuse, gambling, eating disorders, and cults. Causative factors and treatment options are also examined. Each essay features lengthy reference lists. (Also published as *Journal of Adolescent Chemical Dependency*, vol. 1, no. 4, 1991.)

Szasz, Thomas Stephen
Our Right to Drugs: The Case for a Free Market
New York, Praeger, 1992. 199p. ISBN 0275942163

Szasz, a Professor Emeritus of Psychiatry (State University of New York, Syracuse), argues eloquently that restricting the use of drugs creates as much of a problem as it prevents. He believes that it is "a grievous mistake to conceptualize certain drugs as a 'dangerous enemy' we must *attack* and *eliminate*, instead of *accepting* them as potentially helpful...and learning to *cope* with them competently." He examines constitutional law, government intervention, and social issues (including crack use in African American communities). Notes, a bibliography, and name and subject indexes are provided.

Taylor, William N.
Macho Medicine: A History of the Anabolic Steroid Epidemic
Jefferson, N.C., McFarland, 1991. 198p. ISBN 0899506135

Taylor claims credit for the classification of anabolic steroids as a "dangerous narcotic." By interweaving research studies and anecdotal evidence he discusses the discovery, pharmacology, and legal and addiction issues of anabolic steroids. He concludes with a call for cooperation among medical researchers and practitioners in recognizing and treating illicit use. Appendices include "Steroid Trafficking Act of 1990," "Statement of the American Medical Association," and a report to the U.S. Senate, "Drug Misuse, Anabolic Steroids and Human Growth Hormone." Includes notes, tables, and an index.

Thornton, William E.
Codependency, Sexuality and Depression
Summit, N.J., PIA Press, 1990. 139p. ISBN 0929162188

Thornton covers sexuality only in relation to personality development, and there is very little explicit information on sexuality given. Though there is scant coverage of substance abuse, this is a relevant title because of its focus on personality development. The author is interested in getting patients to define more clearly their needs and inadequacies.

Treadway, David C.
Before It's Too Late: Working with Substance Abuse in the Family
New York, Norton, 1989. 215p. ISBN 0393700682

Based on the author's personal experience as a therapist, this book relates the techniques he uses with families coping with substance abuse. Primarily a collection of case studies to illustrate particular techniques, this work is very readable, and is written in a compelling style. Includes a bibliography and an index.

Twerski, Abraham J.
Addictive Thinking: Understanding Self-Deception
San Francisco, Calif., Harper and Row, 1990. 123p.
ISBN 0062553976

In this insightful volume, Twerski discusses the contradictory and fallacious thinking patterns substance abusers exhibit, and the denial system they construct. He defines addictive thinking generally as the "inability to reason with oneself." Many case vignettes are given for illustration. The theoretical is tempered with humor, and practical techniques are given for use in counseling. A select bibliography is provided.

Wadler, Gary I., and Brian Hainline
Drugs and the Athlete
Philadelphia, F. A. Davis, 1989. 353p. ISBN 0803690088

This is a definitive text on the use of performance-enhancing drugs, and the consequent health and addiction problems. Excellent diagrams and references are provided in each chapter. Drugs covered include anabolic steroids, human growth hormone, amphetamines, cocaine, phenylpropanolamine, caffeine, marijuana, narcotics, and other "miscellaneous" (such as beta blockers and diuretics). Blood doping is also addressed. Also included is a chapter on privacy rights of athletes, liabilities of team physicians, drug testing, and other legal issues. Several policy statements are appended, including those from the International and United States Olympic committees, the National Collegiate Athletic Association (NCAA), and major sports organizations. *Contemporary Exercise and Sports Medicine Series*, 2.

Wallace, Barbara C., ed.
The Chemically Dependent: Phases of Treatment and Recovery
New York, Brunner/Mazel, 1992. 364p. ISBN 087630675X

This collection of scholarly essays by treatment professionals and academic researchers is divided into four parts, focusing on current treatment models for different phases of substance abuse; psychoanalytic approaches; behavioral and relapse prevention approaches; and trends in research, especially regarding the disease model of addiction. References are included with each essay. Includes tables, diagrams, and an index.

Wegscheider-Cruse, Sharon
Another Chance: Hope and Health for the Alcoholic Family
2nd ed. Palo Alto, Calif., Science and Behavior Books, 1989. 324p. ISBN 0831400722

This edition is updated with a stronger focus on adult children of alcoholics and co-dependency as a disease. The author outlines her views of the "whole person model," of family roles in an alcoholic dynamic. She addresses the concept of spirituality as a core component of transforming oneself in recovery. The author is the founder of the National Association of Children of Alcoholics. Appendices include list of role playing exercises for family therapy and "The Whole Person Inventory." Illustrated with diagrams and sketches.

Weil, Andrew, and Winifred Rosen
From Chocolate to Morphine: Everything You Need to Know About Mind-Altering Drugs
Rev. and updated ed. Boston, Houghton Mifflin, 1993. 240p. ISBN 0395660793

An updated edition of *Chocolate to Morphine : Understanding Mind-Active Drugs* (Houghton Mifflin, 1983). This readable text is designed for teens, but is comprehensive enough to be appropriate for the undergraduate level. Includes updated information on crack cocaine and AIDS. Well-illustrated with photographs, cartoons, and case studies. Historical background is balanced with contemporary theory in a factual presentation. All major drug groups are discussed, and basic terms are defined in detail. Sources are noted throughout, and a glossary and index are included.

Weinstein, Dava L., ed.
Lesbians and Gay Men: Chemical Dependency Treatment Issues
New York, Harrington Park Press, 1992. 155p. ISBN 1560243937

This collection of wide-ranging essays by treatment professionals covers topics such as heterosexual therapists working with homosexuals; homosexual substance abusers with AIDS; utilizing the spiritual aspects of Twelve Step programs in treating gay males; homosexual adolescent substance abusers; identity formation/ transformation; and family therapy techniques. References are provided in each chapter. (Also published as *Journal of Chemical Dependency Treatment*, vol. 5, no. 1, 1992.)

Weiss, Roger D., Steven M. Mirin, and Roxanne L. Bartel
Cocaine
2nd ed. Washington, D.C., American Psychological Association, 1994. 204p. ISBN 0880485493

In plain language the authors address all aspects of cocaine, including its manufacture, use, effects on the body and brain, dependence, and the effects of addiction on the family. This is a readable, authoritative text featuring the most current in research (including reference to DSM-IV). Treatment modalities include individual and group therapy, behavioral therapy, medication, and relapse prevention. Appendix includes a "Frequently Asked Questions" chapter and a "Self Test for Cocaine Dependence." Also includes a bibliography and an index.

Werry, John S., and Michael G. Aman, eds.
Practitioner's Guide to Psychoactive Drugs for Children and Adolescents
New York, Plenum Medical Book Co., 1993. 440p.
ISBN 0306443899

This ready reference pharmaceutical manual provides introductory information, reading lists, diagrams, graphs, and assessment lists. The contributors are clinicians and medical researchers. A chapter by Alan S. Unis and Jon McClellan covers drugs of abuse. Indexed.

White, William L.
The Culture of Addiction
Bloomington, Ill., Lighthouse Training Institute, 1990. 543p.
ISBN 0938475010

White provides a detailed, sociological analysis of drug use, with guidelines for diagnosis and treatment. The author asserts that what begins as a "person-drug relationship" inevitably moves to "enmeshment in the culture of addiction." He examines each aspect of this culture (for example, "tribal organization," psychosocial functions and key roles), and then applies a parallel structure for the recovery process. Book One addresses "The Culture of Recovery;" Book Two presents "A Travel Guide for Treatment Professionals." Includes a self-assessment checklist, a glossary, and an index.

Windle, Michael T., and John S. Searles, eds.
Children of Alcoholics: Critical Perspectives
New York, Guilford Press, 1990. 244p. ISBN 0898621682

In the introduction the editors review the types of literature available on adult children of alcoholics. Subsequent essays take scholarly views of research on children of alcoholics (adult and younger) from the biopsychosocial perspective. This includes looking at biochemical markers, neurological characteristics, personality attributes, family structure, and the life-span perspective. Each chapter includes extensive reference lists. Indexed.

Wright, Bob, and Deborah George Wright
Dare to Confront!: How to Intervene When Someone You Care About Has an Alcohol or Drug Problem
New York, Master Media, 1990. 210p. ISBN 0942361210

A readable manual for intervention written for family members and coworkers of addicts. As background, the authors clearly outline the process of addiction and describe treatment models. They then describe how to set up an intervention team, and how to prepare to confront an addict. Several checklists are included.

Yesalis, Charles, ed.
Anabolic Steroids in Sport and Exercise
Champaign, Ill., Human Kinetics Publishers, 1993. 325p.
ISBN 0873224019

Contributors to this volume are medical and academic researchers, including specialists in exercise physiology, physical therapy, and sport medicine. All aspects of steroids are covered, including adolescent use. Specific recommendations for treating withdrawal are also given. Photographs, tables, and diagrams supplement the text; chapter references are included.

Yoder, Barbara
The Recovery Resource Book
New York, Simon and Schuster, 1990. 314p. ISBN 0671668730

An extremely comprehensive and well-organized introduction to resources for recovery. Conditions discussed include substance abuse, eating disorders, workaholism, gambling, sex addiction, sexual abuse, and dual disorders. It features annotated bibliographies, directories of organizations, checklists, statistics, and tabulated information. Personal histories are sprinkled throughout the text. Appendices include lists of state agencies and departments of rehabilitation, and directories of publishers and self-help clearinghouses.

Yudofsky, Stuart C., Robert E. Hales, and Tom Ferguson
What You Need to Know About Psychiatric Drugs
New York, Grove Weidenfeld, 1991. 646p. ISBN 0802112811

Covers classes of psychiatric drugs, common disorders for which psychiatric drugs are prescribed, possible side effects, and alternative treatment options. Anti-addiction drugs are also covered. Attention is given to the variance of reaction to specific drugs over time. This is an excellent handbook for patients and caregivers to use when making decisions regarding mental health care. The book is readable and professionally presented. Includes graphs, tables, and references for further reading. An appendix includes mental disorders as listed in the DSM-III-R. A glossary and an index are provided.

Zagon, Ian S., and Theodore A. Slotkin, eds.
Maternal Substance Abuse and the Developing Nervous System
San Diego, Calif., Academic Press, 1992. 377p.
ISBN 0127752250

This scholarly, comprehensive study includes essays by clinicians and clinical researchers on topics such as fetal alcohol syndrome and the effects of prenatal and postnatal exposure to smoking, cocaine, marijuana, benzodiazepines and opioids. Each essay presents a current literature review, discussion of research experiments, animal models, controversies and recommendations. References are included with each essay.

Nonprint Materials

Addiction
New York, WNET/New York, 1988. 1 videocassette (60 min.) VHS

George Page narrates this work on what he calls the vast uncontrolled experiment: the use of psychoactive substances both legal and illicit. Neurological research is presented in addition to basic concepts of addiction, withdrawal, and treatment modalities. *The Mind,* program 4.

Chemical Dependency: A Disease of Denial
Center City, Minn., Hazelden Foundation, 1990. 1 videocassette (25 min.) VHS

Features testimonials of substance abusing persons (of wide-ranging ages and races), and interviews with therapists and medical researchers. The disease model of addiction is emphasized using an upbeat narration. Concludes with a call for seeking treatment.

Confidentiality: Ethical and Legal Considerations
Cleveland, Ohio, Fairview, 1989. 1 videocassette (18 min.) VHS

This video is directed primarily to health care workers, but it provides excellent examples applicable to social service areas as well. Brief vignettes dramatize breaches of confidentiality on topics such as sexual orientation or psychiatric condition. Legal cases are presented, and specific situations examined, e.g., revealing information over the phone or through computer data.

Continued Acts of Sabotage
Skokie, Ill., Gerald T. Rogers Productions, 1988. 1 videocassette (38 min.) VHS and 1 guide.

This dramatization of persons in group and individual counseling sessions illustrates various examples of counterproductive behavior. No narration or instruction is included. Useful for opening discussion on patient compliance and treatment issues. A discussion guide is included.

A Cry for Help: The Fetal Drug and Alcohol Crisis
Northbrook, Ill., MTI Film and Video, 1990. 1 videocassette (33 min.) VHS

Health care professionals, social and child protection workers, policy makers and legislators, and patients share their perspectives on the extent of the problem, and the need for prevention and treatment programs. Structured aftercare is emphasized, as well as keeping mothers currently in treatment with their children.

Drug-Affected Children: The Price We Pay
Evanston, Ill., Altschul Group, 1991. 1 videocassette (18 min.) VHS

Nursing and social service professionals examine the social and economic issues of (primarily) cocaine exposed infants. A call is made for health care availability for at-risk women. *Cocaine Series: The Domino Effect.*

Drugs in the Workplace
Ashland, Ore., CNS Productions, 1990. 3 videocassettes (143 min.) and guide (40p.) VHS

Well-recognized medical researchers, clinicians, and therapists provide various (and sometimes conflicting) perspectives on the following topics: the impaired employee, drug testing, and the effect of drugs on work performance. The presentation does not shy away from acknowledging controversy. Specific techniques for supervisor intervention are demonstrated in the first part. The second part covers technical, legal, and ethical considerations for testing in the workplace. The third video discusses how various classes of drugs affect the body, along with how to meet the requirements of the Drug Free Workplace Act. *Haight-Ashbury Training Series*, 3.

Dual Diagnosis: The Mentally Ill Chemical Abuser
Ashland, Ore., CNS Productions, 1991. 3 videocassettes (160 min.) VHS and 1 training manual (33p.)

The first two tapes are designed for treatment professionals. Tape One provides an overview of the concept of dual diagnosis, and various treatment theories. The second tape demonstrates techniques for assessment and treatment. Tape Three, designed for clients, illustrates fundamental concepts of how psychoactive drugs affect users, and the biological and environmental factors in mental illness. *Haight-Ashbury Training Series*, 4.

Emotional Healing
Boston, Appropriate Media Services, SITE Productions, 1990. 1 videocassette (99 min.) VHS

Terry Hunt directs workshops for adult children of alcoholic and abusive parents with the goals of relating the pain they have experienced and the steps they learn to take control. As the group therapy sessions progress, Hunt describes in detail the various stages taking place.

Family Baggage
Boulder, Colo., Creative Recovery, 1988. 2 videocassettes (112 min.) VHS

Lou Montgomery and Errol Strider act out vignettes illustrating co-dependency issues, dysfunctional family issues, and the experiences of adult children of alcohol and substance abusing parents. Situations are presented in an entertaining, generally humorous way.

Introduction to Street Drug Pharmacology
Park Ridge, Ill., Parkside Pub. Corp., 1988. 1 videocassette (59 min.) VHS

The basic principles of the physiological effect of drugs and substance abuse are presented in lecture format. Terms and concepts being explained are provided on screen. The schedules of controlled substances are discussed, and the forms, names, effects and methods of administering licit and illicit drugs are given.

Mainline
Washington, D.C., U.S. Department of Labor, 1991. 1 videocassette (16 min.) VHS

Presented in a newscast format, this video is the case study of a worker in an industrial warehouse who abuses various drugs and is responsible for a deadly accident due to his impaired judgment. Abstinence is stressed. Substance Abuse Data Sheets (SADS) are discussed and are included in the accompanying trainer's guide and participant's manual.

Marketing Booze to Blacks
Washington, D.C., Center for Science in the Public Interest, 1990. 1 videocassette (17 min.) VHS and 1 discussion guide

Clinicians, public health officials, and policy makers discuss how alcohol advertising is deliberately targeted toward the African American community, particularly young black males. The focus is on billboards, but television and print ads are also examined. This video concludes with an outline of suggestions for public action against such advertising techniques.

Substance Abuse: Causes, Consequences and Treatment Choices: A Videotape Training System
Denver, Colo., Center for Interdisciplinary Studies, 1990.
12 videocassettes (574 min.) VHS, 1 study guide (82p.) and test file (26 sheets)

Topics addressed include an introduction to the concept of addiction; biological factors; societal factors; criminality; assessment and treatment planning; medical considerations; Twelve Step versus cognitive-behavioral therapy; dual diagnosis and self-medication; addictive disease; traffic offenders; maternal drug use and substance abuse among women; and concurrent mental disorders. Presentations are by well-recognized researchers in the field. Presented in a lecture format.

Substance Abuse Disorders
s.l., Alvin H. Perlmutter, 1991. 1 videocassette (59 min.) VHS

Prepared on the behalf of the Annenberg/CPB Project, this instructional presentation discusses basic models of addiction and risk factors for each major class of psychoactive substance, with terms and concepts outlined on screen. Presents information on treatment matching, including a medicine wheel model. A unique prevention method—a "safe drinking class"—is demonstrated in some detail. *The World of Abnormal Psychology*, 6.

Toma
Atlanta, Ga., WXIA-TV, ATI MARK Productions, Inc., distributor, 1989. 1 videocassette (50 min.) VHS

David Toma, a dynamic, confrontational speaker, presents a lecture on his perceptions of the dangers of drug abuse. He stresses abstinence and calls on students and parents to educate themselves about addiction.

Social Work and Human Services

Lynn Namsick
Pima Community College
Tucson, Arizona

Introduction

Social and human service workers attempt to help individuals and families cope with a variety of social problems. Social factors impacting this profession include poverty, homelessness, inadequate housing, unemployment, lack of occupational skills, illnesses, disabilities, chemical dependency, and familial or personality dysfunctions. Employed by private or public agencies, social workers and their assistants perform their duties through individual or family casework, direct counseling, group work, or community organization efforts.

Social and human service programs in community colleges prepare students to enter the workforce as assistants to social workers and psychologists, and in various roles in social service agencies. Most social workers concentrate in specialized fields such as child and family services, school social work, medical or mental health arenas, substance abuse, criminal justice, industrial occupational social work, gerontology, or clinical social work. Assistants in these fields may have a wide range of job titles: social service technician, case management aide, social work assistant, residential counselor, alcohol or drug abuse counselor, mental health technician, child abuse worker, community outreach worker, or gerontology aide, to name a few. In addition, the range of responsibilities and level of supervision may vary significantly based on education and experience.

According to the *Occupational Outlook Handbook*, there are excellent opportunities for qualified applicants. Positions are expected to more than double between 1992 and the year 2005, making it a viable and rapidly growing career option. The most rapid growth is projected in job training programs, residential settings, and private social service agencies. As the mean population ages, positions in such areas as adult daycare and meal delivery programs will increase dramatically. The traditional roles in services to the community are also expected to increase: child welfare, family services, school social work, substance abuse, criminal justice, occupational counseling, geriatric, and clinical work. Some employers may prefer a four-year college degree, while others are willing to hire high school graduates or community college students with preparation in social work. The levels of education may correlate to the levels of responsibilities and positions attained. Employers often provide in-service education for professional growth and development.

This bibliography is designed for undergraduate students pursuing a career in the social work and human services arena, as well as students involved in continuing education and other majors. For alcohol and drug abuse counseling, refer to that chapter in this bibliography.

Selected Associations

Alcohol and Drug Problems Association of North America
444 N. Capitol St. NW, Suite 706
Washington, D.C. 20001

American Association for Marriage and Family Therapy
1100 17th St. NW, 10th Floor
Washington, D.C. 20036

American Association of State Social Work Boards
700 S. Ridge Parkway, Suite 301B
Culpepper, Va. 22701

American Psychological Association
1200 17th St., NW
Washington, D.C. 20036

Council on Social Work Education
1600 Duke St.
Alexandria, Va. 22314

National Association of Social Workers
7981 Eastern Ave.
Silver Spring, Md. 20910

National Board for Certified Counselors
5999 Stevenson Ave., Suite 402
Alexandria, Va. 22304

National Organization for Human Service Education
Box 6257
Fitchburg State College
Fitchburg, Mass. 01420

Selected Journals

Aging.
Raven Press. Irregular. ISSN 0160-2721.

Child Welfare.
Transaction Publishers. Bimonthly. ISSN 0009-4021.

Children Today.
U.S.G.P.O. Bimonthly. ISSN 0361-4336.

Families in Society: The Journal of Contemporary Human Services.
Family Service America. Ten times a year. ISSN 1044-3894.

Federal Probation.
U.S.G.P.O. Quarterly. ISSN 0014-9128.

Journal of Marriage and the Family.
National Council on Family Relations. Quarterly. ISSN 0022-2445.

Journal of Social Work Education.
Council on Social Work Education. Three times a year. ISSN 0022-0612.

Modern Maturity.
American Association of Retired Persons. Bimonthly. ISSN 0747-6302

Public Welfare.
American Public Welfare Association. Quarterly. ISSN 0033-3816.

Social Policy.
Union Institute. Quarterly. ISSN 0037-7783.

Social Problems.
Univ. of California Press. Quarterly. ISSN 0037-7791.

Social Work.
National Association of Social Workers. Bimonthly. ISSN 0037-8046.

Print Materials

◆ Career and Vocational Guidance

Bolles, Richard Nelson
What Color Is Your Parachute? A Practical Manual for Job-Hunters & Career-Changers
1995 ed. Berkeley, Ca., Ten Speed Press, 1995. 446p.
ISBN 0898155335

This practical self-help guide for job-hunters and career-changers discusses self-assessment, career planning, effective job-hunting methods, employment interviews, and avoiding depression. Main concepts in each chapter are supported with the liberal use of statistical data and extensive footnotes. The tone is lighthearted and encouraging. Includes lengthy bibliography and is updated annually.

Collison, Brooke B., and Nancy J. Garfield
Careers in Counseling and Human Development
Alexandria, Va., American Association For Counseling and Development, 1990. 170p. ISBN 1556200722

This career guide describes over ninety career options within counseling and human development fields. Includes the perspectives of professionals on their individual experiences, credentials, education, and salary. Chapters discuss careers in a variety of settings. Appendices provide a list of professional associations (including certification and certifying associations) and a matrix of titles and work settings. Indexed.

DeGalan, Julie, and Stephen Lambert
Great Jobs for Psychology Majors
Lincolnwood, Ill., VGM Career Horizons, 1995. 264p.
ISBN 0844243523

DeGalan and Lambert take students by the hand down the job search and career exploration paths. Includes information on the assessment of strengths and interests, geographic location, standard of living, unusual career options, and the identification of the "best" employers.

DiNitto, Diana M., and Carl Aaron McNeece
Social Work: Issues and Opportunities in a Challenging Profession
Englewood Cliffs, N.J., Prentice Hall, 1990. 400p.
ISBN 0138169438

Attempts to portray a balanced picture of the social work profession and the rewards, difficulties, and struggles in clinical practice. Topics include history of the profession, education and accreditation, social work theory, social work administration, and responses to a variety of social issues. Also presents feminist perspectives in both the social work profession and society.

Eberts, Marjorie, and Margaret Gisler
Careers for Good Samaritans and Other Humanitarian Types
Lincolnwood, Ill., VGM Career Horizons, 1991. 165p.
ISBN 0844281263

Provides descriptions of a variety of career options for those who wish to be involved in humanitarian causes. Listings include jobs in areas such as hunger relief, homelessness, disaster relief, peacekeeping, and medical aid. Professionals working in these fields share their experiences. Appendices include lists of InterAction member agencies, missionary organizations, and state offices of volunteerism. *VGM Careers for You* series.

Eberts, Marjorie, and Margaret Gisler
Careers for Kids at Heart and Others Who Adore Children
Lincolnwood, Ill., VGM Career Horizons, 1994. 137p.
ISBN 0844241113

Careers involving children range from child care to welfare. Eberts and Gisler present a variety of career options within the traditional and nontraditional concepts, such as arts, recreation, and sports. This work is designed to open opportunities for those involved in career exploration.

Emener, William G., and Margaret A. Darrow, eds.
Career Explorations in Human Services
Springfield, Ill., Charles C. Thomas, 1991. 319p.
ISBN 0398057338

An overview of various human service professions including adult and vocational education, communication sciences and disorders, counseling, criminal justice, gerontology, legal professions, nursing, psychology, public administration, public health, rehabilitation counseling, social work, and special education. The introductory chapter discusses the general nature of human service careers and professionals; the conclusion addresses professionalism and ethics in human services.

Garner, Geraldine O.
Careers in Social and Rehabilitation Services
Lincolnwood, Ill., VGM Career Horizons, 1994. 98p.
ISBN 0844241881

This short, readable work discusses the nature of social work, including a description of clients, employment settings, training and other qualifications, possibilities for advancement, and sources of additional information for various social service fields. Includes specific information on rehabilitation counseling; social work; employment; mental health; medical, therapeutic, and educational services; Christian and Jewish clergy; and probation/parole officers. Also discusses career paths and upward mobility, professional licensure and certification, and other issues. Appendices list professional organizations and licensure/certification organizations.

Pierce, Dean
Social Work and Society: An Introduction
New York, Longman, 1989. 277p. ISBN 0582286654

Designed for those people exploring a career in the social work profession, this work includes definitions; social work as a profession and its role in the social welfare system; fields of practice; colleague, client, and public perspectives; the purposes of social work in society; a practice model; contexts for practice; managing change in people, society, policy, and the profession; working with AIDS patients, the elderly, school children, and Hispanics; and commitment to the profession.

Wells, Carolyn Cressy
Social Work Day-to-Day: The Experience of Generalist Social Work Practice
2nd ed. New York, Longman, 1989. 205p. ISBN 080130041X

Employs colorful, detailed case illustrations to exemplify the responsibilities, practical skills, knowledge, and values of a social worker. Sections discuss professional choices and development, practice skills, human behavior in the social environment, and policy issues. The epilogue addresses evaluation and termination as well as organizational, political, and personal changes in the social work profession.

✦ General Works—
Practice, Administration, Ethics, and Theory

Barker, Robert L.
The Social Work Dictionary
2nd ed. Silver Spring, Md., National Association of Social Workers, 1991. 287p. ISBN 0871011905

This specialized reference source, which defines approximately 5,000 terms, also contains organizations, concepts, values, and historical events relating to social work. Definitions include terminology employed in social work administration, research, policy development and planning, human growth and development, micro and macro social work, areas of practice, and practice methods. Provides a historical overview of the development of social work and social welfare, the NASW Code of Ethics, a listing of state boards regulating social work, and a list of NASW state chapter offices.

Corey, Gerald, Marianne Schneider Corey, and Patrick Callahan
Issues and Ethics in the Helping Professions
3rd ed. Pacific Grove, Calif., Brooks/Cole Pub. Co., 1988. 441p.
ISBN 0534080820

Examines the ethical issues in counseling and related helping professions. Contains open-ended case illustrations to stimulate readers to formulate their own opinions. Includes topics such as ethical decision-making, multicultural counseling, marriage and family counseling, informed consent and clients' rights, and legal issues. This work also addresses counseling supervision, the client-counselor relationship, paraprofessional training, group work, and the use of specific therapeutic strategies. Appendices contain codes of ethics of professional counseling and social work organizations.

Corey, Marianne Schneider, and Gerald Corey
Groups: Process and Practice
4th ed. Pacific Grove, Calif., Brooks/Cole Pub. Co., 1992. 439p.
ISBN 0534161227

A basic, readable handbook for beginning group leaders that offers descriptions of group processes, and practical applications of principles in working with various groups. Includes topics such as ethical and professional issues; developmental stages of a group process; and specific group practice with children, adolescents, adults, and the elderly.

Demone, Harold W., and Margaret Gibelman
Services for Sale: Purchasing Health and Human Services
New Brunswick, N.J., Rutgers Univ. Press, 1989. 450p.
ISBN 0813513618

Examines the purchasing of health and social services, including the philosophical and pragmatic reasons underlying growth, past and current sociopolitical bases, and applications to a variety of health and human services. Discusses monitoring, evaluation, current trends, and future developments with respect to purchasing services. Includes case examples from various areas of practice.

Ephross, Paul H., and Thomas V. Vassil
Groups That Work: Structure and Process
New York, Columbia Univ. Press, 1988. 230p. ISBN 0231057385

A practical guide for social workers and other practitioners working with groups. Topics include the roles of members of the group, sociology of group work, a model of working groups, conflict (and problem solving), and a discussion of education requirements.

Garner, Leslie Holland
Leadership in Human Services: How to Articulate and Implement a Vision to Achieve Results
San Francisco, Calif., Jossey-Bass, 1989. 167p.
ISBN 155542144X

Discusses leadership challenges, including articulating a vision, forecasting trends, identifying client needs and ways of meeting those needs, and setting priorities and performance standards. Also discusses budget management, monitoring and evaluating measures of success, focusing agency efforts, responding to new demands, encouraging change, and the art of effective leadership. Emphasizes reliability, accountability, results-oriented management, and translating vision into specific goals and objectives.

Gitterman, Alex, ed.
Handbook of Social Work Practice with Vulnerable Populations
New York, Columbia Univ. Press, 1991. 804p. ISBN 0231070489

A collection of twenty-two readings that concern social work practice with vulnerable people such as AIDS patients, chemically dependent individuals, and the mentally ill and deficient. Analyzes a variety of factors including unique problems, needs, services, and interventions for each group. Includes case examples to illustrate typical problems and situations each client population encounters.

Hull, Grafton H., and others
Building the Undergraduate Social Work Library: An Annotated Bibliography
Rev. ed. Alexandria, Va., Council on Social Work Education, 1993. 53p. ISBN 0872930386

This selective bibliography identifies a core collection of titles for an undergraduate library supporting programs in social work and human services. Two-thirds of the 299 annotated entries are books published after 1988. Subject categories include the history of social welfare, human behavior and the social environment, practice, values and ethics, research and evaluation methodology, specific practice areas, social work education, and human diversity.

Jordan, Bill
Social Work in an Unjust Society
New York, Harvester Wheatsheaf, 1990. 242p. ISBN 0745008968

This work presents a British perspective of the moral and ethical issues that social workers face concerning the conflict between the needs of society and the needs of the individual. Topics include the social worker and the citizen, morality and rules, cooperation, community and democracy, and moral reasoning and judgment. Includes case examples.

Loewenberg, Frank M., and Ralph Dolgoff
Ethical Decisions for Social Work Practice
4th ed. Itasca, Ill., F.E. Peacock Publishers, 1992. 260p.
ISBN 0875813569

Analyzes ethical decisions and issues to consider when handling dilemmas in social work practice. Discussion includes social considerations, diagnosis and interventions, interpersonal responsibilities, and regulation and policy issues.

MacKay, Ruth C., Jean R. Hughes, and E. Joyce Carver, eds.
Empathy in the Helping Relationship
New York, Springer Pub. Co., 1990. 194p. ISBN 0826161405

Edited by three Canadian nursing scholars, this work attempts to clarify the nature of empathy, and to provide greater understanding of scientific validity and theory concerning empathy. The authors suggest practical applications of current knowledge and define empathy from the perspectives of competence, knowledge, and sensitivity. Topics include a definition and the significance of empathy, a historical overview of empathy in the helping professions, empathy as a personality trait and skill, and a conceptual model and measurement of empathy.

McMahon, Maria O'Neil
The General Method of Social Work Practice: A Problem-Solving Approach
2nd ed. Englewood Cliffs, N.J., Prentice Hall, 1990. 356p.
ISBN 0133503801

This practical text focuses on the concerns of beginning social work students and discusses the holistic foundation of social work practice and processes. Incorporates concepts from systems theory, ecological perspectives, and professional wisdom, so that students can focus their practice on people and environment as a unit. Content includes a foundation for general practice, the general method of social work, human diversity in social work practice, and other issues. Encompasses individual, family, group, and community work.

Reamer, Frederic G.
Ethical Dilemmas in Social Service
2nd ed. New York, Columbia Univ. Press, 1990. 266p.
ISBN 0231069685

Examines ethical issues that occur routinely in social work practice, and provides a framework to aid social workers in dealing with ethical problems. Topics concern the nature of ethics as applied to social work, specifically in areas such as practice, service to individuals and families, planning and policy, and collegial relationships. Includes information on the goals of ethical analysis, the use of ethics committees, informed consent, paternalism, and the limits of intervention.

Reamer, Frederic G.
The Philosophical Foundations of Social Work
New York, Columbia Univ. Press, 1993. 219p. ISBN 0231071264

A scholarly examination of political, moral, and ethical formal logic, epistemology, and aesthetics underlying the social work profession. The goal is to lay a foundation for a philosophical statement on the mission, methods, and orientation of the profession. Reamer contends that social workers must learn how to think critically, examine an argument, and link philosophical matters with the practical skills and decisions involved in daily practice.

Reeser, Linda Cherrey, and Irwin Epstein
Professionalization and Activism in Social Work: The Sixties, the Eighties, and the Future
New York, Columbia Univ. Press, 1990. 165p. ISBN 0231067887

Examines data from two surveys that compared the level and type of social activism as well as the professionalization of social workers in the 1960s and the 1980s. Analysis of data shows that social workers have generally resisted the individualistic ideology of the 1980s, and that they believe social problems have their basis in structural forces as was thought in the 1960s, rather than in individualistic explanations. The authors argue that social workers are more committed to activism in the 1980s than they were in the 1960s, but that methods have changed from protest activities to increased involvement in political processes such as lobbying, testifying, and participating in election campaigns.

Reid, Kenneth E.
Social Work Practice with Groups: A Clinical Perspective
Pacific Grove, Calif., Brooks/Cole Pub. Co., 1991. 311p. ISBN 0534148204

This work integrates theoretical knowledge and practical skills needed for daily leadership of groups in clinical practice. Discusses strategies and techniques that increase the chances for a group's success and for the growth of its members. Emphasizes social work group practice in clinical, hospital, social agency, and residential settings. Includes clinical examples to illustrate concepts and techniques.

Richie, Nicholas D., and Diane E. Alperin
Innovation and Change in the Human Services
Springfield, Ill., Charles C. Thomas, 1992. 154p. ISBN 039805763X

A historical overview of social services in the twentieth century, including government, voluntary, and private organizations. Emphasizes recent developments such as hospices, continuing care for the elderly, AIDS services agencies, and domestic violence shelters. Discusses the role of the human services worker in terms of assessment, referral, client advocacy, coordination, and evaluation, Also addressed are future trends of national health insurance, services for the homeless, day care for children, lifelong occupational counseling, and services for the terminally ill. The lack of coordination between agencies and the fragmentation of services that result in a confused system of care, gaps, and unnecessary duplication in services are also discussed.

Rooney, Ronald H.
Strategies for Work with Involuntary Clients
New York, Columbia Univ. Press, 1992. 405p. ISBN 0231067682

Examines the nature and problems of working with involuntary clients in social work practice, both those required to see a practitioner and those pressured to "seek help." Includes socialization strategies, negotiation and contracting, middle-phase intervention and termination, working with involuntary families, group work, the involuntary practitioner, and the system. The appendix lists training videos on strategies for working with involuntary clients.

Rubington, Earl, and Martin S. Weinberg, eds.
The Study of Social Problems: Six Perspectives
4th ed. New York, Oxford Univ. Press, 1989. 301p. ISBN 0195057236

Analyzes major social problems from the angle of sociopathology, social disintegration, value conflict, deviancy, labeling, the critical perspective, and review perspective. Attempts to provide sociological explanations for these problems rather than simply describing them. Each section includes an introductory essay outlining the individuals and major works that influenced that area of sociology, the historical background of American sociology at the time, reprinted readings, a critique of the perspective covered, and an annotated list of references.

Sheafor, Bradford W., Charles R. Horejsi, and Gloria A. Horejsi
Techniques and Guidelines for Social Work Practice
2nd ed. Boston, Allyn and Bacon, 1991. 521p. ISBN 0205127681

Thoroughly discusses the fundamental concepts and techniques common to all areas of social work practice, and covers specific approaches to direct and indirect practice in detail. Includes topics such as the artistic and scientific aspects of social work, the roles and functions of social workers, communication and helping skills, work load and case management, personal and professional development, and the organizational and community contexts of practice.

Specht, Harry, and Mark E. Courtney
Unfaithful Angels: How Social Work Has Abandoned Its Mission
New York, Free Press, 1994. 209p. ISBN 0029303559

This controversial and thought-provoking critique of the social work profession contends that the helping professions in the U.S. have strayed from their original mission of serving the poor and underprivileged by placing too much faith in individualistic solutions to social problems. The authors further state that psychotherapy is not an appropriate method for treating social problems, and that a large number of certified social workers educated with public money have abandoned community service to the poor in favor of private practice tailored to the more affluent middle class. Argues for the return of the social work profession to active community service and proposes a community-based system of social care for the twenty-first century.

Thyer, Bruce A., and Marilyn A. Biggerstaff
Professional Social Work Credentialing and Legal Regulation: A Review of Critical Issues and an Annotated Bibliography
Springfield, Ill., Charles C. Thomas, 1989. 107p. ISBN 0398055882

A detailed, critical review of literature published between 1938-1988 on credentialing and the legal regulation of social workers in the U.S. Sections include a review essay on social work credentialing and legal regulation covering the history, current status, and future trends in the field. Includes bibliographies.

Vourlekis, Betsy S., and Roberta R. Greene, eds.
Social Work Case Management
New York, Aldine de Gruyter, 1992. 199p. ISBN 020236075X

A collection of readings on topics including practice functions and settings, policy and professional contexts, school-based services for adolescent parents, early intervention for disabled infants and toddlers, children with HIV/AIDS, developmentally disabled adults, the chronically mentally ill, employee assistance programs, the elderly, child welfare services, family advocacy in the military, and quality control in case management.

Weinbach, Robert W.
The Social Worker as Manager: Theory and Practice
New York, Longman, 1990. 328p. ISBN 0801300428

This primer on management theory and practice is designed for social work students and others in the human services professions. Using case examples, topics address external social work management; purpose and direction in the planning process; employee growth; personnel actions; work performance; control and leadership; and characteristics of healthy and unhealthy organizations.

Wells, Carolyn Cressy, and M. Kathleen Masch
Social Work Ethics Day to Day: Guidelines for Professional Practice
Prospect Heights, Ill., Waveland Press, 1991. 167p.
ISBN 0881335460

This brief volume emphasizes not only the social workers' accountability to clients in affirming their worth, dignity, and uniqueness, but also their accountability to the public whose tax dollars finance their agencies. Discusses ethical problems that occur in daily social work practice and examines the major principles of the National Association of Social Workers Code of Ethics with case illustrations. Topics include the agency setting; the social worker's conduct and ethical responsibility to clients, colleagues, and employers; and the social work profession and society. The Code of Ethics is appended.

Zastrow, Charles
The Practice of Social Work
4th ed. Belmont, Calif., Wadsworth Pub. Co., 1992. 599p.
ISBN 0534170048

This clearly written text covers the fundamental theoretical knowledge and practical skills required for beginning social workers. Material includes an overview of social work practice and values, assessing, interviewing and counseling individuals, group work, social work with families, community practice, evaluating practice, cultural diversity, and burnout prevention. Discusses client-centered, Gestalt, reality, and rational counseling theories; therapies; transactional analysis; behavior modification; sex counseling and therapy; neurolinguistic programming; specific treatment techniques; and comparative analysis of therapeutic approaches.

✦ Community Action and Empowerment

Betten, Neil, and Michael J. Austin, eds.
The Roots of Community Organizing, 1917-1939
Philadelphia, Temple Univ. Press, 1990. 230p. ISBN 0877226628

This compilation of twelve articles discusses three approaches to community organizing and provides a historical overview of twenty-two years of community organizing. Includes topics such as the roots and intellectual origins of community organizing, the Cincinnati unit experiment, grass roots movements, rural organizing and the Agricultural Extension Service, and the legacy of community organizing at the end of the Great Depression.

Homan, Mark S.
Promoting Community Change: Making It Happen in the Real World
Pacific Grove, Calif., Brooks/Cole Pub. Co., 1994. 442p.
ISBN 0534142508

Aimed at social work students and professionals working on the community level, this text focuses on strengthening communities, promoting community responsiveness, and producing positive changes at the community level. Part One addresses the nature of and the need for community change, and relates change to professional practice and personal life. Part Two delineates the practicalities of knowing the community and its power structures: identifying issues, planning, working with different types of people, funding, publicizing, organizing, and developing special strategies. Part Three discusses major arenas of change such as improving the quality of neighborhoods, increasing organizational effectiveness, and lobbying for change. Includes illustrative vignettes, anecdotal examples, and a running narrative with ordinary characters to facilitate the understanding of key concepts.

Kahn, Si
Organizing: A Guide for Grassroots Leaders
Rev. ed. Silver Spring, Md., National Association of Social Workers, 1991. 345p. ISBN 0871011972

A readable, highly engaging step-by-step handbook for organizing and uniting people to mobilize for social change on a grass roots level. Discusses influencing power structures, becoming successful organizers and fund-raisers, and empowering people to bring about social change in their communities. The author, a renowned civil rights, labor, and community organizer, is also a musician and composer who wrote, recorded, and performed songs about struggle and social change.

Pope, Jacqueline
Biting the Hand That Feeds Them: Organizing Women on Welfare at the Grass Roots Level
New York, Praeger, 1989. 161p. ISBN 0275929221

Discusses various factors related to the social welfare policy of welfare recipients in Brooklyn, and examines the overall effectiveness of the Brooklyn Welfare Action Council (B-WAC, 1967-1973). Examines topics such as the nature and historical development of the welfare bureaucracy; New York City's economic, political, and welfare systems in the 1960s; the philosophy and background of the movement's white, affluent organizers; the relationship between the Catholic Church and the organizers; and the reasons underlying the growth of neighborhood groups and the birth of the B-WAC. Presents a working model for other advocates of grass roots self-help groups and for local community organizing.

Rees, Stuart
Achieving Power: Practice and Policy in Social Welfare
North Sydney, Australia, Allen and Unwin, 1991. 203p.
ISBN 0044423357

Proposes a fresh approach to the notion of empowerment in social services. Draws on insights and contributions of poets, philosophers, political scientists, sociologists, and social work educators in developing a political theory for human service professionals, particularly those involved in community work and organization.

Richan, Willard C.
Lobbying for Social Change
New York, Haworth Press, 1991. 277p. ISBN 1560240792

Introduces effective lobbying: lobbying elected and appointed officials, using the mass media, speaking before audiences, and testifying at hearings. Basic steps covered include assessing one's strengths and limitations; setting goals, priorities, and the agenda; understanding and choosing the right policy-makers; gathering evidence; and preparing the case. Discusses applying practice to policy content. The appendix addresses the various elements of the policy-making machinery—the President, the states, the courts, and fiscal and administrative agencies—and their implications for the policy advocate.

✦ Deviance and Criminality

Endelman, Robert
Deviance and Psychopathology: The Sociology and Psychology of Outsiders
Original ed. Malabar, Fla., Krieger Pub. Co., 1990. 300p.
ISBN 0894643444

A readable, well-documented guide to understanding the varieties of social deviancy, employing research from both sociology and psychology. Includes topics such as crime and delinquency, homosexuality and AIDS, prostitution, substance abuse, mental illness, bohemianism, radicalism, and the persistence of deviance. Attempts to integrate sociological and psychological perspectives concerning each major type of deviancy. Of particular value to social workers who draw on various perspectives in their daily practice.

Goode, Erich
Deviant Behavior
3rd ed. Englewood Cliffs, N.J., Prentice Hall, 1990. 382p.
ISBN 0132040417

Examines the sociologist's theoretical analysis of rule making and rule breaking, and society's reactions to rule-breaking behavior. Defines the notion of deviance and scrutinizes traditional and contemporary sociological theories of deviancy. Explores drug and alcohol abuse; prostitution and pornography; homosexuality; violent behavior such as homicide, family violence, and rape; property crimes; white-collar crime; and mental illness. Includes illustrative case examples.

Koss, Mary P., and Mary R. Harvey
The Rape Victim: Clinical and Community Interventions
2nd ed. Newbury Park, Calif., Sage Publications, 1991. 313p.
ISBN 0803938942

This readable text covers all aspects of rape including legal definitions, types and incidence, causes, the traumatic effects on and responses of rape victims, rape crisis centers, and treatment/ intervention. Includes a chapter on rape as a community issue and discusses treatment approaches for survivors. The chapters are well documented and include reviews of recent clinical and empirical literature concerning rape. Provides case vignettes of victim experiences along with case studies of successful community interventions.

Morris, Norval, and Michael Tonry
Between Prison and Probation: Intermediate Punishments in a Rational Sentencing System
New York, Oxford Univ. Press, 1990. 283p. ISBN 019506108X

Two leading criminologists explore innovative solutions to overcrowded prisons and an overburdened probation system. They argue that a system of intermediate punishments for convicted felons would be more effective than the current limited choice between prison and probation. They stress that the lack of punishment in between these two extremes is one of the fundamental reasons for the current crisis in the American corrections system. Proposes a range of sentencing options that include fines and other financial punishments, community service, house arrest, intensive probation, and electronic surveillance of movement.

✦ Families, Children, Adolescents, the Elderly

Abel, Emily K.
Who Cares for the Elderly? Public Policy and the Experiences of Adult Daughters
Philadelphia, Temple Univ. Press, 1991. 220p. ISBN 0877228140

Examines the experience of caring for elderly relatives from the perspective of adult daughters, through interviews with fifty-one female caregivers for elderly parents. Discusses the existing policy framework of caregiving and services for the aged, and provides an overview of family caregiving in the U.S. since 1800. Argues that the amount of caregiving women devote to elderly relatives is influenced by demographics and deficiencies in the long-term care system in the U.S. Additional content concerns work and leisure, brokering services, social supports, and suggestions for constructive change.

Ackerman, Robert J., and Dee Graham
Too Old to Cry: Abused Teens in Today's America
Blue Ridge Summit, Pa., TAB Books, 1990. 255p.
ISBN 083063407X

Discusses abused adolescents, abusers, and the causes and effects of abuse. Chapters look at teens in abusive families, the value systems of those families, and signs of abuse and neglect. The authors offer coping and prevention strategies, and information on creating change, community needs, and networking. May seem somewhat elementary, but will serve the needs of beginning students in the social/human services curriculum at the community college level.

Adamec, Christine A., and William L. Pierce
The Encyclopedia of Adoption
New York, Facts on File, 1991. 382p. ISBN 0816021082

This highly readable work is one of a series of specialized encyclopedias focusing on social problems. Contains approximately 400 entries, many of which include bibliographies, and covers topics such as adoption placement; the psychological effects of adoption on adoptees; birth parents and adoptive parents; drug-addicted and "boarder" babies; the impact of child abuse; adoptive family support groups; the status of birth fathers; open adoption; gay and lesbian adoption; children with AIDS; adolescent pregnancy; and transracial adoption. The helpful appendices list social service offices and adoption agencies, adoptive parent support groups, statistical data, and periodicals.

Bahr, Stephen J., ed.
Family Research: A Sixty-Year Review, 1930-1990
New York, Free Press, 1991-1992. 2 v. ISBN 0685591794

This scholarly collection of well-documented essays provides a history of family research from 1930-1990. Includes topics such as sexuality, interpersonal communication, family violence, gender roles, marital success and failure, the influence of employment variables on family interaction and well-being, the effects of divorce on children, and other research.

Balcerzak, Edwin A., ed.
Group Care of Children: Transitions Toward the Year 2000
Washington, D.C., Child Welfare League of America, 1989. 423p. ISBN 0878682872

Examines group care of children in contemporary society, including residential group care from 1966-1981, the Casey Family Program experience, characteristics of adolescents and their families in residential treatment, policy issues, family-oriented care, programs for high-risk adolescents, and treatment for emotionally impaired children.

Barth, Richard P., and Marianne Berry
Adoption and Disruption: Rates, Risks, and Responses
New York, Aldine de Gruyter, 1988. 247p. ISBN 0202360490

This work analyzes disrupted adoptions of older and special-needs children, using empirical research to highlight the particular factors associated with successful and unsuccessful adoption placement. The authors delineate risk factors, offer suggestions for improving placement, and attempt to determine those services that might alleviate known risk factors.

Barth, Richard P., and David S. Derezotes
Preventing Adolescent Abuse: Effective Intervention Strategies and Techniques
Lexington, Mass., Lexington Books, 1990. 222p.
ISBN 0669209031

Examines California's high school child abuse programs that reach 350,000 adolescents annually. Reports on the evaluation of physical and sexual abuse programs, and presents a framework for understanding and directing these programs. Additional concerns include methods and outcomes of adolescent abuse prevention, research implications for a K-12 abuse prevention program, a theoretically sound strategy for delivery of abuse prevention programs, and bridging the education, family, and child welfare fields. Concentrates on the social environment as the basis for individual or family pathology.

Bass, Deborah S.
Caring Families: Supports and Interventions
Silver Spring, Md., NASW Press, 1990. 281p. ISBN 0871011859

Identifies the stresses and the rewards experienced by caregivers, mostly women, in balancing the daily care of elderly, chronically ill, or disabled family members along with their other responsibilities. Analyzes both the practical needs of families who provide care to a needy family member, and the social worker's professional role in providing service and support to those families. Content includes the risks and benefits of caregiving, training families to meet needs, stages of caregiving, employment site services and benefits, community response, and broader implications for research and policy. Appendices list state and federal agencies concerned with health care, aging, developmental disabilities, handicapped children, and special education.

Bass, Ellen, and Laura Davis
The Courage to Heal: A Guide for Women Survivors of Child Sexual Abuse
New York, Perennial Library, 1988. 493p. ISBN 0060551054

A practical workbook for victims of childhood sexual abuse that can be used alone, with a professional counselor, or in a support group setting. The content is based on the experiences of survivors of child sexual abuse rather than on psychological theories. Describes the reconciliation of past wounds and includes specific suggestions, case examples, and exercises to aid the survivor in the healing process. Also includes a chapter on healing resources and an annotated bibliography.

Berrick, Jill Duerr, and Neil Gilbert
With the Best of Intentions: The Child Sexual Abuse Prevention Movement
New York, Guilford Press, 1991. 210p. ISBN 0898625645

Based on a study of eight California programs, this work provides an overview and history of child abuse prevention and an examination of contemporary methods of rape prevention. Attempts to illuminate the design, purpose, and effects of sexual abuse training prevention programs given to children in preschool and early elementary grades. Additional topics include program design and content development, age and learning in the early grades, informed parental consent, the role of teachers as protective guardians, prevention policy, and an "alternative protection" model.

Besharov, Douglas J., ed.
Family Violence: Research and Public Policy Issues
Washington, D.C., AEI Press, 1990. 278p. ISBN 0844737070

A collection of fifteen scholarly essays that examines research and issues concerning aspects of domestic violence, including spousal abuse, violence in relationships, child abuse and neglect, marital violence, interventions with men who batter, and legal policy.

Besharov, Douglas J.
Recognizing Child Abuse: A Guide for the Concerned
New York, Free Press, 1990. 270p. ISBN 0029030811

Discusses information that professionals and other involved parties need to know about child abuse, including legal issues such as obligations in reporting, liability for failure to report, protection for those who report, and the reporting process. The section on reporting child abuse includes discussion of sources of suspicion, physical and sexual abuse, neglect, endangerment and abandonment, psychological abuse, mentally disabled parents, and interviews with parents. A final section outlines what to do if parents suspect that their child has been abused by someone else, what to do if they feel they are abusing their own child, and what to do if they have been reported.

Blau, David, ed.
The Economics of Child Care
New York, Russell Sage Foundation, 1991. 192p.
ISBN 0871541181

A compilation of revised papers on the economics of child care presented at the Carolina Public Policy Conference at the University of North Carolina at Chapel Hill on May 16, 1990. This volume analyzes child care costs, working women, the quality and supply of child care services, parental choice, public policy in relation to child care, services, and issues for welfare parents. Includes both the economist's and the developmental psychologist's perspectives on child care policy.

Brody, Elaine M.
Women in the Middle: Their Parent-Care Years
New York, Springer Pub. Co., 1990. 288p. ISBN 0826163807

Discusses caregiving daughters and daughters-in-law who must balance caring for disabled older family members with other equally demanding responsibilities. Includes a historical overview, demographic and socioeconomic trends, the scope of parent care, and the effects of caregiving on caregivers. Addresses subjective experiences, role reversal and control, caregiving life stages, caregivers and their siblings, married and unmarried daughters, and daughters-in-law. Includes case illustrations.

Browne Miller, Angela
The Day Care Dilemma: Critical Concerns for American Families
New York, Insight Books, 1990. 316p. ISBN 0306434350

Provides a detailed examination of contemporary American child care services and their complex problems, discloses the results of a large study of parents of children in day care, and emphasizes the need for a rating system for child care services. Includes an overview of this complex problem, obstacles to a national child care program, dimensions of care, an analysis of six day care programs in California, consumer rating of day care services, and suggestions for change.

Bumagin, Victoria E., and Kathryn F. Hirn
Helping the Aging Family: A Guide for Professionals
New York, Springer Pub. Co., 1990. 310p. ISBN 0826175309

This handbook provides concrete advice written in a readable, engaging style. Part One discusses initial stages of intervention and the interaction approach with the environment; interviewing strategies assessment; and counseling philosophy and techniques. Part Two addresses ongoing work and specific problems of aging such as physical decline, socially distressing symptoms and behaviors, personality changes, and cognitive loss. Part Three deals with termination of treatment and practice evaluation. Principles and intervention strategies are illustrated with vignettes and case examples.

Buzawa, Eve Schlesinger, and Carl G. Buzawa
Domestic Violence: The Criminal Justice Response
Newbury Park, Calif., Sage Publications, 1990. 157p.
ISBN 0803935757

This work is the sixth in a series entitled *Studies in Crime, Law, and Justice*. The first section is on the classic response to domestic violence and discusses the causes and prevalence of domestic violence, the history of intervention by the criminal justice system, the police response to family violence, the police-citizen encounter, and the judicial response. Part Two looks at the process of change: the role of arrests, mandatory arrest policies, training programs, and prosecutorial and judicial changes.

Capuzzi, Dave, and Larry B. Golden
Preventing Adolescent Suicide
Muncie, Ind., Accelerated Development, 1988. 491p.
ISBN 0915202743

Concerned with the stark reality of adolescent suicide, this text includes understanding, predicting, and preventing teen suicide.

Topics include treatment after attempted suicide, societal and psychological factors and trends, personality traits of at-risk teens, depression, assessing lethality, interventions with Native Americans and college students, family therapy and networking, and legal issues of concern for the practitioner.

Carrieri, Joseph R.
Child Custody, Foster Care, and Adoptions
New York, Lexington Books, 1991. 372p. ISBN 0669276383

The author of this work, an attorney specializing in the field of foster care, serves as counsel for two large foster care agencies in New York State. He explains in detail aspects of the foster care system, and discusses legal matters such as custody, abuse and neglect, abandonment, the rights of all parties in foster care custody proceedings, adoption, and the role of the social worker in foster care proceedings. Includes the role of family law attorneys, judges, mental health practitioners, social workers, and educators. Focuses on the laws of New York State, but professionals in other states will find the information useful as well.

Child Welfare League of America
Child Welfare League of America Standards for Adoption Service
Rev. ed. Washington, D.C., The League, 1988. 115p.
ISBN 0878683658

These standards are designed to be employed as goals for the improvement of child welfare and adoption services in local, state, provincial, and national agencies as well as in voluntary and sectarian organizations. Standards address adoption as a child welfare service; services for birth parents, children, and adoptive parents; services during and after adoption placement; organization and administration; and adoption services and other community agencies involved with the adoption process.

Child Welfare League of America
Child Welfare League of America Standards for Service for Abused or Neglected Children and Their Families
Washington, D.C., The League, 1989. 1 v. (loose-leaf)
ISBN 0878683534

Beginning with a historical overview of child protection in the U.S., these standards concern protective services for abused or neglected children, social work in child protection service, the role of the courts in child protection, organization and administration of the child protective service agency, protection of children outside the family, and community responsibility in child protection.

Child Welfare League of America
Child Welfare League of America Standards for Services to Strengthen and Preserve Families with Children
Washington, D.C., The League, 1989. 76p. ISBN 0878682171

This short volume is addressed to all concerned with the improvement of child welfare services: the general public, citizen groups, public officials, child welfare workers, and agency administrators. Based on current knowledge, the developmental needs of children and their families, and proven ways of effectively meeting those needs, these standards can be used as goals for improving child welfare practice. Major sections cover family resources, support, and educational services; family-centered services (casework, crisis services, and staffing evaluation); and service delivery system support.

Cicchetti, Dante, and Vicki Carlson, eds.
Child Maltreatment: Theory and Research on the Causes and Consequences of Child Abuse and Neglect
New York, Cambridge Univ. Press, 1989. 794p.
ISBN 0521364558

This collection of well-documented essays discusses numerous facets of child development and child maltreatment, including neglect. Topics include a historical overview; a definition of child maltreatment; sexual abuse; intergenerational transmission of child abuse; developmental concerns such as attachment, peer relations, and social cognition; adolescent abuse; child abuse and its relation to delinquency and criminality; and prevention of maltreatment. Somewhat technical, with statistics, charts, and tables as supporting data.

Clark, Robin E., and Judith Freeman Clark
The Encyclopedia of Child Abuse
New York, Facts on File, 1989. 328p. ISBN 0816015848

One of a series of specialized encyclopedias focusing on social problems, this source provides a comprehensive, highly readable overview of all aspects of child abuse and contains over 500 concise entries, some with bibliographies. Includes information drawn from clinical cases, official statistics, social surveys, charts, tables, and lists. The introductory overview discusses historical background; the definition, scope, causes, and consequences of child abuse and neglect; and implications for further research. The detailed, helpful appendices lists organizations, child welfare resource centers, state child protection agencies, legislation and state statutes, statistics, the U.N. Declaration of Children's Rights, Canadian child welfare statutes, and provincial definitions of children in need of protection.

Combs-Orme, Terri
Social Work Practice in Maternal and Child Health
New York, Springer Pub. Co., 1990. 311p. ISBN 082616370X

Provides a complete discussion of the role of social work practice in the area of maternal and child health for professionals in medicine, nursing, public health, social work, and nutrition. Topics cover orientation into social work practice in this field, prenatal care, high-risk infants, infant death, young children, children with disabilities, adolescent sexuality and pregnancy, family planning services, and ethical and personal values in practice. Emphasizes a family-centered approach that should include the role of fathers as well as mothers and children.

Dryfoos, Joy G.
Adolescents at Risk: Prevalence and Prevention
New York, Oxford Univ. Press, 1990. 280p. ISBN 0195057716

Covers at-risk youth (aged ten to seventeen) who are likely to experience difficulties in school, at home, or in their communities. Explores four major problem areas of public concern: delinquency, substance abuse, adolescent pregnancy, and school failure. Examines how all four risky behaviors are interconnected and seeks to understand the size and scope of interventions needed for prevention. Concludes with illustrations of the successful application of principles at the local, state, and federal levels.

Faller, Kathleen Coulborn
Understanding Child Sexual Maltreatment
Newbury Park, Calif., Sage Publications, 1990. 251p.
ISBN 0803938411

This text provides a resource and review of case management for child welfare practitioners, mental health professionals, and professionals in allied disciplines such as law, medicine, and education, who must deal with child sexual abuse. Employs research, clinical experience, protocols, and actual case examples to clarify concepts. Defines and describes sexual maltreatment and its prevalence along with discussing diagnosis and assessment, sexual abuse in special situations, and the responses of legal and social welfare institutions.

Fanshel, David, Stephen J. Finch, and John F. Grundy
Foster Children in a Life Course Perspective
New York, Columbia Univ. Press, 1990. 352p. ISBN 0231071809

Examines the findings of a longitudinal-like investigation of several hundred children in foster care in the Casey Family Program, a privately funded agency located in western states. Content analysis includes preplacement life histories of the subjects, their experiences while in foster care, the circumstances surrounding their termination from the program, and their postplacement adjustment. Appendices contain a case reading schedule, follow-up interview, questionnaire, and the follow-up study.

Ferrara, Matthew L.
Group Counseling with Juvenile Delinquents: The Limit and Lead Approach
Newbury Park, Calif., Sage Publications, 1992. 156p.
ISBN 0803938853

This source advocates the use of group work with delinquent youth and attempts to integrate punitive and rehabilitative approaches to juvenile delinquency. Discusses topics such as the characteristics of delinquent youth; the traits of effective group counselors; interaction between the counselor and youth; guidelines for conducting groups; group counselor training; agency and facility administration; treatment services; and supervision, training, and evaluation. Includes a trainee workbook, group leader training post-test, group monitoring form, and personal history work sheet.

Fraser, Mark W., Peter J. Pecora, and David Haapala, eds.
Families in Crisis: The Impact of Intensive Family Preservation Services
New York, Aldine de Gruyter, 1991. 354p. ISBN 0202360695

A collection of essays that examines family preservation programs for high-risk children and families, and analyzes a study of several hundred families who participated in those services. Includes discussion of the family-based intensive treatment project, a literature review of family-based and intensive preservation services, factors contributing to the success and failure of these programs, project methodology, family and child characteristics, therapist perceptions, and implications for practice, policy, and research.

Garbarino, James, and others
Children in Danger: Coping with the Consequences of Community Violence
San Francisco, Jossey-Bass, 1992. 262p. ISBN 1555424163

The aim of this source is to aid teachers, social workers, psychologists, community workers, and other caregivers in helping children cope with violence in their communities and to analyze the negative impact that constant exposure to violence has on child development. Topics include post-traumatic stress disorder, the resilience and coping abilities of at-risk children, early intervention in the school and school-based programs, and the role of art and play in healing. Contains illustrative case examples.

Gartner, Alan, Dorothy Kerzner Lipsky, and Ann P. Turnbull
Supporting Families with a Child with a Disability: An International Outlook
Baltimore, Md., Paul H. Brooks Pub. Co., 1991. 233p.
ISBN 155766059X

Explores the inextricably intertwined issues of disability, culture, and family, and the ways in which they interrelate in the lives of families with disabled children. Discusses developments and trends concerning disability issues and activities, analyzes the manner in which different cultures respond to families with a disabled child, and examines family support systems.

Gelles, Richard J., and Claire Pedrick Cornell
Intimate Violence in Families
2nd ed. Newbury Park, Calif., Sage Publications, 1990. 159p.
ISBN 0803937180

This highly readable, engaging text for undergraduate students discusses various aspects of family violence, including child abuse, marital violence, hidden victims (siblings, adolescents, parents, the elderly), and widespread myths. Also provides profiles of violent homes, with an analysis of family life, prevention, and treatment programs. Each chapter includes discussion questions and suggested assignments.

Ginsburg, Evelyn Harris
School Social Work: A Practitioner's Guidebook: A Community-Integrated Approach to Practice
Springfield, Ill., Charles C. Thomas, 1989. 228p.
ISBN 0398055602

Focuses on integrated short-term, problem-oriented interventions in helping troubled children in their families, the public school, and the larger community. Section One concerns the child, the family, and significant social factors that affect the child. Section Two discusses the school environment, special education, and the school social worker's role. Section Three describes the specific job of the school social worker: procedures, methods, accountability, and utilization of resources. Detailed empirical case studies concern a token-based economy program, modification of a child's fire-setting behavior, a hyperactive learning disabled child, and the effects of reinforcement on disruptive behavior in the classroom.

Hewlett, Sylvia Ann
When the Bough Breaks: The Cost of Neglecting Our Children
New York, Basic Books, 1991. 346p. ISBN 0465091652

Discusses governmental and private sector policies and programs on child neglect and shows how attitudes and policies have worked against children. Hewlett advocates favorable conditions in which all children can flourish, and argues that decreased funding for services and education paired with an increase in single parenting places children in greater danger of being neglected.

Hutchings, Nancy
The Violent Family: Victimization of Women, Children, and Elders
New York, Human Sciences Press, 1988. 201p. ISBN 0898853834

This collection of readings covers all types of domestic violence and the roles that social work practitioners play in working with victims. Includes topics such as the causes of family violence, the economic status of women, battered women, child abuse, elder abuse, rape victims, and pornography's relation to violence against women. The appendix contains the *State* v. *Sheppard* court case concerning the use of video equipment to present the testimony of a ten-year-old victim in a sexual assault case.

Kadushin, Alfred, and Judith A. Martin
Child Welfare Services
4th ed. New York, Macmillan, 1988. 784p. ISBN 0023627107

A comprehensive study of child welfare services in the U.S., this commonly used source emphasizes a broad knowledge of child welfare services rather than practical skills. Content includes orientation, scope, and perspectives on child welfare services; supportive services such as home-based care; supplementary services (homemaker, day care, and protective services); substitutive care (foster family, adoption, residential institutions); and single adolescent parents and their children. Argues that the child welfare system's general orientation towards crisis, rescue, remedy, and response to emergency situations has, in many respects, failed America's children.

Kinney, Jill, David Haapala, and Charlotte Booth
Keeping Families Together: The Homebuilders Model
New York, Aldine de Gruyter, 1991. 235p. ISBN 0202360679

Emphasizes searching for cost-effective means of averting "out-of-home placement of children." Intervention topics include advantages/disadvantages of placement; components of family preservation; safety issues; maintaining progress between visits; assessing strengths and problems; aiding clients in helping themselves; administering family preservation services; staff management; and program evaluation.

Knudsen, Dean D., and JoAnn L. Miller, eds.
Abused and Battered: Social and Legal Responses to Family Violence
New York, Aldine de Gruyter, 1991. 232p. ISBN 0202304132

This collection of sixteen essays analyzes the difficulties that victims of domestic violence encounter with an unresponsive criminal justice system. Includes the perspectives of professionals in medicine, law, criminal justice, and social work, and examines the implications of programs and procedures intended to remedy the effects of family violence. Contends that the prevalence of domestic violence and high recidivism rates are most likely due to the negative ways in which police and the courts respond to victims of abuse, and to domestic violence in general.

Kozol, Jonathan
Rachel and Her Children: Homeless Families in America
New York, Crown Publishers, 1988. 261p. ISBN 051756730X

This analytical narrative, using real people and actual events, discusses the societal forces that cause and sustain homelessness, describes the change in the social characteristics of this population, and examines the inevitable results of becoming homeless. Appendices look at economies of scale concerning death and burial of homeless persons, and future prospects for American society.

Levy, Stephen J., and Eileen Rutter
Children of Drug Abusers
New York, Lexington Books, 1992. 260p. ISBN 0669273325

Examines the dynamics of families in which the parents are drug-addicted and focuses on family-centered clinical practice with these children. Covers topics such as the effects of chemical dependency on women and children; fetal alcohol syndrome; the effects of opiate addiction on newborns; the ways in which drug abuse programs, child welfare agencies, and the courts fail the children of drug abusers; and healing and treating these children. Intended as an impetus for change in attitude and policy toward these less fortunate children, whose needs have been neglected in therapeutic and social work arenas. Includes case illustrations.

Lindley, Mary E.
A Manual on Investigating Child Custody Reports
Springfield, Il., Charles C. Thomas, 1988. 174p. ISBN 0398054878

Written by a child welfare social worker who specializes in conducting court-ordered child custody investigations in Illinois. This detailed, step-by-step manual guides the social work or legal professional through the procedures of obtaining background information; interviewing children, parents, and collaterals; testifying in court; and writing the investigative custody report. Additional content covers the role of the investigator, assessment, relevant custody issues, and children's rights. Although aimed at practitioners in Illinois, others will find this manual useful as well.

Mace, Nancy l., and Peter V. Rabins
The Thirty-Six Hour Day: A Family Guide to Caring for Persons with Alzheimer's Disease, Related Dementing Illnesses, and Memory Loss in Later Life
Rev. ed. Baltimore, Md., Johns Hopkins Univ. Press, 1991. 329p. ISBN 0801840333

A critically acclaimed, comprehensive, practical guide for caregivers of relatives impaired by Alzheimer's Disease and other forms of progressive dementia. Addresses all medical, financial, legal, and emotional factors associated with this type of illness. Concepts are illustrated with specific examples and case vignettes. Appendices list sources for further reading, organizations, suppliers, state agencies and nursing home residents' rights.

Miller, Brent C., and others, eds.
Preventing Adolescent Pregnancy: Model Programs and Evaluations
Newbury Park, Calif., Sage Publications, 1992. 296p. ISBN 0803943903

Analyzes, evaluates, and compares several programs aimed at preventing teenage pregnancy. Specific programs addressed include small-group sex education in the school, improving social and cognitive skills, postponing sexual involvement, an age-phased approach, a teen outreach program that integrates volunteer community service with curriculum, school-linked reproductive health services, and school-based clinics.

Parham, Iris A.
Gerontological Social Work: An Annotated Bibliography
Westport, Conn., Greenwood Press, 1993. 207p.
ISBN 0313285381

This review of the literature provides detailed, evaluative annotations of books, journal articles, curriculum modules, reports, and bibliographies. Includes general works, clinical practice, education, geriatric health services, and other resource materials (journals, audiovisual materials, institutions and associations, and media resource guides).

Patton, Michael Quinn, ed.
Family Sexual Abuse: Frontline Research and Evaluation
Newbury Park, Calif., Sage Publications, 1991. 246p.
ISBN 0803939604

This anthology of fourteen essays examines various aspects of familial sexual abuse and treatment interventions. Subject matter encompasses an overview of child sexual abuse, a description of the Minnesota Family Sexual Abuse Project, the effects of sexual abuseon preschool children, sibling incest, intergenerational transmission, intrafamilial sexual abuse in Native American families, and evaluation of treatment and intervention.

Pecora, Peter J., James K. Whittaker, and Anthony N. Maluccio
The Child Welfare Challenge: Policy, Practice, and Research
New York, Aldine de Gruyter, 1992. 526p. ISBN 0202360814

Addresses the major problems and issues the child welfare field encounters, and how they affect and shape current child welfare practice, policy, and research. Emphasizes service innovations that could help meet these challenges. Topics include the policy context of child welfare; family-centered child welfare services; income support for families; child abuse and neglect; assessment; family-based services; foster care and adoption; group child care; and organizational requirements for child welfare services.

Pfeffer, Cynthia R., ed.
Suicide Among Youth: Perspectives on Risk and Prevention
Washington, D.C., American Psychiatric Press, 1989. 235p.
ISBN 0880481676

Fourteen scholarly essays on assessment and prevention examine methods of psychological autopsy in completed suicide cases of children and teenagers. Addresses epidemiological issues, depression, suicide clusters, effects of media, life stress and family risk factors, genetic factors and biological correlates, and prevention strategies. This work is a multidisciplinary effort involving educators, counselors, sociologists, epidemiologists, physicians, psychiatrists, and psychologists.

Pillari, Vimala
Scapegoating in Families: Intergenerational Patterns of Physical and Emotional Abuse
New York, Brunner/Mazel, 1991. 215p. ISBN 0876306393

Employing case illustrations, this source addresses scapegoating: its nature, the impact that it has on children and adult victims of abuse, and the types of children who become scapegoats (and the ways parents rationalize this behavior). Focuses on both the intrapsychic and interpersonal dynamics involved, and uses insights from psychoanalysis and systems theory.

Plumer, Erwin H.
When You Place a Child—
Springfield, Ill., Charles C. Thomas, 1992. 246p.
ISBN 0398057702

A beginning social worker's guide to placing a child in alternative care outside the home. This handbook, discussing principles and procedures for placement, emphasizes sensitivity to the needs of both children and parents; delineates fundamental considerations and rules; examines the effects of separation on both the child and the parent(s); and outlines the details of the placement process. Also discusses placement resources such as supplemental parenting, foster family care, and group care.

Roberts, Albert R.
Juvenile Justice: Policies, Programs, and Services
Chicago, Dorsey Press, 1989. 376p. ISBN 0256060193

A comprehensive overview of the U.S. juvenile justice system for students in social work and criminal justice. Includes definitions of juvenile justice, statistics, and methods of measuring delinquency. The historical perspective discusses treatment of juveniles in institutional and open settings, community interventions, the origins of juvenile court and probation, and diversion programs. Contemporary treatment approaches concern police work, the juvenile court, preadjudicatory detention, juvenile diversion, wilderness programs, family treatment, and social work advocacy and practice. Concludes with future developments and possibilities in juvenile justice and corrections.

Sachdev, Paul
Unlocking the Adoption Files
Lexington, Mass., Lexington Books, 1989. 247p.
ISBN 0669209759

Based on a study sample of 300 adoptees, adoptive parents, birth mothers, and adoption agency social workers, and covering the adoption years of 1958, 1968, 1978, this random study examines in detail the attitudes of all involved parties concerning major issues surrounding adoption. In order to analyze any response differences of the participants due to the time period of the adoption, the cases are analyzed in historical order. Gender differences in major adoption controversies are also analyzed.

Schneider, Robert L., and Nancy P. Kropf, eds.
Essential Knowledge and Skills for Baccalaureate Social Work Students in Gerontology
Washington, D.C., Council on Social Work Education, 1989. 232p. ISBN 0962313815

This text aims to prepare social work students for clinical work with the aged and their families. Material covers social work in home, health, and community care services, as well as in nursing homes, case management, discharge planning, and social work practice with elderly minorities and women. Includes illustrative case studies and related audiovisual resources.

Sgroi, Suzanne M., ed.
Vulnerable Populations: Evaluation and Treatment of Sexually Abused Children and Adult Survivors
Lexington, Mass., Lexington Books, 1988-1989. 2 v.
ISBN 0669163368 (vol.1); ISBN 0669209422 (vol. 2)

This handbook for practitioners in family therapy examines children's sexual behavior, assessment and treatment of the adult survivor of sexual abuse, treatment of male victims of child sexual abuse in the military, criminal investigation of child sexual abuse cases, videotaping the sexually abused child, using art therapy in treating sex offenders, and the sexual mores of the Yuit Eskimo. Analyzes causes, effects, and interventions with victims and offenders.

Starr, Raymond, and David A. Wolfe
The Effects of Child Abuse and Neglect: Issues and Research
New York, Guilford Press, 1991. 304p. ISBN 0898627591

Studies the effects of child abuse and neglect through the use of longitudinal research, and advocates this approach in order to fully understand the problem. Subject matter includes life-span developmental issues, a longitudinal study of high-risk families, the developmental consequences of child abuse, methodological considerations, qualitative methods, measurement of parental personality traits and psychopathology, child rearing attitudes and belief systems, assessment of parent-child interaction and development in abused children, and the effects of child abuse on child health.

The State of America's Children Yearbook
Washington, D.C., Children's Defense Fund, 1994- .
ISSN 1055-9213

This annual reference source contains statistical data concerning the well-being of America's children. Factors analyzed include family income, health, children and families in crisis, child care and early childhood development, housing and homelessness, hunger and nutrition, teenage pregnancy and youth development, and violence and risk factors associated with violent crime. Appendices cover maternal labor force participation, youth unemployment rates, and state statistics on factors related to child well-being. The introductory essay of the 1994 yearbook addresses the issue of children and guns in the United States.

Stein, Theodore J.
Child Welfare and the Law
New York, Longman Pub. Co., 1991. 213p. ISBN 080130315X

Explains the special relationship between child welfare and law in the United States as it applies to social work practice. Content encompasses a definition and sources of the law, a summary of child welfare and the law, the policy framework surrounding child welfare programs, client rights, litigation, reform measures, and the participation of child welfare practitioners in the legal process. Contains historical examples, legislation, court decisions, and a table of cases.

Stith, Sandra M., Mary Beth Williams, and Karen H. Rosen, eds.
Violence Hits Home: Comprehensive Treatment Approaches to Domestic Violence
New York, Springer Pub. Co., 1990. 363p. ISBN 0826172709

Provides an overview of domestic violence and discusses treatment interventions for spousal abuse, child physical and sexual abuse, adult survivors of incest, and other types of child molestation, elder abuse, and neglect. Treatment methods include crisis intervention, family therapy, multidisciplinary team approaches, Gestalt therapy, and a cognitive-behavioral approach. Concludes with theoretical models and perspectives of assessment and treatment.

Thompson, Travis, and Susan C. Hupp, eds.
Saving Children at Risk: Poverty and Disabilities
Newbury Park, Calif., Sage Publications, 1992. 190p.
ISBN 0803939671

This collection of twelve readings analyzes the correlation between developmental disabilities in children and poverty. Explores developmental risk factors and poverty, prevention measures, early intervention measures, longitudinal interventions, early childhood education for children in inner-city classrooms, promoting cognitive development in at-risk children, education outcomes for Down's syndrome children, adopted disabled children, and national policy issues.

Van Ornum, William, and John B. Mordock
Crisis Counseling with Children and Adolescents: A Guide for Nonprofessional Counselors
New expanded ed. New York, Continuum Pub. Co., 1990. 240p.
ISBN 082640474X

Devoted to teachers, social workers and counselors, medical personnel, attorneys and law enforcement officers, clergy, and others in the helping professions who are involved daily with the crises troubled children face and who may not possess training in child development, this practical, engaging guide covers topics such as definitions of crisis, communication with children, adolescence, empathic focusing, death and divorce, health crises and handicaps, child and sexual abuse, foster care, defiant children, and alcohol issues. Includes illustrative examples, case vignettes, and suggested readings for children.

Wasik, Barbara Hanna, Donna M. Bryant, and Claudia M. Lyons
Home Visiting: Procedures for Helping Families
Newbury Park, Calif., Sage Publications, 1990. 304p.
ISBN 0803935412

Designed for home visitors, trainers, program directors, and policy makers, this work attempts to provide professionals with comprehensive information concerning issues, philosophy, skills, and procedures essential for home visiting with families. Subject matter encompasses a historical overview, philosophy, illustrative home visiting programs, personnel issues, assessment and documentation, and future directions in home visiting. Appendix contains sample home visit report forms used in an infant health and development program.

Whittaker, James K., and others, eds.
Reaching High-Risk Families: Intensive Family Preservation in Human Services
New York, Aldine de Gruyter, 1990. 206p. ISBN 0202360571

Describes intensive family preservation services, emphasizing the homebuilders approach to intervention, and the rationale and implications for social work education. Aspects covered include the implications of family preservation for placement, prevention, effective service delivery, the public policy context and implications, theoretical perspectives, the direct social work practice curriculum, practice research methods, the homebuilders model, community leadership and social change, and prevention of problem behavior.

✦ Human Diversity—Gender, Race/Ethnicity, Sexual Orientation

Applewhite, Steven R., ed.
Hispanic Elderly in Transition: Theory, Research, Policy and Practice
New York, Greenwood Press, 1988. 238p. ISBN 0313244782

Focuses on aging and the elderly from the Hispanic point of view, and provides information related to Hispanic aging: health, economic, social, and cultural factors. Includes discussion of access to and utilization of health and human services, interest group politics and empowerment, income maintenance, and the extended family. Also looks at the diversity of the Hispanic community from areas such as Mexico, Puerto Rico, and Cuba.

Burgest, David R., ed.
Social Work Practice with Minorities
2nd ed. Metuchen, N.J., Scarecrow Press, 1989. 348p.
ISBN 0810822075

This collection of essays discusses the different components of education and the practical skills necessary for clinical social work practice with multiethnic and multiracial minority groups, including those from Third World cultures. Includes information on literature and resources on social work practice with minority clientele; identifying prejudice, racism, and cultural conflict in the therapeutic relationship; the importance of being sensitive to clients of different ethnic/racial backgrounds; and strategies and methods employed by minority scholars and practitioners in the area of social work with minorities.

Chau, Kenneth K. L., ed.
Ethnicity and Biculturalism: Emerging Perspectives of Social Group Work
New York, Haworth Press, 1991. 133p. ISBN 1560240946

Focuses on the nature of social work with multicultural groups and emphasizes the strength perspective in group work practice. Includes information on topics such as facilitating bicultural development and intercultural skills in ethnically mixed groups, group work with transracial foster parents, multiculturalism in school-based adolescent groups, biculturalism in the workplace, working with Vietnamese refugees in single-session groups, and teaching ethnic and racial sensitivity through groups.

Child Welfare League of America
Serving Gay and Lesbian Youths: The Role of Child Welfare Agencies: Recommendations from a Colloquium, January 25-26, 1991
Washington, D.C., The League, 1991. 29p. ISBN 0878684956

This brief overview of research on service to gay and lesbian youths considers the unique psychosocial problems of this high-risk group. Discusses the barriers to service these youths experience and recommends policies, products, and advocacy needed to advance child welfare practice and services for these adolescents. The list of selected readings is particularly helpful.

Cornell, Stephen E.
The Return of the Native: American Indian Political Resurgence
New York, Oxford Univ. Press, 1988. 278p. ISBN 0195037723

Analyzes Native American-white relations from a sociological perspective and explores the reasons why Native Americans did not turn to militant politics before the 1960s and early 1970s. Discusses the patterns of alternating militant and peaceful political responses made by Native Americans, domination and subordination by white Americans.

Devore, Wynetta, and Elfriede G. Schlesinger
Ethnic-Sensitive Social Work Practice
3rd ed. New York, Merrill Pub. Co., 1991. 367p.
ISBN 0675212863

Outlines a conceptual base for ethnic-sensitive social work practice and illustrates assumptions, principles, and approaches with case examples. Ethnic-sensitive practice is defined and explored in terms of the ethnic reality, ethnicity and the life cycle, social work values, and knowledge of human behavior. Provides strategies and procedures for direct and macro practice. The appendix contains a sample community profile.

Dinnerstein, Leonard, Roger L. Nichols, and David M. Reimers
Natives and Strangers: Blacks, Indians and Immigrants in America
2nd ed. New York, Oxford Univ. Press, 1990. 362p.
ISBN 0195057228

The chronological coverage begins with the colonial period, and goes on to discuss topics such as immigrants and nativists, industrialism and population migrations, the development of the West, new immigration and the ethnic character of contemporary America, recent refugees, and immigration legislation and policies. Concentrates on economic growth and the development of the social attitudes of racial, religious, and native minorities. Recounts Native American efforts to maintain independence and manage relations with society at large.

Everett, Joyce, Sandra S. Chipungu, and Bogart R. Leashore, eds.
Child Welfare: An Africentric Perspective
New Brunswick, N.J., Rutgers Univ. Press, 1991. 325p.
ISBN 0813517125

The eleven readings which comprise this work discuss child welfare issues and interventions with African American children and families from a perspective based on cultural strengths. Addresses cultural consciousness, family structures and service options, and examines various aspects of both the policy and the practice arenas of child welfare services. The introductory essay examines the extent of the crisis affecting African American children, the historical background of this situation, and the consequences of ignoring the issue of race and difference in child welfare policy and practice.

Farley, Reynolds, and Walter R. Allen
The Color Line and the Quality of Life in America
New York, Oxford Univ. Press, 1989. 493p. ISBN 0195060296

Provides a detailed, scholarly review of African-American demographic and socioeconomic changes and the progress made in counteracting racism and discrimination in the U.S. Statistical data is drawn from the 1980 census and earlier. Topics discussed include the continuing dilemma of race in America, population growth and distribution, mortality and fertility, housing segregation, and other socioeconomic factors.

Gibbs, Jewell Taylor, and Larke Nahme Huang
Children of Color: Psychological Interventions with Minority Youth
San Francisco, Jossey-Bass, 1989. 423p. ISBN 1555421563

Addresses the unique psychological and behavioral problems and the special mental health needs of minority children and adolescents. Discusses clinical assessment and treatment interventions with Asian, Native American, African American, Hispanic American, and biracial youth. Employs both an ecological perspective in its coverage of specific interventions and a comparative cross-cultural approach to various topics. Outlines implications for research, training, and practice.

Ginzberg, Lori D.
Women and the Work of Benevolence: Morality, Politics, and Class in the Nineteenth-Century United States
New Haven, Conn., Yale Univ. Press, 1990. 230p.
ISBN 0300047045

Provides a detailed historical background of the prominent role of women in formal benevolence and analyzes the ways in which upper- and middle-class interests and religious affiliations were powerful influences in a wide variety of charitable and reform causes in the antebellum years. Argues that these well-to-do women described their benevolence as Christian (Protestant), moral, and uniquely female—reflecting an ideology that combined the concept of femininity with ideas about morality.

Gordon, Linda, ed.
Women, the State, and Welfare
Madison, Wisc., Univ. of Wisconsin Press, 1990. 311p.
ISBN 0299126609

These twelve essays analyze the relationship between gender and the contemporary welfare state. Addresses feminist scholarship in the welfare state; gender bias; women and American political society, 1780-1920; family violence; late-capitalist political culture from a socialist-feminist critical viewpoint; the ineffectiveness of the war on poverty in conquering the feminization of poverty; and African American women and AFDC.

Harrison, Diane F., John S. Wodarski, and Bruce A. Thyer, eds.
Cultural Diversity and Social Work Practice
Springfield, Ill., Charles C. Thomas, 1992. 247p.
ISBN 0398057559

The value of incorporating unique strategies in working with culturally diverse populations is presented in the nine well-documented essays in this book. Includes an overview of cultural diversity and social work practice, and discusses methods for practice, research, and education along these lines.

Herek, Gregory M., and Kevin Berrill, eds.
Hate Crimes: Confronting Violence Against Lesbians and Gay Men
Newbury Park, Calif., Sage Publications, 1992. 310p.
ISBN 0803945418

Surveys the scope of violence and discrimination directed against gay men and lesbians in the U.S. Examines the social context of hate crimes, the psychology of anti-gay violence, and the underlying motivations, mental health consequences, and treatment interventions for hate crime victims. Provides strategies for activists attempting to counteract the hate crime mentality, and presents policy implications.

Ho, Man Keung
Minority Children and Adolescents in Therapy
Newbury Park, Calif., Sage Publications, 1992. 236p.
ISBN 0803939124

Covers treatment interventions with Asian/Pacific American, Native American, African American, and Hispanic American children and adolescents. Part One presents a transcultural framework for assessment and therapy; the second section deals with understanding and assessing the mental health difficulties of these youth; and Part Three looks at culture-specific intervention and treatment approaches in individual, family, and group therapies. The appendix contains an ethnic competence skill model that can be used in psychological interventions with minority youth.

Jacobs, Carolyn, and Dorcas D. Bowles, eds.
Ethnicity and Race: Critical Concepts in Social Work
Silver Spring, Md., National Association of Social Workers, 1988. 242p. ISBN 0871011557

The premise of this anthology is that understanding the dual perspective of people of color is essential in providing adequate, effective, ethnically sensitive treatment to minority clients. Includes material on value systems, family interactions, role assignments, religion, immigration and cultural adjustment, the extended family network, and help-seeking patterns. Discusses practice issues as they relate to specific ethnic minority groups; curriculum issues centered on incorporating cultural diversity studies into social work education; and racism, ethnocentrism, and sexism.

Kismaric, Carole
Forced Out: The Agony of the Refugee in Our Time
New York, Random House, 1989. 191p. ISBN 0679723471

This photographic documentary was inspired by the work of the Human Rights Watch committee groups, who gather facts and provide advocacy and education regarding the cause of human rights. Data included in this work on the plight of displaced people from all over the world reflect the situation as of the fall of 1988. Includes testimony of victims from numerous countries, and provides discussion of government policies and procedures that intensify refugees' problems, including refugee camps.

Kitano, Harry H. L., and Roger Daniels
Asian Americans: Emerging Minorities
Englewood Cliffs, N.J., Prentice Hall, Inc., 1988. 214p. ISBN 0130491640

Provides a historical overview of Asian Americans, describing their experiences in the latter half of the twentieth century, and discusses changes in immigration laws and refugee policy. Specific groups considered include the Chinese, Japanese, Filipinos, Asian Indians, Koreans, Pacific Islanders, and Southeast Asians. Outlines the present status of Asian Americans and examines their responses to prejudice and discrimination, as well as their civic assimilation.

Logan, Sadye Louise, Edith M. Freeman, and Ruth G. McRoy, eds.
Social Work Practice with Black Families: A Culturally Specific Perspective
New York, Longman, 1990. 289p. ISBN 0801300126

The historical overview presented by these essays provides a basic framework for understanding the needs of African American families. Aspects discussed include theoretical perspectives, culturally relevant assessment and interventions, and the future of social work practice and research with those families.

Lum, Doman
Social Work Practice and People of Color: A Process-Stage Approach
2nd ed. Pacific Grove, Calif., Brooks/Cole Pub. Co., 1992. 252p. ISBN 0534170404

This text is designed primarily for minority social work educators and students. Emphasis is on generalist social work practice and process with Africans, Asians, Hispanics, Native Americans and immigrants and refugees. Integrates case studies with recent social science theory and research, and discusses implications for clinical practice such as values, ethnicity, culture, social class, and other socioeconomic factors.

Molnar, Stephen
Human Variation: Races, Types, and Ethnic Groups
3rd ed. Englewood Cliffs, N.J., Prentice Hall, 1992. 354p. ISBN 0134461622

Analyzes how biology influences and determines variations among human beings. Discusses topics such as racial variation and the perception of human differences, the genetic basis for human variation, traits of simple and complex inheritance, and the relation between human variability and behavior. Speculates on the future of the human species in terms of population growth and changes due to the effects of disease, a changing environment, selective reproduction through genetic counseling, and ongoing evolutionary processes. A well-documented study with tabular data, statistics, and solid references.

Muller, Charlotte Feldman
Health Care and Gender
New York, Russell Sage Foundation, 1990. 258p. ISBN 0871546108

Drawing on national statistics and data from medical, health, and behavioral literature, this work examines gender bias and health care service utilization by men and women. Discusses matters such as health and treatment issues, insurance and financial issues, the special needs of the elderly, the role of Medicaid with respect to women's health care, reproductive care and services, the implications of gender in health care, and suggestions for reform and future research.

Notman, Malkah T., and Carol C. Nadelson, eds.
Women and Men: New Perpectives on Gender Differences
Washington, D.C., American Psychiatric Press, 1991. 144p. ISBN 0880481366

Employs well-documented biological and social science findings to present a case for the basis for gender differences. Analyzes the range, extent, and ways that gender differences are affected by social, cultural, and biological influences, and concludes that gender differences are an important reality. Includes specific information on topics such as gender differences in brain structure, hormones and behavior, the psychology of women, and the acquisition of femininity. Also discusses the development of gender.

Rhode, Deborah L., ed.
Theoretical Perspectives on Sexual Difference
New Haven, Conn., Yale Univ. Press, 1990. 315p. ISBN 0300044275

This feminist anthology examines gender from an interdisciplinary perspective, including the areas of feminism, sociobiology, psychology, sociology, anthropology, political science, economics, philosophy, and law. Provides a historical and theoretical overview of feminism and sexual differences and a psychosocial model of gender. Discuses specific topics such as Darwinism, children and gender, ethics (from a female point of view), and sexual equality.

Rogler, Lloyd H., Robert G. Malgady, and Orlando Rodriguez
Hispanics and Mental Health: A Framework for Research
Malabar, Fla., Krieger Pub. Co., 1989. 163p. ISBN 0894642480

Constructs a theoretical framework for organizing clinical-service mental health research pertaining to Hispanic populations. Presents a set of research problems centered on Hispanic mental health concerns, and examines the complex issues regarding specific mental health problems and needs. Includes information on the emergence of mental health problems, help-seeking behavior, evaluating mental health, psychotherapy services, and post-treatment adjustment research problems.

Thompson, Richard H.
Theories of Ethnicity: A Critical Appraisal
New York, Greenwood Press, 1989. 196p. ISBN 0313266360

This comparative study of competing theoretical models by social theorists concerning race and ethnicity analyzes how the state defines ethnicity and the role capitalism plays. Presents cross-national perspectives for the understanding and defining of ethnicity. Examines ethnicity and human nature, the sociobiology of race and ethnicity, the fundamental sentiments of ethnicity and the state, and Marxist approaches to race and ethnicity.

Tinker, Irene, ed.
Persistent Inequalities: Women and World Development
New York, Oxford Univ. Press, 1990. 302p. ISBN 0195059352

Discusses the role of women in economic development, particularly women in Third World countries; income distribution; the sexual division of labor; access to technology and education; and the effects of patriarchy as a persistent barrier to women's equality. This anthology represents a variety of ideologies rather than one specific view of history or of economic development.

✦ Industrial/Occupational Social Work

Gould, Gary M., and Michael Lane Smith
Social Work in the Workplace: Practice and Principles
New York, Springer Pub. Co., 1988. 362p. ISBN 0826153801

Discusses occupational and industrial social work and employee assistance/counseling programs, including both knowledge and skill content. Includes topics such as ethical issues, crisis intervention, substance abuse, workplace accidents, identification and referral of health enhancement programs, shift work problems, child care, affirmative action/equal opportunity, and AIDS in the workplace. Includes case examples.

Roman, Paul M., ed.
Alcohol Problem Intervention in the Workplace: Employee Assistance Programs and Strategic Alternatives
New York, Quorum Books, 1990. 413p. ISBN 0899304591

These twenty-four well-documented essays provide a detailed, scholarly analysis of alcohol problems and treatment interventions in the workplace, focusing on employee assistance programs. Includes an overview of alcohol problems, identifying those who are most at risk for alcohol abuse (the alienated, burned out, and unchallenged); a literature review of jobs, occupations, and patterns of alcohol abuse; the structure, dynamics, and problems of employee assistance programs; and implications for intervention. Provides tables, statistics, and other empirical data.

✦ Psychology, Psychiatry, and Human Behavior

Aguilera, Donna C.
Crisis Intervention: Theory and Methodology
6th ed. St. Louis, Mo., Mosby, 1990. 298p. ISBN 0801600634

A well-documented discussion of crisis intervention theory and methodology. Includes topics such as crisis intervention methods, psychotherapeutic techniques, group work, sociocultural factors affecting intervention, AIDS, and burnout syndrome. Employs case studies to illustrate the techniques used in specific crisis situations.

Bloom, Martin, ed.
Changing Lives: Studies in Human Development and Professional Helping
Columbia, S.C., Univ. of South Carolina Press, 1992. 427p. ISBN 0872497550

This anthology surveys theories concerning the structure and dynamics of human behavior and development, arguing that this is an essential knowledge base for the helping professional. Discusses the role of services in areas such as social work practice, sibling relationships and family size, education, problem solving and stress, depression, homelessness, and intergenerational relations.

Fink, Paul Jay, and Allan Tasman, eds.
Stigma and Mental Illness
Washington, D.C., American Psychiatric Press, 1992. 236p. ISBN 0880484055

This introductory source book for mental health practitioners containing nineteen essays includes both scholarly studies and personal experiences that describe the problem of stigma. Discusses the devastating effects of stigmatization on mentally ill persons, their families, and the professionals who treat them, and the need for more sensitivity on the part of practitioners and the lay public alike. Also covers historical aspects, social issues and stereotypes, and institutional issues.

France, Kenneth
Crisis Intervention: A Handbook of Immediate Person-to-Person Help
2nd ed. Springfield, Ill., Charles C. Thomas, 1990. 261p. ISBN 0398056307

This handbook's interdisciplinary orientation provides the caregiver with a practical framework for furnishing immediate assistance to people in crisis in a variety of social service settings: crisis centers, hot lines, college counseling centers, hospitals, schools, correctional institutions, and youth programs. Material covers crisis theory intervention skills; clients with chronic difficulties; suicidal clients; rape and other crime victims; patients with terminal illnesses; bereaved persons; community relations; and administrative issues.

Gilbert, Paul
Depression: The Evolution of Powerlessness
New York, Guilford Press, 1992. 561p. ISBN 0898628849

Provides an overview of numerous psychological approaches for understanding and treating depression. Part One presents a historical and descriptive analysis; Part Two examines evolution theory, its relation to depression, and the social dimension of this condition; Part Three studies theories of depression and their strengths and weaknesses. Appendices address the measurement of depression, the theory and styles of personal adjustment, endogenous and neurotic depression, and culture and change.

Gilliland, Burl E., and Richard K. James
Crisis Intervention Strategies
Pacific Grove, Calif., Brooks/Cole Pub. Co., 1993. 622p.
ISBN 053419494X

Employing an eclectic approach that incorporates a variety of treatment modalities, this work discusses applied therapeutic counseling generally, and in crisis intervention specifically. Subject matter includes crisis intervention theory; case management; suicide; post-traumatic stress disorder; assaults; chemical dependency; bereavement; hostage crises; human service worker burnout; and trends in crisis intervention. Contains illustrative case studies and examples.

Goldberger, Leo, and Shlomo Breznitz, eds.
Handbook of Stress: Theoretical and Clinical Aspects
2nd ed. New York, Free Press, 1993. 819p. ISBN 0029120357

A collection of thirty-nine scholarly, well-documented essays examining stress research and its history. Includes psychological and biological processes; the measurement of stress and coping; common factors and developmental causes throughout the life cycle; psychiatric, somatic, and psychosomatic conditions; and extreme events such as domestic violence, disasters, the Holocaust, migration, and HIV infection. Presents treatment interventions and social supports. Contains a discussion of the role that personality and emotion play in stress, work-induced stress, and depression.

Greene, Roberta R., and Paul H. Ephross, eds.
Human Behavior Theory and Social Work Practice
New York, Aldine de Gruyter, 1991. 361p. ISBN 0202360717

This overview of human behavior for social work students presents human behavior theories as an aid to understanding people—their functions in various social groups and cultures, their relationship to the environment, and their development throughout the life cycle. Eclectic in nature, the text includes discussion of many types of theories including psychoanalytic, cognitive, symbolic interaction and ecological system theories, and those espoused by Erikson and Rogers. Also surveys the roles of genetics and environment in human development.

Hendricks, James E., ed.
Crisis Intervention in Criminal Justice/Social Service
Springfield, Ill., Charles C. Thomas, 1991. 267p.
ISBN 0398057451

The aim of this anthology is to present theoretical, analytical, and practical information on crisis intervention for preservice and in-service criminal justice and social service crisis intervenors, particularly first responders. Provides a historical background and a theoretical overview of crisis intervention and discusses crisis intervention in relation to spousal abuse, child abuse, elder maltreatment, rape, death notification, suicide intervention, and stress in criminal justice and social service work. Includes models and methods for applying theory to actual situations.

Lewis, Dan A., and others
Worlds of the Mentally Ill: How Deinstitutionalization Works in the City
Carbondale, Ill., Southern Illinois Univ. Press, 1991. 198p.
ISBN 0809314770

Describes and evaluates the care given by the public mental health system, focusing on mental health care in Chicago. Suggests reforms in treatment and care of the mentally ill, and argues that the traditional concepts employed to understand and describe mental illness are no longer relevant in the current mental health care system. Discusses the role of the state hospital after deinstitutionalization, characteristics of the state mental patient, readmission to the hospital, a mental patient's life in the community, criminally insane patients, and mental health policy. The appendix includes sampling and interviewing procedures, and a life events interview.

Nye, Robert D.
Three Psychologies: Perspectives from Freud, Skinner, and Rogers
4th ed. Pacific Grove, Calif., Brooks/Cole Pub. Co., 1992. 160p.
ISBN 053416224X

A brief, straightforward overview of the main ideas and contributions to modern psychology of three major theorists: Sigmund Freud, B.F. Skinner, and Carl Rogers. Compares and contrasts each theory in relation to basic human nature, personality development, society's role, the study of human behavior, views on aggression, and controlling human behavior. Presents evaluative research supporting and criticizing the work of each theorist.

Roberts, Albert R., ed.
Crisis Intervention Handbook: Assessment, Treatment, and Research
Belmont, Calif., Wadsworth Pub. Co., 1990. 341p.
ISBN 0534125107

This handbook is a collaborative endeavor involving clinical social workers, health social workers, counseling and clinical psychologists, and child psychologists. Applies crisis intervention theory, principles, and techniques to youth, victims of violence and their abusers, and both health- and mental health-related crises. Includes an overview of crisis theory and information on topics such as suicidal adolescents, crisis intervention in high schools and colleges, battered women, rape and incest victims, substance abuse, disaster victims, and evaluation of crisis intervention. Includes case illustrations.

✦ Research Methods and Statistics

Alter, Catherine, and Wayne Evens
Evaluating Your Practice: A Guide to Self-Assessment
New York, Springer Pub. Co., 1990. 195p. ISBN 0826169600

Addressed to social work practitioners and students who must concern themselves with accountability to clients, organizations, and communities. This handbook employs case studies to clarify self-assessment research in terms of data collection, statistical analysis, qualitative and quantitative designs, and research validity. Details each step of the evaluative self-assessment process as it is applied to social work practice and uses the empirical approach to measure practice effectiveness and to predict client outcomes. Discusses the moral and ethical dimensions of social work practice.

Blythe, Betty J., and Tony Tripodi
Measurement in Direct Social Work Practice
Newbury Park, Calif., Sage Publications, 1989. 159p.
ISBN 0803930801

Addresses the need for continual, empirical evaluation of the objectives and processes of social work practice. The authors advocate using social workers to define, measure, and evaluate variables objectively in order to improve practice. Applies empirical measurement concepts to an actual "phase model" of clinical practice and illustrates these concepts with concrete examples. Contents include information on measurement, assessment, the planning of interventions, termination, and follow-up.

Grinnell, Richard M.
Social Work Research and Evaluation
4th ed. Itasca, Ill., F. E. Peacock Publishers, 1993. 468p.
ISBN 0875813666

This comprehensive resource discusses research methods and evaluation, including such topics as hypothesis formulation, instrument design, sampling and measurement, validity and reliability, and data gathering methodology. The analysis portion includes information on program evaluation, writing and evaluating research reports, and meta-analysis. Grinnell attempts to dispel the misconception that the nature of social work practice eludes "scientific research" methodologies.

Pilcher, Donald M.
Data Analysis for the Helping Professions: A Practical Guide
Newbury Park, Calif., Sage Publications, 1990. 259p.
ISBN 0803937245

A fundamental guide to statistical methodology for instructors, practitioners, and students of nursing, social work, and education. Includes information on data distribution, statistical computations, and data analysis. A useful glossary and an appendix concerning the use of probability tables are also provided.

Royse, David D.
Research Methods in Social Work
Chicago, Nelson-Hall, 1991. 287p. ISBN 0830412107

Explains the basic concepts and techniques employed in research methods: hypothesis formation and testing; use of variables; design; validity; data collection and analysis; presentation of results; and program evaluation. Also discusses ethical issues in research, and methods for writing professional research reports and journal articles. Appendices contain information on attitudes toward research courses, a table of random numbers, and a sample drug attitude questionnaire.

Weinbach, Robert W., and Richard M. Grinnell
Statistics for Social Workers
2nd ed. White Plains, N.Y., Longman, 1991. 217p.
ISBN 080130413X

A fundamental, readable guide to statistical concepts and procedures for social workers who are not mathematically inclined. Content includes tabular and graphic presentations, measures of central tendency, variability and normal distribution, hypothesis testing, correlations, cross-tabulation, comparison of averages, and various statistical tests. Includes social work practice problems and examples that facilitate the application of statistical methodology to both social work research and daily practice.

Wolcott, Harry F.
Writing Up Qualitative Research
Newbury Park, Calif., Sage Publications, 1990. 94p.
ISBN 080393792X

This readable, straightforward primer concentrates on the special writing problems of social science researchers, the nature of qualitative research, data collection, and data management. Begins with a description of the writing process itself, and then discusses particular difficulties, such as the nature of social understanding, reporting formats, authorial voice, language use, and presentation formats. Discusses organizing data, staying committed to the project, revising and editing, and publishing the final product.

✦ Sexually Transmitted Diseases

Adimora, Adaora A., and others
Sexually Transmitted Diseases: Companion Handbook
New York, McGraw-Hill, 1993. 436p. ISBN 0070003807

This compact handbook is addressed to clinicians and public health professionals. Discusses microbiology, antimicrobial susceptibility, epidemiology, clinical manifestations in men and women, treatment and therapy, and prevention of sexually transmitted diseases. Treatment recommendations conform to the Centers for Disease Control and the National Center for Prevention Services' *Sexually Transmitted Diseases: Treatment Guidelines* (1993). Discusses also patient counseling and sexual assault.

DeVita, Vincent T., and others, eds.
AIDS: Etiology, Diagnosis, Treatment, and Prevention
3rd ed. Philadelphia, Lippincott, 1992. 607p. ISBN 0397512295

Discusses important aspects of AIDS, such as origins, etiology, epidemiology, immune response, clinical manifestations, testing for complications, high-risk sexual practices, and the impact of the HIV infection on health care workers.

Reamer, Frederic G., ed.
AIDS and Ethics
New York, Columbia Univ. Press, 1991. 317p. ISBN 0231073585

This is a collection of well-docmented essays concerning the complex ethical dilemmas engendered from the AIDS crisis. Includes contributions by scholars and professionals in the fields of law, medicine, philosophy, political science, religion, and social work. Considers controversial issues including the conflict between civil liberties and public health, mandatory HIV screening and testing, human subjects research, and the patient-physician relationship.

Sloan, Irving J.
AIDS Law: Implications for the Individual and Society
New York, Oceana Publications, 1988. 154p. ISBN 0379111667

Discusses the evolution of laws concerning AIDS and AIDS victims. Covers topics such as the status of research and testing, civil and criminal liability for the sexual transmission of AIDS, liability for transmission via the blood supply, AIDS discrimination in the workplace, school children and AIDS, the military and AIDS, and AIDS in prisons. Appendices include listings of legal organizations and state laws affecting AIDS patients.

✦ Social Welfare—History and Policy

Abramovitz, Mimi
Regulating the Lives of Women: Social Welfare Policy from Colonial Times to the Present
Boston, South End Press, 1988. 406p. ISBN 0896083306

Critically analyzes the connection between American women and social welfare programs from a feminist perspective. Argues that poor women have been ignored in social welfare literature and that social welfare policy has been shaped by a preoccupation with both the work and family ethics. An additional, weighted factor presented in public policy is the notion that the nuclear family is the only legitimate family structure.

Albert, Vicky N.
Welfare Dependence and Welfare Policy: A Statistical Study
New York, Greenwood Press, 1988. 195p. ISBN 031326175X

Examines the controversial Aid to Families with Dependent Children (AFDC) income-maintenance program in California in comparison to welfare systems across the nation, and presents a method of analyzing welfare dependence and dynamics. Calculates the effects of policy changes, labor market conditions, and population shifts; attempts to explain cause and effect; emphasizes welfare policies initiated by the Reagan administration and measures their impact by forecasting what would have happened otherwise. The results of this study describe the situation in California but can be used as a model for research in other states.

Berkowitz, Edward D.
America's Welfare State: From Roosevelt to Reagan
Baltimore, Md., Johns Hopkins Univ. Press, 1991. 216p. ISBN 0801841275

This historical overview of U.S. social welfare policy traces the development of Social Security from Roosevelt in 1935 through the Reagan administration in the 1980s. Discusses political attitudes towards American social welfare policy and examines the bureaucracies and beneficiaries of the welfare state, and the successes and shortcomings of social welfare programs. Attempts to define the terms "welfare" and "welfare state."

Bernstein, Merton C., and Joan Brodshaug Bernstein
Social Security: The System That Works
New York, Basic Books, 1988. 321p. ISBN 0465079164

Presents a positive outlook toward the social security system and its viability over the long term. Addresses the roles and problems of income-maintenance systems, outlines options, and attempts to account for the strengths and weaknesses of several systems. Argues that private retirement pensions and state or local retirement plans are overrated in terms of reliability.

Brown, Michael K., ed.
Remaking the Welfare State: Retrenchment and Social Policy in America and Europe
Philadelphia, Temple Univ. Press, 1988. 312p. ISBN 0877225419

This collection of essays, written by renowned social scientists, discusses the politics involved in the welfare state and the impact of attempts by conservative and social democratic/liberal governments in Europe and America to decrease the scope of Western welfare states during the Reagan years. Emphasizes the role of budget cuts, disintegrating political coalitions, beneficiaries of the welfare system, welfare system employees, privatization, and other social forces.

Chelf, Carl P.
Controversial Issues in Social Welfare Policy: Government and the Pursuit of Happiness
Newbury Park, Calif., Sage Publications, 1992. 161p. ISBN 0803940424

Analyzes American social welfare policy, discussing the federal government's role, the war on poverty, hunger, homelessness, unemployment, disability, the elderly, aid to poor children and families, and the ongoing search for viable social welfare policies. Illuminates the controversy over whether social programs are more harmful than beneficial, with data evaluating national programs. Also presents the debate between proponents of individualism, self-reliance, and the work ethic, and supporters of egalitarianism, justice, and the well-being of all members of society.

Cherlin, Andrew J., ed.
The Changing American Family and Public Policy
Washington, D.C., Urban Institute Press, 1988. 263p. ISBN 0877664226

This collection of sociological readings examines the changes that American families have undergone in the last twenty-five years. Explores demographic changes; transformations in the workforce; the well-being of children; the implications for public policy; adolescent well-being and family change; and the relationship between government policy and family structure. Concepts are supported with statistical charts, tables, and other numerical data.

Deegan, Mary Jo
Jane Addams and the Men of the Chicago School, 1892-1918
New Brunswick, N.J., Transaction Books, 1988. 352p. ISBN 0887380778

This thoughtful, well-written historical overview discusses Jane Addams and her function as a sociologist and founder of the predominately male Chicago School. It also examines the intellectual foundations of Addams' work, explores her transformation from sociologist to social worker, and looks at discrimination against women in the pioneering years of the social work profession.

Dobelstein, Andrew W.
Social Welfare: Policy and Analysis
Chicago, Nelson-Hall Publishers, 1990. 275p. ISBN 0830411445

Part One of this discussion of social welfare policy examines the relationship between policy analysis and public decision-making in the U.S. Includes information on public policy concepts, including policy analysis methodology, data gathering and analysis. Section Two applies policy analysis theories to specific social welfare areas, such as income maintenance, health and housing policy, child welfare, and the elderly. Concludes with an examination of an unresolved policy agenda, the intellectual perspectives of both the neo-conservative and the neoliberal positions, the philosophical influence of John Locke, and the overall contributions of policy analysis to the policy-making process.

Dougherty, Charles J.
American Health Care: Realities, Rights, and Reforms
New York, Oxford Univ. Press, 1988. 227p. ISBN 0195052714

Analyzes the practical aspects and philosophical bases of the U.S. health care system. Following an introductory section on the realities of access to health care, the quality of care and rising costs, the concept of the right to health care is critically examined from the perspectives of utilitarianism, egalitarianism, libertarianism, contractarianism, and plural foundations. The final section concerns market reforms, the roles of diagnosis-related groups, HMOs, vouchers, and various national health care plans.

Elwood, David T.
Poor Support: Poverty in the American Family
New York, Basic Books, 1988. 271p. ISBN 0465059961

Explores the causes of poverty of families in the U.S. and attempts to reconcile an overall mistrust of the welfare system with the professed desire of most Americans to help the poor. Topics include welfare policy, values (such as individual autonomy, the work ethic, family, and community), changes in American families, poverty among two- and single-parent families, and ghetto poverty. Recommends specific social welfare programs in the conclusion.

Feinberg, Renee, and Kathleen Knox
The Feminization of Poverty in the United States: A Selected, Annotated Bibliography of the Issues, 1978-1989
New York, Garland Publishing, 1990. 317p. ISBN 0824012135

An extensive, annotated bibliography of recent research on the feminization of poverty. Sources include books, journal articles, government documents, and research papers organized by subject and sub-arranged alphabetically by author. Includes discussions on topics such as poverty issues, women heads of households, children, employment issues, Reaganomics, women and social security, adolescent mothers, housing, welfare issues, and family policy. Each chapter contains an essay discussing the major issues as they relate to women and poverty.

Ginsburg, Norman
Divisions of Welfare: A Critical Introduction to Comparative Social Policy
Newbury Park, Calif., Sage Publications, 1992. 228p.
ISBN 0803984405

This critical evaluation of modern social welfare policy in Sweden, Germany, the United States, and Great Britain presents factual data about social policies derived from government statistics. Discusses the contradictory nature of the welfare state in mitigating and exacerbating class, gender, and racial divisions and inequities. Examines Sweden as a social democratic welfare state, Germany as a welfare state in the social market economy, the U.S. as a welfare state in the corporate market economy, and Britain as a liberal collectivist welfare state.

Glazer, Nathan
The Limits of Social Policy
Cambridge, Mass., Harvard Univ. Press, 1988. 215p.
ISBN 0674534433

A compilation of Glazer's work from the 1970s and 1980s that describes the evolution of the national mood of caution and skepticism regarding governmental social welfare programs. Contends that many large-scale governmental social programs have done more harm than good, explores the notion of dismantling these large programs, and proposes more important roles for states and localities, nongovernmental institutions, beneficiaries, and clients. Argues that government cannot effectively solve many of the social problems of the United States.

Hammerle, Nancy
Private Choices, Social Costs, and Public Policy: An Economic Analysis of Public Health Issues
Westport, Conn., Praeger Publishers, 1992. 231p.
ISBN 0275941728

Identifies and analyzes the economic and social costs of unhealthy personal behavioral choices that have public health consequences. Evaluates current and proposed public policy strategies on lessening the incidence of harmful choices, reducing their costs, and promoting those policies that effectively reduce the societal burden. Proposals include increasing access to prenatal care, making health care affordable and accessible to the working poor, legalizing certain drugs, increasing taxes on cigarettes and alcoholic beverages, and providing quality boarding care for abused children.

Handler, Joel F., and Yeheskel Hasenfeld
The Moral Construction of Poverty: Welfare Reform in America
Newbury Park, Calif., Sage Publications, 1991. 269p.
ISBN 0803941978

Analyzes the conflicts, contradictions, and moral ambiguities characteristic of the debate regarding welfare and work in U.S. society. Attempts to define welfare policy in terms of responses to poverty and provides a historical overview from its nineteenth-century origins to the present. Evaluates critical workfare programs for AFDC recipients and takes a dim view of the possibility that the AFDC program can be reformed significantly.

Jansson, Bruce S.
The Reluctant Welfare State: A History of American Social Welfare Policies
Belmont, Calif., Wadsworth Pub. Co., 1988. 278p.
ISBN 0534084907

A well-written and thoroughly researched historical analysis that enables students to understand the evolution of U.S. social welfare policy in terms of five key issues: the morality of providing social services, the essence of social obligation, the controversy surrounding preferential and compensatory policies, various types of intervention, and the role of the federal government. Includes a discussion devoted to the special problems that minorities have experienced throughout U.S. history.

Jencks, Christopher
Rethinking Social Policy: Race, Poverty, and the Underclass
Cambridge, Mass., Harvard Univ. Press, 1992. 280p.
ISBN 0674766784

Reexamines critical American assumptions about race, poverty, crime, heredity, welfare, and the underclass. Argues that neither liberal nor conservative philosophies on these issues stand up to close examination, and contends that less emphasis should be placed on political sentiments and more attention should be paid to the effectiveness of specific programs. Discusses major works on affirmative action and quotas, the safety net, the influence of genetics on learning and criminal tendencies, urban ghetto culture, and the growing underclass. The concluding chapter looks at welfare reform in relation to the work ethic and marriage, proposing the restructuring of Aid to Families with Dependent Children (AFDC) programs based on these two concepts.

Jones, Jacqueline
The Dispossessed: America's Underclasses from the Civil War to the Present
New York, Basic Books, 1992. 399p. ISBN 0465001270

Employing social science data, stories of families, and historical information, this thoroughly documented scholarly work analyzes the underclass debate in U.S. society by examining the situations of migrant workers, black and white field hands, the southern plantation economy, migration to the north, and the underclasses in late-twentieth-century America.

Katz, Michael B.
In the Shadow of the Poorhouse: A Social History of Welfare in America
New York, Basic Books, 1988. 352p. ISBN 0465032265

Presents the history of public and private social welfare in the U.S. from the poorhouse of colonial times through the New Deal, the war on poverty and the war on welfare, to the current crisis of homeless people. Discusses the conflicts in the four goals of social welfare: to relieve misery and want, to maintain social order, to regulate the labor market, and to mobilize political power. Asserts that private, voluntary charity is not an effective response to dependency and that, instead, creative welfare policy on the part of government is the answer to alleviating poverty and its effects.

Katz, Michael B., ed.
The Underclass Debate: Views from History
Princeton, N.J., Princeton Univ. Press, 1993. 505p.
ISBN 069104810X

This collection of essays places the contemporary underclass debate in a historical context. Focus is on the complex set of long-term social processes at work, rather than on individual and family behavior as the causes of chronic poverty and degradation in America's inner cities. Includes information on the origins of ghetto poverty, the transformation of American cities, the ethnic niche of immigrants and minorities, the emergence of underclass familial patterns (1900-1940), family composition and poverty since 1940, African American families, the politics of oppression (1929-1970), urban education and the disadvantaged, the war on poverty in the 1960s, and restructuring of the underclass debate.

Katz, Michael B.
The Undeserving Poor: From the War on Poverty to the War on Welfare
New York, Pantheon Books, 1989. 293p. ISBN 0394534573

This historical overview analyzes the concepts that have influenced contemporary public welfare policy from Lyndon B. Johnson's war on poverty to Ronald Reagan's war on welfare. Katz criticizes the categorizing of poor people into "deserving" and "undeserving"; reconsiders the work of Harrington, Moynihan, and Guilder; and illuminates the bias inherent in terms such as "the culture of poverty" and labeling of the poor as an "underclass." Argues that both liberals and conservatives in America tend to blame poverty on behavioral deficiencies in individuals, rather than on external social forces that cause and perpetuate the situation. Suggests approaches to poverty and its related social problems.

Loewenberg, Frank M.
Religion and Social Work Practice in Contemporary American Society
New York, Columbia Univ. Press, 1988. 176p. ISBN 0231064527

Loewenberg cites the value of religion to social work practitioners and students, and the spiritual dimension it adds to the lives of their clients. Explores the role of religious institutions and clergy, emphasizes empathy and sensitivity to client differences, and looks at the implications of sin and guilt as well as the relationship between religious experience and mental illness.

McCarthy, Michael, ed.
The New Politics of Welfare: An Agenda for the 1990s?
Houndmills, Basingstoke, Hampshire, England, Macmillan, 1989. 274p. ISBN 0333471563

Discusses issues regarding social welfare politics, including personal social services, health, housing, social security, employment, education, criminal justice, community care, and welfare. The introduction addresses the boundaries of welfare and social welfare policy in Great Britain during the 1980s and the 1990s.

Marmor, Theodore R., Jerry L. Mashaw, and Philip Harvey
America's Misunderstood Welfare State: Persistent Myths, Enduring Realities
New York, Basic Books, 1990. 268p. ISBN 0465059694

This analysis examines the myths, fallacies, and realities pertaining to America's social programs. The authors criticize the conservative mentality that influenced social policy for twenty years, and provide a reasonable defense of liberal social policy. Presents a balanced assessment of programs such as Social Security, public assistance, and Medicare, and suggests various reform ideas to handle shortcomings in these programs.

Mead, Lawrence M.
The New Politics of Poverty: The Nonworking Poor in America
New York, Basic Books, 1992. 356p. ISBN 0465059627

Argues that workfare programs requiring employable welfare recipients to work are one of the best ways to combat and possibly overcome chronic poverty and dependency in the U.S. Explores such topics as the current crisis of reform, the costs of nonwork, low wages, employment barriers, human nature and the psychology underlying nonwork, welfare reform, the meaning of dependency, and future prospects.

Moroney, Robert
Social Policy and Social Work: Critical Essays on the Welfare State
New York, Aldine de Gruyter, 1991. 257p. ISBN 020236061X

This collection of readings asserts that social policy in the U.S. is related to macroeconomic needs, and discusses the interplay between our political and economic systems and social welfare policy. Covers topics such as policy analysis, income maintenance, housing, families and dependent children, the elderly, the underclass, and the political economy of social welfare. Maintains that America's emphasis on individualism and freedom has spawned a divided society in which citizens are pitted against one another in competition for resources.

Pampel, Fred C., and John B. Williamson
Age, Class, Politics, and the Welfare State
New York, Cambridge Univ. Press, 1989. 199p.
ISBN 0521372132

This source evaluates popular theories of the welfare state: industrialism, capitalism, social democracy, and interest group politics, by using comparative cross-national data and empirical models. Emphasizes the importance of interest groups, especially the role of the elderly and the middle class in welfare state policy and growth, and contends that public welfare policy may reflect age, ethnic, gender, and other divisions across class boundaries.

Pelton, Leroy H.
For Reasons of Poverty: A Critical Analysis of the Public Child Welfare System in the United States
New York, Praeger, 1989. 203p. ISBN 0275930734

The thesis of this study is that the child welfare system has failed in its intention to serve and implement child welfare policies, seeming to be obstructive rather than constructive in meeting the needs of children and their families. Examines the philosophy, system operation, and reliance on out-of-home placement of children. Argues that social workers should be limited to nonpunitive support and prevention arenas, and that investigation of severe child abuse and neglect cases should be relegated to law enforcement agencies.

Rodgers, Harrell R.
Poor Women, Poor Families: The Economic Plight of Female-Headed Households
Rev. ed. Armonk, N.Y., M.E. Sharpe, 1990. 193p.
ISBN 087332594X

This clearly written, factual study looks at the increase in poverty among female heads of households and its effects on children. Provides a statistical overview of the feminization of poverty, explores the social welfare response to families headed by women, and includes a discussion of the western European approach to social welfare. Presents practical suggestions for reforming the American welfare system.

Sherraden, Michael W.
Assets and the Poor: A New American Welfare Policy
Armonk, N.Y., M.E. Sharpe, 1991. 324p. ISBN 0873326180

Sherraden criticizes the failure of American subsistence-based welfare policy, and outlines liberal (social structure), conservative (individual, behavioral), and Marxist theories, on income and wealth distribution. Asserts that the current American welfare policy has been a grossly ineffective antipoverty measure. Scrutinizes the notion of redirecting welfare policy toward asset accumulation, savings, and investment, a concept which the author believes will encourage risk-taking and give the poor a stake in the system.

Skocpol, Theda
Protecting Soldiers and Mothers: The Political Origins of Social Policy in the United States
Cambridge, Mass., Belknap Press of Harvard Univ. Press, 1992. 714p. ISBN 0674717651

A detailed historical account of the development of American social policies, including a transnational comparative analysis which aids in understanding policy origins. Includes topics such as the origins of modern social provision in America, insurance, trade unions and social legislation, the effects of Progressive Era politics, women's civic involvement and political reforms in the early twentieth century, protection of women workers, the Children's Bureau and the Sheppard-Towner Act. Concludes with the legacies of America's initial social policies and their lessons for the present and the future.

Wenocur, Stanley, and Michael Reisch
From Charity to Enterprise: The Development of American Social Work in a Market Economy
Urbana, Ill., University of Chicago Press, 1989. 327p.
ISBN 0252015568

Reviews the historical development of the social work profession in the U.S. from its origins in the late nineteenth century through the creation of the social welfare industry by 1950. Argues that professionalization, in many respects, has obstructed the primary mission of the social work enterprise. Contends that the social work profession is organized to advance the special interests of its practitioners, and that professionalization and regulation of its practice have been an organized effort on the part of some "occupational entrepreneurs" to gain influence and power.

Zigler, Edward, and Mary E. Lang
Child Care Choices: Balancing the Needs of Children, Families, and Society
New York, Free Press, 1991. 271p. ISBN 0029358213

Examines the relationship between child care and society-at-large in the U.S., analyzing the needs of children in an environment of dual-income, intact families, and single parents who face an inadequate supply of affordable, quality child care. Recommends ways to meet the needs of infants and toddlers, school-age children, and children with special needs. Discusses child care regulation and suggests a comprehensive child care plan for the twenty-first century. Advocates setting up a child allowance trust fund as part of a national child care policy.

Zimmerman, Shirley
Understanding Family Policy: Theoretical Approaches
Newbury Park, Calif., Sage Publications, 1988. 197p.
ISBN 0803927983

A scholarly review of the theoretical frameworks that have influenced values concerning the family. Discusses historical, cultural, and philosophical aspects that have affected public policy. Includes topics related to policy frameworks, such as the game approach under competitive conditions, interest groups and elite theories, exchange and choice theories, conflict theory, symbolic interaction, and family stress theory. Also discusses families as social systems and applies frameworks to areas including welfare reform, family well-being measures, and care for the disabled elderly.

✦ Substance Abuse

Blum, Kenneth, and James E. Payne
Alcohol and the Addictive Brain: New Hope for Alcoholics from Biogenetic Research
New York, Free Press, 1991. 320p. ISBN 0029037018

Addressed to a general audience, this work examines neurophysiological, pharmacological, and biogenetic research on alcoholism performed in the past fifty years. Explains the nature of alcoholism, its human costs, scientific insights providing fresh clues into the causes of alcoholism, and illuminates research that shows the promise of prevention or cure. Recognizes the importance of psychological research and sociological findings in relation to alcoholism, but focuses on alcoholism as a physiological disease process having its origin in genetically-based biochemical imbalances and deficiencies in the brain. Includes evidence that abstinence is the only permanent solution, and that controlled drinking is not an effective choice.

Cermak, Timmen L.
A Time to Heal: The Road to Recovery for Adult Children of Alcoholics
Los Angeles, Jeremy P. Tarcher, 1988. 227p. ISBN 0874774543

This highly readable work is directed to both adult children of alcoholics (ACOAs) and the mental health professionals who work with them. Addresses the trauma associated with growing up in homes controlled by alcohol abuse, and illustrates various self-healing techniques ACOAs can employ in facilitating their own recovery.

Cocores, James
The 800-Cocaine Book of Drug and Alcohol Recovery
New York, Simon and Schuster, 1991. 255p. ISBN 0671744860

This step-by-step guide provides information on how to gain a lifetime of sobriety. Covers recognizing addiction and seeking help, finding and evaluating a treatment center, selecting an appropriate program, developing positive relationships, rebuilding sound health through diet and exercise, re-establishing intimacy and sexuality, avoiding and confronting relapse, finding help for co-dependents, and coping with life outside of the rehabilitation setting.

Collins, R. Lorraine, Kenneth E. Leonard, and John S. Searls, eds.
Alcohol and the Family: Research and Clinical Perspectives
New York, Guilford Press, 1990. 386p. ISBN 0898621690

This scholarly, thoroughly documented overview on alcohol and the family covers adolescent drinking, children of alcoholics, family culture, marital and sexual functioning, the role of bio-genetics, and treatment approaches. Provides family therapy practitioners with a scientific foundation on which to base the therapeutic treatment of alcoholics.

Collins, Stewart, ed.
Alcohol, Social Work, and Helping
New York, Taristock/Routledge, 1990. 176p. ISBN 0415025788

This collection of six essays provides advice on counseling clients whose use of alcohol is problematic. Discusses intervention within the contexts of society and the agency, and deals with the practical issues of working with the problem drinker. Suggests short-term goal setting and assistance in the event of relapse. Focuses on historical perspectives and provides information on new developments in helping problem drinkers, the special needs of women, family contexts, group work, and the development of community services.

Nelson, Jane, Riki Intner, and Lynn Lott
Clean and Sober Parenting: A Guide to Help Recovering Parents
Rocklin, Calif., Prima Publishing, 1992. 264p. ISBN 1559581654

Written by three recovery and parenting experts, this guide helps recovering parents take charge of their lives, create order and consistency, build closeness and trust through emotional honesty, connect with support groups, break patterns and co-dependence, develop healthy communication skills, institute structure and limits, and eliminate fear and guilt in healing the frayed relationship with their children.

O'Brien, Robert, and Morris E. Chafetz
The Encyclopedia of Alcoholism
2nd ed. New York, Facts on File, 1991. 346p. ISBN 081601955X

This highly readable encyclopedia focuses on social problems. More than 600 entries define and explain all aspects of alcoholism, including biological and medical factors, psychological aspects, the social and economic impact, legal issues, and treatment. Also provides information on alcohol use in other countries, and lists organizations connected with alcoholism. Appendices contain statistical tables and other data, sources for further information, and selected English language periodicals of interest to laypeople and professionals in medicine and social services.

O'Brien, Robert, and others
The Encyclopedia of Drug Abuse
2nd ed. New York, Facts on File, 1992. 500p. ISBN 0816019568

More than 1,000 entries define and explain biological, medical, social options, and legal aspects of drug abuse. Includes information on the social and economic impact of drug abuse; helping organizations; statistical data; drug use in other countries; causes of drug abuse; and treatment. Of special interest are the sections which discuss organized crime and drug trafficking, crack babies, and employee assistance programs. Appendices provide street and slang terminology, state alcohol and drug abuse profiles, and sources for further information.

Potter-Efron, Ronald T., and Patricia S. Potter-Efron, eds.
Aggression, Family Violence, and Chemical Dependency
New York, Haworth Press, 1990. 226p. ISBN 0866569642

Concerns the treatment of both the chemically dependent person and co-dependent individuals. Includes information on topics such as the aggression and violence associated with substance abuse, spouse battering and child maltreatment, incest and marital rape, and preventive measures. The introductory chapter contains a literature review and a brief discussion of specific addictive substances and their relation to aggression and violence.

Read, Edward M., and Dennis C. Daley
Getting High and Doing Time: What's the Connection? A Recovery Guide for Alcoholics and Drug Addicts in Trouble with the Law
Washington D.C., St. Mary's Press, 1990. 80p. ISBN 0929310314

This short, easy-to-read book addressed directly to substance abusers involved with the criminal justice system discusses the connection between substance abuse and trouble with the law, the nature and extent of addiction, recovery and twelve-step programs, relapse prevention, the addict's family and recovery, and cooperation with parole and probation officers. Includes a bibliography of suggested readings and a list of self-help organizations and publishers of recovery materials.

Trimpey, Jack
The Small Book: A Revolutionary Alternative for Overcoming Alcohol and Drug Dependence
Rev. ed. New York, Delacorte Press, 1992. 274p.
ISBN 0385305583

Employs the principles of rational-emotive therapy as an alternative to the traditional twelve-step program of Alcoholics Anonymous, in recovery from chemical dependency and the emotional difficulties associated with substance abuse. Discusses the nature of rational recovery and rational recovery self-help groups.

Watts, Thomas D., and Roosevelt Wright, eds.
Alcoholism in Minority Populations
Springfield, Ill., Charles C. Thomas, 1989. 227p.
ISBN 0398055416

This collection of well-documented articles provides a summary introduction to alcoholism among African Americans, Hispanics, Native Americans, and Asian Americans. Addresses the federal role in alcoholism treatment and prevention for minorities, policy and administrative issues in minority alcoholism programs, and alcohol prevention strategies. The introduction notes inadequate services for low-income and minority alcoholics in the United States due to a lack of funding and low priority.

Nonprint Materials

AIDS: Helping Families Cope
Silver Spring, Md., National Association of Social Workers, 1988. 1 videocassette (53 min.) VHS and booklet (26p.)

In this video, social workers cover the major issues associated with AIDS: the psychosocial impact, stigma, daily living requirements, loss and grief, and support groups. Includes an accompanying report with suggestions for meeting the psychosocial needs of AIDS patients and their families.

Community: Program Ten of The Sociological Imagination
New York, Insight Media, 1991. 1 videocassette (30 min.) VHS

In defining the notion of "community," this video examines the ways in which communities meet the emotional, physical, and economic needs of individuals. Addresses American individualism and the search for community in territorial settings (such as neighborhoods, homogenous enclaves, small towns, suburbs, and large cities), as well as in nonterritorial communities (employment sites, professional associations, special interest groups, and churches).

A Crime Never Forgotten
Northbrook, Ill., MTI Film and Video, 1992? 1 videocassette (23 min.) VHS

Originally broadcase as an episode of the television program *20/20*, this video examines the deep-seated hurt that childhood sexual abuse causes in children, and the current tendency of the legal system to permit victims to employ delayed discovery in lawsuits.

Forever Young
Alexandria, Va., PBS Video, 1990. 1 videocassette (58 min.) VHS

Narrated by Dr. C. Everett Koop, this videocassette presents some of the issues and problems of aging: preventive health practices following retirement, frustrations and expectations of older people, recommendations for enhanced insurance coverage, and health programs for the elderly.

Homeless Not Helpless: Opening Doors
Berkeley, Calif., University of California, Extension Center for Media and Independent Learning, 1992. 1 videocassette (45 min.) VHS

Analyzes the organization and empowerment of homeless people, political activism that has taken place on their behalf, and efforts to obtain affordable housing. Also looks at programs for the homeless: Christian missions, church programs, government programs, shelters, and programs for families with children, for the chemically dependent, and for the mentally disabled.

Lobbying Tips for Social Workers
Silver Spring, Md., National Association of Social Workers, 1988. 1 videocassette (26 min.) VHS

Designed as a training tool for social workers who desire to become involved in the lobbying process. Discusses lobbying as an essential part of the legislative process, and emphasizes the role that lobbying plays. Discusses selling an issue and lobbying tactics, compromise and negotiation, and political action.

Race and Racism: Red, White, and Black: Volume Six of The Nineties
Chicago, Subtle Communications, 1991. 1 videocassette (60 min.) VHS

This video combines live action, film transfer, and animation to look at race and racism in contemporary society. Includes a discussion on how whites and blacks view each other, KKK violence, minorities and the Persian Gulf War, and interviews with a Black Panther Party member, Nelson Mandela, a Jewish-Japanese married couple, and Malcolm X, among others. Technically strong with unique content, and highly recommended.

Sex and Sacrifices: What They Did for Love
Fort Worth, Tex., Gladney Center, 1991. 1 videocassette (20 min.) VHS and 1 guide

This production is the result of a career development class project in which sixty young women and adolescents (at the Gladney Center) talk to their peers about the consequences of poor decision-making. Explores the effects of being sexually active, and includes a discussion of adoption as an alternative in coping with unplanned pregnancy. Includes guide.

The World of Abnormal Psychology: Substance Abuse Disorders
South Burlington, Vt., The Annenberg/CPB Collection, 1991. 1 videocassette (60 min.) VHS. ISBN 1559466790

Discusses the treatment and interventions employed to assist people in overcoming a variety of chemical addictions. Includes illustrative case studies.

What Do Social Workers Do?
Silver Spring, Md., National Association of Social Workers, 1992. 1 videocassette (21 min.) VHS

Renowned CBS correspondent Charles Kuralt presents a balanced overview of the social work profession in interviews with practitioners, administrators, and business leaders about society's need for social services. Examines the profession's altruistic and realistic characteristics. and topics such as working with abused and neglected children, the elderly, and the homeless.

Indexes

Name Index 577

Title Index 594

Journal Index 620

Name Index

Abadinsky, Howard, 278
Abbott, Tony, ed., 331
Abel, Emily K., 557
Abel, Ernst, 535
Abercrombie, Stanley, 430
Abling, Bina, 419, 467
Abrami, Patrick F., ed., 78
Abramovitz, Mimi, 570
Abrams, Stan, 287
Accredited Standards Committee Z49, 392
Ackerman, Norman, 524
Ackerman, Robert J., 557
Ackley, Betty J., 77
Acquaviva, Jane D., ed., 98
Ada, Louise, ed., 104
Adamec, Christine A., 557
Adams, Ansel, 445
Adams, Cynthia H., 58
Adams, Lee, 372
Adams, Robert, 447
Adams, Wanda L., 49
Addison, Lois A., 58
Ades, Dawn, 455
Adimora, Adaora A., 569
Adkisson, Mary Ann, 15
Adler, Arlene McKenna, ed., 23
Adler, Bill, 137
Adler, Elizabeth, 236, 397
Adler, Peter, 417
Adler, Susan S., 104
Aglow, Stanley H., 142, 160
Aguilera, Donna C., 567
Ahrens, C. Donald, 358
Aiken, Lewis R., 511
Aiken, Tonia D., ed., 78
Air and Waste Management Association, 367
Aitken, Peter G., 175
Alba, Augusto, 59
Albert, Vicky N., 570
Albrecht, Donna G., 217
Albrecht, Steve, 300
Aldrich-Ruenzel, Nancy, ed., 397, 469
Alexander, Alison, ed., 240
Alfano, R.R., ed., 377
Alfaro-Lefevre, Rosalinda, 84
Alfeld, Louis Edward, 131
Allen, Anne, 467
Allen, Claudia Kay, 98, 102
Allen, David Grayson, 167
Allen, Jeanne, 418
Allen, Judy, 311
Allen, Robert G., 218
Allen, Sam, 137
Alliance For Environmental Education, 500
Allison, E. Jackson, ed., 41
Alpert, Geoffrey P., 288
Alter, Catherine, 568
Althouse, Andrew Daniel, 392
Althouse, Rosemary, 311
Altman, Patricia B., 417
Altschull, J. Herbert, 240
Amaden-Crawford, Connie, 413
Amatayakul, Margaret K., 54
Ambrose, James E., 129, 132
Ambry, Margaret,, 258
Amendola, Joseph, 192
American Association of Cardiovascular and Pulmonary Rehabilitation, 114
American Association of Woodturners, 132
American Bar Association, 298
American College of Legal Medicine, 41
American College of Sports Medicine, 104
American Dental Association, 4
American Dietetic Association, 34

American Health Information Management Association, 55
American Kennel Club, 524
American Library Association, 333
American Medical Record Association, 55
American National Standards Institute, 142
American Occupational Therapy Association, 104
American Society of Architectural Perspectivists, 430
American Welding Society, 389, 390, 391, 392, 393
Ames, Steven E., 247, 460
Amon, Elenore M., 175
Amos, Wally, 258
Anderson, Charles M., ed., 28
Anderson, Douglas A., 252
Anderson, Kenneth, 192
Anderson, Kenneth, ed., 69
Anderson, Kenneth N., 51
Anderson, Laura Killen, 249
Anderson, Patricia, 511
Anderson, Pauline Carter, 5
Anderson, Sandra K., 59
Anderson, Scott R., 397
Anderson, Shauna Christine, 59
Andolina, Valerie, 23
Andreas, John C., 142
Andrews, Lori, P., 367
Andujo, Emily, ed., 5
Ang, K. K., 27
Angeli, Primo, 454
Anglade, Pierre, ed., 518
Angle, Deborah K, 98
Annis, William H., 160
Anthenat, Kathy Smith, 132
Anthony, Robert Newton, 167
Antin, Tony, 234
Anton, Thomas J., 92
Aoki, Byron Y., 41
Aoshima, Hitoshi, 5
Appleby, Kristyn S., 306
Applegate, Elizabeth Ann, 34
Applewhite, Steven R., ed., 564
Apts, David W., 92
Aquaro, Mariann, 98
Arbuckle, J. Gorden, 367
Arcangelo, Virginia Poole, 84
Arends, Mark W., 430
Arick, Martin, 175
Armer, Alan A., 265
Armpriester, Kate, 132
Arntson, Amy E., 458
Arntzen, Charles J., ed., 477
Arozena, Steven, ed., 336
Arwady, Joseph W., 266
Asantewa, Doris, 339
Ascher, L. Michael, ed., 535
Ash, Major M., 5
Ashery, Rebecca Sager, ed., 535
Ashley, Perry J., ed., 249
Ashley, Ruth, 175
Ashton, Floyd M., 487
Asken, Michael J., 41, 492
Askue, Vaughan, 346
Asperheim, Mary Kaye, 54
Atkinson, Henry F., 380
Attenborough, David, 528
Auerbach, Paul S., 41
Austrin, Miriam G., 51
Avery, James K., 5
Avery, Thomas Eugene, 506
Aves, Melanie, 430
Aves, Pirrie B., 430
Aviation Week Group Newsletters, 358
AVO Multi-Amp Institute, 142
AWS Committee on Arc Welding and Cutting, 393
AWS Committee on Definitions and Symbols, 389
Axler, Bruce H., 192
Ayala, Emma, 186
Ayer, Jennifer, 75
Ayres, James J., 275

Ayres, Julia, 398

Baas, Linda S., ed., 72
Babikian, Viken L., 29
Bacha, William J., 525
Bachtler, J. R., 492
Backer, Barbara A., 88
Bacon, Mark S., 223
Baden, John B., ed., 486
Badgett, Tom, 175, 460
Baer, Charold Lee Morris, 84
Baer, John Samuel, ed., 535
Bagust, Harold, 518
Bahr, Stephen J., ed., 557
Bailey, Adrian, 420
Bailey, Susan Pritchard, 55
Bailey, William G., ed., 285
Baird, Ronald J., 393
Baker, Brian L., ed., 279
Baker, Charles J., 492
Baker, Edward, 528
Baker, Georgia O'Daniel, 419
Baker, Kim, 275, 398
Baker, Margaret, 105
Baker, Michael John, 258
Baker, Sharon L., 333
Balay, Robert, ed., 337
Balcerzak, Edwin A., ed., 558
Baldassarre, Guy A., 506
Baldwin, David A., 329
Baldwin, Diane M., 305
Ball, Jeff, 518
Ball, John, 23
Ball, John E., 129
Ball, Marion J., ed., 53
Ball, Vic, ed., 520
Ballast, David Kent, 430
Ballinger, Phillip W., 23
Band, Richard E., 220
Banerjee, Partha P., 377
Bann, David, 398
Barad, Dianne, 311
Baran, Nicholas, 176
Baran, R., 186
Barber, Linda George, 55
Barber, Steve, 276
Barbosa, Pedro, 499
Barbour, William, ed., 241
Barchers, Suzanne I., 311
Bardi, James A., 202
Barker, Robert L., 553
Barker, Thomas, 289
Barlin, Anne Lief, 311
Barnes, Colin, 419
Barnes, Thomas A., 114
Barnes, Thomas A., ed., 114
Barnett, Barbara, 274
Barnett, David W., 311
Barnhurst, Kevin G., 247
Barnum, Barbara Stevens, 81
Baron, Samuel, ed., 59
Barr, Gary K., 220
Barr, Vilma, 430
Barrett, Charles Francis, 209
Barrett, Terry Michael, 450
Barrett, Thomas M., ed., 518
Barron, Marlene, 311
Barry, Patricia D., 86
Barsamian, David, 241
Barth, Richard P., 558
Bartlett, Peggy F., 477
Barton, Barbara J., 519
Barton, Bob, 311
Barton, Richard O., 114
Barton, Roger E., ed., 5
Bartz, Wilfried J., ed., 352
Baskett, Peter J. F., 41
Baskette, Floyd K., 237
Basmajian, John V., ed., 105

Name Index

Bass, Deborah S., 558
Bass, Ellen, 558
Basta, Nicholas, 501
Bastow, Donald, 352
Bates, Barbara, 78
Baty, Gordon B., 223
Bauch, Jerold P., ed., 311
Bauer, Jim, 132
Baumgardner, Jeannette Mahan, 312
Baumgardt, Bill R., ed., 481
Bayer, Patricia, 430
Bayer, Ronald, ed., 92
Bayt, Phyllis Theiss, 54
Beach, Mark, 396, 398, 456
Beals, Melba, 259
Beard, F. Richard, 393
Beard, Tyler, 420
Beardsley, Lyda, 312
Beasley, Joseph D., 535
Beasley, Maurine Hoffman, 254
Beattie, Melody, 535
Beattie, Mollie, 507
Beaty, Janice J., 312
Becan-McBride, Kathleen, 59
Beck, Aaron T., 535
Becklin, Karonne J., 53
Bedworth, David D., 372
Bee, Helen L., 312
Beebe, Jack M., 454
Beech, Jill, ed., 525
Beeman, G. Marvin, ed., 528
Behzadizadeh, Arman, 377
Belanger, Sandra E., comp., 236
Belcher, Anne E., 88
Bell, C. Gordon, 223
Bell, Frank, 384
Bell, John F., 506
Bell, Judith A., 423
Bell, W. Frazier, 218
Bell, Welden E., 5
Bellandi, Robert, ed., 367
Bellardo, Lewis J., comp., 330
Belsey, Richard, 59
Bender, Jim, 486
Benedict, Helen, 254
Benedict, Larry, 266
Benfield, Jack, 143
Benjamin, Martin, 78
Benne, Mae, 331
Benner, Margaret P., 89
Bennett, A. E., 390
Bennett, Gerald, ed., 536
Bennett, Kristina, ed., 525
Benson, K. Blair, 264
Benson, Tedd, 132
Bensussen, Rusty, 413
Bentel, Gunilla Carleson, 27
Bentley, William K., 285
Benton, Mike, 465
Bergant, Kathleen Ann, 186
Berge, James M., 393
Berger, Florence, 202
Berger, Steven N., 304
Bergeron, J. David, 42
Bergman, Thomas, 105
Bergmann, Jorg R., 259
Berk-Seligson, Susan, 278
Berkovitz, B. K. B., 5
Berkowitz, Edward D., 570
Berle, Gustav, 223, 337
Berman, Eleanor, 227
Bermel, Stephanie Nowysz, 16
Bernal, Jose, 377
Bernard, Robert, 266
Bernardo, Barbara, 298
Berndt, Thomas J., 312
Berner, R. Thomas, 237
Berns, Joel M., 5
Bernstein, Merton C., 570

Bernstock, Peter N., 176
Berquist, Thomas H., ed., 28
Berrick, Jill Duerr, 558
Berry, Susan, ed., 458
Berton, Lee, 167
Berutti, Al, ed., 143
Besharov, Douglas J., 558
Besharov, Douglas J., ed., 558
Bethune, James D., 372
Betten, Neil, ed., 556
Betts, Richard M, 215
Betzina, Sandra, 413
Beyer, Jinny, 441
Biagi, Shirley, 238
Bianco, David P., ed., 259, 536
Bick, Rodger L., ed., 59
Bicknell, Treld, ed., 464
Bierals, Gregory P., 143
Bigan, Tammy, 186
Bilbo, Mark K., 176
Bill, Robert, 528
Biller, Henry B., 312
Billings, Paul R., ed., 287
Bingham, Karen Havill, 336
Bininger, Carol J., 87
Binns, Betty, 398, 458
Bintliff, Russell L., 288, 290
Biondo, Charles, 455
Birchard, John, 132
Birchenall, Joan M., 75
Birchfield, John C., 192
Bird, S. Elizabeth, 241
Birmingham, Jacqueline Joseph, 51
Birnbach, Nettie, ed., 81
Birnbaum, Alexandra Mayes, ed., 227
Birnbaum, Stephen, 227
Birren, Faber, 431
Bishop, John Melville, 266
Biswell, Harold H., 503
Bittinger, Gayle, 312
Bivins, Thomas, 236, 398
Bixby, Robert, 398
Black, Alison, 398
Black, Claudia, 536
Black, Henry Campbell, 279, 294, 300
Black, Jay, 245
Black, Roger, 398
Blackburn, Graham, 122, 132
Blackburn, Susan Tucker, 82
Blackman, William C., 367
Blackwell, Lewis, 469
Blandford, Percy W., 122
Blankenbaker, E. Keith, 160
Blatner, David, 399
Blatner, Theodore H., 289
Bledsoe, Bryan E., 42
Blissmer, Robert H., 176
Blitzer, Roy J., 333
Block, Jonathan, 458
Bloom, Martin, ed., 567
Blosser, Fred, 92
Blossom, Bonnie, 105
Blowey, R. W., 525
Blozis, George G., 16
Blum, Eleanor, 236
Blum, Kenneth, 536, 573
Blundell, William E., 239
Bly, Amy Sprecher, 218
Bly, Robert W., 259
Blyth, W. John, 209
Blythe, Betty J., 569
Boardman, R. D., 202
Bobak, Irene M., 82
Bobath, Berta, 98, 105
Bochnak, Peter M., 492
Bogart, Leo, 241
Boggs, Donald L., 528
Bohle, Robert, 252

Bohle, Robert H., 248
Boisset, Caroline, 519
Bokat, Stephen A., ed., 92
Bolander, Verolyn Barnes, ed., 79
Bolles, Richard Nelson, 552
Bolwell, Christine, 70
Bonanno, Alessandro, ed., 484
Bonder, Bette R., 98
Bono, Saverio G., 353
Bontrager, Kenneth L., 23
Boockholdt, James L., 167
Book, Albert C., 234
Boone, Morell D., 333
Borecki, Madeline, 16
Borman, Jami Lynne, 399
Boroson, Warren, 218
Borowsky, Irvin J., 399
Borsenik, Frank D., 203
Bosch, Robert, 353
Bosker, Gideon, 42
Boss, Richard W., 331
Boswell, Suzanne, 414
Botkin, James W., 223
Botts, Jack, 249
Bouchard, Eric A., 23
Bowditch, William A., 392
Bower, John, 221
Bowers, Arden C., 79
Bowers, Charlotte R., ed., 49
Bowes, Anna De Planter, 74, 192
Bowles, Dorothy A., 237
Boyd, Alan, comp., 340
Boyd, Mildred W., 89
Boyd, Wilma, 227
Brackelsberg, Phyllis, ed., 414
Braden, Maria, 254
Bradley, Fern Marshall, ed., 519
Bradley, Jean C., 72
Bradley, Julia Case, 399
Bradshaw, John, 536
Brady, Philip, 469
Brady, Robert N., 363
Brand, Richard W., 5
Brandler, Sondra, 536
Brandt, Mary, 52
Brandt, Peter B., 431
Brannigan, Francis L., 492
Branson, Gary D., 129, 132, 431
Brasch, Walter M., 241
Brauer, Roger L., 92
Brazelton, T. Berry,]312
Breggin, Peter Roger, 536
Brejcha, Mathias F., 353
Brenner, Donald J., 397
Brickell, Christopher, ed., 518
Bricklin, Mark, 34
Bridge, John P., 157
Briggs, Shirley A., 521
Brigham, Nancy, 236
Brill, Jack A., 218
Brillman, Judith C., ed., 42
Brimer, Mark A., 105
Brimson, James A., 167
Bristow, Nicholas, 399
Brittain, James L., 374
Broadus, Jerry R., 384
Brodeur, Paul, 143
Brody, E. W., 259
Brody, Elaine M., 559
Bromberg, Joan Lisa, 377
Bronner, Simon J., ed., 312
Brooker, Donald B., 482
Brookes, Andrew, 503
Brooks, Brian, 246
Brooks, Brian S., 250
Brooks, Charles, ed., 235
Brooks, Jane B., 312
Brooks, Teddar S., 298
Brooks, Terri, 239

Brophy, Maureen, 275, 303
Brottman, May, ed., 331
Brown, Alan, 445
Brown, Alex, 469
Brown, Curtis M., 384
Brown, Faye, 49
Brown, Gary, 506
Brown, Lillian, 259
Brown, Marvin, 259
Brown, Michael K., ed., 570
Brown, Nancy, 446
Brown, Robert Wade, 132
Brown, W. H., 507
Browne Miller, Angela, 559
Browne, David, 399
Browne, Steven E., 267
Browne, William P., ed., 483
Bruce, Jo Anne Czecowski, 52
Brumbaugh, James E., 132
Brundage, Dorothy J., 88
Brunner, Thomas W., 280, 298
Bruno, Carole A., 303
Brunzel, Nancy A., 59
Bryan, John L., 492
Bryant, Ernie, 133, 137
Bryant, Neville J., 59
Bryer, Richard, 264
Bryson, McDowell, 203
Buckland, Michael Keeble, 333
Buckleitner, Warren, 312
Buckley, John, 501
Buckner, R. B., 384
Buehrens, Carol, 375
Bukowski, Richard, 492
Bulechek, Gloria M., ed., 77
Bullivant, Lucy, 431
Bullock, Margaret I., ed., 105
Bullough, Vern L, 81
Bumagin, Victoria E., 559
Bunnell, Peter C., 448
Burchard, Arliss, 313
Burchell, S. C., 431
Burden, Ernest E., 431
Bureau of National Affairs, 59
Burgauer, James, 220
Burger, Jeff, 461
Burgest, David R., ed., 564
Burgett, Gordon, 239, 252
Burke, James F., 227
Burke, Margaret Barton, 83
Burke, Mary M., 75
Burke, Mike, 92
Burns, Elizabeth M., ed., 536
Burns, Evelyn Frank, 23
Burns, Patrick, 209
Burns, Russell M., 499
Burstein, David, 431
Burt, Brian A., 6
Burtis, Grace, 35
Burton, George C., ed., 114
Burton, Gwendolyn R. W., 76
Burton, Kermit, 218
Burton, L. DeVere, 477
Busch, David D., 400
Busché, Don, 400
Bush-Brown, Albert, 431
Bushong, Stewart C., 23, 28, 29
Butcher, Judith, 276
Butler, Arlene Kay, 227
Butler, Brian, 227
Buttel, Frederick H., 478
Buzawa, Eve Schlesinger, 559
Byers, Brenda A., 192
Bygate, J. E., 346
Byrne, Kevin P., 35

Cadogan, Mary, 358
Cahalan, Don, 536
Cahill, Lawrence B., ed., 367

Cailliet, Rene, 105
Calasibetta, Charlotte Mankey, 412
Calbreath, Donald F., 60
Calderone, Robert, 358
Callahan, Bob, ed., 465
Callaway, M. Brett, ed., 487
Calloway, Stephen, 431
Calloway, Stephen, ed., 431
Campbell, Douglas S., 245
Campbell, Hallock C., comp., 390
Campbell, James B., ed., 525
Campbell, John E., 42
Campbell, June Mundy, 60
Campbell, Linda, ed., 53
Campbell, Mary, 209
Campbell, Mary V., 176, 461
Campbell, Reginald L., 92
Campbell, Robert Jean, 281
Campbell, Robin, 313
Campbell, Susan B., 313
Campbell, Suzann K., ed., 105
Campbell, Terry W., 525
Campeau, Frances, 23
Cane, Andries C., 288
Canobbio, Mary M., 88
Capachi, Nick, 363
Capon, Brian, 518
Cappon, Rene J., 246
Caputo, Janette S., 334
Capuzzi, Dave, 559
Carberry, Judith B., 367
Cargill, Jennifer S., ed., 340
Carlebach, Michael L., 450
Carlton, Richard R., 23
Carmichael, D. R., ed., 167
Carnegie, Mary Elizabeth, 81
Carnevali, Doris L., 77
Caroline, Nancy L., 42
Carpenito, Lynda Juall, 77, 78
Carr, Roberta, 414
Carrieri, Joseph R., 559
Carroll, C. Ronald, ed., 478
Carroll, Kevin O., 16
Carroll, Marianne, 400
Carroll, Quinn B., 24
Carroll-Johnson, Rose Mary, ed., 78
Carson, Wayne G., 492
Carter, Alison J., 420
Carter, David E., 461
Carter, David E., ed., 456
Carter, Harry R., 492
Carter, Rob, 469
Caruso, James R., 265, 268
Cary, Howard B., 392
Case-Smith, Jane, ed., 98
Cason, Jeff, 444
Cassidy, Kevin A., 492
Castiglia, Patricia Thorson, 83
Castilaw, Diane, 276
Catalano, Joseph T., 73, 89
Cates, Jo A., 236
Catron, Carol E., 313
Catsberg, C. M. E., 35
Caughman, Joyce L., 216
Cawley, John, 466
Celant, Germano, 448
Cella, June H., 60
Cembrowski, George S., 60
Cermak, Timmen L., 537, 574
Chabner, Davi-Ellen, 51, 60, 281
Chaffin, Don B., 92
Chaganti, Rajeswararao, 223
Chaiken, Craig, 176
Chalmers, Irena, comp., 192
Chaney, Michael, ed., 336
Chapin, June R., 313
Charbeneau, Gerald T., 6
Charest-Papagno, Noella, 186
Charlesworth, Rosalind, 313

Chase, Linda, 432
Chatburn, Robert L., 114, 120
Chau, Kenneth K. L., ed., 564
Chavarria, Lazaro, 16
Chelf, Carl P., 570
Chenevert, Melodie, 70
Chenoune, Farid, 420
Cheo, Peter K., ed., 377
Cheremisinoff, Paul, 92
Cherlin, Andrew J., ed., 570
Cherniak, Reuben, ed., 115
Chernik, Barbara E., 334
Chernow, Barbara A., ed., 337
Chervenak, Frank A., ed., 29
Chesire, David F., 266
Chesky, Sheldon R., 186
Chesser, Jerald W., 192
Chiarenza, Carl, 448
Chiche, Gerard J., 6
Child Welfare League of America, 559, 564
Chilo, V., 6
Chin, S. L., 377
Ching, Frank, 129, 432
Chinn, Peggy L., 82
Chipps, Esther M., 88
Chitty, Kay Kittrell, ed., 77
Choate, Curt, 381
Christ, Mary Ann, 89
Christen, Jenna, 176
Christensen, Gordon J., 16, 17
Christensen, Thomas, H., 368
Christiansen, Charles, ed., 99
Christianson, Elin B., 339
Christy, Joe, 358
Cibbarelli, Pamela, ed., 331
Cicchetti, Dante, ed., 560
Cichy, Ronald F., 192
Ciulla, Anna P., ed., 60
Clancy, Kevin J., 259
Clark, Eric, 259
Clark, F. Bryan, ed., 499
Clark, James F., 400
Clark, Jim, 227
Clark, Julia B., 84
Clark, Robin E., 560
Clark, Roy Peter, 237
Clark, Walter B., ed., 537
Clark, William E., 492
Clayton, George D., ed., 93
Clemen-Stone, Susan, 72
Cliffe, Roger W., 123
Clifton, Merritt, 400
Clochesy, John M., 73
Clough, Richard Hudson, 131
Coaldrake, William Howard, 131
Cobb, Hubbard H., 137
Cochran, Phillip E., 525
Cochrane, Willard Wesley, 478, 483
Cocores, James, 537, 574
Coffin, David Page, 414
Coffin, Michael J., 152
Cohen, Allan R., 209
Cohen, Barbara J., 70
Cohen, Barbara Janson, 51
Cohen, Elaine L., 70
Cohen, Frederick B., 176
Cohen, Howard, 289
Cohen, Morris L., 280, 301
Cohen, Sidney, 537
Cohen, Susan M., 82
Cohen, William A., 223
Colahan, Patrick T., 525
Colchester, Chloë, 423, 432
Coleman, Elizabeth A., 420
Coleman, Ellen, 35
Coleman, Margaret Cecil, 313
Coleman, Ronny J., 493
Coleman, Ronny J., ed., 493
Collett, Iris Weil, 167

Name Index

Colletti, Anthony B., 186
Collier, David, 236, 400
Collin, P. H., 397
Collin, Simon, 461
Collings, George, 133
Collins, Clarence Gerald, 287
Collins, Douglas, 448
Collins, H. Thomas, 313
Collins, N. Mark, ed., 503
Collins, R. Lorraine, ed., 537, 574
Collins, Richard L., 358
Collins, Stewart, ed., 574
Collins, Thomas S., ed., 105
Collins, W. J. N., 6
Collison, Brooke B., 552
Coltman, Michael M., 192, 203, 227
Colville, Joann, 525
Colvin, Raymond, 93
Colvin, Thomas S., 143
Combs, Gerald F., 35
Combs-Orme, Terri, 560
Compesi, Ronald J., 265
Condon, Mary Ann Blum, 88
Confer, Robert G., 93
Confer, Robert G., ed., 93
Conference of Funeral Service Examining Boards, 511
Conner, Floyd, 472
Conner, Jesse R., 368
Conners, C. Keith, 35
Connery, Thomas Bernard, ed., 241
Connolly, Barbara H., ed., 106
Connor, Jane Gardner, 331
Conover, Ernie, 123
Conover, Mary Boudreau, 73
Conover, Theodore E., 248, 469
Conrad, Lynne Hutnik, 89
Conroy, Dennis L., 289
Cook, Alton, 469
Cook, David, 377
Cook, Philip S., ed., 241
Cooklin, Gerry, 414
Cookman, Brian, 400
Cool, Lisa Collier, 239
Cooper, David G, 278
Cooper, Elmer L., 481
Cooper, Jim, 133
Cooper, Kenneth H., 313
Cope, Constantin, 29
Corbett, Jane Vincent, 76
Corbin, John, 301
Corey, Gerald, 553
Corey, Marianne Schneider, 553
Corn, Jacqueline K., 93
Cornell, Stephen E., 564
Cornett, Emily F., 85
Corporation for Public Broadcasting, 264
Corr, Charles A., 511
Cosnett, Jay, 401
Cote, Arthur E., ed., 493
Cottell, Philip G., 167
Cotton, Sherrie L., 42
Cottone, James A., 6
Council of Biology Editors, 468
Counts, David R., ed., 511
Cournoyer, Norman G., 192, 203
Covey, Stephen R., 209
Cowan, Fred F., 6
Cowan, Henry J., 156
Cowell, Rick L., ed., 525
Cowger, James F., 287
Cox, James D., ed., 27
Cox, John, 372
Cozens, Bronwyn, 186, 187
Crafer, R. C., 377
Craig, James, 401, 470, 472
Craig, Marveen, 29
Craig, Robert G., 6
Craig, Robert G., ed., 6
Craig, Stephen R., 203

Cramb, Ian, 157
Crane, Dale, ed., 346
Craven, Ruth F., 79
Crawford, Jane Diehl, 525
Crawford, Walt, 340, 401
Crenshaw, David A., 511
Crocker, Dee, 294
Crosby!, 401
Cross, Carla, 217
Crouch, Tom D., 346
Crouse, David B., 401
Cudleigh, Patricia, 29
Culinary Institute of America, 192
Cullinan, Angeline M., 24
Cullinane, Robert, 470
Cullum, Carolyn N., 313
Cummings, Jack, 215, 228
Cummins, Julie, ed., 465
Cunningham, Rebecca, 416
Cunningham, Stephen, 203
Currid, Cheryl, 209
Curry, Thomas S., 24
Curtis, James, 450
Cushman, Robert Frank, 220
Cutlip, Scott M., 259
Cutting, P., 187
Cypert, Samuel A., 168

Da Silva, Raul, 264
Daffner, Richard H., 24
Dagostino, Frank R., 131
Dahmer, Sondra J., 193
Dailey, Robert H., 42
Dale, Barry G., ed., 6
Dale, John, 519
Daley, Dennis C., 537
Daley, Dennis C., ed., 537
Dallow, Theodore J., 218
Dalton, John W., 187
Damerow, Gail, 478
Damrosch, Barbara, 519
Dancyger, Ken, 268
Daniel, Marilyn S., 99
Daniells, Lorna M., 337
Daniels, Joanne M., 85
Darby, Michele Leonardi, ed., 6
Daria, Irene, 424
Das, Pankai K., 377
Dates, Jannette Lake, 254
Davidoff, Philip G., 228
Davidson, Daniel V., 281
Davidson, Harold, 520
Davidson, Margaret, 246
Davidson, Neil, ed.,]313
Davidson, Sidney, 168
Davies, A. C., 392
Davies, Jacqueline, 401
Davies, Mary E., 203
Davies, Patricia M., 106
Davies, Thomas D., 203
Davis, Anne J., 78
Davis, Brenta G., ed., 60
Davis, Carol M., 106
Davis, Harold, 446
Davis, John P., 358
Davis, Judi Ratliff, 74
Davis, Linda J., ed., 99
Davis, Sally Prince, 401
Dawe, Renee A., 60
Dawes, Christopher, 377, 393
Dawids, S. G., ed., 60
Dawson, Peter E., 6
Day, David A., 131, 363
Day, David E., 209
Day, John, 223
De Chiara, Joseph, ed., 432
De La Haye, Amy, 420
De Lyre, Wolf R., 7
De Marly, Diana, 420

De Vos, Dianne C., 24
De Vries, Mary A., 209
De Vries, Mary Ann, 275
Dearie, Tammy Nickelson, comp., 336
Dearstyne, Bruce W., 330
DeBiase, Christina B., 6
Decker, Celia Anita, 313
Decker, Phillip J., 138
DeCristoforo, R. J., 133
Deegan, Mary Jo, 570
Deem, Bill R., 353
DeGalan, Julie, 552
DeGraaf, Richard M., 506
Delaney, John F., 228
Delone, N. B., 377
Delong, Marilyn Fuller, 282
DeLuca, Chester, 17
Demissie, Ejigou, 483
Demone, Harold W., 553
Demornex, Jacqueline, 420
Dempsey, Paul, 364, 381
Denissen, Harry, 7
Dennerll, Jean Tannis, 51
Dennis, Cynthia A., 24
Dennis, Everette E., 241
Denniston, Lyle W., 252
Dental Education Center, 17
Denton, Frank, 241
Department of Veterans Affairs, 17
DePew, John N., 330
Deppa, Joan, 241
Dern, Daniel P., 176, 332
Dernocoeur, Kate Boyd, 42
Deroche, A. G., 353
Des Jardins, Terry R., 115
Destree, Tom, ed., 401
Devall, Bill, ed., 507
DeVita, Vincent T., ed., 569
Devore, Wynetta, 564
Dewey, Barbara I., 336
Dewey, Patrick R., 397, 470
Dewsnap, Don, 401
Diamant, Lincoln, ed., 264
Diamant, Rachel B., 99
Diamond, Ellen, 423
Diamond, Michael R., 223
Dick, Richard S., ed., 53
Dickason, Elizabeth J., 82
Dickinson, Bradley W., 176
Diehl, Marcy Otis, 282
Diekelman, Donald, 203
Dietrich, William, 507
Dietz, Albert G., 133
Dietz, Albert G.H., 133
Differding, Debra A., 275
Digregorio, Charlotte, 239
Dik, Kees J., 526
Dilenschneider, Robert L., 259
Dimarogonas, Andrew D., 372
DiNitto, Diana M., 552
Dinnerstein, Leonard, 565
Dionne, Raymond A., 7
Dison, Norma, 85
Dittmer, Paul, 193
Ditzel, Paul C., 493
Dix, Mark, 176
DiZazzo, Raymond, 265, 269
Dizik, A. Allen, 432
Dobelstein, Andrew W., 570
Dodson, Laurie, 55
Dodt, Lorette C., 470
Doenges, Marilynn E., 78
Doherty, S. Adele, ed., 7
Doka, Kenneth J., ed., 512
Dolan, Joan T., 73
Dolber, Roslyn, 424
Donaldson, Barry, 133
Donatelli, Robert, ed., 106
Donavin, Denise Perry, ed.,]314

Donovan, Mary Deirdre, ed., 193
Dore, Ian, 193
Dorf, Martin E., 432
Dorris, Michael, 537
Doty, David B., 401
Dougherty, Charles J., 571
Doughty, Dorothy Beckley., 88
Douglas, John E., eds., 287
Douglass, Laura Mae, 70
Douglass, Robert W., 501
Dow, Karen Hassey, ed., 83
Dowd, Steven B., 27
Downey, Ray, 43, 493
Downs, Robert Bingham, 250
Dox, Ida, 51, 70
Doyle, Francis R., 301
Drafke, Michael W., 24, 55
Drake, Ellen, 54
Drake, Margaret, 99
Dreistadt, Steve H., 521
Dreizen, LaVerne, 55
Drpic, Ivo D., 432
Drum, Sue, 528
Drummond, Karen Eich, 35, 193, 203
Dryfoos, Joy G., 560
DuBoff, Leonard D., 444
DuBois, Dorothy, 228
DuBrin, Andrew J., 334
DuBrul, E. Lloyd, 7
Dudek, Susan G., 35, 74
Duff, Jon M., 375
Duffy, James E., 353
Duginske, Gen, 123
Duginske, Mark, 123
Duke, James A., 504
Dulfano, Celia, 537
Duncan, Justin, 160
Dundas, James L., 153
Dunlop, Reginald, 358
Dunn, Vincent, 493
Dunn, Winnie, ed., 99
Dunne, Lavon J., 35
Dupayrat, Jacques, 282
Dupuis, Yvon G., 115
DuPuy, Richard, ed., 353
Durham, Jerry D., ed., 73
Durkin, Kenneth Wayne, 288
Dutton, Wilbur H., 314
Dworetzky, John, 314
Dworin, Lawrence, 218
Dworsky, Alan L., 280
Dyer, W. Gibb, 223
Dykstra, John Jefferson, 203
Dzik, Stanley J., 347

Earley, Mark W., ed., 143
Earnest, Vicki Vine, 79
Eason, Susan, 294
East Central AIDS Education and Training Center, 17, 18
Easterling, K. E., 391
Eberhart, George M., comp., 334
Ebersole, Priscilla, 75
Eberts, Marjorie, 552, 553
Echternach, John L., ed., 106
Eckersley, Pamela M., ed., 106
Eckhardt, Robert C., 402
Eckman, William F., 493
Eddy, Lynne, 106
Eddy, Sandra, 177
Edeling, Joy, 106
Edelman, Robert R., ed., 28
Edge, Valerie, 88
Edgell, David L., ed., 228
Edgerton, Leslie, 187
Educational Foundation of the National Restaurant Association, 193
Edwards, David P., 43
Edwards, Julia, 254

Edwards, Kenneth W., 217
Edwards, Linda L., 278
Effner, Jim, 123
Effros, William G., 218
Eggland, Ellen Thomas, 79
Ehler, R. L., 234
Ehrlich, Ann B., 7, 51
Ehrlich, Jeffrey, 129
Ehrlich, Paul R., 504
Ehrlich, Ruth Ann, 24
Eichel, Margarethe Eichel, 377
Eichenberger, Jerry A., 347
Eirksson, Joann Huang, 89
Eiseman, Leatrice, 458
Eisenberg, Ronald L., 24, 25
Ekhaml, Leticia T., 335
Elam, Kimberly, 470
Eldridge, Burgess C., 307
Eliason, Alan L., 177
Eliopoulos, Charlotte, 75
Elkind, David, 326
Ellefson, Connie Lockhart, 519
Ellis, Barbara W., ed., 521
Ellis, J. Nigel, 93
Ellis, Janice Rider, 77, 79
Ellmore, R. Terry, 250
Elster, Allen D., 28
Elwood, David T., 571
Emanuelsen, Kathy Lynn, 89
Emener, William G., ed., 553
Emery, Michael C., 242
Emery, Richard, 467
Emery, Richard S., 402
Emery, William H., 193
Endelman, Robert, 557
Endicott, Eve, ed., 502
Engelmeier, Regine, ed., 420
Engle, Mary, 332
Engle, Robert P., ed., 106
Engler, Nick, 123
Ephross, Paul H., 554
Erdmann, Robert, 35
Erickson, Duane E., 478
Escott-Stump, Sylvia, 35, 75
Eskew, Mike, 280, 301
Espenschied, Roland F., 387
Essa, Eva, 314
Estes, Jack C., 214
Estrada, Susan, 177
Estrin, Chere B., 298
Ettinger, Stephen J., ed., 526
Ettlinger, Steve, 194
Ettorre, E. M., 537
Eubanks, David H., 115
Euganeo, Kathleen Doran, 32
Evans, Arthur G., 446
Evans, D. H., 30
Evans, G. Edward, 337, 340
Evans, Katie, 538
Evans, Mariwyn, 217
Evantash, Allan B., 32
Everett, Joyce, ed., 565
Evett, Jack B., 384
Ewing, William A., 448
Eyres, Patricia S., 303

Fageria, N. K., 482
Fahey, Victoria, ed., 73
Falcone, Joseph D., 433
Faller, Kathleen Coulborn, 560
Faludi, Susan, 254
Fanshel, David, 560
Fanson, Barbara A., 402
Farewell, Susan, 228
Farley, Reynolds, 565
Farren, Carol E., 433
Fassel, Diane, 538
Fawcett, Marcy Davis, 303
Fay, Jack R., 168

Fay, John J., 285
Fecko, Mary Beth, 340
Fedler, Fred, 246
Feeney, Daniel A., 528
Feeney, Kerri W., 303
Feigenbaum, Harvey, 30
Feinberg, Renee, 571
Feininger, Andreas, 499
Feirer, John Louis, 133
Feldman, Phil, 177
Felici, James, 250, 402
Fellers, Patrick W., 402
Fellows, Jane, 204
Fennelly, Lawrence J., ed., 290
Fensch, Thomas, 239, 252
Ferdico, John N., 285
Ferguson, Donald L., 238
Ferguson, Roy C., 484
Ferl, Terry Ellen, 340
Ferleger, Lou, ed., 478
Ferrara, Matthew L., 560
Ferrari, Mario J., 286
Ferry, Mike, 216
Ferry, Ted S., 93
Fetterman, Roger L., 461
Field, Shelly, 259
Field, Tiffany, 314
Fields, Louis W., 168
Fiell, Charlotte, 433
Fiesta, Janine, 89
Finberg, Howard I., 248
Fine Woodworking Magazine, 123, 124
Fingarette, Herbert, 538
Fink, Joanne, 467
Fink, Paul Jay, ed., 567
Finkbeiner, Betty Ladley, 8
Finkelstein, Karen, 276
Finlayson, Iain, 421
Finley, Paul R., ed., 61
Finn, Molly M., 294
Finn, Patrick J., 314
Finn, Susan Calvert, 35
Finnegan, John, 538
Finnegan, Rita, 50
Finucane, Edward W., 93
Fiore, Mariano S. H. di, 526
Fire, Frank L., 493
Fischbach, Frances Talaska, 72, 76
Fischer, Lionel L., 252
Fisher, Barry A. J., 288
Fisher, James C., 334
Fisher, Scott, 461
Fishman, Stephen, 245
Fitch, Rodney, 433
Fitchen, John, 131, 364
Fitzharris, Tim, 446
Flaskerud, Jacquelyn Haak, 73
Flavell, John H., 314
Fleisher, Beverly, 483
Fleishman, Michael, 472
Fleming, John, 433
Fleming, Michael F., ed., 538
Flesch, R. A., 288
Flexner, Bob, 124
Flight, Myrtle, 52
Flint, Mary Louise, 521
Flocke, Lynne, 254
Flora of North America Editorial Committee, 504
Floyd, Elaine, 234, 402
Floyd, Sally, 276
Flynn, Janet-Beth McCann, 73
Flynn, John C., 538
Flynn, John C., ed., 48, 61
Foa, Linda, 433
Foley, W. Dennis, 30
Folsom, Rose, 467
Fong, Elizabeth, 61
Forbes, C. D., 61
Forbes, Glenn S., 32

Name Index

Foreman, Cindy, 347
Forest Products Laboratory (U.S.), 129
Fornell, David P., 494
Forrest, Lewis C., 194, 204
Fortinash, Katherine M., 86
Fortson, Judith, 334
Foster, David R., 314
Foster, Dennis L., 228
Foster, Jennifer, 209
Foster, Malcolm S., 8
Foster, Robert J., 372
Foster, Steven R., 375
Foth, H. D., 504
Fowler, Cary, 482
Fox, Edward L., 106
Foye, James, 347
Fraase, Michael, 402
France, Kenneth, 567
Frank, James A., 494
Frank, Marjorie, 314
Frank, Robyn C., ed., 194
Fraser, Beverly A., 107
Fraser, James Howard, 260, 455
Fraser, Mark W., ed., 560
Frater, Harald, 461
Frechette, Leon A., 138
Fredericks, Anthony D., 314
Freedman, George A., 8
Freedman, Warren, 93
Freeman, Edith M., ed., 538
Freeman, Harry, ed., 368
Freeman, Yvonne S., 315
Freer, Carolee, 275
Freiberg, Marcos A., 4
Frenay, Agnes Clare, 51
French, Thomas Ewing, 372
Frew, Mary Ann, 48, 61
Friedberger, Mark, 478
Friedes, Harriet, 315
Friedman, Jack P., 214, 216
Friedman, Raymond, 494
Friedmann, Barb, ed., 315
Frisby, Thomas N., 131
Fritz, Edward C., 508
Frome, Michael, 228, 501
Frommer, Herbert H., 8
Frost, Carolyn O., 340
Fry, Edward Bernard, 315
Fry, Prem S., 315
Fryxell, David, 239
Fuller, James Edward, 374
Fuller, Jill, 79
Fuller, Margaret, 501
Fulmer, Terry T., ed., 73
Fulton, Marianne, 448
Funk, Sandra G., ed., 73
Fussell, Betty Harper, 482

Gabbard, Carl, 315
Gadow, Sandy, 218
Gage, Loretta, 528
Gagnan, Patricia J., 228
Gagne, Raymond J., 504
Gahart, Betty L., 85
Gaines, William, 246
Galanter, Marc, 538
Galanti, Geri-Ann, 538
Galassi, Peter, 448
Galaty, Fillmore W., 215
Galeno, Joseph J., 161
Gallagher-Allred, Charlette R., 36
Garbarino, James, 315, 560
Garber, David A., 8
Garcia, Mario, 248
Garcia, Mario R., 248
Gardner, Martha, 8
Garfinkel, Perry, 252
Garner, C. William, 168
Garner, Gerald W., 288

Garner, Geraldine O., 553
Garner, Leslie Holland, 554
Garrett, Linda J., 339
Garrett, T. K., 353
Garrett, Vena, 298
Garrett, Wendell D., 433
Garrison, Bruce, 252
Garrison, Paul, 347
Garrison, Robert H., 36
Garrod, Andrew, ed., 315
Gartner, Alan, 561
Garton-Good, Julie, 218
Garvey, Catherine, 315
Garza, Diana, 61
Gater, D. W., 402
Gearhart, Susan Wood, 187
Geary, Don, 124, 390
Gebhardt, Susan E., 194
Gee, Chuck Y., 204, 228
Geffen, Alice M., 228
Gelenberg, Alan J., ed., 538
Gelijns, Annetine, ed., 61
Gelles, Richard J., 561
Gellman, Stuart, 290
Gendel, Steven M., ed., 481
Genishi, Celia, ed., 315
George, Julia B., ed., 82
George, Susan C., comp., 329
Gerberg, Mort, 466
Gere, Charlotte, 433
Gerken, John M., 393
Gero, David, 359
Gersten, Rita, 419
Gertler, Nat, 461
Gesell, Laurence E., 359
Gesner, Rusty, 374
Gettrust, Kathy V., 78
Getz, Donald, 229
Gibbons, Bob, 445
Gibbs, Jewell Taylor, 565
Gibbs, Mark, 177
Gibilisco, Stan, 144, 377
Gibson, Martin L., 237
Giesecke, Joan, ed., 334
Gifis, Steven H., 280
Giger, Joyce Newman, ed., 79
Gikow, Jacqueline, 464
Gilbert, Charles E.,ed., 368
Gilbert, Paul, 567
Gilbert, Sara, 36
Gill, Bob, 458
Gill, James O., 168
Gillard, Quentin, 229
Gillespie, John Thomas, ed., 315
Gillespie, T. D., 353
Gilliland, Burl E., 568
Gillmor, Donald M., 245
Gilman, Alfred Goodman, ed., 539
Gilman, Barbara R., 89
Gilman, Mary Louise, 277
Gilmore, Gene, 237
Gilster, Paul, 177
Ginsburg, Evelyn Harris, 561
Ginsburg, Herbert, 316
Ginsburg, Norman, 571
Ginzberg, Lori D., 565
Gioia, Louis V., 403
Gisslen, Wayne, 194
Gitterman, Alex, ed., 554
Gittinger, Mattiebelle, ed., 417
Gittleman, Ann Louise, 36
Glantz, Meyer, ed., 539
Glasbergen, Randy, 253
Glassner, Andrew S., 461
Glazer, Nathan, 571
Gleason, Henry A., 505
Gleason, Roger, 445
Gleick, James, 448
Gleim, Irvin N., 359

Glickstein, Joan K., 99
Glondys, Barbara A., 53
Glover, Gary, 403
Gluckman, Dale Carolyn, 417
Goad, Karen, 133
Godish, Thad, 93
Goering, Carroll E., 364, 481
Goetsch, David L., 152, 374
Goffin, Stacie G., 316
Gold, Annalee, 421
Gold, Mark S., ed., 539
Goldberg, Barry B., ed., 30
Goldberg, Vicki, 450
Goldberger, Leo, ed., 568
Goldfarb, Roz, 473
Goldman, Arnold J., 281
Goldmann, Nahum, 332
Goldstein, Avram, 539
Goldstein, Nathan, 458
Goldstein, Norma, ed., 250
Goldstein, Ronald E., 8
Golomb, Claire, 316
Golub, Sharon, 89
Gonsoulin, Sheryl M., 43
Gonzalez-Mena, Janet, 316
Goode, Erich, 539, 557
Goodman, Catherine Cavallaro, 107
Goodman, Danny, 209
Goodman, David, ed., 484
Goodrich, David Lee, 298
Goodrum, Charles A., 260
Goodsell, Don, 354
.Goodwin, Daniel, 220
Gordon, Barbara, 473
Gordon, Elliott, 464
Gordon, Karen Elizabeth, 250
Gordon, Linda, ed., 565
Gordon, Marjory, 78
Gordon, Thomas, 316
Gore, Alan, 433
Gore, Albert, 505
Gorman, G. E., 330
Gorman, J. R., 433
Gorman, Michael, 340
Gorman, Michael, comp., 340
Gorrie, Trula, 82
Gosney, Michael, 403
Goss, Larry D., 375
Goss, Tom, ed., 403, 456
Gots, Ronald E., 93
Gott, Philip G., 354
Gottfried, Adele Eskeles, ed., 316
Gottschall, Edward M., 470
Goulart, Ron, 466
Goulart, Ron, ed., 466
Gould, Gary M., 567
Gould, Toni S., 316
Gowen, Michele C., ed., 274, 298
Grabowski, Ralph, 177, 374
Grace, Rich, 177
Graham, Betsy P., 239
Granatstein, David, 487
Grant, Edgar, 229
Grant, Harvey D., 43
Grant, Ruth, ed., 107
Graphic Artists Guild (U.S.), 473
Graphic Design: USA (Firm), 455
Grauer, Ken, 43
Gray, Mikel, 88
Gray, William S., 204
Great Plains Agricultural Council, 486
Greco, Joseph, ed., 372
Green, William B., 210
Green, William R., 434
Greenberg, Arnold, 229
Greenberg, Arnold E., ed., 368
Greenberg, Naomi Schubin, 99
Greene, Christine Bergren, 305
Greene, Harry L., ed., 61

Greene, Roberta R., ed., 568
Greenough, Sarah, 448
Greenwald, E. K., ed., 144
Gregory, Aryear, 229
Greiman, April, 461
Griffiths, Joan, 282
Griffiths, José-Marie, 339
Grilliot, Harold J., 279
Grimes, Deanna E., 88
Grimm, Nils R, 152
Grinnell, Richard M., 569
Grinspoon, Lester, 539
Grisogono, Vivian, ed., 107
Groenwald, Susan L., ed., 83
Groome, James J., 204
Gross, Kim Johnson, 418
Gross, Larry, ed., 248
Gross, Lynne S., 264
Grossman, William, ed., 29
Grotta, Sally Wiener, 403, 461
Grow, Lawrence, 129
Grundberg, Andy, 457
Grupe, Fritz H., 177
Grzimek, Bernhard, ed., 506
Guenther, Jeff, 374
Guest, J. Russell, 161
Guild, Robin, 434
Guild, Stephen, 177
Guild, Tricia, 434
Guiley, Rosemary, 238
Guiling, Stanley Douglas, 154
Guither, Harold D., 478
Gulanick, Meg, 79, 84
Gunderloy, Mike, 250
Gunn, Clare A., 229
Gunston, Bill, 347
Gunston, Bill, ed., 347, 359
Gurewich, Ori, 462
Gurley, LaVerne Tolley, ed., 25
Gustafson, Robert J., 144
Gutierrez, A, 124
Guttman, Helene N., ed., 526
Guy, Julia F., 48
Guyot, Dorothy, 287
Guyton, Arthur C., 71
Guzzetta, Cathie E., 73

Haas, Francois, ed., 115
Haase, Patricia T., 81
Haass, Richard, 334
Habegger, Jerryll, 434
Haber, Judith, 86
Haessler, Walter M., 494
Hafen, Brent Q., 43
Haga, Michio, 8
Hage, Wayne, 485
Hagen-Ansert, Sandra L., 30
Hahn, Fred E., 235
Hahn, Harley, 177, 332
Haines, Robert G., 194
Haines, Roger W., 152
Halas, John, 466
Hales, T. B., 61
Haley, Allan, 470
Half, Robert, 168
Hall, Daniel, 305
Hall, Dinah, 434
Hall, Lee, 421
Hall, Mary Jo Larkin, 89
Hall, R. J., 348, 359
Hall, Stephen K., 93
Hall, William R., 434
Hallam, Adele, 340
Hallberg, M. C., 483
Halliday, Peter, ed., 467
Halloran, James W., 223
Halperin, Edward C., 27
Halse, Albert O., 434
Hambourg, Maria Morris, 448

Hamilton, Betty, 282
Hamilton, Eva May Nunnelley, 36
Hamilton, Gene, 138
Hamlin, Christopher, 483
Hammer, Nelson, 434
Hammerle, Nancy, 571
Hammett, Carol Totsky, 316
Hampton, Mark, 434
Hamstra, Bruce, 539
Hanak, Marcia, 76
Handee, William R., 25
Handler, Joel F., 571
Haney, Wava G., ed., 478
Hanken, Mary Alice, 51
Hanks, Kurt, 457
Hannah, Kathryn J., 79
Hanowell, Leland H., ed., 115
Hansen, J. Doan, ed., 93
Hanson, A. A., ed., 477
Hanson, Larry, 178
Hansson, Tore, 107
Harbin, Andrew L., 384
Hardesty, Larry L., ed., 331
Hardwick, Phil, 218
Haring, Joen Iannucci, 8
Harl, Neil E., 484
Harlow, William Morehouse, 499
Harms, Thelma, 316
Harper, Victoria, 187
Harpole, Greg, 53
Harr, J. Scott, 289
Harrigan, Jane T., 238
Harrington, Linda, 306
Harrington, Linda, ed., 306
Harris, Cyril M., ed., 434
Harris, Dale, 229
Harris, Jack C., 215
Harris, Jennifer, ed., 412
Harris, Jonathan M., 484
Harris, Norman O., 8
Harris, Peter, 204
Harris, Richard Jackson, 242
Harris, Richard W., 519
Harris, Ruth Roy, 8
Harrison, Annette, 316
Harrison, David, 168
Harrison, Diane F., ed., 565
Harrower, Tim, 248
Harry, Keith, ed., 331
Hart, John Fraser, 478
Hartmann, Hudson, 521
Hartmann, Thom, ed., 403
Hartnett, Johnette, 512
Hartwell, Frederic P., 144
Hartzell, Hal, 499
Hartzog, George B., 501
Harvard Student Agencies Staff, 229
Harvey, Carol J., ed., 82
Harvey, Liz, ed., 445
Harvey, Michael, 468
Harwood, Bruce M., 215
Haskin, David, 462
Hatch, Scott A., 278
Hauser, Barbara A., 368
Hauser-Cram, Penny, 317
Hausman, Carl, 245, 246
Hawkes, Nigel, 131
Hawkey, C. M., 526
Hay, Millicent V., 240
Hayes, Karen W., 107
Haygreen, John G., 508
Haynes, Kathleen J. M.,]340
Hayward, Tom, 178
Hazinski, Mary Fran, ed., 84
Healy, Mary, 187
Heath, Earl D., 94
Heath, Ernie, 229
Heath, Galen G., 115
Heavens, O. S., 377

Hecht, Jeff, 378
Heckman, James D., 43
Hecox, Bernadette, ed., 107
Heddy, Edward J., 157
Hedgecoe, John, 445
Hedman, Glenn, ed., 107
Heilman, Grant, 479
Heilman, Joan Rattner, 229
Hein, Eleanor C., ed., 70
Heinemann, Allen W., ed., 539
Heinze, Art, 102
Heitschmidt, Rodney K., ed., 485
Held, Gilbert, 178
Heldman, Robert K., 178
Heller, Steven, 457, 464, 466
Heller, Steven, comp., 465
Helmuth, John W., ed., 484
Helsel, Marjorie Borradaile, 434
Heltshe, Mary Ann, 317
Helvie, Carol O., 72
Hemphill, Barbara J., 99
Hendler, Sheldon Saul, 36
Hendricks, James E., ed., 568
Hengesbaugh, Mark, 403
Henkins, Shepard, 204
Henn, Harry G., 334
Hennessy, Stephen M., 290
Henricks, Lorraine, 539
Henry, John Bernard, ed., 61
Henry, Mark C., 43
Henry, Paul S., ed., 378
Herberholz, Barbara J., 317
Herbst, Sharon Tyler, 194
Herek, Gregory M., ed., 565
Heriteau, Jacqueline, 519
Herlihy, Barbara, ed., 539
Hermann, R. L., 299
Hernberg, Sven, 94
Herren, Ray V., 477
Herrmann, Tracy, 32
Hershey, Gerald L., 178
Hertfelder, Sarah, ed., 100
Herubin, Charles A., 385
Heward, William L., 317
Hewitt, John, 269
Hewitt, William E., 283
Hewlett, Sylvia Ann, 561
Heyman, Therese Thau, ed., 448
Hiatt, Richard S., ed., 144
Hickey, Joanne V., 77
Hickman, Harold R., 266
Hicks, Roger, 445
Hicks, Tyler Gregory, 218
Highway Users Federation for Safety and Mobility, 364
Hill, Lowell D., 482
Hillam, Christine, ed., 9
Hillard, James M., 337
Hillegass, Ellen A., ed., 107
Hillel, Daniel, 486
Hine, Darlene Clark, 81
Hines, Paul R., 144
Hinitz, Blythe Simone Farb, 317
Hinrichs, Kit, 470
Hirama, Haru, 100
Hiro, John E., 133, 435
Hirons, Jean L., ed., 340
Hisrich, Robert D., 223
Hiss, Stephen S., 25
Hitchcock, Michael, 417
Hitner, Henry, 54
Hixson, Amanda C., 178
Ho, Man Keung, 565
Hoag, Philip M., 9
Hobart Institute, 394
Hobbelink, Henk, 481
Hodges, Carol A., 194
Hodges, Elaine R. S., ed., 468
Hoefler, Patricia A., 87

Name Index

Hoeltke, Lynn B., 61
Hoerner, Harry, 381
Hogg, Min, 435
Hogstel, Mildred O., ed., 76
Holden, Henry M., 348, 359
Holechek, Jerry, 485, 503
Holland, D. K., ed., 455
Holland, Kathleen E., ed., 317
Holloway, Nancy Meyer, ed., 73
Holmes, Alan, 403
Holmes, Gwendolyn, 368
Holroyd, Sam V., ed., 9
Holsinger, Erik, 462
Holsopple, Curtis R., 269
Holt, Francis X., 494
Holtz, Herman, 238
Holtz, Larry E., 287, 288, 290
Holzberlein, Deanne, 340
Holzgang, David A., 404
Holzman, Harvey N., 144
Homan, Mark S., 556
Home Decorating Institute, 435
Hop, Frederick Uhlen, 134
Hope, Augustine, 435
Hopkins, Helen L., 100
Hopkins, Tom, 216
Hordenski, Michael F., comp., 174
Hornbostel, Caleb, 129
Horne, Mima M., 79
Horton, Brian, 248
Horwin, Michael, 270
Hoskins, Johnny D., 528
Hotch, Ripley, 204
Houlbrooke, Ralph A., ed., 512
Houldcroft, P.T., 390, 393
Houston, Alan B., ed., 30
Howard, Jimmy L., ed., 526
Howard, Philip H., ed., 94
Howder, Cynthia L., 115
Howe, Geoffrey L., 9
Howell, David W., 229
Hoy, Frank P., 248
Hoyle, Leonard H., 204
Hubbard, Benjamin Jerome, ed., 253
Hubbard, J.T.W., 238
Hubbard, Robert, 540
Hubin, Wilbert N., 348
Huck, Charlotte S., 317
Hudak, Carolyn M., 73
Hudman, Lloyd E., 229
Hudson, Howard Penn, 404
Hudson, Wendy E., ed., 502
Huffman, Edna K., 53
Hufnagel, James A., 138
Hughes, Howard L., 204
Hughes, James G., 354
Hughes, Kenneth L., 404
Hull, Alastair, 435
Hull, Grafton H., 554
Humphrey, Doris, 53
Humphrey, James Harry, 317
Hunt, Constance Elizabeth, 503
Hunt, Gary C., ed., 107
Hunt, Marsha, 421
Hunt, V. Daniel, 372
Hunter, Chris, 503
Hunter, Fil, 445
Hunter, Malcolm L., 499
Huszar, Robert J., 43
Hutchings, Donald E., ed., 540
Hutchings, Nancy, 561
Huth, Mark W., 134
Hutson, Beverly K., 303
Huws, Ursula, 178
Huxley, Anthony, ed., 518
Hylton, Bil, 125
Hyypia, Erik, 224

Iatridis, Mary D., 317
Ibbs, K. G., ed., 378
Ibsen, Olga A. C., 9
Iga, Kenichi, 378
Ijiri, Yuji, 168
Ikram, Muhammed, 528
Illman, Paul E., 359
Imes, Rick, 519
Inaba, Darryl S., 540
Inciardi, James A., 540
Inciardi, James A., ed., 540
Ingham, Peter, 373
Ingham, Rosemary, 416
Innerarity, Sheryl A., 89
Institute of Medicine (U.S.). Comm. on Dietary Guidelines, 36
Institute of Medicine (U.S.). Comm. on Nutrition Components, 36
Institute of Medicine (U.S.). Food and Nutrition Board, 36
Interagency Scientific Committee to Address the Conservation of the Northern Spotted Owl, 506
International Association of Chiefs of Police, 287
International Association of Plumbing and Mechanical Officials, 161
Ireland, Patrick John, 419
Ireland, Sandra Jones, 25
Irish, Donald P., ed., 512
Irving, Clive, 348
Irwin, Robert, 219
Irwin, Scott, ed., 108
Isenberg, Henry D., ed., 62
Isern, Thomas D., 481
Iserson, Kenneth V., 512
Iverson, Kathleen M., 204
Ivey, Pat, 43
Iyer, Patricia W., 79
Izard, Ralph S., 253

Jablonski, Stanley, 4
Jackman, Dianne R., 435
Jackson, Albert, 138, 161
Jackson, Debra Broadwell, 84
Jackson, Henry L., 391
Jackson, Kenneth T., 512
Jacobi, Peter, 240
Jacobs, Carolyn, ed., 566
Jacobs, David H., 129, 134, 354
Jacobs, David S., ed., 62
Jacobs, Donald Trent, 44
Jacobs, Marvin, 260, 404
Jacobs, Stephen Paul, 373
Jacsó, Péter, 332
Jaderstrom, Susan, 210
Jaffe, Hilde, 414
Jaffe, Lee David, 332
Jaffe, Marie S., 75, 76, 84
Jaffe, Richard, 30
Jakab, Peter L., 348
Jalongo, Mary Renck, 317
James, John W., 512
James, Lynn F., ed., 485
Jamsa, Kris A., 178
Janik, Carolyn, 217
Jankowski, Wanda, 435
Jansma, Paul, 317
Jansson, Bruce S., 571
Jarvis, Carolyn, 79
Jawad, Ibrahim A., 30
Jayne, Charlotte, 187
Jeffus, Larry F., 392
Jencks, Christopher, 571
Jennings, Lucy Mae, 210
Jerde, Judith, 412
Jernigan, Marian H., 423
Jerram, Peter, 462
Jerstorp, Karin, 424
Johnson and Johnson Medical, Inc., 9
Johnson, Barbara L., 187

Johnson, Ben, 251
Johnson, D. Gale, 484
Johnson, G. E., 85
Johnson, Jan L., 53
Johnson, Joan M, 54
Johnson, Lois, 404
Johnson, Mary, ed., 254
Johnson, Peggy, ed., 340
Johnson, Richard C., 194
Johnson, W. E., 134
Johnson, Warren T., 521
Johnson-Maloney, Nancy, 301
Jones, Carroll J., 318
Jones, Chester, 435
Jones, Clarence E., 290
Jones, Frederic H., 435
Jones, Jacqueline, 572
Jones, John C., 18
Jones, Julie Miller, 36, 194
Jones, Marleeta K., 50
Jones, Patrick, 336
Jones, Robert T., 348
Jones, Shirley A., 44
Jones, William E., ed., 526
Jong, Anthony W., ed., 9
Jordan, Bill, 554
Jordan, Cora, 219
Jordan, F. T. W., ed., 528
Jordan, Larry E., 179
Jordan, Ronald E., 9
Jordon, Dorothy Ann, 229
Jorwic, Therese M., ed., 50
Joseph, Joanne M., 318
Joseph, Marjory L., 424
Jouve, Marie-André, 421
Jozwik, Francis X., 520
Julien, Robert M., 540
Juma, Calestous, 481
Juniper, Richard, 9
Jurek, Ken, 270
Justice, Jennifer, ed., 318
Juvinall, Robert C, 373

Kacmarek, Robert M., 116
Kacmarek, Robert M., ed., 116
Kadir, Saadoon, ed., 29
Kadushin, Alfred, 561
Kahn, Joseph, 108
Kahn, Mahmood A., ed., 195
Kahn, Sherry, 89
Kahn, Si, 556
Kallaus, Norman Francis, 210
Kaminer, Wendy, 540
Kanatsu, Takashi, 168
Kane, Peter E., 245
Kaneko, Jiro J., ed., 526
Kaniss, Phyllis C., 242
Kaplan, Dorlene V., ed., 195
Kaplan, Lawrence A., ed., 62
Kapleau, Philip, 512
Karen, Robert, 318
Karnofsky, Brian, ed., 368
Karp, Rashelle S., 334
Kasavana, Michael L., 195, 204
Kase, Donald W., 94
Kasle, Myron J., 9
Kastenbaum, Robert, 512, 529
Katsigris, Costas, 195
Katsuyama, Shigeru, 9
Katz, Michael B., 572
Katz, Michael B., ed., 572
Katz, Stan J., 540
Katz, William A., 337
Katz, William A., comp., 337
Katz, William A., ed., 329
Kaufman, Donald, 435
Kaufman, Donald G., 368
Kaufman, Edward, ed., 540
Kaufman, James A., ed., 62

Page 584

Kaufman, Milton, 269
Kavanagh, Barry F., 385
Kavanaugh, Raphael R., 195, 205
Kavass, Igor I., 280
Kawabe, Seiji, 9
Kaye, Peggy, 318
Kazarian, Edward A., 195
Kearns, Michael P., 306
Keding, Ann, 260, 454
Kee, Joyce LeFever, 76, 85
Keegan, Lynn, 82, 89
Keen, John E., 385
Kehoe, Brendan P., 332
Keir, Lucille, 48, 54, 62
Keith, Michael C., 265, 269
Keller, Leslie, 84
Kelly, Lucie Young, 77
Kelvin, George V., 447, 468
Kendall, Bonnie L., 9
Kendall, Florence Peterson, 108
Kennedy, Danielle, 216
Kennedy, George, 246
Kennedy, Patrick M., 494
Kennedy, Shirley, 421
Kennedy, William V., 242
Kenner, Carole, 82
Kent, Dorothy, 179
Keown, Ian, 230
Kerley, Peggy N., 278, 303
Kerlow, Isaac Victor, 373, 404
Kessler, Lauren, 251
Ketterer, Manfred, 195
Kettlehake, Jane E., 32
Key, Wilson Bryan, 260
Khan, Mahmood A., 195
Khan, Mahmood A., ed., 205
Khantzian, Edward J., 541
Kicklighter, Clois E., 195
Kidd, J. Steven, 44, 494
Kidder, Tracy, 134
Kidwell, Claudia Brush, ed., 418
Kiefer, J., 25
Kieffer, John, 447
Kiehne, H. A., ed., 144
Kiely, Terrence F., 279
Kieran, Michael, 404
Kiernat, Jean M., ed., 100
Kiley, Martin, 140
Killenberg, George M., 239, 253
Kilpatrick, John A., 134
Kim, Tok-su, ed., 30
Kimball, Debi, 368
Kinder, Robin, ed., 337
King, Carol A., 195
King, Jean Callan, 404
King, R. T., 502
Kinn, Mary E., 49, 51
Kinney, Jean, 541
Kinney, Jill, 561
Kinney, John, 37
Kinney, Lisa, 334
Kinney, Marguerite Rodgers, ed., 74
Kirby, Robert R., ed., 116
Kirk, Paul Leland, 494
Kirk, Robert Warren, ed., 526
Kirk, Ruth, 503
Kirkpatrick, James M., 373
Kirsch, A.,, 18
Kirschner, Celeste G., 50
Kisiel, Marie, 299
Kismaric, Carole, 566
Kisslo, Joseph A., 30
Kitano, Harry H. L., 566
Kitchens, James A., 541
Kittle, James L., 152
Klaidman, Stephen, 242
Klatell, Jack, 10
Kleeman, Walter, 436
Klein, Erica Levy, 235

Klein, Ruth, 210
Kleper, Michael L., 236
Kletz, Trevor A., 94
Kloppenburg, Jack Ralph, 482
Kluge, Pamela Hollie, ed., 253
Knackstedt, Mary V., 436
Knapp, Mary H., 274
Knight, John Barton, 37, 195
Knight, Les, 289
Knight, Robert L., 374
Knowlton, Steven R., ed., 245
Knudsen, Dean D., ed., 561
Knuppel, Robert A., ed., 82
Knutson, Ronald D., 483
Kobre, Kenneth, 248
Kocar, Marcella J., 274, 277
Koch, John N., 479
Koch, Marylane Wade, 70
Kochan, Stephen G., 179
Kochmer, Jonathan, 179
Koel, Leonard, 134, 157
Koerselman, Virginia, 295, 299
Koff, Patricia Beck, ed., 116
Kohl, Mary Ann F., 318
Kohn, James P., 94
Kohr, Robert L., 205
Kominars, Sheppard B., 541
Komives, Margaret, 415
Konikow, Robert B., 454
Konner, Melvin, 318
Konz, Stephen A., 94
Kopelow, Gerry, 447
Kopp, Ernestine, 415
Koss, Mary P., 557
Kostelnik, Marjorie J., 318
Kotoski, Gabrielle M., 50
Kotschevar, Lendal Henry, 37, 195, 196
Kovich, Karen M., 100, 108
Kozier, Barbara, 80
Kozol, Jonathan, 561
Kramer, Peter D., 541
Kratz, Kenneth E., 385
Kraynak, Joe, 179
Krebs, Carol A., 30
Krebs, Connie, 49
Krebs, Dennis R., 44
Kremkau, Frederick W., 30
Kricher, John C., 499
Krider, Terrance M., 116
Krigger, John, 152
Kroenke, David, 179
Krol, Ed, 179, 332
Krueger, Robert F., 10
Krumm, Rob, 210
Kubasak, Michael W., 513
Kumar, Vinay, 62
Kunz, Christina L., 280
Kurian, George Thomas, 286
Kurjak, Asim, ed., 30
Kurtz, Howard, 242
Kurzweil, Ray, 179
Kusiak, Andrew, 373
Kusiak, Andrew, ed., 373
Kutscher, Austin, H., ed., 513
Kuwayama, Yasaburo, 456
Kwiterovich, Peter, 37

Laarman, Jan G., 499
L'Abate, Luciano, ed., 541
LaBelle, Dave, 248, 249
Labovitz, Arthur, 30
Labuz, Ronald, 457, 462, 471
Lacayo, Richard, 249
Ladbury, Ann, 415
Ladewig, Patricia A., 83
LaGreca, Gen, 196
Lamar, Laura, 404
LaMay, Craig L, ed., 242
Lambert, Mark, 264

Lambert, Patricia, 459
Lambert, Steve, 332
Lambeth, Edmund B., 245
Lamport, Nancy K., 100
Lancaster, J. F., 391
Landon, Robert P., 385
Lane, Karen, 55
Lane, Karen, ed., 49
Lane, Marc J., 224
Lang, Jovian, ed.,]338
Lang, W. Paul, 18
Langer, R. H. M., 482
Langford, Michael John, 445
Langlais, Robert P., 10
Langland, Olaf E., 10
Langley, Billy C., 153
Langman, Larry, 264
Lanier, Pamela, 205
Lank, Edity, 219
Lankford, Terry T., 359
Lansky, Mitch, 508
Lansky, Vicki, 230
Lanzer, P., 31
LaQuey, Tracy L., 332
Larkin, David, ed., 436
Larsen, Jack Lenor, 436
Larson, George A., 436
Larson, Roy A., 521
Larson, Steven B., 275
Laseau, Paul, 436
Lathrop, James K., 495
Lattin, Gerald W., 196, 205
Lauer, David A., 459
Laughlin, Mildred, 318
Lauriault, Jean, 499
Lauter-Klatell, Nancy, comp., 318
Lauwerys, Robert R, 94
Lavender, Kenneth, 341
Laventhol and Horwath, 196
Lavin, Michael R., 338
Lawrence, Eleanor, 282
Lawrence, Mike, 138
Lawrence, Mike, ed., 157
Lawrie, Robert J., ed., 145
Lawson, Gary W., ed., 541
Layne, Ken, 354
Leach, Noel Johnson, 125
Leach, Penelope, 318
Leach, Robert H., 404
Leach, Sid DelMar, 436
Leary, William M., 359
Leary, William M., ed., 348
Lecca, Pedro J., 541
Leddy, Susan, 77
Lederberg, Joshua, ed., 526
Lee, Genell, 44
Lee, Joseph K.T., 27
Leese, Elizabeth, 421
Lefave, Linda, 25
Leffingwell, Randy, 481
Lego, Suzanne, 89
Lehman, Cheryl R., 169
Lehr, Susan S., 318
Leibrock, Cynthia, 436
Leigh, Ronald W., 179
Leimone, Christine A., 10
Leinfelder, Karl F., 10
Leiter, Richard A., ed., 301
Lem, Dean Phillip, 237, 260, 405, 471
Leming, Michael R., 513
Lenburg, Jeff, 466
Lenz, Hans Peter, 354
Leonard, Peggy C., 51, 52, 62, 282
Leonard, William L., 25
Leong, Carol L. H.,]341
Leopold, Aldo, 505
Lerch, Gregory D., 219
Lerner, Joel J., 169
Lesly, Philip, ed., 260

Name Index

Lester, Paul Martin, 249
Lester, Paul Martin, ed., 249
Leukefeld, Carl E., ed., 541
Levenson, Marc D., 378
Levi, Judith N., 277
Levine, John R., 332
Levine, Kristin Johnson, 100
Levine, Michael, 260
Levison, H., 10
Levitt, Seymour H., ed., 27
Levy, Matthys, 131
Levy, Mitchell A., 219
Levy, Stephen J., 562
Lewis, Carole Bernstein, 108
Lewis, Dan A., 568
Lewis, Greg, 249
Lewis, Judith A., 541
Lewis, Marcia A., 52
Lewis, Richard J., Sr., 94
Lewis, Robert C., 205
Lewis, Sharon Mantik, ed., 76
Li, Xia, 397
Library of Congress, 339, 341
Lichty, Tom, 405
Lieberman, Dan, 219
Lieberman, Judy Serra, 196
Lief, Nina R, 319
Light, William J. Haugen, 542
Lima, Carolyn W., 319
Limpert, Rudolf, 354
Lin, Geraline C., ed., 542
Lindberg, Janice B., 77, 80
Lindheim, Richard D., 266
Lindley, Mary E., 562
Lindsey, Bonnie Joan, 55, 62
Lindstrom, Robert L., 462
Linley, David, 436
Linsley, Leslie, 436
Lipinski, Robert A., 196
Lipkin, Lawrence, 169
Lipkis, Andy, 499
Lipow, Anne Grodzins, ed., 330
Lipp, Markus D. W., 18
Lippert, Lynn, 108
Lipsey, Sally I, 85
Lipson, Ashley, 275
Lipson, Ashley S., 303
Listman, Robert J., 169
Litchfield, Michael W., 138
Littell, Elizabeth H., 108
Litten, Julian, 513
Livingston, Alan, 457
Livingstone, John Leslie, ed., 169
Livo, Norma J., 319
Lobenthal, Joel, 421
Locke, Jim, 134
Locke, Lafe, 466
Lodge, Thomas E., 505
Loebl, Suzanne, 85
Loewenberg, Frank M., 554, 572
Logan, Sadye Louise, ed., 566
Logsdon, Gene, 479
Lohmar, Ceil, 219
Lombardo, David A., 348
London, Barbara, 249
London, Mark, 158
London, Mel, 270
London, P. S., 44
Long, Barbara C., ed., 76
Long, Lynette, 72
Long, Shirley M., 25
Lorenz, Richard, 449
Lose, M. Phyllis, 529
Lotz, Roy, 242
Louderback, Joseph G., 169
Louw, Eric, 180
Lovaas, O. Ivar,]326
Love, T. W., 134
Lovell, Ronald P., 253, 446

Lovell, Ross, 446
Lovern, John D., 542
Lowe, David W., ed., 506
Lowinson, Joyce H., ed., 542
Loyd, Richard E., 145
Lu, Cary, 180
Lu, Frank C., Sr., 94
Lubell, David L., 29
Lucas, George Blauchard, 487
Lucas, Jeffrey, 116
Luening, R. A., 483
Lum, Doman, 566
Lumley, James E. A., 219
Luna, Paul, 405
Lundberg, Donald E., 196, 205, 230
Luscher, Keith F., 260
Luther, Arch C., 462
Lutz, William, 251
Luxon, James T., 378
Lynch, Gerard C. J., 158
Lynton, Jonathan S., 275, 281, 303
Lyons, Gail G., ed., 216

Mabberley, D. J., 518
MacArthur, John R., 242
MacDonald, Linda Brew, 332
Mace, James D., 25
Mace, Nancy I., 562
Machover, Carl, 373
Mack, Daniel, 44
MacKay, Ruth C., ed., 554
Mackenzie, Colin F., ed., 108
MacKenzie, James J., ed., 482, 504
Mackrell, Alice, 421
Madden, Chris Casson, 436
Madden, Margretta M., 81
Maffei, Anthony C., 319
Magdoff, Fred, 486
Mager, N. H, 277
Maggio, Rosalie, 251
Magrill, Rose Mary, 330
Magryta, Leslie, 299
Maguire, Byron W., 134, 135
Mahan, L. Kathleen, 37
Mahan, L. Kathleen, ed., 75
Maher, Ann Butler, 83
Mahoney, Gene, 495
Makarius, Theodore, 405
Makeham, J. P., 483
Makens, James C., 205
Malamed, Stanley F., 10
Malley, William J., 116
Mallory, Bruce L.,ed., 319
Malnar, Joy Monice, 436
Malone, William F. P., 10
Maloney, Roy T., 215
Maloney, Timothy J., 145
Malott, Jack C., 25
Maloy, Otis C., 487
Maltzman, Jeffrey, 230
Manahan, Stanley E., 369
Mancini, Marc, 230
Mancuso, Joseph, 224
Manheim, Carol J., 108
Manion, Paul D., 500
Mann, Jonathan M., 10
Mann, Sally, 449
Mann, Thomas E., ed., 242, 260
Mann, William C., 100
Manning, Richard, 135, 508
Mansfield, Richard, 405
Mantus, Roberta, 237, 405
Marantz, Sylvia S., 465
Marcie, Dorothy Anne, 188
Marieb, Elaine Nicpon, 62, 71
Marini, John J., ed., 116
Marion, Nino, 391
Markell, Edward K., 62
Marler, Jerilyn, 210

Marmor, Theodore R., 572
Marowski, Daniel G., ed., 369
Marquis, Bessie L., 71
Marriner-Tomey, Ann, 71
Marriner-Tomey, Ann, ed., 82
Marschall, Richard, 466
Marshall, David H., 169
Marshall, Evan P., 289
Marshall, George R., 405
Marshall, Jacquelyn, 49, 55, 62
Marshall, John, 417
Marshall, Shelly, 542
Martin, Donald J., 205
Martin, Frances L., 88
Martin, Judy, 468
Martin, Philip L., 479
Martin, Ray, 135
Martin, Richard, 418, 421, 422
Martin, Robert J., 205
Martindale, Carolyn, ed., 254
Martinez, Benjamin, 459
Marvin, Bill, 196
Marvullo, Joe, 446
Marzolf, Marion, 242
Massey, Howard C., 162
Matassarin-Jacobs, Esther, 87
Mathewson Kuhn, Merrily, 85
Matteson, Stephen R., 10
Matthews, Martin, 180
Matunde, Skobi, 513
Maul, Lyle R., 224
Maurer, Daphne, 319
Mauro, John, 251
May, Donald F., 117
May, Katharyn A., 83
Mayer, Jean, 37
Mayer, Martin, 242
Mayer, Robert G., 513
Mayes, Kathleen L., 180
McAfooes, Julie, 90
McAtee, Robert E., 108
McBroom, Michael, 444
McCabe, Gerard B., ed., 330
McCall, Ruth E., 62
McCance, Kathryn L., 71
McCarthy, Frank, 18
McCarthy, Karen, 100
McCarthy, Michael, ed., 572
McClatchey, Kenneth D., ed., 63
McClellan, Dorien Smith, 305
McClelland, Deke, 462
McClintock, Michael, 135
McConnell, Primrose, 479
McCord, Robert R., 230
McCormac Jack C., 385
McCormick, John, 462
McCurnin, Dennis M., 529
McCurnin, Dennis M., ed., 529
McDaniel, George, ed., 174, 276
McDaniel, Julie Ann, 338
McDougall, Angus, 249
McDowell, Colin, 422
McDowell, M. C., 196
McEntee, Kenneth, 526
McFarland, Gertrude K., 86
McFarland, Gertrude K., ed., 78
McFarland, Mary Brambilla, 63
McGarry, Richard M., 437
McGarvey, Charles L., ed., 109
McGivney, Glen P., 11
McGrath, Ellen, 542
McGrath, Norman, 437
McGrew, P. C., 405
McIlwraith, C. Wayne, 527
McIntosh, Robert Woodrow, 230
McKenna, Rosalie Thorne, 449
McKenry, Leda M., 85
McKenzie, Dennis J., 215
McKerns, Joseph P., ed., 251

Name Index

McKinney, William E., 25
McLane, Joan Brooks, 319
McLaughlin, Gerald, 11
McLeish, Robert, 269
McLuhan, Marshall, 260
McMahon, Maria O'Neil, 554
McMillan, Dana, 319
McNair, Carol Jean, 169
McNeil, Charles, ed., 11
McNulty, Tom, 339
McPherson, Steven P., 117
Mead, Lawrence M., 572
Meade, James W., 504
Medina, Pat, 305
Medlik, S., 206
Meehan, Joseph, 446, 464
Meehan, William R., ed., 504
Meers, Gary D., 130
Meggs, Philip B., 405, 457, 459
Meilach, Dona Z., 210
Meine, Curt, 505
Melby, Pete, 520
Meller, Susan, 413
Mellody, Pia, 542
Melvin, Jeanne L., 100
Memmler, Ruth Lundeen, 71
Mench, Joy A., ed., 529
Mencher, Melvin, 247
Merickel, Mark, 374
Merideth, Robert W., 505
Merrill, Ronald E., 224
Merrill-Oldham, Jan, 341
Mertzlufft, Bonnie, 319
Metcalf, Peter, 513
Metcalf, Robert Lee, 487
Metheny, Norma Milligan, ed., 80
Mettling, Stephen R., ed., 216
Metzdorf, Martha, 473
Metzler, Ken, 239
Meyer, Eugene, 495
Meyer, Fred P., 506
Meyer, Karl Ernest, ed., 235
Meyer, Philip, 247
Meyer, Richard E., ed., 513
Meyer, Sylvia, 196
Meyers, Larry F., 169
Microsoft, 210
Middleton, D. H., ed., 348
Middleton, Kent R., 246
Mieczkowski, Zbigniew, 230
Mike Byrnes and Associates, 364
Milbank, Caroline Rennolds, 422
Miles, Dale A., 11
Miles, Martin J., 214
Milkman, Harvey B., 542
Milkman, Harvey B., ed., 542
Mill, Robert Christie, 206
Miller, Arthur P., 502
Miller, Benjamin Frank, 63, 70
Miller, Brent C., ed., 562
Miller, C. D., 44
Miller, Casey, 251
Miller, Charles, 135
Miller, Chris H., 11
Miller, G. Tyler, 505
Miller, Jack E., 197, 206
Miller, Joseph Arthur, 369
Miller, Judith, 437
Miller, Kate, 180
Miller, Norman S., 543
Miller, Peter G., 219
Miller, Phyllis Bell, 416
Miller, Rex, 135, 145, 381
Miller, Robert H., 44
Miller, Rosalind E., 342
Miller, S. Dennis, 290
Miller, Teresa, comp., 319
Millerson, Gerald, 268
Milliron, Robert R., 169

Mills, Heidi, 319
Mills, Irving J., 197
Mills, Kay, 254
Mills, Thomas S., 360
Milonni, Peter W., 378
Minceberg, Mella, 211
Miner, Tom, 197
Minkoff, Kenneth, ed., 543
Minton, David, 348, 349
Miraldi, Robert, 243
Miranti, Paul J., 169
Misanchuk, Earl R., 405
Mitchell International, Inc., 354
Mitchell, Jeffrey T., 495
Mitchell, William J., 450, 463
Mitsch, William J., 504
Mix, Floyd M., 145
Moeller, D. W., 94
Moen, Daryl R., 248
Moffitt, Peggy, 422
Mohr, Carolyn, 320
Molnar, Stephen, 566
Moloney, Margaret M., 77
Monaco, Fred, 270
Monk, Lorraine, 448
Monmonier, Mark S., 251
Montgomery, John H., 369
Montgomery, Patricia C., ed., 109
Montgomery, Paula Kay, 320
Monti, Peter M., 543
Moog, Carol, 261
Moore, Keith L., 83
Moore, Matthew S., 320
Moore, Mike, ed., 243
Moore, Robert J., 29
Moore, Robin, 320
Moore, Ronald E., 44
Moore, Stanley B., 522
Moore, Wayne D., 495
Mora, Gilles, 449
Moran, Michael L., 100, 109
Morehouse, Ward, 206
Morgan, Bradley J., 261
Morgan, Bradley J., ed., 63, 238, 261, 501
Morgan, Donald P., ed., 521
Morgan, Ernest, 513
Morgan, Joe P., ed., 529
Morgan, John D., ed., 320, 514
Morgan, William J., 37
Morin-Spatz, Patrice, 50
Morley, John, 269
Moroney, Robert, 572
Morris, Leslie R., 336
Morris, Norval, 557
Morrison, Alastair M., 197
Morrison, George S., 320
Morrison, Micah, 501
Morrissey, Maria, 375
Morton, Joyce, 295
Morton, Patricia Gonce, ed., 80
Moses, Kathy, 320
Moss, Edna Jean, 54
Moss, Miriam, 418
Moss, Stephen J., 11
Moss-Salentijn, Letty, 11
Motley (pseudonym), 417
Mott, Robert L., 268
Mourad, Leona A., 88
Moyle, Peter B., 507
Mueller, Scott, 180
Muia, Paul J., 11
Muir, William, 527
Muisener, Philip P., 543
Muller, Charlotte Feldman, 566
Mulligan, Nan, 18
Mullin, Dale H., 219
Mullin, Ray C., 145
Mulvagh, Jane, 422
Mulvihill, Mary L., 49

Munger, Richard L., 320
Murfitt, Janice, 197
Murphy, John M., 456
Murray, Donald Morrison, 247
Murray, John J., ed., 11
Murray, Katherine, 180, 408
Murray, Maggie Pexton, 422
Murray, Ruth Beckman, 86
Musberger, Robert B., 266
Mutter, Scott, 449

Nabhan, Gary Paul, 479
Nadler, James, 406
Nailen, Richard L., 146
Nakamura, Sadao, 437
Nance, Barry, 180
Nanda, Navin C., 31
Nanda, Navin C., ed., 31
Nardantonio, Dennis N., 266
Nash, George, 138
Nash, Jay Robert, ed., 286
National Agricultural Library (U.S.), 477
National Association of Emergency Medical
  Technicians, 44
National Association of Legal Assistants, 295
National Association of Legal Secretaries, 295
National Association of Meat Purveyors, 197
National Association of Safety and Health
  Professionals, 369
National Council for Interior Design Qualification, 437
National Council of Examiners for Engineering and
  Surveying, 385
National Institute for Occupational Safety and
  Health, 369
National Museum of American Art (U.S.), 449
National Police Chiefs and Sheriffs Information
  Bureau, 286
National Research Council (U.S.). Subcommittee on the
  10th Edition of RDAs, 37
National Research Council (U.S.), 479, 484, 485, 486
National Research Council (U.S.). Committee on
  DNA Technology in Forensic Science, 288
National Restaurant Association, 197
Nau, Jim, 522
Naylor, Colin, ed., 413, 437, 444, 457
Naylor, W. Patrick, 11
Neal, Margo Creighton, 78
Nebel, Eddystone C., 206
Neely, Keith A., 45
Neese, William A., 349
Neff, Jack, 473
Neff, Janet A., 74
Neilson, James P., ed., 31
Nelson, Jane, 574
Nelson, John A., 125
Nelson, Paul V., 520
Nelson, Roger M., 109
Nelson, Roy Paul, 406, 454
Nelson, Stephen L., 211
Nemeth, Charles P., 299, 303, 305, 307
Netting, Robert, 487
Neuendorff, G., 18
Neuenschwander, Brody, 468
Neville, Brad W., 11
Nevine, Kathleen B., 295
New York Landmarks Conservancy, 138
Newhouse, Jeffrey H., 28
Newkirk, Thomas, 320
Newman, Arnold, 503
Newman, Diane Kaschak, 76
Newman, Jeanne D., 220
Newman, Michael G., 11
Newman, Morton, 157, 158
Newsom, Doug, 261
Newsom, Jayne, 18
Newton, Dennis W., 360
Newton, Randall S., 375
Ng, Joseph, 377
Nicholas, Ted, 224

Name Index

Nichols, David G., ed., 45
Nicoll, Leslie H., ed., 82
Nield, Jill Shiffer, 12
Nielson, Karla J., 437
Ninemeier, Jack D., 197
Nixon, Judith M., 197
Nizel, Abraham E., 12
Noback, Charles Robert, 12
Nolan, Michael J., 406
Norkin, Cynthia C., 109
Norman, John, 495
Norman, Richard B., 463
Noronha, Shonan F. R., 270
Norrgard, Lee E., 514
Norris, Christopher M., 109
Norse, Elliott A., 500
North American Veterinary Technician Association, 531
Northouse, Peter Guy, 72
Norton, Boyd, 447
Norton, Peter, 180
Nosse, Larry J., 109
Notman, Malkah T., ed., 566
Nourie, Alan, ed., 251
Novak, Theodore J., 306
Novoa, Jane, 90
Nowak, Ronald M., 507
Nowinski, Joseph, 543
Noyes, Robert, ed., 369
Nuckols, Cardwell C., 543
Nugent, Patricia Mary, 87
Nunemacher, Greg, 180
Nunn, Richard V., 146
Nye, Robert D., 568
Nyer, Evan K., 369
Nykiel, Ronald A., 206
Nyvall, Robert F., 487

Oakes, Dana F., 117
Obeid, Anis I., 31
Ober, Gary J., 364
Oberg, Erik, 152
Oberschulte, William, 135
O'Block, Robert L., 286, 338
O'Brien, Joseph D., ed., 63
O'Brien, Robert, 574
Occupational Therapy Roles Task Force, 101
O'Connell, Daniel J., 215
O'Connor, Linda J., 109
O'Connor, Walter F., 170
O'Day, Kate, 180
O'Donnell, Lewis B., 270
Odwin, Charles S., 31
Ogawa, Joseph M., 521
Ogawa, Yoko, 418
Ohanian, Thomas A., 268
O'Hara, Shelley, 181
O'Hare, David, 360
Olds, Sally B., 83
Oles, Paul Stevenson, 437
Olian, JoAnne, ed., 422
Oliver, Chadwick Dearing, 500
Olsen, Gary, 463
Olsen, June Looby, 88
Olson, Kent C., 302
Olson, Nancy B., 342
O'Malley, Penelope Grenoble, 360
Omura, George, 373
O'Neill, Paula N., 90
Opie, John, 486
Oran, Daniel, 281
Orenstein, Vik, 447
O'Riley, Ronald P., 146
Orlik, Deborah K., 304
Orlik, Peter B., 265, 269
Orman, Levent V., 181
Orme, Alan Dan, 138
Orr, Lisa, ed., 334
Oster, Clinton V., 360

Ostroff, Harriet, 198
O'Sullivan, Judith, 466
O'Sullivan, Susan B., 109
Othmer, Ekkehard, 543
Otis, Clark E., 349
O'Toole, Marie, 52
O'Toole, Randal, 502
Ots, Peter, 406
Otto, Shirley E., ed., 83
Owen, David, 139
Owen, K., 355

Pace, Steve, 349
Packer, William, 419
Paddock, Bruce T., 406
Padfield, R. Randall, 360
Padgett, JoAnn, ed., 406
Padgett, Mark, 188
Paduano, Joseph, 446
Pagana, Kathleen Deska, 63, 76
Paietta, Ann Catherine, 336
Paige, Grace, 206
Palais, Joseph C., 378
Paley, Vivian Gussin, 320
Palko, Tom, 56
Palladino, Leo, 188
Palmquist, Roland E., 146
Pampel, Fred C., 573
Panarese, William C., 157
Panati, Charles, 514
Pandit, Milind S., 181
Pangrazi, Robert P., 320
Pansini, Anthony J, 146
Panzer, Robert J., ed., 63
Parcel, Guy S., ed., 45
Parelli, R. J., 25
Parenti, Michael, 243
Parham, Iris A., 562
Parish, Justine Limpus, 419
Park, Edwards, 360
Parker, Dana, 181
Parker, Edwin B., 479
Parker, Julie, 424
Parker, Kathryn G., 95
Parker, Roger C., 237, 406, 407, 471
Parmley, Robert O., ed., 153
Parsigian, Elise K., 247
Parsons, Alyce, 418
Pasour, E. C., 483
Patrick, Gay D., 336
Patterson, Marva, 188
Patterson, Nancy, 274
Pattison, Anna Matsuishi, 12
Patton, David R., 507
Patton, Michael Quinn, ed., 562
Paul, J. Ellis, 12
Paul, William E., ed., 63
Pauli, Eugen, 198
Paulsen, Deborah, 407
Pavlik, John Vernon, ed., 243
Paxton, Albert S., 140
Payne, Neil F., 504
Payton, Otto D., 101
Peacock, John, 397, 422
Peacock, John, ed., 407
Pearce, Douglas G., 230
Pearce, J. Malcolm, ed., 31
Pearson, John K., 304
Pecktal, Lynn, 417
Pecora, Peter J., 562
Pedigo, Larry P., 488
Pedretti, Lorraine Williams, 101
Peele, Stanton, ed., 543
Pegler, Martin M., ed., 437
Pelle, Marie Paule, 423
Pellowski, Anne, 321
Peloquin, Albert A., 437
Pelton, Gordon E., 181
Pelton, Leroy H., 573

Pence, Gregory W., 53
Pence, Terry, ed., 78
Penn, Ira A., 342
Pennak, Robert W., 507
Pennell, Allison A., ed., 369
Penney, Edmund F., 264
Perdue, Peggy K., 321
Perel, Azriel V., ed., 117
Perez, Carlos A., ed., 27
Perfect, Christopher, 471
Peringian, Lynda, 109
Perl, Lila, 423
Permar, Dorothy, 12
Perret, Geoffrey, 360
Perrin, David H., 109
Perry, Anne Griffin, 80
Perry, Dorothy A., 12
Perry, Greg M., 181
Perry, Jacquelin, 110
Persing, Gary, 117
Persson, Conrad, 230
Peters, Charles J., 117
Petersen, Dan, 95
Petersen, Jack E., 95
Peterson, Lorna, 334
Peto, Gloriajean, 45
Pettit, Thomas H., 529
Pfeffer, Cynthia R., ed., 562
Pfeifer, Ken, 455
Phagan-Schostok, Patricia A., 12
Phelan, Susan, E., 63
Philbin, Tom, 139
Phillips, Alan, 437
Phillips, Barty, 438
Phillips, Charles, 45
Phillips, Ralph W., 12
Phillips, Steven J., 130
Phipps, William E., 514
Picard, Robert G., 243
Pielou, E. C., 500
Pierce, Dean, 553
Pierce, Sydney J., ed., 342
Pike, Jayna, 373, 463
Pilbeam, Susan P., 117
Pilcher, Donald M., 569
Pile, John F., 438, 457
Pillari, Vimala, 562
Pillitteri, Adele, 83
Pinson, Linda, 170, 224
Piotrowski, Christine M., 438
Pipes, Peggy L. ed., 75
Pittman, Nancy P., ed., 479
Pivar, William H., 216, 217
Place, Stan, 188
Plachy, Sylvia, 449
Plaut, Simone, 26
Plawin, Paul, 230
Plog, Barbara A., ed., 95
Plumer, Erwin H., 562
Plumridge, Andrew, 158
Plunkett, Signe J., 527
Poggenpohl, Sharon Helmer, ed., 473
Poignard, Renee, 188
Pokorny, Elizabeth J., 335
Poleman, Charlotte M., 37
Politis, Jonathan F., 45
Pollan, Stephen M., 139
Pollock, Michael L., 110
Pond, Kathleen Lingle, 230
Pope, Jacqueline, 556
Porter, Donald J., 349
Porter, Thomas, 18
Porter, Tom, 459
Porterfield, James A., 110
Porth, Carol, 71
Portland Cement Association, 158
Postel, Sandra, 486
Potter, Patricia Ann, 80
Potter-Efron, Ronald T., ed., 574

Page 588

Name Index

Powe, L. A. Scot, 246
Powell, Charles C., 521
Powers, Lawrence W., 63
Powers, Thomas F., 198, 206
Powis, Raymond L., 31
Poyner, Rick, 471
Pozgar, George D., 53
Pratsinak, George, ed., 543
Pratt, Pat Nuse, 101
Pratt, Paul W., 529
Pratt, Paul W., ed., 529
Prentice, William E., ed., 110
Preston, Jack D., ed., 12
Preston, Karen, 19
Pribor, Hugo C., 63
Price, W. D., 514
Prince, Mary Miles, 302
Pripps, Robert N., 481
Proctor, Nick H., 95
Professional Training Schools (Dallas, Tex.), 514
Prokohorov, A. M., 378
Protess, David, 243
Provenza, D. Vincent, 12
Prue, Lucinda K., 26
Ptacnik, Donald J., 45
Puccio, Joseph A., 338
Puckett, Craig D., ed., 50
Puette, William, 243
Pulsifer, Nancy, 276, 304
Punch, Walter T., 519
Purtilo, Ruth B., 72
Pyle, Ransford Comstock, 279, 299

Quigley, Christine, 514
Quiller, Stephen, 438
Quillman, Susan M., 89
Quincy, Matthew, 37
Quirt, John, 243

Raabe, William A., 306
Rabinowicz, Tzvi, 514
Rachlin, Harvey, 270
Racine, Drew, ed., 342
Rada, Robert, 19
Raether, Howard C., 514
Ragan, Robert C., 170
Rahn, Arthur O., 13
Railton, W. Scott, 95
Raines, Shirley C., 321
Rajab, Jehan S., 417
Ralston, Anthony, ed., 175
Ramirez, Gonzalo, 321
Ramsay, Linda M., 438
Ramsey, Dan, 135
Rand, Ellen, 139
Randall, Charles T., 438
Randall, Lynn M., 276, 303
Rao, Krishna C.V.G, 28
Rapp, Stan, 261
Rasband, Judith, 415
Raschko, Bettyann Boetticher, 438
Rasmussen, Richard A., 13
Raso, Jack, 38
Rateitschak, Klaus H., 13
Rathbone, Andy, 463
Ratliff, Thomas A., 64
Rau, Joseph L., 117
Ravel, Richard, 64
Rawlins, Ruth Parmelee, 86
Ray, Mary Barnard, 277
Raymond, Eric S., comp., 175
Read, Edward M., 544, 574
Reamer, Frederic G., 554
Reamer, Frederic G., ed., 569
Redden, Kenneth R., 281
Redman, Barbara Klug, 80
Reece, E. Albert, 31
Reed, Kathlyn L., 101
Reed, Lewis, 198

Reed, Maxine K., 270
Reed, Robert M., 265
Reeder, Sharon J., 83
Reeders, Jacques W.A.J., ed., 26
Rees, Stuart, 556
Rees, Yvonne, 438
Reese, Linda E., 88
Reeser, Linda Cherrey, 555
Regan, Joan M., 89
Regens, James L., 505
Regezi, Joseph A., 13
Reichard, David C., 375
Reid, Kenneth E., 555
Reid, Robert D., 198, 206
Reid, William H., 544
Reilly, James P., 386
Reilly, John W., 214
Reinhart, Richard O., 360
Reintzell, John F., 289
Rekus, John, 95
Remington, R. Roger, 457
Renner, Peter Franz, 206
Renner, Robert P., ed., 13
Renshaw, Sharon, 90
Renstrom, Peter G., 295
Requa-Clark, Barbara, 13
Reston, James, 243
Rettinger, Michael, 438
Reynolds, D. W., 287
Reynolds, Monica, 216
Reznikoff, S. C., 439
Rheingold, Howard, 181
Rhode, Deborah L., ed., 566
Rhodes, Richard, 480
Riahi-Belkaoui, Ahmed, 170
Ribar, L. John, 181
Ribbens, William B., 355
Riccardi-Cubitt, Monique, 439
Ricchiardi, Sherry, ed., 255
Rice, Jane, 52, 54, 282
Rice, Laura Williams, 519
Rice-Licare, Jennifer, 544
Rich, Carole, 247
Richan, Willard C., 556
Richard, Eugene, ed., 50
Richards, E. Earl, 19
Richardson, Alan R., ed., 267
Richardson, Janice K., 110
Richie, Nicholas D., 555
Rickert, Colleen M., 321
Rieman, Timothy D., 439
Riffe, Tanya S., 19
Riggs, J. Rosemary, 439
Riley, Sam G., ed., 251
Rimmer, James Hunter, 110
Rimmer, Steve, 407
Ripka, L. V., 162
Ritchin, Fred, 249, 450
Ritter, Merrill A., 110
Rivers, William L., 240
Rivinus, Timothy M., ed., 544
Roach, William H., 53
RoAne, Susan, 262
Robak, Rostyslaw, 544
Roberts, Albert R., 563
Roberts, Albert R., ed., 568
Robertson, Bozena-Eva, 544
Robertson, Nan, 255
Robertson, Paul B., 13
Robillard, Walter G., 386
Robinson, Corinne H., 38, 75
Robinson, Gordon, 500
Robinson, Judith Schiek, 335
Robinson, Julian, 419
Robinson, Leigh, 220
Robinson, N. E., 527
Roblee, Charles L., 495
Rocabado Seaton, Mariano, 110
Rodale, Robert, 484

Roderick, Wanda Walker, 281
Rodgers, Harrell R., 573
Rodriguez, Walter, 373
Roemer, Joan, 321
Rogers, Ronald, 544
Rogers, Vickie, 50
Rogler, Lloyd H., 566
Rogondino, Michael, 463, 471
Rollock, Barbara, 465
Roman, Paul M., ed., 567
Romano, Deborah, 19
Roncarelli, Robi, 466
Roney, Raymond G., ed., 335
Rooney, Joe, 373
Rooney, Ronald H., 555
Roopnarine, Jaipaul L., ed., 321
Roper, Brent D., 303
Rose, Carla, 181, 407
Rose, Jay, 181
Rose, Walter, 131
Roselle, Laurie P., ed., 304
Rosellini, Gayle, 544
Rosen, Ben, 471
Rosen, Robert J., 29
Rosenberg, Jerry Martin, 175
Rosenberg, Paul, 147
Rosenblatt, Paul C., 480
Rosenbluth, Hal F., 230
Rosenborg, Victoria, 181
Rosenthal, Lawrence, 170
Rosentswieg, Gerry, ed., 459
Rosini, Neil J., 246
Ross, E. Betsy, 514
Ross, George R., 544
Ross, John, 407
Ross, Mildred, 101
Rossbach, Sarah, 439
Rossotti, Hazel, 495
Roth, Alan W., 386
Roth, Alfred C., 381
Roth, Cliff, 267
Roth, Eli, 38
Roth, Paula, ed., 544, 545
Roth, Wendy, 231
Rothenberg, David, 502
Rothmyer, Karen, 235
Rothstein, Jules M., 110
Rotunda, Ronald D., 295
Rouf, A., 95
Rouse, Elizabeth, 418
Roush, James K., 530
Rowan, Carl Thomas, 243
Rowel, Jo Ann C., 54
Rowh, Mark, 392
Rowland, I. R., 38
Roy, Greg, 135
Roy, Robert L., 135
Royse, David D., 569
Royster, Judith Doswell, 95
Rubash, Joyce, 198
Rubel, G. Alexander, 527
Rubington, Earl, ed., 555
Rubinstein, Hilary, ed., 231
Rudin, Claire, 339
Rudman, Jack, 13, 19, 38, 56, 95, 188, 274, 355, 381, 386
Ruff, Thomas P., 321
Ruggles, Philip Kent, 407
Rundback, Betty R., 231
Runge, Val M., 28
Runnells, Robert R., 13
Rupp, William E., 439
Ruppel, Gregg, 118
Rush, George E., 286
Russel, Charlie, 182
Russell, Beverly, 439
Russell, Dale, 459
Russell, David L., 321
Russell, Harold E., 290

Name Index

Rust, Frances O'Connell, 321
Rutledge, Devallis, 287, 290
Ryan, Chris, 231
Ryan, Sally E., ed., 101

Saari, David J., 276
Sabato, Larry J., 243
Sabbagh, Karl, 136
Saberin, Gloria, 125
Sabin, William A., 211, 295
Sachdev, Paul, 563
Sacher, Ronald A., 64
Sader, Marion, ed., 338
Safford, Edward L., 378
Saia, D. A., 26
Salaman, R. A., 125
Salamon, Sonya, 480
Salser, Carl Walter, 295
Salter, Charles A., 337
Salter, Jeffrey L., 339
Salvadori, Mario George, 131
Salway, J. G., ed., 64
Sample, V. Alaric, 502
Sampson, Carol A., 439
Sanders, Barbara, ed., 110
Sanders, Catherine M., 515
Sanders, Danielle M., 321
Sanders, Linda, 407
Sanders, Roger C., ed., 31
Sanders, Rosanne Bryce, 182, 211
Sandler, Stanley, 360
Sangster, Alan, 170
Santos, Virginia, 4
Santrock, John W., 322
Sapp, Gregg, ed., 330
Saracho, Olivia N., ed., 322
Sato, Sadakatsu, 13
Satterlund, Donald R., 504
Satterwhite, Joy, 447
Saunders, Dave, 447
Saunders, Laverna M., ed.,]333
Sauselein, Theodore B., 147, 153
Savage, Craig, 136
Savage, Peter, 231
Sawin, Philip V., comp., 206
Sax, Nancy, 170
Saxton, Dolores F., ed., 87
Saye, Jerry D., 342
Sayer, Chloe, 417
Scales, Barbara, ed., 322
Scali-Sheahan, Maura T., 188
Scanlon, Craig L., ed., 118
Schaaf, T. W. van der, ed., 95
Schactman, Tom, 267
Schaef, Anne Wilson, 545
Schaefer, John Paul, 446
Schaeffer, Dorcas O., 530
Schaetzing, Edgar E., 206
Scharff, Robert, 355
Scharff, Robert, ed., 355
Schatt, Stanley, 182
Schell, Frank R., 391
Schenck, Mary, 407
Schiff, David, 125
Schiff, Kenny, 407
Schihl, Robert J., 267
Schildt, Herbert, 182
Schilit, Rebecca, 545
Schiller, Pamela Byrne, 322
Schinke, Steven Paul, 545
Schirrmacher, Robert, 322
Schlemmer, Richard M., 471
Schlessinger, Bernard S., ed., 338
Schmidgall, Raymond S., 198, 207
Schmidt, Arno, 198
Schmidt, Karen A., ed., 330
Schneeman, Angela, 306
Schneider, Arthur, 268
Schneider, Madelin, 207

Schneider, Robert L., ed., 563
Schoeff, Larry E., 64
Schoeneck, Therese S., 515
Schrank, Jeffrey, 442
Schroeder, Dirk G., 231
Schroll, R. Craig, 495
Schuckit, Marc Alan, 545
Schulman, Mark, 182
Schulte, Henry H., 247
Schultz, Judith M., 86
Schultz, Morton J., 355
Schulz, Marjorie Rittenberg, 207
Schumacher, Michael, 239
Schumann, Gail L., 488
Schuttner, Scott, 136
Schutzer, Daniel, 182
Schwab, James, 480
Schwartz, Carol A., ed., 224
Schwarz, Barb, 217, 221
Schwarzrock, Shirley Pratt, 14
Schwenke, Karl, 158
Sciarra, Dorothy June, 322
Scorer, Richard Segar, 360
Scott, Gini Graham, 231
Scott, Randall W., 467
Scott, Ronald W., 14, 111
Scriven, Carl, 198
Scully, Crispian, 14
Scully, Rosemary M., ed., 111
Seaberg, Albin G., 198
Seale, Arthur C., 147
Sears, Jean L., 335
Sedan, Paul S., 136
Seeburger, Francis F., 545
Seefeldt, Carol, 322
Seefeldt, Carol, ed., 322
Seeley, Rod R., 64
Seeram, Euclid, 27
Seibel, Rainer M. N., 29
Seidel, Henry M., 80
Selekman, Janice, 89
Self, Charles R., 136, 158
Selman, Joseph, 28
Sennett, Ted, 467
Sgroi, Suzanne M., ed., 563
Shaeffer, Claire B., 413, 415
Shannon, Margaret T., 85
Shapiro, Barry A., 118
Shapiro, Fred R., comp., 302
Shapiro, Howard I., 364
Shapiro, Jacob, 26
Shapiro, Paul D., 45
Sharp, Bill, 501
Sharp, Craig W., 182
Shaw, Susan, 445
Shea, Thomas M., 322
Sheafor, Bradford W., 555
Sheal, Peter R., 335
Sheda, Constance, 101
Sheehy, Susan Budassi, ed., 74
Sheldon, Thomas, 182
Shell, Ellen Ruppel, 322
Shelov, Steven P., 322
Sher, Paula, 473
Sherding, Robert G., ed., 527
Sherlock, Paul, 224
Sherman, Karen M., 89
Shernoff, Michael, ed., 545
Sherraden, Michael W., 573
Sherry, John E. H., 199, 207
Sherwood, Gerald E., 136
Shevell, Richard Shepherd, 349
Shiba, Akihiko, 14
Shibukawa, Ikuyoshi, 459
Shields, J., ed., 369
Shigo, Alex L., 500
Shim, Jae K., 170
Shirato, Linda, ed., 331, 333
Shirkhoda, Ali, ed., 31

Shivers, Natalie W., 139
Shives, Louise, R., 86
Shock, Patti J., 207
Shoemaker, Pamela J., 244
Shoemark, Pete, 381
Sholly, Dan R., 502
Shook, Frederick, 265
Shoup, Cynthia A., 118
Shugart, Grace Severance, 38, 199
Shumaker, Terence M., 374
Shurr, Donald G., 111
Shushan, Ronnie, 408
Shuster, William A., 381
Sides, Maria B., 87
Siegel, Eric S., 224
Siegel, Joel G., 170, 171
Sierra, Judy, 322, 323
Signey, Phillip J., 304
Sills, James R., 118
Silver, Gerald A., 408
Silver, Linda, ed., 408, 454
Silver, Susan, 211
Silverman, Norman H., 31
Silverman, Sol, 4
Simini, Joseph Peter, 171
Simmons, Arthur, 64
Simmons, H. Leslie, 139
Simmons, John V., 188
Simone, Charles B., 38
Simone, Luisa, 408
Simpson, Alan, 182
Sine, Charlotte, ed., 477
Sing, Bill, ed., 255
Singer, Eleanor, 244
Sirmans, C. F., Jr., 216
Siropolis, Nicholas C., 224
Sirota, David, 220
Siuru, William D., 360
Skale, Nedra, 84
Skocpol, Theda, 573
Skoll, Geoffrey R., 545
Slater, Jeffrey, 171
Slavin, Robert E., ed., 323
Sloan, Annie, 439
Sloan, Irving J., 569
Sloan, W. David, comp., 236
Sloan, William David, 235
Sloan, William David, ed., 235
Sloane, David Charles, 515
Slote, Stanley J., 342
Small, Jacquelyn, 545
Smead, David, 147
Smedinghoff, Thomas J., 463
Smeltzer, Suzanne, ed., 76
Smit, Kornelis, 130
Smith, Bradford P., ed., 527
Smith, Carol Cox, 546
Smith, Charles A., 323
Smith, Coleman Cecil, 266
Smith, David, 343
Smith, Donna Phillips, ed., 84
Smith, Gary A., ed., 118
Smith, Genevieve Love, 52
Smith, Hubert, 349, 360
Smith, Jeannette, 261
Smith, L. Murphy, 171
Smith, Lee, 162
Smith, Myron J., 360
Smith, Robert Charles, 408, 459
Smith, Robert L., 147
Smith, Robert Leo, 505
Smith, Robert Sellers, 306
Smith, Robin, 343
Smith, Sandra Fucci, 80, 87
Smith, Susan, 72
Smith, Valene L., ed., 231
Smith, W. J., 527
Smith, Warren J., 378
Smoot, E. Clyde, 20

Smutny, Joan F., 323
Snider, Nancy, ed., 199
Snyder, John M., 373
Snyder, Mariah, ed., 77
Sobell, Mark B., 546
Sobieszek, Robert A., 449
Sobon, Jack A., 136
Sochats, Ken, 175
Society of Automotive Engineers, 355
Sokoll, Gary J., 515
Solbrig, Otto T., 480
Soley, Lawrence C., 244
Sollock, Tom, 189
Solomon, Charles, 467
Solomon, Eldra Pearl, 71, 283
Solomon, Robert E., ed., 496
Solvberge, Arne, 182
Somer, Elizabeth, 38
Somers, Martha F., 111
Sonderegger, Theo, ed., 546
Song, Jizhong, 343
Sonnenschmidt, Frederic H., 199
Sonsino, Steven, 455
Sonsthagen, Teresa F., 530
Sormunen, Carolee, 52, 283
Sosinsky, Barrie A., 408
Soule, Judith D., 487
Southard, Samuel, comp., 515
Southern Weed Science Society (U.S.)., 488
Spacek, Leonard, 171
Spahn, Edwin J., 496
Sparks, Sheila M., 78
Sparks, Shirley N., 546
Specht, Harry, 555
Spector, Rachel E., 80
Speer, Kathleen Morgan, 84
Spence, William Perkins, 136
Spencer, Donald D., 397
Spencer, Dorothy, 440
Spencer, Roberta Todd, 86
Spenser, Donald D., 175
Sperry, Neil, 520
Spiller, David, 330
Spock, Benjamin, 323
Spodek, Bernard, 323
Spodek, Bernard, ed., 323
Sprawls, Perry, 26
Spurling, David, 171
Spurr, Stephen Hopkins, 500
Squibb, Sharon, 454
Squires, James D., 244
Stachowitsch, Michael, 527
Staebler, Wendy W., 440
Stafford, Peter G., 546
Stallcup, James G., 147
Stallings, William, 182
Stamper, Anita A., 418
Stamps, Georgiana M., 88
Stanfield, Peggy, 283
Stanhope, Marcia, ed., 72
Stankus, Bill, 125
Starck, Patricia L., ed., 74
Stare, Fredrick J, 38
Stark, David D., 28
Stark, John, 373
Starr, Raymond, 563
Starr, Richard, 323
Statkiewicz-Sherer, Mary Alice, 26
Statsky, William P., 300, 304, 305
Staub, Norman C., 118
Stauffer, H. Brooke, 147
Steel, G. Gordon, ed., 28
Steele, Valerie, 413
Steele, Victoria, 335
Steele, Virginia, comp., 337
Stefanelli, John M., 199
Steffy, Gary R., 440
Stegelin, Dolores, ed.,]323
Stegemeyer, Anne, 413

Stein, Alice M.,ed., 87
Stein, Theodore J., 563
Steingress, Frederick M., 153
Steinhaus, Nancy Hoover, 416
Stem, Seth, 440
Stephen, George, 139
Stephens, Diane, ed., 323
Stephens, Mitchell, 244
Stephens, Richard C., 546
Stephenson, Peter, 182
Stern, Daniel N., 323
Stern, Ron, 289
Sternloff, Robert E., 502
Stevens, Lawrence L., 220
Stevens, Mark, 171
Stewart, Bernice, 324
Stewart, Charles E., 45
Stewart, Darlene L., ed., 111
Stewart, Kenneth L., 14
Stewig, John W., ed., 324
Stinchcomb, Craig, 393
Stith, Sandra M., ed., 563
Stits, Don, 349
Stitt-Gohdes, Wanda, 277
Stone, Elaine, 423
Stone, Janet I., 324
Stonecipher, Harry W., 240
Stopke, Judy, 408
Stoskopf, Michael K., ed., 527
Stout, Chris E., ed., 546
Stramler, James H., 95
Strange, Marty, 480
Strangelove, Michael, ed., 333
Streitmatter, Rodger, 255
Strentz, Herbert, 244
Strianese, Anthony J., 199
Strobel, Caroline D., ed., 171
Stroebel, Leslie, 445
Stryker, John A., 28
Stuart, Gail Wiscarz, ed., 86
Stubbendieck, James L., 485
Stucker, Steve, 270
Stueart, Robert D., 335
Stueve, Thomas F., 515
Stufflebeam, Charles E., 530
Stutts, Alan T., 231
Sublette, Kathleen, 515
Suddarth, Doris Smith, 80
Sudjic, Deyan, 423
Sullivan, Constance, ed., 449
Sullivan, Thomas F. P., ed., 369
Sultan, William J., 199
Sunal, Cynthia S., 324
Sundeen, Sandra J., 72
Supon Design Group, 457
Surette, Ray, 244
Surkin, Howard B., 118
Sutherland, Martha, 468
Sutherland, Zena, 324
Swain, Steven F., 527
Swann, Alan, 408, 459, 460
Swanson, Janice M., 72
Swartz, Mark H., 80
Swearingen, Pamela L., 80
Swearingen, Pamela L., ed., 74
Sweeney, Mary Ann, 90
Sweet, Eileen Smith, ed., 546
Swenson, S., 153
Syvanen, Bob, 136
Szabo, John F., 515
Szarkowski, John, 449
Szasz, Thomas Stephen, 546

Taintor, Jerry F., 14
Takamura, Zeshu, 420
Tamparo, Carol D., 55, 64
Tanaka, Paul K., 50
Tanenbaum, Andrew S., 183
Tanke, Mary L., 199

Tannenbaum, Jerrold, 530
Tappan, William T., 217
Tarrant, John, ed., 480
Tate, Philip, 71
Tate, Sharon Lee, 424
Taylor, Barbara J., 324
Taylor, Cecelia Monat, 86
Taylor, David, 530
Taylor, Dean, 373
Taylor, Don, 355
Taylor, Lawrence, 279
Taylor, Margaret, 338
Taylor, Pamela, 189
Taylor, Patrick J., 416
Taylor, Richard L., 349, 361
Taylor, Stephen, 136
Taylor, Thomas D., 14
Taylor, V., 125
Taylor, William N., 546
Tebbel, John William, 244
Tecklin, Jan Stephen, ed., 111
Teele, Bruce W., ed., 496
Teele, Rita L., 31
Tempkin, Betty Bates, 32
Tenenbaum, David, 153, 156
Tennant, Roy, 183, 333
Tennenhouse, Dan J., 283
Terrell, Margaret E., 199
Terrill, Richard J., 286
Testa, Laura J., 408
Tetsch, Peter, 20
Thames, Bill, 420
Thelan, Lynne A., 74
Thibodeau, Gary A., 71
Thiel, Richard, 381
Thiessen, F. J., 355
Thomas, Bob, 467
Thomas, Clayton L., ed., 70
Thomas, Glyn V., 324
Thomas, James L.,]324
Thomas, Kas, 349
Thomas, Marilyn H., 408
Thomas, Michael, 361
Thomas, Regina Dreyer, 14
Thomas, Steve, 139
Thomen, James R., 95
Thompson, Charles L., 324
Thompson, Douglas, 231
Thompson, George J., 290
Thompson, Jo Ann Asher, ed., 440
Thompson, June M., 80
Thompson, Michael A., 26
Thompson, Richard H., 567
Thompson, Travis, ed., 563
Thomsett, Michael C., 214
Thomson, Ellen Mazur, comp., 457
Thomson, Linda Kohlman, 101
Thoreau, Henry David, 505
Thorell, L. G., 463
Thornton, Peter, 440
Thornton, Richard S., 460
Thornton, William E., 547
Thorp, James H., ed., 507
Thumann, Albert, ed., 148
Thurston, Stephen E., 20
Thyer, Bruce A., 555
Tietz, Norbert W., ed., 64
Tilkian, Ara G., 81
Tilley, Alvin R., comp., 440
Tilton, Richard C., ed., 64
Timby, Barbara, Kuhn, 81
Timmons, Jeffrey A., 225
Tinder, Richard F., 374
Tinker, Irene, ed., 567
Titchener, Campbell B., 253
Toale, Bernard, 408
Tober, James A., 503
Tobin, Joseph Jay, 324
Tobin, Martin J., ed., 118

Name Index

Tobler, Rita, 89
Todd, Ken, 137
Toenjes, Leonard P., 130
Toglia, Joan P., 101
Tolhurst, William A., 183
Tolpin, Jim, 126
Tomal, Daniel R, 148
Toman, Walter, 324
Tomosy, Thomas, 446
Toogood, Alan R., 520
Torres, Hazel O., 14
Torres, Lillian S., 26
Torrice, Antonio F., 440
Tortorici, Marianne R., 26
Tosh, Dennis S., Jr., 214
Touissant-Samat, Maguelonne, 199
Toula, Nicholas, 50
Tovey, Priscilla, ed., 231
Townsend, James J., 183
Towsley, Doreen, 26
Trachtenberg, Alan, 450
Traister, John E., 130, 148
Trawick-Smith, Jeffrey W., 324
Treadway, David C., 547
Trelease, Jim, 324
Treshow, Michael, 505
Trimpey, Jack, 575
Trofino, Rita Bolek, ed., 74
Trombly, Catherine A., ed., 111
Trotta, Marcia, 339
Troy, Leo, 225
Tschihold, Jan, 409
Tucker, Susan M., 71
Tucker, Susan Martin, 81
Tuckman, Diane, 424
Tumility, Thomas, 374
Turbide, David A., 374
Turecki, Stanley, 325
Turgeon, Mary Louise, 64
Turnbull, George I., ed., 111
Twerski, Abraham J., 547

U. S. Public Health Service. Office of the Surgeon General, 38
U.S. Department of Energy. Office of Environment, Safety, and Health, 95
Ubassy, Gerald, 14
Ullman, David G., 374
Ullmann, John, ed., 247
Umphred, Darcy Ann, 111
Underhill, Roy, 126
Underhill, Sandra L., 90
Underwood, Doug, 244
United States. Bureau of the Census, 477
United States. Dept. of Agriculture, 477, 480
United States. Federal Aviation Administration, 361
United States. National Park Service, 231
United States. Bureau of Land Management, 480
United States. Environmental Protection Agency, 367
United States. Federal Bureau of Investigation, 286
Upp, E.L., 370
Urbonas, Ben, 370
Utterback, Ann S., 265
Utz, Peter, 267

Vallentine, John F., 485
Vallerand, April Hazard, 89
Van Buren, Chris, 183
Van Harssel, Jan, 232
Van Hise, James, 467
Van Ornum, William, 564
Van Sickle, Jan, 386
Van Wyk, Christopher J., 183
Vanden Heuvel, Nichole, comp., 286
Vander Woude, Judith, 325
VanDyke, Scott, 460
VanEgmond-Pannell, Dorothy, 199
Varcarolis, Elizabeth M., 86
Vardaman, James M., 508

Varner, Dickson D., 527
Vartabedian, Roy .E., 39
Vasey, Daniel E., 480
Vasquez-Nutall, Ena, 325
Vasta, Ross, 325
Vasta, Ross, ed., 325
Vaughan, Tay, 463
Vazquez, Moya, ed., 74
Velarde, Giles, 455
Venit, Sharyn, 409
Ventolo, William L., 215
Ventura, Judy, 189
Vernon, Julia, ed., 300
Viera, John David, 268
Villarosa, Riccardo, 418
Vladimir, Andrew, ed., 232
Vlcek, Charles W., 336
Voisinet, Donald D., 374
Volk, Wesley A., 65
Voth, Eric R., 225
Vourlekis, Betsy S., ed., 555

Wadler, Gary I., 547
Wagner, Andrea, 300
Wagner, John D., 137, 480
Wagner, Willis H., 126
Walden, Joyce, 307
Walker, Aidan, 130
Walker, John R., 390, 393
Walker, Lisa, 409
Walkin, Colin, 409
Walkowski, Debbie, 183
Wallace, Barbara C., ed., 547
Wallace, Henry Agard, 482
Wallace, Patricia E., 211
Wallace, Thomas J., 131
Wallace, Wanda A., 171
Wallis, L. W., ed., 472
Walman, David, 220
Walraven, Gail, 45
Walstad, John D., ed., 501
Walter, J. B., 65
Walter, Richard, 269
Walters, Norma J., 65
Walters, Roger L., 269
Wang, C. C., 28
Wang, Shan K., 152
Wanger, Jack, 119
Wanttaja, Ron, 350
Ward, Hiley H., 240, 247
Ward, Kory M., ed., 65
Warda, Mark, 216
Warfield, Gerald, 253
Warner, John W., 289
Warner, Mickey, 39
Warner, Ralph, 300
Warren, Jean, 325
Washer, Doug, 126
Wasik, Barbara Hanna, 564
Wasserman, Dick, 261
Wasserman, Paul, 296
Wasserman, Paul, ed., 338
Wasserman, Steven R., 302
Wasserman, Steven R., ed., 338
Waterman, Laura, 503
Watkins, A. M., 137
Watkins, Christopher, 409
Watkins, Gary L., 26
Watkins, Paul R., ed., 183
Watterson, Andrew, 521
Watts, Thomas D., ed., 575
Watts, Tim J., 335, 515
Wearing, Robert, 126
Weaver, David H., 244
Webb, Steve, 27
Webber, B. A., ed., 111
Weber, H., 20
Weber, Janet, ed., 81
Webster and Associates (N.S.W.), 409

Webster, Carrie, 409
Wegscheider-Cruse, Sharon, 547
Weidensaul, Scott, 530
Weil, Andrew, 547
Weil, Roman L., 171
Weiler, Betty, ed., 232
Weinbach, Robert W., 556, 569
Weinberg, Robert, 464
Weiner, Richard, 252, 265
Weingand, Darlene E., 336
Weinhold, Virginia Beamer, 440
Weinstein, Bruce D., ed., 15
Weinstein, Dava L., ed., 547
Weinstein, Philip, 15
Weis, Lois, ed., 325
Weisman, Leslie, 440
Weiss, Irwin, 277
Weiss, Roger D., 547
Welch, William C., 520
Weldy, Norma Jean, 81
Wellisch, Hans H., 342
Wells, Alexander T., 361
Wells, Carolyn Cressy, 553, 556
Welsh, Heidi J., 530
Welsh, Pat, 520
Wenke, Robert A., 279
Wenocur, Stanley, 573
Wentz, Gini, ed., 27
Werry, John S., ed., 548
Wesley, Ruby L., 82
West, Bessie Brooks, 199
West, John B., 119
West, Patricia, 86
West, Suzanne, 472
Westbrook, Catherine, 28
Westcott, Cynthia, 521
Westfall, Patricia Tichenor, 240
Westman, Randall Perry, 15
Weyman, Arthur E., 32
Whaley, Lucille F., 84
Wheeler, Patricia R., ed., 306
Whelan, Tensie, ed., 232
Whitaker, Jerry C., 148
Whitaker, Kent B., 119
White, Alex, 472
White, Antony, 441
White, Gary C., 119
White, George R., 157
White, Graham E., 15
White, Jan V., 409, 460, 472
White, John H., 356
White, William L., 548
Whitehead, G. Kenneth, 507
Whitford, Ronald, 527
Whitson, Tom D., 488
Whittaker, James K., ed., 564
Whittaker, Ron, 267
Widman, Rudy, 333
Wiener, Eric, 214
Wiese, Michael, 267
Wigger, G. Eugene, 199
Wiggins, Joanne L., 189
Wigglesworth, David T., ed., 370
Wild, Alan, 370
Wilde, Judith, 460
Wildman, Iris J., 335
Wilhide, Elizabeth, 441
Wilkie, Bernard, 268
Wilkins, Esther M., 15
Wilkins, Robert L., 119
Wilkins, Robert L., ed., 119
Wilkoff, Wm. L., 441
Wille, Christopher M., 501
Willett, Norman D., ed., 15
Williams, David F., ed., 4
Williams, Elizabeth, 137
Williams, Michael, 364, 500
Williams, Ric, 276
Williams, Robin, 237, 409

Williams, Sue Rodwell, 39, 75
Williams, Sue Rodwell, ed., 39
Williamson, Darcy D., 304
Willins, Michael, ed., 445
Willis, Barry Donald, 331
Willis, Barry Donald, ed., 331
Willis, William James, 245
Willis-Thomas, Deborah, 450
Williston, Ed M., 508
Willson, Jane Wynne, 515
Willumson, Glenn Gardner, 450
Wilson, A. Bennett, 101
Wilson, Clint C., 255
Wilson, Diana Hardy, 468
Wilson, George, 325
Wilson, Holly Skodol, 87
Wilson, J., 379
Wilson, James F., 530
Wilson, Jim, 520
Wilson, John M., 240
Wilson, R. Dean, 153
Wilson, Steven, 139
Wilson, Susan Fickertt, 88
Wilson, Thomas G., ed., 15
Winckler, Suzanne, 232
Windle, Michael T., ed., 548
Winick, Myron, ed., 39
Winship, D. R., 515
Winslow, Taylor F., 149
Winter, Ruth, 39
Wischnitzer, Saul, 4
Wistreich, George, 52
Witek, Theodore J., 120
Witt, Leonard, ed., 240
Wodaski, Ron, 464
Wodrich, David L., 325
Woelfel, Julian B., 15
Wojciechowski, Gene, 253
Wolbarst, Anthony B., 27
Wolcott, Harry F., 569
Wold, Gloria, 76
Wolf, Paul R., 386, 506
Wolf, Zane Robinson, 89
Wolfe, David, 183

Wolfe, Mary Gorgen, 424
Wolfelt, Alan, 516
Wolfgram, Douglas E., 464
Wolseley, Roland Edgar, 255
Wolverton, Van, 183
Wong, Donna L., 84
Wong, Wucius, 460
Wood, Ernest, 441
Wood, John, ed., 450
Wood, Peter R., 15
Wood, Phyllis, 468
Wood, Robert J., 261
Wood, Robert W., 149
Wood, Roy C., 207
Woodall, Irene R., ed., 15
Woods, Mary Ann, 49
Woods, Robert H., 200
Woodson, R. Dodge, 162, 163
Wootton, R., ed., 27
Workman, M. Linda, 74
World Chamber of Commerce, 232
World Wildlife Fund, 370
Worsing, Robert, ed., 45
Worthington-Roberts, Bonnie S., 75
Worthley, John Abbott, 54
Wren, Christopher G., 302
Wrey, Lady Caroline, 441
Wright, Bob, 548
Wright, Kieth C., 339
Wright, Orville, 350
Wright, R. Gerald, 507
Wyatt, Allen, 184
Wylde, Margaret A., 137, 441
Wynar, Bohdan S., 342

Yariv, Amnon, 379
Yarwood, Doreen, 423
Yeager, Jan, 441
Yee, Martha M., comp., 342
Yenne, Bill, 350
Yesalis, Charles, ed., 548
Yew, Wei, ed., 456
Yoder, Barbara, 548
Yogev, Ram, ed., 84

York, Kenneth J., 496
Young, Donald S., 65
Young, Doyald, 456
Young, James A., 500
Young, Jay A., ed., 96
Young, Raymond Allen, ed., 500
Yudd, Ronald A., 200
Yudofsky, Stuart C., 548
Yvorra, James G., ed., 45

Zabel, Mary Kay, 325
Zablotney, Sharon L., ed., 65
Zaccerelli, Herman E., 200
Zadai, Cynthia Coffin, ed., 111
Zagon, Ian S., ed., 548
Zaid, Barry, 207
Zakus, Sharron M., 49, 54, 65
Zalewski, Dianne Dupre, 305, 306
Zaloga, Gary P., 75
Zamkoff, Bernard, 416
Zangrillo, Frances Leto, 416
Zastrow, Charles, 556
Zawid, Carole Israeloff, 89
Zeid, Ibrahim, 374
Zelanski, Paul, 460
Zelezny, John D., 246
Zellers, Margaret, 232
Zeman, Frances J., 39
Zerwekh, JoAnn Graham, ed., 77
Ziegler, Shirley Melat, ed., 82
Zigler, Edward, 573
Zimbardo, Philip G., 261
Zimmerman, Shirley, 573
Zinsser, William Knowlton, 240
Zobel, Louise Purwin, 253
Zoller, Bettye Pierce, 262
Zubrowski, Bernie, 325
Zuckerman, Jim, 446
Zulawski, David E., 288
Zvoncheck, Juls, 232
Zweifel, Frances W., 469
Zwemer, Thomas J., ed., 5
Zwiebel, William J., ed., 32

Title Index

1-2-3 Math, 325
1-2-3 Reading and Writing, 325
100 Years of Western Wear, 420
101 Desktop Publishing and Graphics Programs, 397
1001 Solved Surveying Fundamentals Problems, 386
1991 ASHRAE Handbook, 151
1991 Masonry Codes and Specifications, 156
1992 ASHRAE Handbook, 151
1993 ASHRAE Handbook: Fundamentals, 151
3000 Questions and Answers for HVAC/R Licensing Examinations, 153
The 479 Best Public Golf Courses in the United States, Canada, the Carribean, and Mexico, 230
6,000 Soundalikes, Lookalikes, and Other Words Often Confused, 277
The 800-Cocaine Book of Drug and Alcohol Recovery, 574
The 800-COCAINE Book of Drug and Alcohol Recovery, 537

The A to Z's of Clipper Cutting, 189
A to Zoo: Subject Access to Children's Picture Books, 319
AACD Ethical Standards Casebook, 539
AACN'S Clinical Reference for Critical Care Nursing, 74
ABC's of Landscape Pruning, 522
Absolute Beginner's Guide to C, 181
Absolute Beginner's Guide to Multimedia, 464
Absolute Beginner's Guide to Networking, 177
Absolute Beginner's Guide to Programming, 181
Abused and Battered, 561
AC Power Systems, 148
Accepting Ourselves, 541
Access EPA, 367
Access Services: A Handbook, 336
Access Services: Organization and Management, 337
Access Services in Libraries, 330
Access to Information, 339
Accident Prevention for Hotels, Motels, and Restaurants, 205
Accident Prevention Manual for Business and Industry, 92
Accosting the Golden Spire, 167
Accountancy Comes of Age, 169
Accountant's Handbook, 167
Accountant's Handbook of Formulas and Tables, 169
Accounting: The Language of Business, 168
Accounting and Auditing Careers Passbook, 167
Accounting and Budgeting in Public and Nonprofit Organizations, 168
Accounting and Office Systems, 172
Accounting and Taxation, 170
Accounting Certification, Educational and Reciprocity Requirements, 168
Accounting Ethics: A Practical Guide for Professionals, 167
Accounting for Business Combinations and Restructurings, 171
Accounting for Success, 167
Accounting Handbook, 171
Accounting Information Systems, 167
Accounting Made Simple, 171
Accounting Research Methods, 171
Accounting's Changing Roles in Social Conflict, 169
Achieving Power: Practice and Policy in Social Welfare, 556
ACI Manual of Concrete Inspection, 156
The Acid Rain Controversy, 505
ACJS/Anderson Monograph Series, 291
ACLS, 43
Acquisitions Management and Collection Development, 330
Activities of Daily Living, 100

Activity Accounting: An Activity-Based Costing Approach, 167
Activity Analysis Handbook, 100
Adams' Guide to Coding and Reimbursement, 49
Adapting the Home for the Physically Disabled, 102
Addiction, 548
Addiction: From Biology to Drug Policy, 539
Addiction and Responsibility, 545
Addiction and the Vulnerable Self, 541
An Addictions Curriculum for Nurses and Other Helping Professionals, 536
Addictive Behaviors Across the Life Span, 535
Addictive Disorders, 538
The Addictive Organization, 545
Addictive Thinking: Understanding Self-Deception, 547
Administering Medication, 530
Administering Medications, 54
Administrative Office Management, 210
Administrative Procedures for the Electronic Office, 211
Administrative Procedures in the Electronic Office, 182
Adolescents at Risk: Prevalence and Prevention, 560
Adoption and Disruption: Rates, Risks, and Responses, 558
Adult Hemiplegia: Evaluation and Treatment, 98, 105
Advance Fashion Sketchbook, 467
Advanced Aircraft Systems, 348
Advanced Composites: An IAP, Inc. Training Manual, 347
Advanced Court Reporting Technology: Computer Concepts, 276
Advanced Emergency Care for Paramedic Practice, 44
Advanced Excel for the PC, 178
Advanced Fashion Sketchbook, 419
Advanced Framing, 137
Advanced Home Wiring, 142
Advanced Life Support Skills, 41
Advanced Professional Cooking, 194
Advanced Radiographic Techniques Part I, 16
Advanced Routing: Techniques for Better Woodworking, 123
Advanced Skills, 87
Advances in Echo Imaging Using Contrast Enhancement, 31
Advances in Periodontics, 15
Adventures in Virtual Reality, 178
Advertise! An Assessment of Fundamentals for Small Business, 260
Advertising Career Directory, 261
Advertising from the Desktop, 234, 402
Advertising in America: The First 200 Years, 260
Advertising's Ten Best of the Decade, 1980-1990, 258
Affairs in Order: A Complete Resource Guide to Death and Dying, 511
African Majesty: The Textile Art of the Ashanti and Ewe, 417
After Suicide: A Ray of Hope, 514
After the Smoke Clears, 290
Aftermath, 358
Age, Class, Politics, and the Welfare State, 573
The Age of Intelligent Machines, 179
The Age of Manipulation, 260
Agency: Choices, Challenges, and Opportunities, 220
Aggression, Family Violence, and Chemical Dependency, 574
Agrarian Policies and Agricultural Systems, 484
Agricultural and Food Policy, 483
Agricultural Bioethics: Implications of Agricultural Biotechnology, 481
Agricultural Biotechnology, 481
Agricultural Mechanics: Fundamentals and Applications, 481
Agricultural Plants, 482
Agricultural Risk Management, 483
Agricultural Statistics, 477
Agricultural Wiring Handbook, 144
Agriculture: America's Most Crucial Industry, 488
Agriculture: Illustrated Search Strategy and Sources, 479
Agriculture and National Development, 478

Agriculture and the State: Market Processes and Bureaucracy, 483
Agriculture and the Undergraduate: Proceedings, 484
Agriculture and Water Quality in the Great Plains, 486
Agriculture and Water Quality: International Perspectives, 486
The Agriculture Dictionary, 477
The Agriculture Fact Book, 477
Agriscience and Technology, 477
Agroecology, 478
AIDS: Etiology, Diagnosis, Treatment, and Prevention, 569
AIDS: Helping Families Cope, 575
AIDS: Identifying Community Resources, 17
AIDS: Vignettes for Dental Professionals, 16
AIDS and Ethics, 569
AIDS and HIV Infection, 88
AIDS in the Dental Office?, 13
AIDS in the World, 10
AIDS Law: Implications for the Individual and Society, 569
AIDS Prevention for Laboratory Professionals, 65
AIM/FAR, 346
Air Conditioning and Refrigeration Toolbox Manual, 153
Air Emissions, Baselines, and Environmental Auditing, 369
Air Pollution Engineering Manual, 367
Air Pollution's Toll on Forests and Crops, 482, 504
Air Words: Writing for Broadcast News, 269
Aircraft Detail Design Manual, 347
Aircraft Electrical Systems: Single and Twin Engine, 346
Aircraft Gas Turbine Engines of the World and Dictionary of the Gas Turbine, 349
Aircraft Hydraulic Systems, 349
Aircraft Ignition and Electrical Power Systems, 346
Aircraft Performance: The Forces Without, 361
Aircraft Rescue and Firefighting, 492
Aircraft Systems: Really Knowing Your Airplane, 349
Aircraft Systems: Understanding Your Airplane, 348
Aircraft Technical Dictionary, 347
The Airline Industry, 359
Airman's Information Manual/Federal Aviation Regulations, 358
The Airway: Emergency Management, 42
AJN/Mosby-Yearbook Nursing Boards Review, 87
ALA Handbook of Organization and Membership Directory, 333
Alba's Medical Technology: Board Examination Review, Volume II, 59
Alcohol, 543
Alcohol and Drugs are Women's Issues: Volume One, 544
Alcohol and Drugs are Women's Issues: Volume Two, 545
Alcohol and the Addictive Brain, 536, 573
Alcohol and the Family: Research and Clinical Perspectives, 537, 574
Alcohol Disabilities Primer, 544
Alcohol in America: Drinking Practices and Problems, 537
Alcohol Problem Intervention in the Workplace, 567
Alcohol, Social Work, and Helping, 574
Alcoholism and Substance Abuse in Special Populations, 541
Alcoholism and Women, Genetics, and Fetal Development, 542
Alcoholism in Minority Populations, 575
Aldo Leopold: His Life and Work, 505
Aldo Leopold's Wilderness, 505
Alfred Stieglitz, 448
All About Cotton: A Fabric Dictionary and Swatchbook, 424
All About Escrow, 218
All About Mortgages: Inside Tips to Finance the Home, 218
All About Silk: A Fabric Dictionary and Swatchbook, 424

Title Index

All Suite Hotel Guide, 205
All You Wanted to Know About Day Care Centers, 311
Almanac of Business and Industrial Financial Ratios, 225
Alternative Agriculture, 479
Alternative Housebuilding, 135
Ambulance Calls: Review Problems for the Paramedic, 42
America and the Daguerreotype, 450
America the Bountiful, 488
American Aviation: An Illustrated History, 358
American Bench: Judges of the Nation, 300
The American Billboard, 260
The American Billboard: One Hundred Years, 455
American Children's Folklore, 312
American Dreams, Rural Realities, 477
The American Farm Crisis, 478
The American Farm Tractor: A History of the Classic Tractor, 481
American Graphic Design: A Guide to the Literature, 457
American Health Care: Realities, Rights, and Reforms, 571
The American Horticultural Society Encyclopedia of Garden Plants, 518
The American Horticultural Society Encyclopedia of Gardening, 518
American House: A Guide to Architectural Styles, 441
American Image: 150 Years of Photography, 450
American Journalism History: An Annotated Bibliography, 236
The American Journalist, 244
American Jurisprudence Desk Book, 300
The American Law Dictionary, 295
American Library Directory, 333
American Magazine Journalists, 1741-1850, 251
American Magazine Journalists, 1900-1960, 251
American Mass-Market Magazines, 251
American Newspaper Publishers, 249
American Nursing: A Biographical Dictionary, 81
American Nursing Review for NCLEX-RN, 87
American Photography Showcase, 446
American Reference Books Annual, 337
American Tree Houses and Play Houses, 132
American Typography Today, 469
Americans and Their Forests: A Historical Geography, 500
Americans with Disabilities Act Handbook, 339, 430
America's Great Comic-Strip Artists, 466
America's Misunderstood Welfare State, 572
America's Welfare State: From Roosevelt to Reagan, 570
America's Wild Turkey, 508
Amiga Animation Video, 271
Anabolic Steroid Abuse, 542
Anabolic Steroids in Sport and Exercise, 548
Anatomy and Physiology, 64
Anatomy and Physiology, 71
Anatomy of a Business Plan, 224
Anatomy of Orofacial Structures, 5
Ancient Egyptian Construction and Architecture, 130
Ancient Forests, 508
Ancient Forests of the Pacific Northwest, 500
And Access for All: ADA and Your Library, 343
Andre Kertesz, 448
Angiography and Interventional Radiology, 29
Animal Handling and Restraint, 530
Animal Testing and Consumer Products, 530
Annual Editions: Criminal Justice, 291
Annual Review of Microbiology, 59
Annual Review of Nursing Research, 77
Annual Review of Nutrition, 34
Anonymous Sources, 255
Another Chance: Hope and Health for the Alcoholic Family, 547
Anthony's Color-Coded ICD-9-CM, 50
Antibiotic/Antimicrobial Use in Dental Practice, 11
A.P.E. Surfshop Accounting Videos, 171
Aperture Masters of Photography, 448

The Apple Macintosh Book, 180
Appleton and Lange's Review for the Dental Assistant, 5
Appleton and Lange's Review for the Medical Assistant, 56
Appleton and Lange's Review for the Radiography Examination, 26
Appleton and Lange's Review for the Ultrasonography Examination, 31
Application of Nursing Process and Nursing Diagnoses, 78
Applied Ergonomics Handbook, 92
Applied Foodservice Sanitation: A Foundation Textbook, 193
Applied Laboratory Medicine, 64
Applied Nutrition and Diet Therapy, 35
Applied Nutrition and Diet Therapy for Nurses, 74
Applied Pharmacology for the Dental Hygienist, 13
Applied Radiographic Calculations, 24
Applying the 1993 National Electrical Code, 143
The Appraisal of Apartment Buildings, 215
The Appraisal of Real Estate, 215
Approaches to Literature through Subject, 320
Approaches to Moral Development, 315
APTA Resource Guide: Audiovisual Catalog 1992-1993, 104
Aquaculture: Farming the Waters, 508
Aquaculture Management, 504
Arboriculture, 519
Arc Welding: Basic Fundamentals, 393
Architectural Drawing: Options for Design, 436
Architectural Delineation: Professional Shortcuts, 432
Architectural Detailing in Contract Interiors, 440
Architectural Drawing and Design: Principles and Practices, 433
Architectural Lighting Design, 440
Architectural Rendering, 434
Architecture and Design, 1970-1990: New Ideas in America, 439
Architecture and Ornament: A Visual Guide, 441
Architecture in Perspective:, 430
The Archival Enterprise, 330
Are They Selling Her Lips? Advertising and Identity, 261
Are You Buying a Home? A Professional Guide to Home Inspection, 221
The Art and Craft of Feature Writing, 239
Art and Creative Development for Young Children, 322
The Art and Science of Culinary Preparation, 192
The Art and Science of Medical Radiography, 25
The Art and Science of Professional Makeup, 188
Art Deco Architecture, 430
Art Deco Interiors, 430
Art of Bedside Care: ABC's of Nursing Procedures, 90
The Art of Editing, 237
The Art of Facial Reconstruction, 515
The Art of Fashion Draping, 413
The Art of Hanna-Barbera: Fifty Years of Creativity, 467
The Art of Infrared Photography, 446
The Art of Outdoor Photography, 447
The Art of Persuasion: A History of Advertising Photography, 449
The Art of Selecting a Jury, 279
The Art of the Cabinet: Including a Chronological Guide to Style, 439
The Art of the Stonemason, 157
The Artist's Friendly Legal Guide, 472
Artists of the Page: Interviews with Children's Book Illustrators, 465
As It Should Be Done, 20
ASA Aviation Mechanic's Handbook, 346
ASHRAE Handbook, 151
Asian Americans: Emerging Minorities, 566
Asian Pacific Americans, 255
ASID, American Society of Interior Designers Professional Practitioners Manual, 440
Aspects of the Computer-Based Patient Record, 53
Assessing and Screening Preschoolers, 325

Assets and the Poor: A New American Welfare Policy, 573
Assistant Land Surveyor, 386
Assistive Technology for Persons with Disabilities, 100
Associate Degree Nursing Education, 81
The Associated Press Guide to News Writing, 246
The Associated Press Photo-Journalism Stylebook, 248
The Associated Press Stylebook and Libel Manual, 250
At Nature's Pace: Farming and the American Dream, 479
At Twelve: Portraits of Young Women, 449
Atlas of Color Doppler Echocardiography, 31
Atlas of Correlative Imaging Anatomy of the Normal Dog, 528
Atlas of Dental Radiographic Anatomy, 9
Atlas of Diagnostic Radiology of Exotic Pets, 527
Atlas of Diagnostic Radiology of the Horse: Part 2, 526
Atlas of Diagnostic Radiology of the Horse: Part 3, 526
Atlas of Interventional Radiology, 29
Atlas of Mammographic Positioning, 26
Atlas of Normal Histology, 526
Atlas of Porcelain Restorations, 7
Atlas of Ultrasonography in Obstetrics and Gynecology, 30
Attorney's Medical Deskbook, 283
Atypical Orthopedic Radiographic Procedures, 26
The Audubon Society Guide to Nature Photography, 446
Auscultation of Normal Breath Sounds, 89
Authentic French Fashion of the Twenties, 422
Authoring Interactive Multimedia, 462
Auto Electricity, Electronics, Computers, 353
Auto Technology: Theory and Service, 353
AutoCAD, 179
AutoCAD: Drafting and 3D Design, 374
AutoCAD: Methods and Macros, 374
AutoCAD and its Applications: Release 12 for Windows, 374
The AutoCAD Book: Drawing, Modeling, and Applications, 373
AutoCAD for Electronics: A Tutorial, 374
AutoCAD for the Apparel Industry, 416
The AutoCAD Productivity Book for Releases 10, 11, 12, 374
AutoCAD Quick Reference, 182, 374
AutoCAD Reference Guide, 179
The AutoCAD Technical Reference, 177, 374
Automatic Sprinkler Systems Handbook, 496
Automatic Transmissions, 356
Automatic Transmissions and Transaxles, 353
Automating the Welding Process, 393
Automotive Air-Conditioning Refrigerant Service Guide, 354
Automotive Electrical and Electronic Systems, 353
Automotive Engine Performance: (Vol 1), 354
Automotive Engine Performance: (Vol 2), 354
Automotive Engine Repair and Rebuilding, 355
Automotive Fuels and Fuel Systems, 353
Automotive Fuels Handbook, 355
Automotive Handbook, 353
Automotive Principles and Service, 355
Auxiliary Oriented Diagnostic Appointment, 16
Avian Hematology and Cytology, 525
The Aviation and Aerospace Almanac, 358
Aviation and the Law, 359
Aviation Disasters, 359
Aviation's Golden Age: Portraits from the 1920s and 1930s, 348
Awakening the Hidden Storyteller, 320
AWS Film Directory, 390

Back Injury Prevention Handbook, 92
Backlash: The Undeclared War Against American Women, 254
Backyard Brickwork, 157
Backyard Composting, 522
The Bakers' Manual, 192
Balanced Nutrition: Beyond the Cholesterol Scare, 38
Balenciaga, 421

Title Index

Ball Culture Guide: The Encyclopedia of Seed Germination, 522
Ball Pest and Disease Manual, 521
Balloons: Building and Experimenting with Inflatable Toys, 325
Band Saw Basics, 123
Band Saw Handbook, 123
Bankruptcy and Collections: The Paralegal Perspective, 304
Bankruptcy Law and Procedure, 304
Bankruptcy Rules and Forms Handbook, 304
The Bar and Beverage Book: Basics of Profitable Management, 195
Barrier-Free Residential Design, 437
Barriers to Intimacy: For People Torn by Addiction, 544
Barron's Guide to Medical and Dental Schools, 4
Barron's Real Estate Handbook, 215
Basic Accounting Video, 172
Basic Aerodynamics, 350
Basic Alternating Current Control Diagrams, 144
Basic Anatomy of the Head and Neck, 8
Basic Arrest and Prisoner Control Tactics, 288
The Basic Business Library: Core Resources, 338
Basic Clinical Lab Competencies for Respiratory Care, 119
Basic Clinical Radiobiology: For Radiation Oncologists, 28
Basic Clipper Techniques, Vol. 1, 189
Basic Concepts of Psychiatric-Mental Health Nursing, 86
Basic Desktop Design and Layout, 236
Basic Emergency Care of the Sick and Injured, 45
Basic Filing for Health Information Management, 53
Basic Food and Beverage Cost Control, 197
Basic Geriatric Nursing, 76
Basic Graphic Design, 408, 459
Basic Guide to Pesticides: Their Characteristics and Hazards, 521
Basic Hazardous Waste Management, 367
Basic Hotel Front Office Procedures, 206
Basic ICD-9-CM Coding Handbook, 50
Basic Keyboarding for the Medical Office Assistant, 54
Basic Knife Defense for Criminal Justice, 288
Basic Medical Laboratory Techniques, 65
Basic Medical Techniques and Patient Care for Radiologic, 26
Basic Microbiology, 65
Basic Neuroscience for the Health Professions, 108
Basic Nutrition and Diet Therapy, 39, 75
Basic Oral Anatomy for the Dental Health Team, 16
Basic Pathology, 62
Basic Pharmacology for Health Occupations, 54
Basic Physics of Radiation Therapy, 28
Basic Plumbing Skills, 160
Basic Plumbing with Illustrations, 162
Basic Principles of Oral and Maxillofacial Radiology, 11
Basic Principles of Radiographic Exposure, 24
Basic Printmaking Techniques, 409
Basic Real Estate Appraisal, 215
Basic Reference Sources: A Self-Study Manual, 338
Basic Rescue and Emergency Care, 45
Basic Respiratory Physiology, 118
Basic Stairbuilding, 136
Basic System, 172
Basic Techniques of Photography: An Ansel Adams Guide, 446
Basic Toxicology, 94
Basic Typography: A Design Manual, 470
Basic Wiring, 146
Basic Wiring and Electrical Repairs, 143
Basics for Builders, 131
Basics of Dental Implantology, 20
Basics of GMAW and GTAW, 393, 394
The Basics of Paralegal Studies, 298
Bathroom Remodeling, 138
Battery Technology Handbook, 144
Battling for the National Parks, 501

The BBS Construction Kit, 183
Beatrix Potter: Artist, Storyteller and Countrywoman, 326
Beautiful Barrier-Free: A Visual Guide to Accessibility, 436
Becoming a Fundraiser, 335
Becoming Attached, 318
Becoming Naturally Therapeutic, 545
Bed and Breakfast USA, 231
Before It's Too Late, 547
Before the Story, 239
Before We Are Born, 83
Beginner's Guide to Producing TV, 265
Beginning Bulletin Boards: Basic Skills, 319
Behavior Disorders: Theory and Practice, 313
Behavior Problems in Preschool Children, 313
Behavioral Engineering Through Safety Training, 94
Behavioral Treatment of Autistic Children, 326
Behind the Scenes: The Advertising Process at Work, 473
Being a Medical Admissions Clerk, 55
Being a Medical Clerical Worker: An Introductory Core Text, 55
Being a Medical Information Coder, 55
Bench Tools, 123
Benfield Conduit Bending Manual, 143
Bereavement: Counseling the Grieving Throughout the Life Cycle, 511
Best: From the Interior Design Magazine Hall of Fame, 430
The Best 1001 WordPerfect Tips Ever, 209
The Best American Sports Writing, 1992, 235
Best Books for Children: Preschool through Grade 6, 315
Best Books for Public Libraries, 336
The Best Comics of the Decade, 465
Best Editorial Cartoons of the Year, 235
The Best in Lobby Design, Hotels, and Offices, 437
Best Newspaper Writing, 235
The Best of AMJ Maintenance Tips, 346
The Best of Newspaper Design, 235
The Best of Photojournalism: The Year in Pictures, 235
Best of the Best for Children, 314
The Best of the Desktop Publishing Forum on Compuserve, 403
Best's Safety Directory: Industrial Safety, Hygiene, Security, 92
Better Documentation, 87
Better Said and Clearly Written, 236
Better Trout Habitat, 503
Between Home and Heaven, 449
Between Prison and Probation, 557
Beyond Cholesterol, 37
Beyond Intuition: A Guide to Writing and Editing Magazine Nonfiction, 240
Beyond Maximarketing: The New Power of Caring and Daring, 261
Beyond the Beauty Strip: Saving What's Left of Our Forests, 508
Beyond the Desktop, 408
Beyond the Inverted Pyramid, 246
Bibliographic Instruction in Practice, 331
Bieber's Dictionary of Legal Abbreviations, 279, 294
Bieber's Dictionary of Legal Citations, 294, 302
The Big Fix-Up, 139
The Big Six: The Selling Out of America's Top Accounting Firms, 171
The Bilingual Courtroom, 278
Bilingual Dictionary of Dental Terms, 4
Biographical Dictionary of American Journalism, 251
A Biographical Dictionary of Science Fiction and Fantasy Artists, 464
Biological Radiation Effects, 25
Biology and Agriculture Courseware, 488
Biomedical Ethics for Radiographers, 26
Biosphere 2000: Protecting Our Environment, 368
Biotechnology and the Future of World Agriculture, 481
The Birder's Miscellany, 530
The Birds of North America, 506

Birnbaum's Disneyland, 227
Birnbaum's Eastern Europe, 227
The Bishop's Boys: A Life of Wilbur and Orville Wright, 346
Biting the Hand That Feeds Them, 556
Black Authors and Illustrators of Children's Books, 465
Black Journalists in Paradox, 255
The Black Press U.S.A, 255
Black Women in White, 81
Black's Law Dictionary, 279, 300
Bleaching Vital Teeth at Home and in the Dental Office, 16
Blessed are the Brood Mares, 529
Bloodborne Pathogens, 16, 65
The Bluebook: A Uniform System of Citation, 279, 294, 300
The Blueprint Dictionary of Printing and Publishing, 397
Blueprint of a LAN, 176
Blueprint Reading for Plumbers: Residential and Commercial, 161
Blueprint Reading for Welders, 390
Blueprint Reading for Welders and Fitters, 394
Blueprint Reading Made Easy, 160
The BOCA National Building Code, 129
Body Fluids and Electrolytes: A Programmed Presentation, 81
Boeing 737, 349
The Boeing 747, 349
Boiler Operator's Workbook, 153
Book Repair: A How-To-Do-It Manual for Librarians, 341
Book Selection: Principles and Practice, 330
Bookkeeping Made Simple, 168
Books for Children to Read Alone, 325
Botany for Gardeners: An Introduction and Guide, 518
Boucher's Clinical Dental Terminology, 5
Bowes and Church's Food Values of Portions Commonly Used, 7, 192
Bowker Annual: Library and Book Trade Almanac, 337
The Boy Who Would Be a Helicopter, 320
Bradshaw On The Family: A Revolutionary Way of Self-Discovery, 536
Brady Emergency Care, 43
Brady EMT Review Manual, 43
Brady Paramedic Emergency Care, 42
Braiding, 188
Braiding Beautiful: Basic and Advanced Techniques, 189
Braiding Made Easy, 189
Brake Design and Safety, 354
Branemark System of Oral Reconstruction: A Color Atlas, 13
Brazing Handbook, 389
Breaking Barriers: A Memoir, 243
Bricklaying: A Homeowner's Illustrated Guide, 158
Brickwork: Architecture and Design, 158
Bringing Computers to the Hospital Bedside, 78
Bringing Us Together, 339
Broadcast Voice Handbook: How to Polish Your On-Air Delivery, 265
Broadcast Voice Performance, 265
Broadcast Writing: Dramas, Comedies, and Documentaries, 268
Broadcast Writing: Principles and Practice, 269
Broadcast/Cable Copywriting, 269
Broadcasting and Cable Yearbook, 264
Brodovitch, 457
The Broken Cord, 537
Brunner and Suddarth's Textbook of Medical-Surgical Nursing, 76
BTLS: Basic Prehospital Trauma Care, 42
Build a Classic Timber-Framed House, 136
Build Your Own Stone House: Using the Easy Slipform Method, 158
The Builders: Marvels of Engineering, 130
Building a Medical Vocabulary, 51, 282
Building a Multi-Use Barn for Garage, Animals, Workshops, Studio, 480
Building and Designing Decks, 136

Title Index

Building Construction Before Mechanization, 131, 364
Building Construction for the Fire Service, 492
Building Construction Illustrated, 129, 432
Building Construction: Site and Below-Grade Systems, 132
Building Doors, Windows, and Skylights, 133
Building for a Lifetime, 137, 441
Building Garden Furniture, 135
Building Security and Personal Safety, 336
Building Soils for Better Crops, 486
Building Structures, 129
Building the Reference Collection, 336
Building the Undergraduate Social Work Library, 554
Building Thoreau's Cabin, 136
Building Trades Dictionary, 130
Buildings for Hospitality, 204
Bull Threshers and Bindlestiffs, 481
Bumper to Bumper, 364
Business Administration for the Dental Assistant, 7
Business and the Environment: A Resource Guide, 369
Business Cards: Dynamic Graphic Design, 454
The Business Chef, 197
Business Decisions with Computers: New Trends on Technology, 182
Business Dictionary of Computers, 175
Business English: Quick and Easy, 277
The Business Guide to Local Area Networks, 182
Business Information: How to Find It, How to Use It, 338
Business Information Sourcebook, 337
Business Information Sources, 337
Business Information Systems: An Introduction, 179
Business Law, Principles and Practices, 281
Business Law: Principles and Cases, 281
The Business of Hotels, 206
Business Software Companion, 175
The Business Travel Survival Guide, 228
Business Week Guide to Multimedia Presentations, 462
Buying a Home When You're Single, 217

The C4 Handbook, 373
Cabinetmaking, 122
Cabinets and Bookcases, 122
Cables and Wiring, 142
The CAD Design Studio, 373
CAD Systems in Mechanical and Production Engineering, 373
CAD/CAM Theory and Practice, 374
CADKey Light, 375
The Caldecott Video Library, Volume Four, 326
Calibrating, Mixing, and Applying Pesticides, 522
The Calligrapher's Dictionary, 467
Calligraphy in the Graphic Arts, 468
Calligraphy Masterclass, 467
Camera Maintenance and Repair, 446
Cancer and Nutrition, 38
Cancer Nursing, 88
Cancer Nursing: Principles and Practice, 83
Canine Research Environment, 529
Car Suspension and Handling, 352
Carbusters! (Series), 46
Cardiac Catheterization, Angiography, and Intervention, 29
Cardiac Doppler Ultrasound: A Clinical Perspective, 30
Cardiopulmonary Pharmacology, 115
Cardiopulmonary Physical Therapy, 108
Cardiovascular Disorders., 88
Cardiovascular Nursing: Holistic Practice, 73
Cardiovascular Resource Videodisc, 90
The Care and Use of Amphibians, Reptiles and Fish in Research, 530
Care of the Older Adult, 75
Career Decisions for Dental Hygienists, 14
Career Explorations in Human Services, 553
The Career Legal Secretary, 295
Career Opportunities for Writers, 238
Career Opportunities in Advertising and Public Relations, 259

Career Opportunities in Television, Cable, and Video, 270
Careers by Design, 473
Careers for Good Samaritans and Other Humanitarian Types, 552
Careers for Kids at Heart and Others Who Adore Children, 553
Careers for Travel Buffs and Other Restless Types, 230
Careers in Beauty Culture, 187
Careers in Clothing, 425
Careers in Counseling and Human Development, 552
Careers in Film and Video Production, 270
Careers in Power Engineering and Boiler Operation, 154
Careers in Social and Rehabilitation Services, 553
Careers in the Forest Service, 508
Careers in Video: Getting Ahead in Professional Television, 270
Caribbean 1994, 232
Caribbean Hideaways, 230
Caring Families: Supports and Interventions, 558
Caring for Patients from Different Cultures, 538
Caring for Your Baby and Young Child, 322
The Carpenter's Manifesto, 129
The Carpenter's Toolbox Manual, 130
Carpentry, 134
Carpentry: Tools, Shelves, Walls, Doors, 122
Carpentry and Construction, 135
Carpentry and Exterior Finish: Some Tricks of the Trade, 136
Carpentry and Interior Finish, 136
Carpentry in Commercial Construction, 134
Carpentry Layout, 137
Cartooning: The Art and the Business, 466
Cases In Hospitality Marketing and Management, 205
The Cat: Diseases and Clinical Management, 527
Cataloging and Classification for the Small Library, 340
Cataloging Nonbook Resources, 340
Cataloging Rules for the Description of Looseleaf Publications, 340
Cataloging Sound Recordings: A Manual with Examples, 340
Catering to Every Whim, 200
Cath Lab: An Introduction, 29
CD-ROM Software, Dataware, and Hardware, 332
CDT-1: Current Dental Terminology, A User's Manual, 4
Celebrations of Death: The Anthropology of Mortuary Ritual, 513
Cement and Concrete Terminology, 156
Cemeteries and Gravemarkers: Voices of American Culture, 513
Censorship: Opposing Viewpoints, 334
Census of Agriculture, 477
Central Hardwood Notes, 499
Ceramic Tile Setting, 157
Certification: The Measure of a Professional, 441
Certified Dental Technician (CDT), 13
Certified PLS Resource Manual, 294
Chamber of Commerce Directory, 232
Change Your Smile, 8
The Changing American Family and Public Policy, 570
Changing Children's Behavior Quickly, 320
Changing Economics of Medical Technology, 61
Changing Gears, 354
Changing Lives, 567
Changing Styles in Fashion: Who, What, Why, 422
Changing Teaching, Changing Schools, 321
Changing Workplace, 211
Channelized Rivers: Perspectives for Environmental Management, 503
The Chef's Book of Formulas, Yields, and Sizes, 198
The Chef's Guide to Practical Restaurant Cookery, 193
The Chemical Brain: The Neurochemistry of Addictive Disorders, 537
Chemical Dependency: A Disease of Denial, 549
Chemical Fixation and Solidification of Hazardous Wastes, 368
Chemical Safety Data Sheets, 92
Chemical Safety in the Laboratory, 65, 93

The Chemically Dependent: Phases of Treatment and Recovery, 547
Chemistry: A Teaching Aid for the Cosmetology Student, 189
Chemistry of Hazardous Materials, 495
Chemotherapy Care Plans: Designs for Nursing Care, 83
Chesney's Radiographic Imaging, 23
Chest Physiotherapy in the Intensive Care Unit, 108
Chest Tube Therapy, 89
Chest Tubes; Patient and System Management, 120
Chiarenza: Landscapes of the Mind, 448
Chic Simple: Clothes, 418
Chicago Architecture and Design, 436
The Chicago Manual of Style, 397
Child Care Choices, 573
Child Custody, Foster Care, and Adoptions, 559
Child Development, 312, 322
Child Development: One to Three, 326
Child Health Care: Process and Practice, 83
Child Health Nursing, 84
Child Maltreatment, 560
Child Psychology: The Modern Science, 325
Child Welfare: An Africentric Perspective, 565
Child Welfare and the Law, 563
The Child Welfare Challenge: Policy, Practice, and Research, 562
Child Welfare League of America Standards for Adoption Service, 559
Child Welfare League of America Standards for Service for Abused or Neglected Children and Their Families, 559
Child Welfare Services, 561
Childhood, 318
Children and Books, 324
Children in Danger, 560
Children of Alcoholics: Critical Perspectives, 548
Children of Chemically Dependent Parents, 544
Children of Color, 565
Children of Drug Abusers, 562
Children of Prenatal Substance Abuse, 546
Children's Book Illustration and Design, 465
Children's Books in Print, 331
Children's Library Services Handbook, 331
Children's Literature in the Elementary School, 317
Children's Play in Diverse Cultures, 321
Children's Psychological Testing, 325
Childrenswear Design, 414
Child's Creation of a Pictorial World, 316
The Child's Developing Sense of Theme, 318
A Child's Place, 322
Chilton Small Engine Repair: 2 Hp to 12 Hp, 380
Chilton Small Engine Repair: 13 Hp to 20 Hp, 381
Chisels on a Wheel: A Comprehensive Reference to Modern Woodworking Tools and Materials, 123
Christensen's Physics of Diagnostic Radiology, 24
Christian Dior, 420
Chronicle of Aviation, 347, 359
Circular Work in Carpentry and Joinery, 133
Circulation, 343
City Surveyor, 386
Civil Litigation for the Paralegal, 278, 303
Civilizing Voices: American Press Criticism, 1880-1950, 242
CLA Review Manual, 299
CLA Review Manual: A Practical Guide to CLA Exam, 295
Clark on Surveying and Boundaries, 386
Class I Occlusal and Buccal Pit Amalgam Restoration, 16
Classic Cases in Medical Ethics, 53
Classic Woodworking Woods: And How to Use Them, 122
Classical Cooking the Modern Way, 198
Classical Furniture, 436
Classification of Nursing Diagnosis, 78
Clean and Sober Parenting: A Guide to Help Recovering Parents, 574

Page 597

Title Index

Clearcut: The Tragedy of Industrial Forestry, 507
Clearcutting: A Crime Against Nature, 508
Clinical Application of Blood Gases, 118
Clinical Application of Respiratory Care, 118
Clinical Applications of Ventilatory Support, 116
Clinical Assessment in Respiratory Care, 119
Clinical Biochemistry of Domestic Animals, 526
Clinical Blood Gases: Application and Noninvasive Alternative, 116
Clinical Calculations: A Unified Approach, 85
Clinical Chemistry: A Fundamental Textbook, 60
Clinical Chemistry: Concepts and Applications, 59
Clinical Chemistry: Theory, Analysis, and Correlation, 62
Clinical Diagnosis and Management by Laboratory Methods, 61
Clinical Electrotherapy, 109
Clinical Guide to Laboratory Tests, 64
Clinical Imaging: An Atlas of Differential Diagnosis, 24
The Clinical Interview Using DSM-III-R, 543
Clinical Kinesiology for the Physical Therapist Assistants, 108
Clinical Laboratory Handbook, 61
Clinical Laboratory Instrumentation and Automation, 65
Clinical Laboratory Manual Series: Phlebotomy, 61
Clinical Laboratory Medicine, 63, 64
Clinical Laboratory Science: Strategies for Practice, 60
Clinical Laboratory Tests: Values and Implications, 76
Clinical Magnetic Resonance Imaging, 28
Clinical Manifestations of Respiratory Disease, 115
Clinical Manual of Health Assessment, 79
Clinical Manual of Pediatric Nursing, 84
Clinical Manual of Psychiatric Nursing, 86
The Clinical Medical Assistant, 49
Clinical Microbiology, 49
Clinical Microbiology Procedures Handbook, 62
Clinical Nursing Skills, 80
Clinical Nursing Skills and Techniques, 80
Clinical Nutrition and Dietetics, 39
Clinical Oncology for Students of Radiation Therapy Technology, 28
Clinical Orthopaedic Physical Therapy, 110
Clinical Pharmacology and Nursing, 84
Clinical Pharmacology and Nursing Management, 86
Clinical Pharmacology in Dental Practice, 9
Clinical Practice of Neurological and Neurosurgical Nursing, 77
Clinical Practice of the Dental Hygienist, 15
Clinical Procedures for Medical Assistants, 49
Clinical Procedures for Medical Assisting, 48
Clinical Radiology: The Essentials, 24
Clinical Removable Partial Prosthodontics, 14
Clinical Restorative Materials and Techniques, 10
Clinical Simulations: The Right Way! The Complete Guide for Respiratory Care Providers, 117
Clinical Skillbuilders, 87
Clinical Skills in Nursing Practice, 79
Clinical Sonography: A Practical Guide, 31
Clinical Textbook for Veterinary Technicians, 529
Clip Art: Image Enhancement and Integration, 403
Clipper Maintenance and Blade Care, 531
Coaching Writers: The Essential Guide for Editors and Reporters, 237
Coated Abrasives Reference Manual, 126
Cocaine, 538, 547
Cocaine: From Dependency to Recovery, 543
Cocaine Solutions: Help for Cocaine Abusers and Their Families, 544
The Codependency Conspiracy, 540
Codependency, Sexuality and Depression, 547
Codependent's Guide to the Twelve Steps, 535
Coding for Prospective Payment, 50
Cognitive Development, 314
A Cognitive Psychology of Mass Communication, 242
Cognitive Rehabilitation, 101
Cognitive Style and Early Education, 322
Cognitive Therapy of Substance Abuse, 535
Coherent Lightwave Communications, 378

Colefax and Fowler: The Best in English Interior Decoration, 435
Collection Development for Libraries, 330
Collection of Ceramic Works, 5
Collier's Rules for Desktop Design and Typography, 400
Color, 460
Color: Natural Palettes for Painted Rooms, 435
Color Atlas and Text of Clinical Medicine, 61
Color Atlas and Textbook of Oral Anatomy, Histology, and Embryology, 5
Color Atlas of Clinical Oral Pathology, 11
Color Atlas of Common Oral Diseases, 10
Color Atlas of Comparative Veterinary Hematology, 526
Color Atlas of Diseases and Disorders of Cattle, 525
Color Atlas of Diseases and Disorders of the Pig, 527
Color Atlas of Oral Manifestations of AIDS, 4
Color Atlas of Porcelain Laminate Veneers, 8
Color Atlas of the Hair, Scalp, and Nails, 186
Color Atlas of Tooth Whitening, 11
Color Atlas of Veterinary Histology, 525
Color Choices, 438
The Color Compendium, 435
Color Confidence!, 441
Color Doppler Flow Imaging, 30
Color Doppler Imaging in Obstetrics and Gynecology, 30
Color for the Electronic Age, 409, 460
Color in Decoration, 439
Color in Everyday Life, 441
Color in Fashion, 418
The Color Line and the Quality of Life in America, 565
Color on Color, 402
Color Publishing on the Macintosh, 398
The Color Source Book for Graphic Designers, 437
Color Vision, 446
Colorworks, 459
Colour Atlas of Diagnosis After Recent Injury, 44
The Columbia Encyclopedia of Nutrition, 39
The Columbia-Knight Bagehot Guide to Economics and Business, 253
Combined Fixed/Removable Prosthesis with the SAE Spark Erosion System, 20
Come Fly With Me, 358
The Comic Book in America: An Illustrated History, 465
The Comic Book Reader's Companion, 466
Comic Books and Strips: An Information Sourcebook, 467
The Coming Crisis in Accounting, 170
Commercial Aviation Safety, 361
Committed Journalism: An Ethic for the Profession, 245
Common Threads: A Parade of American Clothing, 421
The Common-Sense Mortgage, 219
Commonsense Cataloging: A Cataloger's Manual, 342
Communication and Image in Nursing, 89
Communication in the Nursing Context, 72
Communication Skills for Cosmetologists, 186
Communications and Networking for the IBM PC and Compatibles, 179
Communications in Nursing, 72
Communications Law: Liberties, Restraints, and the Modern Media, 246
Community: Program Ten of The Sociological Imagination, 575
Community College Reference Services, 329
Community Dental Health, 9
Community Health Nursing, 72
Community Health Nursing: Theory and Practice, 72
Community Living Skills Workbook for the Head Injured Adult, 98
Compact Disc Packaging and Graphics, 455
Compendium of Veterinary Products, 525
Competency in Cosmetology: A Professional Text, 186
Comping Techniques, 472
Complete Automotive Engine Rebuilding and Parts Machining, 355
The Complete Aviation/Aerospace Career Guide, 358
The Complete Beverage Dictionary, 196
The Complete Book of Comprehensives, 470

Complete Book of Cordwood Masonry Housebuilding, 135
The Complete Book of Curtains and Draperies, 441
The Complete Book of Decorative Paint Techniques, 439
The Complete Book of Feature Writing, from Great American Feature Writers, Editors, and Teachers, 240
The Complete Book of Kitchen Design, 139
The Complete Book of Shaker Furniture, 439
The Complete Book of Silk Painting, 424
The Complete Cataloging Reference Set, 342
Complete Confined Spaces Handbook, 95
The Complete Court Reporter's Handbook, 274
The Complete Dog Book, 524
The Complete DUAT Book, 358
The Complete Guide to Barrier-Free Housing, 431
The Complete Guide to Decorative Landscaping with Brick and Masonry, 157
The Complete Guide to Factory-Made Houses, 137
The Complete Guide to Fashion Illustration, 419
The Complete Guide to Log and Cedar Homes, 132
The Complete Guide to Lumber Yards and Home Centers, 129
The Complete Guide to Magazine Article Writing, 240
The Complete Guide to Residential Deck Construction, 135
The Complete Home Renovation Manual, 138
The Complete Home Restoration Manual, 138
The Complete Idiot's Guide to Multimedia, 462
The Complete Manual of Corporate and Industrial Security, 290
The Complete Off-Premise Caterer, 196
The Complete Polygraph Handbook, 287
The Complete Printmaker: Techniques, Traditions, Innovations, 407
The Complete Scanner Handbook for Desktop Publishing, 400
Complete Secretary's Handbook, 209
Complete Siding Handbook: Installation, Maintenance, Repair, 132
Complete Textbook of Phlebotomy, 61
The Complete Travel Marketing Handbook, 232
The Complete Typographer, 471
Composite Materials in Aircraft Structures, 348
The Comprehensive Book of Photographic Lenses, 446
Comprehensive Cardiac Care, 74
Comprehensive Child and Family Nursing Skills, 84
Comprehensive Dental Hygiene Care, 15
Comprehensive Family and Community Health Nursing, 72
Comprehensive Medical Assisting, 48
Comprehensive Neonatal Nursing: A Physiologic Perspective, 82
Comprehensive Perinatal and Pediatric Respiratory Care, 119
Comprehensive Psychiatric Nursing, 86
Comprehensive Radiographic Pathology, 24
Comprehensive Respiratory Care: A Learning System, 115
Computed Body Tomography with MRI Correlation, 27
Computed Tomography, 27
Computer-Aided Design, 373
Computer-Aided Drafting with AutoCAD, 374
Computer-Aided Engineering Design Graphics, 373
Computer Aided Machine Design, 372
The Computer Animation Dictionary, 466
The Computer-Based Patient Record, 53
Computer Color: 10,000 Computer-Generated Process Colors, 463
Computer Dictionary, 175
Computer Graphics for Designers and Artists, 373, 404
The Computer in Graphic Design: From Technology to Style, 462
Computer-Integrated Design and Manufacturing, 372
Computer-Integrated Manufacturing Handbook, 372
Computer Literacy for Health Care Professionals, 59
Computer Networks, 183
Computer Principles for Physical and Occupational Therapists, 100, 109

Title Index

Computer Shorthand: Skillbuilding and Transcription, 275
Computer Shorthand: Speedbuildng and Transcription, 275
Computer Type: A Designer's Guide to Computer-Generated Type, 471
Computerized Aviation Publications, 350, 361
Computerized Litigation Management for Paralegals, 303
Computerized Medical Office Management, 54
Computers in Accounting: Buyers Guide, 167
Computers in Manufacturing, 374
Computers in the Fashion Industry, 416
Computers in the Law, 303
Concepts in Medical Radiographic Imaging, 26
Concepts of Foodservice Operations and Management, 195
Conceptual Basis of Professional Nursing, 77
The Concierge: Key to Hospitality, 203
The Concise AACR2, 1988 Revision, 340
A Concise Dictionary of Interior Design, 435
Concise Encyclopedia of Interior Design, 432
Concise Encyclopedia of Medical and Dental Materials, 4
Concrete Formwork, 157
Concrete Masonry Handbook for Architects, Engineers, Builders, 157
Concrete Technology, 157
Concurrent Engineering, 373
Condemnation of Property, 306
Confessions of a PR Man, 261
Confidentiality in the Use of Library Materials, 335
Confidentiality: Ethical and Legal Considerations, 549
Conical Double-Crown Telescopic Removable Periodontic Prosthesis, 14
Connecting to the Internet: A Buyer's Guide, 177
Connecting Young Adults and Libraries, 336
CONSER Cataloging Manual, 340
CONSER Editing Guide, 341
The Conservation Atlas of Tropical Forests, 503
A Conservation Strategy for the Northern Spotted Owl, 506
Considerations of Discipline, 326
Construction Electrical Contracting, 148
Construction Equipment Guide, 131, 363
Construction Litigation for Paralegals, 306
Construction Manual: Finish Carpentry, 134
Construction Manual: Rough Carpentry, 134
Construction Materials, 129
Construction Materials for Interior Design, 439
Construction Productivity: On Site Measurement and Management, 131
Construction Project Management, 131
Construction Revisited: An Illustrated Guide to Construction Details of the Early Twentieth Century, 129
Construction Technology, 134
A Consumer's Dictionary of Food Additives, 39
The Contemporary Animator, 466
A Contemporary Approach to Permanent Waving, 188
Contemporary Criminal Procedure, 1993-1994, 287
Contemporary Dental Hygiene Practice, 12
Contemporary Designers, 413, 437, 457
Contemporary Graphic Design, 457
Contemporary Leadership Behavior: Selected Readings, 70
Contemporary Medical Office Procedures, 53
Contemporary Newspaper Design: A Structural Approach, 248
Contemporary Photographers, 444
Contemporary Sports Reporting, 252
Continued Acts of Sabotage, 549
Contract Interior Finishes, 434
The Contrary Farmer, 479
Contrary Investing for the 90s, 220
Control Optics Optical Engineering Series, 377
Controlling Color, 459
Controlling Movement, 105
Controversial Issues in Social Welfare Policy, 570

The Conversion and Seasoning of Wood, 507
Cooperative Learning in Mathematics, 313
Coping with the Final Tragedy, 511
Cops: The Men and Women Behind the Badge, 290
Copy-Editing, 276
The Copyright Handbook: How to Protect and Use Written Works, 245
Core List for an Environmental Reference Collection, 367
Core Textbook of Respiratory Care Practice, 114
Corel Draw! Version 2.0, 410
Corn and Its Early Fathers, 482
Corporate and Organizational Video, 267
Corporate Magazines of the United States, 251
Corporate Scriptwriting: A Professional's Guide, 269
Corporate Television: A Producer's Handbook, 265
Corpus Juris Secundum, 301
Corrective Haircoloring: A Hands-On Approach, 189
Cosmetologist, 186
Cosmetologists State Board Exam Review, 186
Cost Control for the Hospitality Industry, 203
Costume Design: Techniques of Modern Masters, 417
Costume Design in the Movies, 421
The Costume Designer's Handbook, 416
The Costume Technician's Handbook, 416
COTA Supervision Information Packet, 99
Counseling Chemically Dependent People with HIV Illness, 545
Counseling Children, 324
The Courage to Heal, 558
The Court and Free-Lance Reporter Profession, 276
Court Hearing Reporter, 274
Court Law Stenographer, 274
Court Reporter, 274
Court Reporter I, 274
Court Reporter II, 274
Court Reporters and Stress: How to Find the Time to Live, 274
Court Reporter's Language Arts Workbook, 277
Court Reporting: Grammar and Punctuation, 276
Courtroom Survival: The Officer's Guide to Better Testimony, 287
Couture: The Fine Art of Sewing, 414
Couture Sewing Techniques, 415
Covers and Jackets!, 465
CPR '93, 46
CP "Teach" Expert Coding Made Easy!, 50
CPT 1994: Physician's Current Procedural Terminology, 50
CPT Coding Made Easy, 50
The Craft of Corporate Journalism, 252
Crafting the Successful Business Plan, 224
Crafts in Therapy and Rehabilitation, 99
Cranes and Derricks, 364
Create Excellent Video, 267
Create Your Own Multimedia System, 462
Creating a Sense of Place: Photographs by Joel Meyerowitz., 447
Creating Multimedia on Your PC, 460
Creating Multimedia Presentations, 464
Creative Editing for Print Media, 237
Creative Fashion Skills for Fashion Design, 416
Creative Interviewing, 239
Creative Sewing Ideas, 414
The Creative Stroke, 467
Creative Techniques for Photographing Children, 447
The Creative Traveler: A Guidebook for All Places and All Seasons, 231
Creative Window Treatments, 435
Cremation and the Funeral Director, 513
Cremation Concerns, 514
Crime and the American Press, 242
The Crime Classification Manual, 287
A Crime Never Forgotten, 575
Criminal Evidence for Law Enforcement Officers, 288
Criminal Justice Abstracts, 285
Criminal Justice Periodicals Index, 285
Criminal Justice Research Sources, 286, 338
Criminal Law and Procedure, 305

Crisis Counseling with Children and Adolescents, 564
Crisis Drugs, 87
Crisis Intervention, 567
Crisis Intervention: Theory and Methodology, 567
Crisis Intervention Handbook: Assessment, Treatment, and Research, 568
Crisis Intervention in Criminal Justice/Social Service, 568
Crisis Intervention Strategies, 568
Crisis of Conscience: Perspectives on Journalism Ethics, 245
Critical Care Nursing: A Holistic Approach, 73
Critical Care Nursing: Diagnosis and Management, 74
Critical Care Nursing, 73, 74
Critical Care Nursing of the Elderly, 73
Critical Care Obstetrical Nursing, 82
Critical Perspectives on Early Childhood Education, 325
Criticizing Photographs, 450
Critiquing Radio and Television Content, 265
Crop Improvement for Sustainable Agriculture, 487
Cross Infection Control in Dentistry, 15
Crossing the Great River, 513
Crossing the Internet Threshold: An Instructional Handbook, 183, 333
Crucial Skills for Legal Secretaries—and Receptionists, 294
Cruises: Selecting, Selling, and Booking, 232
A Cry for Help: The Fetal Drug and Alcohol Crisis, 549
Culinary Nutrition for Food Professionals, 194
Cultural Diversity and Social Work Practice, 565
Cultural Diversity in Health and Illness, 80
The Culture of Addiction, 548
Cumulative Analysis of Uniform Plumbing Code Changes, 161
Cumulative Trauma Disorders, 95
Current Practice of Interventional Radiology, 29
Current Respiratory Care, 116
Current Therapy in Equine Medicine, 3, 527
Current Therapy of Respiratory Disease, 115
Current Veterinary Therapy: Food Animal Practice, 3, 526
Current Veterinary Therapy XI: Small Animal Practice, 526
Curriculum Models and Early Childhood Education, 316
Curtains, Draperies, and Shades, 432
The Customer Comes Second, 230
Customer Service: More Than a Smile, 343
Cycling the U.S. Parks, 227
Cytology and Hematology of the Horse, 525

D.V.M. Series, 527
Daily Activities After Your Hip Surgery, 104
Dancers: Photographs by Annie Liebovit, 447
Danger Signs and Symptoms, 87
Daralism: Phases of Hair Care for the Bedridden and Disabled, 186
Dare to Confront!, 548
Data Analysis for the Helping Professions: A Practical Guide, 569
Data Analysis in Hotel and Catering Management, 203
Data Sources, 174
Data Structures and C Programs, 183
The Day Care Dilemma: Critical Concerns for American Families, 559
Deadline: A Memoir, 243
Dealing Creatively with Death, 513
Dealing with Individual Differences, 323
Death and Dying: A Bibliographical Survey, 515
Death and Dying, Life and Living, 511
Death and Grief: A Guide for Clergy, 516
Death Dictionary, 514
Death, Ritual, and Bereavement, 512
Death, Society, and Human Experience, 512
Death to Dust: What Happens to Dead Bodies?, 512
Deciphering Difficult ECGs, 87
Decision Making in Medicine, 61
The Decision-Making Process in Journalism, 246
Deck Planner: Twenty-Five Outstanding Decks You Can Build, 132

Title Index

Deep Disagreement in U.S. Agriculture, 483
Definitions, Conversions, and Calculations for Occupational Safety and Health Professionals, 93
The Deluxe Transitive Vampire, 250
Demystifying Media Technology, 243
Denim: An American Legend, 421
Dental Anatomy: Its Relevance to Dentistry, 15
Dental Anatomy: Mandibular Incisors, 17
Dental Anatomy: Maxillary and Mandibular Canines, 17
Dental Anatomy: Maxillary Incisors, 17
Dental Anatomy: Maxillary Premolars, 17
Dental and Oral Tissues: An Introduction, 11
Dental Assistant, 5
The Dental Assistant, 5
Dental Assistant Boards (NDAB), 19
Dental Assisting Exam Preparation, 14
Dental Assisting: Basic and Dental Sciences, 10
Dental Cements: Selecting the Best Type, 16
Dental Ethics, 15
Dental Health Education: Theory and Practice, 6
Dental Hygiene: A Profession of Opportunities, 17
Dental Hygiene Employment Reference Guide, 7
Dental Hygienist As Change Agent, 7
Dental Hygienist Boards (NDHB), 19
Dental Implants: Are They For Me?, 14
Dental Maintenance for Patients with Periodontal Disease, 15
Dental Materials: Properties and Manipulation, 6
Dental Pharmacology, 6
Dental Radiology, 10
Dental Science in a New Age, 8
Dentistry, Dental Practice, and the Community, 6
The Deposition Handbook, 278
Deposition Manual for Paralegals, 305
Depression: The Evolution of Powerlessness, 567
Desairology: Hairstyling of the Deceased, 186
Design and Composition, 458
Design and Layout of Foodservice Facilities, 192
Design and Repair of Residential and Light Commercial Foundation, 132
Design Apparel Through the Flat Pattern, 415
Design Basics, 459
Design Essentials: A Handbook, 458
Design for Hospitality, 203
Design Guidelines for Desktop Publishing, 237, 405
The Design of Advertising, 454
Design of Reinforced Concrete Structures, 156
Design Presentation, 431
Designed Right!, 459
Designer Primer: For Architects, Graphic Designers, and Artis, 459
Designer's Guide to Color 5, 459
Designer's Guide to Making Money with Your Desktop Computer, 473
The Designer's Guide to PostScript Text Type, 404
Designer's Guide to Print Production, 397, 469
Designing and Making Stage Costumes, 417
Designing and Planning Bedrooms, 432
Designing Exhibitions, 455
Designing Furniture from Concept to Shop Drawing, 440
Designing Interventions for Preschool Learning, 311
Designing To Sell, 430
Designing with Color, 458
Designing with Illustration, 464
Designing with Light: Residential Interiors, 435
Designing with Two Colors, 398, 458
Designing with Type: A Basic Course in Typography, 470
Designs for Marketing: Number One, 454
Desktop Design, 404
Desktop Design: Getting the Professional Look, 400
The Desktop Design Workbook, 400
Desktop Multimedia Bible, 461
Desktop Publisher's Easy Type Guide, 401
Desktop Publisher's Survival Kit, 399
Desktop Publishing: Design Basics, 403
Desktop Publishing: Dollars and Sense, 397

Desktop Publishing By Design, 408
Desktop Publishing for Librarians, 401
Desktop Publishing Handbook, 404
Desktop Publishing in Color, 404
Desktop Publishing Secrets, 402
Desktop Publishing Sourcebook, 399
Desktop Publishing Using PFS, First Publisher (Version 3.0), 408
Desktop Publishing Using Publish It!—IBM Version, 399
Desktop Publishing with Word for Windows Through Version 6, 405
Desktop Publishing with WordPerfect 6, 405
The Desktop Style Guide, 250, 402
Destructive and Useful Insects: Their Habits and Control, 487
Developing and Administering a Child Care Center, 322
Developing Child, 326
The Development of American Agriculture, 478
Developmental Motor Activities for Therapy, 101
Deviance and Psychopathology, 557
Deviant Behavior, 557
DHAT: Dental Hygiene Aptitude Test, 13
Diagnosing and Managing Chemical Dependency, 535
Diagnostic and Statistical Manual of Mental Disorders: DSM-IV, 86
Diagnostic and Surgical Arthroscopy in the Horse, 527
Diagnostic Cytology of the Dog and Cat, 525
Diagnostic Hematology: Clinical and Technical Principles, 63
Diagnostic Imaging of AIDS, 26
Diagnostic Parasitology for Veterinary Technicians, 525
Diagnostic Reasoning and Treatment Decision Making in Nursing, 77
Diagnostic Strategies for Common Medical Problems, 63
Diagnostic Test Implications., 88
Diagnostic Testing and Nursing Implications, 76
Diagnostic Tests, 60, 88
Diagnostic Ultrasound: Physics, Biology and Instrumentation, 29
Diagnostic Ultrasound: Principles and Instruments, 30
Diary of a Baby, 323
Dictionary and Thesaurus of Environment, Health, and Safety, 95
Dictionary for Human Factors/Ergonomics, 95
Dictionary of Architecture and Construction, 434
Dictionary of Automotive Engineering, 354
Dictionary of Aviation, 348, 359
The Dictionary of Bias-Free Usage, 251
Dictionary of Biomedical Acronyms and Abbreviations, 282
Dictionary of Broadcast Communications, 264
Dictionary of Clinical Tests, 60
The Dictionary of Criminal Justice, 286
Dictionary of Gardening, 518
Dictionary of Legal Terms, 280
Dictionary of Marketing and Advertising, 258
Dictionary of Printing and Publishing, 397
Dictionary of Real Estate Terms, 214
Dictionary of Travel, Tourism, and Hospitality, 206
Dictionary of Twentieth Century Design, 438, 457
Dictionary of Woodworking Tools, c. 1700-1970, and Tools of Allied Trades, 125
Diet and Nutrition Guide, 37
Diet Counseling Procedure for the Dental Clinic, 17
Diet Right!, 37
Differential Diagnosis in Physical Therapy, 107
Differential Leveling, 387
Difficult Child, 325
Digital Engineering Design: A Modern Approach, 374
Digital Imaging for Visual Artists, 403, 461
Digital Nonlinear Editing, 268
Digital Type Specimens: The Designer's Computer Type Book, 471
Dimensions of Professional Nursing, 77
Dimensions of the Hospitality Industry: An Introduction, 193
Dining Room and Banquet Management, 199

Direct Digital Control for Building HVAC Systems, 152
Directing Television and Film, 265
Directory of Criminal Justice Information Sources, 286
Directory of Educational Software, 70
Directory of Electronic Journals, Newsletters, 333
Directory of Electronic Journals, Newsletters and Academic Discussion Lists, 174
Directory of Environmental Information Sources, 369
Directory of Food and Nutrition Information for Professionals and Consumers, 194
Directory of Institutions Offering Programs, 335
Directory of Library Automation Software, Systems, and Services, 331
Directory of Print Media Advertising Resources, 234
Directory of Women's Media, 250
Disaster Planning and Recovery, 334
Discipline That Works, 316
Discipline: Appropriate Guidance for Young Children, 326
Discipline: What Lily Learned, 326
Discover Bookkeeping and Accounts, 171
Discovering AutoCAD Release 12, 176
Discreet Indiscretions: The Social Organization of Gossip, 259
Discrimination by Design, 440
Diseases, 88
Diseases and Management of Breeding Stallions, 527
Diseases of the Temperate Zone: Tree, Fruit, and Nut Crops, 521
Disenfranchised Grief: Recognizing Hidden Sorrow, 512
Disney's Art of Animation, 467
Display Lighting, 426
The Dispossessed, 572
Distance Education: A Practical Guide, 331
Distance Education: New Perspectives, 331
Distance Education: Strategies and Tools, 331
Diversity and Developmentally Appropriate Practices, 319
Divisions of Welfare, 571
DNA on Trial: Genetic Identification and Criminal Justice, 287
DNA Technology in Forensic Science, 288
Do-It-Yourself Advertising, 235
Do-It-Yourself Direct Marketing: Secrets for Small Business, 223
Do-It-Yourself Investment Analysis, 220
The Doctors' Vitamin and Mineral Encyclopedia, 36
Documentation Requirements for the Acute Care Patient Record, 53
Documenting Care, 72
Documenting Functional Outcomes in Physical Therapy, 111
Doing Ethics in Journalism: A Handbook with Case Studies, 245
Domestic Violence: The Criminal Justice Response, 559
Doors, Windows, and Skylights, 135
Doppler Color Flow Imaging, 30
Doppler Echocardiography, 31
Doppler Echocardiography: The Quantitative Approach, 30
Doppler Ultrasound, 30
Doppler Ultrasound: Principles and Instruments, 30
Doppler Ultrasound in Perinatal Medicine, 31
Dorland's Electronic Medical Speller, 65
Dorland's Hematology/Oncology Speller, 60
Dorland's Medical Speller, 60
Dorland's Pocket Medical Dictionary, 60
DOS, 175
DOS: The Complete Reference, 178
Dosage Calculations, 88
Dosages and Solutions, 85
Double Duty, 536
Double Your Income in Real Estate Sales, 216
Doublespeak, 251
The Douglas DC-3, 348
The Dover Electronic Clip Art Library, 406
Down by the River, 503

Dr. Spock's Baby and Child Care, 323
Draping for Fashion Design, 414
Drawing, A Creative Process, 432
Drawing Fashion, 420
Drawing the Fashion Body, 419
Drawing the Future, 437
A Dream of Wings, 346
Dressing with Color, 418
Drug Abuse Treatment: A National Study of Effectiveness, 540
Drug Abuse Treatment in Prisons and Jails, 541
Drug and Alcohol Abuse, 545
Drug Handbook: A Nursing Process Approach, 84
Drug, Alcohol and Other Addictions, 537
Drug-Affected Children: The Price We Pay, 549
Drug-Test Interactions Handbook, 64
Drugs and Behavior: A Sourcebook for the Helping Professions, 545
Drugs and the Athlete, 547
Drugs in American Society, 539
Drugs in the Workplace, 549
Drunk Driving Defense, 279
Drying and Storage of Grains and Oilseeds, 482
DSM-III-R Training Guide, 544
Dual Diagnosis: Counseling the Mentally Ill Substance Abuser, 538
Dual Diagnosis: The Mentally Ill Chemical Abuser, 549
Dual Diagnosis in Substance Abuse, 539
Dual Diagnosis of Major Mental Illness and Substance Disorder, 543
Dual Disorders, 537
Dwelling House Construction, 133
Dwelling Requirements of the Uniform Plumbing Code, 161
Dying, Death, and Bereavement, 511
Dynamic Physical Education for Elementary School Children, 320

Early Childhood Art, 317
Early Childhood Curriculum, 313, 322
Early Childhood Education, 323
Early Childhood Education and the Elementary School, 314
Early Childhood Education in the Schools, 311
Early Childhood Education Today, 320
Early Childhood Language Arts, 317
Early Childhood Social Studies, 324
Early Childhood Teacher Preparation, 323
Early Defibrillation, 43
Early Education in the Public Schools, 317
Earth In the Balance: Ecology and the Human Spirit, 505
Easy Access to National Parks, 231
Easy Coder, 50
Easy DOS, 181
Easy Multimedia: Sound and Video for the PC Crowd, 462
Easy-To-Tell Stories for Young Children, 316
Eat For Life, 36
Eat Smart, 39
Eating for Endurance, 35
ECG Interpretation, 88
Echocardiography, 30
Echocardiography in Clinical Practice, 31
An Ecological History of Agriculture, 480
Ecology and Classification of North American Freshwater Invertebrates, 507
Ecology and Field Biology, 505
Economics for Hotel and Catering Students, 204
The Economics of Child Care, 558
Ecotours and Nature Getaways, 228
Editing in the Electronic Era, 237
Editor and Publisher International Yearbook, 250
Editorial and Persuasive Writing, 240
The Editorial Eye, 238
Education for the Earth, 500
The Educational Annotation of ICD-9-CM, 50
Edward Weston, 448

Effective Dental Assisting, 14
Effective Documentation for Occupational Therapy, 98
The Effective Nurse: Leader and Manager, 70
Effective Public Relations, 259
The Effects of Child Abuse and Neglect: Issues and Research, 563
Effects of Drugs on Clinical Laboratory Tests, 65
Egan's Fundamentals of Respiratory Care, 118
EGT Systems: Basic and Advanced Light Plane Maintenance, 349
Electric Woodwork: Power Tool Woodworking, 122
Electrical Blueprint Reading, 149
Electrical Course for Apprentices and Journeymen, 146
Electrical Grounding: Bringing Grounding Back to Earth, 146
Electrical Hazards and Accidents: Their Cause and Prevention, 144
Electrical Raceways and Other Wiring Methods, 145
Electrical Systems: Compact Equipment, 381
Electrical Wiring, 147
Electrical Wiring, Commercial, 145
Electrical Wiring, Industrial, 147
Electrical Wiring, Residential, 145
Electrical Wiring: Residential, Utility, Service Areas, 143
Electrical Wiring, Industrial, 147
Electricity: Fundamental Concepts and Applications, 145
Electronic Access to Information, 333
Electronic Color, 463
The Electronic Invasion, 209
Electronic Post-Production and Videotape Editing, 268
Electronic Post-Production Terms and Concepts, 268
Electronic Publishing: Evaluating, Procurement, and Management, 402
Electronic Style: A Guide to Citing Electronic Information, 397
Electronic Troubleshooting, 148
Electronics Math, 353
The Elegant Man: How to Construct the Ideal Wardrobe, 418
Elegant Small Hotels: A Connoisseur's Guide, 205
Elementary School Child Health: For Parents and Teachers, 317
Elementary Social Studies: A Practical Guide, 313
Elementary Surveying, 386, 506
Elements of Dental Materials, 12
Elements of Information Systems: Components and Architecture, 181
Elements of Newspaper Design, 247, 460
Elements of Paediatric Physiotherapy, 106
Elements of Style, 431
The Embalmer's Guide to Cardiovascular Anatomy, 514
Embalming Notes, 515
Embalming: History, Theory, and Practice, 513
Emergencies in Dental Practice, Diagnosis and Management, 9
Emergency Care and Transportation of the Sick and Injured, 43
Emergency Care in the Streets, 42
Emergency Communications Management, 494
Emergency Drugs, Devices, and Procedures, 18
Emergency Medical Technician Paramedic, 45
Emergency Medical Treatment, 42
Emergency Nursing: Principles and Practice, 74
Emergency Planning and Management in College Libraries, 329
Emergency Procedures, 88
Emergency Procedures for the Small Animal Veterinarian, 527
Emergency Services Stress, 495
Emotional Healing, 549
Empathy in the Helping Relationship, 554
Employee Rights Handbook for Dental Hygienists, 7
Employment Savvy, 7
EMS Driving: The Safe Way, 45
EMS Training Software, 46
EMT: Basic National Standards Review Self Test, 44
EMT: Beyond the Lights and Sirens, 43

EMT: Prehospital Care, 43
EMT Certification Preparation and Review, 44
The EMT Review Manual, 45
Enchanted Drawings: The History of Animation, 467
Encyclopedia and Dictionary of Medicine, Nursing, and Allied Health, 63, 70
The Encyclopedia of Adoption, 557
Encyclopedia of Agricultural Science, 477
The Encyclopedia of Alcoholism, 574
Encyclopedia of American Comics, 466
The Encyclopedia of Animated Cartoons, 466
The Encyclopedia of Calligraphy Techniques, 468
The Encyclopedia of Child Abuse, 560
Encyclopedia of Computer Science, 175
Encyclopedia of Criminology, 285
The Encyclopedia of Drug Abuse, 574
Encyclopedia of Legal Information Sources, 279, 296, 301, 338
Encyclopedia of Microbiology, 526
Encyclopedia of Physical Sciences and Engineering, 338
Encyclopedia of Police Science, 285
Encyclopedia of Television, Cable, and Video, 265
Encyclopedia of Textiles, 412
The Encyclopedia of Window Fashions, 438
The Encyclopedia of Wood, 130, 507
Encyclopedia of World Crime, 286
Encyclopedic Dictionary of Accounting and Finance, 170
The Endless Quest: Helping America's Farm Workers, 479
Endotracheal Intubation, 120
Enduring Seeds, 479
Energy and Pollution Control Opportunities to the Year 2000, 368
Energy Efficient Electric Motors: Selection and Application, 142
Engine and Tractor Power, 364, 481
Engine Oils and Automotive Lubrication, 352
Engine Troubleshooting Guide, 347
Engines: Compact Equipment, 381
English Interiors: An Illustrated History, 433
The English Way of Death, 513
Entomology and Pest Management, 488
Entourage, 431
The Entrepreneur and Small Business Problem Solver, 223
The Entrepreneurial Experience, 223
The Entrepreneurs, 225
The Entrepreneur's Guide to Starting a Successful Business, 223
Entrepreneur's Road Map to Business Success, 224
Entrepreneurship, 223
Entrepreneurship for the Nineties, 223
Entry-Level Respiratory Care Review: Study Guide and Workbook, 117
Environmental Audits, 367
Environmental Career Directory, 501
The Environmental Career Guide, 501
Environmental Health, 94
Environmental Law Handbook, 367
Environmental Science: Sustaining the Earth, 505
Environmental Systems and Engineering, 367
Environmental Viewpoints, 369
The Environmentalist's Bookshelf, 505
Equine Anesthesia: Monitoring and Emergency Therapy, 527
Equine Medicine and Surgery, 525
Equine Respiratory Disorders, 525
Equine Sports Medicine, 526
Equipment Theory for Respiratory Care, 119
Ergonomics: The Physiotherapist in the Workplace, 105
Eric Carle: Picture Writer, 326
The Ernst and Young Business Plan Guide, 224
Errors, Lies, and Libel, 245
Eruption of Permanent Teeth: A Color Atlas, 13
Essential Dental Microbiology, 15
The Essential Feature: Writing for Magazines and Newspapers, 240

Title Index

Essential Knowledge and Skills for Baccalaureate Social Work Students in Gerontology, 563
The Essential PageMaker 5.0, 407
Essentials for the Small Laboratory and Physician's Office, 59
Essentials of Accounting, 167, 294
Essentials of Cardiopulmonary Physical Therapy, 107
Essentials of Cardiovascular Nursing, 72
Essentials of Dental Assisting, 7
Essentials of Dental Radiography for Dental Assistants and Hygienists, 7
Essentials of Fire Fighting, 493
Essentials of Human Anatomy and Physiology, 62
Essentials of Maternal-Newborn Nursing, 83
Essentials of Meteorology: An Invitation to the Atmosphere, 358
Essentials of Oral Histology and Embryology, 5
Essentials of Paralegalism, 300
Essentials of Periodontics, 9
Essentials of Psychiatric Nursing, 86
Essentials of Real Estate Economics, 215
Essentials of Real Estate Investment, 220
The Essentials of Respiratory Care, 116
Essentials of Skull Radiography, 32
Estate Planning and Administration for Paralegals, 307
Esthetic Composite Bonding: Techniques and Materials, 9
Esthetic Dentistry, 6
Esthetic Restorations, 11
Esthetics of Anterior Fixed Prosthodontics, 6
Estimating for Interior Designers, 439
Estimating in Building Construction, 131
Ethical and Legal Aspects of Nursing, 89
Ethical Decisions for Social Work Practice, 554
Ethical Dilemmas and Legal Issues in Care of the Elderly, 90
Ethical Dilemmas and Nursing Practice, 78
Ethical Dilemmas in Social Service, 554
Ethics Conference I, 8
Ethics for the Legal Assistant, 304
Ethics in Nursing, 78
Ethics in Nursing: An Anthology, 78
The Ethics of Photojournalism: A Special Report, 249
Ethnic Interiors: Decorating with Natural Materials, 434
Ethnic Variations in Dying, Death, and Grief, 512
Ethnic-Sensitive Social Work Practice, 564
Ethnicity and Biculturalism, 564
Ethnicity and Race: Critical Concepts in Social Work, 566
Ethnicity and the American Cemetery, 513
Europe for Free, 227
Europe's Wonderful Little Hotels and Inns: The Continent, 231
Evaluating Apparel Quality, 418
Evaluating Radiographs, 24
Evaluating Your Practice: A Guide to Self-Assessment, 568
Evaluation, Diagnosis, and Treatment of Occlusal Problems, 6
Evaluation, Stabilization, and Transport of the Critically Ill Child, 41
The Everglades Handbook: Understanding the Ecosystem, 505
Everyday Things: Photographs by Neil Winokur, 447
Everyone's Guide to Successful Publications, 236, 397
Everything You Wanted to Know About the Mac, 178
Evidence and Procedure for Boundary Location, 384
Evidence Handbook for Paralegals, 305
Examination for Purchase, 528
Examination Guide, 437
Examples, the Making of Forty Photographs, 445
Excavation and Grading Handbook, 363
Excel for Windows: The Complete Reference, 180
Exceptional Children, 317
Exercise in Health and Disease, 110
Exercise Physiology, 110
Expanding Dental Practice with Computer Technology, 8
Experiments in Wave Optics, 377

Expert Systems in Business and Finance, 183
Exploring CADKey 3, 375
Exploring Careers in Accounting, 170
Exploring Careers in the Travel Industry, 229
Exploring Sand and the Desert, 312
Exploring the Mysteries of Film, 451
Exploring the UNIX System, 179
Explosion Investigation and Analysis: Kennedy on Explosions, 494
Expose Yourself, 259
Expressive Typography: The Word as Image, 470
Exterior Tree and Shrub Pruning, 522
Exterior Wall Systems, 133
An Eye for Type, 408
Eyes on the News, 248
Eyewitness: 150 Years of Photojournalism, 249

F-117A Stealth Fighter, 349
FAA Statistical Handbook of Aviation, 361
Fabric Sewing Guide, 413
Fabrics and Wallpapers: Sources, Design, and Inspiration, 438
Fabulous Fit, 415
Fachbuch für Hotellerie und Gastronomietronomie, Reiseburo und Reiseveranstalter, 206
Facilitated Stretching, 108
Facing Codependence, 542
The Factory-Crafted House, 136
Facts on File Dictionary of Film and Broadcast Terms, 264
Fairchild's Dictionary of Fashion, 412
Fairly Familiar Phrases, 277
Faith in a Seed, 505
Families: Crisis and Caring, 312
Families, Alcoholism and Recovery, 537
Families in Crisis, 560
Family Baggage, 549
Family Constellation, 324
Family Farming: A New Economic Vision, 480
Family Law, 305
Family Pictures: Photographs by Nicholas Nixon, 447
Family Research: A Sixty-Year Review, 1930-1990, 557
Family Sexual Abuse: Frontline Research and Evaluation, 562
Family Therapy of Drug and Alcohol Abuse, 540
Family Violence: Research and Public Policy Issues, 558
FAR '93 Handbook for Aviation Maintenance Technicians, 347, 359
Farm, 479
Farm: A Year in the Life of an American Farmer, 480
Farm Buildings Wiring Handbook, 143
Farm Chemicals Handbook, 477
The Farm Debt Crisis of the 1980s, 484
Farm Families and Change in Twentieth-Century America, 478
The Farm Management Handbook, 483
Farming and Food, 480
The Farming Game Now, 483
Farming in Nature's Image, 487
Farming Is in Our Blood: Farm Families in Economic Crisis, 480
Fashion! A Study of Clothing Design and Selection, Textiles, the Apparel Industries, and Careers, 424
Fashion Art for the Fashion Industry, 419
The Fashion Cycle, 424
Fashion Design for the Plus-Size, 416
Fashion Display Skills, 426
Fashion Drawing: Basic Principles, 467
Fashion Drawings in Vogue: Carl Erickson, 419
Fashion Drawings in Vogue: René Bouët-Williamuez, 419
Fashion in Film, 420
Fashion in the Western World, 1500-1990, 423
Fashion Merchandising: An Introduction, 423
Fashion Merchandising: Concepts and Careers, 425
Fashion Merchandising and Marketing, 423
Fashion Retailing, 423

Fashion Sketchbook, 419
Fashion Source Book, 420
Fathers and Families: Paternal Factors in Child Development, 312
Fats, Nutrition, and Health, 35
Fear and AIDS/HIV, 89
Fear of Sewing, 413
Fear WordPerfect No More, 209
Federal Jobs in Law Enforcement, 289
Federal Rules of Appellate Procedure: December 1, 1993, 278
Federal Rules of Civil Procedure: With Forms: December 1, 1993, 278
Federal Rules of Criminal Procedure: December 1, 1993, 278
Federal Rules of Evidence: December 1, 1993, 278
Feeding Frenzy, 243
Feeding the Brain: How Foods Affect Children, 35
The Feminization of Poverty in the United States, 571
Fences for Pasture and Garden, 478
Fences, Decks, and Other Backyard Projects, 135
Ferdico's Criminal Law and Justice Dictionary, 285
Festivals, Special Events, and Tourism, 229
Fetal Alcohol Syndrome, 535
Fiber Optic Communications, 378
Fiber Optics Laboratory Experiments and Projects, 377
Fiberoptics and Laser Handbook, 378
Fibers: Manufactured and Natural, 425
Field Crop Diseases Handbook, 487
A Field Guide to Eastern Forests, North America, 499
Field Guide to Microsoft Word 6 for Windows, 211
Field Manual for the Investigation of Fish Kills, 506
Fighters: The World's Greatest Aces and Their Planes, 360
Film and Video Financing, 267
Film Animation Techniques: A Beginner's Guide and Handbook, 466
Final Celebrations, 515
Final Choices: Making End-Of-Life Decisions, 514
The Final Forest, 507
Financial Aid for Minorities in Journalism/Mass Communications, 238
Financial Control for Your Hotel, 203
Find the Law in the Library: A Guide to Legal Research, 301
Finding the Law: An Abridged Edition of How to Find the Law, 280
The Fine Art of Fashion: An Illustrated History, 419
Fine Homebuilding on Finish Carpentry, 133
Fine Homebuilding on Foundations and Masonry, 157
Fine Motor Dysfunction, 100
Fine Woodworking Design Book Six: 266 Photographs of the Best, 124
Fine Woodworking on More Proven Shop Tips, 124
Fingerprint Science, 287
Finish Carpentry Illustrated, 137
Finishes and Finishing Techniques, 124
Fire, 495
Fire: Fundamentals and Control, 494
Fire Alarm Signaling Systems, 492
Fire Department Aerial Operations, 493
Fire Department Company Officer, 493
Fire Department Occupational Safety, 93, 493
Fire Department Pumping Apparatus, 493
Fire Department Water Supply Handbook, 493
Fire in Paradise, 501
Fire Instructor's Training Guide, 492
Fire Loss Control: A Management Guide, 492
Fire Officer's Handbook of Tactics, 495
Fire Protection for Industry, 494
Fire Protection Guide on Hazardous Materials, 494
Fire Protection Handbook, 493
Fire Protection Systems, 492
Fire Safety and Loss Prevention, 492
Fire Service Instructor, 494
Fire Service Radio Communications, 496
Fire Stream Management Handbook, 494
Fire Stream Practices, 494
Fire Suppression and Detection Systems, 492

Title Index

Fire Suppression Practices and Procedures, 495
Fireboats, 493
Firefighter's Handbook of Hazardous Materials, 492
Firefighting Principles and Practices, 492
Firewall Forward—Maintaining Power, 347
The First Book of Excel 5.0 for Windows, 183
The First Book of Lotus 1-2-3, Release 2.4, 182
The First Book of Lotus 1-2-3 for Windows, 175
The First Book of Microsoft Works for the PC, 183
The First Book of MS-DOS 6, 179
The First Book of Personal Computing, 179
The First Book of the Mac, 181
The First Book of Word for Windows 6, 177
The First Book of WordPerfect 6, 180
First Class: An Introduction to Travel and Tourism, 228
First Responder, 42
First Steps in Cake Decorating, 197
First the Seed, 482
First Things First, 209
First Words, 81
Fish: An Enthusiast's Guide, 507
Fish and Shellfish Quality Assessment, 193
Fish Medicine, 527
Fit to Fly: A Pilot's Guide to Health and Safety, 360
Fitness and Rehabilitation Programs for Special Populations, 110
Fix It Fast, Fix It Right, 138
Flair: Fashion Collected by Tina Chow, 421
Flight Testing Homebuilt Aircraft, 346
Flightdeck Performance: The Human Factor, 360
Floors, Stairs, and Carpets, 433
Floors, Walls, and Ceilings, 132
Flora of North America: North of Mexico, 504, 518
Flora Photographica: Masterpieces of Flower Photography, 448
Fluid and Electrolyte Balance: Nursing Considerations, 80
Fluid, Electrolyte, and Acid-Base Balance, 79
Fluid Flow Measurement, 370
Fluids and Electrolytes, 89
Fluoride: Professionally Applied and Home Use, 16
Flying IFR, 358
Fodor's Paris, 228
Following the Money, 168
The Font Problem Solver, 400
Food and Beverage Service, 192
Food Equipment Facts: A Handbook for the Foodservice Industry, 198
Food Features Video, 39
Food for Fifty, 38, 199
Food Handbook, 35
Food Lover's Companion, 194
The Food Professional's Guide, 192
Food Safety, 36, 195
Food Service in Institutions, 199
Food Service Management by Checklist, 200
Food Service Manual for Health Care Institutions, 192
Foodservice Facilities Planning, 195
Foodservice Numbers, 197
For Enquiring Minds: A Cultural Study of Supermarket Tabloids, 241
For Hearing People Only, 320
For Reasons of Poverty, 573
For Sale by Owner, 219
The For Sale by Owner Kit, 219
For the Bereaved: The Road to Recovery, 513
For Yourself or for Investments, 138
Forbes MediaGuide 500, 250
Forced Out: The Agony of the Refugee in Our Time, 566
Forerunners of Revolution, 241
Forest and Rangeland Birds of the United States, 506
The Forest and the Trees: Guide to Excellent Forestry, 500
Forest Ecology, 500
Forest Fires, 501
Forest Measurements, 506
Forest Products and Wood Science: An Introduction, 508

Forest Recreation, 501
Forest Resource Management in the Twenty-First Century, 502
Forest Stand Dynamics, 500
Forever Young, 575
The Form of the Book: Essays on the Morality of Good Design, 409
Format Integration and Its Effect on the USMARC, 341
Forty-Seven Printing Headaches (And How to Avoid Them), 407
Foster Children in a Life Course Perspective, 560
Fostering Children's Cognitive Competence, 315
Foundations of Law for Paralegals, 279, 299
Foundations of Maternal-Newborn Nursing, 82
Foundations of Psychiatric-Mental Health Nursing, 86
The Fourth Estate and the Constitution, 246
Fracture Radiography, 32
Frame Carpentry, 133
Franchising, 225
Frank Lloyd Wright: The Masterworks, 436
The Free Life of a Ranger, 502
Free-Floating Subdivisions: An Alphabetical Index, 341
Freeing Someone You Love from Alcohol and Other Drugs, 544
Fresh-Water Invertebrates of the United States, 507
Friction Ridge Skin, 287
Friendly Multimedia, 461
Friends and Relations: Photographs by Tina Barney, 447
From Buyer Beware to Broker Beware, 221
From Charity to Enterprise, 573
From Chocolate to Morphine, 547
From Design to Finish, 134
From Line to Design: Design Graphics Communication, 460
From Milton to McLuhan: The Ideas Behind American Journalism, 240
From News to Newsprint: Producing a Student Newspaper, 252
From The Land, 479
From the Watching of Shadows, 27
From Top Hats to Baseball Caps, from Bustles to Blue Jeans, 423
From Wonder to Wisdom: Using Stories to Help Children Grow, 323
Frommer's Budget Travel Guide: South America on $40 a Day, 229
Frozen Food Book of Knowledge: A Foodservice Reference, 199
Fuchs's Radiographic Exposure, Processing and Quality Control, 24
Fuel and Oil Systems, 347
The Functions of Mass Communication, 255
Fundamental Immunology, 63
Fundamental Nursing Procedures, 89
Fundamental Nursing Skills, 79
Fundamental Skills and Concepts in Patient Care, 81
Fundamental Skills for the Clinical Laboratory Professional, 62
Fundamentals I: Introduction to Dental Terminology, Charting, and Procedures, 7
Fundamentals II: Infection Control, Local Anesthesia and Oral Surgery, 7
Fundamentals of CAD with CADKey for Engineering Graphics, 375
Fundamentals of Copy and Layout, 234
Fundamentals of Dental Hygiene Instrumentation, 12
Fundamentals of Electricity for Agriculture, 144
Fundamentals of Flight, 349
Fundamentals of Hazardous Materials Incidents, 92
Fundamentals of Immunohematology, 64
Fundamentals of Industrial Hygiene, 95
Fundamentals of Laser Optics, 378
Fundamentals of Laser Optoelectronics, 377
Fundamentals of Machine Component Design, 373
Fundamentals of Nonlinear Optics of Atomic Gases, 377
Fundamentals of Nursing, 88
Fundamentals of Nursing: Concepts, Process and Practice, 80

Fundamentals of Nursing: Human Health and Function, 79
Fundamentals of Obstetric and Gynecologic Ultrasound, 31
Fundamentals of Oral Histology and Embryology, 12
Fundamentals of Soil Science, 504
Fundamentals of Successful Newsletters, 236
Fundamentals of Urine and Body Fluid Analysis, 59
Fundamentals of Vehicle Dynamics, 353
The Funeral: An Endangered Tradition, 512
Funeral Directing Investigator, 512
The Funeral Director's Practice Management Handbook, 514
The Funeral Industry: Regulating the Disposition of the Dead, 515
Funeral Rites and Customs, 514
Funeral Service: A Historical Perspective, 514
Funerals Without God, 515
Furniture, Modern and Postmodern: Design and Technology, 438
Furniture for the Workplace, 433
Future Flight: The Next Generation of Aircraft Technology, 360
Future Harvest: Pesticide-Free Farming, 486
The Future Is Now: The Changing Face of Technical Services, 343
The Future of News, 241

Gait Analysis: Normal and Pathological Function, 110
Gale Directory of Publications and Broadcast Media, 250, 336
Games for Learning, 318
Garages: Complete Step-By-Step Building Plans, 133
Garden Primer, 519
The Garden Trees Handbook, 520
Gardeners Dictionary of Horticultural Terms, 518
The Gardener's Palette, 519
Gardening by Mail: A Source Book, 519
Gas Tungsten Arc Welding, 393
Gastrointestinal Disorders, 88
Gauged Brickwork: A Technical Manual, 158
The Gene Hunters: Biotechnology and the Scramble for Seeds, 481
General Aviation Law, 347
The General Method of Social Work Practice, 554
General Reference Books for Adults, 338
General Training Air Conditioning—Fundamentals, 154
Generative Models for Computer Graphics and CAD, 373
Genetics of Domestic Animals, 530
Genitourinary Disorders, 88
The Genius of Japanese Carpentry, 130
Geography of Travel and Tourism, 229
Geometrical Optics Experiments, 377
Geriatric Care Plans, 76
Geriatric Nursing, 75
Geriatric Nutrition and Diet Therapy, 75
Geriatric Physical Therapy: A Clinical Approach, 108
Gerontologic Nursing, 89
Gerontologic Nursing: Care of the Frail Elderly, 75
Gerontological Nursing, 75
Gerontological Social Work: An Annotated Bibliography, 562
Get Ready to Read, 316
Getting at the Source, 370
Getting High and Doing Time, 544, 574
Getting into Video: A Career Guide, 270
Getting It Printed, 398
Getting Started Drawing and Selling Cartoons, 253
Getting Started in Computer Graphics, 463
Getting the Most for Your Home in a Down Market, 219
Getting the Story, 247
Ghost and I: Scary Stories for Participatory Telling, 318
Giants of the Sky: The Biggest Aeroplanes of All Time, 347
Giorgio Armani: Images of Man, 422
The Girls in the Balcony: Women, Men and the New York Times, 255
Giving Drugs by Advanced Technique, 87

Title Index

Giving Your Child a Smile: Correcting Cleft Lip and Palate, 20
Glass Ionomer Dental Cement, 9
Global Forests: Issues for Six Billion People, 499
Global Primer: Skills for a Changing World, 313
Global Telecommunications: Layered Networks' Layered Services, 178
Glossaries for Court Reporters, 277
A Glossary for Archivists, Manuscript Curators and Records Managers, 330
Glossary of Healthcare Terms, 51
Glossary of Prosthodontic Terms, 4
Golden Hour: The Handbook of Advanced Pediatric Life Support, 45
Good Cholesterol, Bad Cholesterol, 38
Good Day Bad Day, 312
A Good House: Building a Life on the Land, 135
Goodman and Gilman's The Pharmacological Basis of Therapeutics, 539
Goodnight Toes!, 311
Gotcha!: The Art of the Billboard, 456
Govoni and Hayes Drugs and Nursing Implications, 85
Grain Grades and Standards, 482
Grammar for Court Reporters, 277
Granpa, 326
The Graphic Artist's Guide to Marketing and Self-Promotion, 401
Graphic Artists Guild Handbook, 473
Graphic Artists Guild Handbook: Pricing and Ethical Guidelines, 403
Graphic Arts Production, 470
Graphic Communications Today, 248, 469
Graphic Design: A Career Guide and Education Directory, 473
Graphic Design: Los Angeles, 459
Graphic Design, New York, 458
Graphic Design Basics, 458
Graphic Design Career Guide, 472
Graphic Design for Desktop Dummies, 260, 404
Graphic Design for the Electronic Age, 472
Graphic Design in America: A Visual Language History, 456
Graphic Design Made Difficult, 458
Graphic Design One: Demonstrating the Values of Good Design, 473
The Graphic Design Portfolio: How to Make a Good One, 473
Graphic Design School, 460
Graphic Design Two: Demonstrating the Values of Good Design, 473
The Graphic File Toolkit: Converting and Using Graphic Files, 407
Graphic Illustration in Black-And-White, 464
The Graphic Spirit of Japan, 460
Graphic Style: From Victorian to Post-Modern, 457
Graphic Thinking for Architects and Designers, 436
Graphical Communication Principles: A Prelude to CAD, 372
Graphically Speaking, 396, 456
Graphics for the Desktop Publisher, 406
Graphics Master 4, 237, 471
Graphics Master 5, 260, 405
The Gray Book: Designing in Black and White on Your Computer, 403
Grazing Management, 485
Grazing Management: An Ecological Perspective, 485
The Great American Comic Strip, 466
The Great Bear Almanac, 506
Great Editorials: Masterpieces of Opinion Writing, 235
Great Jobs for Psychology Majors, 552
Great Newspaper Graphics, 235
Great Package Design, 455
The Great Picture Hunt, 248
The Great Power-Line Cover-Up, 143
Great Print Advertising, 234
The Great Reporters, 235
Great Vacations with Your Kids, 229
The Greenhouse and Nursery Handbook, 520
Greenhouse Operation and Management, 520

The Gregg Reference Manual, 211, 295
Grief: The Mourning After; Dealing with Adult Bereavement, 515
The Grief Recovery Handbook, 512
Groundwater Chemicals Field Guide, 369
Groundwater Treatment Technology, 369
Group Care of Children: Transitions Toward the Year 2000, 558
Group Counseling with Juvenile Delinquents, 560
Group Work: Skills and Strategies for Effective Interventions, 536
Groups: Process and Practice, 553
Groups that Work: Structure and Process, 554
Growing Up Cavity Free, 11
Growth and Mineral Nutrition of Field Crops, 482
The Growth of Arthur Andersen and Company, 1928-1973, 171
Grzimek's Encyclopedia of Mammals, 506
Guardians of Yellowstone, 502
Guerrilla PR, 260
A Guide for Legal Assistants, 274, 298
A Guide for Newspaper Stringers, 246
The Guide to Cooking Schools, 1994, 195
Guide to ECG Analysis, 73
Guide to Hardware and Software for Desktop Publishing, 401
A Guide to Life: Jewish Laws and Customs of Mourning, 514
A Guide to Multimedia, 181
A Guide to Neurological and Neurosurgical Nursing, 77
Guide to Nursing Management, 71
Guide to Physical Examination and History Taking, 78
A Guide to Preparing a Restaurant Business Plan, 197
Guide to Reference Books, 337
Guide to Residential Carpentry, 133
Guide to Services of the National Agricultural Library, 477
Guide to Surviving Nursing School, 89
Guide to Technical Services Resources, 340
The Guide to Textiles for Interior Designers, 435
Guide to the 1993 National Electrical Code, 144
Guide to the Automobile Certification Examination, 354
Guide to the Library Binding Institute Standard, 341
Guide to U.S. Government Publications, 335
Guidebook to Successful Safety Programming, 93
Guidelines and Procedures for Obtaining ABA Approval, 298
Guidelines for Exercise Testing and Prescription, 104
Guidelines for Pulmonary Rehabilitation Programs, 114
Guidelines for the Well-Being of Rodents in Research, 526
Guiding Children's Social Development, 318
The Guild Handbook of Scientific Illustration, 468

HAACP Reference Book, 193
Hair Additions: The Fourth Dimension, 187
Haircutting for the Now Nineties, A Family Affair, 189
Haircutting with Clippers: Basic Techniques, 189
Hairdressing: Theory, Science, and Practice, 187
Hand Pain and Impairment, 105
The Hand Scanner Handbook, 400
Hand Tools, 124
Handbook for Assessing and Treating Addictive Disorders, 546
Handbook for Beginning Legal Assistants and Receptionists, 295
Handbook for Dental Hygienists, 6
The Handbook for Microcomputer Technicians, 176
Handbook for Proofreading, 249
Handbook for Public Playground Safety, 316
Handbook for the Hospital Medical Secretary, 282
Handbook of Advertising Art Production, 471
Handbook of Air Conditioning and Refrigeration, 152
Handbook of Architectural Acoustics and Noise Control, 438
A Handbook of Biological Illustration, 469
Handbook of Clinical Dietetics, 34
A Handbook of Costume Drawing, 419

Handbook of Differential Treatments for Addictions, 541
Handbook of Edible Weeds, 504
Handbook of Emergency Response and Toxic Chemical Releases, 369
Handbook of Environmental Fate and Exposure Data for Organic Chemicals, 94
Handbook of Environmental Management and Technology, 368
Handbook of Forensic Science, 288
Handbook of General and Modified Diets, 36
Handbook of HVAC Design, 152
Handbook of Joinery, 125
Handbook of Local Anesthesia, 10
Handbook of Loss Prevention and Crime Prevention, 290
Handbook of Mammography, 25
Handbook of Mechanical Ventilatory Support, 117
Handbook of Medical Surgical Nursing, 76
The Handbook of Nonsexist Writing, 251
Handbook of Nursing Diagnosis, 77
Handbook of Photography, 446
Handbook of Real Estate Terms, 214
Handbook of Respiratory Care, 114
Handbook of Social Work Practice with Vulnerable Populations, 554
Handbook of Solid-State Lasers, 377
Handbook of Stress: Theoretical and Clinical Aspects, 568
Handbook of Structural Welding, 391
Handbook of Wood and Wood-Based Materials for Engineers, Architects, and Builders, 129
Handgun Stopping Power: The Definitive Study, 289
Hands On! Skills Series, 46
Hands-On Math: Manipulative Math for Young Children, 324
Hardwood Floors, 135
Hate Crimes: Confronting Violence against Lesbians and Gay Men, 565
Hats: Status, Style, and Glamour, 422
The Haynes Small Engine Repair Manual, 381
Hazard Communication for the Dental Health Team, 18
Hazardous Chemicals Desk Reference, 94
Hazardous Materials Emergency Response: Pocket Handbook, 92
Hazardous Materials/Waste Handling for the Emergency Responder, 496
Hazardous Waste Chemistry, Toxicology, and Treatment, 369
Hazardous Waste Management Compliance Handbook, 368
Hazardous Waste Minimization, 368
Hazardous Waste Site Remediation: The Engineer's Perspective, 367
HazMat for First Responders, 494
Head and Neck Screening Examination Procedures for the Dental Hygienist, 18
Head Injury, 100, 108
Healing the Planet, 504
Healing Yourself, 89
Health and Safety in Small Industry, 94
Health and Safety of Workers, 92
Health Assessment: A Nursing Approach, 79
Health Assessment in Nursing, 80
Health Care and Gender, 566
Health Care Malpractice, 111
Health Care Management in Physical Therapy, 105
Health Communication: Strategies for Health Professionals, 72
Health in the Headlines, 242
Health Information Management, 53
Health Professional and Patient Interaction, 72
Health Risks and the Press, 243
Healthcare for Older Adults, 90
Hearing Conservation Programs, 95
Heat and Massage for the Lower Back, 112
Heat, Massage, and Exercise for the Neck and Upper Back, 112

Title Index

Heat, Massage and Quad Setting Exercise for the Knee, 112
Heating Systems Troubleshooting Handbook, 153
Heavy Drinking: The Myth of Alcoholism as a Disease, 538
Heavy-Duty Truck Fuel Systems, 363
Heavy-Duty Truck Power Trains, 363
Heavy-Duty Truck Suspension, Steering, and Braking Systems, 363
Help Yourself to Travel: Travel Planning Made Easy, 229
Helping Children Learn to Read, 314
Helping the Aging Family: A Guide for Professionals, 559
Hematology: A Combined Theoretical and Technical Approach, 64
Hematology: Clinical and Laboratory Practice, 59
Henderson's Dictionary of Biological Terms, 282
Henn on Copyright Law: A Practitioner's Guide, 334
Herbicide Effect on Plants, 488
Hey—We're Being Audited!, 172
Hey, We're In Business, 172
The Hidden Dimension of Illness: Human Suffering, 74
Hidden Stories in Plants, 321
High Heels and Ground Glass: Pioneering Women Photographers, 451
High Performance Management Strategies for Entrepreneurial Companies, 223
High Pressure Boilers, 153
High Tech Illustration, 468
High-Performance CAD Graphics in C, 372
High-Risk Patrol: Reducing the Danger to You, 288
High-Risk Pregnancy: A Team Approach, 82
High-Tech Ventures: The Guide for Entrepreneurial Success, 223
High/Scope Buyer's Guide to Children's Software 1993, 312
Hiring Legal Staff: Determining Cost and Value, 298
Hispanic Elderly in Transition, 564
Hispanics and Mental Health: A Framework for Research, 566
Historic Homes of America, 441
A History of Food, 199
A History of Furniture, 431
A History of Graphic Design, 405, 457
A History of Men's Fashion, 420
A History of News: From the Drum to the Satellite, 244
HIV/AIDS : A Guide to Nursing Care, 73
HIV/AIDS: Epidemiology for Primary Care Health Professionals, 18
HIV/AIDS: Testing and Risk Assessment, 18
Holography: An Instructional Lab Manual, 377
Home Electrical Wiring Made Easy, 149
Home Furnishings Merchandising and Store Design, 437
Home Heating and Air Conditioning Systems, 152
Home Improvement Tools and Equipment, 129
Home Ownership: The American Myth, 219
Home Plumbing Illustrated, 162
Home Plumbing Projects and Repairs, 161
The Home Remodelers' Combat Manual, 137
Home Respiratory Care, 116
Home Visiting: Procedures for Helping Families, 564
Home Workshop, 125
The Homebuyer's Kit, 219
Homecoming: Reclaiming and Championing Your Inner Child, 536
Homeless Not Helpless: Opening Doors, 575
The Homeowner's Guide to Building with Concrete, Brick, and Stone, 158
The Homeowner's Guide to Carpentry and Cabinetry, 132
Horror Comics: The Illustrated History, 465
Hose Practices, 494
Hospitable Design for Healthcare and Senior Communities, 431
Hospital Administration for Veterinary Staff, 529
Hospital Laboratory, 61
Hospitality and Recreation, 207
Hospitality and Travel Marketing, 197

Hospitality Index: An Index for Hotel, Foodservice, and Travel Industries, 202
Hospitality Industry, 207
Hospitality Industry Managerial Accounting, 198, 207
Hospitality Management Accounting, 192, 203
Hospitality Marketing Management, 198, 206
Hotel and Catering Accounts, 202
Hotel and Motel Management and Operations, 204
The Hotel and Restaurant Business, 205
Hotel and Restaurant Industries: An Information Sourcebook, 197
Hotel Catering: A Handbook for Sales and Operations, 207
Hotel Front Office Management, 202
Hotel Room with a View: Photographs by Bruce Weber, 447
The Hotel Sales and Marketing Planbook, 205
Hotel, Restaurant and Travel Law: A Preventive Approach, 192, 203
Hotel/Motel Front Desk Personnel, 206
Hotshot, 501
House, 134
House Wiring, 146
House Wiring Simplified, 145
Housebuilding: A Do-It-Yourself Guide, 133
Housekeeping Management in the Hospitality Industry, 203
Housing Interiors for the Disabled and Elderly, 438
How About a Career in Real Estate?, 217
How Computers Really Work, 181
How Multimedia Works, 462
How Therapists Diagnose, 539
How to Buy Advertising Like the Pros—and Save 15% to 50%, 259
How to Buy Foreclosed Real Estate for a Fraction of Its Value, 218
How to Check and Correct Color Proofs, 398
How to Create Animation, 466
How to Deal with Difficult Patrons, 343
How to Design Logos on Your Computer, 461
How to Design Trademarks and Logos, 456
How to Develop a Six-Figure Income in Real Estate, 216
How to Do Leaflets, Newsletters and Newspapers, 236
How to Do Your Own Accounting for a Small Business, 169
How to Draft Basic Patterns, 415
How to Draw Art for Comic Books, 467
How to Find a Job As a Paralegal: A Step-By-Step Job, 299
How to Find the Law, 280
How to Form Support Groups and Services for Grieving People, 515
How to Get a Job in Radio, 270
How to Get a Mortgage in Twenty-Four Hours, 219
How to Get the Best Home Loan, 218
How to Incorporate, 223
How to Land Your First Paralegal Job, 300
How to Lie With Maps, 251
How to List and Sell Real Estate in the 90s, 216
How to List Residential Real Estate Successfully, 217
How to Make a Living as a Travel Writer, 228
How to Make Money Growing Trees, 508
How to Make One Million Dollars in Real Estate in Three Years Starting with No Cash, 218
How to Make the Perfect Hotel Deal, 204
How to Manage a Successful Catering Business, 195
How to Master the Art of Listing and Selling Real Estate, 216
How to Negotiate a Lease on Your Terms, 220
How to Negotiate Real Estate Contracts, 216
How to Overcome Fear of Dentistry, 10
How to Photograph Buildings and Interiors, 447
How to Prepare Your Home for Sale...so It Sells, 221
How to Produce Creative Advertising, 260, 454
How to Produce Creative Publications, 236, 398
How to Read and Understand the Financial News, 253
How to Repair and Restore Bodywork, 354
How to Repair Diesel Engines, 364

How to Restore Your Farm Tractor, 481
How to Run a Small Business, 224
How to Sell Your Home in Five Days, 218
How to Sell Your Home When Homes Aren't Selling, 219
How to Sell Your House, Condo, Co-Op, 218
How to Sell Your Photographs and Illustrations, 464
How to Start and Run a Writing and Editing Business, 238
How to Start and Run Your Own Bed and Breakfast Inn, 204
How to Survive in a Law Firm: Client Relations, 276, 304
How to Teach Reading: For Teachers, Parents, Tutors, 315
How to Understand and Use Grids, 408
How to Use Ami Pro for Windows, Release 2 and 3, 407
How to Work a Room, 262
How to Write a Police Report, 290
How to Write and Illustrate Children's Books and Get Them Published!, 464
How to Write Fast (While Writing Well), 239
How To Write Irresistible Query Letters, 239
How We Invented the Airplane: An Illustrated History, 350
Human Anatomy and Physiology, 71
Human Behavior Theory and Social Work Practice, 568
Human Body, 65, 66
Human Diseases: A Systemic Approach, 49
Human Nervous System: Introduction and Review, 12
Human Relations for the Hospitality Industry, 205
Human Resource Management for the Hospitality Industry, 193, 199
Human Variation: Races, Types, and Ethnic Groups, 566
Hungry Hearts: On Men, Intimacy, Self-Esteem, and Addiction, 543
HVAC Controls and Systems, 153
HVAC Field Manual, 153
Hybrid Imagery: The Fusion of Technology and Graphic Design, 461

I CAD Can You?, 416
I Learn to Read and Write the Way I Learn to Talk, 311
IBM Dictionary of Computing, 174, 276
ICD-9-CM Basic Coding Handbook, 50
ICD-9-CM Coding Handbook for Physician Practices, 50
ICD-9-CM Coding Handbook, with Answers, 49
Ice Pack, 9
The Idealizing Vision: The Art of Fashion Photography, 447
Identification Guide to the Trees of Canada, 499
Identifying Wood: Accurate Results with Simple Tools, 508
If Wishes Were Horses: The Education of a Veterinarian, 528
Illustrated Changes in the 1993 National Electrical Code, 144
Illustrated Changes of the 1993 NEC, 147
Illustrated Computer Graphics Dictionary, 397
Illustrated Dictionary for Building Construction, 130
Illustrated Dictionary for Electrical Workers, 148
Illustrated Dictionary of Microcomputers, 174
Illustrated Encyclopedia of General Aviation, 347
The Illustrated Encyclopedia of Woodworking Handtools, Instruments, and Devices, 122
The Illustrated Guide to Aerodynamics, 349, 360
Illustrated Guide to the National Electrical Code, 1993, 148
The Illustrated Handbook of Desktop Publishing and Typesetting, 236
The Illustrated Real Estate Dictionary, 214
Illustrating for Science, 447, 468
Illustrating Science: Standards for Publication, 468
I'm Dysfunctional, You're Dysfunctional, 540
I'm Tired of a Messy Desk, 211
Image Ethics, 248
Imaging Principles of Cardiac Angiography, 29

Title Index

Immunology and Serology in Laboratory Medicine, 64
Imogen Cunningham: Ideas without End: A Life in Photographs, 449
Improving Access to Justice, 300
Improving America's Diet and Health, 36
Improving Safety in the Chemical Laboratory, 96
Improving Your Field Procedures, 386
IMZ Implant System, Part I: Clinical Aspects, 18
IMZ Implant System, Part II: Laboratory Procedure, 18
In My Room: Designing for and with Children, 440
In Our Own Image, 249, 450
In Print: Text and Type, 469
In the Romantic Style, 432
In the Shadow of the Poorhouse, 572
In-House Publishing in a Mainframe Environment, 405
The Independent Paralegal's Handbook, 300
Index to Legal Periodicals, 301
Indexing from A to Z, 342
Indonesian Textiles, 417
Indoor Action Games for Elementary Children, 314
Indoor Air Pollution Control, 93
Industrial Chemical Exposure, 94
Industrial Fire Protection Handbook, 495
Industrial Health, 95
Industrial Hygienist Trainee, 95
Industrial Lasers and Their Applications, 378
Infancy, 314
Infant/Toddler Environment Rating Scale, 316
Infants, Toddlers, and Caregivers, 316
Infection Control, 11
Infection Control Card File, 9
Infection Control for Lodging and Food Service Establishments, 203
Infection Control for the Dental Health Team, 18
Infection Control in the Dental Branch Operatory, 15
Infection Control in the Dental Environment, 17
Infection Control Made Cost Effective!, 16
Infectious Disease in Emergency Medicine, 42
Infectious Diseases, 88
Infectious Medical Waste: Safe Handling and Disposal, 66
Influence Without Authority, 209
Influences of Forest and Rangeland Management on Salmonid Fishes and Their Habitats, 504
Information Plus: The Information Series on Current Topics, 291
Information Systems Engineering: An Introduction, 182
INNovation: Creativity Techniques for Hospitality Managers, 202
Innovation and Change in the Human Services, 555
Innovative Approaches in the Treatment of Drug Abuse, 540
The Ins and Outs of Interfacings, 415
Insects That Feed on Trees and Shrubs, 521
Inside AutoCAD, 374
Inside Fashion Design, 424
Inside Generic CADD, 375
Inside Television Producing, 266
Insight into Optics, 377
Installation Requirements of the 1993 National Electrical Code, 147
Instant Multimedia for Windows 3.1, 178
Instrument Transfer for the Dental Assistant, 19
Integrated Curriculum, 314
Integrated Mathematics of Radiographic Exposure, 25
Integrated Pest Management in Agriculture, 489
Integrated Quality Management, 71
Integrative Group Therapy, 101
Intellectual Freedom Manual, 333
Intelligent Design and Manufacturing, 373
Intelligent LAN Management with Novell NetWare, 175
Interactions in the Classroom, 324
An Interactive Approach to Radiographic Anatomy and Positioning, 32
Interior Construction and Detailing for Designers and Architects, 430
Interior Design, 438
The Interior Design Business Handbook, 436
Interior Design of the Electronic Office, 436

Interior Design Reference Manual, 430
Interior Design with Feng Shui, 439
The Interior Designer's Drapery, Bedspread, and Canopy Sketchfile, 434
Interior Designers' Showcase of Color, 430
Interior Dimension: A Theoretical Approach to Enclosed Space, 436
Interior Finish Materials for Health Care Facilities, 440
Interior Landscape Design, 434
Interior Lighting: Bringing Rooms to Life, 442
Interior Presentation Sketching for Architects and Designers, 430
Interiors, 435
Interlibrary Loan Policies Directory, 336
Interlibrary Loan Trends, 336
International Agricultural Trade and Market Development Policy in the 1990s, 484
International Brand Packaging Awards, 455
International Classification of Diseases, 9th Revision, Clinical Modification, 50
The International Farm Crisis, 484
International Furniture Design for the 90s, 435
International Interior Design, 431
International Logotypes (vol. 1), 456
International Logotypes (vol. 2), 456
The International Menu Speller, 192
The International Typebook, 470
International Video Graphic Design, 454
International Women in Design, 457
Internet Basics, 332
The Internet Companion, 332
The Internet Complete Reference, 177, 332
Internet Connections, 332
The Internet for Dummies, 332
The Internet Guide for New Users, 176, 336
The Internet Navigator, 177
The Internet Passport, 179
Internet World's on Internet, 331
The Internet Yellow Pages, 177, 332
Internship Program Policies and Procedures, 288
Interpersonal Skills and Health Professional Issues, 58
Interpersonal Skills Training, 516
Interventional Computed Tomography, 29
Interviews That Work: A Practical Guide for Journalists, 238
Intimate Violence in Families, 561
Intra-Oral Radiographic Technique, 16
Intravenous Medications, 85
Intravenous Therapy, 90
Introducing Computers: Concepts, Systems and Applications, 176
Introducing the Internet: A Trainer's Workshop, 332
Introduction to Basic Concepts in Dental Radiography, 18
Introduction to Cataloging and Classification, 342
Introduction to Child Development, 314
An Introduction to Computer Graphics Concepts, 373, 463
Introduction to Computerized Accounting, 172
Introduction to Contracts for Paralegals, 305
Introduction to Critical Care Skills, 73
Introduction to Databases, 183
Introduction to Electronic Document Management Systems, 210
Introduction to Fall Protection, 93
Introduction to Fashion Design, 419
Introduction to Fashion Merchandising, 425
Introduction to Floriculture, 521
Introduction to Forest and Shade Tree Insects, 499
Introduction to Forest Science, 500
Introduction to Hospitality Management, 204
Introduction to Human Anatomy and Physiology, 71, 283
Introduction to Immunohematology, 59
Introduction to Law and the Legal System, 279
Introduction to Legal Assisting, 298
Introduction to Library Public Services, 337
Introduction to Library Services, 334
Introduction to Metal Ceramic Technology, 11

Introduction to Networking, 180
Introduction to Nonlinear Laser Spectroscopy, 378
Introduction to Nursing: Concepts, Issues and Opportunities, 77
Introduction to Nursing Informatics, 79
Introduction to Occupational Epidemiology, 94
An Introduction to Paralegal Studies, 278
Introduction to Paralegal Studies: A Skills Approach, 299
Introduction to Paralegalism, 300
Introduction to Personal Computers, 180
Introduction to Personal Computers: Self-Teaching Guide, 182
Introduction to Plant Diseases: Identification and Management, 487
Introduction to Radiography and Patient Care, 23
Introduction to Radiologic Technology, 25
An Introduction to Radiometry, Photometry, and Colorimetry, 377
Introduction to Reference Work, 337
Introduction to Street Drug Pharmacology, 549
Introduction to Technical Services, 340
Introduction to the Hospitality Industry, 198, 206
Introduction to the Physical Metallurgy of Welding, 391
Introduction to the Principles of Disease, 65
Introduction to the Psychology of Children's Drawings, 324
Introduction to Travel and Tourism: An International Approach, 227
Introduction to Ultrasonography and Patient Care, 29
Introduction to UNIX, 182
Introduction to Vascular Ultrasonography, 32
Introduction to VersaCAD Macintosh, 375
Inventing Reality: The Politics of News Media, 243
The Invertebrates: An Illustrated Glossary, 527
Investigating Science with Young Children, 311
Investigation of Fires, 495
Investigative and Operational Report Writing, 290
Investigative Reporting for Print and Broadcast, 246
Investing from the Heart, 218
Irish Castles, 207
Is It Painful to Think? Conversations with Arne Naess, 502
Island Press Bibliography of Environmental Literature, 369
Isokinetic Exercise and Assessment, 109
Issues and Ethics in the Helping Professions, 553
The Italian Renaissance Interior, 1400-1600, 440
It's a Child's World, 320
"It's Not Gonna Happen to Me": [Tractor Accidents], 489
IV Catheter Placement, 531
I.V. Solutions, 89
I.V. Therapy, 88

Jablonski's Dictionary of Dentistry, 4
Jane Addams and the Men of the Chicago School, 1892-1918, 570
Janes' All the World's Aircraft, 359
J. K. Lasser's Real Estate Investment Guide, 220
J. K. Lasser's Your Income Tax, 168
Jobs In Paradise, 230
Jocks and Nerds: Men's Style in the Twentieth Century, 418
Joinery: Methods of Fastening Wood, 136
Joining In, 319
Joint Structure and Function, 109
Joseph's Introductory Textile Science, 424
Journalism: A Guide to the Reference Literature, 236
Journalism: State of the Art, 245
Journalism and the Aging Population, 254
The Journalism of Outrage, 243
The Journalist's Moral Compass: Basic Principles, 245
Journalists of the United States, 250
Journalist's Road To Success: A Career and Scholarship Guide, 238
Journeying: Children Responding to Literature, 317
Juvenile Justice: Policies, Programs, and Services, 563

Title Index

Karl Imhoff's Handbook of Urban Drainage and Wastewater Disposal, 369
Karsh: The Art of the Portrait, 449
Kawabe's Complete Dentures, 9
Keep Your Outboard Motor Running, 381
Keeping Eden: A History of Gardening in America, 519
Keeping Families Together: The Homebuilders Model, 561
Keeping the Books, 170
Key Aspects of Recovery, 73
Key Issues in Neurological Physiotherapy, 104
Keyguide to Information Sources in CAD/CAM, 372
Keys to Mortgage Financing and Refinancing, 216
Kid Fitness, 313
Kids' Stuff Book of Math for the Primary Grades, 314
Kids Who Do, Kids Who Don't, 539
Kilim: The Complete Guide, 435
Kiplinger's Buying and Selling a Home, 219
Kirk's Fire Investigation, 494
Kitchen Cabinet Construction, 126
Kitchens and Bathrooms, 138
The Kitchenware Book, 194
Kitplane Construction, 350
Knee Ligament Rehabilitation, 106
Knock It Down, Break It Up, 139
Kohlman Evaluation of Living Skills: KELS, 101
Krause's Food, Nutrition, and Diet Therapy, 37, 75

Lab Safety, 66
Lab Safety: Handling Hazardous Chemicals, 66
Lab Safety: The Chemical Hygiene Plan, 66
Laboratory and Diagnostic Tests with Nursing Implications, 76
Laboratory Consultant, 63
Laboratory Exercises in Respiratory Care, 118
Laboratory Immunology and Serology, 59
Laboratory Mathematics: Medical and Biological Applications, 60
Laboratory Procedures for Veterinary Technicians, 529
Laboratory Quality Assurance System, 64
Laboratory Quality Management: QC [and] QA, 60
Laboratory Safety: Containing HIV and HBV Barriers for Your Protection, Laboratory Test Handbook, 62
Laboratory Tests and Diagnostic Procedures with Nursing Diagnosis, 76
Ladybirds: The Untold Story of Women Pilots in America, 359
LAN Primer: The Definitive Guide To Networking Fundamentals, 180
Land Conservation Through Public/Private Partnerships, 502
Land Measurement, 387
Land Stewardship in the Next Era of Conservation, 502
Land Survey Review Manual, 384
Land Surveying Computations, 384
Land Surveying Law, 385
Land Surveyor, 386
Land Surveyor Reference Manual, 384
Land Surveyor Trainee, 386
The Land That Feeds Us, 478
Landlording, 220
Landlording as a Second Income: The Survival Handbook, 220
The Landlord's Handbook, 220
Landscape Equipment Safety, 522
Landscape Linkages and Biodiversity, 502
Landscaping with Container Plants, 520
Language and Literacy in Early Childhood Education, 323
Language Development, 326
Language in the Judicial Process, 277
Language Lessons for the Curriculum, 311
Language Lessons for the Curriculum: Level 1, 325
The Language of Computer Publishing, 397
The Language of Medicine, 281
The Language of News: A Journalist's Pocket Reference, 249
The Language of Real Estate, 214
Large Animal Internal Medicine, 527

Large Quantity Recipes, 199
Larousse Gardening and Gardens, 518
Laser Chemical Processing for Microelectronics, 378
The Laser Guidebook, 378
Laser Heating of Metals, 378
The Laser in America, 1950-1970, 377
Laser Pioneers, 378
Laser Processing and Manufacturing, 377
Laser Welding, 377, 393
Lasers, 378
Lasers and Optical Engineering, 377
The Last Great Necessity: Cemeteries in American History, 515
The Last Rain Forests: A World Conservation Atlas, 503
Last Stand: Logging, Journalism, and the Case for Humility, 508
The Lathe Book: A Complete Guide for the Wood Craftsman, 123
Lathes and Turning Techniques, 124
Law and Ethics of the Veterinary Profession, 530
Law and Justice: An Introduction to the American Legal System, 278
Law and Legal Information Directory, 302
Law and Occupational Injury, Disease, and Death, 93
Law Enforcement Employment Guide, 289
Law, Liabilities, and Ethics for Medical Office Personnel, 52
The Law of Corporations, Partnerships, and Sole Proprietorships, 306
The Law of Public Communication, 246
The Law of Real Property, 306
Law Office Automation for Paralegals, Administrators, and Legal Secretaries, 275, 303
Law Office Management for Paralegals, 275, 303
Law Office Software: Attorney's Guide to Selection, 275
Law Office Transcription, 275
The Laws of Innkeepers: For Hotels, Motels, Restaurants, and Clubs, 199, 207
The Lawyer's Almanac, 1994, 301
Lawyer's Desk Book, 281
Layout, Design, and Typography for the Desktop Publisher, 408
Leadership and Management of Volunteer Programs, 334
Leadership in Human Services, 554
Leadership in Safety Management, 95
Learjets: The World's Executive Aircraft, 349
Learn Ami Pro In a Day: For Versions 2.0 and 3.0, 406
Learn CAD Now, 373
Learners with Disabilities, 322
Learning Desktop Publishing with PageMaker 4.0, 403
Learning from Accidents in Industry, 94
Learning Human Anatomy: A Laboratory Text and Workbook, 48
Learning Medical Technology: A Worktext, 51
Learning to Fly Helicopters, 360
Legal Aspects of Documenting Patient Care, 14
Legal Aspects of Health Care Administration, 53
Legal Assistant I, 299
Legal Assistant Trainee, 299
Legal Assistants: Are You Ready for the 90s, 299
Legal Assistant's Handbook, 280
The Legal Assistant's Handbook, 298
Legal Assistants, 1988, 306
Legal Assistants, 1989, 306
Legal Assistants, 1990, 306
Legal Clerk, 299
Legal, Ethical and Political Issues in Nursing, 78
Legal Ethics and Professional Responsibility, 281
Legal Handbook for Small Business, 224
Legal Medicine: Legal Dynamics of Medical Encounters, 41
The Legal Problem Solver for Foodservice Operators, 194
Legal Procedures and Terminology for, 301
Legal Procedures and Terminology for Court Reporters and Paralegals, 280

Legal Research in a Nutshell, 280, 301
Legal Research Made Easy, 307
Legal Secretarial Procedures, 295
Legal Secretary's Complete Handbook., 275
Legal Studies, to Wit, 281
Legal Writing for Paralegals, 276
Legal Writing—Getting It Right and Getting It Written, 277
Legendary Decorators of the Twentieth Century, 434
Lesbians and Gay Men: Chemical Dependency Treatment Issues, 547
Lesly's Handbook of Public Relations, 260
Lessons in Death and Life, 249
Let's Go: The Budget Guide to the USA and Canada, 229
Letterhead and Logo Design: Creating the Computer Image, 409
Lettering Arts, 467
Lettering for Architects and Designers, 468
Letterwork: Creative Letterforms in Graphic Design, 468
Levitt and Tapley's Technological Basis of Radiation Therapy, 27
Library and Information Center Management, 335
Library and Information Services for Handicapped Individuals, 339
Library Management and Technical Services, 340
The Library Manager's Guide to Automation, 331
A Library, Media, and Archival Preservation Glossary, 330
A Library, Media, and Archival Preservation Handbook, 330
Library of Congress Rule Interpretations for AACR2, 340
Library of Congress Subject Headings, 341
Library Paraprofessionals: A Bibliography, 334
Library Resources for the Blind and Physically Handicapped, 339
Library Safety and Security, 336
A Life in Photography, 449
Life Safety Code Handbook, 495
Life Threatening Emergencies in Dentistry, 6
Lifespan Development, 312
Light: Science and Magic, 445
Light, Color and Environment, 431
Lighting Efficiency Applications, 148
Lighting for Film and Electronic Cinematography, 268
Lighting for Video, 268
Lighting Secrets for the Professional Photographer, 445
Lights! Camera! Advertising!, 447
Limited Radiography, 23
The Limits of Social Policy, 571
Lippincott Manual of Nursing Practice, 80
Lippincott's Review Series, 88
The LIRT Library Handbook, 331
List of Available Publications of the US Department of Agriculture, 477
Listening In: Children Talk About Books (and Other Things), 320
Listening to Prozac, 541
Literacy and the Library, 339
Literature-Based Social Studies, 318
Literature for Children: A Short Introduction, 321
A Literature Guide to the Hospitality Industry, 206
Litigation Handbook for the Lawyer's Assistant, 294
Litigation Organization and Management for Paralegals, 276, 303
Litigation Paralegal, 304
Litigation, Pleadings, and Arbitration, 303
Live Animal Carcass Evaluation and Selection Manual, 528
Living By Design, 442
Living on 12 Volts with Ample Power, 147
Living the Life of a Pre-Schooler: A Day in the Life, 326
Living with OSHA, 516
Loader-Backhoe Safety: Part I Operator Safety, 364
Loader-Backhoe Safety: Part II Worker Safety, 364
Lobby for Your Library: Know What Works, 334

Title Index

Lobbying for Social Change, 556
Lobbying Tips for Social Workers, 575
Local Anesthesia, 18
Local Anesthesia in Dentistry, 9
The Lodging and Food Service Industry, 196, 205
Lodging and Restaurant Index, 202
Log Homes Made Easy, 133
Log Scaling and Timber Cruising, 506
Logos of Major World Corporations, 456
Logotypes and Letterforms, 456
Looking Closely, 319
Looking Good in Print, 237, 406, 471
Low Budget Video Bible, 267
Lumber Manufacturing, 508
Lung Sounds: A Practical Guide, 119

Mac Online! Making the Connection, 181
Machinery's Handbook, 152
Macho Medicine, 546
The Macintosh CAD/CAM Book, 372
Macs for Beginners, 176
Madeleine Vionnet, 420
Magazine and Feature Writing, 240
The Magazine Article: How to Think It, Plan It, Write It, 240
Magazine Article Writing, 239
Magazine Editing for Professionals, 238
The Magazine in America, 1741-1990, 244
Magazines Career Directory: A Practical, One-Stop Guide to Getting a Job in Newspaper Publishing, 238
The Magic Garment: Principles of Costume Design, 416
Magnetic Resonance Angiography, 28
Magnetic Resonance Imaging: Physical and Biological Principles, 28
Magnetic Resonance Imaging, 28
Magnetic Resonance Imaging of the Brain, 28
Mainline, 549
Mainstream Multimedia: Applying Multimedia in Business, 461
Maintenance, Disclosure, and Redisclosure of Health Information, 52
Make Your Own Handcrafted Doors and Windows, 132
Make Your Own Japanese Clothes, 417
The Makeover Book, 237
The Makeover Book: 101 Design Solutions for Desktop Publishing, 406
Making Accurate, Easy, Alginate Impressions, 17
Making Basic Plumbing Repairs, 163
Making Built-In Cabinets, 123
Making Desks and Bookcases: Techniques for Better Woodworking, 123
Making It in Video, 266
Making Local News, 242
Making Money in Films and Video, 264
Making News, 242
Making Space: Remodeling for More Living Area, 137
Making Television Programs: A Professional Approach, 264
Making the Record, 277
Mammographic Imaging: A Practical Guide, 23
Mammography: Pretest Self-Assessment and Review, 25
Mammography for Radiologic Technologists, 27
Man with No Name Turns Lemons into Lemonade, 258
Management, 225
Management by Menu, 195
Management Decision Making for Nurses: 118 Case Studies, 71
Management in the Fire Service, 492
Management of Food and Beverage Operations, 197
Management of HIV Infection in Infants and Children, 84
The Management of Maintenance and Engineering Systems in the Hospitality Industry, 203
Management of Pain and Anxiety in Dental Practice, 7
The Management of People in Hotels and Restaurants, 196, 205
Management Principles for Physical Therapists, 109

The Manager's Guide to Desktop Electronic Publishing, 405
Managing Bar and Beverage Operations, 196
Managing CAD/CAM, 373
Managing Computer Viruses, 180
Managing Computers in Health Care: A Guide for Professionals, 54
Managing Computers in the Hospitality Industry, 195
Managing Controls, 146
Managing Conventions and Group Business, 204
Managing Fire Services, 493
Managing for Productivity in the Hospitality Industry, 206
Managing for Profit in Commercial Agriculture, 484
Managing Front Office Operations, 204
Managing Hospitality Human Resources, 200
Managing Hotels Effectively: Lessons from Outstanding General Managers, 206
Managing Library Outreach Programs, 339
Managing Media Services: Theory and Practice, 336
Managing Motors, 146
Managing Pilot Stress, 361
Managing Quality Child Care Centers, 322
Managing Technical Services in the 90s, 342
Managing the Experience of Labor and Delivery, 89
Managing the Small Law Enforcement Agency, 287
Managing Today's Public Library, 336
Managing Your Business: Milady's Guide to the Salon, 187
Managing Your Time, 211
Mancuso's Small Business Resource Guide, 224
Manheimer's Cataloging and Classification: A Workbook, 342
Manual for Functional Training, 108
Manual for Physical Agents, 107
Manual for the Lawyer's Assistant, 295
Manual of Critical Care, 74
A Manual of Laboratory and Diagnostic Tests, 76
Manual of Pediatric Nursing Procedures, 84
Manual of Prehospital Emergency Medicine, 44
Manual of Psychiatric Nursing Care Plans, 86
Manual of Pulmonary Function Testing, 118
A Manual of the Principal Instruments Used in American Engineering and Surveying, 385
Manual of Vascular Plants of Northeastern US and Adjacent Canada, 505
A Manual on Investigating Child Custody Reports, 562
Manual Therapy for Chronic Headache, 106
MARC for Library Use, 340
Maria: Photographs by Lee Friedlander, 447
Marianne Carroll's Super Desktop Documents with Microsoft Word 5.0 for the Macintosh, 400
Marihuana: The Forbidden Medicine, 539
Marketing Accounting Services, 169
Marketing and Selling the Travel Product, 227
Marketing Booze to Blacks, 549
Marketing Hospitality, 206
Marketing in a Global Economy, 489
Marketing in the Hospitality Industry, 206
Marketing Leadership in Hospitality, 205
Marketing Myths That Are Killing Business, 259
Marketing Strategy for Small Business, 225
Marketing Tourism Destinations: A Strategic Planning Approach, 229
Martindale-Hubbell International Law Directory, 301
Mary Ellen Mark, 25 Years, 448
MASA: Medical Acronyms, Symbols, and Abbreviations, 282
Masonry: How to Care for Old and Historic Brick and Stone, 158
The Mason's Toolbox Manual, 156
Mass Media: Opposing Viewpoints, 241
Mass Media Bibliography, 236
Mass Media Writing, 247
Master Dictionary of Food and Wine, 198
Master Guide for Passing the Respiratory Care Credentialing Exams, 116
Master Photographs: Master Photographs from PFA Exhibitions, 449

Mastering Advanced Assessment, 87
Mastering Real Estate Mathematics, 215
Mastering Television Technology: A Cure for the Common Video, 266
Material Wealth: Living with Luxurious Fabrics, 436
Materials and Components of Interior Design, 439
Maternal and Child Health Nursing, 83
Maternal and Neonatal Nursing: Family-Care, 83
Maternal Drug Abuse and Drug Exposed Children, 542
Maternal, Fetal, and Neonatal Physiology, 82
Maternal-Infant Nursing Care, 82
Maternal, Neonatal, and Women's Health Nursing, 82
Maternal-Neonatal Nursing, 89
Maternal-Newborn Nursing, 88
Maternal Newborn Nursing: A Family-Centered Approach, 83
Maternal Substance Abuse and the Developing Nervous System, 548
Maternity and Gynecologic Care, 82
Maternity Nursing: Family, Newborn and Women's Health Care, 83
Math and Dosage Calculations for Health Occupations, 60
Math for Nurses: A Problem Solving Approach, 85
Math for Welders, 391
Math Principles for Food Service Operations, 194
Math Workbook for Foodservice/Lodging, 196
Mathematics Children Use and Understand, 314
Mathematics for the Heating, Ventilating, and Cooling Trades, 152
Mathematics, the Gas Laws and the Respiratory Practitioner, 117
Matters of Light and Depth, 446
Maurice Sendak Library, 327
Mayo Clinic Diet Manual, 37
McBroom's Camera Bluebook, 444
McCracken's Removable Partial Prosthodontics, 11
The McGraw-Hill Thirty-Six Hour Real Estate Investing Course, 215
McGraw-Hill's Compound Interest and Annuity Tables, 214
McGraw-Hill's Interest Amortization Tables, 214
Means Concrete and Masonry Cost Data, 156
Means Electrical Change Order Cost Data, 145
Means Electrical Cost Data, 145
Means Mechanical Cost Data, 1994, 151
Means Plumbing Cost Data, 162
Means Plumbing Estimating, 161
The Measure of Man and Woman: Human Factors in Design, 440
The Measurement and Evaluation of Library Services, 333
Measurement in Direct Social Work Practice, 569
The Meat Buyers Guide, 197
The Mechanical Design Process, 374
Mechanical Design Using CADD, 374
Mechanical Drawing: CAD-Communications, 372
Mechanical Low Back Pain, 110
Mechanical Ventilation, 120
Mechanical Ventilation: Physiological and Clinical Applications, 117
Mechanics of Small Engines, 380
Media Access and Organization, 340
The Media and Disasters: Pan Am 103, 241
Media and the Environment, 242
Media Circus, 242
Media, Crime, and Criminal Justice: Images and Realties, 244
Media Polls in American Politics, 242, 260
Media Portrayals of Terrorism, 243
Mediating the Message, 244
Medical Acronyms and Abbreviations, 282
Medical Assistant, 56
Medical Assistant: 800 Multiple-Choice Questions with Explanatory Answers, 55
The Medical Assistant: Administrative and Clinical, 49
The Medical Assistant Examination Guide, 55
Medical Assisting, 66

Title Index

Medical Assisting: Administrative and Clinical Competencies, 48, 54,
Medical Assisting: Clinical Competencies, 49
Medical Device Packaging Handbook, 63
Medical Emergencies in the Dental Office, 10
Medical Imaging Physics, 25
Medical Laboratory Technician, 63
Medical Law, Ethics, and Bioethics in the Medical Office, 52
Medical Microbiology, 59
Medical Office Procedures, 53
Medical Parasitology, 62
Medical Record, 55
Medical Records and the Law, 53
Medical Records Assistant, 56
Medical Records Clerk, 56
Medical Records Technician, 56
Medical, Surgical and Anesthetic Nursing For Veterinary Technicians, 529
Medical-Surgical Nursing, 88
Medical-Surgical Nursing, 89
Medical-Surgical Nursing: A Nursing Process Approach, 76
Medical-Surgical Nursing: Assessment and Management of Clinical Problems, 76
Medical-Surgical Nursing Care Plans, 76
Medical Technologists and Technicians Career Directory, 63
Medical Technology Examination Review and Study Guide, 60
Medical Terminology: A Programmed Text, 52
Medical Terminology: A Self-Learning Text, 51
Medical Terminology: A Short Course, 51, 60
Medical Terminology: An Illustrated Guide, 51, 70
Medical Terminology: Building Blocks for Health Careers, 51
Medical Terminology: Principles and Practices, 283
Medical Terminology for Health Professionals, 51
Medical Terminology in Action, 52
Medical Terminology Made Easy, 51
Medical Terminology with Human Anatomy, 52
Medical Testing: The Complete Resource Guide to Good Laboratory Practice and HFCA Compliance, 59
Medical Transcription: Fundamentals and Practice, 53
Medical Transcription Guide: Do's and Don'ts, 282
Medical Typing and Transcribing: Techniques and Procedures, 282
Medical Waste Handling, 66
Medication Administration: Module I, 90
Medication Errors, 89
Medications, 88
Medicine for the Outdoors, 41
Medicolegal Issues for Physical Therapists, 105
Medicolegal Issues for Radiographers, 25
The Medium Is the Massage, 260
Melloni's Illustrated Medical Dictionary, 51, 70
Men and Women: Dressing the Part, 418
Men's Hairstyling and Beard Design, 189
Menswear: Suiting the Customer, 414
Mental Health and Mental Illness, 86
Mental Health and Psychiatric Nursing, 89
Mental Health and Psychiatric Nursing, 88
Mental Health Psychiatric Nursing, 88
Menu Design, Merchandising, and Marketing, 198
Merck Index, 282
Merck Manual of Diagnosis and Therapy, 282
Merriam-Webster's Secretarial Handbook, 210
Merrill's Atlas of Radiographic Positions and Radiologic Procedures, 23
Metallurgy of Welding, 391
Metric Practice Guide for the Welding Industry, 391
Mexican Patterns: A Design Source Book, 417
Michael Langford's 35mm Handbook, 445
Microbiology for Health Care Careers, 61
Microbiology for the Health Sciences, 76
Microbiology for Veterinary Technicians, 528
Microcomputer Applications, 177
Microcomputers on the Farm: Getting Started, 478
The Micrografx Designer Companion, 398

Microsoft C: Secrets, Shortcuts and Solutions, 178
Microsoft Publisher by Design, Version 2, 408
Microsoft Word for Windows: The Complete Reference, 176
Midwest Wildflowers: Wildflowers of Woodlands and Wetlands, 522
Milady's Art and Science of Nail Technology, 188
Milady's Color Crazy, 187
Milady's Guide to Owning and Operating a Nail Salon, 189
Milady's Makeup Techniques, 189
Milady's Salon Receptionist Training, 189
Milady's Standard Textbook of Cosmetology, 188
Milady's Standard Textbook of Professional Barber-Styling, 188
The Military and the Media, 242
Miller-Keane Encyclopedia and Dictionary of Medicine, Nursing, and Allied Health, 52
Millwork Handbook, 133, 435
Mind's Eye, Mind's Truth: FSA Photography Reconsidered, 450
Minor League: Photographs by Andrea Modica, 447
Minor White: The Eye That Shapes, 448
Minority Children and Adolescents in Therapy, 565
Mitchell Automechanics, 354
Mixing Glass Ionomer Cements, 16
Mixture Formation in Spark-Ignition Engines, 354
Mobility Series, 90
Model for Oral Hygiene in Long Term Care Facilities, 19
The Modeling of Design Ideas, 373
Modern Accident Investigation and Analysis, 93
Modern Automotive Mechanics, 353
Modern Dental Assisting, 14
Modern Dictionary for the Legal Profession, 281
Modern Drafting: An Introduction to CAD, 372
Modern Electronic and Electrical Drafting with Computers, 372
Modern Encyclopedia of Typefaces, 1960-90, 472
Modern Furniture Classics Since 1945, 433
Modern Furniture Projects, 124
Modern Image Processing, 409
Modern Masonry: Brick, Block, Stone, 157
Modern Newspaper Editing, 237
Modern Office Procedures, 209
Modern Optical Engineering, 378
Modern Plumbing, 160
Modern Radio Production, 270
Modern Real Estate Practice, 215
Modern Real Estate Practice Study Guide, 215
Modern Residential Financing Methods: Tools of the Trade, 216
Modern Residential Wiring, 144
Modern Tort Liability: Recovery in the '90s, 279
Modern Welding, 392
Modern Welding Technology, 392
Modern Wood Finishing Techniques, 125
Modern Woodworking Workbook: Tools, Materials and Processes, 126
Modular House Design, 134
Modules for Basic Nursing Skills, 79
Momentum Accounting and Triple-Entry Bookkeeping, 168
Monitoring in Respiratory Care, 116
The Moral Construction of Poverty, 571
More Power Sewing, 413
Mort Schultz's Electronic Fuel Injection Repair Manual, 355
Mortgage Payment Handbook, 214
Mortuary Law, 515
Mortuary Science: A Sourcebook, 515
Mosby Comprehensive Review of Nursing, 87
Mosby-Yearbook's Guide to Physical Examination, 80
Mosby-Yearbook's Pharmacology in Nursing, 85
Mosby's Clinical Nursing, 80
Mosby's Clinical Nursing Series, 88
Mosby's Comprehensive Review of Dental Hygiene, 6
Mosby's Diagnostic and Laboratory Test Reference, 63
Mosby's Emergency Dictionary, 45

Mosby's Fundamentals of Medical Assisting, 54, 65
Mosby's Medical, Nursing and Allied Health Dictionary, 51, 69
Mosby's Nursing Drug Reference, 85
Mosby's Paramedic Study Guide, 42
Moss' Radiation Oncology: Rationale, Technique, Results, 27
Mostly Windows with Just Enough DOS, 182
Motivational Interviewing, 543
Motor Auto Body Repair, 355
Motor Control and Physical Therapy, 109
Motorcycle Basics Manual, 381
Mount Sinai Medical Center Family Guide to Dental Health, 10
Movement Activities for Early Childhood, 316
Moving Image Materials: Genre Terms, 342
MRI and CT of the Spine, 28
MRI in Practice, 28
MRI of the Musculoskeletal System, 28
Muckraking and Objectivity: Journalism's Colliding Traditions, 243
Mudworks: Creative Clay, Dough, and Modeling Experiences, 318
Multicultural Explorations: Joyous Journeys with Books, 317
Multicultural Folktales: Stories to Tell Young Children, 322
Multicultural Issues in Child Care, 316
Multiethnic Children's Literature, 321
Multilingual Dictionary of Electricity, Electronics and Telecommunications, 146
Multimedia: Making It Work, 463
Multimedia and CD-ROMs for Dummies, 463
Multimedia Authoring: Building and Developing Documents, 461
Multimedia Illustrated, 461
Multimedia Madness, 464
Multimedia Mania, 461
Multimedia Power Tools, 462
Multimedia Solutions, 264
Multiply Your Success with Real Estate Assistants, 217
Murphy's Laws of DOS, 182
Muscles, Testing and Function: With Posture and Pain, 108
Musculoskeletal Approach to Maxillofacial Pain, 110
My Child Has a Disability, 327
Myofascial Release Manual, 108
Mystical Diets, 38

Nail Art and Design, 186
The Nail File, 188
NALA Manual for Legal Assistants, 295
Nantucket Style, 436
National Arboretum Book of Outstanding Garden Plants, 519
National Association of Legal Assistants Manual for Legal Assistants, 301
National Board Examination Study Guide for Applicants for Certification in the Funeral Service Profession, 511
National Construction Estimator, 140
National Dental Assistant Boards (NDAB), 13
National Dental Hygiene Boards (NDHB), 13
National Directory of Law Enforcement Administrators, 286
National Electrical Code: 1993, 146
National Electrical Code Handbook, 143
National Electrical Code Illustrated Changes Deskbook, 143
National Electrical Estimator, 146
National Electrical Safety Code, 142
National Park Guide, 228
The National Parks Camping Guide, 231
National Plumbing and HVAC Estimator, 162
National Plumbing Codes Handbook, 162
National Repair and Remodeling Estimator, 140
National Restaurant Association's Current Issues Reports, 197
National Survey of State Laws, 301

Title Index

The National Wildflower Research Center's Wildflower Handbook, 519
Natives and Strangers, 565
Natural and Prescribed Fire in Pacific Northwest Forests, 501
Nature Tourism: Managing for the Environment, 232
Nature's Chaos, 448
NCA Review for Clinical Laboratory Sciences, 65
NCLEX-RN Exam: Easy Steps to Passing, 87
Near Miss Reporting as a Safety Tool, 95
Needlestick Prevention: Five Steps to Safety, 66
Negotiating International Hotel Chain Management Agreements, 206
Neighbor Law: Fences, Trees, Boundaries and Noise, 219
Neil Sperry's Complete Guide to Texas Gardening, 520
Neonatal and Pediatric Respiratory Care, 116
Neonatal/Pediatric Respiratory Care, 117
Network Therapy for Alcohol and Drug Abuse, 538
Networking and Communications Desk Reference, 175
Neurologic Disorders, 88
Neurological Rehabilitation, 111
New American Home, 442
The New Comics Anthology, 465
New Complete Do-It-Yourself Manual, 138
The New Complete Guide to Environmental Careers, 501
New Dental Assistant: Impressions and Molds, 18
New Hacker's Dictionary, 175
New Life for Old Houses, 139
The New Owner, 225
The New Police Report Manual, 290
The New Politics of Poverty: The Nonworking Poor in America, 572
The New Politics of Welfare: An Agenda for the 1990s?, 572
The New Precision Journalism, 247
The New Professional Chef, 193
The New Range Wars, 489
New Read-Aloud Handbook, 324
New Rudman's Questions and Answers on the OCE, Occupational Competency Examination in Auto Body Repair, 355
New Rudman's Questions and Answers on the OCE, Occupational Competency Examination in Auto Mechanics, 355
New Rudman's Questions and Answers on the OCE, Occupational Competency Examination in Cosmetology, 188
New Rudman's Questions and Answers on the OCE, Occupational Competency Examination in Quantity Food Preparation, 38
New Rudman's Questions and Answers on the OCE, Occupational Competency Examination in Small Engine Repair, 381
New Television Technologies, 264
The New Textiles: Trends and Traditions, 423, 432
New Tree Biology, 500
The New Video Encyclopedia, 264
The New Vision: Photography Between the World Wars, 448
The New Well-Tempered Sentence, 250
New York Fashion: The Evolution of American Style, 422
News Reporters and News Sources, 244
News Reporting and Writing, 246, 247
The News Shapers: The Sources Who Explain the News, 244
Newsletters from the Desktop, 407
The Newspaper Designer's Handbook, 248, 460
Newspaper Layout and Design, 248
Newspaper Wars, 255
Newspapers Career Directory, 238
Newspapers, Diversity, and You, 254
NFDA Satellite Seminar 1994 Mandatory OSHA Staff Training, 516
NFPA 1500 Handbook, 496
NFPA Inspection Manual, 495
Night and Low-Light Photography, 445

Nine Pioneers in American Graphic Design, 457
Nineteenth-Century Decoration: The Art of the Interior, 433
Ninety Years of Fashion, 421
NIOSH Bookshelf, 369
NIOSH Pocket Guide to Chemical Hazards, 369
No-Fail Art Projects, 313
The Non-Lawyers A-B Trust Kit, 218
Normal and Therapeutic Nutrition, 38
Normal Infant Reflexes and Development, 102
North American Horticulture: A Reference Guide, 518
North American Range Plants, 485
Notes in the Catalog Record Based on AACR2, 342
Nothing Down for the 90s, 218
Noxious Range Weeds, 485
NSNA, NCLEX-RN Review, 87
NTC's Mass Media Dictionary, 250
The Nurse as Healer, 82, 89
Nurse-Client Interaction: Implementing the Nursing Process, 72
NurseReview, 88
Nurse's Drug Handbook, 85
Nurse's Guide to Successful Test-Taking, 87
Nurses' Handbook of Health Assessment, 81
Nurse's Manual of Laboratory Tests, 60
Nurse's Photolibrary, 88
Nurse's Ready Reference, 88
Nursetest, 88
Nursing: A Human Needs Approach, 79
Nursing Assessment and Health Promotion, 86
Nursing Bottle Caries Simulation, 16
Nursing Care in Radiation Oncology, 83
Nursing Care Management: From Concept to Evaluation, 70
Nursing Care of Elderly Patients with Acute Cardiac Disorders, 89
Nursing Care of Infants and Children, 84
Nursing Care of the Burn-Injured Patient, 74
Nursing Care of the Cancer Patient with Compromised Immunity, 90
Nursing Care of the Critically Ill Child, 84
Nursing Care of the Elderly Patient with Chronic Obstructive Pulmonary Disease, 89
Nursing Care of the Immunocompromised Patient, 74
Nursing Care of the Older Adult, 76
Nursing Care Plans: Nursing Diagnosis and Intervention, 79
Nursing Care Plans for Newborns and Children, 84
Nursing Data Review 1993, 70
Nursing Diagnoses and Process in Psychiatric Mental Health Nursing, 86
Nursing Diagnosis: Application to Clinical Practice, 78
Nursing Diagnosis: Process and Application, 78
Nursing Diagnosis and Intervention: Planning for Patient Care, 78
Nursing Diagnosis Care Plans for Diagnosis-Related Groups, 78
Nursing Diagnosis Handbook: A Guide to Planning Care, 77
Nursing Diagnosis in Clinical Practice, 78
Nursing Diagnosis Reference Manual, 78
Nursing Documentation: A Nursing Process Approach, 79
Nursing Documentation: Charting, Recording and Reporting, 79
Nursing Fundamentals, 89
Nursing Implications of Laboratory Tests, 63
Nursing in Today's World: Challenges, Issues and Trends, 77
Nursing Interventions: Essential Nursing Treatments, 77
Nursing Pharmacology, 89
Nursing Procedures, 80
Nursing Student's Guide to Drugs, 85
Nursing the Critically Ill Adult, 73
Nursing Theories: The Base for Professional Nursing Practice, 82
Nursing Theories and Models, 82
Nursing Theorists and Their Work, 82
Nursing Theory: Analysis, Application, Evaluation, 81

Nursing Today: Transition and Trends, 77
Nutripoints, 39
Nutrition, 39
Nutrition: Concepts and Controversies, 36
Nutrition Almanac, 35
Nutrition and Diagnosis-Related Care, 35, 75
Nutrition and Diet Therapy, 75
Nutrition and Metabolism in Patient Care, 37
Nutrition and Respiratory Care, 39
The Nutrition Desk Reference, 36
Nutrition Essentials and Diet Therapy, 37
Nutrition for the Foodservice Professional, 35, 193
Nutrition for Women, 38
Nutrition Handbook for Nursing Practice, 35, 74
Nutrition in Clinical Dentistry, 12
Nutrition in Critical Care, 75
Nutrition in Infancy and Childhood, 75
Nutrition in Pregnancy and Lactation, 75
Nutrition Labeling: Issues and Directions for the 1990s, 36
Nutrition Throughout the Life Cycle, 39
Nutrition, Toxicity, and Cancer, 38
Nutritional Care of the Terminally Ill, 36
Nutritive Value of Foods, 194

Observing Development of the Young Child, 312
Obstetric and Gynecologic Care in Physical Therapy, 109
Obstetric Ultrasound: How, Why and When, 29
Obstetric Ultrasound One, 31
Occlusal Splints, 17
Occupational Biomechanics, 92
Occupational Hazards to Dental Staff, 14
Occupational Health and Safety Terms, Definitions and Abbreviations, 93
Occupational Safety and Health Law, 92
Occupational Safety and Health Management, 92
Occupational Therapy: Overcoming Human Performance Deficits, 99
Occupational Therapy: Practice Skills for Physical Dysfunction, 101
Occupational Therapy and the Older Adult: A Clinical Manual, 100
Occupational Therapy Assistant: A Primer, 100
Occupational Therapy Assistant Career Profile, 99
Occupational Therapy for Children, 101
Occupational Therapy for Physical Dysfunction, 111
Occupational Therapy Protocol Management in Adult Physical Dysfunction, 99
Occupational Therapy Roles and Career Exploration and Development, 101
Occupational Treatment Goals, 98
Of Media and People, 241
Office Automation, 211
Office Design, 431
Office Laboratory, 58
Office on the Go, 275
Office Smarts: 222 Tips for Success in the Workplace, 333
Officers at Risk: How to Identify and Cope with Stress, 289
The Official Guide to the American Marketplace, 258
The Official New Print Shop Handbook, 407
The Official World Wildlife Fund Guide to Endangered Species of North America, 506
Ogallala: Water for a Dry Land, 486
Old Furniture: Understanding the Craftsman's Art, 439
Old-House Dictionary, 130
The Olympic Rain Forest: An Ecological Web, 503
On Assignment: Photographs by Jay Maise, 447
On Base! The Step-By-Step Self-Esteem Program, 315
On Nursing: A Literary Celebration: An Anthology, 81
On Our Own Terms, 105
On the Art of Fixing a Shadow, 448
On the Frontlines, 337
On Writing Well: An Informal Guide to Writing Nonfiction, 240
Oncologic Nursing, 89
Oncology Nursing, 83

Title Index

One Hundred and One Desktop Publishing and Graphics Programs, 470
One Word, Two Words, Hyphenated?, 277
Online Business Computer Applications, 177
Online Information Hunting, 332
Operating Techniques for the Tractor-Loader-Backhoe, 364
Operation of the Offset Press, 405
Operator's Manual for Oxyfuel Gas Cutting, 393
Opportunities in Beauty Culture Careers, 187
Opportunities in Commercial Art and Graphic Design Careers, 473
Opportunities in Dental Care Careers, 9
Opportunities in Desktop Publishing Careers, 407
Opportunities in Fashion Careers, 424
Opportunities in Fire Protection Services, 493
Opportunities in Forestry Careers, 501
Opportunities in Hotel and Motel Careers, 204
Opportunities in Journalism Careers, 238
Opportunities in Printing Careers, 399
Opportunities in Real Estate Careers, 217
Opportunities in Welding Careers, 392
Opposing Viewpoints Series, 291
Optical Electronics, 379
Optimizing Radiographic Positioning, 24
Optoelectronics: An Introduction, 379
The Opulent Era: Fashions of Worth, Doucet, and Pingat, 420
Oral Hygiene Procedures for Bedridden Patients, 19
Oral Pathology, 5
Oral Pathology: Clinical-Pathologic Correlations, 13
Oral Pathology for the Dental Hygienist, 9
Oral Report, 14
Oral Self Care: Strategies for Preventive Dentistry, 15
Oran's Dictionary of the Law, 281
The Organic Gardeners Handbook of Natural Insect and Disease Control, 521
Organized to Be the Best!, 211
Organizing: A Guide for Grassroots Leaders, 556
Origins and Rise of Associate Degree Nursing Education, 81
The Origins of Photojournalism in America, 450
Ornamental Horticulture as a Vocation, 522
Ornamental Plant Diseases, 522
Orofacial Pains, 5
Orthopaedic Nursing, 83
Orthopaedic Physical Therapy, 106
Orthopedic Disorders, 88
OSHA Compliance Handbook, 95
OSHA's Bloodborne Pathogens Standard, 66
Osseointegrated Dental Technology, 15
OT GOALS, 98
An Ounce of Prevention, 536
Our Right to Drugs: The Case for a Free Market, 546
Out of the Earth: Civilization and the Life of the Soil, 486
Outdoor Building Projects, 135
Outside the IBM PC and PS/2: Access to New Technology, 180
Outstanding Special Effects Photography on a Limited Budget, 446
Overexposure: Health Hazards in Photography, 445
The Oxford Dictionary of American Legal Quotations, 302
Oxyacetylene Welding: Basic Fundamentals, 393

Packaging Design: Graphics, Materials, Technology, 455
Packaging Design 4: PDC Gold Awards, 455
Page Design with QuarkXPress 3.2 for the Mac, 402
Page Maker 5: Expert Techniques for Macintosh, 406
PageMaker 5 by Example, 409
PageMaker 5.0 for Macintosh: Techniques and Applications, 401
PageMaker 5.0 for the Mac Page Design, 408
PageMaker 5.0 for Windows: Visual Quickstart Guide, 409
PageMaker for the IBM PC, 401
Pain Control, 112

Paint and Body Handbook, 355
Painting on the PC, 462
Palestinian Costume, 417
Panati's Extraordinary Endings of Practically Everything and Everybody, 514
Panoramas of the Far East: Photographs by Lois Conner, 447
The Pantone Book of Color, 458
Papers for Printing, 398
Paralegal, 298
Paralegal Aide, 299
Paralegal Career Guide, 298
Paralegal Discovery, 305
Paralegal Discovery: Procedures and Forms, 305
Paralegal Drafting Guide, 305
Paralegal Employment, 300
Paralegal Ethics and Regulation, 304
Paralegal Management Handbook, 275, 303
Paralegal Medical Records Review, 306
Paralegal Procedures and Practices, 278
The Paralegal Resource Manual, 299
Paralegal Trial Handbook, 303
The Paralegal's Desk Reference, 300
The Paralegal's Guide to U.S. Government Jobs, 299
Paralegal's Litigation Handbook, 303
The Paralegal's Role at Trial, 303
Paramedic, 45
Paramedic Examination Review Manual, 43
Paramedic National Standards Review Self Test, 44
Paramedic Pocket Reference, 42
Paramedic Review Guide, 45
The Paramedic Review Manual, 45
Paramedic Skills Manual, 45
Parenting Special Children, 327
Park and Recreation Maintenance Management, 502
Park Ranger Guide to Rivers and Lakes, 502
Park Ranger Guide to Seashores, 502
Park Ranger Guide to Wildlife, 502
Partial to Home: Photographs by Birney Imes, 447
Passbooks for Career Opportunities Series, 274
Passenger Airliners of the United States, 1926-1991, 360
The Passion for Fashion, 420
Passport: An Introduction to the Travel and Tourism Industry, 229
Paste-Up 1: Pre-Press Fundamentals for Professional Production, 410
Paste-Up 2: Practical Examples of Production Art, 410
Paste-Up 3: Complex Examples of Production Art, 410
Pat Welsh's Southern California Gardening, 520
Path We Tread: Blacks in Nursing, 1854-1990, 81
Pathophysiology, 71
Pathophysiology: Concepts of Altered Health States, 71
Pathways to Pleasure, 542
Pathways to Reality, 542
Patient Billing: A Computerized Simulation Using Medisoft, 53
Patient Care in Radiography, 24
Patient Care Standards, 81
Patient Care Standards: Nursing Process, Diagnosis, and Outcome, 71
Patient Communication for First Responders and EMS Personnel, 44
Patient Participation in Program Planning, 101
Patient Practitioner Interaction, 106
Pattern Grading for Children's Clothes, 414
Pattern Grading for Men's Clothes, 414
Pattern Grading for Women's Clothes, 414
Patty's Industrial Hygiene and Toxicology, 93
Paul Poiret, 421
The PC Is Not a Typewriter, 237, 409
Peachtree Complete II: Accounting Made Easy, 170
Peachtree Complete III, 170
Pediatric Care Planning, 84
Pediatric Dental Care: An Update for the 90s, 12
Pediatric Echocardiography, 31
Pediatric Neurologic Physical Therapy, 105
Pediatric Nursing, 84, 88, 89
Pediatric Nursing Care Plans, 84
Pediatric Occupational Therapy, 99

Pediatric Occupational Therapy and Early Intervention, 98
Pediatric Physical Therapy, 111
Pediatric Radiation Oncology, 27
Pelvic Ultrasound, 31
The Penguin Dictionary of Decorative Arts, 433
People First: Serving and Employing People with Disabilities, 343
Percy Blandford's Favorite Woodworking Projects, 122
Perennial Garden Color: For Texas and the South, 520
Perfect WordPerfect 6 Documents: A Visual Step-By-Step Guide, 400
Performance Analysis and Appraisal, 335
Performing Advanced Procedures, 87
Perinatal and Pediatric Respiratory Care, 119
Perinatal Substance Abuse, 546
Period Finishes and Effects, 437
Periodontal Instrumentation, 12
Periodontology, 13
Peripheral Vascular Imaging and Intervention, 30
Permar's Oral Embryology and Microscopic Anatomy, 12
Persistent Inequalities: Women and World Development, 567
The Person with AIDS: Nursing Perspective, 73
Personal Injury Paralegal, 307
Personal Injury Paralegal Forms and Procedures, 307
Personal Safety for Real Estate Professionals, 221
A Personnel and Operations Manual for Travel Agencies, 231
Perspectives in Dental Ceramics, 12
Perspectives on Nursing Theory, 82
Perspectives on Oral Manifestations of AIDS, 13
Pesticide Users' Health and Safety Handbook, 521
Pesticides in Agriculture, 489
Pests of Landscape Trees and Shrubs, 521
Pests of the Garden or Small Farm, 521
Peter Norton's Inside the PC, 180
Pharmacologic Basis of Nursing Practice, 85
Pharmacology: A Nursing Process Approach, 85
Pharmacology: An Introductory Text, 54
Pharmacology and the Nursing Process, 85
Pharmacology and Therapeutics in Respiratory Care, 120
Pharmacology for Veterinary Technicians, 528
Pharmacotherapeutics: A Nursing Process Approach, 85
The Philosophical Foundations of Social Work, 554
A Philosophy of Interior Design, 430
Phlebotomy Essentials, 62
Phlebotomy Handbook, 61
Phlebotomy Techniques: A Laboratory Workbook, 63
Photo Atlas of Nursing Procedures, 80
Photo Business Careers, 446
Photo Essay: Photographs by Mary Ellen Mark, 447
The Photo Gallery and Workshop Handbook, 444
Photographed by Bachrach: 125 Years of American Portraiture, 448
The Photographer's Assistant, 447
Photographers at Work: A Smithsonian Series, 447
The Photographer's Business and Legal Handbook, 444
Photographer's Publishing Handbook, 446
Photographic Perspective Drawing Techniques, 436
Photographing Buildings Inside and Out, 437
Photographing People for Stock, 446
Photographs That Changed the World, 449
Photography, 249
Photography at the Bauhaus, 449
Photography for Graphic Designers, 464
Photography Until Now, 449
Photoguide to Drug Administration, 85
Photojournalism: An Ethical Approach, 249
Photojournalism: Content and Technique, 249
Photojournalism: The Professionals' Approach, 248
Photojournalism: The Visual Approach, 248
Physical Agents: A Comprehensive Text for Physical Therapists, 107
Physical and Occupational Therapists' Job Search Handbook, 109

Title Index

Physical Education for Children: Building the Foundation, 315
Physical Examination and Health Assessment, 79
Physical Management of Multiple Handicaps, 107
Physical Rehabilitation: Assessment and Treatment, 109
Physical Therapy, 111
Physical Therapy for the Cancer Patient, 109
Physical Therapy in Craniomandibular Disorders, 107
Physical Therapy in Public Schools: A Related Service, 105
Physical Therapy Management of Parkinson's Disease, 111
Physical Therapy of the Cervical and Thoracic Spine, 107
Physical Therapy of the Foot and Ankle, 107
Physical Therapy of the Hip, 106
Physical Therapy of the Shoulder, 106
Physical Therapy Pharmacology, 106
Physicians' Desk Reference: PDR, 282
Physician's Office Laboratory, 59
Physician's Office Laboratory Guidelines, 63
Physics of Radiology, 27
The Physiological Basis for Exercise and Sport, 106
Physiotherapy for Respiratory and Cardiac Problems, 111
Piaget's Developmental Theory: An Overview, 326
Piaget's Theory of Intellectual Development, 316
Picture Editing and Layout, 249
The Pilot's Guide to Weather Reports, Forecasts, and Flight Plans, 359
The Pilot's Handbook of Aeronautical Knowledge, 359
The Pilot's Radio Communications Handbook, 359
The Pilot's Reference to ATC Procedures and Phraseology, 360
Pinning and Flying Display, 426
Pit and Fissure Sealants, 18
A Place in the News: From the Women's Pages to the Front Page, 254
Plains States, 232
Planning and Administering Early Childhood Programs, 313
Planning and Control for Food and Beverage Operations, 197
Planning and Financing the New Venture, 225
Planning and Forming Your Company, 223
Planning and Implementing End-User Information Systems, 178
Planning and Managing Interior Projects, 433
Plant Disease Control: Principles and Practice, 487
Plant Diseases: Their Biology and Social Impact, 488
The Plant Growth Planner, 519
Plant Propagation Principles and Practices, 521
Plant Stress from Air Pollution, 505
The Plant-Book, 518
The Plant-Feeding Gall Midges of North America, 504
Planting and Staking Landscape Trees, 522
Plaster and Drywall Systems Manual, 433
Play, 315
Play and the Social Context of Development in Early Care and Education, 322
Play, Learn, and Grow, 324
Playing It Safe, 186
Pleasures and Terrors of Domestic Comfort, 448
The Plumber's Troubleshooting Guide, 162
Plumbing, 162
The Plumbing Apprentice Handbook, 162
Plumbing Contractor: Start and Run a Money-Making Business, 163
Plumbing Technology, 160
Plumbing Technology: Design and Installation, 162
Pluralizing Journalism Education, 254
The PMA Fresh Produce Reference Manual for Food Service, 198
PNF in Practice: An Illustrated Guide, 104
Pocket Reference to Plumbing and Pipe Fitting Calculations, 162
Police Deviance, 289
The Police Dictionary and Encyclopedia, 285

The Police Officer's Guide to Survival, Health, and Fitness, 289
Police Vehicles and Firearms: Instruments of Deadly Force, 288
Policing as Though People Matter, 287
Policy for American Agriculture: Choices and Consequences, 483
The Politics of Trees, 508
Politics, Privacy, and the Press. Ethics in America, 255
Pollution Prevention, 370
Pollution Prevention Technology Handbook, 369
Pond Scum and Vultures, 253
Poor Support: Poverty in the American Family, 571
Poor Women, Poor Families, 573
Popular Mechanics Home Answer Book, 139
Popular Mechanics Home How-To: Plumbing and Heating, 161
¿Por Que Los Impuestos?, 172
¿Por Que Nosotros, Los Garcia?, 172
Porcelain and Composite Inlays and Onlays, 8
Porcelain Laminate Veneers, 8
Porcelain Laminate Veneers: Preparation and Placement, 19
Portable MBA in Finance and Accounting, 169
Portrait of Imogen, 451
Positioning for Play, 99
Poultry Diseases, 528
Power and Influence: Mastering the Art of Persuasion, 259
Power and Restraint: The Moral Dimension of Police Work, 289
Power Foods, 34
Power of FrameMaker 4.0 for Windows, 401
The Power of Photography: How Photographs Changed Our Lives, 450
Power, Publicity and the Abuse of Libel Law, 245
Power Publishing with Ventura 3.2, 400
Power Real Estate Letters, 217
Power Real Estate Selling, 216
Power Saws and Planers, 124
Power Sewing: New Ways to Make Fine Clothes Fast, 413
Power Sewing Volume 1: Fear of Sewing, 425
Power Sewing Volume 2: Pattern Sizing and Alteration, 425
Power Sewing Volume 3: Fitting Solutions, 425
Power Sewing Volume 4: Foolproof Pants Fitting, 425
Power Sewing Volume 5: Construction Difficulties, 425
Power Sewing Volume 6: Easy Linings, 426
Power Sewing Volume 7: Hassle-Free Designer Jackets, 426
Power Sewing Volume 12: Handwoven and Quilting, 426
Power Shortcuts: Word 2.0 for Windows, 210
Power Technology, 147
The Power to Persuade, 334
Power Trains: Compact Equipment, 381
PR News Casebook: 1,000 Public Relations Case Studies, 259
Practical Approaches to Legal Research, 302
Practical Aspects of Interview and Interrogation, 288
Practical Baking, 199
Practical Doppler Ultrasound for the Clinician, 31
A Practical Guide to Echocardiography and Cardiac Doppler Ultrasound, 30
The Practical Guide to Libel Law, 246
Practical Guide to Power Distribution Systems for Computers, 145
Practical Guide to Quality Power for Sensitive Electronic Equipment, 143
Practical Guide to Solving Preschool Behavior, 314
Practical Handbook of Agricultural Science, 477
Practical Help for New Supervisors, 334
Practical Horticulture, 519
Practical Hydraulics Handbook, 368
Practical Infection Control in Dentistry, 6
Practical Problems in Mathematics for Welders, 391
Practical Radiation Protection and Applied Radiobiology, 27

Practical Surveying for Technicians, 385
Practical Surveying with GPS, 386
Practical Videography: Field Systems and Troubleshooting, 266
Practice Issues in Occupational Therapy, 101
Practice Management for the Dental Team, 8
The Practice of Social Work, 556
Practicing Universal Design: An Interpretation of the ADA, 441
The Practitioner's Guide to Psychoactive Drugs, 538
Practitioner's Guide to Psychoactive Drugs for Children and Adolescents, 548
Prairie Patrimony, 480
Pre-Hospital Trauma Life Support Committee PHTLS, 44
Pre-School Resource Guide, 315
Prehospital Drug Therapy, 43
Prehospital Emergency Care and Crisis Intervention, 43
Prenatal Abuse of Licit and Illicit Drugs, 540
Prentice Hall Encyclopedic Dictionary of English Usage, 277
Preparation for Credentialing in Radiography, 26
Preparing for the FCC General Radiotelephone Operator's License, 270
Preparing for the RPR and CM Written Knowledge Test, 274
Preparing Instructional Text, 405
Preschool in Three Cultures, 324
Preschool Power 4!, 327
Preschoolers and Substance Abuse, 541
Prescribed Burning in California Wildlands Vegetation Management, 503
Preserving the Press, 241
Press and Public, 241
The Press and the World of Money, 243
The Press in America, 242
Preventing Adolescent Abuse, 558
Preventing Adolescent Pregnancy, 562
Preventing Adolescent Suicide, 559
Preventing Early School Failure, 323
Prevention and Treatment Considerations for the Dental Patient with Special Needs, 19
Prevention Magazine's Complete Nutrition Reference Handbook, 34
Prevention of Dental Disease, 11
The Preveterinary Planning Guide, 525
Primary Preventive Dentistry, 8
A Primer for Today's Substance Abuse Counselor, 544
A Primer of Drug Action, 540
Primrose McConnell's The Agricultural Notebook, 479
Principles and Practice of Echocardiography, 32
Principles and Practice of Electrotherapy, 108
Principles and Practice of Operative Dentistry, 6
Principles and Practice of Psychiatric Nursing, 86
Principles and Practice of Radiation Oncology, 27
Principles of Applied Optics, 377
The Principles of Auto Body Repairing and Repainting, 353
Principles of Children's Services in Public Libraries, 331
Principles of Computer-Aided Design, 373
Principles of Extrication, 495
Principles of Fire Protection Chemistry, 494
Principles of Food, Beverage, and Labor Cost Controls for Hotels and Restaurants, 193
Principles of Imaging Science and Protection, 26
Principles of Laboratory Instruments, 64
Principles of Pharmacology for Medical Assisting, 54
Principles of Radiographic Imaging: An Art and a Science, 23
Principles of Radiography for Technologists, 26
Principles of Surveying, 385
Principles of Two-Dimensional Form, 460
The Print and Production Manual, 407
Printing Basics for Non-Printers, 410
Printing Estimating: Principles and Practices, 407
The Printing Ink Manual, 404
Printmaking Techniques, 398
Print's Best Letterheads and Business Cards 3, 403, 454
Print's Best Logos and Symbols, 456

Title Index

Print's Best Typography, 408
PRISM: A Computerized Tutorial for IBM PC and Compatibles, 343
Prison Slang: Words and Expressions Depicting Life Behind Bars, 285
Privacy and Confidentiality of Health Care Information, 52
Private Choices, Social Costs, and Public Policy, 571
Private Pilot and Recreational Pilot FAA Written Exam, 359
Probate Handbook for the Lawyer's Assistant, 295
Problem Drinkers: Guided Self-Change Treatment, 546
Problems and Solutions: Sources of Error in Chemistry Tests, 66
Procedures for the Automated Office, 210
Procedures in Phlebotomy, 48, 61
The Process of Editing, 237
The Process of Legal Research: Successful Strategies, 280
Process of Parenting, 312
Process of Patient Education, 80
Processing with DacEasy 4.4, 172
Proctor and Hughes' Chemical Hazards of the Workplace, 95
Producing a First-Class Newsletter, 402
Producing Quality Radiographs, 24
Production for the Graphic Designer, 401
Professional Advertising Photography, 447
Professional and Occupational Licensing Directory, 536
Professional Baking, 194
Professional by Choice: Milady's Career Development Guide, 187
The Professional Chef's Art of Gardé Manger, 199
The Professional Chef's Techniques of Healthy Cooking, 193
Professional Cooking, 194
The Professional Cosmetologist, 187
Professional Dining Room Management, 195
The Professional Guide: Dynamics of Tour Guiding, 230
Professional Guide to Alcoholic Beverages, 196
The Professional Housekeeper, 207
Professional Land Surveying Candidate Handbook, 385
Professional Management of Housekeeping Operations, 205
The Professional Medical Assistant: Clinical Practice, 55, 62
Professional Nursing: Concepts and Challenges, 77
Professional Practice for Interior Designers, 438
Professional Practice Standards for Health Information Management Services, 55
Professional Practice Standards for Health Information Management Services in Ambulatory Care, 55
Professional Practice Standards for Health Information Management Services in Long Term Care, 55
Professional Practice Standards for Mental Health Information, 55
Professional Responsibility, 295
Professional Secretaries International Complete Office Handbook, 210
Professional Social Work Credentialing and Legal Regulation, 555
Professional Table Service, 196
Professional Travel Agency Management, 228
Professionalization of Nursing: Current Issues and Trends, 77
The Professionals' Guide to Commercial Property Development, 220
Profile Leveling, 387
Profit Planning, 204
Profits in Buying and Renovating Homes, 218
Programming in C, 179
The Programming Primer: A Guide To Programming Fundamentals, 181
Progress and Issues in Case Management, 535
Project Management for the Design Professional, 431
Projecting a Professional Image, 211
Projects for the Home Craftsman, 125
Promoting Community Change, 556

Promotional Strategy for Small Business, 225
Prosthetics and Orthotics, 111
Protecting Our Children's Teeth, 8
Protecting Soldiers and Mothers, 573
Protecting Yourself Against Bloodborne Pathogens, 66
Psychedelics Encyclopedia, 546
Psycheresponse, 41, 492
Psychiatric and Mental Health Nursing with Children and Adolescents, 86
Psychiatric Dictionary, 281
Psychiatric Nursing, 87
Psychiatric Nursing Care Plans, 86
Psychological Services for Law Enforcement, 289
The Psychology of Attitude Change and Social Influence, 261
The Psychology of Death, 529
Psychology of Dieting, 36
Psychopathology and Function, 98
Psychosocial Crisis, 88
Public Affairs Reporting, 253
Public Affairs Reporting: The Citizens News, 253
Public Broadcasting Directory, 264
The Public Mind, 255
Public Relations Career Directory, 261
Public Relations Writing, 259
Public Relations Writing: Form and Style, 261
Publication Design, 406
Publication Design for Editors, 248
The Publicity Kit, 261
Publishing Newsletters, 404
Pucci: A Renaissance in Fashion, 421
Pulmonary Care of the Surgical Patient, 115
Pulmonary Function Testing: A Practical Approach, 119
Pulmonary Management in Physical Therapy, 111
Pulmonary Rehabilitation Homecare: From Paper to Practice, 114
Pulmonary Therapy and Rehabilitation: Principles and Practice, 115
Pundits, Poets, and Wits, 235
Purchasing: Selection and Procurement for the Hospitality Industry, 199
Pure Invention—The Tabletop Still Life: Photographs by Jan Groover, 447
Purposeful Play with Your Preschooler, 319

Quantity Food Production, Planning, and Management, 37, 195
Quantity Food Purchasing, 196
The QuarkXPress Book, 399
The QuarkXPress for Windows Companion, 401
Questions and Answers in Magnetic Resonance Imaging, 28
Quick and Easy Medical Terminology, 52, 62
Quick E.C.G. Interpretation, 88
Quick Emergency Care Reference, 44
Quick Reference to Occupational Therapy, 101
Quicken 7.0 Quick and Easy, 172
Quicken for Windows, Quick and Easy, 172
Quintessence of Dental Technology: QDT, 13

Rabies, 525
Race and Racism: Red, White, and Black, 575
Rachel and Her Children: Homeless Families in America, 561
Racism in the News, 255
Radial Arm Saw Basics, 123
Radiation Protection: A Guide for Scientists and Physicians, 26
Radiation Protection in Medical Radiography, 26
Radiation Protection in the X-Ray Department, 26
Radiation Protection of Patients, 27
Radiation Therapy for Head and Neck Neoplasms, 28
Radiation Therapy Planning, 27
Radical Rags: Fashions of the Sixties, 421
Radio Operator's License Q and A Manual, 269
Radio Production: Art and Science, 269
Radiographic Exposure and Technique, 32
Radiographic Imaging, 23

Radiographic Imaging for Dental Auxiliaries, 11
Radiographic Interpretation for the Dental Hygienist, 8
Radiographic Pathology for Technologists, 25
Radiographic Positioning, 24
Radiographic Processing and Quality Control, 25
Radiographic Techniques and Safety, 19
Radiography Exam Review, 23
Radiography Examination Review, 25
Radiologic Atlas of Brain Tumors, 32
Radiologic Science for Technologists, 23
Radiology: An Illustrated History, 25
Radiology and Ultrasound of Urogenital Diseases in Dogs and Cats, 524
Radiology for Dental Auxiliaries, 8
Radiology for Dental Hygienists and Dental Assistants, 10
Radiology Management, 23
Radiotherapy for Head and Neck Cancers, 27
Rain Forests: A Part of Your Life, 509
Raising Her Voice, 255
Raising Less Corn and More Hell, 480
Raising Money: Venture Funding and How to Get It, 224
Range Development and Improvements, 485
Range Management: Principles and Practices, 485, 503
Rangeland Health, 485
Rangeland Reform '94, 485
The Rape Victim: Clinical and Community Interventions, 557
Rapid Assessment, 88
Rapid Reference Guide to Adobe Illustrator, 402
Rapid Viz: A New Method for the Rapid Visualization of Ideas, 457
Rational Manual Therapies, 105
Reaching High-Risk Families, 564
Read All About It! The Corporate Takeover of America's Newspapers, 244
The Reader's Advisor: A Layman's Guide to Literature, 337
Reader's Digest Book of Skills and Tools, 130
Reader's Digest Complete Guide to Sewing, 415
Reading American Photographs, 450
Reading Real Books, 313
Readings and Problems in Accounting Information Systems, 171
Readings in Child Development, 318
Real Estate: A VGM Career Planner, 217
Real Estate Agent's Business Planning Guide, 217
Real Estate Careers: Twenty-Five Growing Opportunities, 217
Real Estate Exchange and Acquisition Techniques, 217
Real Estate Finance, 216
Real Estate Principles, 215
Real Estate Prospecting: Strategies for Farming Your Markets, 216
Real Estate Quick and Easy, 215
Real Estate Sales Handbook, 216
The Real Life Nutrition Book, 35
Real Nursing Series, 88
Real-Time Writing, 276
Real World Scanning and Halftones, 399
Recognition and Management of Pesticide Poisonings, 521
Recognizing Child Abuse: A Guide for the Concerned, 558
Recommended Dietary Allowances, 37
Recommended Practices for Gas Metal Arc Welding, 393
Recommended Reference Books for Small and Medium-Sized Libraries and Media Centers, 338
The Reconfigured Eye: Visual Truth in the Post-Photographic Era, 450, 463
Recording Field Notes, 387
Records Management: Integrated Information Systems, 211
Records Management Handbook, 342
Recovery at Work: A Clean and Sober Career Guide, 546
Recovery from Addiction, 538

Title Index

The Recovery Resource Book, 548
Recreational Foodservice Management, 39
Recreational Tourism: A Social Science Perspective, 231
Recycling in America: A Reference Handbook, 368
Red Flags, 172
Redefining Families, 316
Redesigning Library Services: A Manifesto, 333
Reference and Information Services, 337
Reference Books Bulletin, 338
Reference Sources for Small and Medium Sized Libraries, 338
Reforming Farm Policy: Toward a National Agenda, 483
Reforming the Forest Service, 502
Refrigerant Management, 153
Regents/Prentice Hall Medical Assistant Kit, 49
Regents/Prentice Hall Textbook of Cosmetology, 187
Regreening the National Parks, 501
Regulating Drinking Water Quality, 368
Regulating the Lives of Women, 570
Rehabilitation and Continuity of Care in Pulmonary Disease, 117
Rehabilitation in Mental Health, 99
Rehabilitation Nursing for the Neurological Patient, 76
The Rehabilitation Specialist's Handbook, 110
Rehabilitation Technology, 107
Rei Kawakubo and Commes des Garçons, 423
Reinventing the Newspaper: Essays, 241
The Reliable Source for Occupational Therapy, 102
Relief Printmaking: A Manual of Techniques, 409
Religion and Social Work Practice in Contemporary American Society, 572
The Reluctant Welfare State, 571
Remaking the Welfare State, 570
Remodeling, 139
Remodeling and Repairing Kitchen Cabinets, 137
Renal Disorders, 88
Renovating Old Houses, 138
Renovation: A Complete Guide, 138
Repairing and Extending Doors and Windows, 139
Repairing Old and Historic Windows, 138
The Reporter and the Law: Techniques of Covering the Courts, 252
Reporter's Handbook, 247
Reporters Reference Manual, 277
Reporting for the Print Media, 246
Reporting in Depth, 247
Reporting on Disability: Approaches and Issues, A Sourcebook, 254
Reporting on Risk, 244
Reporting Public Affairs: Problems and Solutions, 253
Reporting Religion: Facts and Faith, 253
Reproductive Pathology of Domestic Animals, 526
Rescue, 46
The Rescue Company, 43, 493
Research Manual for the Health Information Management Profession, 54
Research Methods in Social Work, 569
Researching Copyright Renewal, 335
Reshaping the Bottom Line, 487
Residential Electrical Design Revised, 148
Residential Framing: A Homebuilder's Construction Guide, 136
Resilient Child, 318
Resort Development and Management, 204
Respiratory Care: A Guide to Clinical Practice, 114
Respiratory Care: Evolution of a Profession, 118
Respiratory Care: Know the Facts, 115
Respiratory Care Certification Guide, 118
Respiratory Care Pharmacology, 117
Respiratory Care Principles, 114
The Respiratory Care Workbook, 118
Respiratory Disease: Principles of Patient Care, 119
Respiratory Disorders, 88
Respiratory Monitoring, 119
Respiratory Physiology—The Essentials, 119
Respiratory Support, 88
Respiratory Therapy Equipment, 117

Respiratory Therapy Examination Review, 115
Response to Occupational Health Hazards, 93
Restaurant Basics, 196
The Restaurant Operator's Manual, 198
Restaurants That Work: Case Studies of the Best in the Industry, 432
Restorative Dental Materials, 6
Restraint of Domestic Animals, 530
Resuscitation Handbook, 41
Retail Design, 433
The Retail Store: Design and Construction, 434
Rethinking Business to Business Marketing, 224
Rethinking Reference in Academic Libraries, 330
Rethinking Social Policy: Race, Poverty, and the Underclass, 571
The Return of the Native: American Indian Political Resurgence, 564
Review Questions and Answers for Veterinary Technicians, 529
Reviewing the Arts, 253
Reviving Old Houses, 138
Rheumatic Disease in the Adult and Child, 100
The Right Fit, 223
Right in the Middle, 106
The River of the Mother of God and Other Essays, 505
RMA Annual Statement Studies, 170
Robert Half's Success Guide for Accountants, 168
Robert Morris Associates (RMA) Annual Statement Studies, 224
Rodale's All-New Encyclopedia of Organic Gardening, 519
Rodale's Landscape Problem Solver: A Plant-By-Plant Guide, 518
Roger Black's Desktop Design Power, 398
Roger Haines on HVAC Controls, 152
The Role of Occupational Therapy with the Elderly, 99
Roofers Handbook, 134
Roofing Research and Standards Development, 131
Rooms with a View, 436
The Roots of Community Organizing, 1917-1939, 556
Roots of Dentistry, 9
Rope Rescue Manual, 494
Rough Carpentry Illustrated, 137
Routing and Shaping, 125
The Rudi Gernreich Book, 422
Ruffed Grouse, 509
Rules of Practice Before the United States Court of Appeals for the Federal Circuit, 278
Running MS-DOS: Covers Version 6.0, 183
Rural America in the Information Age, 479

Sacred Cows and Hot Potatoes, 483
SAE Transactions, 355
Safe Crane Operation and Practices, 365
The Safe Travel Book, 231
Safety and Health for Engineers, 92
Safety and Survival on the Fireground, 493
Safety Assessment: A Quantitative Approach, 95
Safety Auditing: A Management Tool, 94
Safety Before Selling: A Survival Guide, 221
Safety in Welding and Cutting: Superseding ANSI Z49.1-83, 392
Safety Management, 95
Sales Promotion Design, 454
The Samisdat Method: A Do-It-Yourself Guide to Printing, 400
A Sampling Strategy Guide for Evaluating Contaminants in the Welding Environment, 392
Sandra Smith's Review for NCLEX-RN, 87
Sanitary Landfilling, 368
Sanitation Management, 192
SARA Title III, 493
Saunders Manual of Medical Assisting Practice, 49
Saunders Pharmaceutical Word Book, 1995, 54
Saunders Review for NCLEX-RN, 87
The Savage Mirror: The Art of Contemporary Caricature, 466
Save Thousands on Your Mortgage, 218
Save Three Lives: A Plan for Famine Prevention, 484

Saving Children at Risk: Poverty and Disabilities, 563
Sax's Dangerous Properties of Industrial Materials, 94
Scapegoating in Families, 562
Schaum's Outline of Theory and Problems of Bookkeeping and Accounting, 169
Schematic Wiring: A Step-By-Step Guide, 142
Scholarships and Loans for Nursing Education, 70
School Foodservice Management, 199
The School Librarian's Sourcebook, 339
School Social Work, 561
The Science and Practice of Welding, 392
Science and the Beauty Business, Vol. 1, 188
Science and the Beauty Business, Vol. 2, 188
Science Everywhere, 324
Science Fiction Comics: The Illustrated History, 465
Science in Your Salon, 187
The Science of Flight: Pilot-Oriented Aerodynamics, 348
Scientific Illustration, 468
Screen Printing: Design and Technique, 399
Screenwriting: The Art, Craft, and Business of Film and Television Writing, 269
Scriptwriting for High-Impact Videos, 269
Sculptured Nails, 189
Searching the Law, 301
Second Front: Censorship and Propaganda in the Gulf War, 242
Secrets of Success for Today's Interior Designers and Decorators, 438
Security and Crime Prevention in Libraries, 336
Sedation: A Guide to Patient Management, 10
Seeds of Woody Plants in North America, 500
Seeing for Yourself, 445
Seeing Straight: The f.64 Revolution in Photography, 448
Seeing the Newspaper, 247
Seeking Employment in Law Enforcement, Private Security, and Related Fields, 289
Segregated Skies: All-Black Combat Squadrons of WW II, 360
Self-Contained Breathing Apparatus, 495
Self-Defense for Criminal Justice, 289
Selling Destinations, 230
Senior Court Reporter, 274
Senior Shorthand Reporter, 274
Serials and Reference Services, 337
Serials Cataloging Handbook, 341
Serials Reference Work, 338
Services for Sale: Purchasing Health and Human Services, 553
Serving Gay and Lesbian Youths, 564
Setting Up Your Own Woodworking Shop, 125
The Seventh Old House Catalogue, 129
Severe Weather Flying, 360
Sewing ABCs, 426
The Sewing Book: A Complete Practical Guide, 415
Sewing Pants That Fit, 415
Sex and Sacrifices: What They Did for Love, 575
Sexual Harassment on the Job, 212
Sexual Health, 89
Sexually Transmitted Diseases: Companion Handbook, 569
SFPE Handbook of Fire Protection Engineering, 495
The Shadow World: Life Between the News Media and Reality, 245
Shape and Color: The Key to Successful Ceramic Restorations, 14
Shaper Handbook, 123
Shattering: Food, Politics, and the Loss of Genetic Diversity, 482
She Said What? Interviews With Women Newspaper Columnists, 254
Shepard's Acts and Cases by Popular Names, Federal and State, 302
Shepard's United States Citations. Cases, 302
Shirtmaking: Developing Skills for Fine Sewing, 414
Shirtmaking Techniques, 426
Shoes: Fashion and Fantasy, 422
A Short Course on Computer Viruses, 176

Shortcuts to a Perfect Sewing Pattern, 413
Shorthand Reporter, 274
Showcase of Interior Design, 439
Sicher and DuBrul's Oral Anatomy, 7
Sign Design: Environmental Graphics, 455
Sign Language for the Dental Team, 18
Silent Cities: The Evolution of the American Cemetery, 512
Silent Selling, 423
Silvics of North America, 499
The Simple Act of Planting a Tree, 499
Simple, Fast, High Quality Dental Radiographs (Intraoral), 19
Simple Surgery for Everyday Practice, 17
Simple Temporary Restorations for Fixed Prosthodontics, 17
Simplified Design of Building Foundations, 132
Simplified Drugs and Solutions for Nurses, Including Mathematics, 85
Simplified Irrigation Design, 520
Simplified Management of Medical Emergencies, 18
Simplifying Accounting Language: Don't Lose Your Balance!, 171
Simply DOS, 178
Simulations in Medical Evaluation of Geriatric Patients, 18
Singer Sewing Step-By-Step, 415
Single Camera Video: From Concept to Edited Master, 267
Single-Camera Video Production, 266
Six Theories of Child Development, 325
Sixty Art Projects for Children, 312
The Sixty Second EMT, 42
Sketching and Rendering Interior Spaces, 432
Skills for Radio Broadcasters, 269
Skills of Daily Mouth Care: A Caregiver's Guide, 19
Skyscraper, 136
Small AC Generator Service Manual, 147
Small Animal Allergy: A Practical Guide, 528
Small Animal Physical Diagnosis and Clinical Procedures, 529
Small Animal Wound Management, 527
The Small Book, 575
Small Business Course, 224
The Small Business Encyclopedia, 224
Small Business in Tough Times: How to Survive and Prosper, 223
Small Business Information Handbook, 223
Small Business Management: A Guide to Entrepreneurship, 224
Small Business Sourcebook: The Entrepreneur's Resource, 224
Small Electric Motors, 145
Small Engine Technology, 381
Small Format Television Production, 265
Small Gas Engine Repair, 381
Small Gas Engines, 381
Small Gasoline Engines: Service and Repair, 381
Small Gasoline Engines, Operation, and Maintenance, 381
Small Houses, 136
Small Scale Agriculture in America, 483
Small Wonders, 321
The Small Woodshop, 124
Small Woodworking Projects, 124
The Smaller Academic Library: A Management Handbook, 330
Smallholders, Householders, 487
Smart House Wiring, 147
Smart Litigating with Computers, 303
Smart Vacations, 231
Smoking Cessation Intervention Techniques, 120
So Shall You Reap: Farming and Crops in Human Affairs, 480
So...You Want to Be an Innkeeper, 203
Social and Emotional Development of Exceptional, 318
Social Issues Resources Series, 291
Social Policy and Social Work: Critical Essays on the Welfare, 572

Social Security: The System that Works, 570
Social Studies for the Preschool-Primary Child, 322
Social Welfare: Policy and Analysis, 570
Social Work, 552
Social Work and Society: An Introduction, 553
Social Work Case Management, 555
Social Work Day-To-Day, 553
The Social Work Dictionary, 553
Social Work Ethics Day to Day, 556
Social Work in an Unjust Society, 554
Social Work in the Workplace: Practice and Principles, 567
Social Work Practice and People of Color, 566
Social Work Practice in Maternal and Child Health, 560
Social Work Practice with Black Families, 566
Social Work Practice with Groups: A Clinical Perspective, 555
Social Work Practice with Minorities, 564
Social Work Research and Evaluation, 569
The Social Worker as Manager: Theory and Practice, 556
The Sociology of Agriculture, 478
Software Digest Ratings Report, 175
The Software Publishers Association Legal Guide to Multimedia, 463
Software Reviews on File, 175
Soil and Water Quality: An Agenda for Agriculture, 486
Soils and the Environment: An Introduction, 370
The Solutions Manual for the Land Surveyor Reference Manual, 384
Sorenson and Luckmann's Basic Nursing, 79
Sound Effects: Radio, TV, and Film, 268
Sound Studio Production Techniques, 266
A Sourcebook of American Literary Journalism, 241
Sourcebook of Criminal Justice Statistics, 1992, 286
Sourcebook of Modern Furniture, 434
Southwestern Country Classics, 125
Spacious Skies, 360
Spanish for Dentistry, 4
Special Libraries: A Guide for Management, 339
Special Libraries: Increasing the Information Edge, 339
Special Physical Education, 317
Special Problems in Counseling the Chemically Dependent Adolescent, 546
Special-Interest Tourism, 232
Specialty Cookbooks: A Subject Guide, 198
Specifications for Commercial Interiors, 439
Specs, 198
Speedbuilding for Court Reporting, 276
SPELLRIGHT: A Medical Wordbook, 282
Spinal Cord Injury: Functional Rehabilitation, 111
Splash! A History of Swimwear, 422
Split Image: African Americans in the Mass Media, 254
Sports Injuries, 107
Sports Injuries: Diagnosis and Management for Physiotherapists, 109
Sports Physical Therapy, 110
Sports Reporting, 252
The Sports Writing Handbook, 252
Spreadsheet Style Manual, 168
Springhouse Notes Series, 89
St. Anthony's Clinical Reference to Diagnostic Coding, 50
St. Anthony's DRG Optimizer: 1993, 50
St. Anthony's ICD-9-CM: Questions and Answers, 49
The Staff Development Handbook, 335
Standard Aircraft Workers' Manual, 349
Standard Consumer Safety Playground, 323
Standard for Accreditation of Test Facilities for AWS Certified Welder Program, 391
Standard for AWS Certification of Welding Inspectors, 391
Standard for Health Care Facilities, 147
Standard Handbook of Hazardous Waste Treatment and Disposal, 368
Standard Methods for the Examination of Water and Wastewater, 368
Standard Methods of Mechanical Testing of Welds, 391

Standard Symbols for Welding, Brazing and Soldering, 389
Standards, Principles and Techniques in Quantity Food Production, 37, 195
The Stanley Complete Step-By-Step Book of Home Repair and Improvement, 138
Start Your Own Desktop Publishing Business, 406
Start Your Own Interior Design Business and Keep It Growing, 438
Starting Your Own Small Graphic Design Studio, 472
STAT, 70
State-Approved Schools of Nursing L.P.N./L.V.N., 70
State-Approved Schools of Nursing—R.N., 1993, 70
State Board Review Questions, 187
The State of America's Children Yearbook, 563
Stationary Engineering for Boiler Operations, 153
Statistical Deception at Work, 251
Statistics for Social Workers, 569
Staying Healthy in Asia, Africa, and Latin America, 231
Stenographers to Power: Media and Propaganda, 241
Step-by-Step Bookkeeping, 170
Steps to Small Business Start-Up, 224
Sterilization and Sanitation, 189
Stifle Surgery, 530
Stigma and Mental Illness, 567
Stores of the Year, 440
Stories in the Classroom, 311
Storm over Rangelands: Private Rights in Federal Lands, 485
Stormwater, 370
The Story of Corn, 482
Storytelling Folklore Sourcebook, 319
Storytime Sourcebook, 313
Strategic Planning Basics for Special Libraries, 339
Strategies for Work with Involuntary Clients, 555
The Street Addict Role: A Theory of Heroin Addiction, 546
Street Dancer, 45
Street Fashion, 418
Street Medicine (Video Series), 46
Streetsense: Communication, Safety, and Control, 42
Stress and Burnout in Library Service, 334
Structural Details for Concrete Construction, 157
Structural Details for Masonry Construction, 158
Structural Welding Code Reinforcing Steel, 390
Structural Welding Code Sheet Steel, 390
Structural Welding Code Steel, 390
Structure and Function of the Human Body, 71
Structured Publishing from the Desktop, 402
Structures: The Way Things are Built, 131
A Student's Guide to Am Jur 2d, ALR, and USCS, 281
Studio Drama Processes and Procedures, 267
Studio Techniques for Advertising Agencies and Graphic Designers, 454
The Study of Social Problems: Six Perspectives, 555
Style Traditions: Recreating Period Interiors, 431
Styles of American Furniture, 442
Styling Competition: A Guide to Winning Technique, 186
Subject Cataloging: A How-To-Do-It Workbook, 340
Substance Abuse: A Comprehensive Textbook, 542
Substance Abuse: Causes, Consequences, and Treatment Choices, 600
Substance Abuse: Pharmacologic, Developmental, and Clinical Prospectives, 536
Substance Abuse and Physical Disability, 539
Substance Abuse Counseling: An Individualized Approach, 541
Substance Abuse Disorders, 550
Substance Abuse in Adolescents and Young Adults, 543
Substance Abuse in Children and Adolescents, 545
Substance Abuse in Dentistry, 7
Substance Abuse Treatment: A Family Systems Perspective, 538
Success with Bedding Plants, 522
Successful Aging: Overcoming Barriers to Nutrition and Health, 39
Successful Black and White Photography, 445
Successful Buffet Management, 200

Title Index

The Successful CAD Manager's Handbook, 177
Successful Field-To-Office Automation, 386
Successful Sign Design: Number 2., 455
Suction Tip Placement, 19
Suctioning: Nasotracheal, Oropharyngeal and Endotracheal Techniques, 120
Suicide Among Youth: Perspectives on Risk and Prevention, 562
Sunset Western Garden Book, 520
Super Nutrition for Women, 36
The Supercontinuum Laser Source, 377
Superhero Comics: The Illustrated History, 465
Supervising Student Employees in Academic Libraries, 329
Supervision and Management of Quantity Food Preparation, 37
Supervision Handbook, 286
Supervision in the Hospitality Industry, 195, 197, 205, 206
Supporting Families with a Child with a Disability, 561
The Supreme Court and the Mass Media, 245
Surgeon General's Report on Nutrition and Health, 38
Surrational Images: Photomontages, 449
Survey Crew Manual, 385
Survey of Accounting, 169
Surveying, 384
Surveying: Principles and Applications, 385
Surveying and Setting Out Procedures, 384
Surveying Fundamentals, 385
Surveying Skills, 387
Surveying with Construction Applications, 385
The Surveyor and Property Law, 384
Survival Spanish for the Hospitality Industry, 203
Sustainable Agriculture, 489
Sustaining America's Agriculture: High-Tech and Horse Sense, 489
Sweet's Contract Interiors, 440
Sylvia Plachy's Unguided Tour, 449
Synthetic Fabric Coverings, 349
Systems: Analysis, Design, and Computation, 176

Taber's Cyclopedic Medical Dictionary, 52, 70
Tabletop Presentations: A Guide for the Foodservice, 197
Takeoffs Are Optional, Landings Are Mandatory, 360
Taking Sides: Clashing Views on Controversial Issues in Crime and Criminology, 291
Taking Sides: Clashing Views on Controversial Issues in Mass Media and Society, 240
Taking Their Place, 254
Talk Show and Entertainment Program Processes and Procedures, 267
Tapping the Government Grapevine, 335
Targeted Public Relations, 259
Tax Tips Strategies: Income Tax Hints Every Taxpayer Should, 169
The TCP/IP Companion: A Guide for the Common User, 175
Teach Yourself Electricity and Electronics, 144
Teach Yourself PageMaker 5.0 for Windows, 399
Teacher Tips for Action Time, 319
Teacher Tips for Circle Time, 319
Teacher Tips for Quiet Time, 319
Teaching Behavior to Infants and Toddlers, 324
Teaching Deaf Children: Techniques and Methods, 321
Teaching Language Arts: An Integrated Approach, 311
Teaching Library Skills in Middle and High School, 339
Teaching Reading in the Social Studies, 321
Teaching Science to Children, 317
Teaching Social Studies in Grades K-8, 321
Teaching Social Studies to the Young Child, 317
Teaching Technologies in Libraries, 332
Teaching Young Children with Behavioral Disorders, 325
Teaching Your Child Basic Body Confidence, 315
Team Dentistry, 12
Team Power: Making Library Meetings Work, 336
Technails: Extensions, Wraps, and Nail Art, 186
Technical Services Today and Tomorrow, 340

The Technique of Radio Production: A Manual for Broadcasters, 269
The Technique of Special Effects in Television, 268
Techniques and Guidelines for Social Work Practice, 555
Techniques and Theory of Periodontal Instrumentation, 12
Techniques for Porcelain Laminate Veneers, 8
Techniques for Wildlife Habitat Management of Wetlands, 504
Techniques in Clinical Nursing, 80
Techniques of Crime Scene Investigation, 288
Techniques of Veterinary Radiography, 529
Technology Edge: A Guide to CD-ROM, 181
Teenage Addicts Can Recover: Treating the Addict, Not the Age, 542
Telecommunications: Concepts, Development, and Management, 209
Television and Audio Handbook for Technicians and Engineers, 264
Television and Cable Factbook, 265
Television and Video: A VGM Career Planner, 270
Television Commercial Processes and Procedures, 267
Television Directing, 266
Television Field Production and Reporting, 265
Telework: Towards the Elusive Office, 178
Temporomandibular Disorders, 11
The Tenant's Leasing Handbook, 220
Terminology for Allied Health Professionals, 52, 283
Test Procedures for the Blood Compatibility of Biomaterials, 60
Test Success, 87
Textbook for Dental Nurses, 10
Textbook of Abdominal Ultrasound, 30
Textbook of Complete Dentures, 13
Textbook of Dendrology, 499
Textbook of Diagnostic Ultrasonography, 30
Textbook of Medical Physiology, 71
Textbook of Physical Diagnosis: History and Examination, 80
Textbook of Radiographic Positioning and Related Anatomy, 23
Textbook of Veterinary Internal Medicine, 526
The Textile Design Book, 424
Textile Designs, 413
Textiles, 425
Textiles for Residential and Commercial Interiors, 441
Textiles, 5,000 Years, 412
The Thames and Hudson Encyclopedia of Graphic Design and Designers, 457
Thats Our New Ad Campaign?, 261
Thematic Units, 314
Theoretical Perspectives on Sexual Difference, 566
Theories of Ethnicity: A Critical Appraisal, 567
Theory and Application of Neonatal Ventilation, 120
Theory and Nursing: A Systematic Approach, 82
Theory-Directed Nursing Practice, 82
Therapeutic Communication, 90
Therapeutic Communications for Allied Health Professions, 55, 64
Therapeutic Exercise in Developmental Disabilities, 106
Therapeutic Interventions in Alzheimer's Disease, 99
Therapeutic Modalities in Sports Medicine, 110
Therapeutic Paradox, 535
Thinking Activities for Books Children Love, 320
Thinking Cop, Feeling Cop: A Study in Police Personalities, 290
Thinking Finance, 170
The Third Year of Life, 319
The Thirty-Six Hour Day, 562
This is PR: The Realities of Public Relations, 261
This Old House Kitchens, 139
Threads of Identity, 417
Three Hundred and Forty-Seven Woodworking Patterns, 125
Three Psychologies, 568
Three-D Computer Graphics, 461
Threshers, 481

Through Jaundiced Eyes: How the Media Views Organized Labor, 243
The Timber-Frame Home: Design, Construction, Finishing, 132
Timber-Frame Houses, 137
A Time to Heal, 537, 574
Time-Life Books Complete Home Improvement and Renovation Manual, 139
Time-Saver Standards for Interior Design and Space Planning, 432
Tips and Techniques for Builders, 135
Tips and Traps When Buying A Home, 219
To Listen, to Comfort, to Care, 88
To Make It Home: Photographs of the American West, 447
To Speak with Cloth: Studies in Indonesian Textiles, 417
Toddlers and Parents: A Declaration of Independence, 312
Toma, 550
Too Old to Cry: Abused Teens in Today's America, 557
Tools of the Profession, 335
Tooth Desensitization, 17
Toothbrushing: The Bass Method, 17
Toothbrushing: The Circular Scrub Method, 17
Topical Reference Books, 338
Tort Law for Legal Assistants, 278
Total Data Quality for the Coding Manager, 50
Total Design: Objects by Architects, 440
Touchpoints, 312
Tourism: An Exploration, 232
Tourism: Principles, Practices, Philosophies, 230
Tourism Alternatives, 231
Tourism Geography, 228
Tourism Geography Workbook, 228
Tourism Marketing, 227
Tourism Planning: Basics, Concepts, Cases, 229
The Tourist Business, 230
Tourist Organizations, 230
Toward Healthy Aging, 75
Toxic Psychiatry, 536
Toxic Risks: Science, Regulation, and Perception, 93
TQC for Accounting: A New Role in Companywide Improvement, 168
Tracheostomy Care, Tube Change, and Artificial Airway Cuff Management, 120
Tracing File for Interior and Architectural Rendering, 437
Tractors Since 1889, 364
Training Foodservice Employees, 196
Training for the Hospitality Industry, 194, 204
Training in the Workplace, 94
Training Manual for Law Enforcement Officers, 288
Training of Technical Services Staff, 342
Training Paraprofessionals for Reference Service, 338
Training Student Library Assistants, 333
Training Tape for Living with F.A.S. and F.A.E., 327
Transcranial Doppler Ultrasonography, 29
Transcultural Nursing: Assessment and Intervention, 79
Trauma and Mobile Radiography, 24
Trauma Nursing: The Art and Science, 74
Travel Agent, 227
The Travel Agent: Dealer in Dreams, 229
Travel Career Development, 228
Travel Geography Handbook, 229
The Travel Safety Handbook, 231
The Travel Writer's Guide, 252
The Travel Writer's Handbook, 253
Travel Writing for Profit and Pleasure, 252
Traveler's Hotline Directory, 228
Traveling on Your Own, 227
Traveling with Children and Enjoying It, 227
Travels in the American West: Photographs by Len Jenshel, 447
Travelwise, 228
Treating Adolescent Substance Abuse, 544
Treating Alcohol Dependence: A Coping Skills Training Guide, 543

Title Index

Treating the Chemically Dependent and Their Families, 537
Treatment Choices for Alcoholism and Substance Abuse, 542
Treatment Technologies, 370
Tree Disease Concepts, 500
Trees, 499
Trenchers: Stay Alert Stay Alive, 365
The Trials of Life: A Natural History of Animal Behavior, 528
Tricia Guild on Color: Decoration, Furnishing, Display, 434
Trim Carpentry Techniques, 136
Tropical Rainforest, 503
Trouble-Free Travel with Children, 230
Trouble Shooting Boundary Line Problems, 385
The Trouble with Fonts, 209
Troubleshooting and Servicing Air Conditioning Equipment, 153
Truck Driver's Guide to CDL: Commercial Driver License, 364
Trust, AIDS, and Your Dentist, 15
The Truth, the Whole Truth, and Nothing But, 287
The TV and Movie Business, 270
TV and Video Technology, 264
TV Newscast Processes and Procedures, 267
Twentieth-Century Fashion: The Complete Sourcebook, 422
The Twentieth-Century Poster: Design of the Avant-Garde, 455
Twentieth-Century Type, 469
Twenty Legal Pitfalls for Nurses to Avoid, 89
Twice Upon a Time, 323
Two Hours of Real Estate: One Minute at a Time, 218
Two to Four from 9 to 5, 321
Tylman's Theory and Practice of Fixed Prosthodontics, 10
Type and Color: A Handbook of Creative Combinations, 469
Type and Image: The Language of Graphic Design, 459
Type and Typography, 471
Type in Use, 472
Type Wise, 470
Typefaces for Desktop Publishing: A User Guide, 398
Typo-Graphic Design, 472
Typographic Communications Today, 470
Typographic Design: Form and Communication, 469
Typographic Milestones, 470
Typography and Typesetting, 471
Typography for Desktop Publishers, 403
Typography Now: The Next Wave, 471

U.S. and Worldwide Travel Accommodations Guide for $12-$24 per Day, 231
U.S. Government Documents, 335
U.S. Government Publications for the School Library, 335
Ugo Mulas, 448
The Ultimate Portfolio, 473
Ultrasonography of Infants and Children, 31
Ultrasound Atlas of Disease Processes, 30
Ultrasound Exam Review, 29
Ultrasound in Obstetrics and Gynecology, 29
Ultrasound Scanning: Principles and Protocols, 32
Unbelievably Good Deals and Great Adventures That You Absolutely Can't Get Unless You're Over Fifty, 229
The Underclass Debate: Views from History, 572
Undergrounding Electric Lines, 146
Understanding and Managing Cholesterol, 35
Understanding and Treating Adolescent Substance Abuse, 543
Understanding and Treating Codependence, 541
Understanding Automotive Electronics, 355
Understanding Baking, 192
Understanding Child Development, 313
Understanding Child Sexual Maltreatment, 560
Understanding Color, 442

Understanding Data Communications, 178
Understanding Dying, Death, and Bereavement, 513
Understanding Electrocardiography, 73
Understanding Family Policy: Theoretical Approaches, 573
Understanding Fashion, 418
Understanding Financial Statements, 168
Understanding Heart Sounds and Murmurs with an Introduction to Lung Sounds, 81
Understanding House Construction, 134
Understanding Human Behavior for Effective Police Work, 290
Understanding Lasers, 377
Understanding Lasers: An Entry Level Guide, 378
Understanding Local Area Networks, 182
Understanding Medical Insurance: A Step-By-Step Guide, 54
Understanding Medical Terminology, 51
Understanding MRI, 28
Understanding MS-DOS, 180
Understanding NE Code Rules on Grounding and Bonding, 144
Understanding NE Code Rules on Hazardous (Classified) Location, 148
Understanding NE Code Rules on Medium Voltage Power Systems, 144
Understanding NE Code Rules on Transformers, 144
Understanding Periodontal Diseases, 5
Understanding Radiography, 25
Understanding Regulations on—OSHA Electrical Design Safety, 148
Understanding Regulations on OSHA Electrical Safety, 148
Understanding Regulations on—OSHA Electrical Work Rules, 148
Understanding Substance Abuse and Treatment, 543
Understanding the Business of Library Acquisitions, 330
Understanding the Human Body, 71
Understanding Type for Desktop Publishing, 405
Understanding Wood Finishing, 124
Understanding/Responding, 72
Underwear: The Fashion History, 420
The Undeserving Poor, 572
Unfaithful Angels: How Social Work Has Abandoned Its Mission, 555
Uniform Crime Reports for the United States, 286
Uniform Fire Code Standards, 496
Uniform Plumbing Code, 161
Uniform Plumbing Code Illustrated Training Manual, 161
Uniform Plumbing Code Interpretations Manual, 161
Uniform Solar Energy Code, 161
Uniform Swimming Pool, Spa, and Hot Tub Code, 161
Uniform System of Accounts for Restaurants, 196
Unit Method of Clothing Construction, 414
United States Code, Containing the General, 302
Universal Precautions, 14, 20
Universal Style, 418
Unlocking the Adoption Files, 563
Unlocking WordPerfect 6.0, 210
Upgrading and Repairing PCs, 180
Uppers, Downers and All Arounders, 540
The USA Travel Phone Book, 230
The Use of Markers in Fashion Illustrations, 420
The Use of Upper Extremity Prostheses, 102
User's Guide to the Bluebook, 280
Using 1-2-3 Release 3.4, 176
Using 1-2-3 Release 4 for Windows, 183
Using Assembly Language, 184
Using AutoCAD, 374
Using BASIC, 177
Using Computer Color Effectively: An Illustrated Reference, 463
Using Computerized Accounting, 172
Using Computers in Legal Research, 302
Using DOS in Court Reporting, 275
Using Government Information Sources, 335
Using Literature in the Elementary Classroom, 324
Using Microsoft Publisher 2, 406

Using PageMaker 5 for Windows, 409
Using the Band Saw: Techniques for Better Woodworking, 123
Using the Internet, 183
Using the New AACR2, 343
Using the Table Saw: Techniques for Better Woodworking, 123
Using Type Right, 469
Using Video: Interactive and Linear Designs, 266
USMARC Code List for Countries, 341
USMARC Code List for Geographic Areas, 341
USMARC Code List for Languages, 341
USMARC Format for Authority Data, 341
USMARC Format for Bibliographic Data, 341

Valentino: Thirty Years of Magic, 423
Valuing Diversity, 343
VanDerZee: Photographer, 1886-1983, 450
Vascular Imaging by Color Doppler and Magnetic Resonance, 31
Vascular Nursing, 73
Vehicle Extrication: A Training Manual, 44, 494
Vehicle Rescue and Extrication, 44
Ventilators: Theory and Clinical Application, 115
Ventilatory Failure, 116
Verbal Judo: The Gentle Art of Persuasion, 290
The Verbum Book of Digital Typography, 403
Versacad Corporation's Training Guide 2D and 3D Tutorials, 375
VersaCAD: A Practical Approach to Computer-Aided Design, 375
Vest-Pocket CPA, 171
The Vest-Pocket Real Estate Advisor, 214
Vet on the Wild Side, 530
Veterinary Ethics, 530
Veterinary Pediatrics: Dogs and Cats from Birth to Six Months, 528
Victorian America: Classical Romanticism to Gilded Opulence, 433
The Victorian House Book, 434
Video Commuication, 267
Video Court Reporting: A Bibliography of Recent Information Sources, 283
Video Court Reporting: A Prime, 283
Video Court Reporting: A Review, 283
Video Court Reporting: Tips for Court Managers, 283
The Video Demo Tape, 266
Video Editing and Post Production, 268
Video Field Production, 267
Video Lighting and Special Effects, 268
Video Power, 267
Videography: The Successful Home Video Series, 271
Videoguide to Birds of North America, 508
Videotape Editing: A Postproduction Primer, 267
Videotaped Trial Records: Evaluation and Guide, 283
The Village Carpenter, 131
Violence Hits Home, 563
The Violent Family, 561
Virgin or Vamp: How the Press Covers Sex Crimes, 254
The Virtual Library: Visions and Realities, 333
Virtual Reality, 181
Visions of a Flying Machine, 348
Visions of Addiction, 543
Visual Editing: A Graphic Guide for Journalists, 248
Visual Forces: An Introduction to Design, 459
Visual Guide to Dental Care, 20
The Visual Handbook of Building and Remodeling, 130
Visual Literacy, 460
Visual Merchandising, 426
The Vitamins: Fundamental Aspects in Nutrition and Health, 35
VNR's Encyclopedia of Hospitality and Tourism, 195, 205
Vogue and Butterick's Designer Sewing Techniques, 415
Vogue History of Twentieth Century Fashion, 422
The Vogue/Butterick Step-By-Step Guide to Sewing Techniques, 415

Title Index

Voice Processing, 181
Volunteers in Libraries, 334
Vulnerability to Drug Abuse, 539
Vulnerable Populations, 563
W. Eugene Smith and the Photographic Essay, 450
The Waiter and Waitress Training Manual, 193
The Waldorf-Astoria: America's Gilded Dream, 206
Walk the Walk and Talk the Talk, 545
Walker Evans, 448
Walker Evans: The Hungry Eye, 449
Walker's Mammals of the World, 507
The Wall Street Journal on Accounting, 167
Walls and Molding, 139
The Walls Around Us, 139
The Wantmakers: Inside the World of Advertising, 259
The War on Drugs II, 540
Waste Disposal in Academic Institutions, 62
Wastewater Engineering: Treatment, Disposal, and Reuse, 370
Water-Based Inks, 404
Water for Agriculture: Facing the Limits, 486
Water Quality I: Introduction to Water Quality Management, 509
Water Quality II: Water Quality Dynamics, 509
Water Quality III: Procedures for Water Quality Management, 509
Water Supplies for Fire Protection, 496
Waterfowl Ecology and Management, 506
Waxing Made Easy: A Step-By-Step Guide, 188
The Way Multimedia Works, 461
The Way of the Carpenter: Tools and Japanese Architecture, 131
The Way We Wore, 421
Ways of Assessing Children and Curriculum, 315
Ways of Living: Self-Care Strategies for Special Needs, 99
We Care...The Effective Veterinary Receptionist, 531
Weaning and Extubation, 120
Weaver and Koehler's Programmed Mathematics of Drugs and Solutions, 84
Web Offset Press Operating, 401
Webster's New World Dictionary of Media and Communications, 252, 265
Weed Identification Guide, 488
Weed Science: Principles and Practices, 487
Weeding and Maintenance of Reference Collections, 342
Weeding Library Collections, 342
Weeds of the West, 488
The Weekend Woodworker: 101 Easy-To-Build Projects, 125
Welcome Home: A Consumer's Guide to Home Buying, 221
Welcome to Concepts in Graphic Design, 404
Welcome to Desktop Publishing, 399
The Welder's Bible, 390
Welding, 393, 394
Welding: Principles and Applications, 392
Welding and Cutting, 393
Welding and Fabricating Data Book, 390
Welding as a Career, 394
Welding Certification Questionnaire, 391
Welding Handbook, 390
Welding Print Reading, 390
Welding Technology Fundamentals, 392
Welding Technology Today: Principles and Practices, 393
Welfare Dependence and Welfare Policy: A Statistical Study, 570
Well-Being of Nonhuman Primates in Research, 529
The Well-Built House, 134
Westcott's Plant Disease Handbook, 521
Western Ranching: Culture in Crisis, 489
Westlaw Reference Manual, 301
West's Federal Tax Research, 306
West's Legal Desk Reference, 302
West's Reporter Series, 302
West's Tax Law Dictionary, 306
Wetlands, 504

Wetlands of North America, 504
What Children Can Tell Us, 315
What Color Is Your Parachute?, 552
What Do Social Workers Do?, 575
What Is Good Instruction Now? Library Instruction, 331
What Matters? A Primer for Teaching Reading, 323
What You Need to Know About Psychiatric Drugs, 548
The Wheel of Life and Death: A Practical and Spiritual Guide, 512
Wheelchairs: A Prescription Guide, 101
Wheeler's Dental Anatomy, Physiology, and Occlusion, 5
When Art Became Fashion: Kosode in Edo-Period Japan, 417
When Feeling Bad Is Good, 542
When Good Journalists Do Bad Things, 256
When MBAs Rule the Newsroom, 244
When the Bough Breaks: The Cost of Neglecting Our Children, 561
When Violence Erupts: A Survival Guide for Emergency Responders, 44
When Words Collide, 251
When You Place a Child, 562
Where Did the Time Go?, 210
Where to Find What: A Handbook to Reference Service, 337
Which Process? An Introduction to Welding and Related Processes and a Guide toTheir Selection, 390
The Whitehead Encyclopedia of Deer, 507
Who Cares for the Elderly?, 557
The Whole Internet User's Guide and Catalog, 179, 332
Whole Language for Second Language Learners, 315
Whole Language Kindergarten, 321
Whole Library Handbook, 334
Wholesale in Fashion Merchandising, 426
Who's What and Where, 251
Who's Who in American Law, 302
Who's Who in Fashion, 413
Who's Who in Interior Design, 441
Why Airplanes Crash: Aviation in a Changing World, 360
Why Buildings Fall Down: How Structures Fail, 131
Why Buildings Stand Up: The Strength of Architecture, 131
Why Do These Kids Love School?, 327
Why Occupational Therapists Use Crafts, 102
Why You Should File, 172
Wide-Body: The Triumph of the 747, 348
Widmann's Clinical Interpretation of Laboratory Tests, 64
The Wild West: Photographs by David Levinthal, 447
Wildflowers, 519
Wildland Watershed Management, 504
Wildlife and the Public Interest, 503
Wildlife, Forests, and Forestry, 499
Wildlife Habitat Relationships in Forested Ecosystems, 507
Wildlife Research and Management in the National Parks, 507
Willard and Spackman's Occupational Therapy, 100
William Morris: Decor and Design, 441
Window Style, 438
Window Treatments, 437
Windows 3.1: The Complete Reference, 182
Windows from the Keyboard, 176
Windows, Word, and Excel Office Companion, 209
Wing Theory, 348
Winged Victory: The Army Air Forces in World War II, 360
Winning Combinations, 223
Winning Office Politics: DuBrin's Guide for the 90s, 334
Winning Pulitzers, 235
Wiring, 148
Wiring 12 Volts for Ample Power, 147
Wish You Were Here, 207
With a Little Help from My Friends, 327

With the Best of Intentions, 558
Wolves, 509
Women and Farming: Changing Roles, Changing Structures, 478
Women and Men: New Perpectives on Gender Differences, 566
Women and Substance Use, 537
Women and the Work of Benevolence, 565
Women in the Middle: Their Parent-Care Years, 559
Women in Veterinary Medicine: Profiles of Success, 528
Women of Design: Contemporary American Interiors, 439
Women of Fashion: Twentieth-Century Designers, 413
Women of the World: The Great Foreign Correspondents, 254
Women on Deadline: A Collection of America's Best, 255
Women Photographers, 449
Women with Wings: Female Flyers in Fact and Fiction, 358
Women, the State, and Welfare, 565
Women's Health Care, 88
Wood: Basic Woodworking Tips and Techniques, 126
Wood Frame House Construction, 136
The Wood Worker's Dictionary, 125
Wood-Frame House Construction, 135
Woodall's The Campground Directory for North America, 232
Woodworker's Essential Shop Aids and Jigs, 126
The Woodworkers Guide to Making and Using Jig. .Fixtures and Setups, 125
Woodworking Machines, 126
Woodworking Projects for the Great Outdoors, 124
Woodworking Projects Yearbook, 124
Woodworking with the Router, 125
Woodworking with Your Kids, 125
The Woodwright's Eclectic Workshop, 126
Word 6.0 for Windows Resource Kit, 210
Word for Windows 6: Quick Reference, 177
Word for Windows 6: Self-Teaching Guide, 177
Word Processing for Business Users, 210
WordPerfect 5.1 for Windows, 180
WordPerfect 6 for Windows Solutions, 175
WordPerfect 6 for Windows: The Complete Reference, 184
WordPerfect 6 Made Easy, 211
WordPerfect 6: The Complete Reference, 184
WordPerfect for Windows: Wiley Command Reference, 209
Words' Worth: A Handbook on Writing and Selling Nonfiction, 239
Work Design: Industrial Ergonomics, 94
Work Environment, 93
Work in Progress: Occupational Therapy in Work Programs, 100
Workbenches and Shop Furniture: Techniques for Better Woodworking, 123
Workbook for Clinical Procedures for Medical Assisting, 61
Workbook of Accounting Standards, 170
Worker Protection During Hazardous Waste Remediation, 367
Working at Woodworking, 126
Working in Health Care: What You Need to Know to Succeed, 55
Working in Hotels and Catering, 207
Working Ourselves to Death, 538
The Working Press of the Nation, 252
Working with Faculty in the New Electronic Library, 333
Working with Graphic Designers, 472
Working with Pesticides: A Video Safety Course, 522
Working with Words, 250
Working with Your Woodland: A Landowner's Guide, 507
Workplace Health Protection, 93
Workshop for Legal Assistants: Basic Litigation ..., 304

Title Index

Workshops and Outbuildings, 134
Workshops for Legal Assistants: Bankruptcy, 1992, 304, 306
Workshops for Legal Assistants: Employee Benefits, 306
Workshops for Legal Assistants: Litigation, Legal Research ..., 279, 304, 306
World Agriculture and the Environment, 484
World Agriculture in Disarray, 484
World Aviation Directory, 361
World-Class Accounting and Finance, 169
World Criminal Justice Systems: A Survey, 286
World Dictionary of Legal Abbreviations, 280
World Encyclopedia of Police Forces and Penal Systems, 286
World of Abnormal Psychology, Program 11, 327
The World of Abnormal Psychology: Substance Abuse Disorders, 575
World of Storytelling, 321
World of the Newborn, 319
The World of the Northern Evergreens, 500
The World of the Veterinary Technician, 531
The World of Zines, 250
World Press Photo 1992, 235
World Tourism at the Millennium, 228
World Trends in Tourism and Recreation, 230
Worlds of the Mentally Ill, 568
The World's Worst Aircraft, 350
Write Great Ads: A Step-By-Step Approach, 235
The Writer's Complete Guide to Conducting Interviews, 239
The Writer's Guide to Query Letters and Cover Letters, 239
Writer's Market 1994: Where and How to Sell What You Write, 252
Writing and Reporting News: A Coaching Method, 247
Writing for Your Readers, 247
Writing Opinion, Editorials, 240
Writing Solutions: Beginnings, Middles, and Endings, 239
Writing Up Qualitative Research, 569

Xeriscape Gardening, 519

The Yearbook of Agriculture, 480
Years 3-5: What Lily Learned, 327
The Yew Tree: A Thousand Whispers; Biography of a Species, 499
You and Your Clients: Human Relations for Cosmetology, 187
You Can Be a Columnist: Writing and Selling Your Way to, 239
You Can Double Your Income and Build Clientele as a Stylist, 188
You Can't Say You Can't Play, 320
Young People and Death, 320, 514
Your Baby and Child: From Birth to Age Five, 318
Your Barn House, 137
Your Career as a Court Reporter, 274
Your Gifted Child, 323
Your Home: A Home Buying, Selling, Building, Remodeling Guide, 219
Your Home Cooling Energy Guide, 152
Your House, Your Health: A Non-Toxic Building Guide, 221
Your Injury: A Common Sense Guide to Sports Injuries, 110
Your Public Best, 259
Your Successful Real Estate Career, 217

Zen and the Art of the Internet, 332

Journal Index

AANA Journal, 68
AAOHN Journal, 68
AARCTimes, 114
Aberdeen's Concrete Construction, 128, 156
Abstracts of Clinical Care Guidelines, 58
Access, 3
Accounting Review, 166
Advance for Medical Technologists, 58
Advances in Nursing Science, 68
Advances in the Biology of Disease, 58
Advertising Age, 258
Adweek, 258
Ag Consultant, 476
Ag/Innovator, 476
Aging, 552
The Agricultural Education Magazine, 476
Agricultural Outlook, 476
Agricultural Research, 476
AIDS Patient Care, 68
AIGA Journal of Graphic Design, 396
Air Conditioning, Heating and Refrigeration News, 151
Air Line Pilot, 358
Air Transport World, 358
Aircraft Maintenance Technology, 346
Aircraft Technician, 346
Airport/Facility Directory, 358
Alcohol Health and Research World, 535
Alcoholism Treatment Quarterly, 535
Aldus Magazine, 396
American Bar Association Journal, 274
American Fire Journal, 491
American Forests, 498
American Funeral Director, 511
American Geriatrics Society Journal, 103
American Horticulturist, 517
American Industrial Hygiene Association Journal, 91
American Journal of Alternative Agriculture, 476
American Journal of Clinical Nutrition, 34
American Journal of Infection Control, 68
American Journal of Nursing, 68
American Journal of Occupational Therapy, 98
American Journal of Physical Medicine and Rehabilitation, 103
American Journal of Roentgenology, 22
American Journal of Sports Medicine, 103
American Journalism Review, 234
American Laboratory News, 58
American Libraries, 328
American Machinist, 376
American Midland Naturalist, 498
American Naturalist, 498
American Nurse, 68
American Nurseryman, 517
American Photo, 444
American Printer, 396
American Review of Respiratory Disease, 114
American Salon, 186
American Woodworker, 122
Anesthesia and Analgesia, 114
Anesthesiology, 114
ANNA Journal, 68
Annals of Emergency Medicine, 41
AORN Journal, 68
Aperture, 444
Application Technology, 476
Applied Optics, 376
Applied Radiology, 22
The Appraisal Journal, 214
Appraiser News, 214
Architectural Digest, 429
Archives of Pathology and Laboratory Medicine, 58
Archives of Physical Medicine and Rehabilitation, 104
Art Culinaire, 191
Art Direction, 454

ASBA Today, 222
ASHRAE Journal, 151
ASTA Agency Management, 227
Audubon, 498
The Auk, 498
Auto Service Today, 352
Automotive Engineering, 352
Automotive News, 352
AutoWeek, 352
AV Video, 263
AVC Presentation Development and Delivery, 263
Aviation Consumer, 346
Aviation Equipment Maintenance, 346
Aviation Mechanics Bulletin, 346
Aviation Tradescan, 358
Aviation Week and Space Technology, 358

BBW: Big Beautiful Woman Magazine, 412
Beauty Education, 186
Before and After, 396
Better Homes and Gardens Wood, 122
Bio/Technology/Diversity Week, 476
Biocycle, 366
Blue Seal, 352
Body Language, 352
BodyShop Business, 352
The Book of Video Photography, 266
Booklist, 310
British Journal of Occupational Therapy, 98
Builder, 128
The Bulletin of the American Society of Newspaper Editors, 234
Business Marketing, 258
Business Publishing, 396
Byte, 174, 396

Cadalyst, 372
Cadence, 174, 372
Camera and Darkroom, 444
The Canadian Field-Naturalist, 498
Canadian Hotel and Restaurant: CH&R., 202
Canadian Journal of Medical Radiation Technology, 22
Canadian Journal of Occupational Therapy/Revue, 98
Canadian Journal of Zoology, 498
Canadian Nurse, 68
Cancer Nursing, 68
Car and Driver, 352
Cat Fancy, 524
Catering Today, 191
CEE News [Contractors Electrical Equipment News], 142
Chest, 114
Child and Youth Care Forum, 310
Child Care Information Exchange, 310
Child Development, 310
Child Welfare, 552
Childhood Education, 310
Children Today, 310, 552
Children's Literature Association Quarterly, 310
Choice, 328
Clinical Chemistry (Reference Edition), 58
Clinical Lab Letter, 58
Clinical Laboratory Reference, 58
Clinical Laboratory Science, 58
Clinical Microbiology, 62
Clinical Nurse Specialist, 68
Clinical Pharmacology and Therapeutics, 58
Coding Clinic for ICD-9-CM, 48
Collection Building, 329
Collection Management, 329
College and Research Libraries, 329
College and Research Libraries News, 329
Colonial Homes, 429
Columbia Journalism Review, 234
Commercial Investment Real Estate Journal, 214
Communication Arts, 396, 454
Communication Briefings, 258
Communications of the ACM, 174
Community and Junior College Libraries, 329

Compute, 174
Computer Artist, 454
Computer Graphics, 396
Computer Graphics World, 372, 396
Computer Language, 174
Computers in Nursing, 68
Computerworld, 174
Condé Nast's Traveler, 227
The Condor: A Journal of Avian Biology, 498
Constructor, 128
Consumer Magazine and Agri-Media Rates and Data, 258
Contractor, 160
The Cornell Hotel and Restaurant Administration Quarterly, 202
Corporate Travel, 227
The Counselor, 535
Creative Camera, 444
Critical Care Medicine, 114
Critical Care Nursing Clinics of North America, 68
Critical Care Nursing Quarterly, 68
Critical Reviews in Clinical Laboratory Sciences, 58
Cruise Travel Magazine, 227
Custom Builder, 128

Darkroom and Creative Camera Techniques, 444
Data Based Advisor, 174
Datamation, 174
Day Care and Early Education, 310
Defenders, 498
Dental Abstracts, 3
Dental Assistant Journal, 3
Dental Clinics of North America, 3
Dental Lab Products, 3
Dental Teamwork, 4
Dentistry Today, 4
DES (Diesel Equipment Superintendent), 363
Design News, 372
Desktop Communications, 396
Details, 412
Developmental Medicine and Child Neurology, 104
Developmental Psychology, 310
Diagnostic Imaging, 22
The Director, 511
DNR: The Magazine, 412
The Docket, 294, 298
Dog Fancy, 524

E: The Environmental Magazine, 366, 498
EAP Digest [The Voice of Employee Assistance Programs], 535
Earth Journal, 498
Ecology, 498
Editor and Publisher, 234
Electrical Code Watch, 142
Electrical Construction and Maintenance [ECM], 142
Electrical Contractor, 142
Electronic Composition and Imaging, 396
Electronic Design, 372
Elle, 412
Emergency, 41
Emergency Librarian, 329
Emergency Medicine, 41
EMS Insider, 41
Entrepreneur, 222
Environment, 498
Environment Abstracts, 367
Environmental Science and Technology, 367
EPA Journal, 366
Equine Practice, 524
Equipment Today, 363
Equus, 524
ERIC/EECE Newsletter, 310
Esquire, 412
Exceptional Children, 310
Exercise and Sport Sciences Review, 104
Existing Home Sales, 214
Families in Society: The Journal of Contemporary Human Services, 552

Journal Index

Farm Chemicals, 476
Farm Journal, 476
Farm Smart, 476
Farmer's Digest, 476
FarmFutures, 476
FBI Law Enforcement Bulletin, 285
Federal Probation, 552
Fine Homebuilding, 128, 156
Fine Woodworking, 122
Fire Chief, 491
Fire Engineering, 491
Fire Safety Journal, 491
Fire Technology, 491
Firefighter's News, 491
Firehouse, 491
FIU Hospitality Review, 202
Five Owls, 310
Fleet Equipment, 363
Flying, 358
Food and Wine, 191
Food Animal, 524
Food Arts, 191
Food Management, 191
Food Technology, 191
Forest Science, 498

General Aviation News and Flyer, 346
General Dentistry, 4
Geriatric Nursing, 69
Glamour, 412
Gourmet, 191
GQ, 412
Graphic Arts Monthly, 396
Graphic Design: USA, 396
Graphis, 454
Grounds Management Forum, 517
Growing Child, 310

Harper's Bazaar, 412
Heart and Lung, 69, 114
Heating, Piping, and Air Conditioning, 151
Home, 429
Home Mechanix, 122
Home Office Computing, 174
Horn Book Guide to Children's and Young Adult Books, 310
Horn Book Magazine, 310
Horticulture, 517
Hospitality and Tourism Educator, 202
Hospitality Design, 202
Hospitality Law, 202
Hospitality Research Journal, 202
Hot Line Farm Equipment Guide, 476
Hot Rod, 352
Hotel and Motel Management, 202
Hotel and Resort Industry, 202
Hotel and Travel Index, 202
Hotels, 202
House Beautiful, 429
How: the Magazine of Ideas and Techniques in Graphic Design, 396, 454

IAEI News, 142
Image—Journal of Nursing Scholarship, 69
Impact, 352
Implement and Tractor, 476
In Confidence, 48
In House Graphics, 396
Inc.: The Magazine for Growing Companies, 208, 222
Industrial Fire Safety, 491
Industrial Fire World, 492
Industrial Hygiene News, 91
Infants and Young Children, 104
Information Technology and Libraries, 329
Instructor, 310
Interior Design, 429
Interiors, 429
International Aviation Mechanics Journal, 346
International Journal of Early Childhood, 310

International Journal of Hospitality Management, 202
International Journal of the Addictions, 535
International Textiles, 412
International Wildlife, 498
Internet World, 174
IRE Journal, 234

JAMA: The Journal of the American Medical Association, 114
JEMS: A Journal of Emergency Medical Services, 41
Job Safety and Health Quarterly: JS and HQ, 91
Journal of Academic Librarianship, 329
Journal of Accountancy, 166
Journal of Accounting Research, 166
Journal of Advertising, 258
Journal of Advertising Research, 258
Journal of AHIMA, 48
The Journal of Bone and Joint Surgery, 104
Journal of Cardiovascular Nursing, 69
Journal of Child and Youth Care, 310
Journal of Clinical Investigation, 58
Journal of Clinical Laboratory Analysis, 58
Journal of Community Health Nursing, 69
Journal of Computing in Childhood Education, 310
Journal of Construction Engineering and Management, 128
Journal of Court Reporting, 274
Journal of Criminal Law and Criminology, 285
Journal of Dental Education, 4
Journal of Dental Hygiene, 4
Journal of Diagnostic Medical Sonography, 22
Journal of Drug Issues, 535
Journal of Emergency Nursing, 69
Journal of Environmental Education, 367
Journal of Environmental Engineering, 367
Journal of Forestry, 498
The Journal of Hand Therapy, 104
Journal of Interlibrary Loan and Information Supply, 329
Journal of Laboratory and Clinical Medicine, 58
Journal of Learning Disabilities, 310
Journal of Mammology, 498
Journal of Marketing, 258
Journal of Marriage and the Family, 552
Journal of Natural Resources and Life Sciences Education, 476
Journal of Neuroscience Nursing, 69
Journal of Nursing Administration, 69
Journal of Nursing Education, 69
Journal of Nutrition, 34
Journal of Obstetric, Gynecologic and Neonatal Nursing, 69
Journal of Occupational Therapy Students: JOTS, 98
The Journal of Orthopaedic and Sports Physical Therapy, 104
Journal of Pediatric Health Care, 69
Journal of Periodontology, 4
Journal of Professional Nursing, 69
Journal of Prosthetic Dentistry, 4
Journal of Psychosocial Nursing and Mental Health, 69
Journal of Public Health Dentistry, 4
Journal of Range Management, 498
Journal of Real Estate Taxation, 214
Journal of Small Business Management, 222
Journal of Social Work Education, 552
Journal of Soil and Water Conservation, 476, 498
Journal of Sports Medicine and Physical Fitness, 104
Journal of Substance Abuse Treatment, 535
Journal of Surveying Engineering, 384
Journal of Taxation, 166
Journal of the American Animal Hospital Association, 524
Journal of the American College of Dentists, 4
Journal of the American Dental Association, 4
Journal of the American Dietetic Association, 34
Journal of the American Society of Echocardiography, 22
Journal of the American Veterinary Medical Association, 524

Journal of the Optical Society of America, 376
Journal of Wildland Fire Management, 498
Journal of Wildlife Management, 498
Journal of X-Ray Science and Technology, 22
Journal of Zoo and Wildlife Medicine, 524
Journal/National Association of Desktop Publishers, 396
Journalism Quarterly, 234

Kids Fashions Magazine, 412
Kitplanes, 346

Lab Animal, 524
Lab Report, 58
Labmedica, 58
Laboratory Equipment, 58
Laboratory Investigation, 58
Language Arts, 310
Laser Focus World, 376
Law and Order, 285
Law Technology Product News, 294, 298
Learning, 310
The Legal Assistant Today, 294, 298
Library Acquisitions: Practice and Theory, 329
Library Administration and Management, 329
Library Hotline, 329
Library Journal, 329
Library Mosaics, 329
Library Personnel News, 329
Library Resources and Technical Services, 329
Light Plane Maintenance, 346
Lincoln Electric Stabilizer, 389
Living Bird, 498
LMT: Lab Management Today, 4
Lodging Hospitality, 202
Lodging, 202

Machine Design, 372
MacUser, 174, 394
MacWorld, 174, 396
Mademoiselle, 412
Management Accounting, 166
Managing Office Technology, 208
Marketing News, 258
The Masonry Society Journal, 156
The Masthead, 234
MCN, The American Journal of Maternal Child Nursing, 69
Media Week, 258
Medical Laboratory Observer, 58
Medicine and Science In Sports and Exercise, 104
Meetings and Conventions, 202
Metropolitan Home, 429
Modern Maturity, 552
Modern Salon, 186
Mortuary Management, 511
Motor Age Mechanics Newsletter, 363
Motor Magazine, 363
Motor Service, 352
Motor Trends, 352
MT Today, 58

NACTA Journal, 476
Nails, 186
National Culinary Review, 191
National Fire Codes, 492
National Gardening, 518
National Parks, 498
National Real Estate Investor, 214
National Wildlife, 498
Nation's Restaurant News, 191
Nature Study, 498
NBIAP News Report, 476
New Directions for Child Development, 310
New England Journal of Medicine, 114
The New Farm, 476
New Homes Magazines, 214
New Scientist, 377
News Photographer, 234
Newspaper Rates and Data, 258

Journal Index

Newspaper Research Journal, 234
NFPA Journal, 142, 492
No-Till Farmer, 476
Northwest Science, 498
NTSB [National Transportation Safety Board Reporter], 358
Nurse Practitioner, 69
Nursing, 69
Nursing and Health Care, 69
Nursing Clinics of North America, 69
Nursing Diagnosis, 69
Nursing Outlook, 69
Nursing Research, 69
Nursing Times, 69
Nutrition Action Health Letter, 34
Nutrition Reviews, 34
Nutrition Today, 34

OAG Business Travel Planner, 202
Occupational Health and Safety, 91
Occupational Hygiene, 91
Occupational Injuries and Illnesses in the U.S. by Industry, 91
Occupational Safety and Health Reporter, 91
Occupational Safety and Health Statistics of the Federal Government, 91
Office Systems: The Magazine for Small and Medium Offices, 208
Office Technology Management, 208
Official, 160
Oncology Nursing Forum, 69
Optical Engineering Report, 377
OSHA Week, 91
OT Week, 98
Overdrive, 363
Owner Operator, 363

Page, 396
Parent and Preschooler, 310
Parents Magazine, 310
Parks and Recreation, 498
PC Magazine: The Independent Guide to IBM-Standard, 208
PC Magazine, 174, 396
PC World, 174, 208, 396
PC/Computing, 174
Pediatric Nursing, 69
Pediatric Physical Therapy, 104
Pediatrics, 104
Perspectives, 524
Pest Control Technology, 476
Petersen's Photographic, 444
Photo Design, 444
Photo Electronic Imaging, 444
Photonics Spectra, 377
Physical and Occupational Therapy in Pediatrics, 104
Physical Therapy, 104
Physiotherapy Canada, 104
Plants and Gardens, 518
Plumbing Engineer, 160
Plumbing-Heating-Cooling Business, 160
Point of Beginning-P.O.B, 384
The Police Chief, 285
Pollution Engineering, 367

Popular Mechanics, 122, 377
Popular Photography, 444
Popular Science, 122, 377
Power Talk: Standard American English, Your Ladder to Success, 262
PR Reporter, 258
Prehospital and Disaster Medicine, 41
Print Media Production Data, 258
Print, 396, 454
Proceedings of the Nutrition Society, 34
Professional Counselor Magazine, 535
Professional Medical Assistant, 48
Professional Safety, 91
Professional Surveyor, 384
Progress in Tourism and Recreation and Hospitality, 202
PT: The Magazine of Physical Therapy, 104
Public Health Nursing, 69
Public Relations Journal, 258
Public Relations Quarterly, 258
Public Relations Review, 258
Public Welfare, 552
Publish!, 396
Publishers Weekly, 329

Quill, 234

Radiologic Clinics of North America, 22
Radiologic Technology, 22
Radiology, 22
Rangelands, 476
RDH: [The National Magazine for Dental Hygiene Professionals], 4
Reading Teacher, 310
Ready, Set, Go! In-Depth, 396
Real Estate Finance Journal, 214
Reference Librarian, 329
Reference Services Review, 329
Regan Report on Nursing Law, 69
Rehabilitation Nursing, 69
Report—Student Press Law Center, 234
Rescue, 41
Research in Nursing and Health, 69
Research Quarterly for Exercise and Sport, 104
Research Strategies: RS, 329
Respiratory Care, 114
Restaurant Hospitality, 191
Restaurants and Institutions, 191
Restaurants USA, 191
RMA Vital Signs, 48
RN, 69
RQ, 329
Rural Heritage, 476

Safety and Health, 91
Sales and Marketing Management, 202
Scholastic Pre-K Today, 310
School Food Service Journal, 191
The Secretary, 208, 294
Serials Librarian, 329
Serials Review, 329
Service Edge, 202
Sesame Street Parents' Guide, 310
Seventeen, 412
Seybold Report on Desktop Publishing, 396

Sierra, 498
Small Animal Practice, 524
Social Policy, 552
Social Problems, 552
Social Work, 552
Special Libraries, 329
Sports Medicine, 104
Spot Television Rates and Data, 258
Step-By-Step Graphics, 396, 444, 454
Successful Farming, 476
Surveying and Land Information Systems, 384

Teaching Exceptional Children, 310
Teaching K-8, 310
Technicalities, 329
Threads Magazine, 412
Top Producer, 476
Topics in Health Information Management, 48
Topics in Veterinary Medicine, 524
Totline, 310
Toward an Electronic Patient Record, 48
Travel Agent, 227
Travel Trade, 227
Trends and Techniques in the Contemporary Dental Laboratory, 4
Trial, 274
TypeWorld, 396

U & LC, 454
UA Journal, 160
Ultrasonic Imaging, 22
UNIX Review, 174
UNIX/World, 174

Veterinary Clinics of North America, 524
Veterinary Technician, 524
Veterinary Update Clinical Abstract Service, 524
Videography, 263
Vogue, 412

Wastetech News, 367
Weatherwise, 499
Welding Design and Fabrication, 389
The Welding Innovation Quarterly, 389
Welding Journal, 389
Wilderness, 499
Wildlife Society Bulletin, 499
The Wilson Bulletin, 499
Wilson Library Bulletin, 329
Windows Magazine, 208, 396
The Wine Spectator, 191
Women's Wear Daily, 412
Wood and Wood Products, 128
Wood Digest, 122
Woodshop News. Northeast, 122
Woodsmith, 122
Woodworker's Journal, 122
WordPerfect: The Magazine, 208
Writer's Digest, 234

Young Children, 311

Zero to Three, 311